Memories OF LIFE ON THE FARM

THROUGH THE LENS OF PIONEER PHOTOGRAPHER *J. C. Allen*

FREDERICK WHITFORD • NEAL HARMEYER

PURDUE UNIVERSITY PRESS
West Lafayette, Indiana
press.purdue.edu

Cataloging-in-Publication data on file at the Library of Congress
Hardback ISBN: 978-1-55753-866-6

Unless otherwise noted, all photographs in this book are from the J. C. Allen and Son Inc. Photographs and Negatives Collection at the Purdue University Archives and Special Collections, West Lafayette, Indiana. While photographs from this collection included in this book are presented as John C. Allen's work, it is possible that a few of the images may have been the work of his son, Chester Allen.

For all photographs, information about individuals, organizations, places, events, and dates has been provided to the extent it is known. The abbreviation *n.p.* indicates that no place is known, while *n.d.* indicates that no date is known.

All places mentioned are in Indiana unless otherwise noted.

As needed, some minor grammatical changes were made to improve the readability of quoted material; these changes did not affect content.

Photo of John C. Allen on opposite page: Tippecanoe County, 1930; courtesy John O. Allen

Designer: Dawn L. Minns
Editor: Carolyn A. McGrew
Indexer: Marilyn Augst

Memories
OF LIFE ON THE FARM

Purdue University Press Founders Series

Books by Frederick Whitford

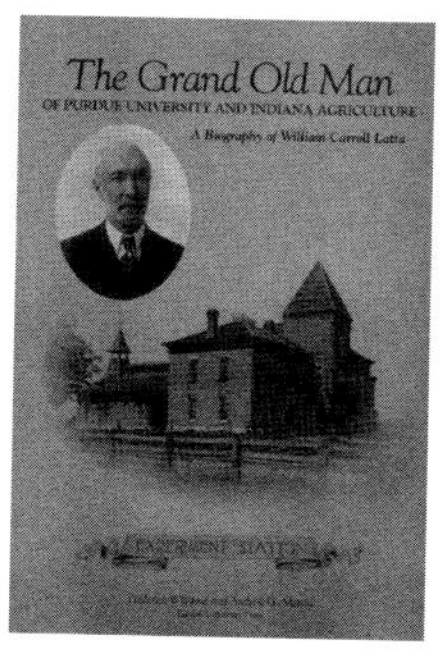

The Grand Old Man of Purdue University and Indiana Agriculture: A Biography of William Carroll Latta

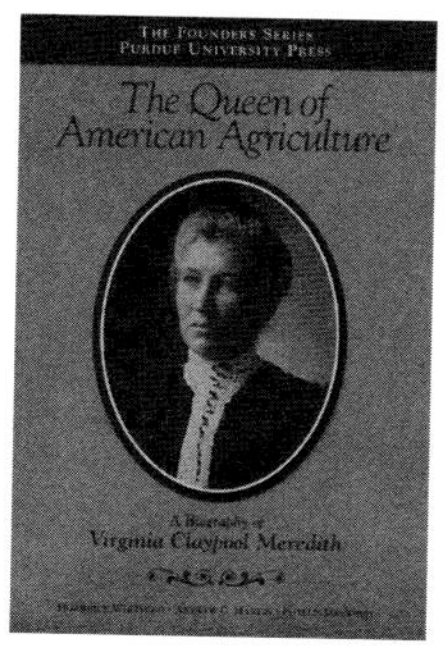

The Queen of American Agriculture: A Biography of Virginia Claypool Meredith

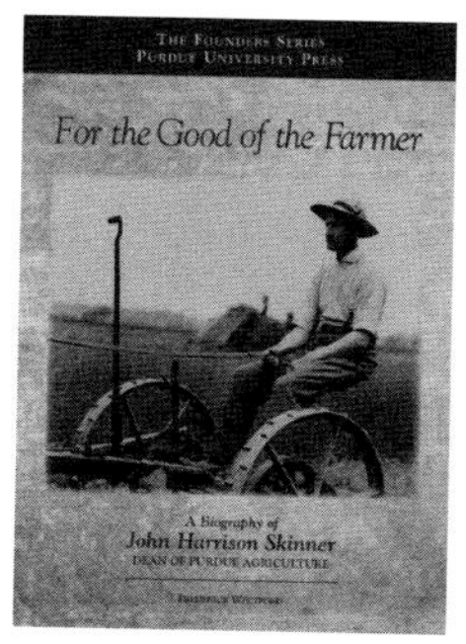

For the Good of the Farmer: A Biography of John Harrison Skinner, Dean of Purdue Agriculture

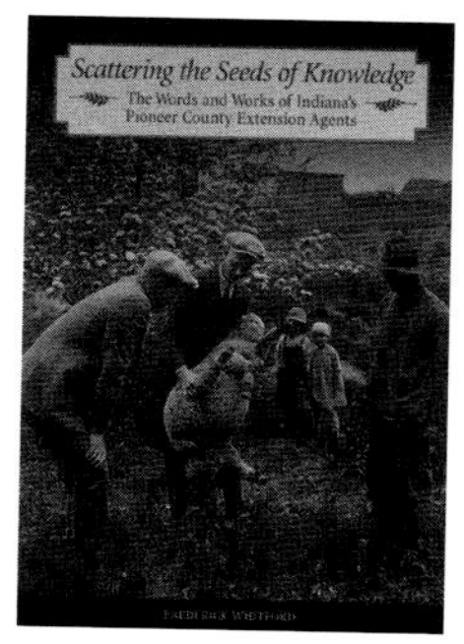

Scattering the Seeds of Knowledge: The Words and Works of Indiana's Pioneer County Extension Agents

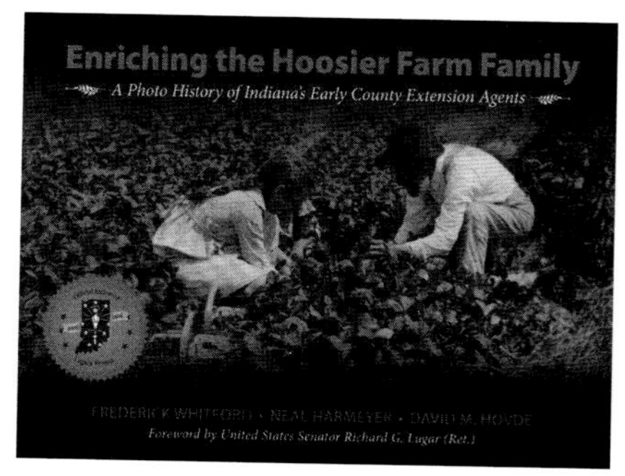

Enriching the Hoosier Farm Family: A Photo History of Indiana's Early County Extension Agents

CONTENTS

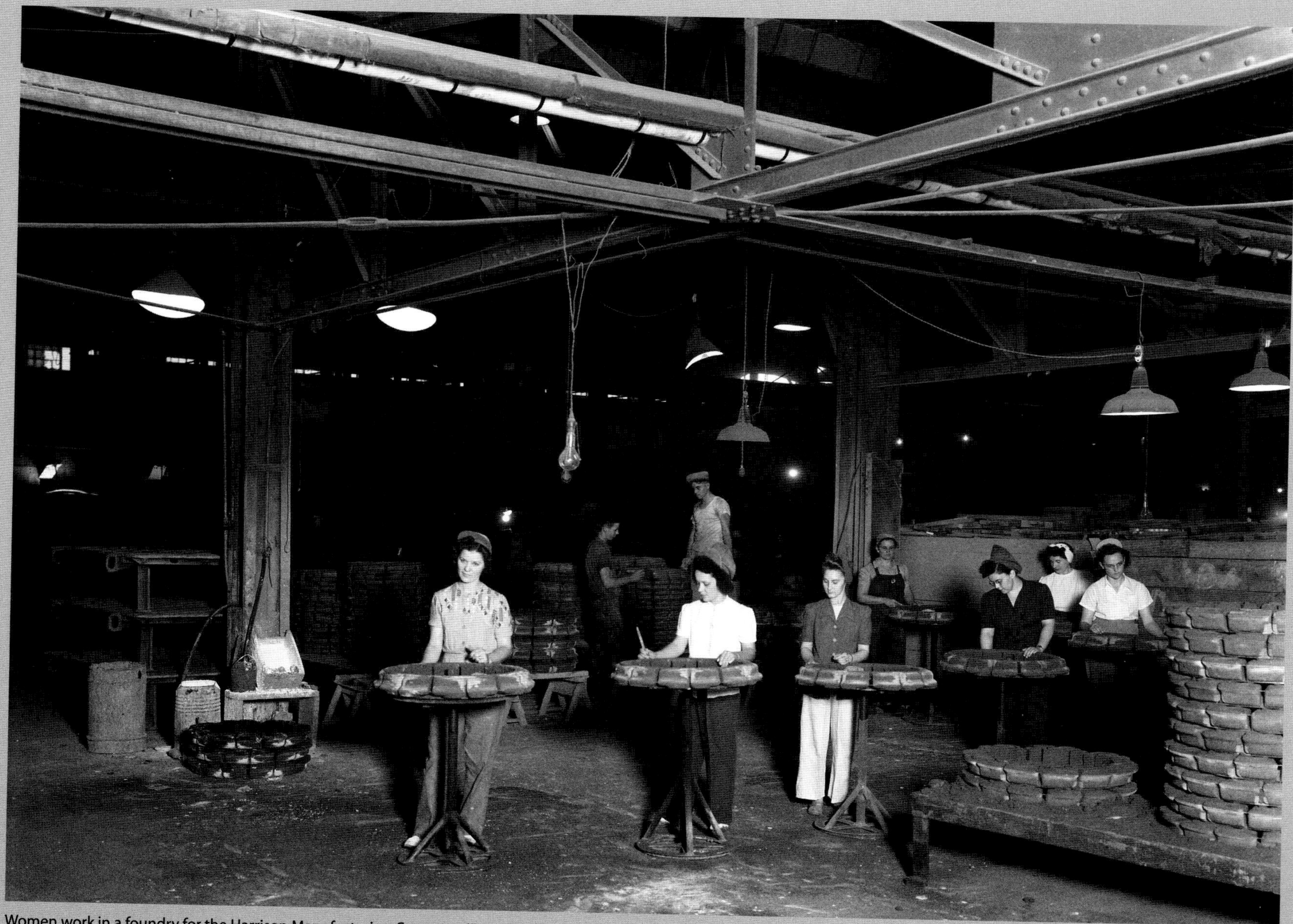

Women work in a foundry for the Harrison Manufacturing Company. (Fountain County, 1944)

FOREWORD

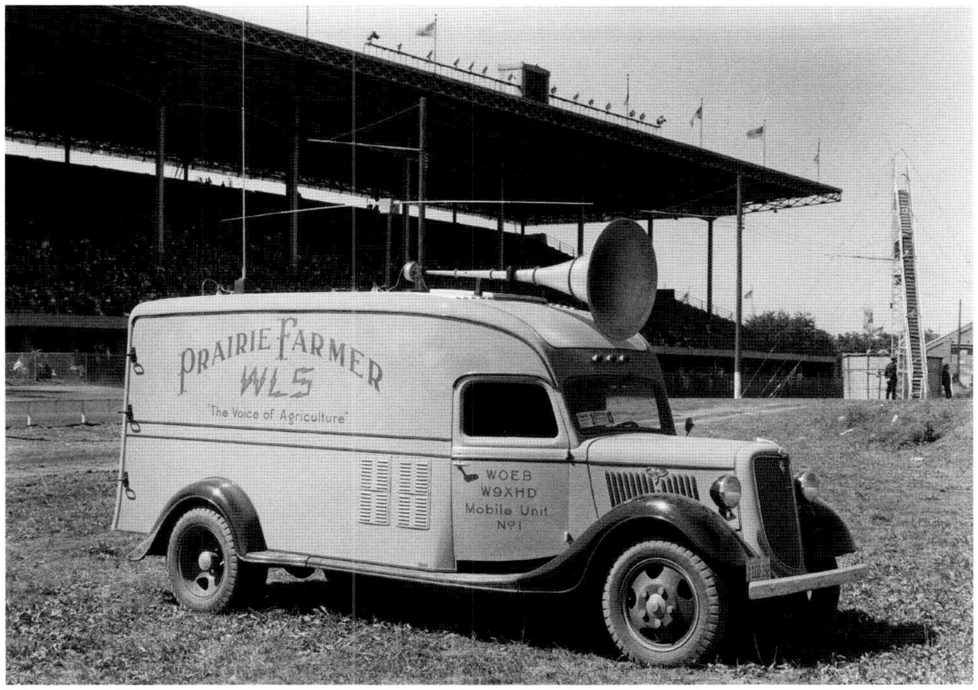

The *Indiana Prairie Farmer* broadcasting truck on location at the Indiana State Fair on September 6, 1935. (Marion County)

The farm magazine that attracted me the most growing up on a small dairy farm in Johnson County, Indiana, in the 1960s was *Prairie Farmer*. It certainly wasn't because I had a premonition that I would someday write for that magazine. Instead, I was 100 percent farm boy, and the big, front-page pictures of a farmer plowing or cows in a pasture always said "Indiana agriculture" to me.

Once I began writing for the magazine as a field editor in 1981—working alongside Tom Budd, the editor, and Carl "Indiana Ike" Eiche—I learned that many of those photos I liked so much were taken by a freelance photographer and later by his son. *Prairie Farmer* and, eventually, *Indiana Prairie Farmer* purchased many photos from J. C. Allen and Son for magazine covers during the mid- and latter part of the twentieth century.

Little did I know that Allen was a legend in his own right. After I had the chance to view hundreds of his photos, going back into the days of black-and-white photography, I understood why. Based in West Lafayette, Indiana, he traveled farm country with his camera equipment, taking photos of real farm people and real farm scenes. He captured everything from a young boy driving a very early tractor to farmers plowing, disking, combining, and working on equipment.

Many years ago, I served on a judging committee for the American Agricultural Editors' Association charged with picking out the best photos submitted by agricultural photographers that year. I soon learned that what I considered good photos and what many others caught up in modern journalism thought were good photos were quite different. They were into artsy-type photos. I was looking for genuine photos that tell a story, such as a herd of cows standing under a tree on a hot day or a farmer cultivating corn with an old Massey-Harris 44 tractor and two-row mounted cultivator with a young boy along for the ride.

J. C. Allen and Son captured hundreds of images that were award winning in my book. Taken together as a body of work over decades, they paint the story of U.S. agriculture during a large part of the twentieth century—a story that includes the people who sweated and worked hard to grow crops and raise livestock, and the technological changes they encountered along the way.

My hat is off to the authors who spent time reviewing hundreds of photos and piecing together another story: that of an unsung hero of American agriculture, John C. Allen. He preserved this history in photos as it unfolded before him. Now the authors have preserved it for you. Enjoy every page!

Tom J. Bechman
Editor, *Indiana Prairie Farmer*

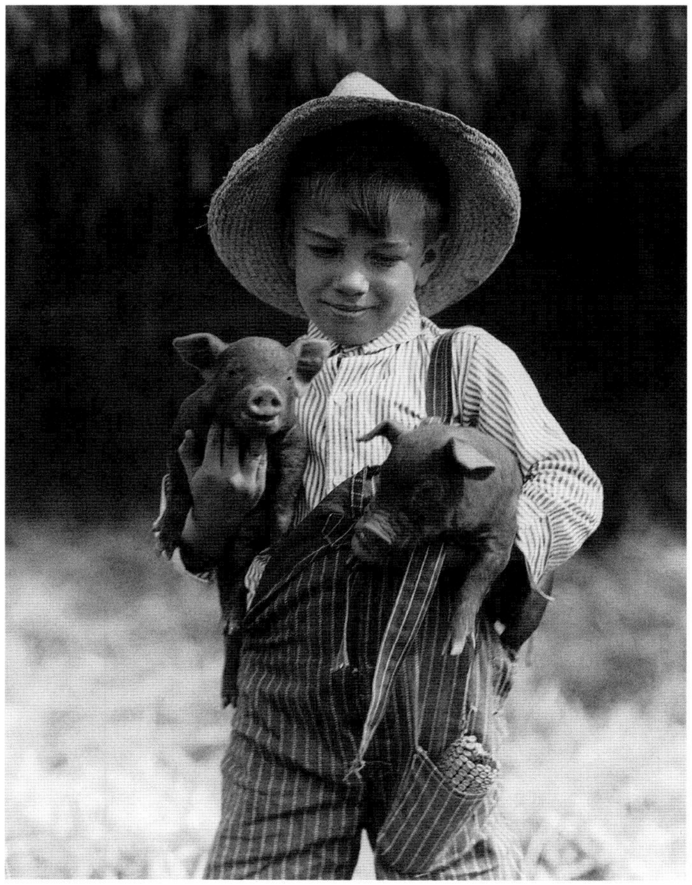

This picture of Chester Allen at approximately four years old holding two pigs was one of John Allen's favorites. (N.p., ca. 1911; courtesy John O. Allen)

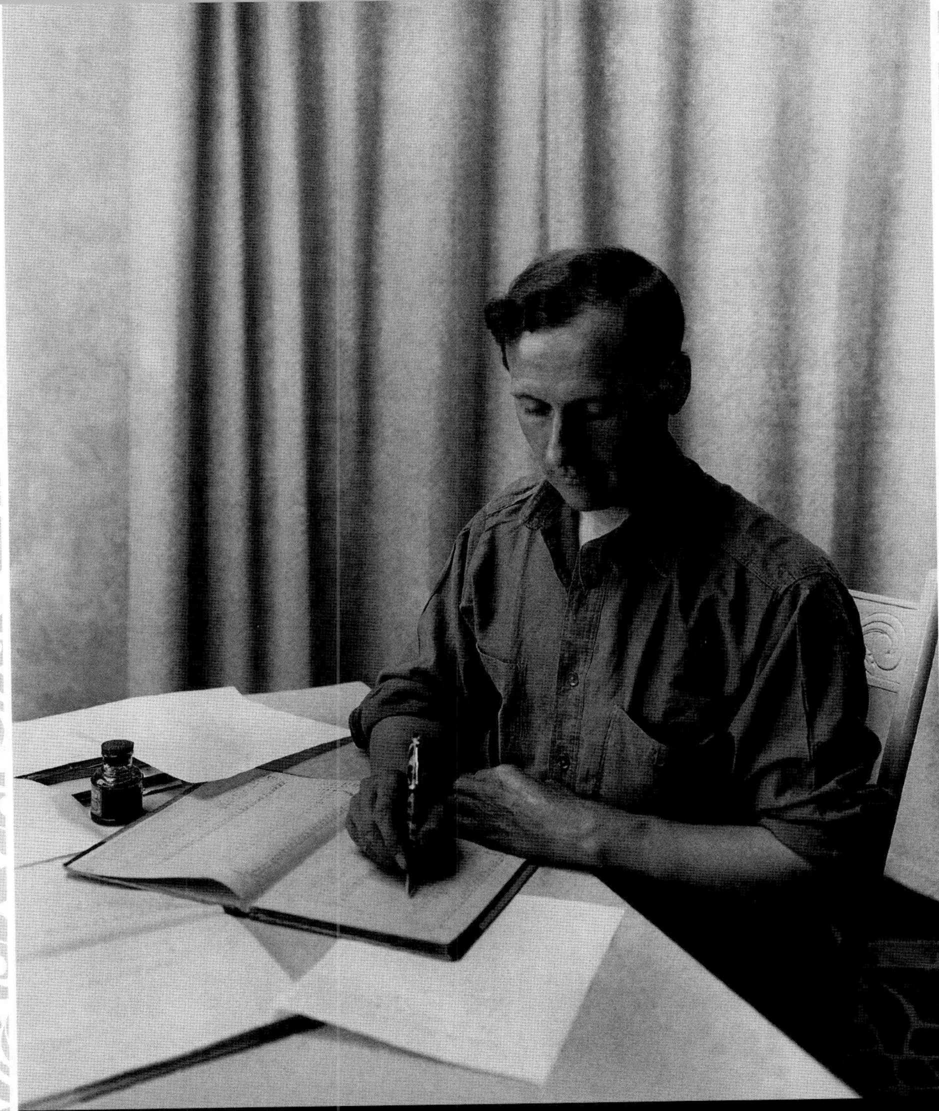

To be most successful in this work, I think one should be a perfectionist and satisfied with nothing less than the best he can do. My work has not all been easy, but it has been pleasant much of the time.

—JOHN CALVIN ALLEN

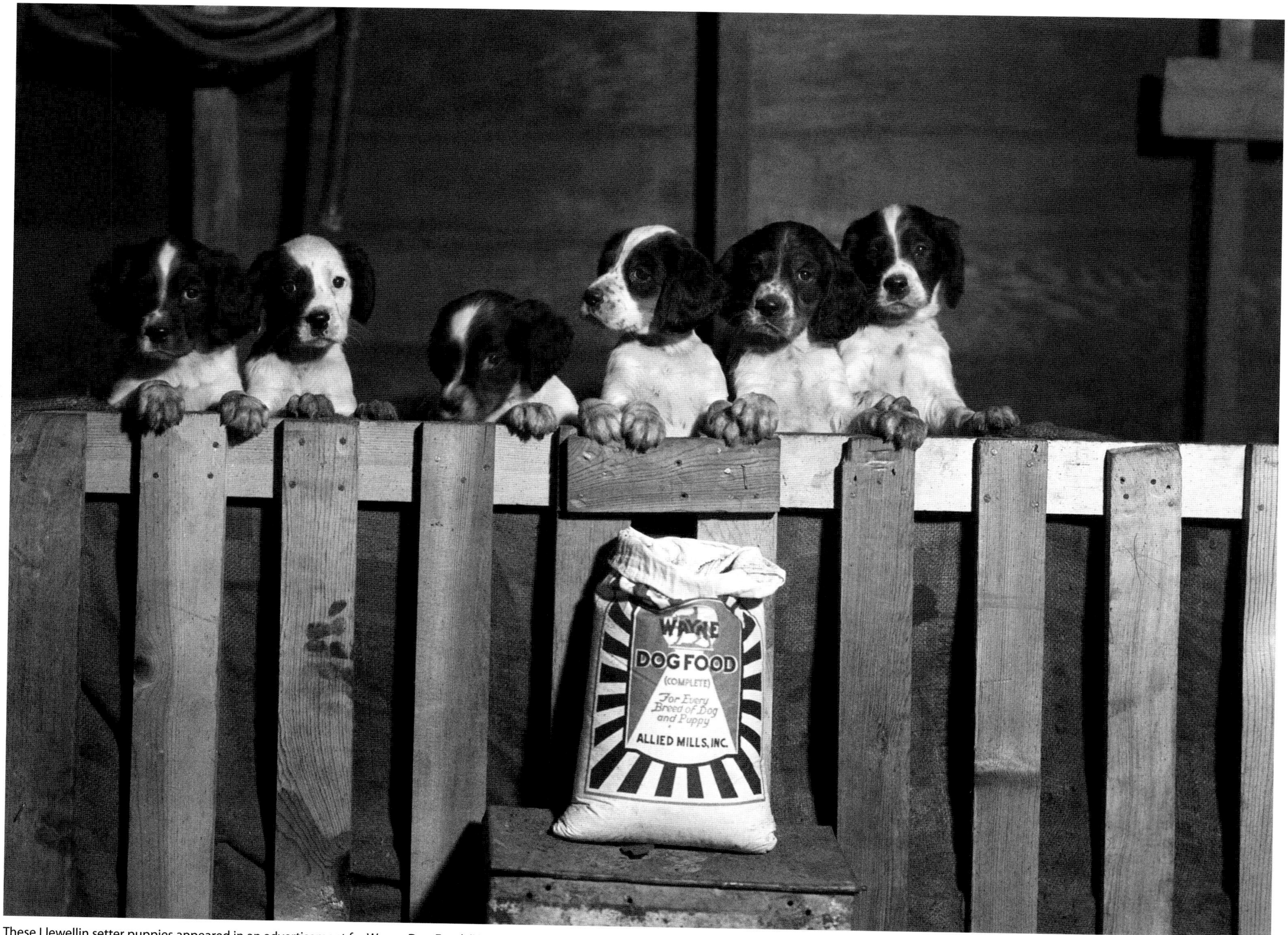

These Llewellin setter puppies appeared in an advertisement for Wayne Dog Food. (N.p., ca. 1933)

Introduction

Who among us has not spent hours studying photographs of family and friends or old sections of our town and countryside? Images transport us to another time, another place, another state of being. Photographs of young children playing in the yard, cradling yellow chicks, or starting school stir something deep within us.

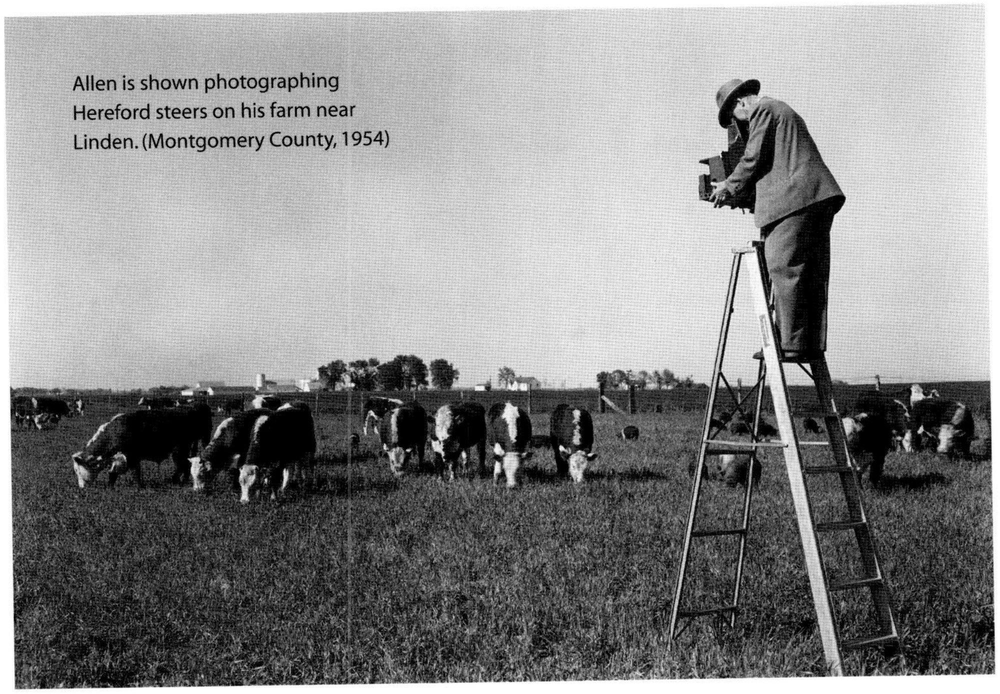

Allen is shown photographing Hereford steers on his farm near Linden. (Montgomery County, 1954)

Photographs capture events, memories, and even emotions, connecting us to those times, places, and moments. When the spoken or written word is unable to convey depth or significance, a photograph speaks. Photographs evoke wonderment, happiness, inspiration, pain, nostalgia, and loss—facets of the human sense and desire for links to our collective memory.

Images provide the tool to look into the eyes of people long gone and places long abandoned. The photographer documents an exact moment, fixing the person, place, or event within his or her scope. Indeed, though the photographer is not visible in the image, that person is a participant. The photographer perceives the scene and then defines it through the lens. The camera acts as a documentary instrument, with the resulting image serving as a proxy to the camera lens, binding the view of all who reminisce or study over a photograph to a fixed perspective. The photographer is a creator, and the photograph, a creation that is redefined with each new viewer.

John Calvin Allen, an amateur-turned-professional photographer, captured a rich collection of photographs depicting life during much of the twentieth century in Indiana—on the farm, in the cities, and at Purdue University—through his work. As a self-taught but highly successful photographer, his more than 100,000 images span all facets of Hoosier life. However, Allen consistently focused his lens upon farms and rural communities, resulting in a rare glimpse of agriculture and rural life at that time.

The J. C. Allen photographs represent a historical account of the transition from pioneer practices to scientific methodologies in agriculture and rural communities from 1909 until the early 1970s. During this major transitional period for agriculture, tractors replaced horses, hybrid corn supplanted open-pollinated corn, and soybeans changed from a novelty crop to one placed in regular rotation on most farms. It was also a time when purebred animals with better genetic pedigrees replaced run-of-the-mill livestock, and livestock producers adopted practices to promote systematic disease prevention in cattle, swine, and poultry. Looking closely at the tractors, livestock, wagons, planters, sprayers, harvesting equipment, and crops Allen photographed gives us a sense of the changing and fast-paced world of agriculture, captured then but experienced now.

But the reach of the J. C. Allen collection goes well beyond men working in fields, animals grazing on green and lush pastures, and participants who excelled at competitions such as the International Live Stock Exposition in Chicago or the Indiana State Fair in Indianapolis. Allen's photographs document clothing

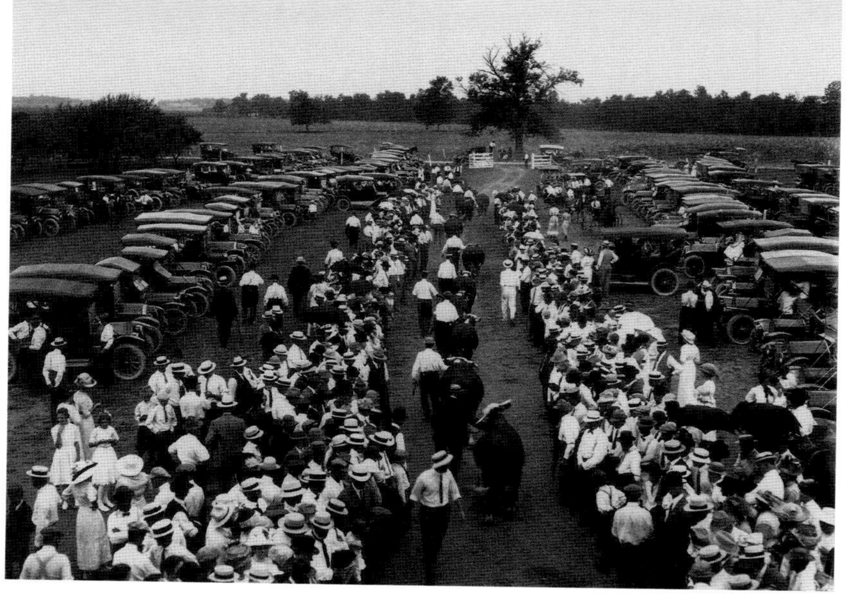

Allen had an eye for capturing crowds in action, such as this one gathered to view Hereford cattle on the Van Natta farm at Battle Ground in 1912. In order to capture this perspective, Allen would have ascended to a higher place, such as a grain bin. This vantage point shows that attendees were wearing their Sunday best. It also indicates that the automobile had supplanted the horse and buggy on many farms by this time. (Tippecanoe County)

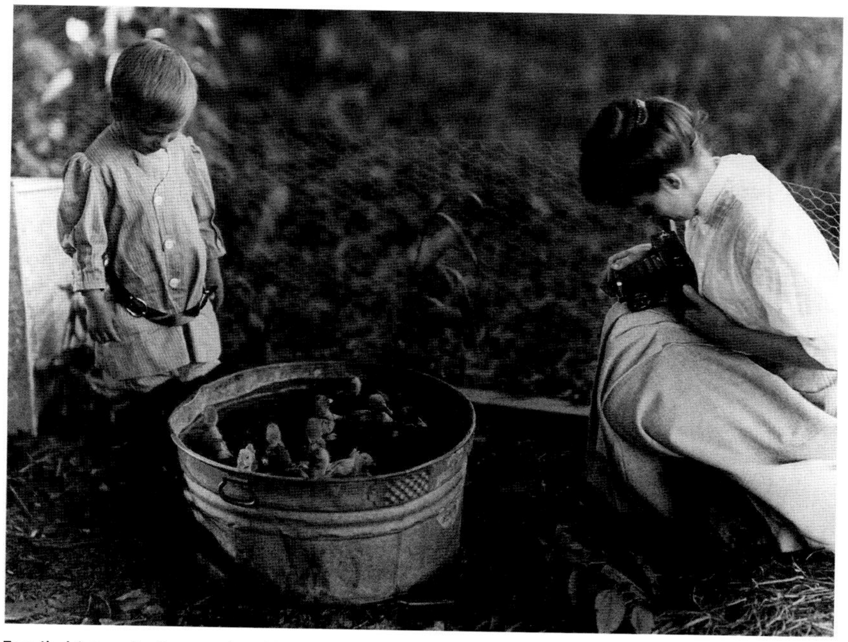

Family history indicates that Mary Abbie Allen taught her husband basic photography skills. Here, John Allen shows his wife taking a photograph of ducklings swimming in a washtub while Chester stands nearby. This photo-within-a-photo documents many things: an intimate family moment, Abbie as a photographer, and the greater scope of the image captured by her. (N.p., ca. 1912)

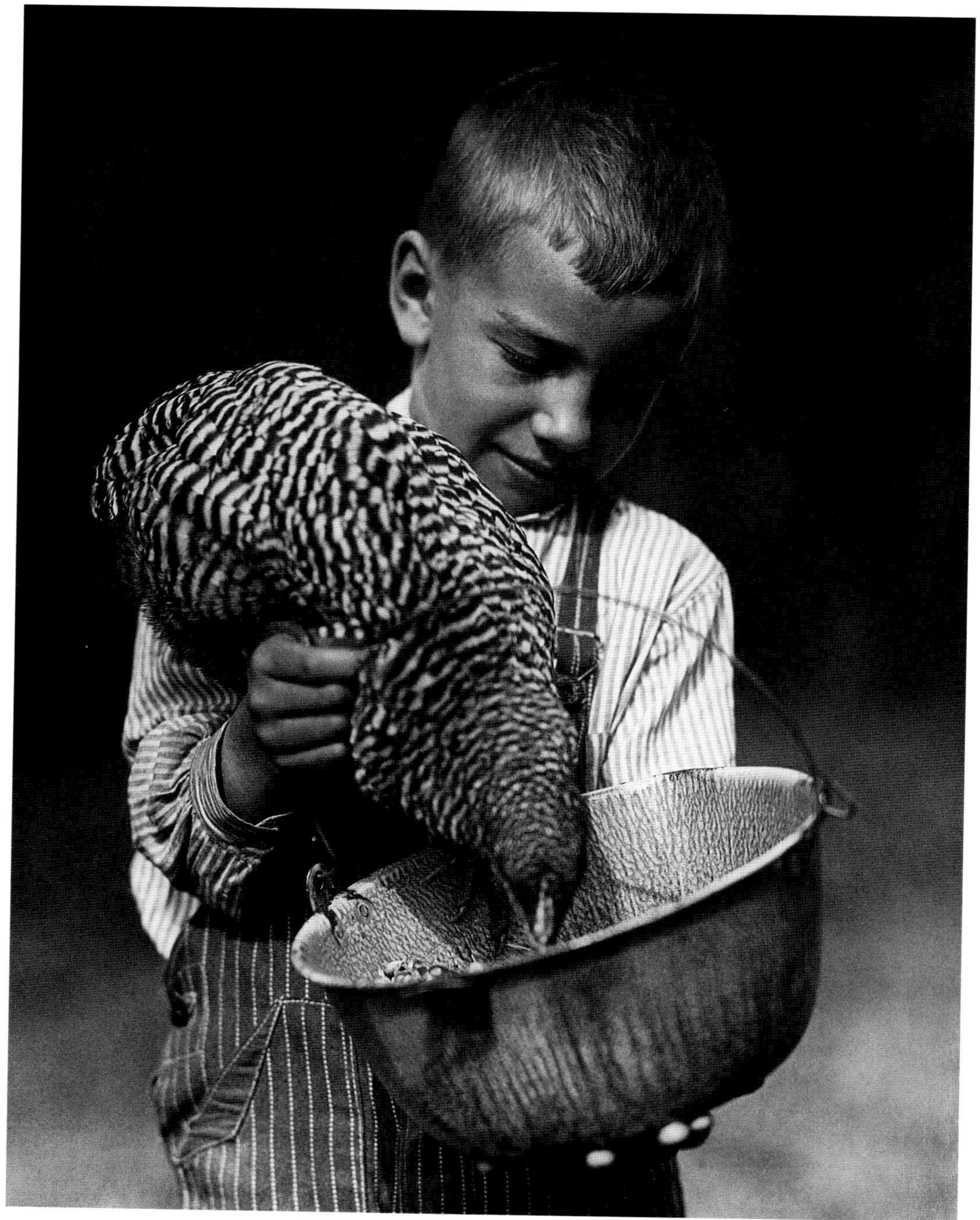

Chester feeds a Barred Plymouth Rock hen at his grandmother's farm near Clay City. (Clay County, ca. 1914)

styles, home furnishings, and the items people thought important as they went about their daily lives. They also record the transformation of the rural landscape and life there. Growing infrastructure would replace muddy and rutted roads with gravel or concrete, making it easier for horses and buggies—and later, automobiles—to travel to and from communities across the state. And the expansion of electric power beyond the city and into the most remote townships of Indiana allowed families to listen to the latest news, sports, and entertainment on their radios, rendering rural isolation a thing of the past.

As part of his work for Purdue University, Allen was able to document and preserve farms and homes, peoples and animals, and machinery and nature for decades to come. Using a heavy wooden camera, he captured the Hoosier and American experience in thousands of images, first on glass-plate negatives and later on film. His earliest images were in black and white but eventually shifted to color by the late 1950s. Little to nothing was off-limits to photograph. His snapshots of rural life depict men, women, and children doing their chores, but they also capture families playing and praying together, celebrating weddings and the birth of children, and honoring deceased loved ones.

Allen recognized that his photographs were more than pretty pictures. He described and documented events and places in detail, storing negatives and prints in specialized sleeves with meticulously typewritten labels. Print

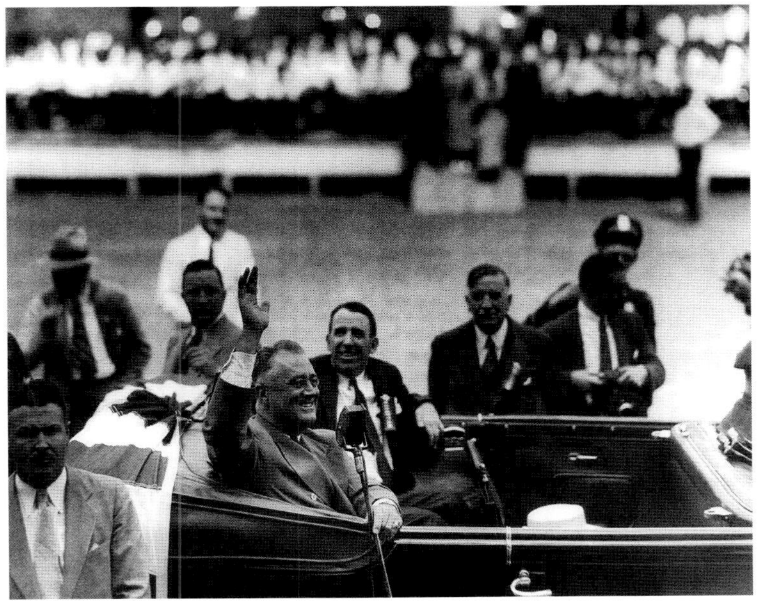

Allen photographed President Franklin D. Roosevelt in Indianapolis as he spoke into a microphone from his car in 1936. (Marion County)

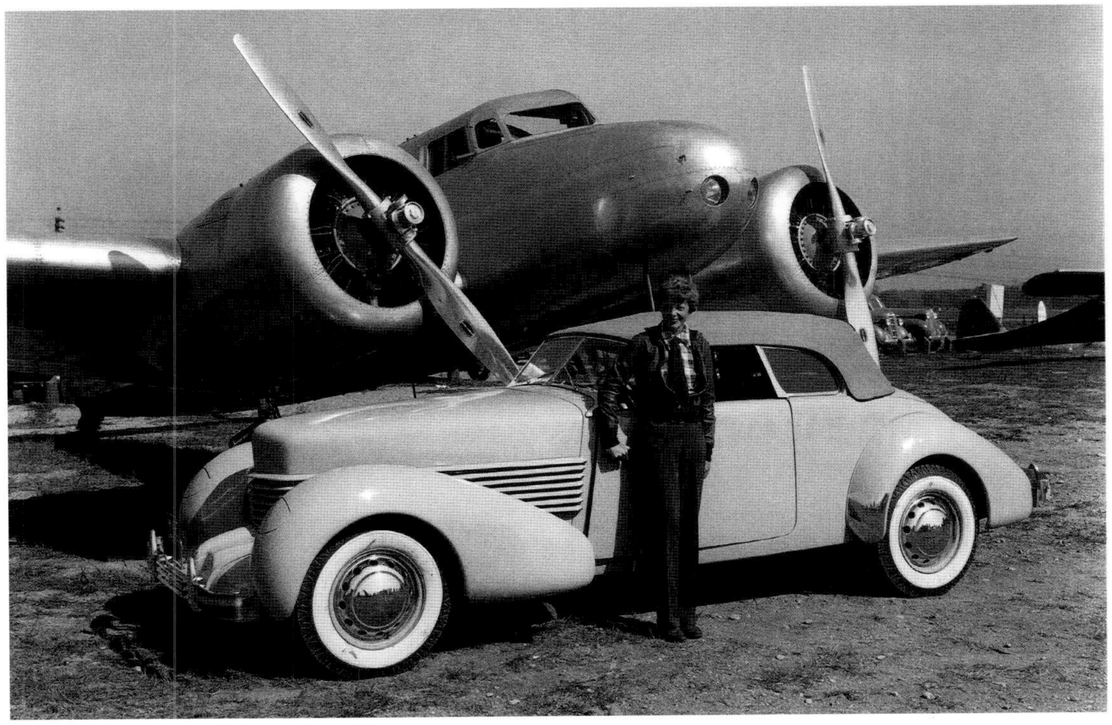

Aviator Amelia Earhart stands in front of her new Cord automobile and Lockheed Electra airplane at the Purdue University airport. Allen would have had the opportunity to photograph Earhart during the time she spent on the Purdue campus in West Lafayette. (Tippecanoe County, 1936; courtesy John Allen Benham)

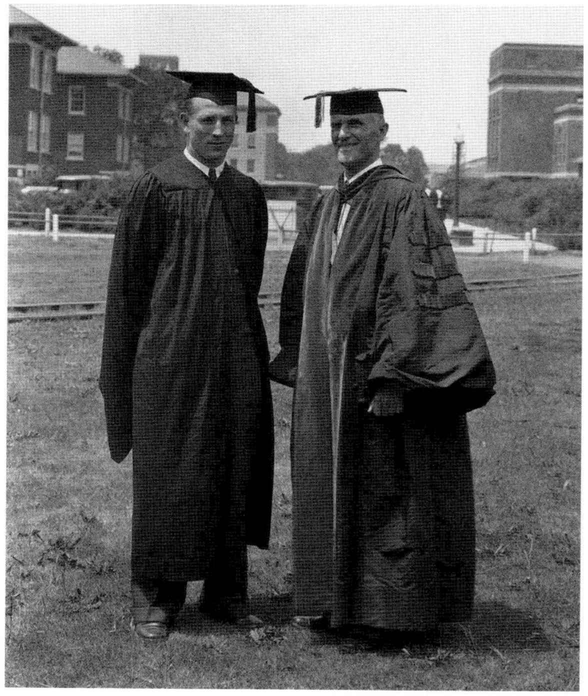

In this photo from the J. C. Allen Collection, Purdue President Edward C. Elliott (right) poses with John Wooden, who received the Big 10 Medal for scholarship and athletic achievement at his June 1932 graduation. Wooden would go on to become the legendary basketball coach for the University of California at Los Angeles. (Tippecanoe County)

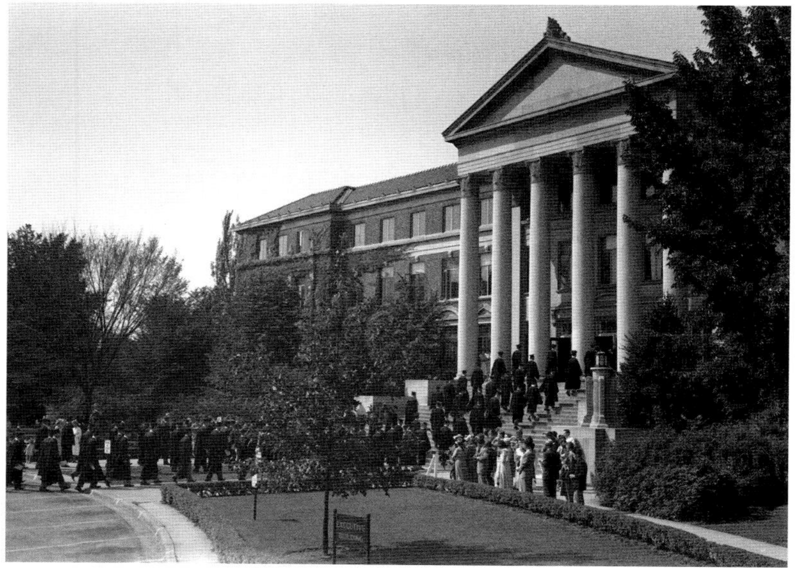

The spring 1949 graduating class at Purdue University marches into the Executive Building—today known as the Frederick L. Hovde Hall of Administration—for its commencement exercises. (Tippecanoe County)

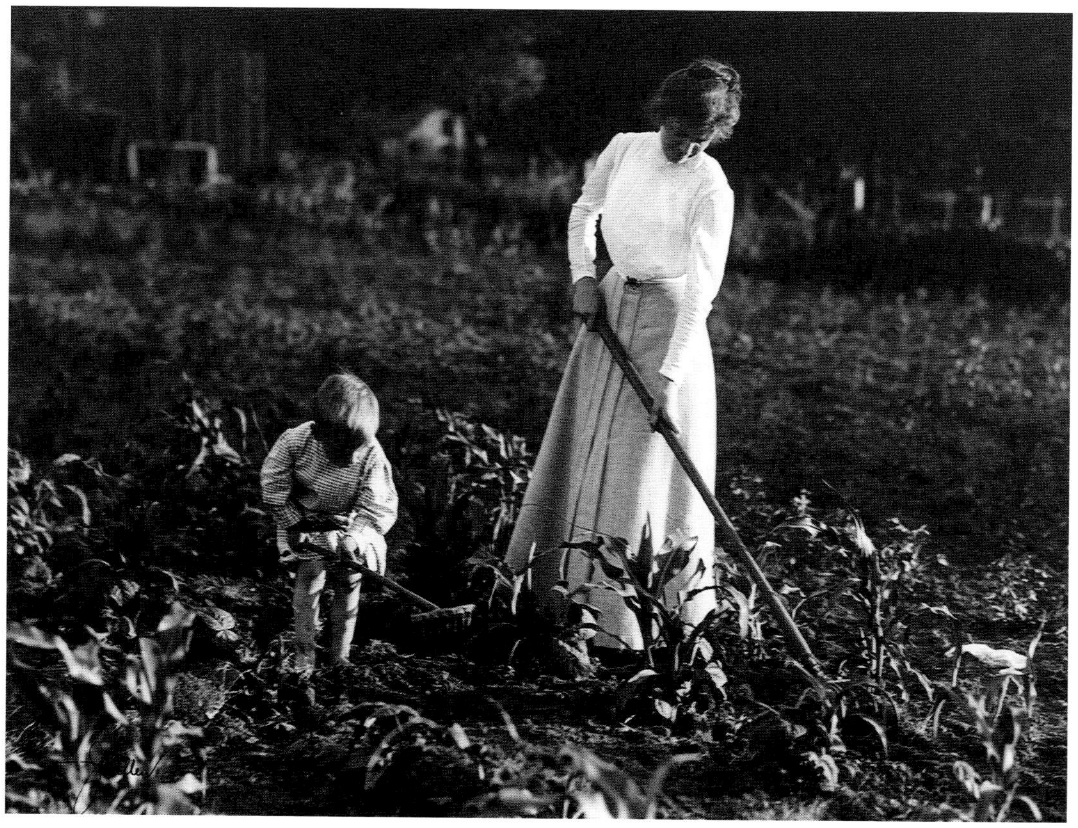

Due to the popularity of family photographs among magazine editors, Allen liked taking pictures of his own family members. Here, Chester Allen (left) works alongside his mother, Abbie Allen, in a garden. (N.p., ca. 1911)

photographs were often labeled with handwritten notes that included individual names, dates, and extra details thought pertinent. A historical photograph without accompanying information may grab one's attention, but that same photograph accompanied by the who, what, when, where, why, and how provides knowledge and evidence. The latter makes each photograph priceless.

There is no indication that Allen intended for his photographs to carry historic significance while he was taking them, but near the end of his life, he realized his collection had captured agricultural history and rural life like no one individual before or after had done.

A Photographer's Record

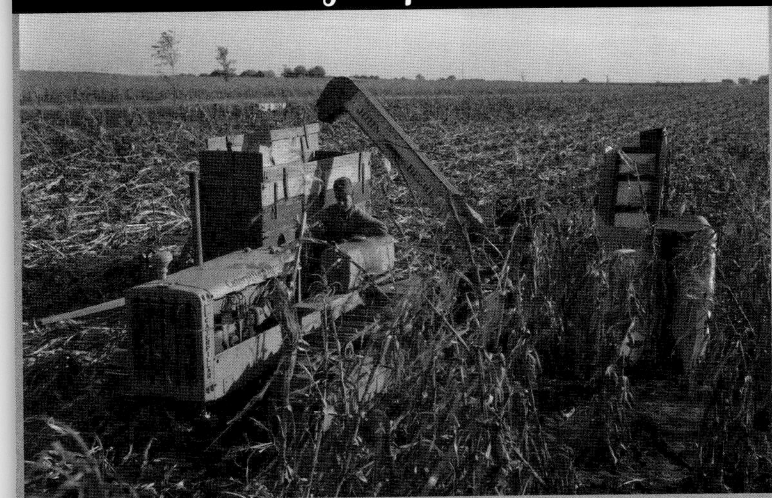

John Allen kept meticulous notes on his photographs. He recorded the usual information, such as the date, location, equipment, and name of the person pictured. But as this particular data sheet shows, he often documented additional information, which here included details about how the equipment had been used, the amount and type of fuel required to run it, and even how the owner/operator had paid for it. When taken as a whole, his photographs and detailed records provide a valuable historical account of rural life during the early to mid-twentieth century. (Tazewell County, IL; 1936)

Photographer's Data Sheet

J. C. Allen and Son, West Lafayette, Indiana
(NAME AND ADDRESS OF PHOTOGRAPHER)

DATE TAKEN Oct. 19, 1936 "CATERPILLAR" MODEL RD 4

EQUIPMENT ~~Two row~~ John Deere Corn picker PTO
(GIVE MAKE AND MODEL, CAPACITY, ETC.)

LOCATION J. P. Werner farm, Tremont, Ill.
(GIVE NAME AND LOCATION OF JOB)

OWNER Schurter Bros. NAME OF OPERATOR Henry Schurter

P. O. ADDRESS Box 21, Tremont, Illinois.

DESCRIPTION OF WORK Picking corn

NUMBER OF HOURS TRACTOR HAS OPERATED 1600 hours

PRODUCTION (Rate of doing work) Picking (900 to 100 bu.) a day 16–20 acres

FUEL Deisel Fuel Oil COST PER GAL. 7¢ GALLONS PER HOUR 10 gal for 15 acres

OWNER'S OR OPERATOR'S REMARKS (Write on reverse side)

DO NOT WRITE IN SPACE BELOW

PASSED BY B. O. P. R. DATE SUBMITTED ACCEPTED

TESTIMONIAL? YES ☐ NO ☐

RELEASED FOR: NEWS ☐ PUBLICITY ☐ LITERATURE ☐ GENERAL ☐ (SEE OTHER SIDE FOR REASON)

NUMBER ASSIGNED

The owner of this tractor is a Farm Power Contractor. He plows, discs, and does all kinds of farm work and also does some road work. He plowed about 1000 acres in 1936. Henry Schurter, who is only 20 years old said he had paid for the outfit in 6 months of custom work.

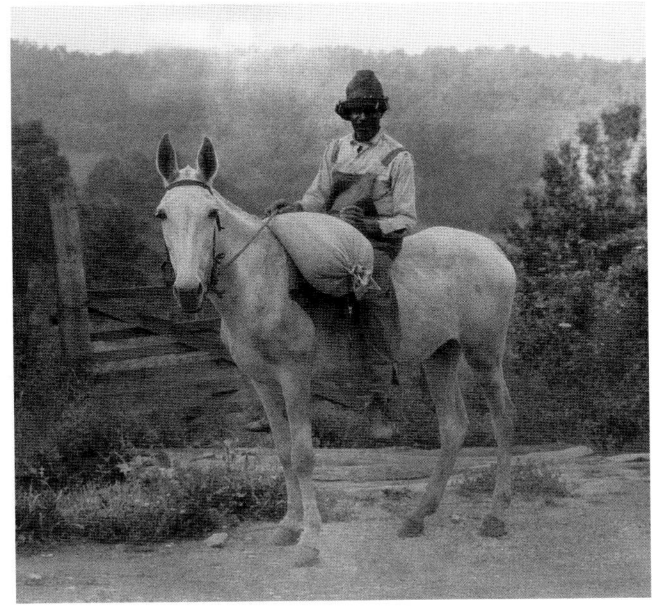

In 1928, Allen took this photograph of Lewis Gardner from Horse Creek, Kentucky, carrying grain to a local mill on the back of mule. (Hart County, KY)

His legacy is forever linked to the tens of thousands of images he amassed during his fifty years of traveling the back roads of Indiana and the Midwest. We, the authors of this work, present these selected images from the eyes, mind, and camera of John Calvin Allen. We hope our readers enjoy this journey to moments long since passed but readily reimagined into existence in all of our minds, hearts, and souls.

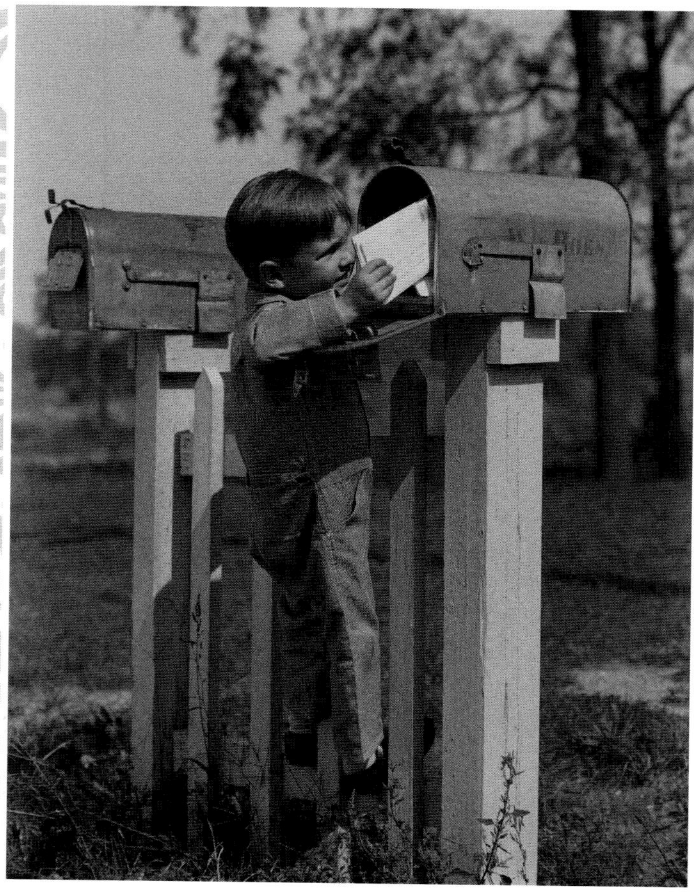
John Bullard works hard to get the day's mail. (Tippecanoe County, 1934)

The Early Life of John C. Allen

In his youth, John Calvin Allen did not aspire to become a photographer, but the twists and turns of life led him there nevertheless. Influences included having a father who was physically unable to work, growing up in an orphanage, developing a genuine interest in livestock, drawing a meager salary as a Purdue University clerk and stenographer, and living through advances in printing technologies that allowed magazines and newspapers to reproduce photographs in print. Collectively, these events changed the personal and professional life of Allen from farmer to photographer.

Even though the American Civil War predated Allen's birth by decades, its effect on his father, James Thomas Allen (1844–97), would directly impact John's young life. Twenty years before John was born, his father was a working farmer and practicing harness maker. Like most of his friends, James joined the Union Army, enlisting in June 1862 with the Fifty-Fourth Indiana Infantry Regiment as a private for a three-month stint. By the end of September, eighteen-year-old James returned to farming in Indiana.

Most Northerners expected the southern Confederates to fall quickly to the well-armed Union army, but the war would continue until 1865. By 1864, the high number of casualties had taken a heavy toll on troop numbers for both armies, and each side needed new recruits to replace the tens of thousands of men who had died or been wounded on the battlefield. The United States offered recruits an additional $100 for one year's service in the Union army. Whether it was patriotism or the money or both, in October 1864 James once again became a soldier.

During this second tour of duty, James and the Thirty-Third Indiana Infantry Regiment were assigned to General William T. Sherman's army, where they would become accustomed to long marches and hard fighting.[1] As a low-ranking foot soldier, James marched to Nashville, Tennessee, where the Union army engaged and defeated the Confederate forces of General John Bell Hood at the Battle of Nashville in the second week of December 1864. As the year drew to an end, James recalled, "[W]e suffered great exposure in forced marches through swamps and sloughs wading until we went in camp at a little town in Alabama called Courtland about New Year's Day of 1865."[2] By the time they arrived, he had marched 130 miles.

During his days in Alabama, James became chronically ill. He was sent from the front lines to a hospital in Chattanooga on January 6, 1865, and soon thereafter to Cumberland Hospital in Nashville, Tennessee. He was diagnosed with hepatitis, rheumatism, and chronic diarrhea. With the Civil War drawing to an end, he was honorably discharged in April 1865.[3]

Returning home, he found work as a clerk in a dry goods store. A railroad accident on December 27, 1877, resulted in his lower right leg being amputated below the knee. Poor in health and now disabled, James went back to making harnesses as well as barrels, buckets, and churns.[4] By 1878, thirty-four-year-old James was no longer physically able to work and earn a living to support his family.[5] In 1893, at age forty-nine, he was admitted to the U.S. National Home for Disabled Volunteer Soldiers—today part of the U. S. Department of Veterans Affairs—at Marion, Indiana. He would eventually die of heart failure in 1897 at age fifty-three while visiting his son in Clay City, Indiana.[6]

John Calvin Allen was born near Darwin, Illinois, on September 11, 1881. His father by this time had remarried and, with John's arrival, now had six children between his first and second marriages. In 1883, tragedy struck the Allen family when John's mother, Joan (1851–83), died at the age of 32. Years later, he would write, "My mother died when I was 18 months old. Father broke up housekeeping and we . . . children were never together again as a family."[7] John's paternal grandmother and an aunt and uncle, Albert and Jennie Lane, took him in to live on an eighty-acre farm roughly eight miles south of Sullivan, Indiana.[8]

The young boy adapted well to his adopted family. He said he loved his uncle like a father and explained:

> I had not lived with my father since I could remember. . . . Those were mostly trouble-free, enjoyable days although we were much poorer than most anyone at the present time. I do not remember being very hungry. It seems we had plenty of cornbread and milk. I do remember getting a half-stick of candy in my shoe at Christmas time. If we went to town we drove two farm work-horses [hitched] to the big wagon over dirt roads, often very muddy and sometimes very dusty.[9]

He and the others slept in a one-room log cabin heated by a fireplace to keep them warm during cold winters. Adjacent to the living quarters was an adjoining kitchen, where meals were eaten on a long wooden table. When he was old enough to attend school, he went to a nearby one-room schoolhouse, where the teacher was Jennie Lane's brother, William A. Curtis.

Then the still-young John Allen experienced a life-changing event. He recounted, "During my last year on the farm . . . I was a happy boy on the farm, then 'the roof fell in.' I was almost six years old when my father, who had been a Northern soldier, came and took me to the Indiana Soldiers' and Sailors' Orphans' Home near Knightstown, Indiana."[10] Why his father took such a drastic measure may never be known.

The institution cared for orphaned, homeless, or impoverished children of Civil War Union soldiers. By the time John was admitted in 1888, the home had become the guardian of boys and girls whose parents were unable or unwilling to care for their children.[11]

John's father wrote on the May 1888 application, "Said applicant [J. C. Allen] is destitute of means of support and education, and the father of said applicant is a permanent cripple having lost his leg by accident since the war."[12] Just a few weeks later, the six-year-old boy was brought to the home, where he told his father good-bye and remained for the next ten years.

John Allen slowly adapted to life at Knightstown, writing, "At first this was a very bitter experience. I was but a small boy and very homesick among larger boys. Somehow I gradually got over it."[13] The boys were issued military uniforms and marched everywhere they went. Allen noted, "At the Orphans' Home we were divided into groups of about 30 and each group marched most everywhere they went, to breakfast, dinner and supper, to school and to chapel. We wore blue uniforms with knee pants and brass buttons and we lined up two and two with the smaller boys in front and had to keep in step."[14]

The children attended classes, where Allen began learning the three Rs: reading, writing, and arithmetic. He observed, "The first year in school I was at the top of the class. The teacher, Miss Laura Walkstetter, showed me the only affection that I can remember when I handed her my slate with what must have been the correct answer."[15] It seems what the young man needed most in his life were kind words and any signs of affection. He recalled, "One nice spring day it was my job to plow part of the home garden with two fine horses and a breaking plow. I was proud of my job, the soil was turned over beautifully and I tried to see how smooth and even I could make it. I got my reward when the old Dutch gardener told my boss, '[I]t don't take the big boys to do a good job.'"[16]

Allen recounted another story about his stay from 1893:

> When I was about 12 years old I had an experience that changed my life. A Mr. and Mrs. John Allen from Illinois visited the orphan's home and because of the same name came to see me. Mrs. Allen especially was a very fine

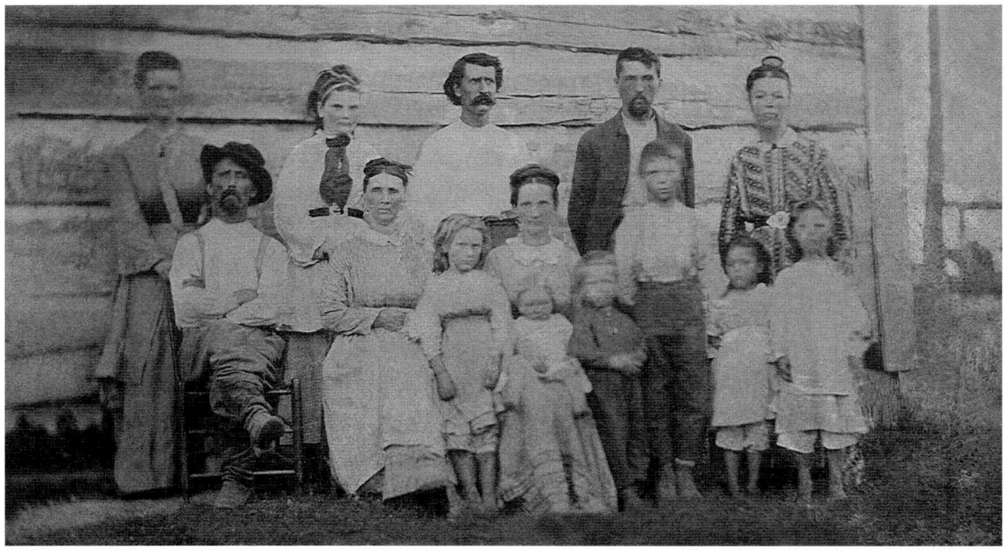

This image is one of the few surviving ferrotypes, also known as tintypes, that include John Allen's parents and older siblings prior to his birth in 1881. John wrote on the back that his mother, Joan, was seated in the center holding daughter Alma in her lap. Another daughter, Cora, stands to her right, while son James stands to her left. Directly behind Joan is Allen's father, also named James, who has a mustache and is wearing a white shirt with no jacket. (N.p., ca. 1874; courtesy John O. Allen)

> person and as I remember it was she who gave me Mr. Allen's own personal Bible. Swearing was a common language among many of the boys, including myself; but gradually I broke the habit. I began going to Christian Endeavor [a non-denominational evangelical society] and took the pledge, a part of which is, "[T]o the best of my ability I will lead a Christian life."[17]

This promise in 1893 that he made to himself would be a trait that his own children and grandchildren would reiterate—swearing and drinking were never allowed in the Allen home.

The older children were tasked with learning a trade they could use when they left the orphanage. Allen recounted his experience with this:

> At the age of 13 each boy or girl was required to learn some kind of a trade by going to school half-day and working half-day. I was either selected or chose (on my own) to work on the dairy farm where in the afternoons I helped feed the dairy herd and I milked two of the cows morning and evening. To stick one's head into the flank of a good dairy cow on a cold winter day, with a bucket between your knees and milking with both hands helps to develop the wrists and is an excellent experience for a growing boy.[18]

It wasn't all work at the school. Allen enjoyed recreational activities at the Knightstown school, too, saying:

> There were fun and recreation periods at the home. A reservoir or lake nearby often froze over in winter and we had excellent skating. I was a good skater and enjoyed playing cross-tag and shiny [ice hockey]. A very beautiful clear creek with rock bottom, not too far away was where we went swimming when we could. The better swimmers often went to Blue River almost two miles away. It was a beautiful stream in those early days. Both places were "out of bounds" but we somehow managed to get to them. I was not too good at baseball, but at marbles I was one of the best. We played for keeps and I usually had a pocket full of "chalkies" [unglazed clay marbles] to prove it.[19]

Allen enjoyed some time away from the school visiting family members in the summer, recalling, "One pleasant redeeming feature during those ten years was that each summer I spent a two-month vacation either with my sister in Terre Haute or with my uncle and aunt who had moved to a small farm near Shelburn, Indiana."[20] Based on where he took his vacations, it does seem that father and son had permanently severed their bond. The young boy enjoyed spending time with his uncle's black rat terrier dog, noting that they were "very close friends." He also described a ride he took on the Pennsylvania Railroad, saying that "late June showed the woods full of wild roses in bloom."[21]

Allen did well in school and was a diligent worker on the farm. During his last year at the home, he was able to take a course in business, where he learned typing and shorthand, and became a proficient stenographer. Nearly sixteen years old now, he received his high school diploma in 1897. He shook hands with teachers and told his friends good-bye. As an adult, he would write, "I was in the Orphans' Home for ten years and although I didn't realize it at the time, this was without a doubt a very good preparation for later life. I learned at a very early age to depend upon my own efforts."[22]

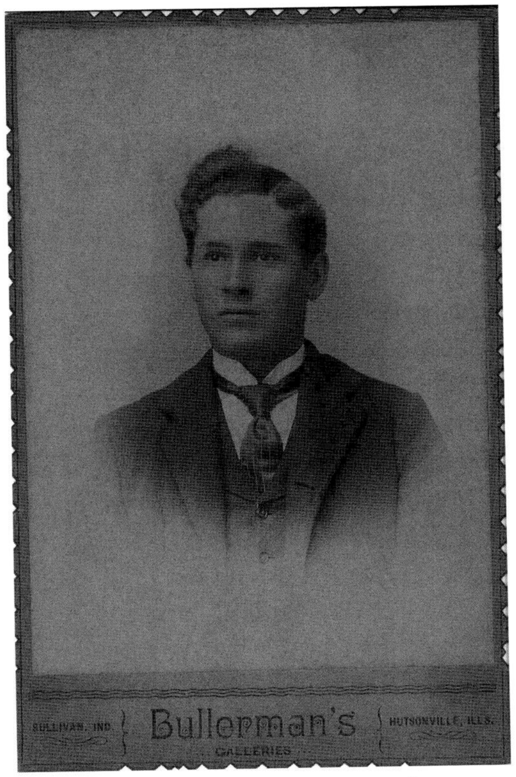

A rare photograph believed to be John C. Allen as a young man of sixteen years old in 1897. That year, he left the Indiana Soldiers' and Sailors' Orphans' Home near Knightstown, Indiana, to start life on his own. (Sullivan County; courtesy John Allen Benham)

He would return in 1929 to take photographs of the dairy herd at the orphanage.

As the teenager left school with a set of new clothes and fifteen dollars in his pocket, he must have wondered what he would do with the rest of his life.[23] He first went to live on his uncle's farm. After the crops were harvested that fall, he took a job on a crew that built railroads. For his effort, he earned $1.35 a day. He started as a nipper [who worked with the rails as they were being assembled on the track] and water boy, and recounted, "[B]efore I finished I was driving the spikes that hold the rails to the new wooden ties—a man's job."[24]

With cold weather halting construction, he went to work in a coal mine, using mules to haul carts through the mine to an area where the coal would be brought to the surface. He recalled, "After one trip, I remember going back and finding a pile of rock on the track big enough to have buried me. I don't remember being afraid. We just cleaned up the rock and went on hauling coal as usual. The man who had the mine leased failed to make a success of it and I lost my last week's wages."[25]

Allen spent much of 1898 working near his uncle's property for a farmer who provided him room and board and thirteen dollars per month in wages.[26] Eventually, his older brother, James Russell Allen, who at the time was practicing dentistry in Clay City, told his younger sibling that he had a spare room in his home and the possibility of a job. John Allen wrote about that opportunity, which was with the Evansville and Indianapolis Railroad, saying:

> [I would be] sweeping the floors, handling the freight, carrying the mail to the post office, delivering telegrams, etc. I would be given the opportunity to learn the telegraph and learn other business connected with running the railroad office. There were no trucks in those days and everything that came to Clay City and the surrounding country came in over the railroad. The railroad office was a busy place.[27]

Eventually, he also served as a relief telegraph operator and agent at multiple locations on the line.

By age 20, Allen sought out and took a job as relief telegraph operator with the Baltimore and Ohio Southwestern Railroad in Washington, Indiana.

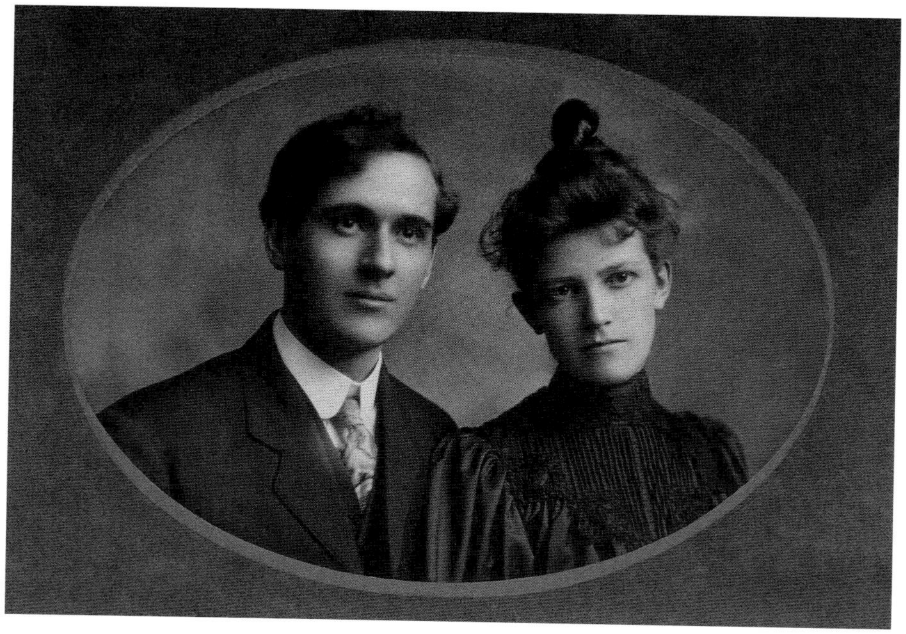

(Above) John Allen and Mary Abbie Peavey Allen around the time of their marriage in 1904. (N.p.; courtesy John Allen Benham)

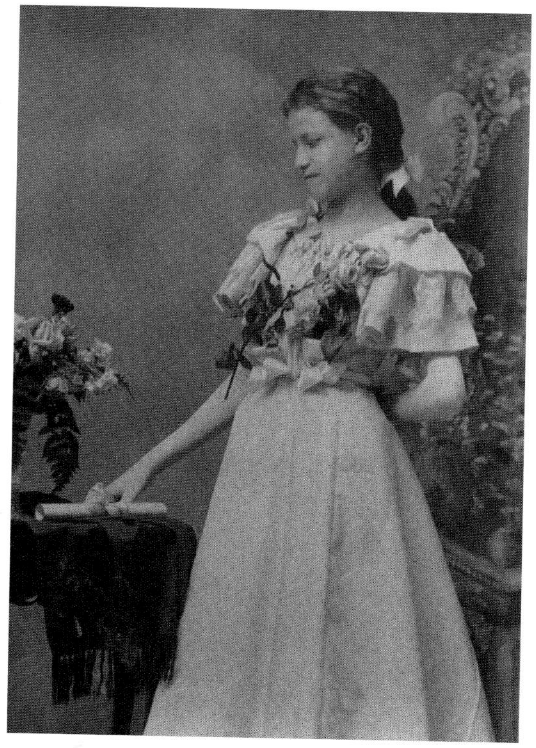

(Right) Mary Abbie Peavey stood for this photograph during her youth. The posed nature of the photograph was common during the late nineteenth century. (N.p., n.d.; courtesy John Allen Benham)

Eventually, he served as a telegraph operator in Oakdale and Rivervale, Indiana, where he assumed additional duties as ticket agent, express manager, and Western Union manager.

Allen distinguished himself at the Baltimore and Ohio Southwestern and by 1902 was made a train dispatcher, which he called "the best telegraph job on the railroad."[28] He described himself as "probably the youngest dispatcher in the USA" and noted, "There probably is nothing better than a job bigger than you are to help to make a young man develop. You either make good or else, and I wasn't looking for the 'else.'"[29]

Allen's good fortune continued when Mary Abbie Peavey (1881–1980), who was called Abbie, agreed to marry him in 1904.[30] As the then twenty-three-year-old Allen recalled, he had married "the best girl in Clay City, Indiana, and we spent our honeymoon at the St. Louis World's Fair, much of the time with my bride and her cousin, a red head. I do not recommend this kind of honeymoon."[31]

At least one account indicates that Allen's lifelong passion with photography had its origins in his visit to the fair.[32] He had purchased for his wife a four-by-five Eastman Kodak camera, which was not common at the time.[33] After learning how to use it, she would teach him how to use it and print his own photographs.[34] By experimenting with the camera and subjects, "he gradually improved his technique and became an expert at the art," taking photographs that captured the attention of those who saw them.[35] Little did he know that this hobby would soon turn into a successful business.

The Allens lived for a short time in Washington, Indiana, where John managed the 3 to 11 P.M. shift as the telegraph operator. He recalled that "things went smoothly for some time. I had a very good salary for those days and passes on the railroad that we often used."[36] But following the arrival of his son, Chester, in 1907, John wrote, "[I]t did not take me many years to decide that I didn't want to work nights all my life seven days a week, and when my father-in-law offered me an opportunity to join with my brother-in-law and operate a 600-acre farm in Eel River bottoms, I decided to try it."[37]

This farm—which included chickens, pigs, horses, cattle, wheat, hayfields, orchards, and watermelons—would have been one of the larger farms in the state at the time.[38] Chester later wrote that he was surprised about this change, saying, "I don't know how they did it but Mother's family talked Dad into leaving his railroad job and taking over the family farm which my Grandmother's parents had left to her."[39] But Chester also noted that "Dad loved agriculture and entered the farming project with big hopes."[40]

Unfortunately for the Allens, poor drainage meant sections of their farm were often underwater when the nearby river flooded, making John's reentry into farming more or less short-lived. He observed, "The farm was not well tiled, we had a wet season, and as one might guess, the farming arrangement did not work out well."[41] With no levees or dams to manage the water, fields often flooded.

When a third year of flooding struck the farm and money was hard to come by, John became discouraged and worried about taking care of Chester: "I gave the farm trial credit for getting me out of the train dispatcher's job. It took something rather unusual to cause me to give up a good interesting and exciting job with a good salary. As a lifetime occupation, it did not have the possibilities that I [had] later on in the photo-illustrating business."[42]

Allen returned to the railroads, working first for a railroad in Hymera, Indiana, and later in the freight office on the Chicago and Eastern Illinois Railroad at Hoopeston, Illinois.[43] It seemed that he was destined to be a railroad man for the rest of his life.

Two years passed quickly. While he enjoyed the work and the pay that went with it, he hated working evenings. He wrote, "After a few years of railroad work I decided I didn't want to sit up nights and dispatch trains all my life, so I went to Purdue University as a clerk and secretary in the agricultural experiment station" in 1909.[44] At twenty-eight years old, Allen missed the outdoors and being on a farm, so he accepted the work in the Animal Husbandry Department at Purdue, saying, "I thought this might give me a chance to learn what others knew about agriculture and farm livestock."[45] Unbeknownst to him, the twists and turns of life were still pushing him toward photography.

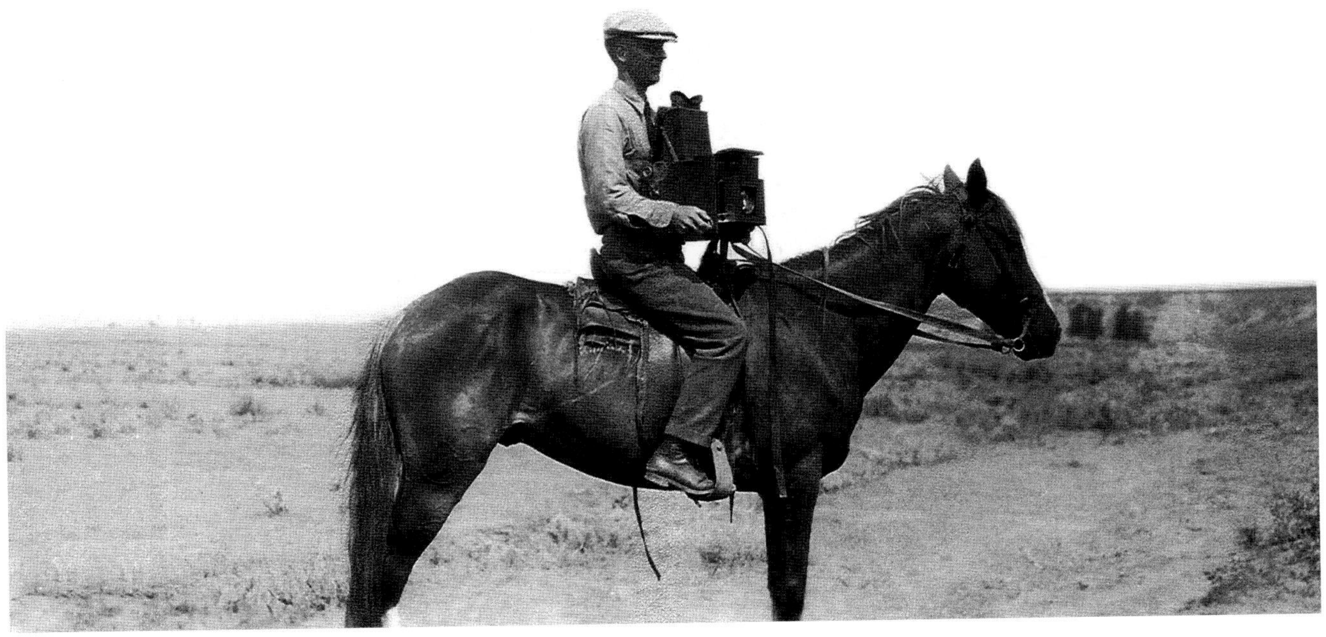

John C. Allen taking pictures on horseback (N.p., n.d.; courtesy John O. Allen)

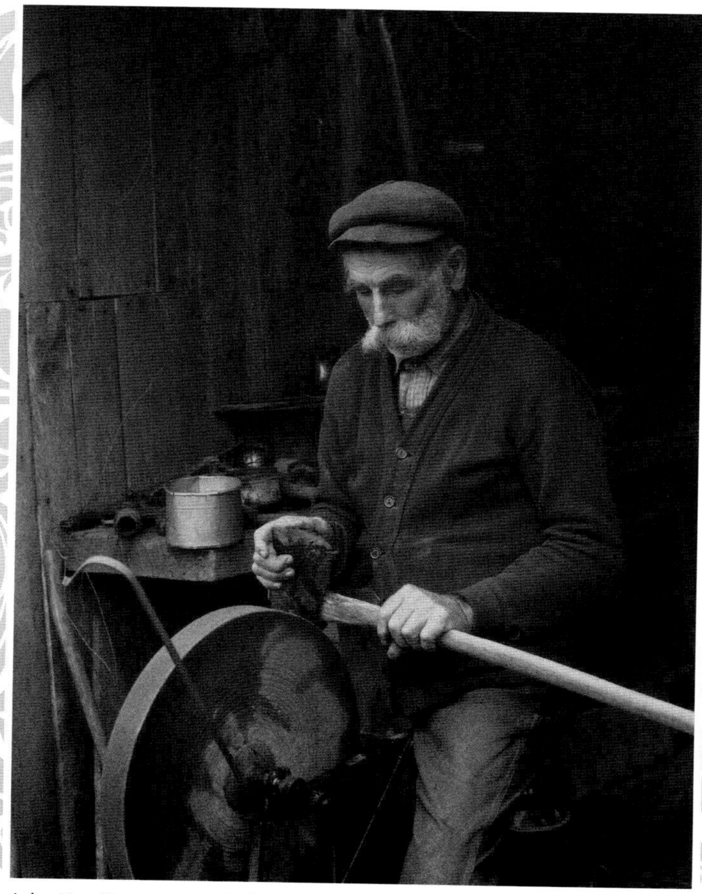

John Hamilton uses a grindstone to sharpen an axe in 1931. At the time of the photograph, he was one of the oldest residents in Bobo, Indiana. (Adams County)

The Photographer Emerges

After working on the farm and for the railroads, John Allen made a pragmatic choice in 1909 to take the position as a clerk and stenographer in the Agricultural Experiment Station at Purdue University on the West Lafayette campus.[1] In addition to his office work, Allen helped select uniform animals for nutritional studies, but he soon became responsible for weighing the livestock at specific intervals and tracking the data.[2]

At the time, Purdue agricultural programs were divided into three distinct and equal units: the Agricultural Experiment Station, which conducted agricultural research; the Extension Service, which oversaw outreach efforts between the university and the people and communities in Indiana; and the School of Agriculture, which was responsible for academic programs for students. Allen was hired into the first unit under the supervision of John H. Skinner, dean of the School of Agriculture and chief of the Department of Animal Husbandry. Skinner had developed a national reputation for his research on feeding balanced diets to cattle, hogs, and sheep in order to ensure maximum and profitable gain.

To Allen's credit, Skinner quickly learned that his new hire could be trusted when it came to collecting, analyzing, and summarizing large amounts of livestock research data.[3] Skinner told another livestock researcher, "Mr. Allen is a first-class calculator"—a high compliment on Allen's research and data-analysis acumen.[4]

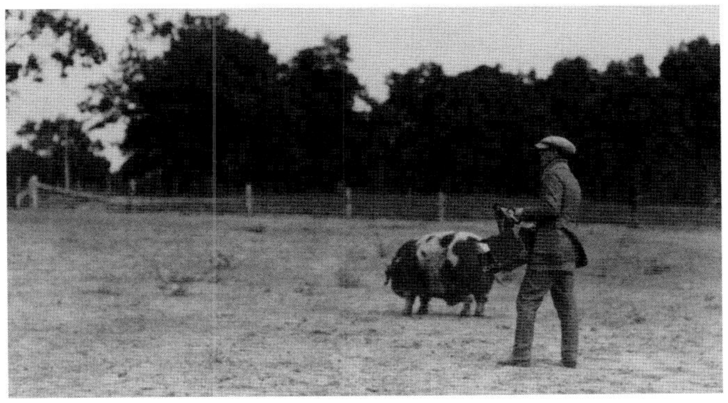

John Allen tries to find the best angle for photographing a hog. (Newton County, 1918; College of Agriculture Extension Reports Collection, Newton County)

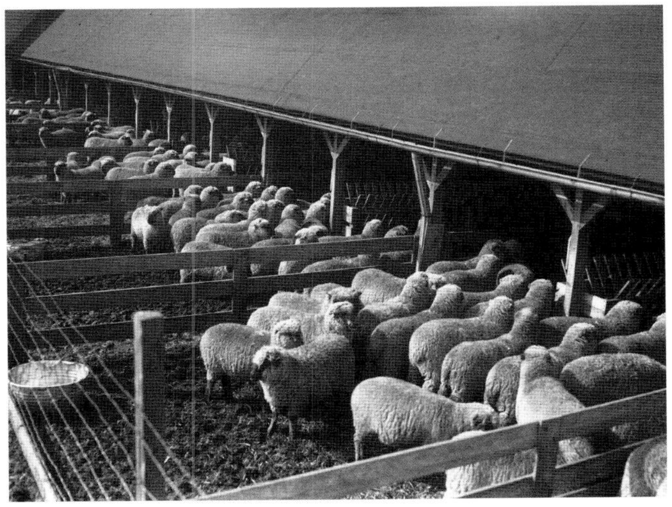

Allen became well known as a photographer when he began documenting the results of Purdue livestock feeding studies, such as this one conducted in 1917. His photos were published in agricultural magazines and newspapers along with articles about the research. (Tippecanoe County)

While Allen had gained plenty of practical working experience with livestock at the orphanage and on his family's farm, he lacked an in-depth understanding of agriculture.[5] Allen's own admission was that he needed to take college classes to increase his fundamental knowledge of livestock.[6]

Allen applied for and was accepted as a special student in the School of Agriculture from September 1911 to June 1912.[7] The thirty-year-old worked his full-time job while taking six college-level classes, including Livestock Judging, Livestock Management, Poultry Housing and Breeding, Advance Livestock Judging, and Animal Nutrition and Principles of Feeding. Allen's official transcripts from Purdue indicated he earned three As, one B, and two Cs in his coursework.

During this time, Allen's interest in photography conjoined with his newfound career and education. Fortunately for him, he had access to a Press Graflex camera that had been purchased for the agricultural programs. This camera provided the opportunity for him to hone his photography skills.[8]

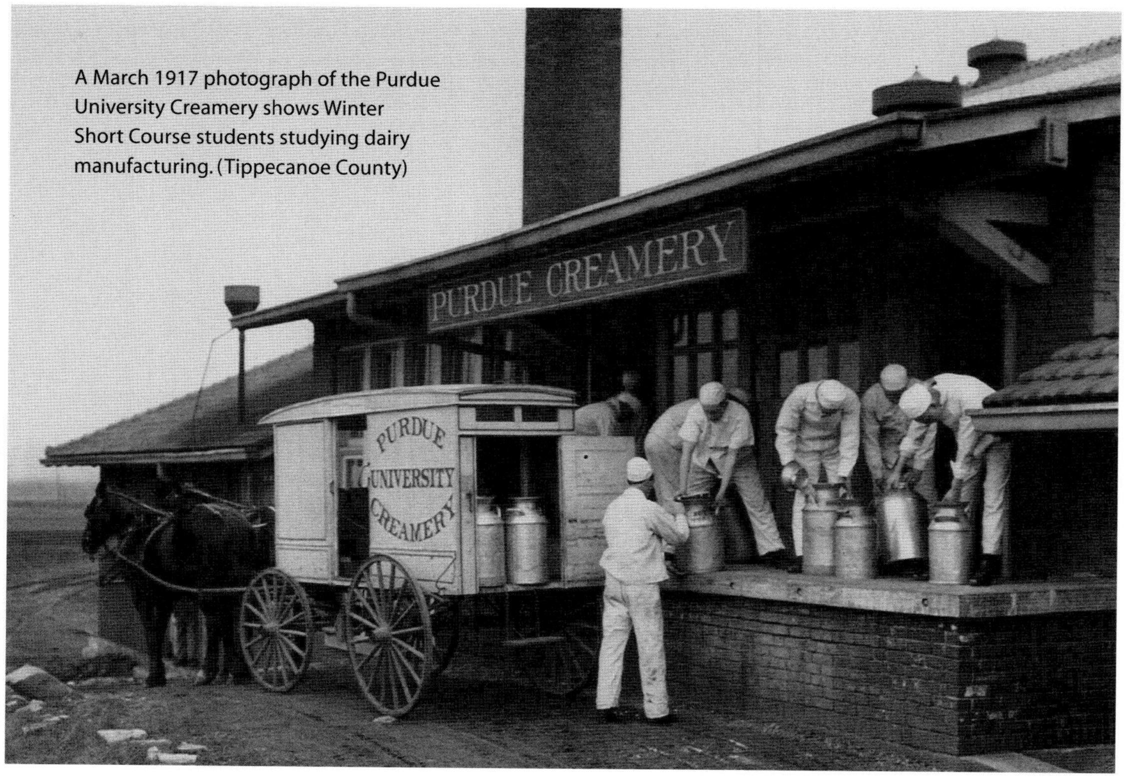

A March 1917 photograph of the Purdue University Creamery shows Winter Short Course students studying dairy manufacturing. (Tippecanoe County)

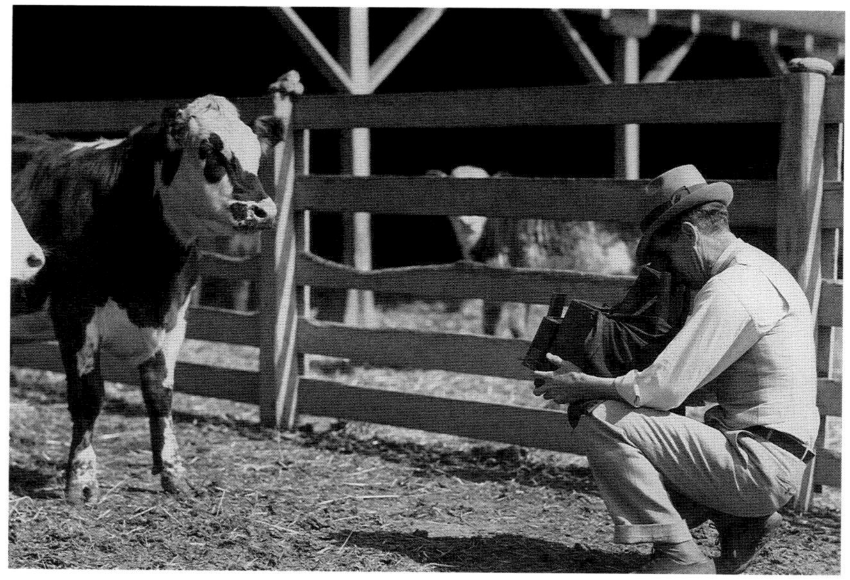

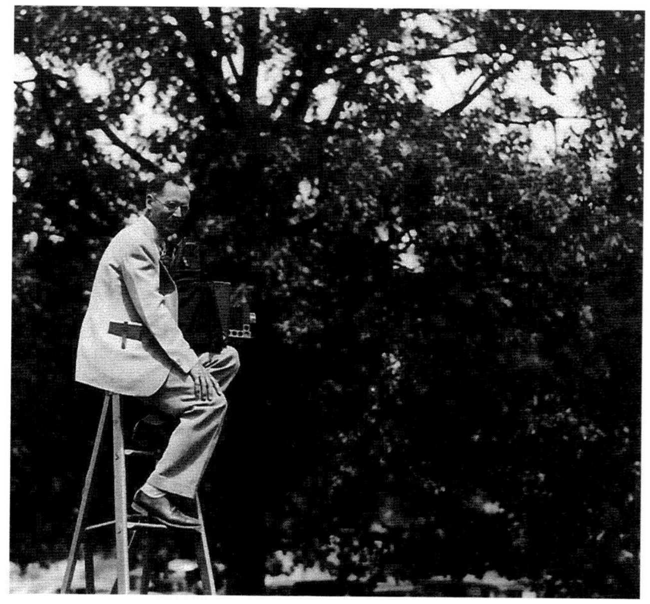

(Top right and bottom) John Allen would do just about anything to get the best angle for the shot he had in mind. (N.p., ca. 1940; courtesy John O. Allen)

He began to explore visually documenting the livestock in experiments as a novel way to complement the numerical data. His photographs immediately and irrefutably demonstrated research outcomes, causing Skinner and other station researchers to take note. Allen began exploring different methods for taking the photographs, changing the distance and angles to ascertain the best approach for documenting research results. As a result, he began developing an eye for what constituted a good photograph.

Allen used the Graflex to capture differences between well-bred and inferior specimens of hogs, horses, sheep, and dairy and beef cattle, with the resulting photographs used for instructing students. Within a few years, he expanded his subjects to include corn, soybeans, strawberries, tomatoes, apples, peaches, and just about anything found in farm fields, barns, and homes. Allen's photographs began appearing within Agricultural Experiment Station and Extension publications. Researchers saw the value of sending photographs with their articles to newspapers and trade publications, and soon his images began to appear in periodicals such as *Hoard's Dairyman*, *Prairie Farmer*, and *Breeder's Gazette*. The photographs published in state and national publications raised the visibility of Purdue's agricultural programs. As Allen's grandson recounted, "Purdue liked the attention that the pictures that were being published in several agricultural magazines brought to Purdue."[9]

Allen himself also received a fair share of positive press from these published photographs. As his renown expanded beyond Indiana's borders, Allen's images caught the attention of Ohio State University administrators. In 1915 Ohio State offered Allen a twofold opportunity: a contract to photograph 10,000 images of rural life in Ohio and an offer to become a full-time photographer for the university. For Allen, this must have been a dream come true. By this time, Allen knew he had both the talent and interest to make photography a successful and profitable career.

Allen told Purdue administrators about the offer, and Purdue quickly matched the Ohio State proposal, offering Allen a full-time position as the official photographer for the Agricultural Experiment Station and Extension Service.[10] Instead, Allen negotiated to work half-time at Purdue so he could concurrently start his own photography business. Purdue agreed to Allen's terms, with Skinner writing in a May 1915 letter to his boss in the Agricultural Experiment Station that "Mr. J. C. Allen be employed on the basis of one-half time, with the understanding that he is to devote the other half to his own private business, his services to the university to distributed as follows; Salary to be paid Mr. Allen to be $900.00 per year and to be understood that his personal business is not to interfere with his experiment station duties."[11] The agreement with Purdue also stated that all photography equipment had to be purchased by Allen's business.

Later in life, when asked about why he became a farm photographer, Allen said, "Probably by necessity and accident. Then with a wife and two children (as with many young people today) the wages were barely enough to meet expenses. So, I had to find some way to supplement my income and I turned to photography. I soon found there was a place for good photographs."[12]

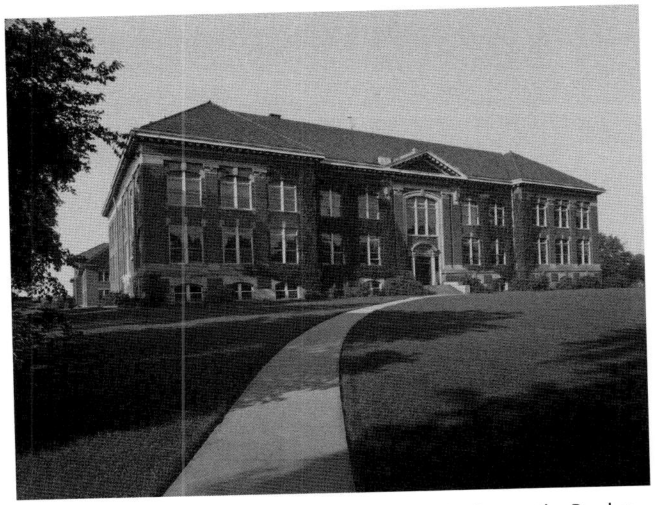

This photograph of the Agricultural Experiment Station on the Purdue campus in West Lafayette was taken in June 1926. Allen's office was on the ground floor. (Tippecanoe County)

Indeed, in 1912 the Allen family had welcomed a daughter, Martha Charlene Allen. With a growing family to support, why would Allen decide to forgo a full-time position, cutting a regular paycheck in half if the family had been experiencing financial difficulties? As early as 1910, Allen had perceived an opportunity to generate a second income by selling high-quality photographs to the agricultural press, and by 1915, he had developed a savvy mind for business. Simply put, he believed he could earn much more by selling photographs through his business than by working full-time for Purdue.

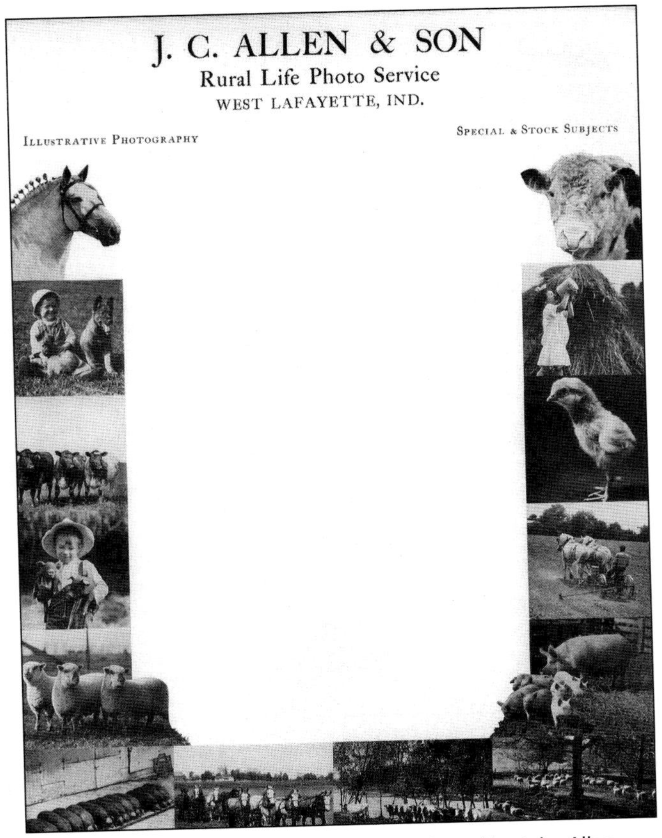

Letterhead (above) and business card (below) used by John Allen (N.d.; courtesy John O. Allen)

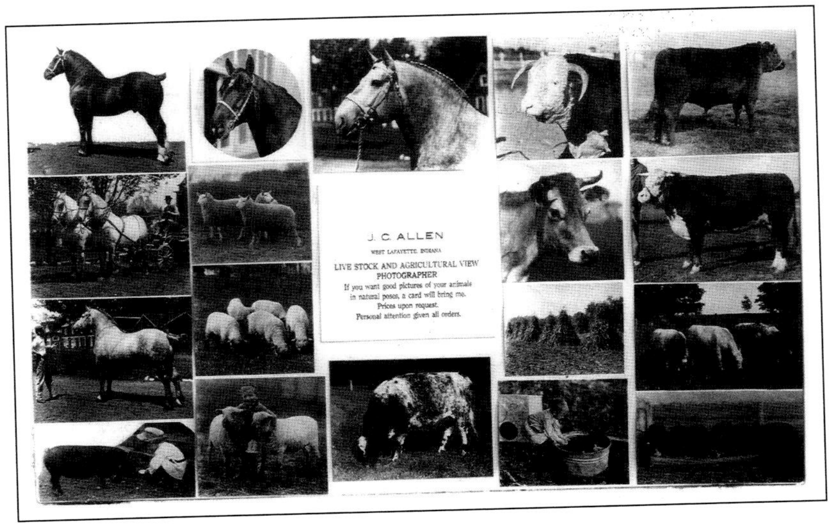

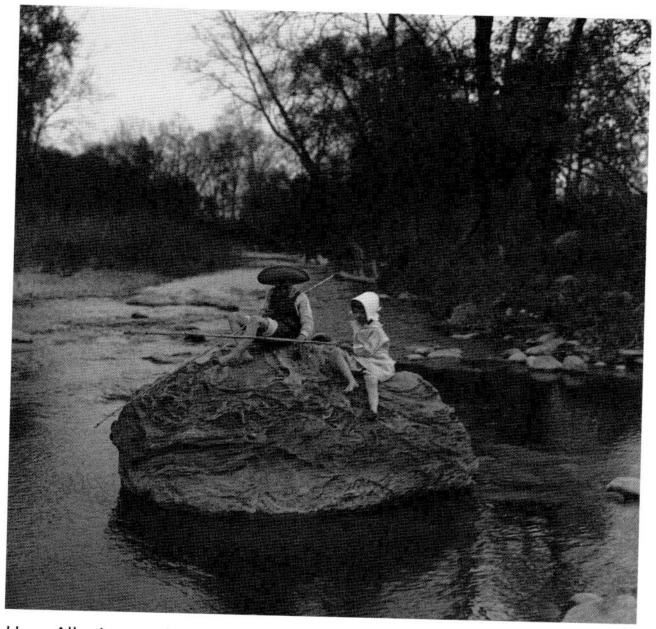

Here, Allen's son, Chester (left), and daughter, Martha, fish on the Wildcat Creek in 1918. John Allen frequently photographed his children as models for his commercial work. (Tippecanoe County)

Allen specifically pointed to advice he had received from a well-respected agricultural editor.

> From Art Page, former editor of *Orange Judd Farmer* and from the *Breeder's Gazette*, I discovered that there was some market for good farm pictures, so I decided to try to fill some of that demand. This gradually developed into a photo-illustrating business. Without attempting to do so I had by accident prepared myself during the past years for just such a business. First, my boyhood experiences on the farm gave me some practical knowledge of the farm, the course in livestock judging at Purdue Univ[ersity], was particularly valuable for making good livestock photos and my former business training on the railroad helped to fit me for success in this kind of enterprise. It still required determination and plenty of hard work.[13]

A Photograph and a Poem

It is often said that a photograph is worth a thousand words, but in the case of one of Allen's, it was worth a poem. In 1934, Dean John Skinner of the Purdue University School of Agriculture received the following letter from John Ashton, managing editor of the *Texas Grower and Valley Farmer* headquartered in Corpus Christi, Texas.

> I write you now mainly on account of an Indiana boy whose name I do not know. He is probably a man by now—or pretty near it, and his picture is on the front cover of our magazine, copies of which are going to you under separate cover. Mr. Allen, your photographer, sold me this picture along with others, about 1928 or 1929. Please give Mr. Allen one of the copies I am sending you, and ask him to identify this boy. It is such a striking picture that, after I had decided to use it, I went home one evening with the image of that Indiana boy in my mind and wrote the little poem about him, on page 9, of the February number of *The Texas Grower and Valley Farmer*, symbolizing him as typical of the American farm boy. Already people are writing me quite enthusiastically about the poem, and the biggest daily newspaper in Texas—*The Dallas News*—published the poem in its issue of Sunday, Feb. 18, 1934. . . . I wish Mr. Allen would bring this matter to the attention of the parents of the boy in question. . . . I wish to say, also, that Mr. Allen's skill as a photographer of rural subjects is outstanding and "The Farmer Boy" is one of his very best compositions.[14]

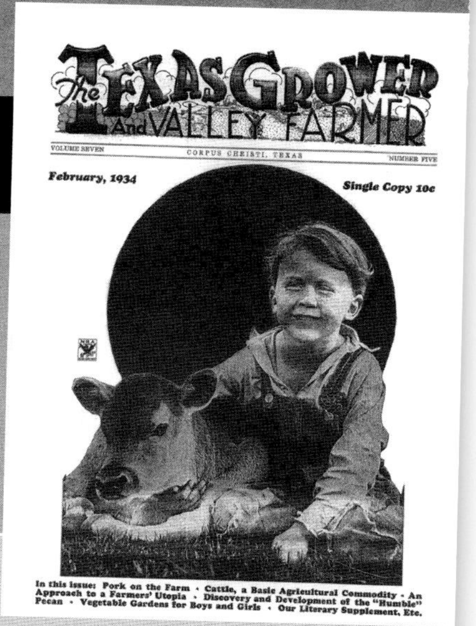

This poem, which Ashton had written after being inspired by Allen's photograph, became popular in agricultural circles.

The Farmer Boy

He's just a country urchin, as happy as can be;
A child of Mother Nature,—'tis plain enough to see.
He's roamed the fields for miles around, he's romped on ev'ry hill,
And woods and streams have had to yield their secrets to his will.

He knows the haunts of 'coon and fox, he's watched the eagle soar;
He's trailed the bobcat to its lair, and heard the cougar roar.
He's climbed the tallest poplar tree, and many a bee's nest found.
And once, when seeking arrow-heads, he struck an Indian mound.

The names of all the trees he knows, and many weeds can name;
And ev'ry season brings its joys to one who knows his game.
He's trained as pets both birds and beasts, for kindness ever pays;
In fact, like Daniel Boone of old, he loves God's creatures' ways.

His freckled face and tousled hair, his smile that won't come off;
His ragged jeans and shoeless feet proclaim a lad who's "tough."
"O, would I were a boy like him!" in fancy comes a wail
From pampered weakling, richly clad, whose health's begun to fail.

He's just a lusty farmer boy, contented as can be;
Nor would he change his humble home for palace, no, not he!
No city wiles his mind beguile; he's made of sterner stuff:
He's reared in Nature's own front yard—a diamond in the rough![15]

Skinner replied to Ashton, stating, "I have your letter and copy of your magazine and poem on the farm boy. I will be glad to refer this to Mr. John Allen, and I think he can identify the lad for you and for the boy's parents. I am delighted with the poem. It has a lot of fine sentiment in it."[16]

Allen described one of his first photographs, saying:

> [It was of] a small lamb taken with a borrowed 5 x 7 Press Graflex. This was back in the days when drawings were the chief means of illustrating farm publications. I sold the picture to a leading agricultural publication for the great sum of fifty cents, and, while the fifty cents was never received, my imagination was fired with the possibilities of agricultural illustrations for the farm press."[17]

Allen worked long hours to get his photography business off the ground, knowing that he had to have the photographs and a market for them in order to make money.[18] He described going out into the countryside during his first years to take photographs of images he thought others would purchase: "All of my off days were spent making photographs with this borrowed [Graflex] camera on farms that lay within walking distance of town, and on occasions I would splurge enough to rent a horse and buggy and make short photographic trips into the country."[19] In time, that gamble would pay off.

Allen credited *Hoard's Dairyman* as among the first to compensate him for one of his photographs, which showed the head of a dairy calf. Allen noted, "In actual cash, I received 50 cents for the picture, but it was worth dollars to me to see this first picture in print."[20]

An earlier photograph was described in a *Purdue Alumnus* article: "It all began in 1910, when John Allen, then employed by the Purdue Animal Husbandry department, borrowed a camera from the Purdue Extension Service and took a picture of his Buster Brown-Clad son, Chester, '29, with a lap full of Purdue prize piglets. A national farm magazine paid the munificent sum of $1.50 for the publication rights."[21]

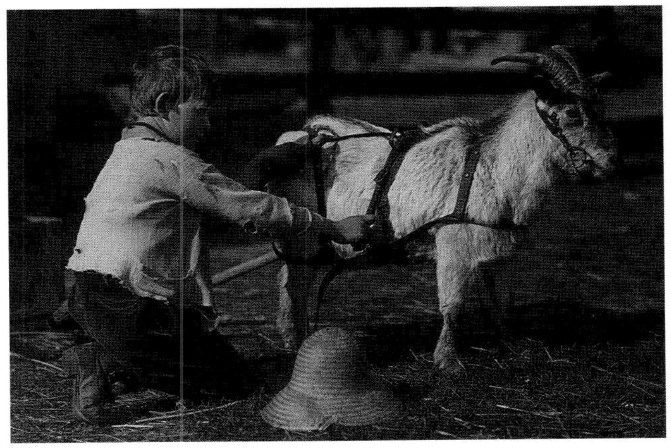

An early photograph sold by John Allen captured this boy with his goat harnessed to a sled. (N.p., ca. 1925)

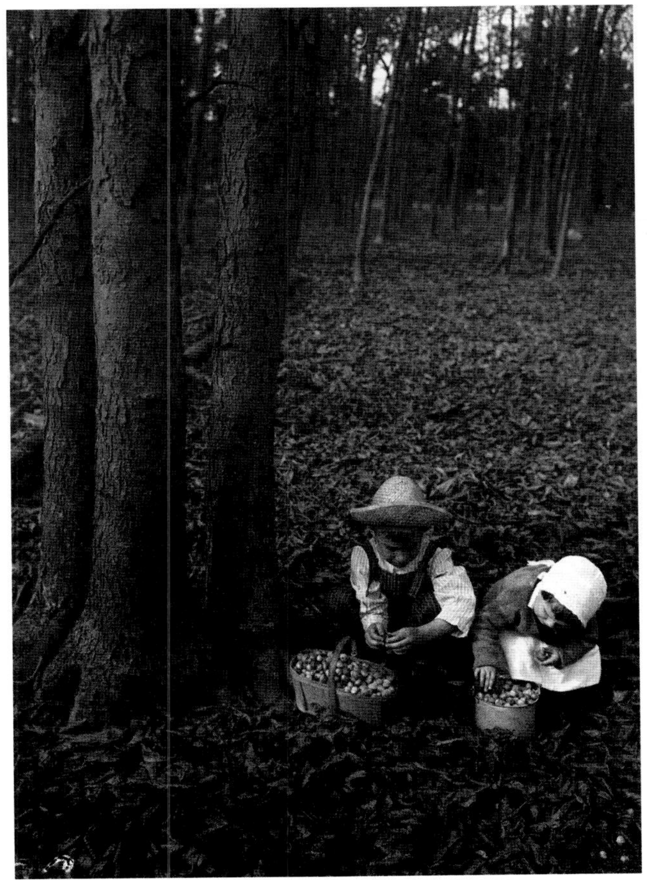

Chester (left) and Martha Allen hull hickory nuts. (N.p., 1913)

Another photograph taken early in Allen's career of a child holding a small goat was sold to *Country Gentleman*.

One of Allen's trademarks throughout his professional career was taking pictures of children—including his own children and grandchildren—interacting with young animals. He strongly believed that a picture needed to quickly capture the viewer's attention, and children always worked to make the viewer take a second look. Reflecting back on his career in 1976, Allen said he was "most proud of his work with children, where animals and children could be brought together."[22]

Ralph Reeder, who worked in agricultural communications at Purdue, noted that soon after Allen's photographs began appearing in popular magazines such as the *Breeder's Gazette* and *Prairie Farmer*, "his photographs were showing up regularly in other national farm magazines."[23] Allen was now a businessman in possession of his own company, which he named J. C. Allen Rural Life Photo Service.[24] His work through that business and for Purdue would eventually earn him a reputation for being one of the country's best rural life photographers.

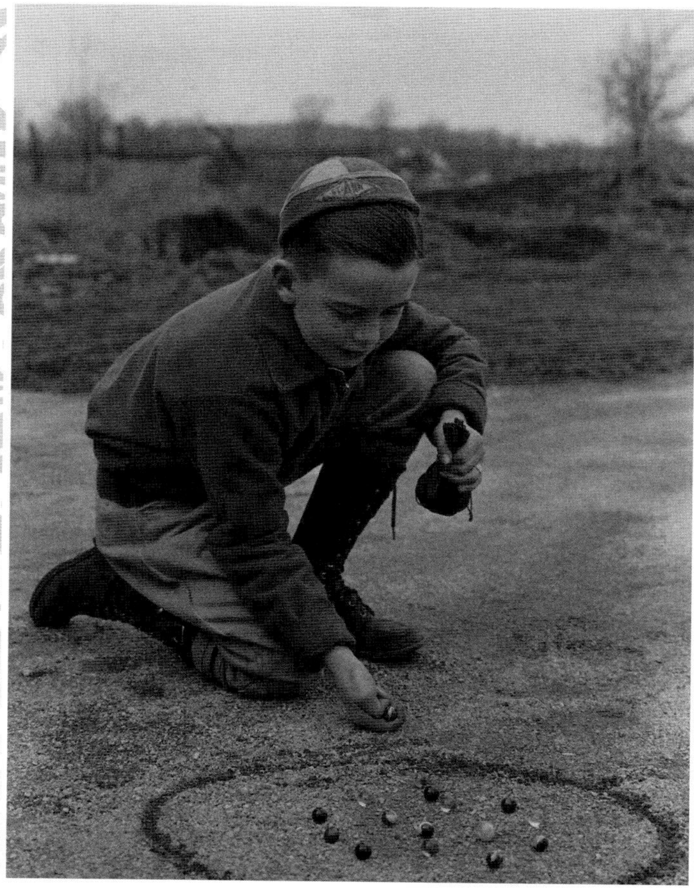

Steve Hurley plays marbles near Lafayette. (Tippecanoe County, 1932)

Crafting a Business One Photograph at a Time

Allen's interest in rural photography was well-timed. He had embarked on his new career just as the print media had acquired the technology to print black-and-white photographs. With the Purdue deal now in place, he began taking images for the university and commercial clients.

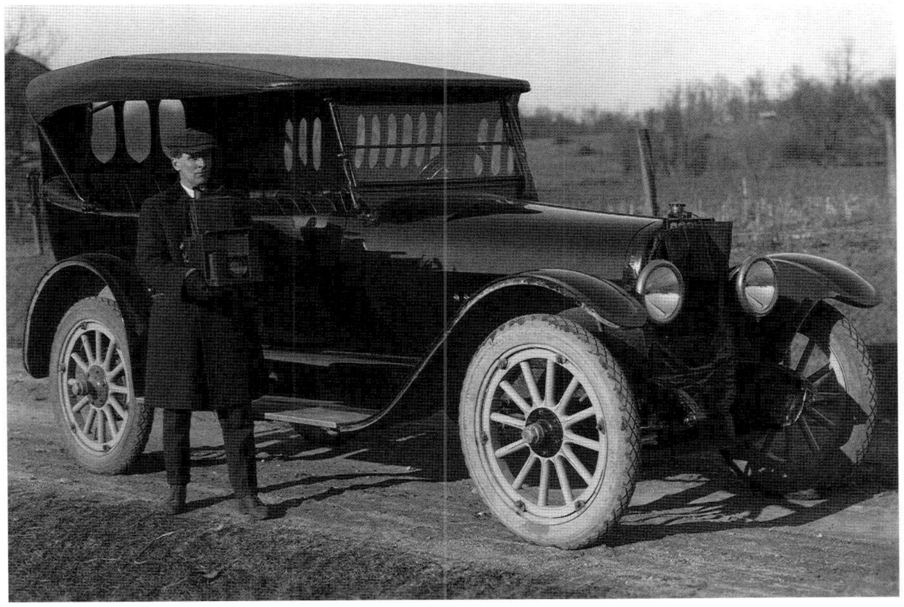

John Allen holds his Graflex camera as he stands next to his Mitchell Touring sedan in 1916. (N.p.; courtesy John O. Allen)

Allen's duties as a Purdue photographer varied. One moment he was taking pictures for the Agricultural Experiment Station researchers and the next, for campus Extension specialists. He also took photographs for the *Debris*, which was the Purdue student yearbook, and the *Exponent*, the student-run campus newspaper. During a career that would last many decades, Allen would take thousands of photographs for Purdue agriculture and campus activities, especially many sporting events.[1] His subjects included famous people, registration, graduation ceremonies, campus buildings, 4-H events, university life, livestock association meetings, classroom lectures, Winter Short Courses, and various state and national conferences.

Most all other subjects were shot to support his ever-expanding personal business. Allen's straightforward motto was simply, "Have camera, will travel."[2] While his strongest focus was on rural photography, he also did photo shoots for calendars, weddings, group club photos, events, funerals, and even accident claims for insurance. He also served as the official photographer for the Indiana State Fair from 1914 into the 1960s.

His ability to capture life at its most earnest, particularly in rural places, charmed people from all

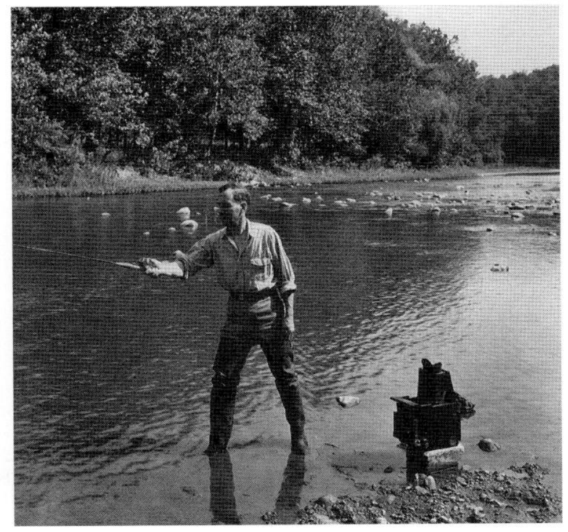

Allen fishes in a river in July 1931. As a spokesperson for the Graflex camera, he made certain the camera was prominently displayed in this photograph. (N.p.; courtesy John O. Allen)

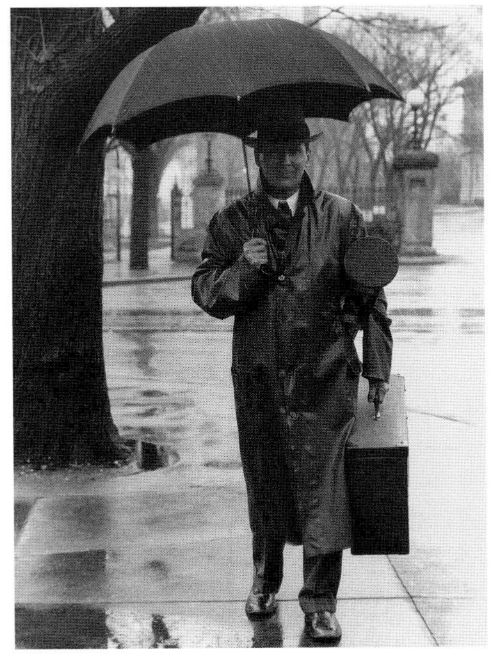

John Allen heading out for a photography shoot. (Tippecanoe County, n.d.; courtesy John O. Allen)

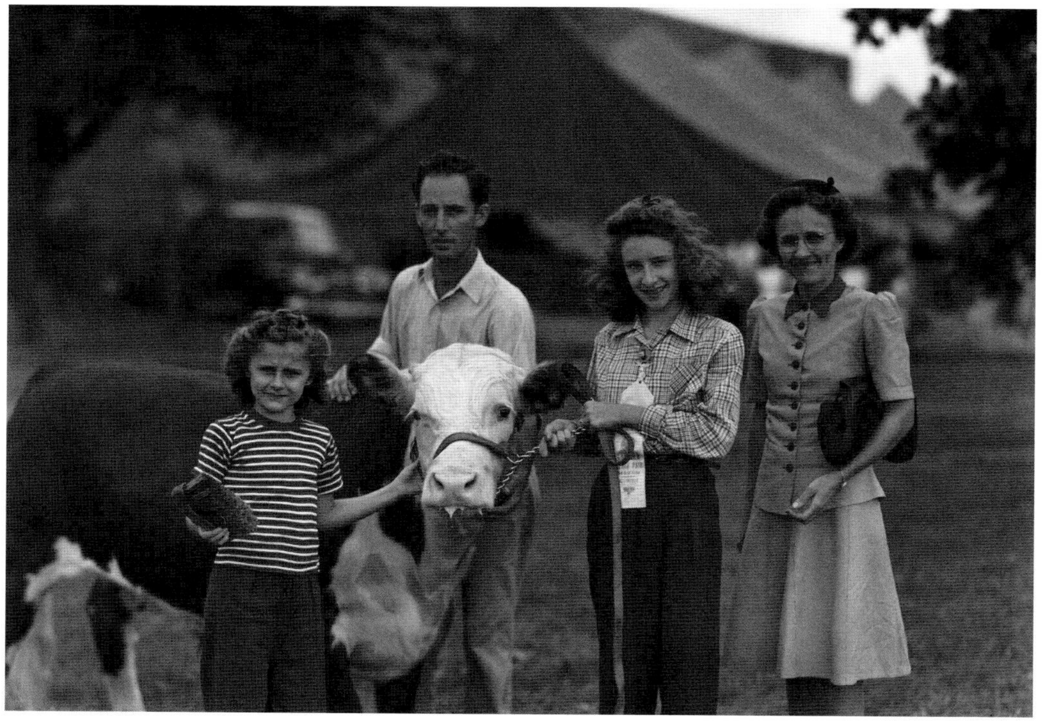

Patty (left) and Nadene Royer (second from right) of New Richmond exhibit a Hereford steer in 1944 alongside their parents. (Montgomery County)

walks of life. Allen inherently understood which images held eye appeal and which would find wide interest in commercial advertising. Writing for *Hoard's Dairyman* in 1959, Allen shared his business philosophy.

> To be successful in photography, you must know what constitutes a good picture and how to select subjects which are best for pictures. . . . Photography, especially of farm livestock and other farm subjects, is and has been my life's work. . . . These subjects of most of our photos, therefore, are farm people and their work, farm livestock and poultry, farm buildings, crops and fruits. We do some commercial photography for special clients and some assignment work, both of which fit in nicely with the farm illustrating business and give us an opportunity to travel. We have been very careful, however, not to get into the portrait business in our studio.[3]

When choosing subject matter, Allen had a twofold approach. First, he wanted to capture the adaptations by farmers that made their jobs easier or more efficient. Second, he looked for impromptu photographic opportunities to create unique images. He stated that the "farm community is loaded with photographic possibilities if the photographer has patience and the eye to locate pleasing subject matter."[4]

As a photographer, Allen saw opportunities throughout the four very distinct seasons of the year.

> Personally spring is my favorite season for farm picture hunting. It is then that new life is in abundance; little pigs, colts that before many years will be plodding along doing the heavy farm labor, lambs and calves frisking about the pastures and baby chicks filling every chick brooder house. It is then that every farmer capable of work is in the field with team or tractor plowing, working the seed bed, planting seed. It is then that blossoms of apple, peach, pear and plum trees form beautiful frames for your many farm photographs. During the summer months come the pictures of growing crops, threshing of small grain and haying at its best. Fall brings the harvest and winter either snow or inside scenes showing the farmer milking, repairing tools for another year or around the radio with his family.[5]

Allen had the foresight to adapt to changing times across rural landscapes. The subjects he photographed shifted over time to reflect the advances in farming practices, introduction of new technologies, and implementation of agricultural research. These included:

- field work done by hand and with horses,
- improved efficiency of farm buildings,
- promotion of purebred livestock herds,
- advantages in fence construction,
- adaptations to reduce hand labor inside poultry houses and dairy barns,
- spraying equipment for apple orchards,
- replacement of horses with steel-wheeled tractors,
- use of rubber tires on tractors,
- switch from open-pollinated corn to hybrid corn,
- introduction of self-propelled equipment,
- use of electricity on farms,
- children and young animals,
- nature scenes, and
- flower gardens and individual flowers.

While there were few subjects that Allen would not photograph, one subject is conspicuously absent: the struggle of Depression–era families during the 1920s and 1930s. Despite a demand for photographs documenting the financial and emotional toll during this time period, Allen made a personal decision to avoid such work. Indeed, he was fortunate to be successfully positioned to make decisions with his conscience at a time of great hardship for so many. He had lived through difficulties and poverty as a young man and chose not to make money off those finding themselves unemployed and struggling to make ends meet.

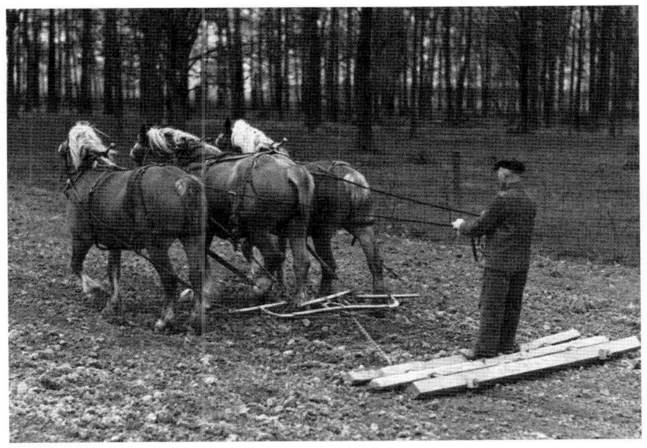

Harvey Hoewischer from Sidney, Ohio, works the soil using Belgian mares in 1943. (Shelby County, OH)

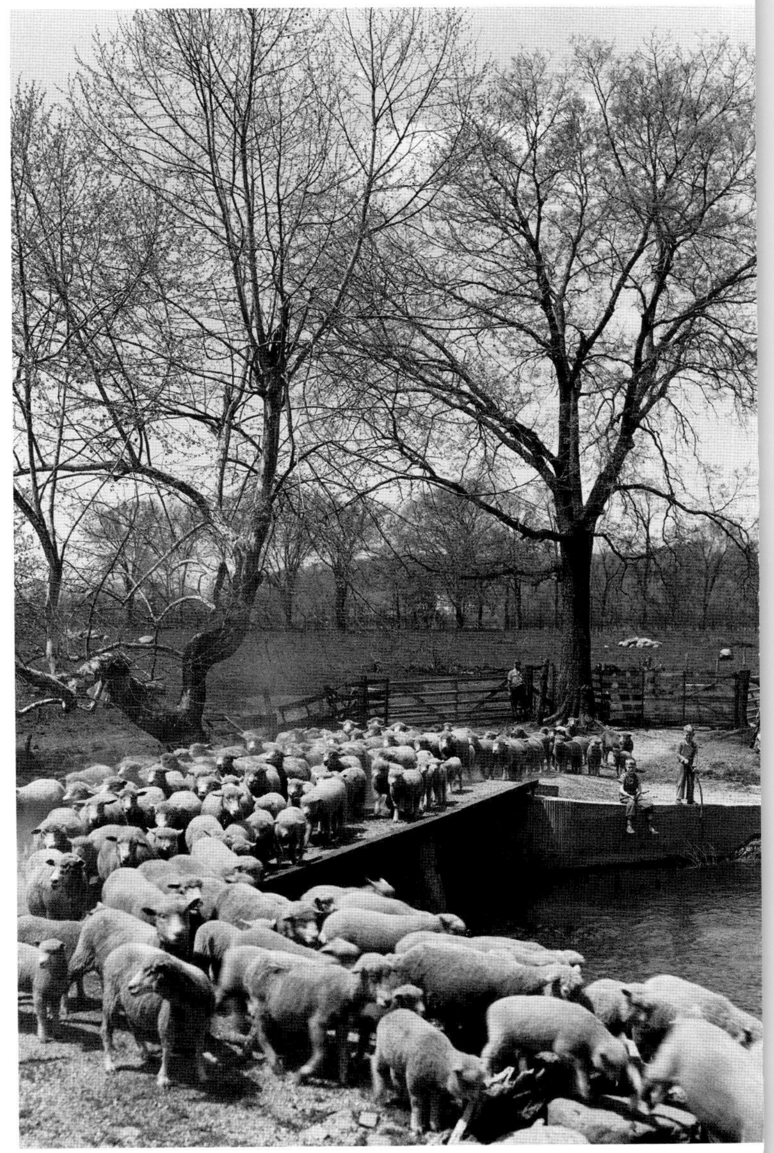

One hundred Corriedale ewes with lambs cross a bridge on the Tow Path Farm belonging to C. L. Reedy in Logansport. (Cass County, 1950)

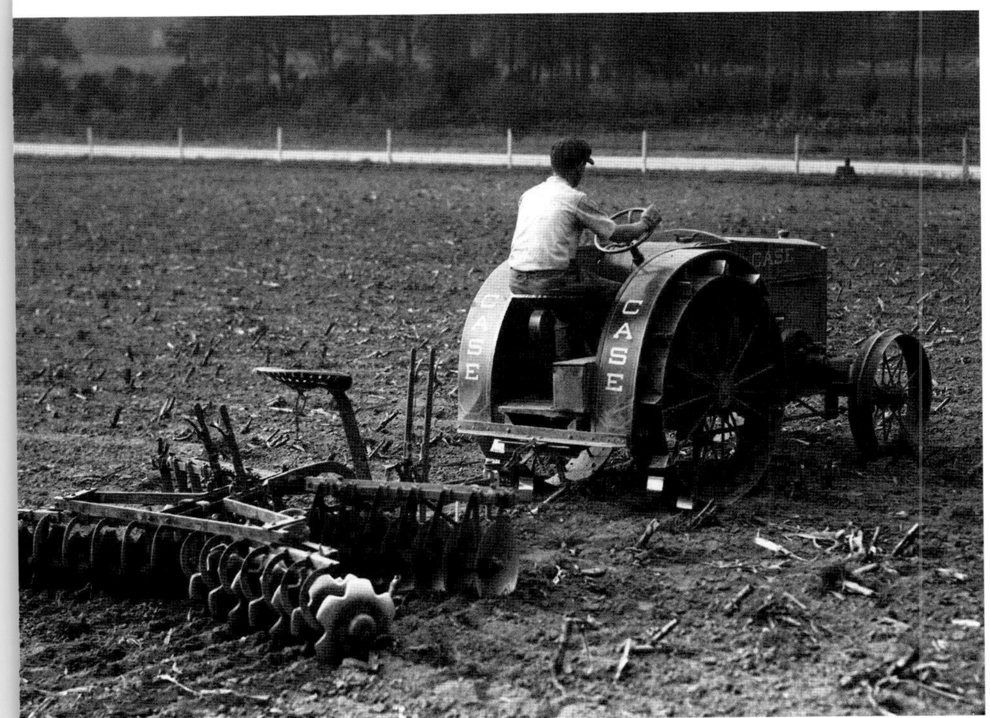

A worker prepares ground at a Purdue University farm using a Case tractor pulling a disc harrow in the fall of 1917. (Tippecanoe County)

Allen's grandson, John O. Allen, said his grandfather "watched for the pictures to appear to him," noting, "I don't think he tried changing what they were doing as much as he tried going along."[6] Indeed, John Allen himself explained that "after making farm photographs for almost thirty years I can assure you that one of the prime requisites of a livestock photographer is patience and more patience—patience with the animals, patience with the animal handlers and owners, and patience with would-be helpers who often swarm like bees to honey when a picture is being made."[7]

Allen described the process he used to get his images: "At least my aim has been to turn out completed photographs that appear as though the subject or subjects didn't realize there was a camera within the state boundaries. However, practically every shot, at least the best ones, were posed to some degree—posed to make them look unposed."[8]

He went on to say that "taking farm pictures as in making any other type, it isn't so much the kind of camera you have, though some have their limitations, but it is your ability to use your camera quickly and skillfully."[9] He explained in an undated and unpublished manuscript how he set up scenes.

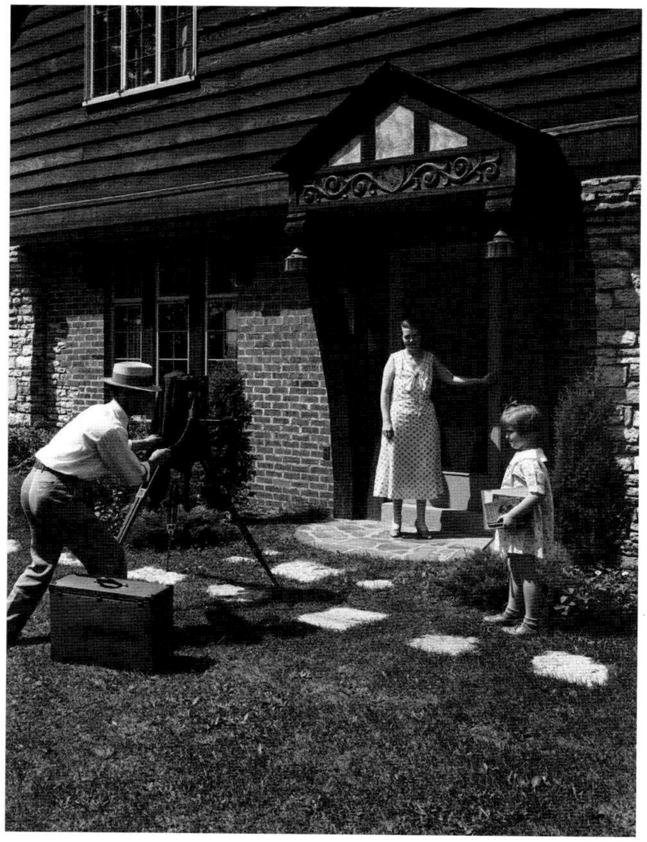

John Allen takes a photograph of Mary Snyder starting out to school in July 1931 as her mother looks on. (Tippecanoe County)

> Usually when photographing farm animals I prefer to go about my work alone or with the help of my assistant, but now and then the farmer can be of great assistance in herding the livestock if you can make him understand just the goal toward which you are working. As stated before, it is essential to obtain pictures that appear natural, but after selecting your background, correct light angles and possibly a tree to frame the photograph you may need to move the animals across the pasture into the camera's line of vision. . . .
>
> When you ask permission to make photographs the farmer may want to place himself or part of his family into the picture for they are usually mighty proud of their livestock. Of course you can't insult the man by saying no, but be sure to keep any individuals to the edge of your picture so that they may be cropped off in turning out your finished print. It isn't safe to crop them off in making the exposure for they may insist on seeing prints. If the farmer alone wants in the picture and is dressed in his work clothes, you can give him a feed bucket or hay fork and he may add atmosphere to your shot. Don't let him look at the camera, but insist that he take a position to indicate that he is actually at work. . . .
>
> I would say: learn the mechanical part of your camera first so that all attention may be directed toward the picture at hand, solicit the cooperation of the farm resident so that you will not be charged with trespassing, decide in advance the general composition and best lighting, then work quietly toward this goal being ready at all times to release the shutter when your selected subjects have taken a nature pose that produces a pleasing composition. . . .
>
> Like other types of activities, each photograph of individual livestock has its individual problems. Therefore it is well to know your camera and film thoroughly before attempting livestock pictures so that all thought may be devoted to your subject. Animals don't understand much about pictures so once you have them set you won't have an opportunity to work out your exposure and play with the various gadgets that may be incorporated on your camera. They may take a perfect pose for a fleeting moment and it is up to you to be ready to release the shutter when that moment comes.[10]

As soon as Allen received word that Purdue had agreed to his part-time status, he began traveling intensively throughout the Hoosier State and beyond. From 1916 to 1922, his excursions also took him to Illinois, Kentucky, Minnesota, Missouri, Nebraska, Ohio, West Virginia, and Wisconsin. He once estimated that he had driven more than 400,000 miles and taken 60,000 photographs by the 1940s.[11] Eventually he traveled to every state within the continental United States for his work.

While traveling across the country in the twenty-first century is common, during the first half of the twentieth century it was a major undertaking that required advanced and detailed planning. There were few hotels, restaurants, and gas stations, so it was necessary to carry extra tires, fuel, oil, provisions, and camping equipment. Allen liked traveling in touring cars such as the Mitchell sedan. It was big enough to get across the larger ruts and spacious enough to carry all of his supplies, such as the hundreds of glass plates needed for lengthy countryside trips.[12]

Allen remained constantly accessible to magazine editors, who requested photographs to highlight stories, and to advertisers, who requested images showcasing their products. The open dialogue Allen maintained with his customers and availability of images on hand was instrumental to Allen's success. Editors frequently contacted him by either letter or telegram. Over time, the telephone became his conduit to his customers. By late in his career, he said, "Today approximately 60 percent of the requests come by General Telephone and almost everyone is in a rush."[13]

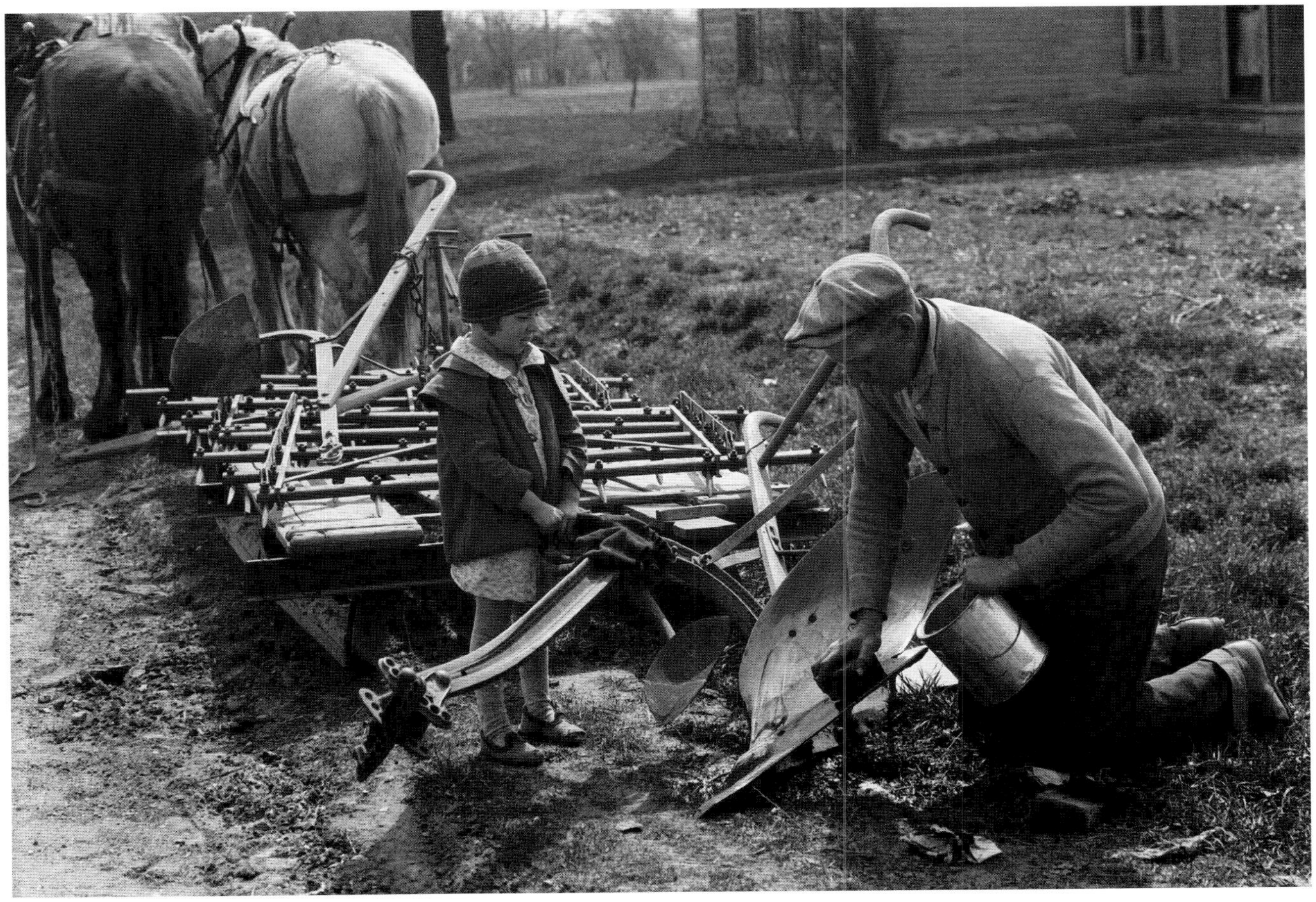

W. E. Bieghler of Russiaville scours rust from a plow while his daughter, Joan, watches. (Howard County, 1932)

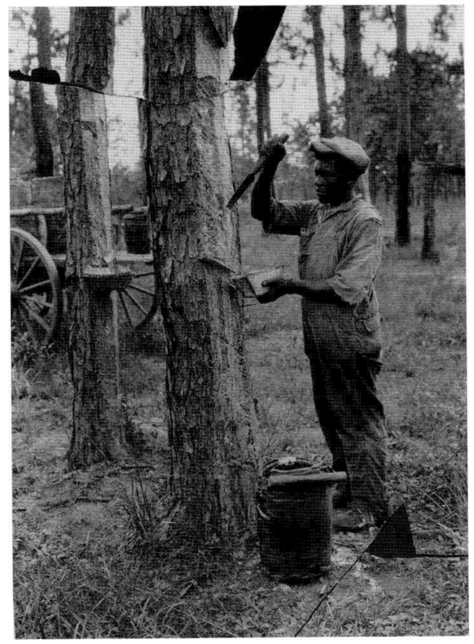

A man collects sap from a pine tree to make into turpentine in Lakeland, Georgia. (Lanier County, GA; 1959)

Whether through experience or reputation, editors knew they could depend on Allen to fulfill their needs quickly and professionally. If he didn't have the specific image in his files that a customer needed, Allen could always find a nearby farm or location to get the images others wanted. In a 2017 interview, John O. Allen indicated that his grandfather "travelled so much and came to know so many people. Grandfather became popular with farmers. When he started out for the day he had an idea where he was going to be and what he was going after."[14]

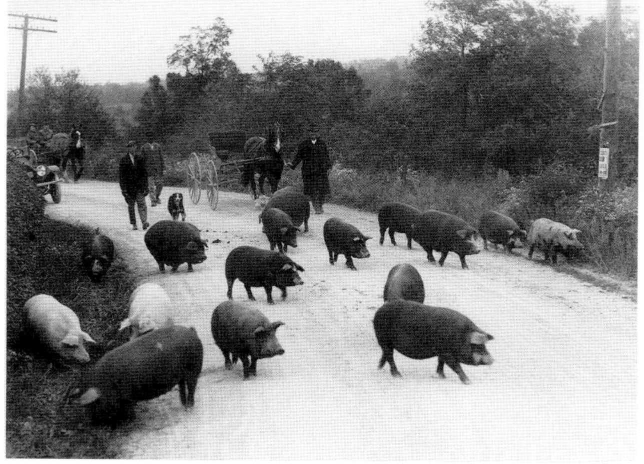

Here, hogs are herded down the road, possibly heading to market. The J. C. Allen and Son photographs captured a way of life that no longer exists, thus preserving for history what once was. (N.p., n.d.)

Allen respected the rights of property owners and did not step onto their property without asking first, explaining: "I make a practice of asking permission before going onto a farm and in my years of travel I can count on the fingers on one hand the times that I have been refused this permission. When refusal was made it was easy to gather that it was due to the farmer's unpleasant contacts with some other camera carrying person who had taken advantage of him."[15]

Occasionally farmers would contract with Allen to take photographs of their property, but typically the landowners were not his customers. He noted, "One kind of business we avoid is that of asking to take pictures on farms in the hope of selling to the farm people themselves. If we have good cooperation from farm folks in taking pictures, we usually send them prints free of charge for their picture albums or collections."[16] Allen gave farmers the free prints in exchange for his right to have unrestricted use of those photographs. Farmers were asked to sign a release so Allen could sell them whenever and to whomever. In a few instances, he sweetened the deal by paying the farmer one dollar for the right to sell the images.

The House on the Hill

When John Allen moved to take the Purdue position in 1909, he and his family lived at several different places around West Lafayette. In 1925, the Allens bought a nearly six-acre property at 1341 Northwestern Avenue. John Allen Benham, one of Allen's grandsons, said his grandfather had every intention of remodeling and renovating the home on the property. However, after receiving estimates for remodeling versus rebuilding, the Allens decided to pursue the latter option.[17]

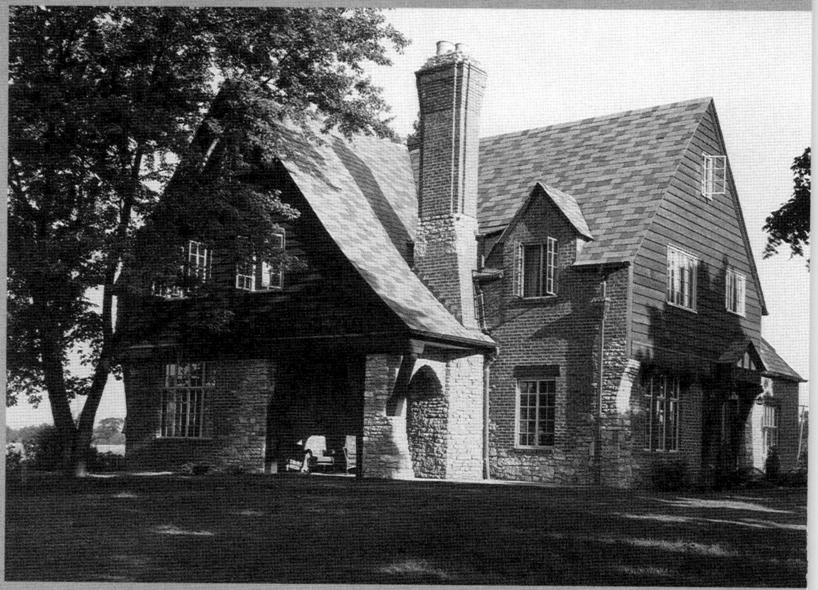

John and Abbie Allen built this beautiful home on Northwestern Avenue in West Lafayette, Indiana. When John left Purdue, he moved his office here. (Tippecanoe County, 1930)

To design his new home, Allen worked with locally prominent architect Walter Scholer, who was well known in the Purdue community as the designer of many campus buildings. In 1930, the Allens moved into their Tudor Revival home.[18] Surrounded with expansive grounds, flower gardens, walking paths, and a large vegetable garden, it became known to the family as the "house on the hill."

As long as Allen remained employed by Purdue, he kept his office in the Agricultural Experiment Station. Eventually, he established a studio in his Northwestern Avenue home to store the overflow of materials from the campus office. The complete inventory of photographs and processing equipment were not transferred to his house on the hill until Allen's retirement from Purdue in 1952.

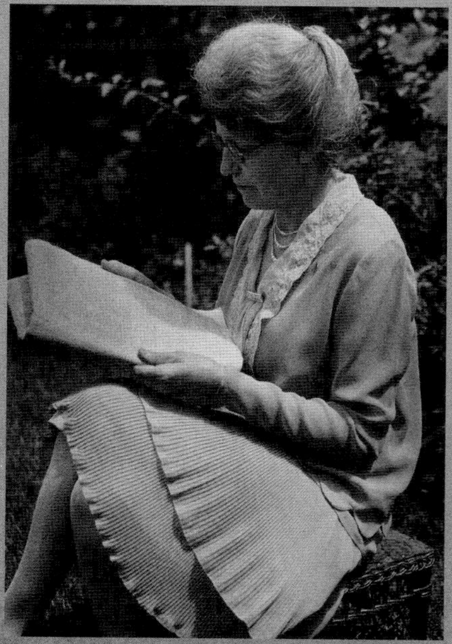

Abbie Allen reads the local newspaper in her garden. (Tippecanoe County, n.d.)

Allen thought that a photographer needed talent, but he also believed a photographer needed "the business ability to market the pictures after they've been taken."[19] From the time he took the risk to set up his own business, he proved to be a capable businessman.

Sometimes, Allen quoted a specific price for a photograph, such as in this 1950 exchange: "We hope one of these pictures [Grand Champion Red Poll Cow at the National Dairy Show] is what you want. You may select from them at $5.00 each for publication in the *Breeders' Gazette* or advertising."[20] At other times, magazine editors offered a flat payment, which Allen had to agree to if he wanted to do business with the publication. But for many would-be customers, he remained quiet when asked about prices. Allen let the person set the price, expecting the offer often to be advantageous to him. If the offer was too low, though, Allen and the customer could negotiate until they agreed on a price.

In the very early years of the business, exclusive rights for using photographs were not as important as they would be in later decades, so Allen would sell a popular photograph multiple times. However, he was savvy enough to make adjustments to suit individual customers, so he often maintained multiple versions of an original—cropped or processed differently—in order to supply similar but slightly different results of the same image.

Allen could even sell photographs of Purdue events, as John Skinner indicated in this 1928 correspondence: "I have the photographs of the Marshall cattle and I appreciate this very much. Undoubtedly some of these will be wanted from time to time for publicity purposes. . . . I would suggest that you do not use the photographs showing picture of myself and Mr. [Henry] Marshall [Purdue trustee and farmer] without my approval. Any of the other pictures can be used wherever you see fit."[21]

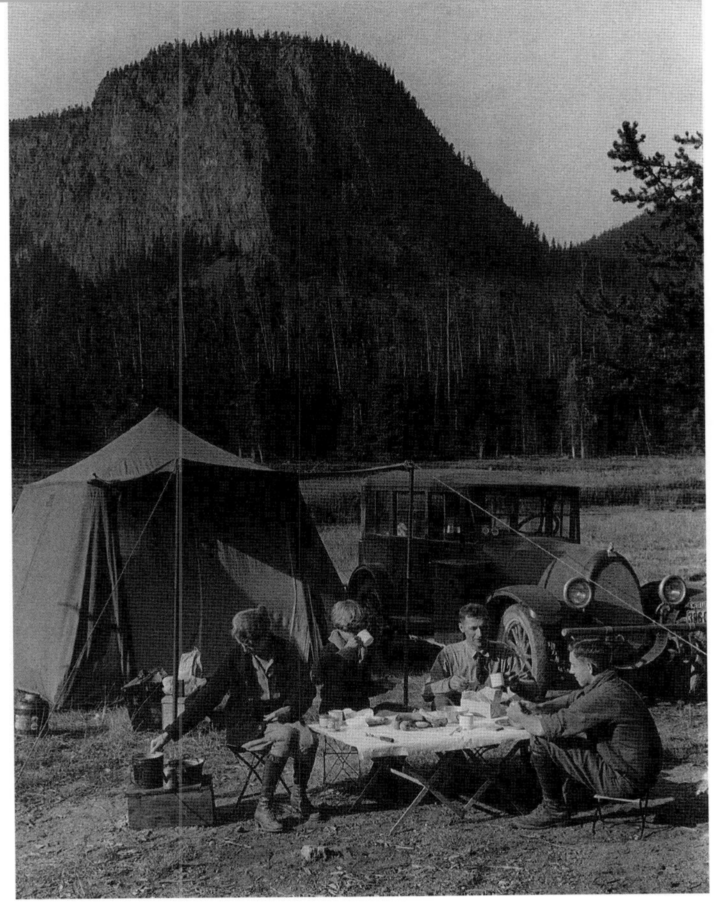

The Allen family relaxes during their annual camping expedition. Here, they visit Yellowstone National Park in the shadow of Mount Haynes. (Ca. 1922; courtesy John O. Allen)

Even when the Allens took their summer vacations, John managed to mix some business with pleasure, taking photographs along the way. In one account, he explained:

> Our son and daughter were now at an age to get the most out of our annual camping trips. We would spend the month traveling, taking pictures of interesting things along the way and fishing. We traveled through Michigan, Wisconsin, and Minnesota and one summer we visited the Yellowstone National Park, camping all the way and taking pictures with which to pay our expenses. Later we made these vacation trips into Canada and spent most of the time fishing.[22]

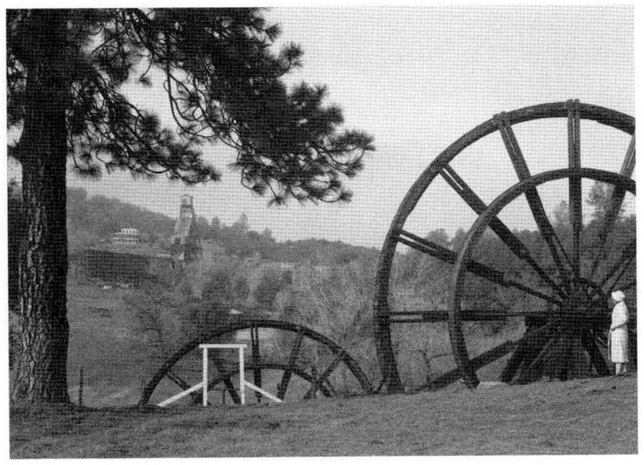

While on vacation in 1961, Allen took this photograph of a gold mine and the Kennedy Tailing Wheels in Jackson, California. (Courtesy John O. Allen)

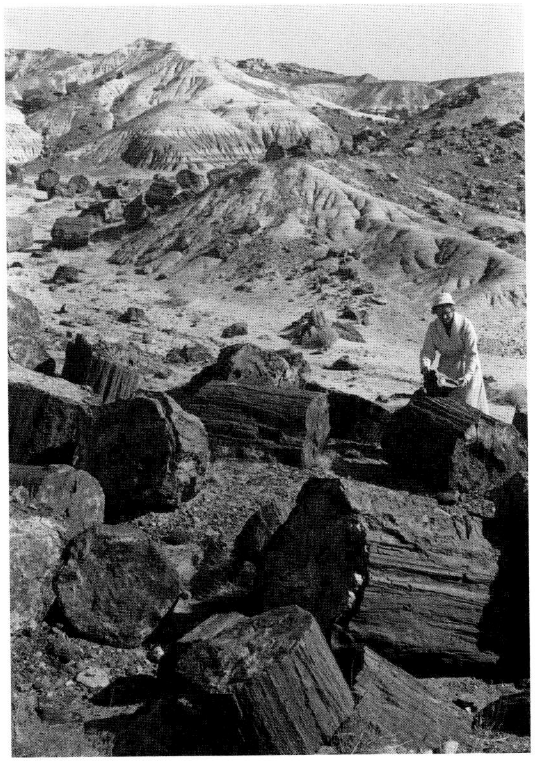

In 1961, this image in the Petrified Forest National Park near Holbrook, Arizona, caught Allen's attention. (Courtesy John O. Allen)

He astutely determined that others would enjoy—and pay for—photographs of these beautiful vacation landscapes. So in addition to taking photographs of the family campsites and other activities, he took hundreds of photographs of monuments, wildlife, flowers, mountains, and canyons.

Even when conditions for taking photographs weren't ideal, Allen was resourceful, noting, "One summer, when on vacation, we ran into a very wet rainy week while we were in a log cabin on the Flambeau River in Wisconsin. We spent the week fishing and picking wild red raspberries. I posed a picture of myself sitting on a big rock along the river bank with some very nice walleye pike and a South Bend bait. The picture won first prize of $500 in gold in the South Bend Bait [Company] contest of that year."[23]

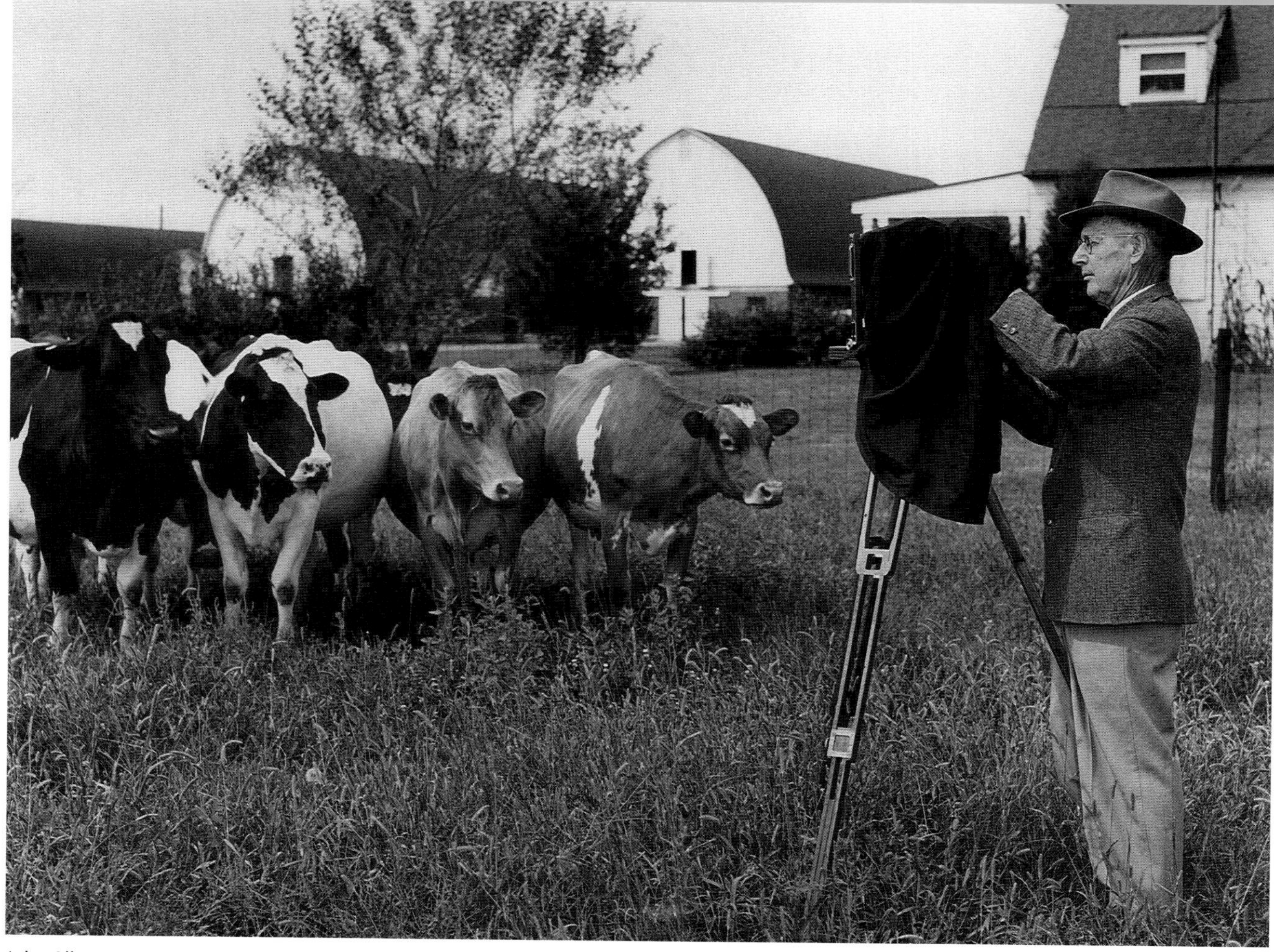

John Allen captures the attention of dairy cows. (N.p., 1958; courtesy John O. Allen)

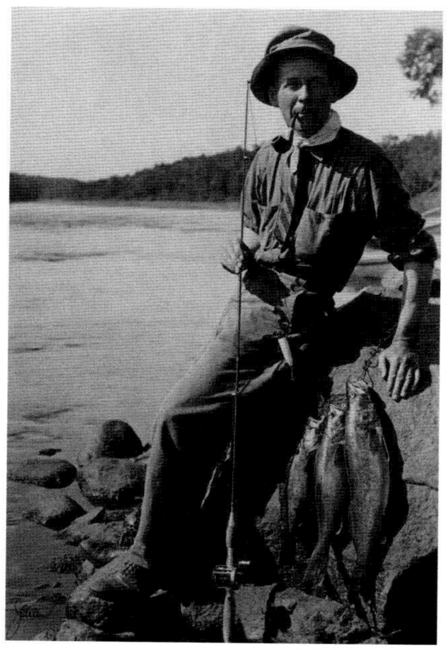

Around 1928, Allen won first place in a photo contest and received $500 in gold from the South Bend Bait Company for this image of himself following a successful fishing expedition. (Flambeau River in Wisconsin)

Allen relied on two Kodak-brand cameras for most of his career: a lighter-weight, wooden view camera and a heavier steel Graflex. In 1933, the *Purdue Alumnus* reported that his scientific approach to photography "enabled him to become an expert and his expanded business forced him to add new types of cameras to his collection and at the present time he has a full battery of view cameras, a Graflex speed camera, and a Cinema outfit."[24]

While he had many cameras in his photographic arsenal, his trademark was the wooden view camera, which sat on top of a wooden tripod. This camera used George Eastman five-by-seven-inch glass-plate negatives but was eventually converted to film.[25] His grandson would later report, "He continued to use these wood view cameras many years after more modern and smaller cameras became available because of the excellent quality of the pictures the large wooden view cameras made."[26]

But from the time he began taking pictures, Allen had taken photos with a Graflex, which could use either celluloid or glass negatives.

He once recalled, "I was able to purchase my own 5 x 7 Press Graflex and though I have used many cameras since then, I still have two of these now-out-of-production cameras and still prefer to use them for many types of livestock pictures."[27] At one time, he even became a spokesperson for the Kodak camera.

In a 1959 interview that appeared in *Hoard's Dairyman*, Allen described the progression of equipment he used.

> The first professional camera I used was a 5- x 7-inch Press Graflex with a 10- or 12-inch lens, Zeiss Tessar f4.5. This is a very heavy outfit and most people wouldn't want it for that reason, but I still use it for certain livestock groups where there is action. This camera is focused through a reflector and the subjects can be viewed in the camera until the exposure is made.
>
> Our next professional equipment was a 5- x 7-inch view camera equipped with a Protar lens in units which give various focal lengths. This lens, used on a tripod, gives a very sharp image and the depth of field at a given stop is greater.
>
> For interiors we use 8- x 10-inch view cameras equipped with [a] special wide-angle lens. Our 4- x 5-inch Speed Graphic is very convenient, especially for color where action must be stopped.
>
> We may have the only 5- x 7-inch Naturalists Graflex, which we had made to order for use in photographing individual beef animals where a long focus lens must be used to avoid distortion. The camera will accommodate a 23-inch lens and is focused though a reflector so the animal is seen on ground glass until exposure is made. This is quite an advantage with restless animals, especially in fly time. . . . We, however, prefer the larger cameras because we think the larger negative, when properly done, makes a better print.[28]

By the 1920s, photographers switched to cheaper and lighter-weight film, and glass-plate negatives fell out of favor.[29] Yet Allen preferred the glass plates over celluloid film and continued using them long after most people had changed over to film. In fact, his devotion to glass plates motivated him to purchase as many as he could find. Even though glass was more expensive, Allen thought they made better prints and did not deteriorate like celluloid film.[30]

When using a camera that required the glass-plate negatives, Allen would load a film holder in the view camera with two plates that were taken from a bag draped over his shoulder. After taking two pictures, he would place the now-used glass-plate negatives in the outside portion of his shoulder bag, and refill the film holder with two new plates from inside the bag. Allen continued putting them in this order until all of the glass plates in the bag were used. Upon finishing, he placed the negatives in a box in the order they were taken. Throughout the process, Allen was a meticulous note taker, writing down information such as the subject's name, location, date, and anything he considered of importance in a small notebook.

With negatives in a box and details in a notebook, Allen would return to his Purdue campus office on the ground floor of the Agricultural Experiment Station, which is known today as the Agricultural Administration Building. Allen had been placed in charge of a newly established university photographic service located there in 1920. In addition to himself, there were normally two employees on staff, with one paid for by the university and one, by the photography business.[31]

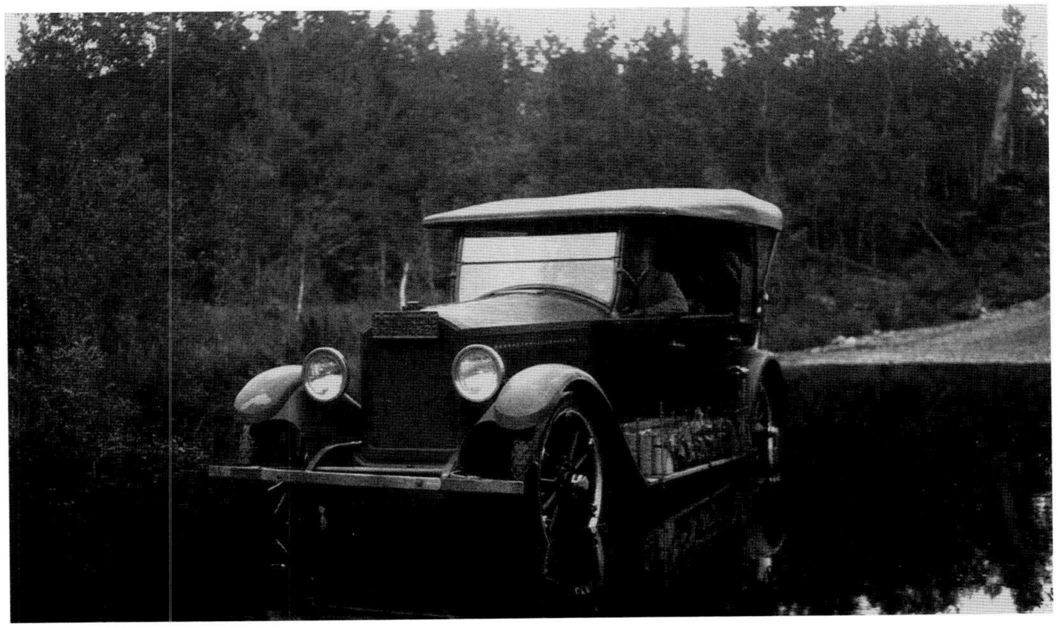

Allen traveled far and wide to obtain images for clients and his stock photo library. Here, he navigates a flooded road as he carries photographic and camping supplies on the running board of his car. (N.p., n.d.; courtesy John O. Allen)

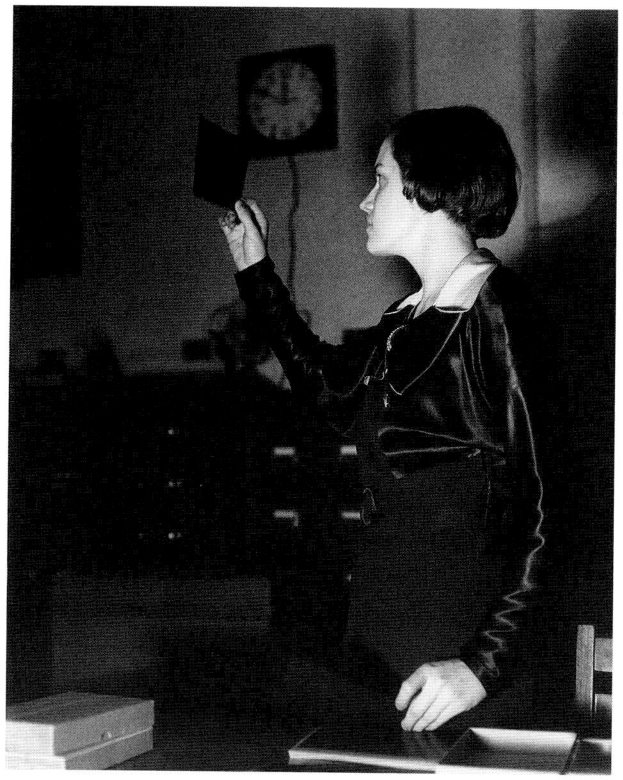

Here, Martha Allen works in the office of J. C. Allen and Son in the Agricultural Experiment Station Building on the Purdue University campus. (Tippecanoe County, ca. 1935)

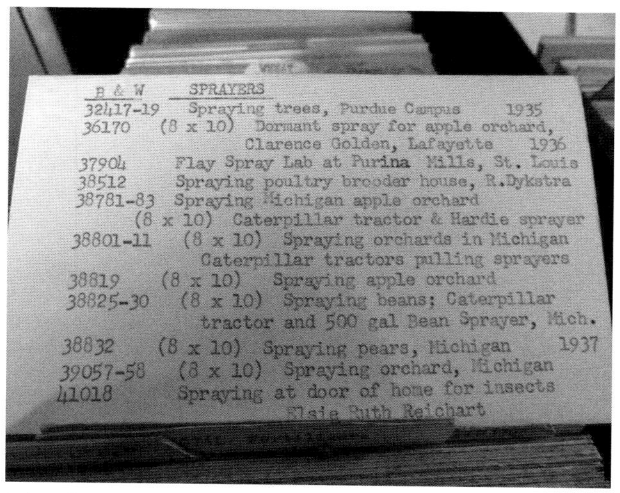

John Allen designed an elaborate card cataloging system to locate a photograph among the many thousands he had stored. Images were numbered, and the unique image identification number placed under the subject matter. (Tippecanoe County, 2017; photo by Frederick Whitford)

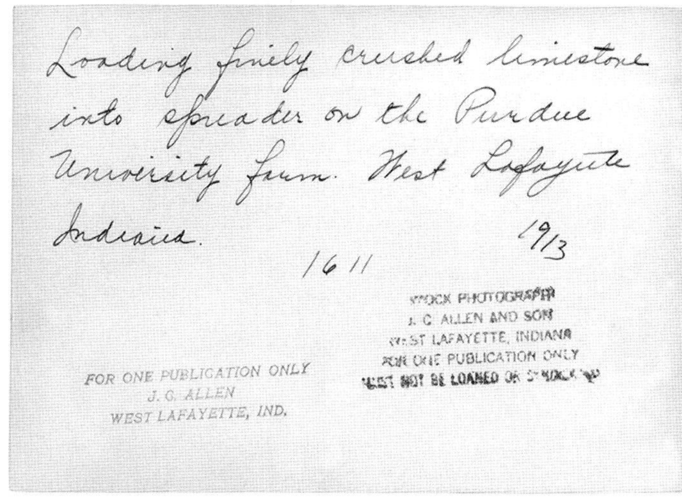

John Allen understood that a photograph loses its meaning without a description, date, or some type of identifying information. This example shows his fastidious attention to recording the details of his photo shoots, a practice which served his business well and continues to add tremendous value to the Allen photographs and negatives. (Tippecanoe County, 2017; photo by Frederick Whitford)

Black-and-white prints were produced in the office. One staff member worked in the darkroom to develop the negatives into prints, while the second sorted the negatives and prints into envelopes. Since the cameras Allen used made five-by-seven-inch prints, enlargements were needed when customers requested bigger sizes, such as eight-by-tens.[32] After the prints were developed, they were washed to remove any leftover chemicals from the darkroom processing. Then, each print was placed on a board and run through a dryer. Color negatives were sent to Kodak for processing.[33]

In the middle of the office was a large wooden table where the prints were sorted. Here, they also received an identification number and an identifying stamp, which read, "Photo By J. C. Allen, W. Lafayette, Ind." Over time, the wording of the stamp would be changed to "Stock Photograph, J. C. Allen and Son, West Lafayette, Indiana, For One Publication Only, Must Not Be Loaned or Syndicated." Allen or his staff would then write a brief description in pencil on the back of the print. Occasionally he typed a very detailed description on paper and glued it to the back of the photograph.

Prints were placed in envelopes that had the same number as the print and notes that had been transferred from Allen's notebook. This photo sleeve also contained the film or glass-plate negative. Some of Allen's first photographs, which were images from Purdue's beef cattle experiments dated 1910, were stored in this manner.

As his Purdue and personal work increased, the number of photographs he had to manage also grew, so he needed a way to quickly retrieve specific photographs. How would he find the one photograph amongst the tens of thousands—that needle in the haystack? To solve that problem, he created a filing system with individual subject cards labeled as horses, hay, electricity, and so on. Each subject was further divided into subcategories, then the identification number on the photo sleeve would be written under the appropriate subject and subcategory. As a result, he could look in his card file, retrieve identification numbers, and select the exact image he wanted. It's been estimated by his grandson that, at one time, the card catalog had 77,000 subject categories that categorized several hundred thousand photographs.[34]

While the employees went about their daily work, Allen could be seen at his rolltop desk. There he maintained the accounting books for his Purdue and personal business, took requests from university and other clients, and looked through the files to fill orders. Years later, he recounted:

> Many of my office hours are spent in assembling a suitable selection of pictures to send in response to telegrams, telephone calls, and mail requests. These selections are made up from our stock prints, which are numbered, cataloged, and filed according to subject matter. We have a stock collection of some 25,000 farm and livestock pictures. As time goes by, some of these may become more valuable as a history of early farming operations.[35]

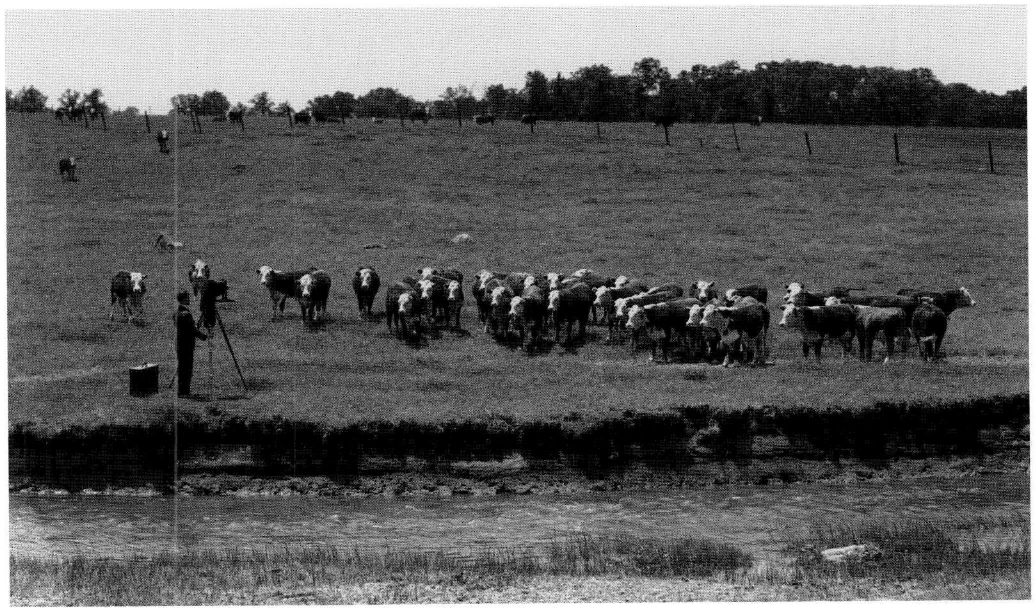

Like father, like son. Chester Allen photographs a herd of Herefords along a creek on the Ira Bordner farm in Brookston. John Allen changed the name of his company when Chester joined the business in 1929. (White County, 1938; courtesy John O. Allen)

As Allen's reputation grew, J. C Allen Rural Life Photo Service began receiving more requests from around the country. In time, Allen's son, Chester, began working part time as a photographer while he was a student at Purdue. When he graduated in 1929 with a bachelor's degree in horticultural science, he became full-time photographer, and the name of the business was officially changed to J. C. Allen and Son Rural Life Photo Service.[36] According to John O. Allen, a grandson to John C. Allen and son of Chester's, the two had "a friendly competition of who could get out in the field and shoot the most pictures. Sometimes they traveled together, but many times they did not."[37]

John C. Allen wrote that Chester "was great help in further developing the photo business I had started. I had put a lot of effort into getting started, but by this time business was going very well and we could spend more of our time in production, hunting new picture material and finding new markets."[38] He also noted that Chester "was quite good with writing stories to go with the pictures."[39] The *Purdue Alumnus* became a steady customer of the pair, stating, "More than half of all the photographs which have appeared in the *Alumnus* since the magazine began using illustrations (shortly after World War I) have come from the same source—John and Chester Allen, official photographers for the Purdue Agricultural Experiment Station and for the Purdue Extension Service."[40]

While John preferred taking images in black and white, Chester converted to color film. John told Purdue agricultural communicator Ralph Reeder that this was "because the magazines and advertising agencies want color even though it is more expensive."[41]

According to John O. Allen, "Chester Allen entered the business during a time when tractors were rapidly replacing horses for power, and the tractor companies wanted photos of their new models. There was a great demand for truck and tire pictures, since that was the era of change from metal wheels on farm equipment to rubber tires."[42] Until his death in 1996 at the age of 89,

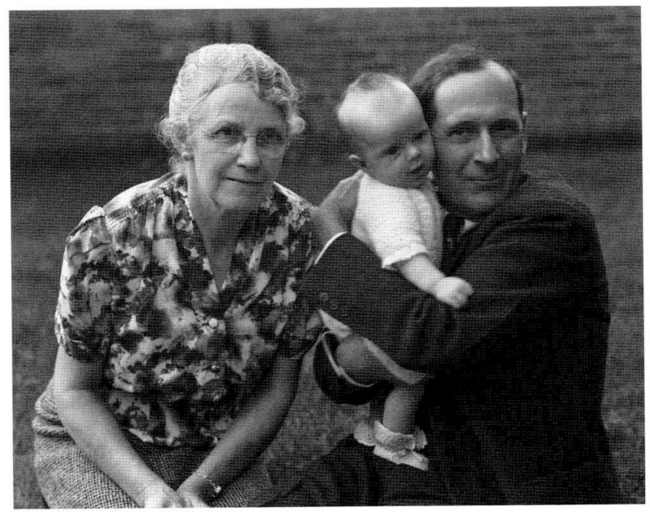

The Allens with their granddaughter, Martha Joan Benham, in 1939. (Tippecanoe County; courtesy John O. Allen)

Farmer for Life

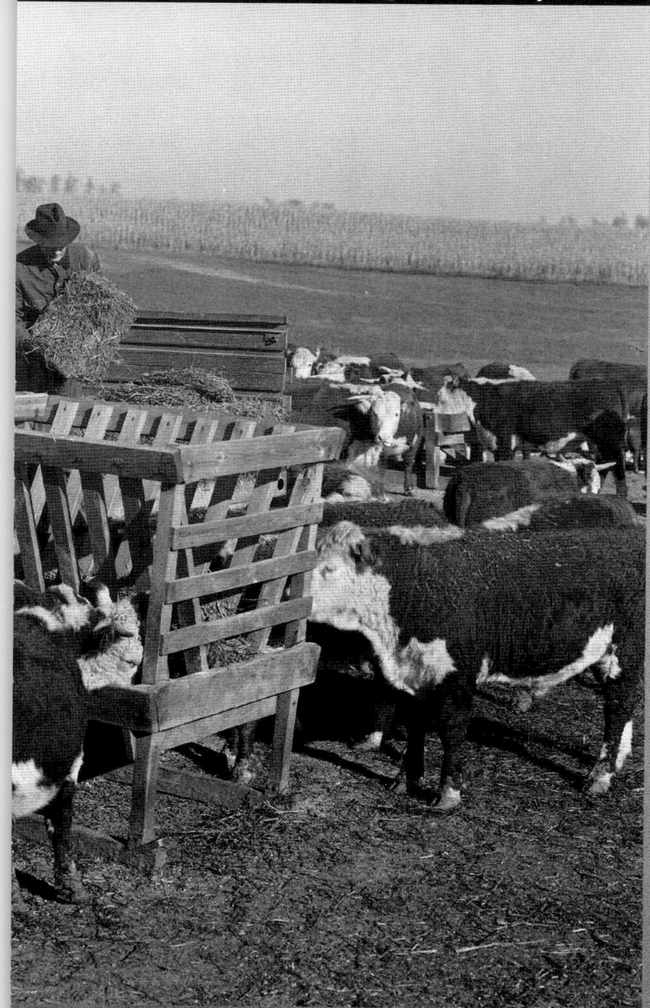

Not only did Allen (shown here) enjoy photographing farm life, but he also owned a number of working farms. (Warren County, 1948)

John Allen and his wife, Abbie, owned a number of farms during their lifetimes. They obtained the land through a variety of methods, including as a gift, an inheritance, and purchases. Abbie received an 80-acre parcel of farm ground in Greene County from her mother in 1911. Over the next thirteen years, the Allens would purchase two additional 40-acre tracts and increase that farm to 160 acres. In 1945, Abbie received 40 more acres of land in Clay County through an inheritance. The following year, the Allens sold their 200 acres in Greene and Clay Counties.

From 1937 to 1960, the Allens bought four farms and used tenants to manage the properties.[43] In 1937, Allen purchased a 346-acre farm in nearby Montgomery County. A few years later, he purchased a 240-acre farm in Warren County and a 320-acre farm in Benton County. Allen purchased his last farmland—238 acres in Tippecanoe County—in 1960.[44]

In 1955, the Allens began divesting themselves of some of the farm property. They gave their daughter, Martha, the Benton County farm and son Chester the farm in Warren County. They raised hogs and cattle between the two remaining farms.[45]

Grandson John Allen Benham noted, "At the time of my grandparents' death, 346 acres in Montgomery County went to me and my sisters. The 238 acres in Tippecanoe County went to our cousins, John and Jimmy [Allen, Chester's sons]."[46]

Chester dedicated his career to photographing rural life, just like his father had done before him.

John C. Allen would eventually live to see yet a third generation join the family business. In 1967, grandson John O. Allen began working with his grandfather and father after receiving a bachelor's degree in agricultural economics from Purdue. Eventually, Daniel G. Allen, who was John C. Allen's great-grandson and John O. Allen's son, would become the fourth-generation photographer with the family business.

John O. Allen explained, "The business went from a company name to a corporation around 1970 in the name of J. C. Allen and Son, Inc. Grandfather was still with us at that time and he was very complimented that we wanted the business to remain permanently with his name starting the title. We would not have had it any other way."[47] Today, the grandson continues what his grandfather started early in the twentieth century.

John C. Allen remained on staff at Purdue until his official retirement in 1952.[48] He had wanted to retire in 1951, but for unknown reasons, the retirement was delayed for a year, with the Purdue Board of Trustees announcing it in June 1952.[49] After leaving the university, Allen moved his business to his home on Northwestern Avenue, noting, "With opportunity to put our entire time into the illustrating business, it continued to grow."[50]

Even as Chester took on more duties and responsibilities, the elder Allen continued to be active in his business.[51] One of his last photography trips was in 1969, when the eighty-eight-year-old went out on

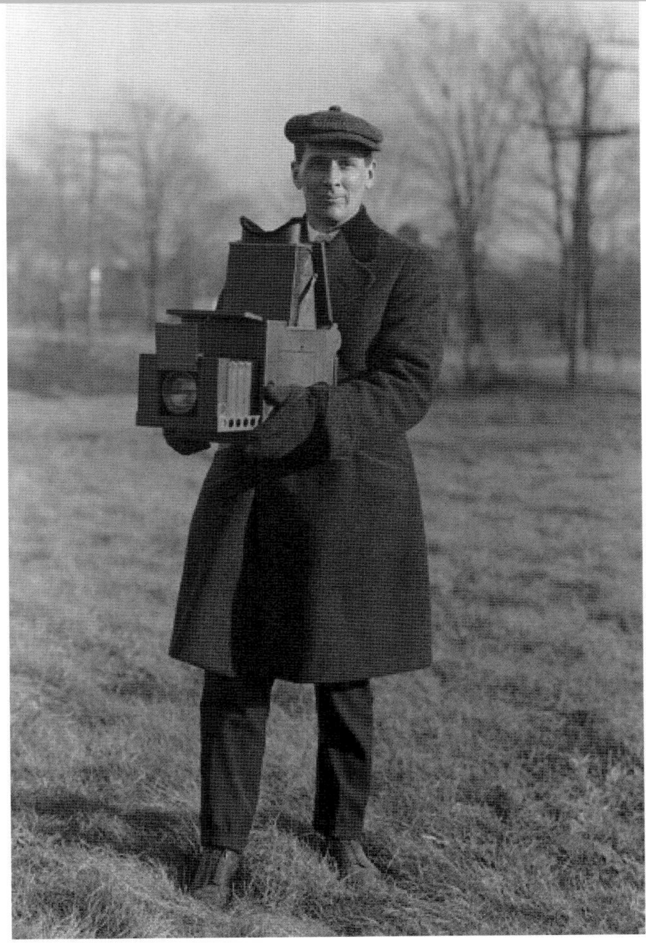

John C. Allen with his Graflex camera (N.p., 1919; courtesy John Allen Benham)

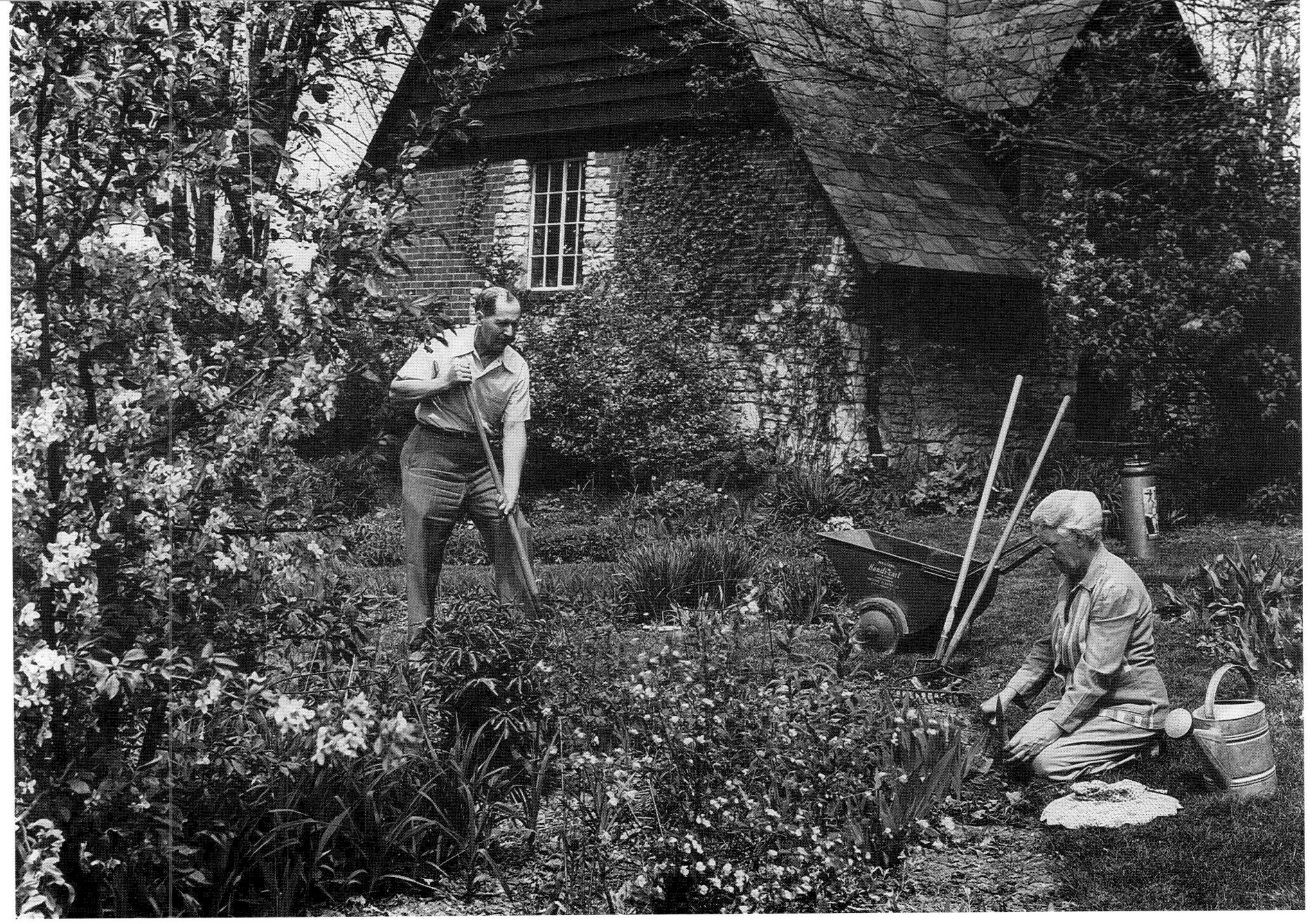

Abbie Allen (right) loved to garden on the grounds of her home on Northwestern Avenue. Her flowers were often photographed by John, who is shown here helping his wife. (Tippecanoe County, 1948; courtesy John O. Allen)

assignment for *Farm Journal*. Chester carried the larger, heavier cameras, while his father carried a lighter Rollei camera. John captured an image of Hereford cattle looking through the slats of a farm gate, which appeared on the cover of the January 1977 issue of the publication.[52]

In retirement, John C. Allen continued to receive recognition for his work. In 1963, the Purdue Agricultural Alumni Association honored him with its Certificate of Distinction, which is the highest honor the organization bestows. In 1977, the American Agricultural Editors' Association named him as an honorary member, summarizing his work by saying, "It's doubtful that any other name has been so widely recognized in agricultural photography. The J. C. Allen and Son mark has probably appeared with more photographs, in more farm publications, than any in our nation."[53]

Even at age ninety-four, Allen kept regular office hours, answering the phone and occasionally venturing out to take photographs.[48] His time during the latter years was also spent enjoying gardening, fishing, golf, bowling, bridge, and Purdue football.[55]

On July 21, 1976, John C. Allen died at home at the age of ninety-four. He was laid to rest at Grand View Cemetery in West Lafayette.[56] Four years later, his wife, Mary Abbie Allen, died on August 19, 1980, at age ninety-eight and was buried next to her husband.

Allen's work continued to receive accolades following his death. In 1996, his business received the Agricultural Science and Heritage Award from the Indiana State Fair. And in 2012, the Indiana Historical Society gave it a Centennial Business Award for its 100 years in business.[57] On April 14, 2004, a permanent display to honor the legacy of John C. Allen was installed inside Pfendler Hall on the Purdue campus at West Lafayette.

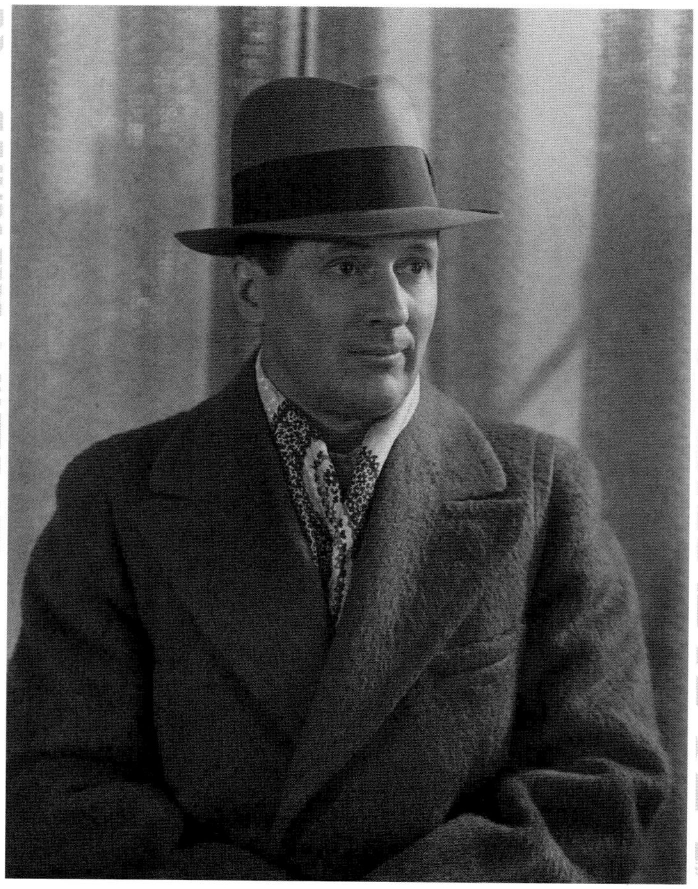

John C. Allen, January 1, 1931 (N.p.; courtesy John Allen Benham)

A Lasting Legacy

Picturesque images of men, women, and children working on the farm remain powerful reminders of life in rural America at the turn of the twentieth century and beyond. While those people and times no longer exist today, their memory is kept alive and preserved by the work of the photographer. A camera in his hands and an eye for photography allowed John Calvin Allen to create indelible visual histories that continue to tell the story of agriculture and rural life from a long ago time.

Allen had a talent for capturing on film the feelings and personalities of the people and what it meant to live in rural areas. Through his work, he showed the very essence of the farmer working the family farm, a woman caring for her home and family, or a child playing and working around the home and farm. Looking at these photographs, one can almost step into the image and know exactly what the person was thinking and doing during that brief moment when Allen focused his camera and snapped the shutter.

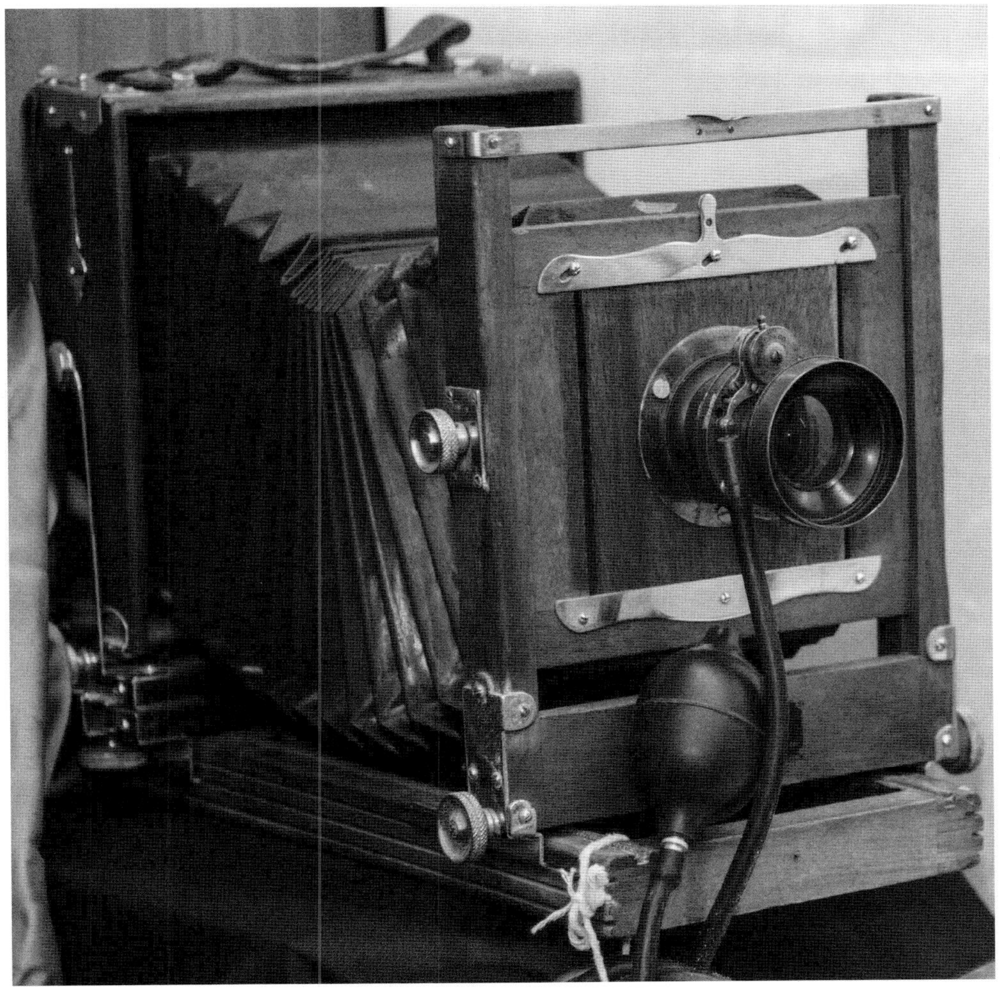

A wooden Kodak camera used by John Allen is displayed in a case in Pfendler Hall on the Purdue campus in West Lafayette, Indiana. (Tippecanoe County, 2018; courtesy John Obermeyer)

When presented with the opportunity for a good photograph, Allen was sure not to pass up that moment, for he understood that exact image might never show itself again. Throughout the United States, his customers such as magazine editors, businessmen, and farmers knew that, whether Allen already had the photo on hand or would seek it out upon request, he would get what they needed and fill their orders quickly. He understood his customers so well that he often took photographs of what they needed before they even seemed to know what they wanted.

John C. Allen's efforts captured rural American lives, livelihoods, and places during a majority of the twentieth century. To him, the collection represented many decades of hard work and an appreciation for how well the photography business had provided for him and his family.

But to the historians of agriculture and of American life who study these photographs, Allen's photographs document moments in time and provide insights into the past.

As John O. Allen reminisced about his grandfather's work, he noted how the collection has taken on this new meaning with time: "In the earlier years we were really not aware of how valuable the record of the development of agriculture and rural life were. The changes seemed to take place slowly. After a period of time passes, you can look back and see the great change that had taken place. That's when it really hits you what the photographs captured."[1]

While the interpretation, meaning, and relevance of these images may shift over time, they will always provide an account of that moment, fixed and unchanging, when the photographer and his camera found their focus.

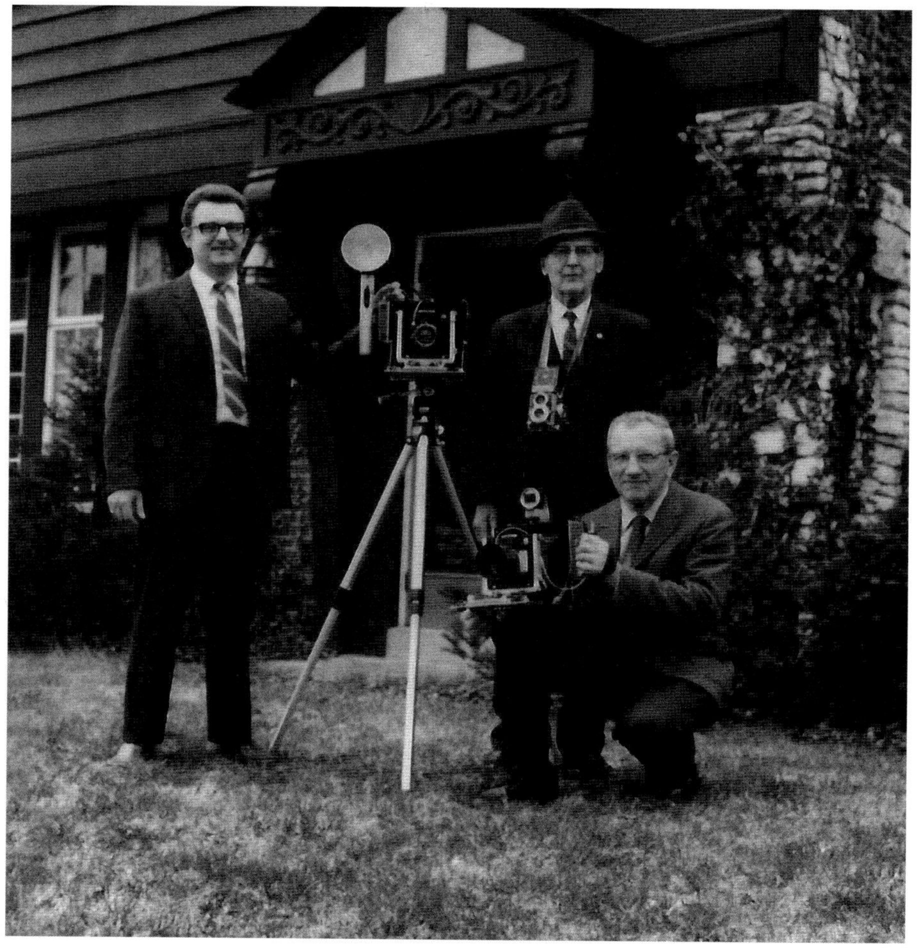

The J. C. Allen and Son photography business would span three generations and more than six decades. Here, the photographers become the subjects as John C. Allen (standing at the back right) poses with son Chester (right foreground) and grandson John. (Tippecanoe County, 1968; courtesy John O. Allen)

Public Collections of J. C. Allen and Son Images

There are two major public holdings of J. C. Allen and Son photographs. A substantial collection of Allen prints and negatives is preserved by the Purdue University Archives and Special Collections in West Lafayette, Indiana. The Indiana Historical Society also has a large collection of J. C. Allen and Son photographs from John C. Allen's work as the official photographer for the Indiana State Fair from 1914 into the 1960s.

THROUGH THE LENS OF PIONEER PHOTOGRAPHER

J. C. Allen

PART I

Life on the Farm

The Farm Homestead

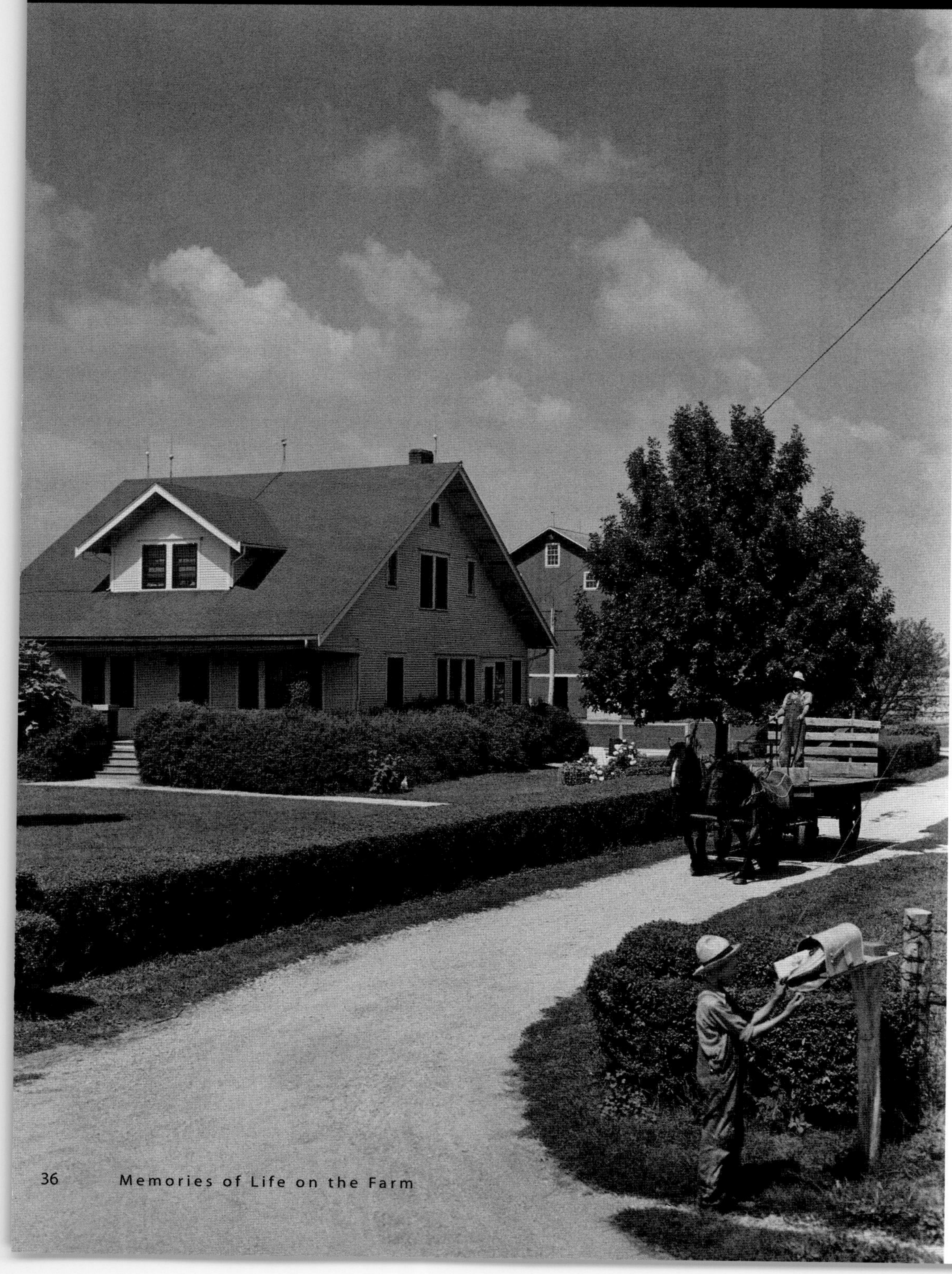

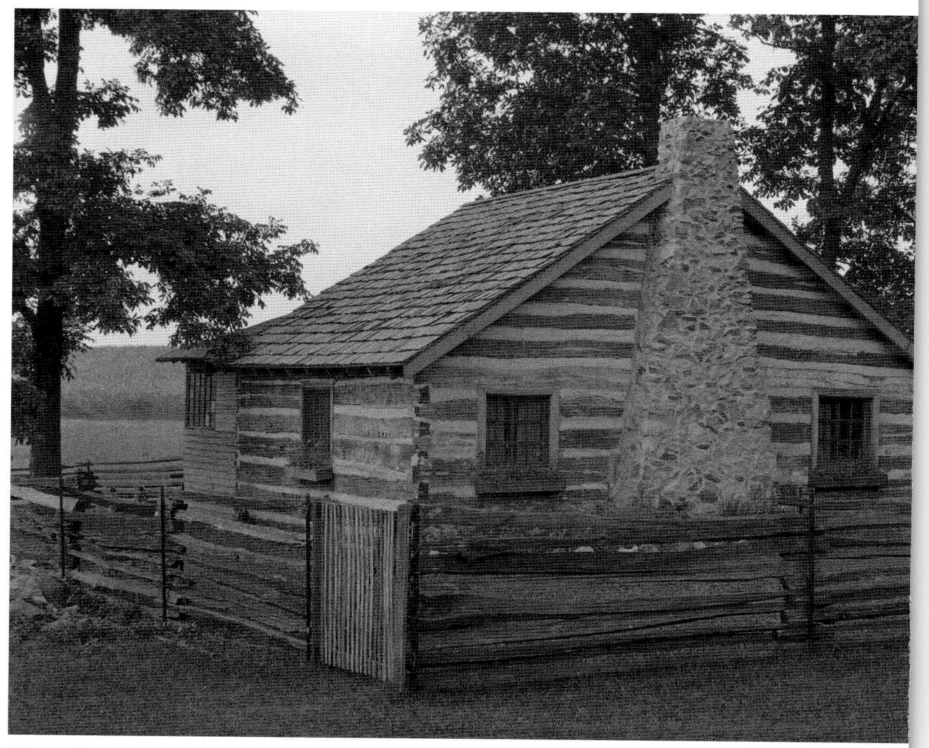

A log cabin near Pendleton (Madison County, 1935)

The well-cared-for farmstead of Ed Woldorf in Mendota, Illinois (LaSalle County, IL; 1937)

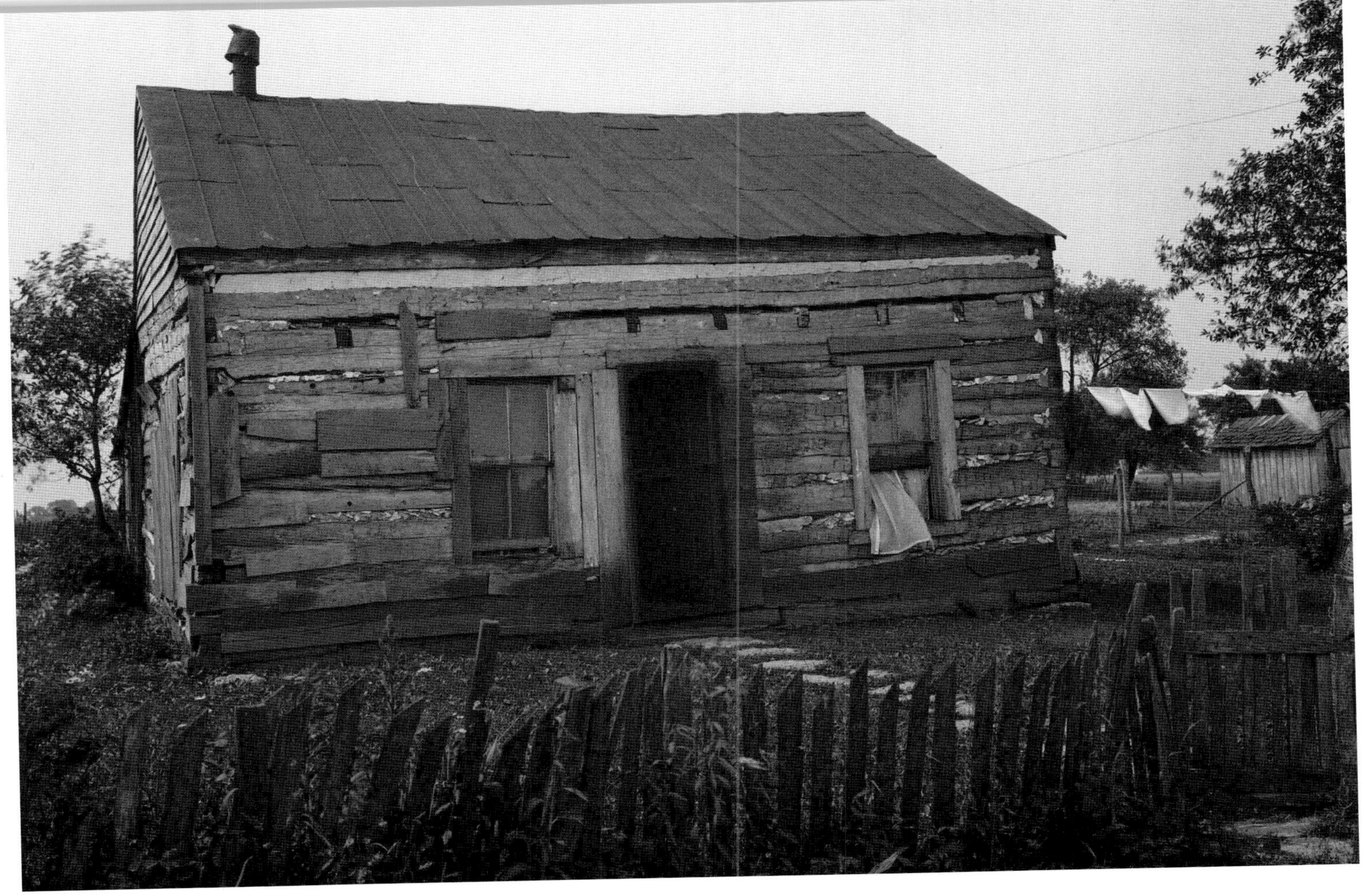

(Top) A log cabin on the John Ballard farm (Switzerland County, 1922)

(Bottom left) A tenant loads a wagon with possessions. (Tippecanoe County, 1916)

(Bottom right) This log cabin with its sod roof was photographed in Medora, North Dakota. (Billings County, ND; 1925)

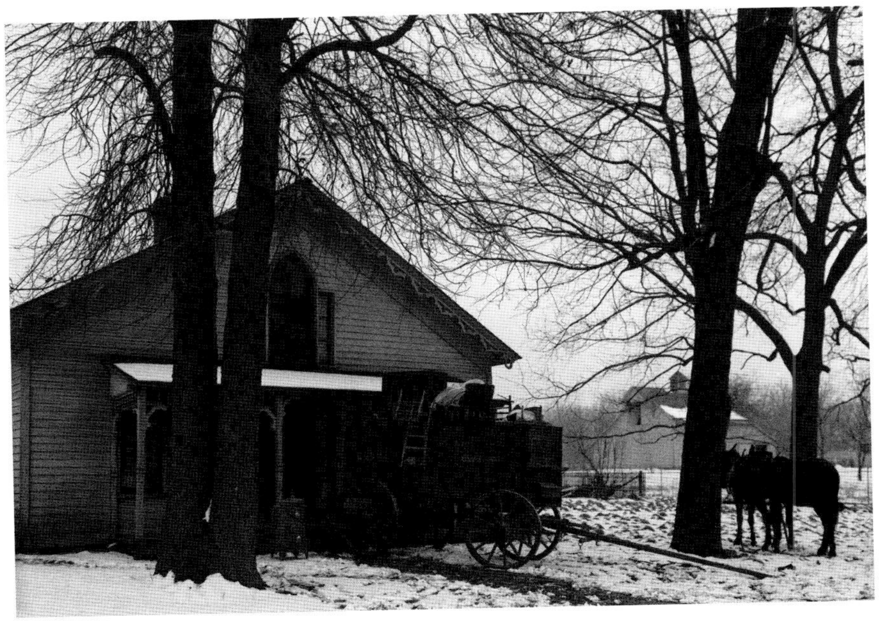

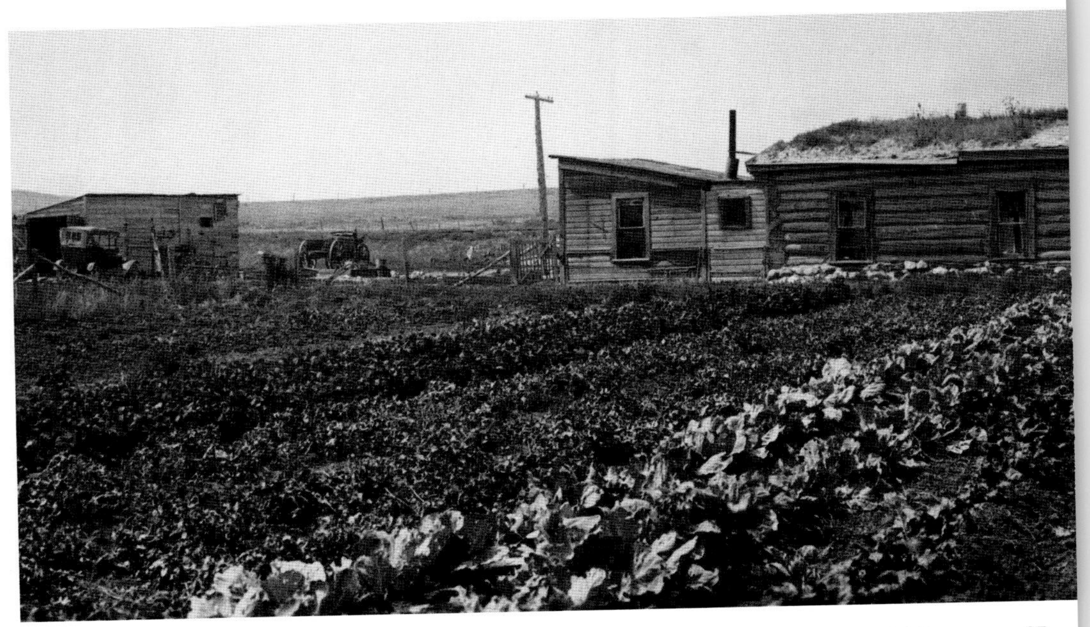

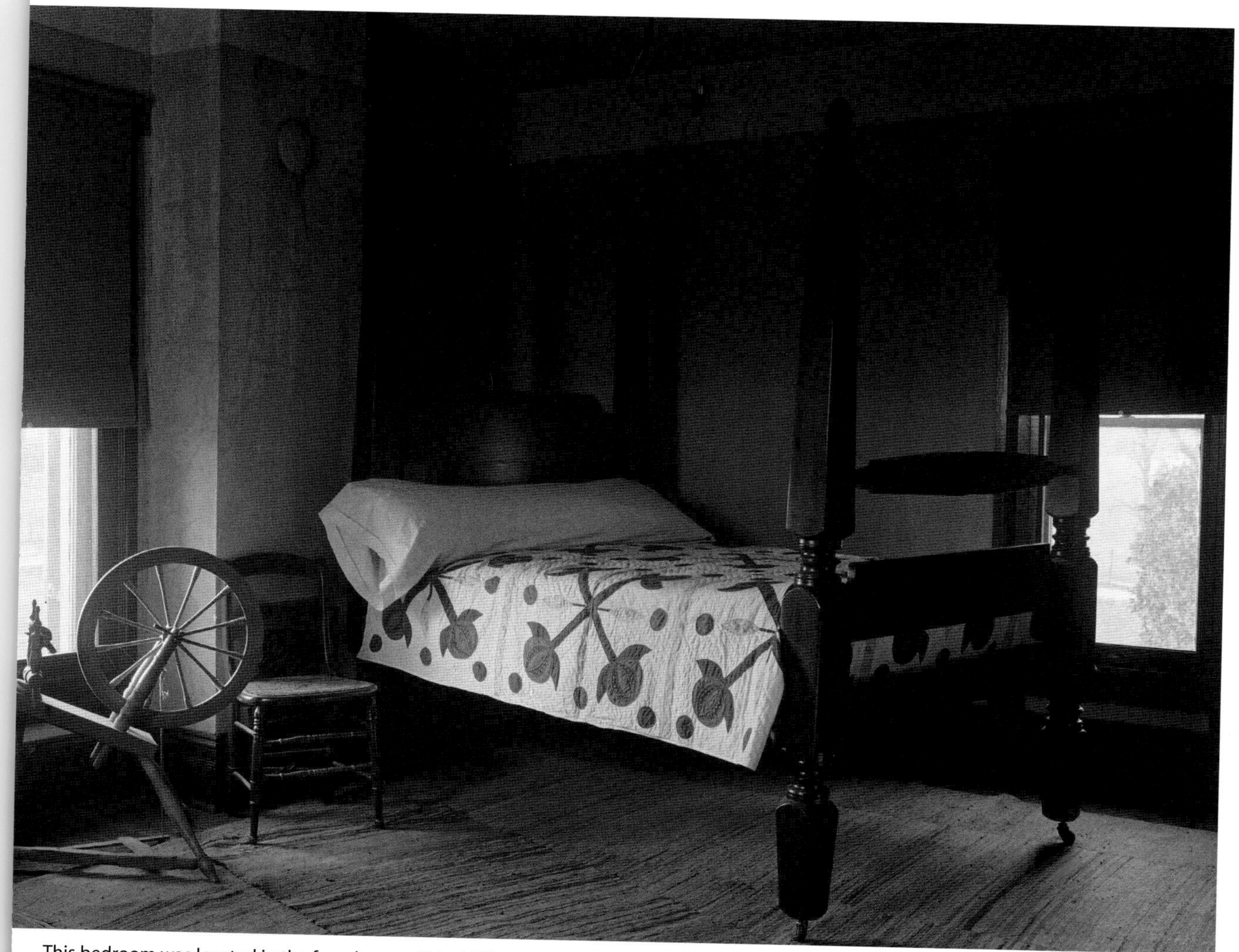

This bedroom was located in the farm home of Mrs. William V. Riggs of Trafalgar. Allen noted, "The quilt on the bed is 75 years old and was made by her grandmother. The bed is of cherry and was made by her grandfather." (Johnson County, 1925)

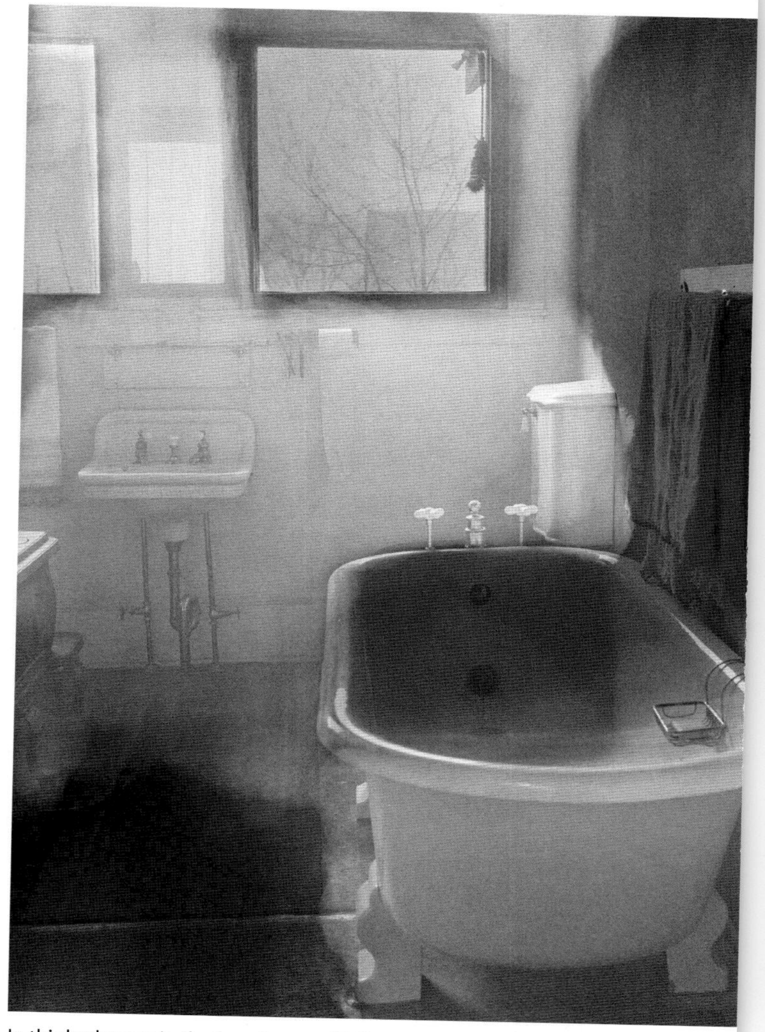

In this bathroom in the farm home of William V. Riggs of Trafalgar, the water for the bath and toilet was pumped from a spring some distance from the house. (Johnson County, 1925)

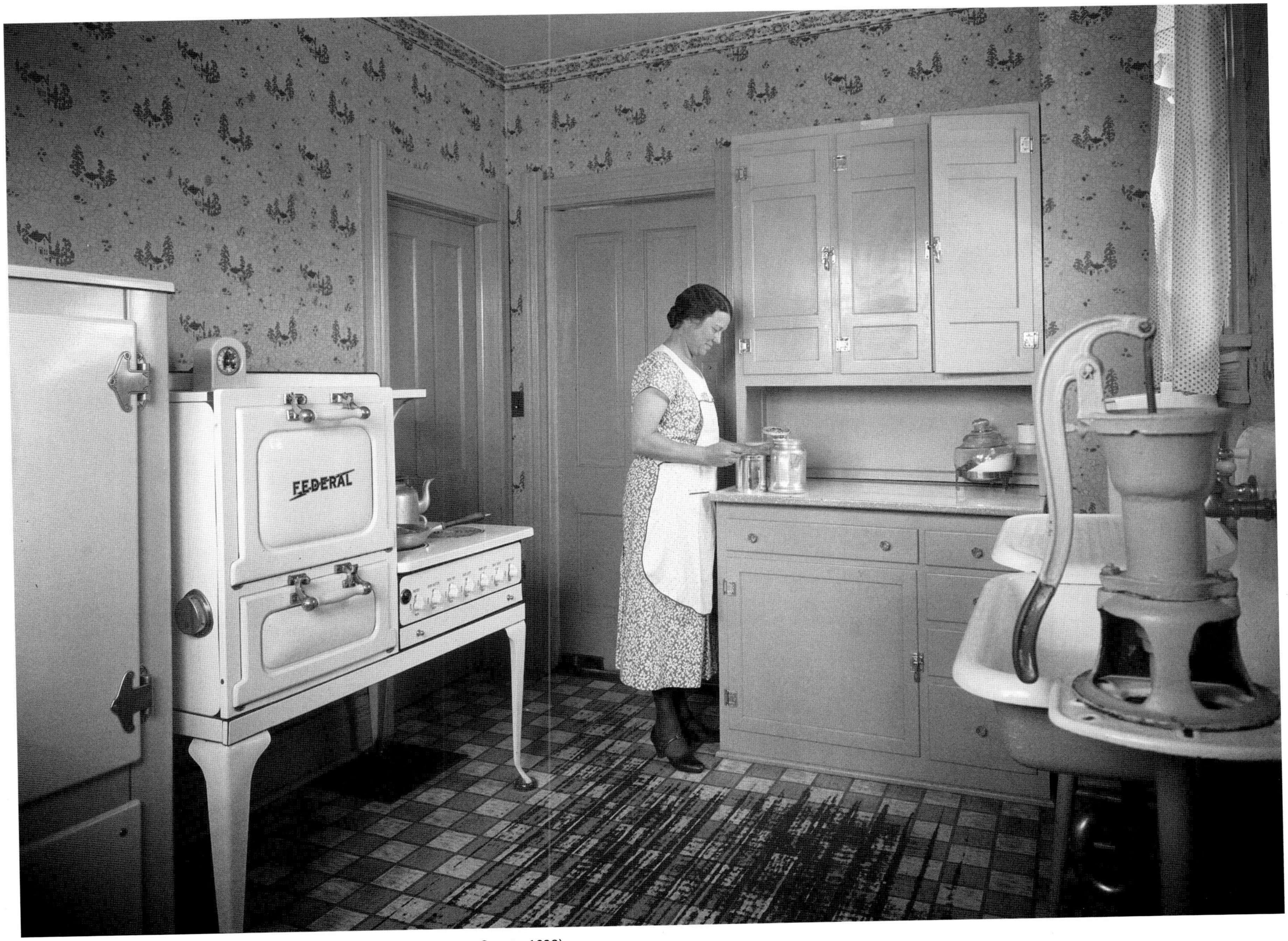

Mrs. C. H. Smalley's modern farm kitchen near West Lafayette (Tippecanoe County, 1932)

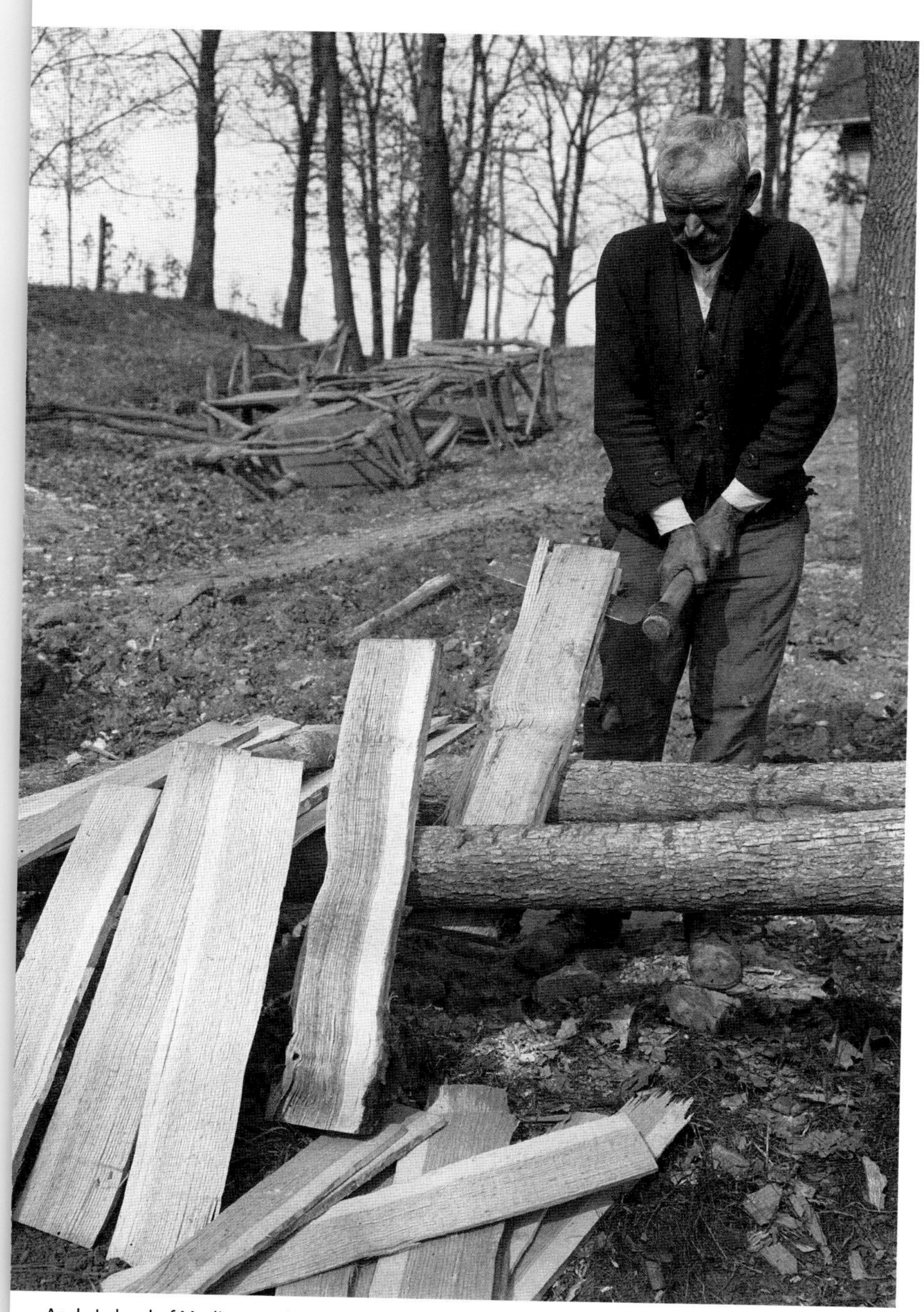

Andy Leland of Madison makes clapboards by splitting white oak logs. (Jefferson County, 1925)

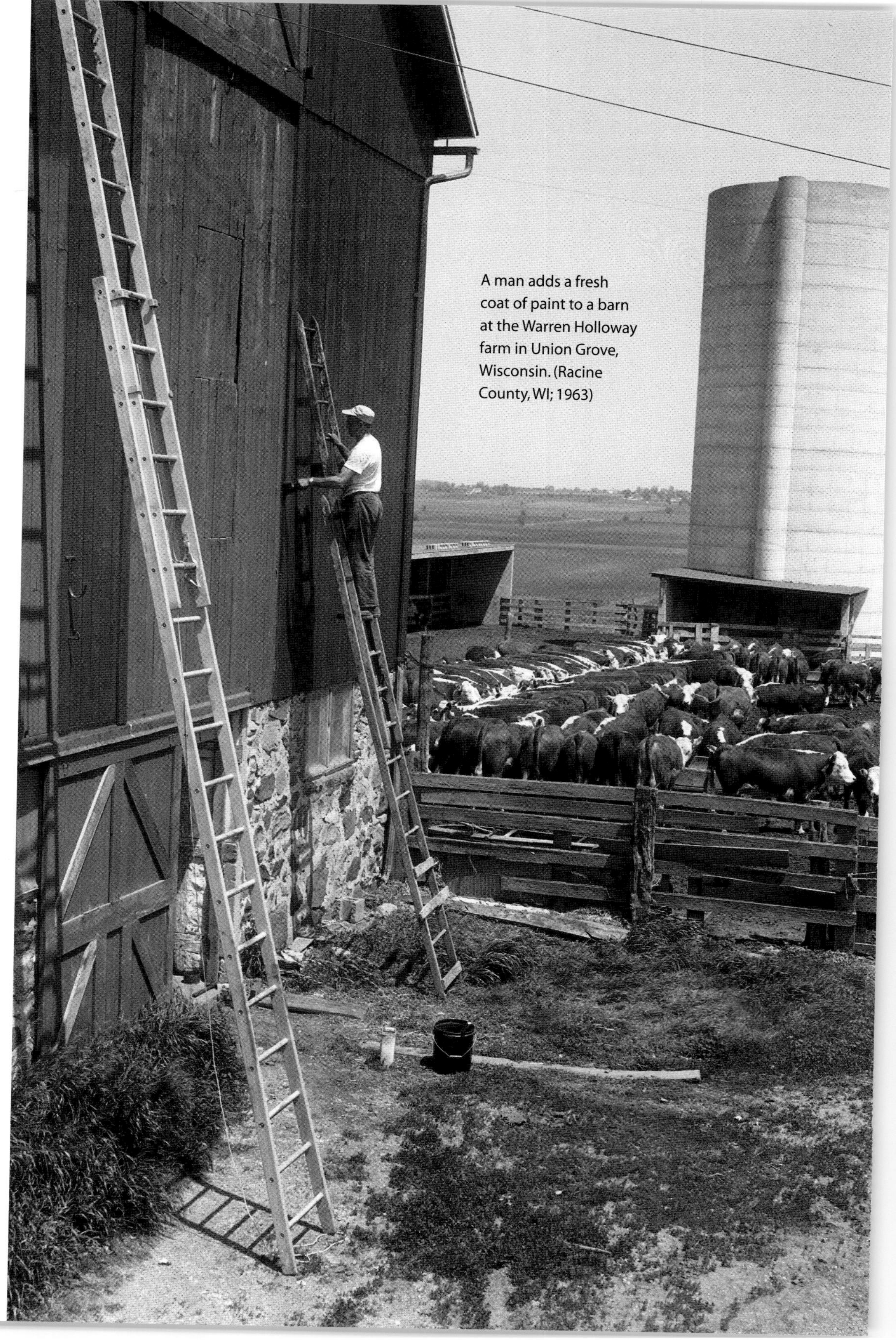

A man adds a fresh coat of paint to a barn at the Warren Holloway farm in Union Grove, Wisconsin. (Racine County, WI; 1963)

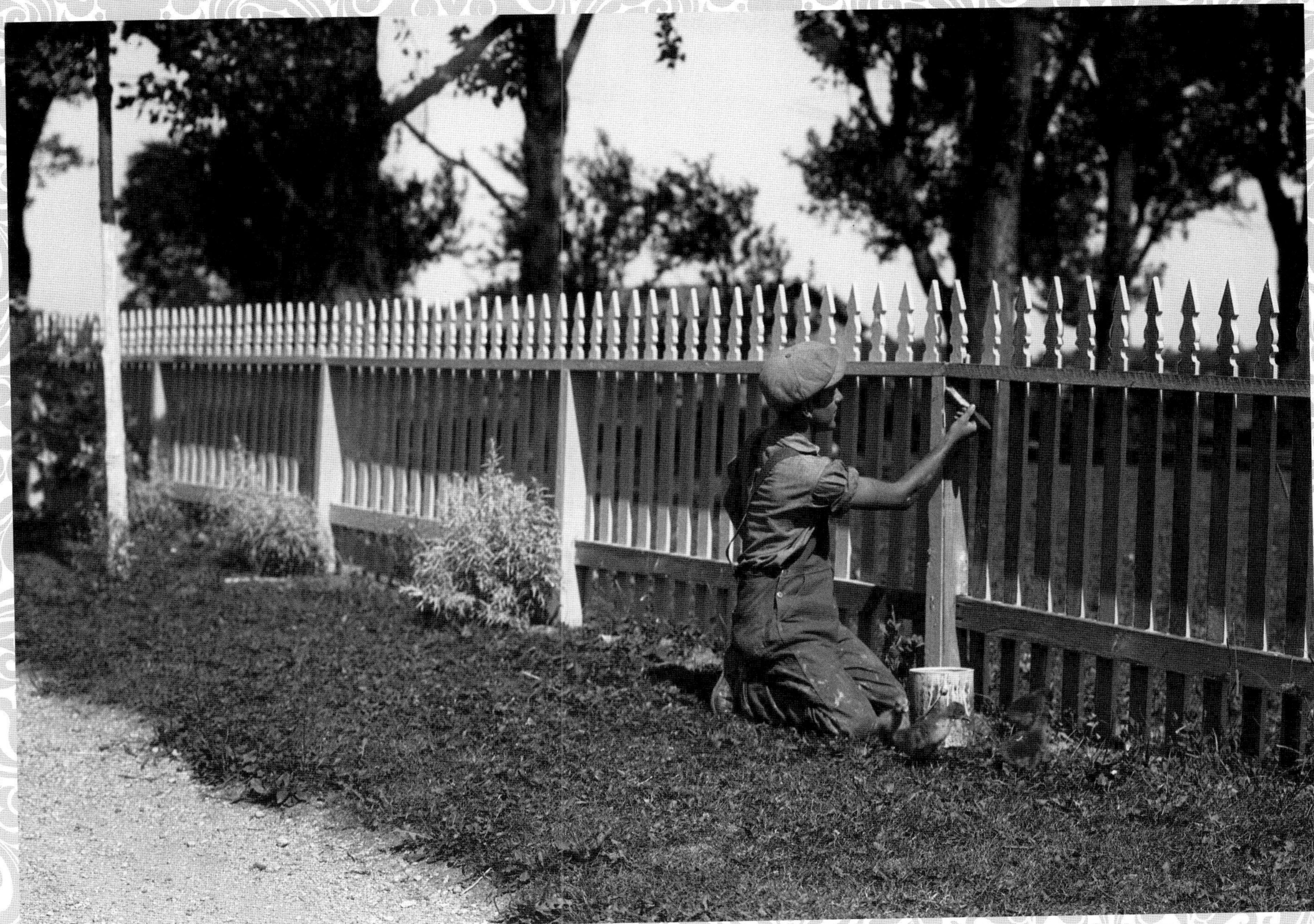

Painting the fence around the farmstead of W. V. Polk, whose property was located in Burlington (Carroll County, 1932)

Fueling the Farm

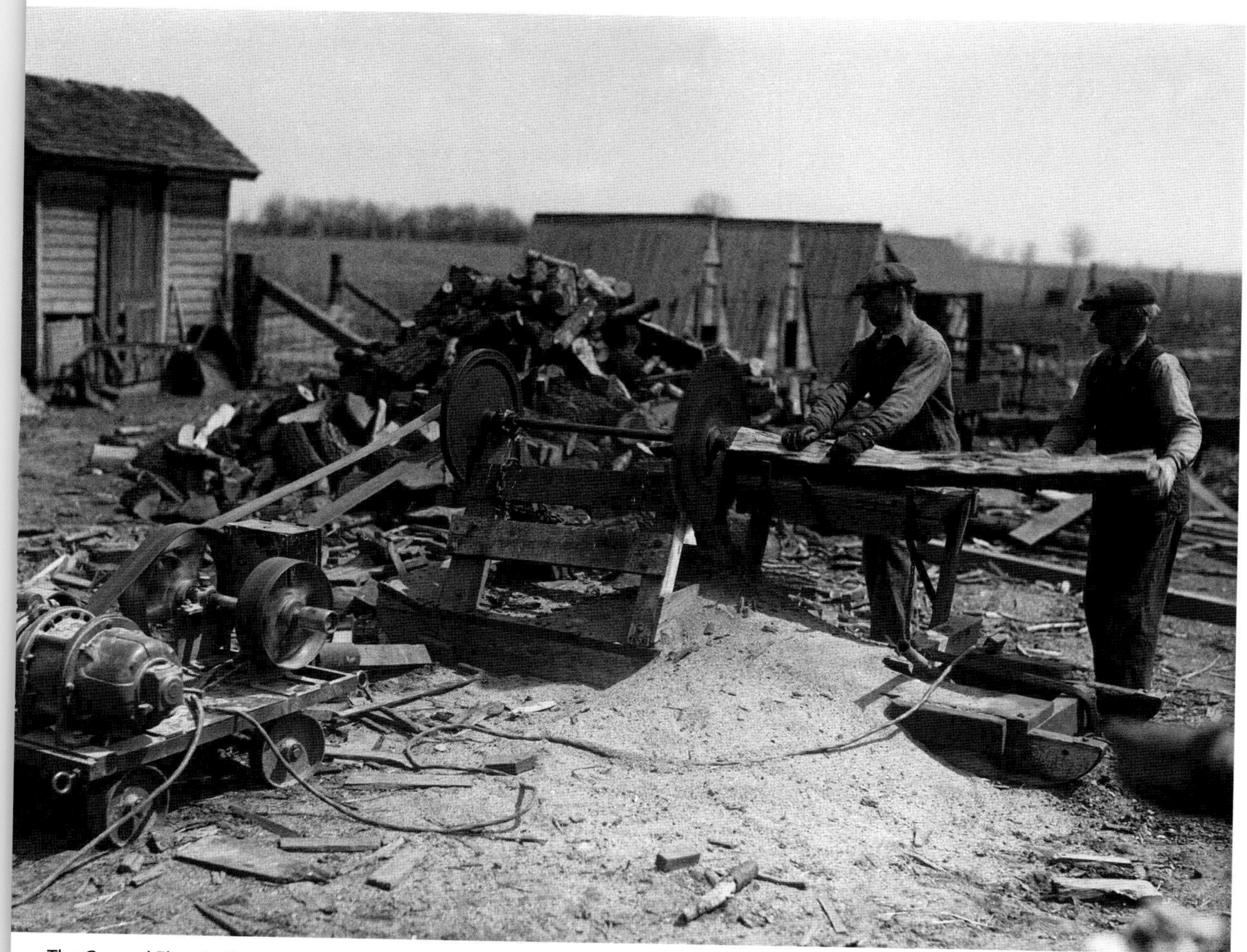

The General Electric Company developed this five-horsepower farm utility motor for small power operations, such as sawing wood at the Pinney Purdue Farm in Wanatah. (LaPorte County, 1926)

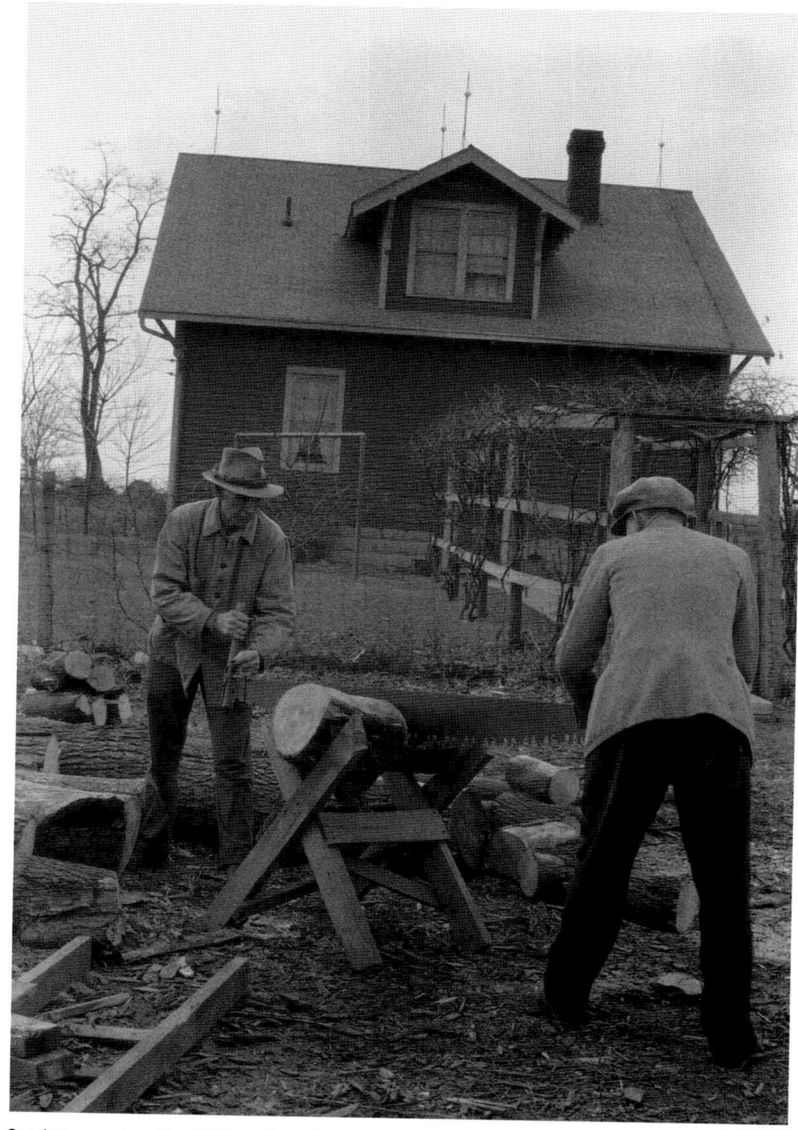

Sawing wood on the William Boes farm near West Lafayette (Tippecanoe County, 1931)

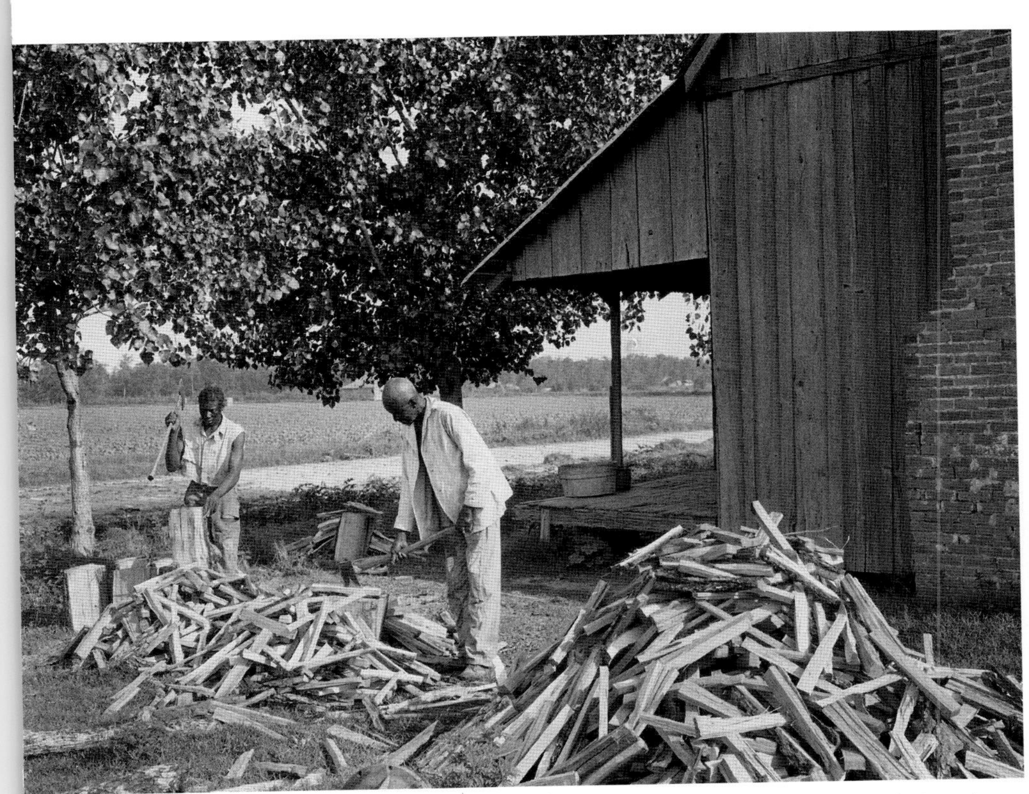

E. R. Brown and his son prepare their winter supply of wood near Scott, Arkansas. (Lonoke and Pulaski Counties, AR; ca. 1935)

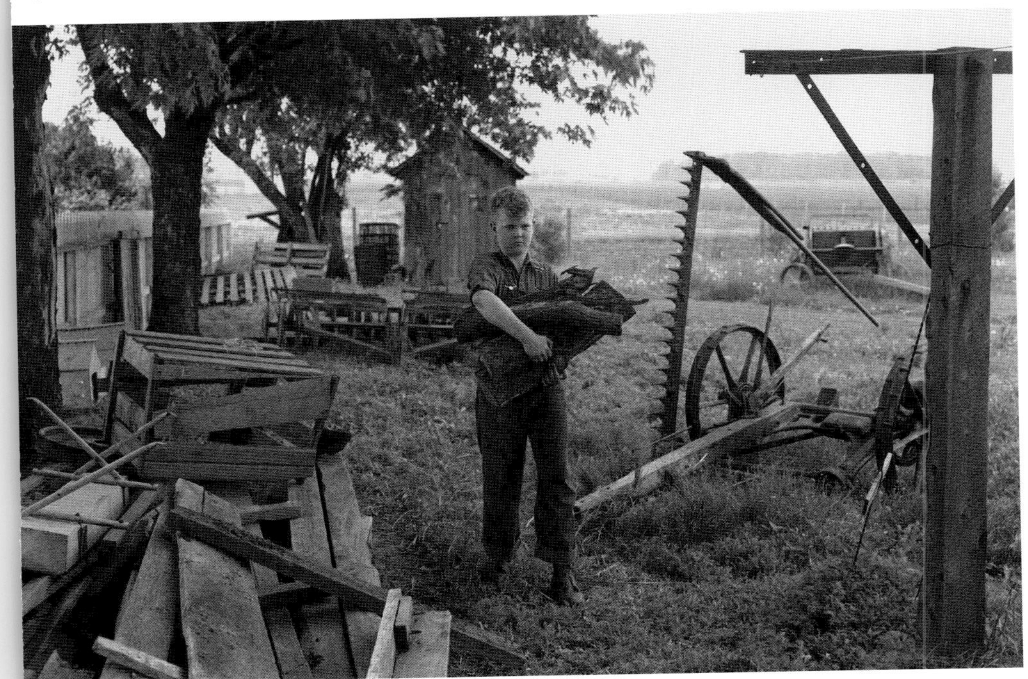

Allan Gustavel carries a load of wood to his family's house. (Tippecanoe County, 1940)

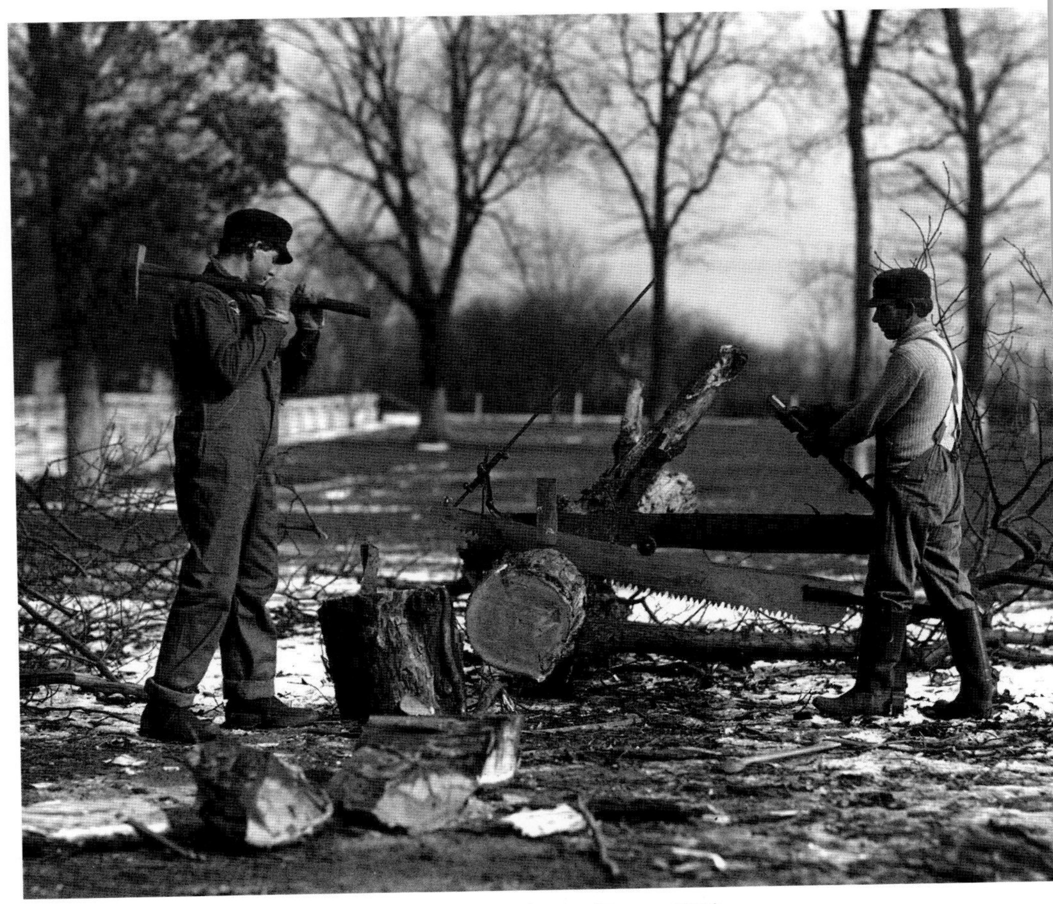

Two men cut wood in the summer before the fall harvest begins. (N.p., ca. 1915)

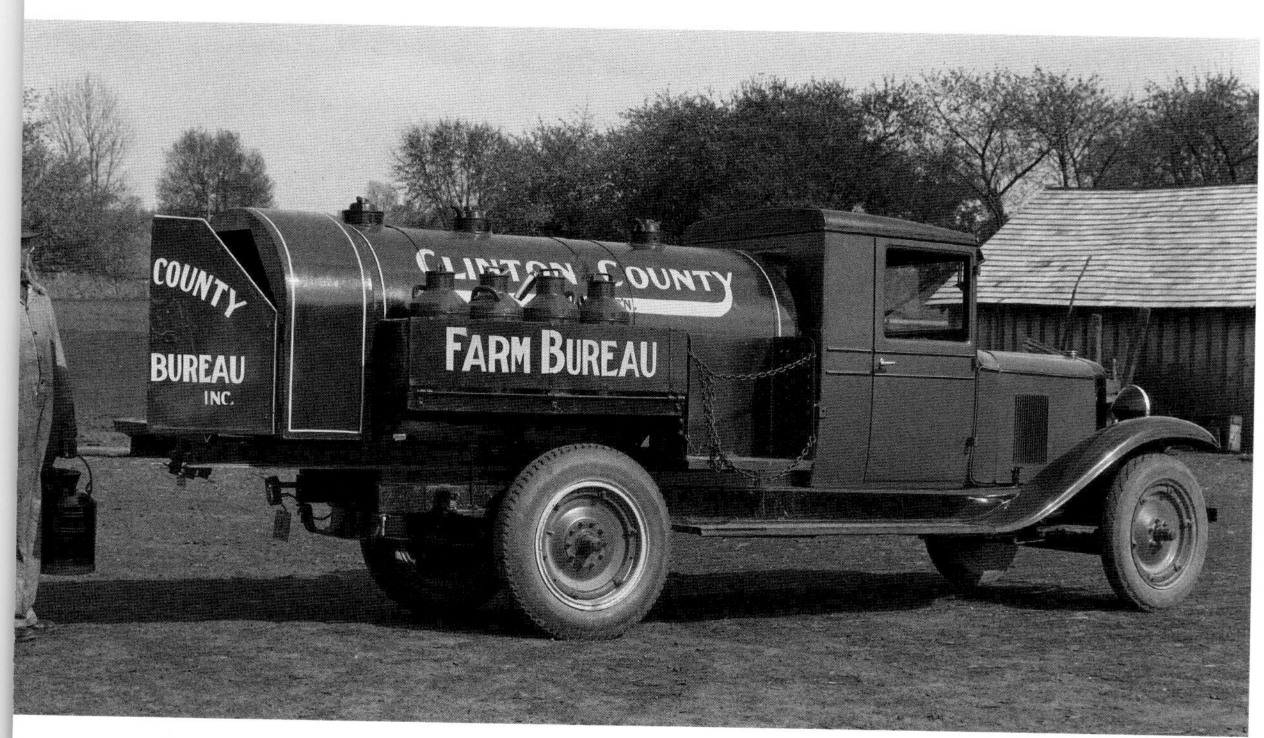

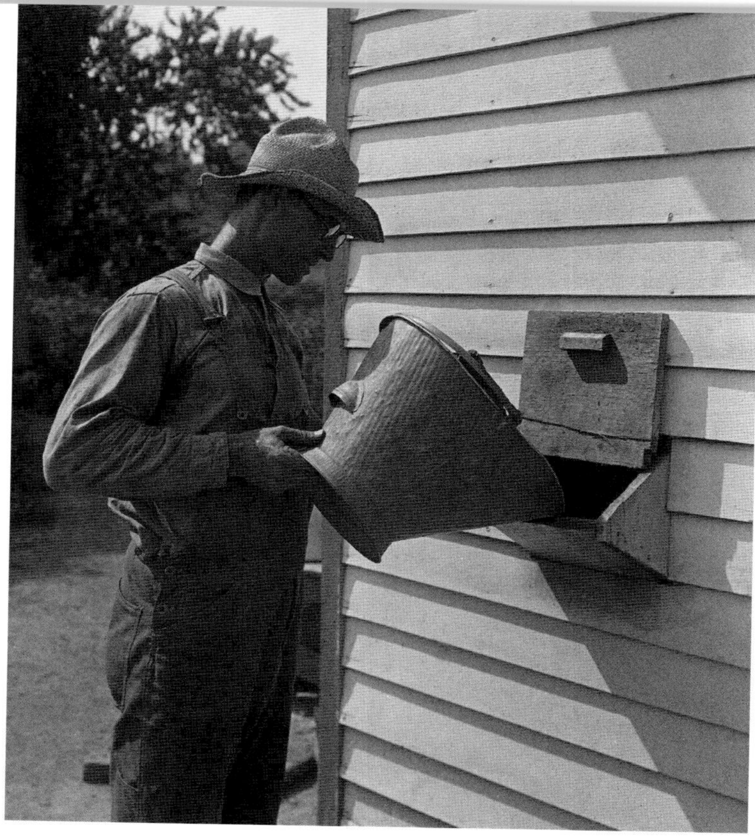

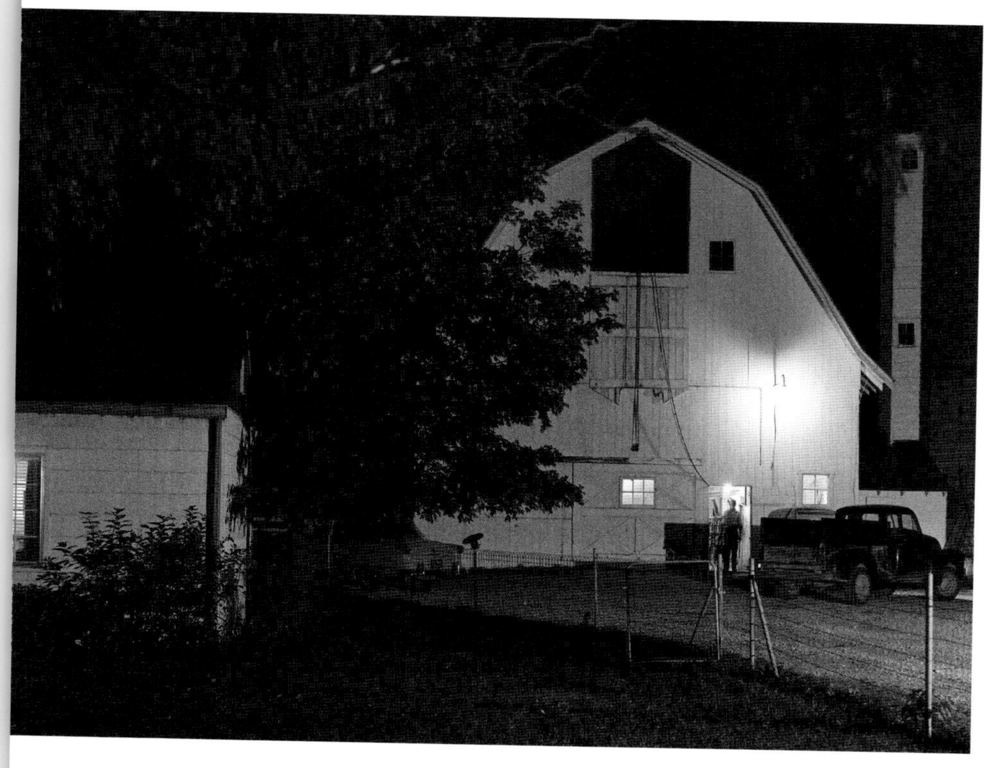

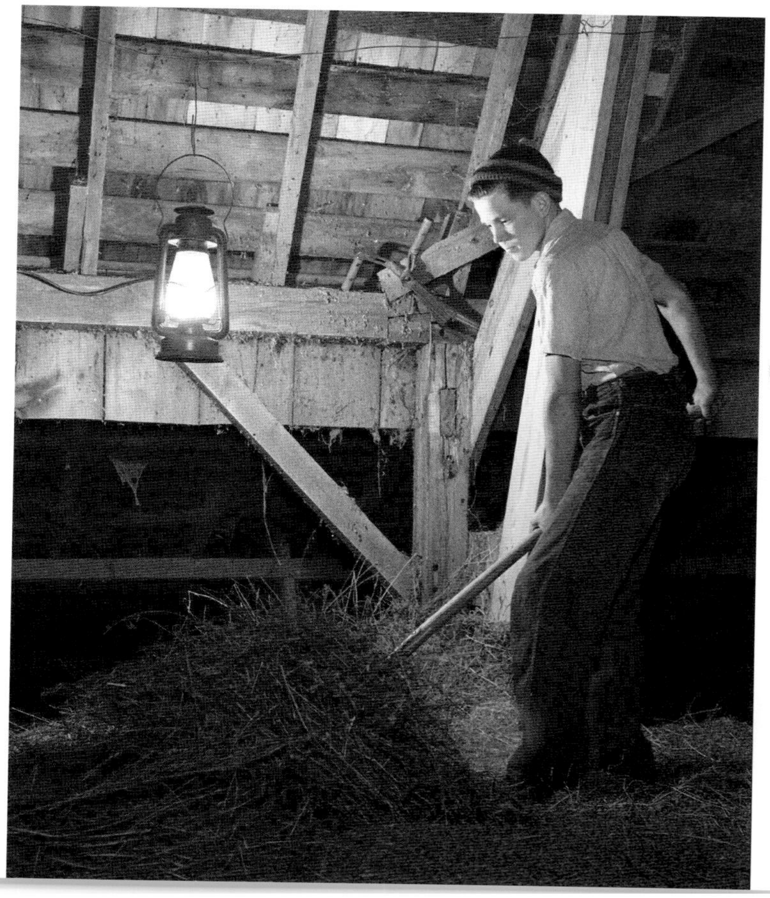

(Top left) This Clinton County Farm Bureau truck was used to deliver gas and oil to customers. (Clinton County, 1929)

(Top right) Raymond Clutter fills a kitchen coal box from the outside. (Vanderburgh County, 1925)

(Bottom left) When electricity came to the farm of Jewell Rosenbloom, he could extend the working hours at his dairy in Plainfield. (Hendricks County, 1954)

(Bottom right) A young man on the farm of C. R. Gustavel gets hay from a haymow by a lantern, which hangs from a wire for safety. (Tippecanoe County, 1940)

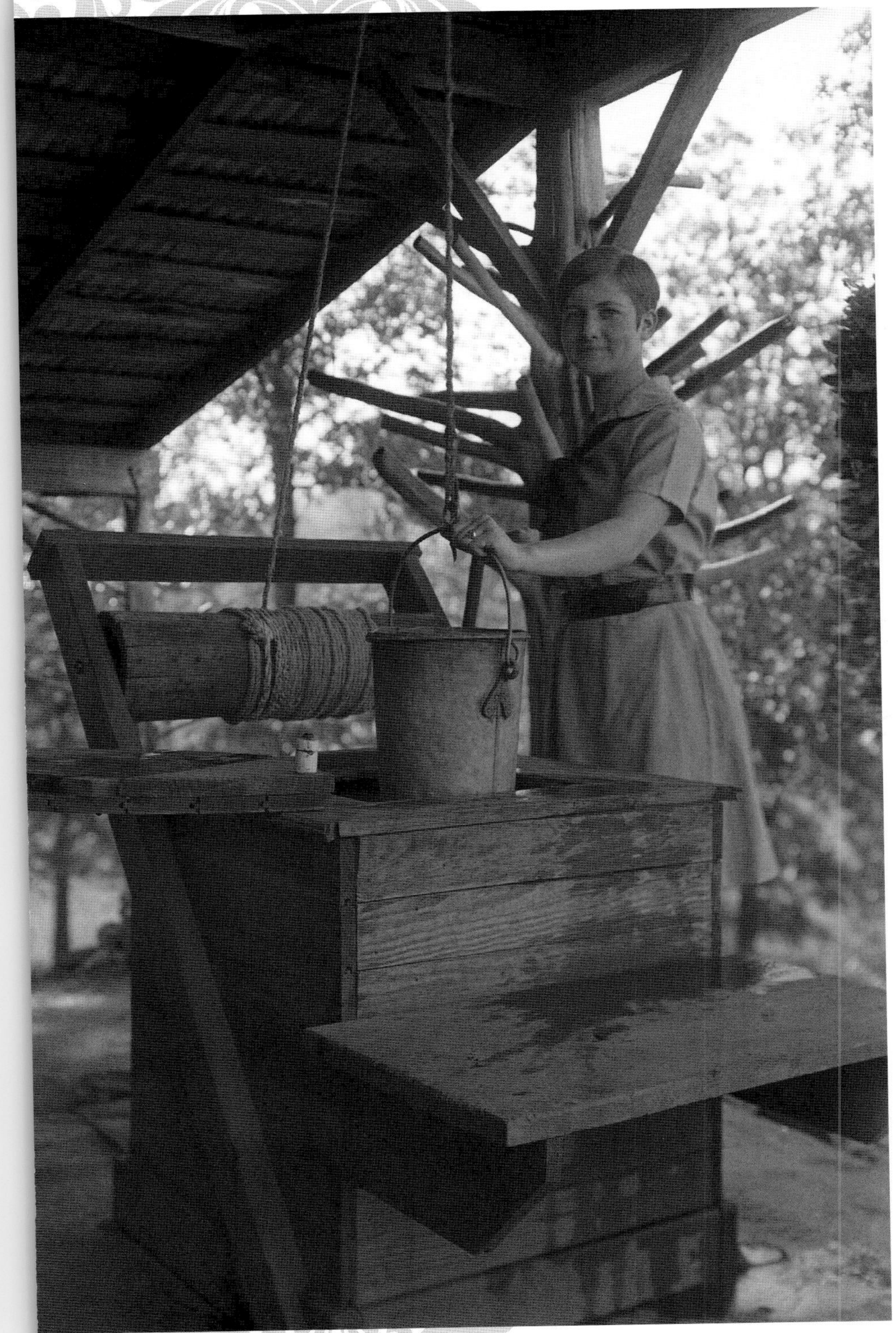

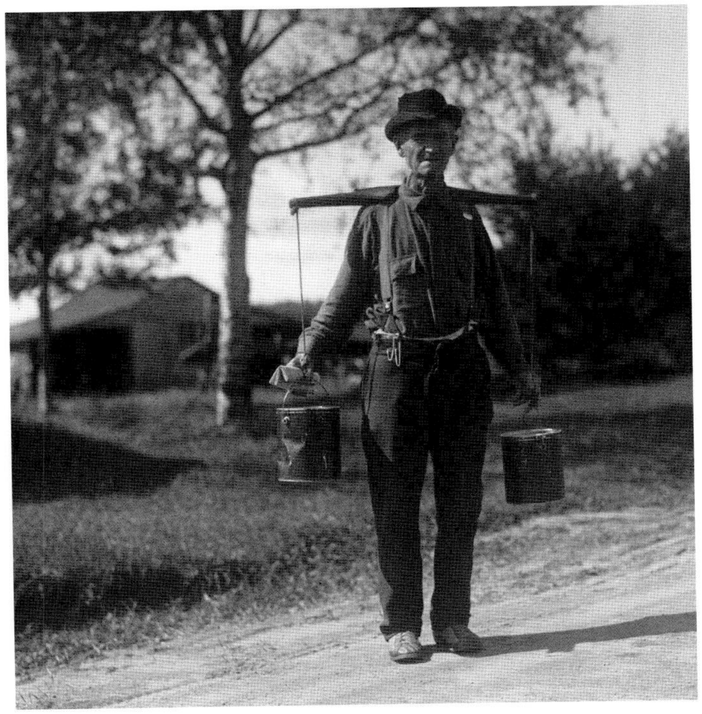

(Far left) Anna Reed Chapman from Sandy Springs, South Carolina, draws water at a well house. (Anderson County, SC; 1927)

(Top right) H. H. Flemming from Cable, Wisconsin, is photographed carrying water with a yoke, a system he had used for 20 years. (Bayfield County, WI; 1924)

(Bottom right) A waterwheel in a small stream pumps water from a nearby spring to George W. Smith's farm home, located 90 feet up the hill. (Anderson County, SC; 1927)

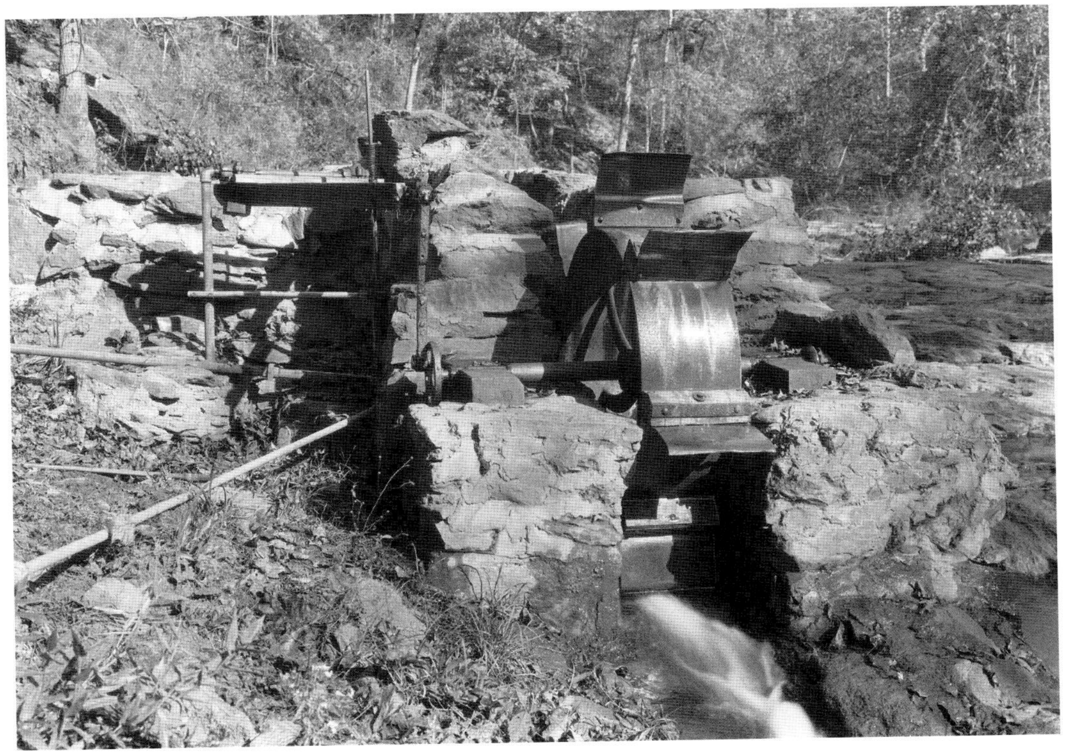

Chores around the Home

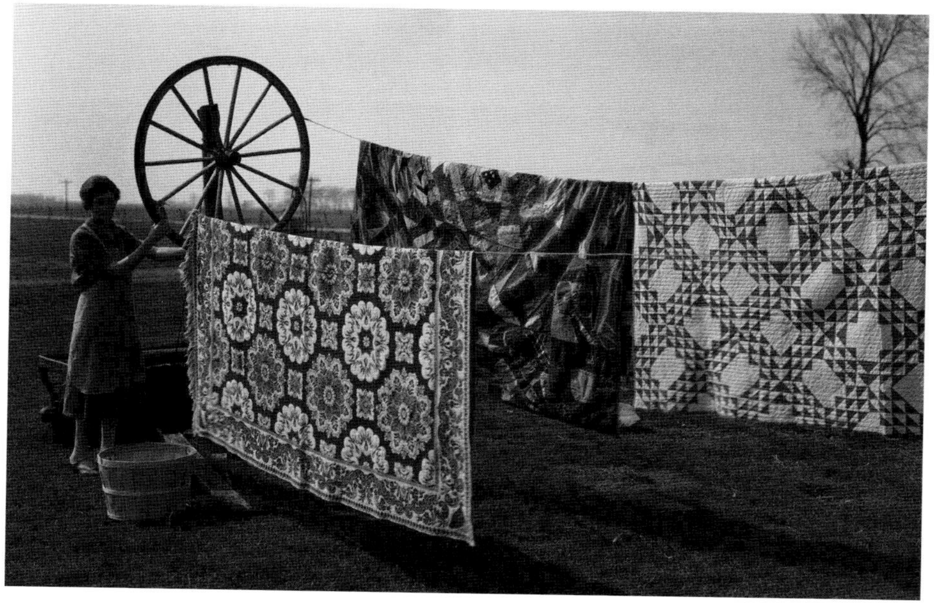

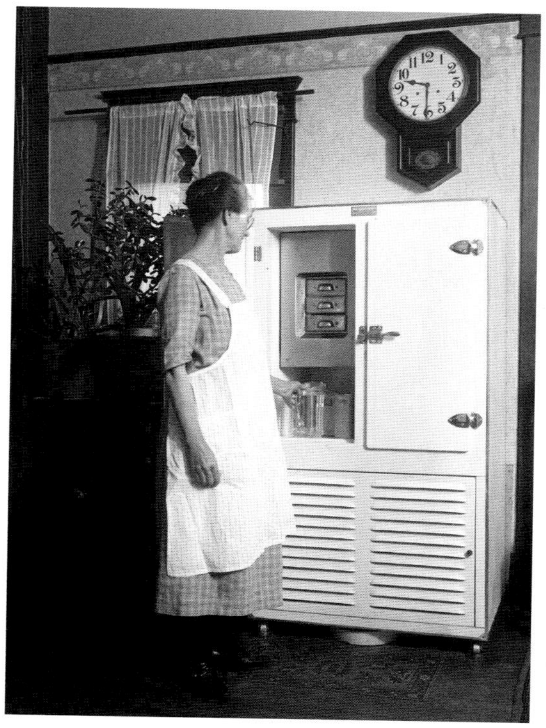

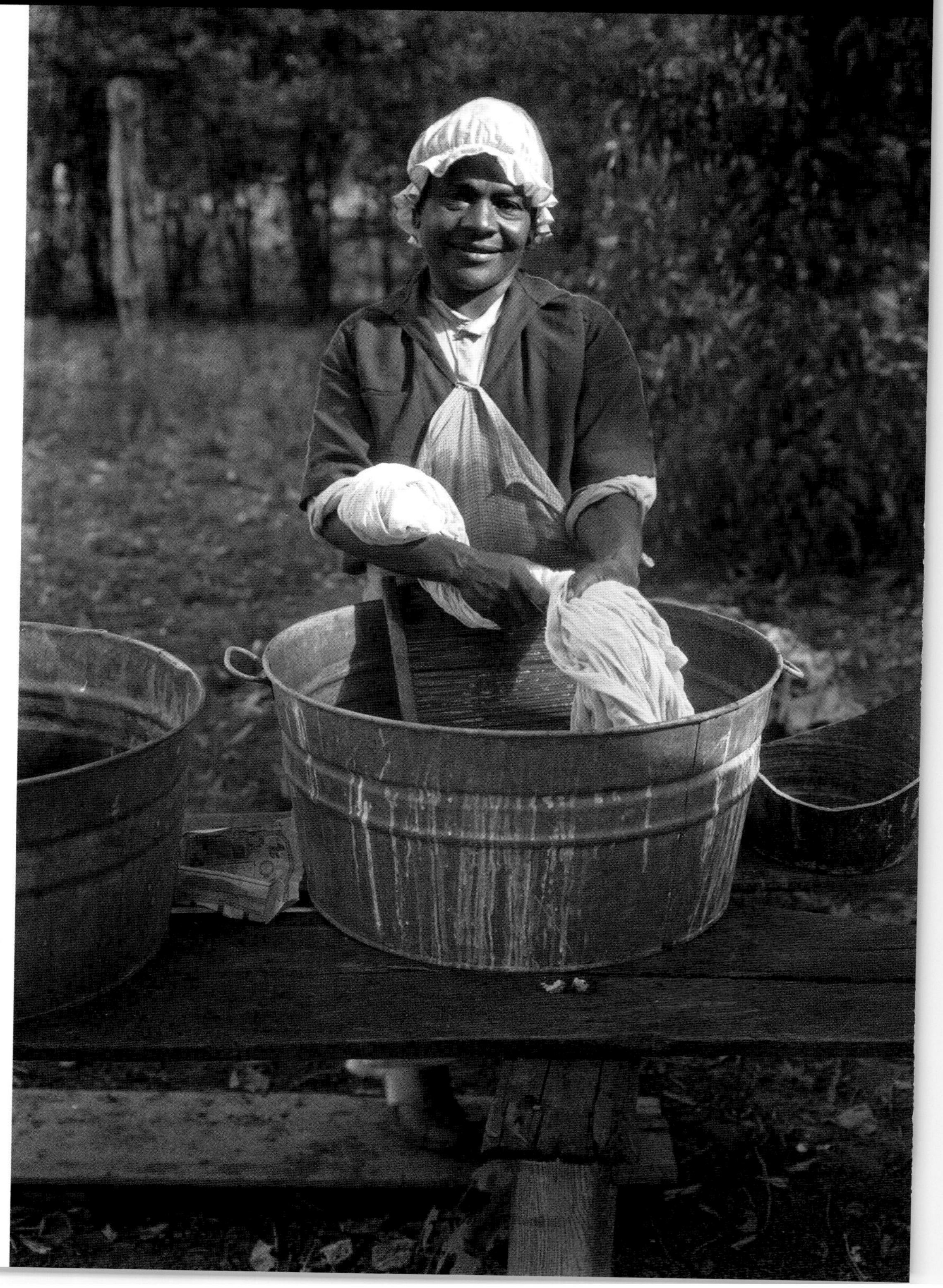

(Top left) A woman uses this mechanism to raise and lower a clothesline on the farm of Carl K. Crist near Lafayette. (Tippecanoe County, 1931)

(Bottom left) The farm home of Colver P. Ryan near Marion featured this Frigidaire electric refrigerator. (Grant County, 1928)

(Far right) A woman washes laundry on the farm of Mr. and Mrs. R. H. Kennedy near Harmony, North Carolina. (Iredell County, NC; 1927)

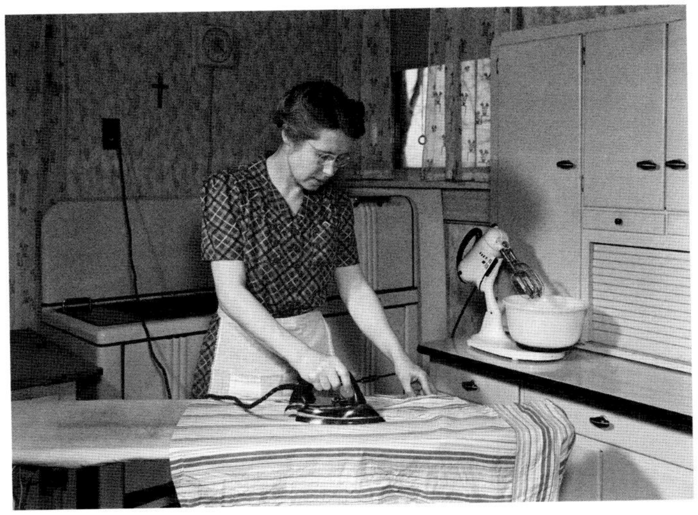

Mrs. W. O. Kirkpatrick of Lafayette presses clothing with an electric iron. (Tippecanoe County, 1941)

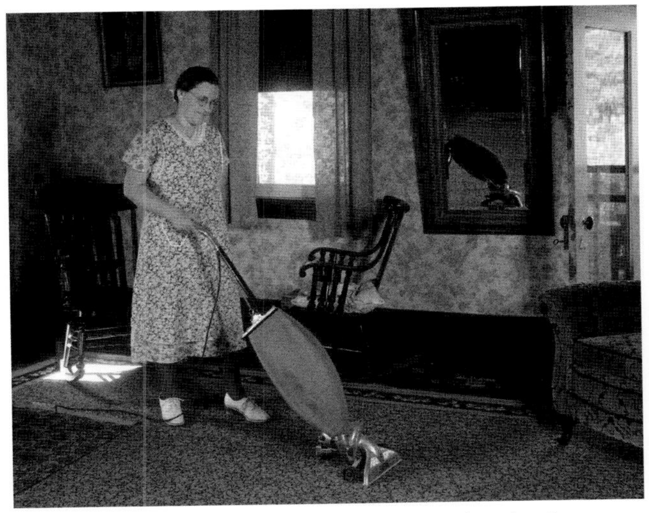

Mrs. William Y. Hartsough of North Manchester uses her electric sweeper. (Wabash County, 1934)

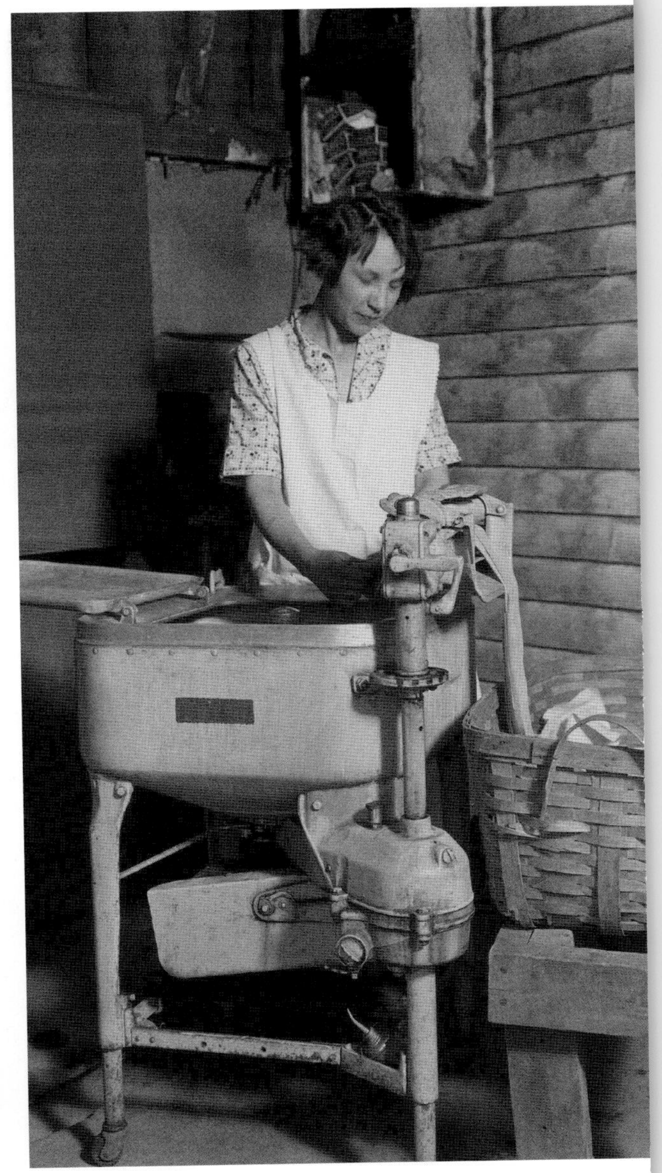

(Above) To do laundry, Mrs. T. V. Carter of Seymour uses a Maytag electric washing machine. (Jackson County, 1928)

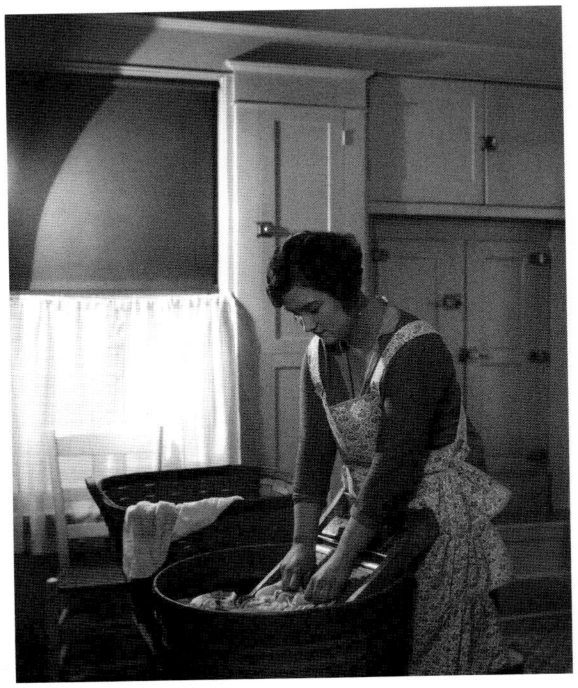

(Above) Helen Ryan of West Lafayette uses a washboard to do laundry. (Tippecanoe County, 1930)

(Right) In this farm kitchen, Mrs. Chalmer Miller builds a fire in a coal range. (Clinton County, 1928)

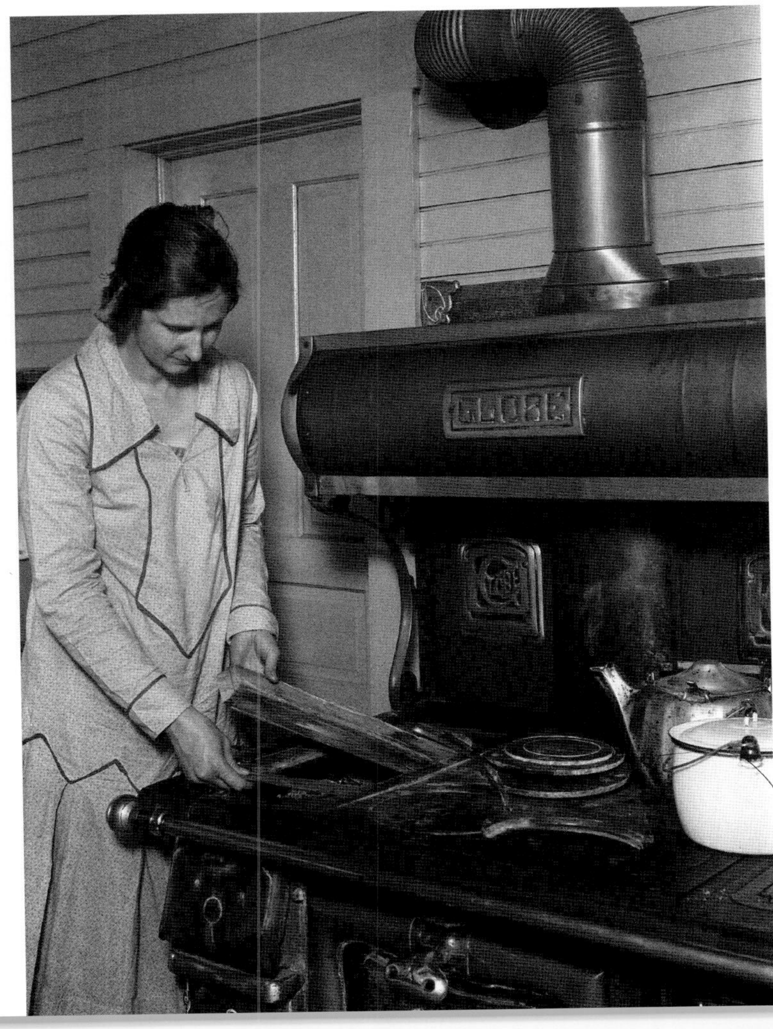

We Grow What We Eat

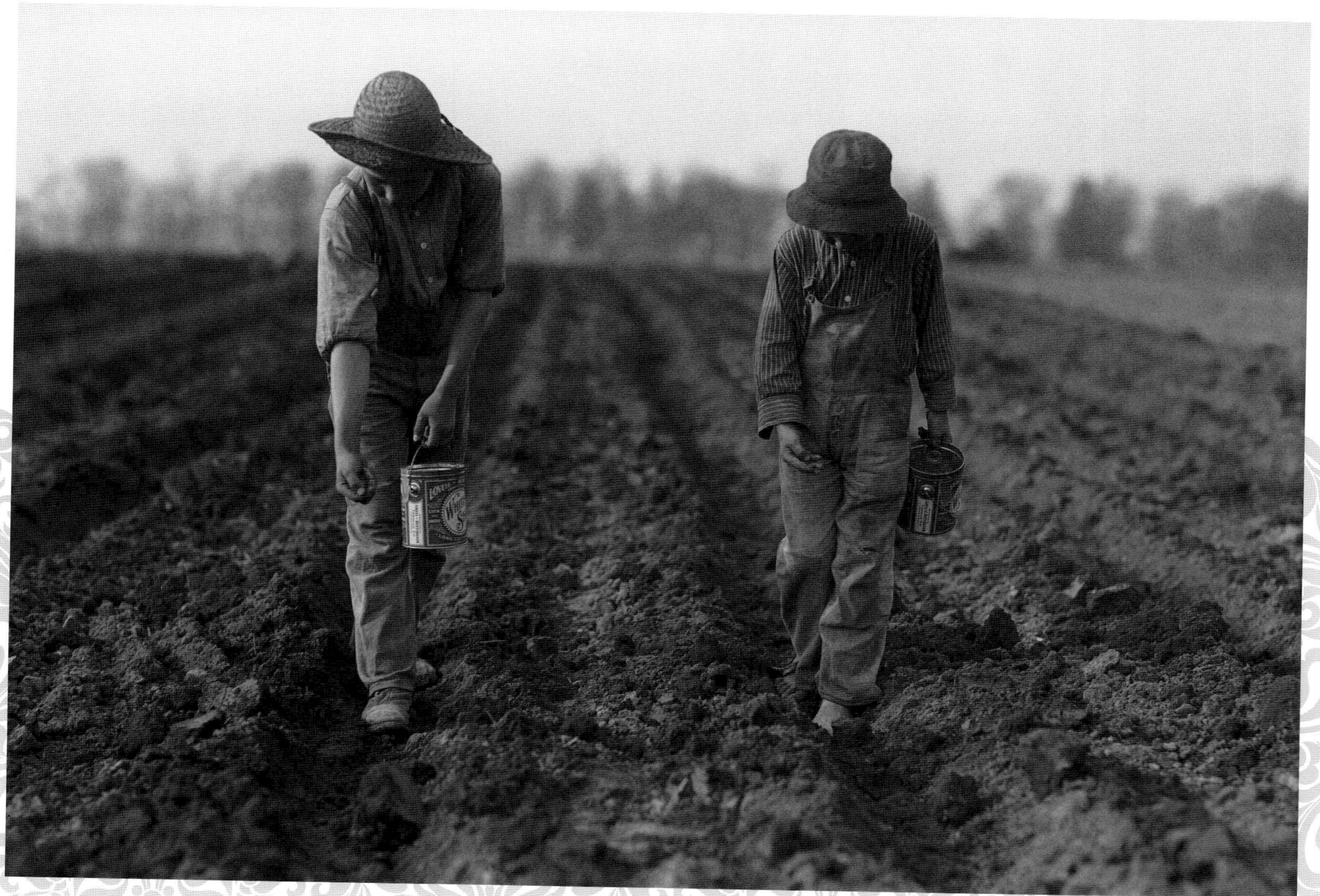

Farm boys plant beans near Lafayette. (Tippecanoe County, 1915)

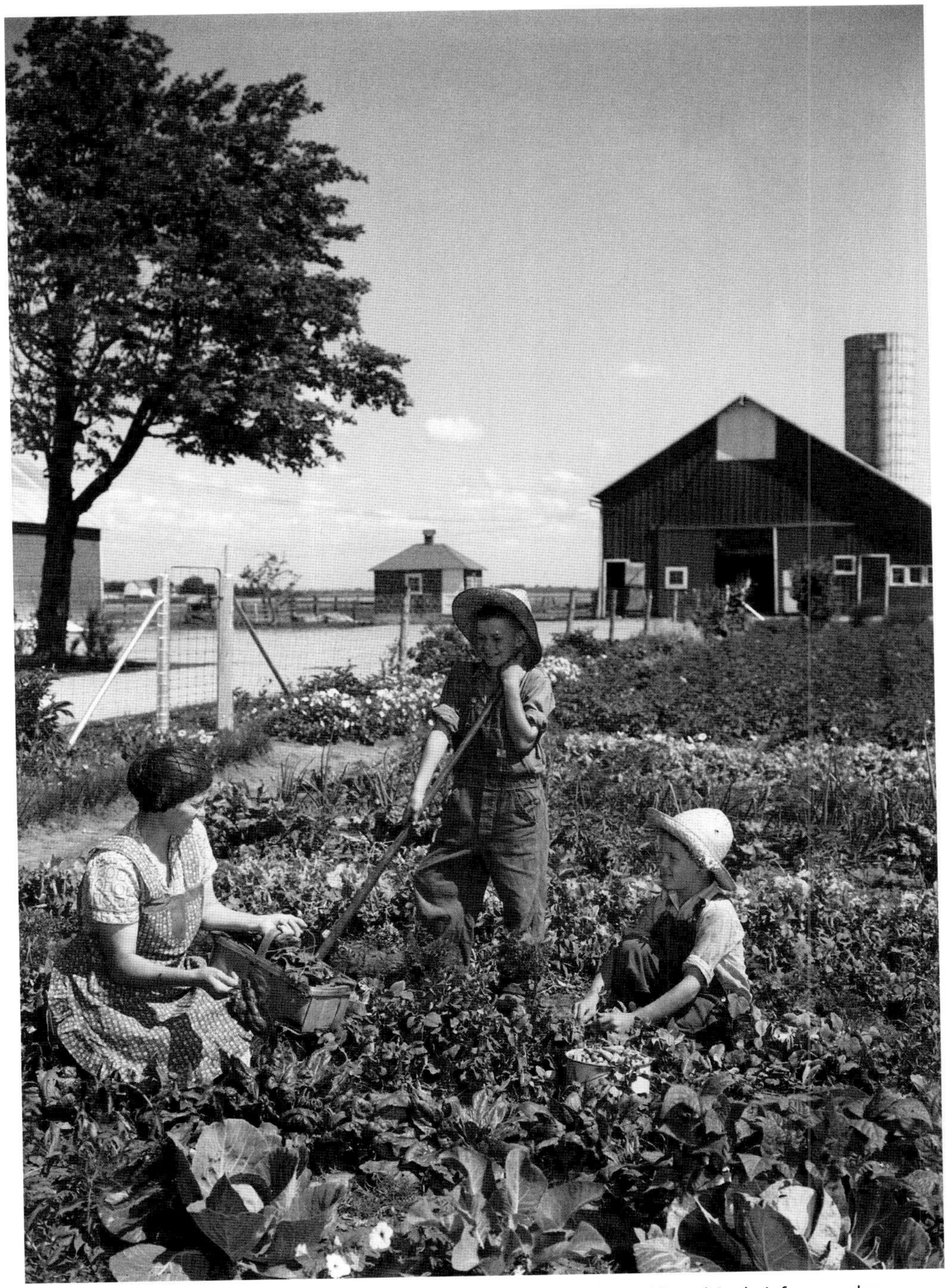

Mrs. Herman Erb of Francesville and her two sons, George (center) and Donald, work in their farm garden. (Pulaski County, 1940)

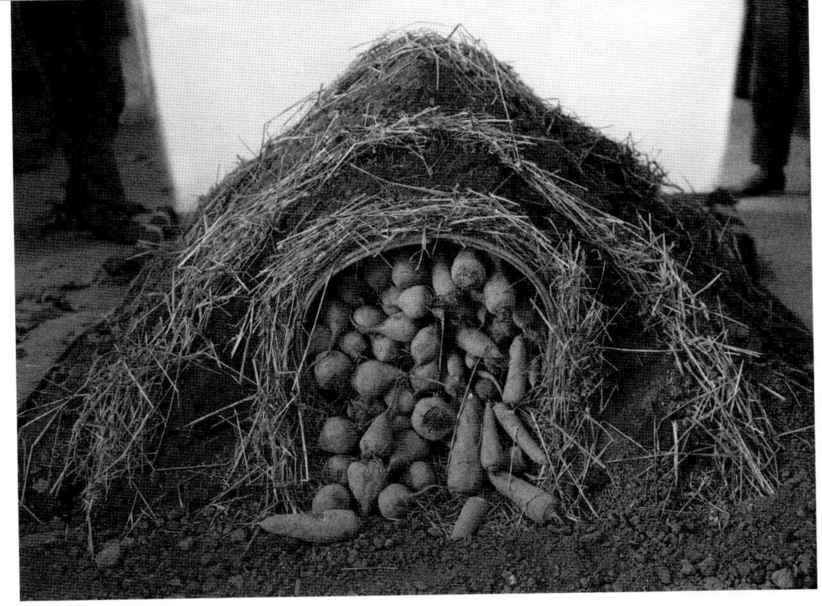

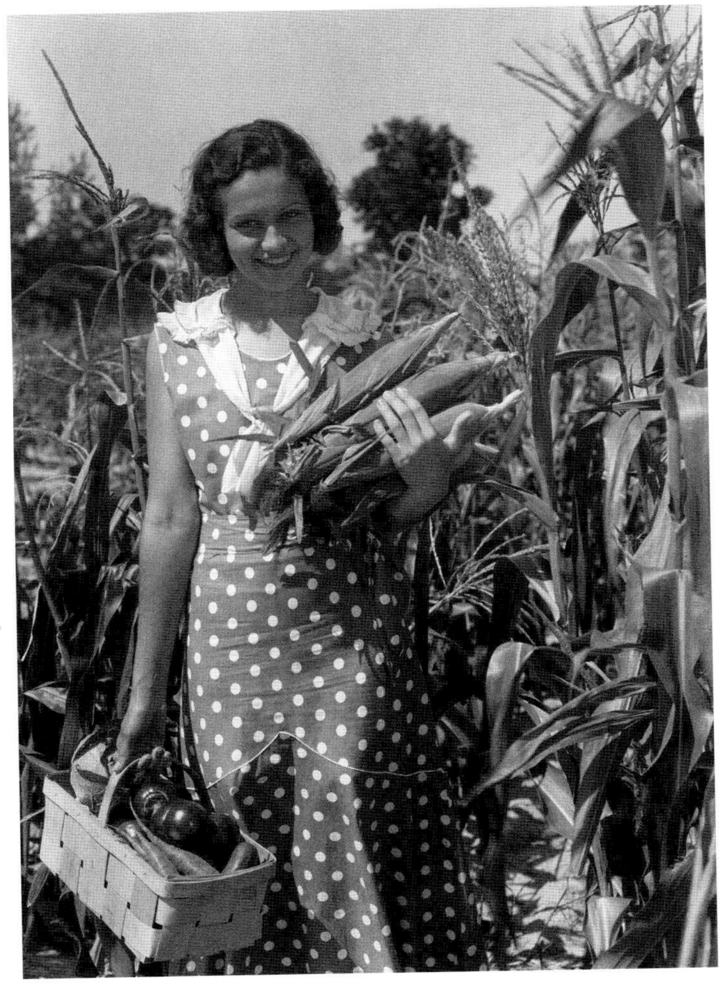

(Top right) Burying beets and carrots preserved them for winter use. (N.p., 1917)

(Bottom right) Geraldine Tudor of West Lafayette harvests corn and other fresh vegetables from a home garden. (Tippecanoe County, 1933)

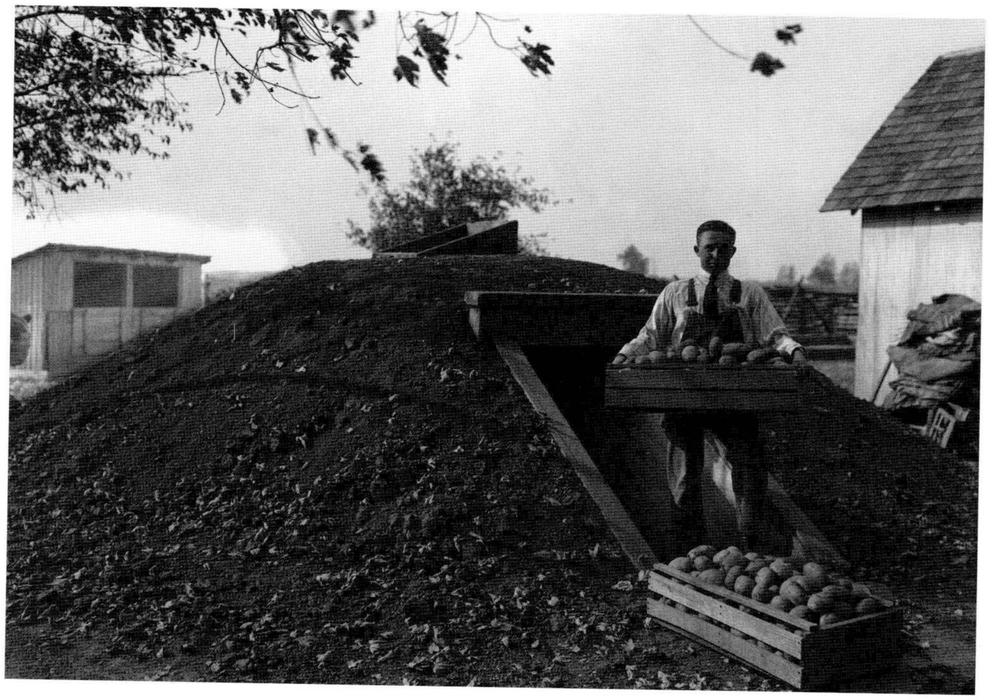

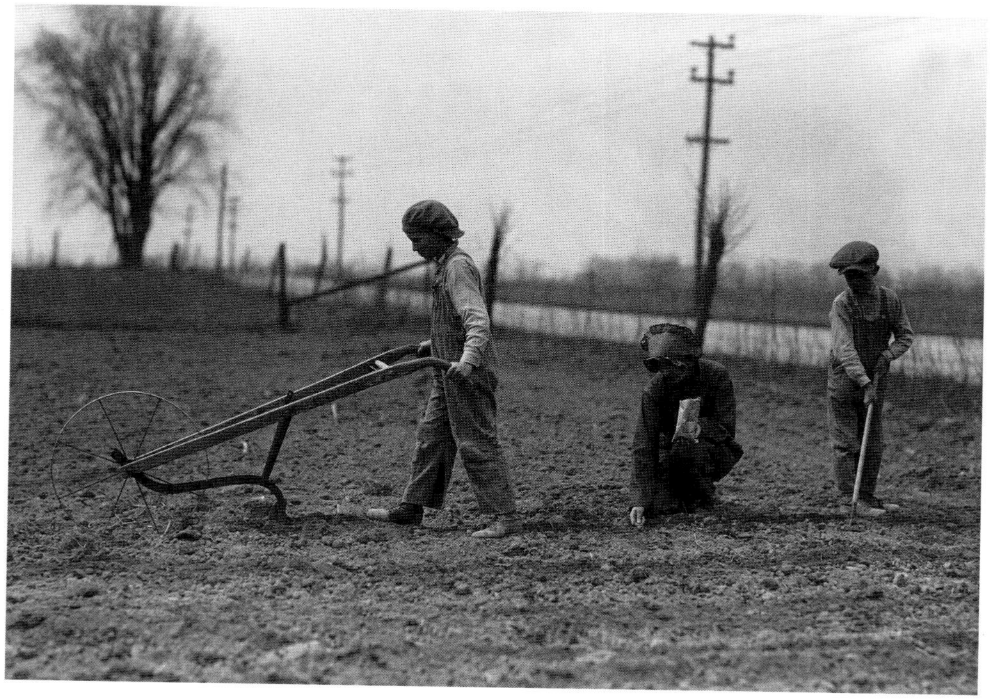

(Top left) Russell Brown of Elwood carries potatoes into a vegetable cellar. (Madison County, 1920)

(Top right) Mrs. Elmer Chasteen (center) and sons Howard and Harold plant beans in their farm garden near Franklin. (Johnson County, 1925)

(Bottom) Mrs. E. E. Brown of Veedersburg uses this coal stove and a pressure cooker to can fresh food in her farm kitchen. (Fountain County, 1924)

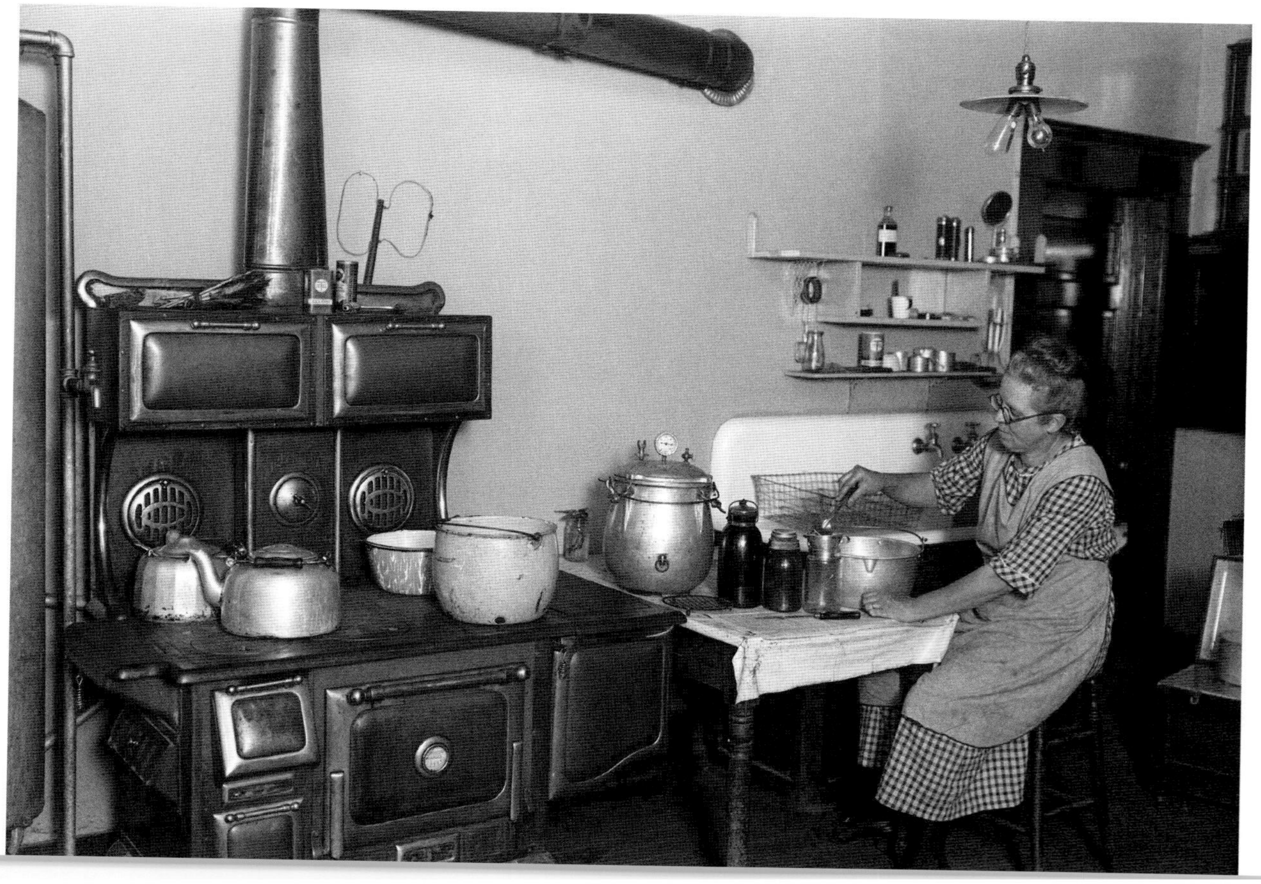

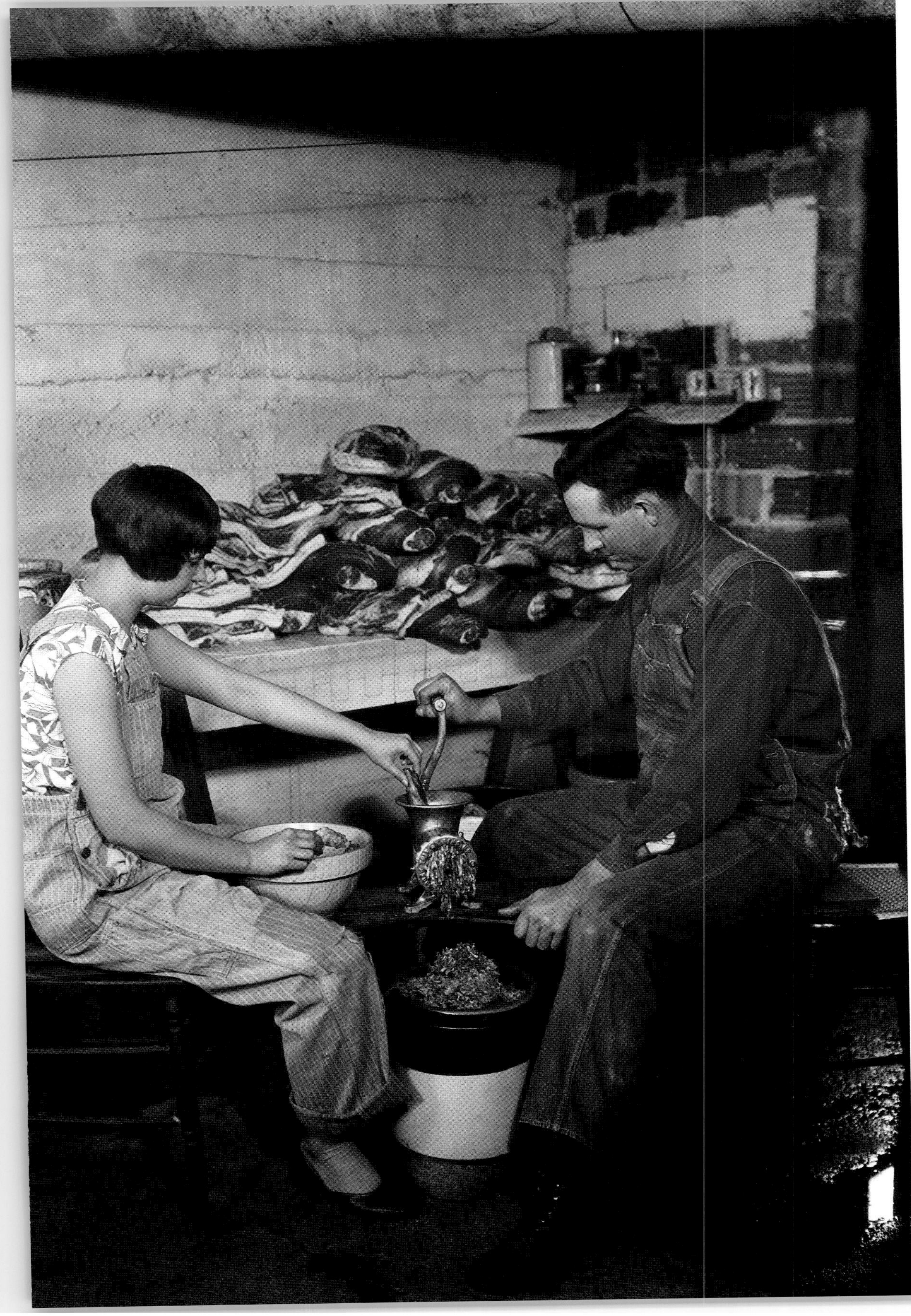

(Left) Mr. and Mrs. Wilbur Wilkey of Kentland grind sausage and trim bacon in their farm basement. (Newton County, 1930)

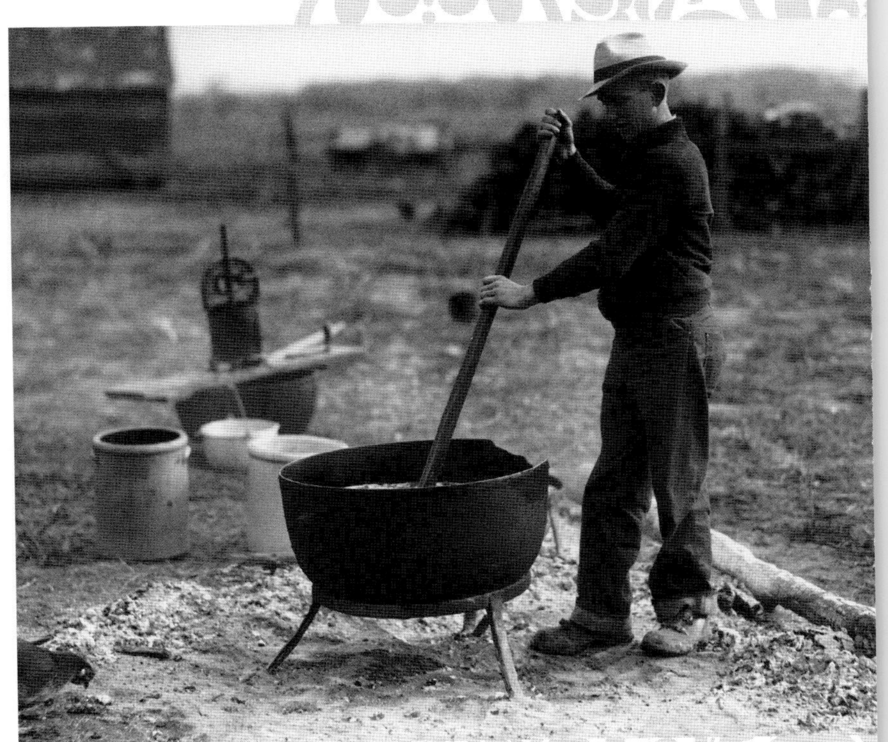

(Above) Floyd Munsen of Delphi renders lard in an open kettle. (Carroll County, 1933)

(Left) Mrs. Loren McGibben of Swayzee operates a cream separator powered by an electric motor on a portable tripod. (Grant County, 1928)

(Right) On the W. A. Smith dairy farm near Terre Haute, Martha Rose Smith pours milk into an electric churn. (Vigo County, 1933)

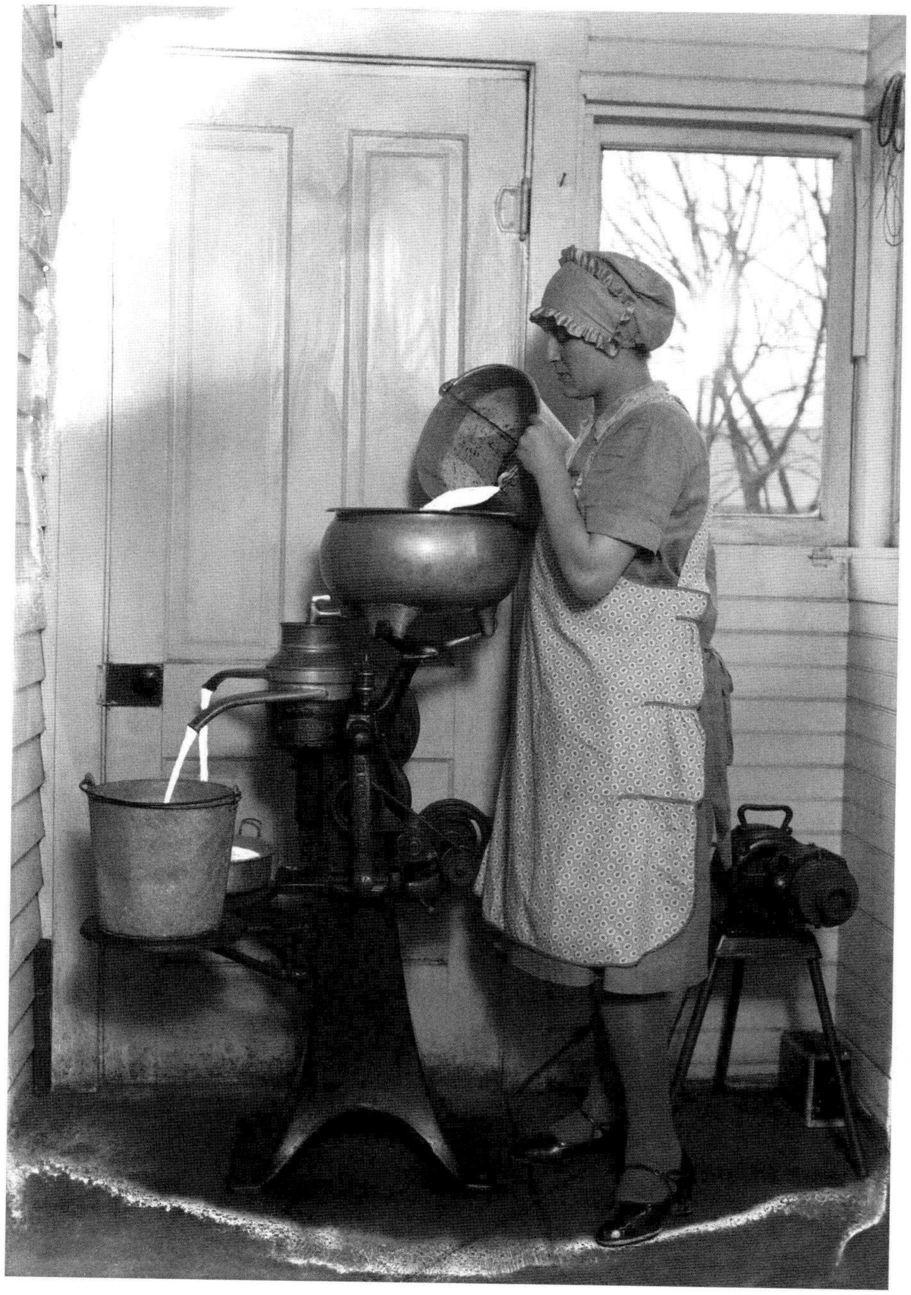

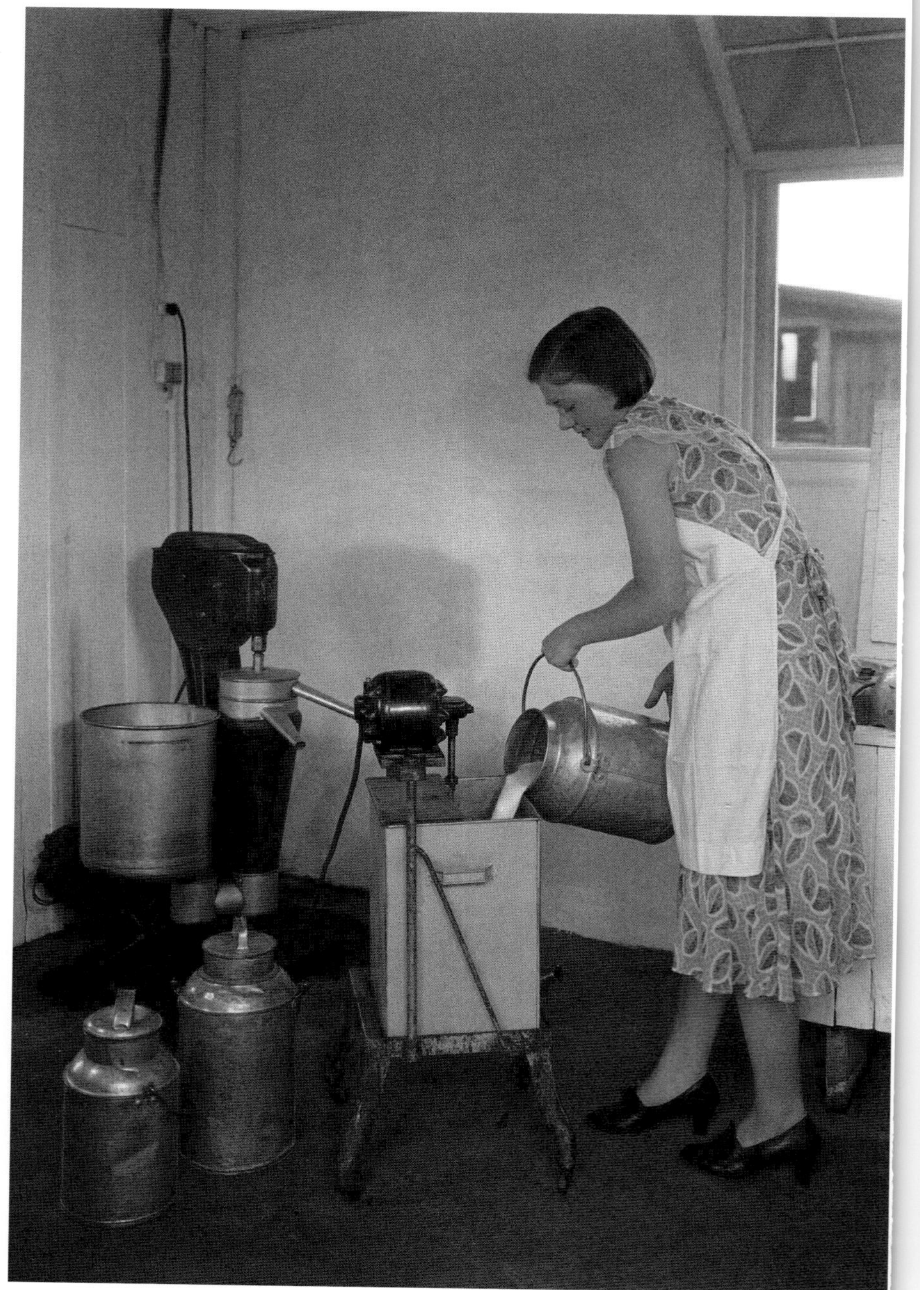

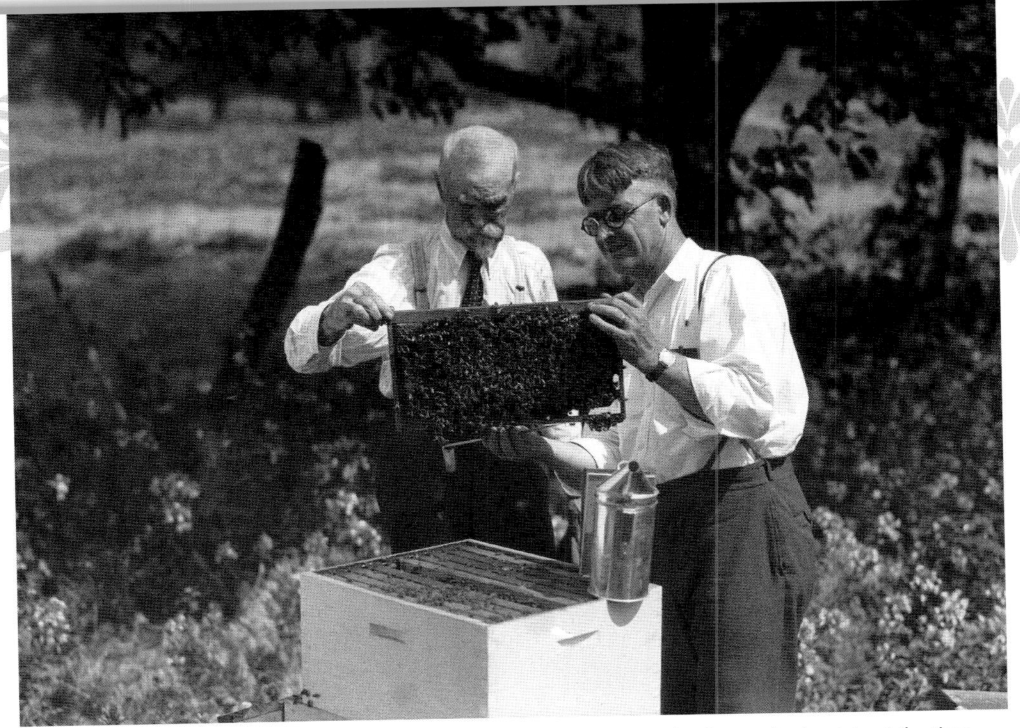

Near Lafayette, Amos Myers (left), a Civil War veteran and the oldest active beekeeper in the state at the time, works with another man to inspect a frame of honey and bees. (Tippecanoe County, 1933)

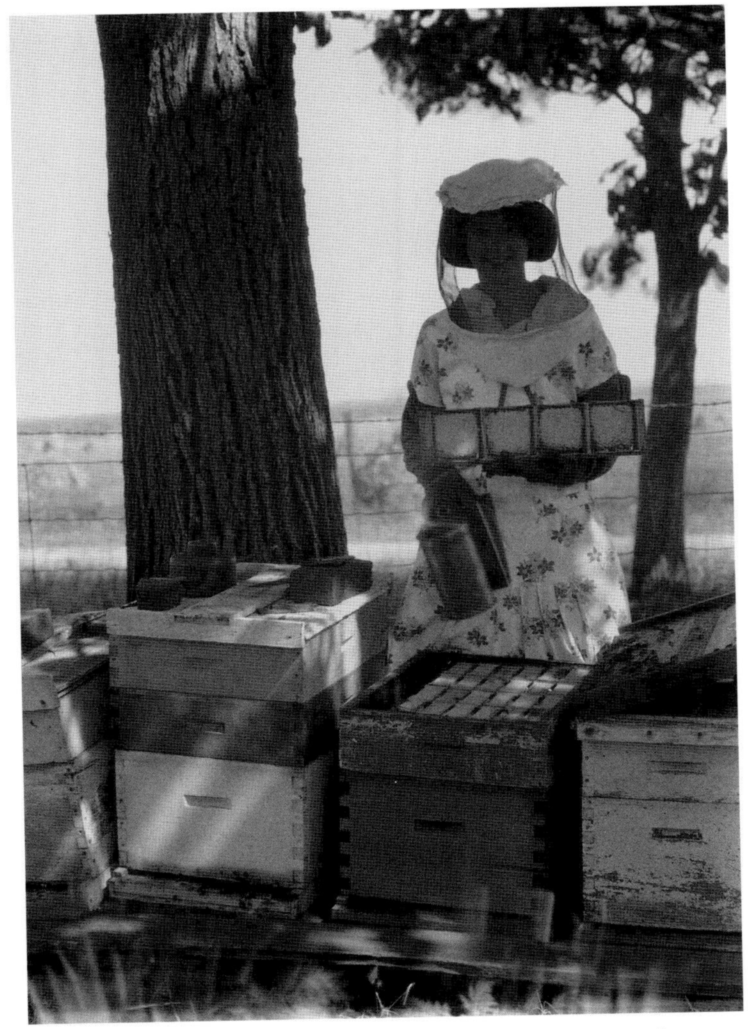

Elise Barton of Goodland takes honey from hives. (Newton County, 1927)

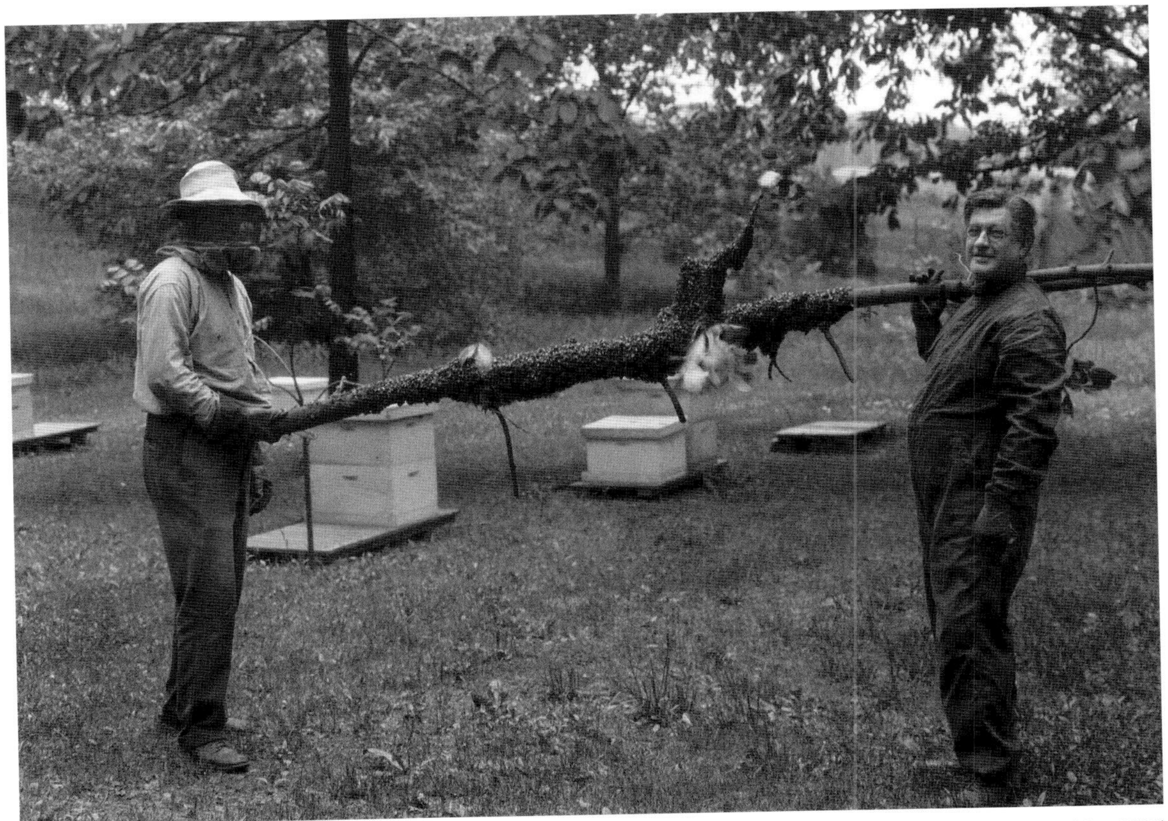

Professor G. A. Young (right) of Purdue University and another man relocate a swarm of bees from a nearby tree to a hive. (N.p., 1929)

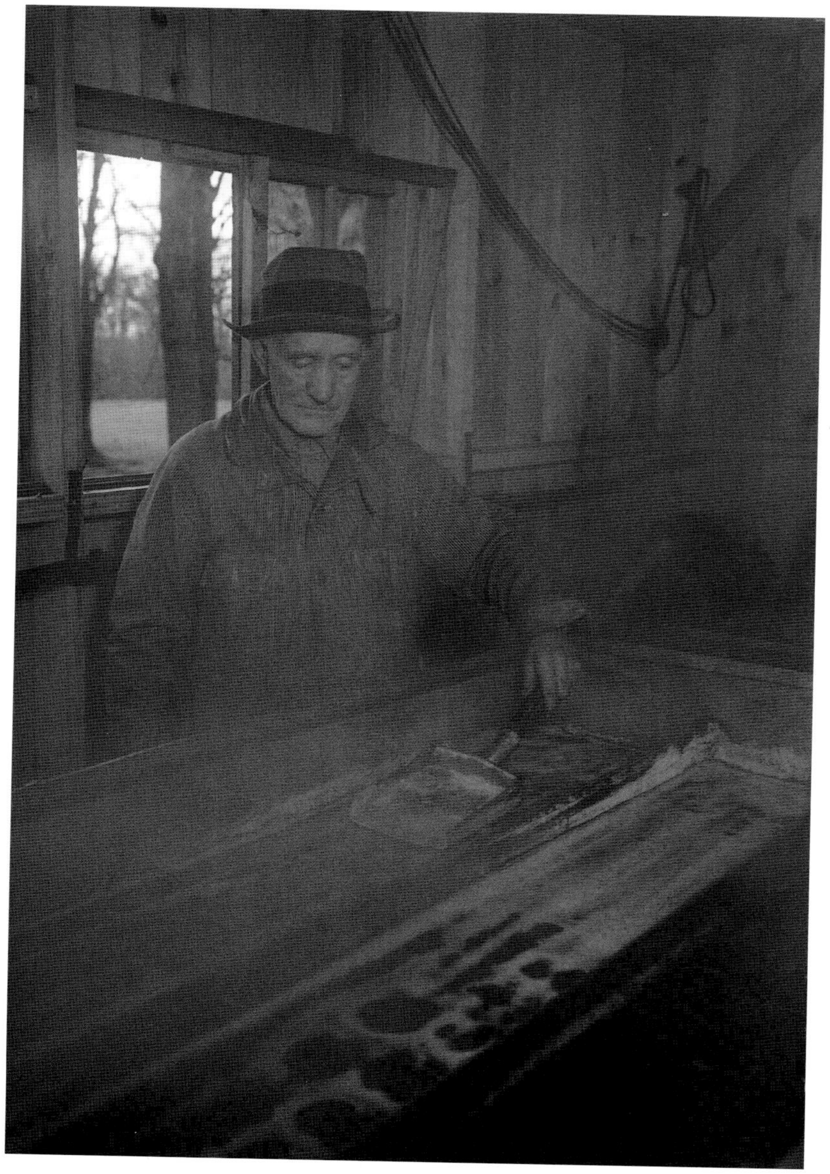

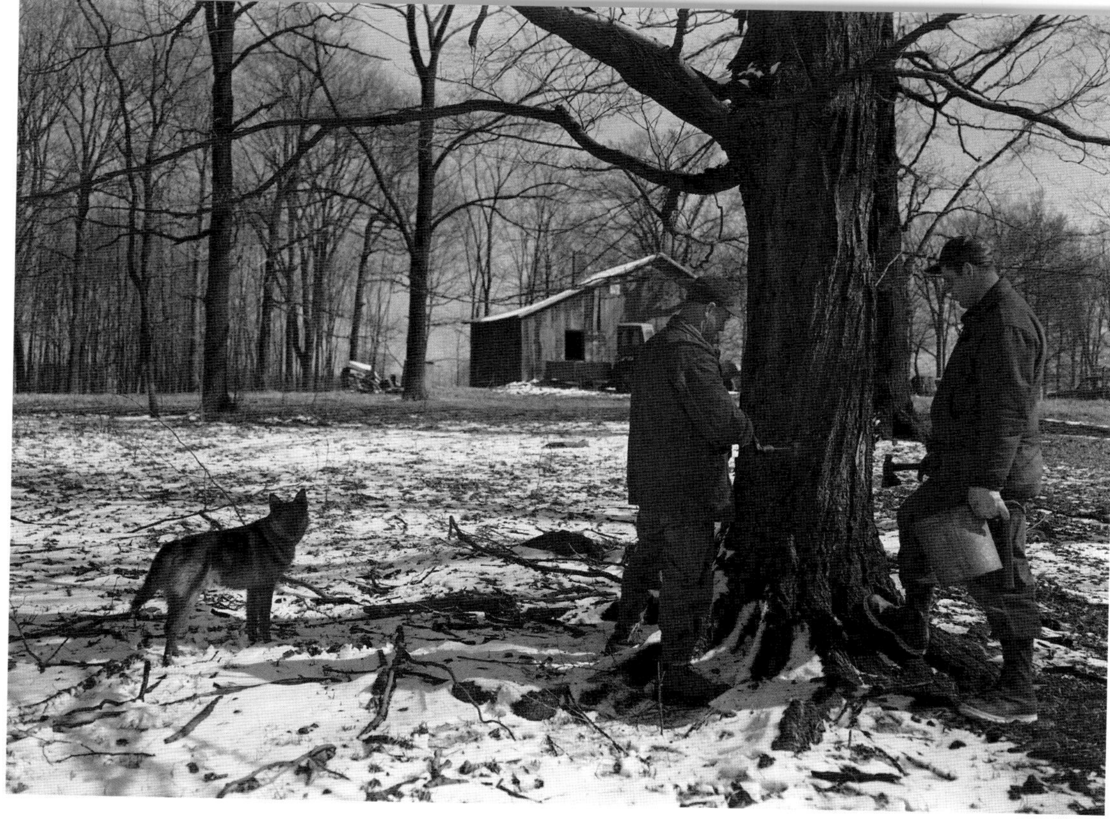

(Left) E. I. Miller skims boiling maple syrup in pans in his maple sugar camp near Lafayette. (Tippecanoe County, 1931)

(Top right) A maple sugar camp on the Adam Yoder farm near Frankfort (Clinton County, 1953)

(Bottom right) Two men tap a maple tree for sap on the John Milnar farm near Sedalia. (Clinton County, 1937)

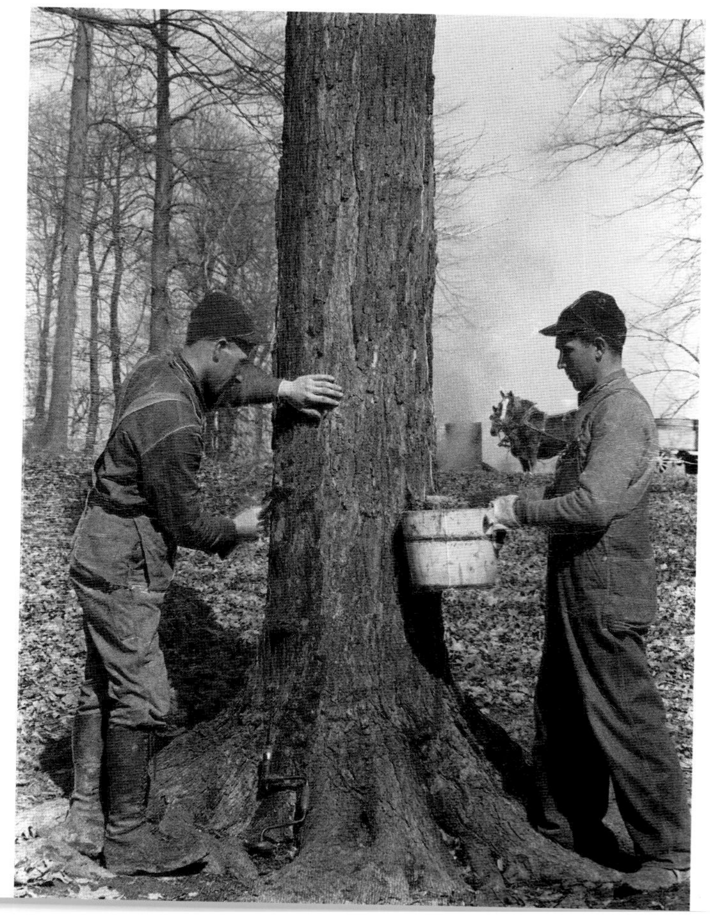

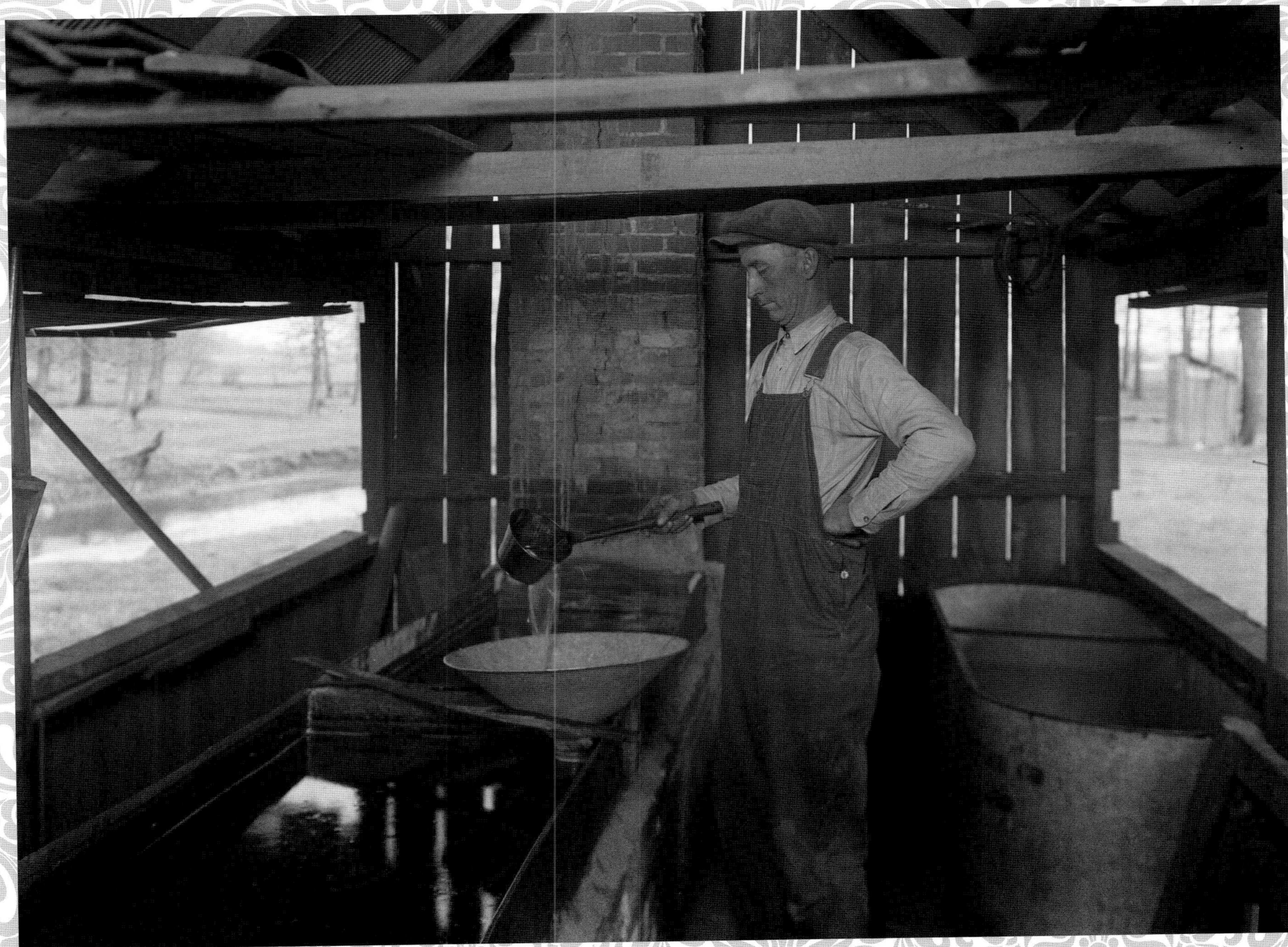

Austin Cochran of Frankfort boils sap to produce maple syrup. (Clinton County, 1927)

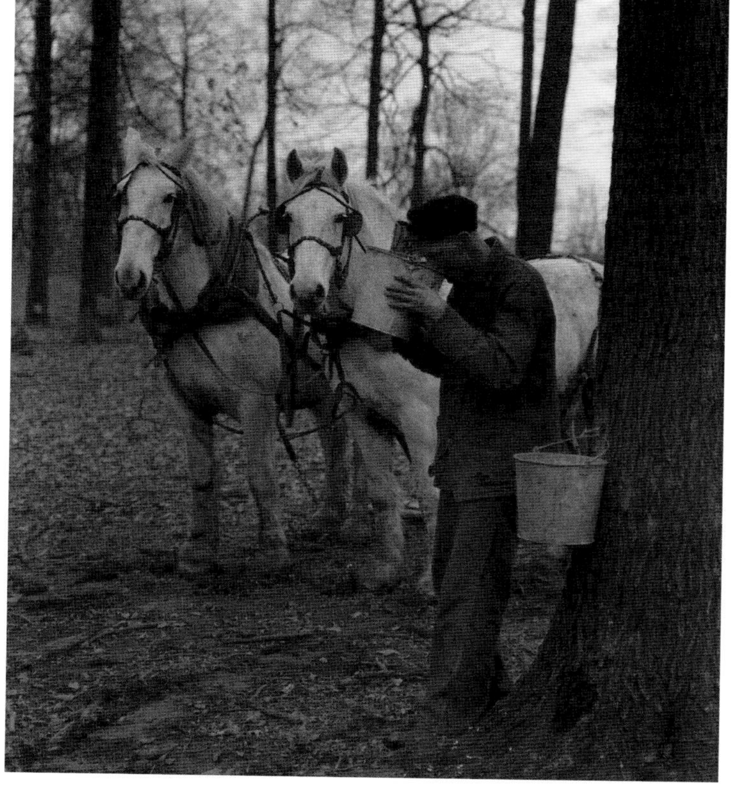

(Above) W. H. Todd of Flora samples the sugar water in his maple sugar camp. Todd, who typically harvested between 75 to 100 barrels of sugar water annually from 230 trees, would boil down each barrel to make one gallon of pure maple syrup. (Below) At the time of this photo, Todd had been operating this camp continuously for 35 years. (Carroll County, 1927)

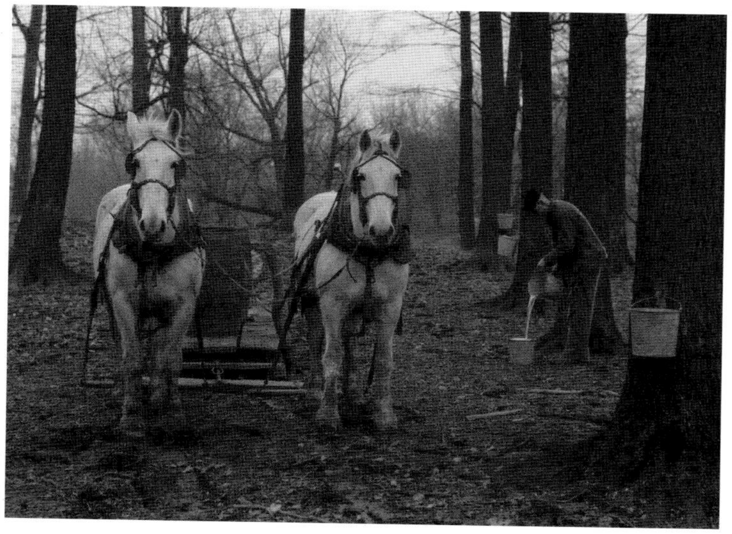

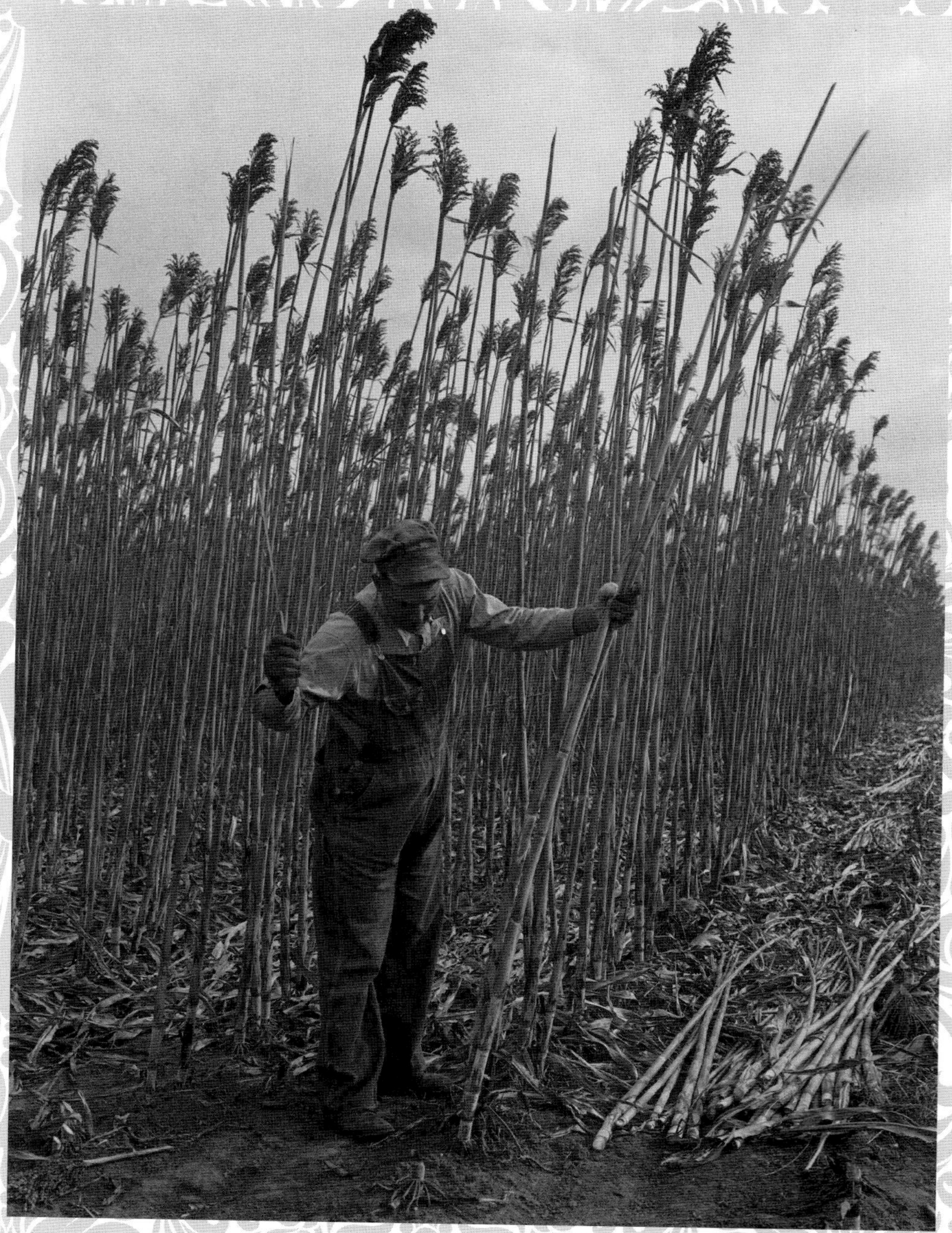

Nelson Hollingsworth cuts a field of sorghum cane on his farm near Russiaville. (Howard County, 1934)

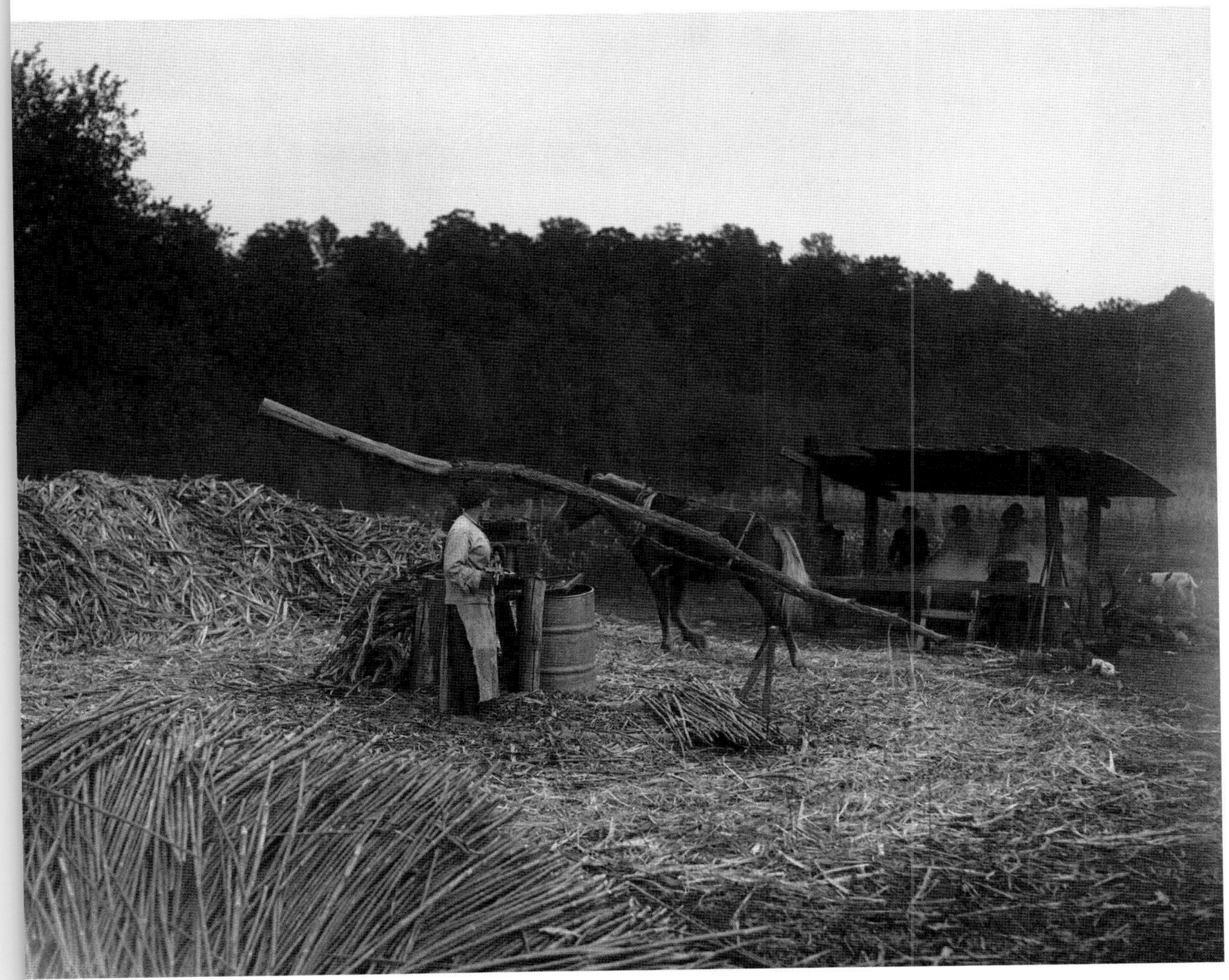

Using a horse to grind sorghum to make molasses (Putnam County, 1918)

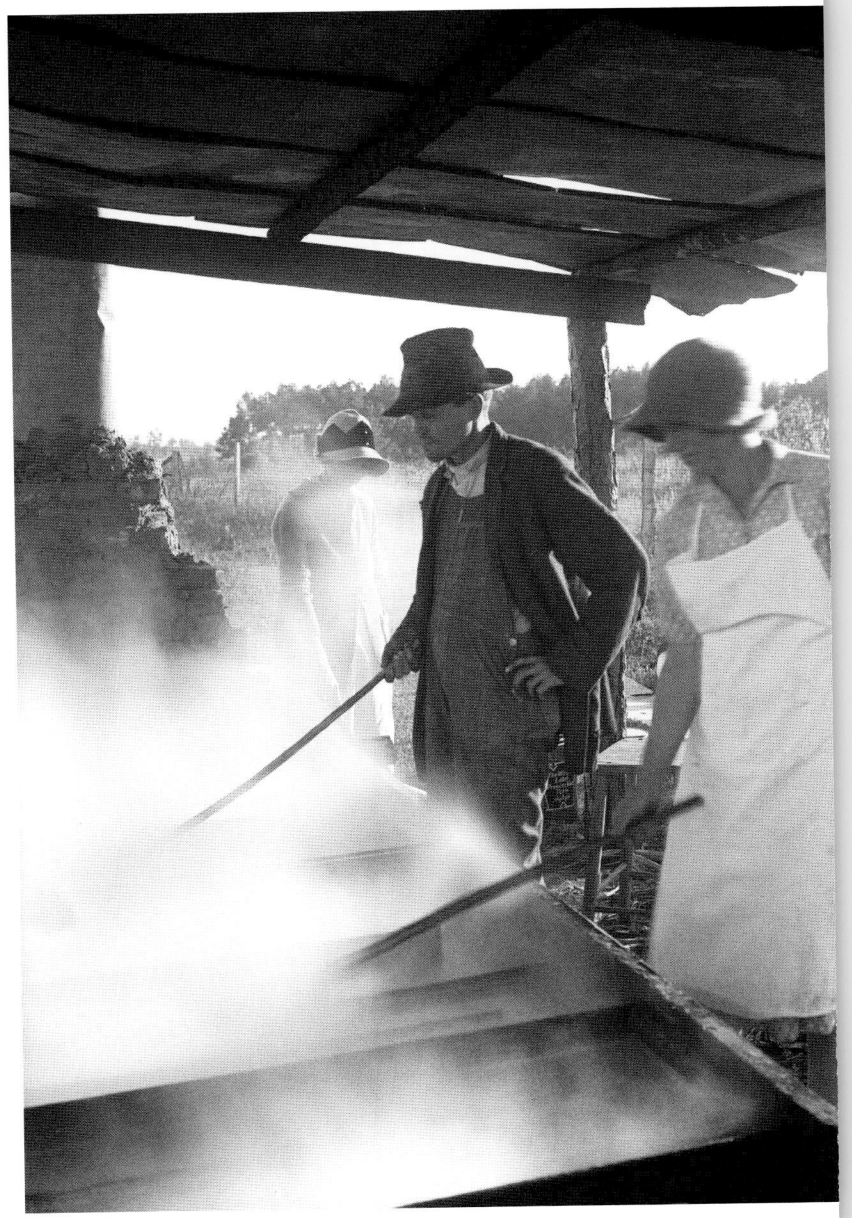

Boiling down sorghum molasses on the farm of W.W. Hughes in Hartselle, Alabama (Morgan County, AL; 1939)

Farm Fresh Products for Sale

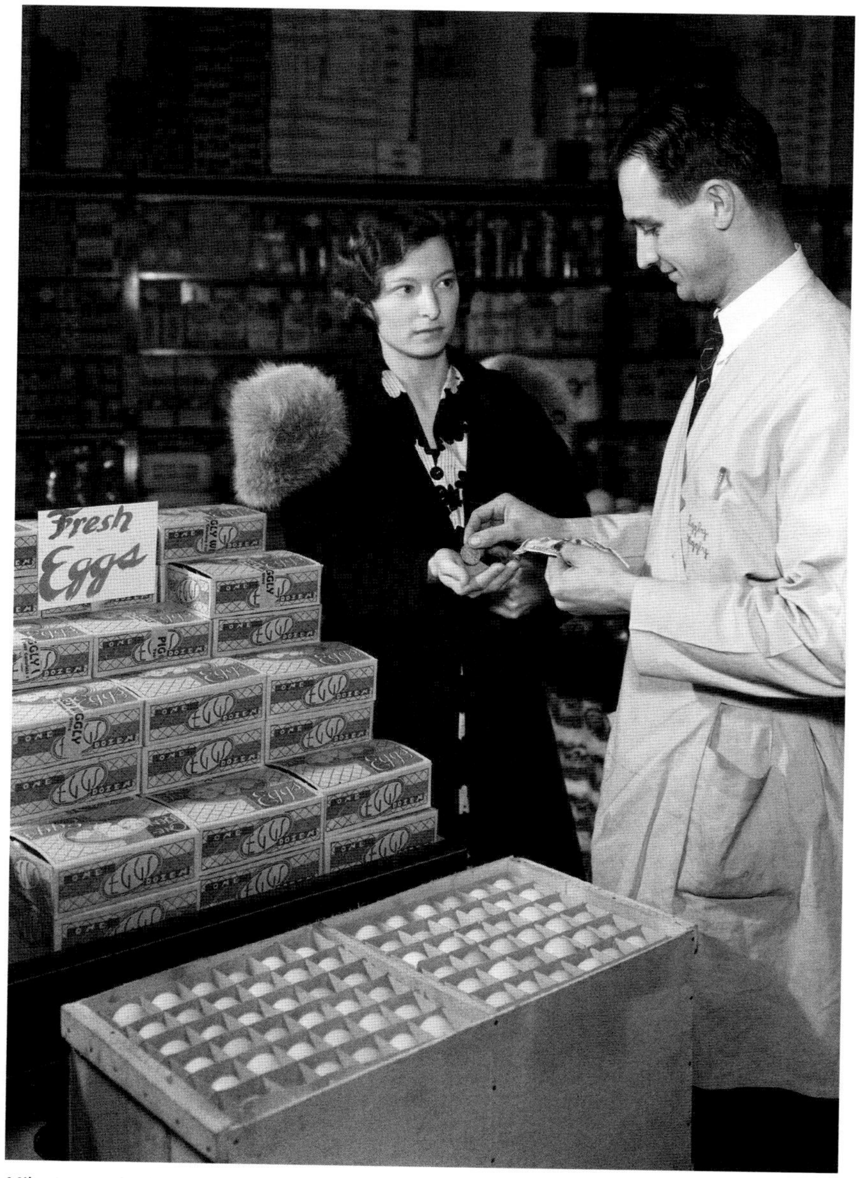

Mike Jester of Lafayette's Piggly Wiggly pays Ella Zink for a case of eggs. (Tippecanoe County, 1939)

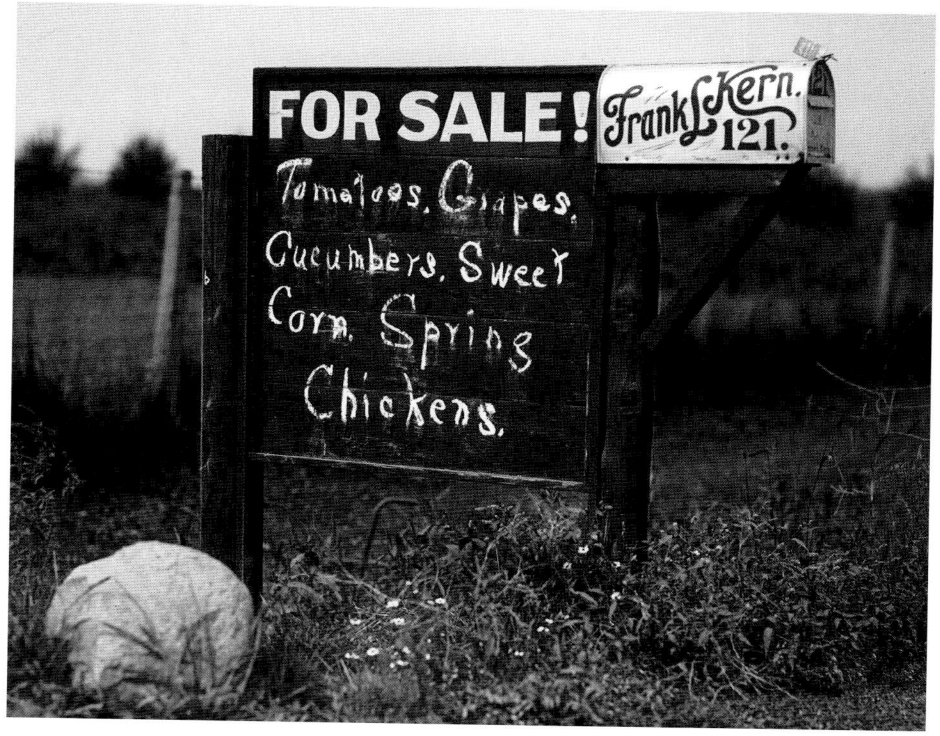

This sign on the Frank L. Kern farm in Cedar Rapids, Iowa, advertises farm products to the public. (Linn County, IA; 1919)

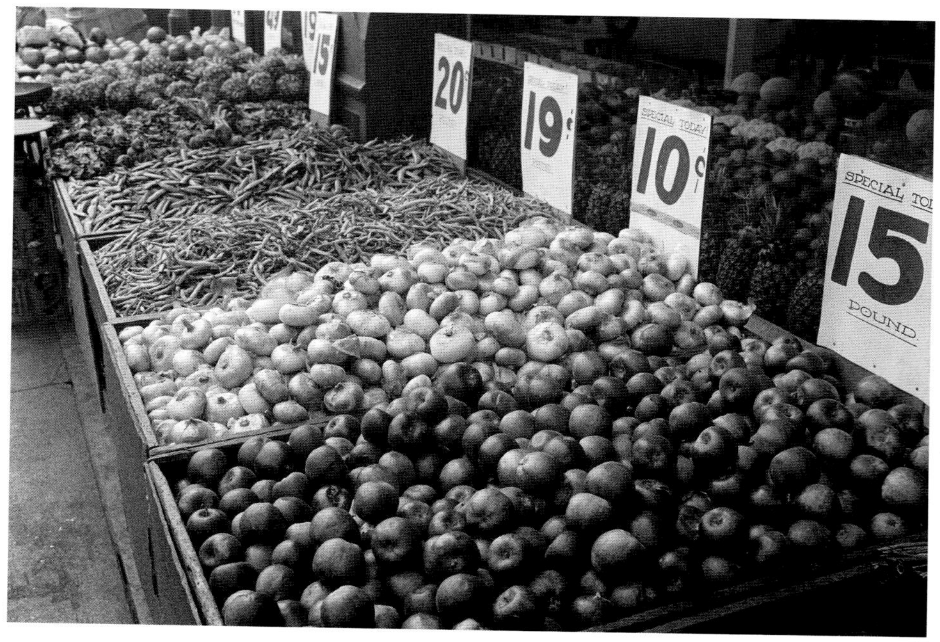

A Richmond-area grocery store sells apples and other locally grown fruits and vegetables. (Wayne County, 1925)

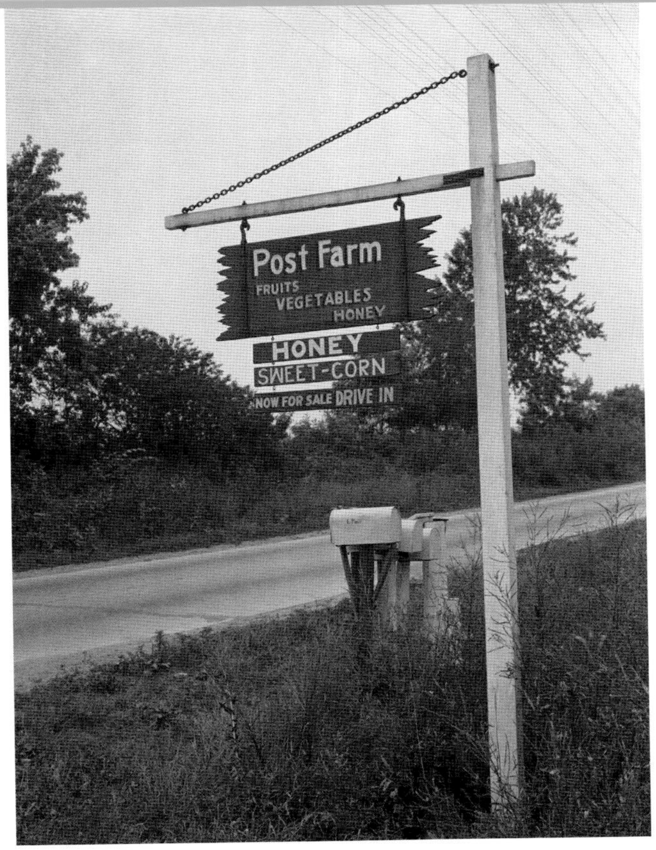

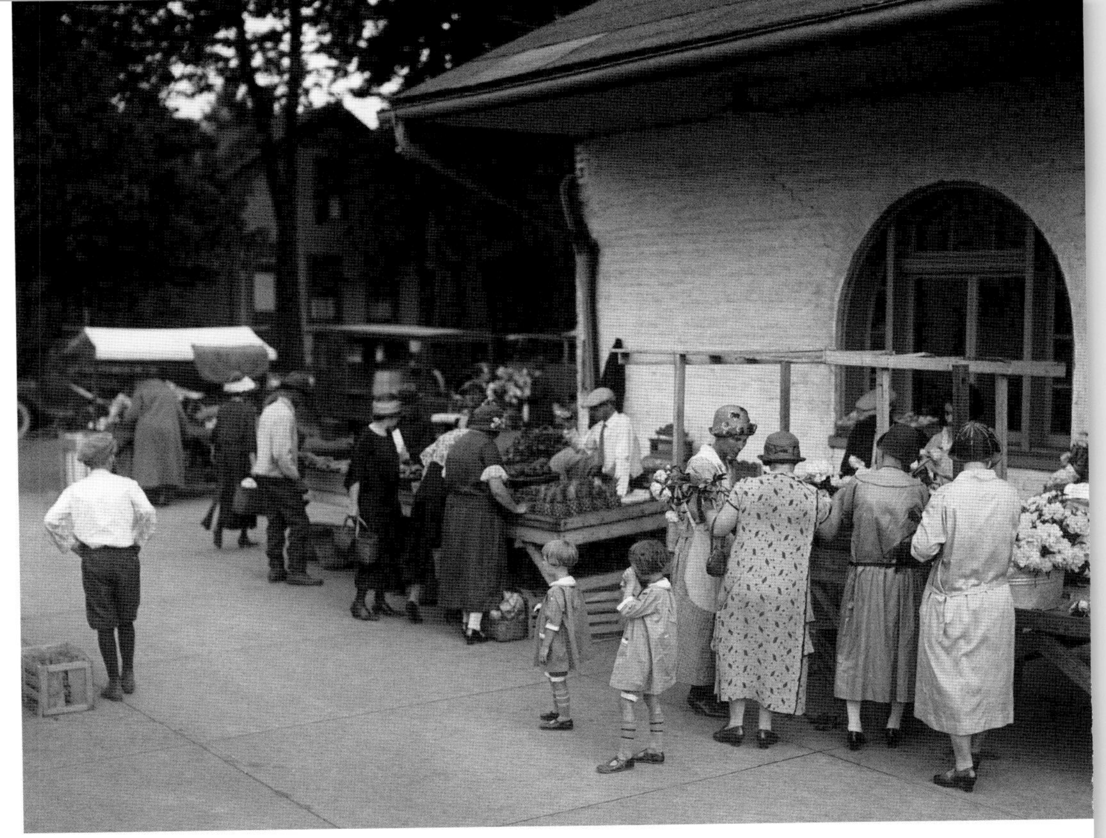

(Top left) A farm advertising sign at the Post Farm in Madison, Wisconsin (Dane County, WI; 1923)

(Top right) A farmers market at Richmond (Wayne County, 1925)

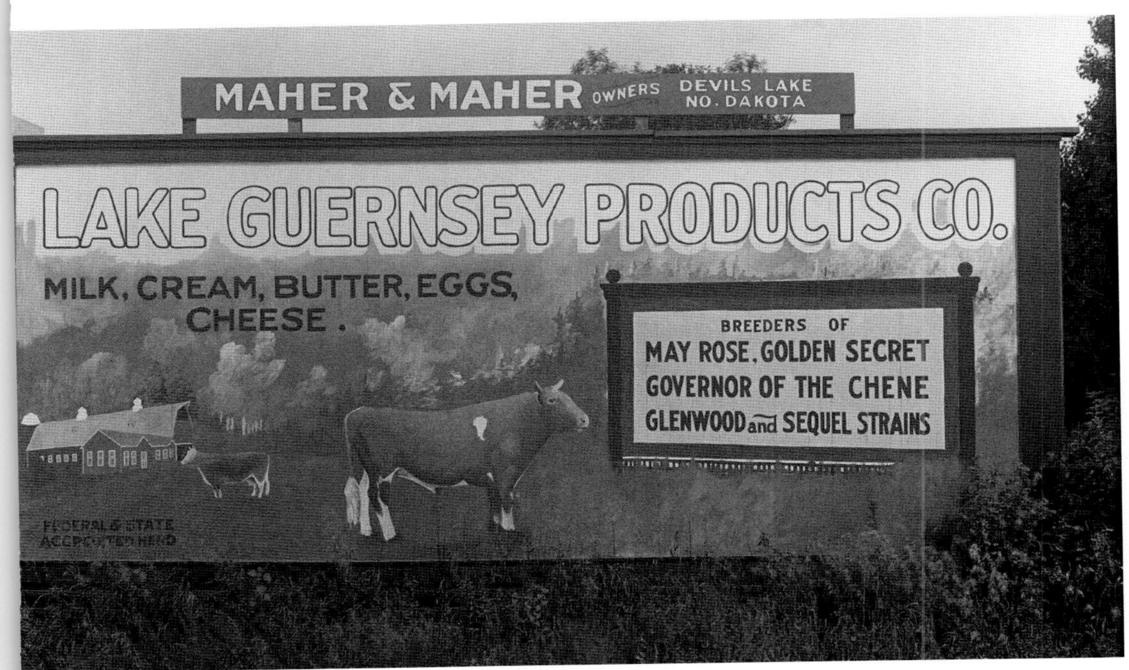

This eye-catching sign promotes products available from the Maher and Maher Farm in Devils Lake, North Dakota. (Ramsey County, ND; 1925)

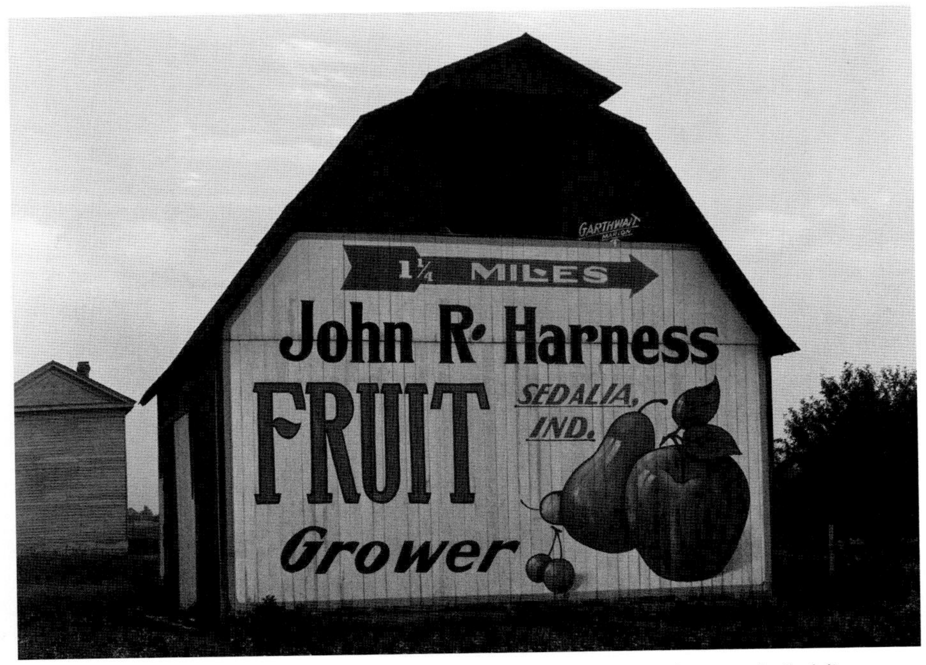

A sign covers one side of a barn along a public highway. John R. Harness, who owned a Sedalia-area fruit farm, believed it paid to advertise his produce. (Clinton County, 1931)

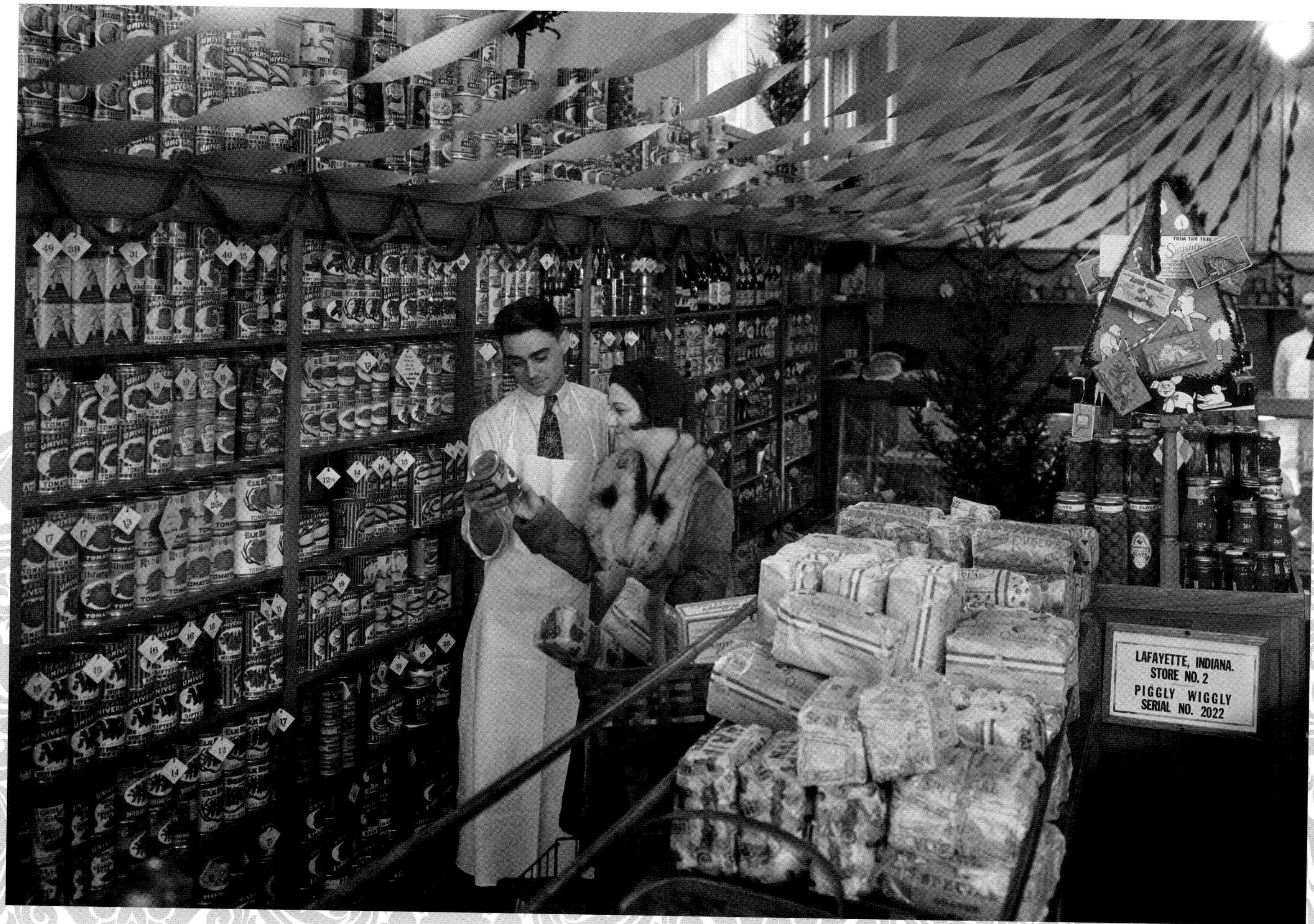

An inside view of a Piggly Wiggly Store (Tippecanoe County, 1931)

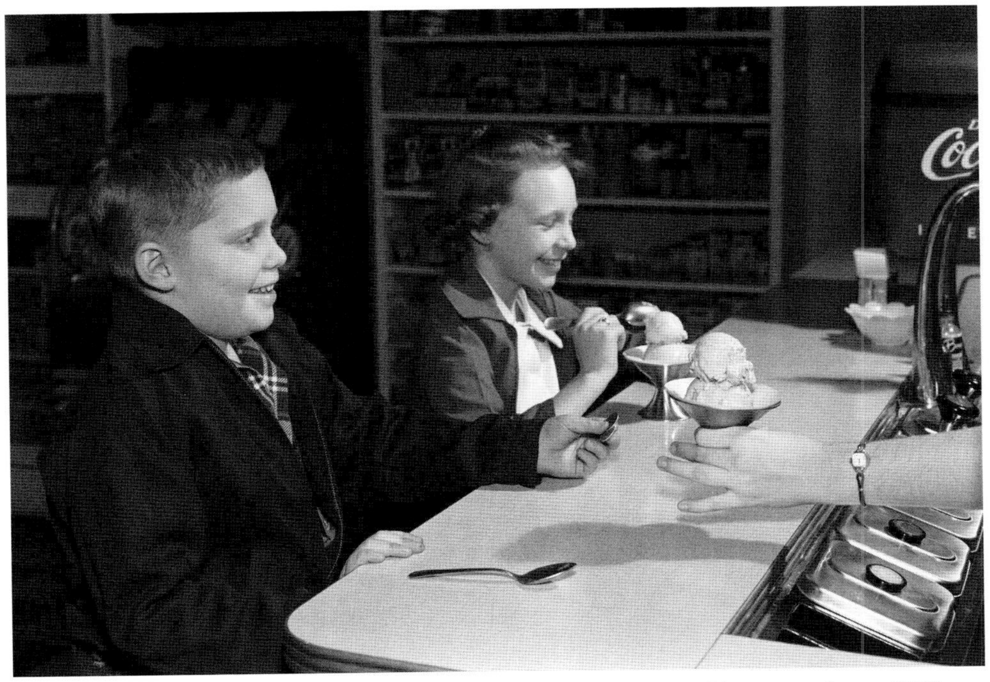

Larry (left) and Terry Parker of West Lafayette enjoy a treat at a local drugstore. (Tippecanoe County, 1954)

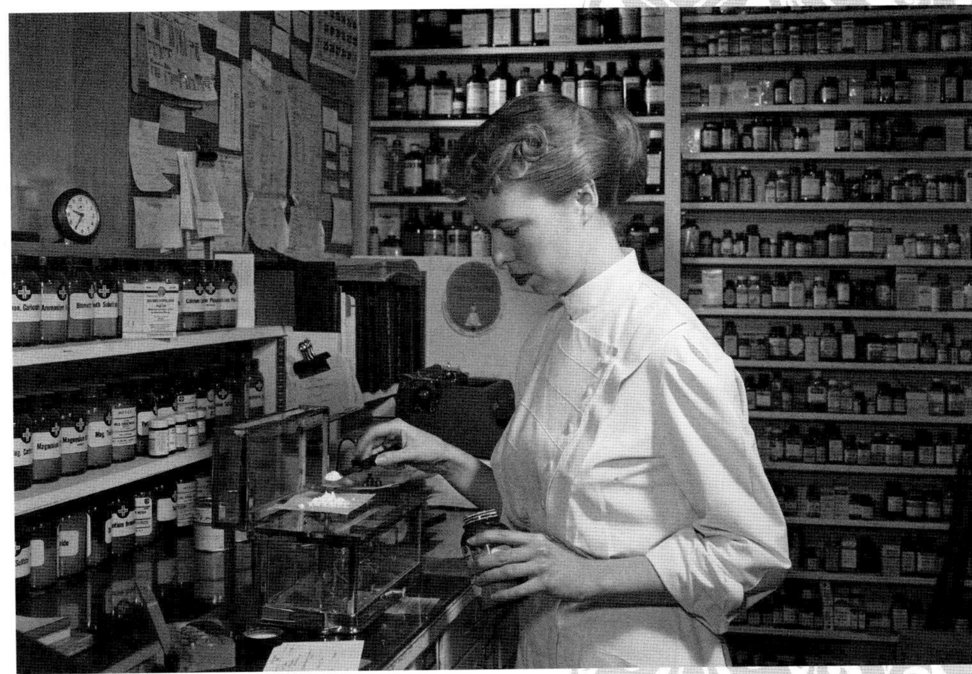

A pharmacist weighs ingredients for a prescription at Arth Drug Store. (Tippecanoe County, 1953)

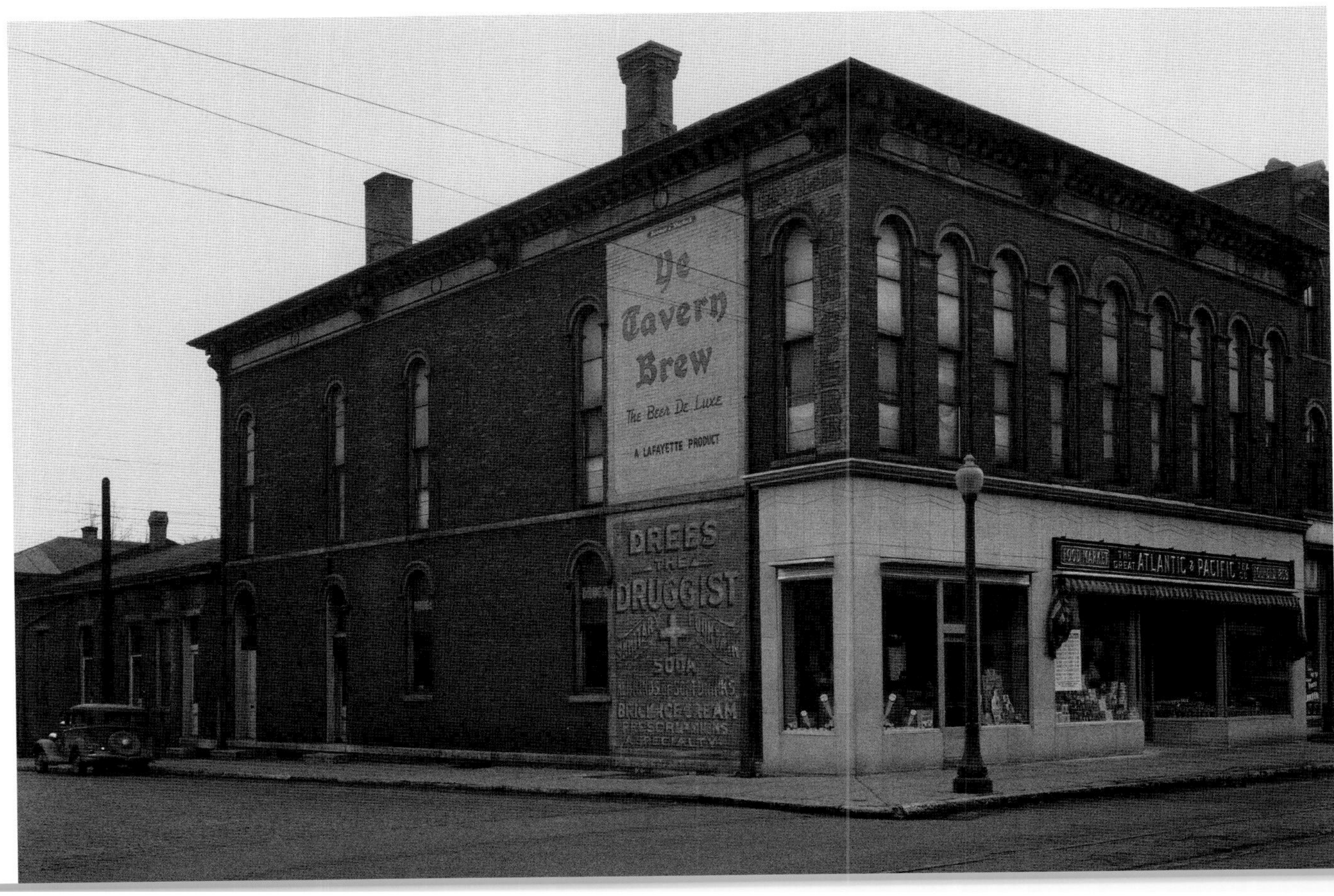

A storefront at the corner of Eighth and Main Streets in Lafayette (Tippecanoe County, 1935)

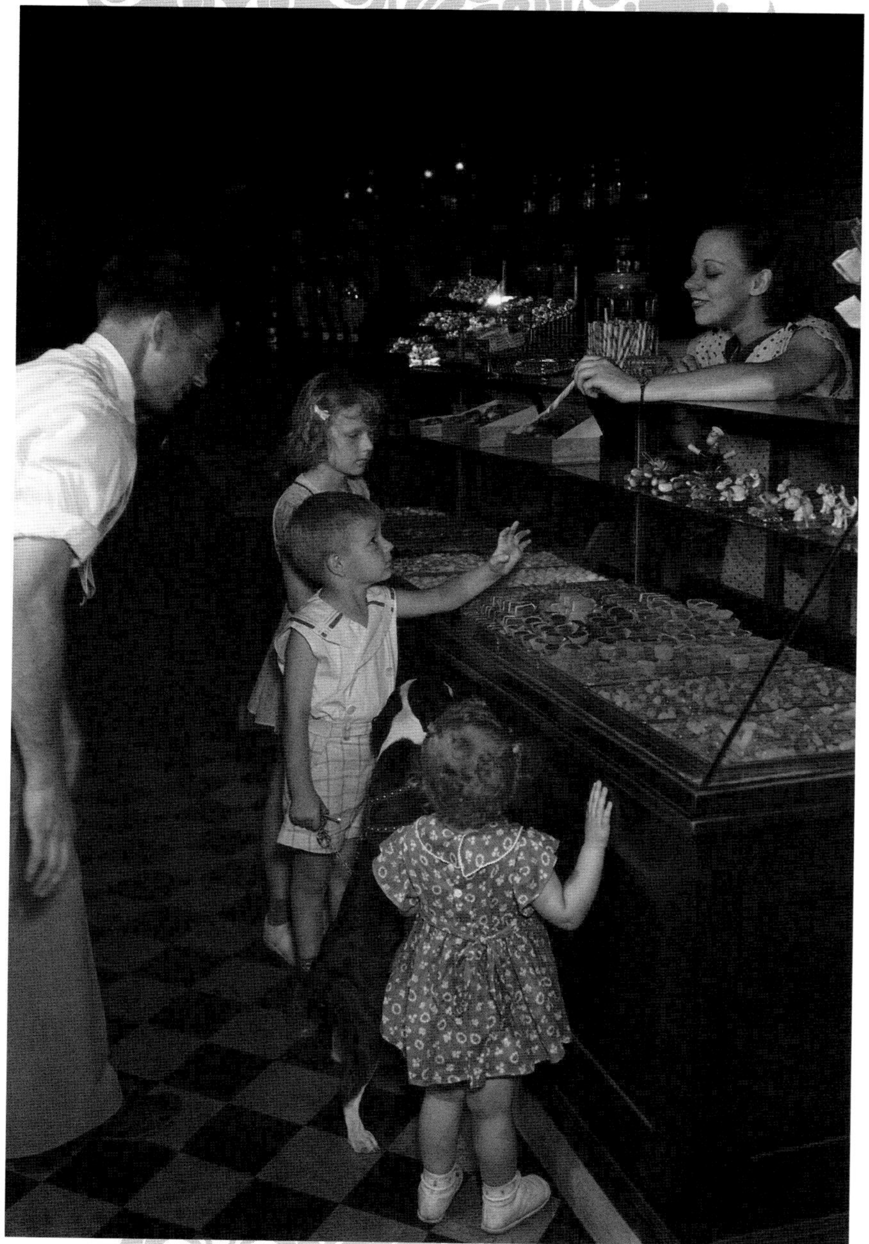

Children and their dog wait their turn for a piece of candy inside Glatz Candies and Luncheonette in downtown Lafayette. (Tippecanoe County, 1938)

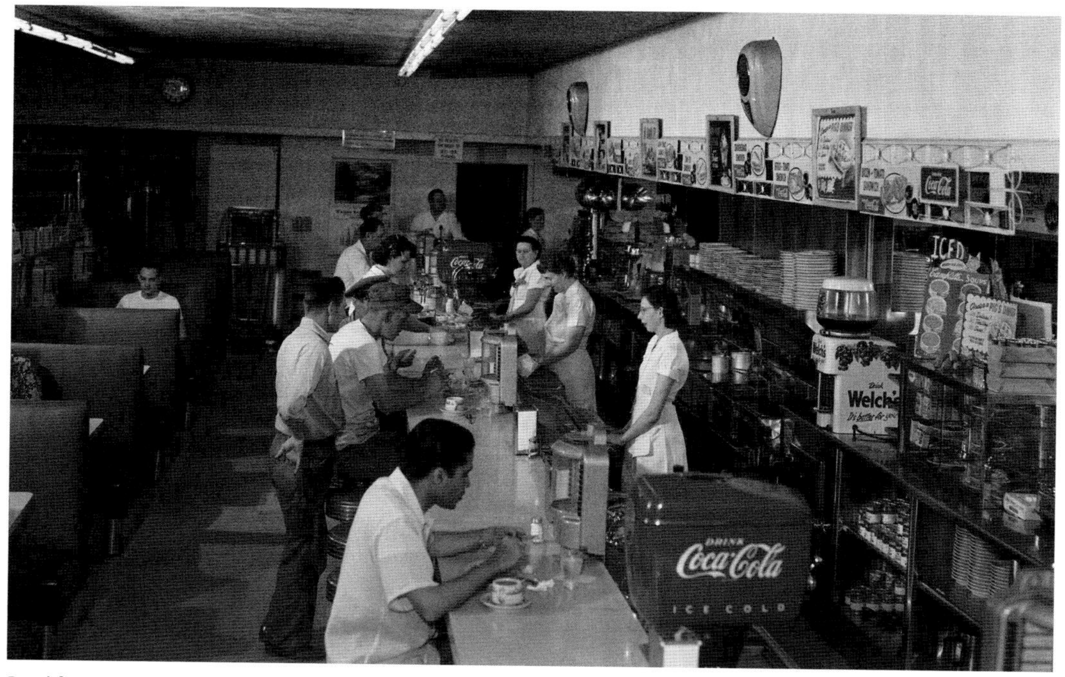

Breakfast is served at a local diner. (Tippecanoe County, 1953)

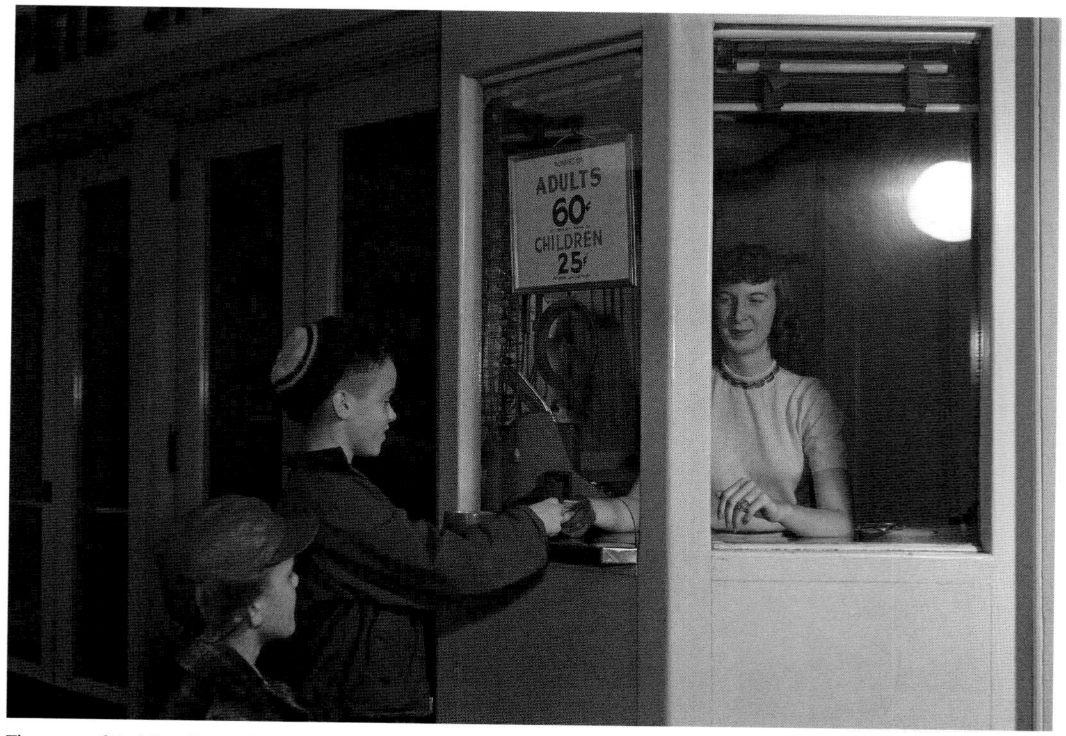

The sons of Earl Butz buy tickets to a show at the Mars Theater in Lafayette. (Tippecanoe County, 1954)

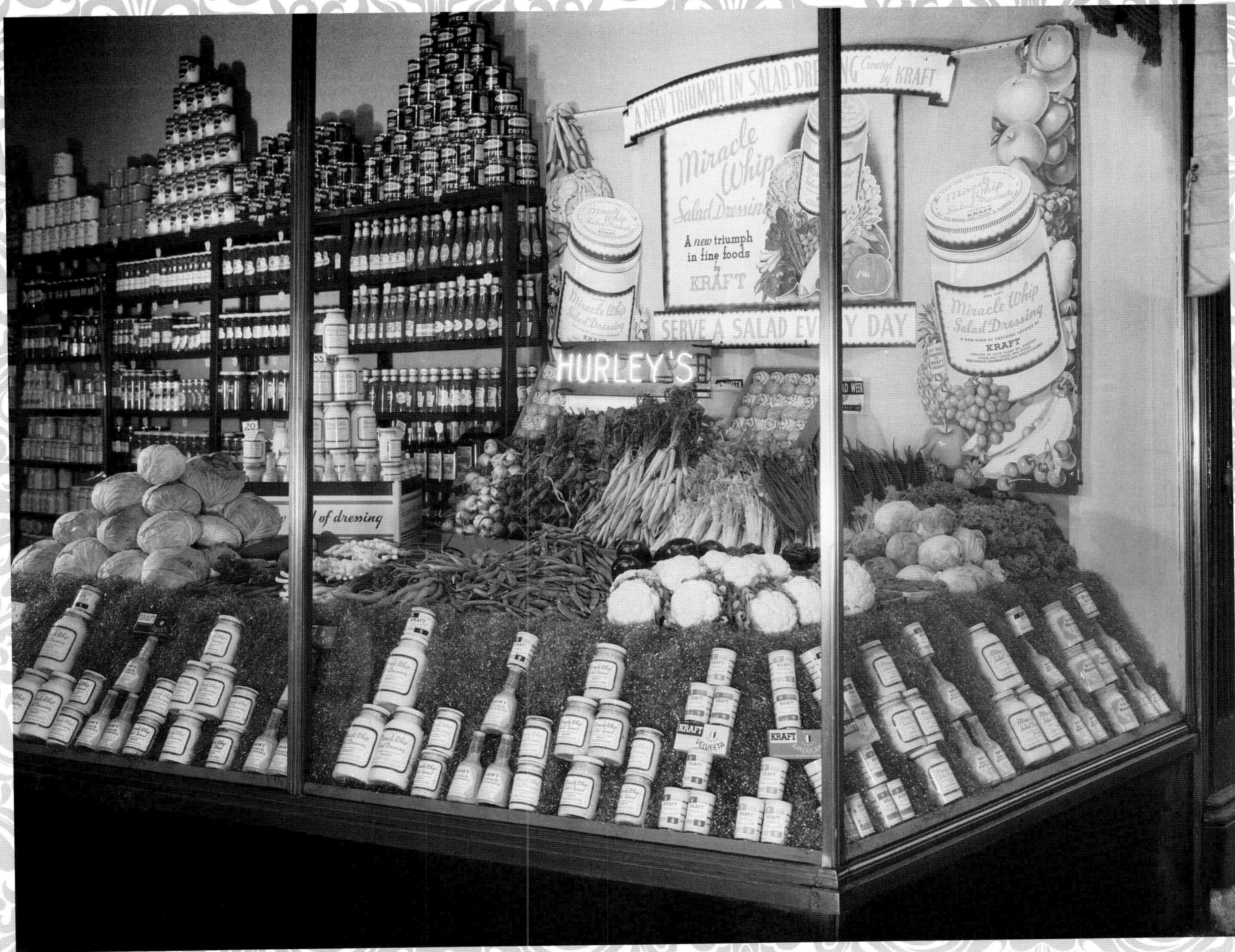

Hurley Grocery Store in Lafayette displays a new product, Kraft Miracle Whip Salad Dressing, in its window. (Tippecanoe County, 1934)

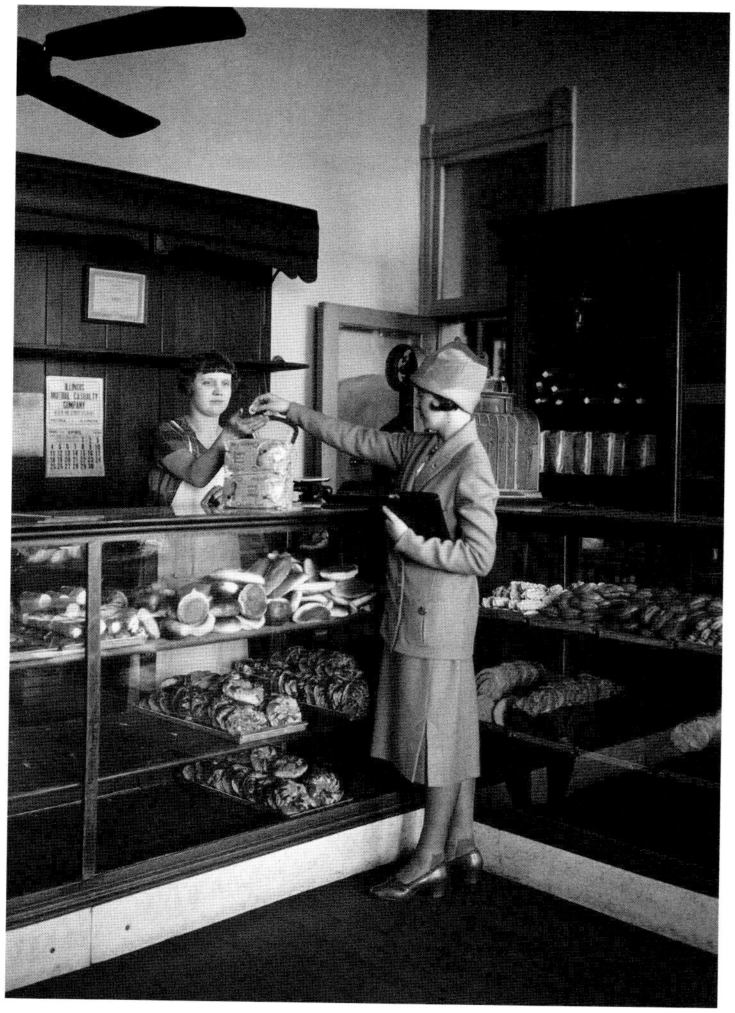

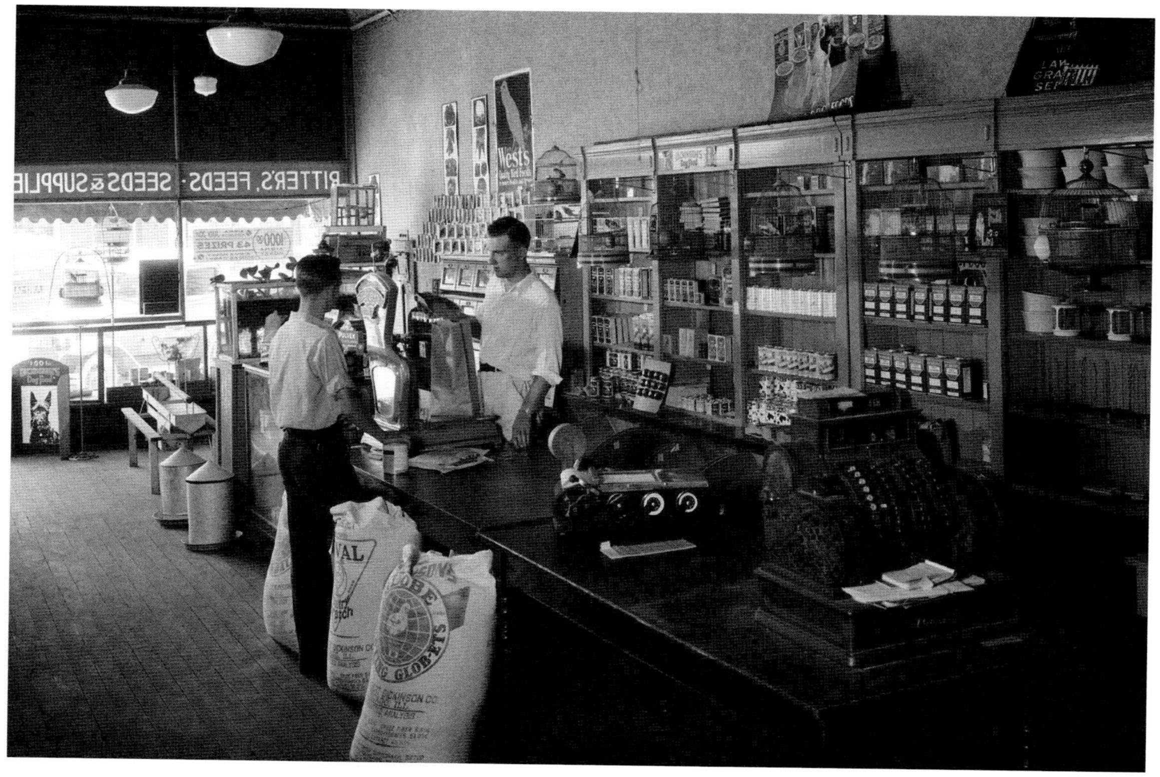

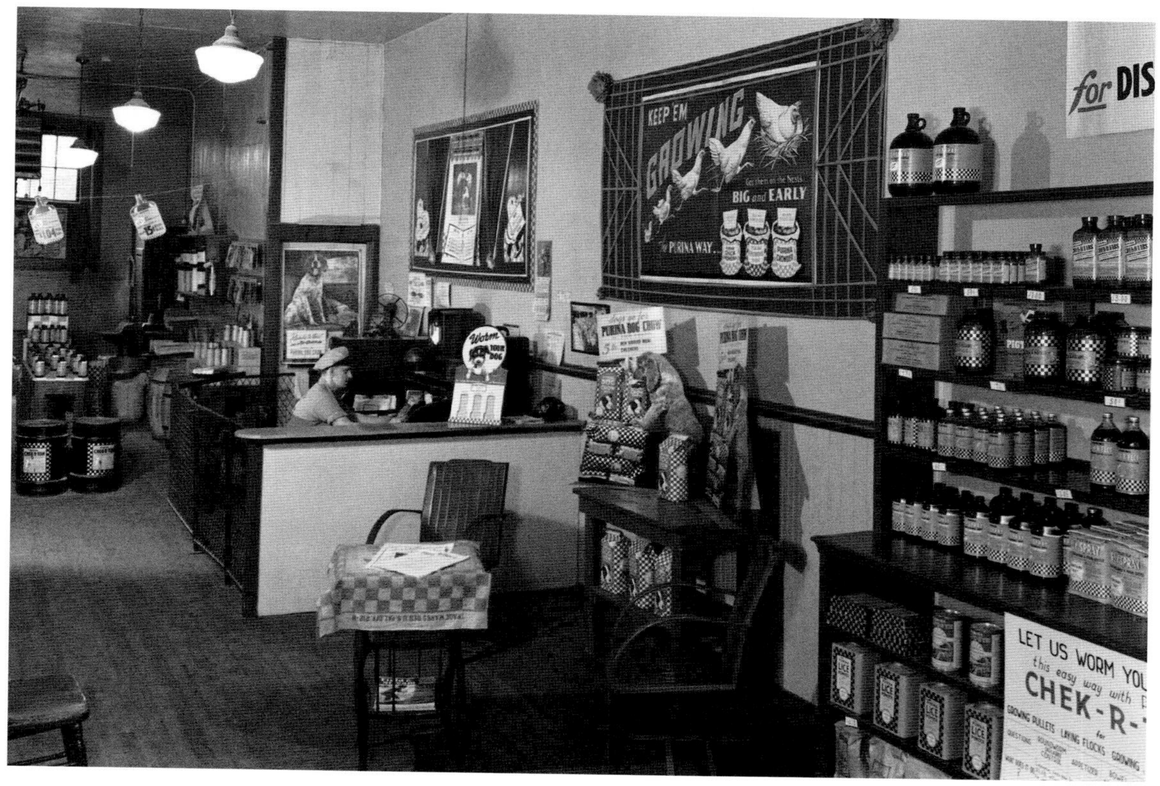

(Above) Sid Keller buys bread at the Ruger Bakery in Lafayette. (Tippecanoe County, 1936)

(Top right) A customer purchases poultry feed at C. E. Ritter's Feeds, Seeds, and Supplies in Anderson. (Madison County, 1933)

(Bottom right) Charles Shuman works in a Lafayette-area Purina Store. (Tippecanoe County, 1946)

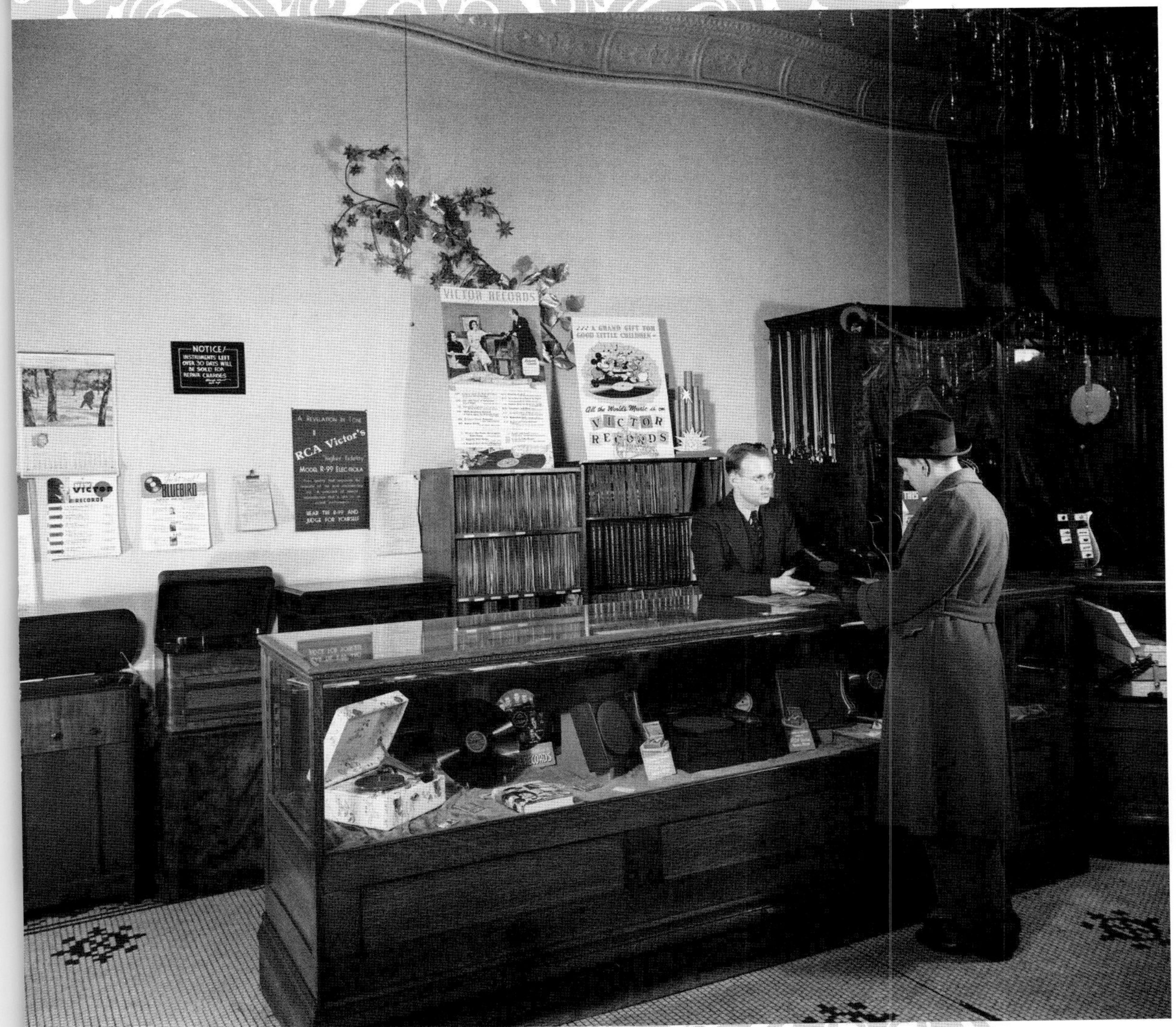

(Left) The McLean Music Store in Lafayette sold record albums and players. (Tippecanoe County, 1937)

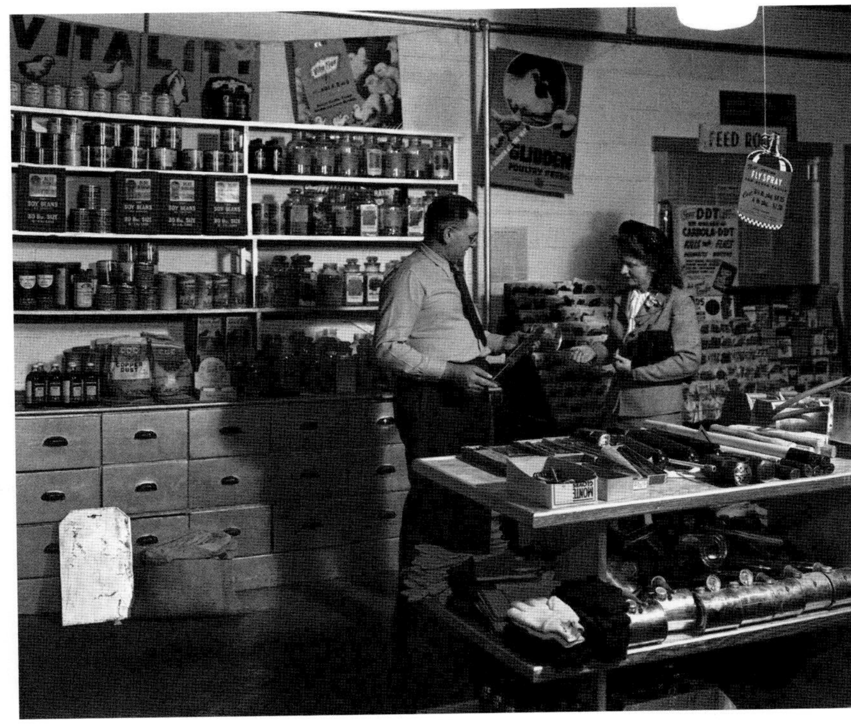

(Above) The Bashore Feed Store and Hatchery in North Manchester sold farm supplies. (Wabash County, 1946)

Connecting to the Outside World

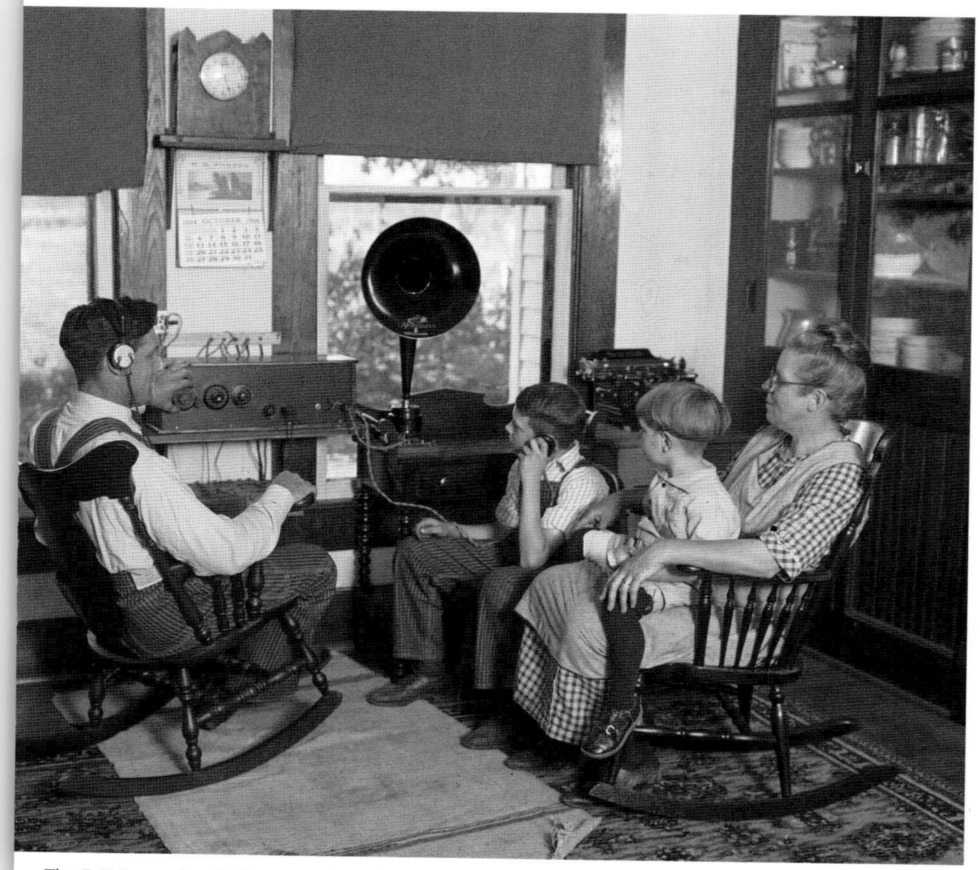

The E. E. Brown family listens to the radio in Veedersburg. (Fountain County, 1924)

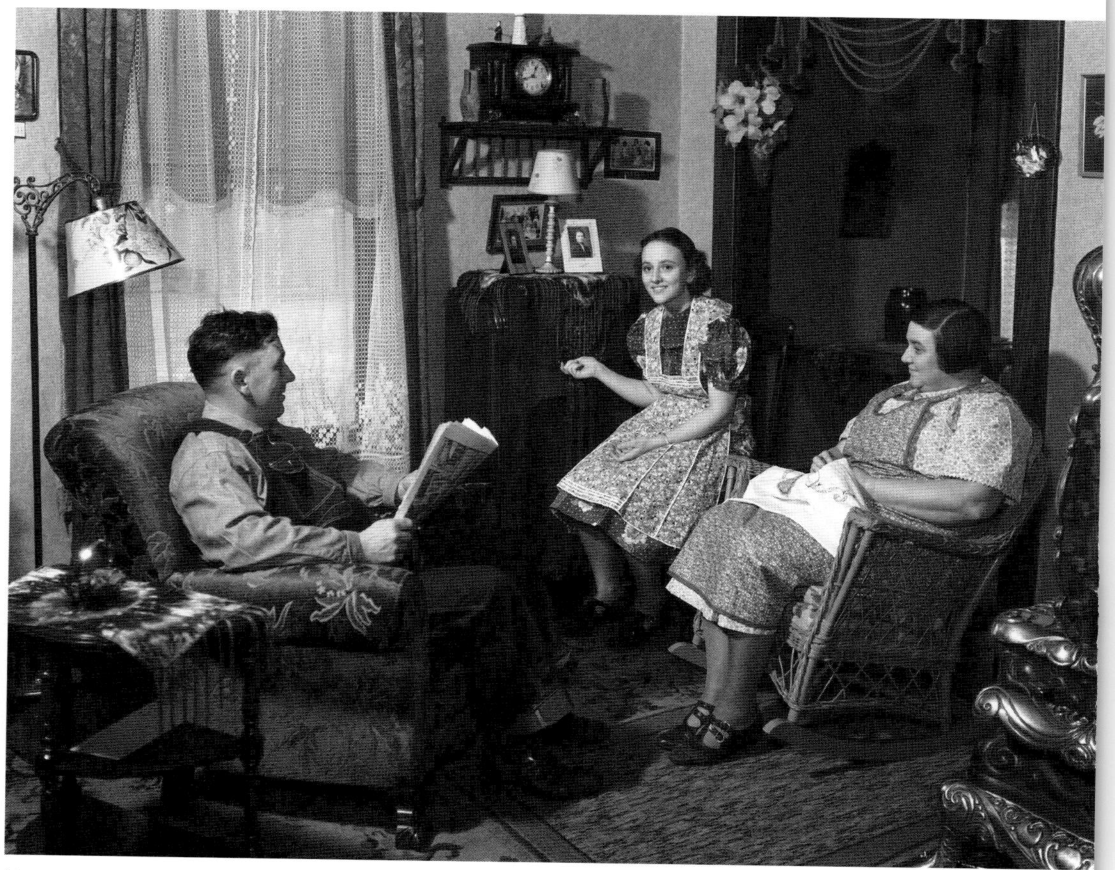

Herman Heaton of Russiaville enjoys a radio program along with his wife and daughter. (Howard County, 1939)

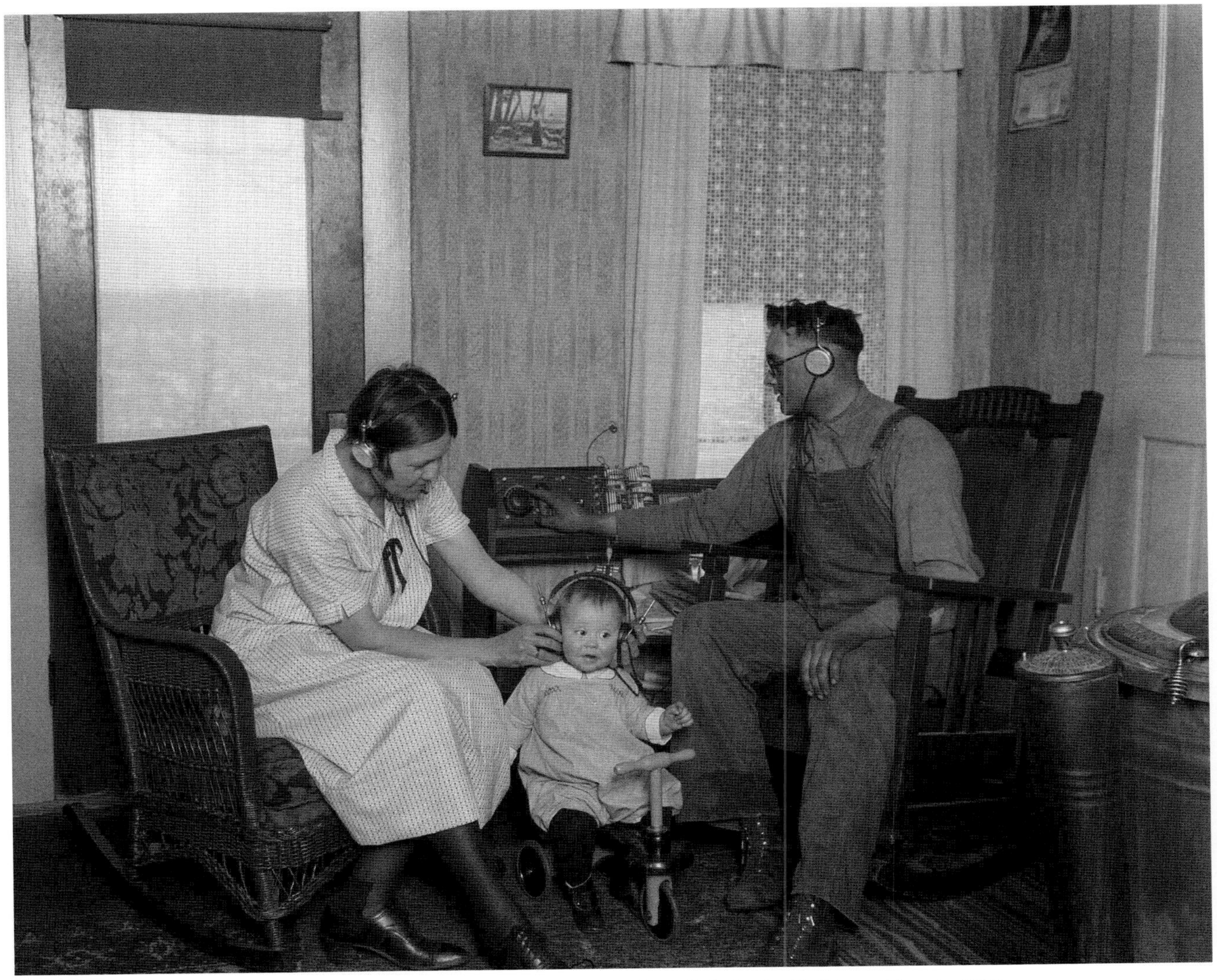

Mr. and Mrs. Hurshel Jacobs and their young son use headphones to listen to the radio in Russiaville. (Howard County, 1925)

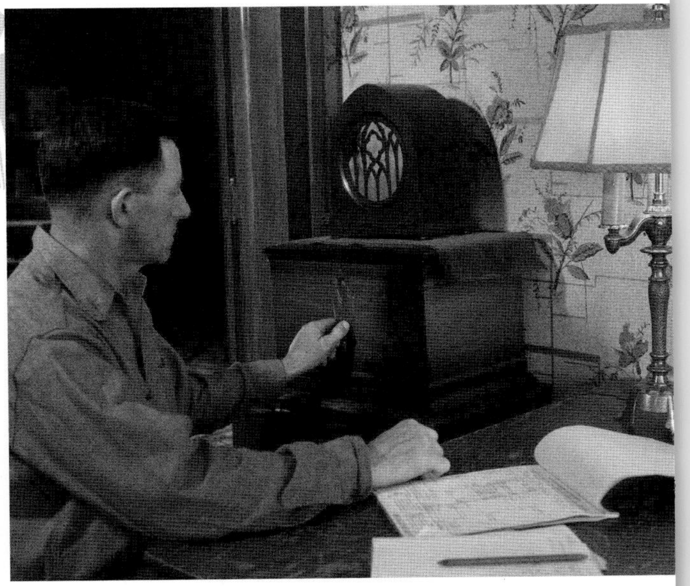

Ray A. Glass of the Kendallville area tunes his radio to a farm program. Allen wrote of Glass, "One morning when he had sheep to sell, he made $60 by obtaining the market report over his radio." (Noble County, 1934)

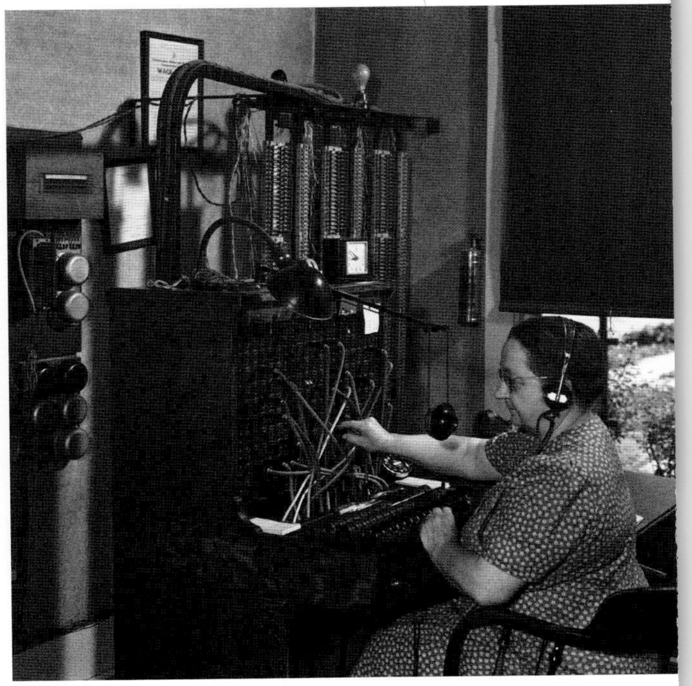

At the time of this photo, Adeline C. Hawkins had served 28 years as a telephone switchboard operator, connecting one party to another in Galveston. (Cass County, 1946)

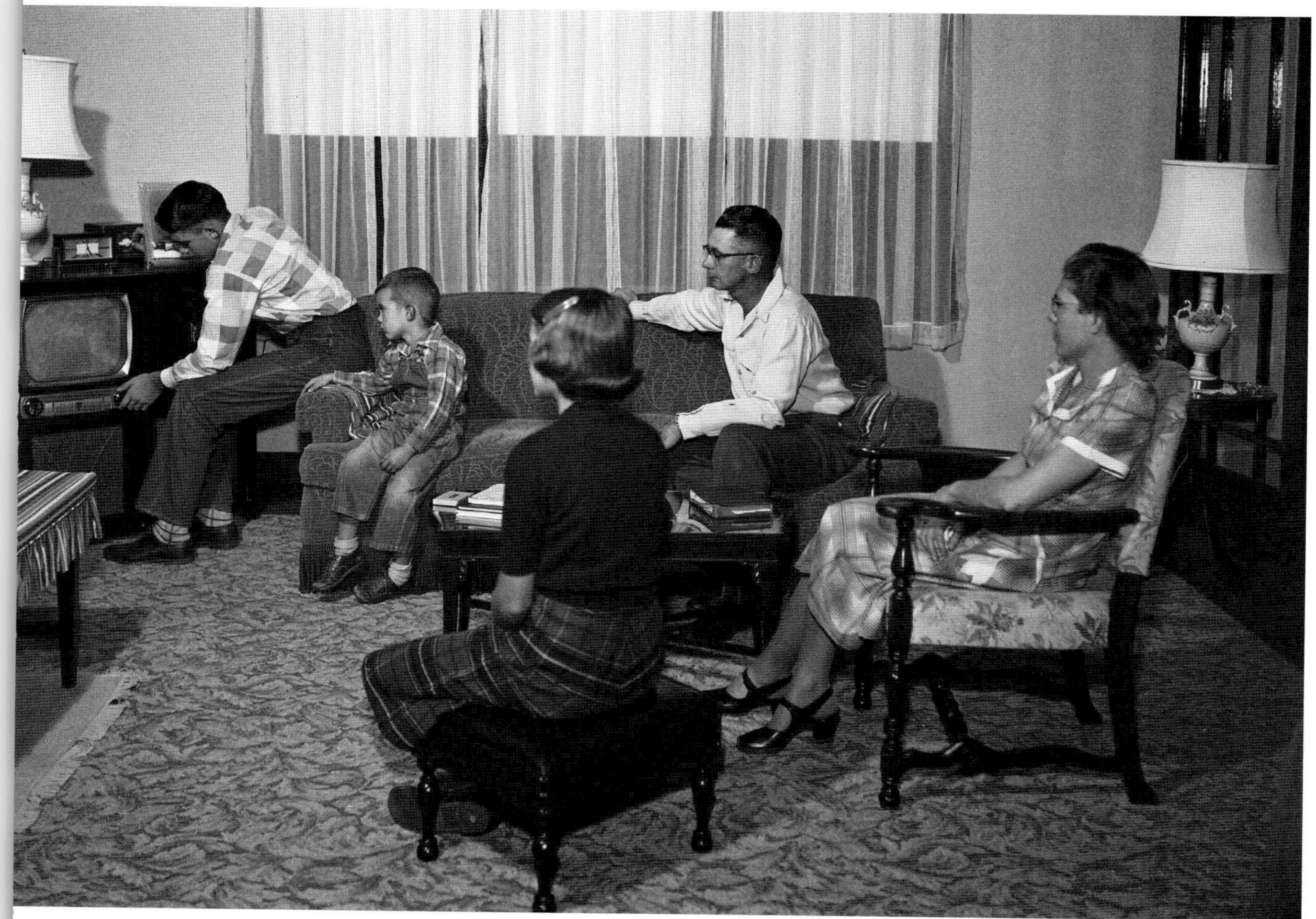

The Leon Johnson family in Orleans prepares to watch a TV show in their living room. (Orange County, 1951)

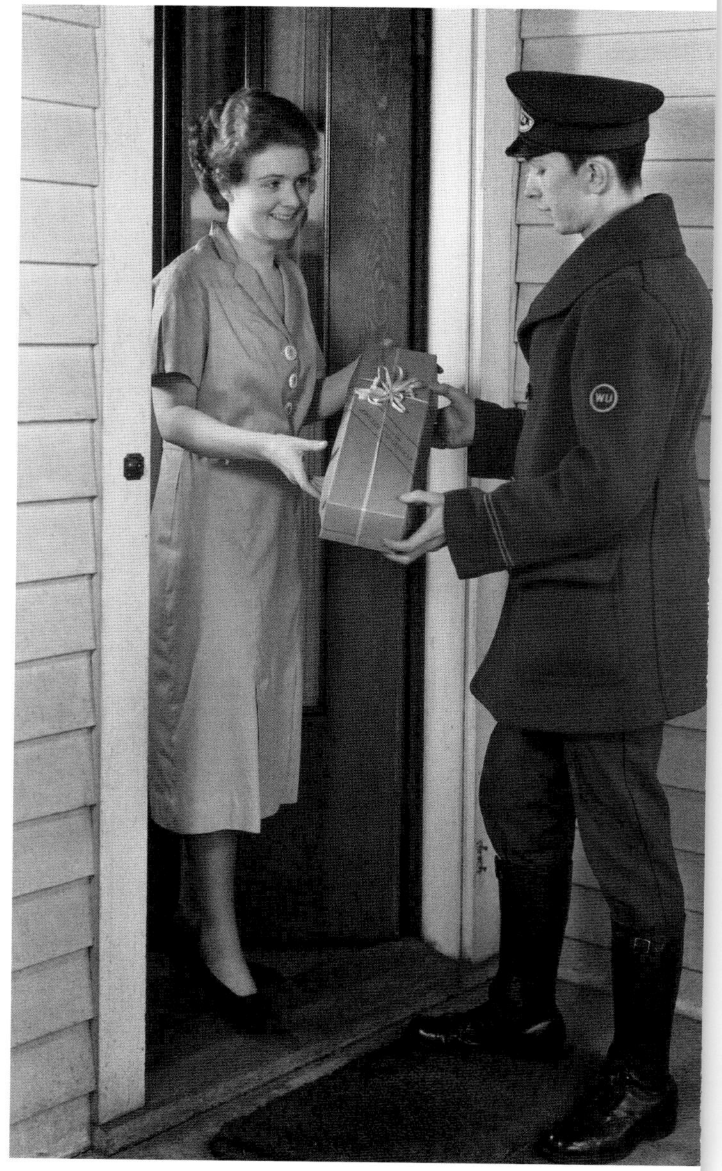

A Western Union employee delivers a box of flowers. (Warren County, 1935)

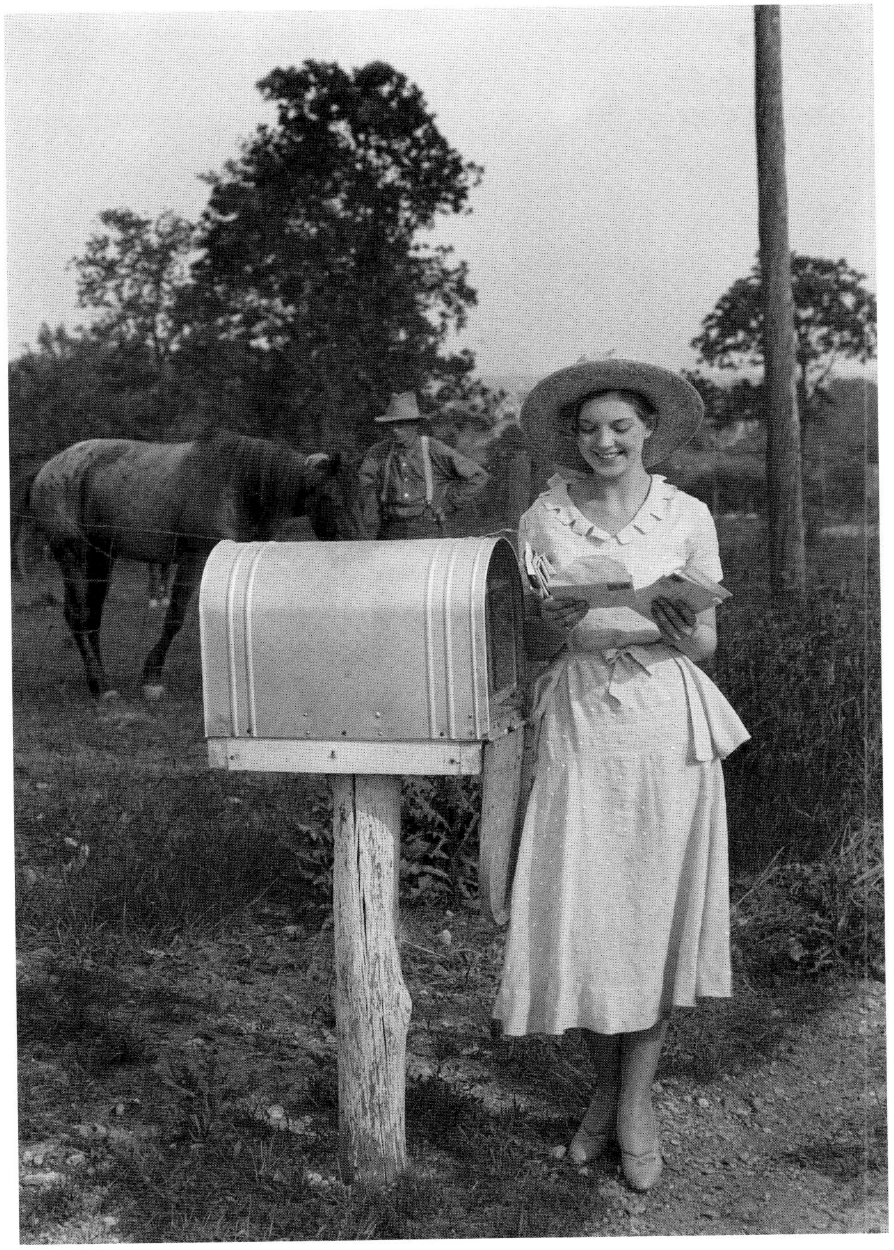

(Above) Priscilla Ramsay reads a letter at her mailbox. (N.p., 1931)

(Right) Morris McCarty of Lafayette looks at the August 1931 issue of the *Progressive Farmer*. (Tippecanoe County, 1931)

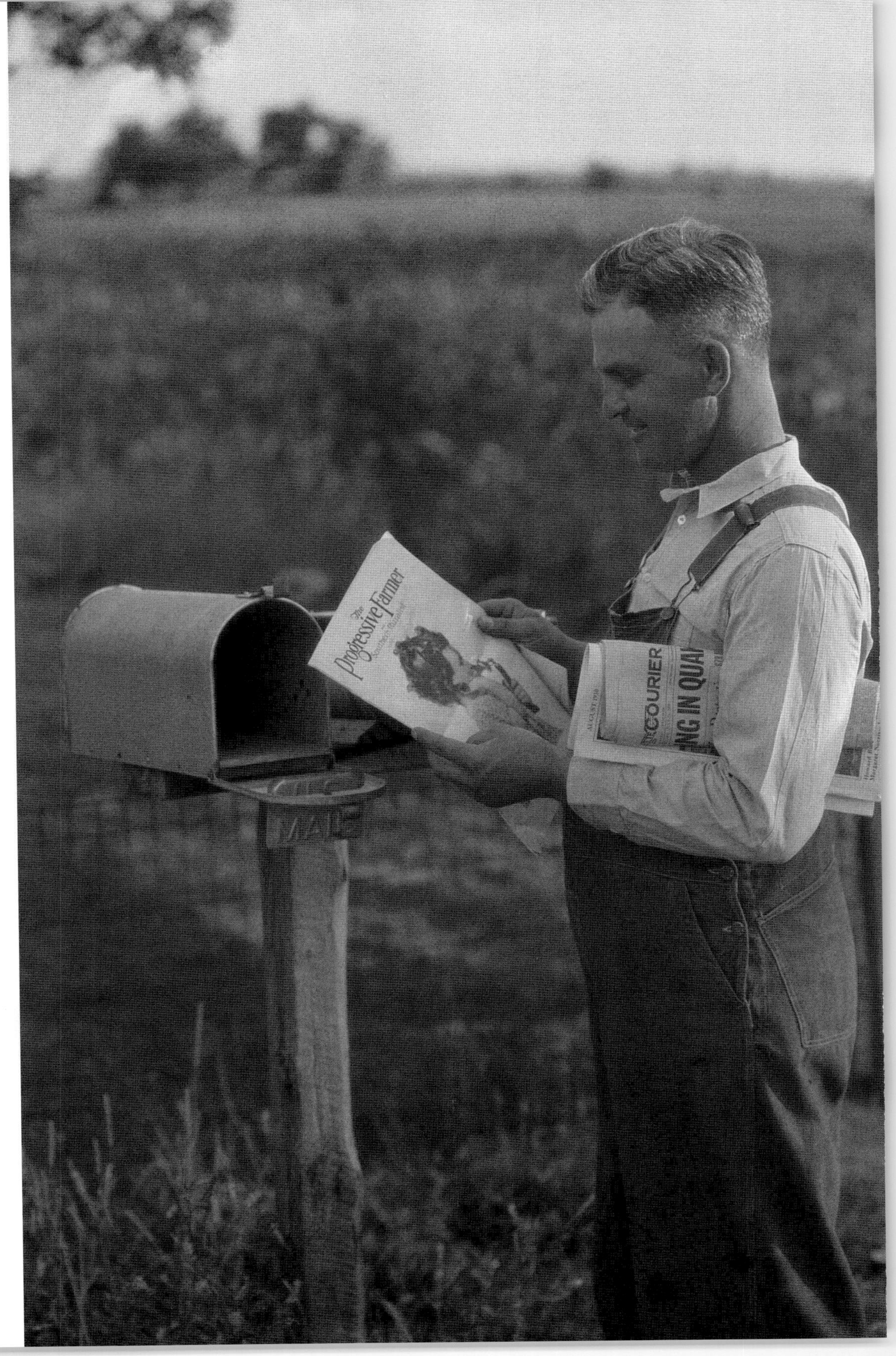

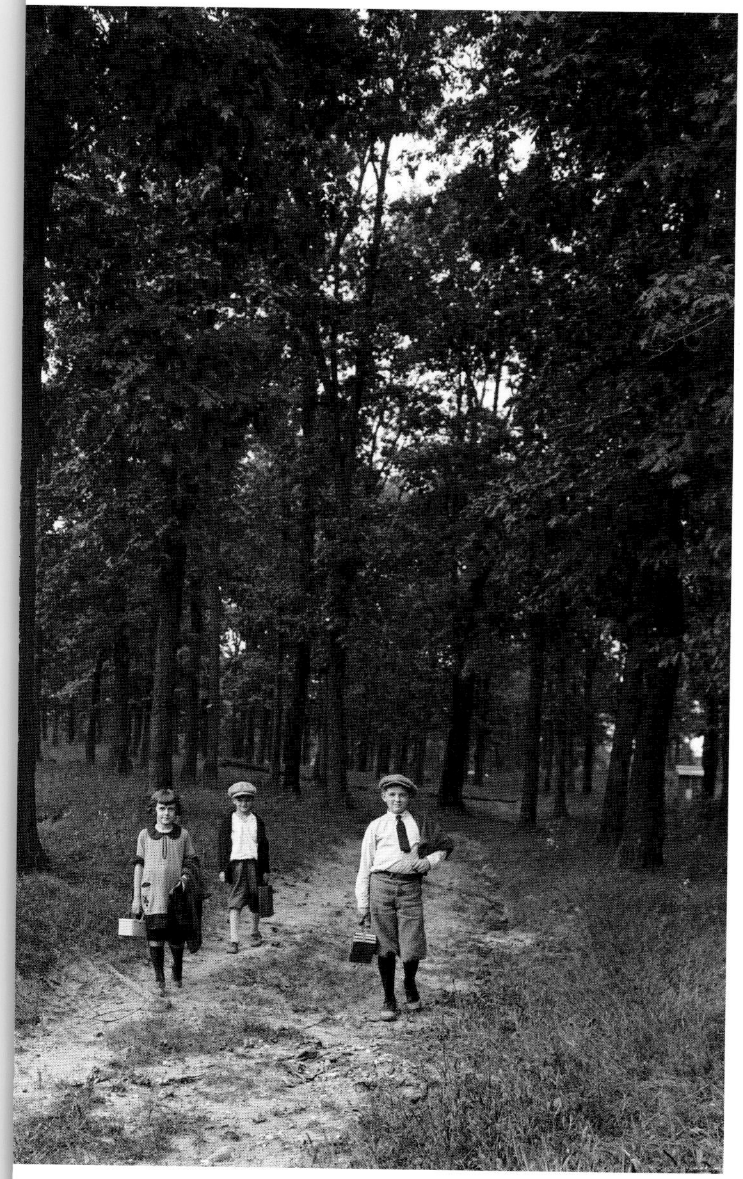

Children with lunch buckets walk to school down a lane in the woods. (N.p., 1923)

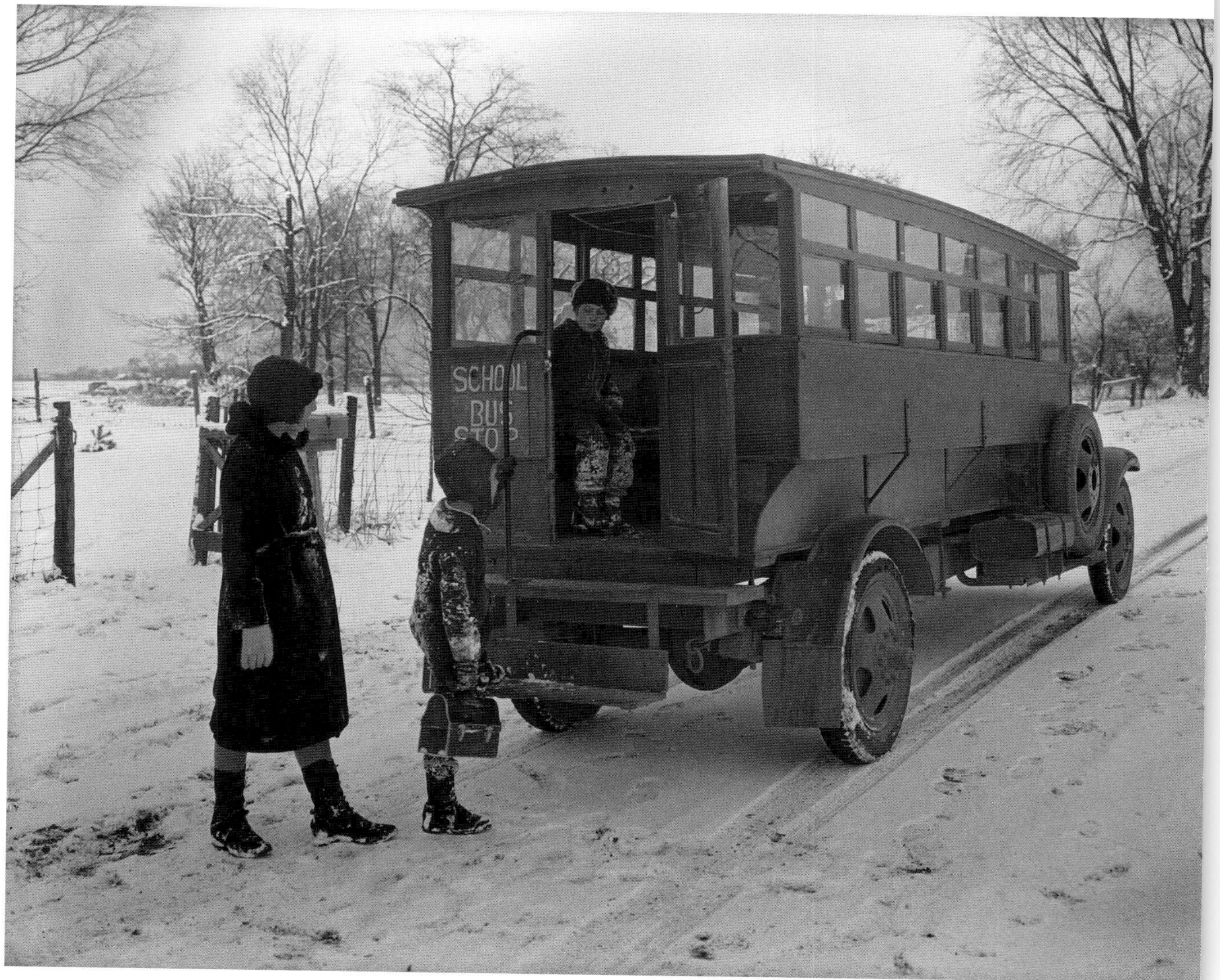

A school bus stops for the children of John Trost near West Lafayette. (Tippecanoe County, 1933)

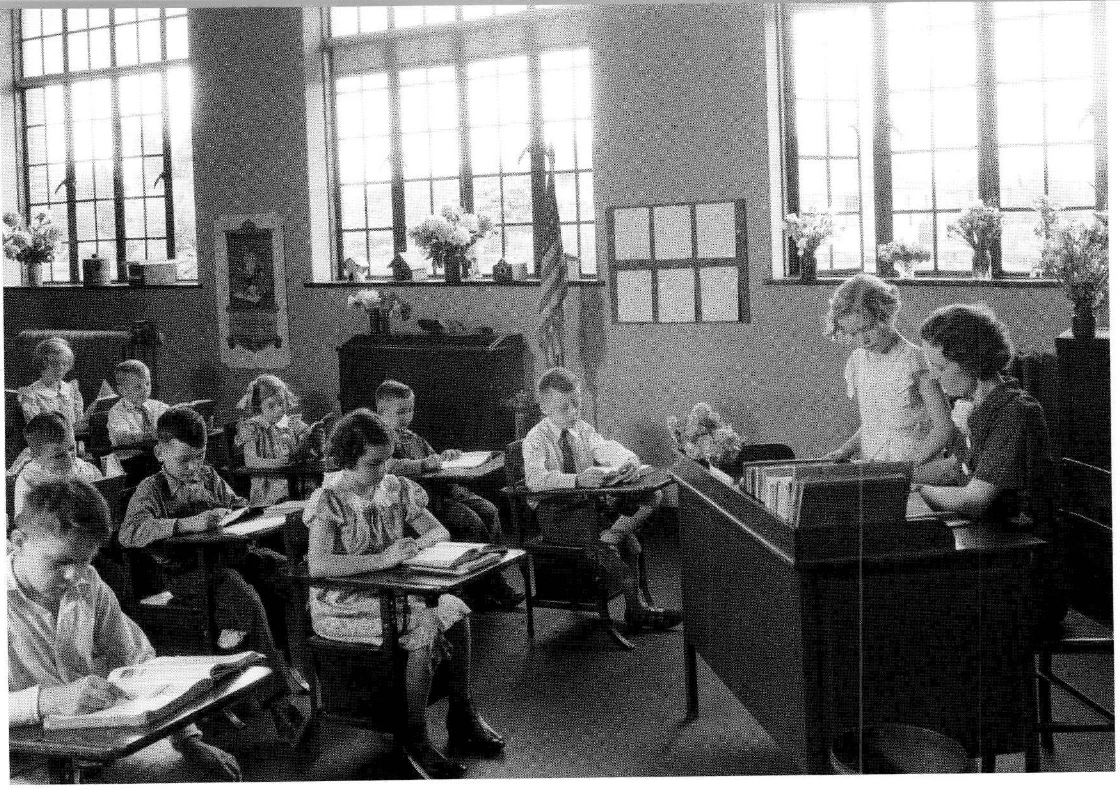

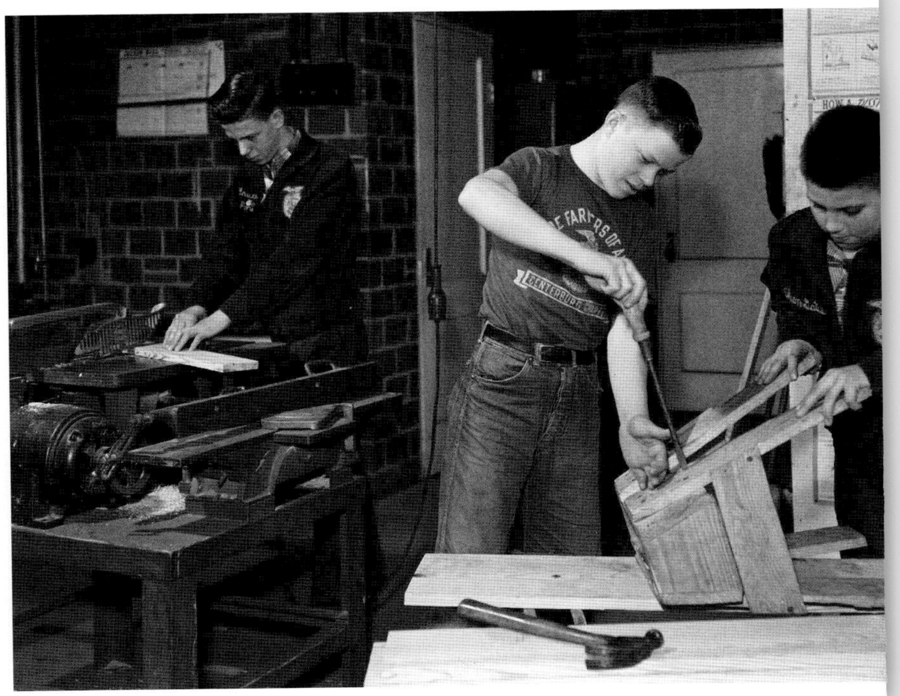

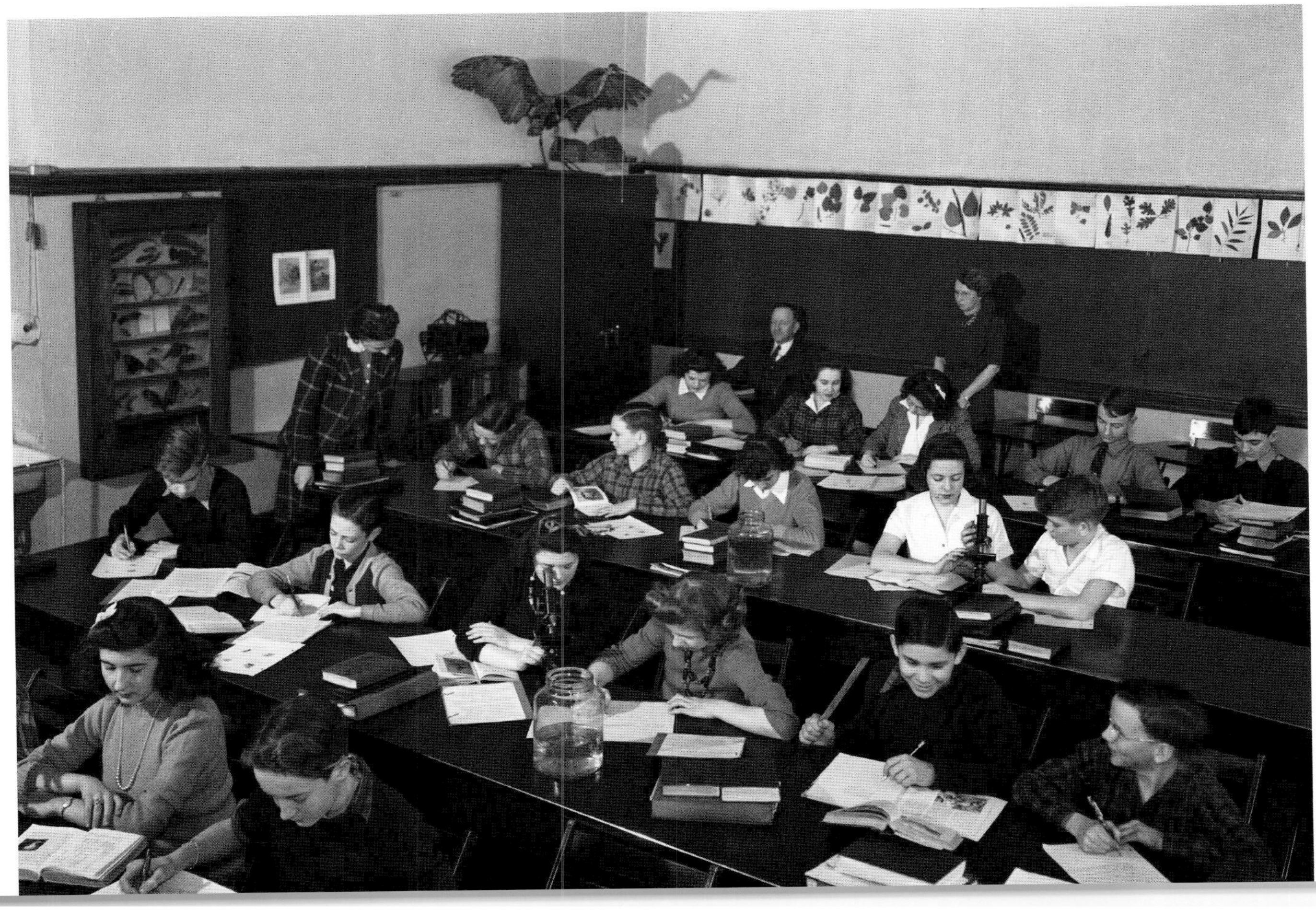

(Top left) Miss Henderson teaches a fourth-grade class at Longlois School in Lafayette. (Tippecanoe County, 1935)

(Top right) Members of the Future Farmers of America—today known as the National FFA—practice their carpentry skills at the Centerburg High School. (Knox County, OH; 1952)

(Bottom) Students work on various projects during a biology lab in the junior high school at West Lafayette. (Tippecanoe County, 1943)

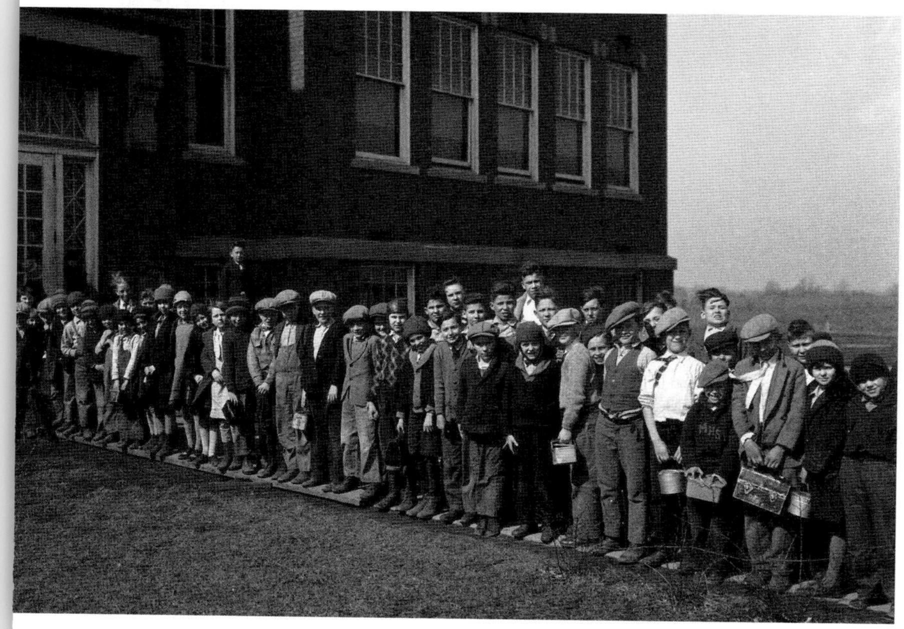

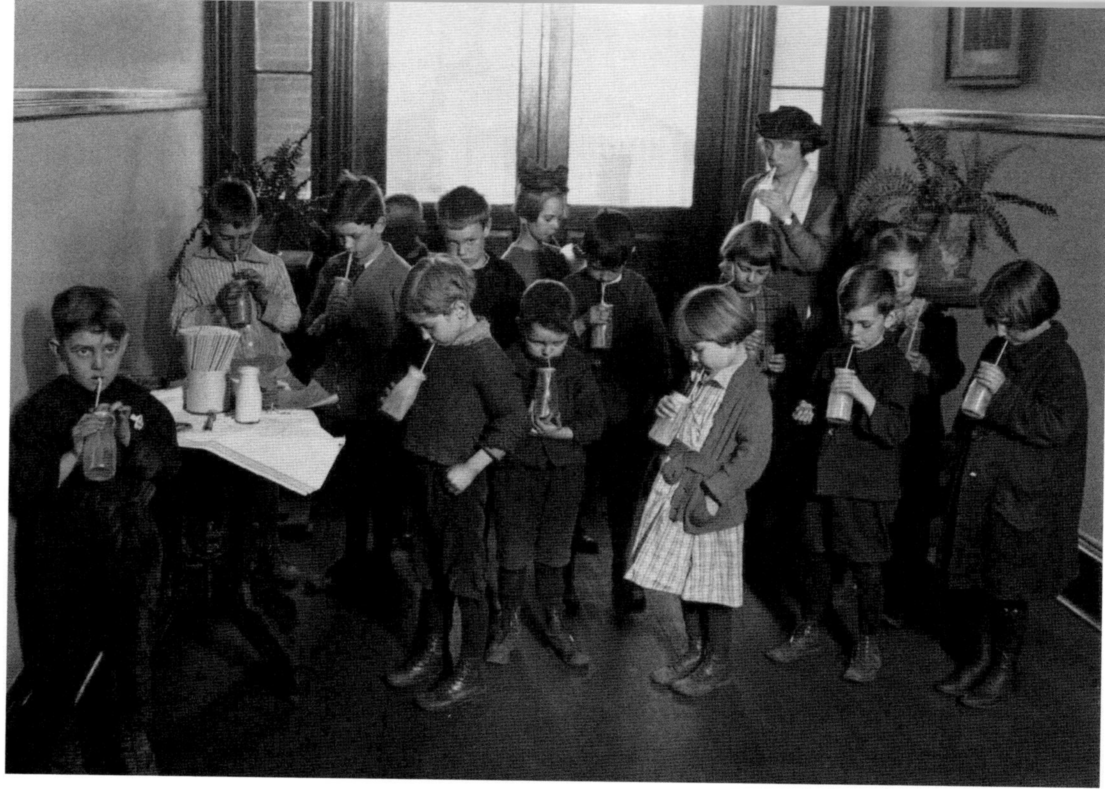

(Top left) Schoolchildren with their lunch buckets at Monitor School (Tippecanoe County, 1928)

(Top right) Children take a break for lunch. (Tippecanoe County, 1919)

(Bottom) Serving hot lunch at Monitor School (Tippecanoe County, 1928)

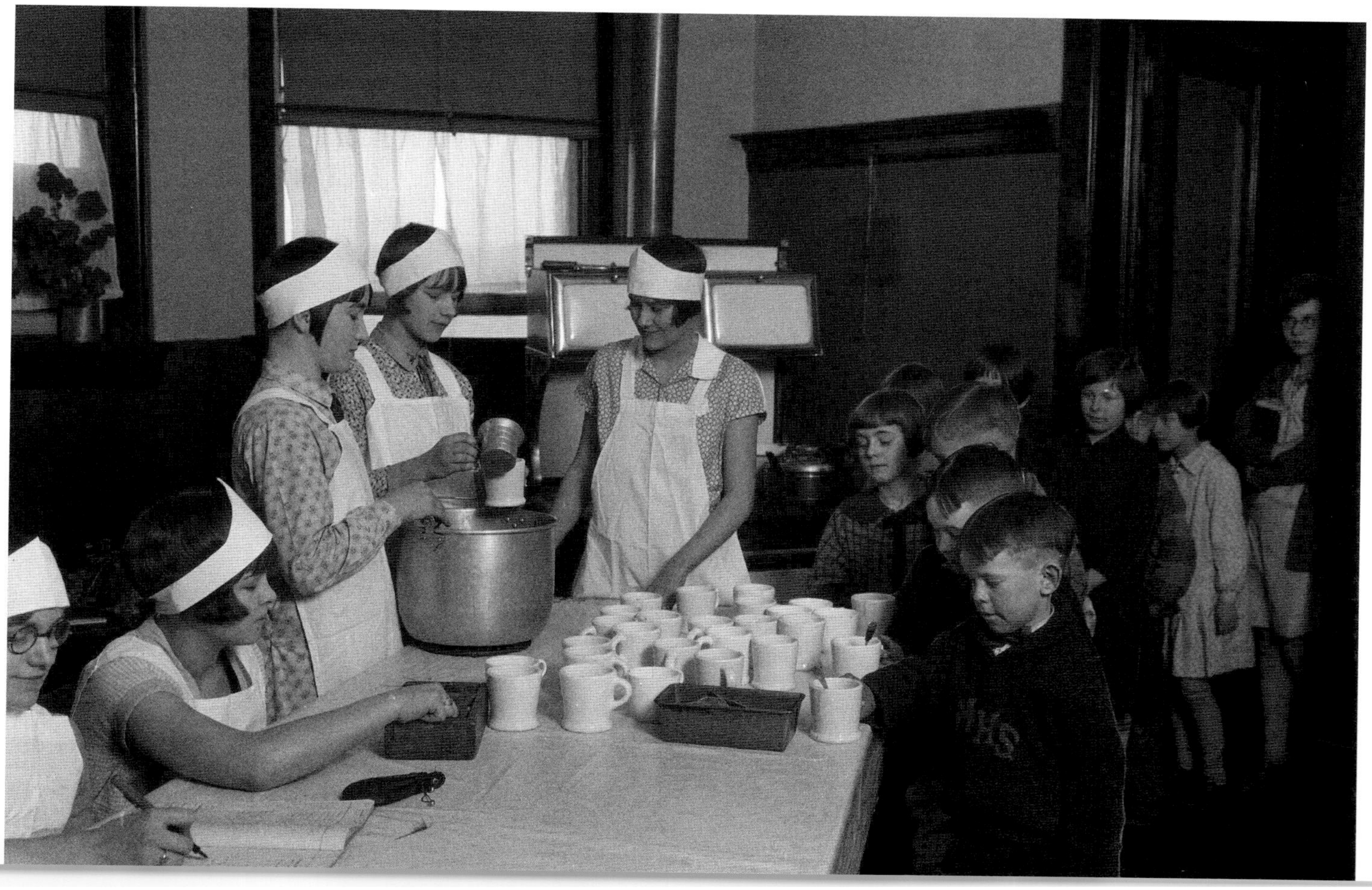

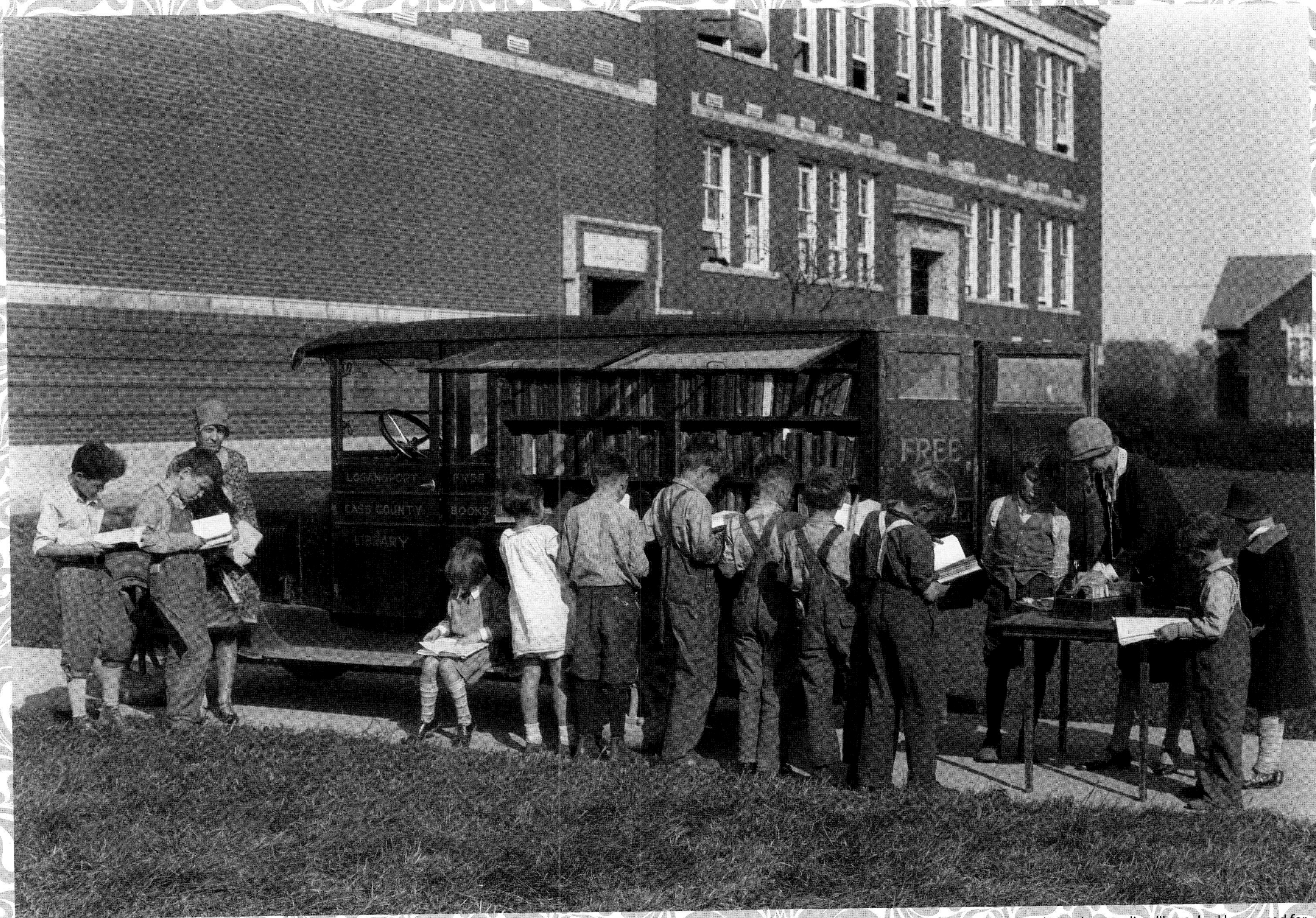

Children from a Washington Township school near Logansport gather around the Bibliobus, a library on wheels operated free of charge by Cass County. At the time of the photo, the traveling library had been used for nine years, with the vehicle traveling 25 miles a day for five days each week. Here, librarians Edna Holden and Lorene Senn assist the children. (Cass County, 1929)

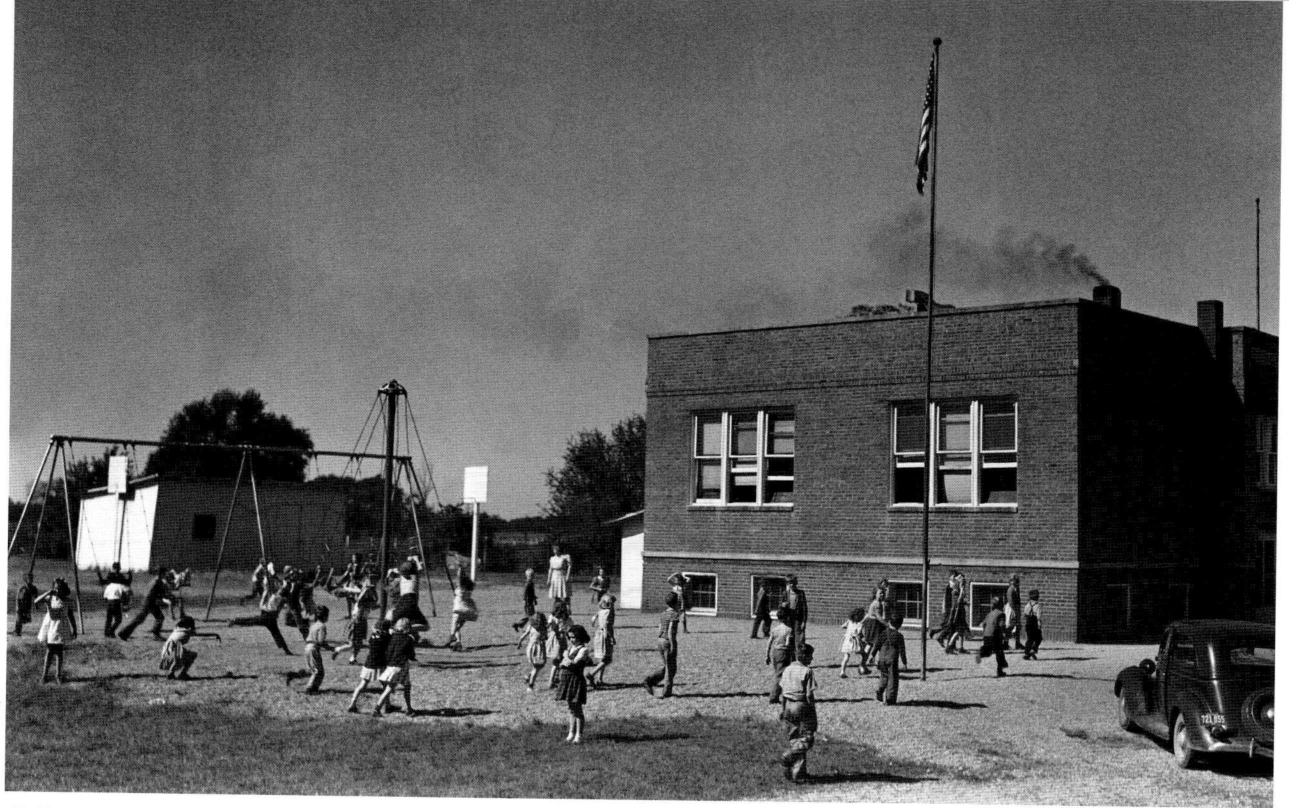

Children play in the school yard at Federal Hill School in Noblesville. (Hamilton County, 1945)

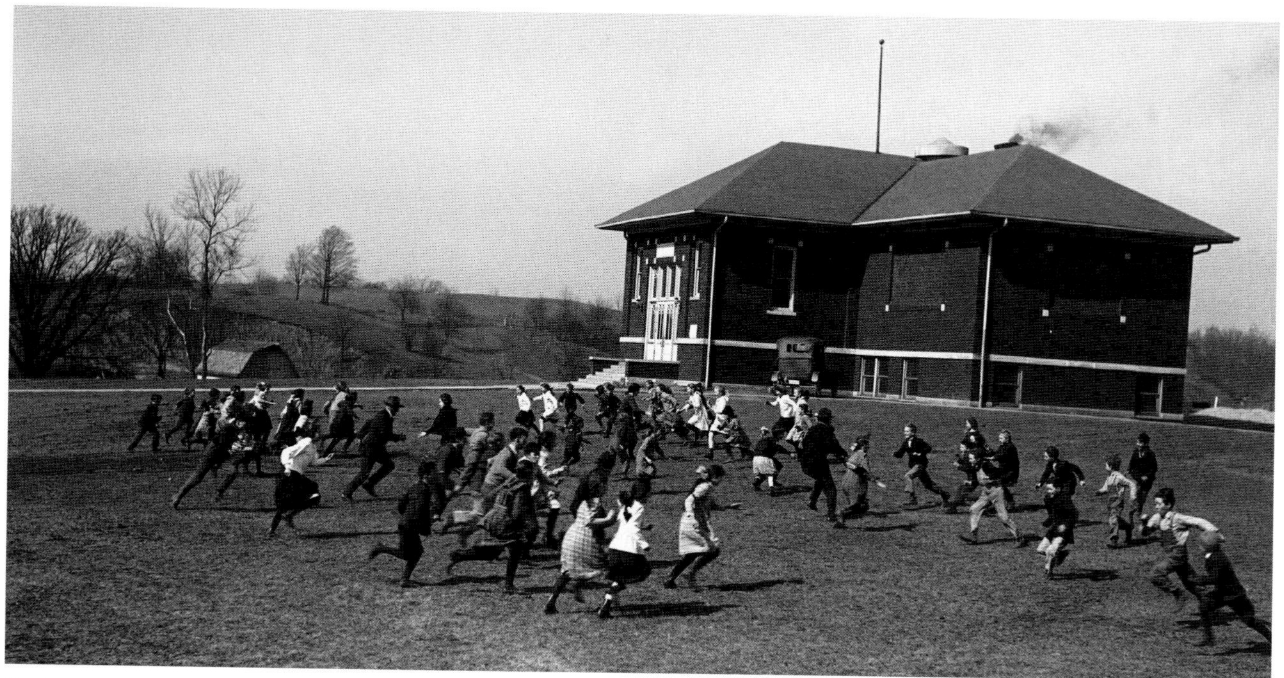

Students participate in a game at their school. (Tippecanoe County, 1919)

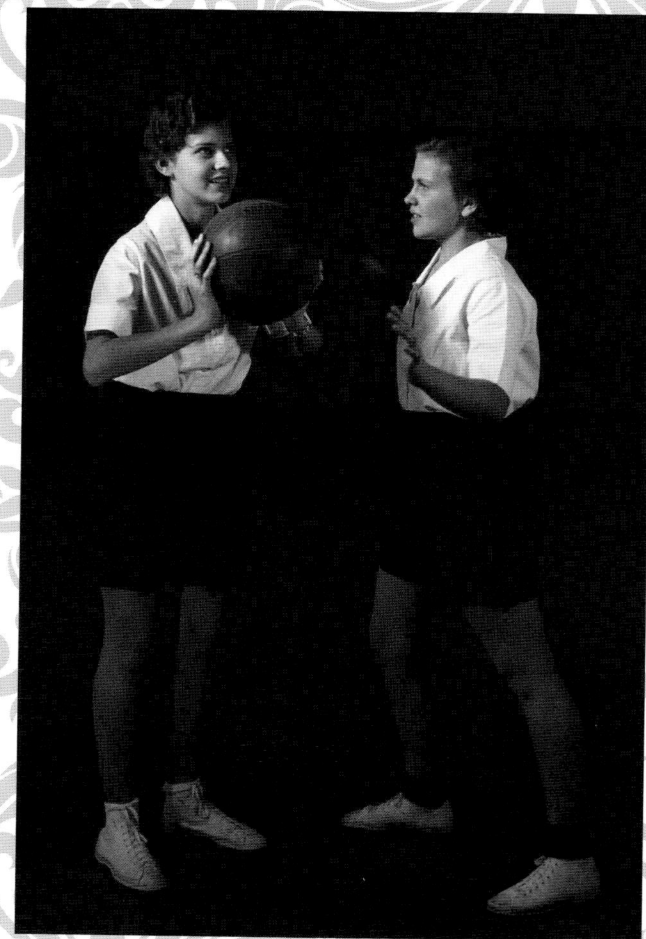

Margaret Young of West Lafayette prepares to throw a basketball while another student tries to block the shot inside a school gym. (Tippecanoe County, 1933)

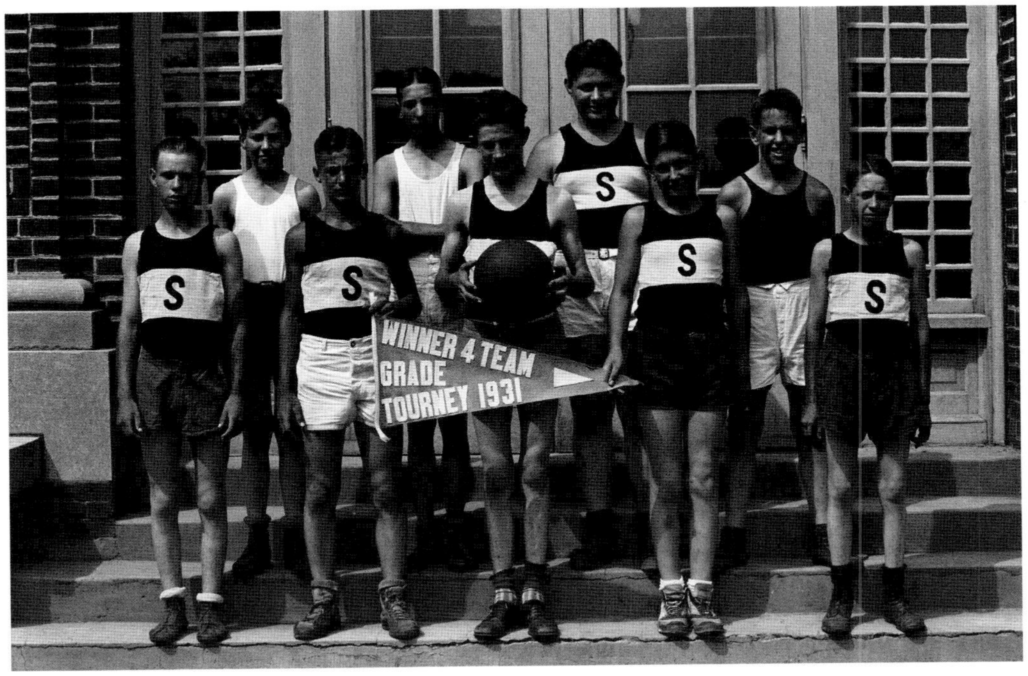

The Shadeland High School basketball team (Tippecanoe County, 1931)

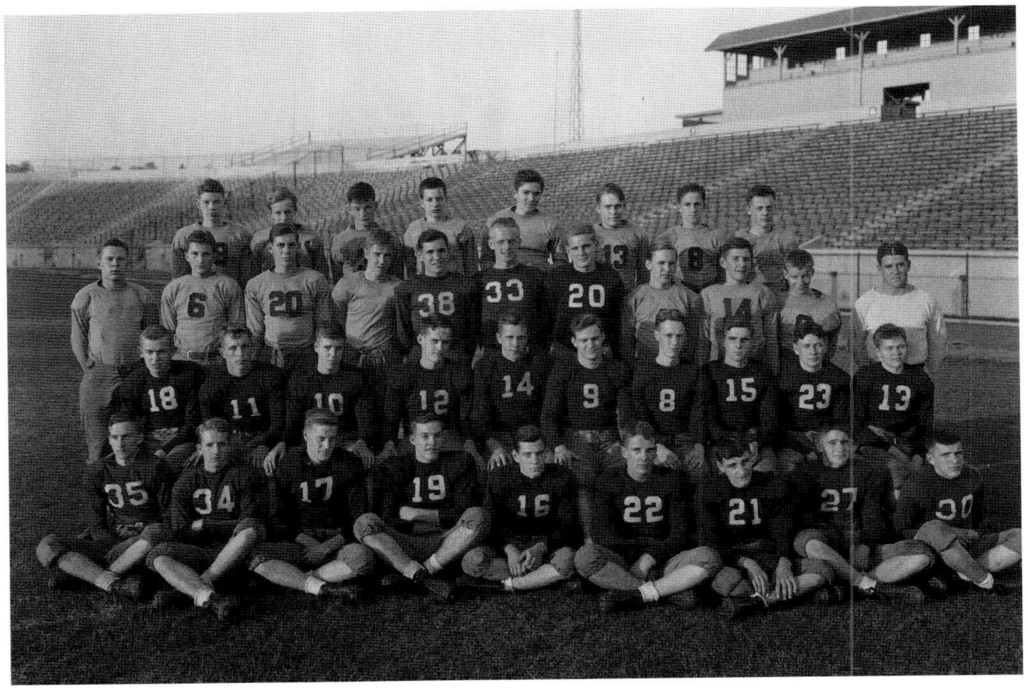

Coach Doxie Moore (third row, far right) poses with the West Lafayette High School football squad. (Tippecanoe County, 1934)

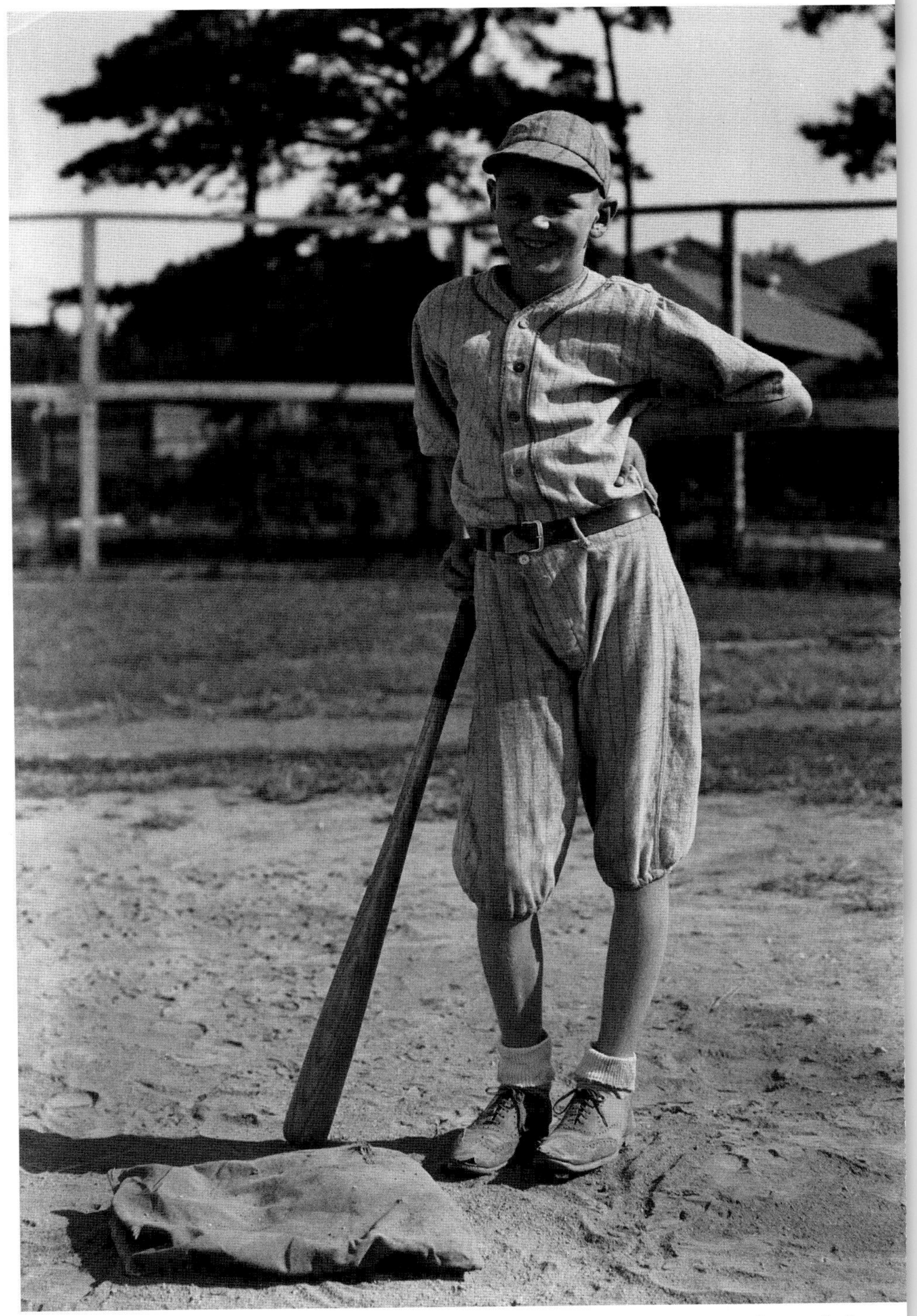

Bill Lux of West Lafayette leans on a baseball bat. Allen's notes included the comment, "Excellent smile." (Tippecanoe County, 1935)

Time for Fun

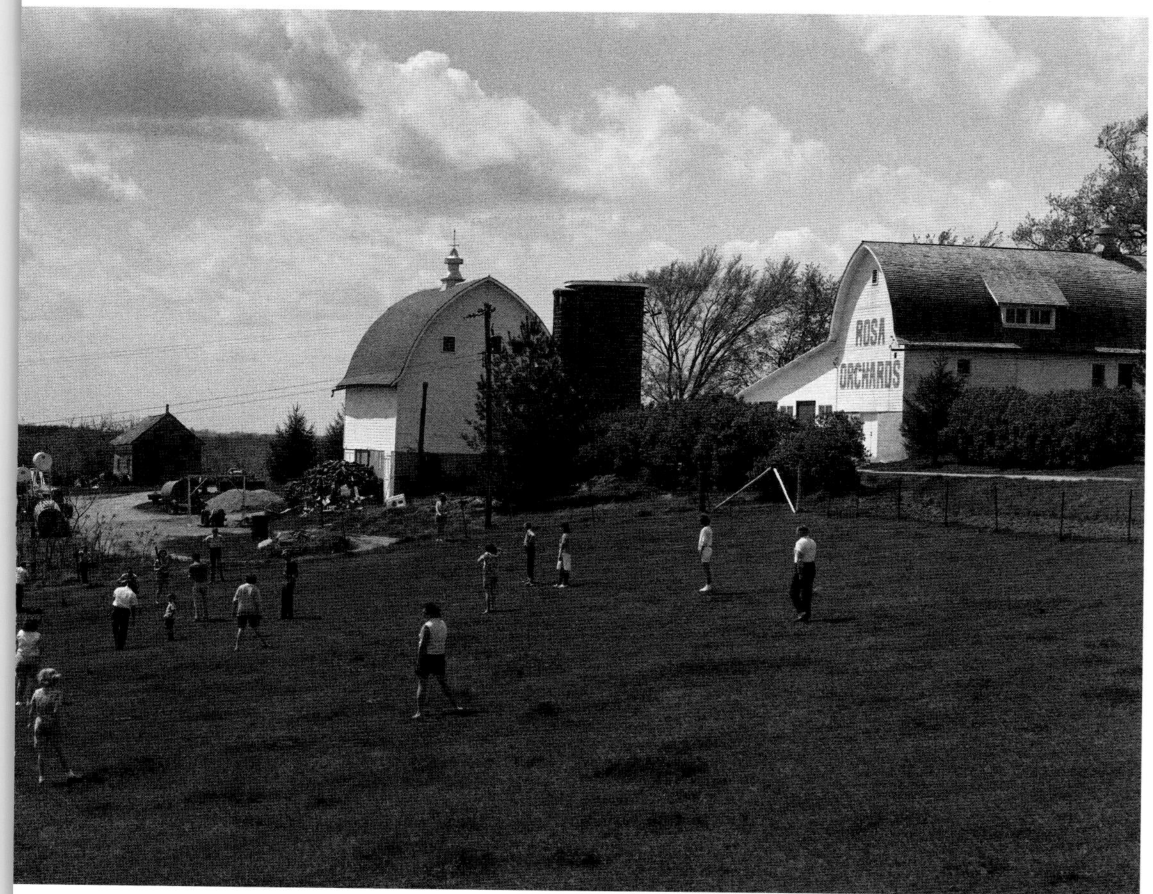

This family reunion at Rosa Orchards near Gays Mill, Wisconsin, included a baseball game. (Crawford County, WI; 1964)

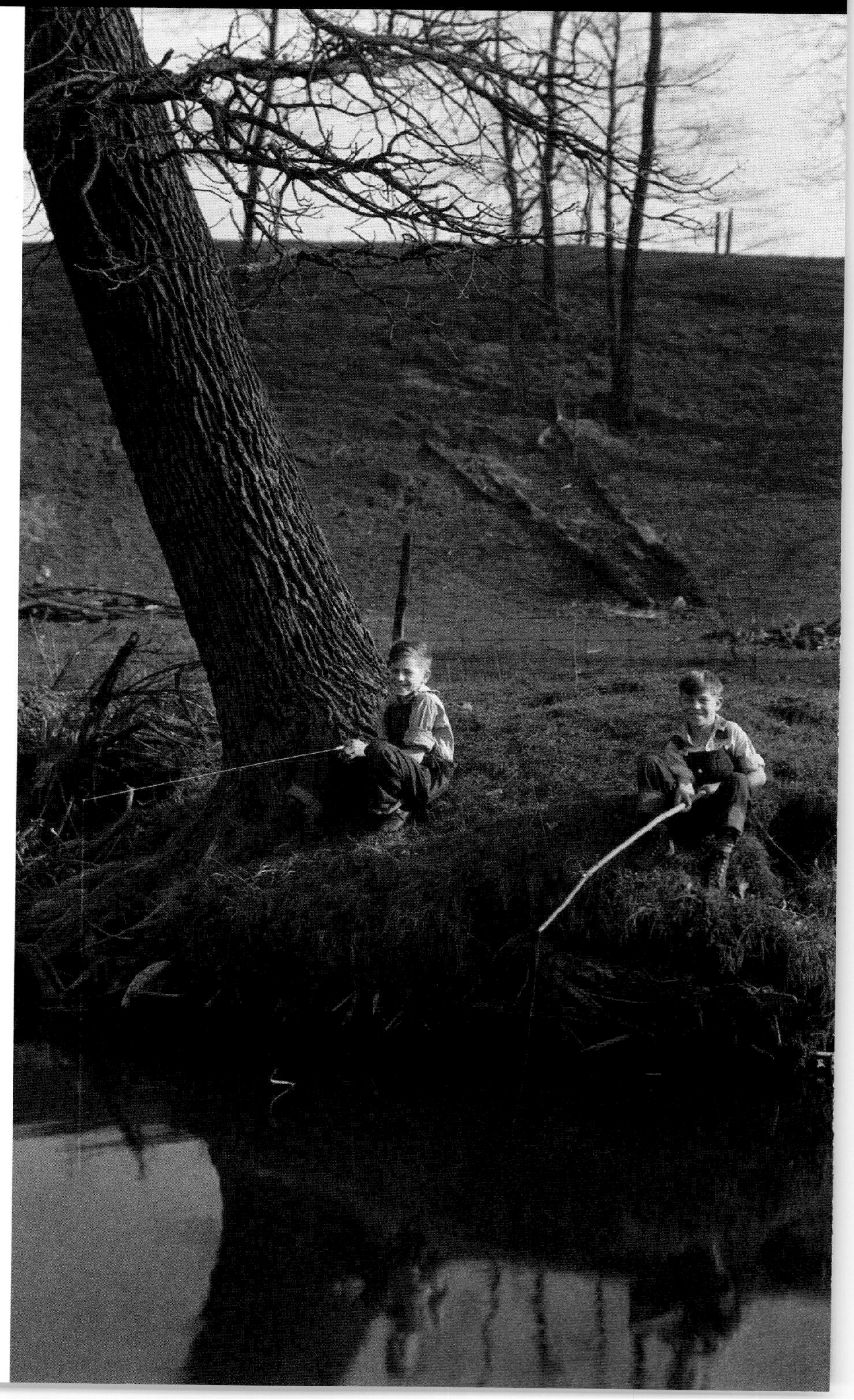

Selby Tarter and Donald Brooks, both of Stockwell, fish in Laramie Creek. (Tippecanoe County, 1931)

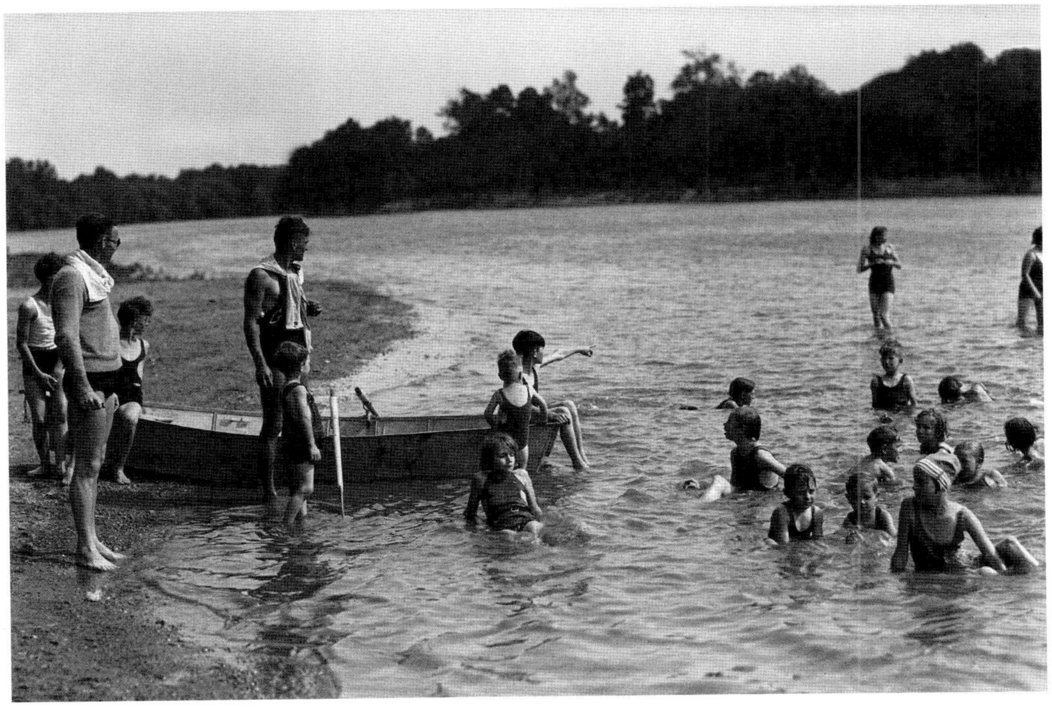

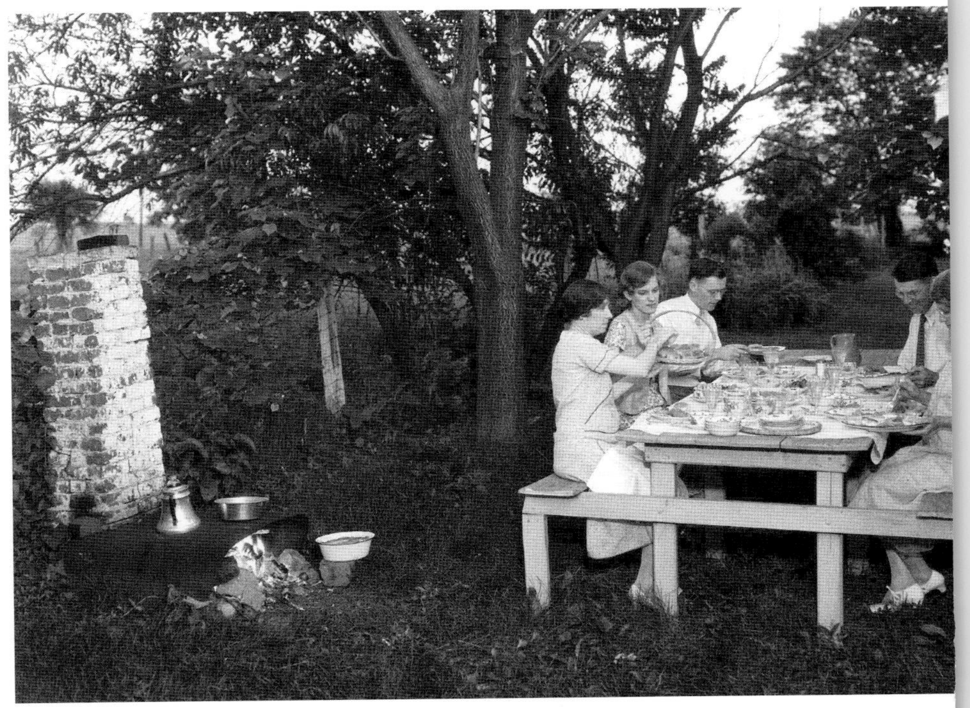

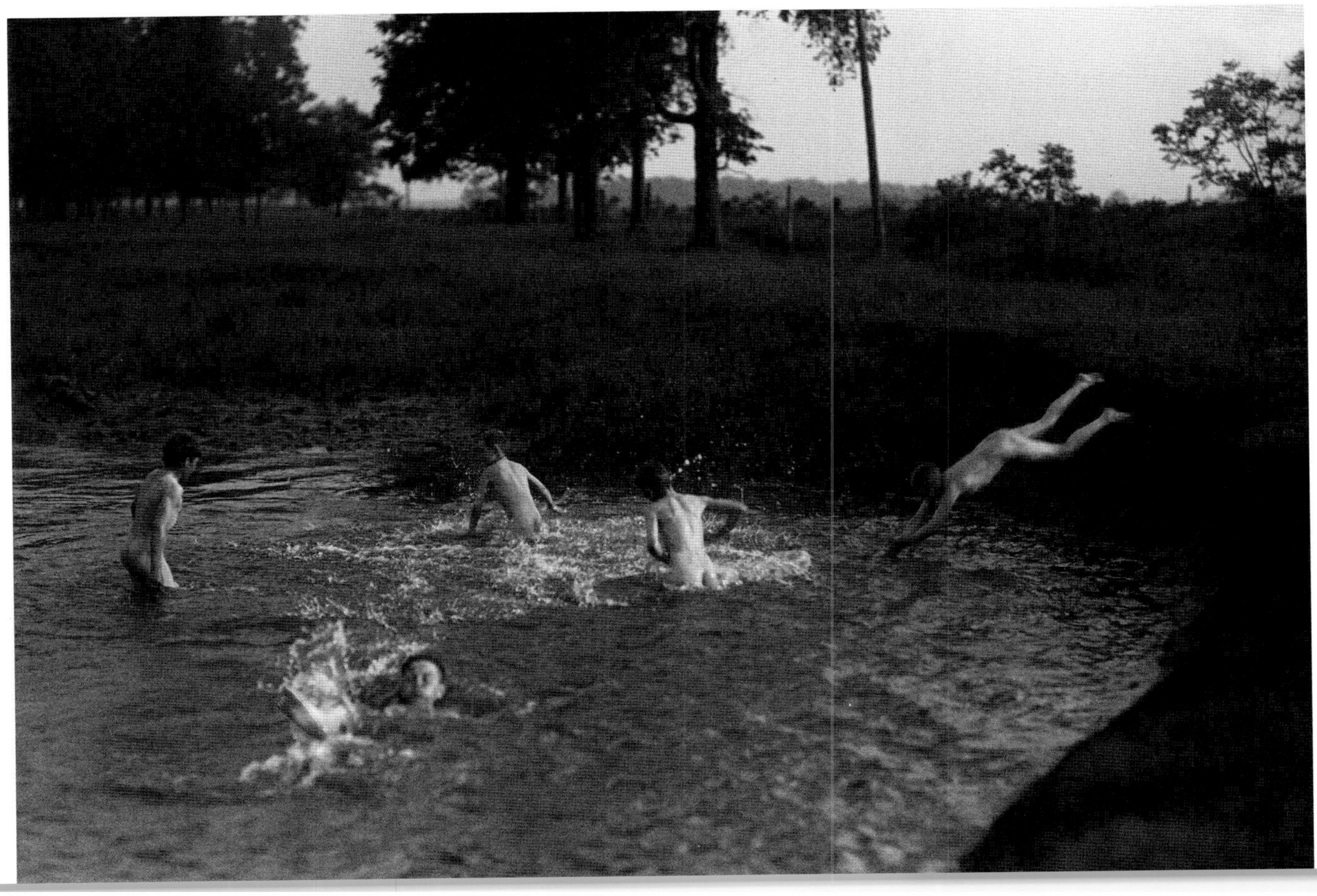

(Top left) A group of eight-year-olds swim in the river at Indianapolis. (Marion County, 1935)

(Top right) These people enjoy an evening picnic in Montmorenci. (Tippecanoe County, 1931)

(Bottom) Boys swim on the farm of W. W. Washburn of New Richmond. (Montgomery County, 1917)

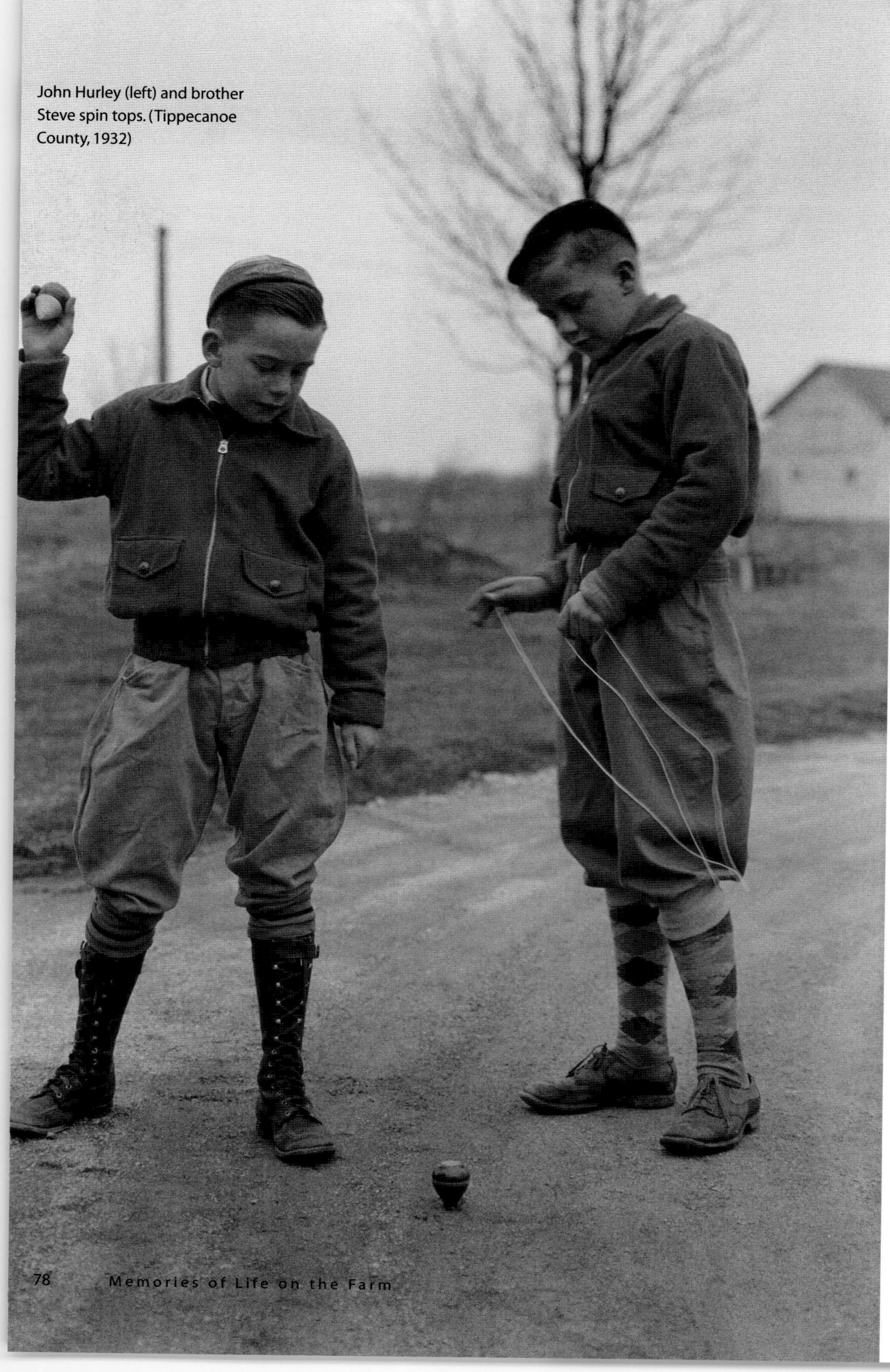

John Hurley (left) and brother Steve spin tops. (Tippecanoe County, 1932)

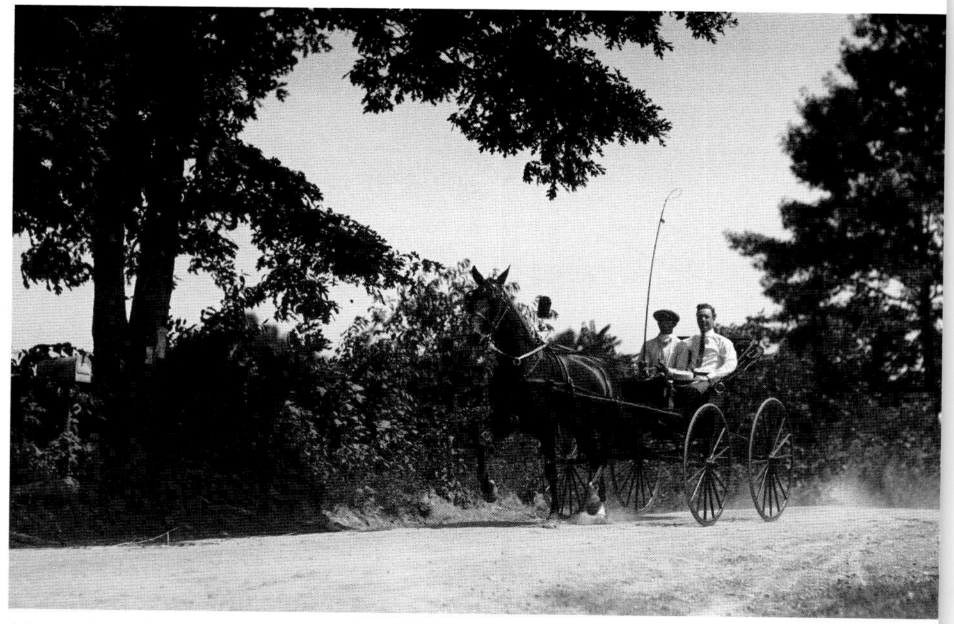

A horse-drawn buggy on a country road (N.p., 1921)

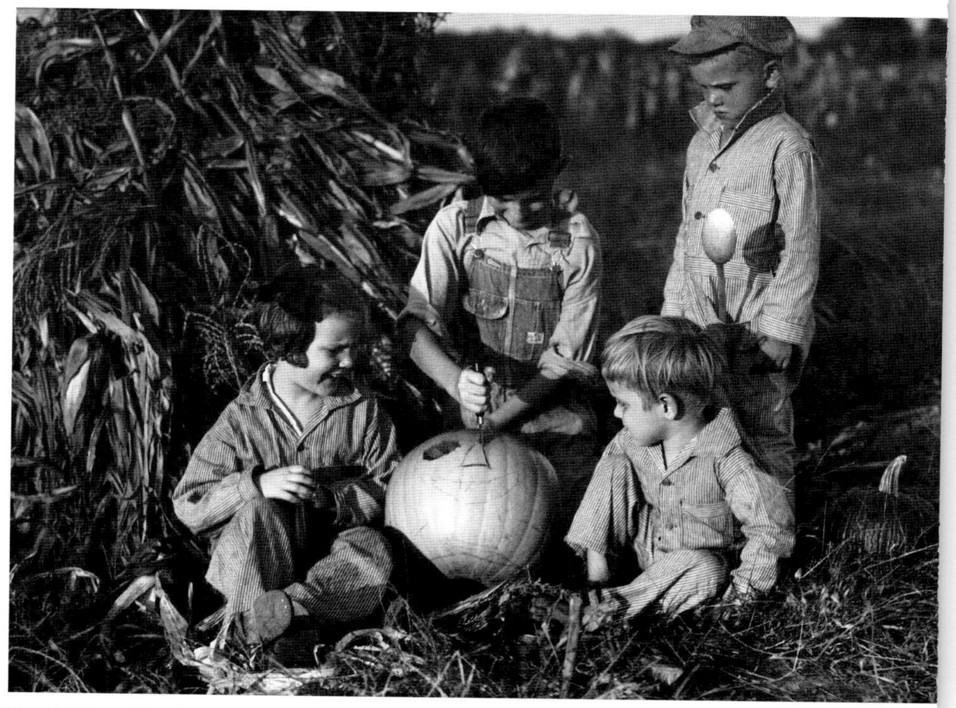

The children of Paul Mitchell of Battle Ground along with a neighbor boy make a face on a pumpkin. (Tippecanoe County, 1933)

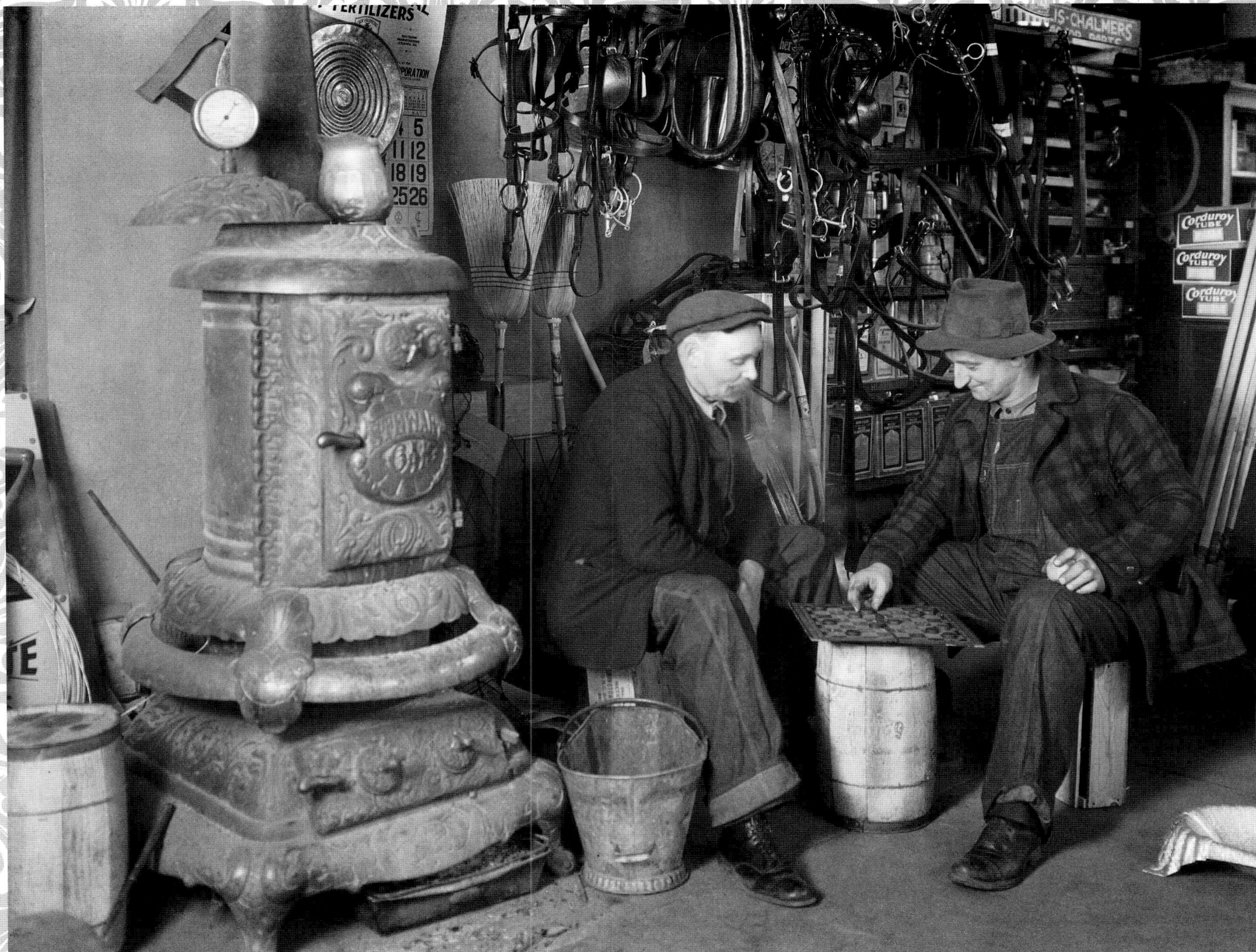

Two men play checkers by a stove in the store of George J. Wettschurack of Montmorenci. (Tippecanoe County, 1932)

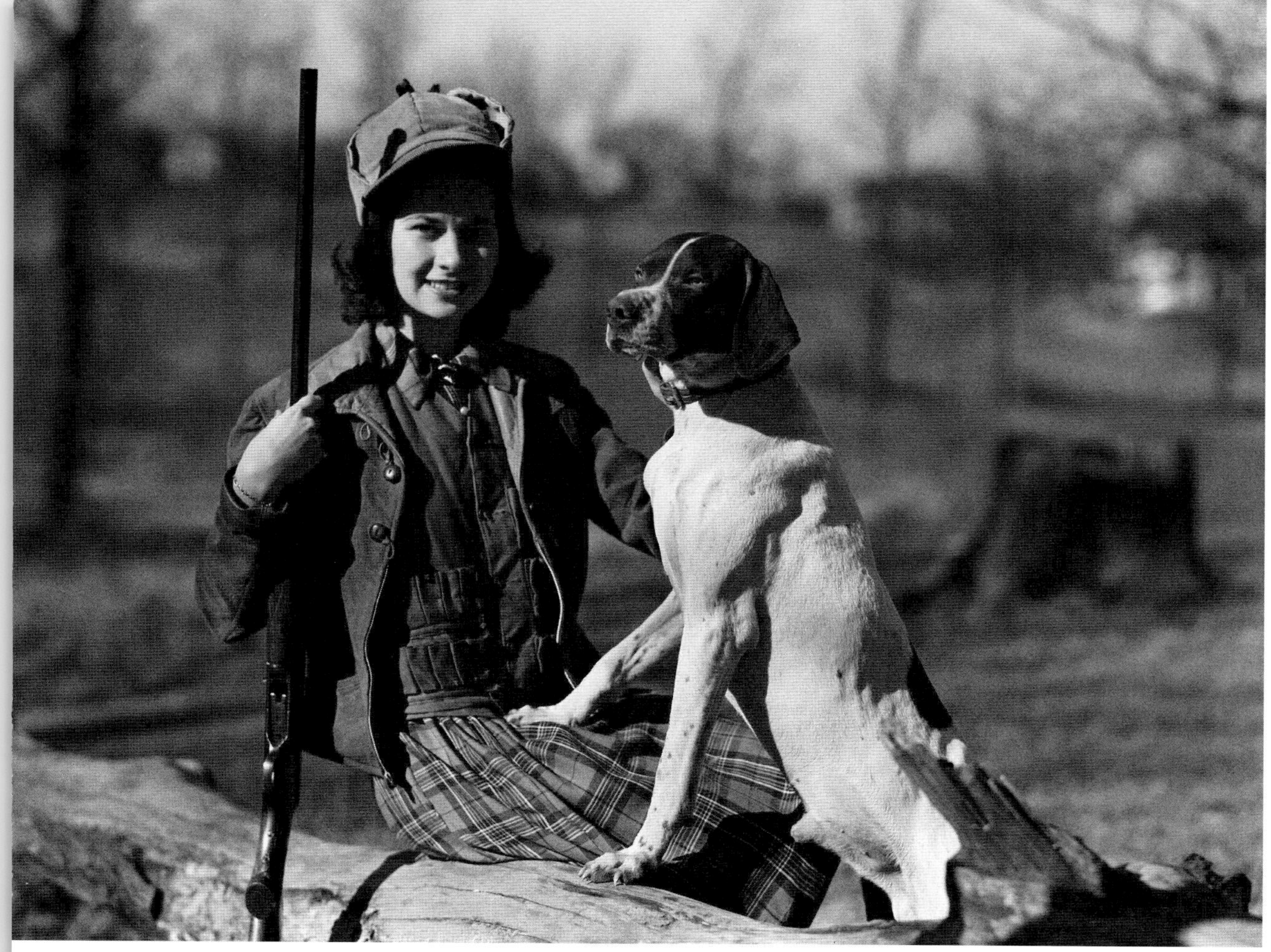

Ann Ostrander of West Lafayette with a dog and gun (Tippecanoe County, 1941)

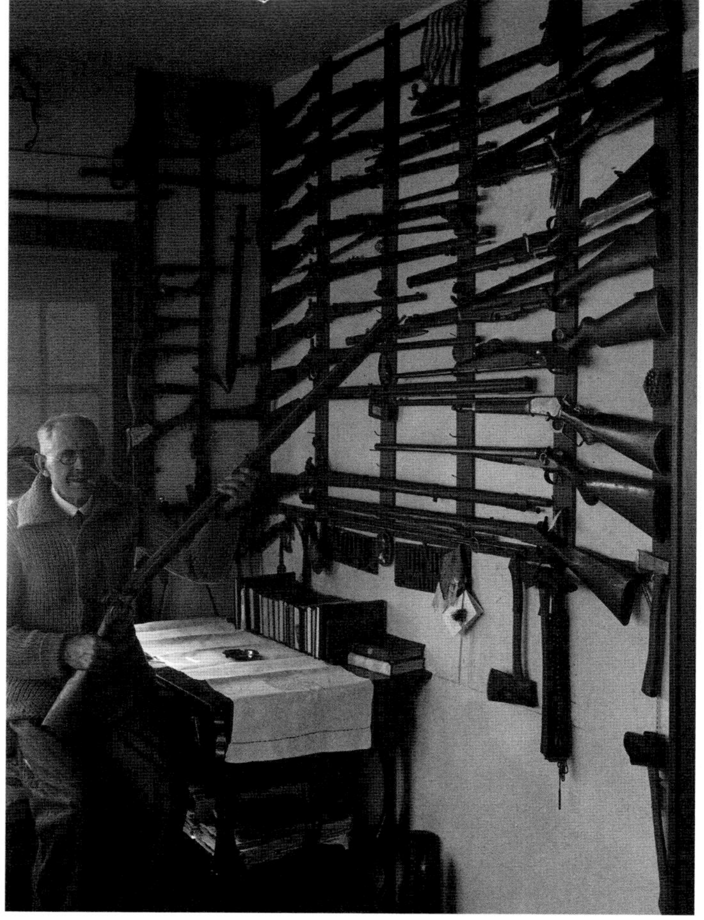

Elmer R. Waters of South Raub shows off his collection of guns in his den. He is handling an old flintlock rifle formerly owned by John Ritchey, who settled in Tippecanoe County in 1834. (Tippecanoe County, 1926)

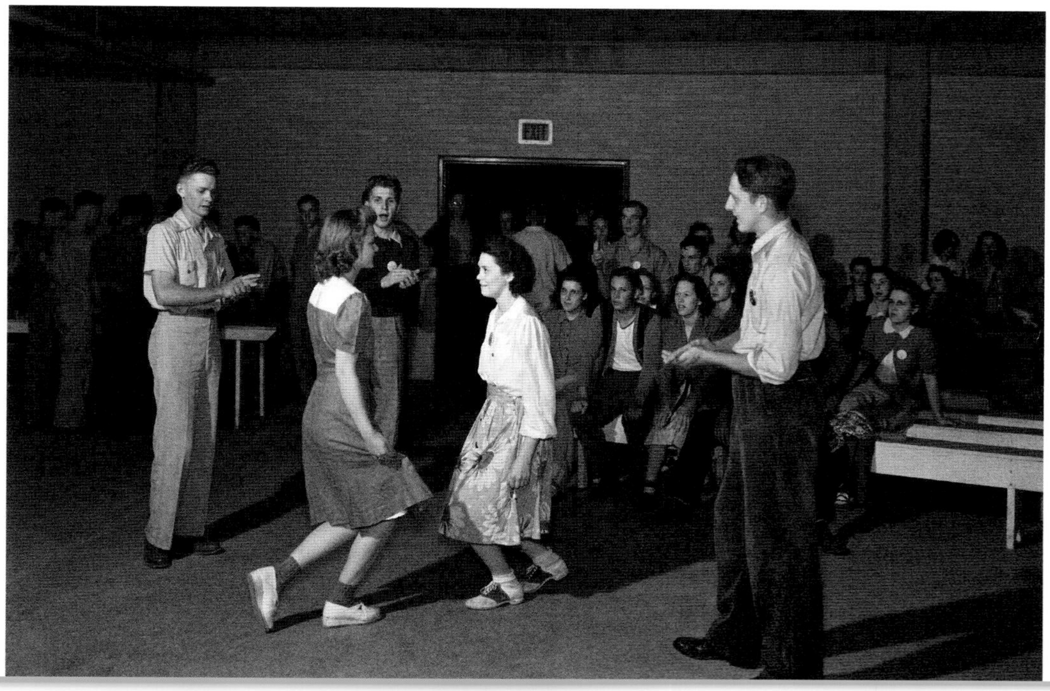

Participants at a 4-H Junior Leadership Conference in Indianapolis take part in a folk dance study. (Marion County, 1941)

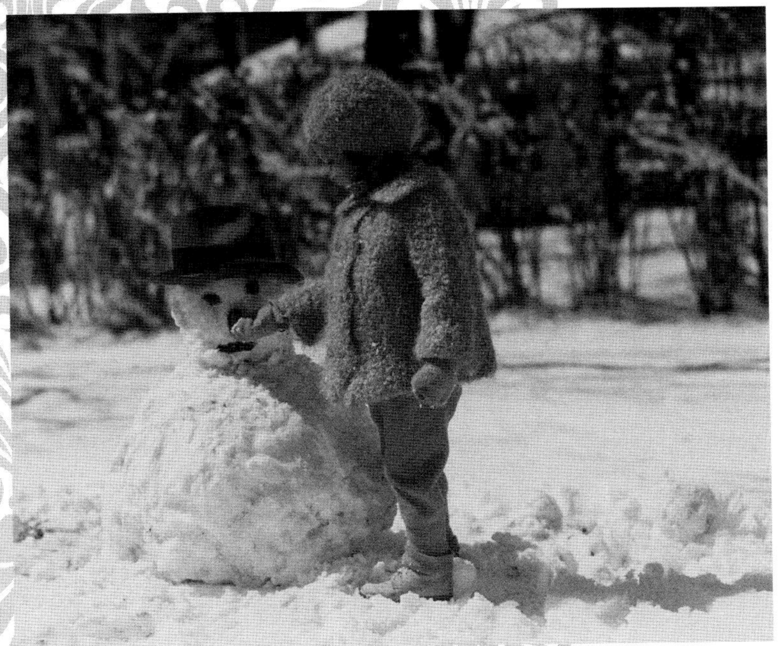

Fanchon Hinrichs, daughter of Mr. and Mrs. F. A. Hinrichs of West Lafayette, looks at a snowman. (Tippecanoe County, 1934)

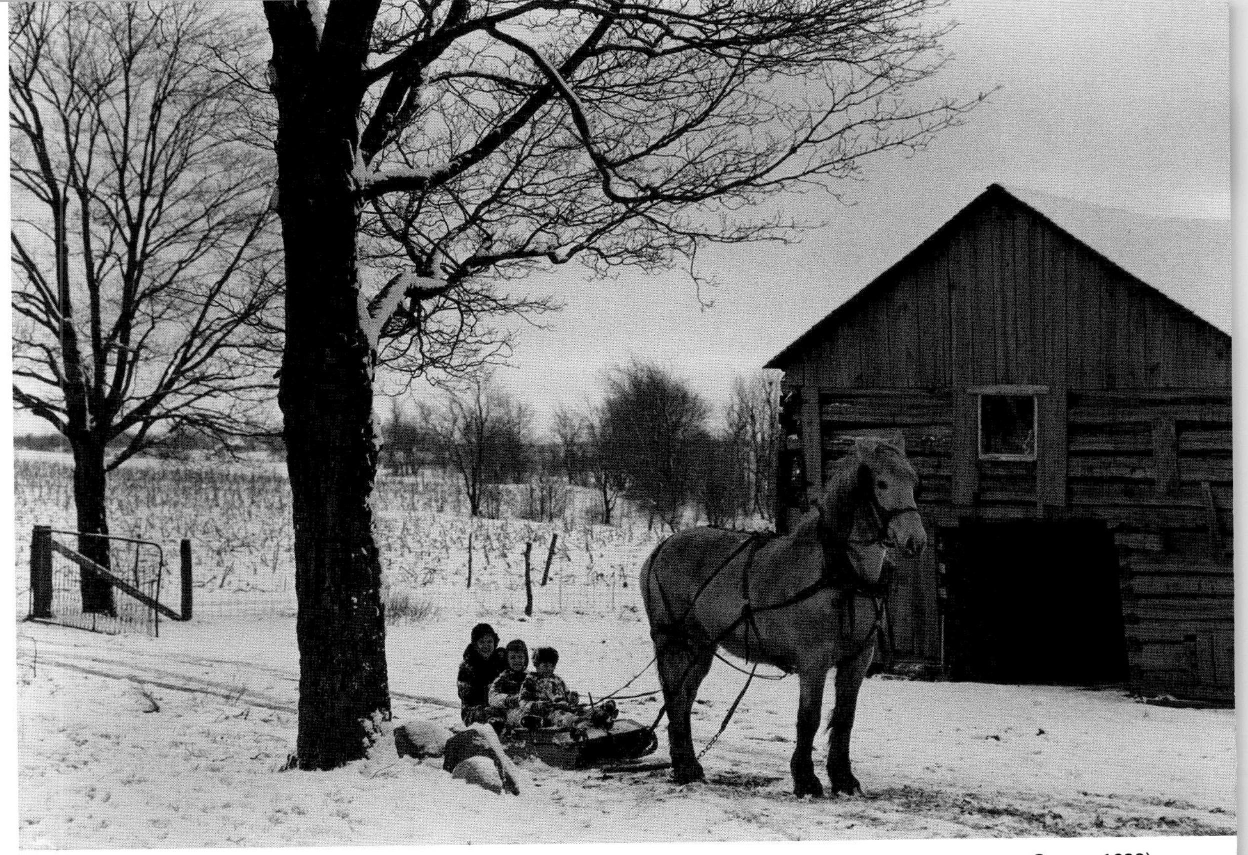

Martha Jane, Fredrick Albert, and Julia May Trost ride on a horse-drawn sled on the farm of John Trost. (Tippecanoe County, 1933)

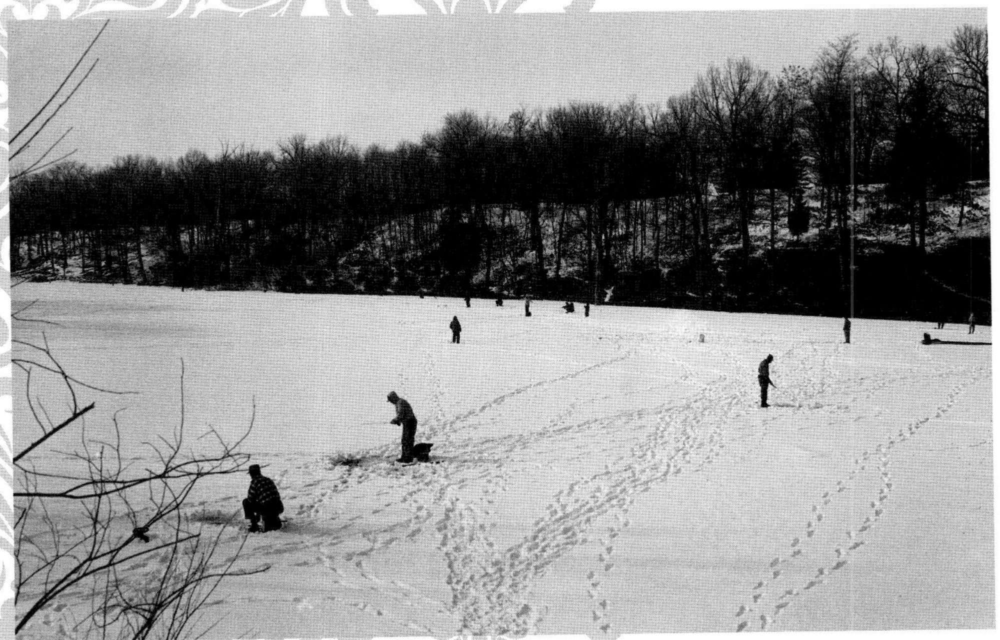

Ice fishing on a small lake on the Jack Van Natta farm in Battle Ground (Tippecanoe County, 1961)

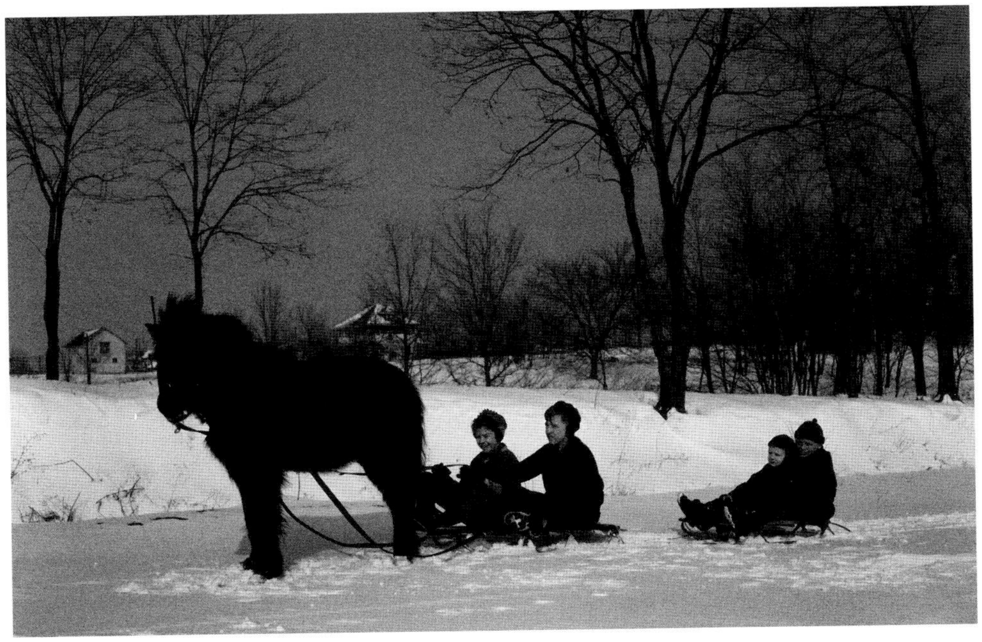

Here, a pony pulls two sleds loaded with boys. (N.p., 1939)

The Beauty of Children

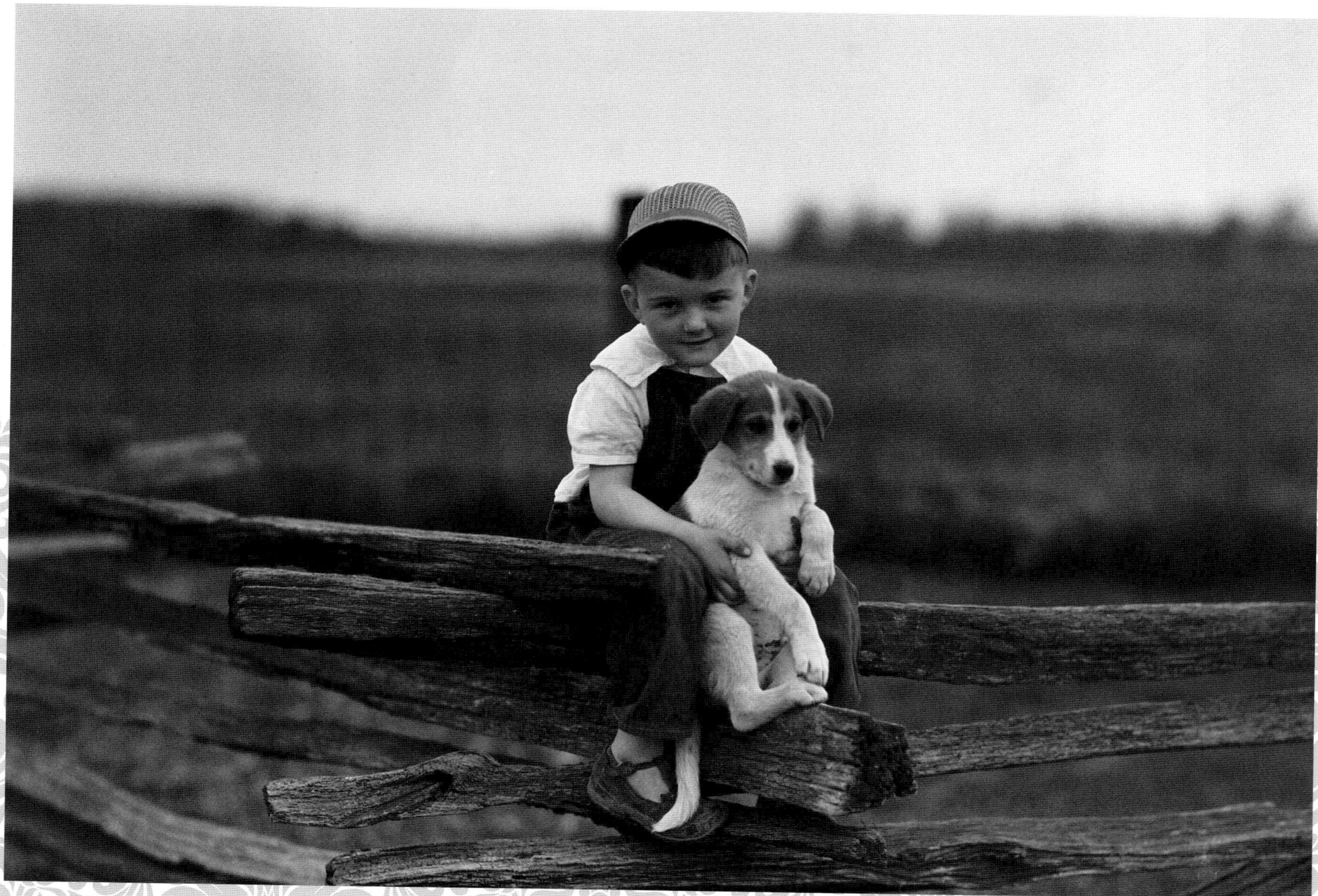

Leonard Schwartz from Berne sits on a split-rail fence while holding his buddy. (Adams County, 1933)

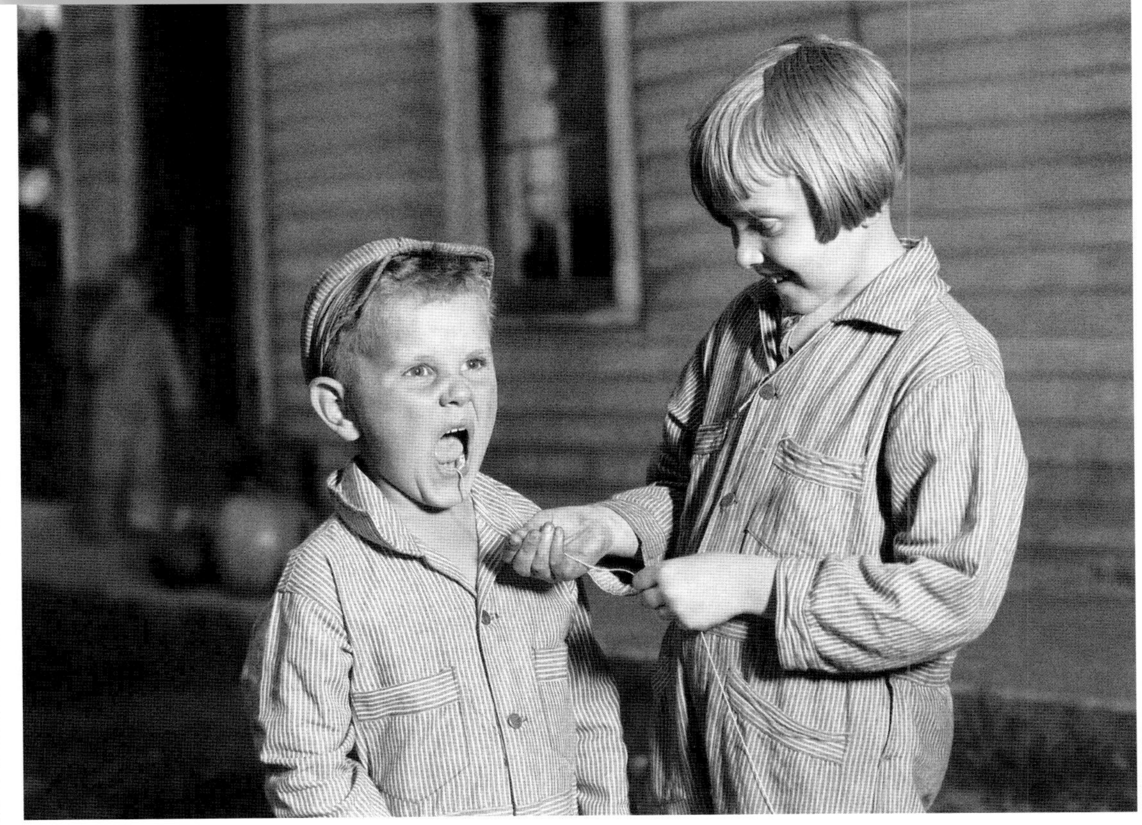

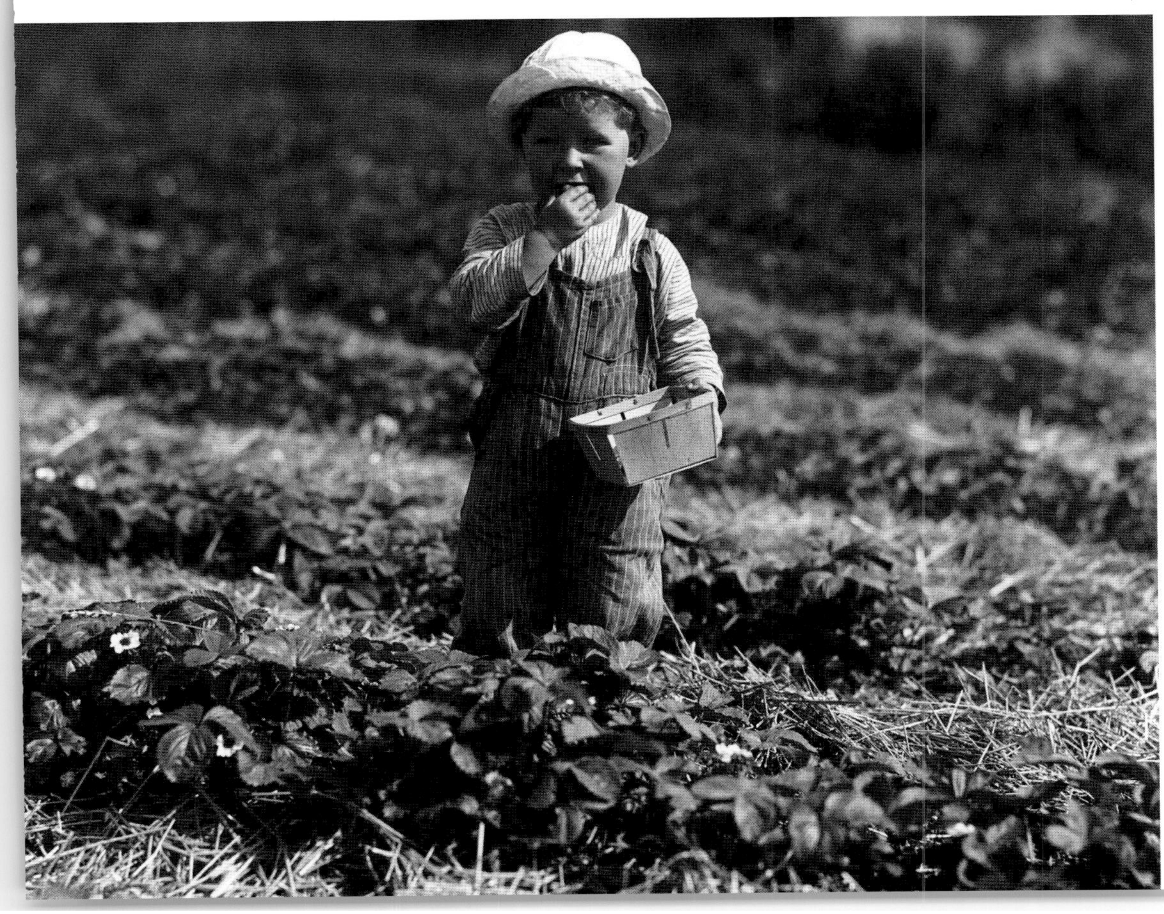

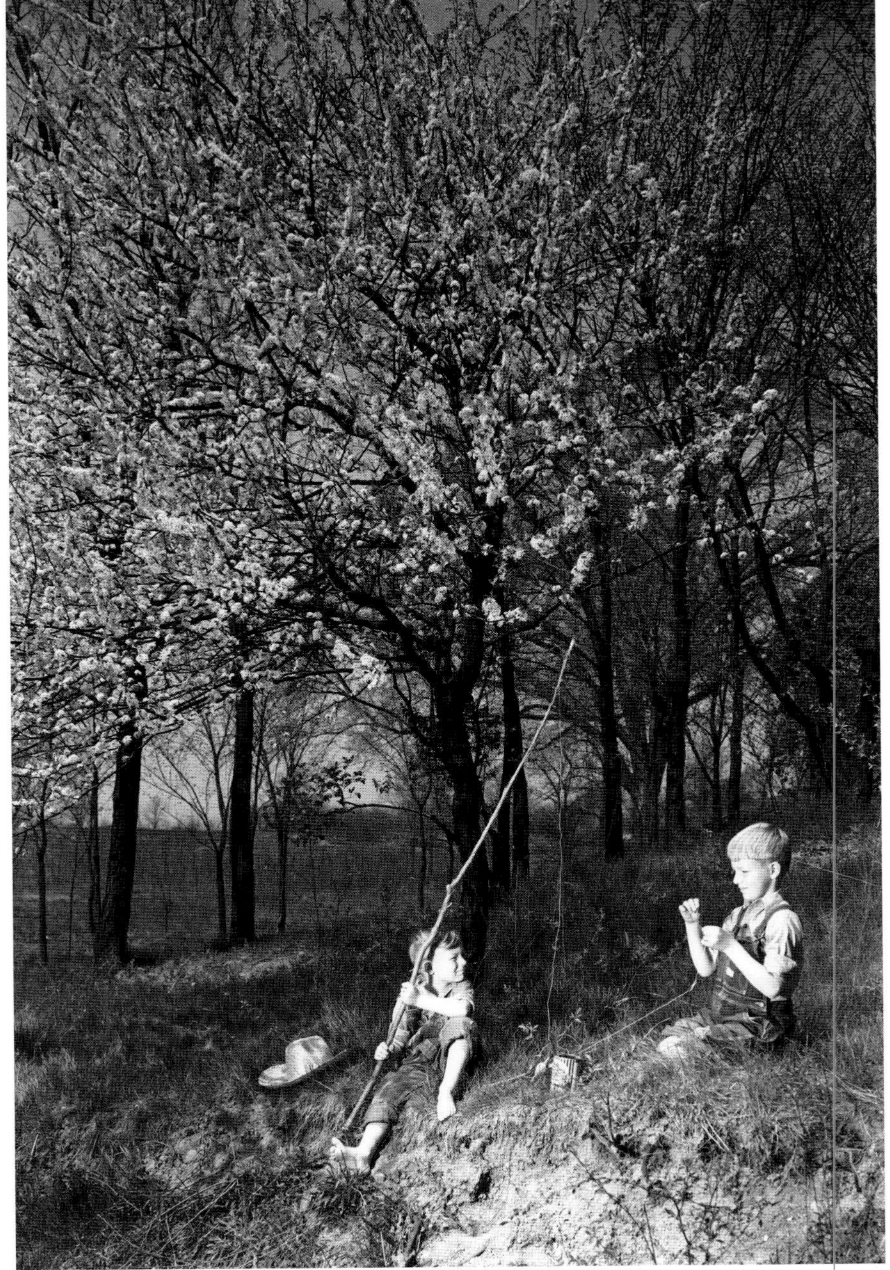

(Top left) Frances Mitchell from Battle Ground uses a string to pull her brother David's tooth. (Tippecanoe County, 1933)

(Bottom left) Two-year-old Gordon Ray Myers fills up his stomach with strawberries from his father's field before filling up his box. (Tippecanoe County, 1924)

(Right) While sitting under a blossoming wild plum tree, John (left) and Jim Allen make a day out of fishing. (Tippecanoe County, 1942)

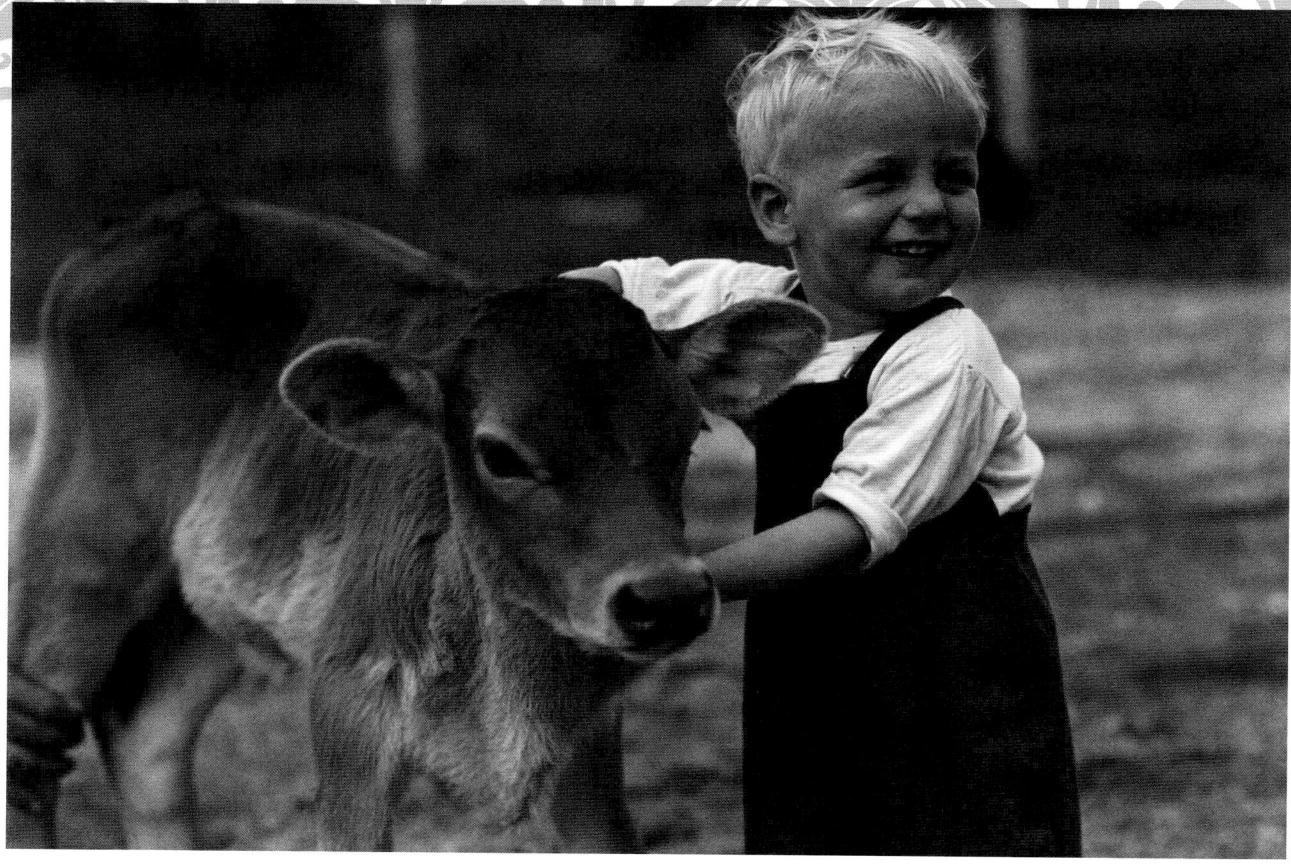

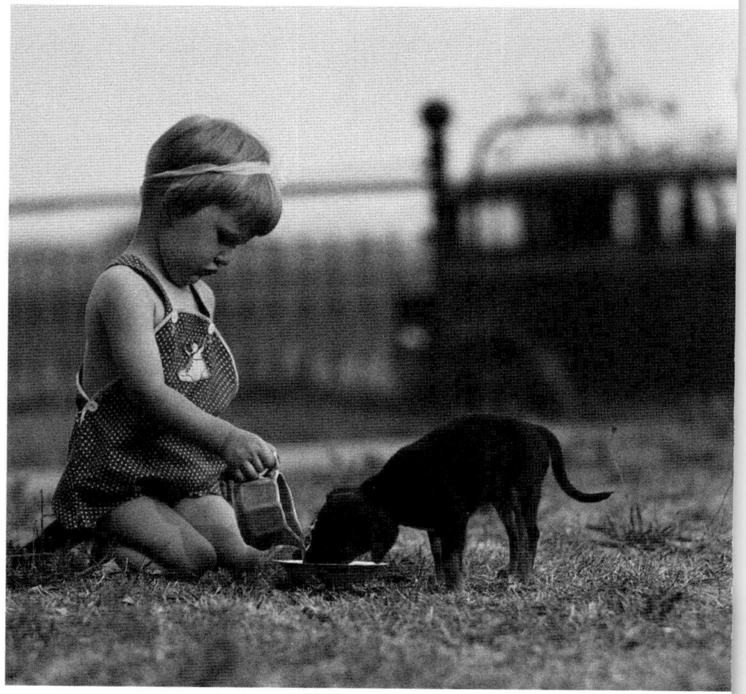

(Top left) A young boy poses with a Jersey calf on the farm of J. C. Piper and Sons near Sumner, Illinois. (Lawrence County, IL; 1947)

(Top right) Vera Joan Taylor feeds her puppy. (Tippecanoe County, 1934)

(Bottom) Ruth and Helen Crumbaker from Romney crack walnuts. (Tippecanoe County, 1929)

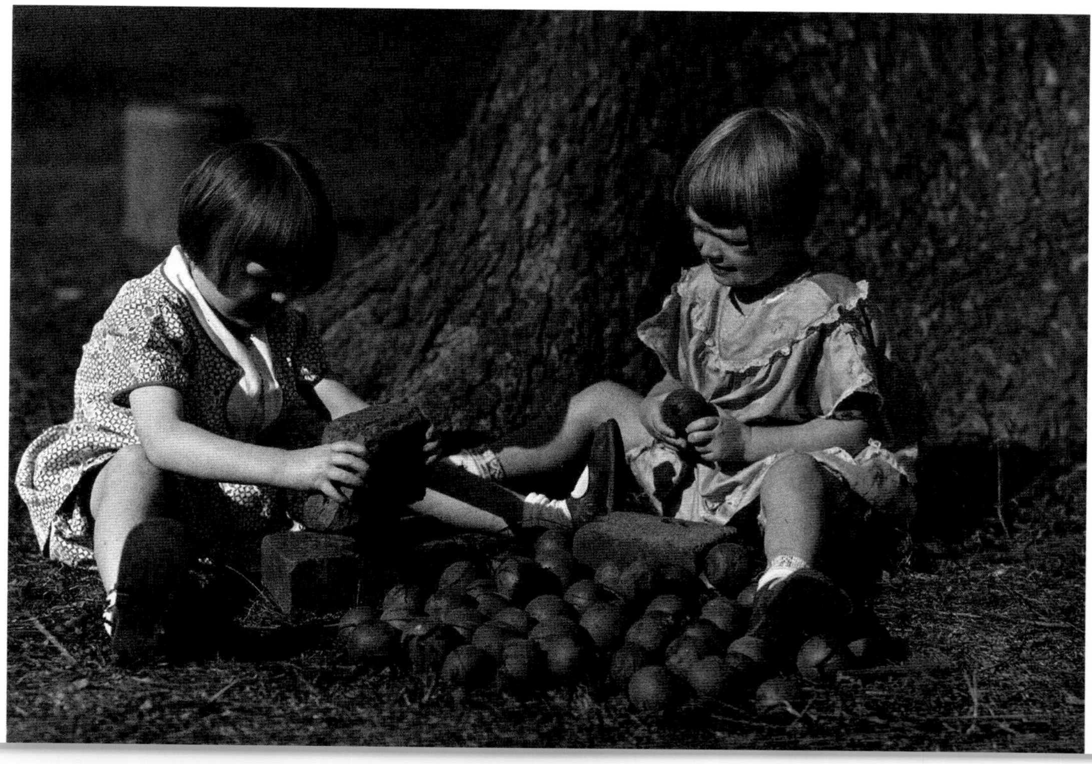

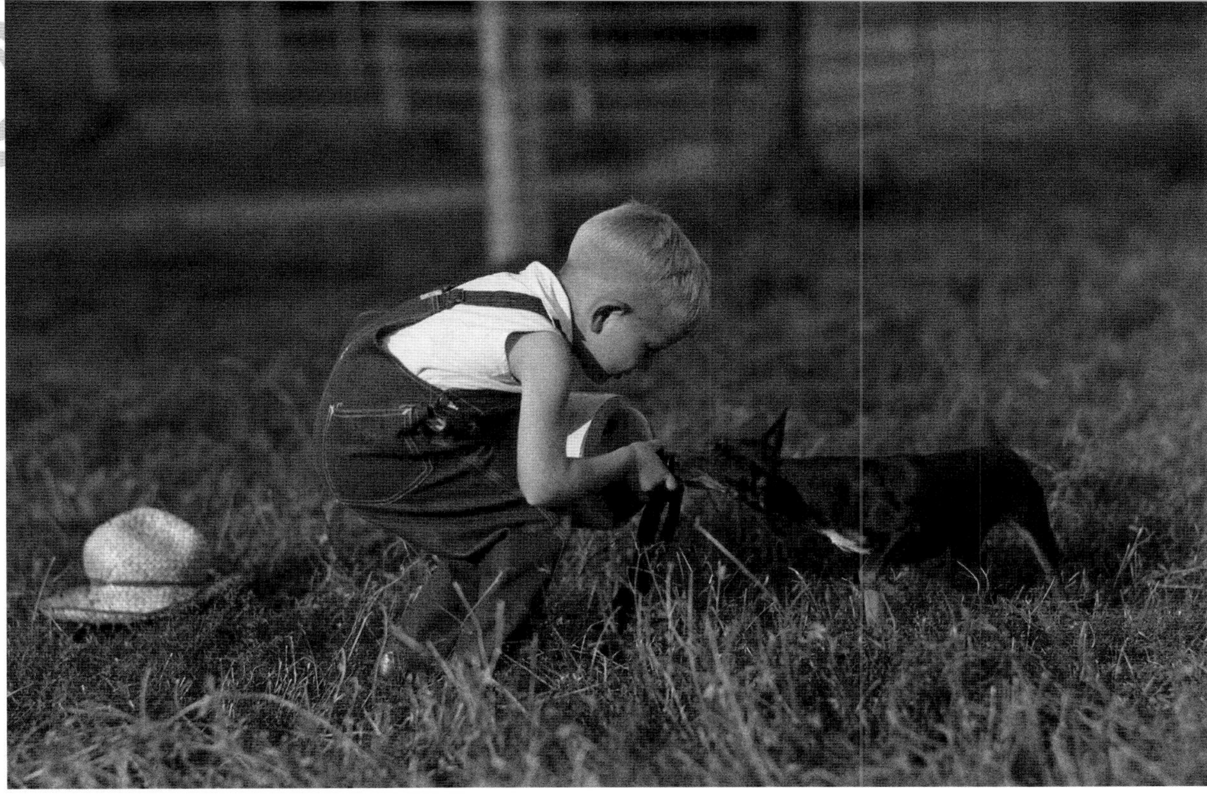

Jimmie Scott from Burrows gives his dog a drink directly from a jug. (Carroll County, 1934)

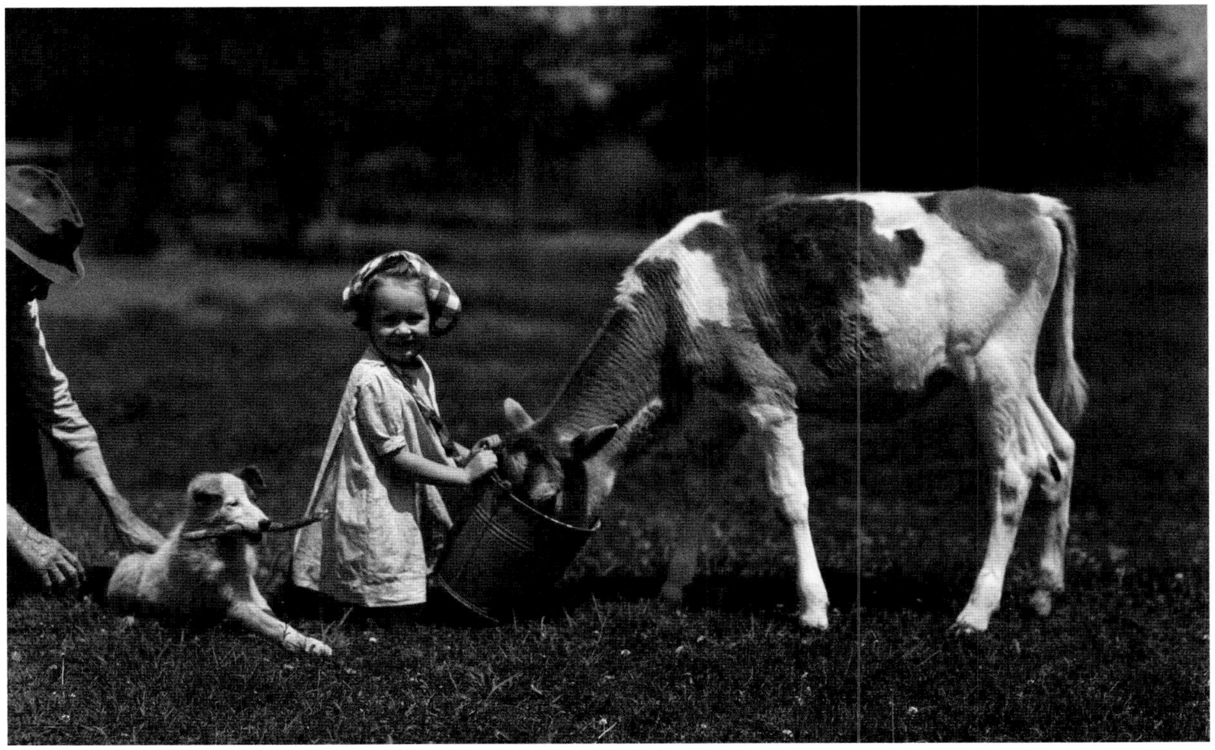

Martha Ann Aikman feeds her calf. (Daviess County, 1935)

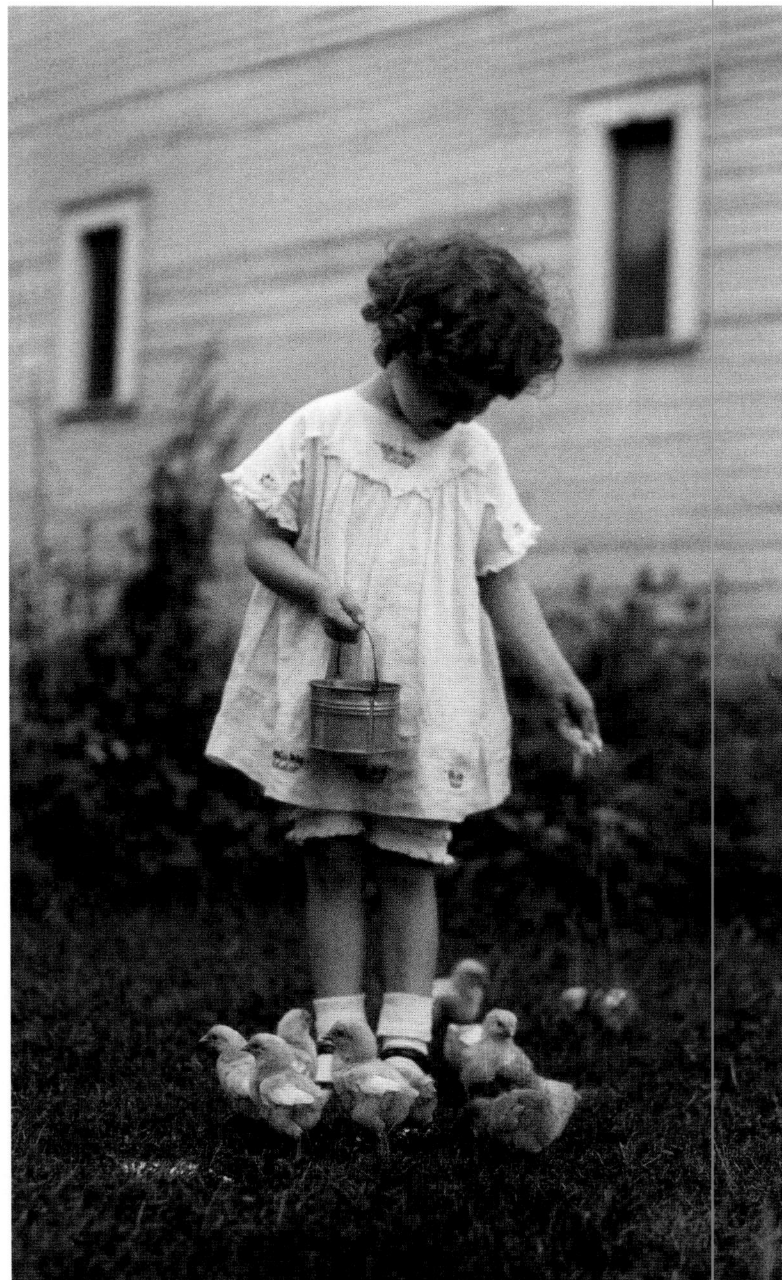

Joan Peavey feeds chicks at her grandmother's farm. (Marion County, 1927)

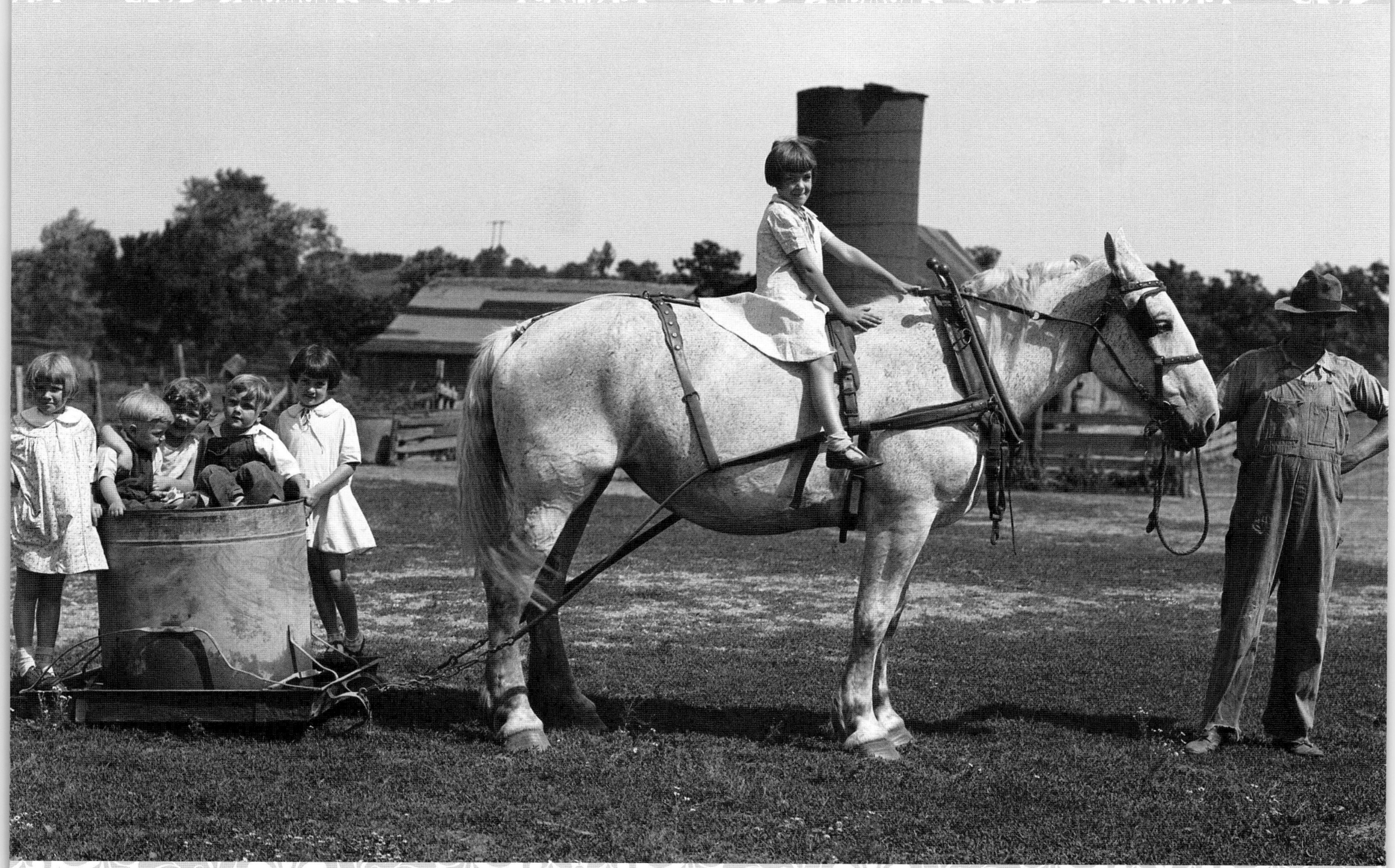

Paul Mitchell leads a horse carrying his daughter Marjorie while his other children—Frances, Floyd Glen, Carolyn, David, and Jean—catch a ride on the water tank. (Tippecanoe County, 1932)

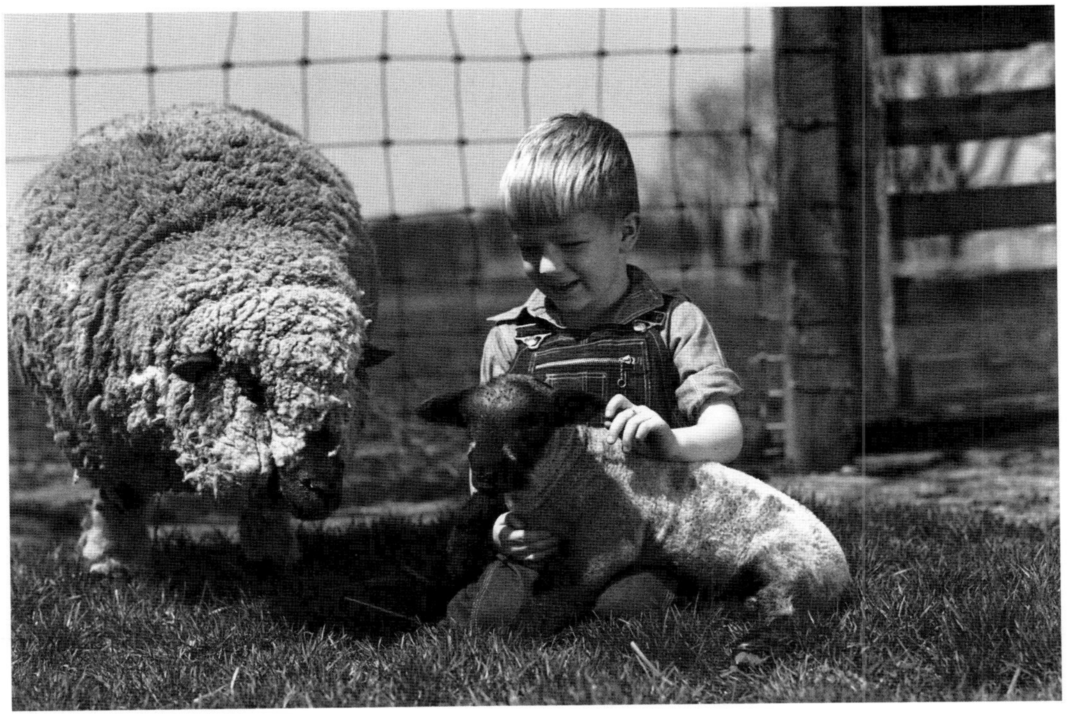

Jim Allen holds a baby lamb under the watchful eye of a Shropshire ewe. (Tippecanoe County, 1939)

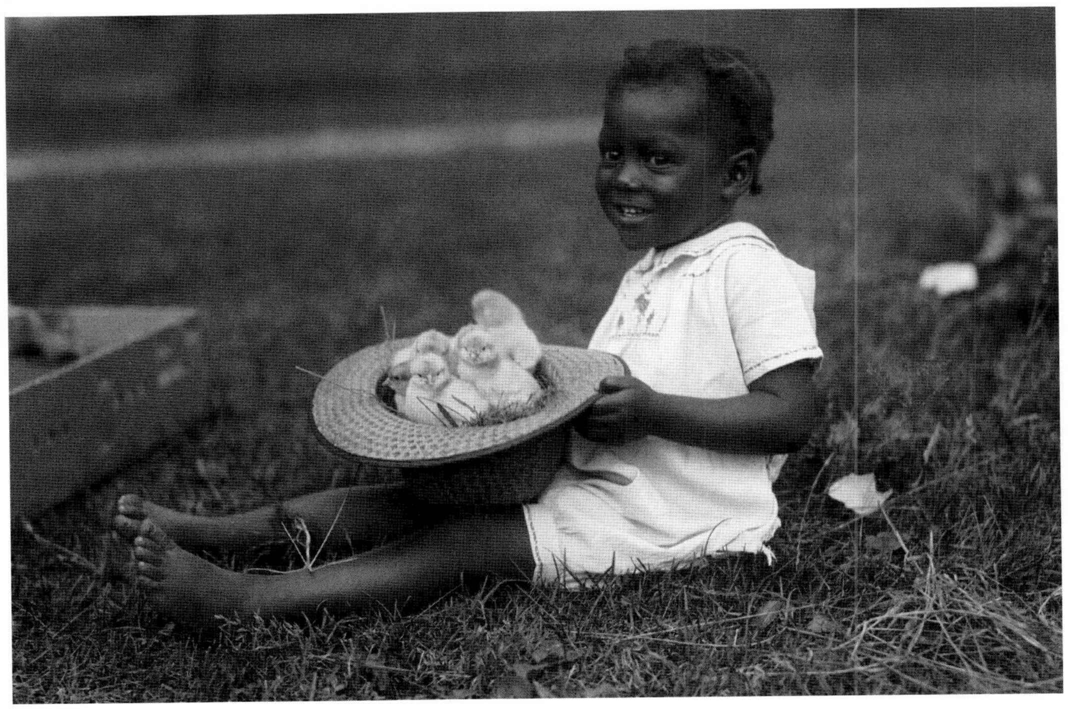

A hat filled with baby chicks makes this little girl smile. (N.p., 1931)

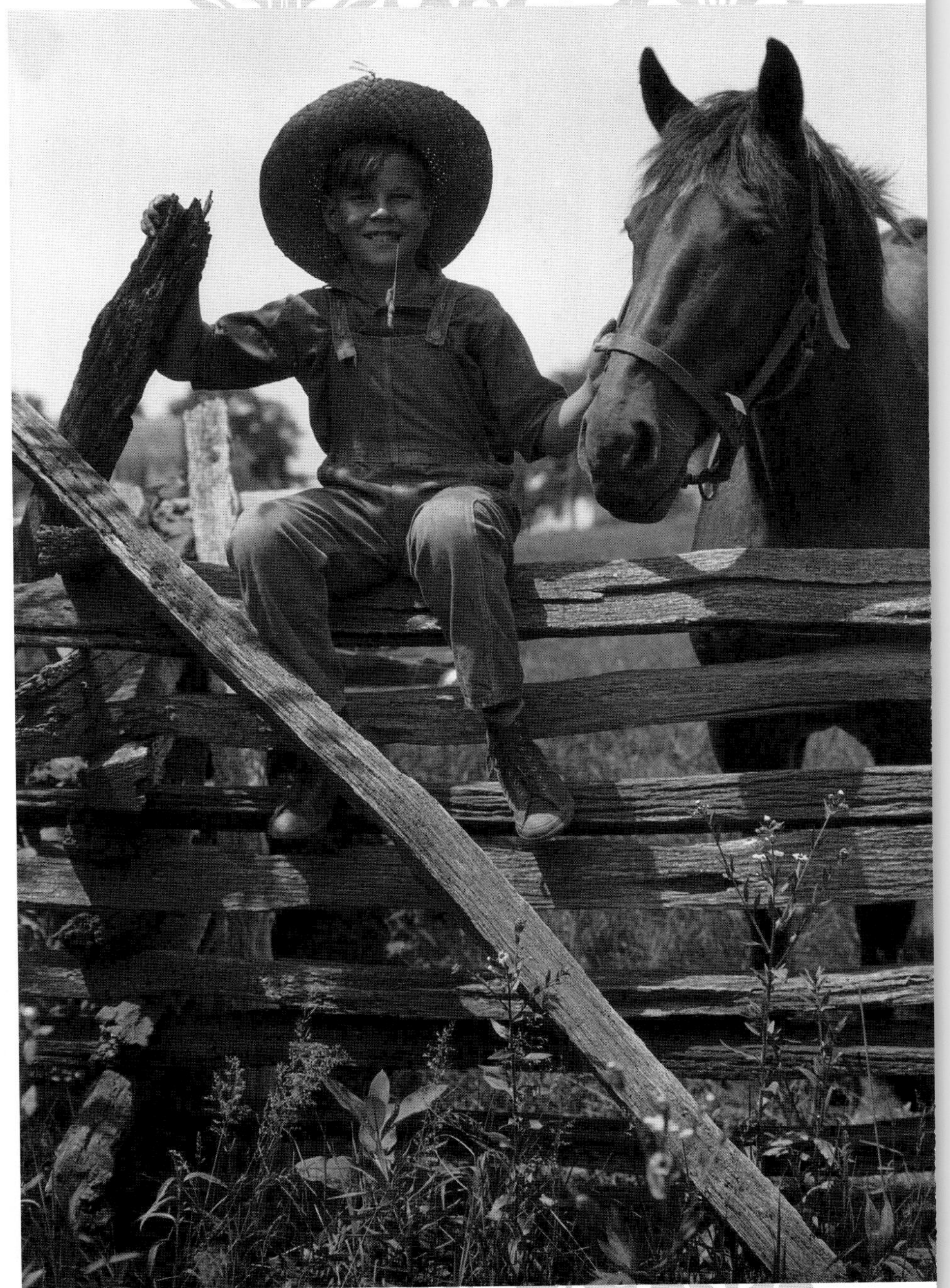

Herman Dale Roth III from West Lafayette proudly shows off his horse. (Tippecanoe County, 1929)

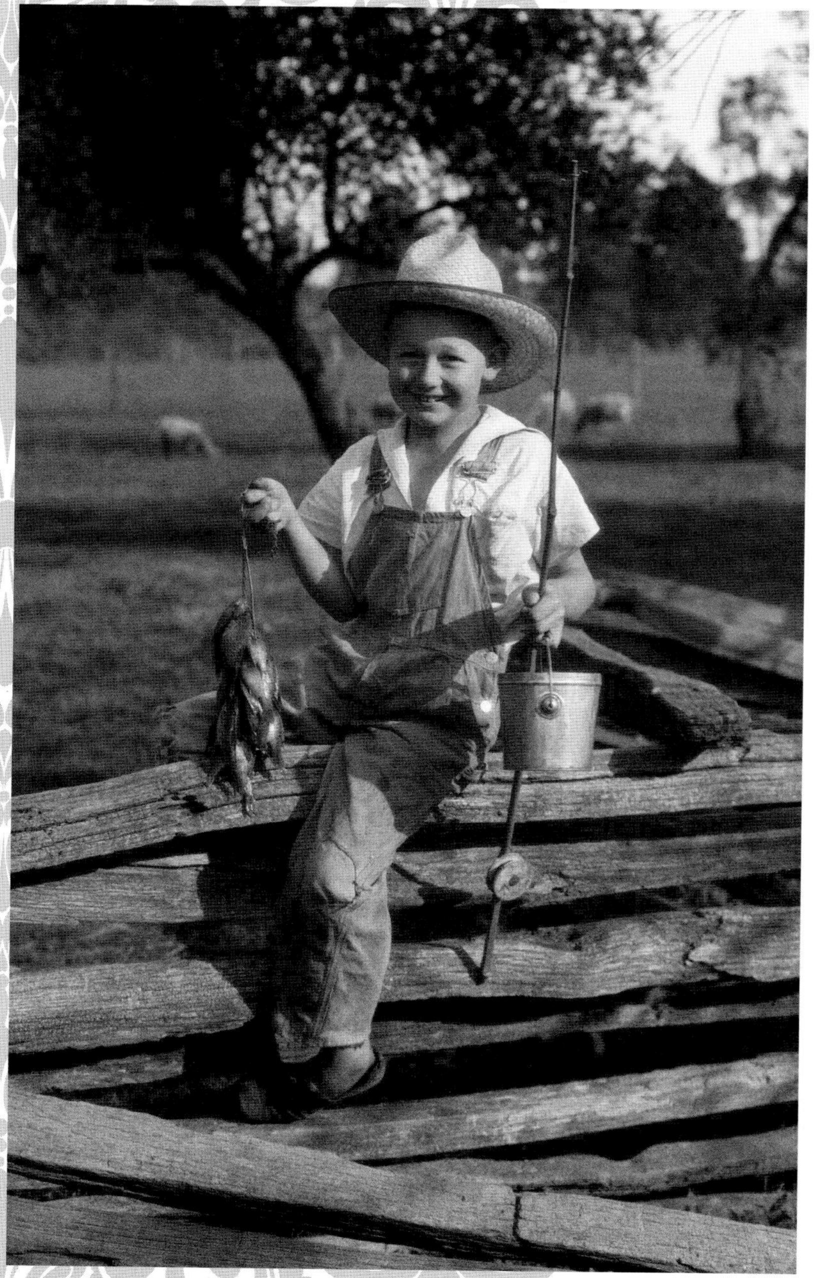

Eldon Robeson displays a string of fish that he caught while fishing in Camden. (Carroll County, 1930)

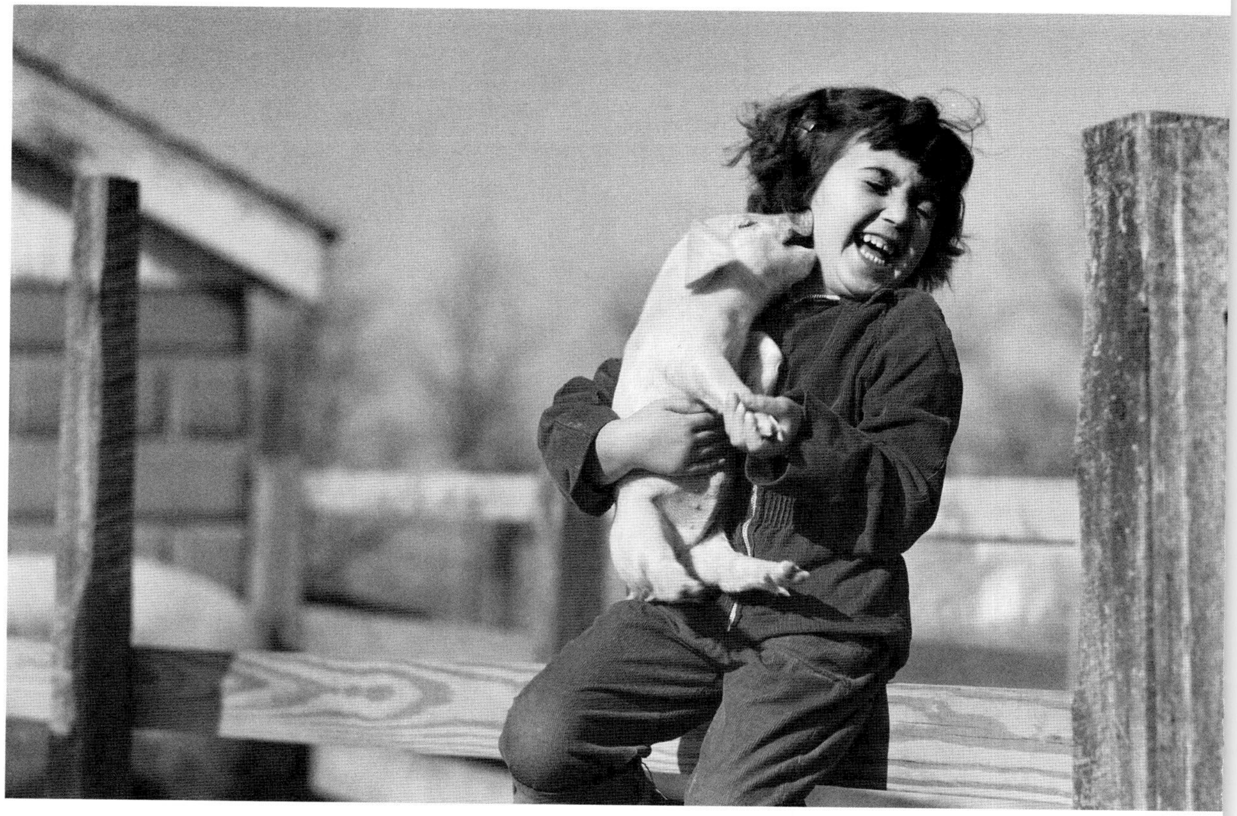

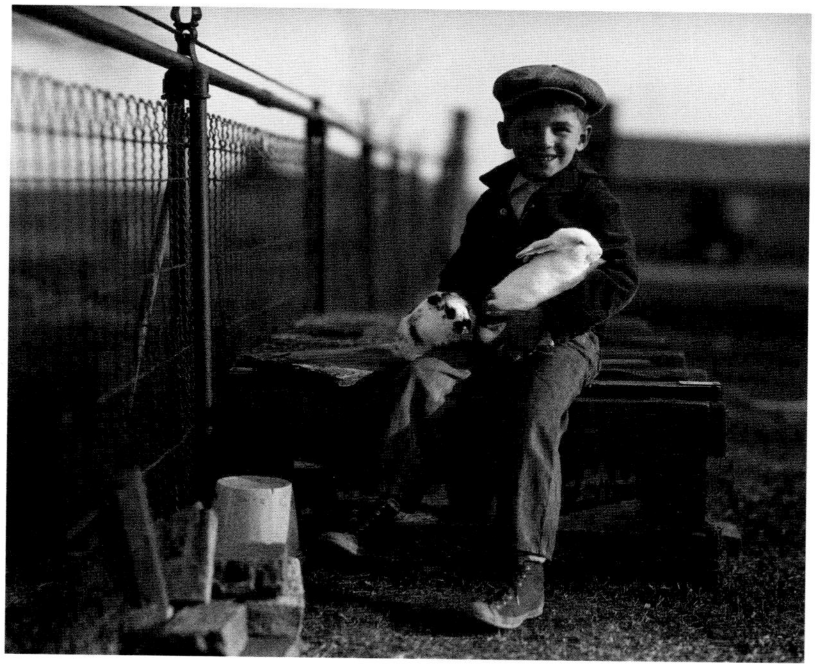

(Above) A seemingly happy Rose Moore receives a kiss from a pig on a farm in Noblesville. (Hamilton County, 1955)

(Bottom right) Bob Caldwell proudly shows off two rabbits. (Fayette County, 1932)

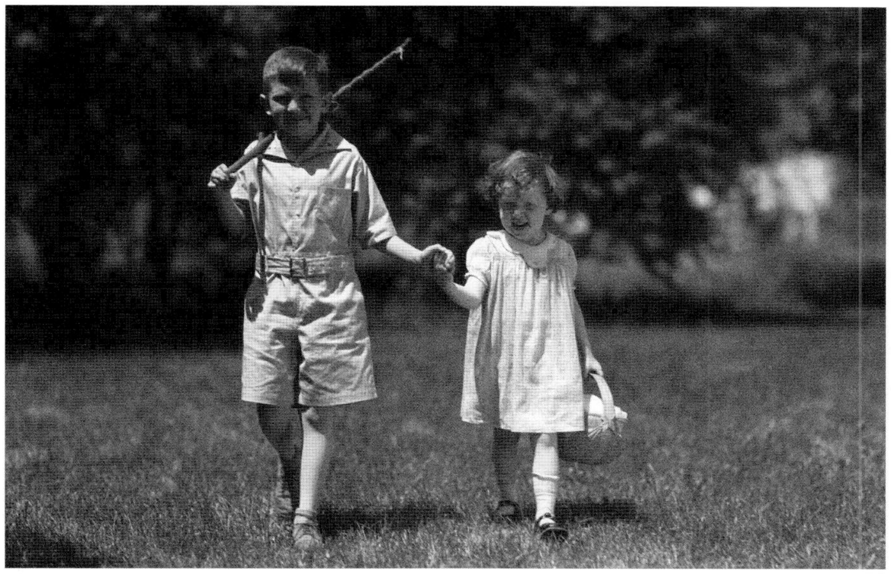

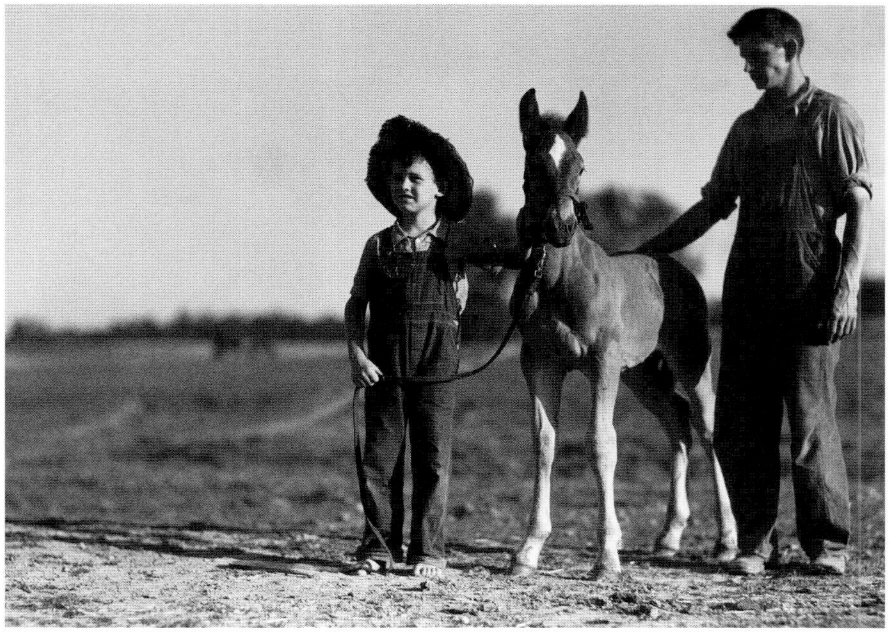

(Top left) Bill Graham (left) and Virginia Jones tote a fishing pole and a lunch basket. (Tippecanoe County, 1933)

(Bottom left) Howard Richards Jr. holds onto his young colt on a farm near New Castle. The unidentified man at the right is likely Howard's father. (Henry County, 1934)

(Right) Eldon Robeson of Camden fills his family's woodbox. (Carroll County, 1931)

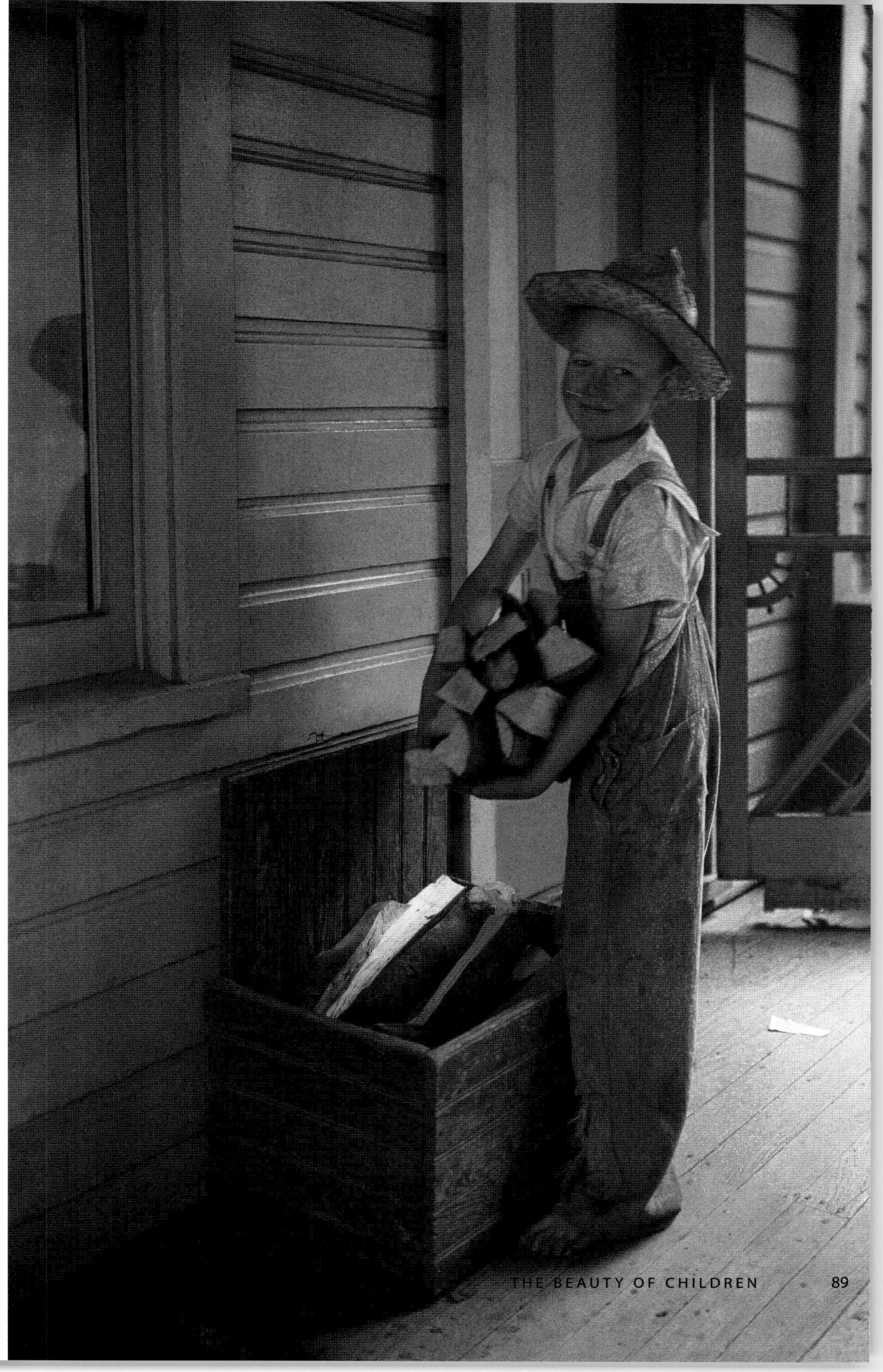

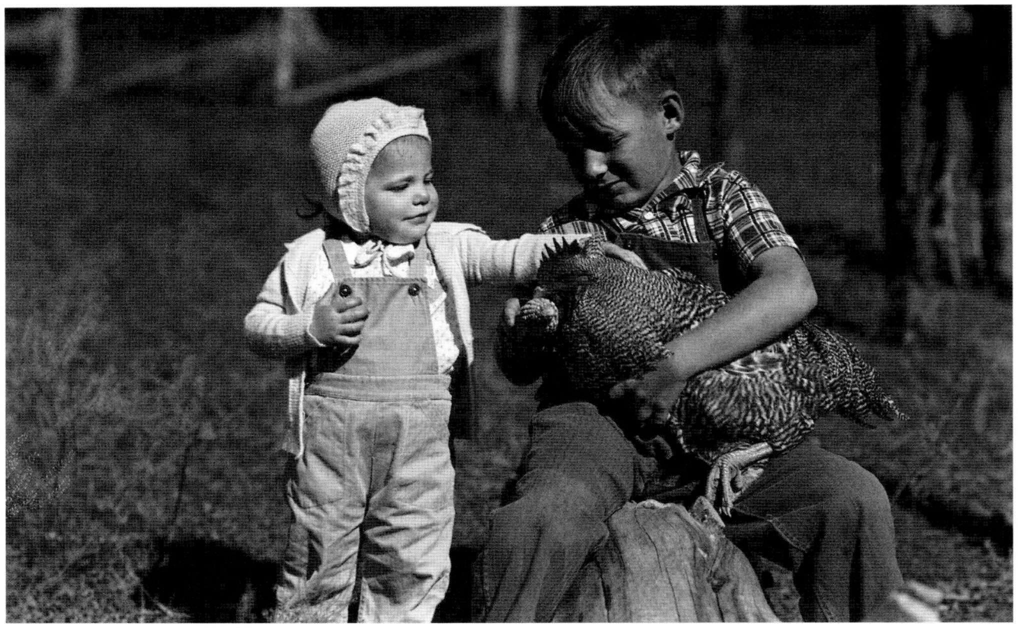

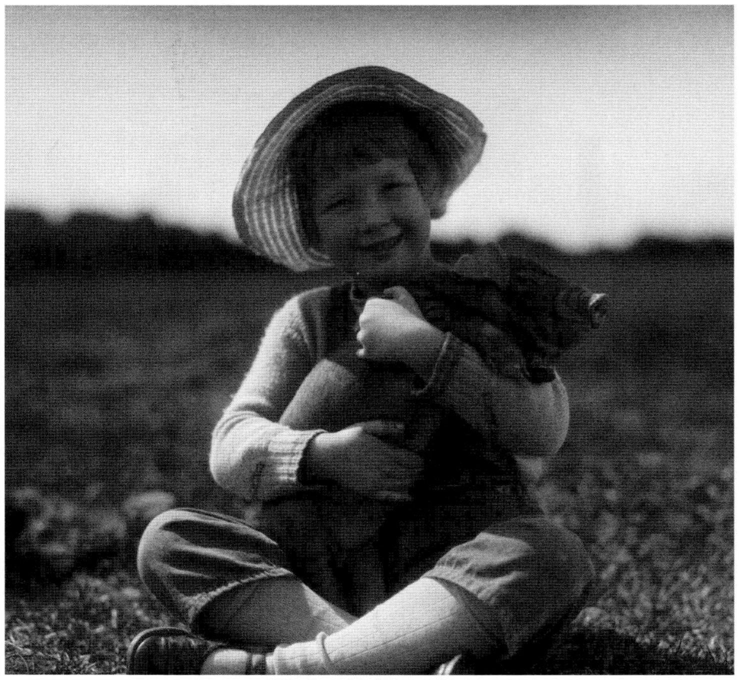

(Top left) A boy holds a rooster while a little girl pets it in Fulton, Missouri. (Callaway County, MO; 1946)

(Bottom left) Beth Schaupp from West Lafayette holds a Duroc Jersey pig. (Tippecanoe County, 1930)

(Right) Steve Hurley hands sister Aileene some pussy willow branches as he climbs over a split-rail fence. (Tippecanoe County, 1933)

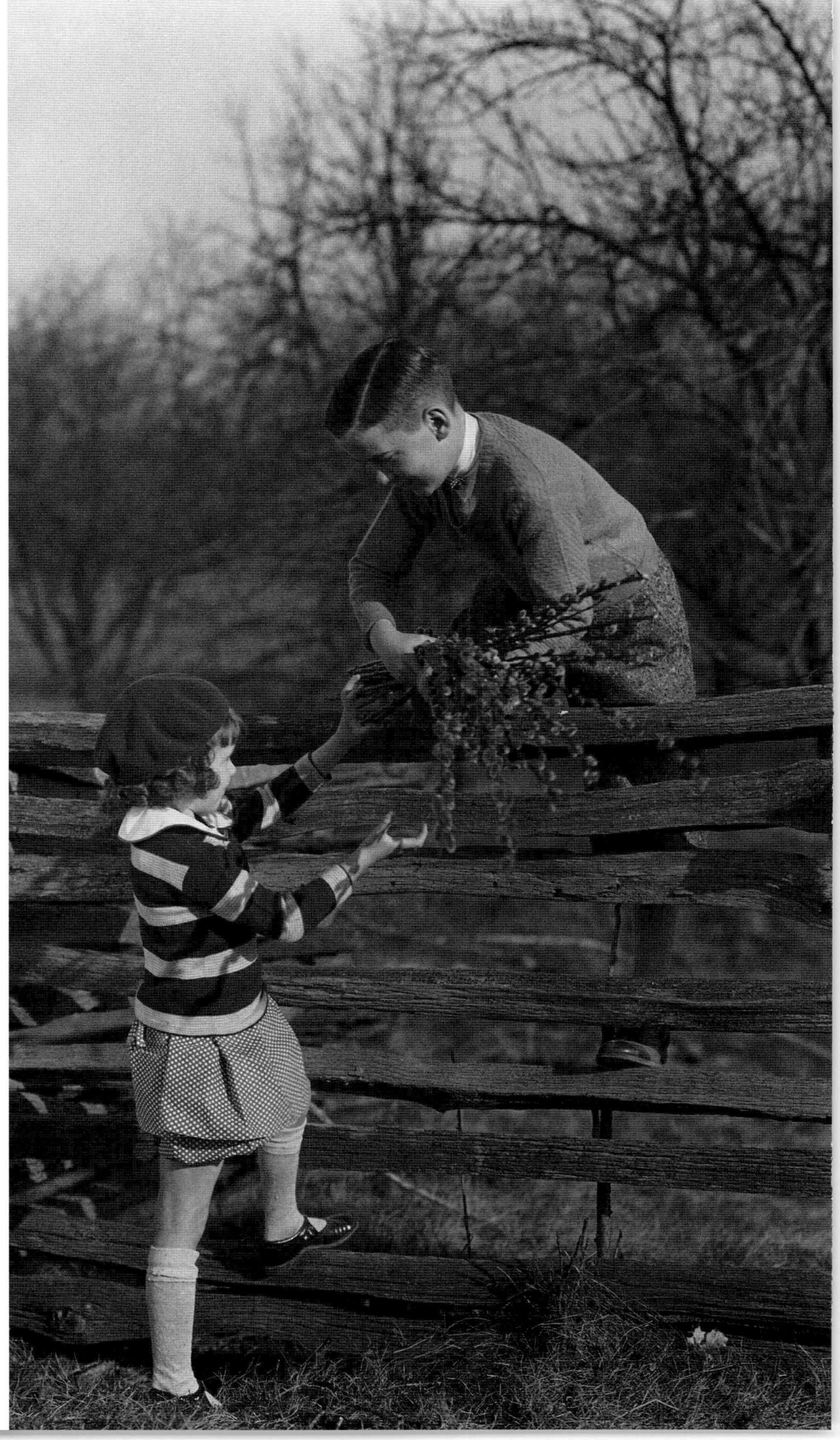

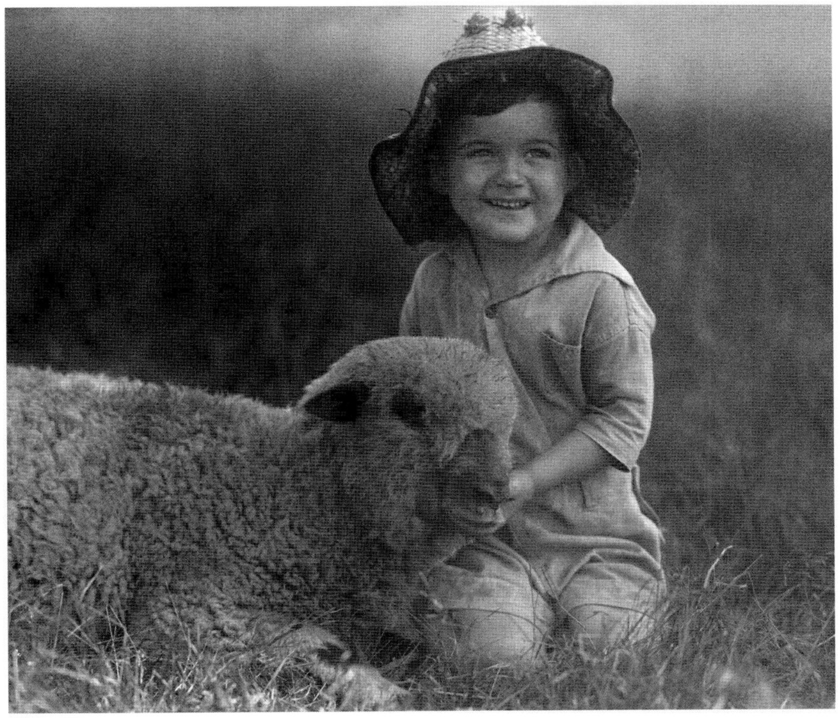

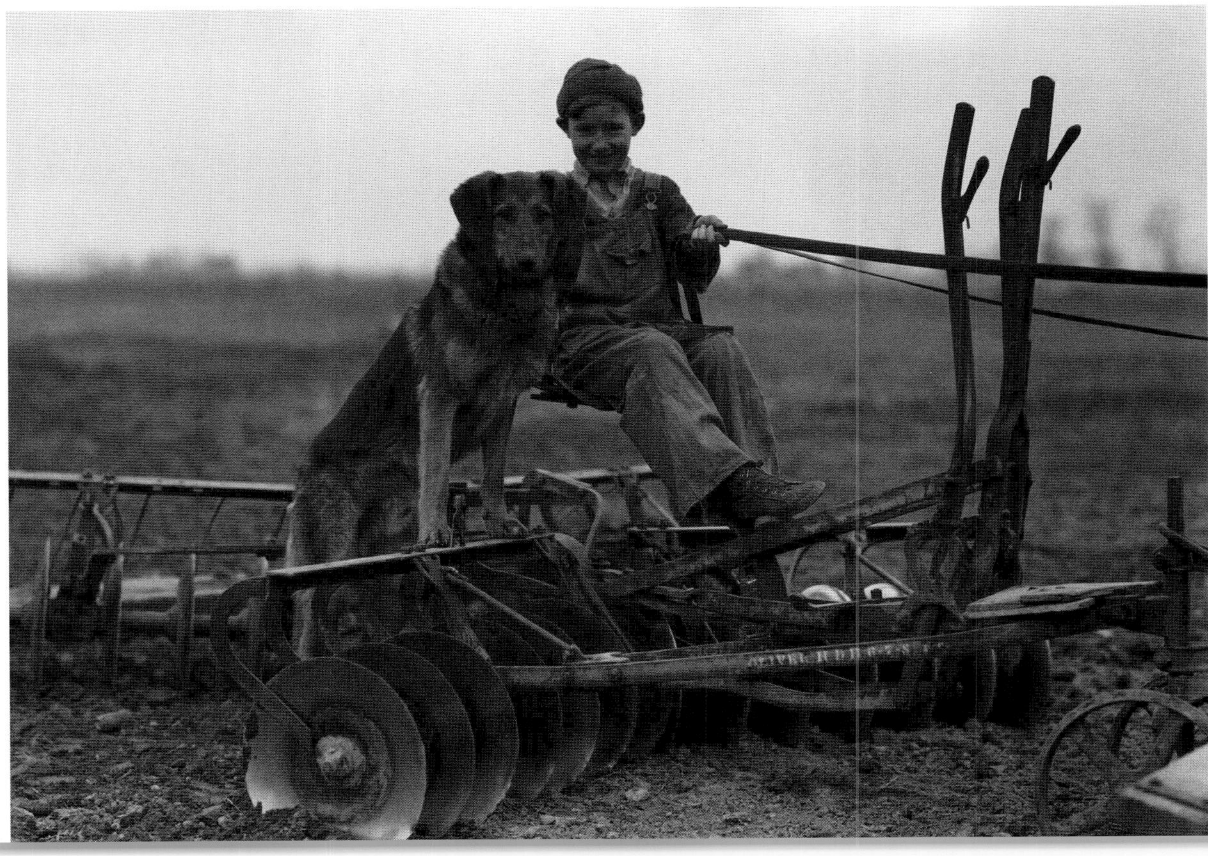

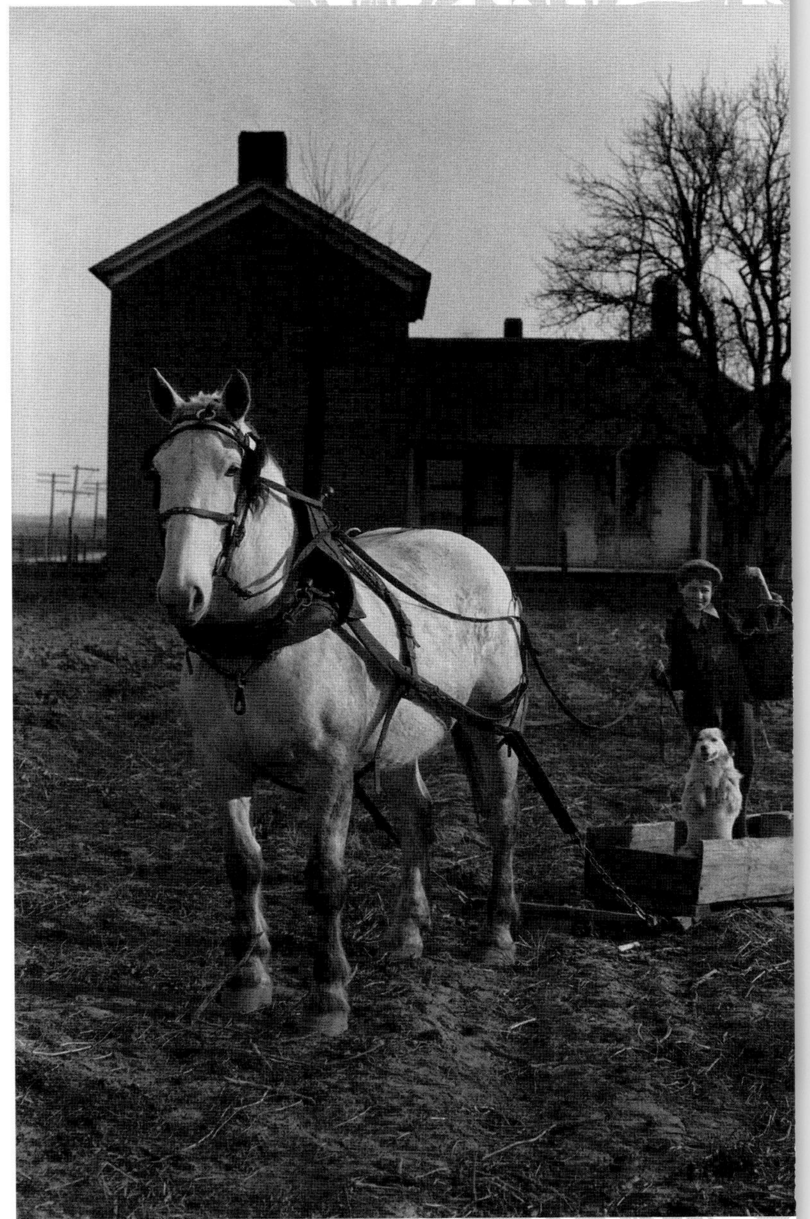

(Top left) Three-year-old Harold Stoner from Ladoga poses with his sheep. (Montgomery County, 1927)

(Bottom left) Junior Lasley from Lebanon stops for a picture on a cultivator with his partner, Jack. (Boone County, 1932)

(Right) E. W. Boram of Noblesville plants his potato patch accompanied by his dog. (Hamilton County, 1932)

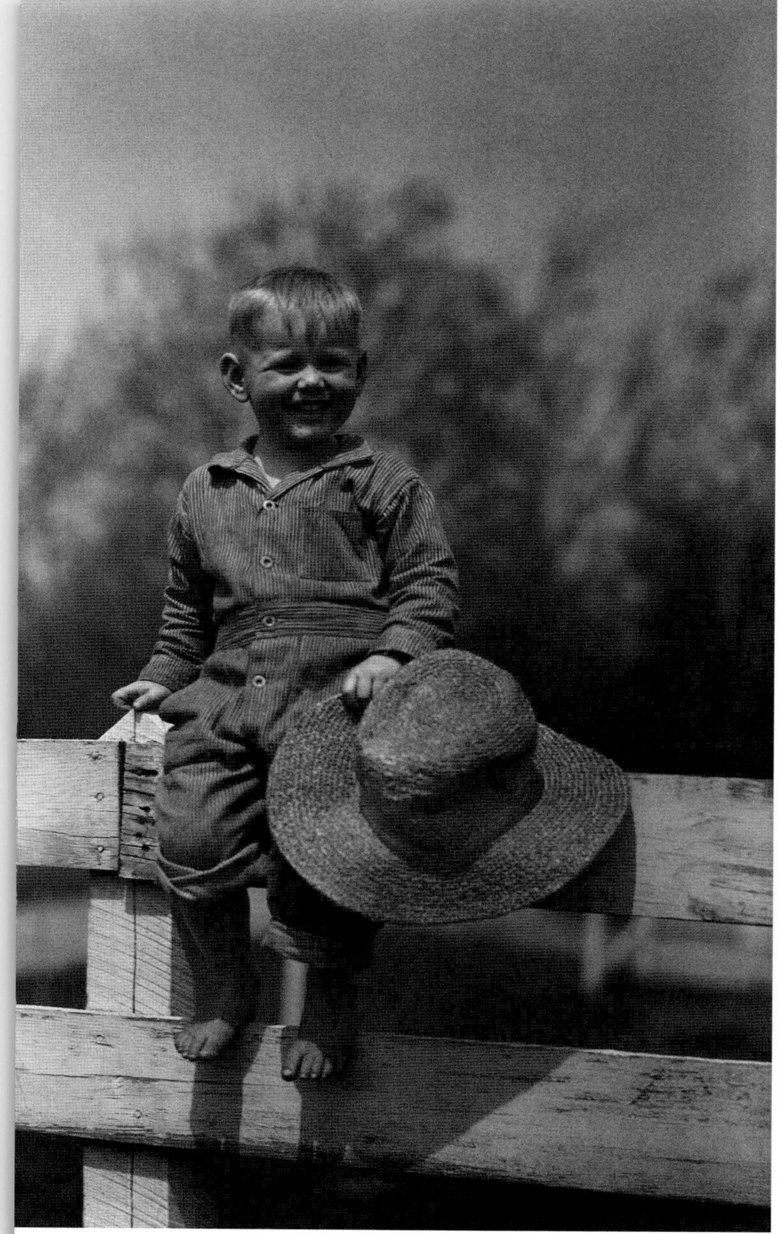

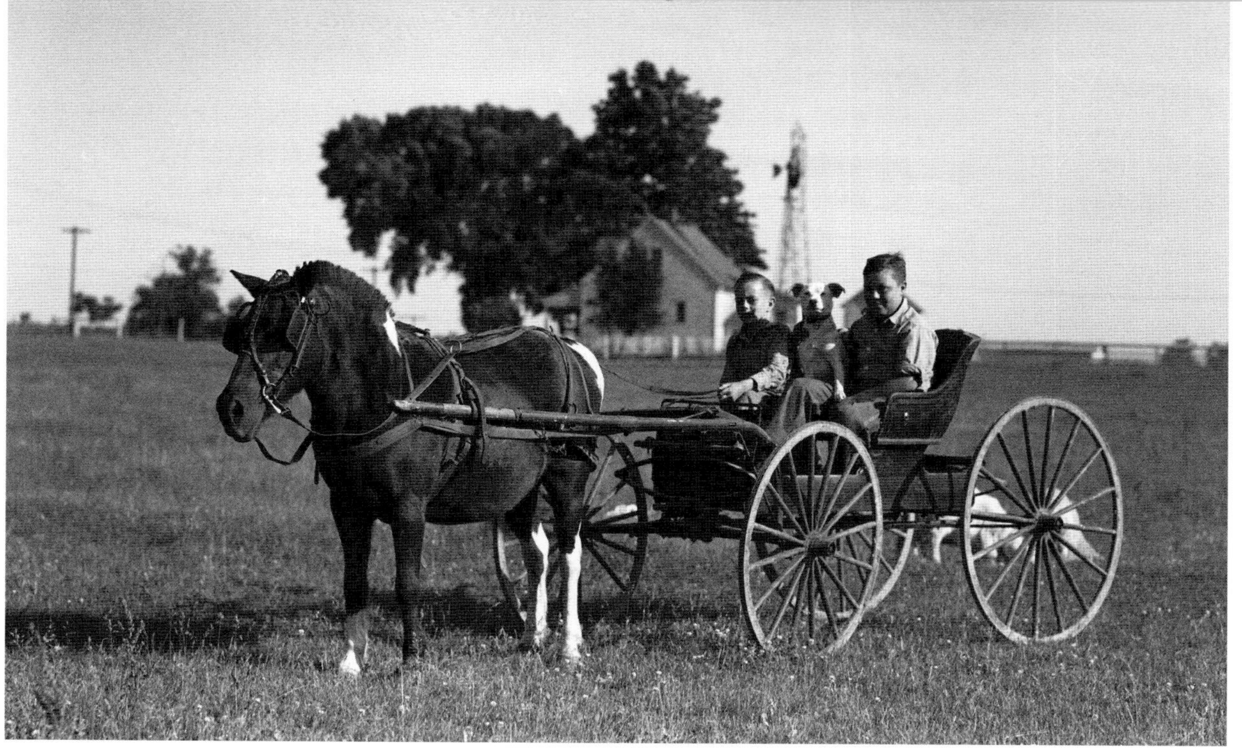

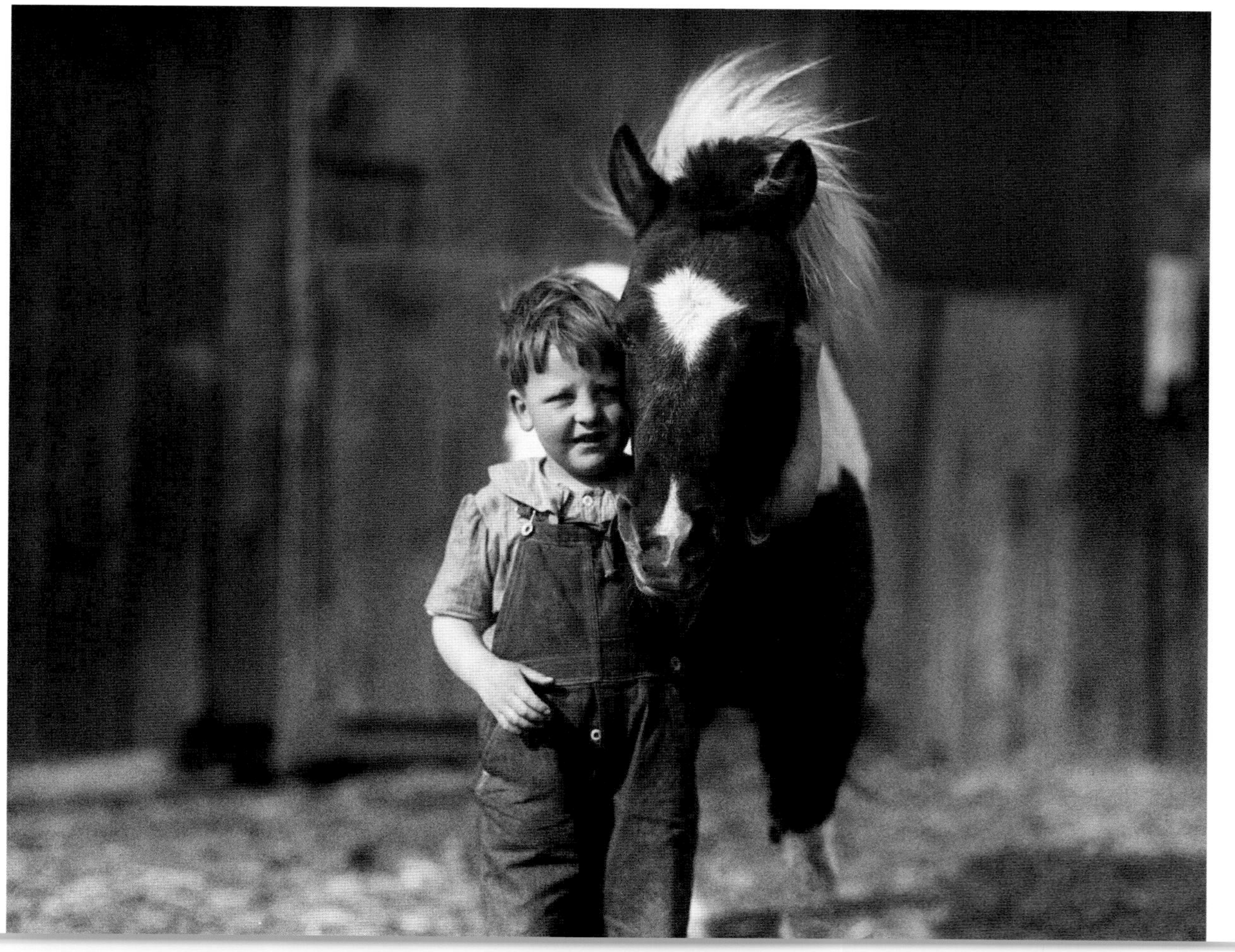

(Above) Jerry Caldwell perches on fence holding his straw hat at his family's farm near Colburn. (Tippecanoe County, 1931)

(Top right) Robert and Charles Wilson ride in a pony cart with their sidekick near Kennard. (Henry County, 1940)

(Bottom right) Harold Goodwin of Wanatah poses beside his pony. (LaPorte County, 1926)

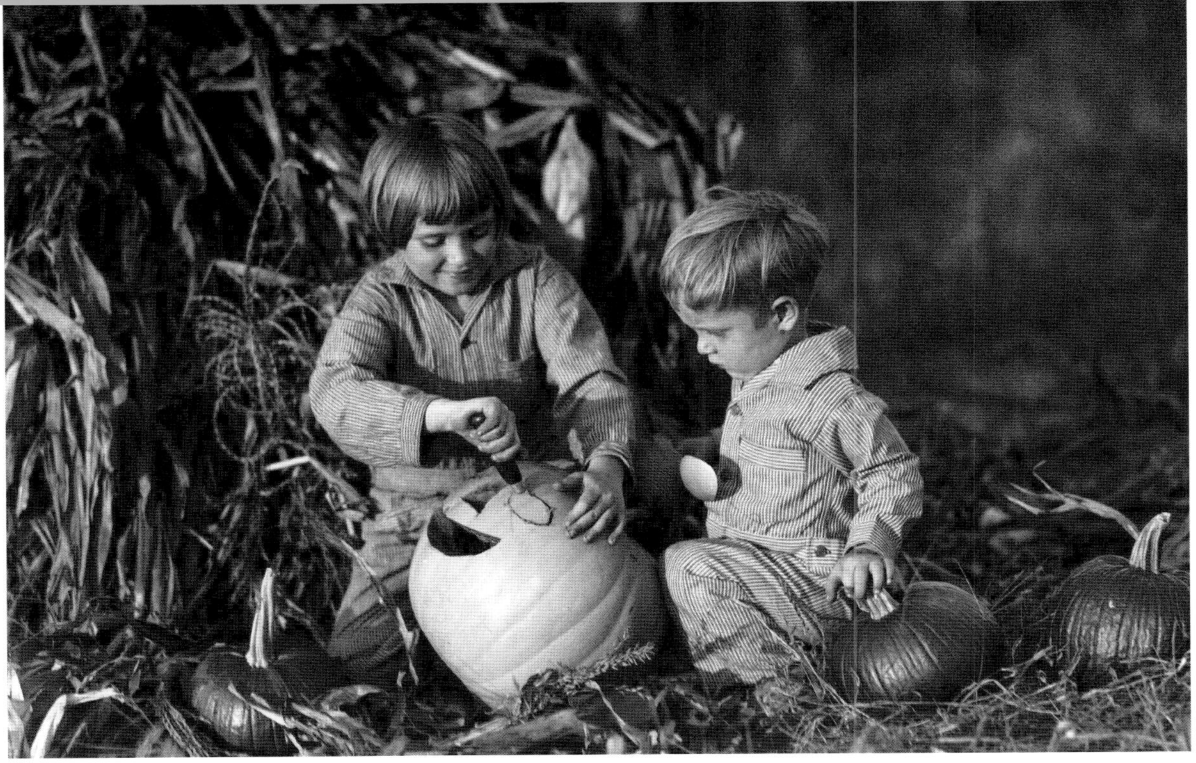

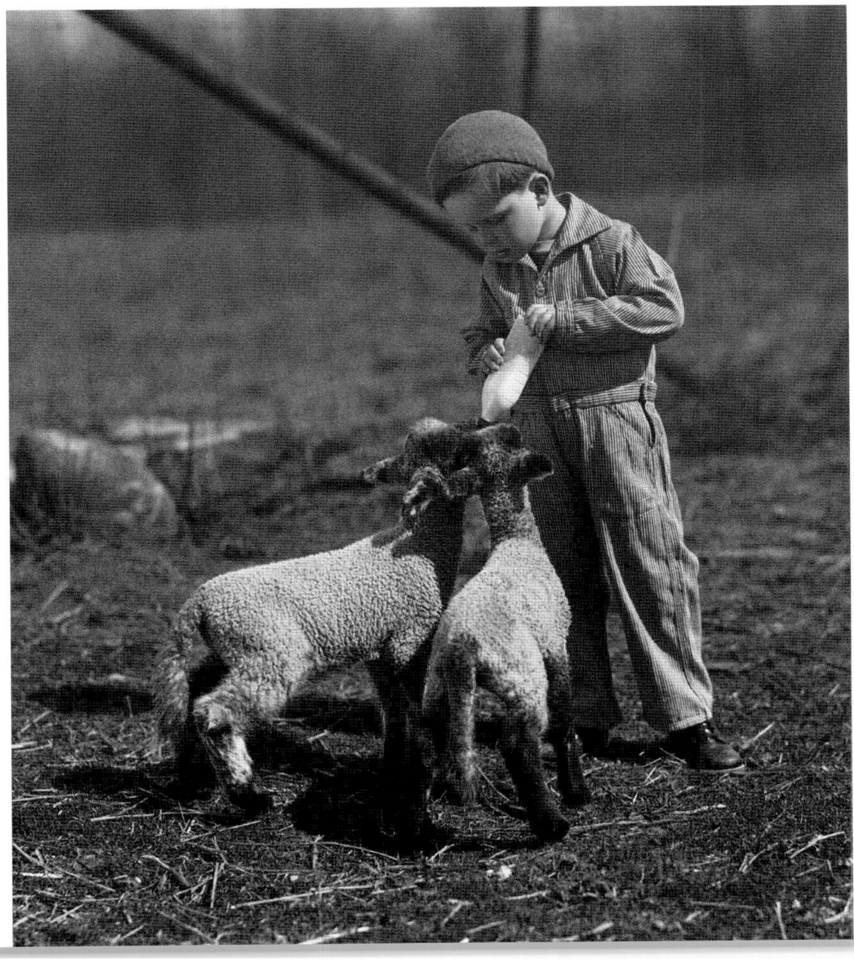

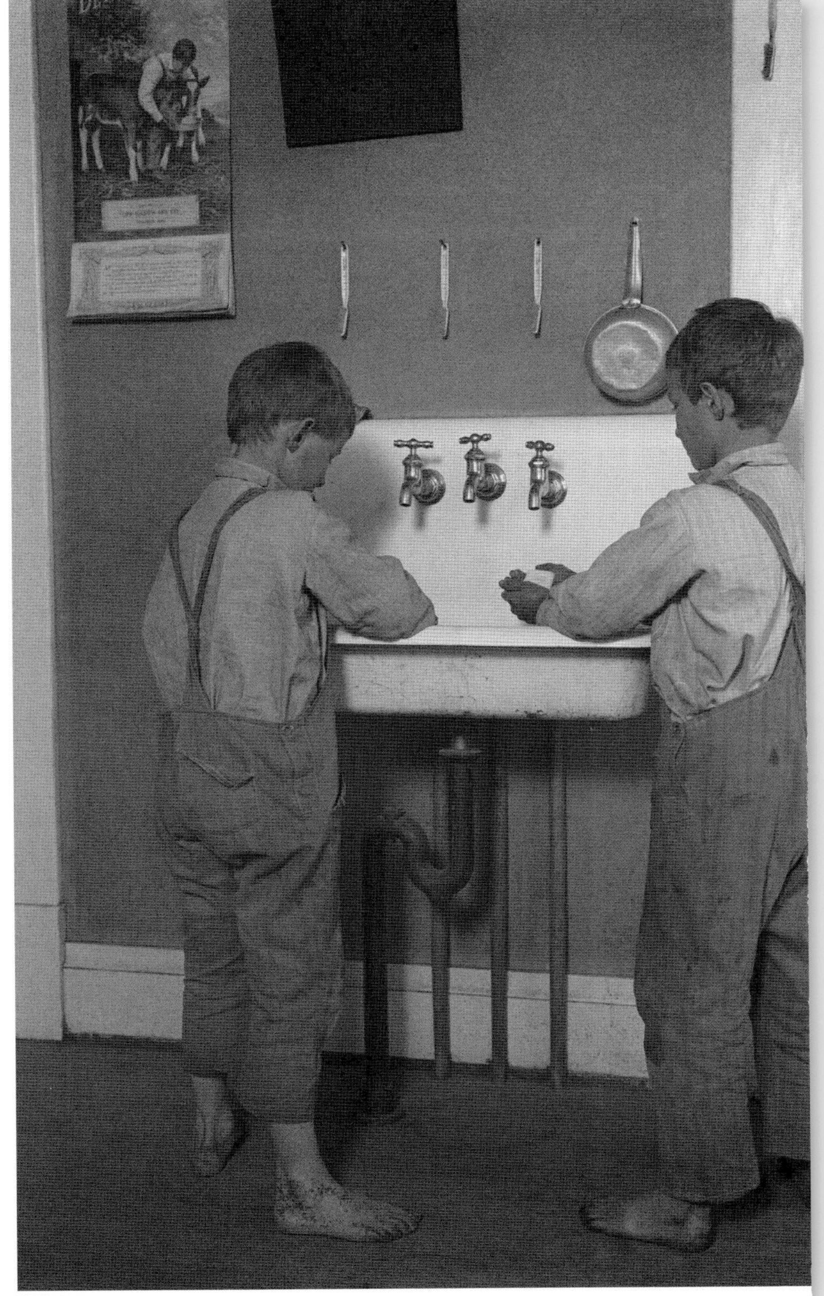

(Top left) Carolyn and David Mitchell cut a jack-o'-lantern face in Battle Ground. (Tippecanoe County, ca. 1932)

(Bottom left) West Lafayette's Buddy Fisher feeds milk to two orphan lambs on the farm of Paul Riley. (Tippecanoe County, 1933)

(Right) Two barefoot farm boys wash up at a kitchen sink. (Benton County, 1915)

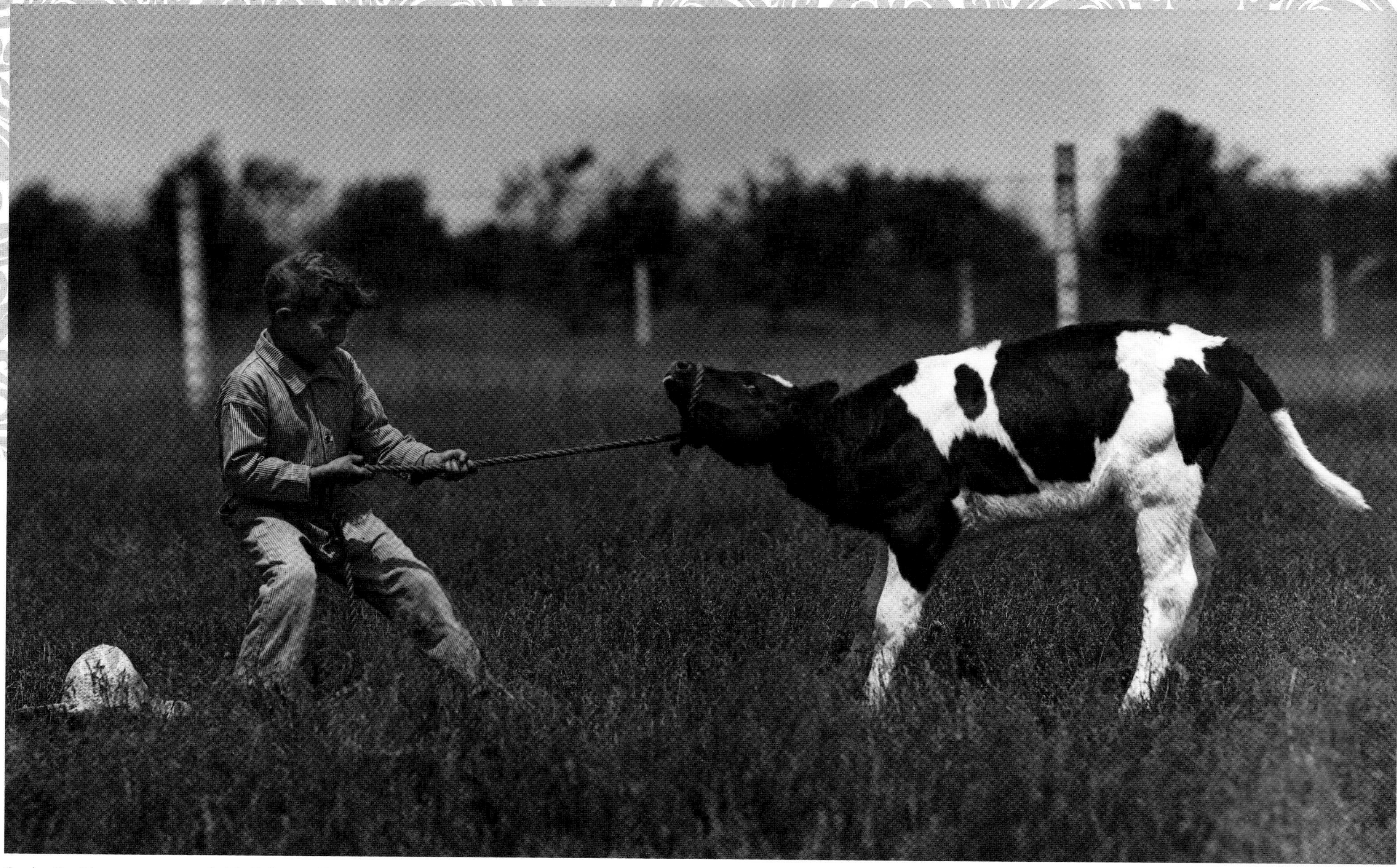

Gordon Ray Myers uses a rope to hold a Holstein calf. (Tippecanoe County, 1930)

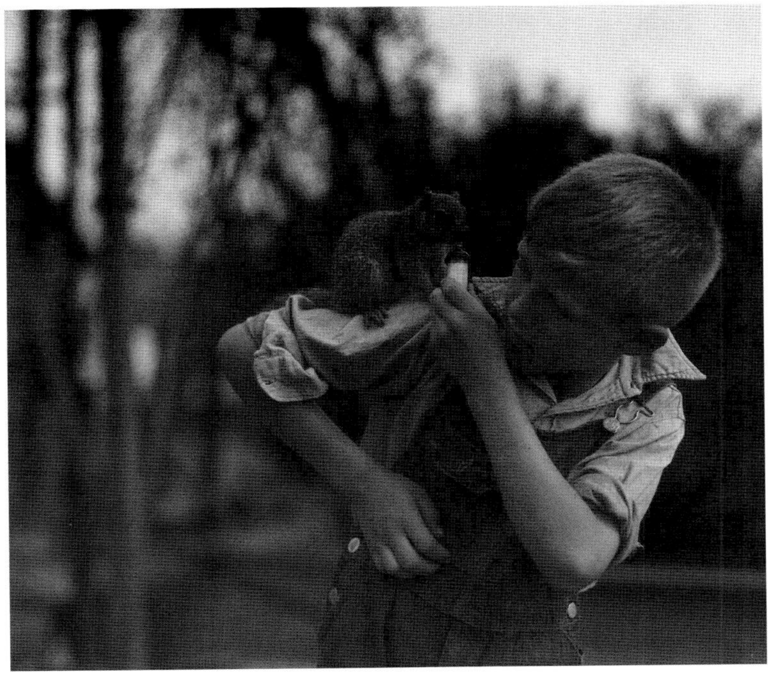

(Top left) LeRoy Brutus feeds a pet squirrel from a bottle containing milk. (Tippecanoe County, 1932)

(Top right) Farmer Fred Guinn's son and two sons of his tenant share an afternoon riding a horse in Noblesville. (Hamilton County, 1935)

(Bottom left) Woods Caperton III is photographed in a cart pulled by his Great Dane. (Hamilton County, 1940)

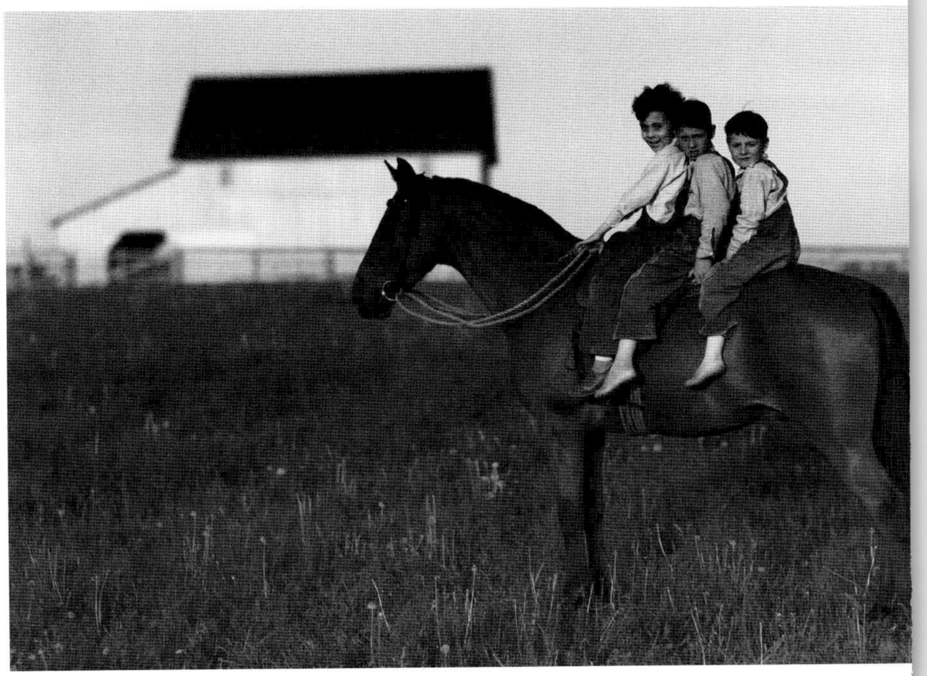

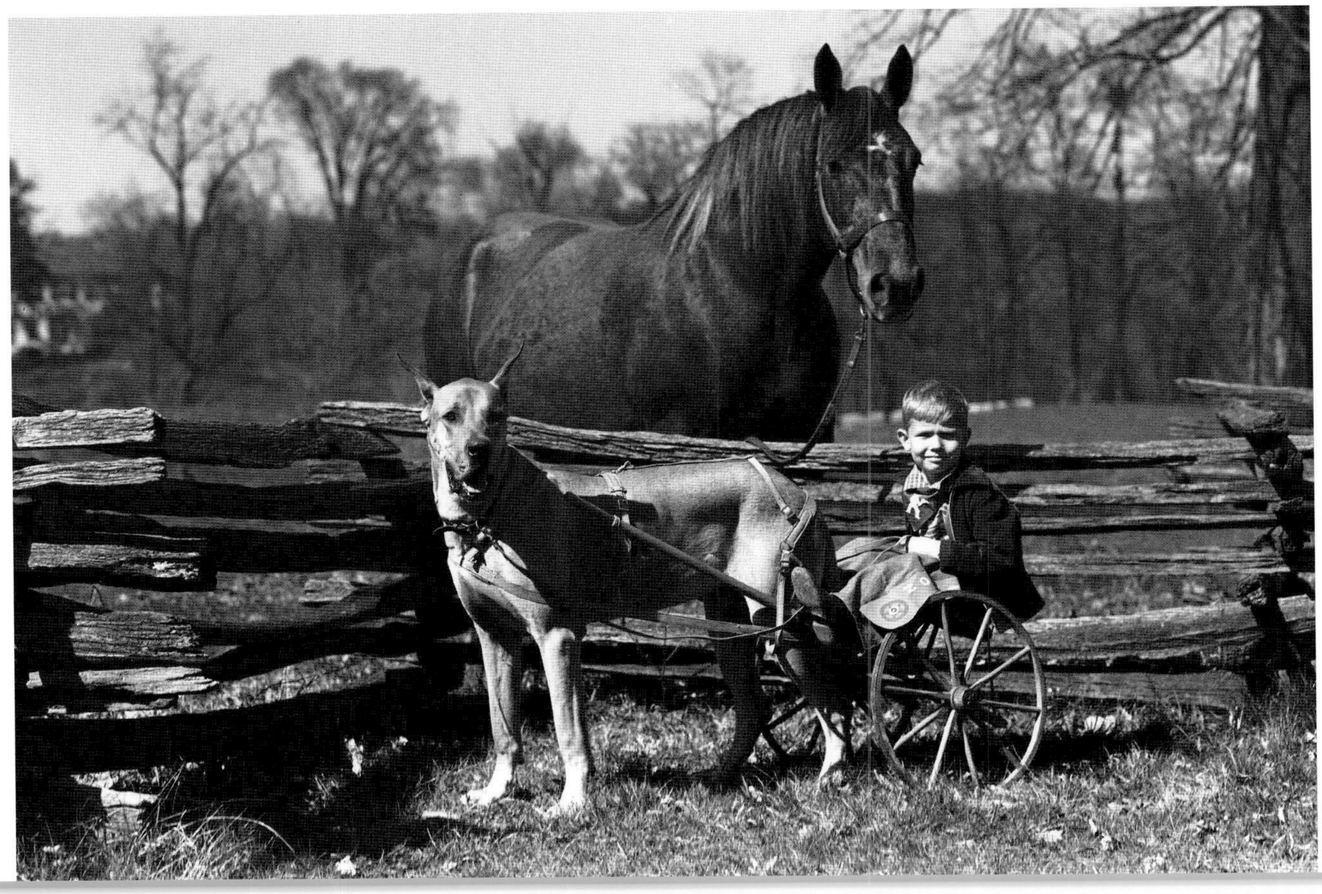

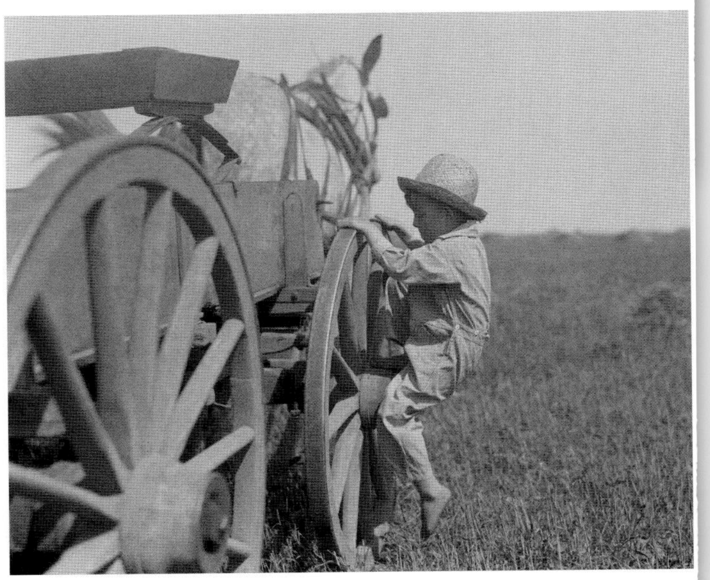

(Above) Dick Vanderhoff uses a wheel to climb onto a wagon on the Henry Marshall farm. (Tippecanoe County, 1932)

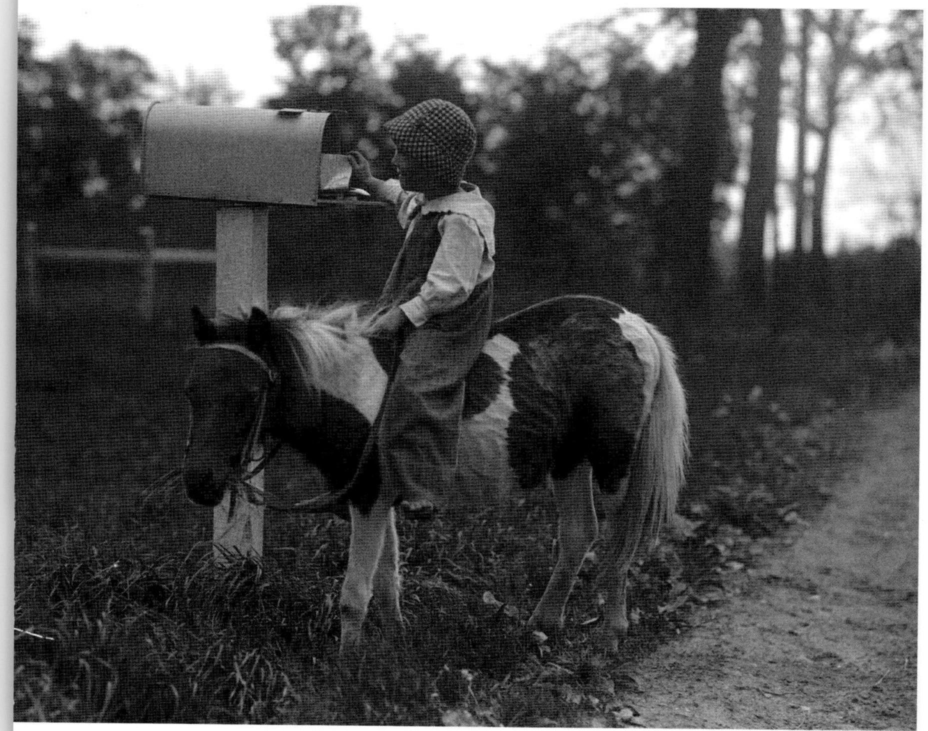

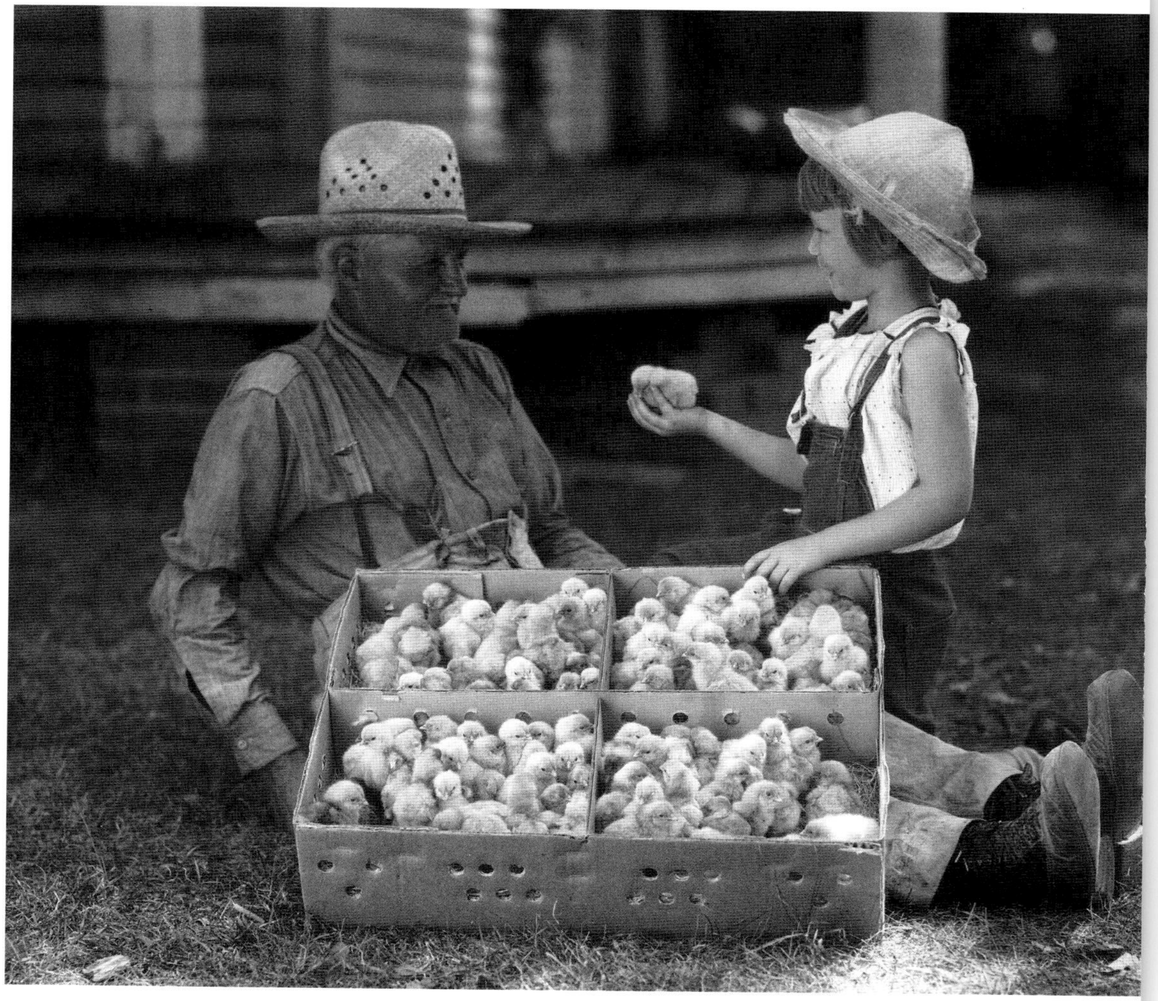

(Above) Bascom O'Hair of Greencastle uses his pony to get the day's mail. (Putnam County, 1916)

(Right) Beth Schaupp holds a little chick while Mr. Cartmill from Pine Village looks on. (Warren County, 1930)

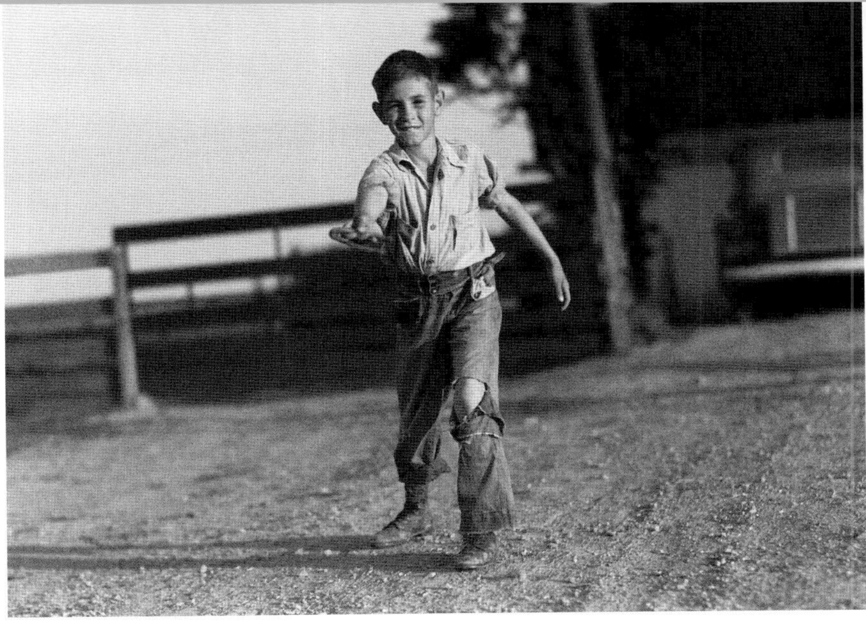

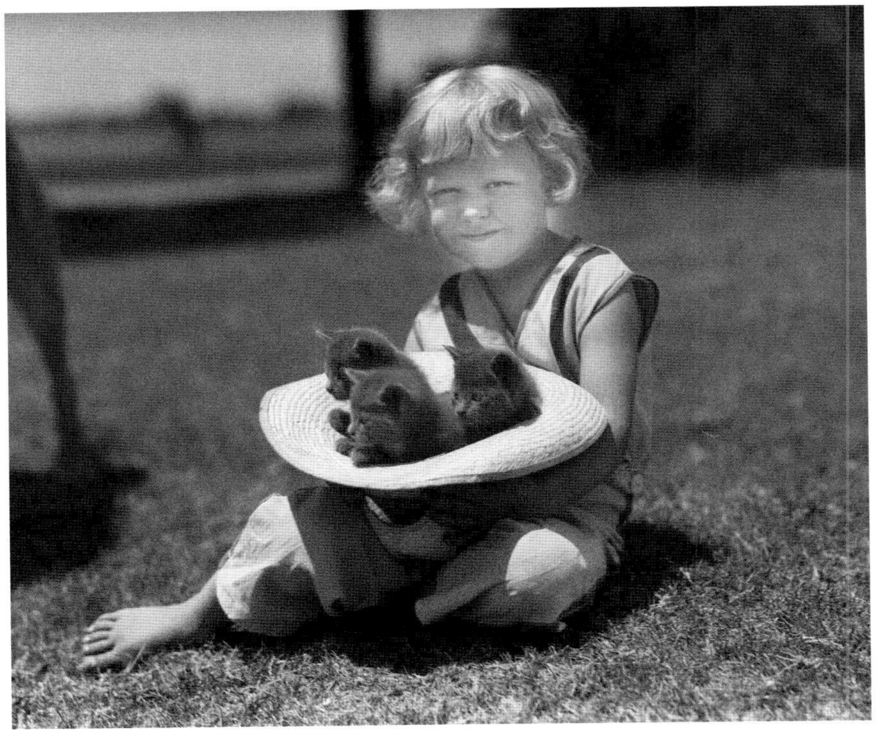

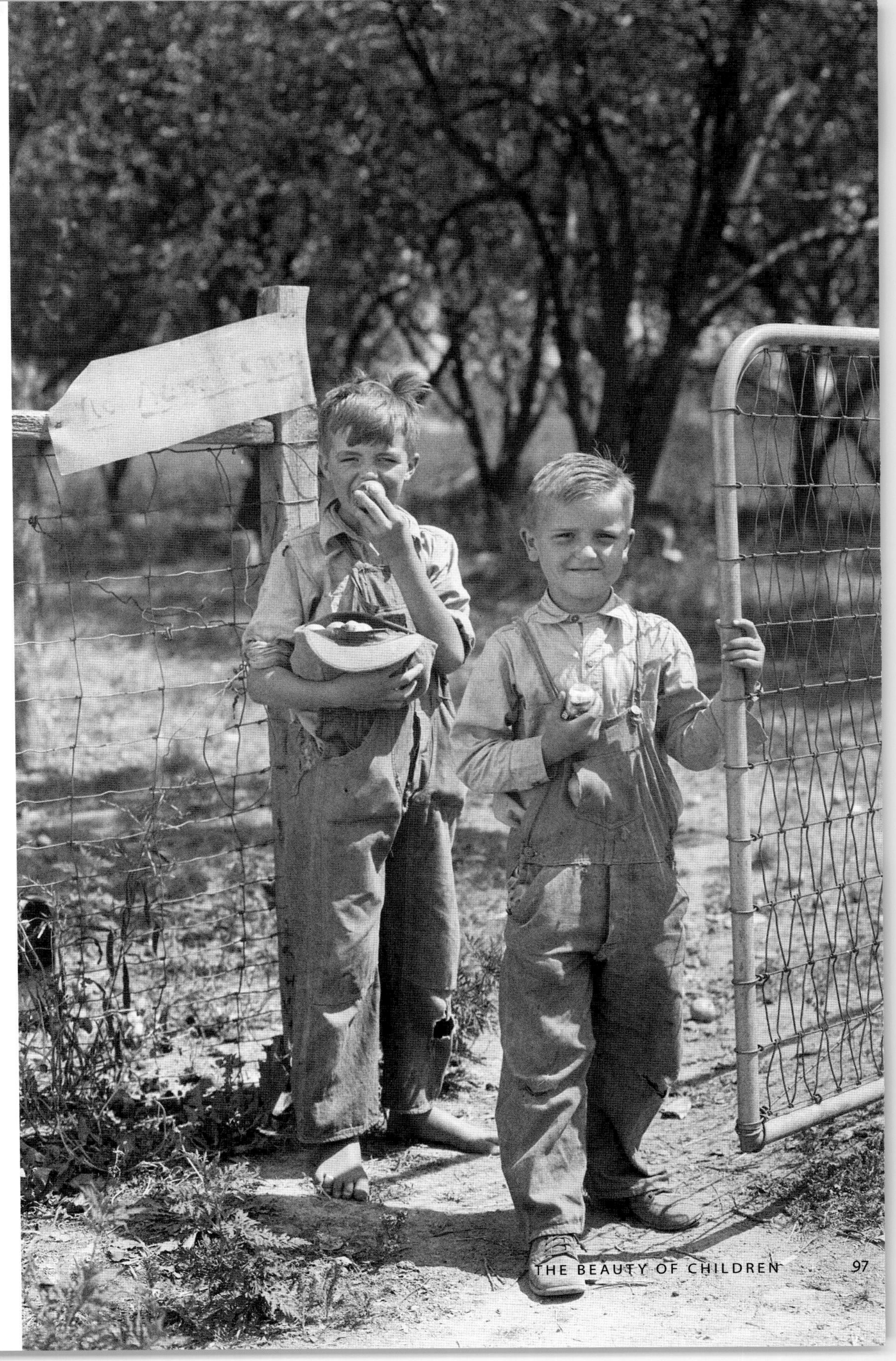

(Top left) Donald Bowman from Rockfield pitches a horseshoe. (Carroll County, 1934)

(Bottom left) Arda Berle Cochran from Forest shows off her three little kittens. (Clinton County, 1934)

(Right) Kenneth Gray and Bobby Wise of Rossville come through the orchard gate toting apples in a hat and their pockets. (Clinton County, 1933)

An advertisement for special baby milk adorns this dairy barn near Greenwood. At night, electric lights illuminated the sign for the public. The dairy owners produced their high-quality milk for the Weber Milk Company in Indianapolis. (Marion County, 1924)

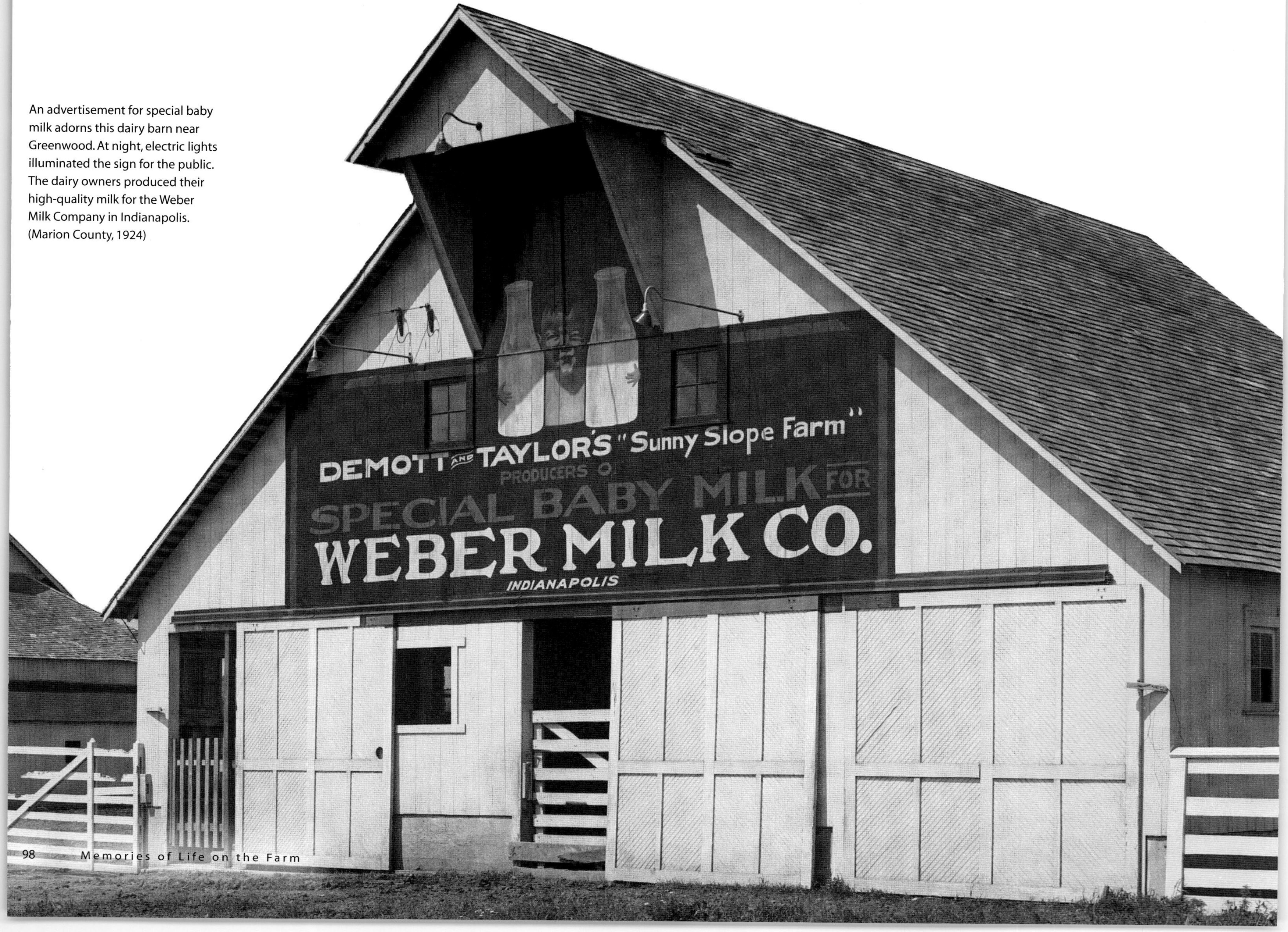

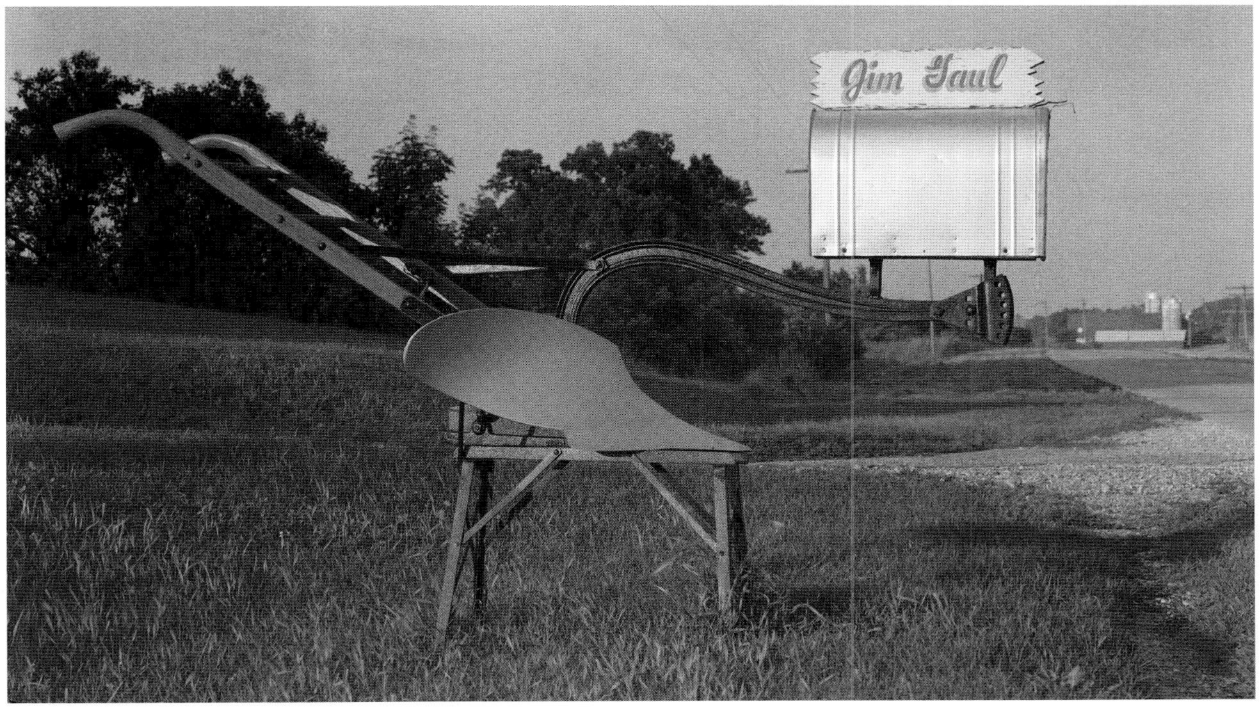

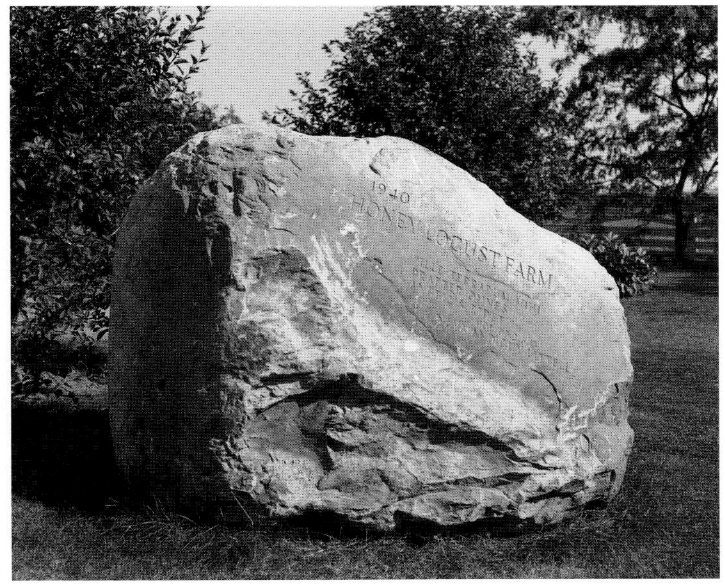

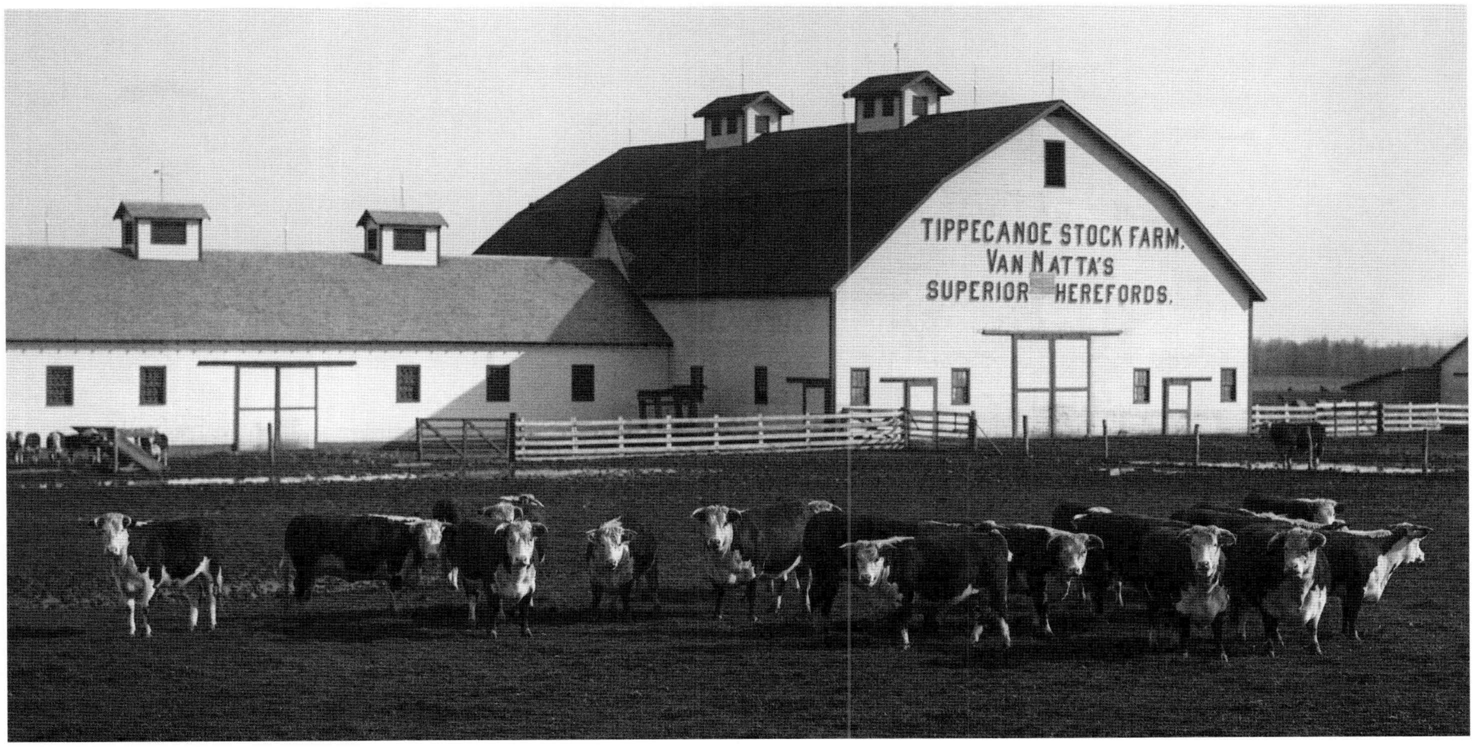

(Top left) An old walking plow holds a mailbox on the farm of Jim Gaul near Harvard, Illinois. (McHenry County, IL; 1957)

(Top right) A large rock on the Honey Locust Farm in Fowler carries a Latin phrase penned by the Italian poet Horace. Translated into English, it states, "That corner of the world smiles in my eye beyond all others." (Benton County, 1955)

(Bottom) Hereford cattle stand in front of a large barn on the farm of J. W. Van Natta in Battle Ground. (Tippecanoe County, 1920)

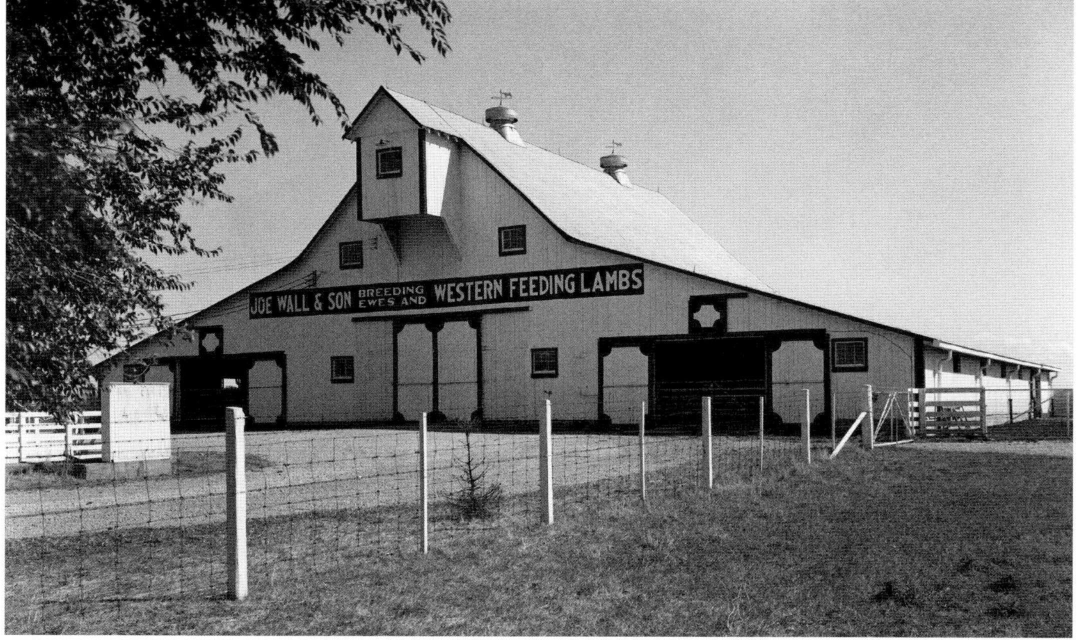

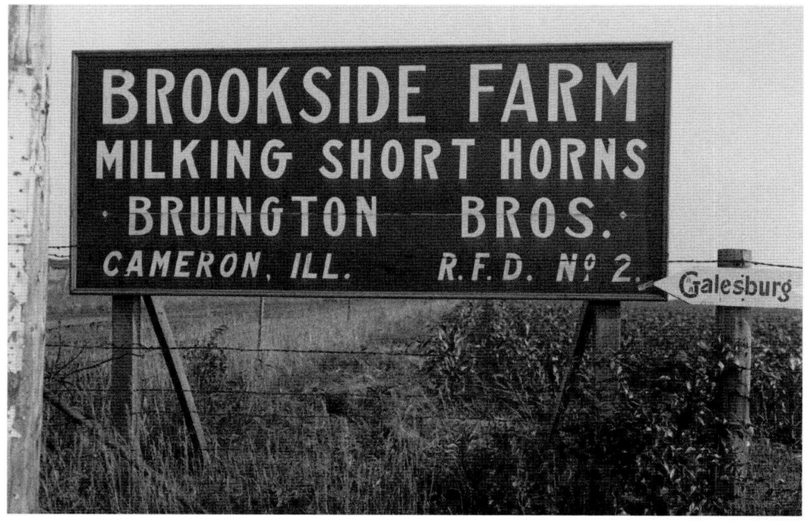

(Top left) Joe Wall's large barn advertises western feeding lambs in Advance. (Boone County, 1953)

(Top right) A farm advertising sign in Cameron, Illinois (Warren County, IL; 1920)

(Bottom) A herd of Angus cattle stand in the feedlot of S. W. Larmore of Tuscola, Illinois, with his cattle feeding barn in the background. The sign on the silo reads, "Indiana Silo Co., Anderson, Ind., S. W. Larmore Agt." (Douglas County, IL; 1920)

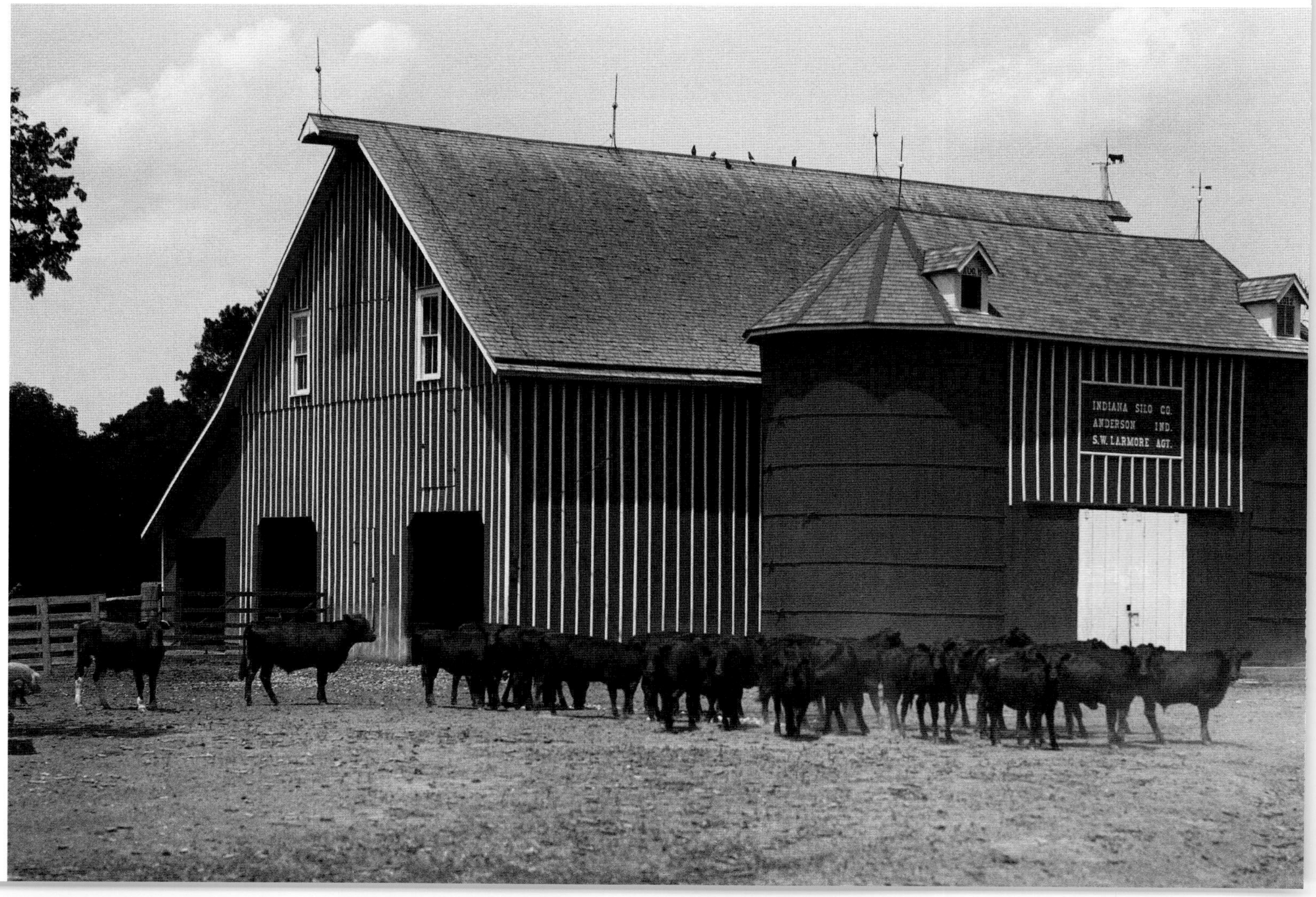

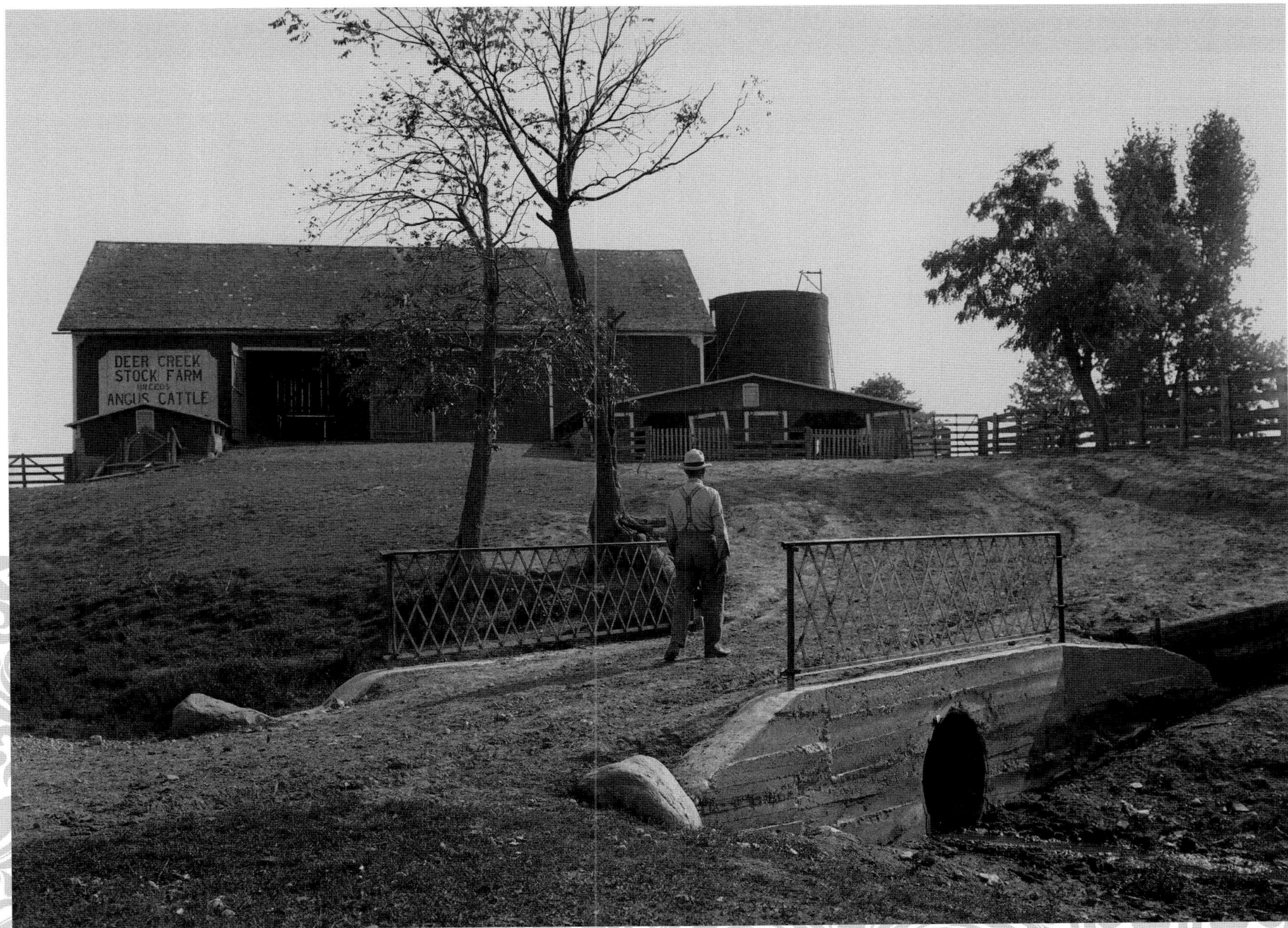

A sign for the Deer Creek Stock Farm owned by Taylor Fouts promotes his Angus cattle. (Carroll County, ca. 1910s)

The Farmer's Workshop

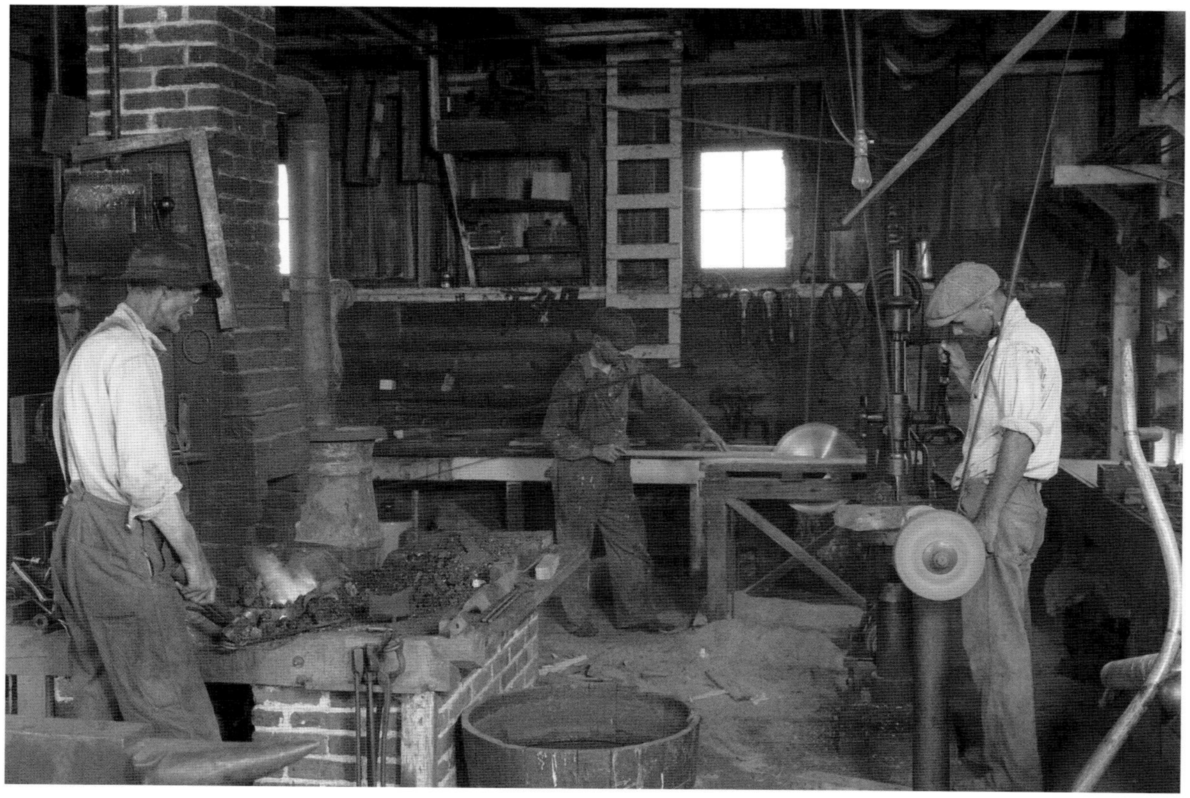

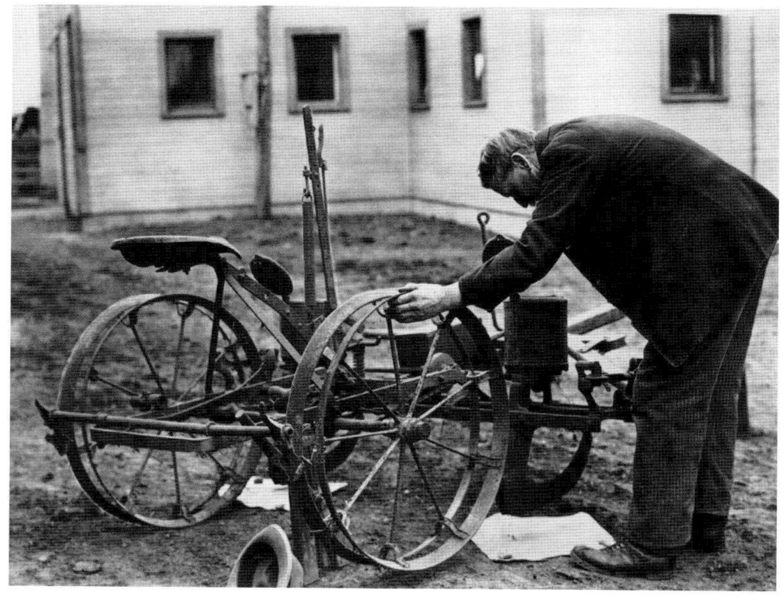

A farmer checks the seed drop for a two-row corn planter. (N.p., ca. 1914)

(Top left) In this farm workshop belonging to former Indiana Governor Warren T. McCray of Kentland, the drill, emery wheel, and saw all operated by motors receiving current from a rural high-power electric line. (Newton County, 1929)

(Bottom left) Placing the head on a Caterpillar tractor in 1930 (Tippecanoe County, 1930)

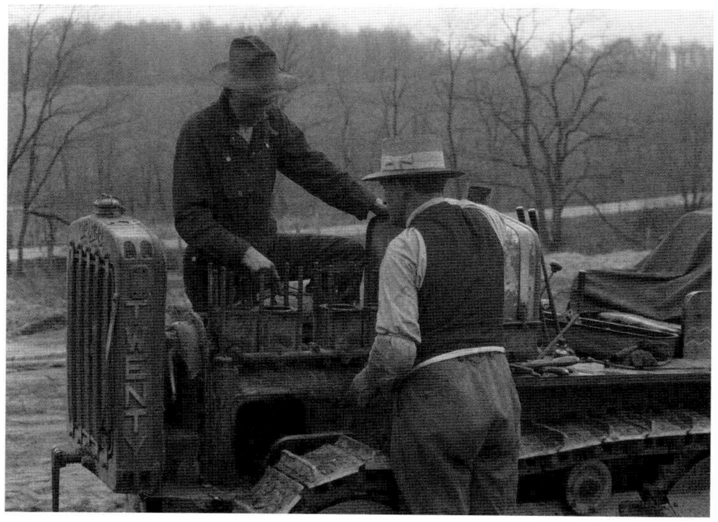

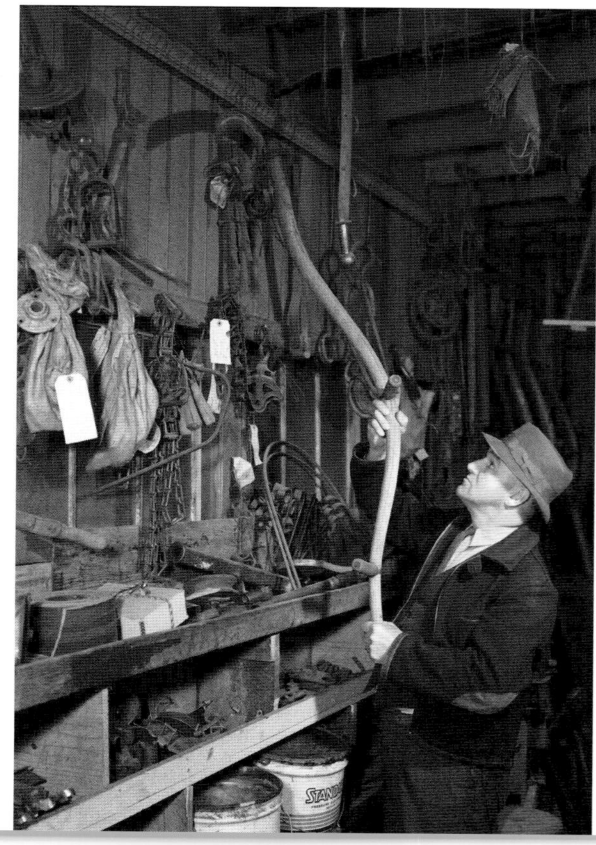

Sam Breese hangs up tools at a Purdue University farm in West Lafayette. (Tippecanoe County, 1940)

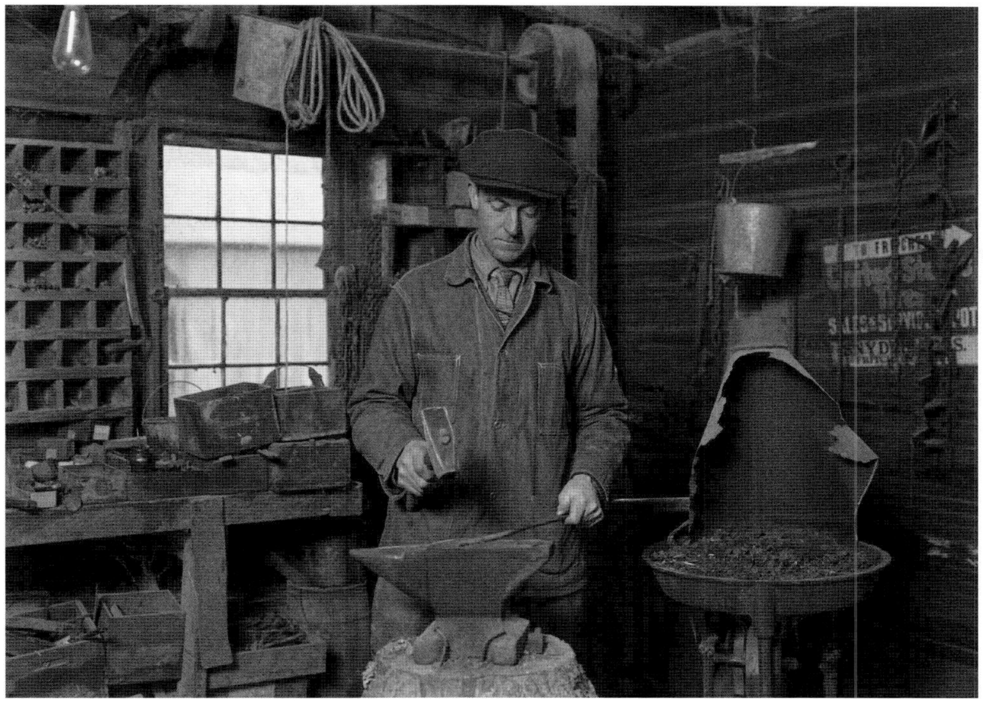

P. L. Johnson works on an anvil in his farm workshop at Vincennes. (Knox County, 1926)

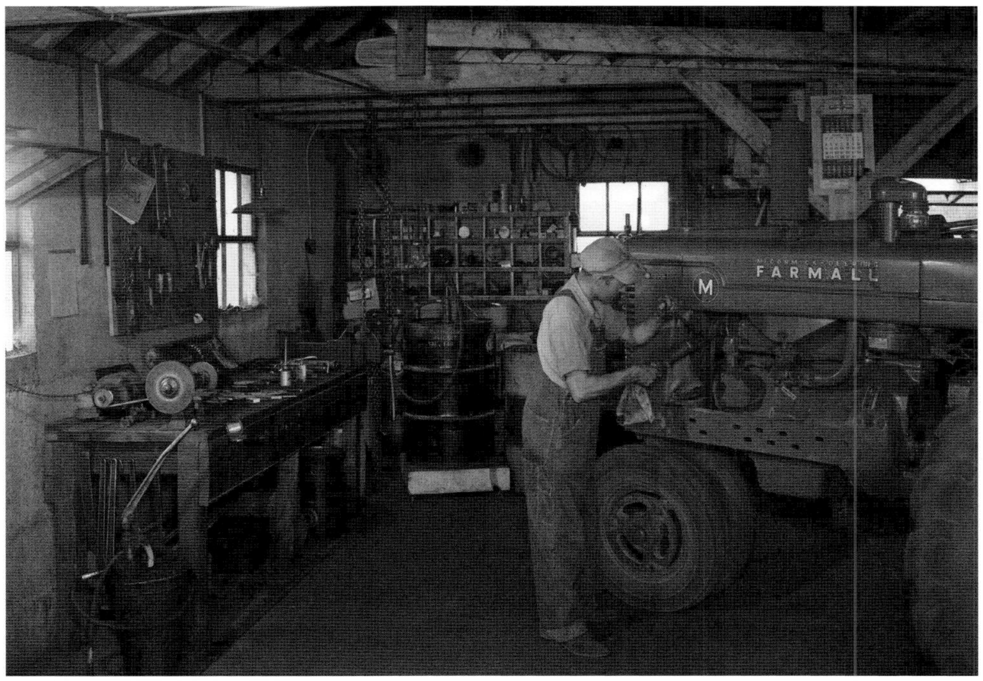

Inside the shop of Major Rafferty of Morocco, Paul E. Merriman repairs a Farmall tractor. (Newton County, 1944)

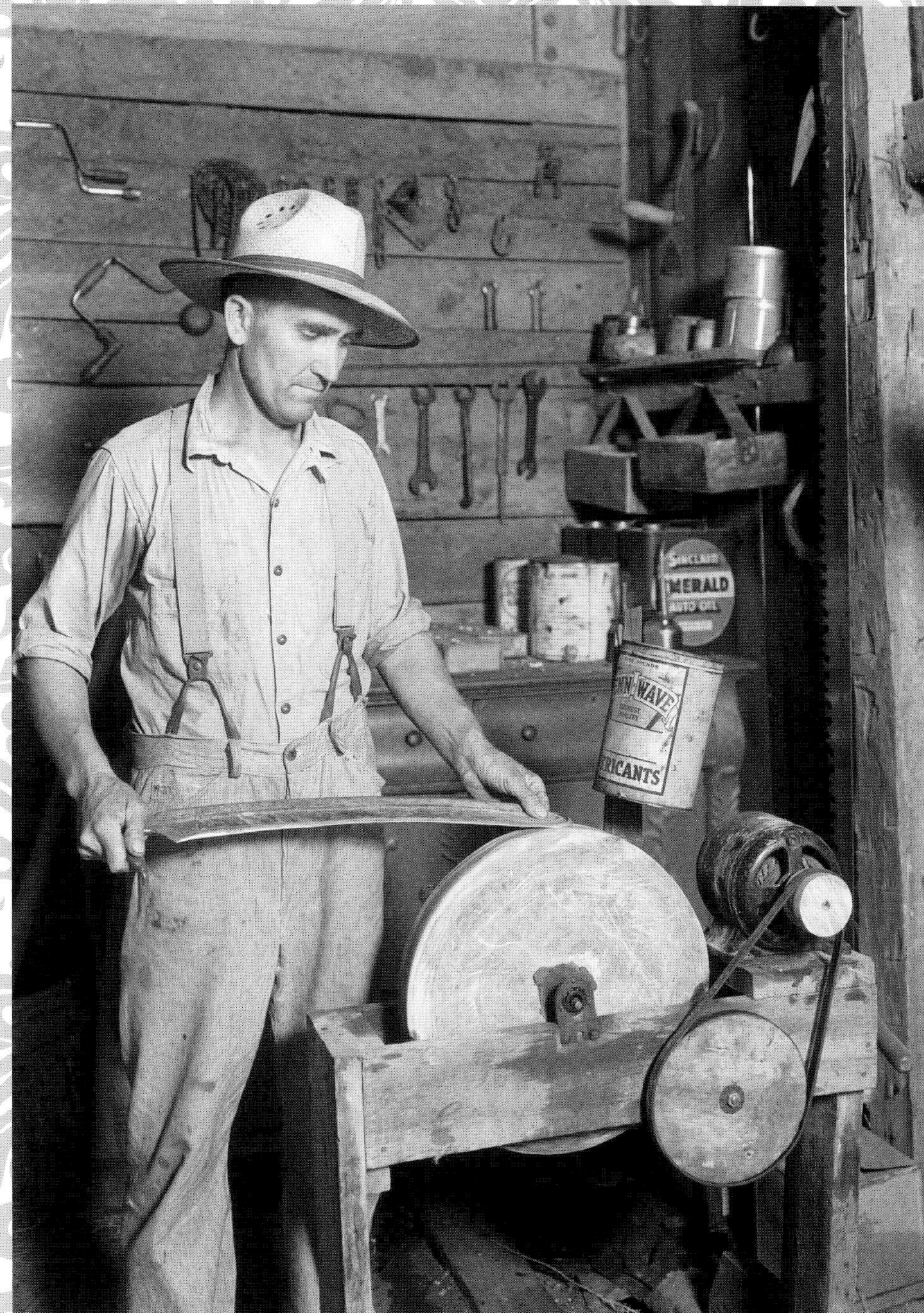

Forest Richards from Converse uses an electric-driven grindstone. (Grant and Miami Counties, 1934)

A Drive through the Countryside

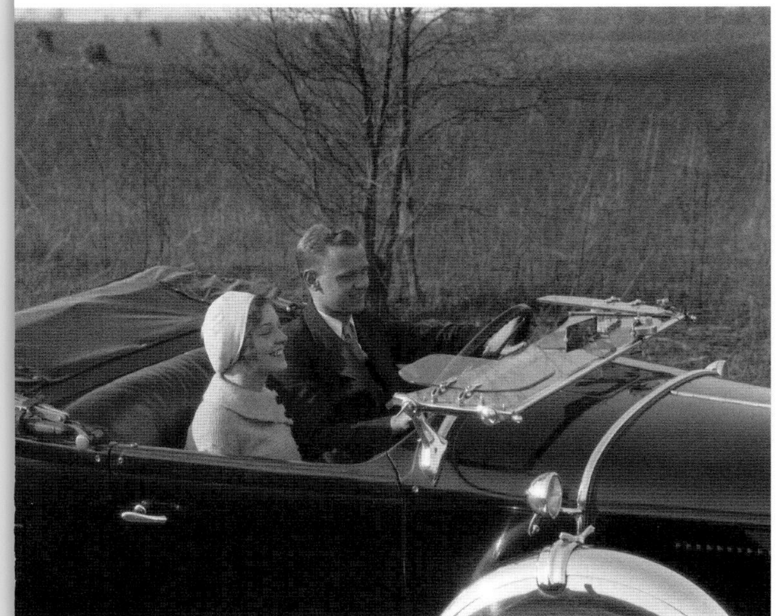

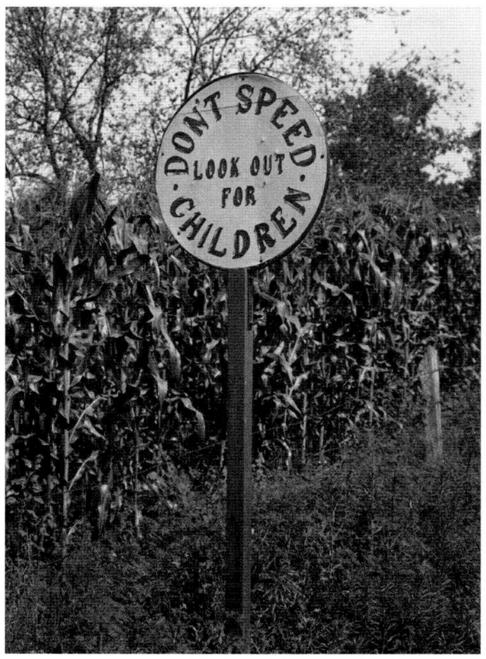

(Top left) Priscilla Ramsay of West Lafayette and John Dils of Indianapolis enjoy a spring drive in a roadster. (Tippecanoe County, 1931)

(Bottom left) A traffic sign in Brookston warns motorists, "Don't Speed. Look Out for Children." (White County, 1915)

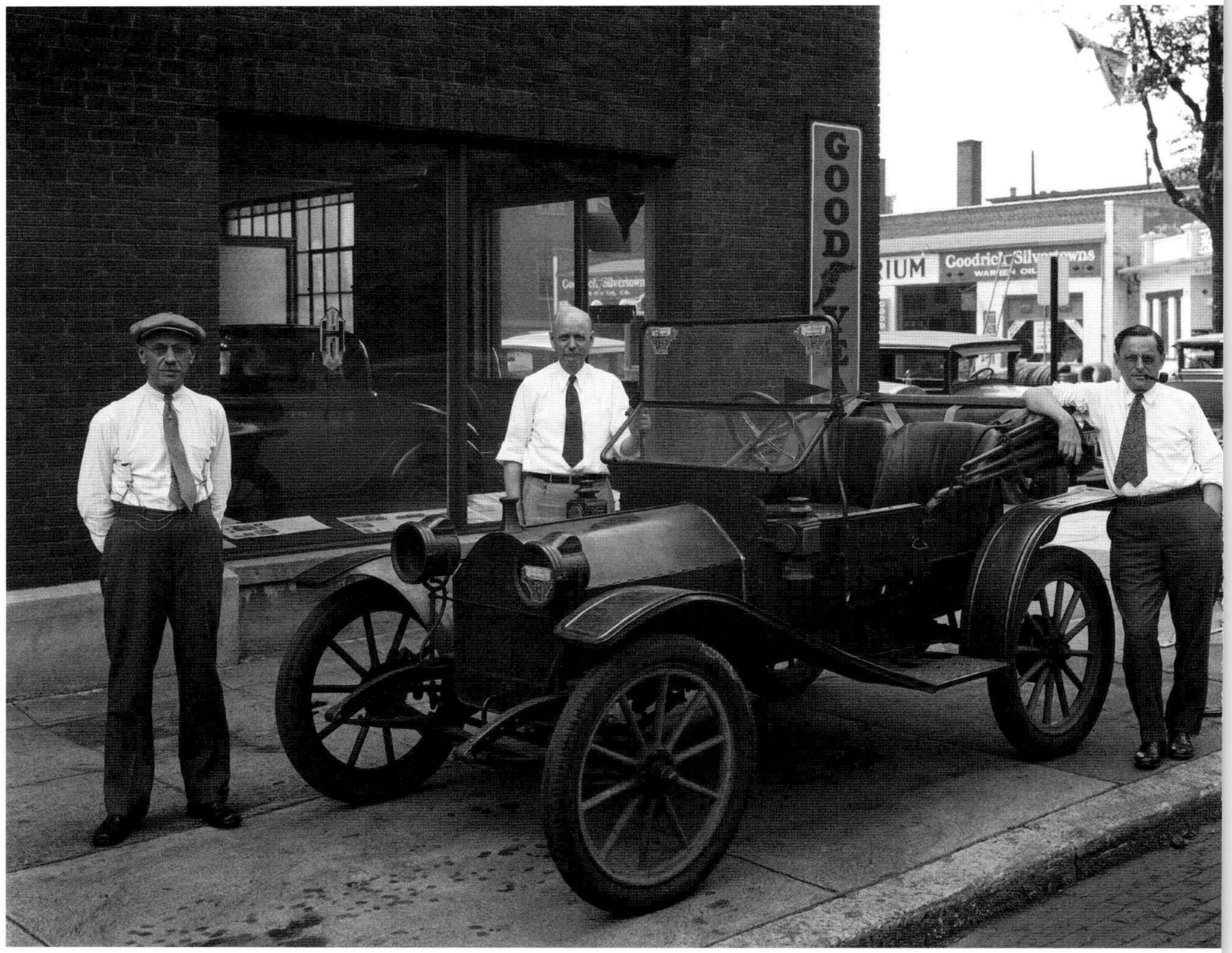

A 1908 Hupp 4 is photographed at a dealership in Warren, Ohio. (Trumbull County, OH; n.d.)

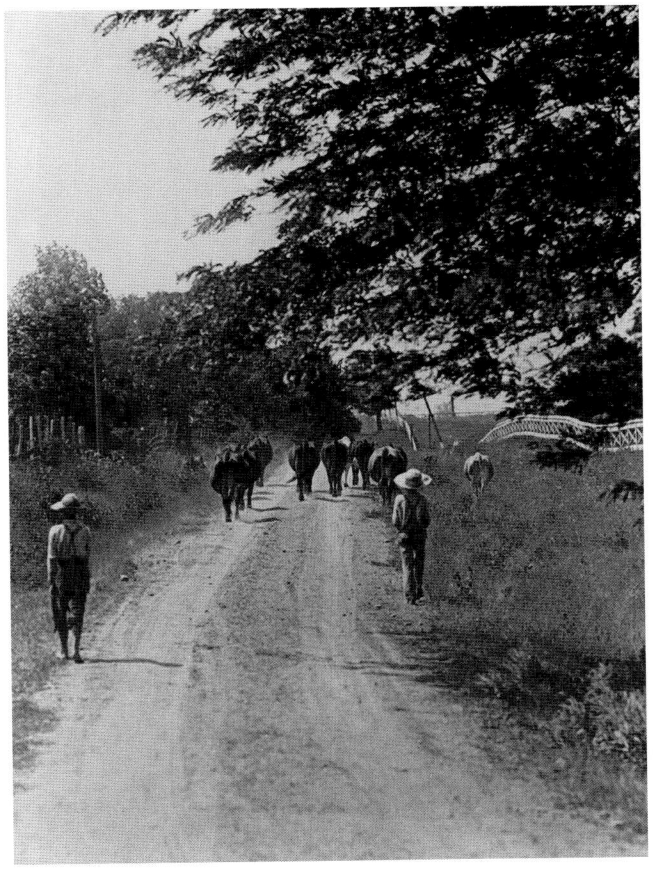

In 1913, Allen photographed these young boys driving cattle down a country road. (Tippecanoe County, 1913)

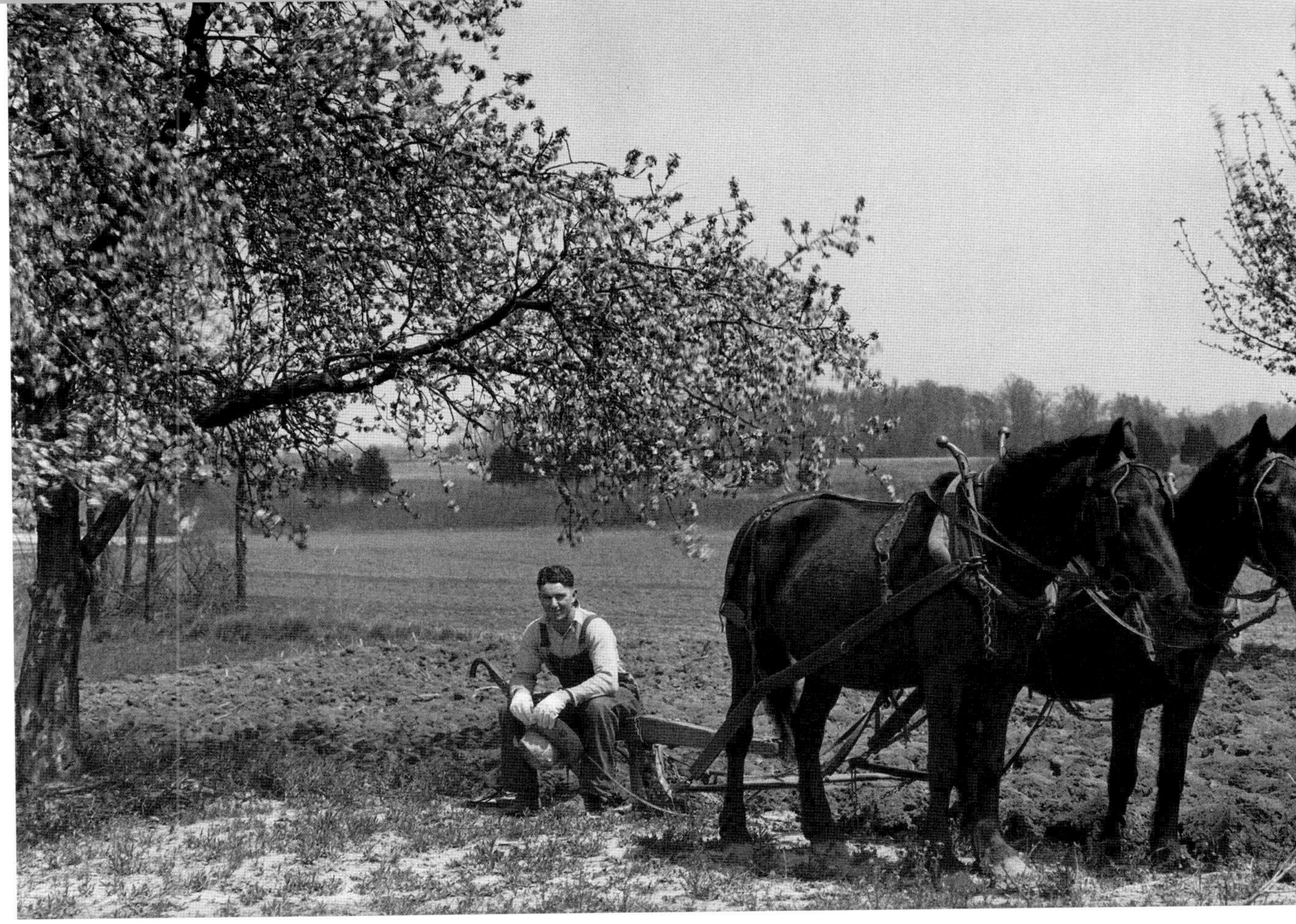

Leland Boxton from Blocher takes a break from plowing under a blossoming apple tree. (Scott County, 1933)

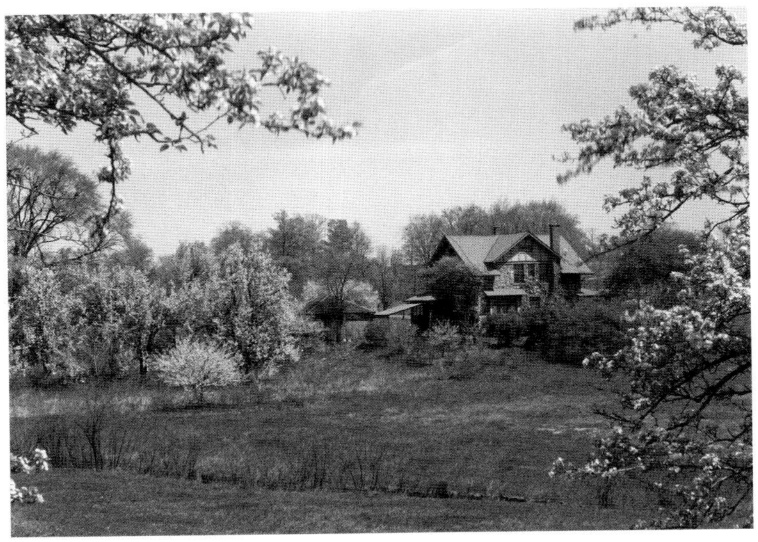

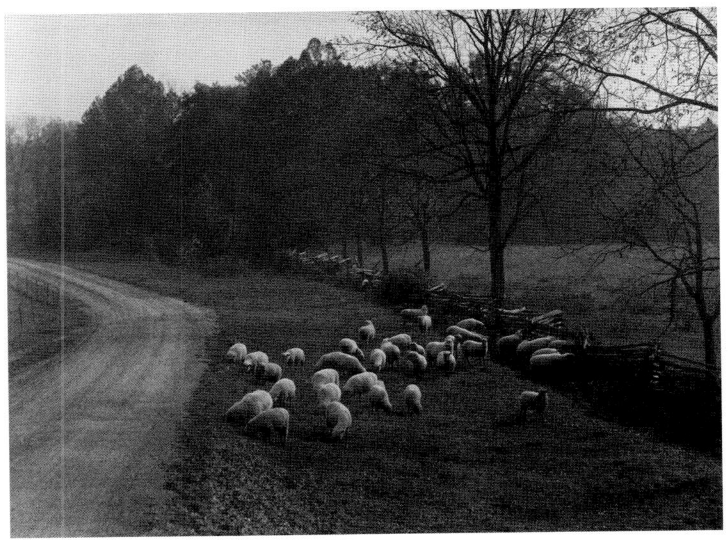

(Bottom left) The West Lafayette home of Harry G. Leslie, governor of Indiana from 1929 to 1933 (Tippecanoe County, 1931)

(Bottom right) While driving near Harmony, North Carolina, Allen stopped to take this photograph of R. H. Kennedy's sheep, which were lazily grazing along a split-rail fence. (Iredell County, NC; 1927)

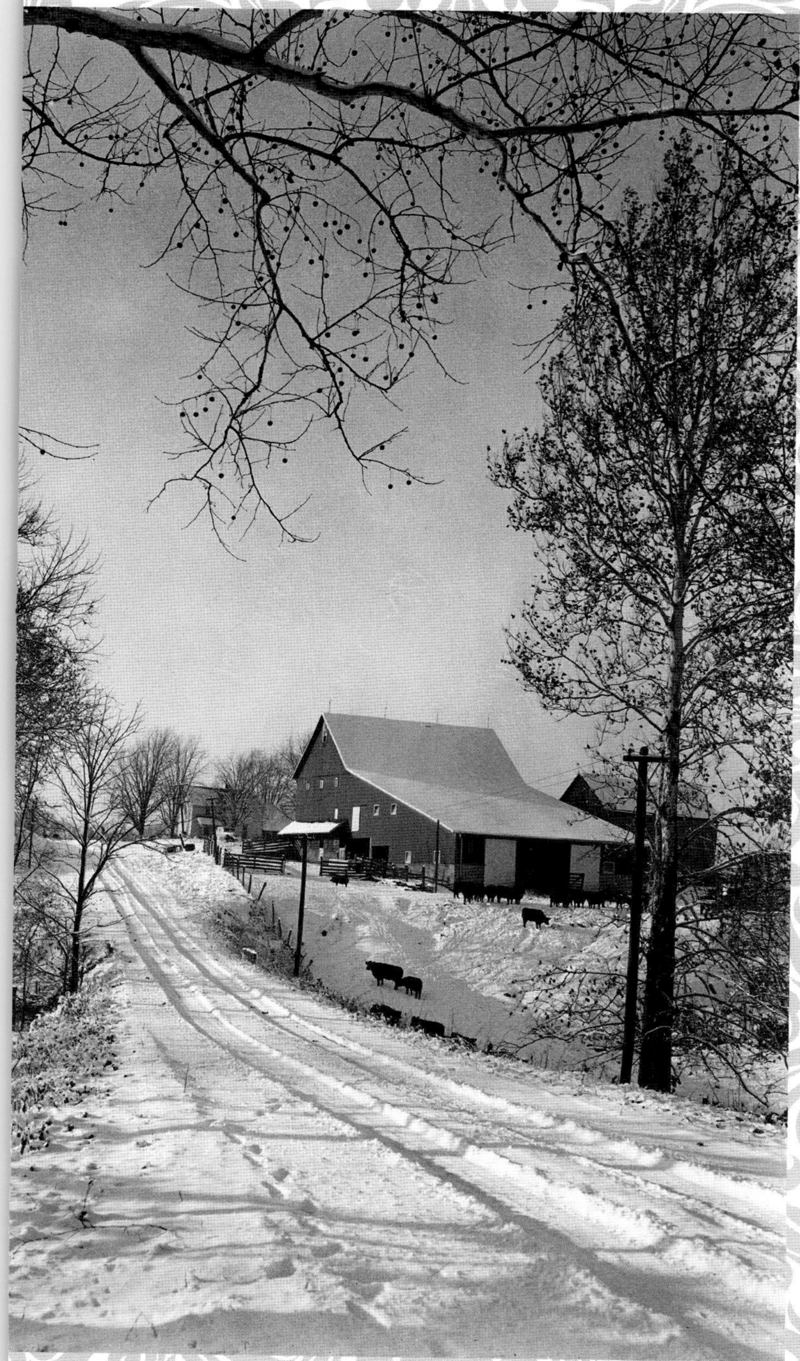

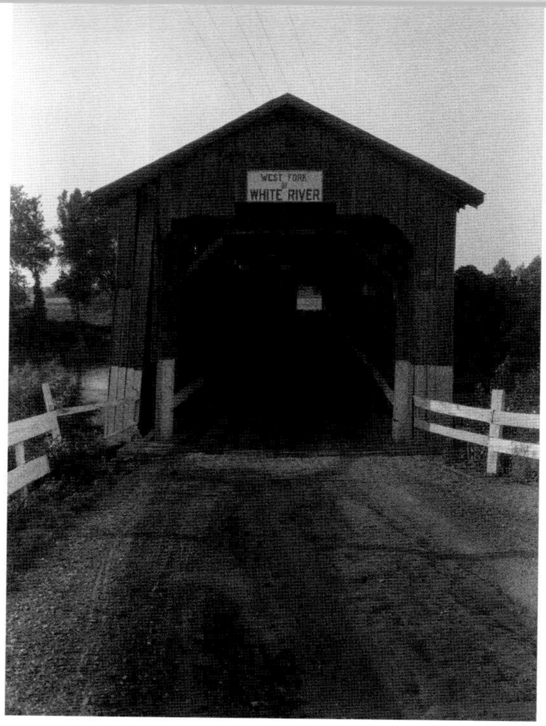

(Left) A winter scene along a road that passed in front of the Harshman Farm (Tippecanoe County, 1947)

(Top right) A covered bridge over the West Fork of the White River near Newberry (Greene County, 1929)

(Bottom right) Cataract Falls on the Eel River (Owen County, 1927)

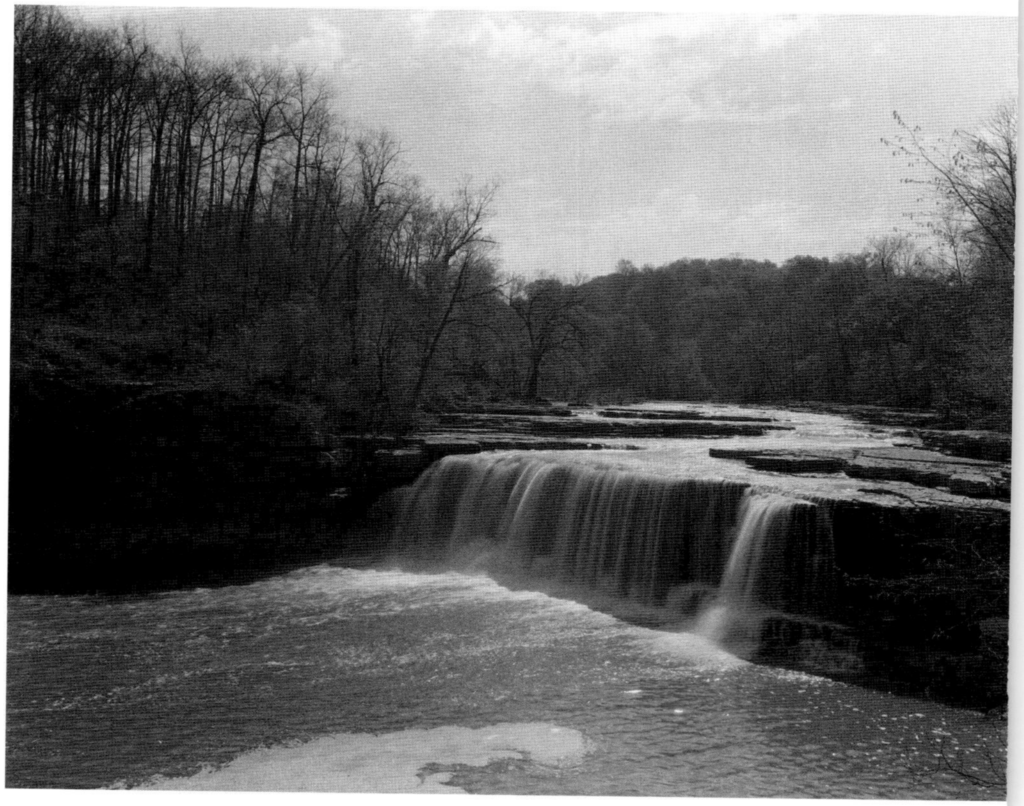

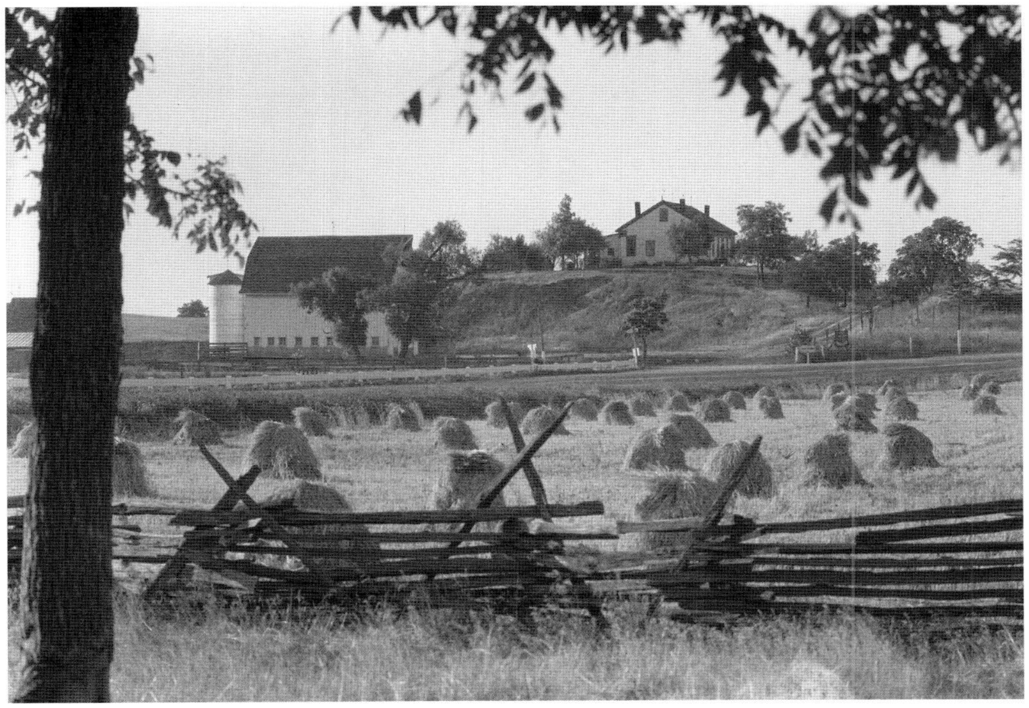

The Albert Ball farm near Muncie during fall harvest (Delaware County, 1933)

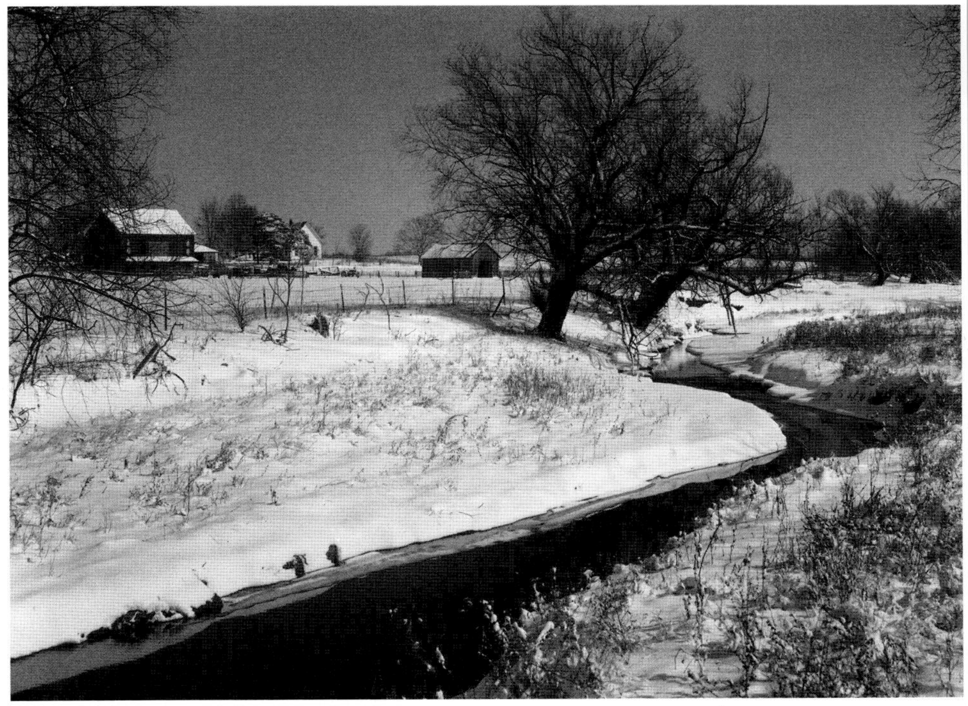

A winter scene near a creek in Rossville includes a farmstead in the distance. (Clinton County, 1963)

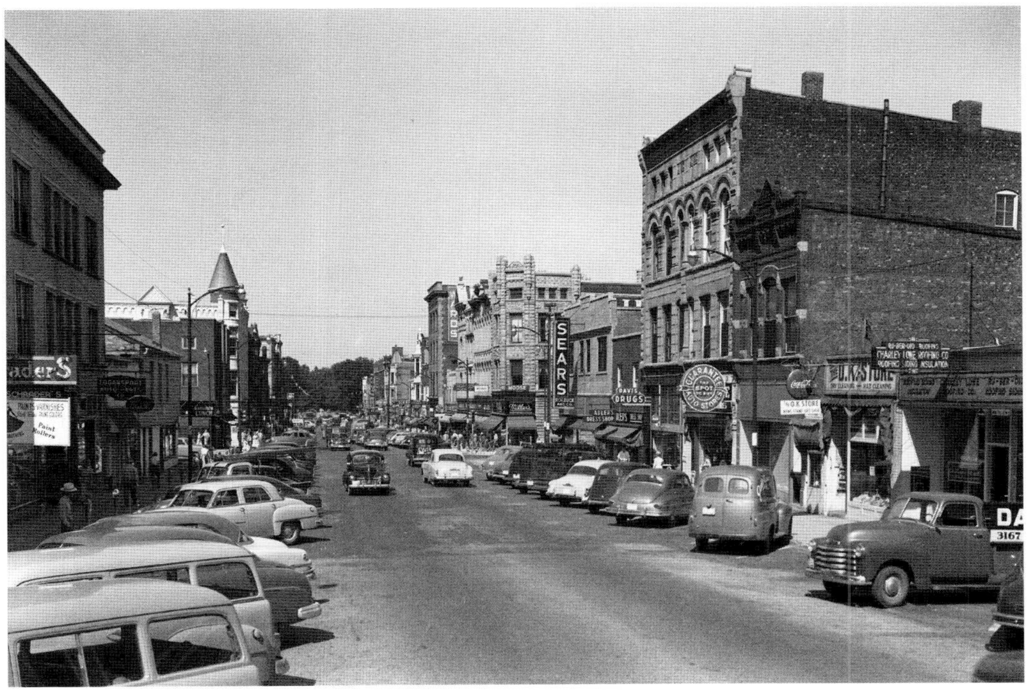

A view of downtown Logansport on a July day (Cass County, 1952)

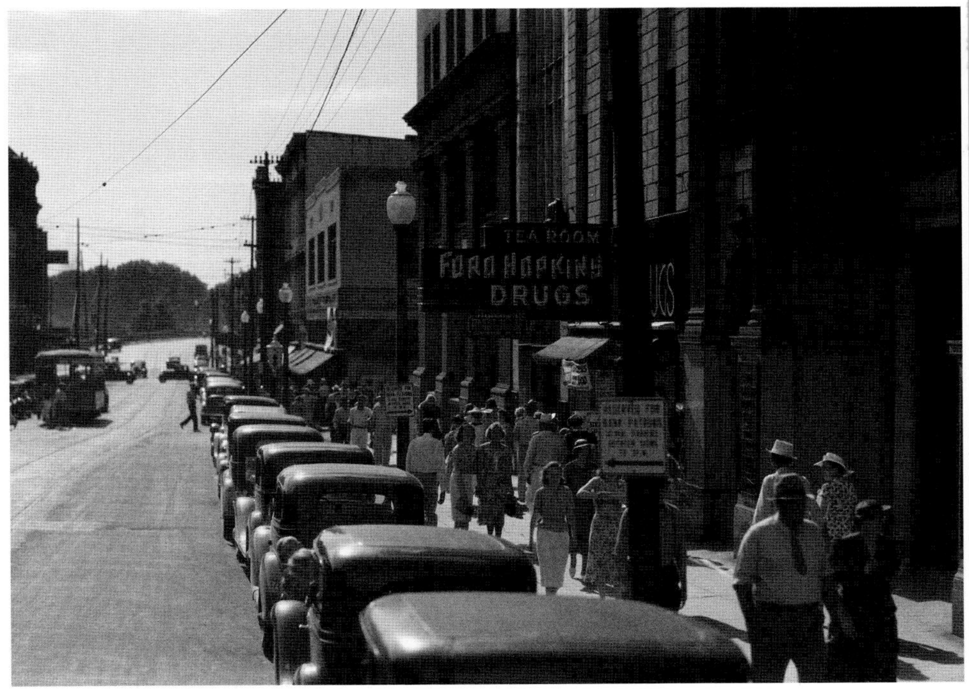

Allen took this photograph on August 6, 1938, capturing a busy Saturday shopping day in downtown Lafayette. (Tippecanoe County)

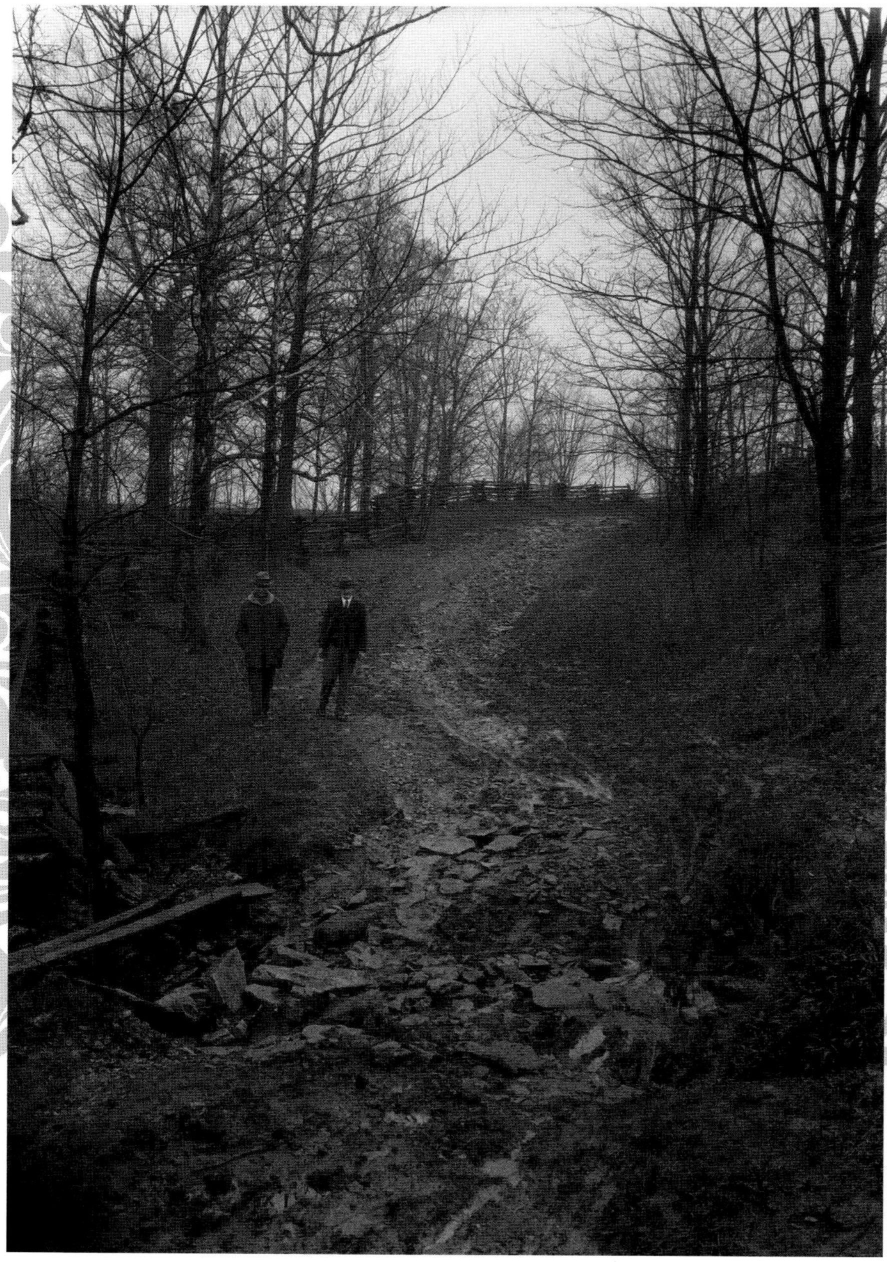

A road in poor condition with a badly washed-out ditch (Monroe County, 1927)

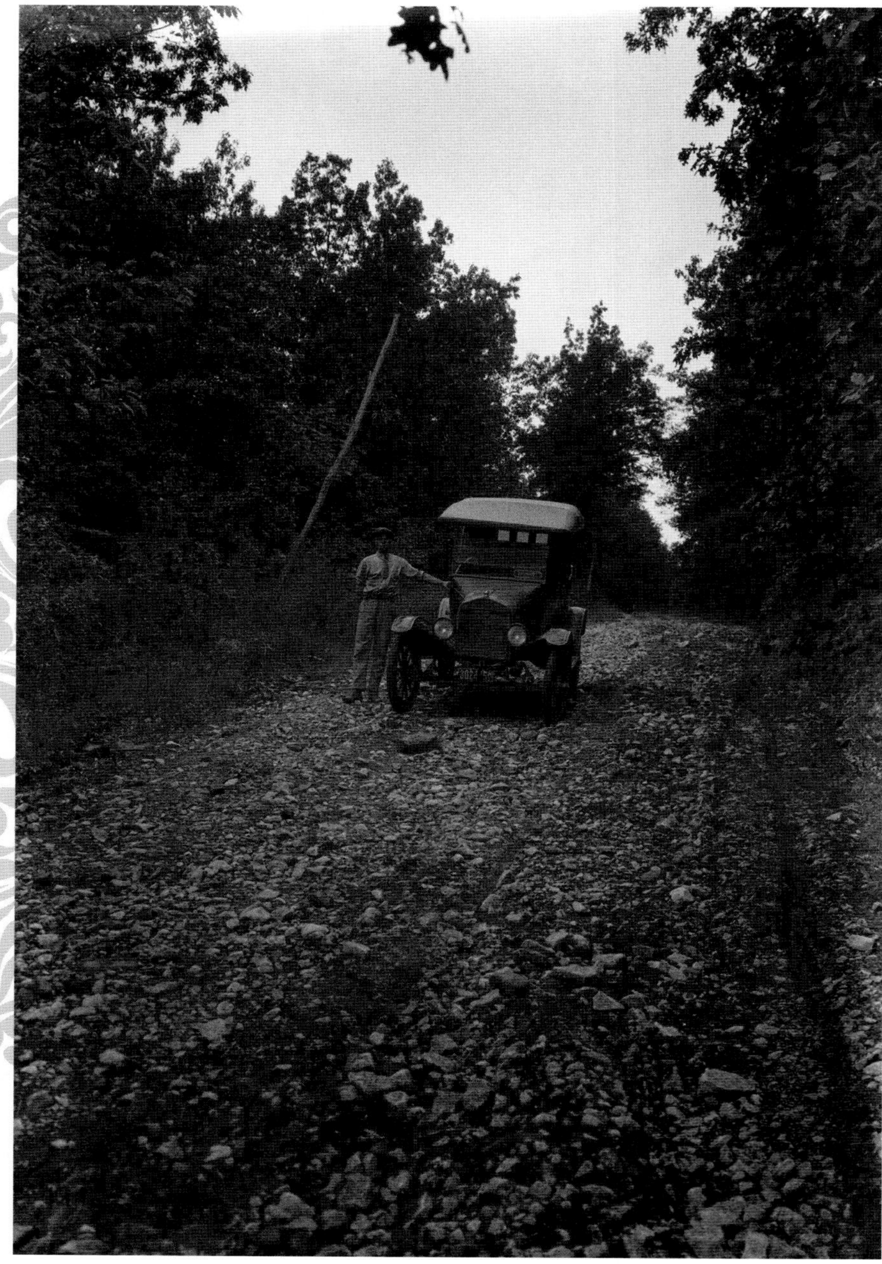

A Ford Model T on a country road covered with large stones (Crawford County, 1922)

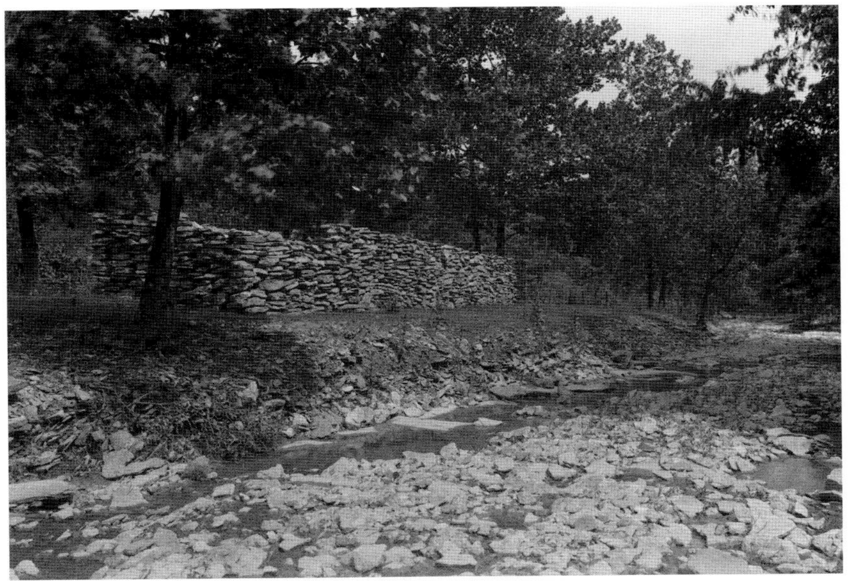

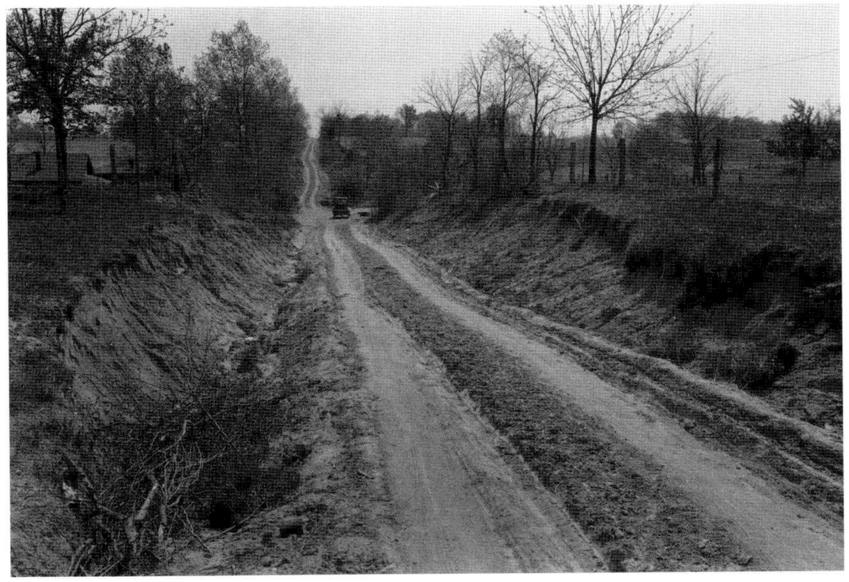

(Top left) This limestone, which was piled beside a creek, was removed from the creek bed, possibly to be used in road building or for agricultural purposes. (Ohio County, 1923)

(Top right) Allen captured this view of a typical Daviess County road in 1928.

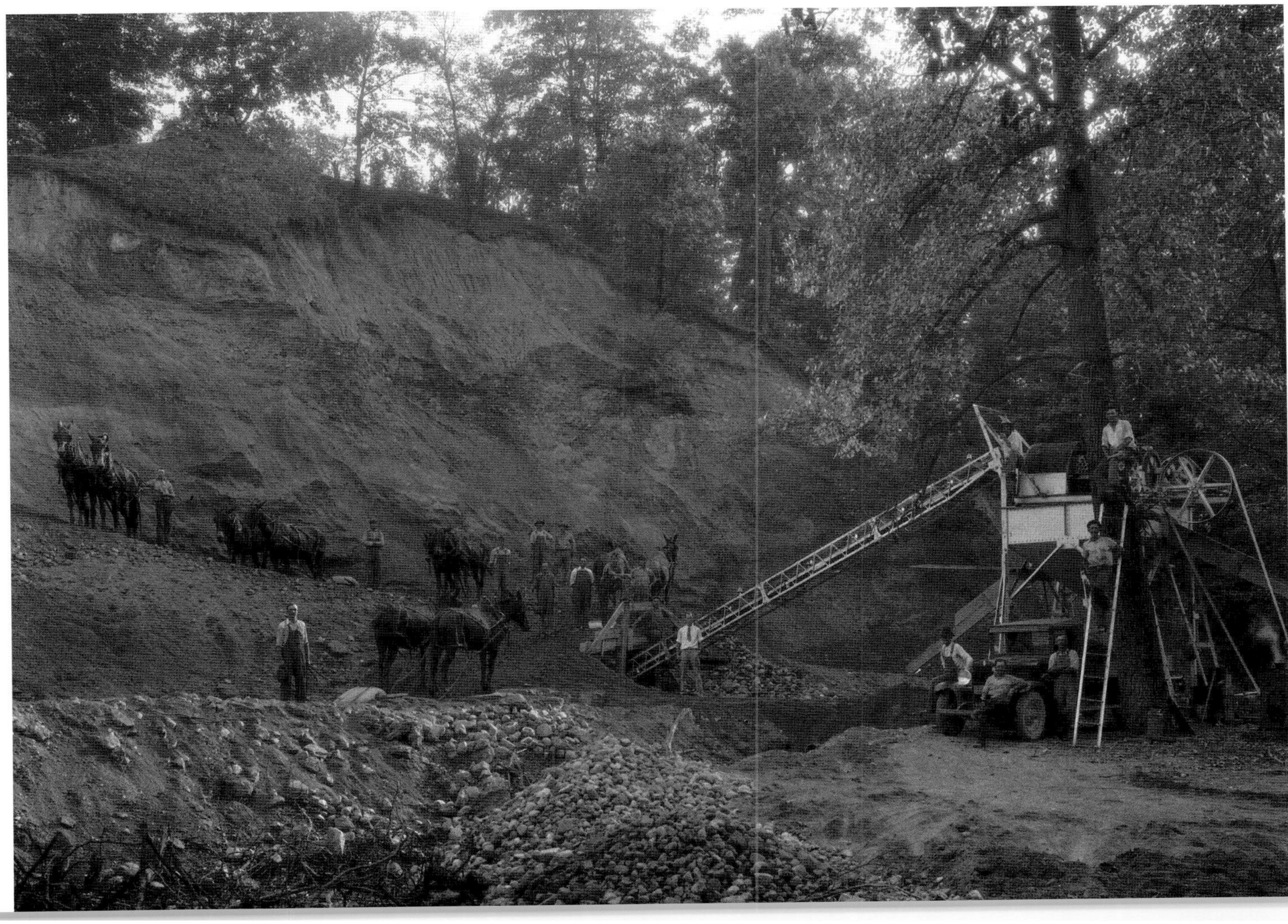

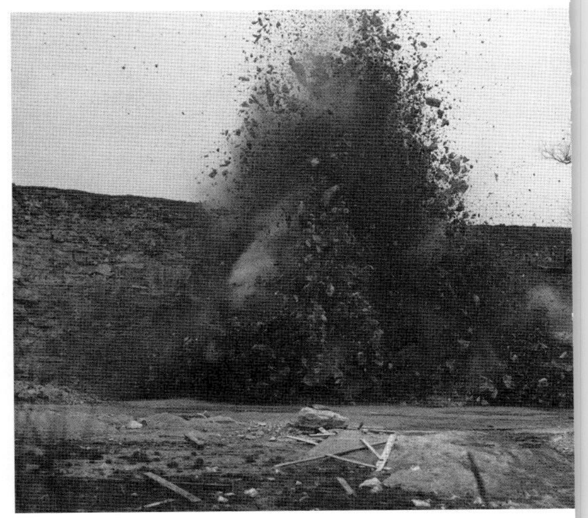

(Bottom left) A Williamsport gravel pit operated by David Stigler (Warren County, 1930)

(Bottom right) At the Monroe Crushed Stone Company quarry near Bloomington, workers blast limestone for road construction. The charge was placed in 31 drill holes, with 14 sticks of dynamite used per hole. Allen recorded in his notes, "Approximately 300 yards of stone are in the air at one time, and large stones were thrown at least 100 feet beyond where the cameraman stood in taking the picture." (Monroe County, 1927)

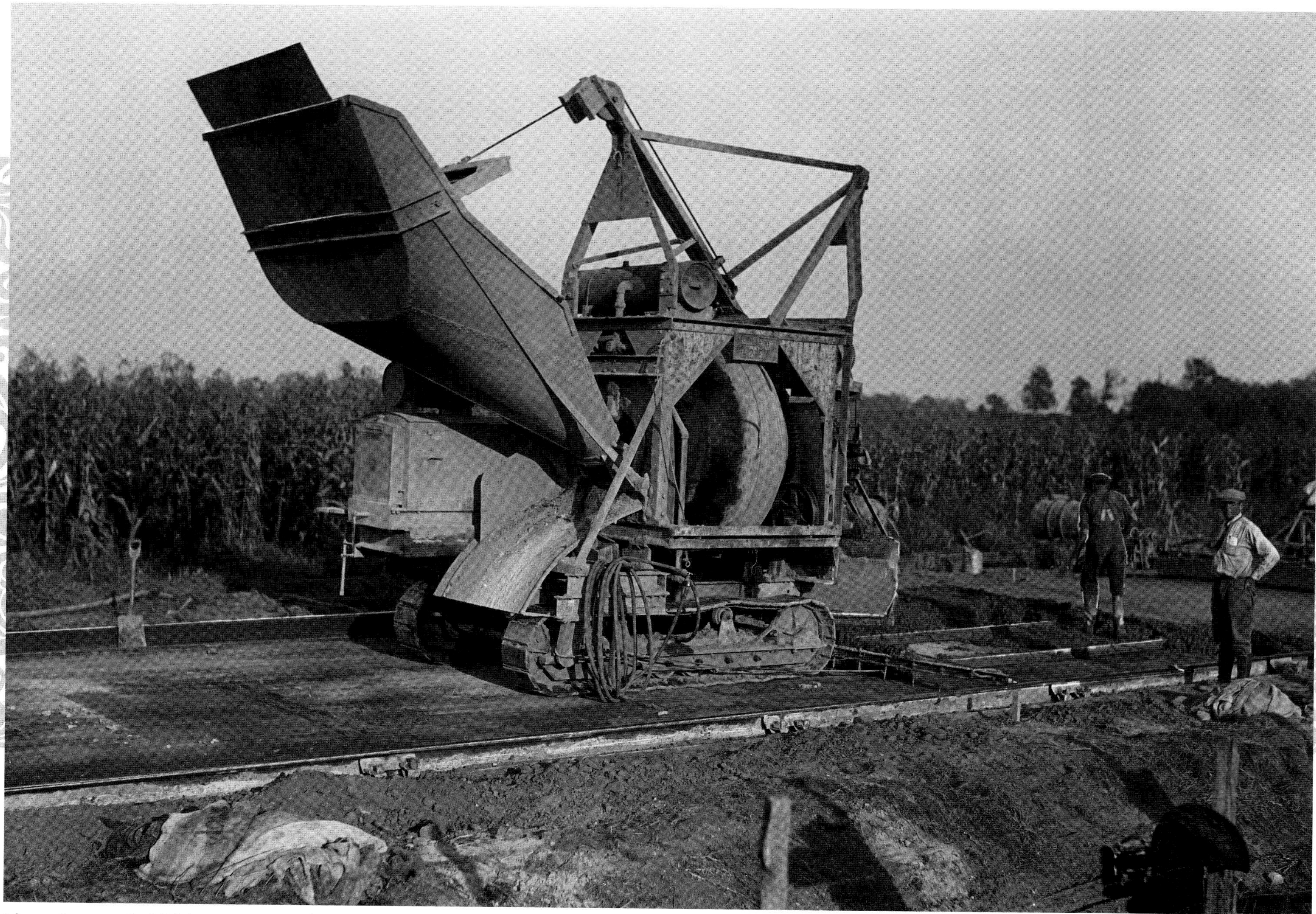

A large mixer owned by S. S. Palma pours concrete for a new road. (Tippecanoe County, 1924)

(Top left) This man breaks rocks with a hammer to fill holes in the road. (Switzerland County, 1922)

(Bottom left) While building a road, S. S. Palma uses a turntable to turn the trucks that haul in the gravel, sand, and cement. (Tippecanoe County, 1924)

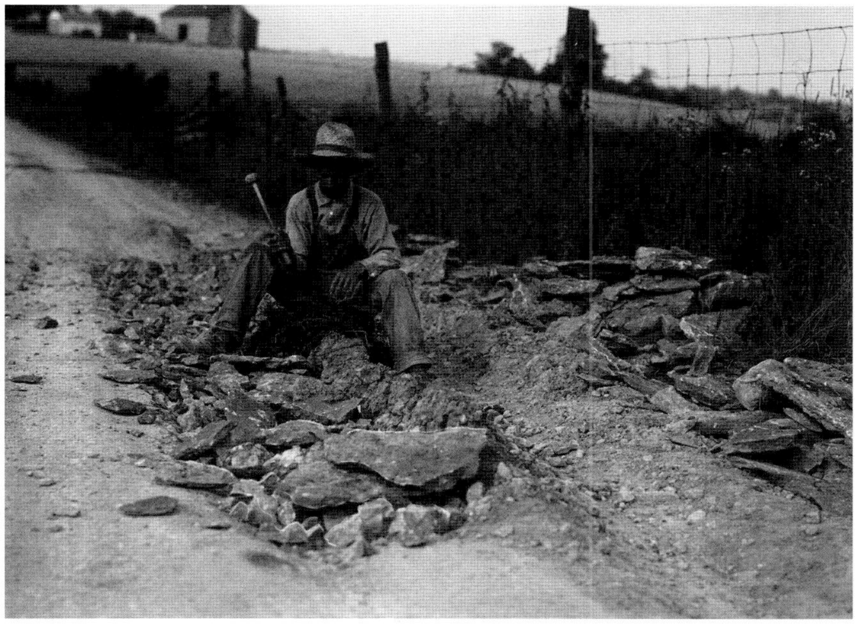

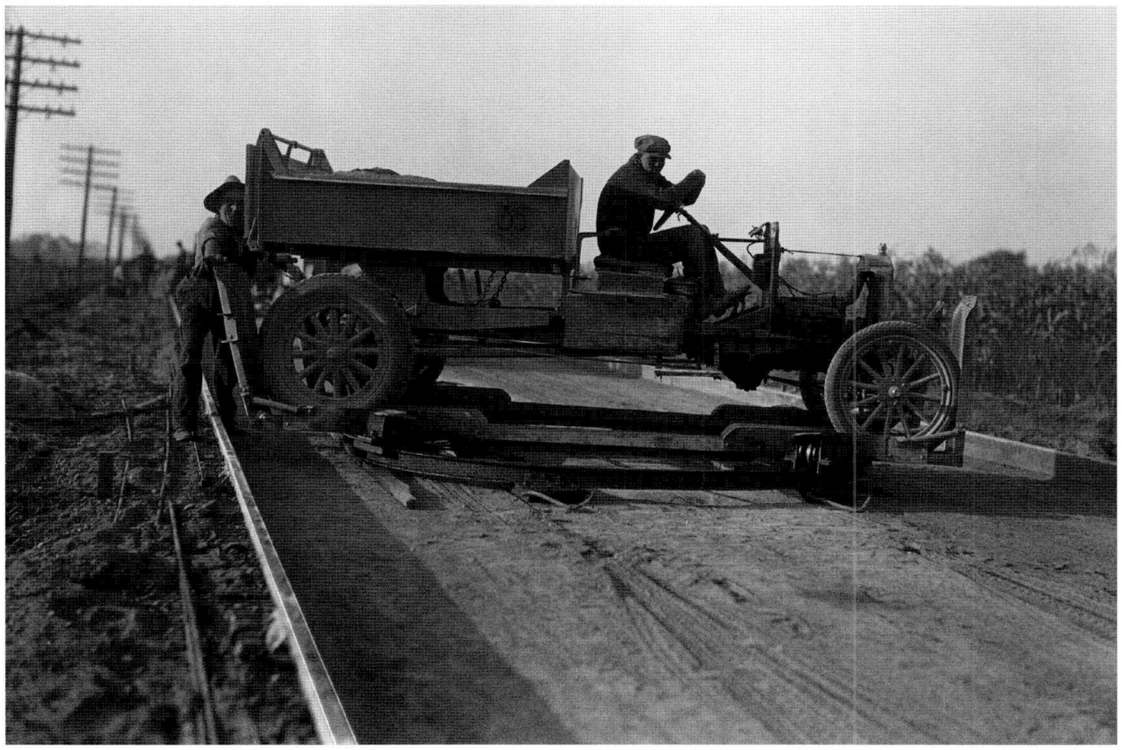

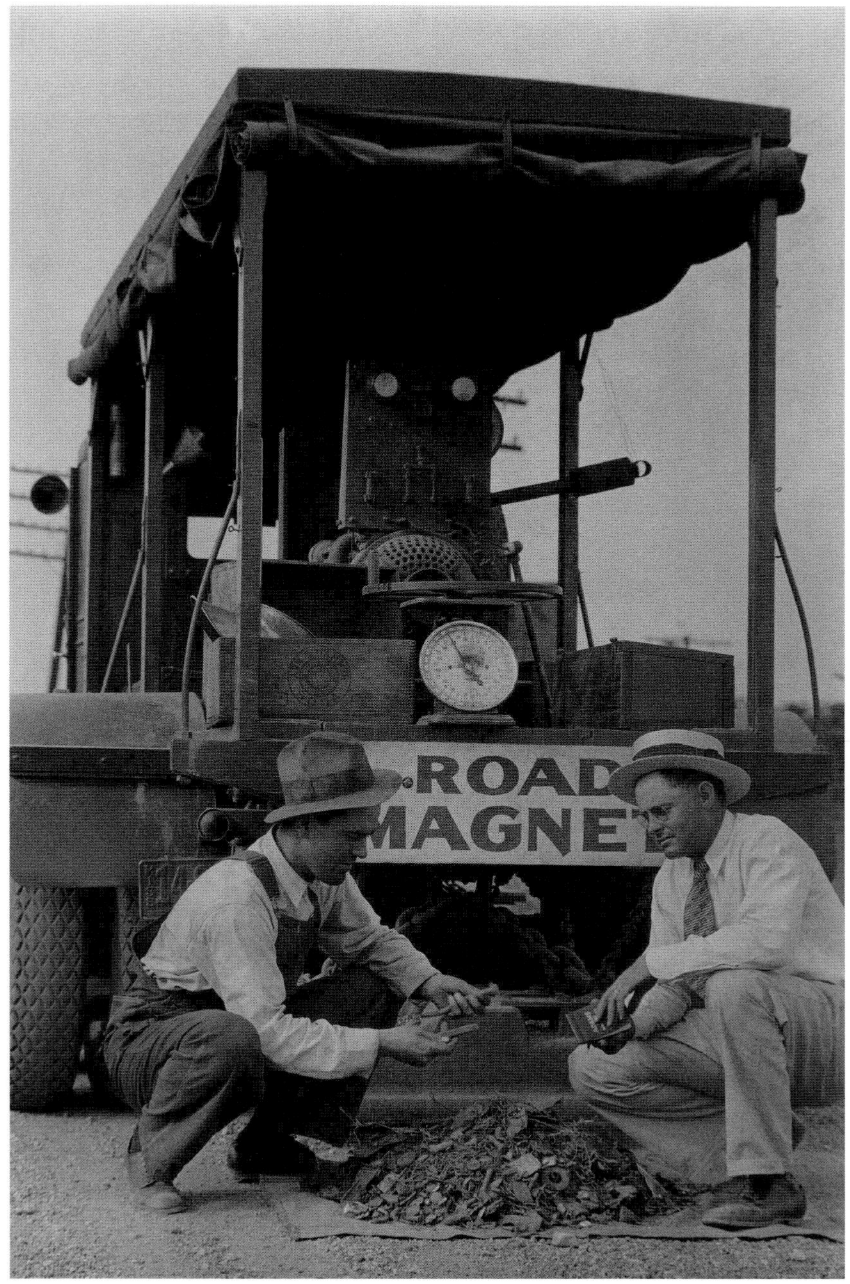

(Above) W. H. Wicker (left), a truck driver from Jackson, Mississippi, and Dan R. Ferguson, superintendent of maintenance for the Mississippi State Highway Commission of Batesville, Mississippi, examine 150 pounds of nails, bolts, springs, horseshoes, and screws picked up in three miles between Clarksdale, Mississippi, and the Tennessee state line. (ca. 1930s)

Barns and Silos of the Midwest

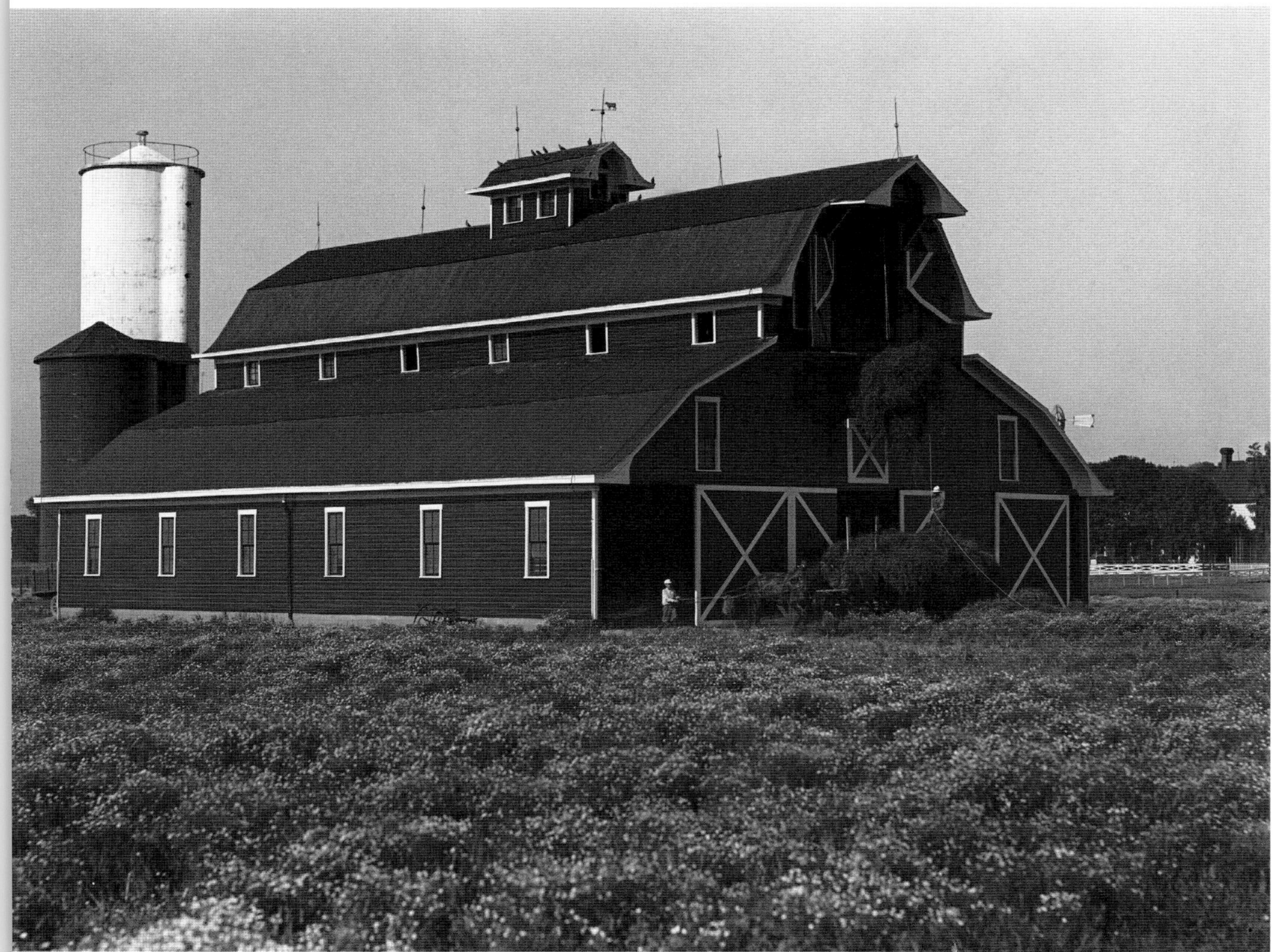

Clover hay is lifted into the loft of a beef cattle feeding barn at the Julius Berlovitz farm near Lafayette. Tenant Ed Coleman managed the 600-acre farm. (Tippecanoe County, 1927)

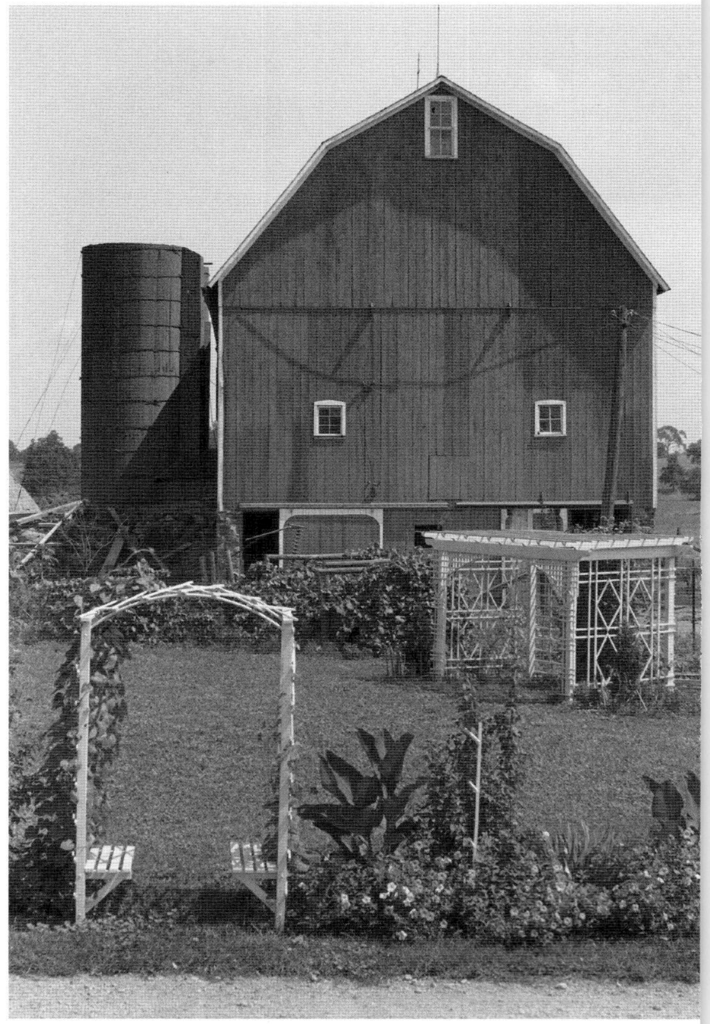

The farm of W. J. Schneider in Ann Arbor, Michigan, caught Allen's attention with its showy flowers and benches. (Washtenaw County, MI; 1933)

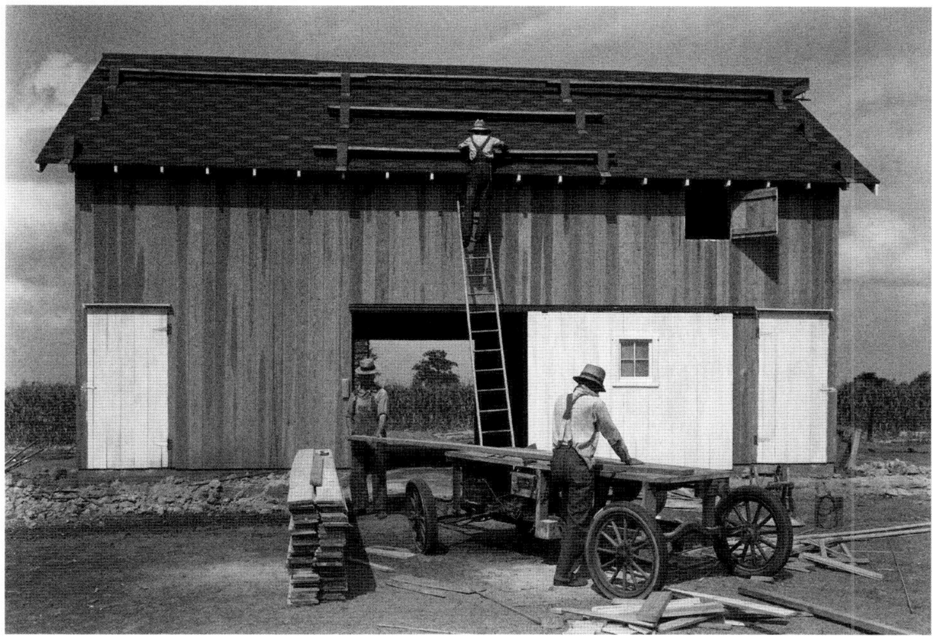

Contractors at the Stoup farm near Rushville use a homemade saw operated by an old Ford motor. The 24-by-36-foot structure cost $830 in 1931. (Rush County, 1931)

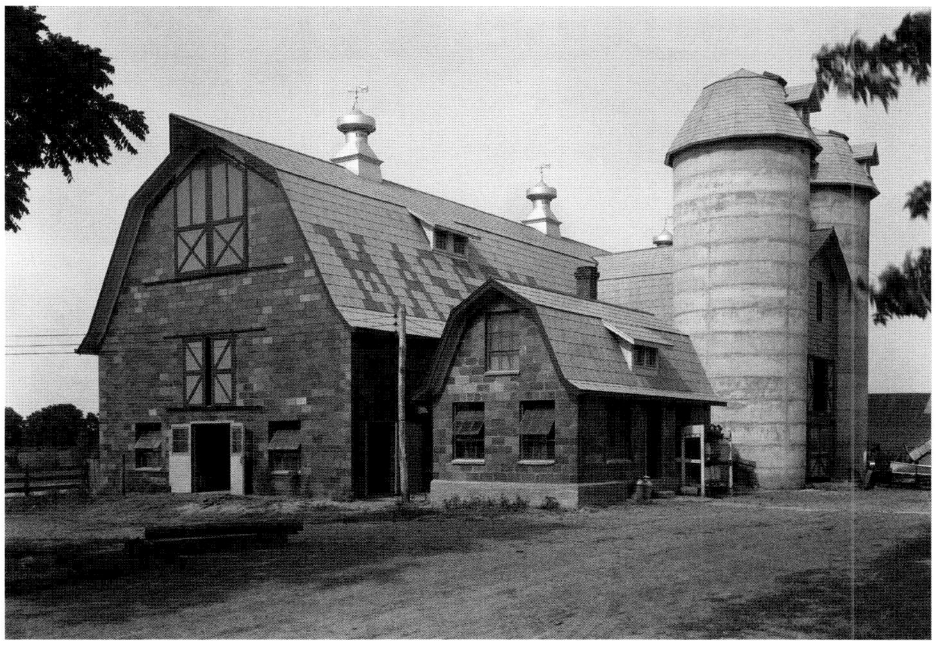

A new dairy barn in 1931 located on the farm of Massey Brothers in Jerseyville, Illinois (Jersey County, IL)

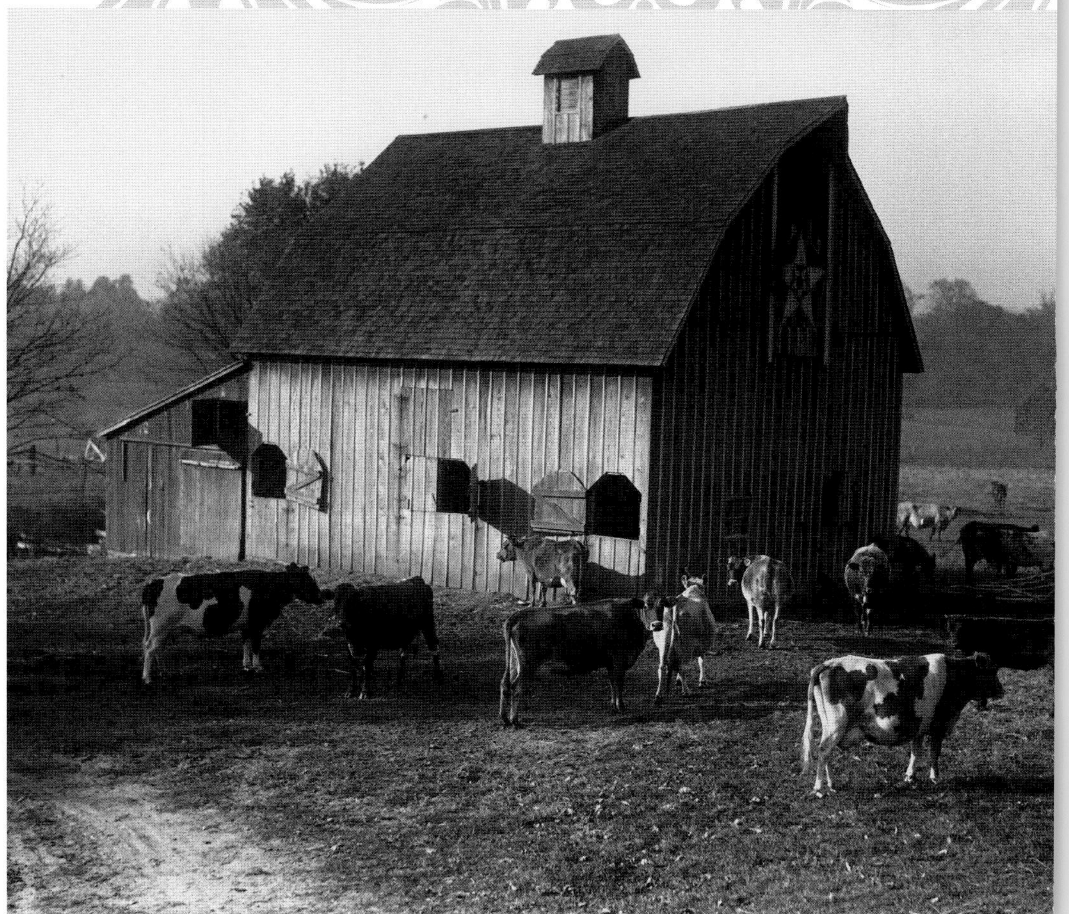

Dairy cattle surround an old barn on the farm of Fred Wichman of Lafayette. (Tippecanoe County, 1928)

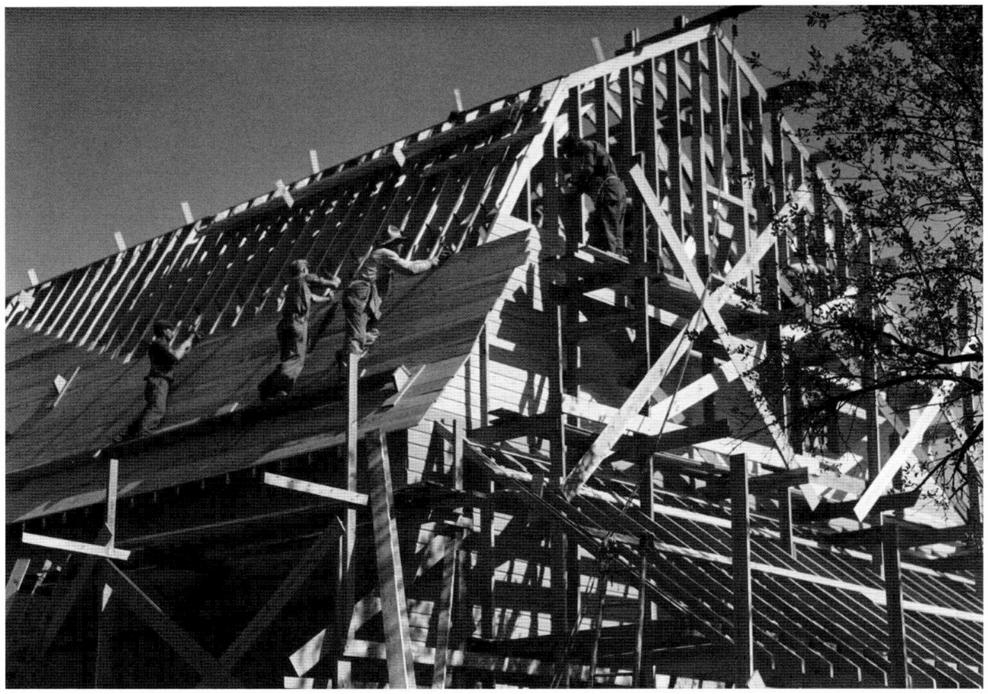

Workers build a horse barn at the Purdue University Livestock Experimental Farm. (Tippecanoe County, 1933)

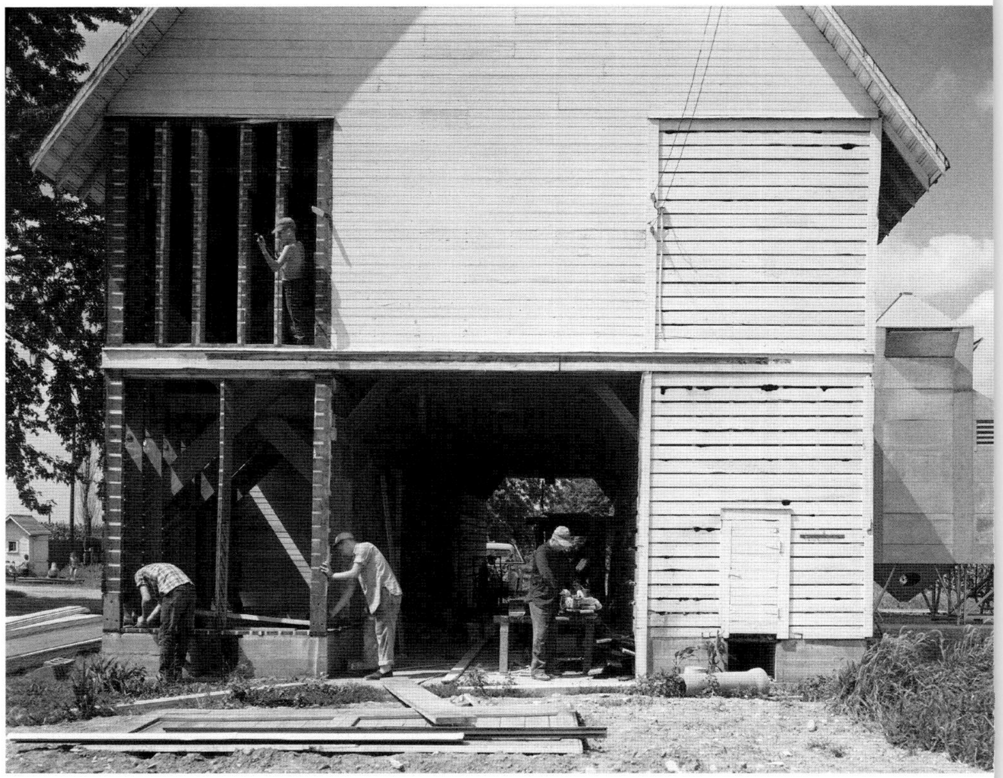

A corn crib is remodeled to hold shelled corn at the Lewis Strasburger farm in Earl Park. (Benton County, 1956)

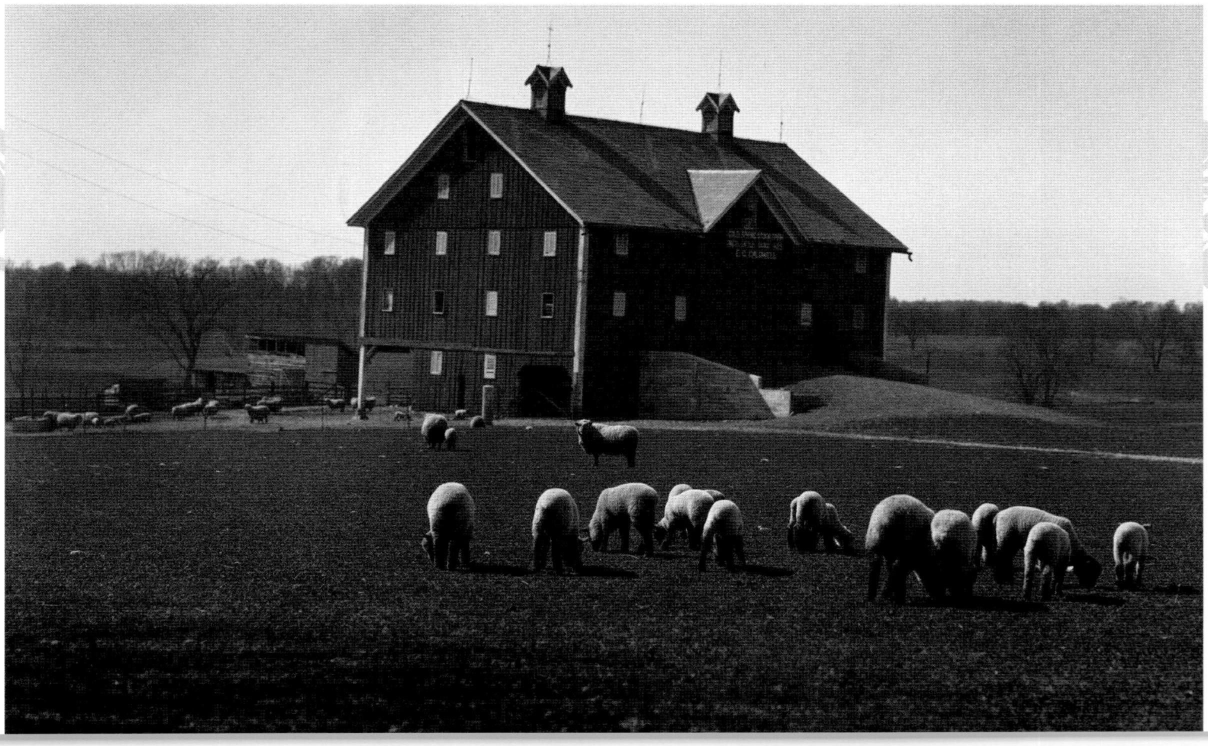

Young lambs graze in a pasture with a beautiful barn as the backdrop on the Cold Spring Stock Farm owned by E. C. Caldwell. (Fayette County, 1932)

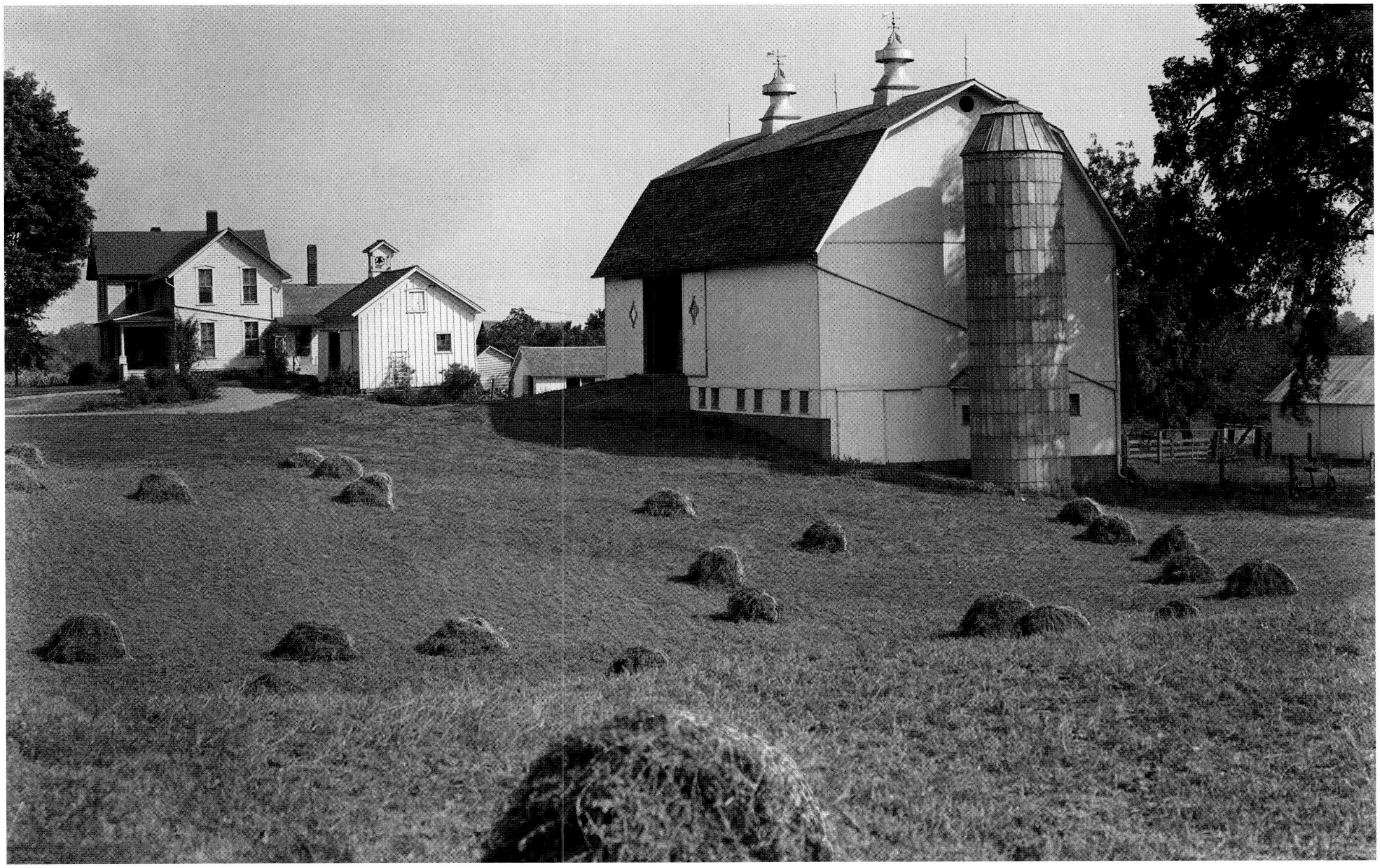

A neat and well-painted farmstead near Fenton, Michigan (Genesee County, MI; 1933)

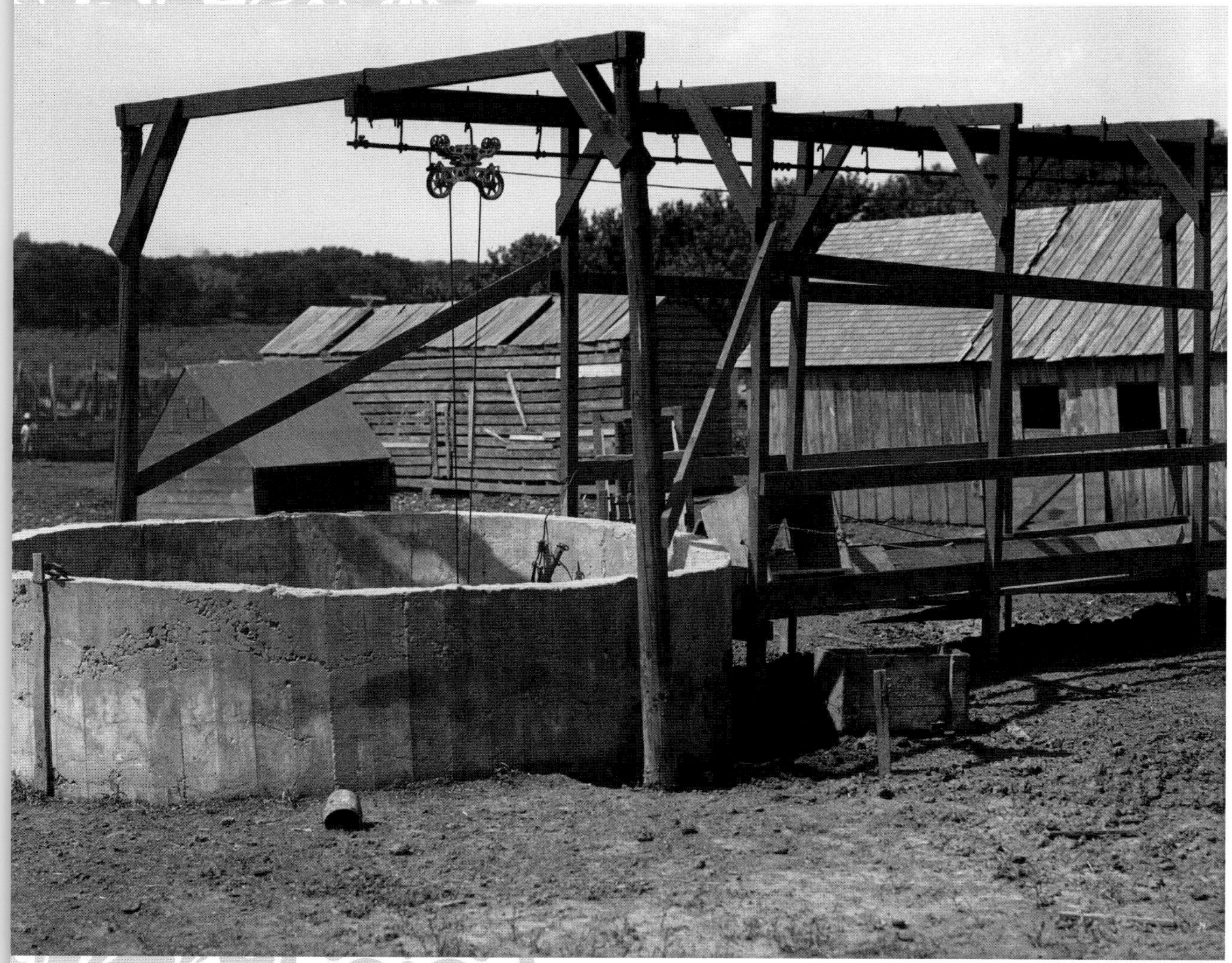

This unusual pit silo stood on the Clyde McElrath farm in Moville, Iowa. (Woodbury County, IA; 1919)

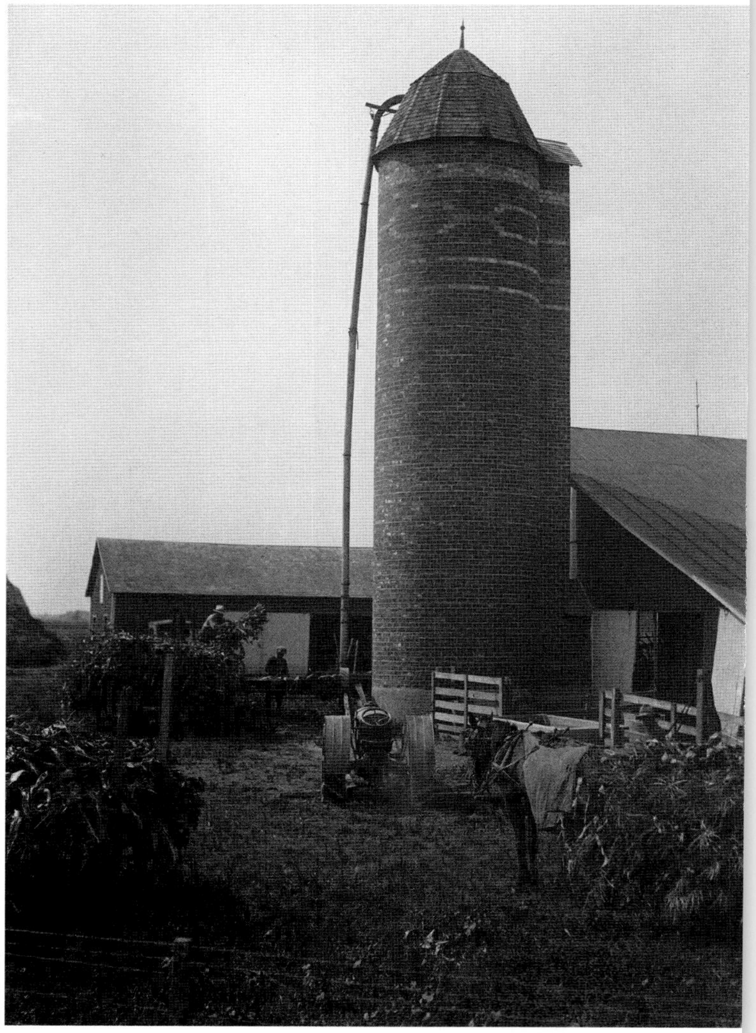

Tom Ruch, a tenant farmer on the Andrew Johnson farm in Mulberry, uses a tractor to fill a silo. (Clinton County, 1923)

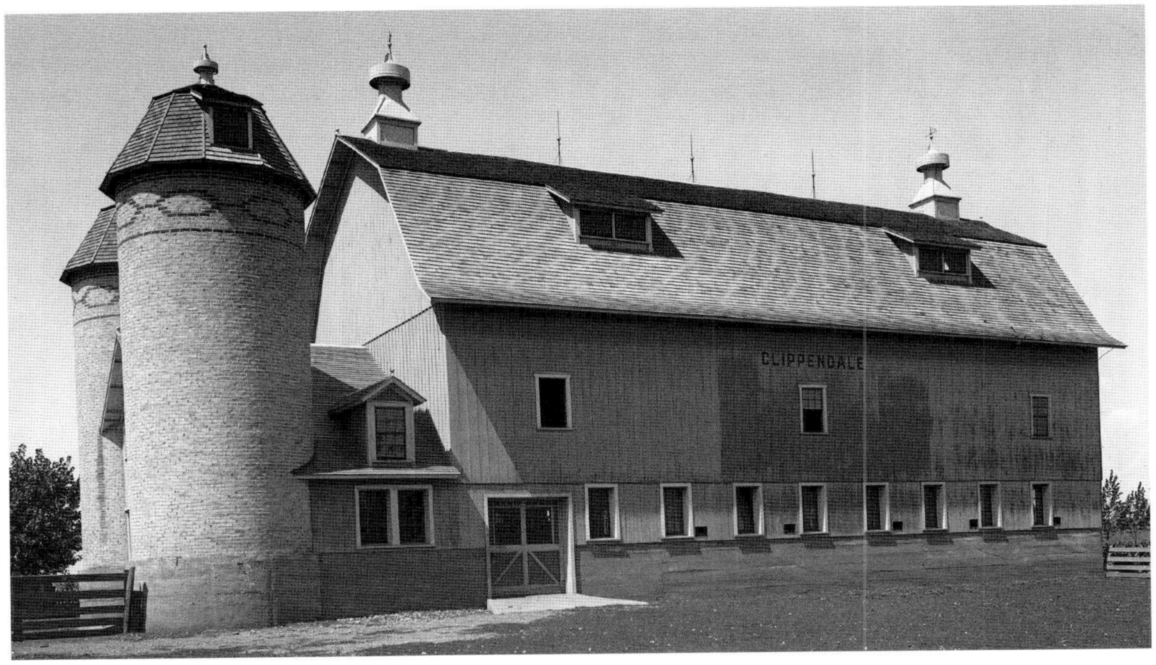

George F. Clark from Janesville, Wisconsin, owned this barn and the brick silos in 1919. (Rock County, WI)

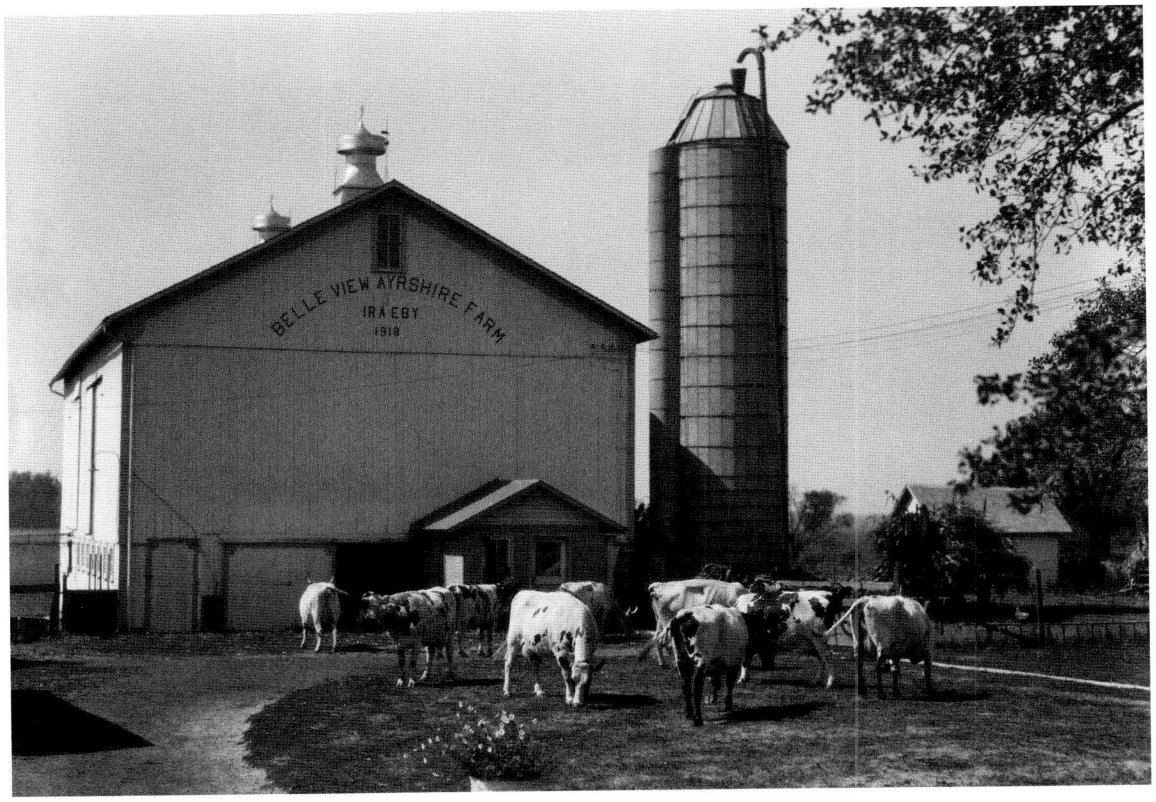

A herd of purebred Ayrshire cattle stands in front of the dairy barn of Ira Eby. (Elkhart County, 1930)

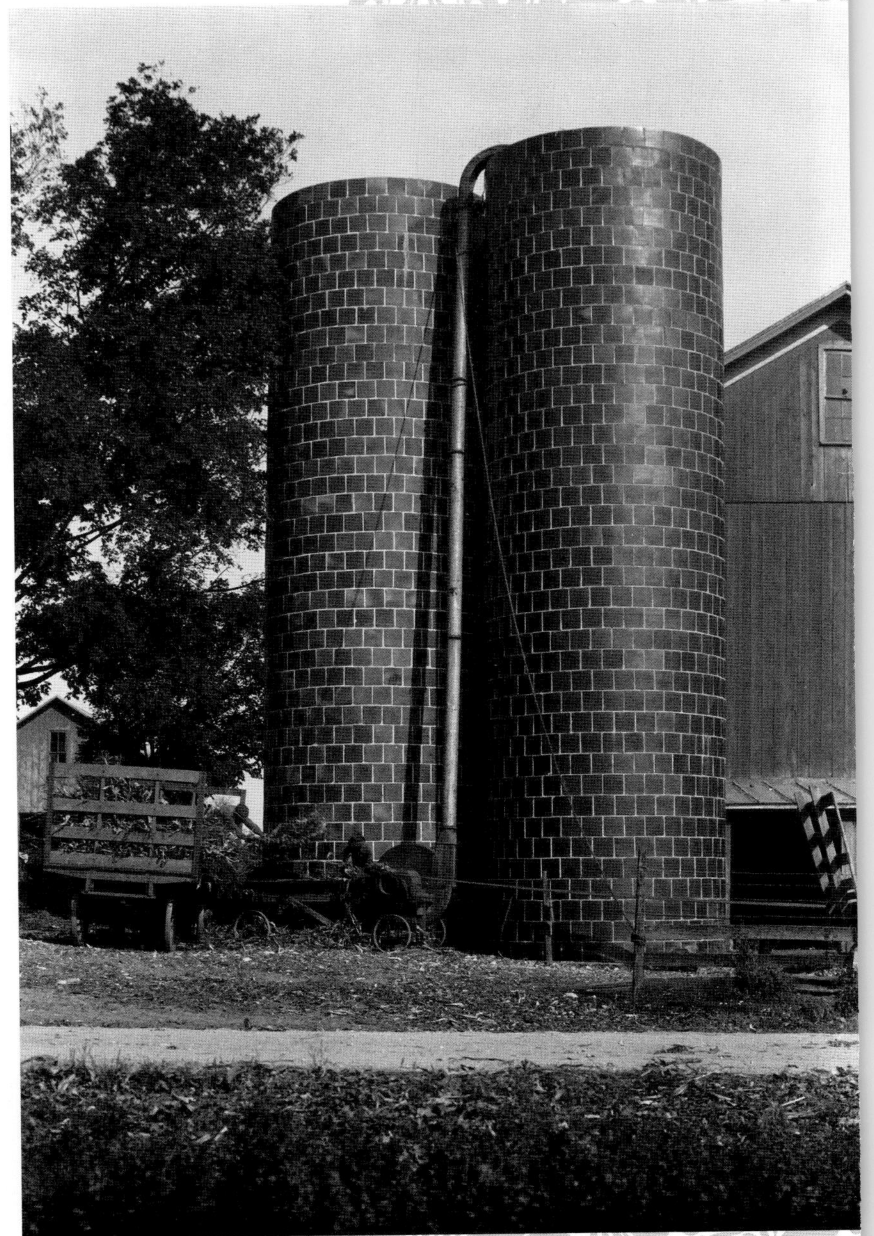

Filling silos on the farm of Glen Gilbert (Steuben County, 1929)

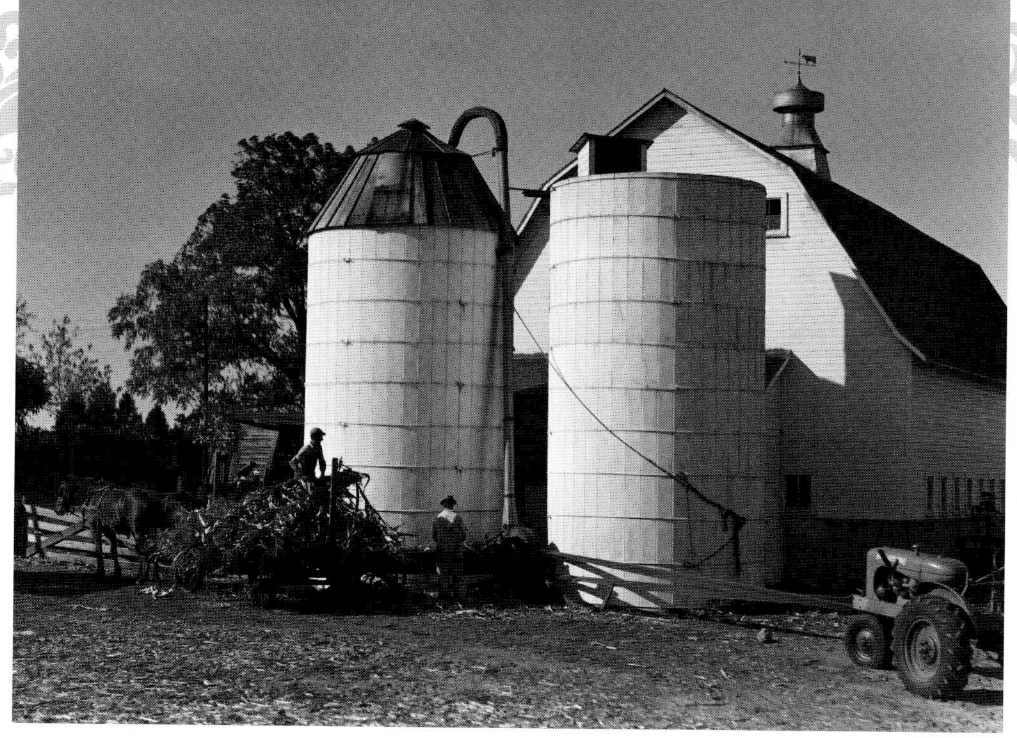

Farmers fill a silo on the Albert and George Waite farm near North Judson. (Starke County, 1944)

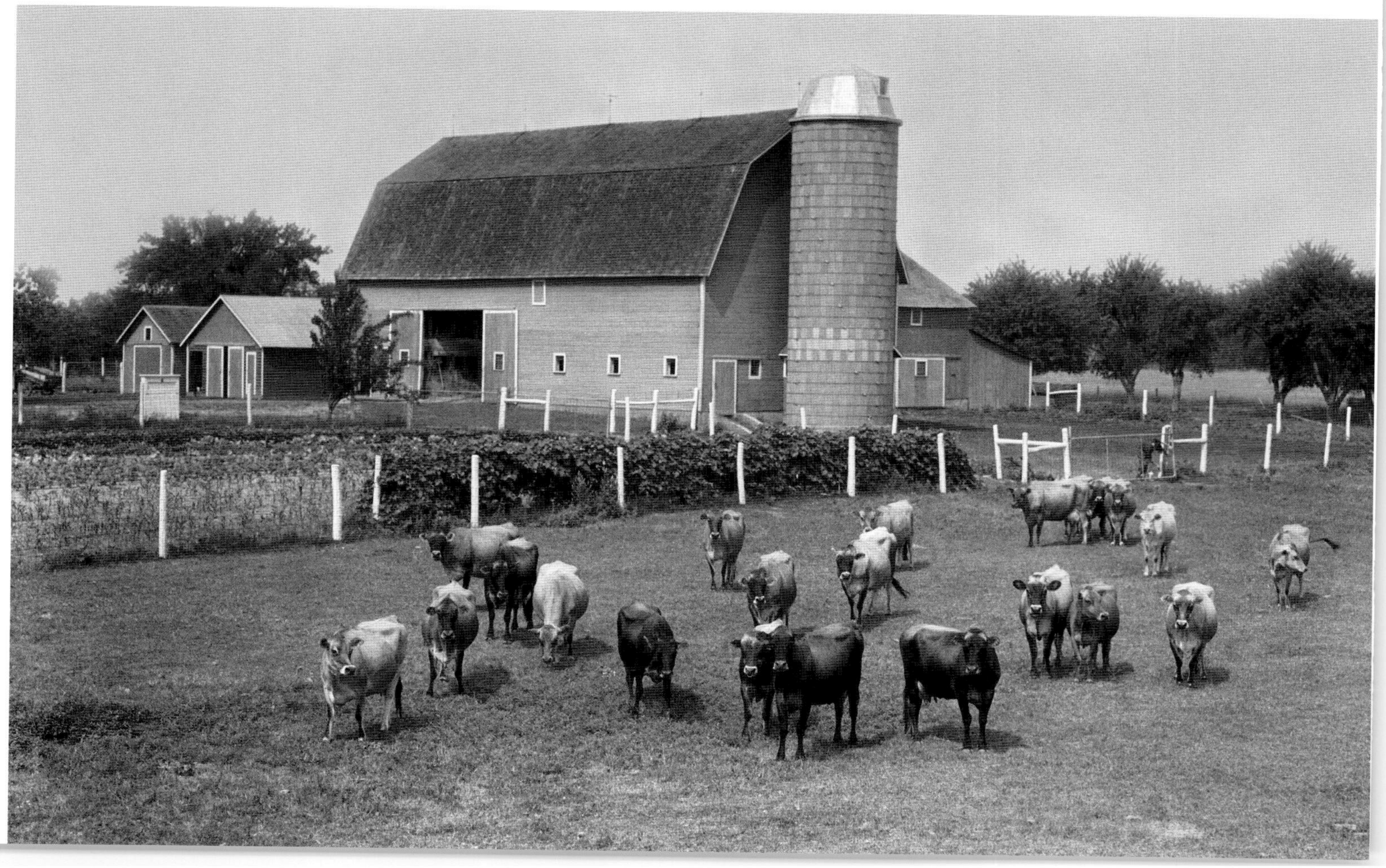

A Jersey herd in front of a barn on the farm of Earl Hodges in Wauseon, Ohio (Fulton County, OH; 1931)

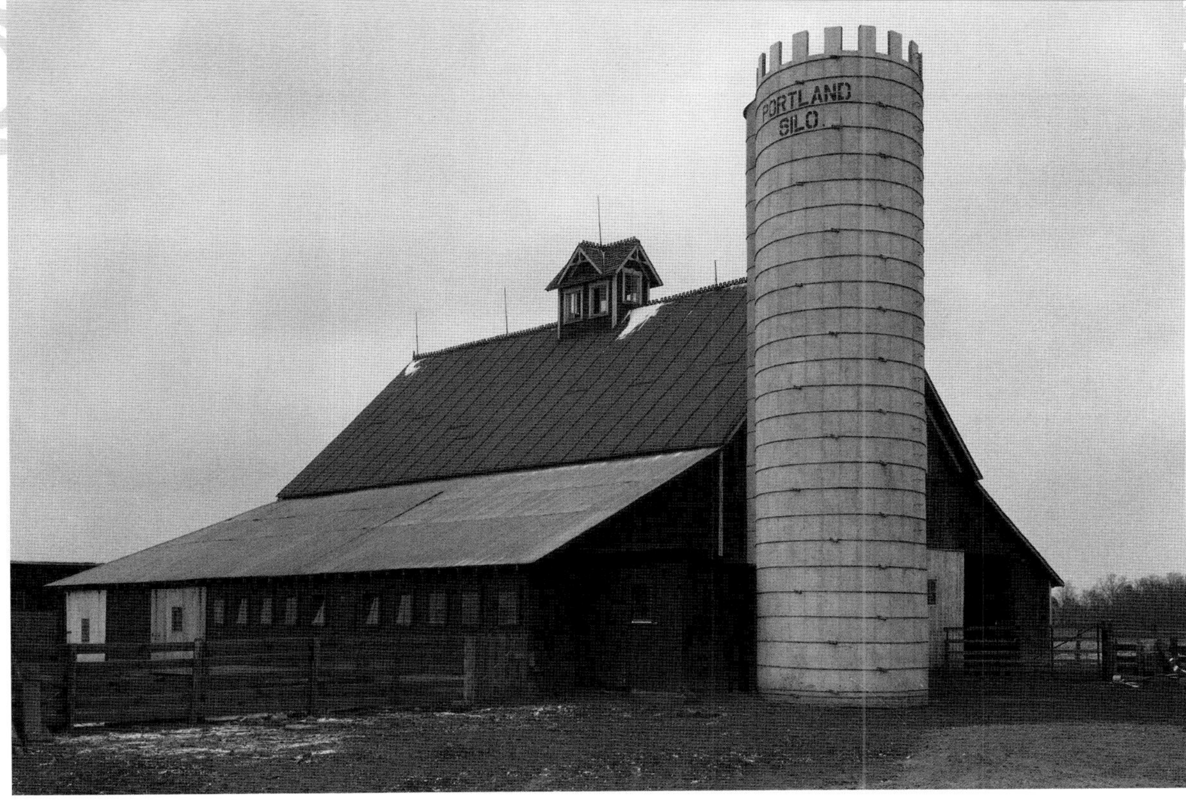

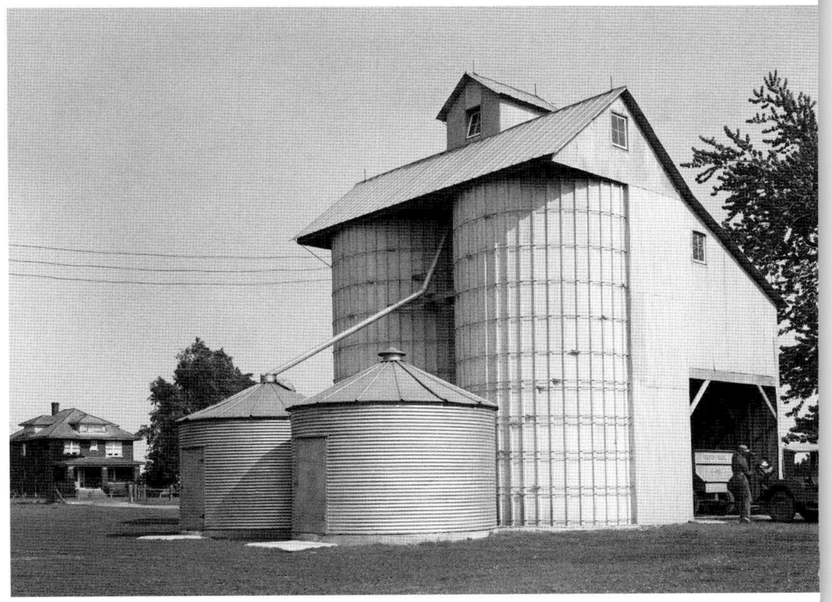

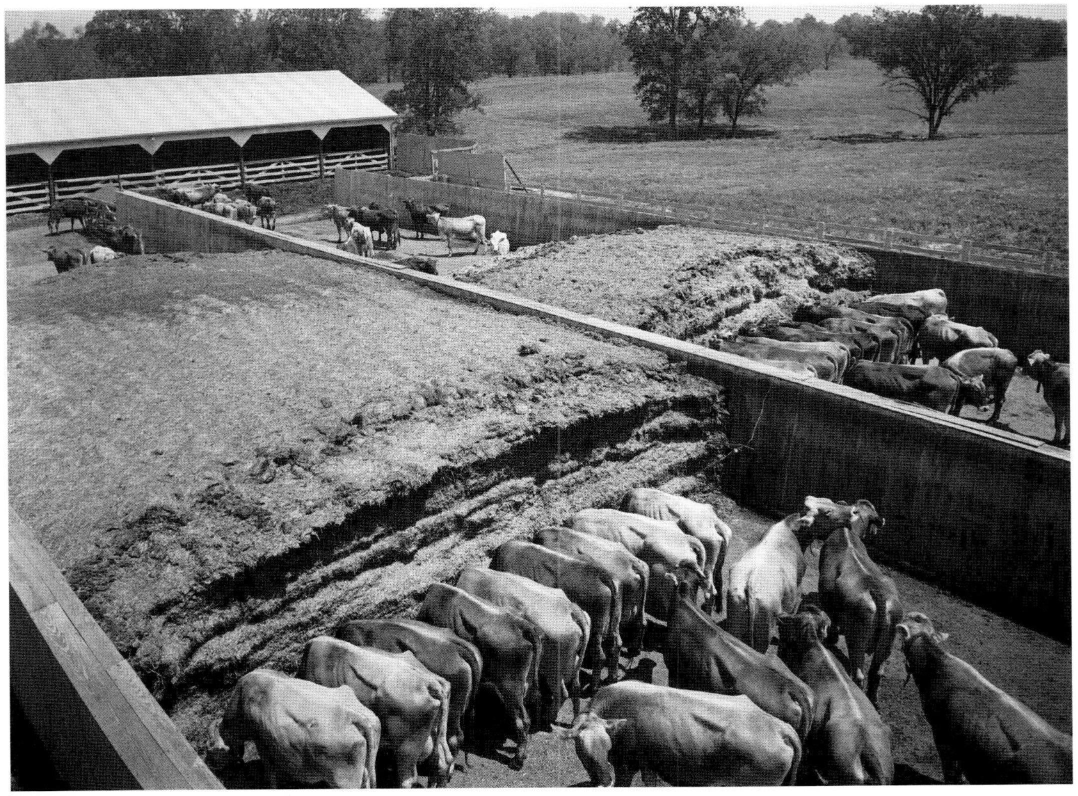

(Top left) A remodeled dairy barn on the farm of William Blickenstaff of Monticello (White County, 1930)

(Top right) Grain storage on the farm of Ralph Yergler in LaCrosse (LaPorte County, 1958)

(Bottom) Dairy cattle eat from the ends of two bunker silos at the same time at the Conner Prairie Farm located in Noblesville. (Hamilton County, 1956)

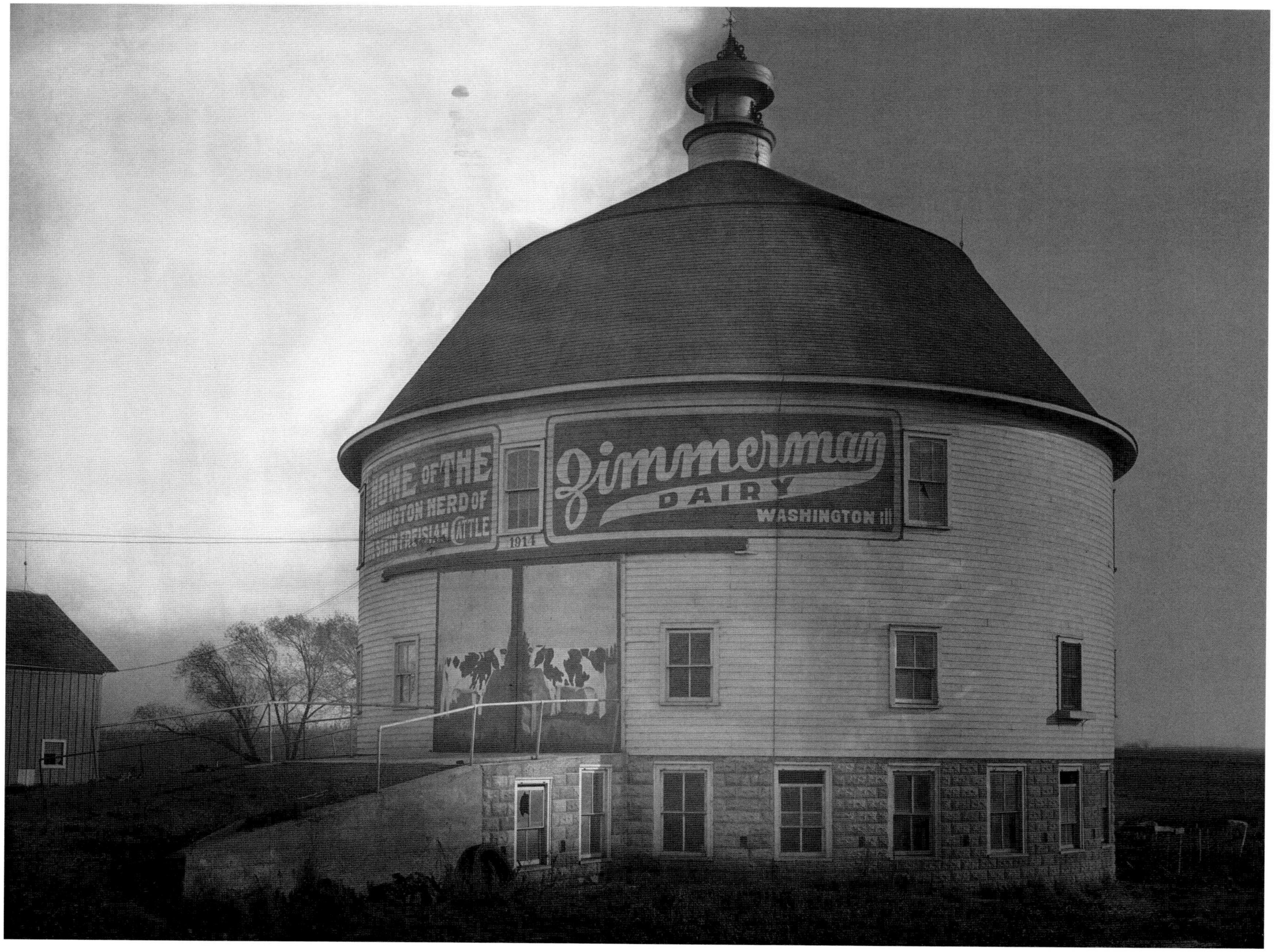

Fred Zimmerman of Washington, Illinois, owned this round dairy barn. (Tazewell County, IL; 1918)

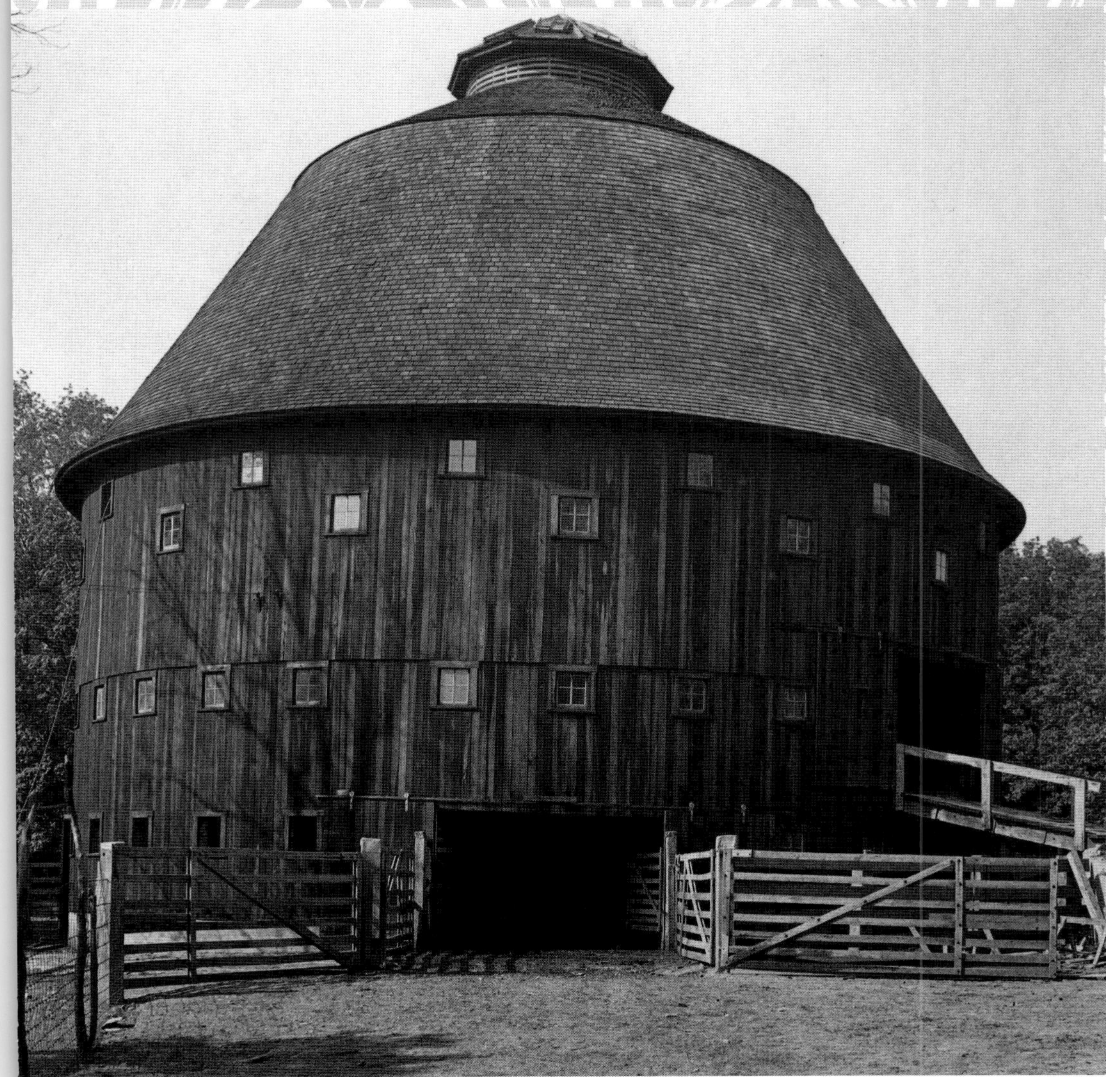

A round barn with wooden siding on the E. O. White farm in Dana (Vermillion County, ca. 1913)

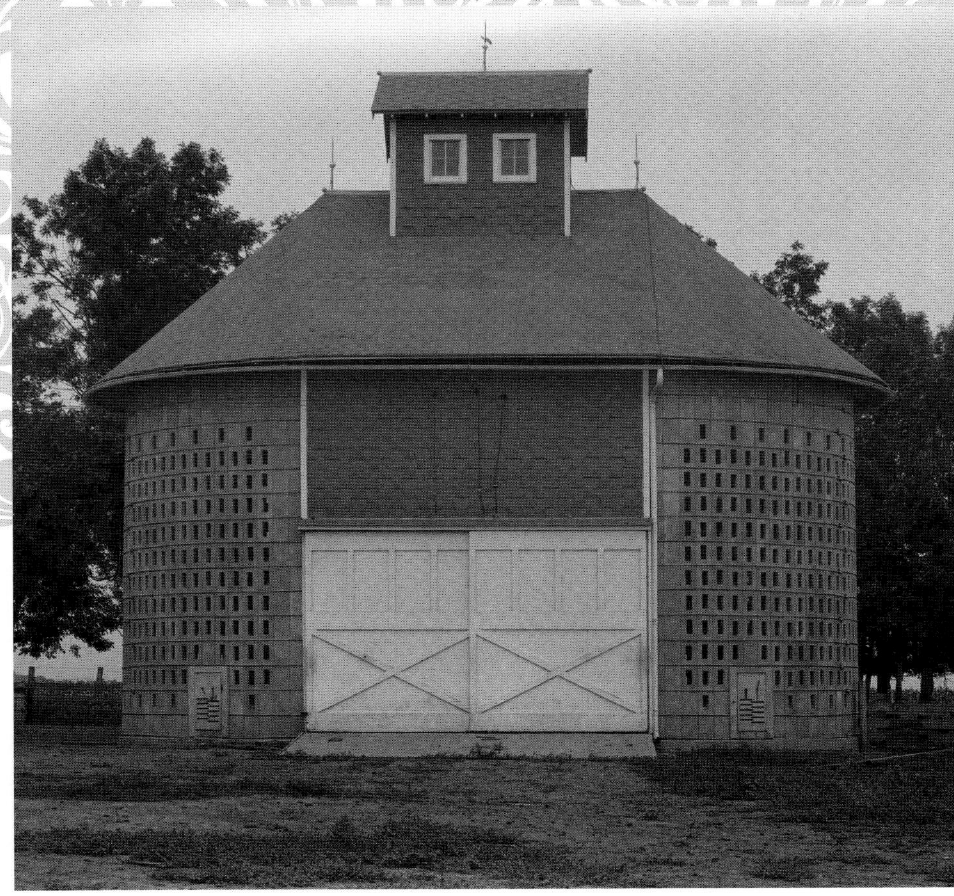

A hollow tile block granary on the farm of G. L. Frost from Grundy Center, Iowa (Grundy County, IA; 1929)

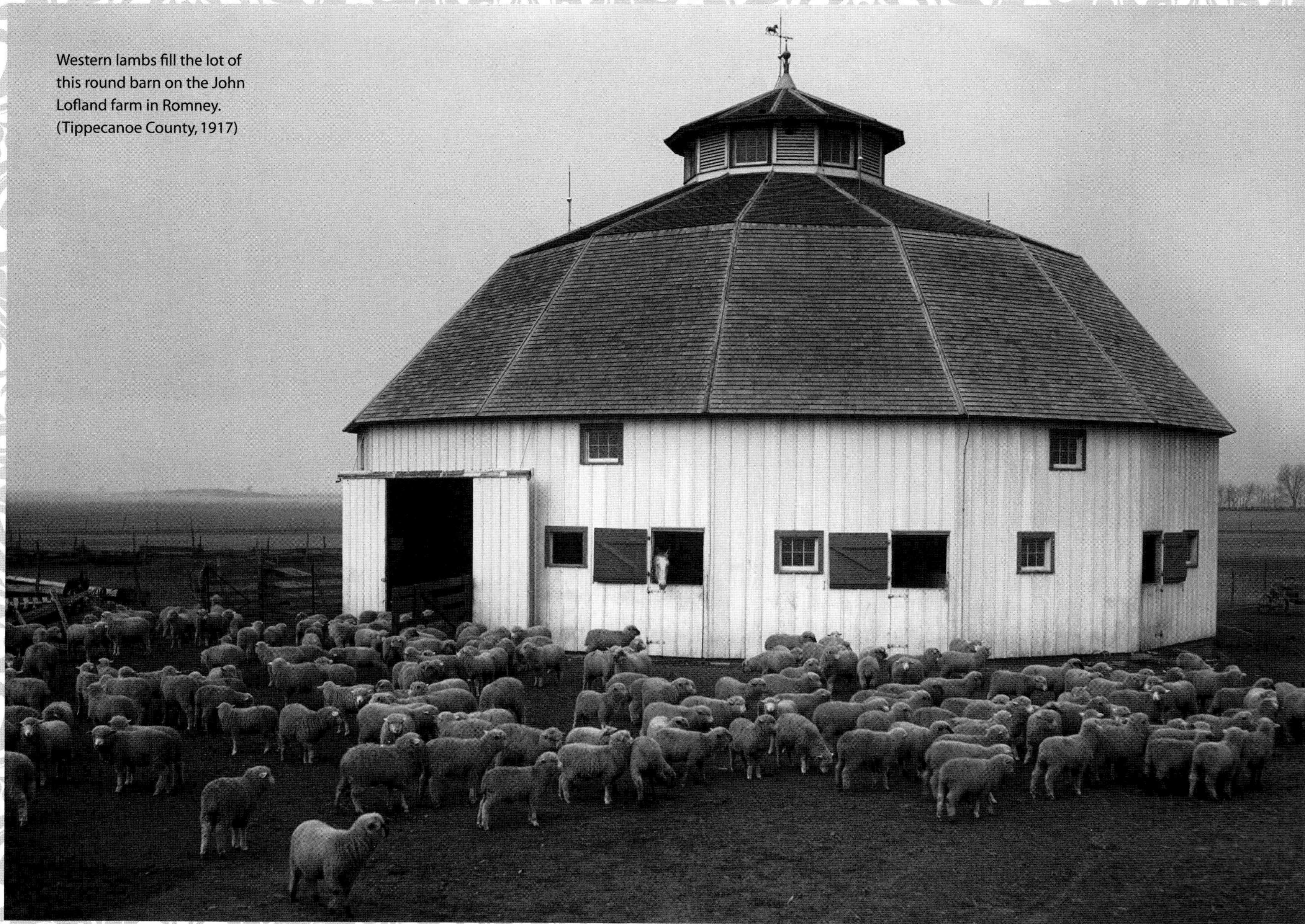

Western lambs fill the lot of this round barn on the John Lofland farm in Romney. (Tippecanoe County, 1917)

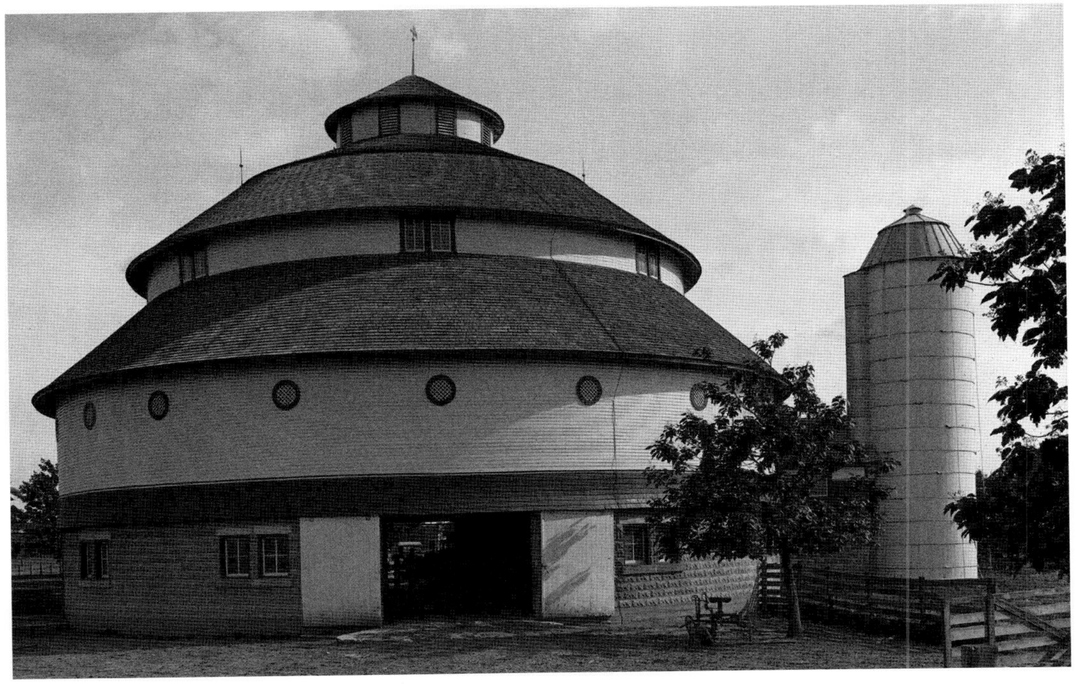

A Clinton County round barn on the farm of T. G. Johnson in Kirklin (Clinton County, 1917)

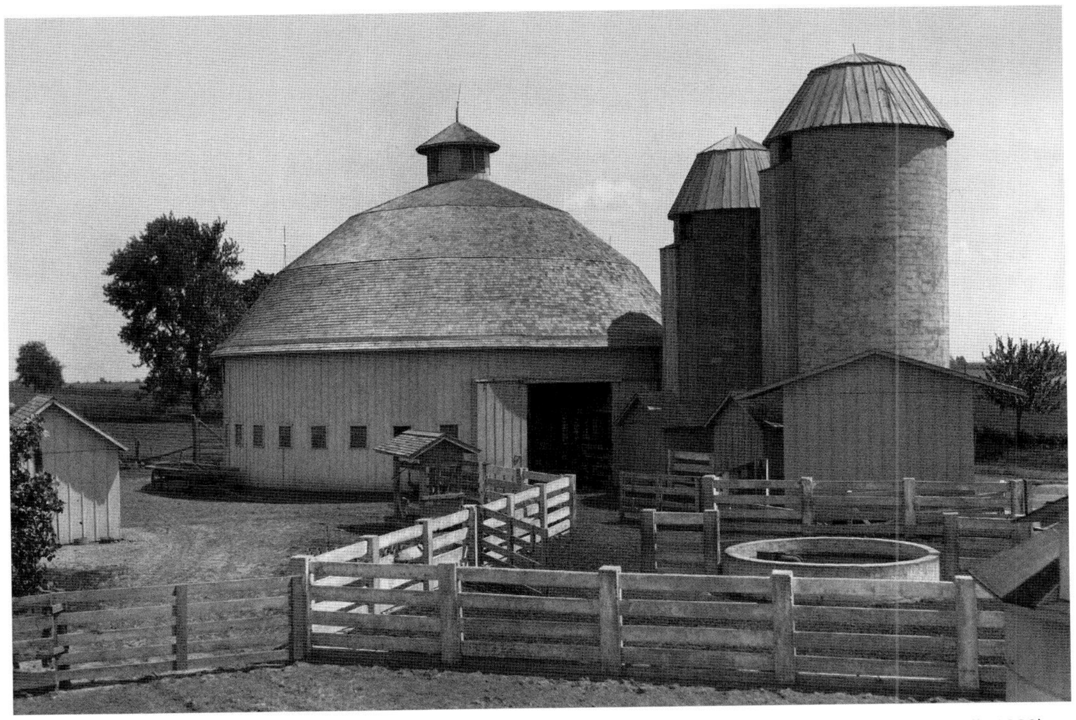

A round barn with hollow tile block silos located on the Russell Davis farm in Clayton, Illinois (Adams County, IL; 1920)

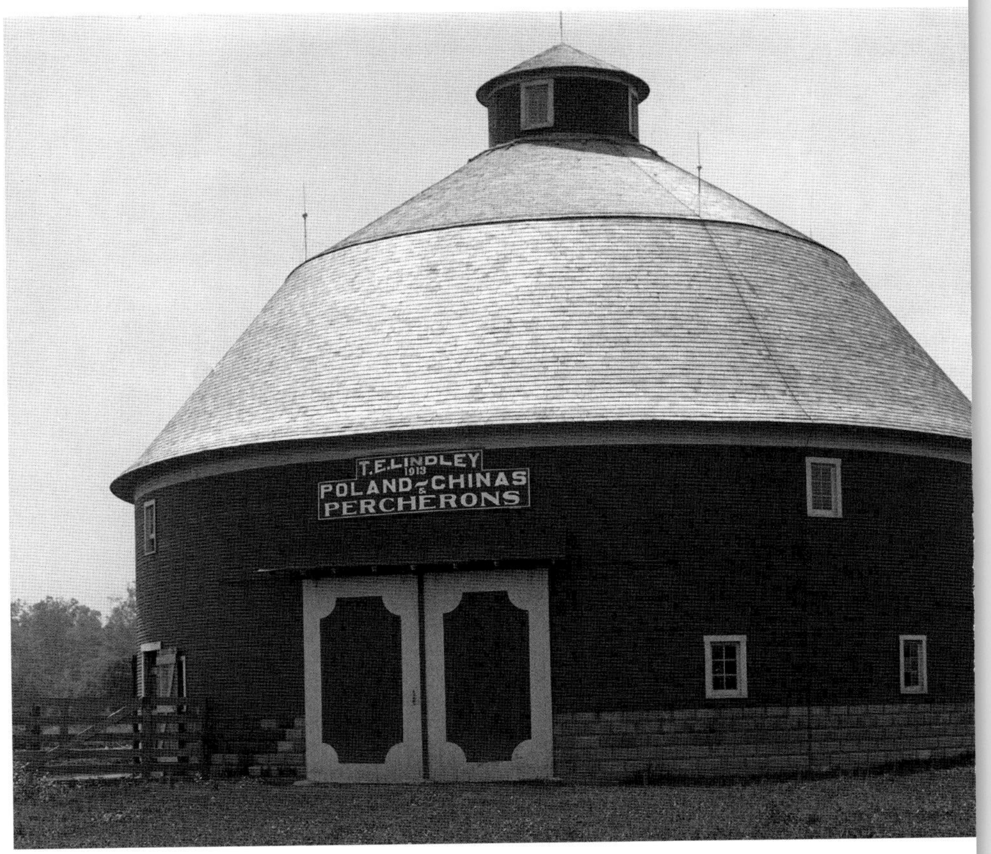

T. E. Lindley owned this round barn in Russiaville. (Howard County, 1916)

Swinging on the Pasture Gate

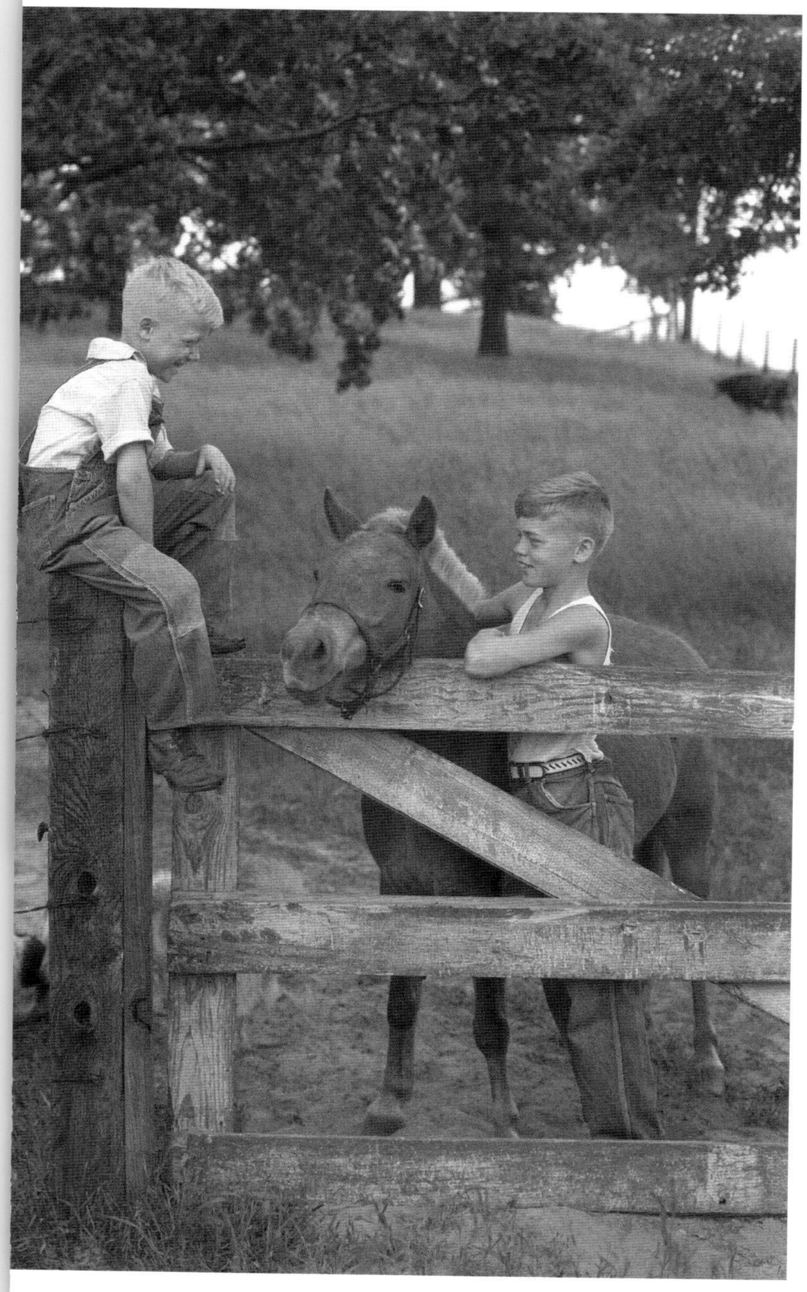

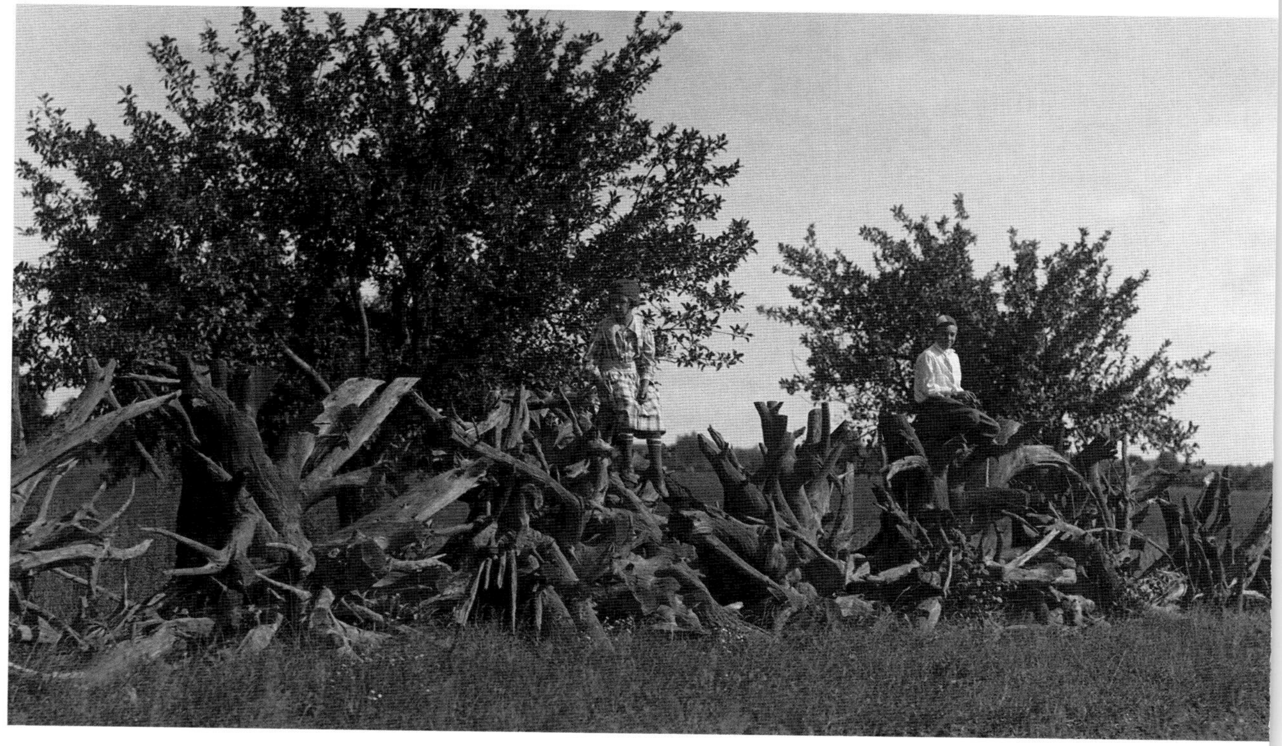

(Left) Robert (seated) and Albert Vandergraff Jr. at a gate with their pony (Tippecanoe County, 1935)

(Top right) Pine stumps were pulled out of a field and used to build this fence in northern Michigan. (1921)

(Bottom right) These fences were built from stones taken off land near Wilkes-Barre, Pennsylvania, in 1926. (Luzerne County, PA)

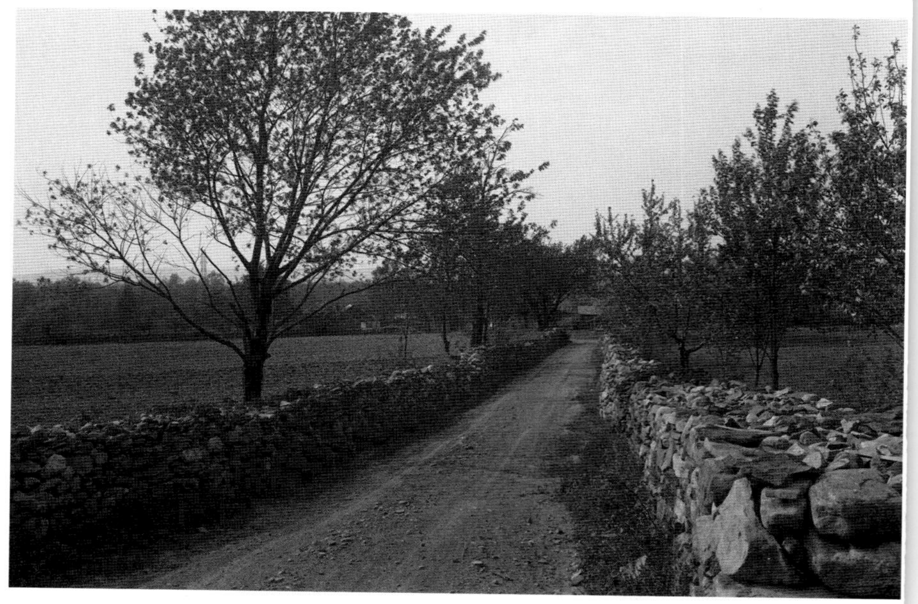

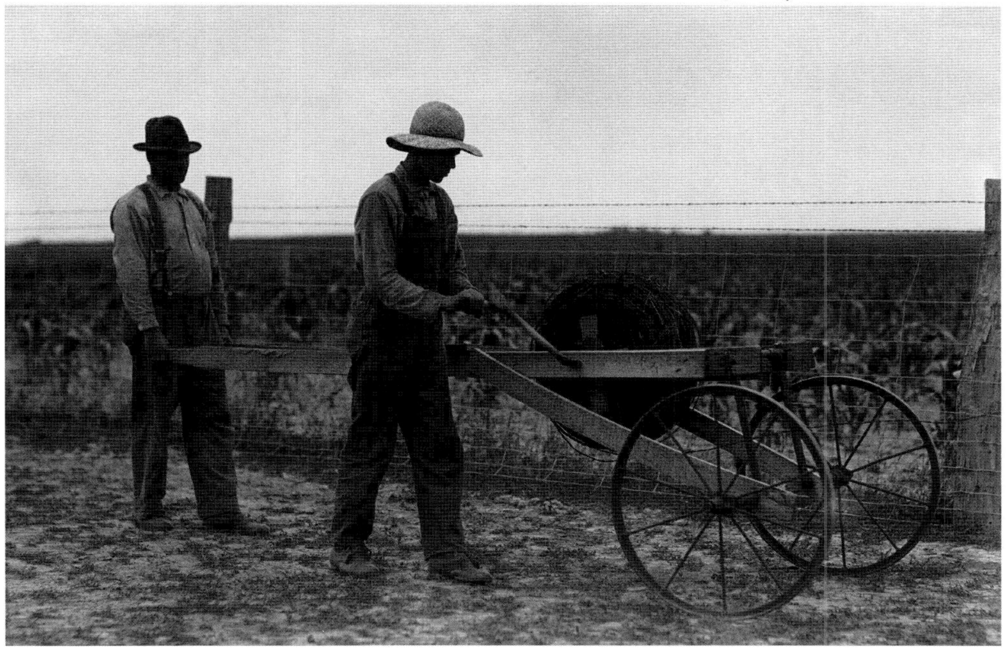

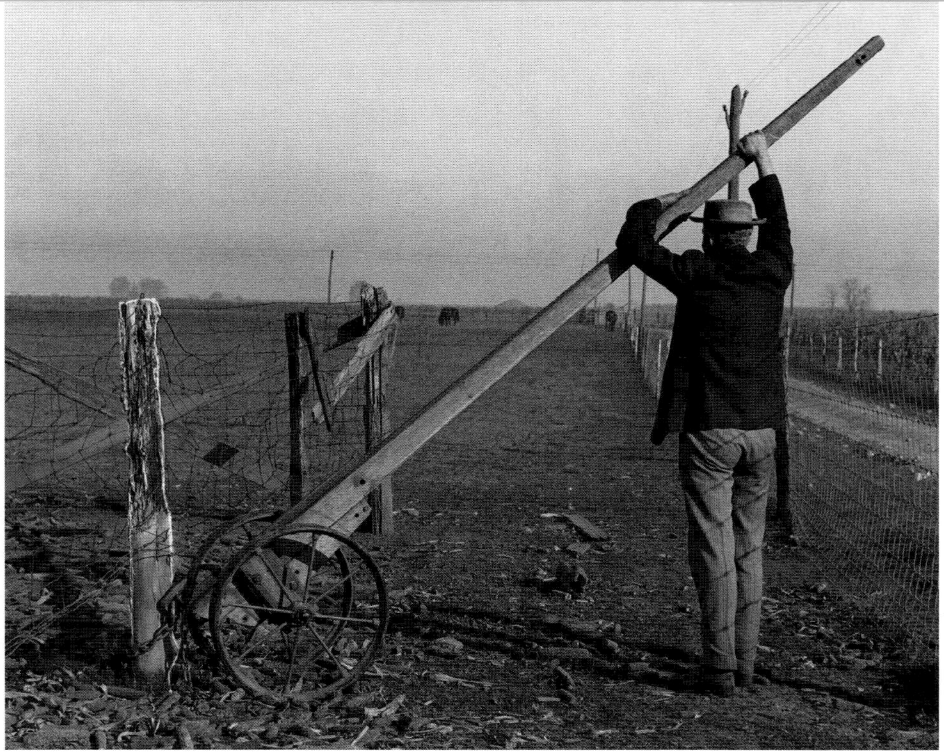

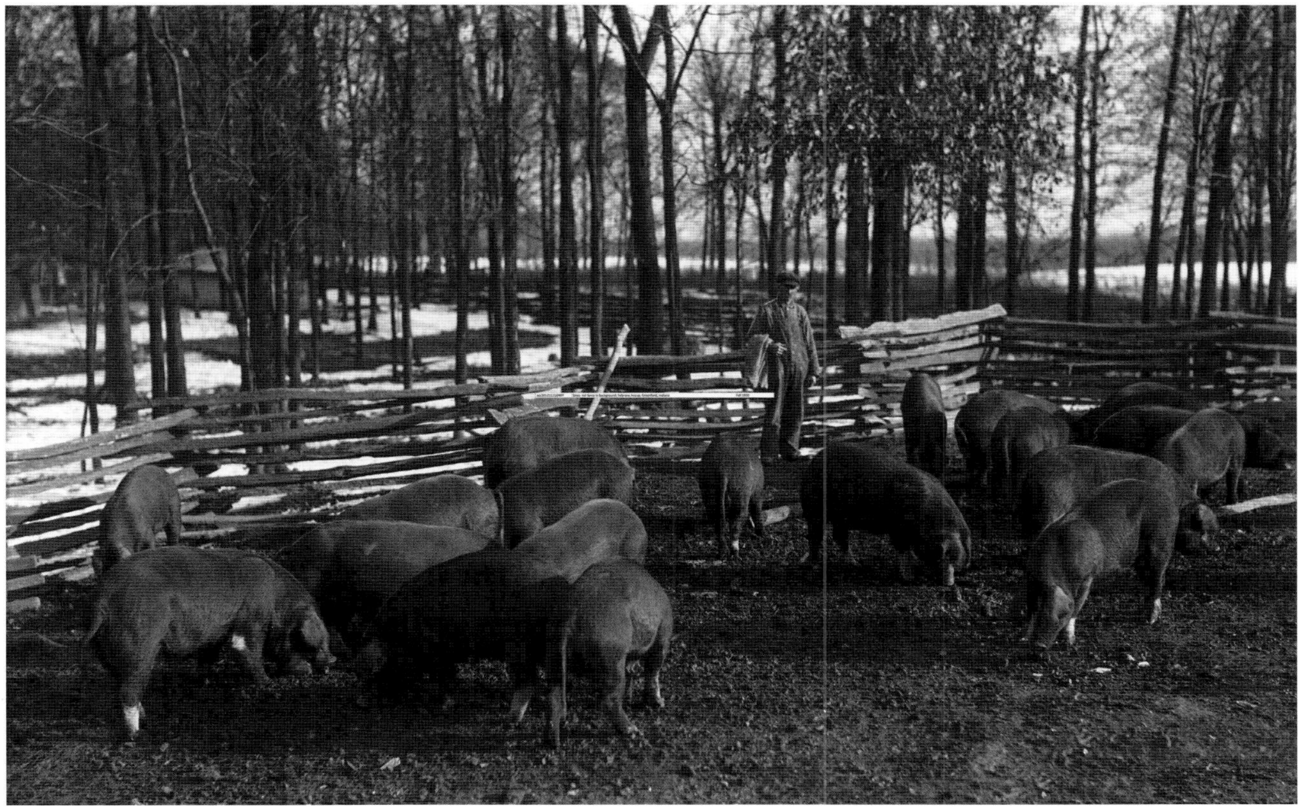

(Top left) This photograph depicting John Greenwalt rolling up barbed wire fencing with a cart and reel in Delphi was sold to the *Breeder's Gazette*. (Carroll County, 1921)

(Top right) This man uses a handy device for pulling fence posts on a farm just southwest of Lafayette. (Tippecanoe County, 1914)

(Bottom) Fattening Duroc Jersey hogs on wheat pasture within a split-rail fence enclosure (Tippecanoe County, 1931)

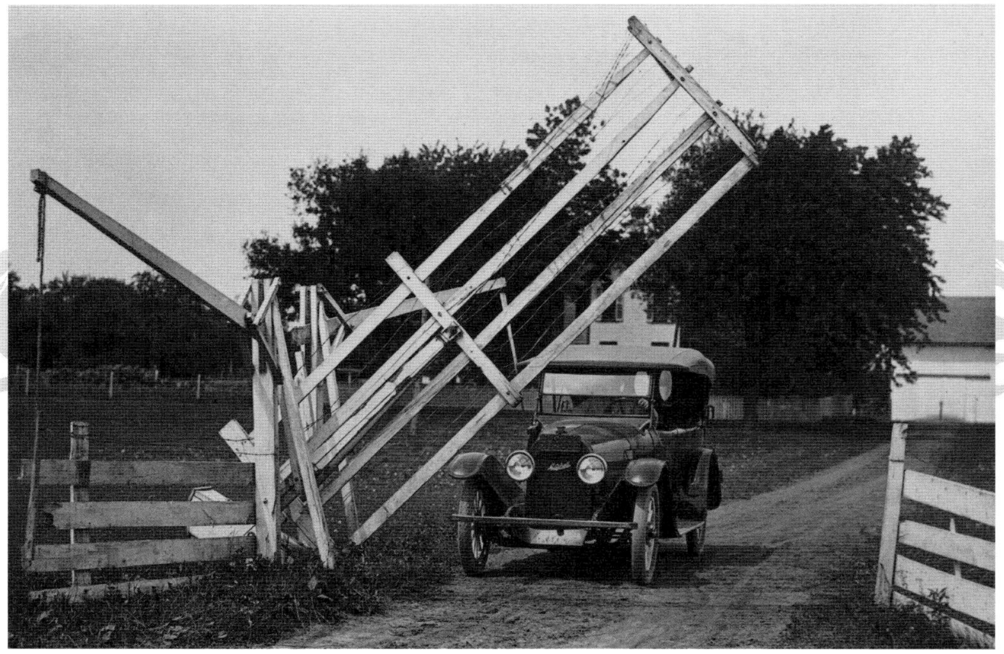

This farm gate could be opened from inside this Mitchell Touring sedan. (McLean County, IL; 1917)

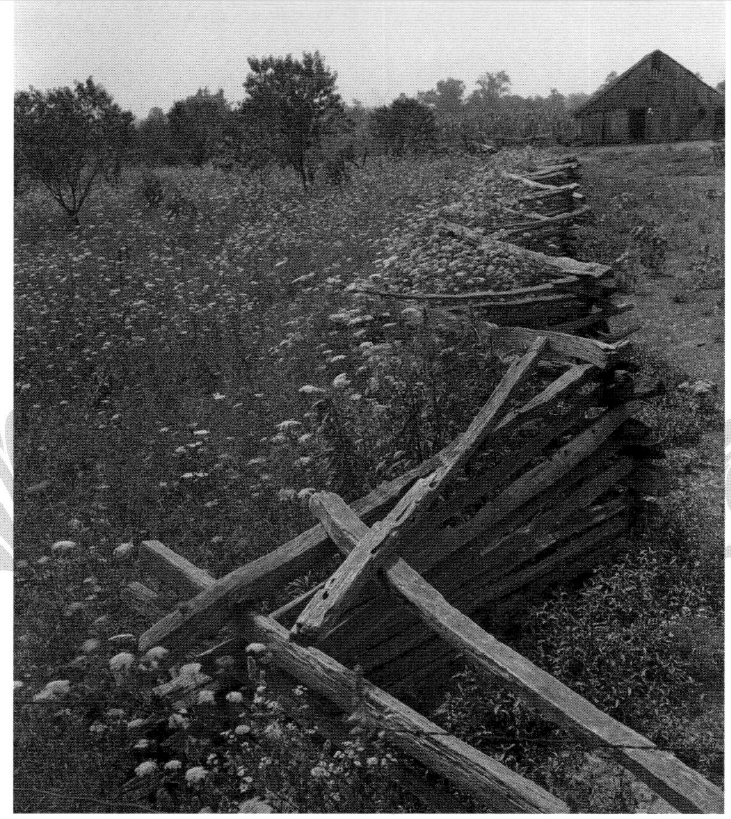

An old fencerow borders a field grown up with wild carrots. (Putnam County, 1915)

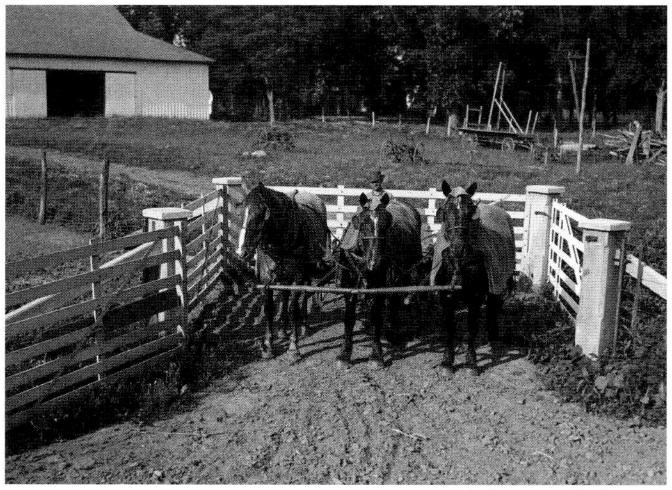

(Above) This group of four gates connected three fields with the barn lot on the farm of William Harman of Oxford. (Benton County, 1919)

(Bottom right) A passenger opens this gate without leaving the automobile on the farm of W. L. Woodrow, Green Valley, Illinois. (Tazewell County, IL; 1918)

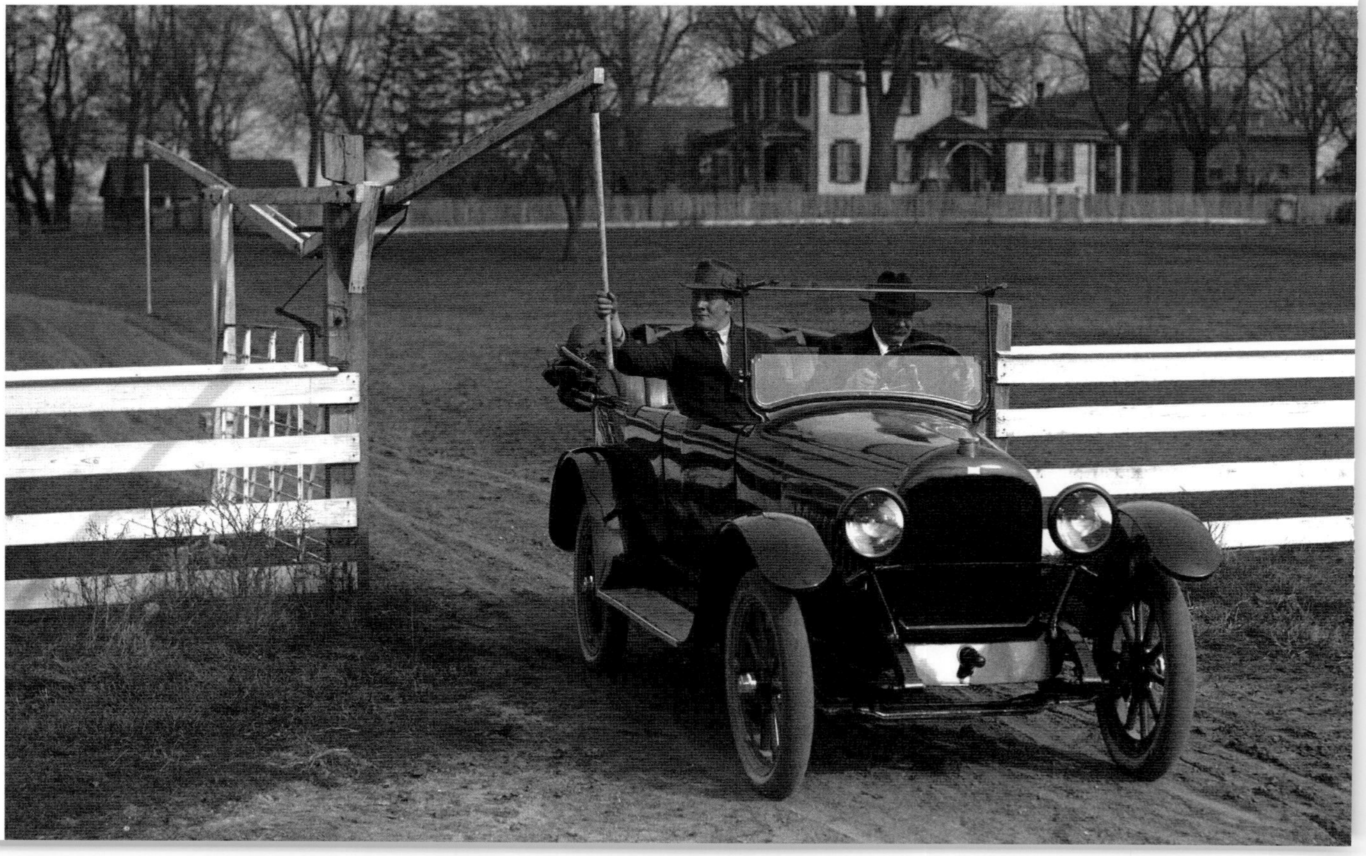

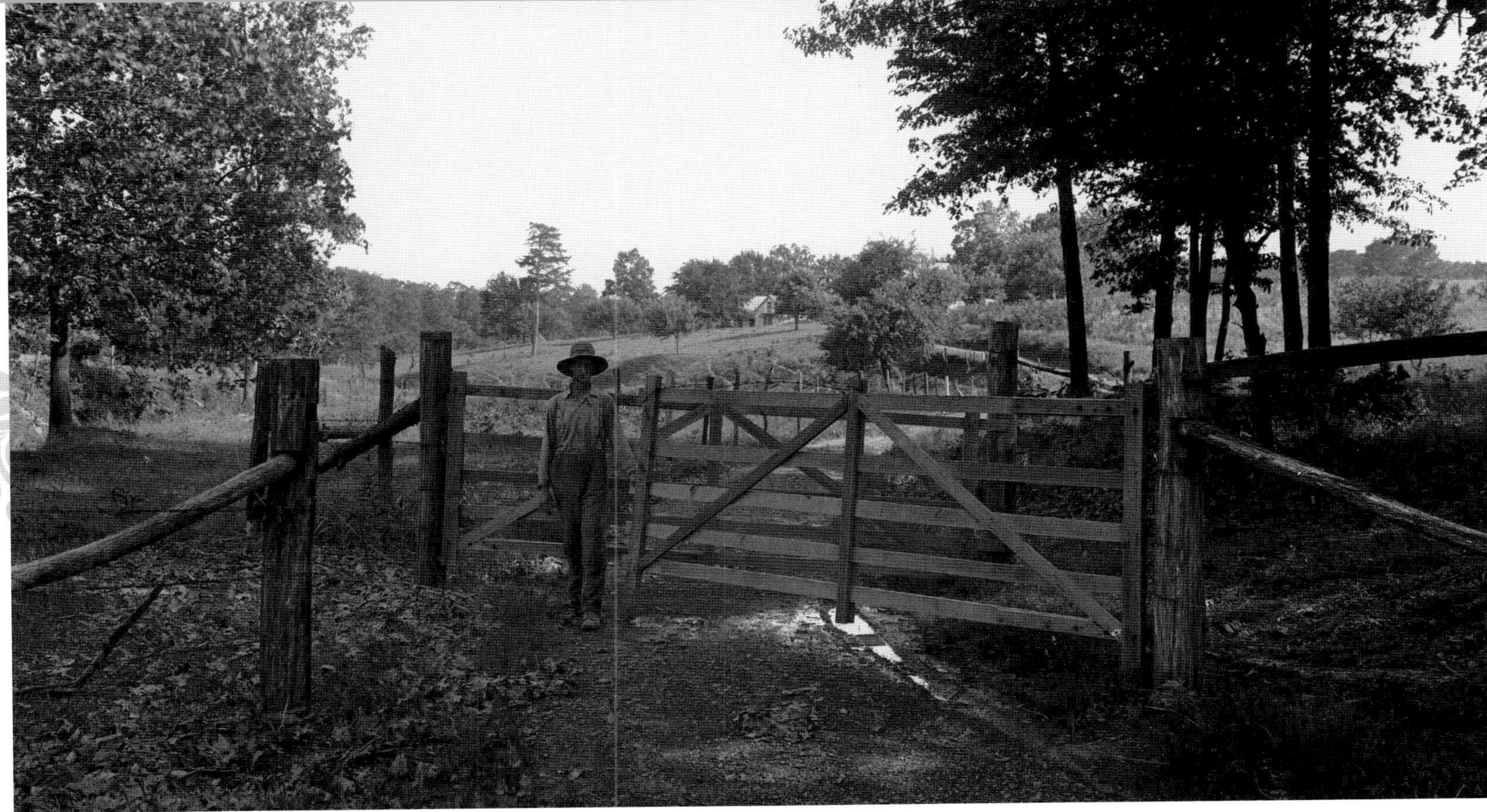

Otto Sperzel from New Albany arranged these two gates to take care of four openings, connecting the wood's pasture with the bottom pasture, or connecting the pastures with the lane of the barn. (Floyd County, 1923)

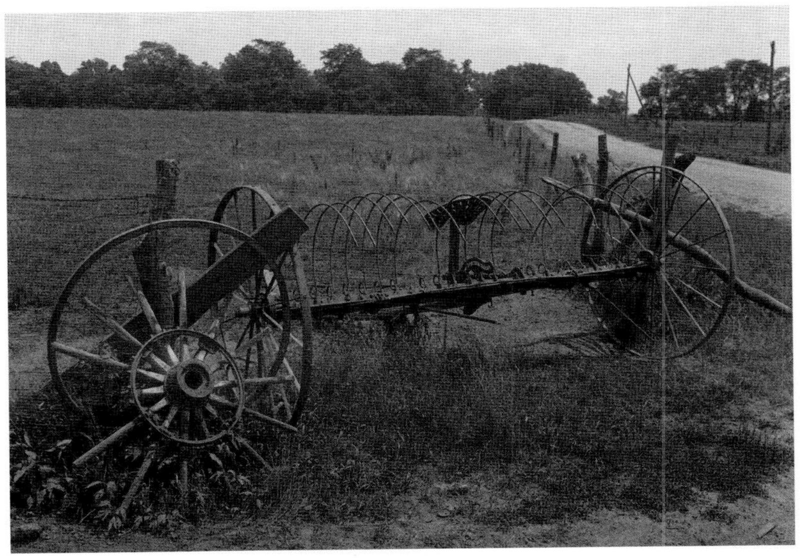

An old hay rake is turned into a farm gate. (Benton County, 1922)

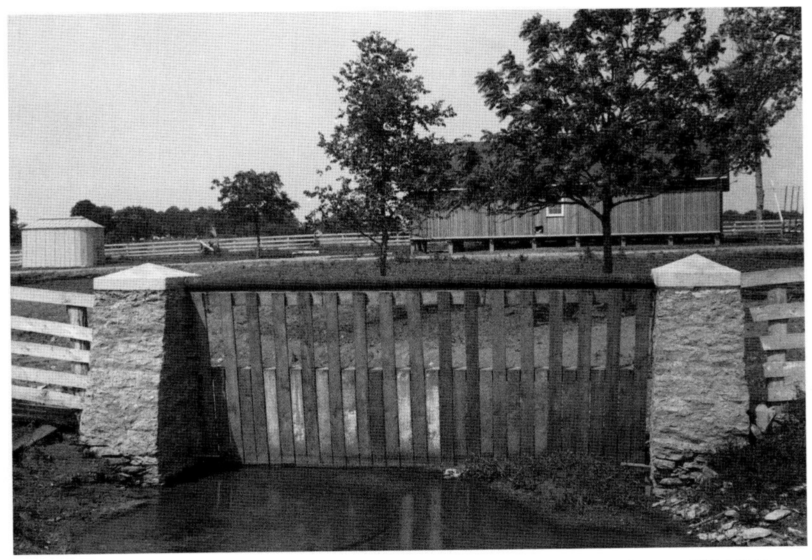

A water gate at the Coldstream Farm in Lexington, Kentucky (Fayette County, KY; 1917)

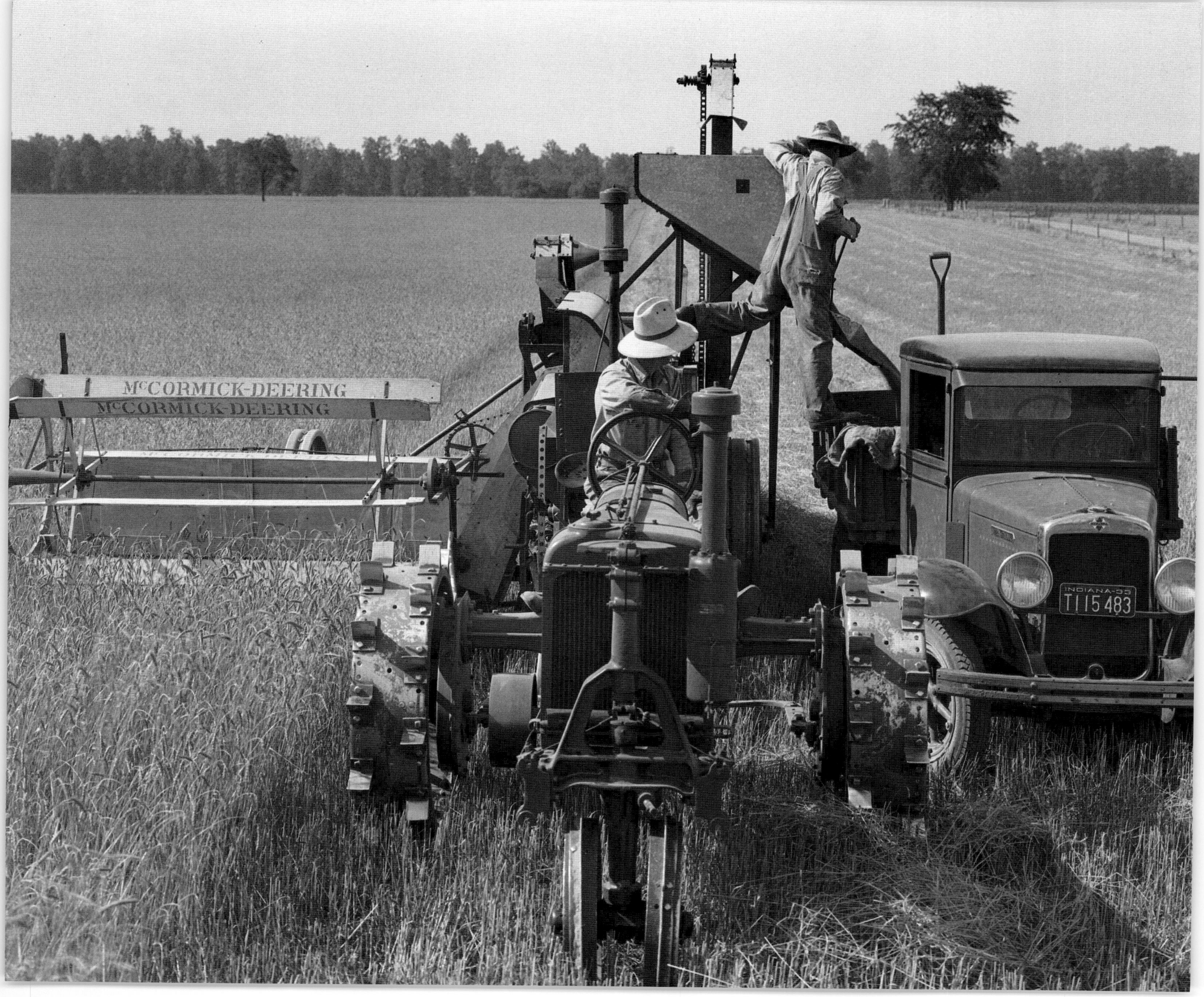
McCORMICK-DEERING
McCORMICK-DEERING
INDIANA-33
T115 483

THROUGH THE LENS OF PIONEER PHOTOGRAPHER
J. C. Allen

PART 2 *Working the Ground*

Clearing Land and Managing Water

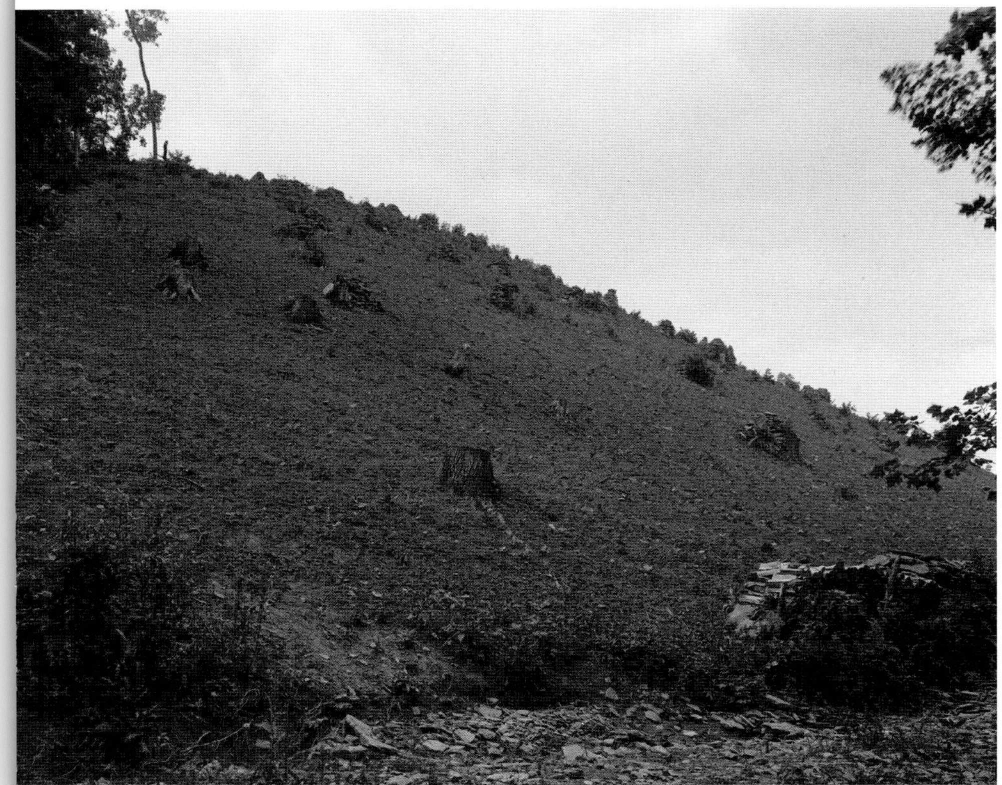

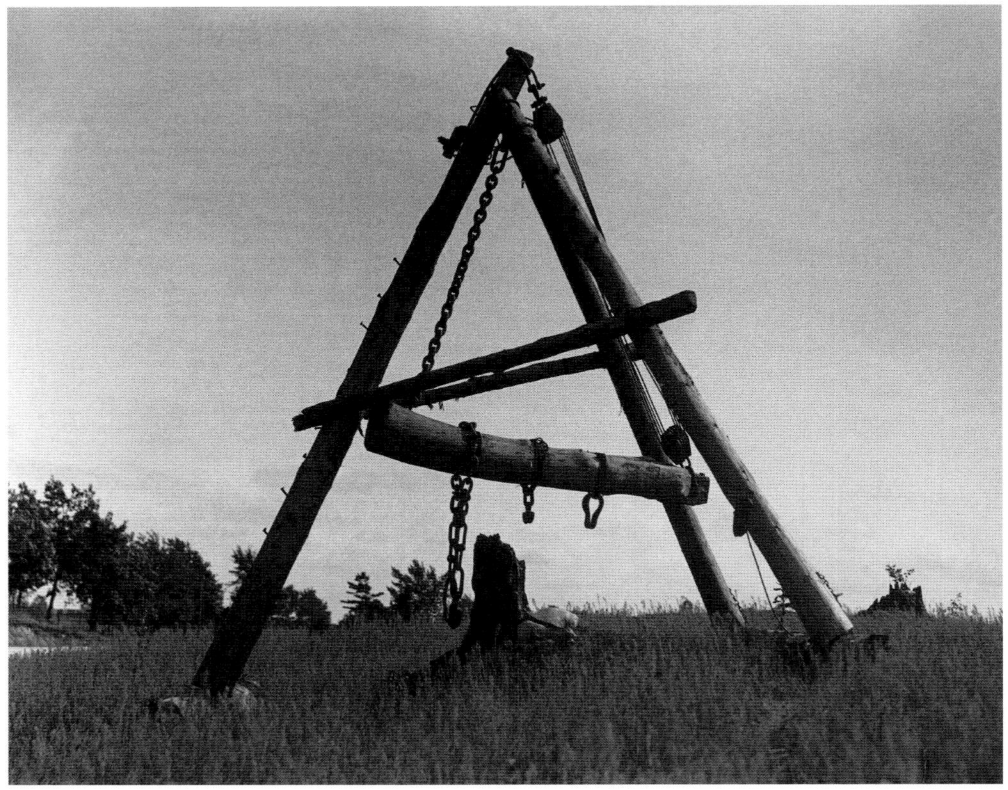

(Top left) This steep hill in southern Indiana had to be cleared so it could be used for planting tobacco. (Ohio County, 1923)

(Top right) A derrick powered by four horses is used for pulling stumps on the farm of Will Force in Lakeview, Michigan. (Montcalm County, MI; 1921)

(Bottom) Clearing hilly land in southern Indiana for tobacco (Ohio County, 1923)

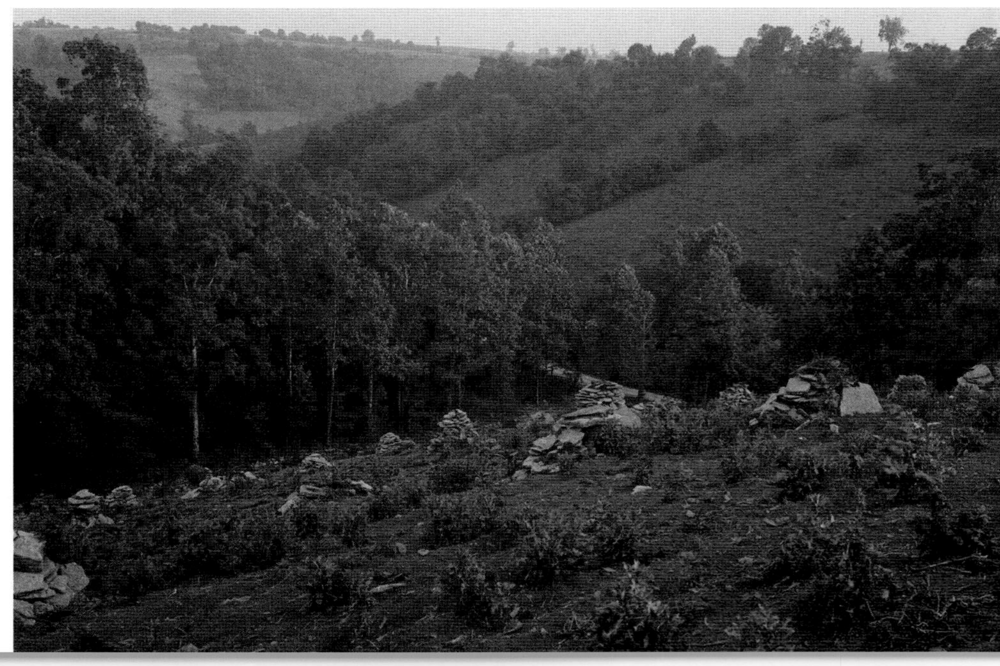

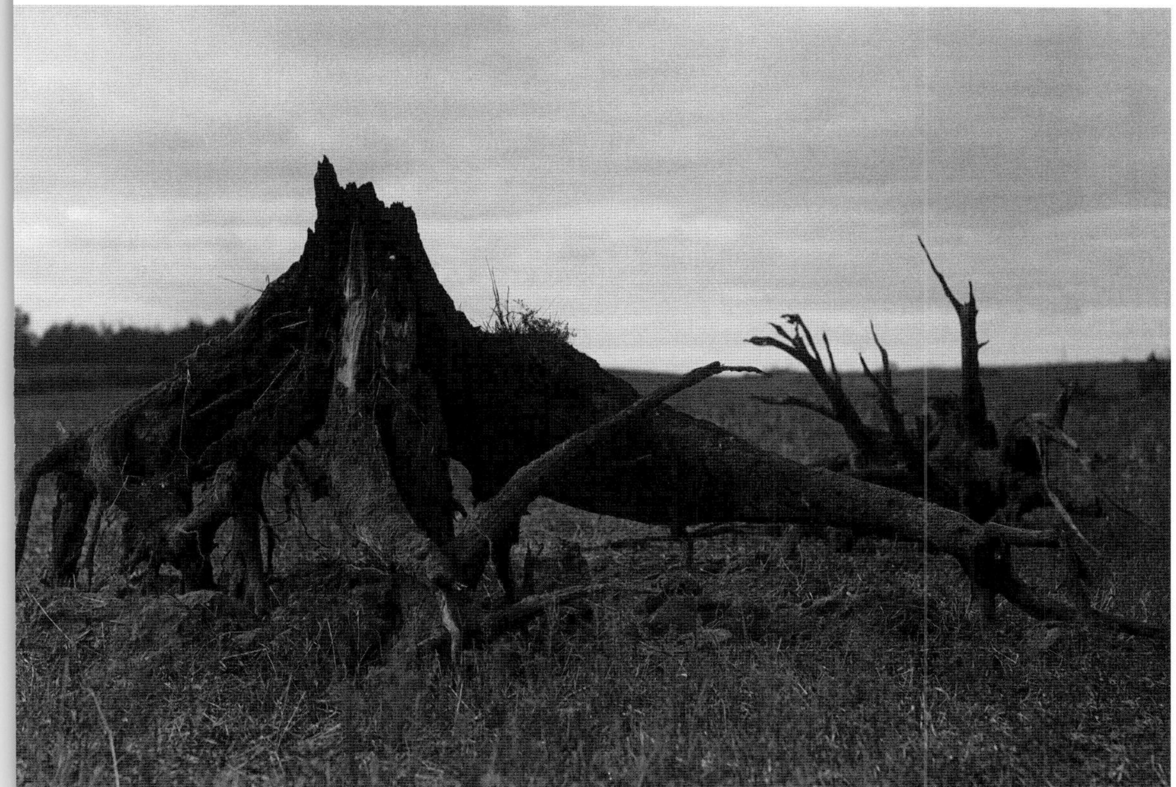

This stump was pulled from a field in Lakeview, Michigan. (Montcalm County, MI; 1921)

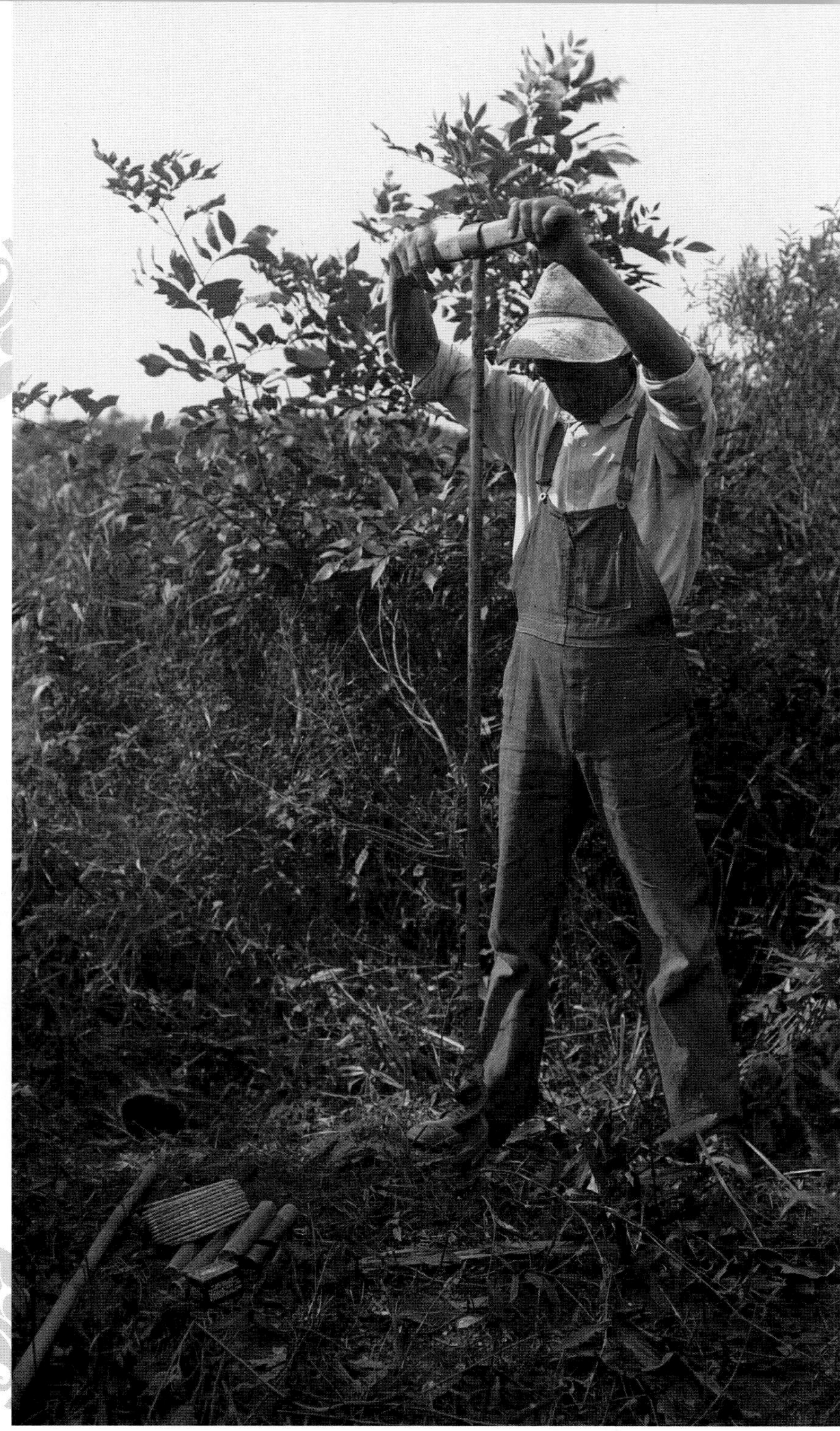

A man uses an auger to make a hole so he can insert dynamite to clear dirt from a tile outlet on the Frank Peavey farm. (Greene County, 1920)

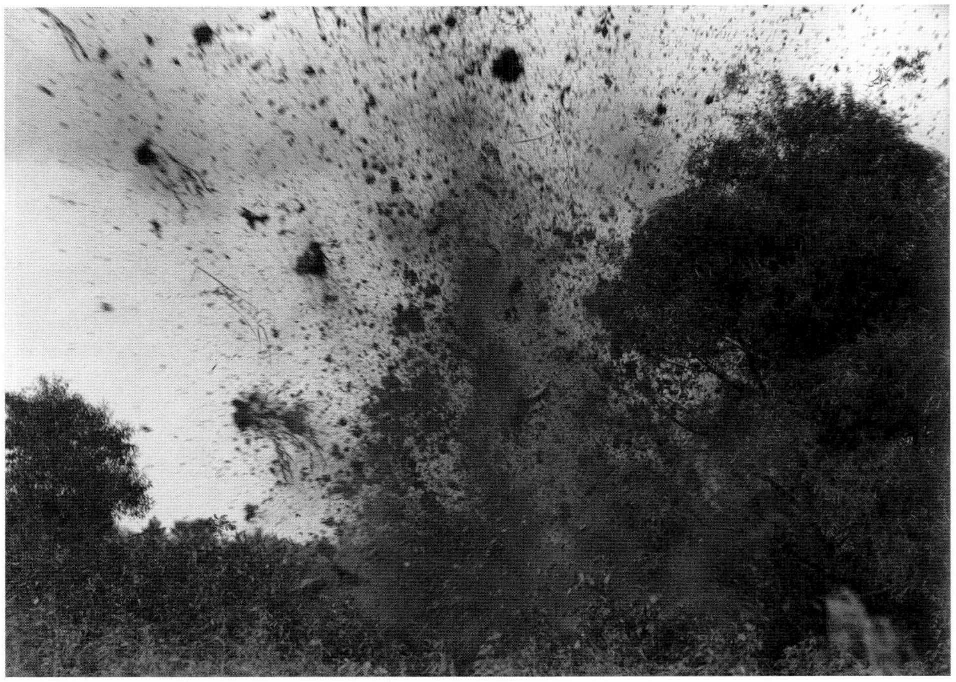

The dynamite blast that sent dirt flying from a tile outlet (Greene County, 1920)

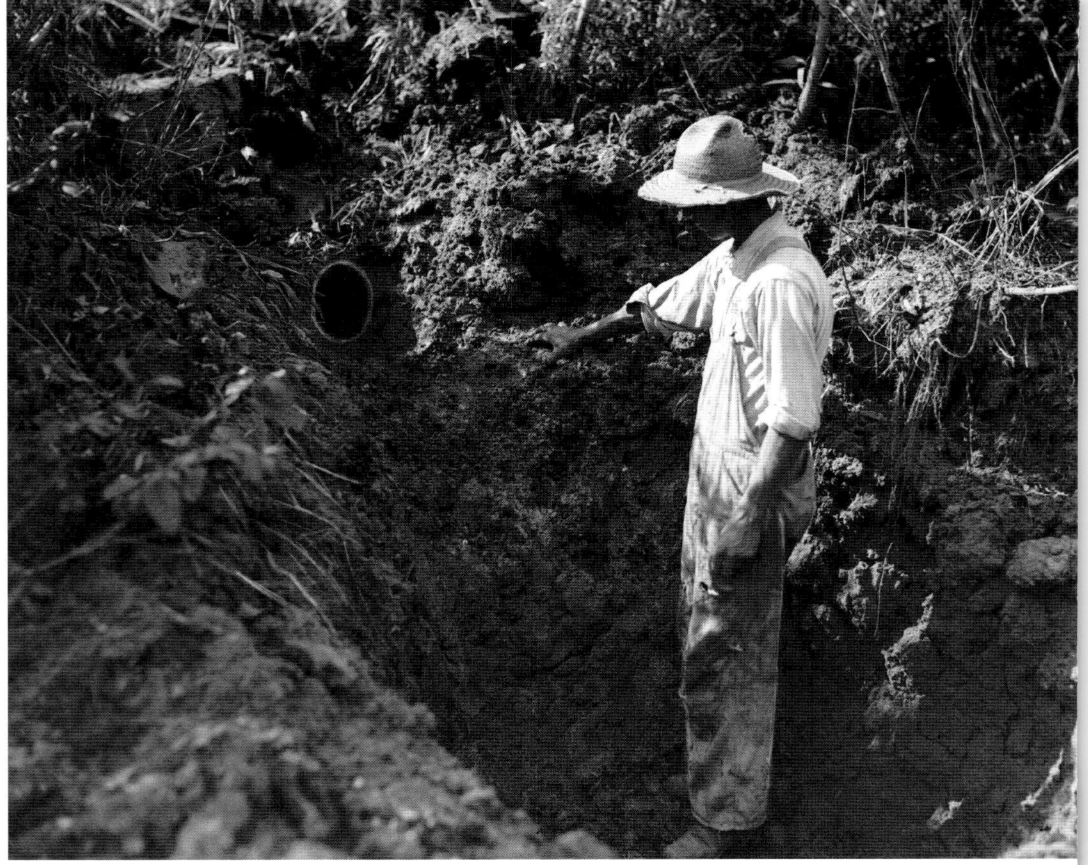

A view of the tile outlet after it was cleared by the dynamite blast (Greene County, 1920)

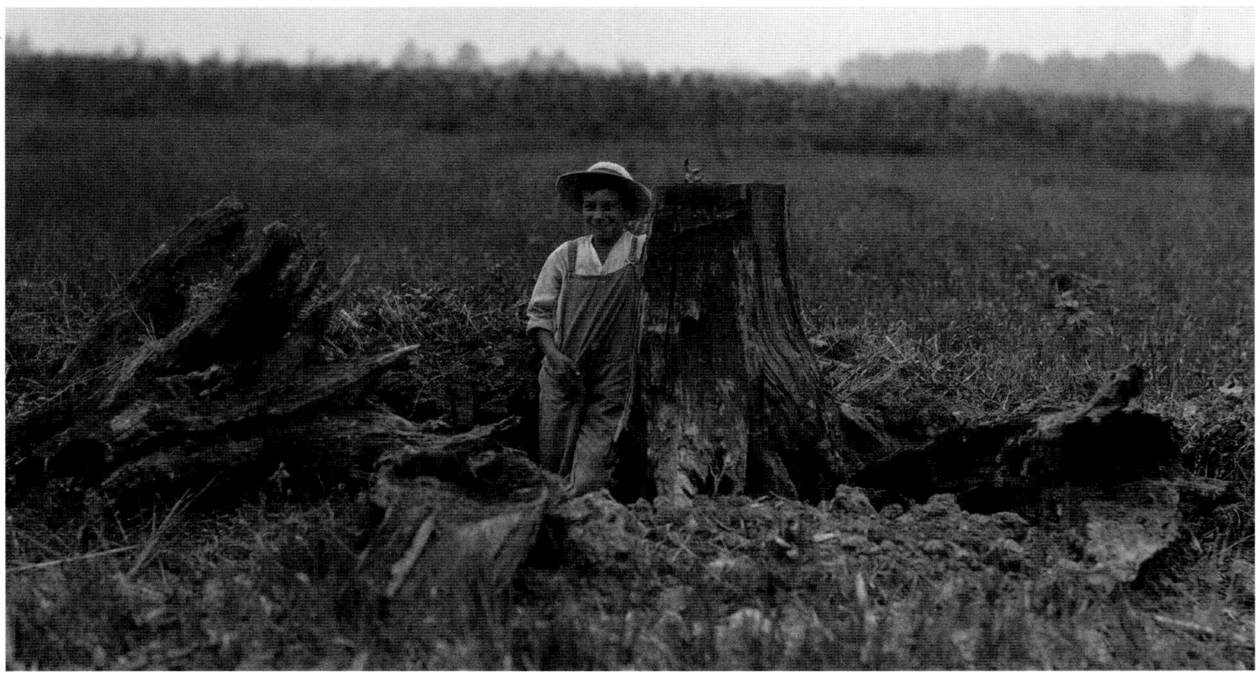

The remnant of a big white oak stump after blasting with dynamite on the farm of Frank Peavey (Greene County, 1920)

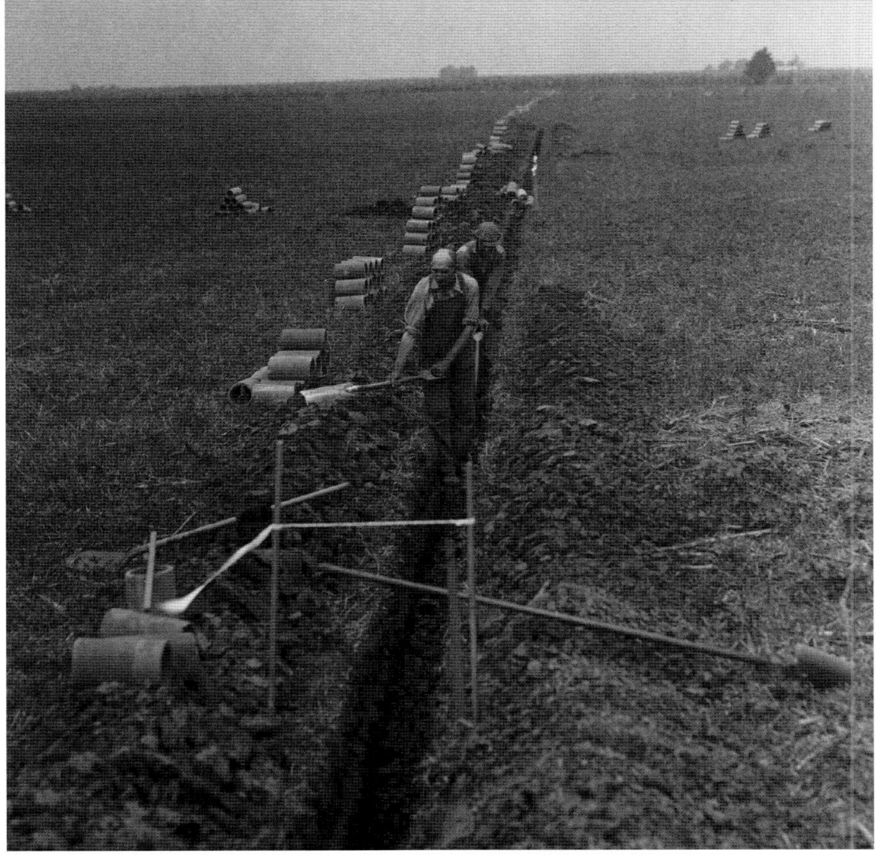

(Top left) These workers lay field tile in a hand-dug trench on the farm of Walter Fols in Potomac, Illinois. (Vermilion County, IL; ca. 1919)

(Bottom left) This tile ditching machine was owned and operated by John Royer of Clarks Hill on the Jim McCabe farm near Otterbein. (Tippecanoe County, 1955)

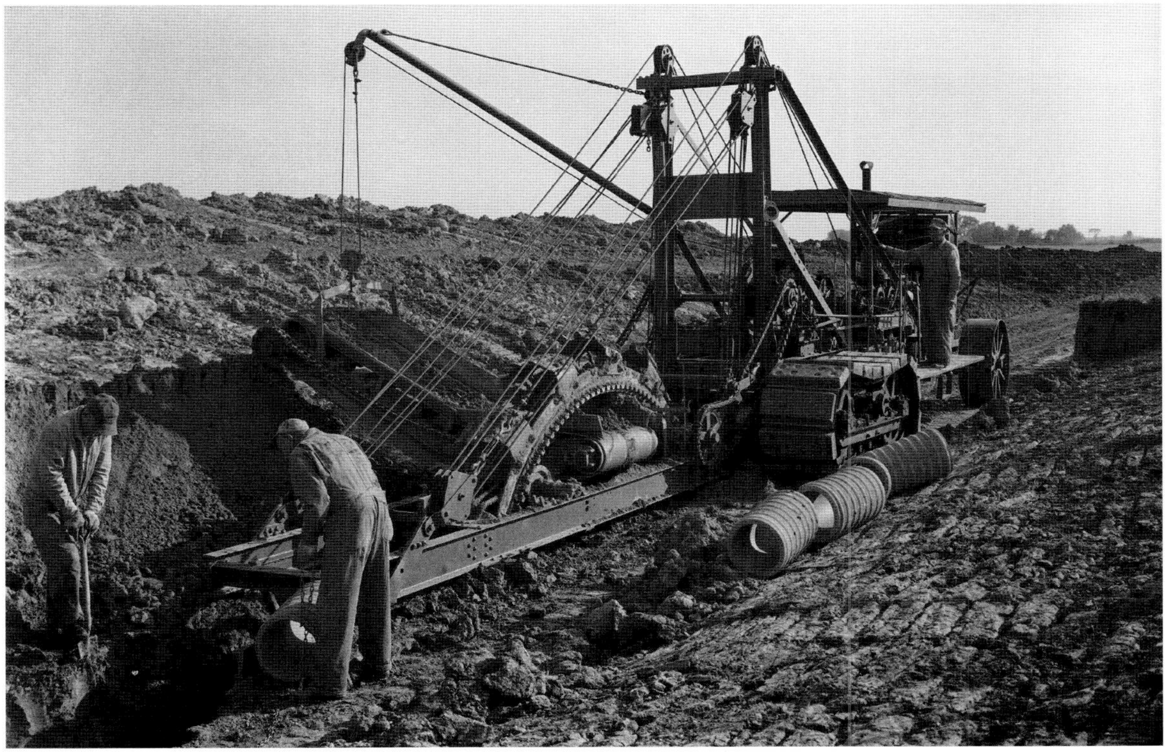

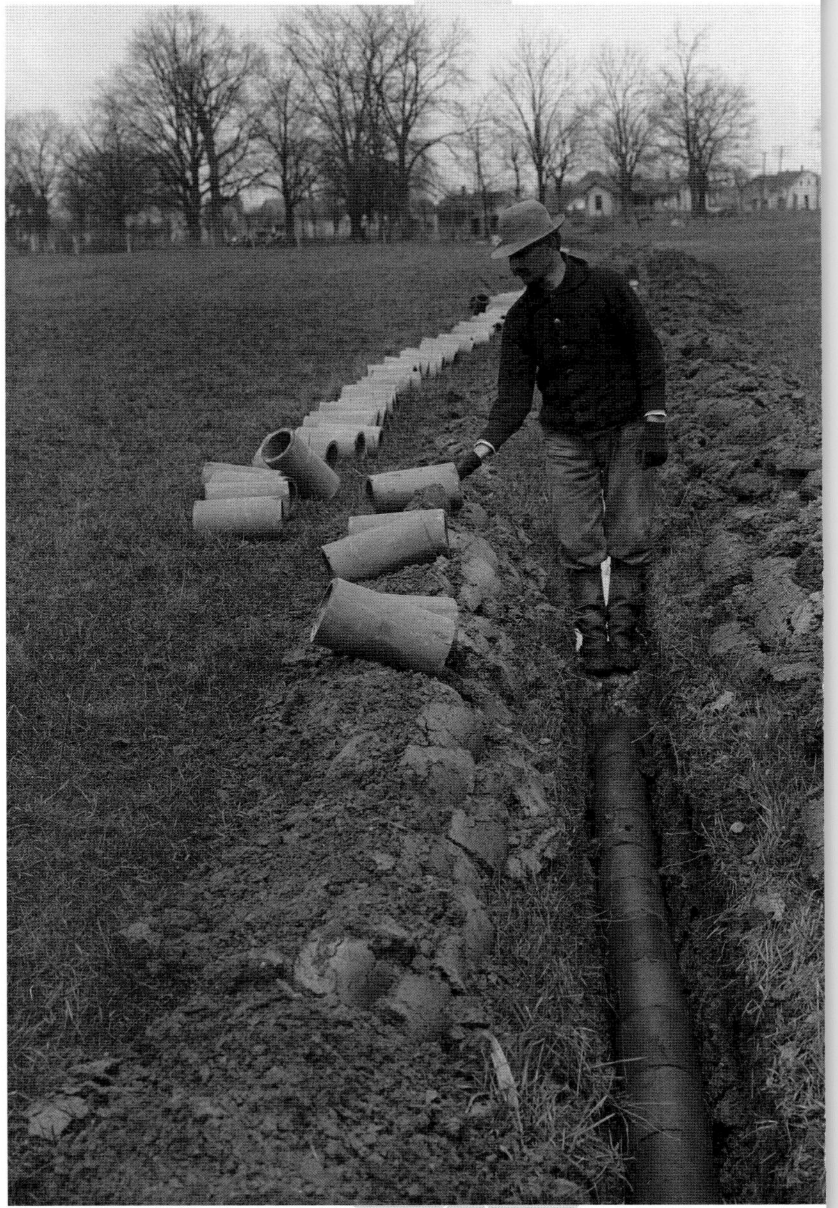

(Above) Laying field tile in a trench (Tippecanoe County, early 1910s)

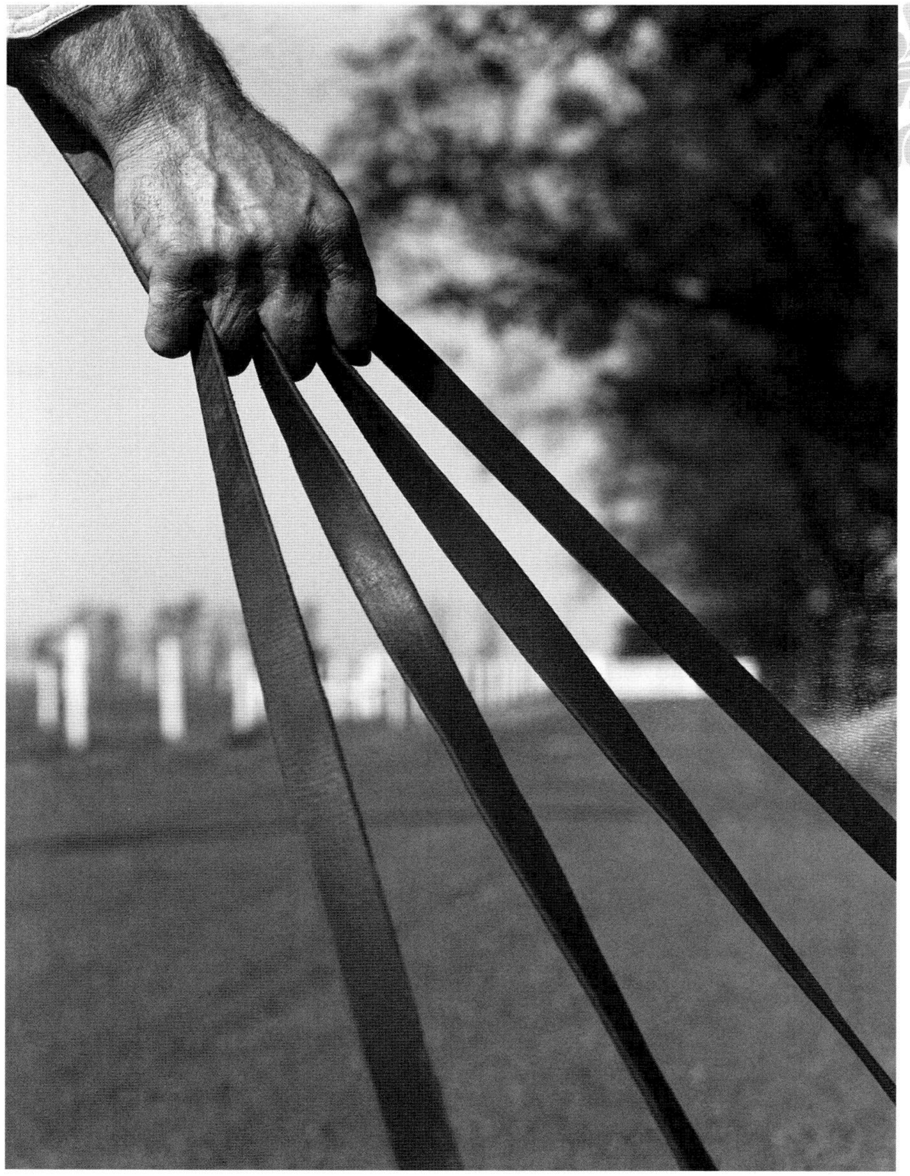

Allen took this picture of a hand holding four lines especially for the *Indiana Farmer's Guide*. (N.p., 1937)

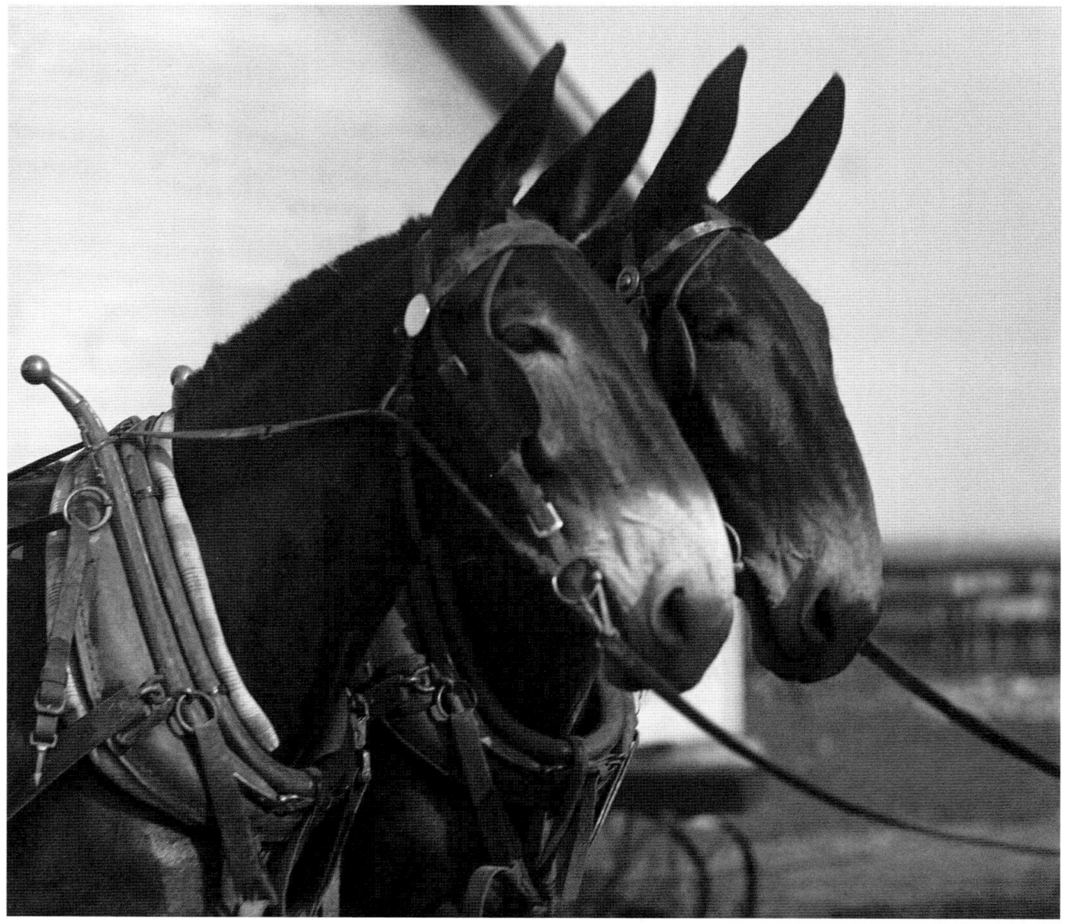

Mules at attention (Indiana, ca. 1920)

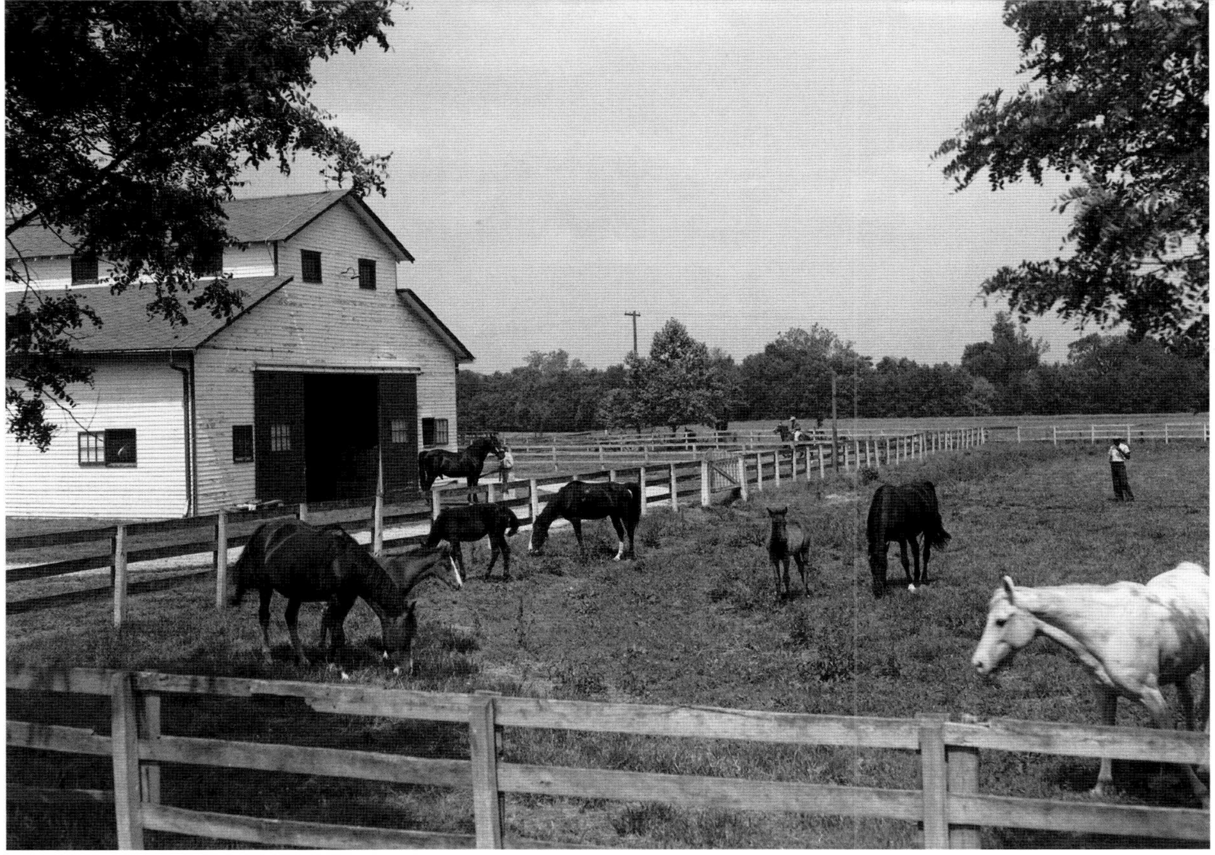

Tennessee Walking Horses at the Harlinsdale Farm in Franklin, Tennessee (Williamson County, TN; 1961)

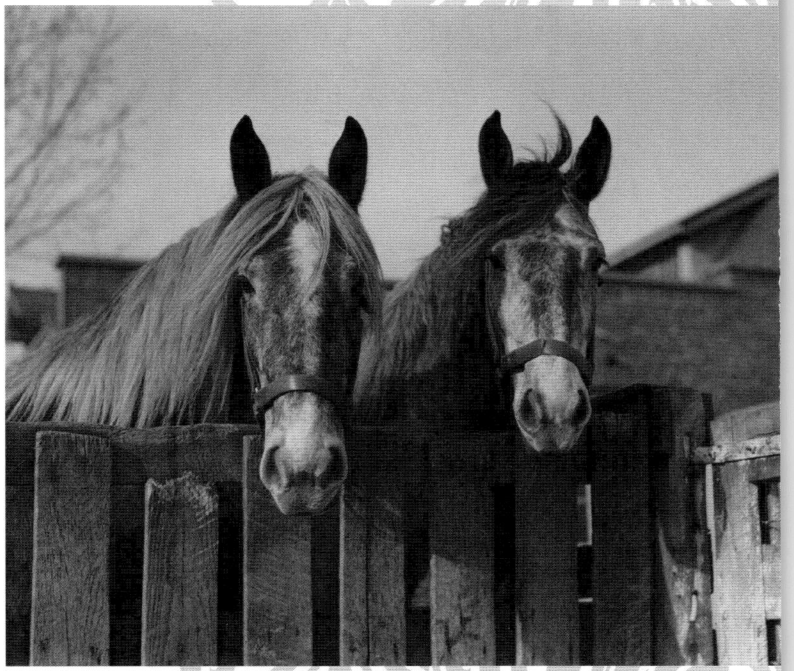

A pair of Percherons look over a fence. (N.p., 1938)

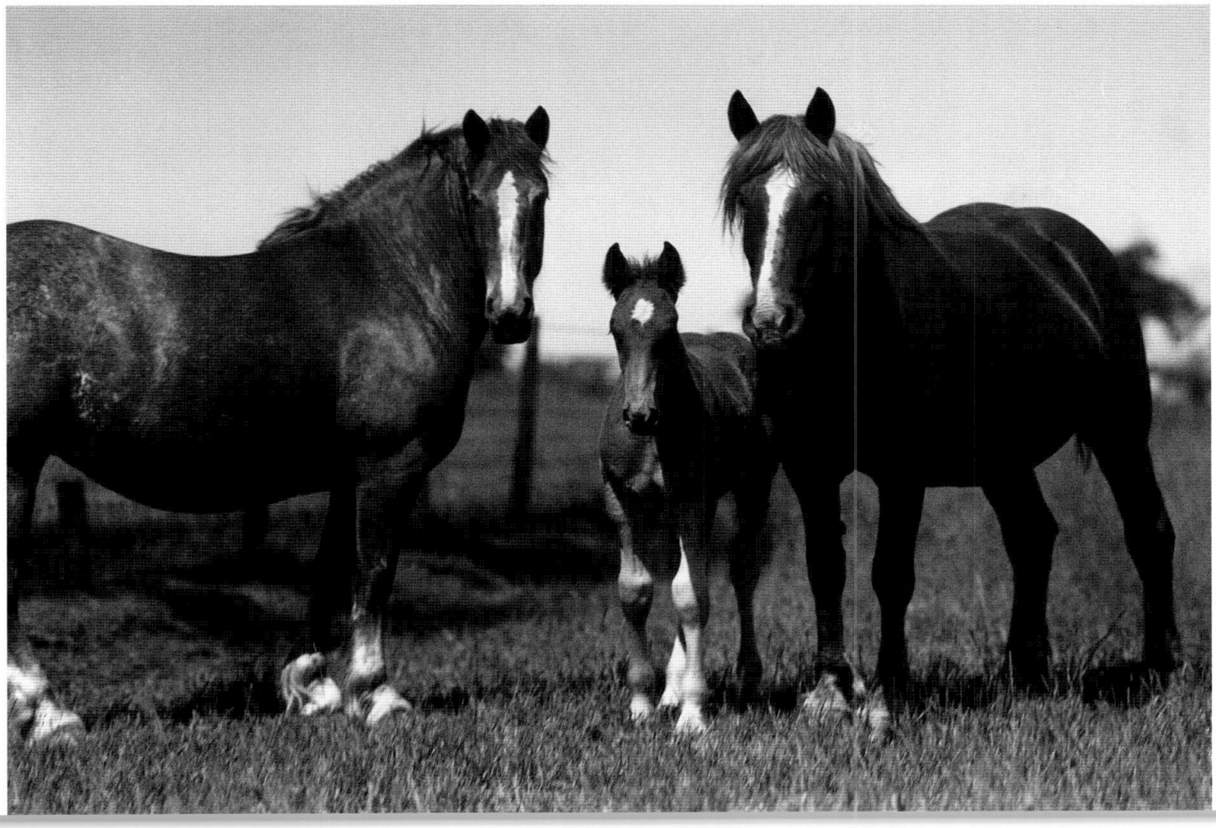

Belgian mares and a foal on pasture at the Purdue University campus farm (Tippecanoe County, 1935)

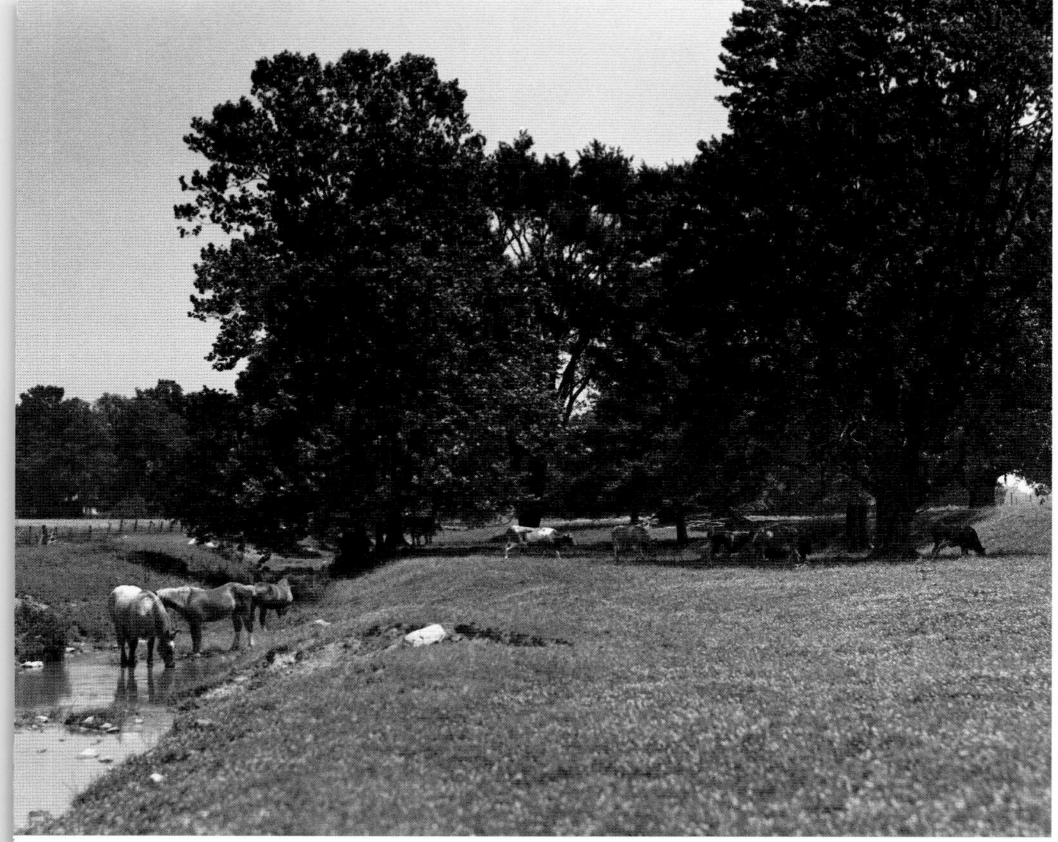

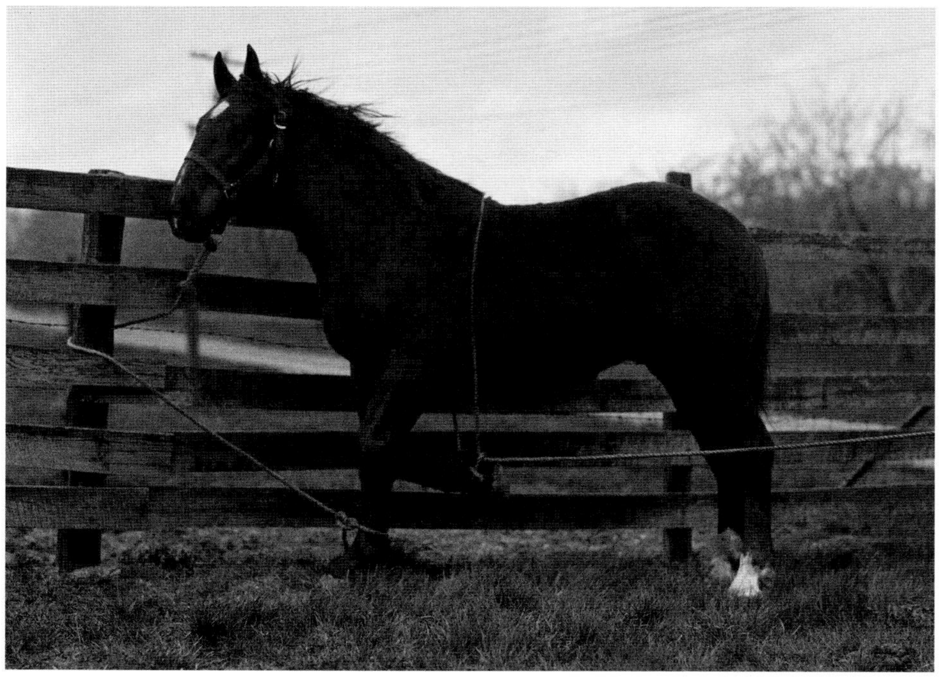

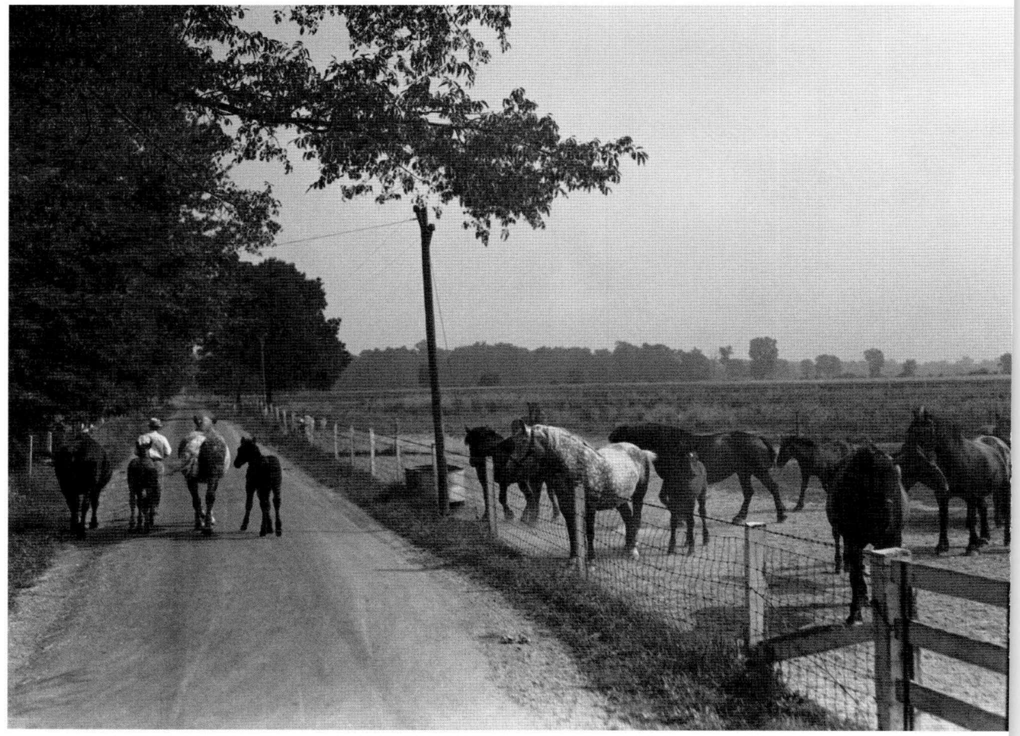

(Top left) Horses drink from a stream on the Cliff Eller farm in Arcadia while cattle graze nearby. (Hamilton County, 1943)

(Bottom left) The Scotch hobble method of restraint works to hold this unruly colt while trimming its feet. (Indiana, 1933)

(Right) Elmer Taft brings Percheron mares and colts to pasture at Purdue University's Lynnwood Farm in Carmel. (Hamilton County, 1943)

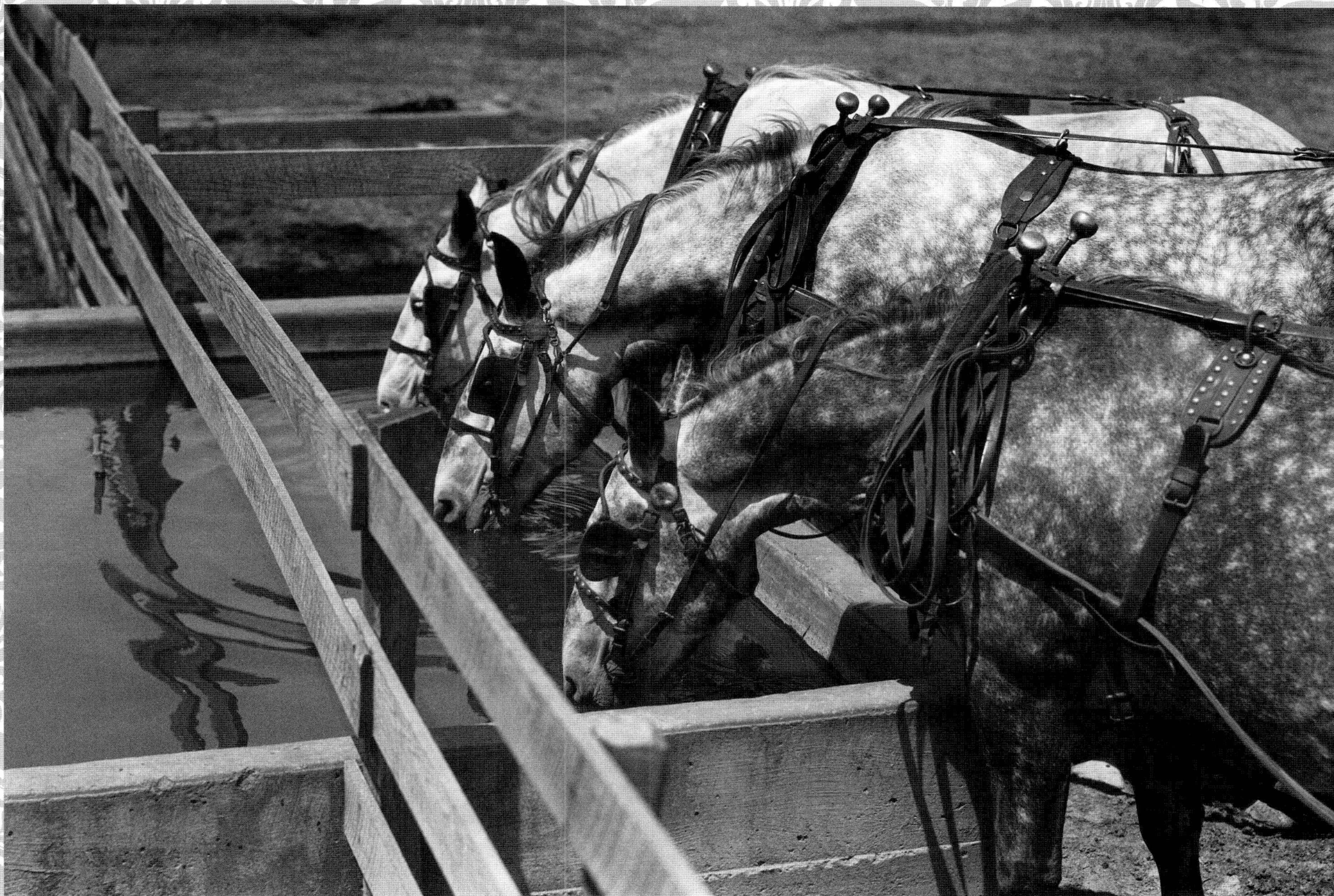

Percherons drink from a concrete watering trough on the farm of C. Orr near Forest. (Clinton County, 1931)

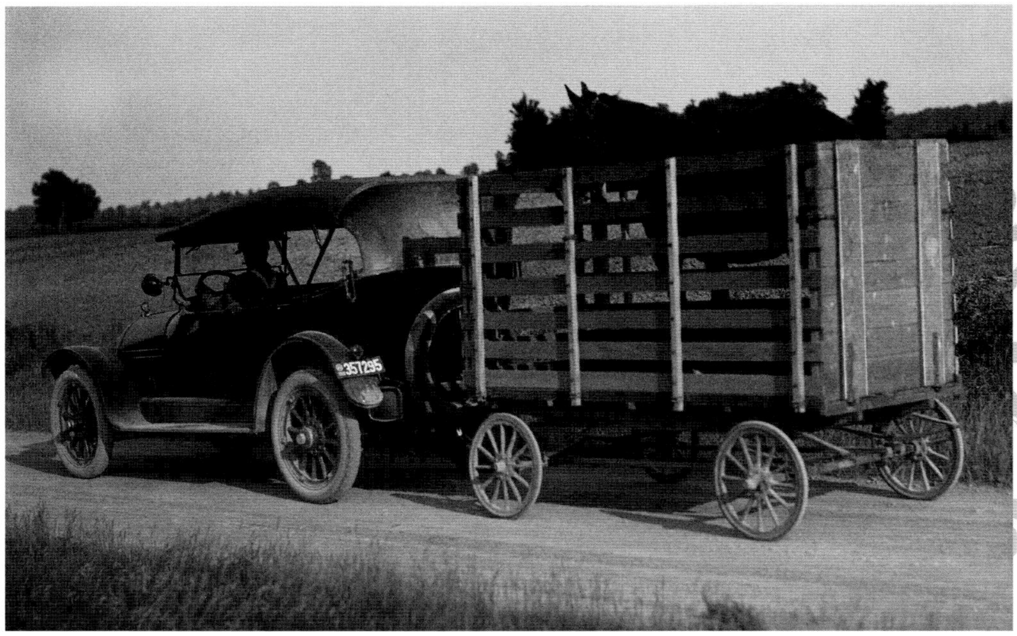

A touring car driven by T. M. Miller pulls a horse trailer in Mendon, Ohio. (Mercer County, OH; 1919)

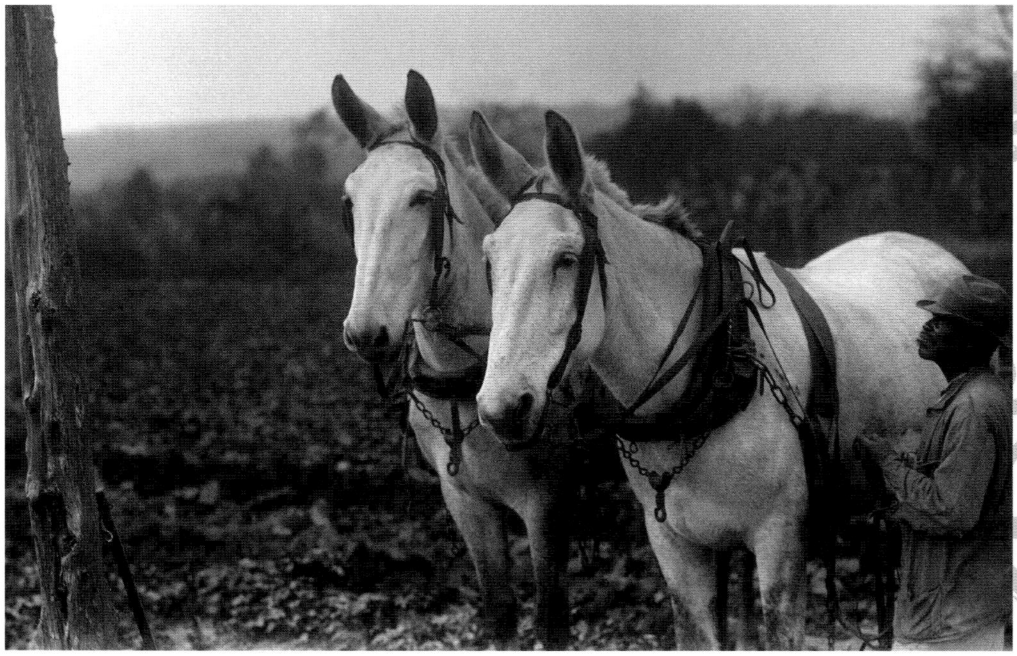

Harnessing a team of mules on the A. R. Anthony farm in Newton, North Carolina (Catawba County, NC; 1927)

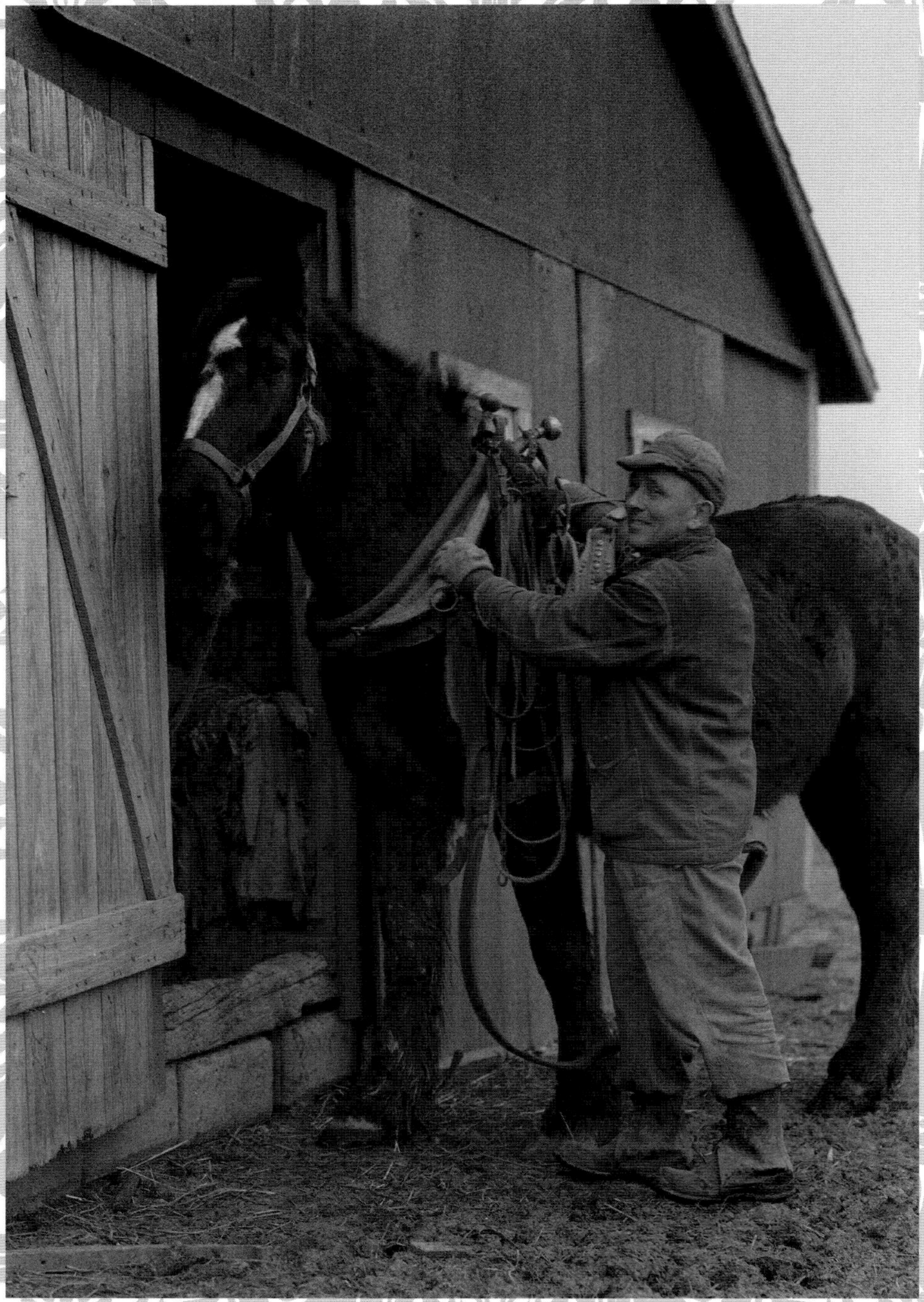

Herman Heaton harnesses his horse in Russiaville. (Howard County, 1939)

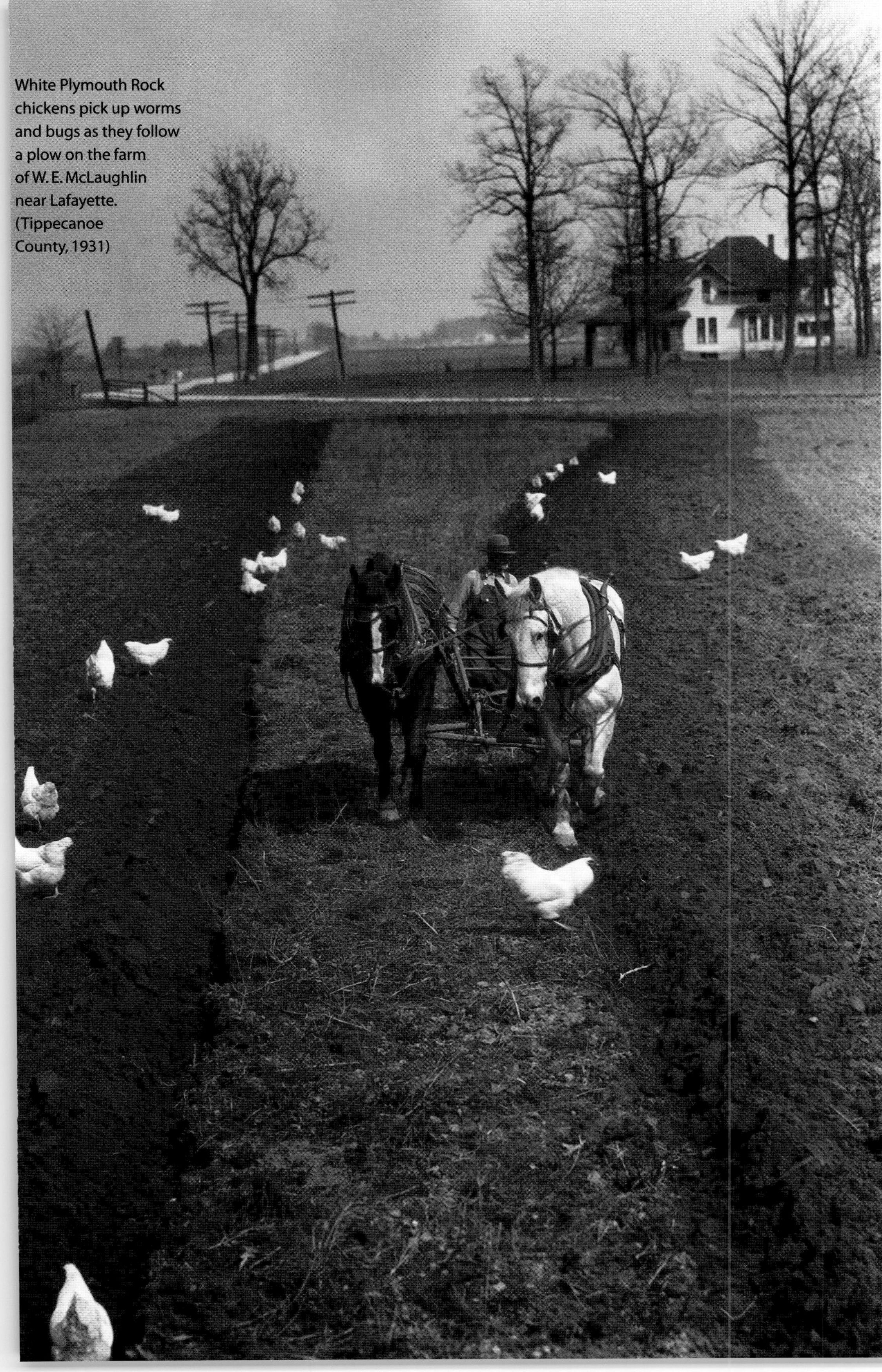

White Plymouth Rock chickens pick up worms and bugs as they follow a plow on the farm of W. E. McLaughlin near Lafayette. (Tippecanoe County, 1931)

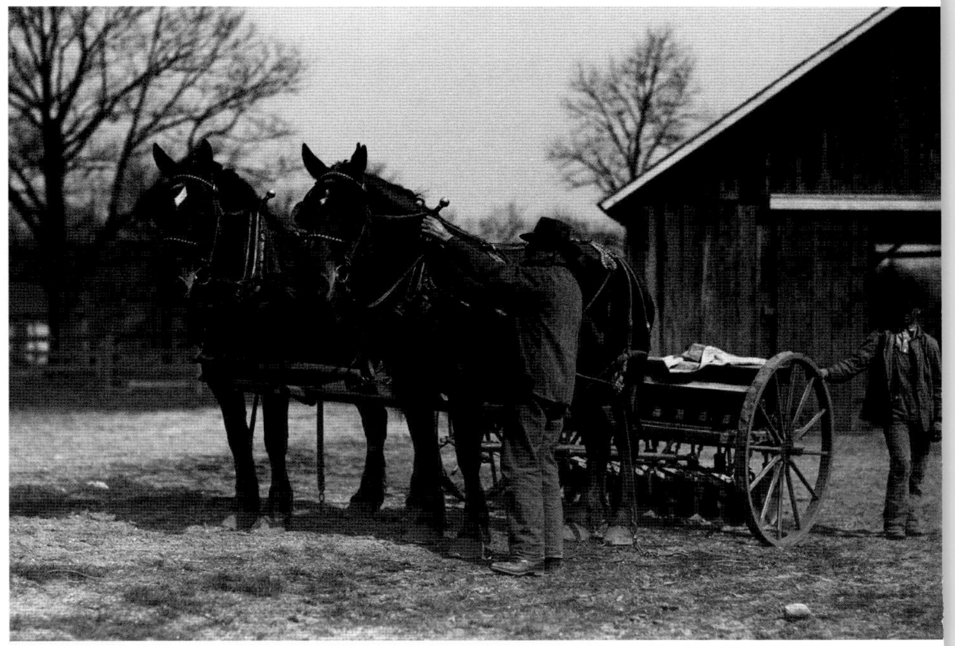

A farmer unhitches a horse team from a grain drill on the Shadeland Stock Farm. (Tippecanoe County, ca. 1910s)

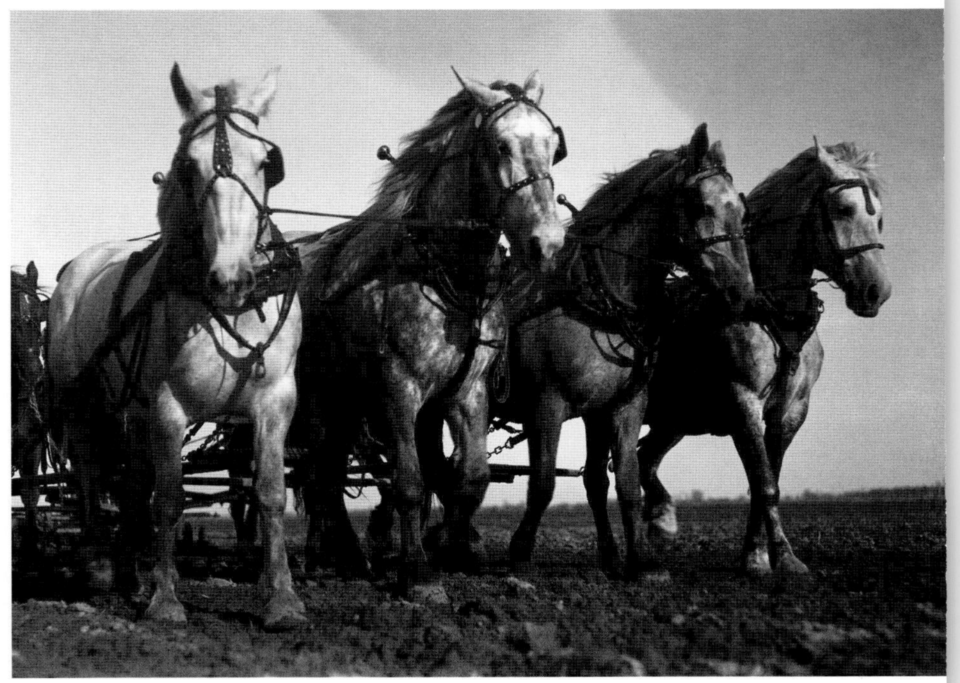

Front view of four Percheron horses in action on the farm of Harry F. Caldwell near Connersville (Fayette County, 1932)

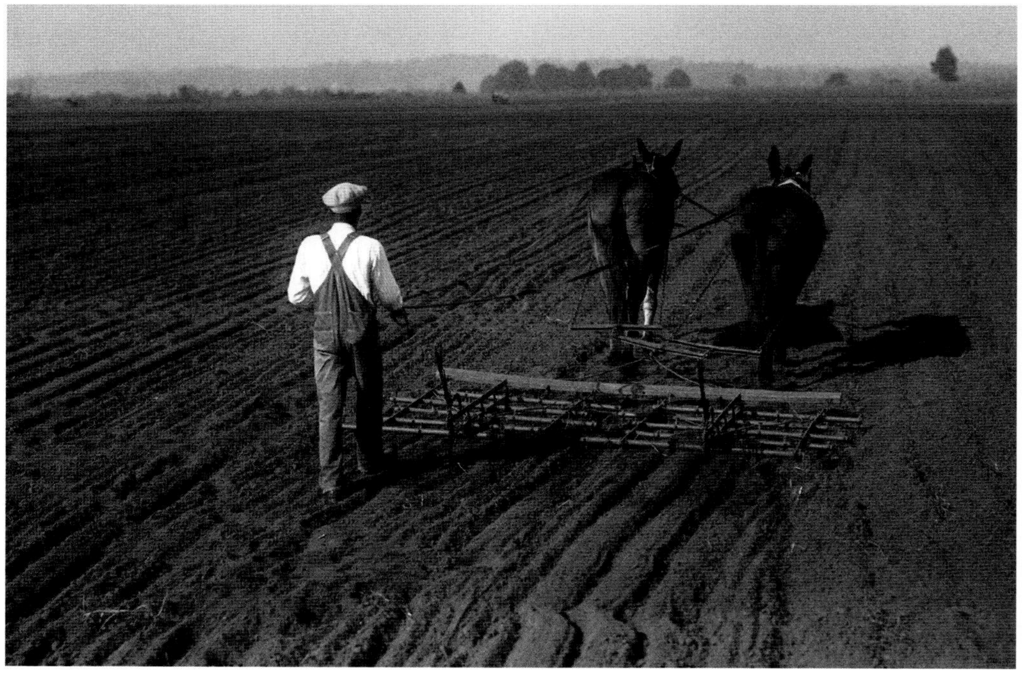

Harrowing with two mules on the farm of L. C. Young and Son in Montgomery, Alabama (Montgomery County, AL; 1927)

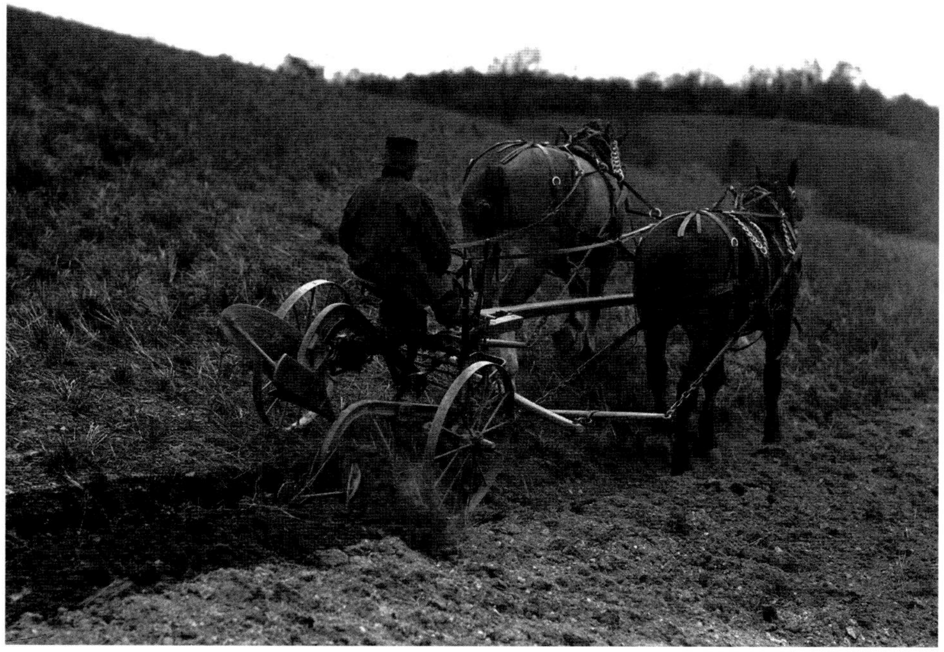

A farmer guides a horse-drawn plow along a hillside on the farm of A. M. McGaughey. (Jefferson County, PA; 1926)

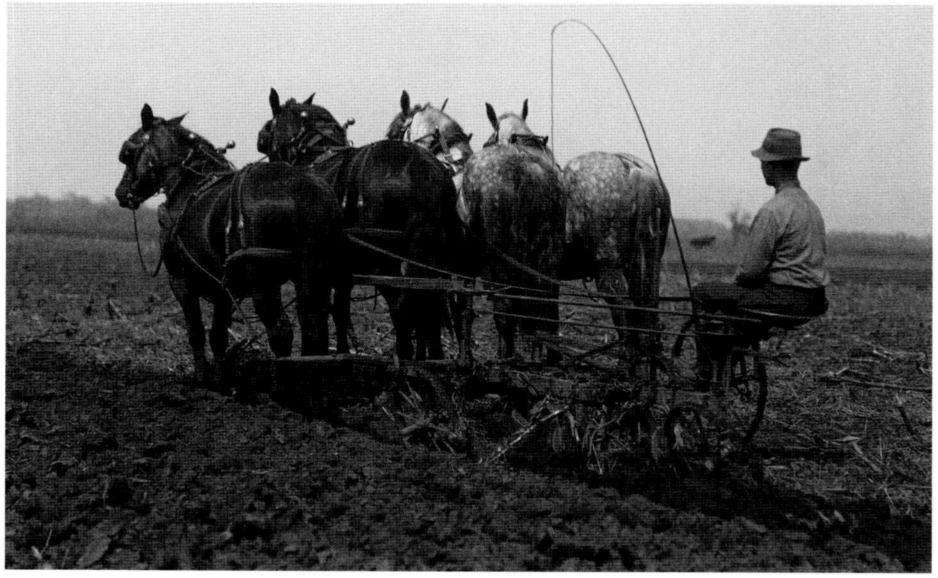

A four-horse team pulls a plow. (N.p., 1915)

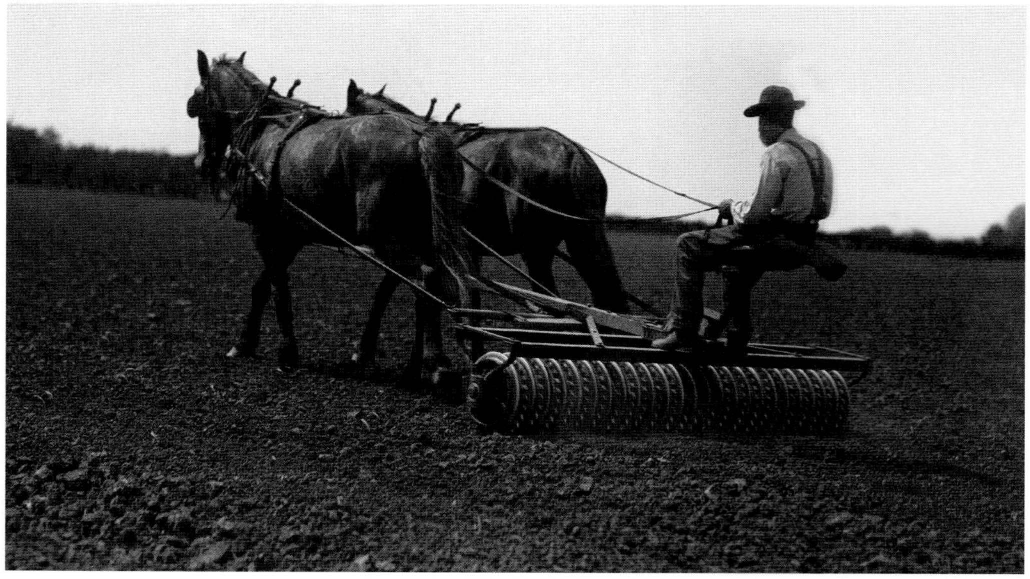

This man operates a horse-drawn roller and clod crusher on the Frank Collins farm in Arthur, Illinois. (Douglas and Moultrie Counties, IL; 1920)

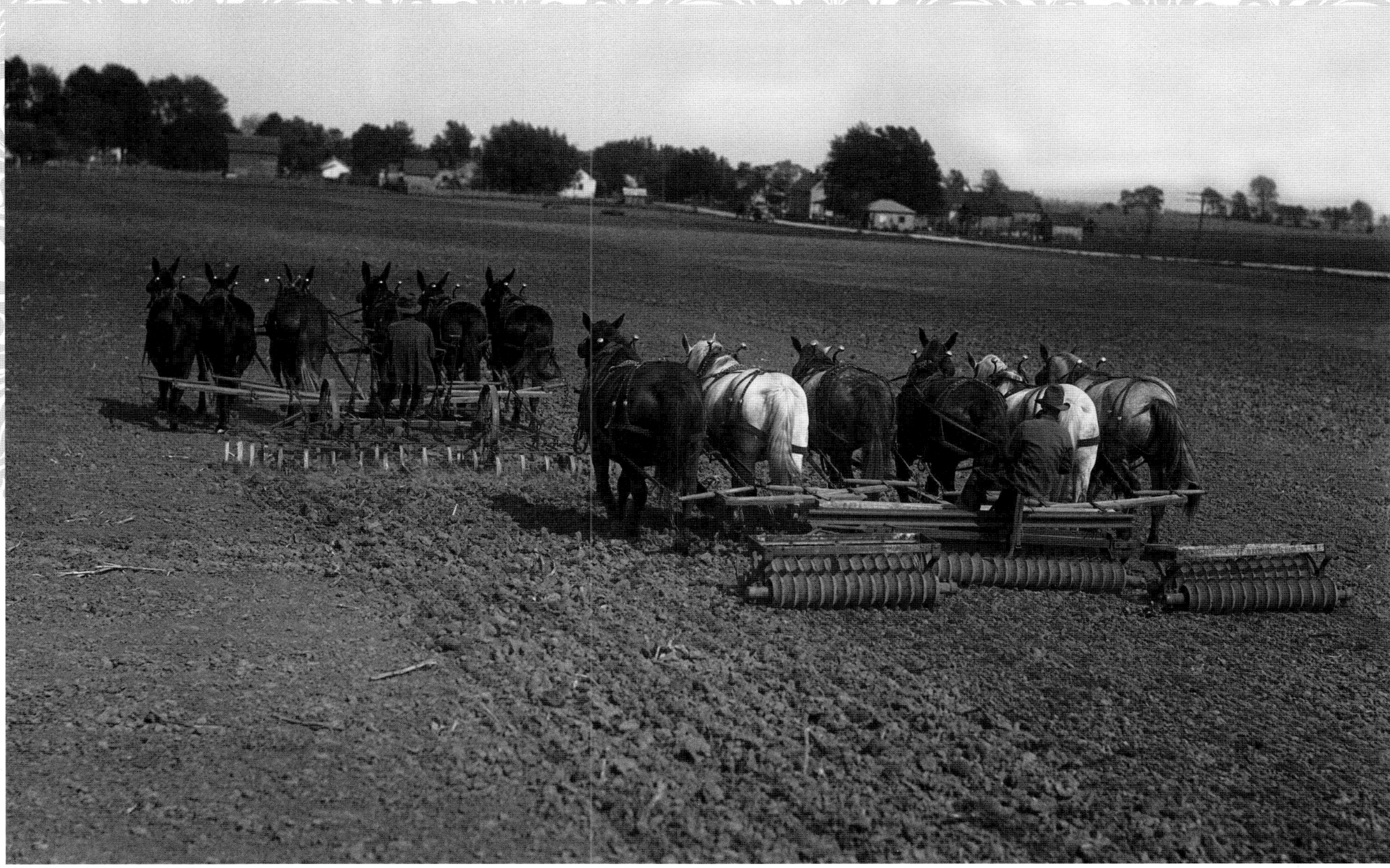

Two men prepare a soybean seedbed using six mules pulling a spring-tooth harrow followed by six horses pulling a corrugated roller on the Charles Meharry farm in Attica. (Fountain County, 1925)

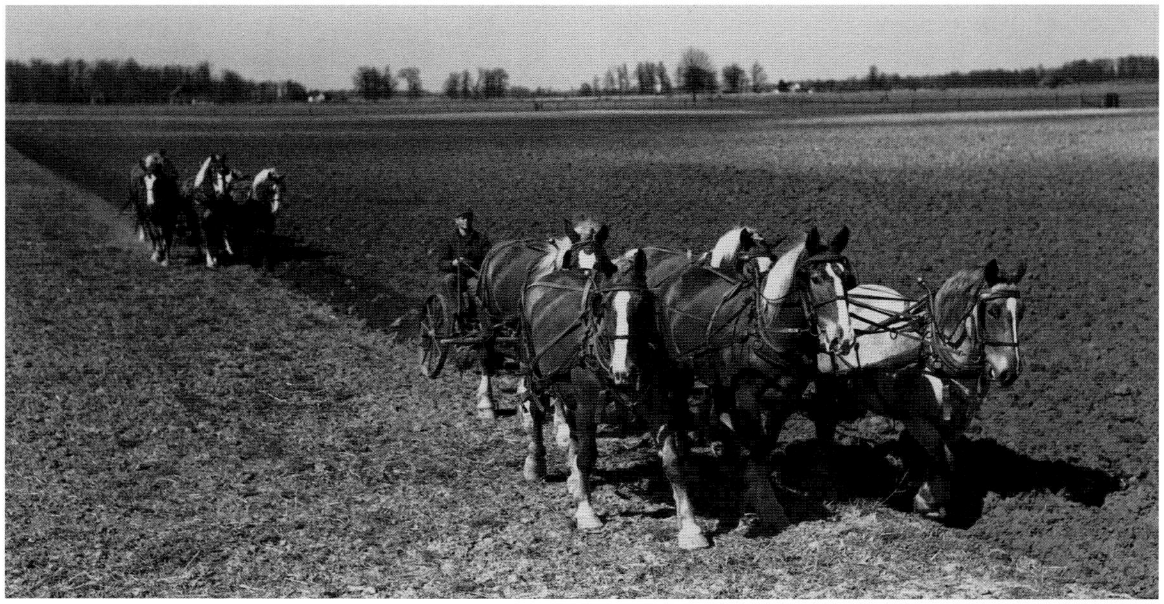

Two teams with five Belgian horses each plow on the farm of Roy Cox of Sheridan. (Hamilton County, 1943)

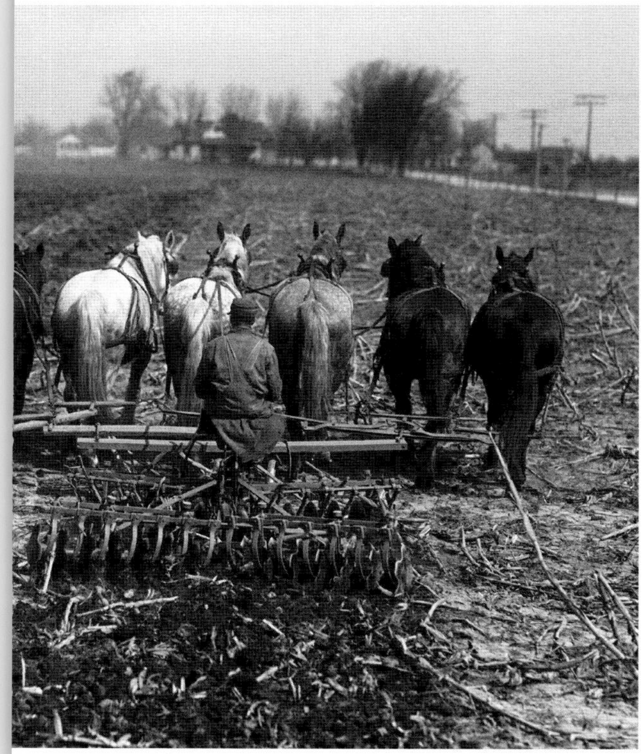

Disking stalks using a six-horse Percheron team on a Frankfort-area farm (Clinton County, 1925)

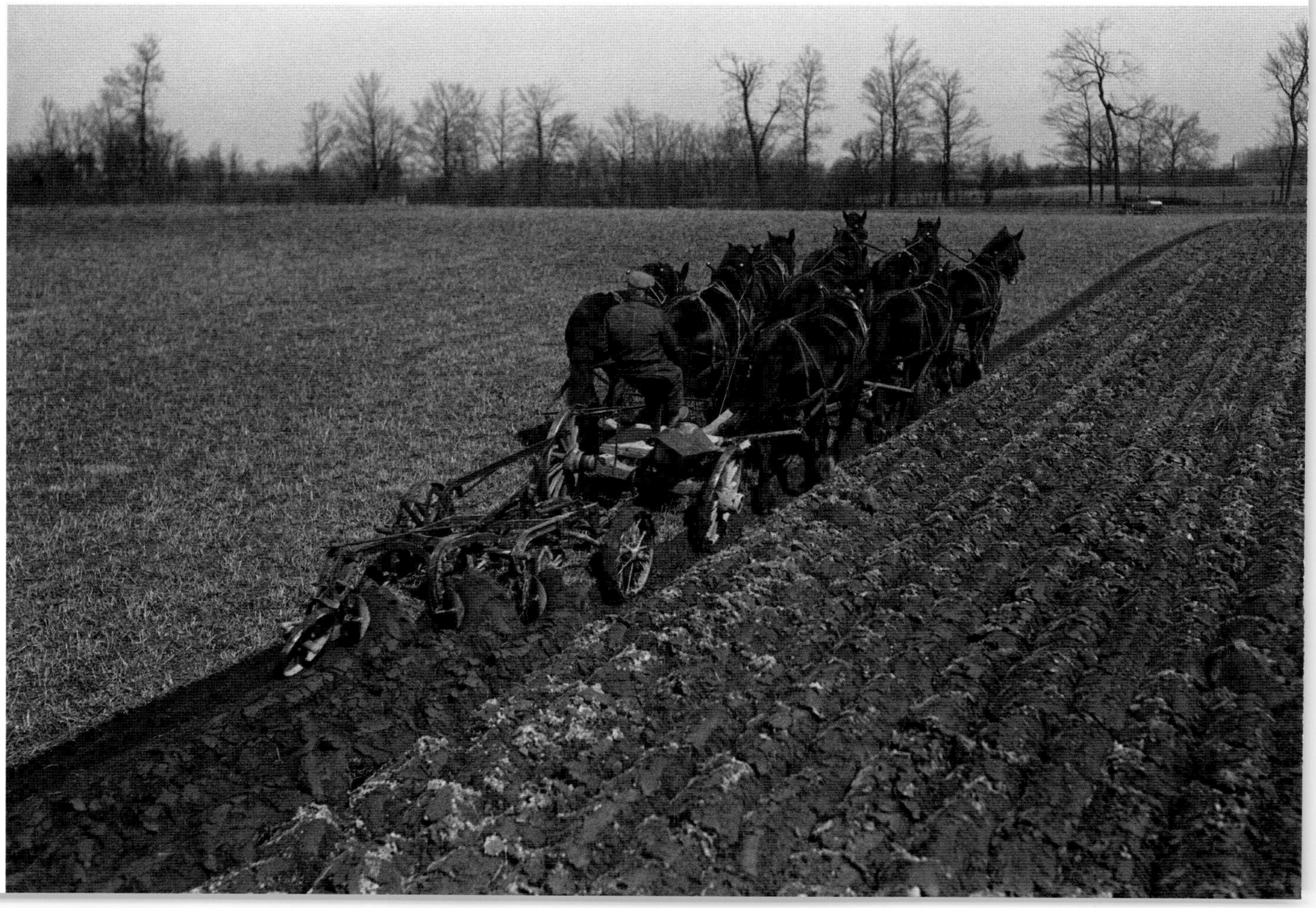

Cleveland Combs from Mulberry plows with nine horses hitched to a three bottom plow. (Clinton County, 1930)

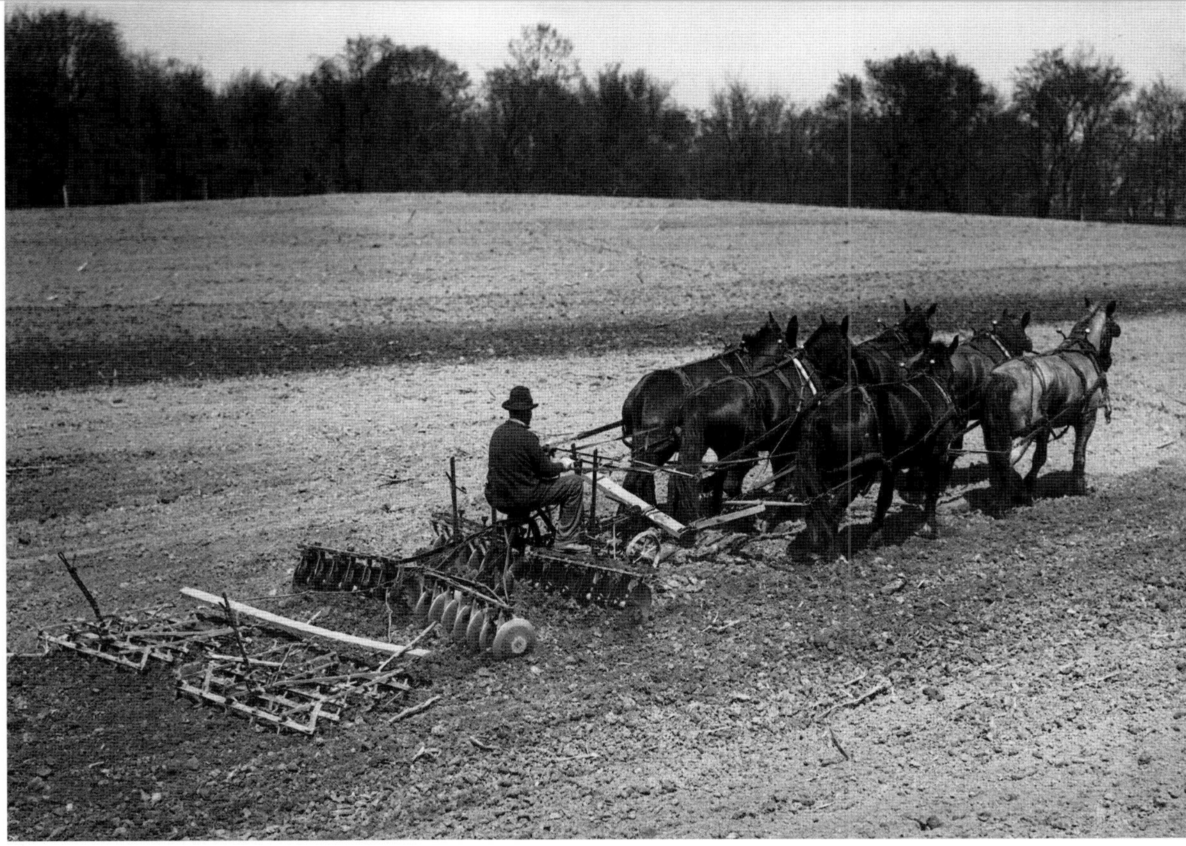

Charles Rutan of Frankfort pulls a large double disc harrow with a team of six horses. (Clinton County, 1929)

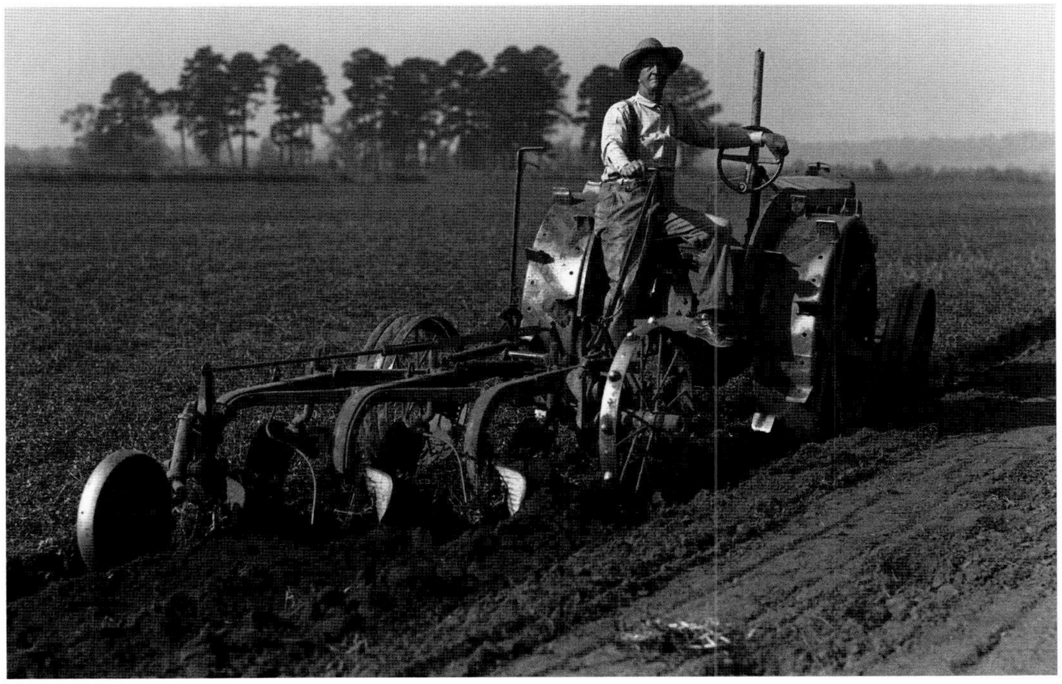

L. C. Young plows with a tractor in Montgomery, Alabama. Allen noted, "The ground was so hard in the fall of 1927 that he could hardly keep his plow in the ground." (Montgomery County, AL; 1927)

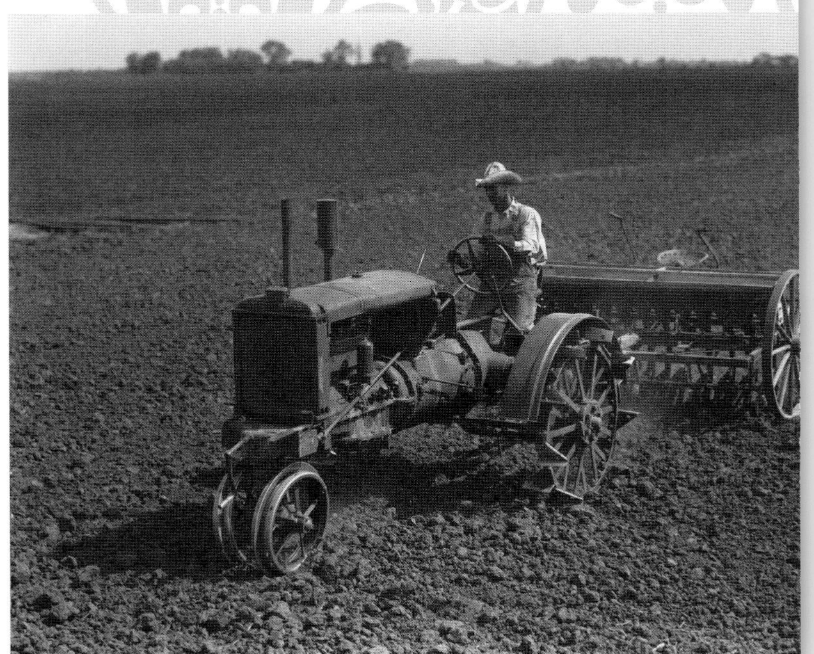

A man uses an Allis-Chalmers all-purpose tractor to work a field. (Indiana, 1932)

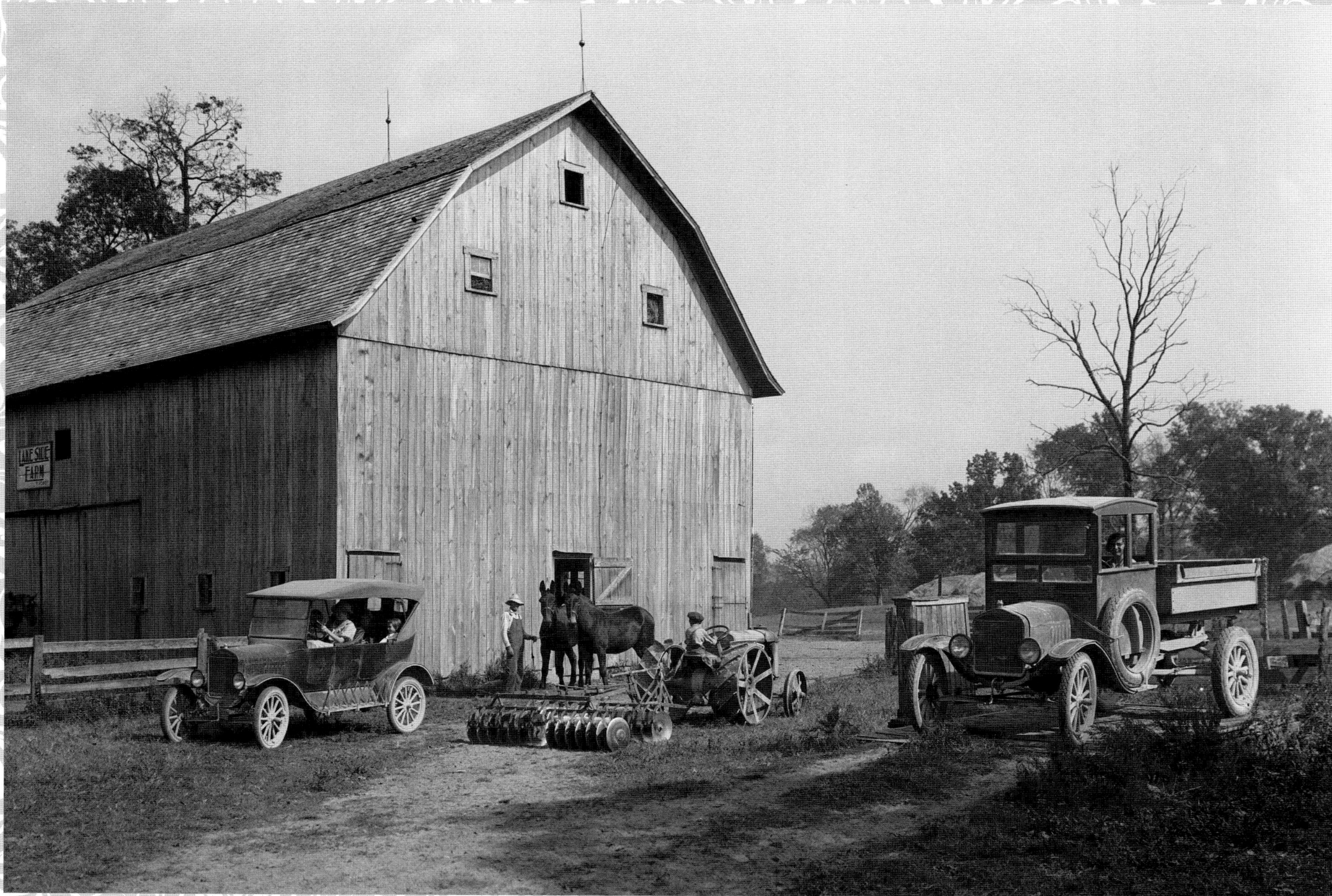

An automobile, tractor, and Ford truck on the farm of Frank Peavey from Jasonville (Greene County, 1924)

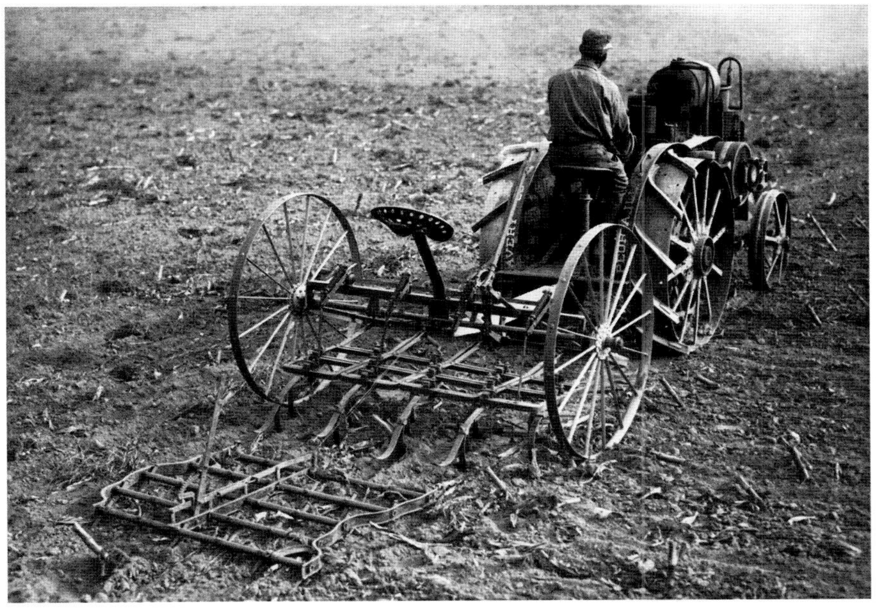

Preparing a seedbed using an Avery tractor to pull a field cultivator and harrow (Tippecanoe County, 1917)

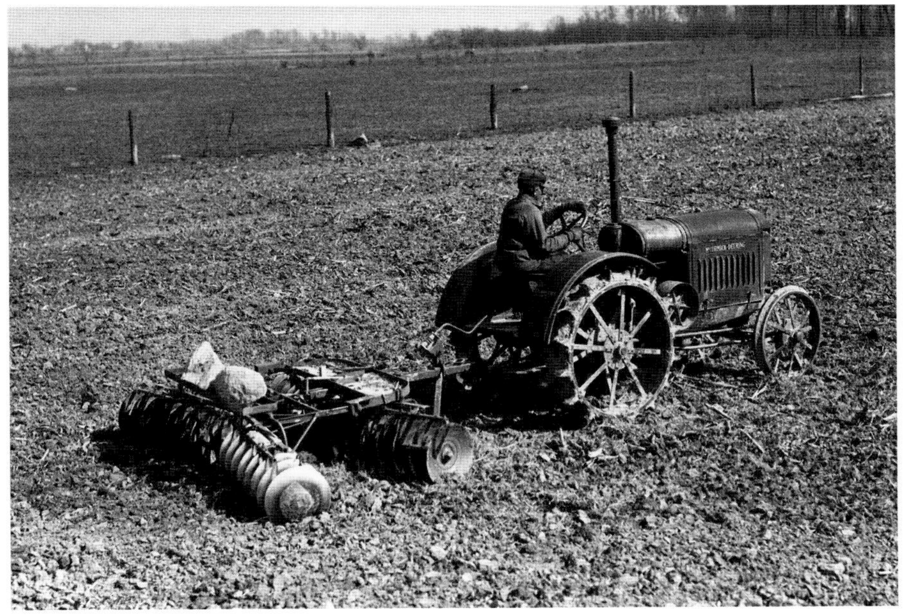

A farmer uses a steel-wheeled McCormick-Deering tractor with added weight from rocks placed on this disc on the George Bremer farm in Lafayette. (Tippecanoe County, 1930)

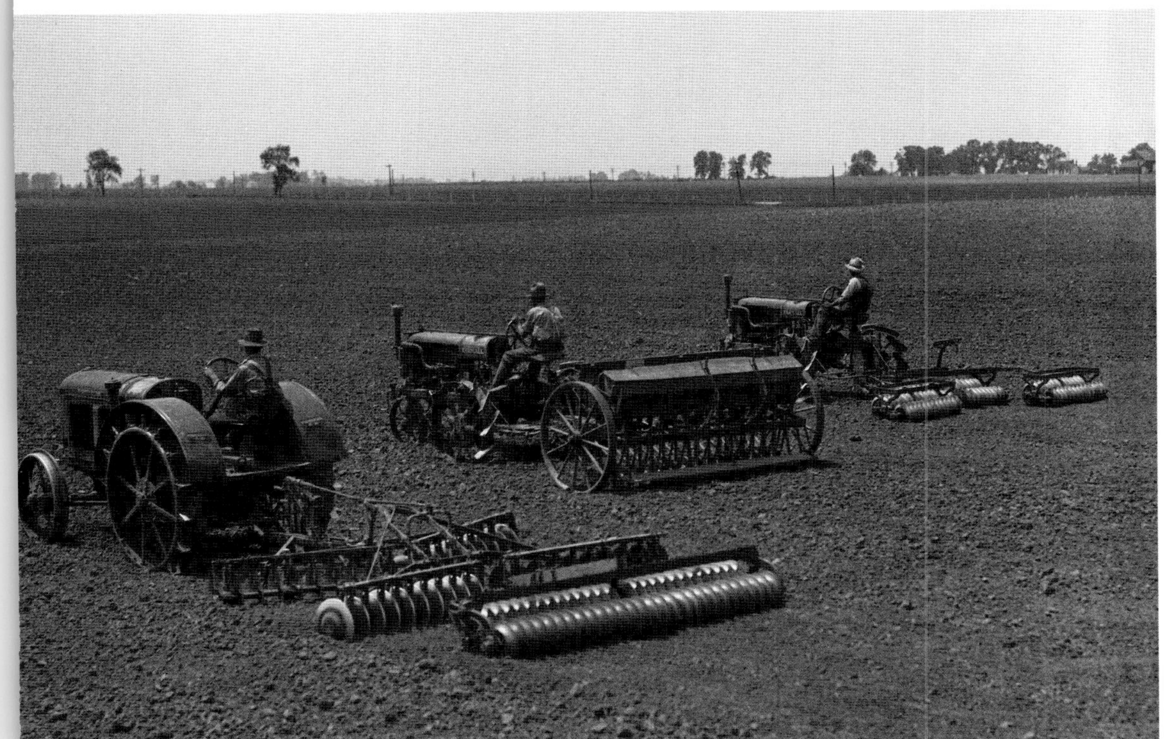

Three steel-wheeled tractors pull a double disc, soybean drill, and corrugated roller on the Henry Marshall farm. (Tippecanoe County, 1930)

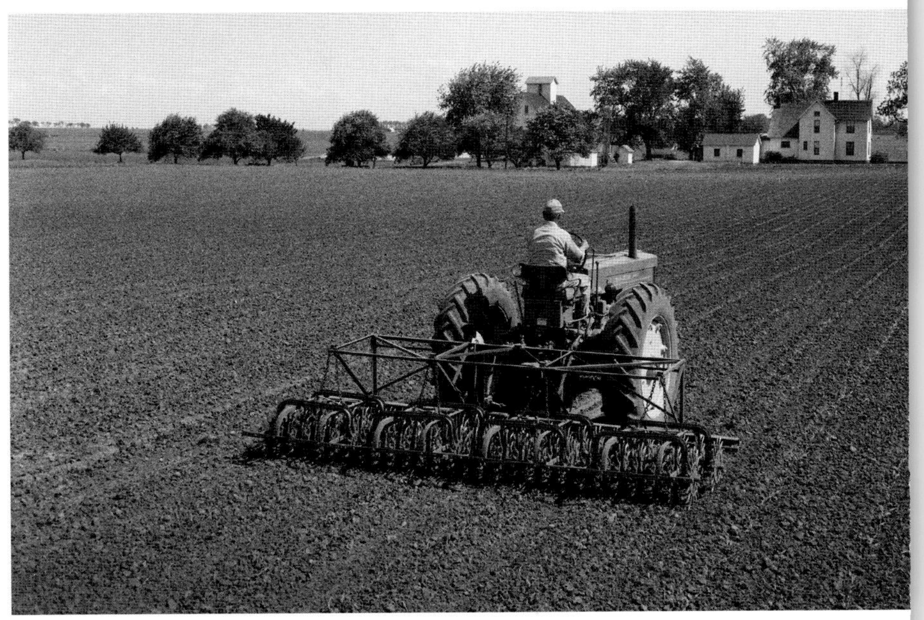

Lewis Strasburger from Earl Park uses a John Deere rotary hoe that he mounted on his tractor to make it easier to lift up and move by highway when going between fields. (Benton County, 1956)

Conserving Soil

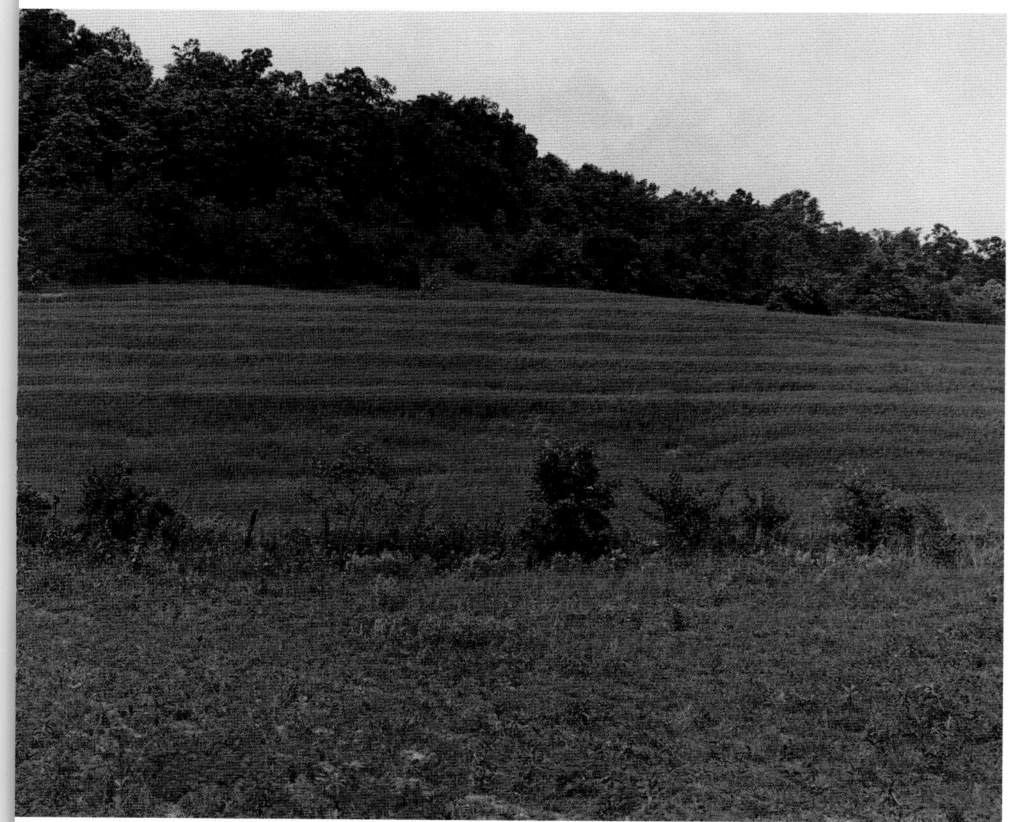

Alfalfa grows in a terraced field on the Arthur Bentley farm in Paoli. The first terraced field was built in Indiana in 1916. (Orange County, 1923)

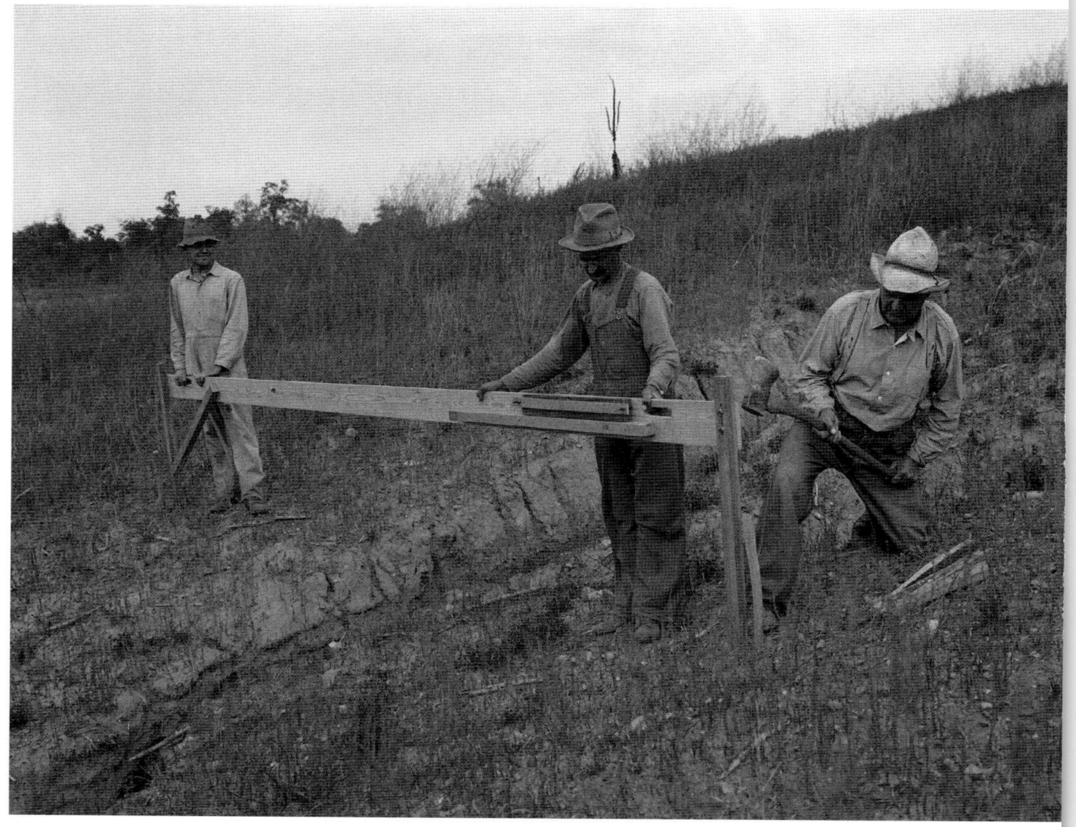

These men use a carpenter level to determine the terrace construction grade on the Russell Thompson farm in Frankfort. (Clinton County, 1922)

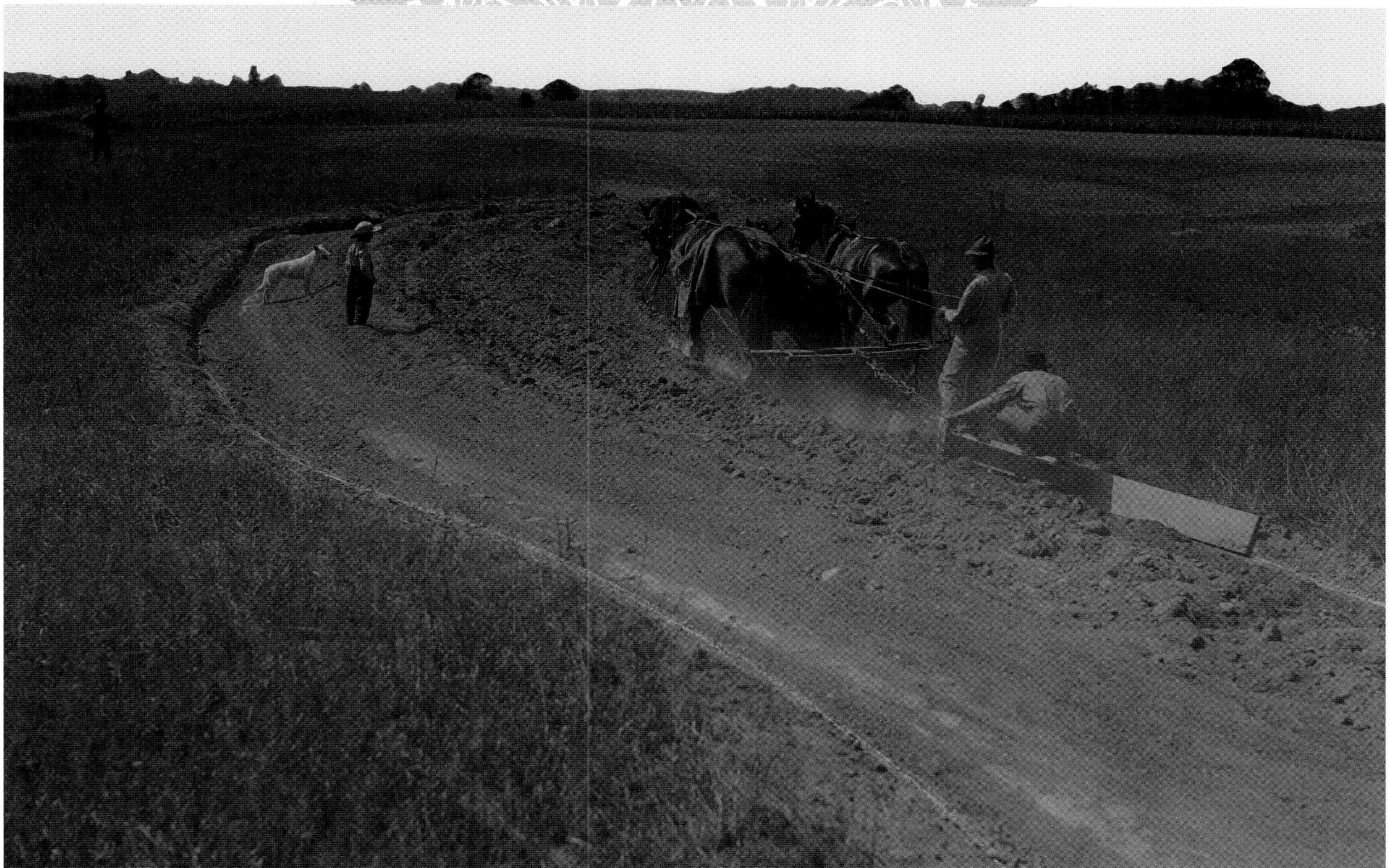

Using a team of three horses to create a ditch and mound dirt for the terrace on the Russell Thompson farm in Frankfort (Clinton County, 1922)

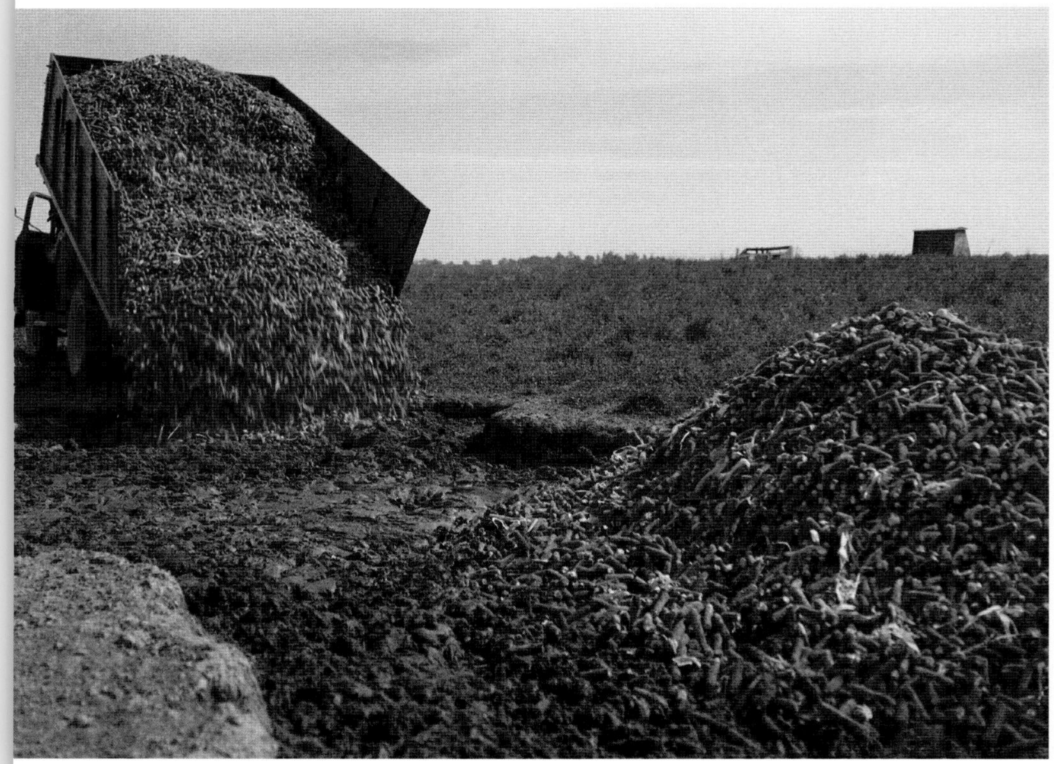

A truckload of corncobs dumped into a hog wallow prevents further soil erosion. (N.p., 1957)

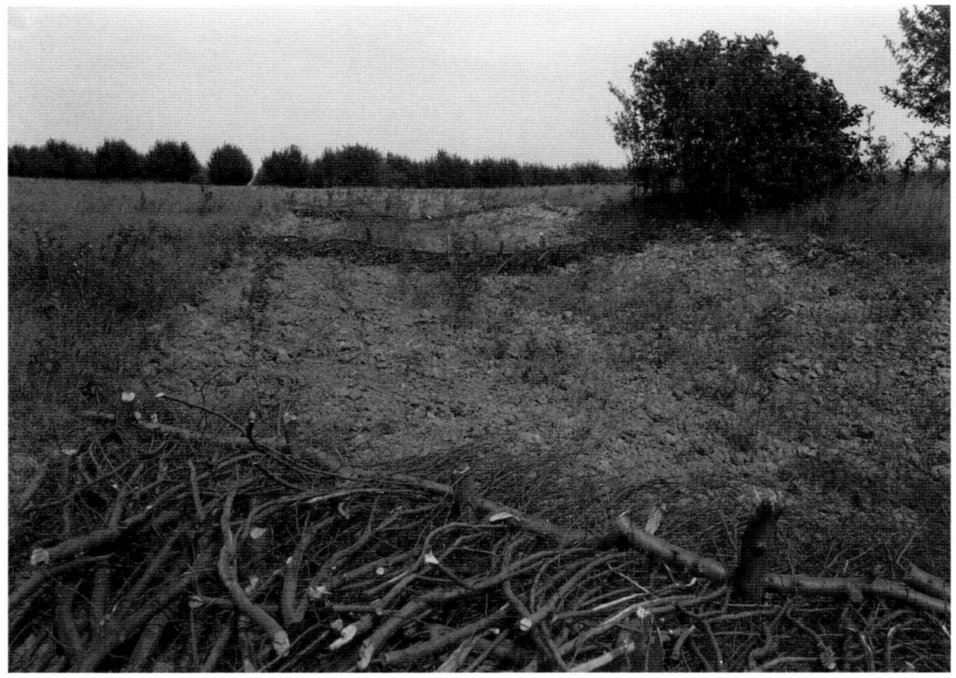

Here, brush dams and young plantings of black locust trees prevent soil erosion at the Purdue University Horticulture Farm near West Lafayette. (Tippecanoe County, 1934)

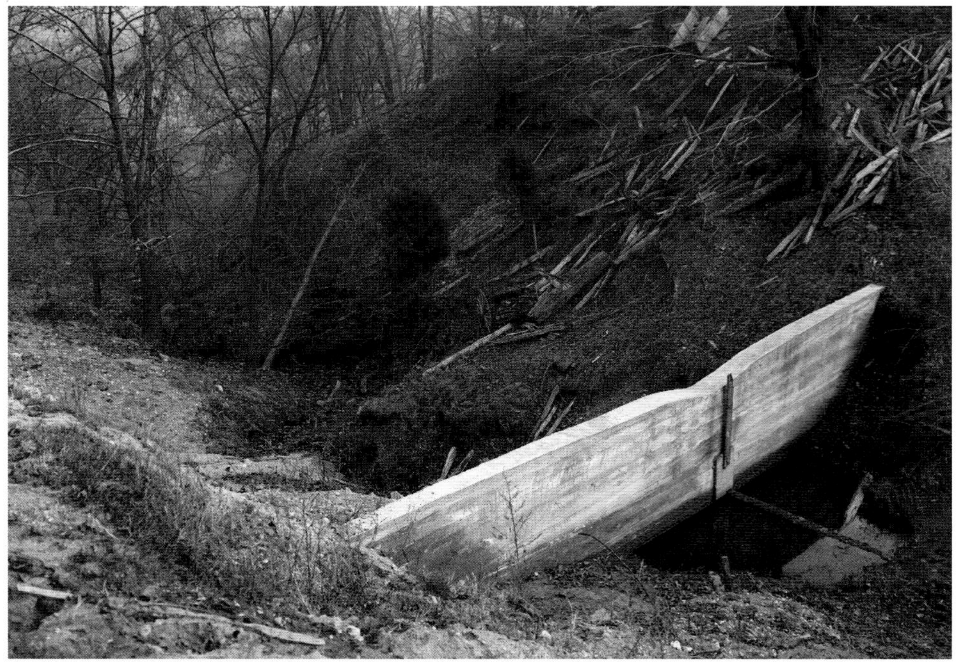

On the Jack Van Natta farm in Battle Ground, a concrete dam stops soil from washing away. (Tippecanoe County, 1917)

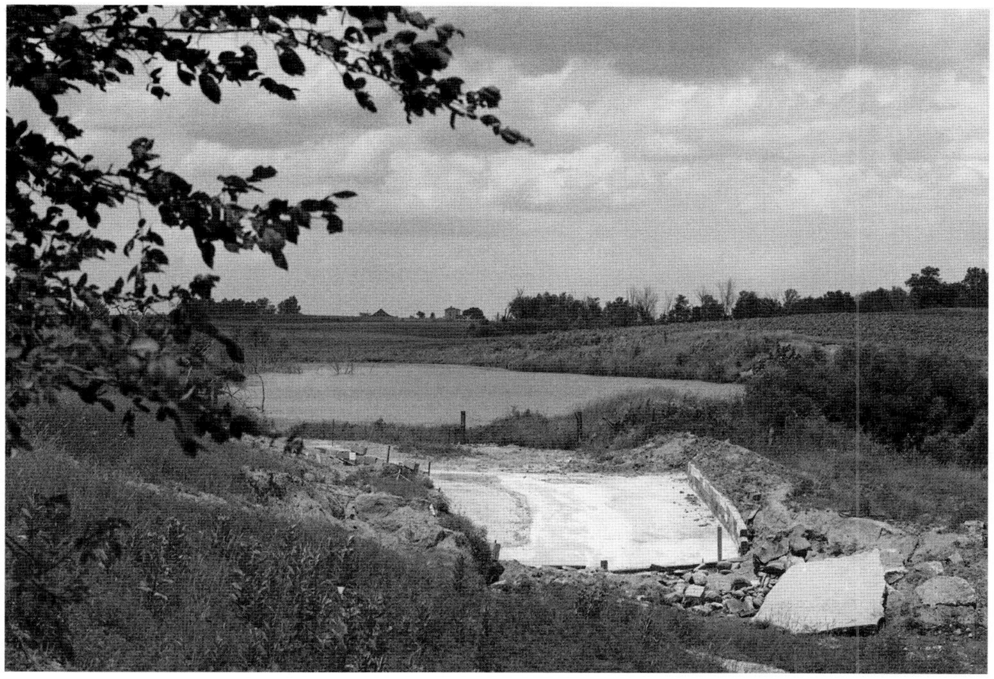

A farm pond showing a spillway (Indiana, 1959)

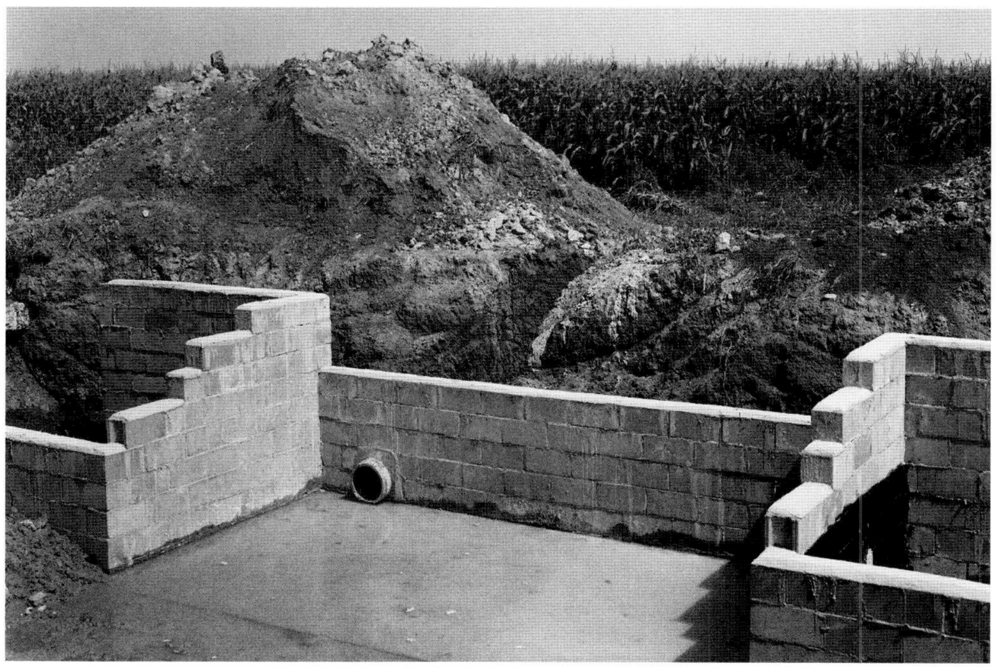

A tile outlet and concrete dam at the end of a sodded waterway tackles an erosion issue. (Benton County, 1960)

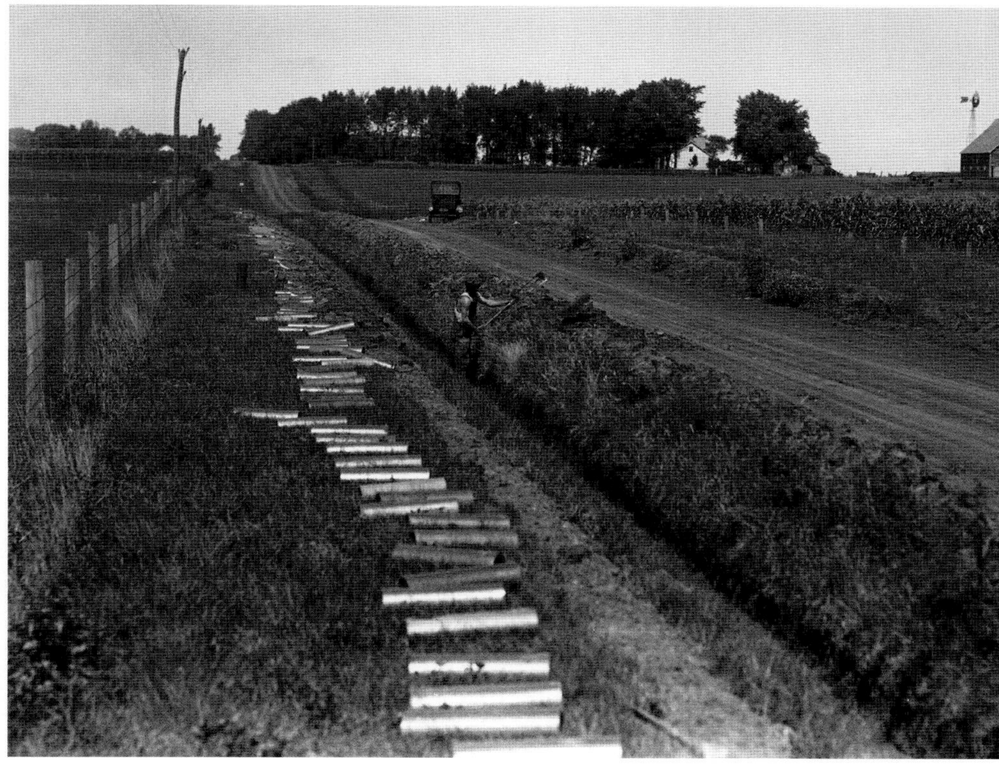

Laying tile for drainage along a country road (Hamilton County, IA; 1923)

Sweetening a Sour Soil

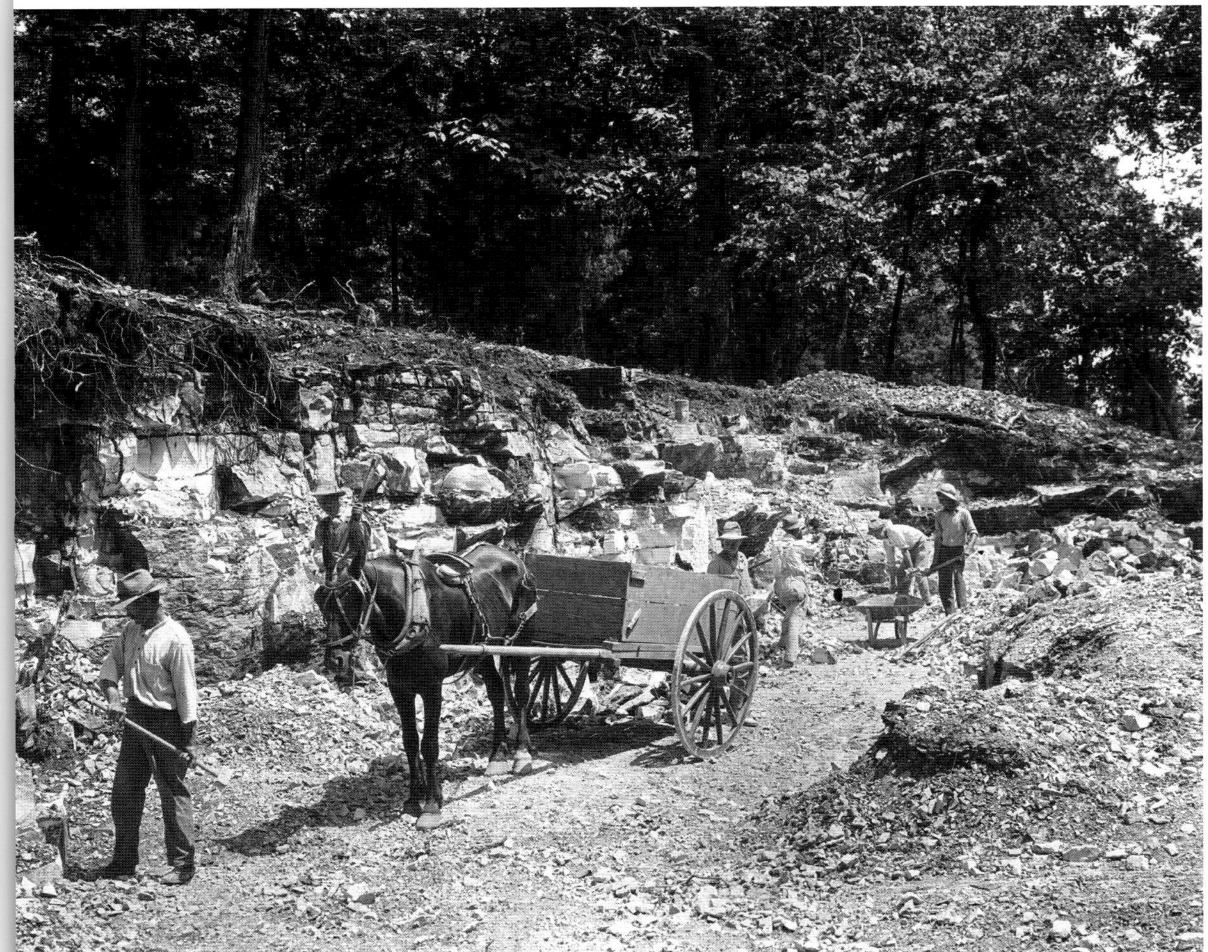

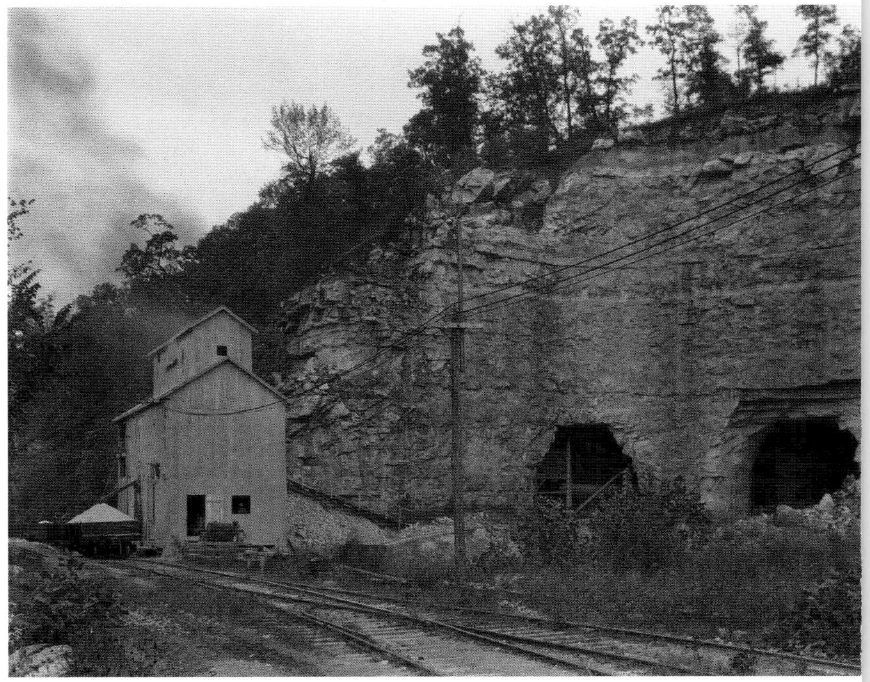

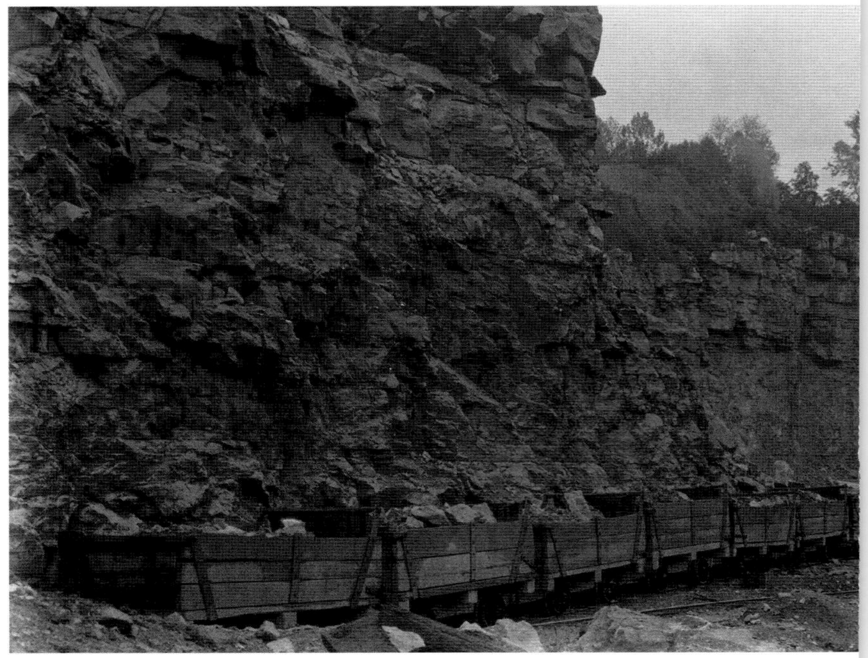

(Above) Workers cut lime in a quarry. (Harrison County, 1921)

(Top right) Loading agricultural limestone onto flatcars in Milltown (Crawford County, 1922)

(Bottom right) Carts filled with limestone at Marengo (Crawford County, 1922)

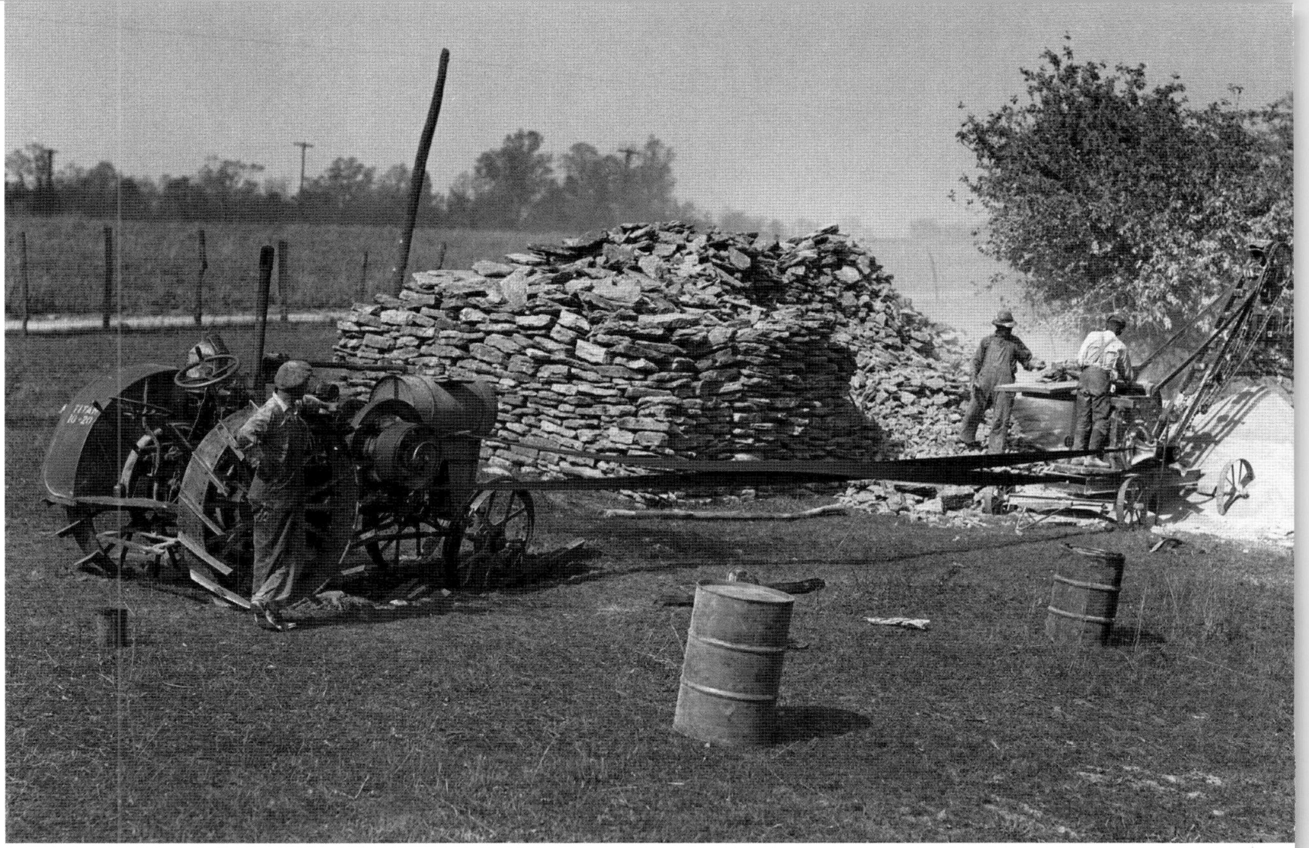

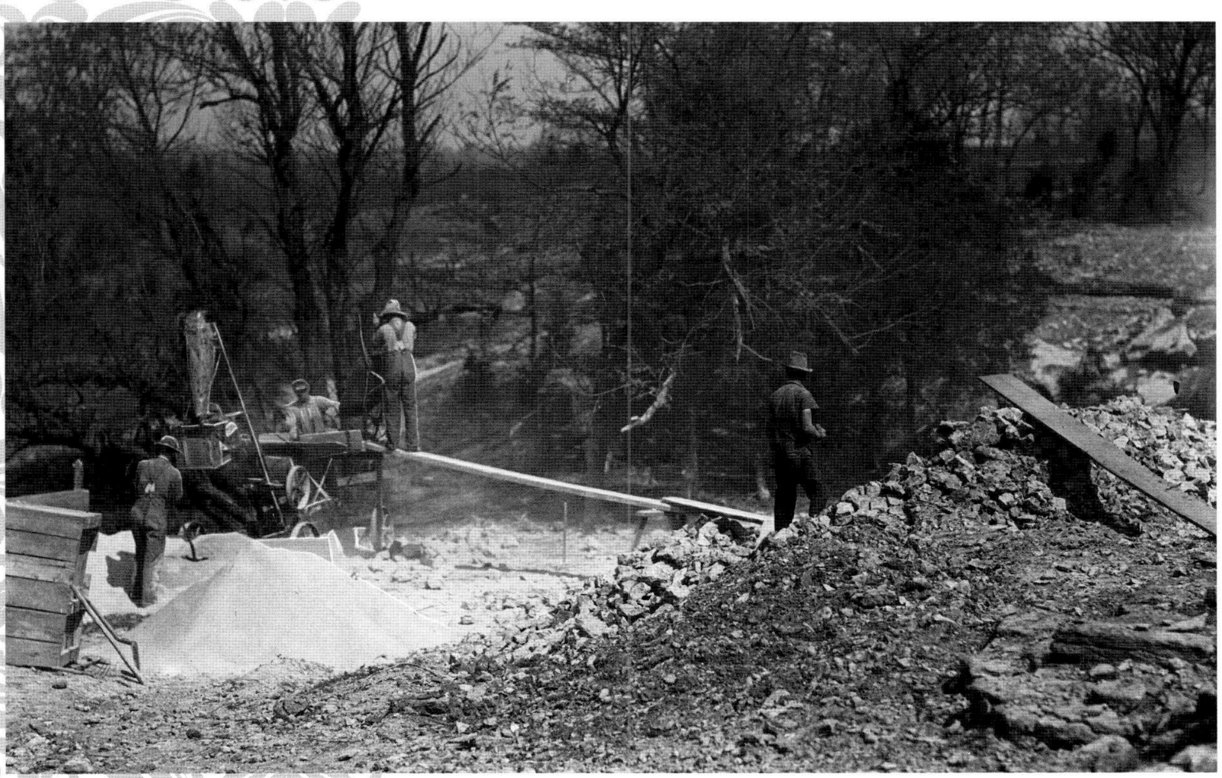

(Top) A crew crushes approximately 120 tons of limestone on the Bert Scott farm in Madison. Scott applied six tons of lime per acre to change his soil pH so he could grow alfalfa. (Jefferson County, 1925)

(Bottom) For this photo, Allen wrote, "The crew of Wagner and Fall from Dupont, Indiana, doing custom limestone pulverizing on the farm of Mr. Schnabelt near Madison, Indiana. It costs the farmer $1.50 per ton to have the stone pulverized if he quarries and delivers the stone himself, or if he prefers to have the crew quarry the stone, it costs him $2.50 per ton pulverized and ready to be applied to his farm. One hundred and fifty tons are being crushed for three different farmers at this location, where the stone is plentiful. During this spring of 1925 up to the first of May, this crew has pulverized 1,500 tons of limestone for farmers in this community, and they have contracted 1,200 tons more in the same community for this fall." (Jefferson County, 1925)

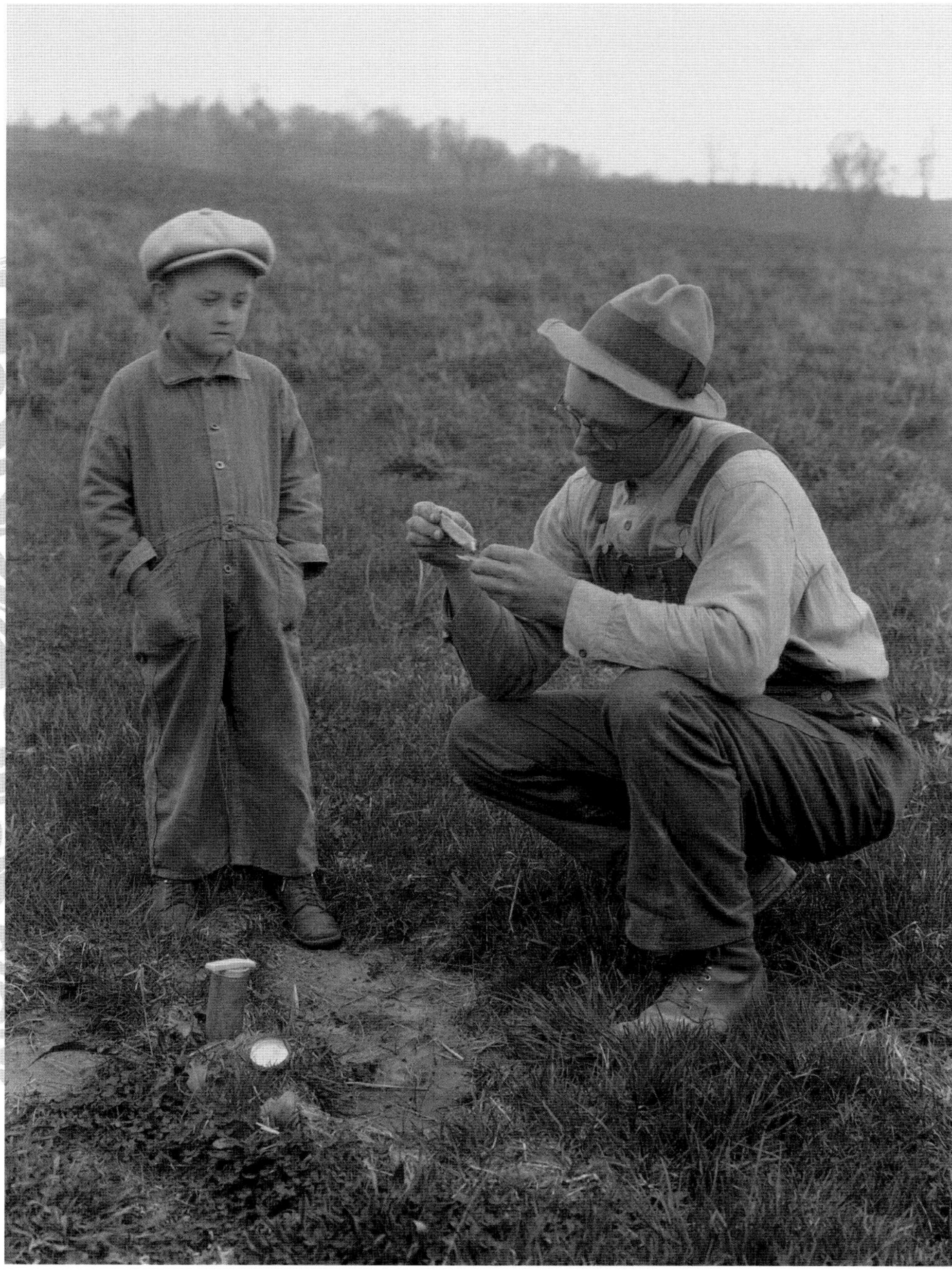

Dairyman Ray Fyffe near Bloomington uses the Soiltex system to test for soil acidity while his son, Warren, looks on. The test would turn yellow for acidic soil, light green for neutral soil, and dark green to blue for alkaline soil. (Monroe County, 1927)

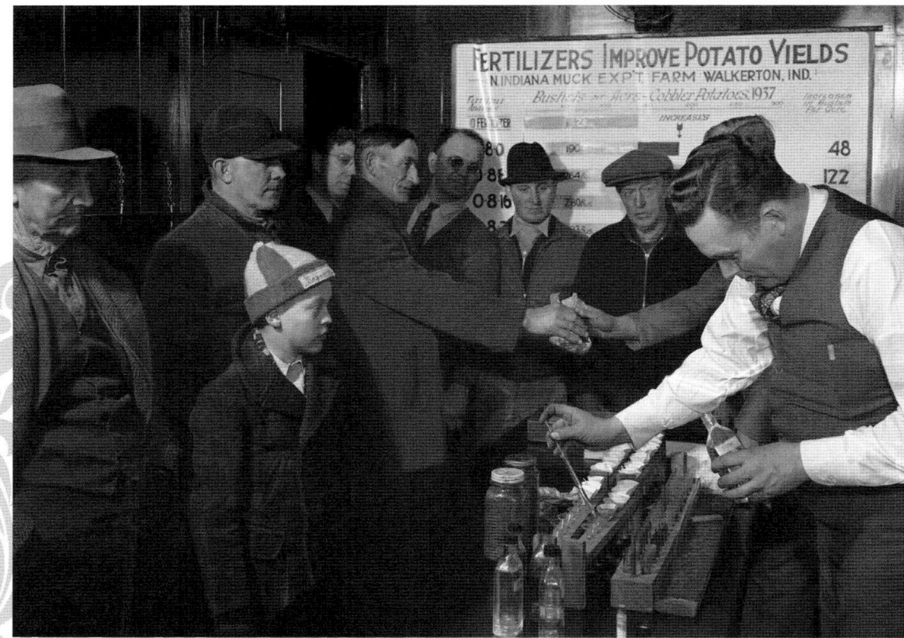

Farmers have their soils tested aboard the Muck Crops Special, an education train from Purdue University that traveled the state via the Baltimore and Ohio Railroad. (Indiana, 1939)

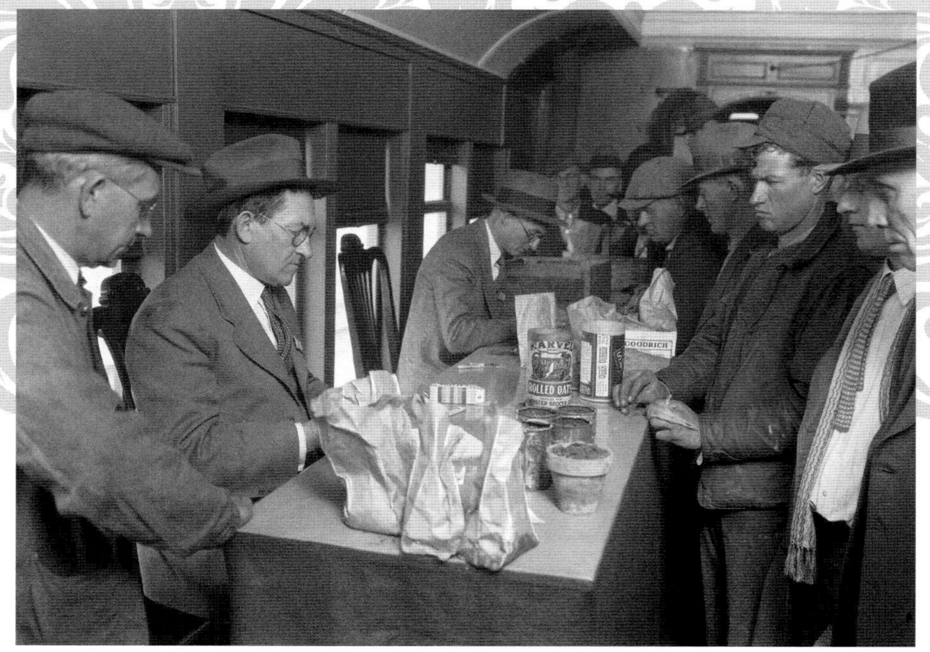

Purdue University Extension staff members test soil samples for acidity on a limestone education train stopped at Cayuga. (Vermillion County, 1929)

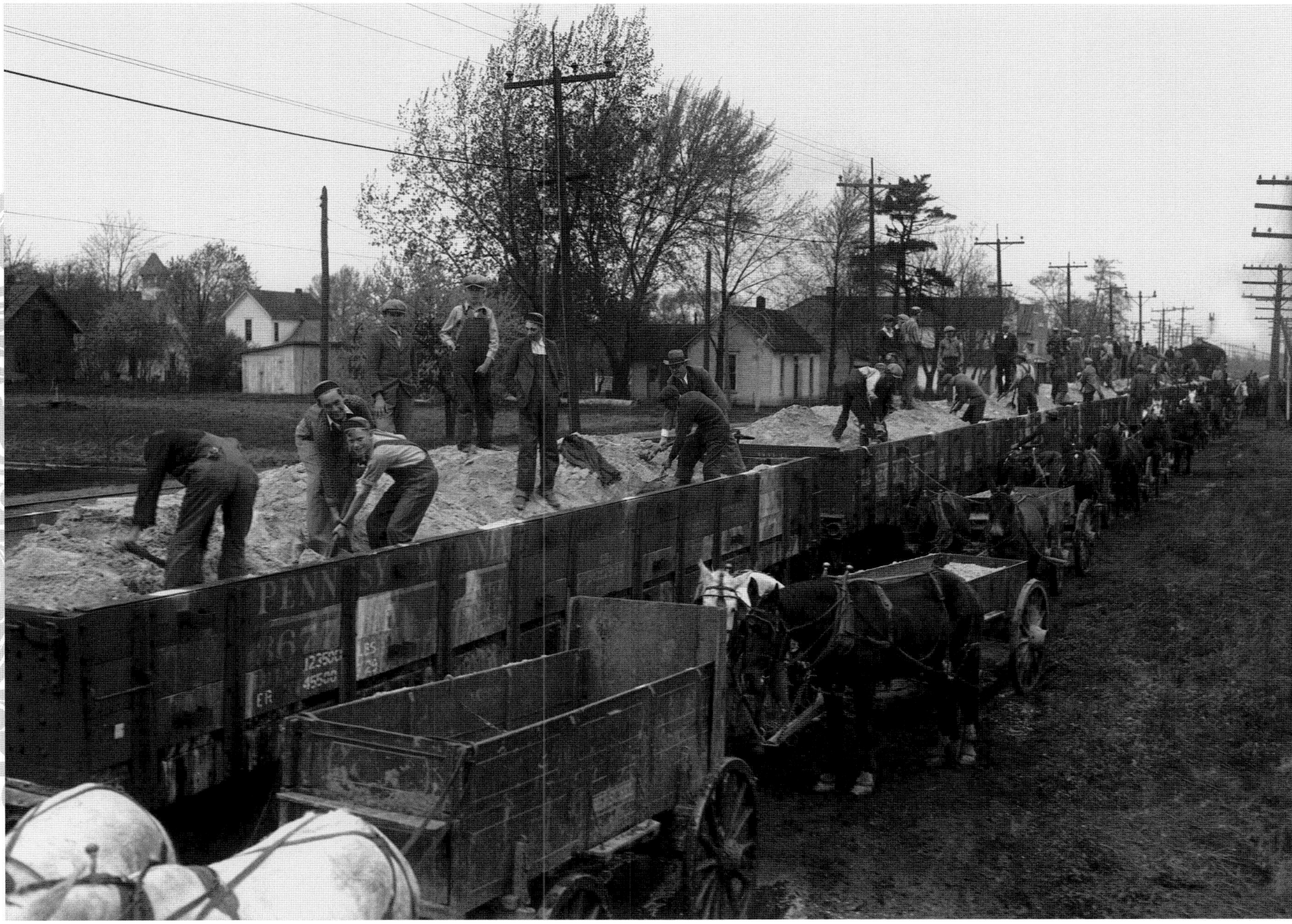

At Pimento, farmers unload lime that was delivered via railcars. (Vigo County, 1929)

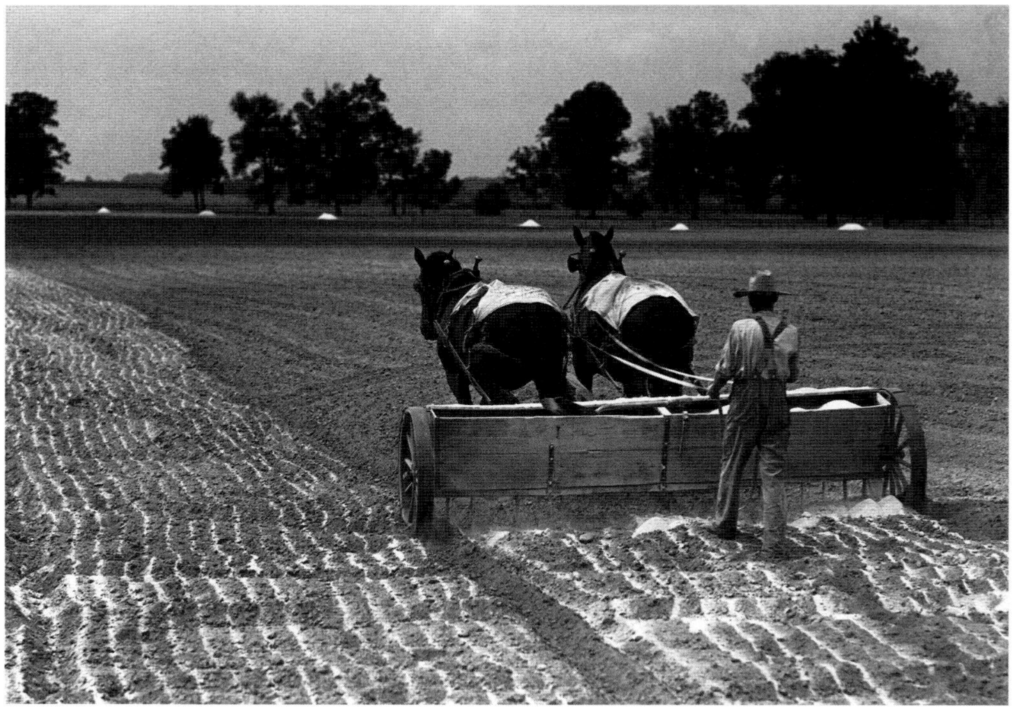

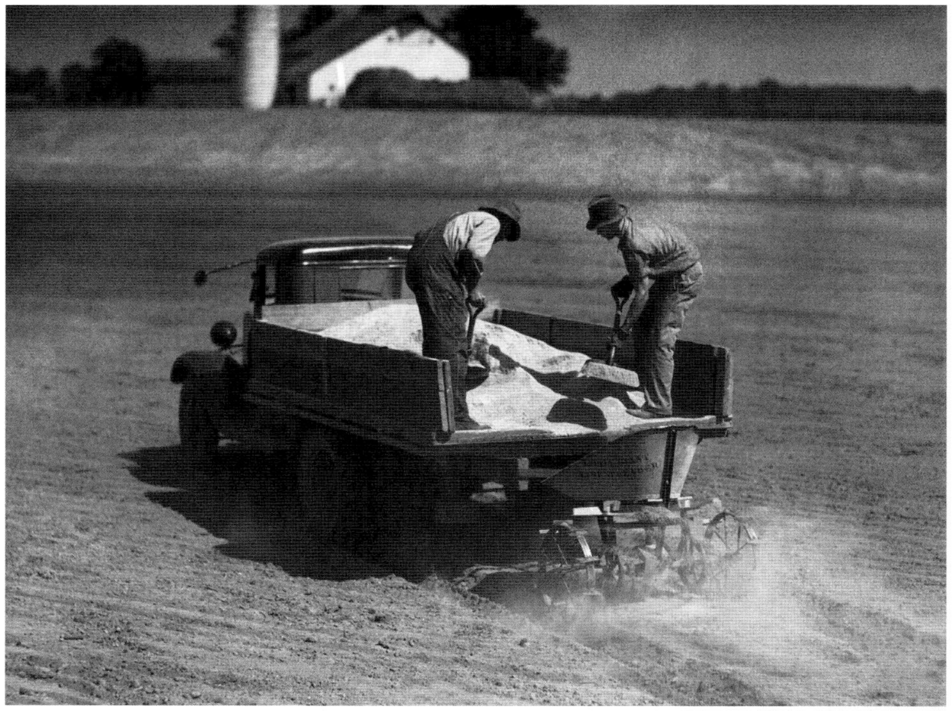

(Top left) A horse-drawn limestone spreader applies three tons of limestone per acre before the field is planted with alfalfa on the farm of Dr. J. H. Black in Lebanon. (Boone County, 1926)

(Top right) Two men spread lime from a truck bed. At the time, the Marshall County Farm Bureau sold lime for $1.65 per ton if delivered by railcar or $2.40 per ton if delivered and spread. Most farmers chose the latter option. (Marshall County, 1935)

(Bottom) John M. McDonald spreads limestone on the W. R. Wiggins farm near Muncie. (Delaware County, 1951)

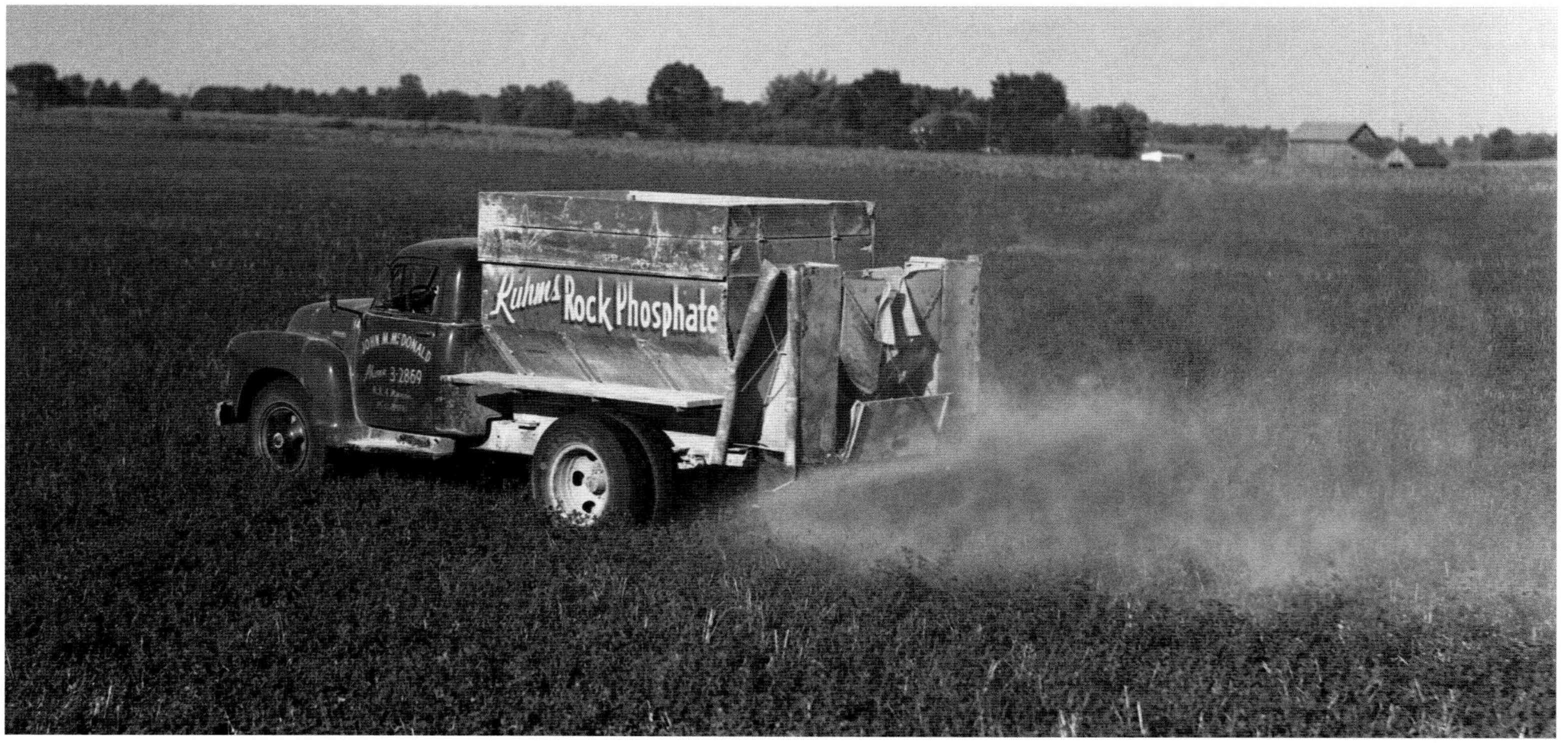

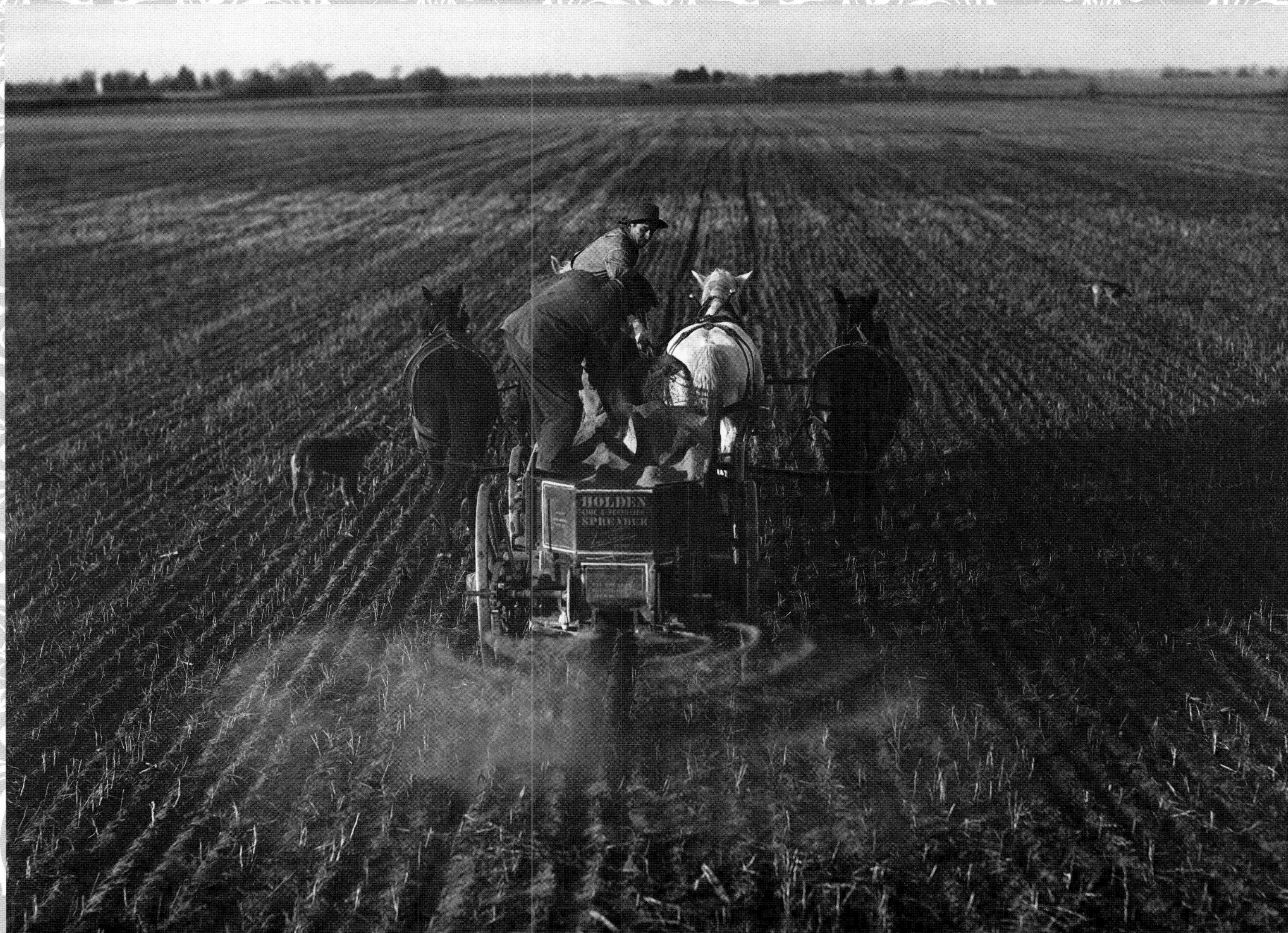

These men use four horses hitched to a wagon equipped with a Holden Spreader attachment to apply lime on the farm of Dr. John S. Morrison in Lafayette. (Tippecanoe County, 1929)

Adding Nutrients to Depleted Soils

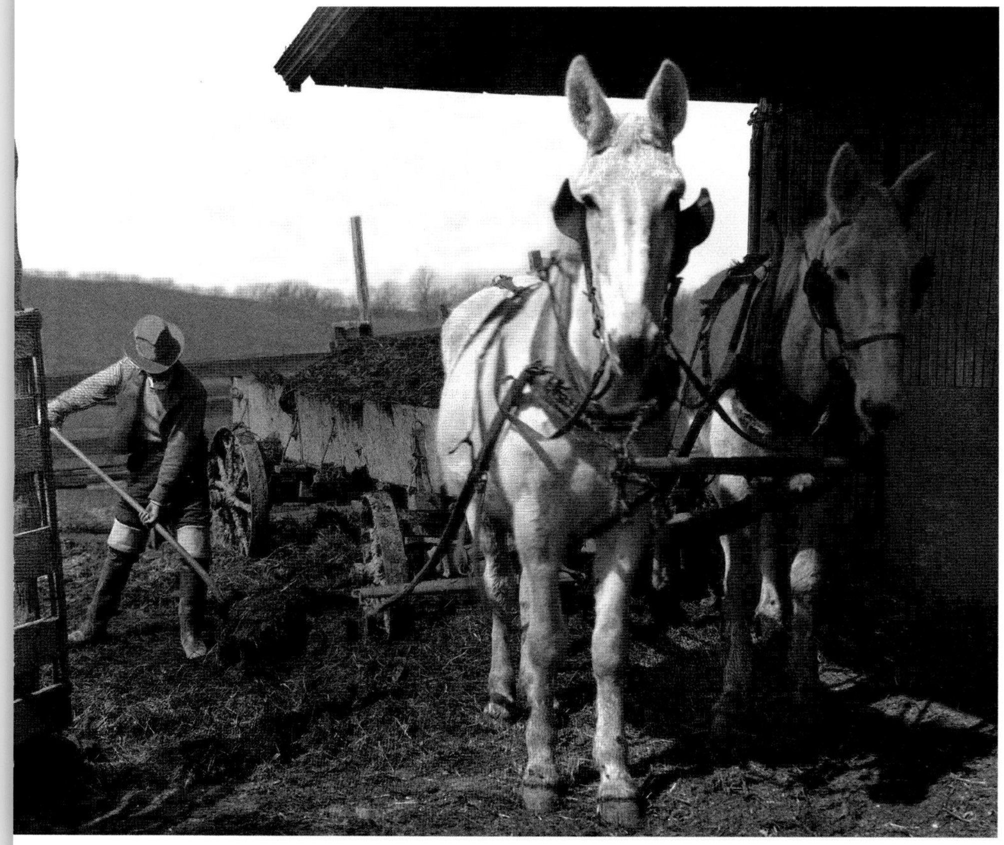

A man loads manure into a spreader on the farm of Garrett Tolan from Pleasant Plains, Illinois. (Sangamon County, IL; 1936)

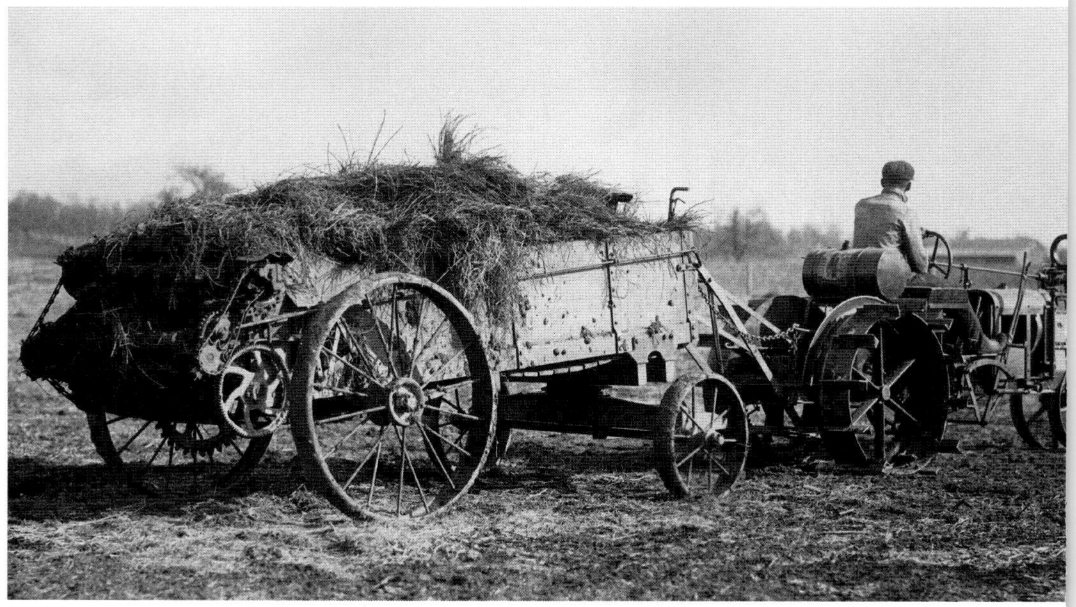

Hauling and spreading manure from a low-wheeled wagon (Wisconsin, 1928)

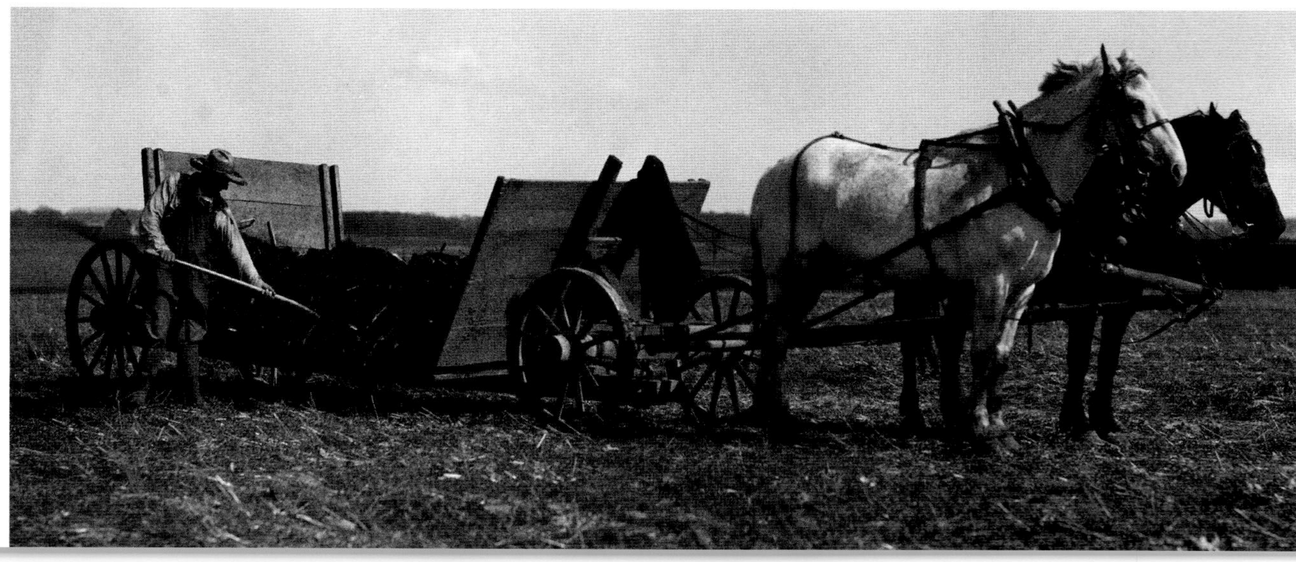

Spreading manure in Carmel at the Lynnwood Farm owned by Purdue University (Hamilton County, 1945)

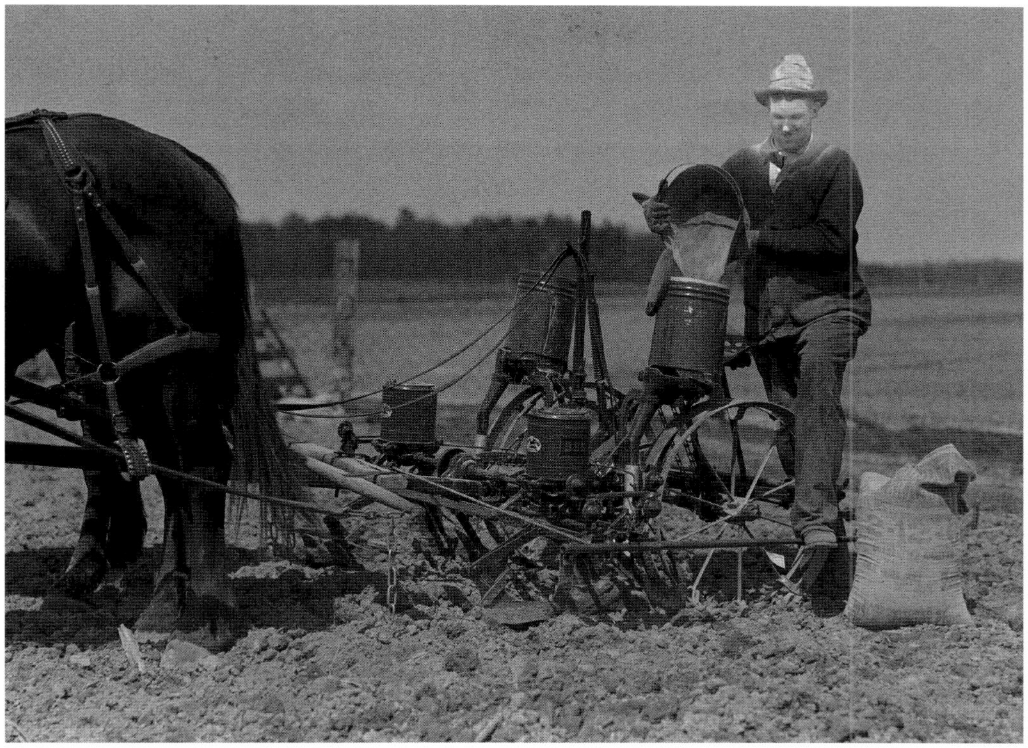

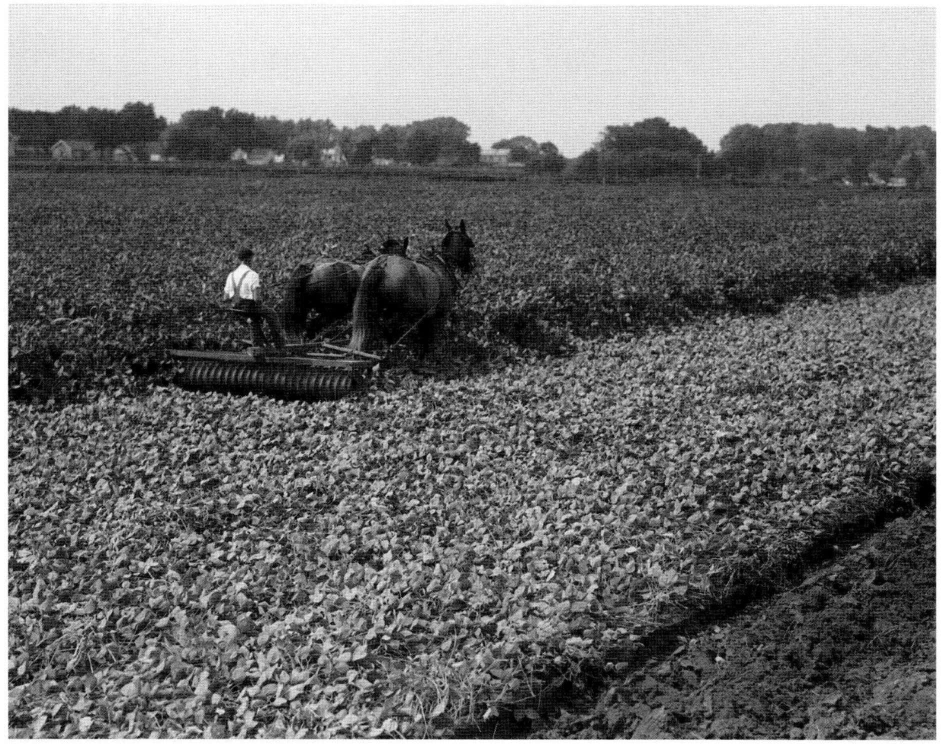

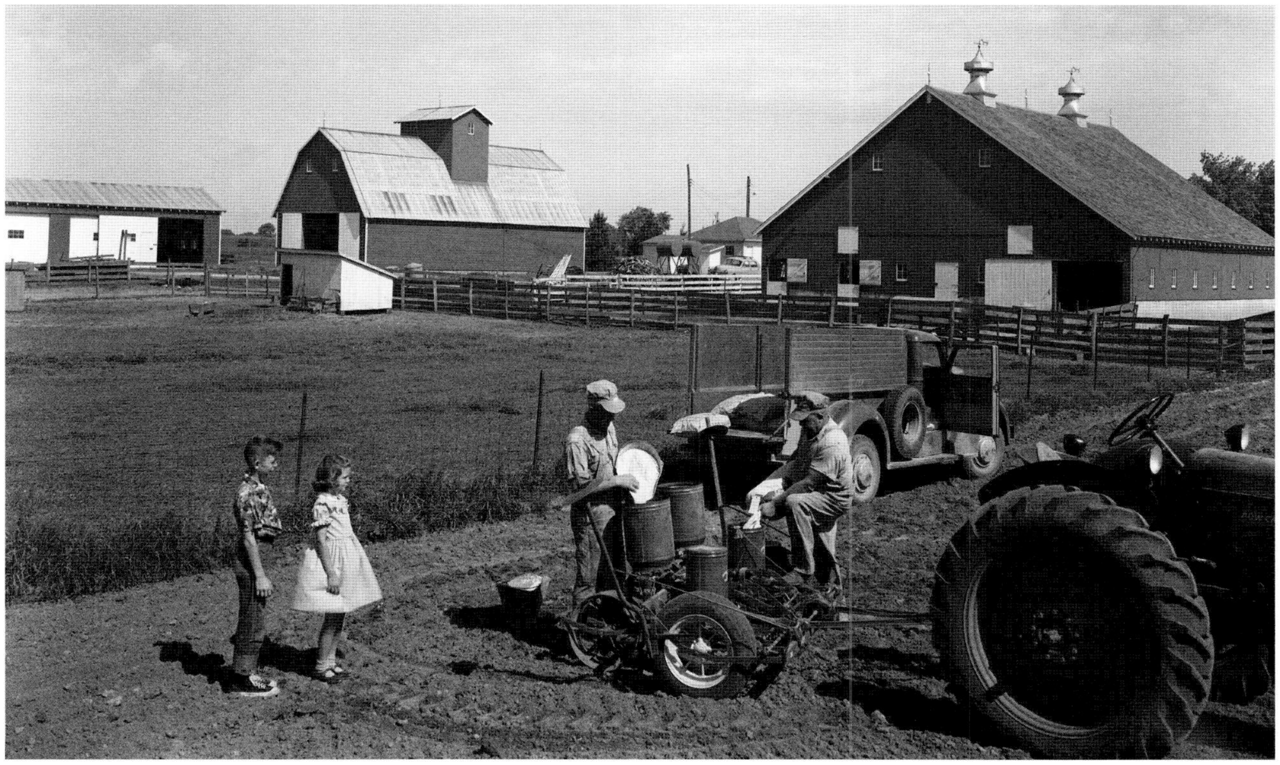

(Top left) Charles Rutan fills fertilizer boxes on his corn planter in Frankfort. (Clinton County, 1929)

(Top right) By rolling down a heavy crop of soybeans, Leonard Hill of Douglas, Illinois, could more easily plow under the crop as a green manure. (St. Clair County, IL; 1932)

(Bottom) Two men fill the fertilizer and seed boxes on a corn planter on the farm of Samuel Voss and Son. (Tippecanoe County, 1952)

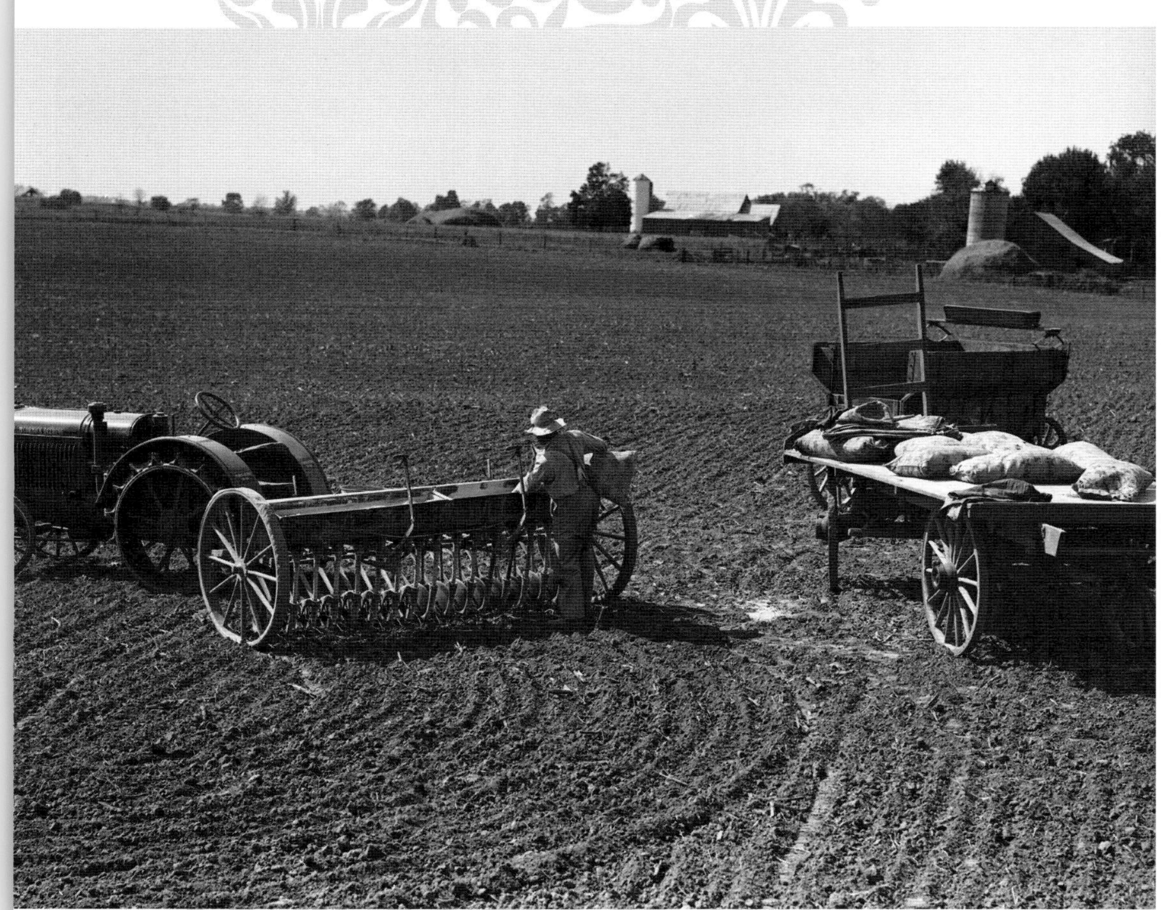

To improve the soil on his Wingate-area farm, Griff Quirk used a wheat drill to spread 200 pounds of 2-12-6 fertilizer. (Montgomery County, 1929)

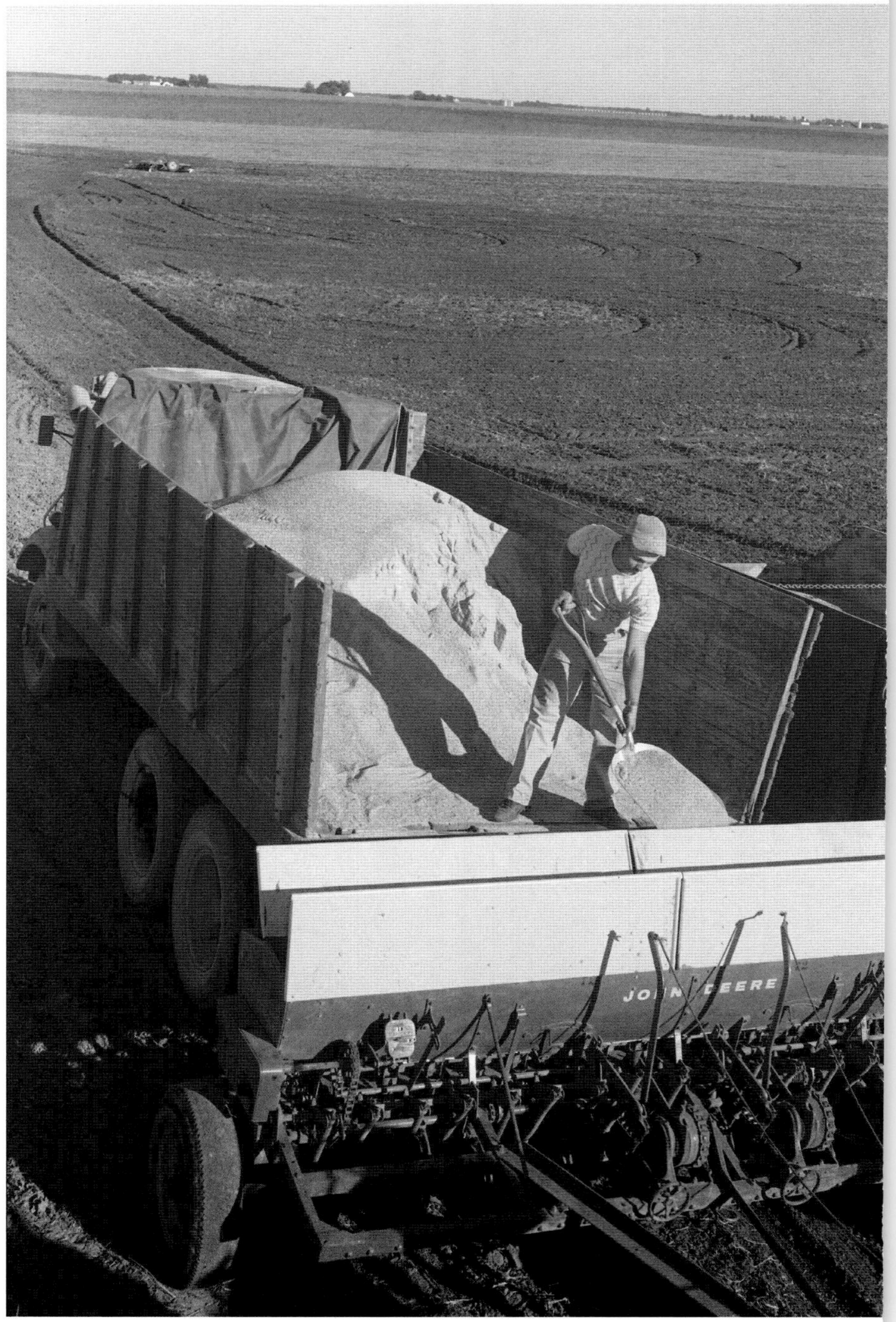

A man shovels bulk fertilizer into a spreader on the Overmeyer farm in Francesville. (Pulaski County, 1960)

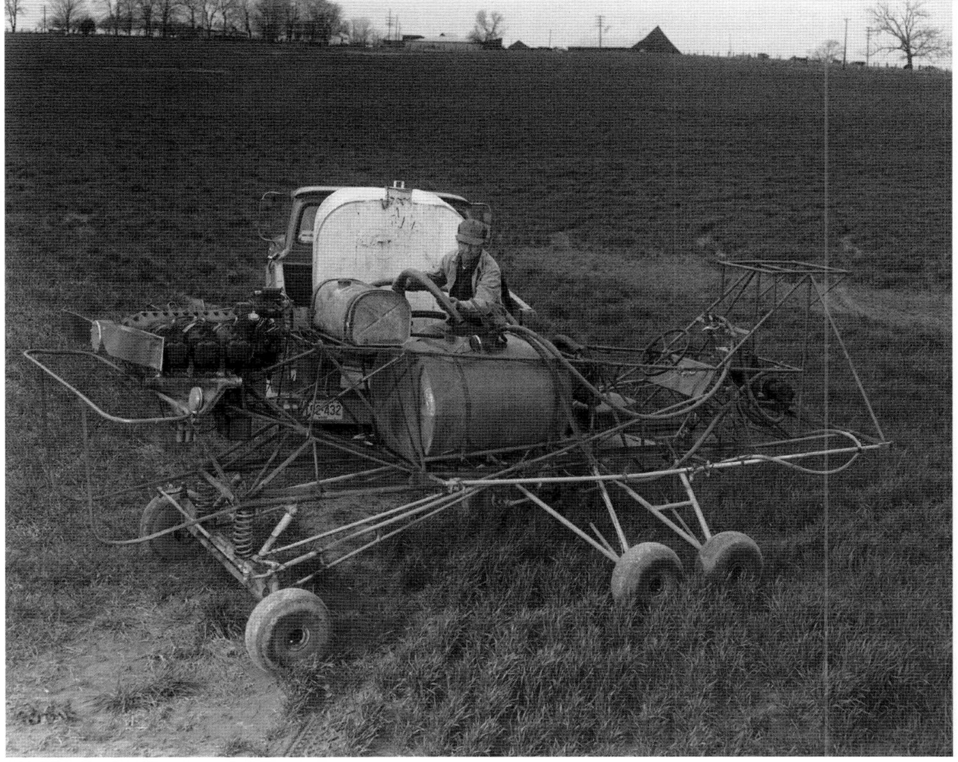

A farmer fills the tank of a spray rig with liquid fertilizer for top-dressing wheat. (N.p., 1964)

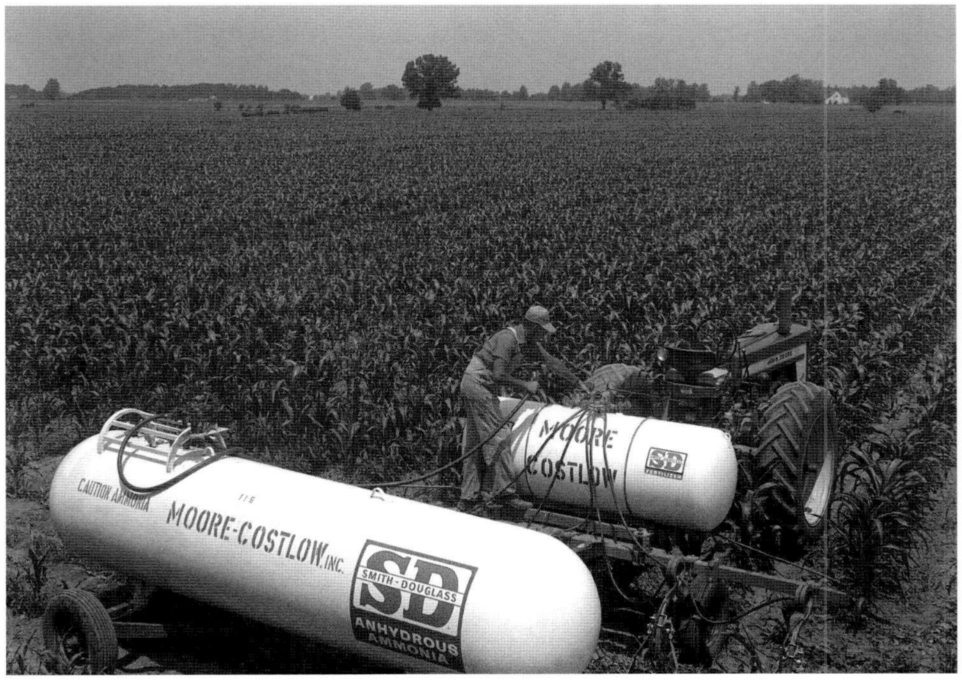

Harold Amos from Kirklin fills an anhydrous ammonia tank from a nurse tank. (Clinton County, 1962)

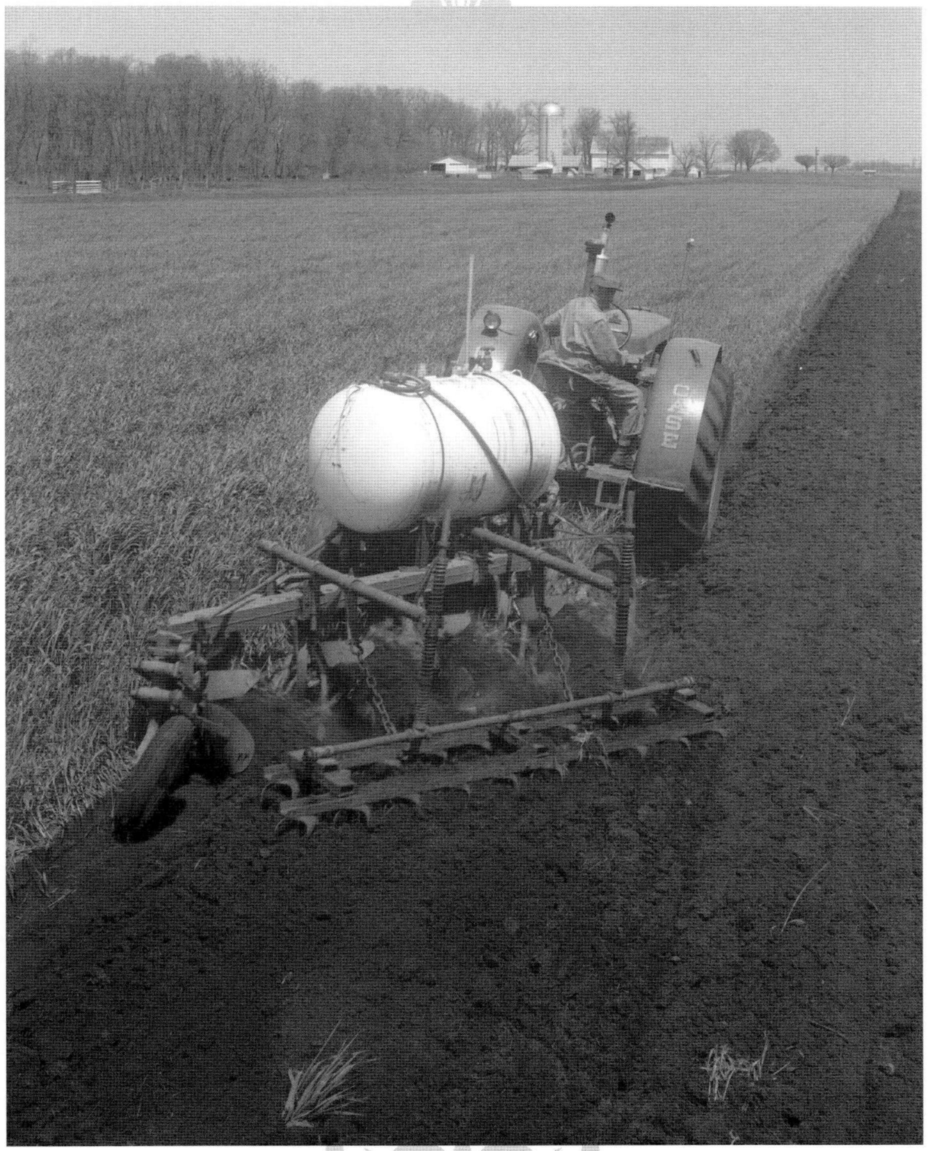

Plowing under rye and applying anhydrous ammonia on the Ward Wilkins farm in Linden (Montgomery County, 1965)

Corn on the Cob

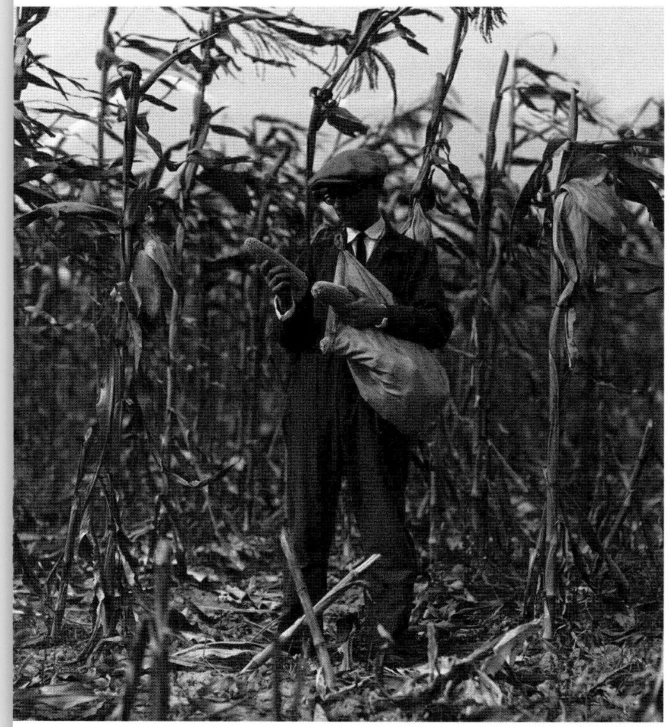

Homer Crandall collects seed corn in Mount Summit. (Henry County, 1924)

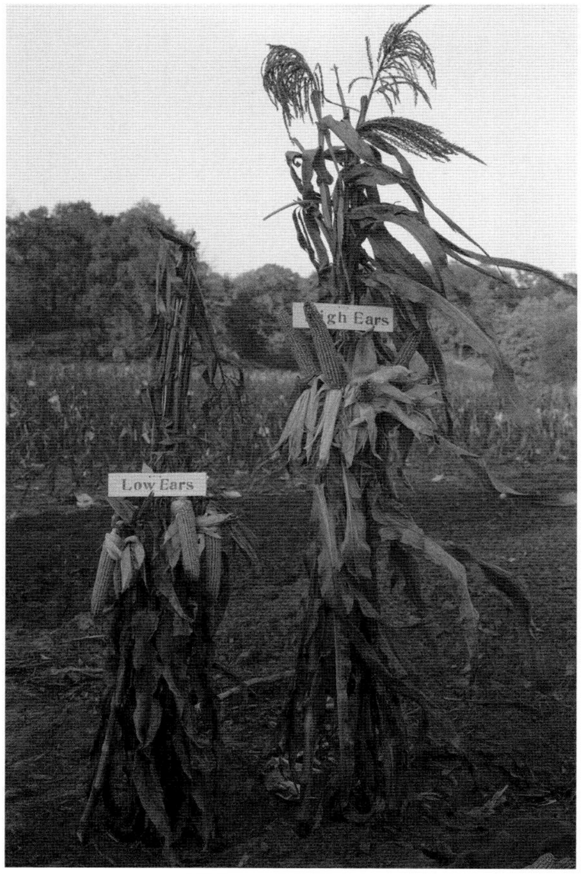

These stalks of corn show low ears and high ears, which was a rather important attribute when corn was picked by hand. (Indiana, 1934)

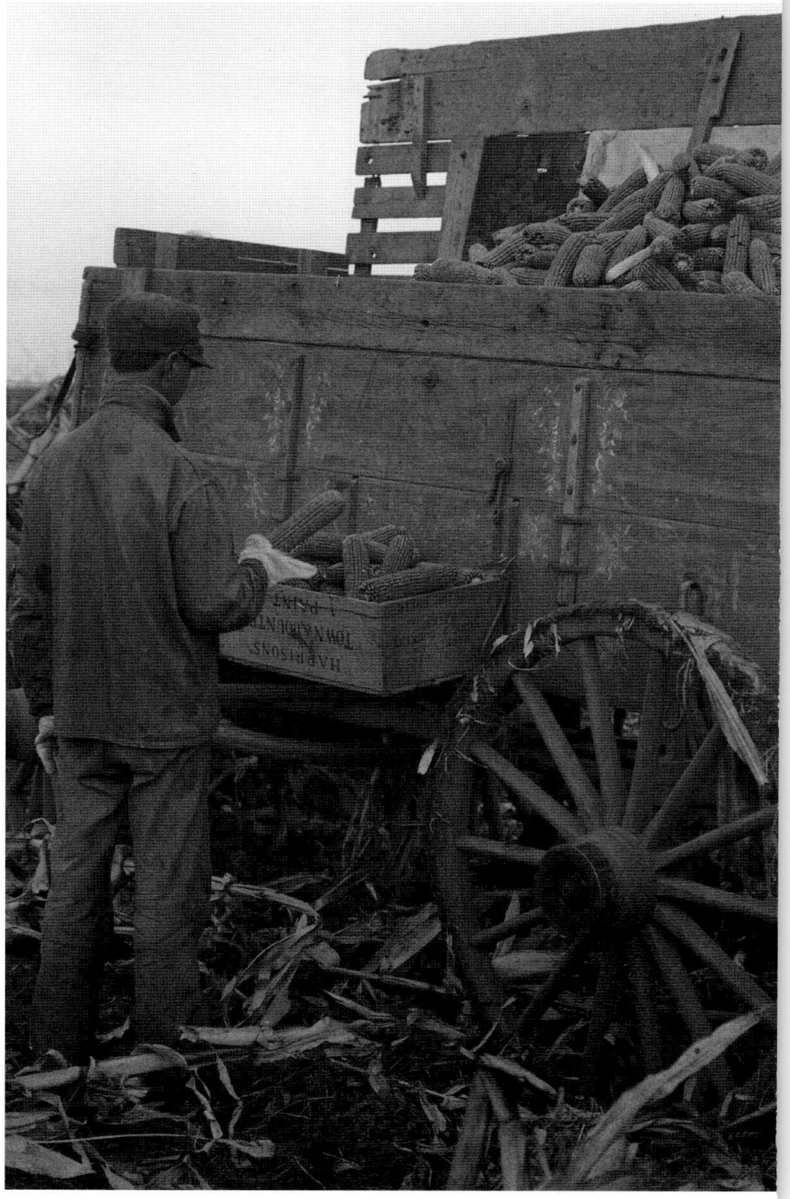

Allen Lewis from Wingate chooses next year's seed corn by placing it in a box on the side of his wagon at harvest. (Montgomery County, 1917)

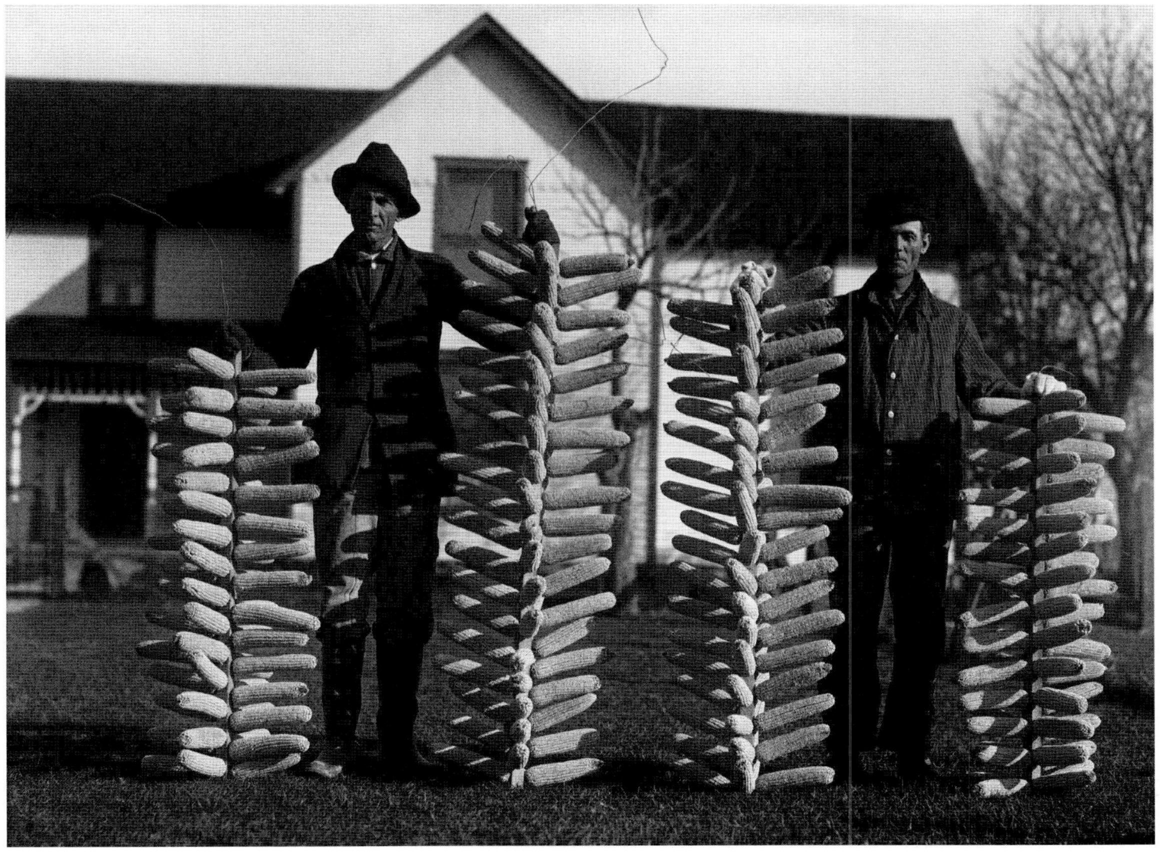

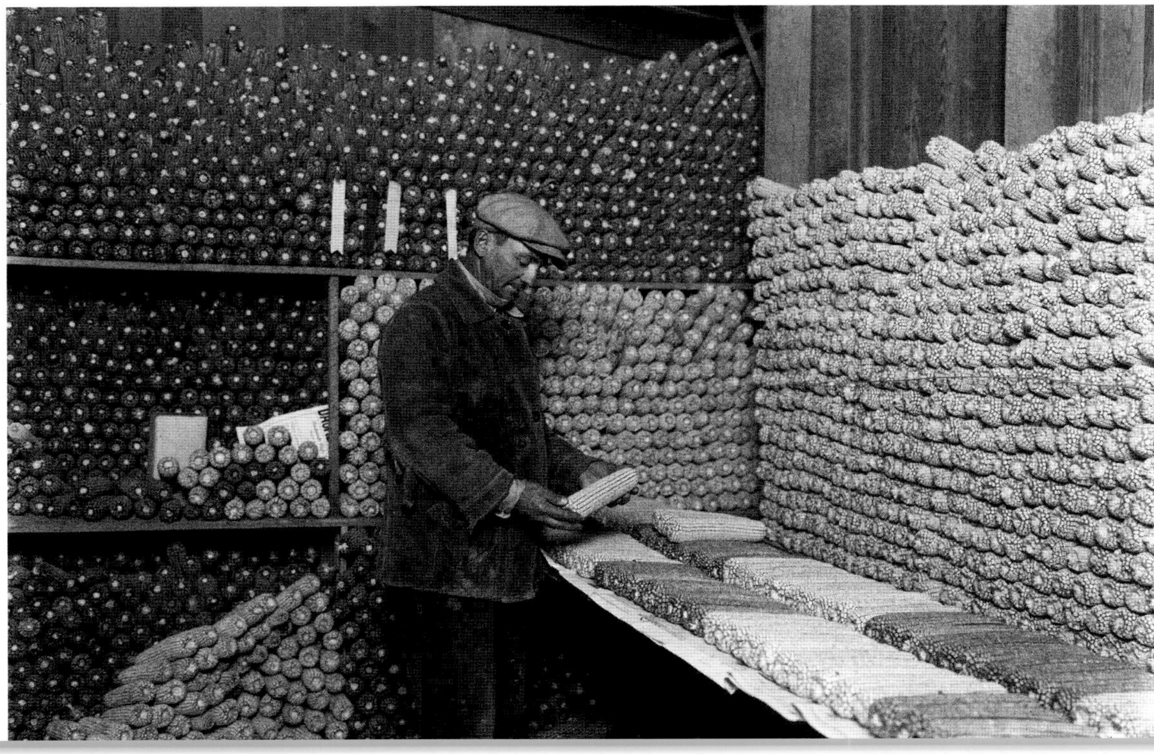

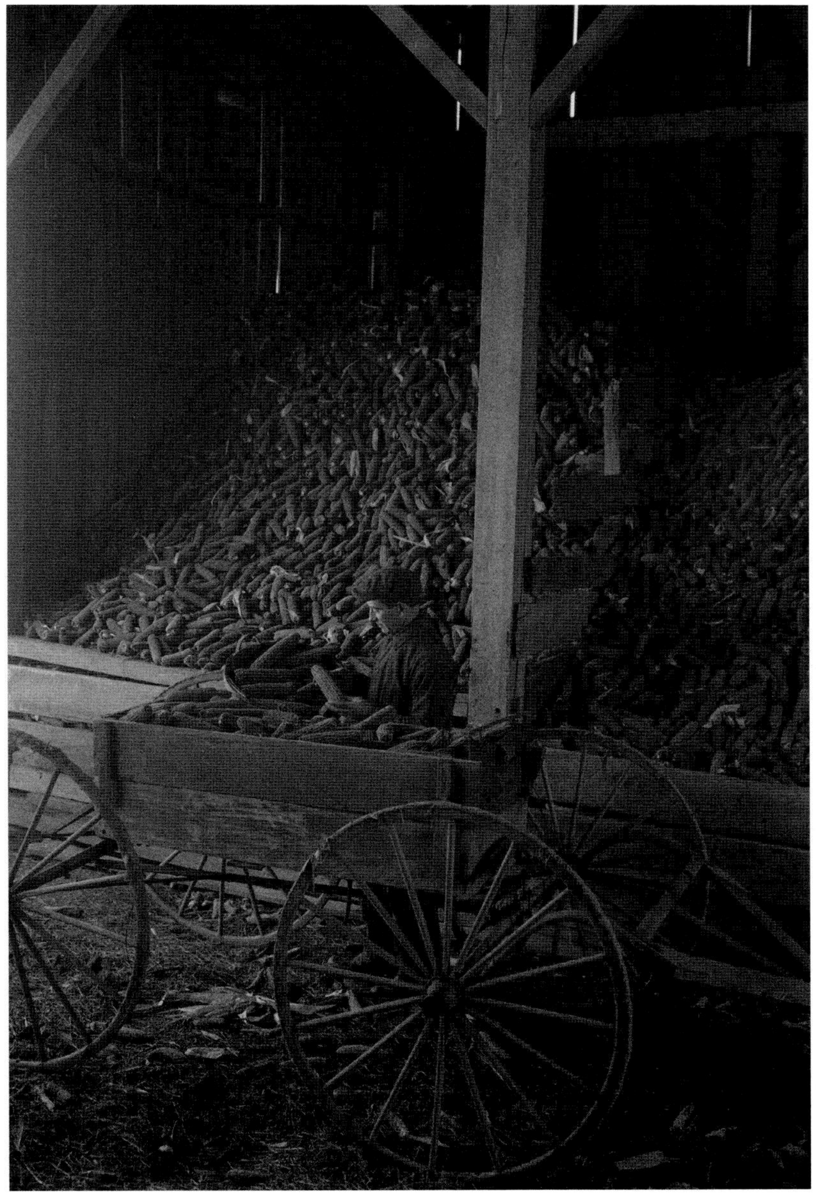

(Top left) Chester and William Joyce, two brothers who farmed in Logansport, display seed corn on hangers for drying. (Cass County, 1921)

(Bottom left) Lawrence Lutz of Haubstadt works in his storage area to select ears of corn to use for seed. (Gibson County, 1926)

(Right) C. J. Ternet Jr., a New Haven–area farmer, used a seed corn cart fastened behind the wagon during husking to collect his best ears for next year's seed. (Allen County, 1923)

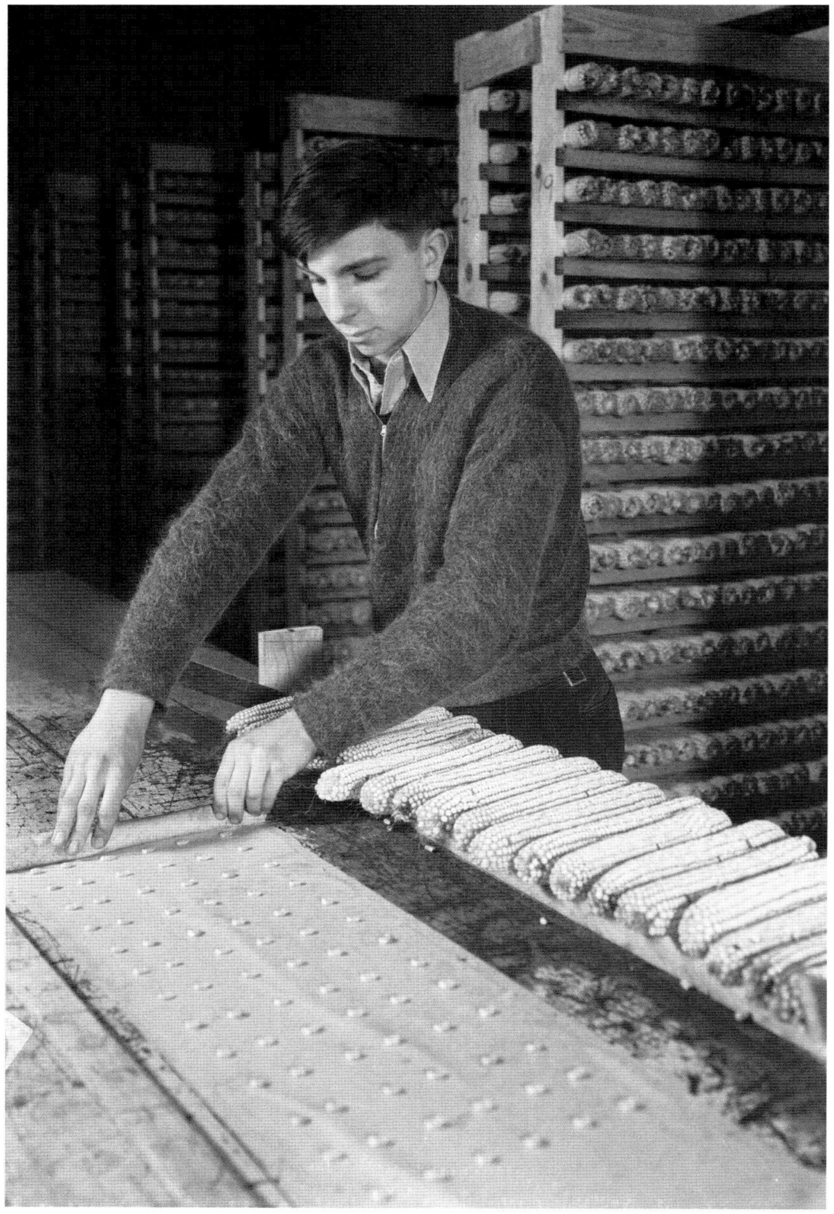

(Above) Robert Keim of West Lafayette rolls seed corn into a rag doll, which was a cloth wrap used to test corn seeds for germination. (Tippecanoe County, 1936)

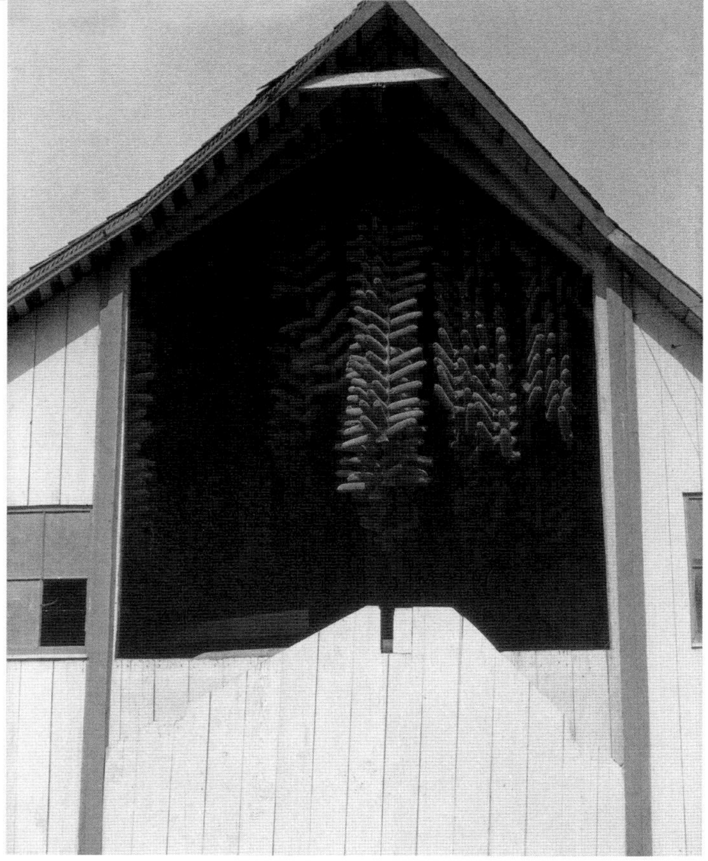

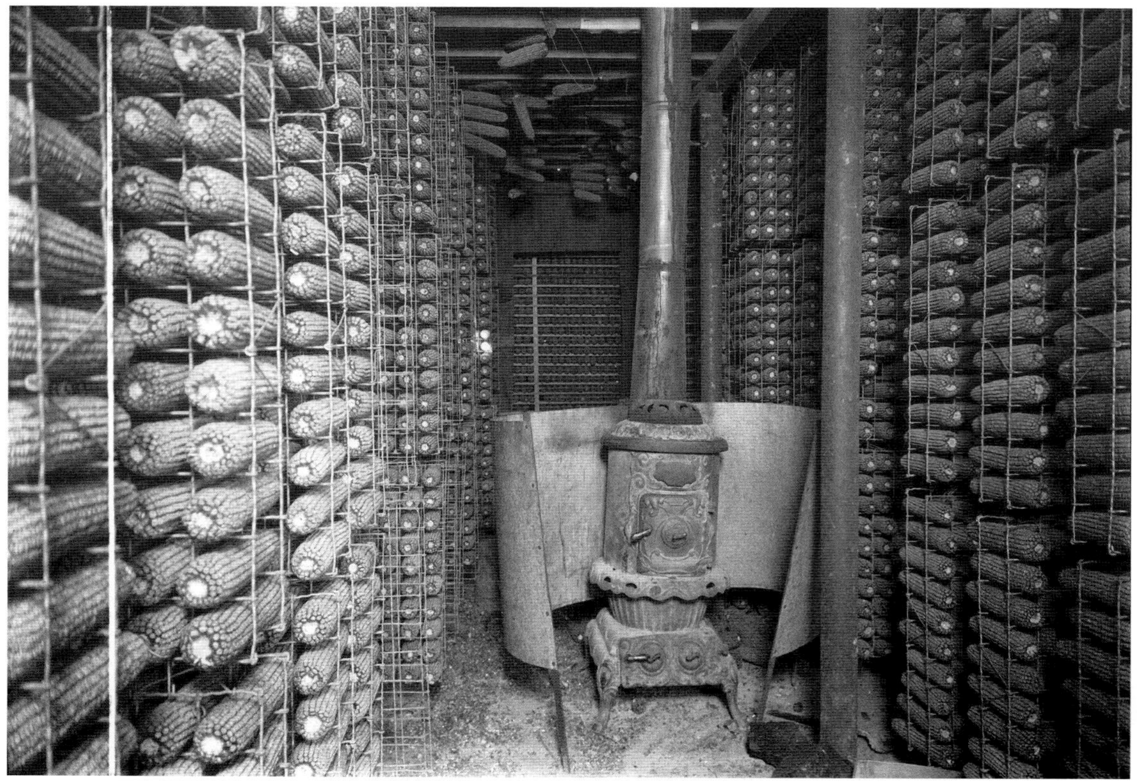

(Top right) Seed corn hangs in a barn loft on the farm of Taylor Fouts in Camden. (Carroll County, 1929)

(Bottom right) Inside the corn storage house of Lowell L. Moffitt of Knightstown (Henry County, 1928)

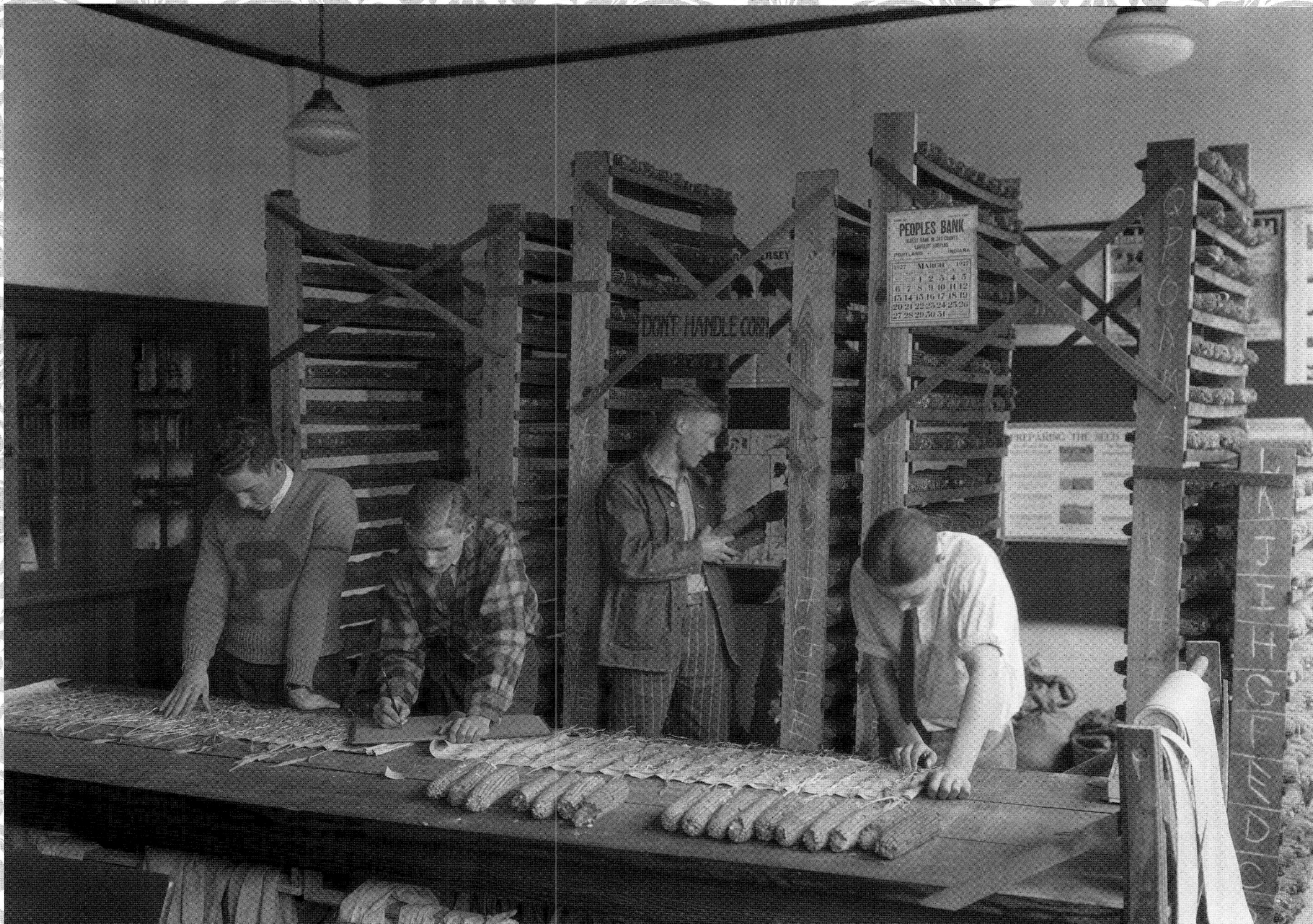

Members of the vocational agriculture class in Portland examine rag doll testers to check seed corn for germination percentage and evidence of disease. (Jay County, 1927)

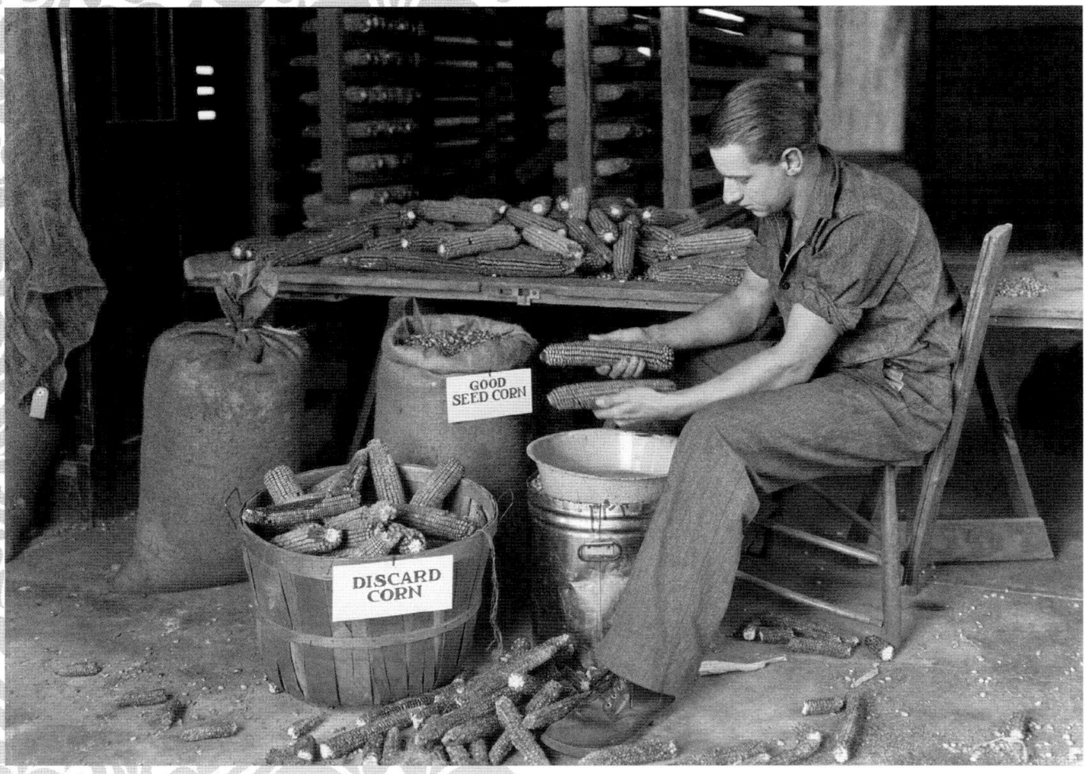

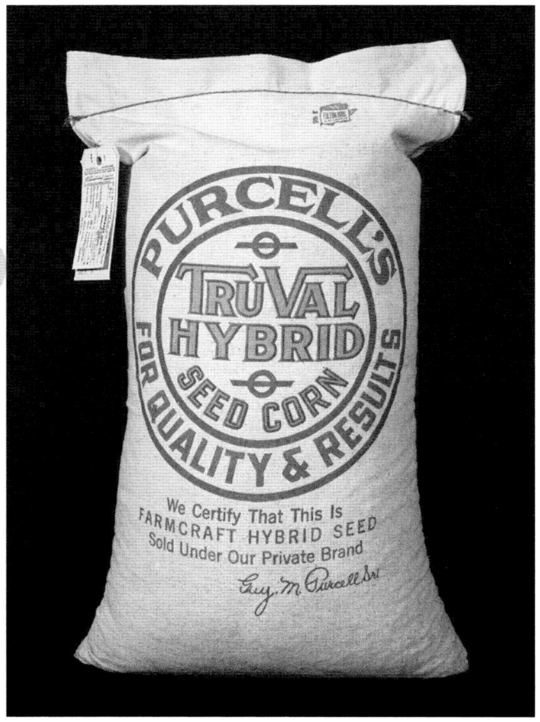

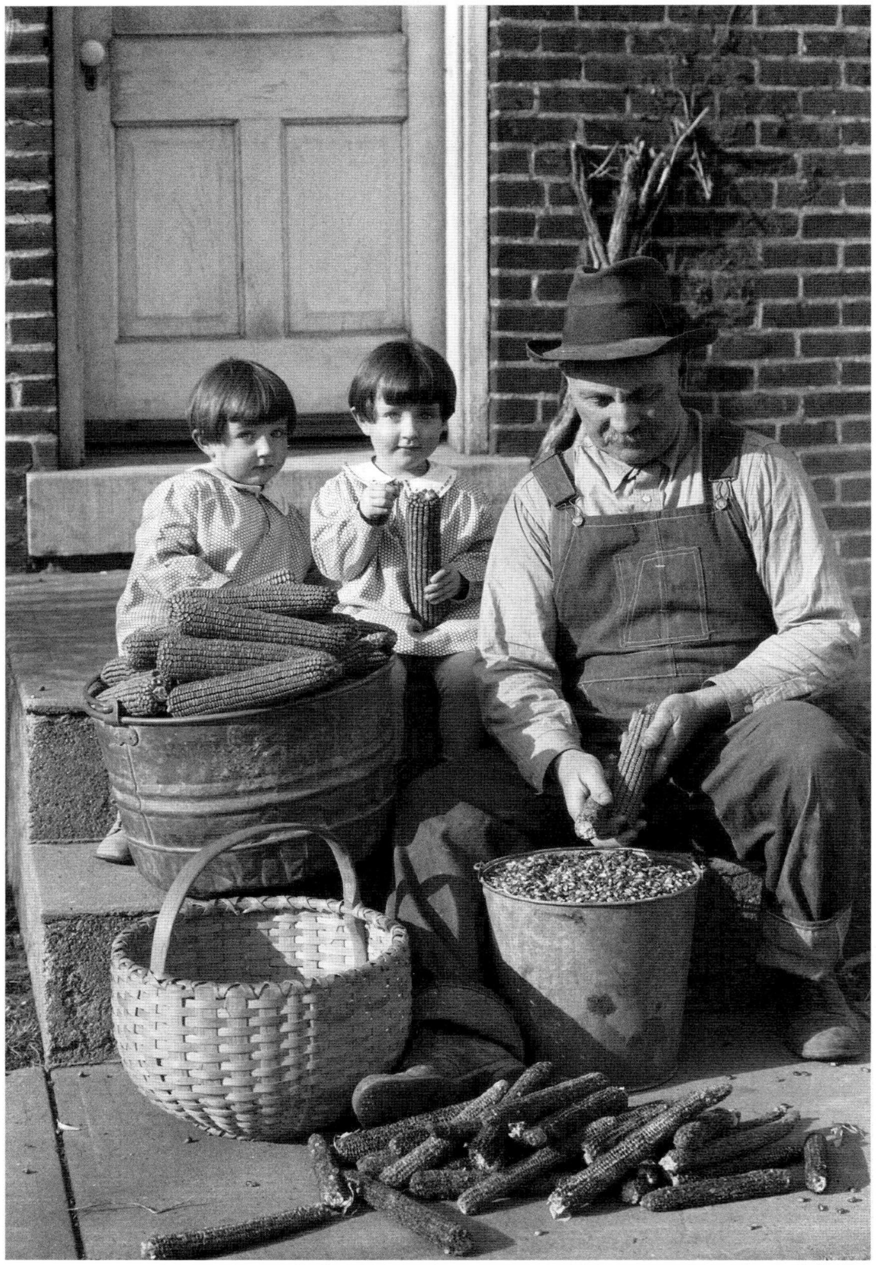

(Top left) After testing the ears, corn was either kept for seed or fed to livestock. (Indiana, 1933)

(Bottom left) Planting commercially grown hybrid seed corn, such as this bag of Farmcraft seed from Oxford, allowed farmers to achieve higher yields and better stands of corn. (Benton County, 1946)

(Right) The daughters of E. E. Baker look on as he shells seed corn in Dayton. (Tippecanoe County, 1930)

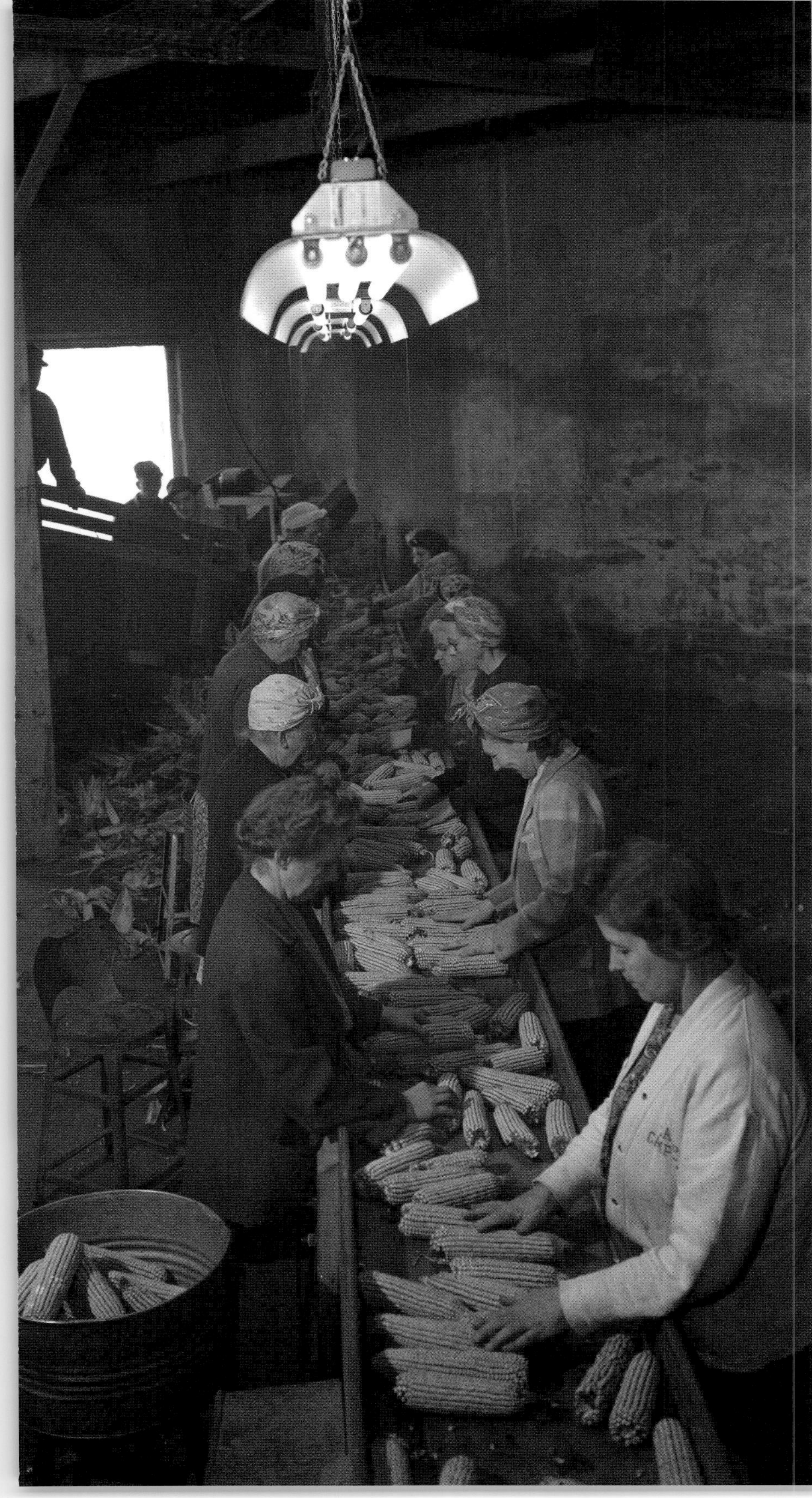

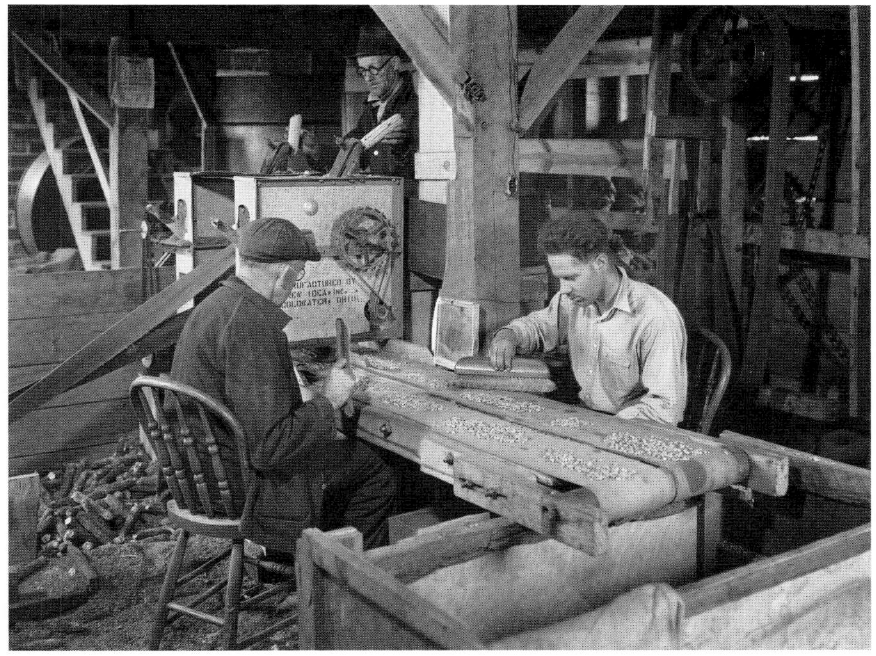

(Far left) Grading corn at the Wickersham Seed Company plant at Brookston (White County, 1946)

(Top right) Shelling and inspecting corn at Brodbeck Brothers new seed house near Wabash (Wabash County, 1937)

(Bottom right) An exterior view of the Oxford Farmcraft Company (Benton County, n.d.)

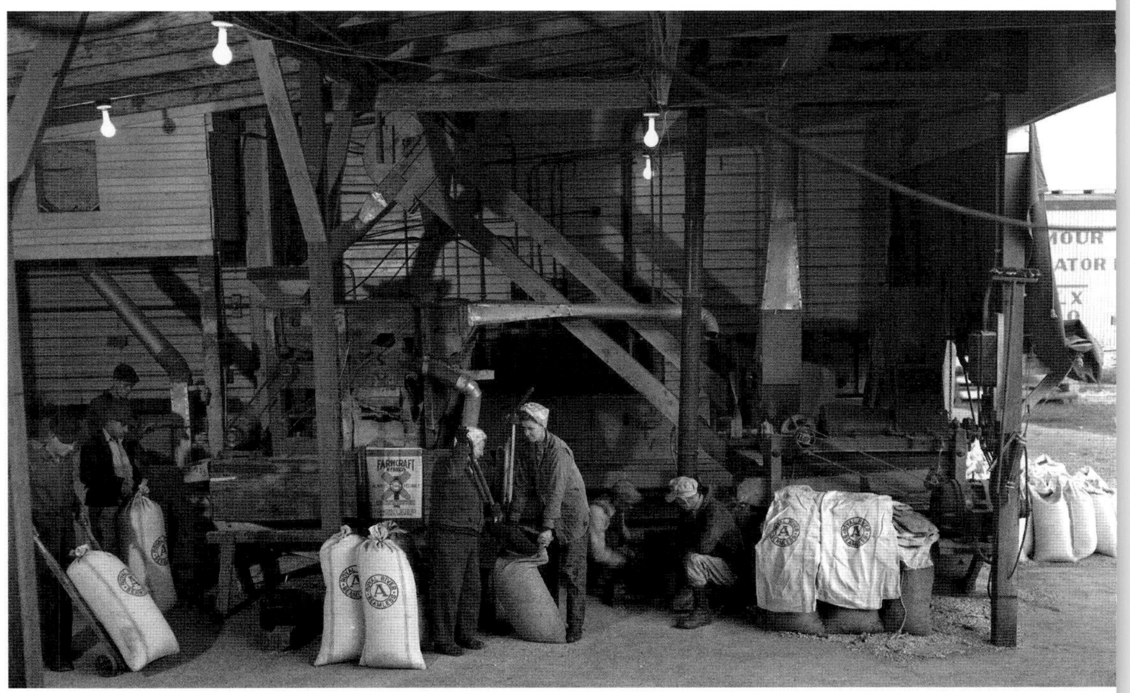

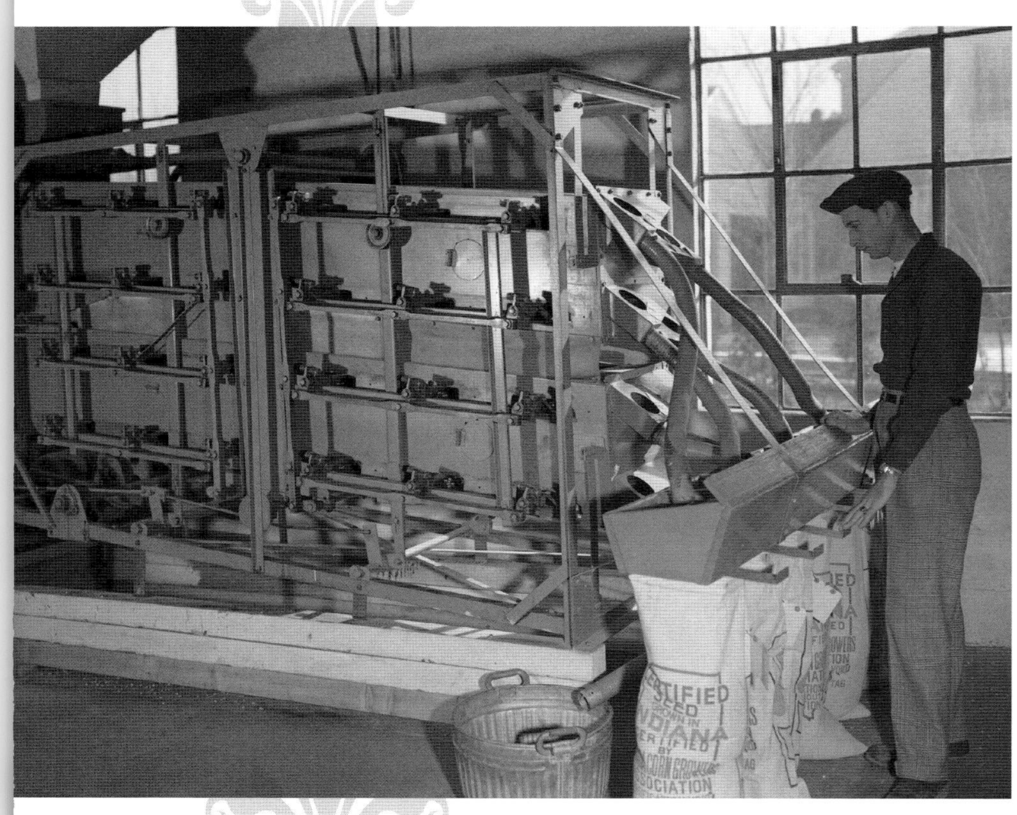

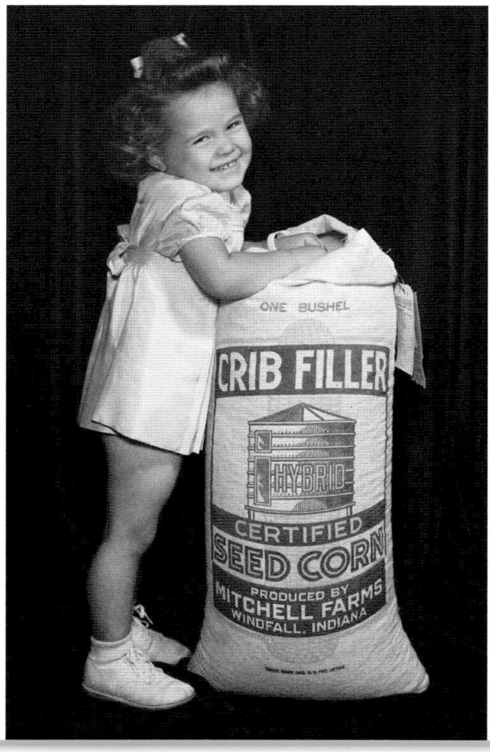

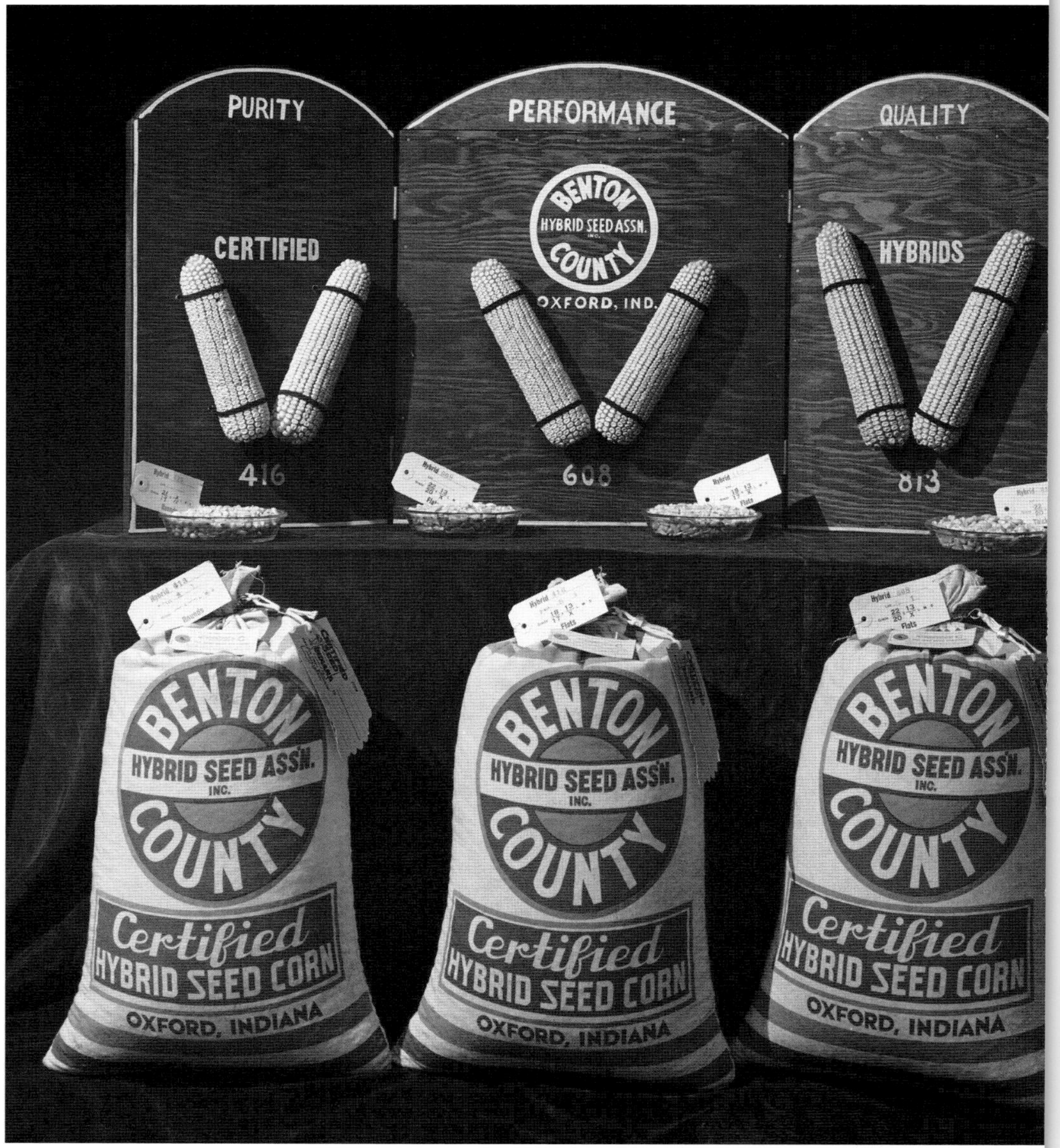

(Top left) A worker uses a machine to grade and sack corn at the Purdue Agricultural Alumni Seed Improvement Association plant. (Tippecanoe County, 1948)

(Bottom left) In this photo taken in Lafayette, a little girl poses next to a bag of certified seed corn produced by Mitchell Farms in Windfall. (Tippecanoe County, 1944)

(Right) Sears and Roebuck stores used this exhibit of certified seed corn from the Benton County Hybrid Seed Association. (Benton County, 1942)

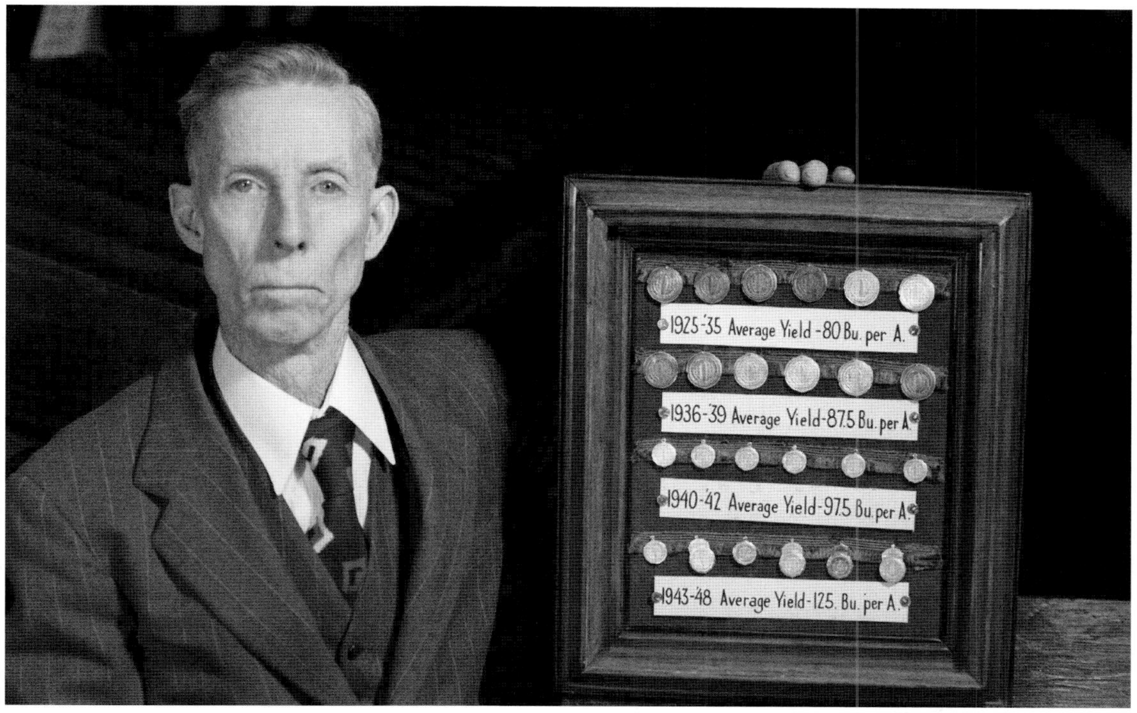

J. H. Bone proudly displays medals he won as a corn grower at the Corn School and Stockmen's Convention at Purdue University. (Tippecanoe County, 1948)

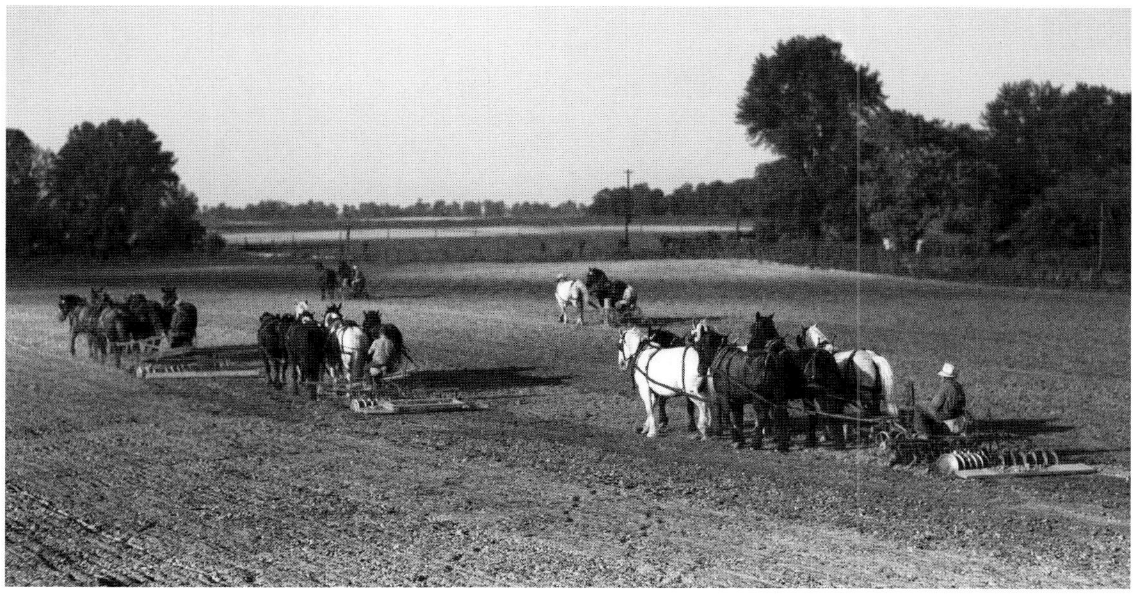

Three 6-horse teams of Percheron mares pull discs while two 2-horse teams drag corn planters on the Penney-Gwinn Farm in Noblesville. (Hamilton County, 1935)

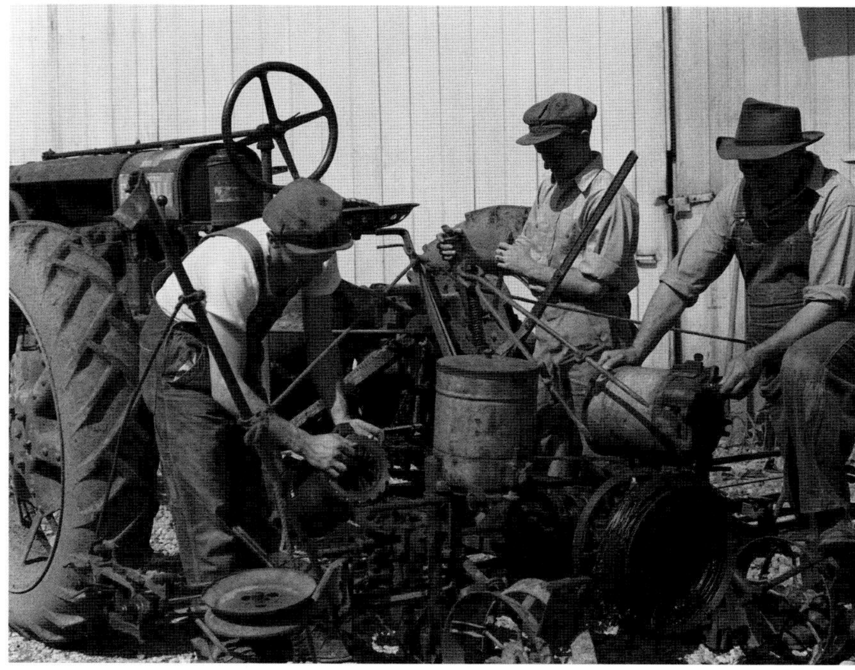

Merle Skinner and his sons, William and Byron, prepare the corn planter for the planting season on their 400-acre farm. (Indiana, 1945)

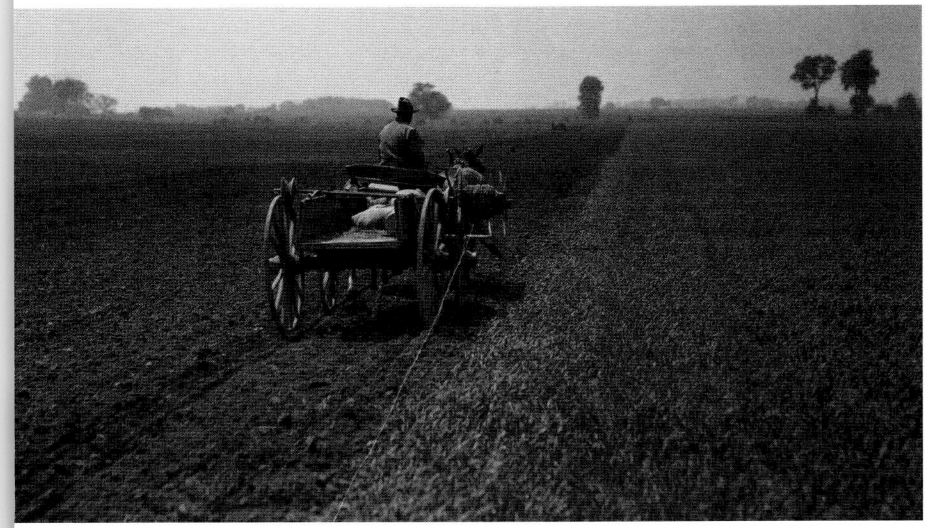

An attachment on this wagon lays the corn planter check wire on the James Bailey farm in Delavan, Illinois. (Tazewell County, IL; 1921)

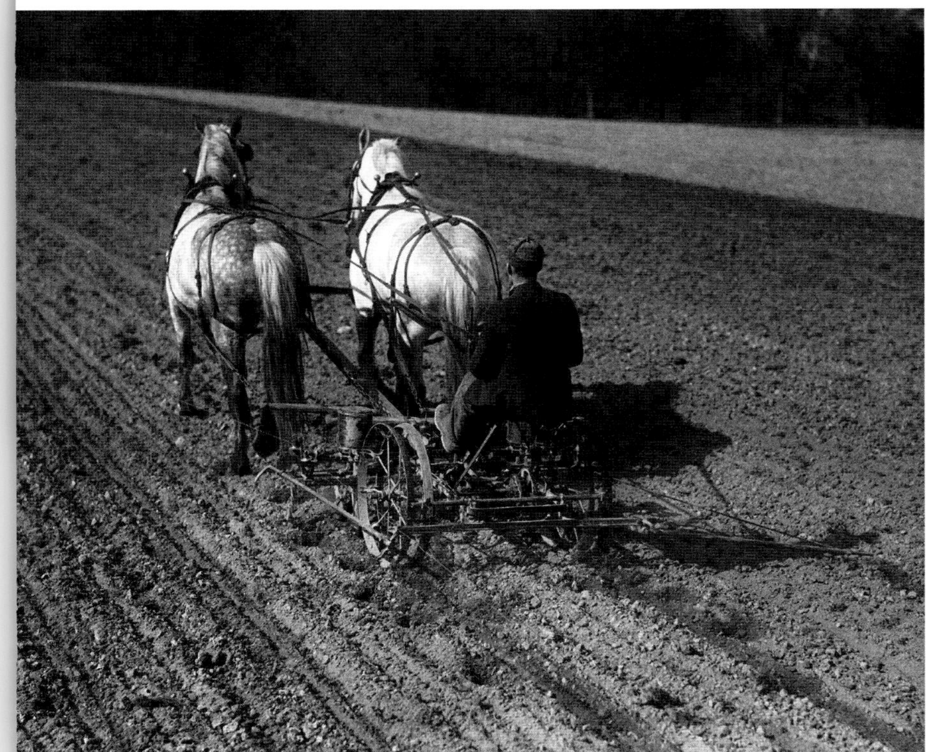

Horses pull a planter using a check wire to keep the rows straight on the Lauros Rutan farm in Mulberry. (Clinton County, 1929)

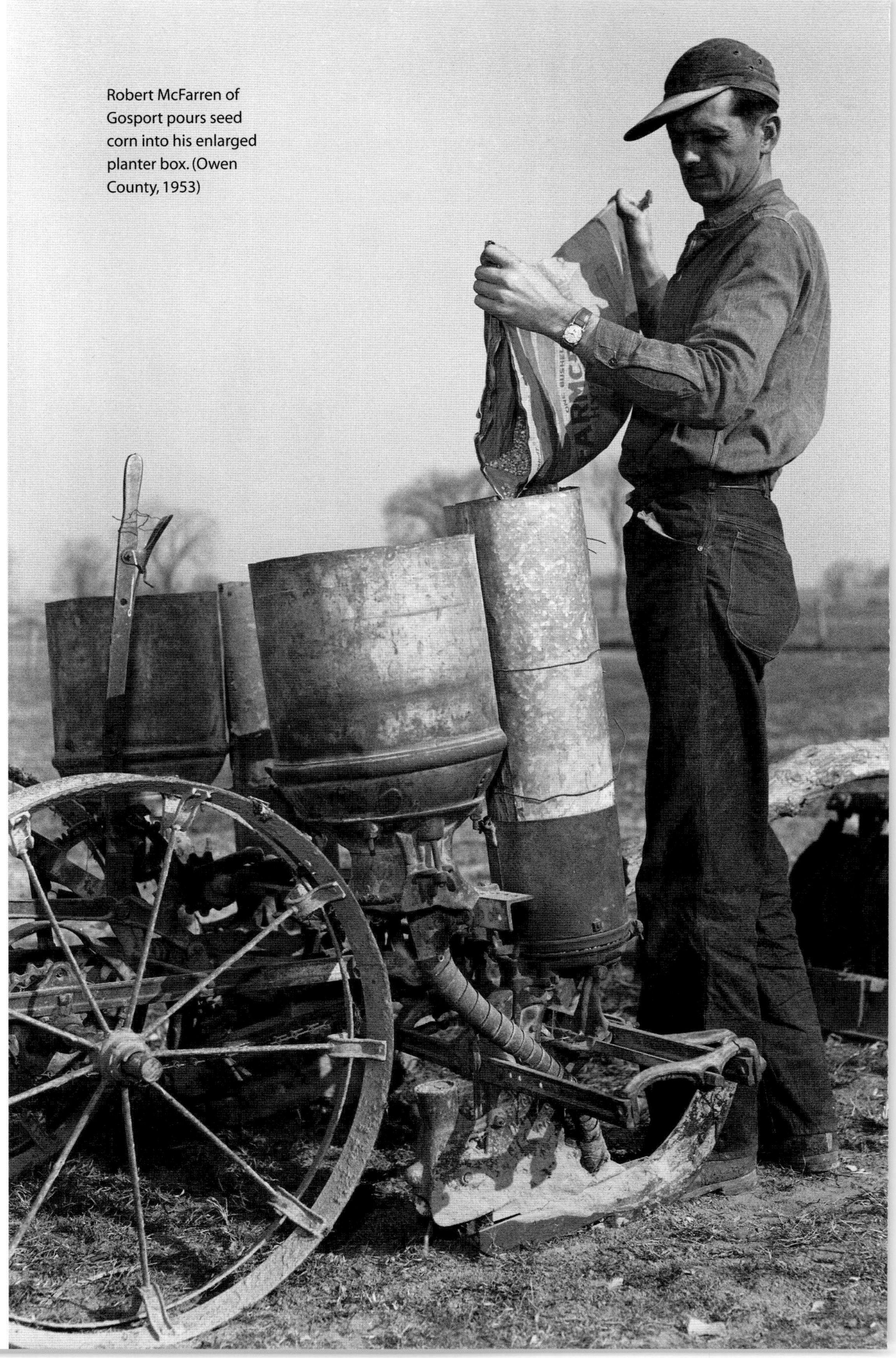

Robert McFarren of Gosport pours seed corn into his enlarged planter box. (Owen County, 1953)

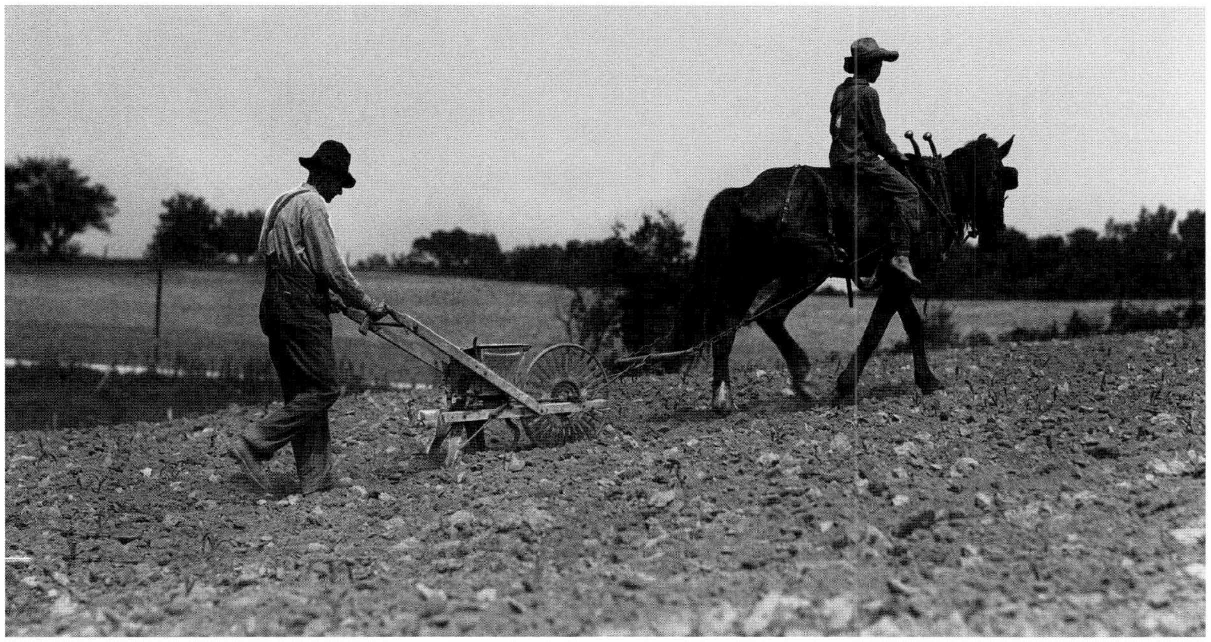

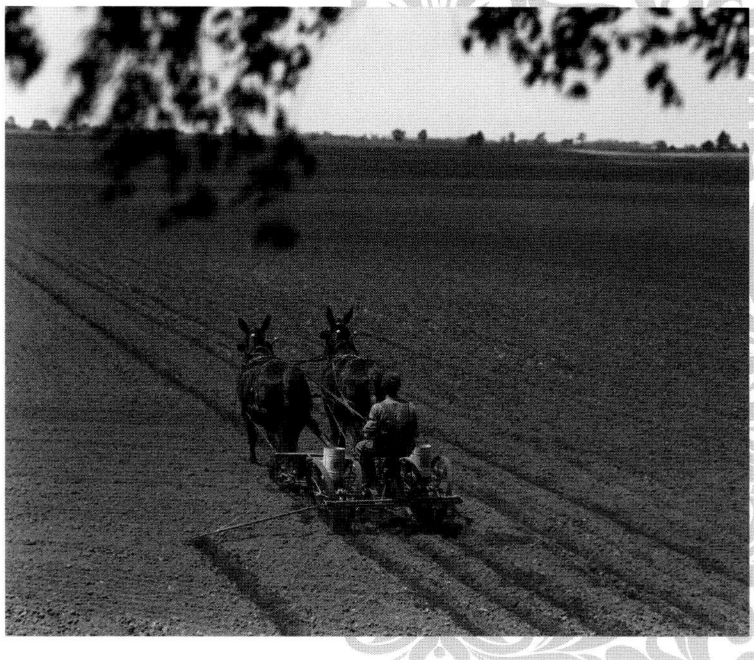

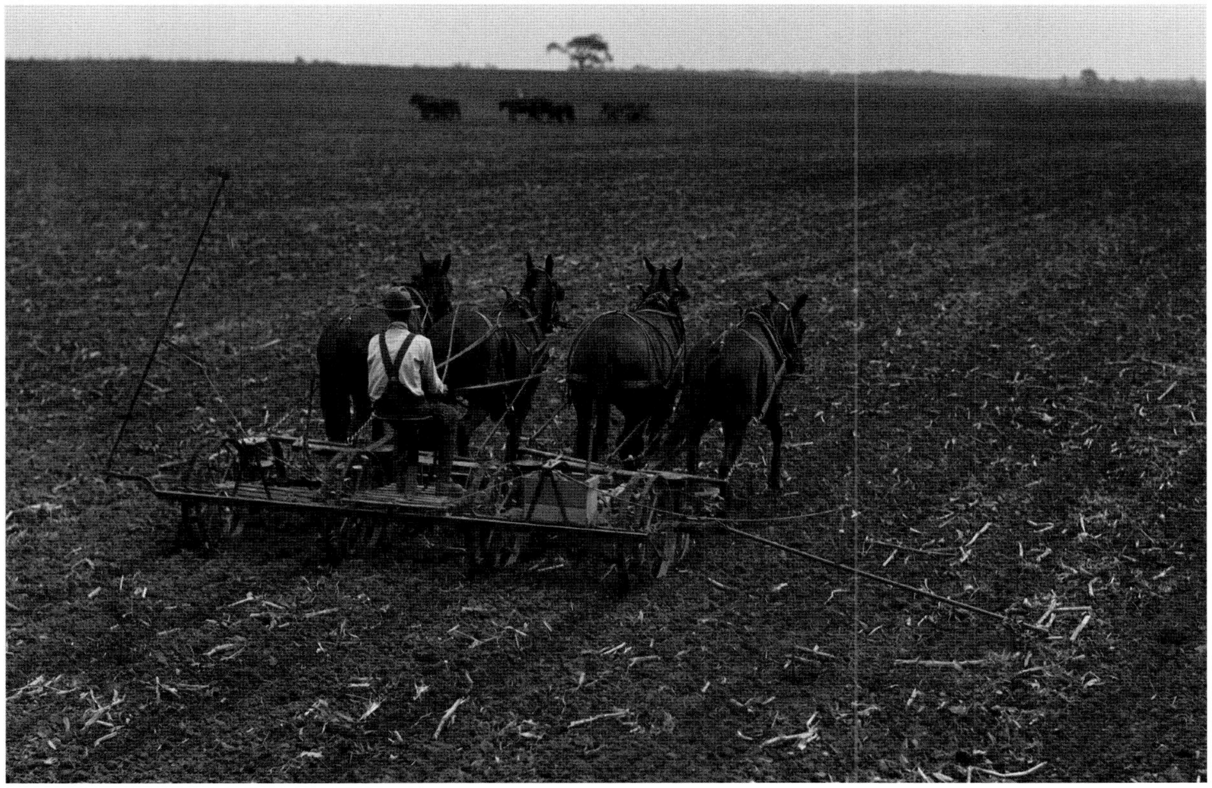

(Top left) Homer Young (walking) replants corn that was washed out on his Bloomington-area farm with the help of his son, Everett. (Monroe County, 1923)

(Top right) A mule team plants corn on the Henry Marshall farm in Lafayette. (Tippecanoe County, 1936)

(Bottom) Using a four-horse planter, George G. Seaman from Taylorville, Illinois, could plant 40 acres of corn per day. (Christian County, IL; 1929)

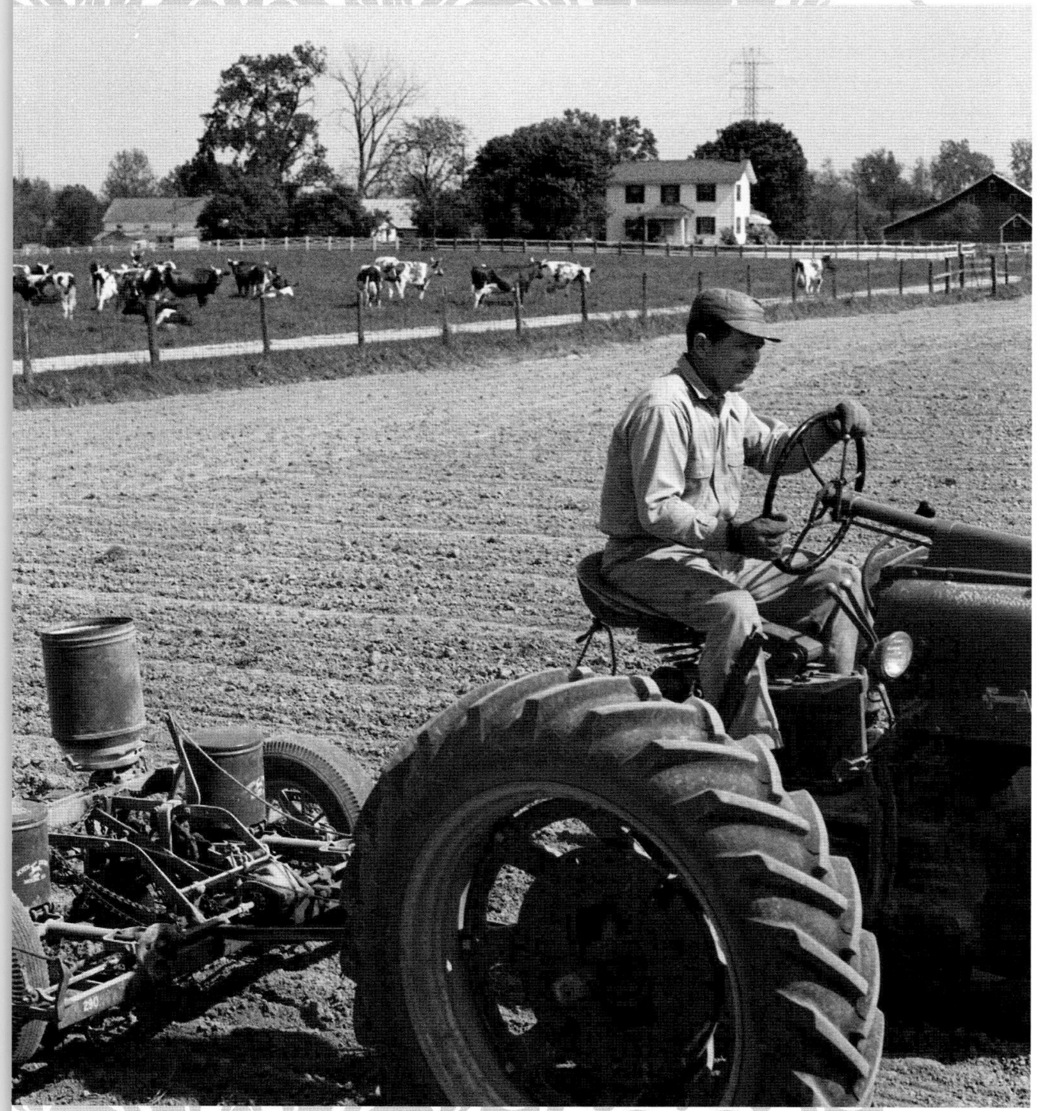

Planting corn on the N. D. Sommers and Son Farm in Grove City, Ohio (Franklin County, OH; 1961)

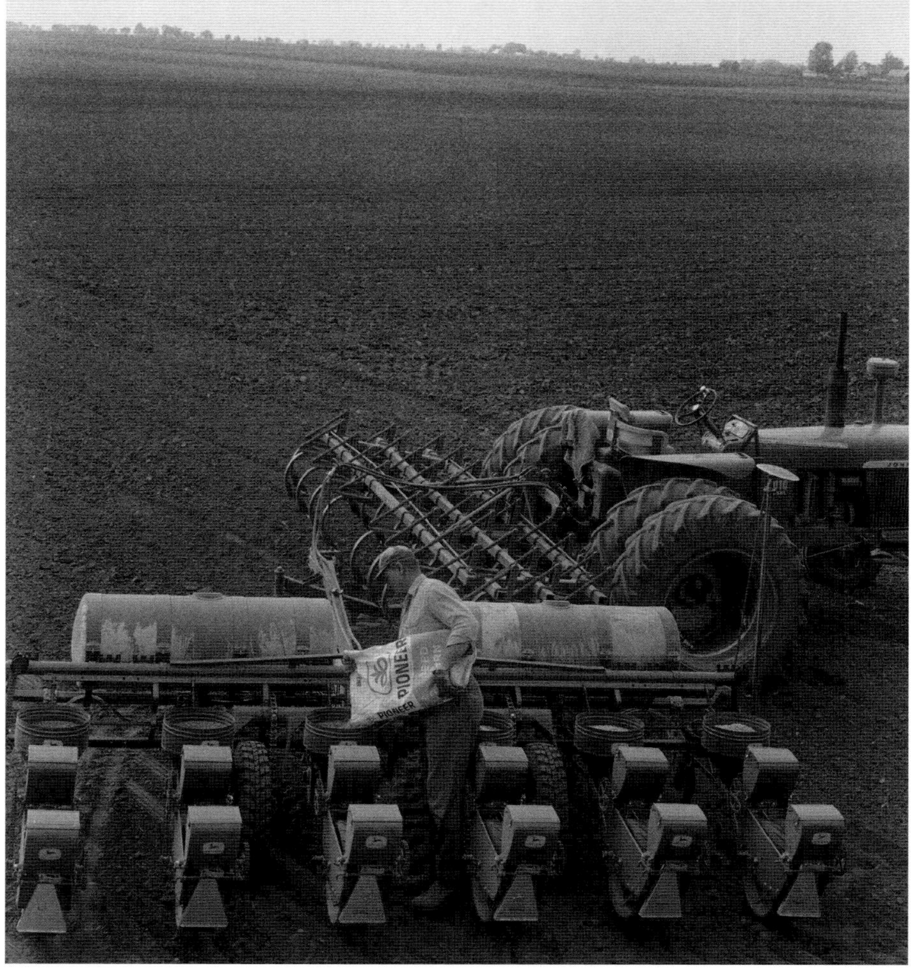

Preparing to plant corn on the Elmer Janssen farm (Tippecanoe County, 1965)

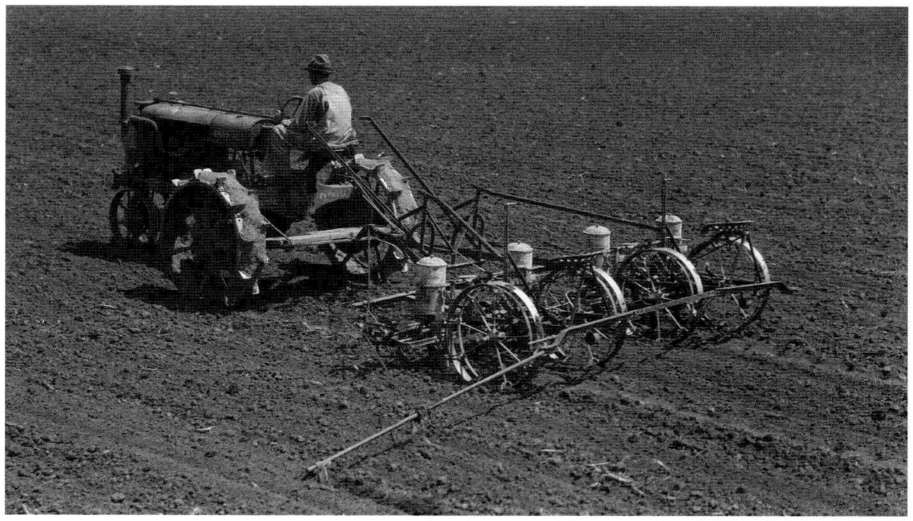

A farmer uses a four-row corn planter on the Kentland farm of Warren T. McCray, governor of Indiana from 1921 to 1924. (Newton County, 1931)

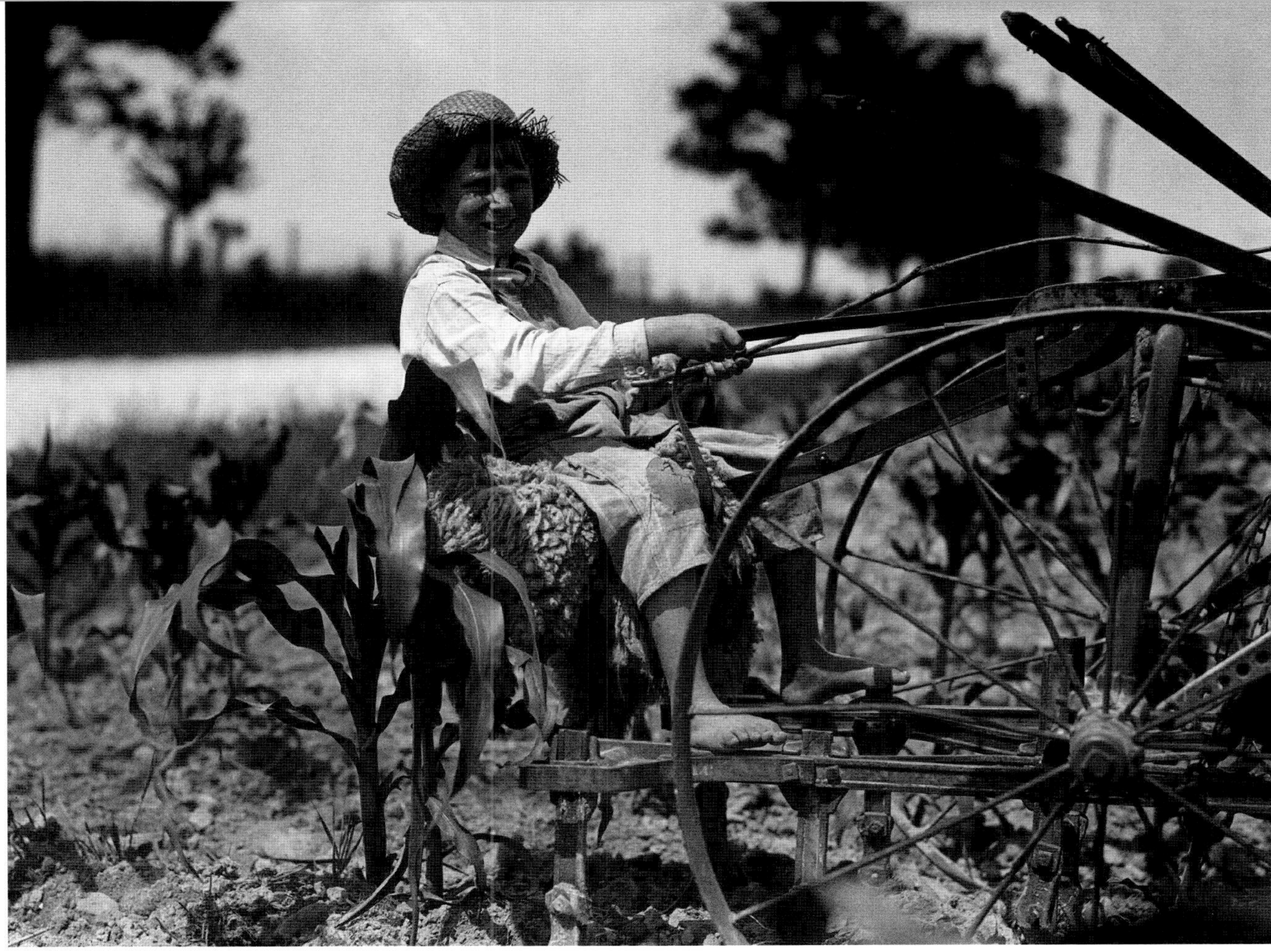

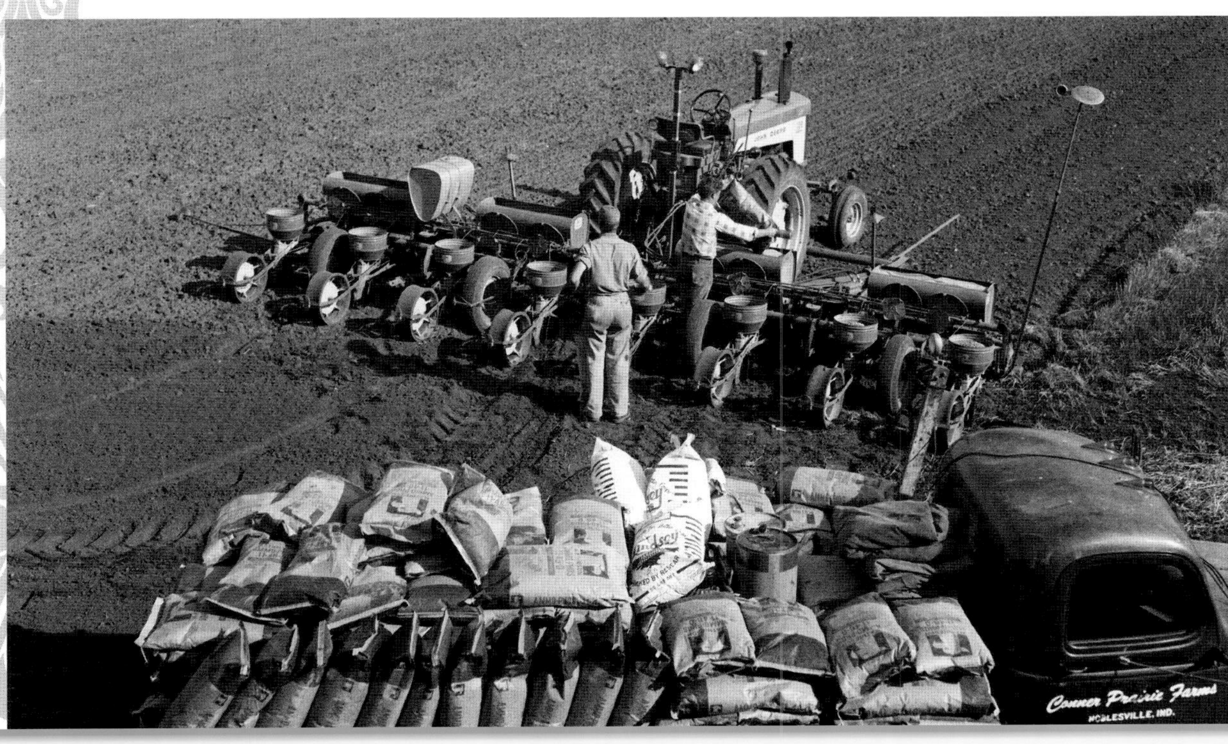

(Top) Eight-year-old George Leo Elrod from Dillsboro poses on a corn cultivator. (Dearborn County, 1923)

(Bottom) Two men fill an eight-row corn planter with seed and fertilizer at the Conner Prairie Farm in Noblesville. (Hamilton County, 1960)

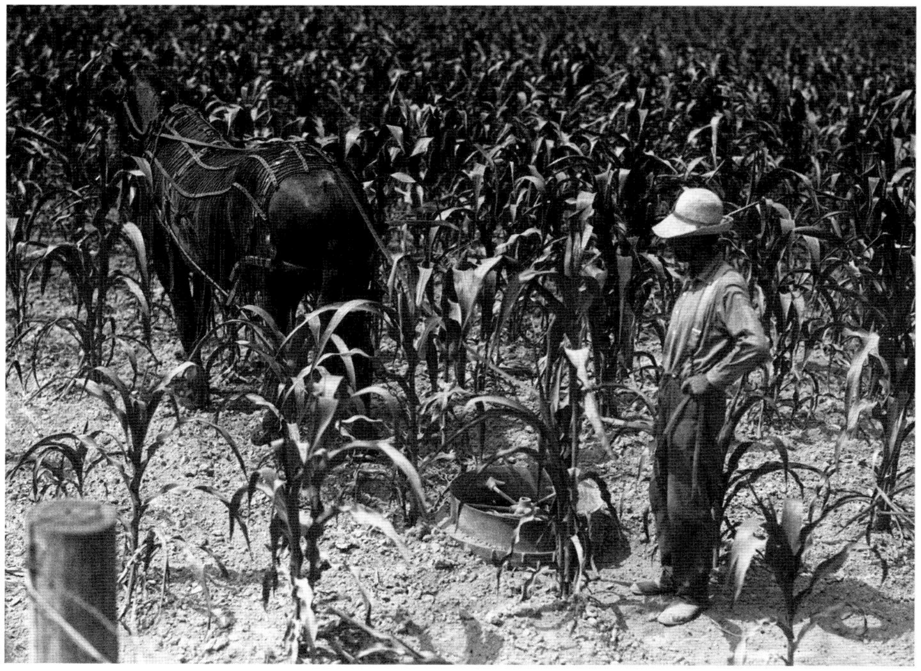

This corn is cultivated by dragging a mower wheel down the rows. (Indiana, ca. 1910s)

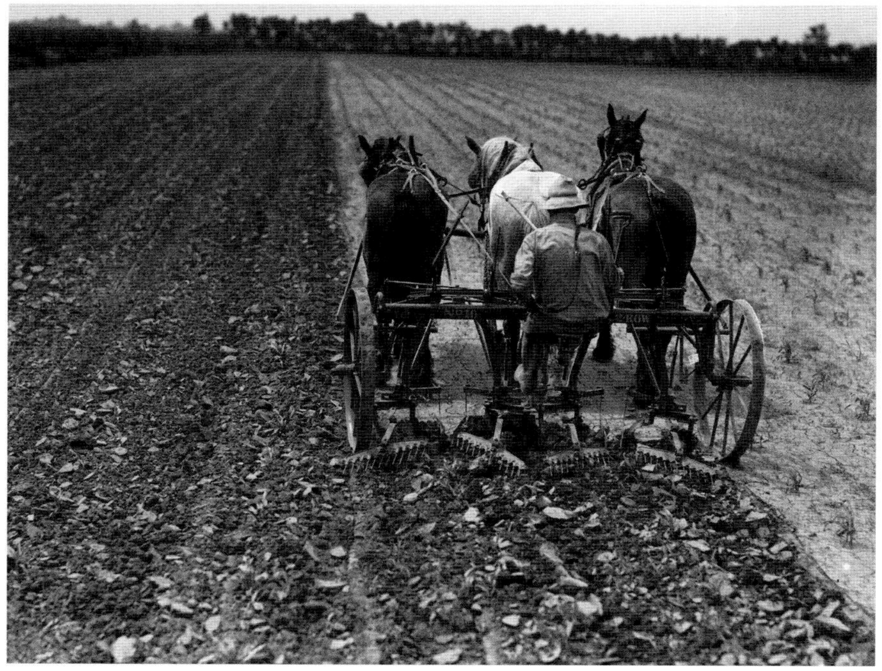

Cultivating corn with a three-horse team on the John Stout farm in Chatham, Illinois (Sangamon County, IL; 1920)

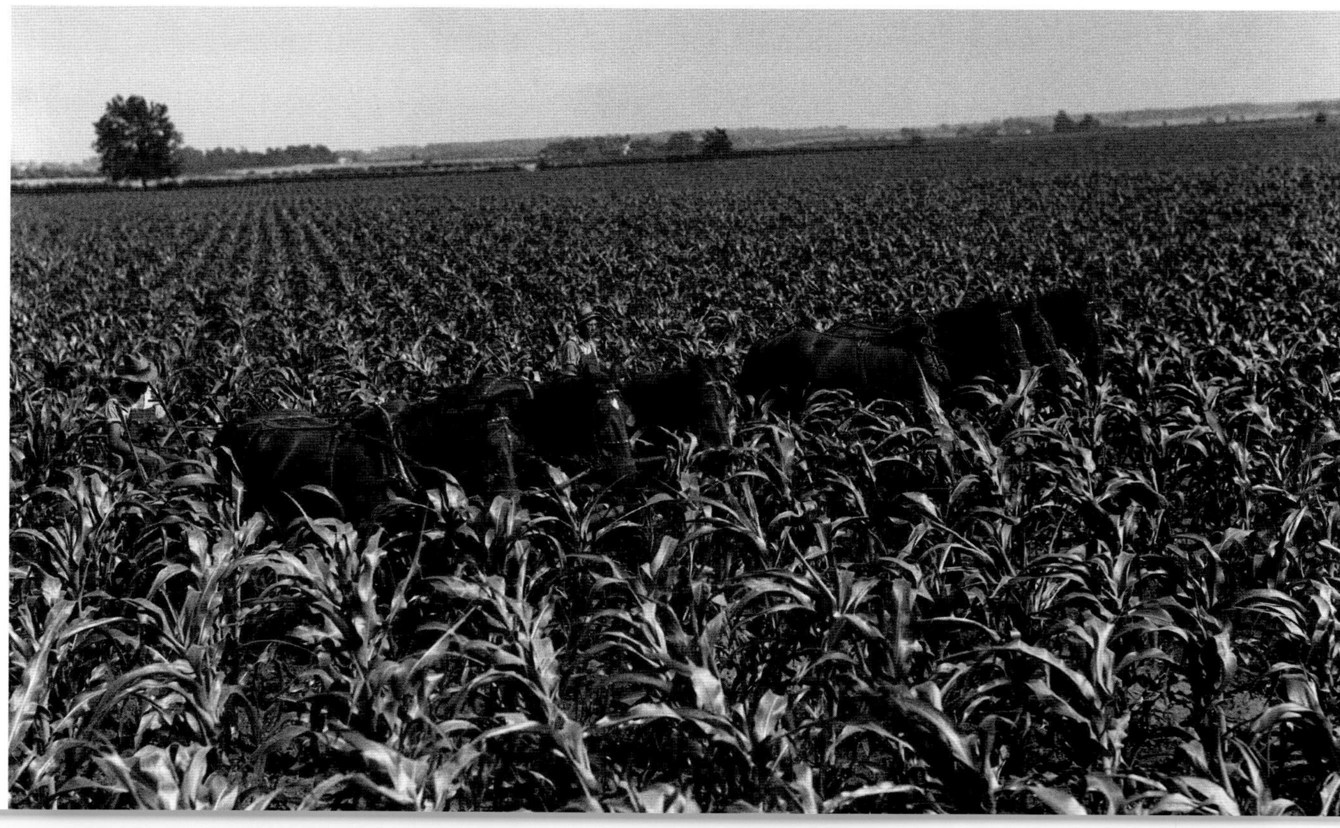

Two 3-horse teams cultivate corn on the Richard Goris farm near Lafayette. (Tippecanoe County, 1934)

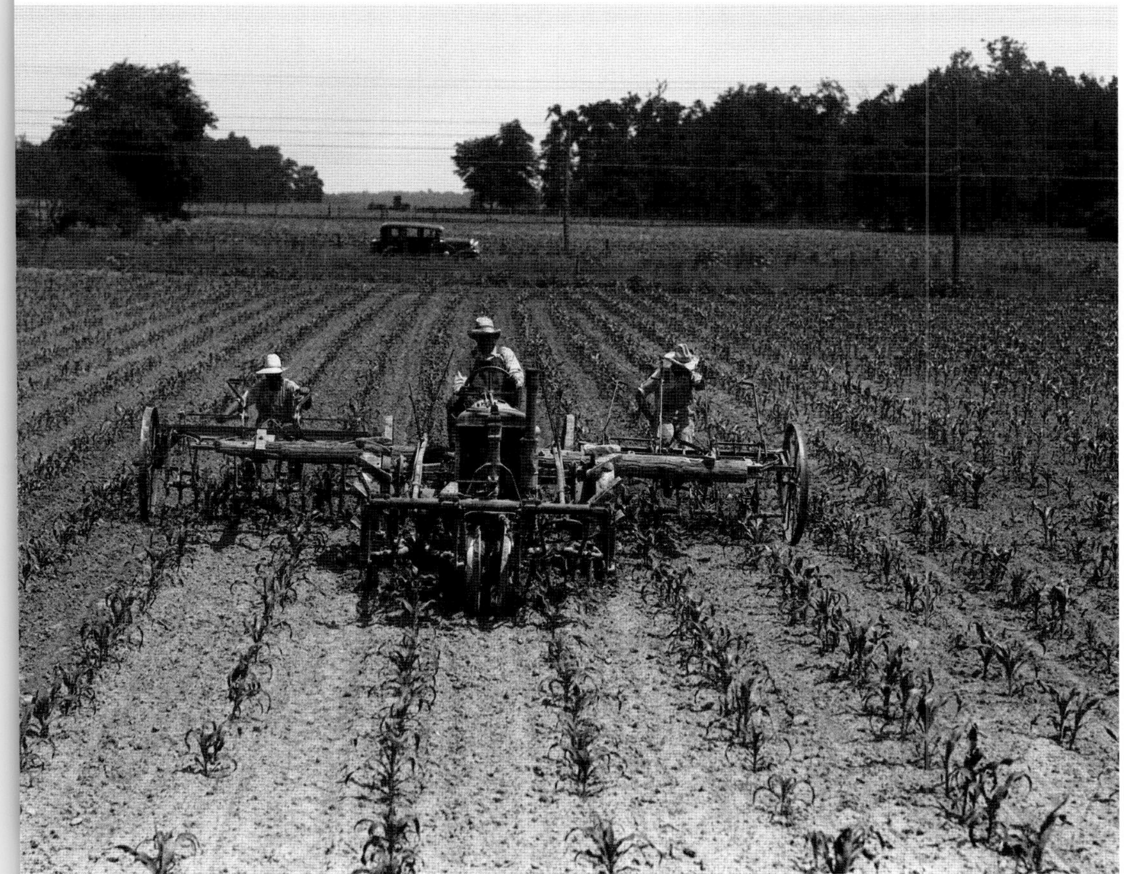

M. S. Steckel from Mulberry uses a six-row cultivator in his cornfield. Allen wrote, "Steckel had his two-row Farmall and decided he could get more efficient use from it. A log was fastened to the drawbar after the drawbar braces were shortened to the proper length of about 12 inches and two 2-row cultivators attached." (Clinton County, 1931)

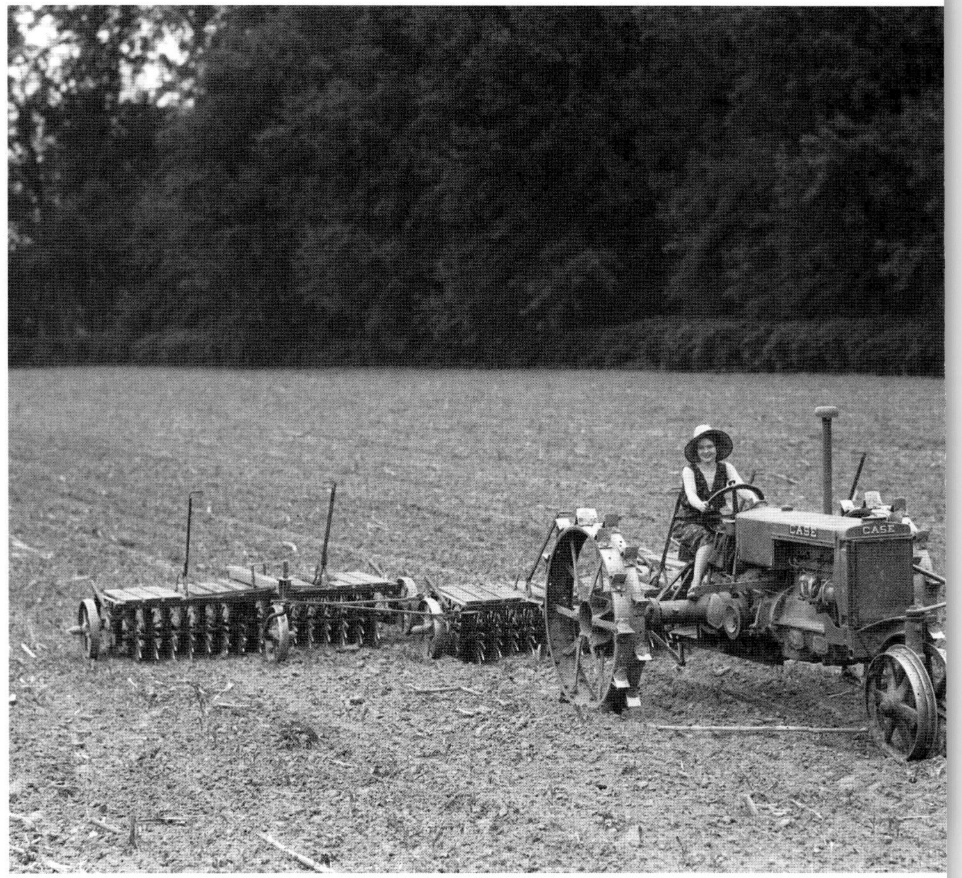

A woman uses a Case tractor to pull three rotary hoes in a cornfield on the *Breeder's Gazette* Farm located at Spencer. (Owen County, 1932)

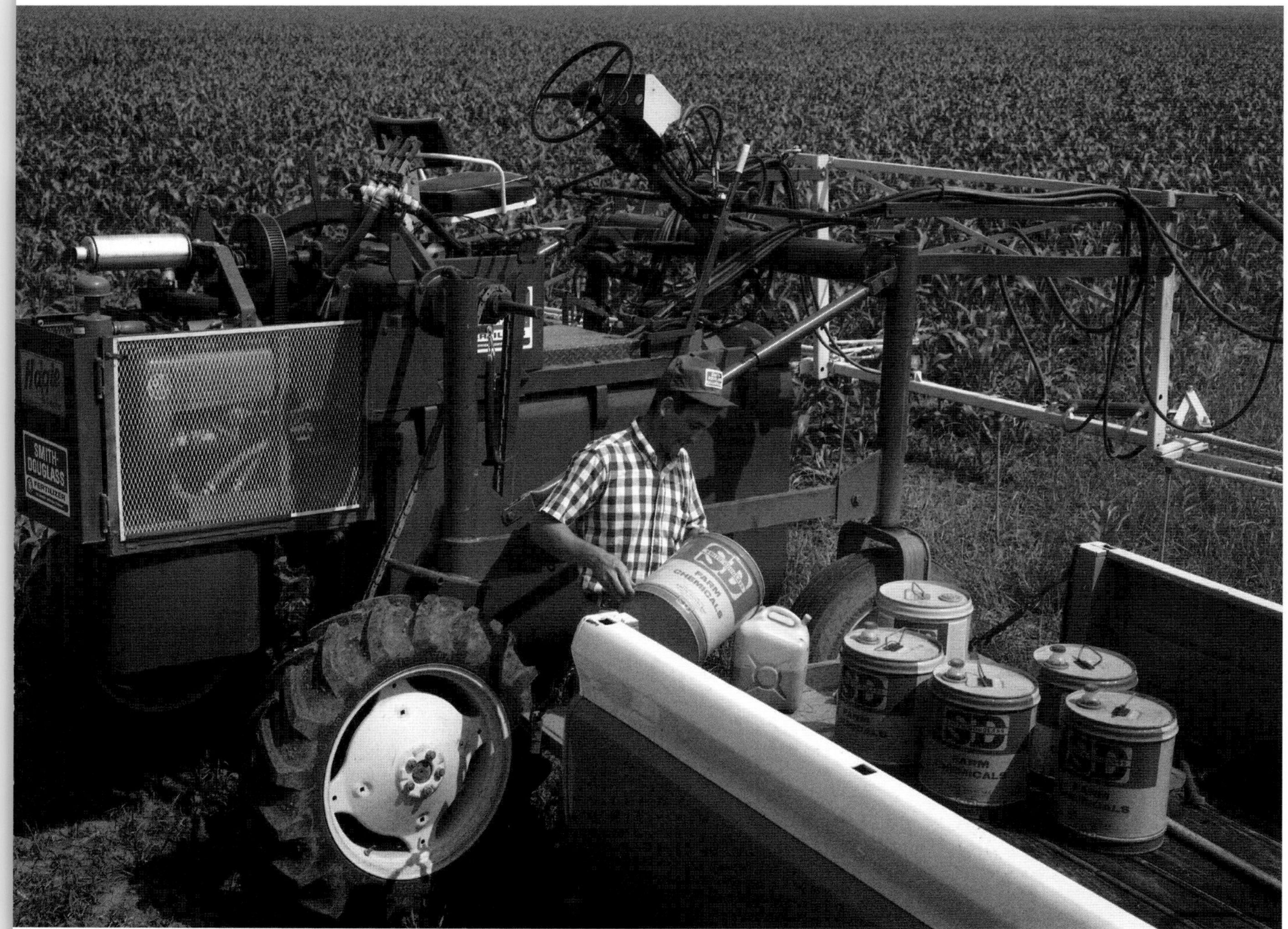

Ed Smith adds a herbicide to a spray rig on the Jess Smith and Son Farm in Urbana, Ohio. (Champaign County, OH; 1966)

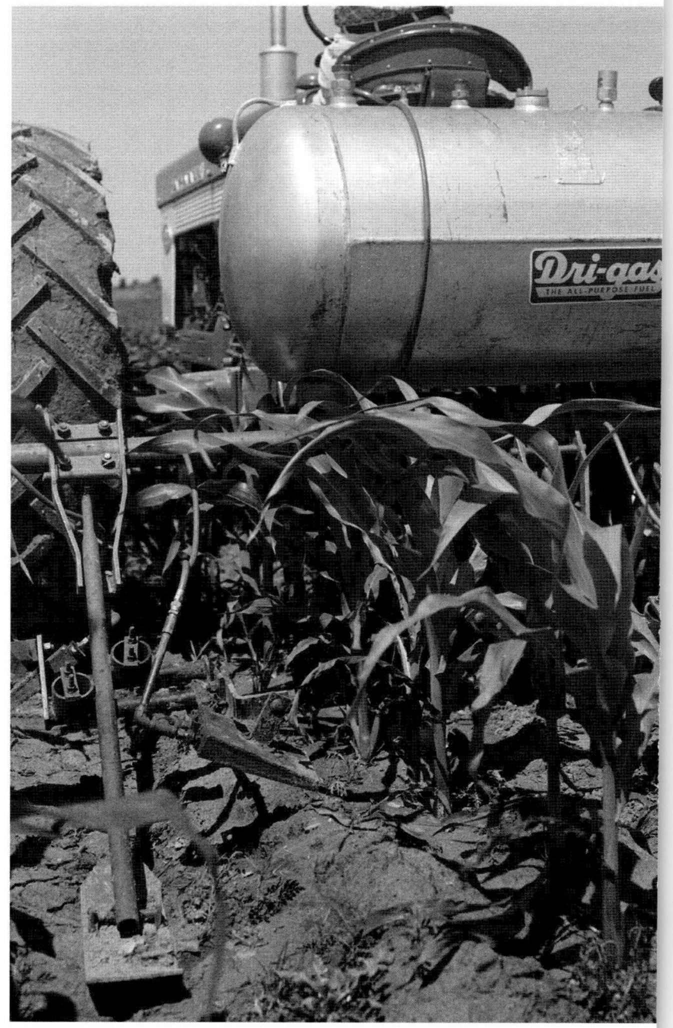

Using an LP gas weed burner to control weeds in corn (Indiana or Illinois, 1960)

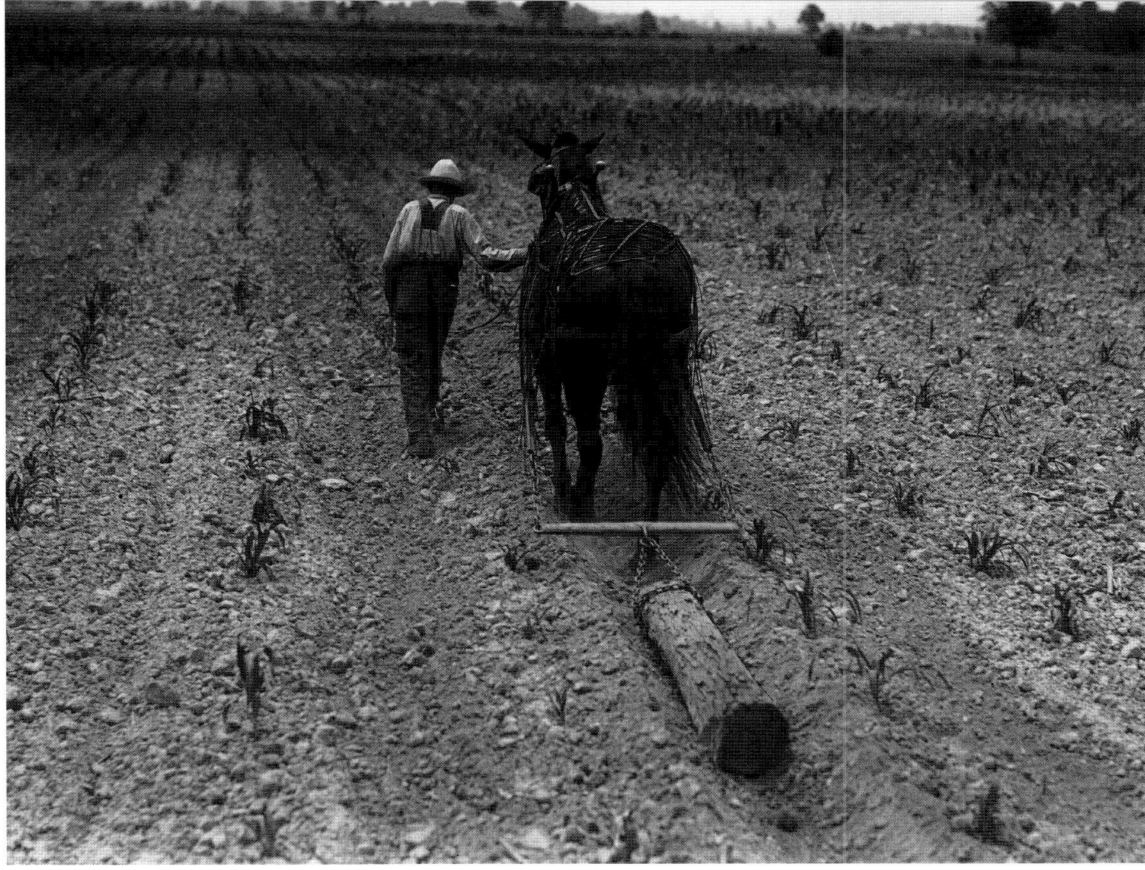

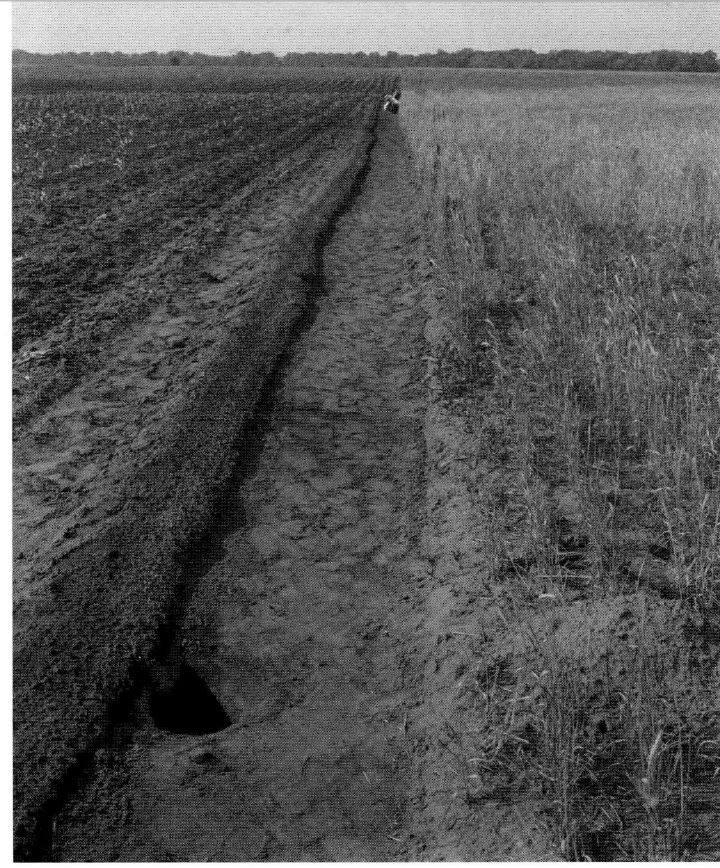

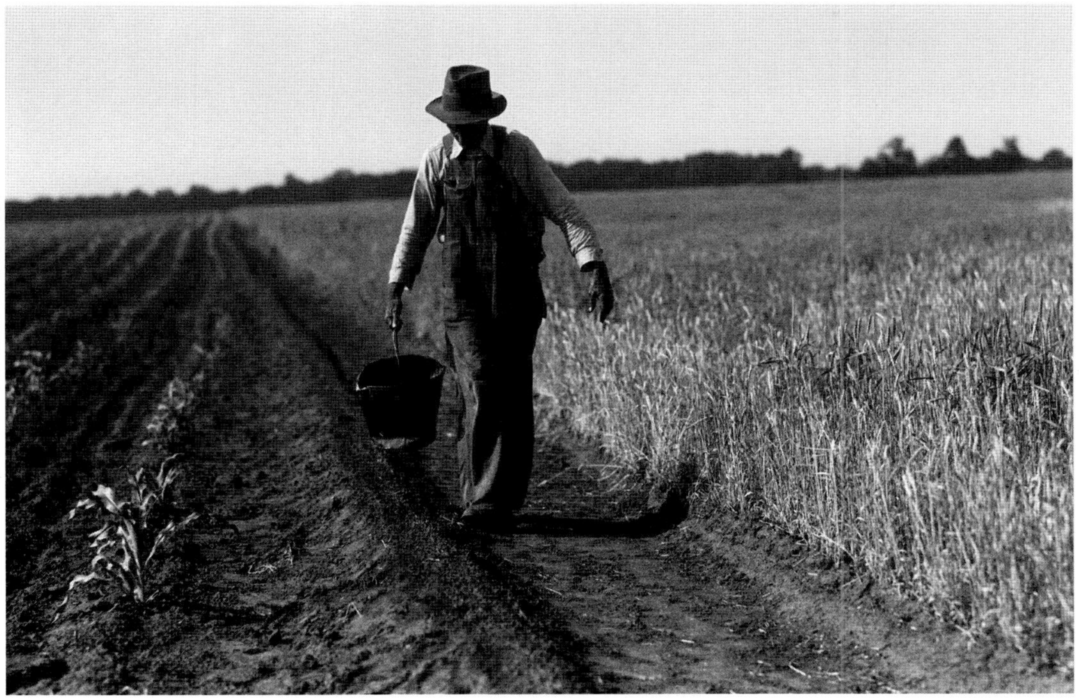

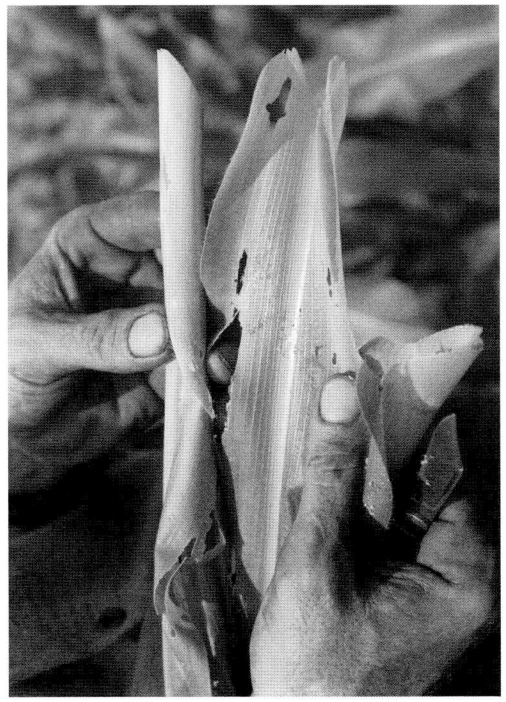

(Top left) A man leads a horse pulling a log between furrows to control chinch bugs on the farm of Jo Hoffman in Bluffton. (Wells County, 1921)

(Top right) A furrow with a creosote barrier and holes for traps keeps chinch bugs from moving out of an infested wheat field into an adjacent corn crop. (Indiana, 1934)

(Bottom left) This man uses a bucket to apply creosote as a barrier for stopping chinch bugs. (Indiana, 1934)

(Bottom right) Small corn borer larvae damage to corn leaves in Paxton, Illinois. (Ford County, IL; 1949)

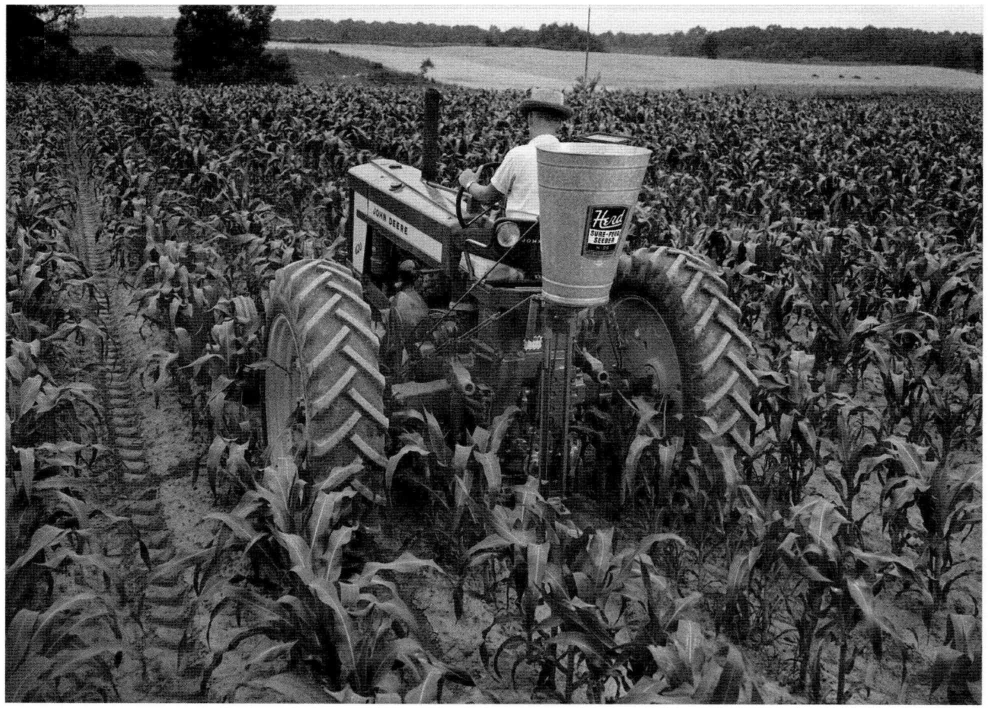

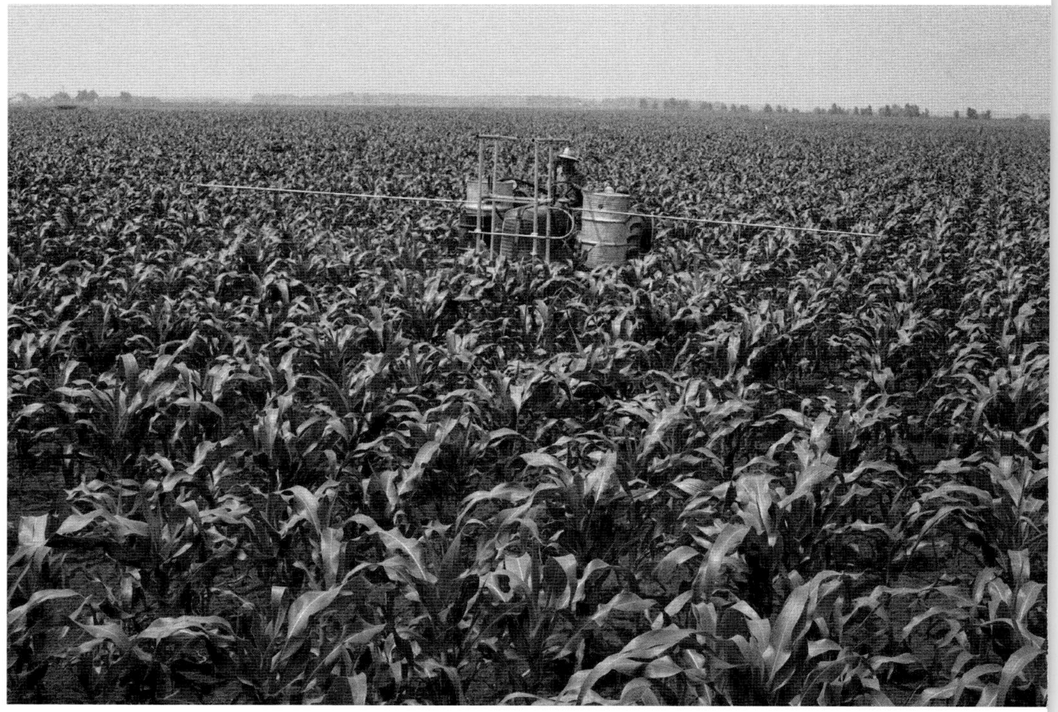

(Top left) This man treats corn for corn borer using a Herd Seeder to spread the insecticide. (Cass County, 1959)

(Top right) Silas Hill from Paxton, Illinois, sprays for corn borer using a 25-percent solution of the insecticide DDT at the rate of two quarts in 10 gallons of water per acre. (Ford County, IL; 1949)

(Bottom) Filling tanks as a crop duster prepares for takeoff (Grant County, 1950)

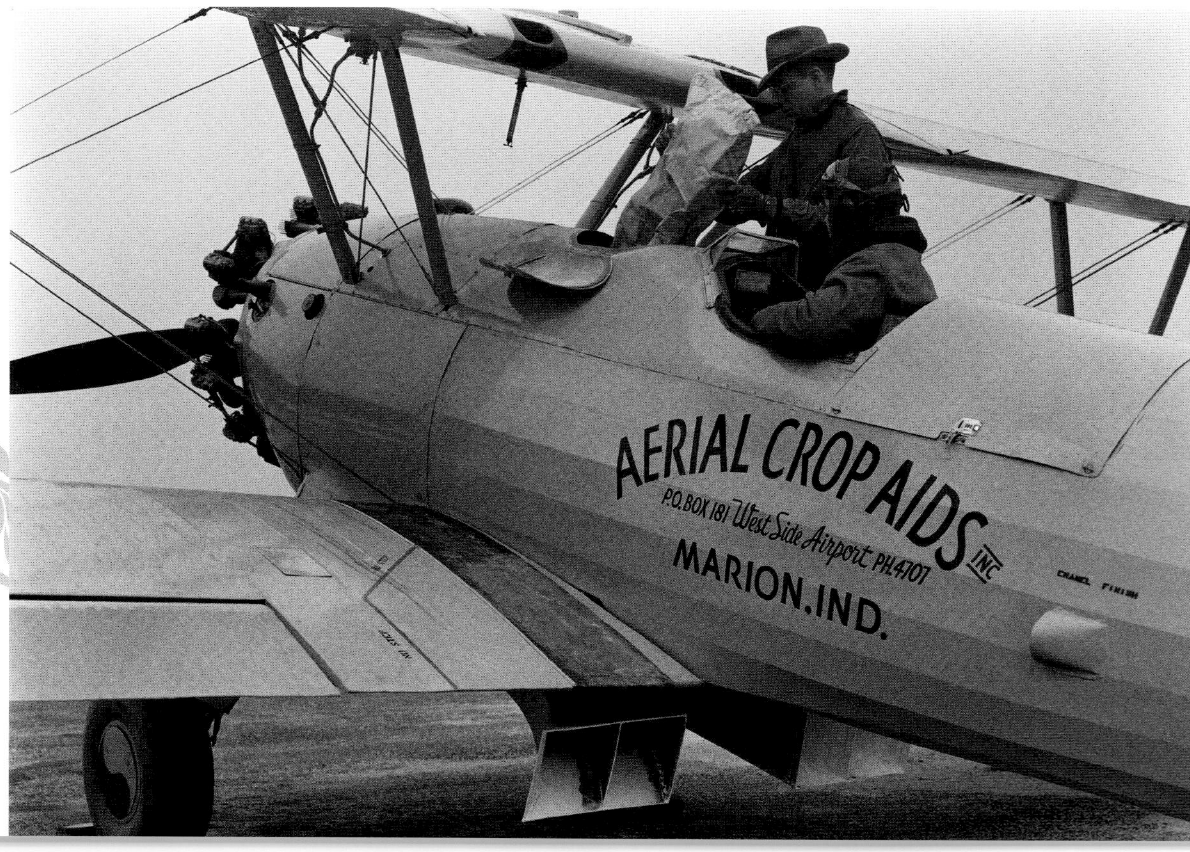

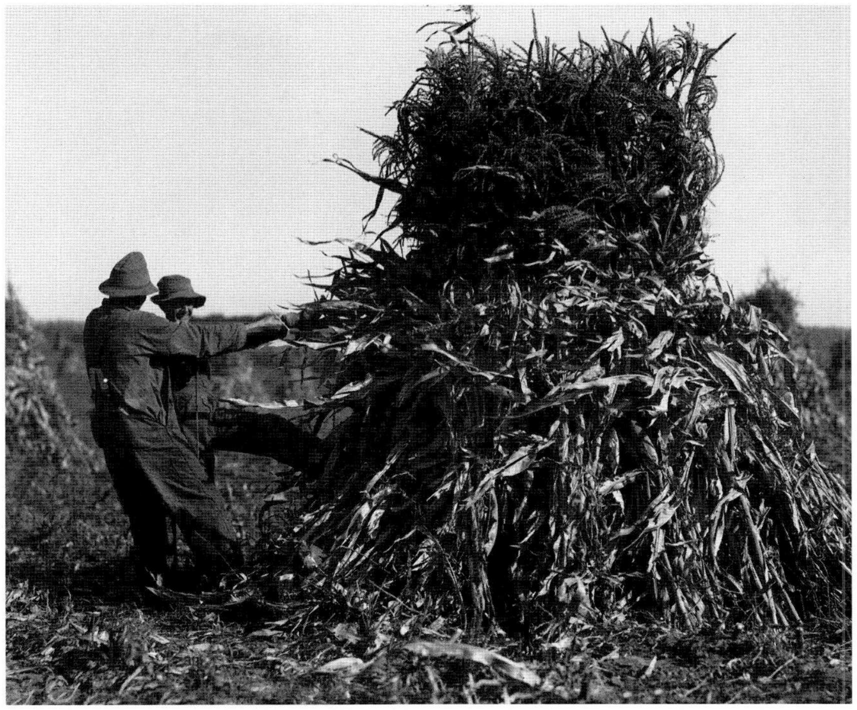

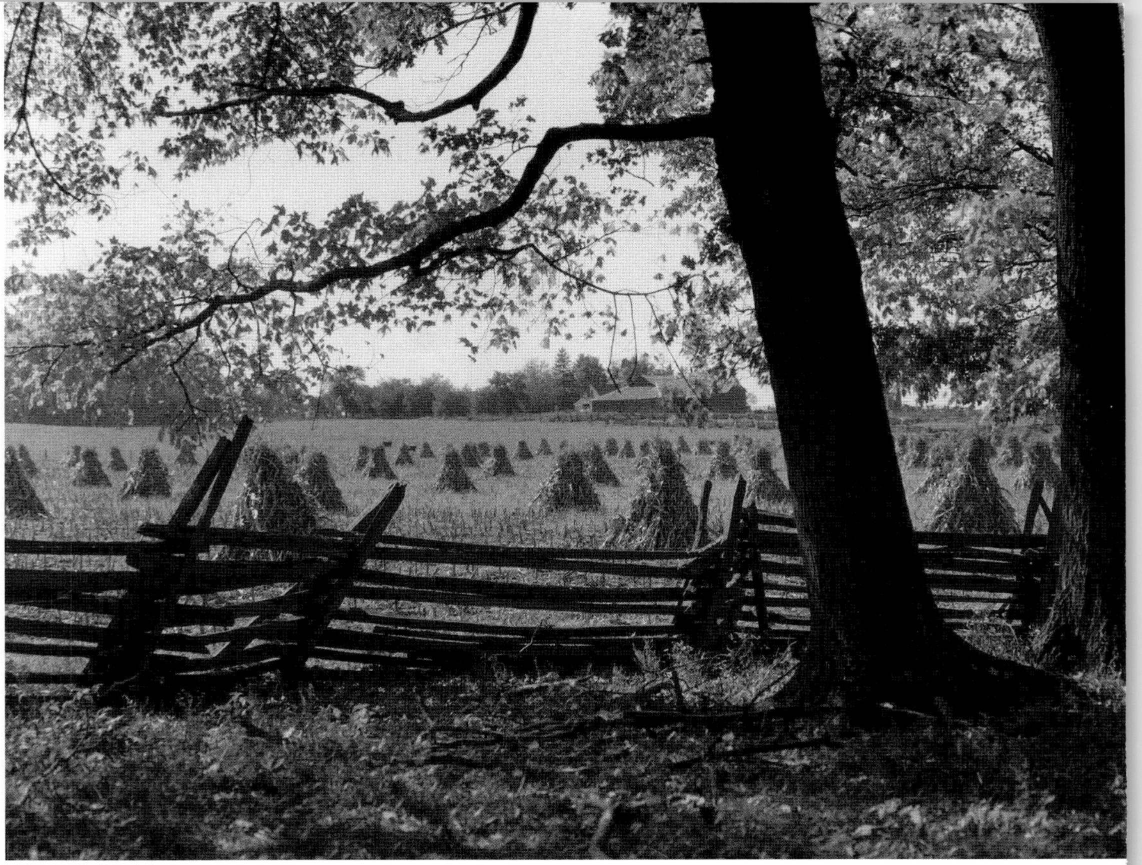

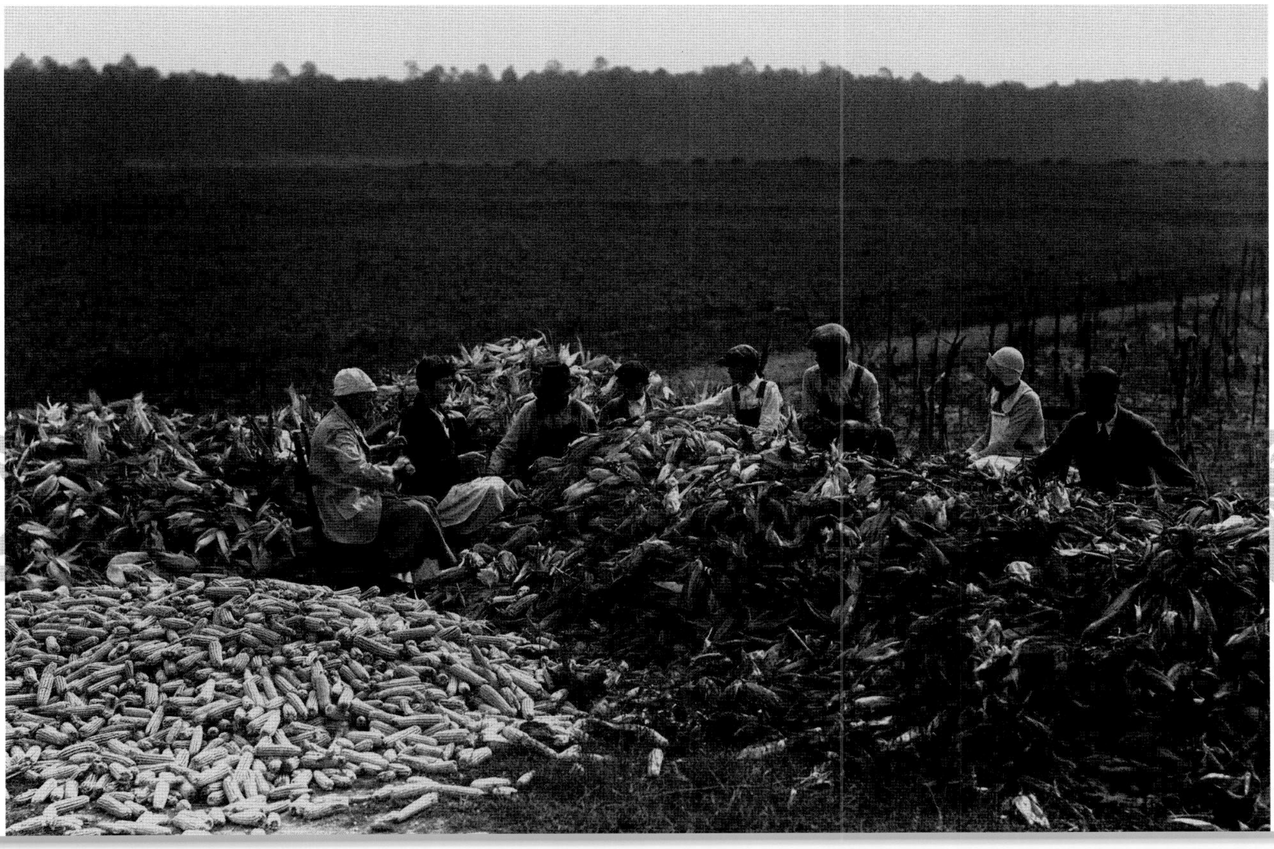

(Top left) Tying a shock of corn on the farm of W. A. Withrow from New Richmond (Montgomery County, 1917)

(Top right) An autumn scene on the farm of Noah Eaton near Mentone shows the shocked corn in the field. (Kosciusko County, 1934)

(Bottom) Husking corn by hand on the W. E. Campbell farm in Olin, North Carolina (Iredell County, NC; 1927)

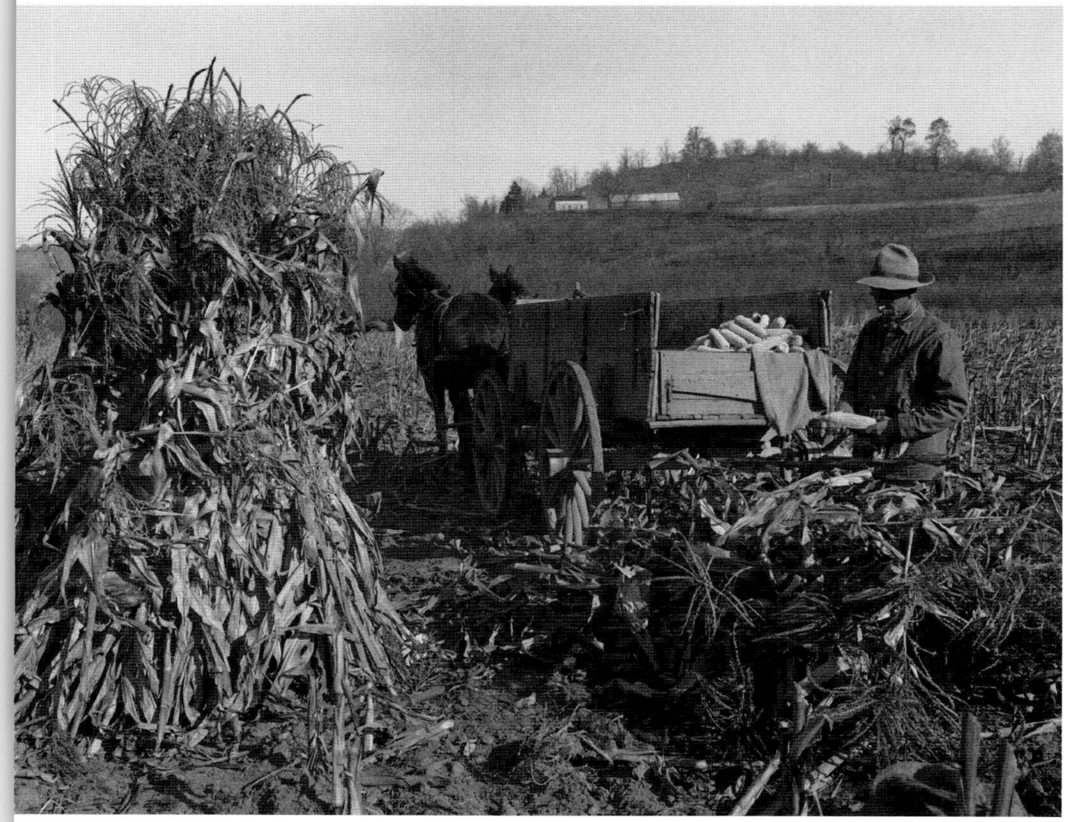

A farmer husks corn from a shock on the Charles A. Martin farm in Brooksburg. (Jefferson County, 1922)

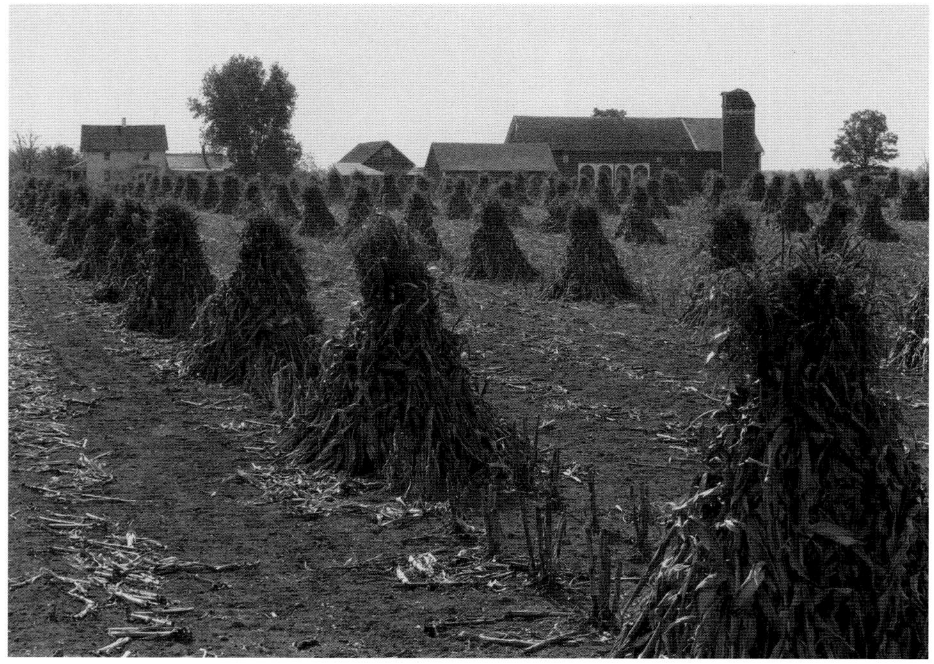

A large field of shocked corn on the farm of Lawrence Braun in Decatur (Adams County, 1929)

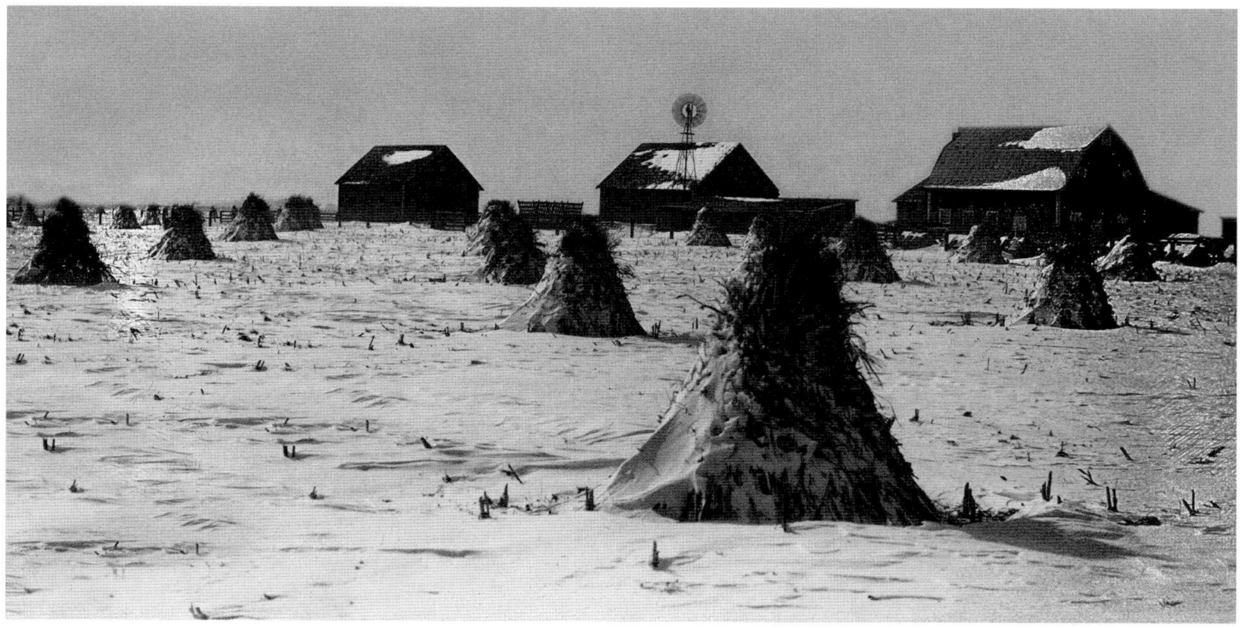

A snow scene showing shocked corn and farm buildings (White County, 1944)

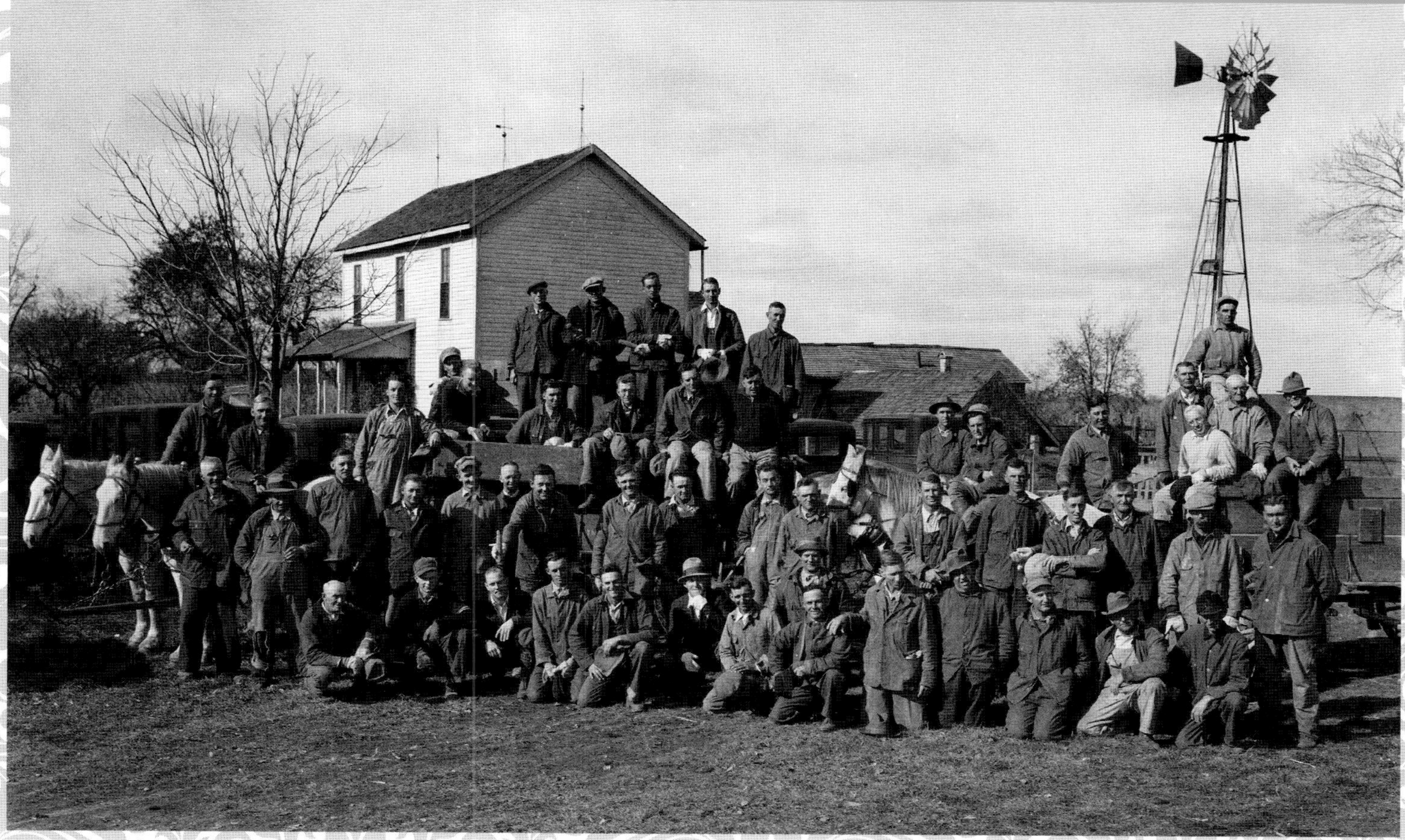

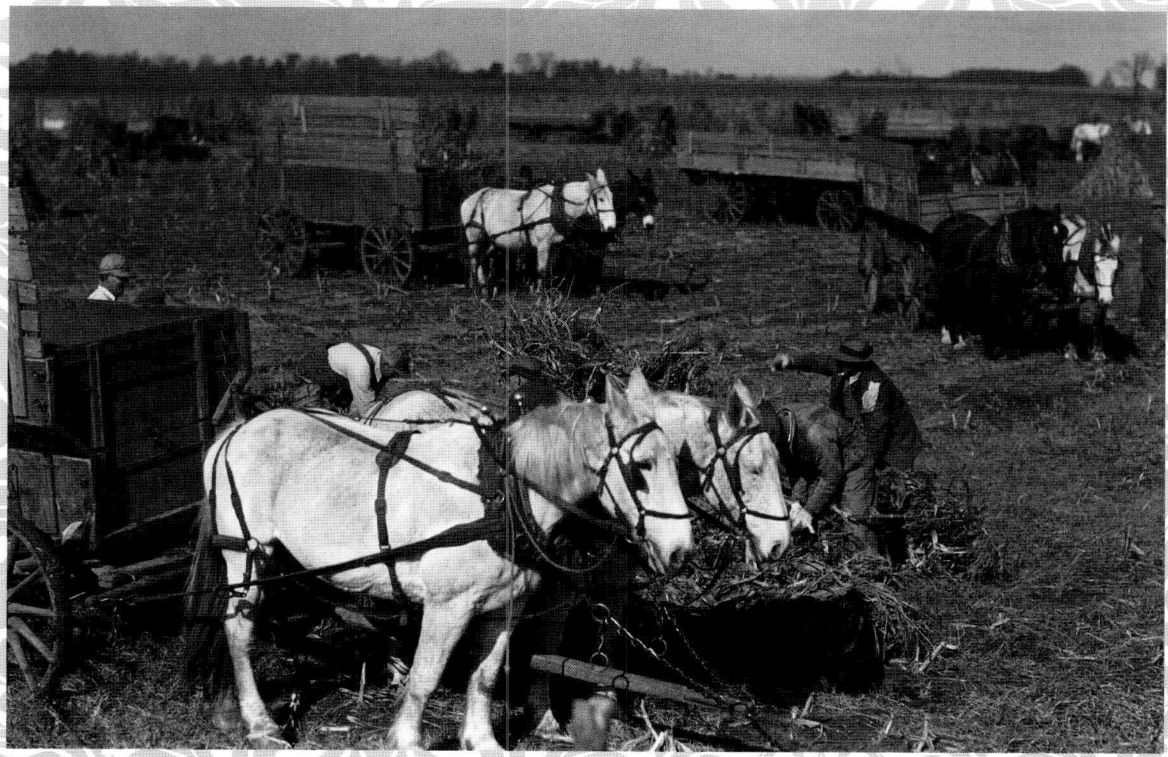

(Top and bottom) Farmers from Stockwell pitch in to harvest a woman's corn crop after the death of her husband. (Tippecanoe County, 1934)

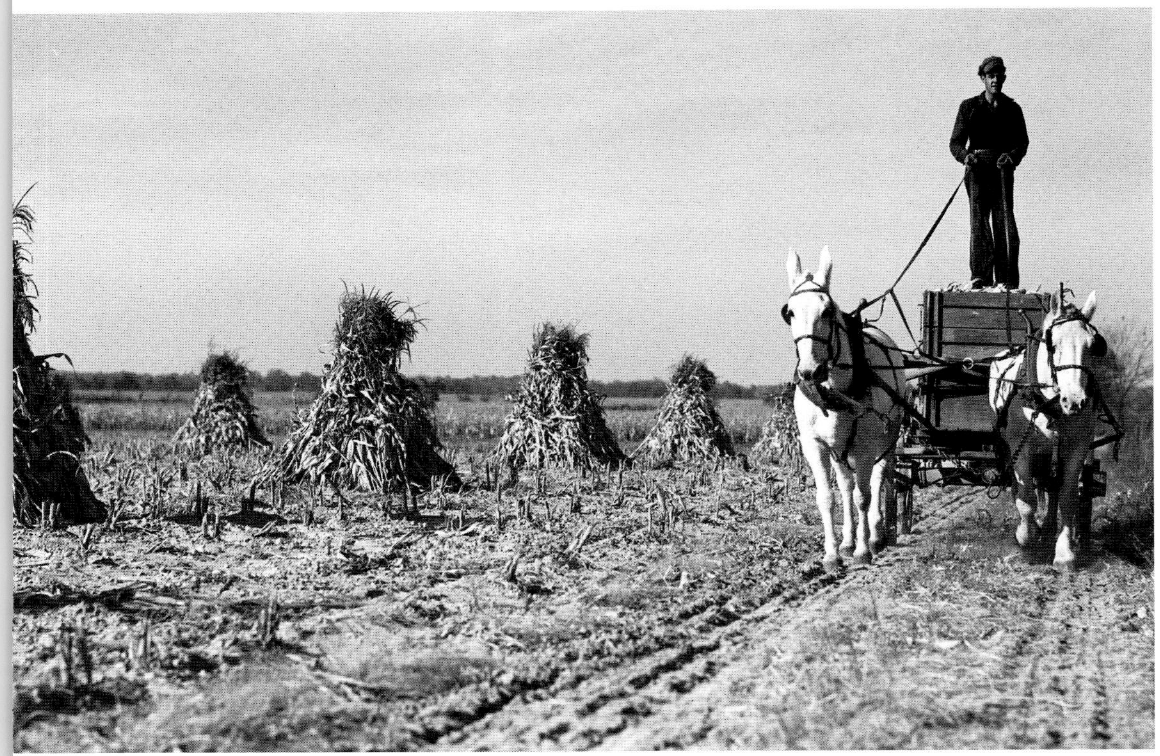

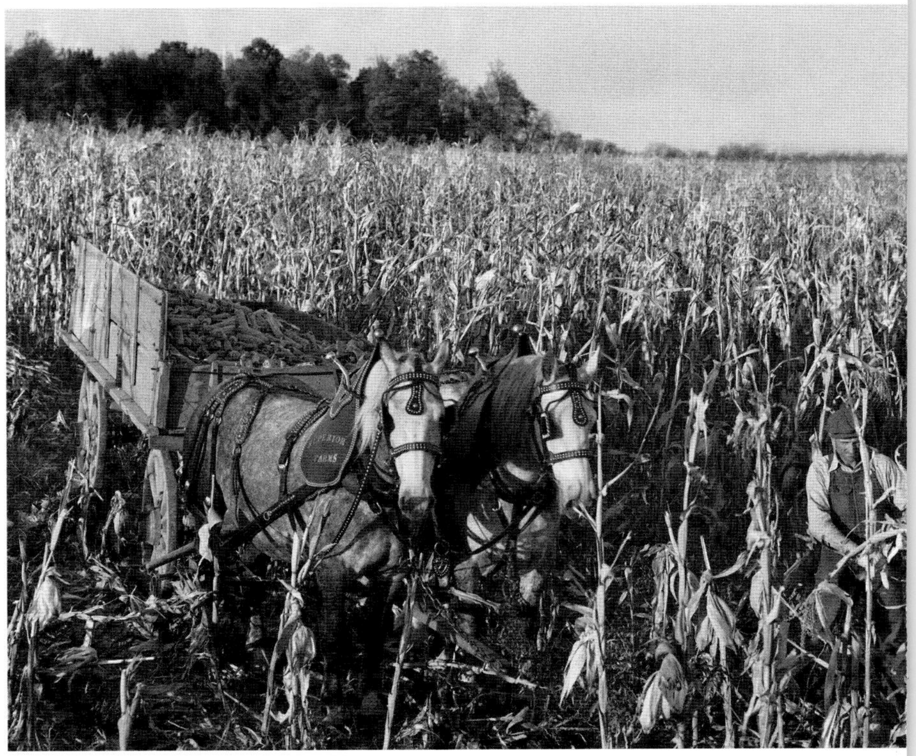

(Top left) A man hauls a wagonload of corn on the Chester Meek farm in Galveston. (Cass County, 1938)

(Top right) This man places husked corn in a wagon pulled by Percheron mares on the Penney-Gwinn Farm in Noblesville. (Hamilton County, 1938)

(Bottom) A farmer hand-husks corn and throws the ears into a wagon. (Indiana, 1936)

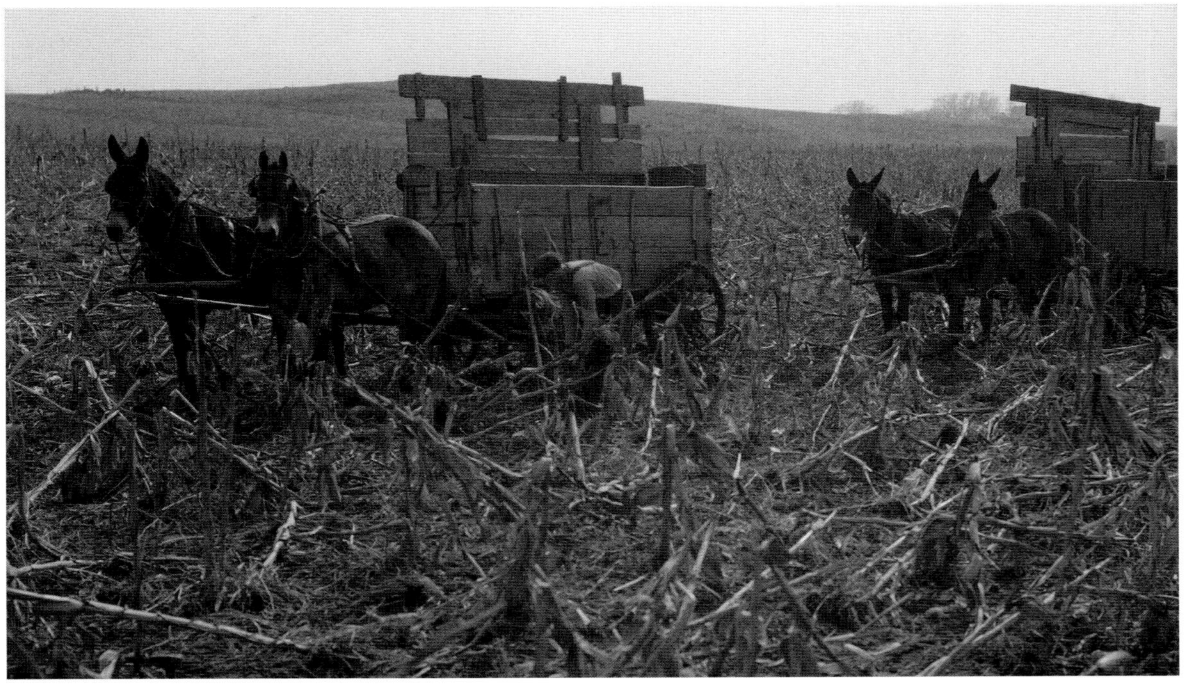

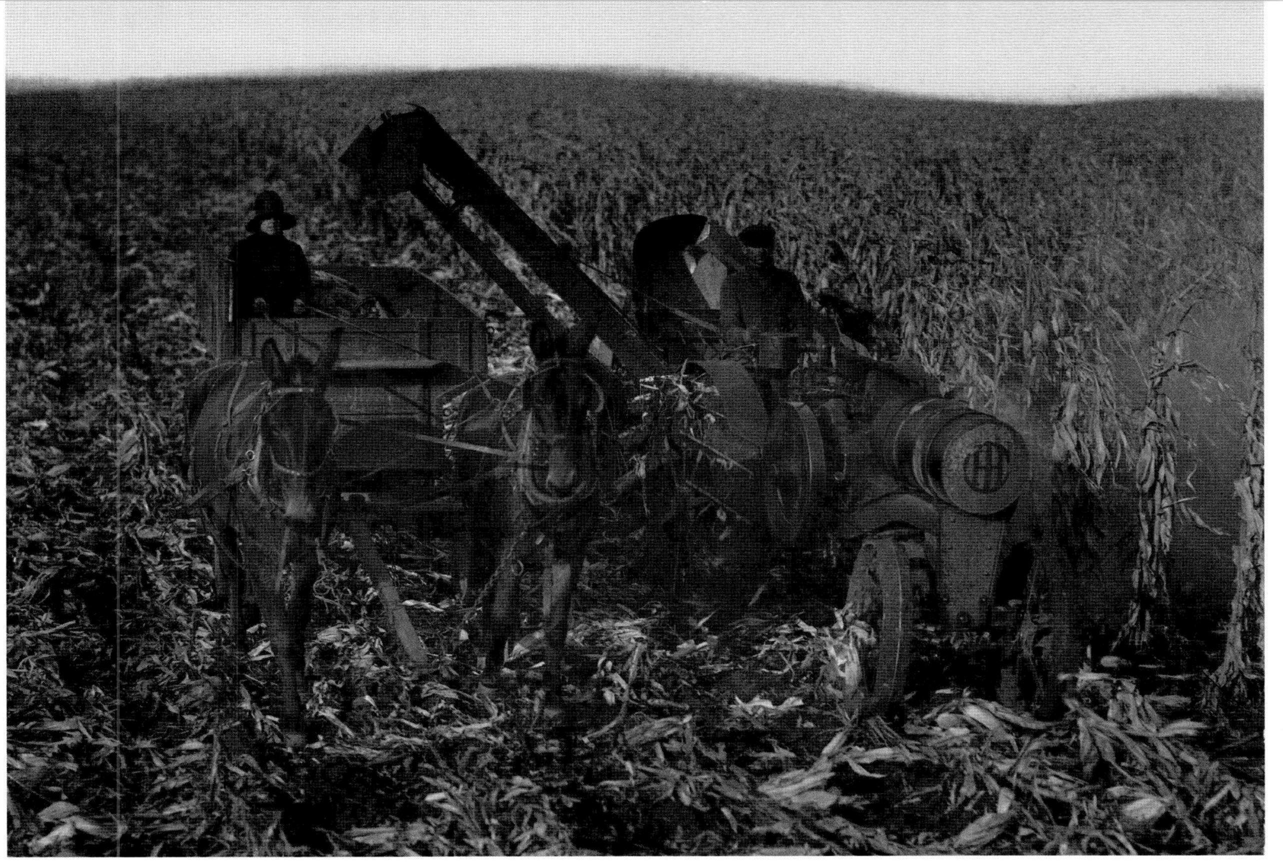

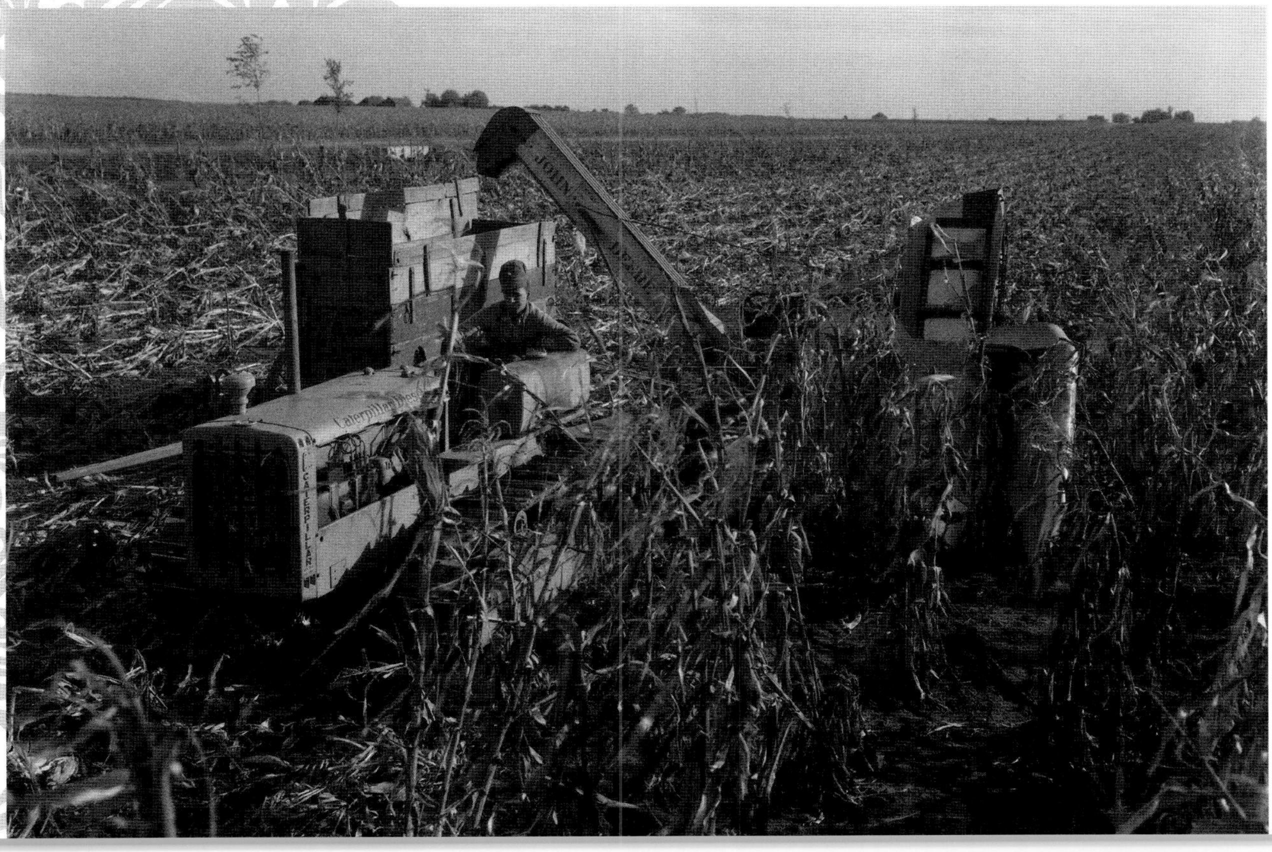

(Top) A. L. Henderson of Linden operates a tractor pulling a corn picker while his wife drives a wagon alongside. (Montgomery County, 1917)

(Bottom) Henry Schurter drives a Caterpillar RD4 pulling a John Deere two-row picker to harvest corn on the J. P. Werner farm near Tremont, Illinois. (Tazewell County, IL; 1936)

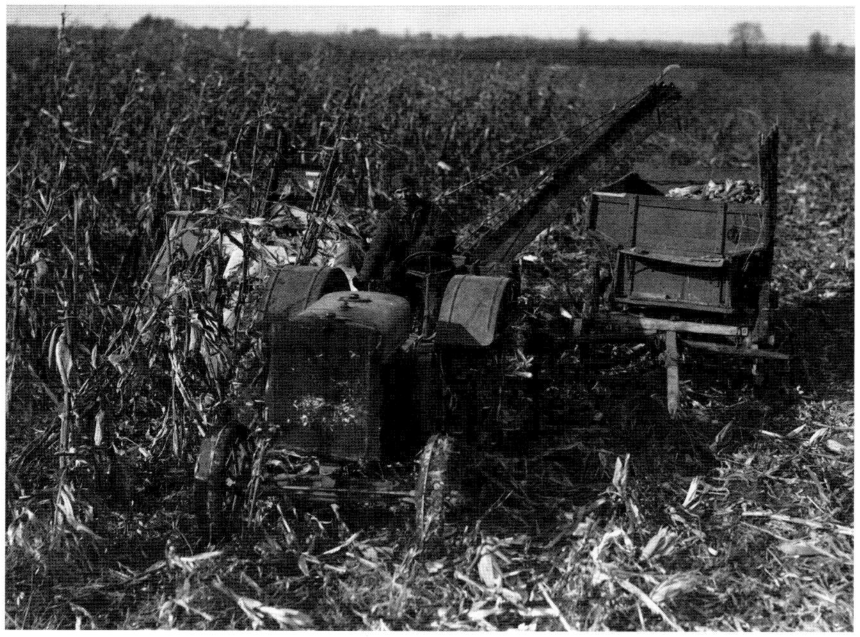

A two-row picker pulled by a John Deere tractor on the farm of Louis Withrow near Romney (Tippecanoe County, 1929)

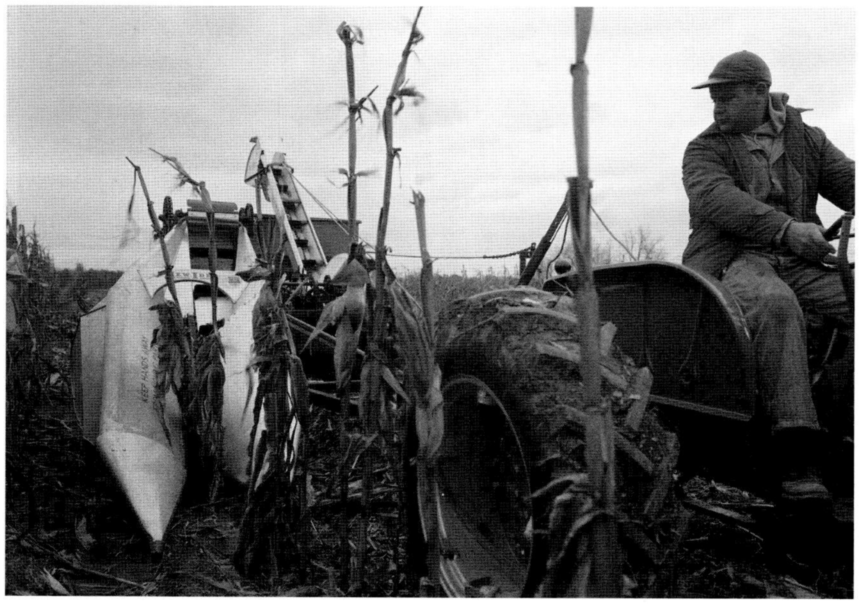

Harvesting corn with a New Idea one-row, pull-type picker on a farm near New Castle (Henry County, 1959)

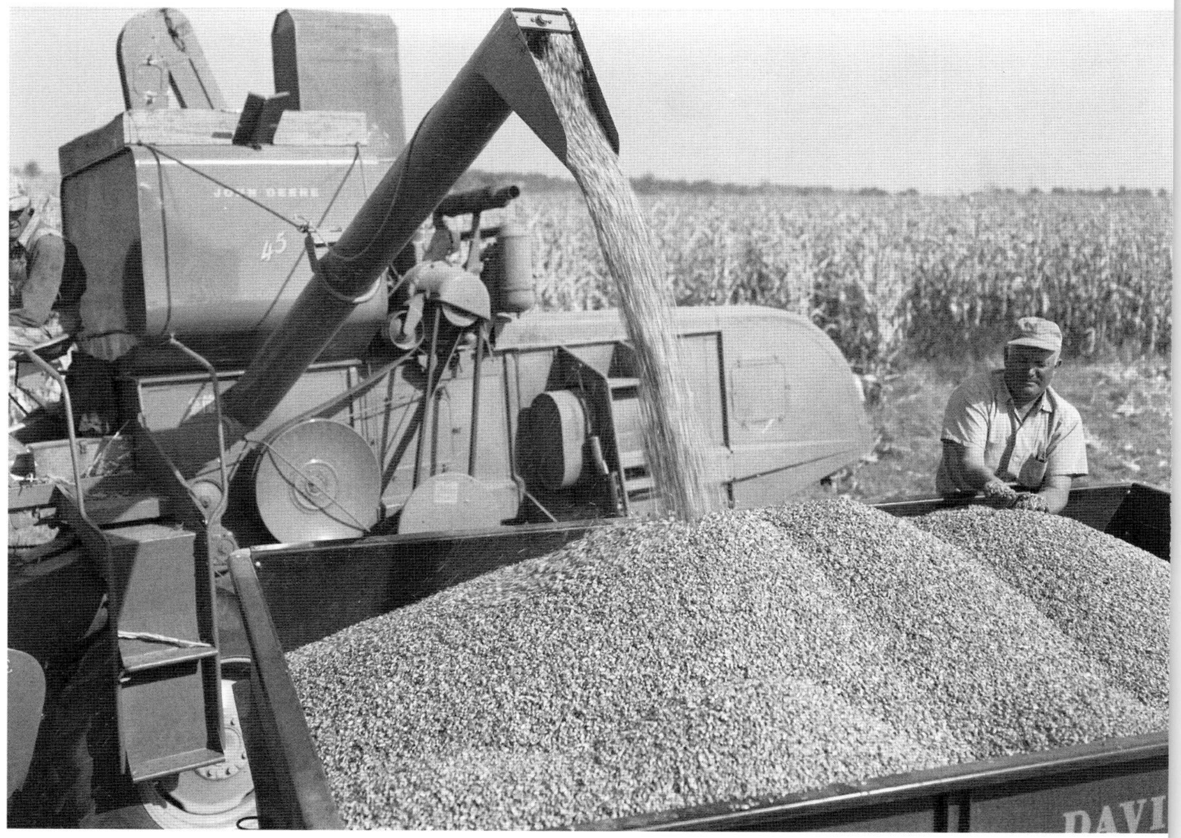

Dumping corn from a picker-sheller into a wagon on the farm of James Kellerman near Romney (Tippecanoe County, 1956)

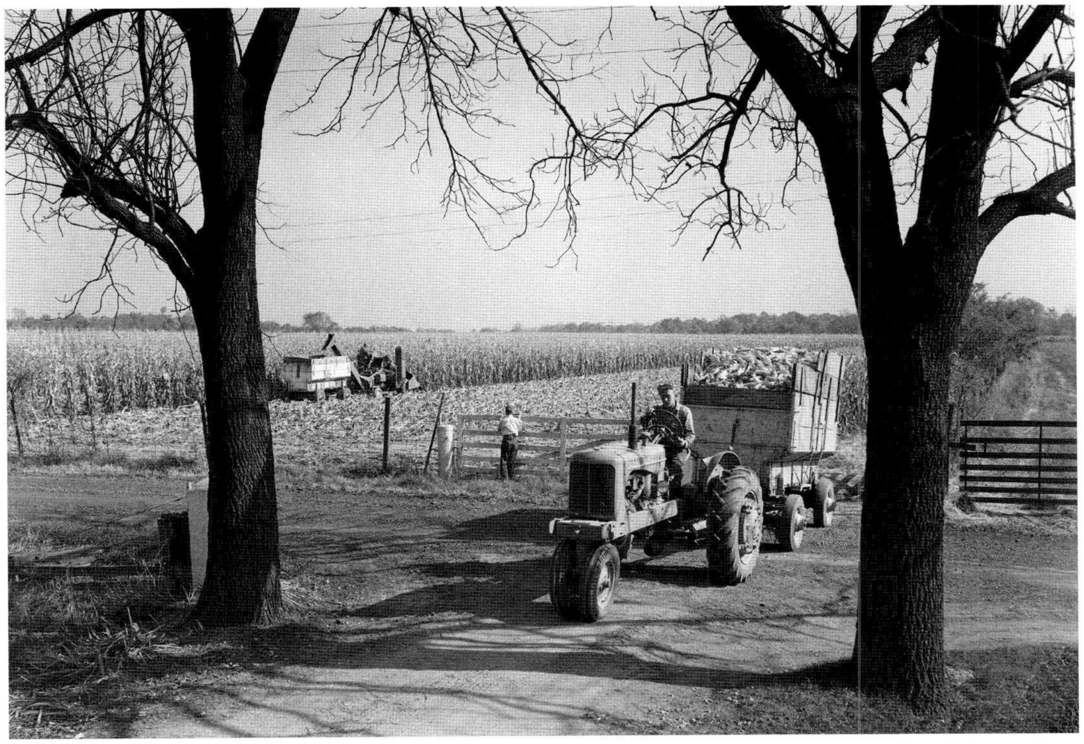

Paul Boes from Lafayette brings corn out of the field. (Tippecanoe County, 1948)

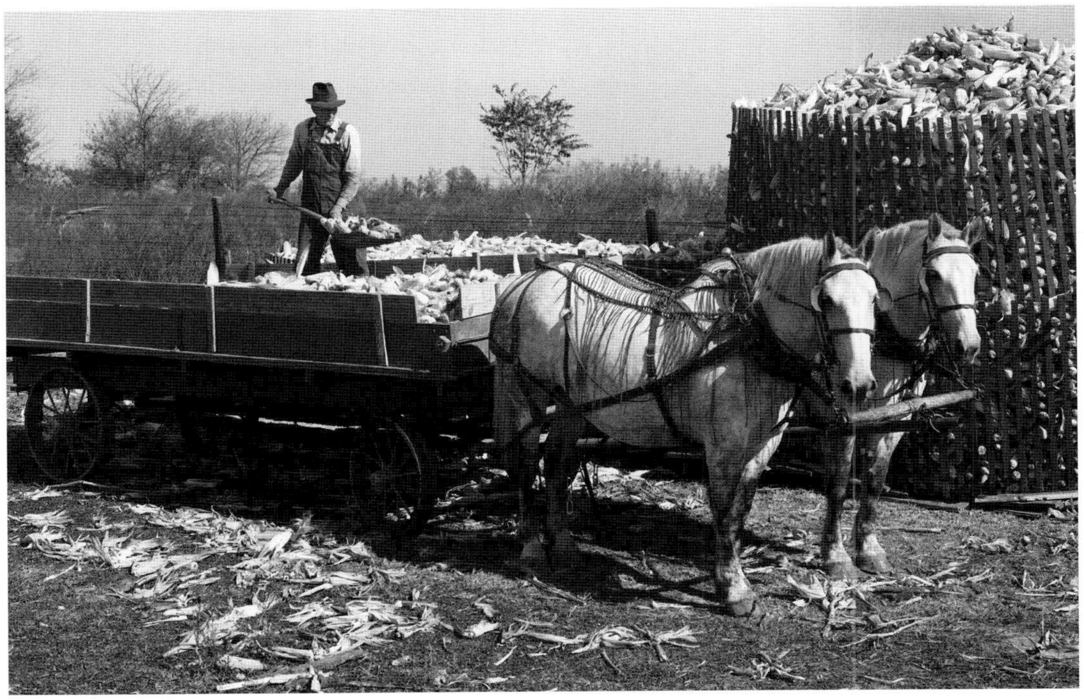

Joe Fisher shovels ear corn from a horse-drawn wagon into a temporary crib on the David S. Wise farm in Flora. (Carroll County, 1938)

This man uses an unusual method to tip a wagon so that popcorn can be unloaded and stored in the elevator on the farm of R. O. Snelling. (Tippecanoe County, 1944)

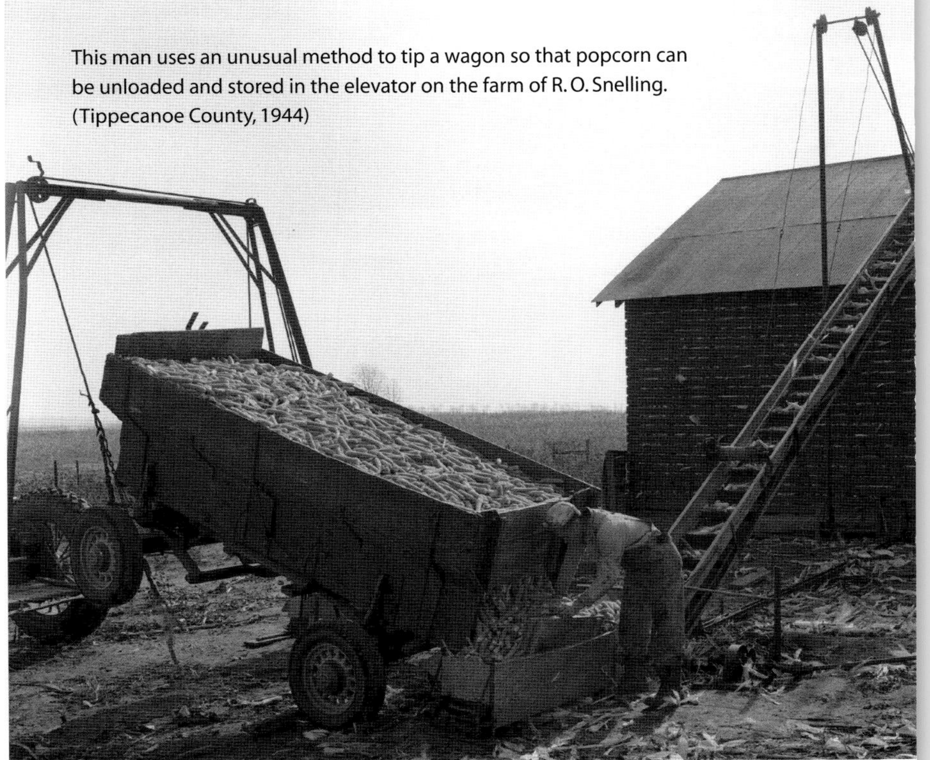

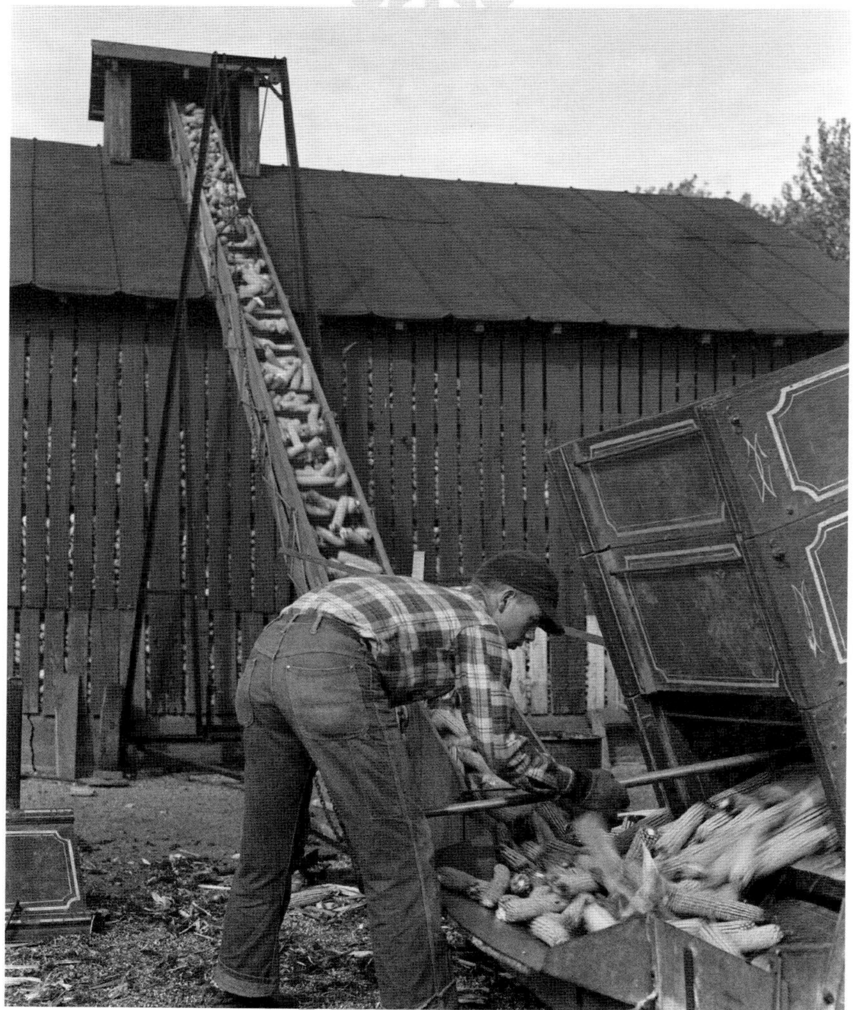

Emptying corn into a crib on the farm of Marvin Yoder (Tippecanoe County, 1951)

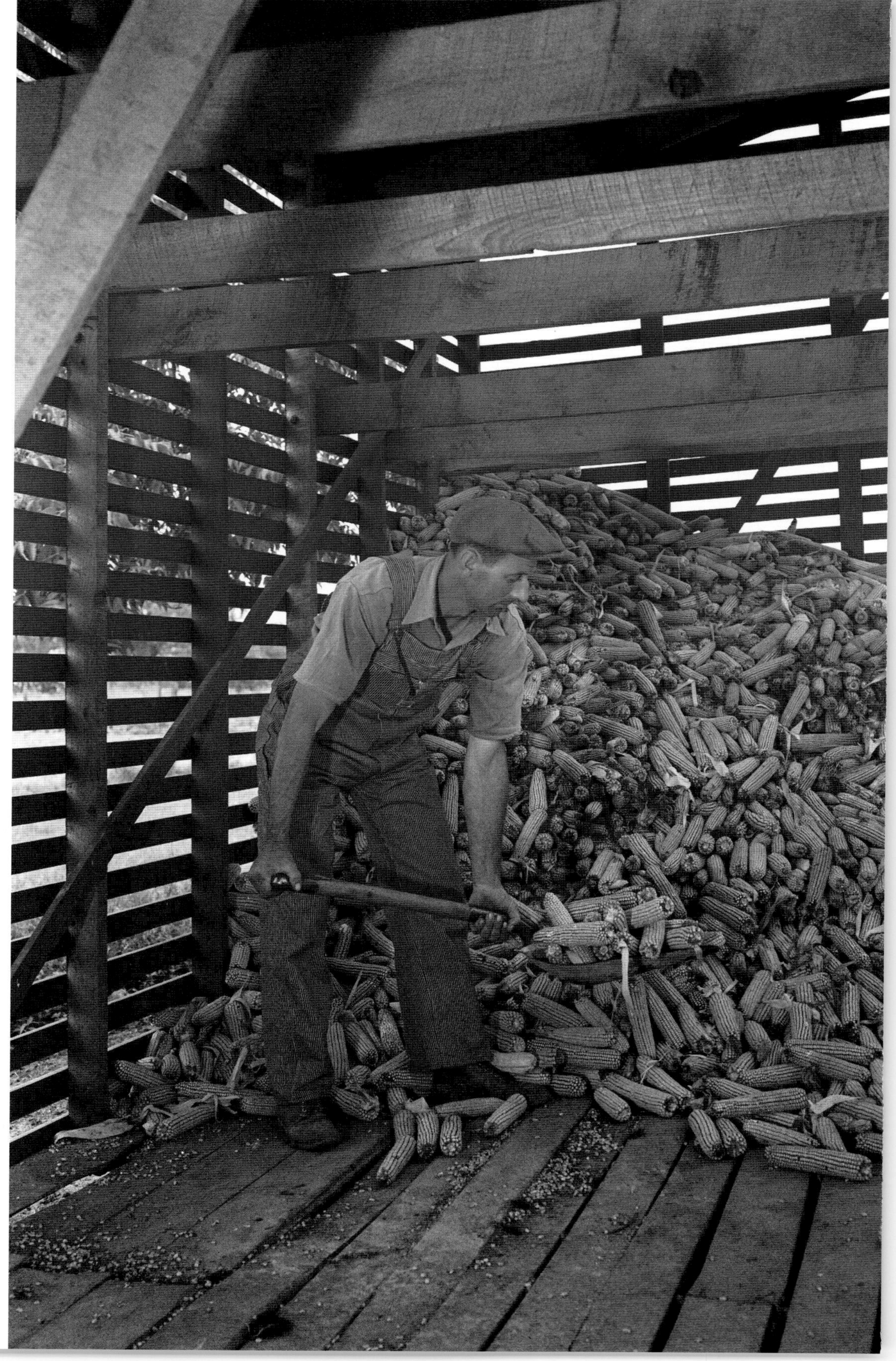

Alonzo Richards shovels corn from a crib in Plymouth. (Marshall County, 1942)

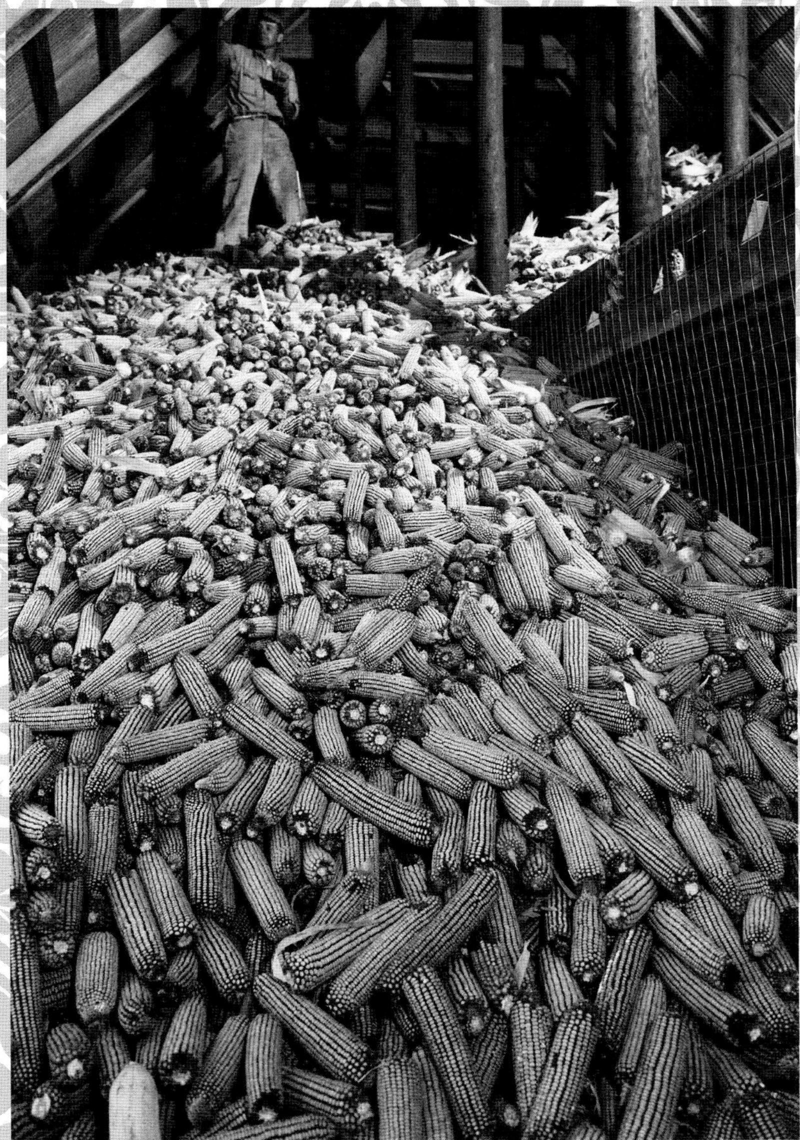

Dave White from Morristown adjusts the spout from the elevator to drop corn where he wants it. (Shelby County, 1955)

After filling this temporary crib on the Russ Skinner farm in Lafayette, a man takes wagonloads of corn to another crib. (Tippecanoe County, 1952)

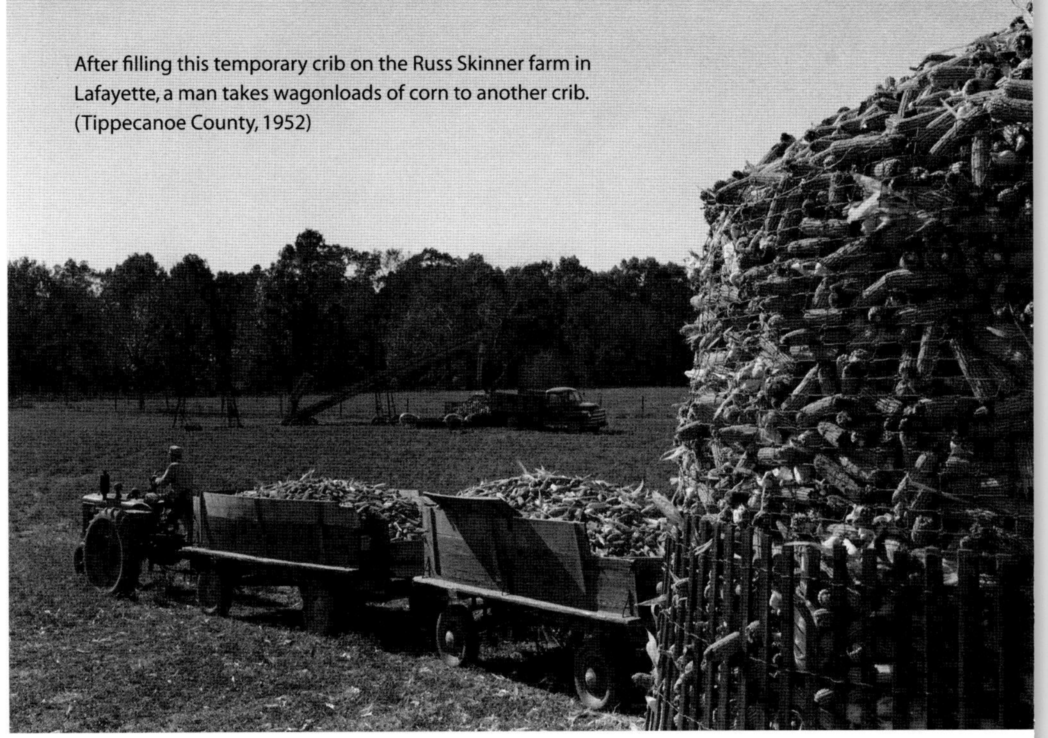

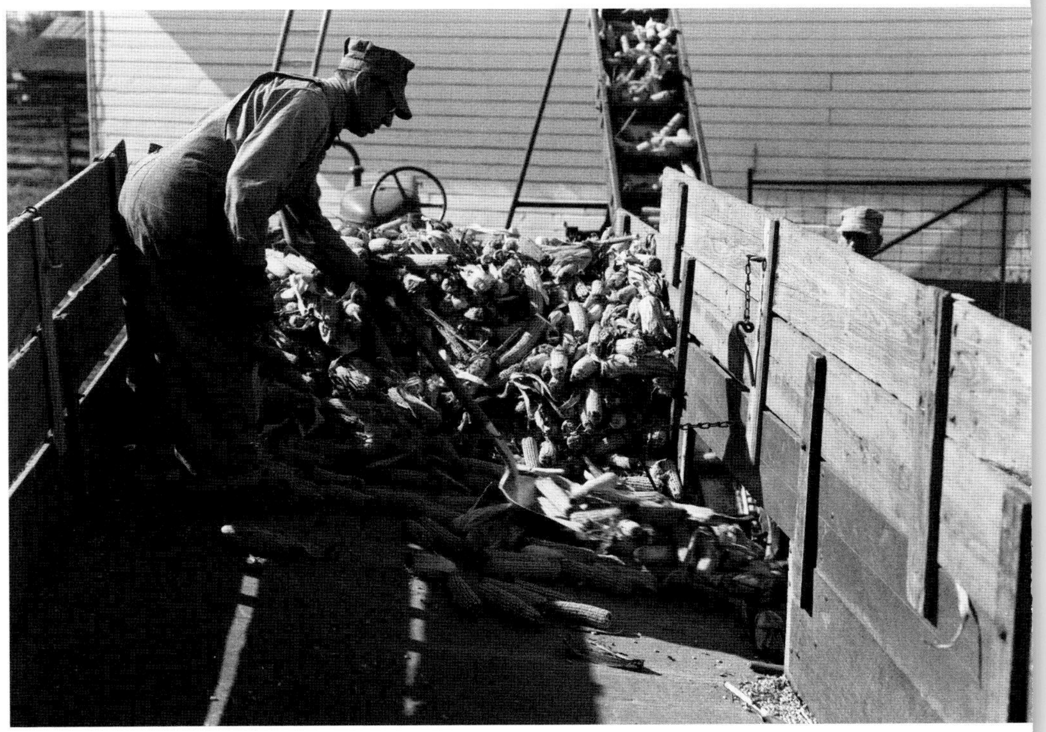

This wagon bed has a sloping floor to make it easier to push corn into the elevator on the farm of O. M. Michel in Tipton. (Tipton County, 1949)

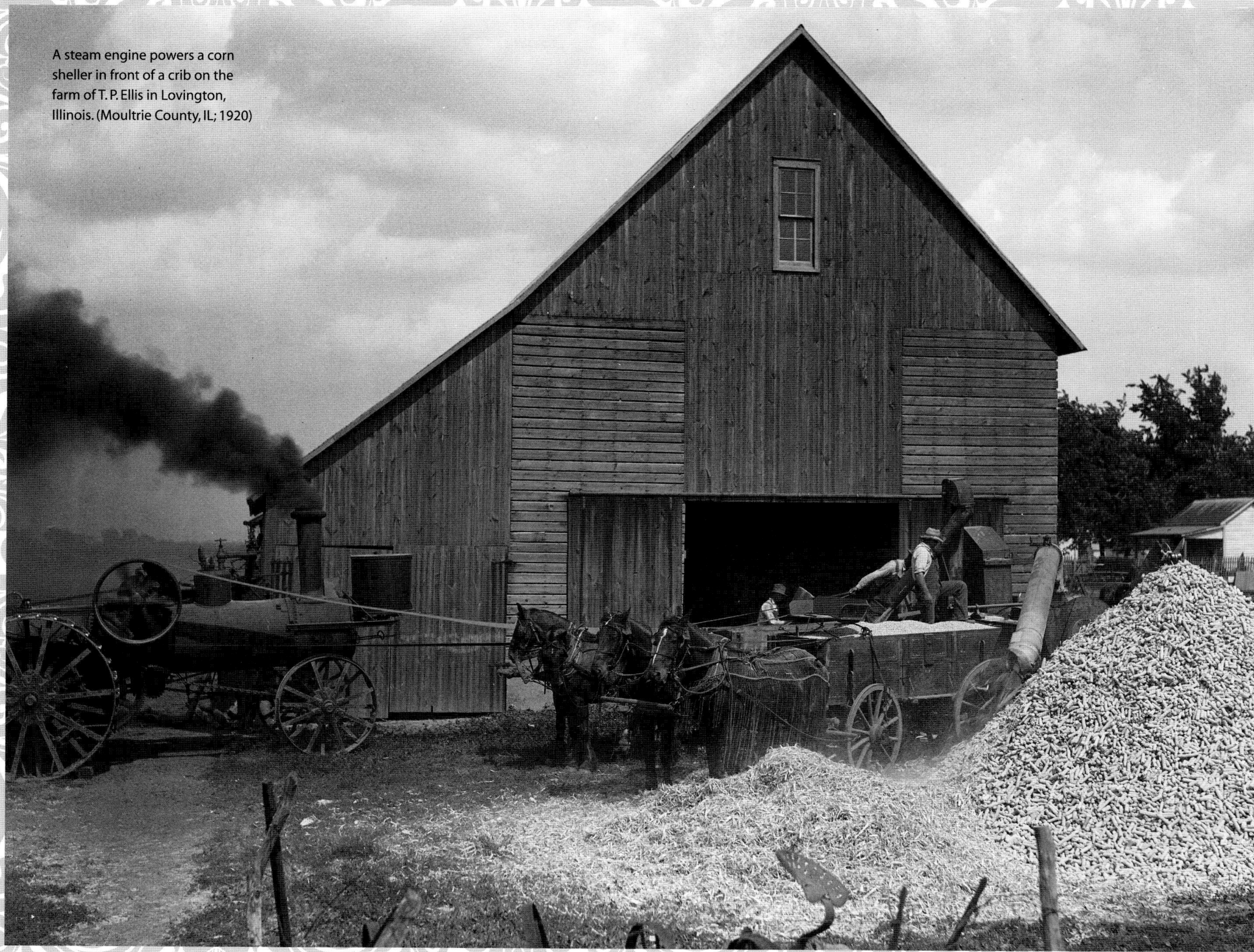

A steam engine powers a corn sheller in front of a crib on the farm of T. P. Ellis in Lovington, Illinois. (Moultrie County, IL; 1920)

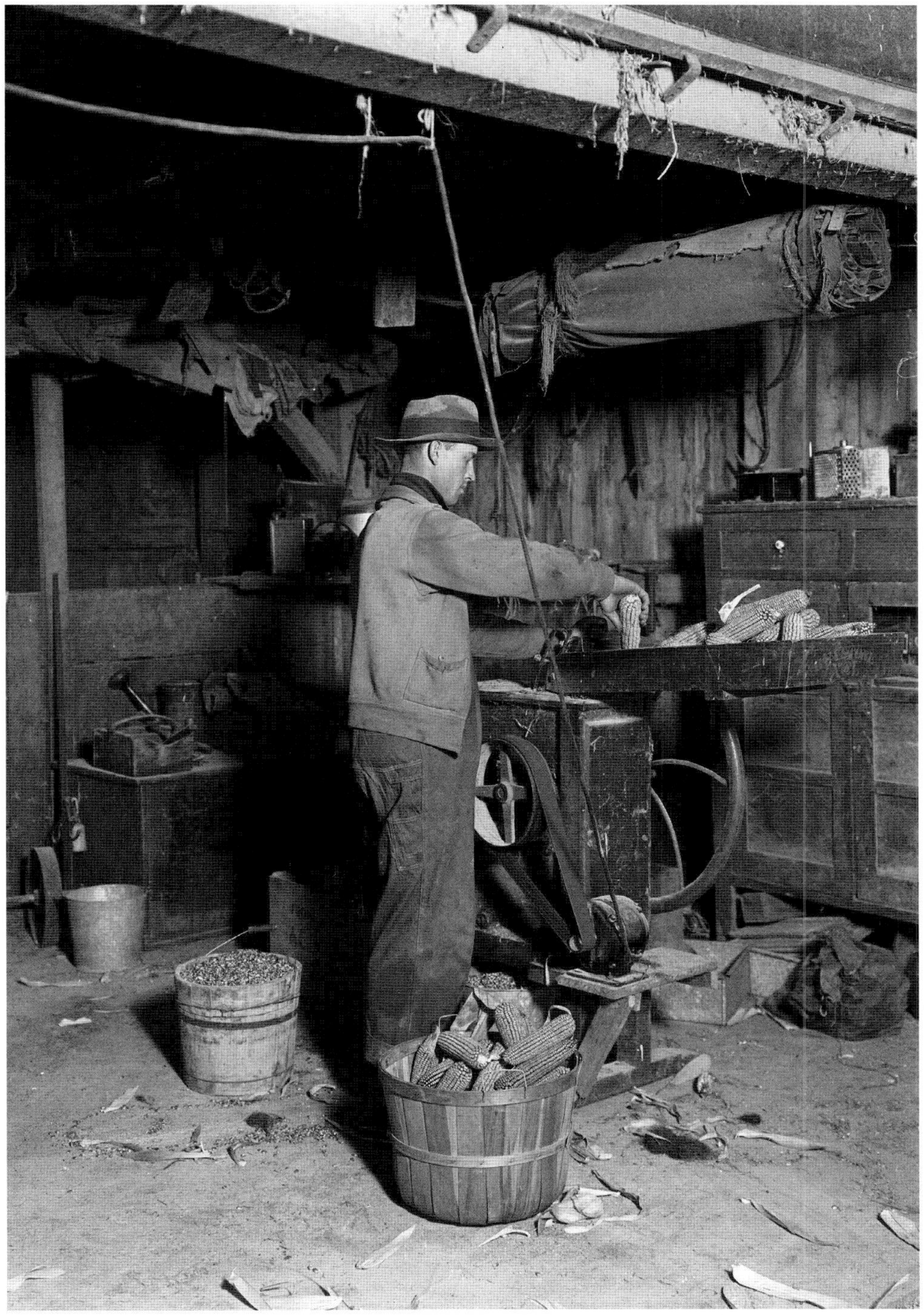

An electric motor runs a corn sheller on the Joe Yundt and Son Farm in Frankfort. (Clinton County, 1928)

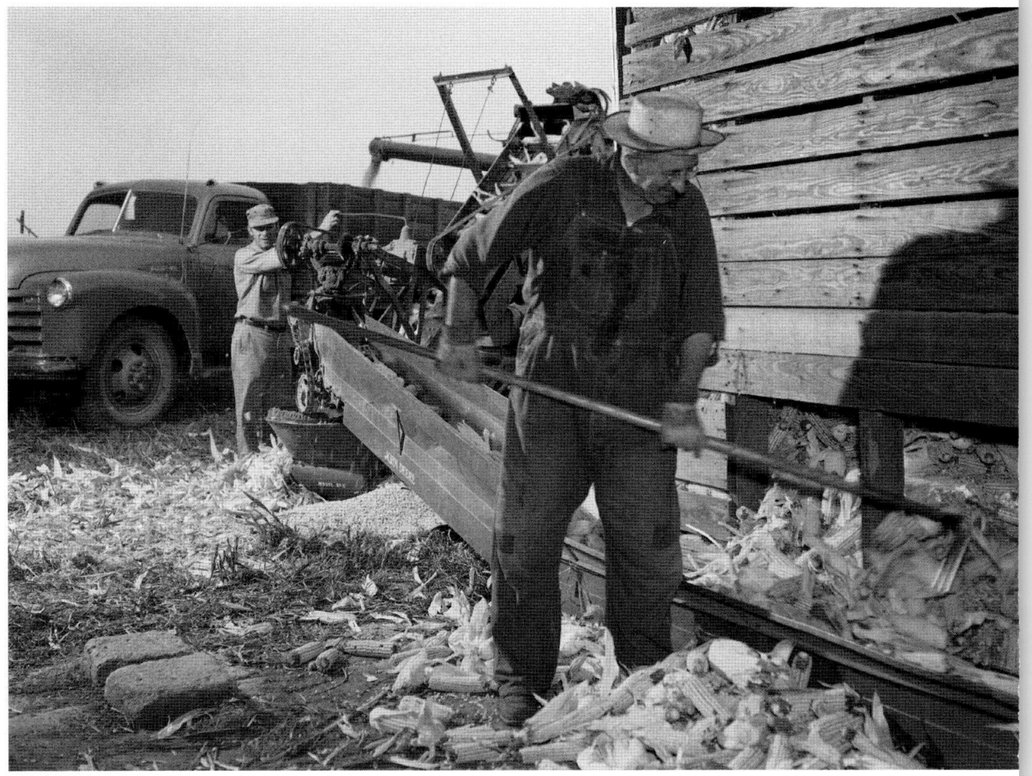

Shelling corn from a crib using a sheller owned by Ken Brummet of Otterbein (Benton County, 1957)

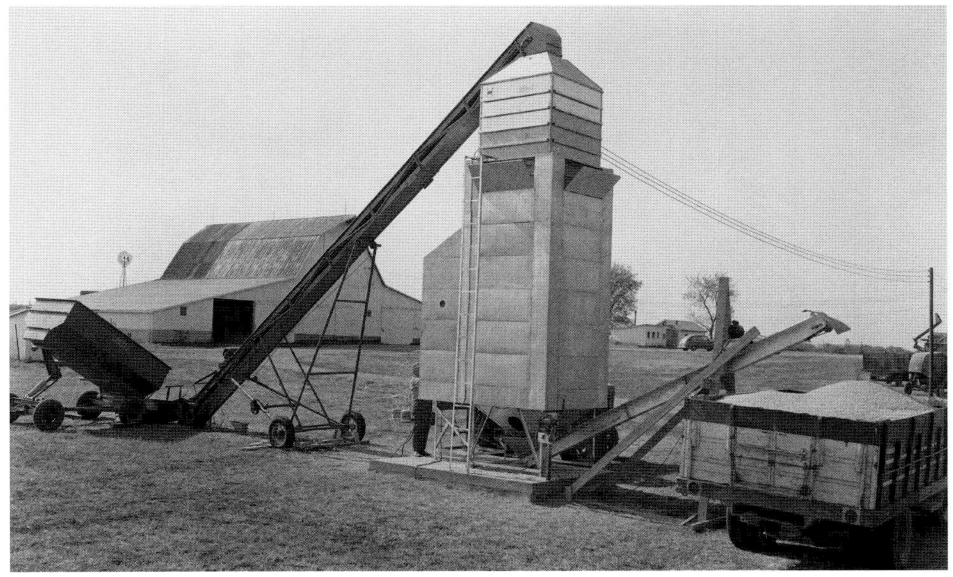

Corn is placed into a dryer on the Jim Kellerman farm in Romney. (Tippecanoe County, 1956)

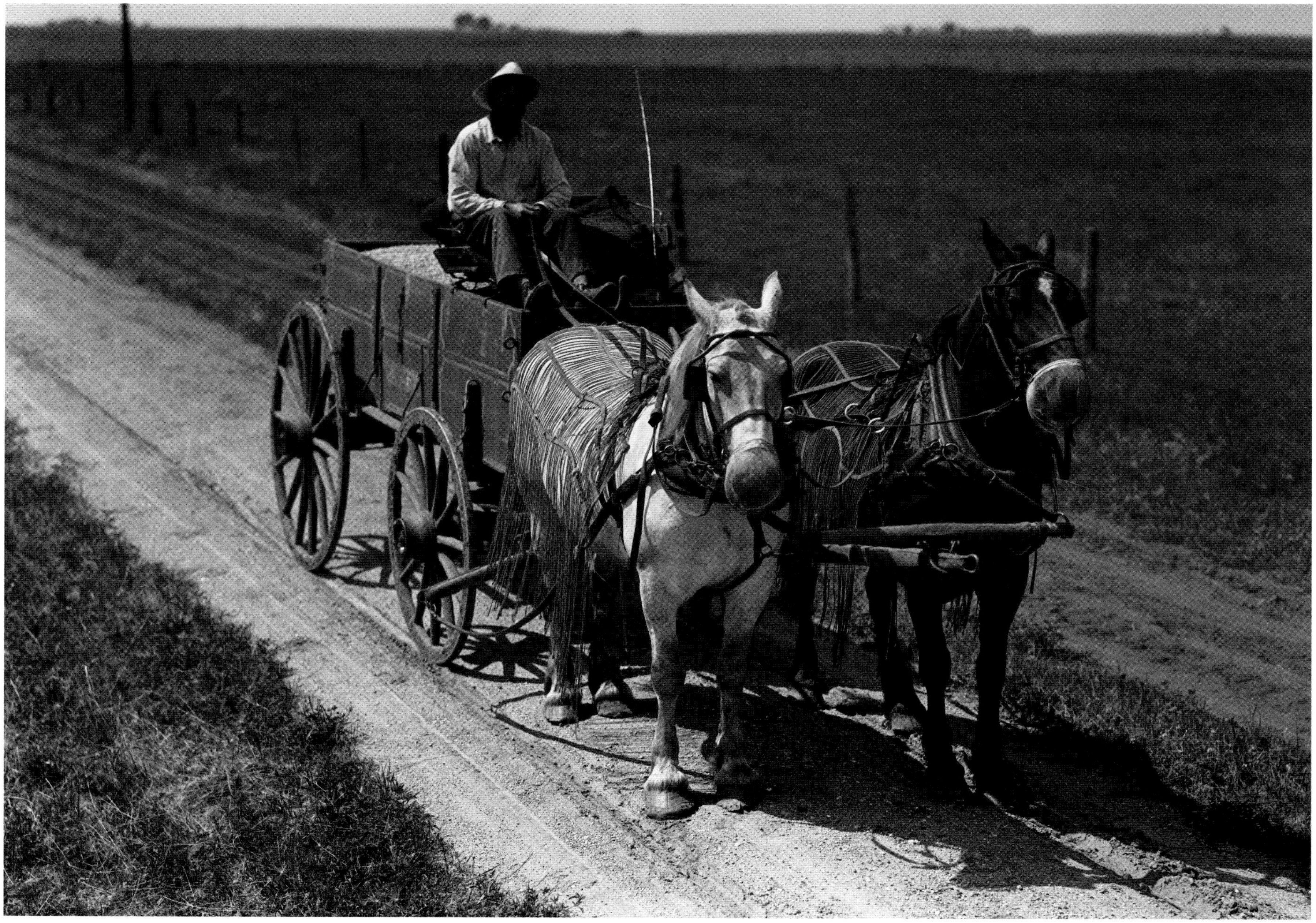

Elvin Wakey from Grand Ridge, Illinois, hauls corn to town with a team of horses and a wagon. (LaSalle County, IL; 1918)

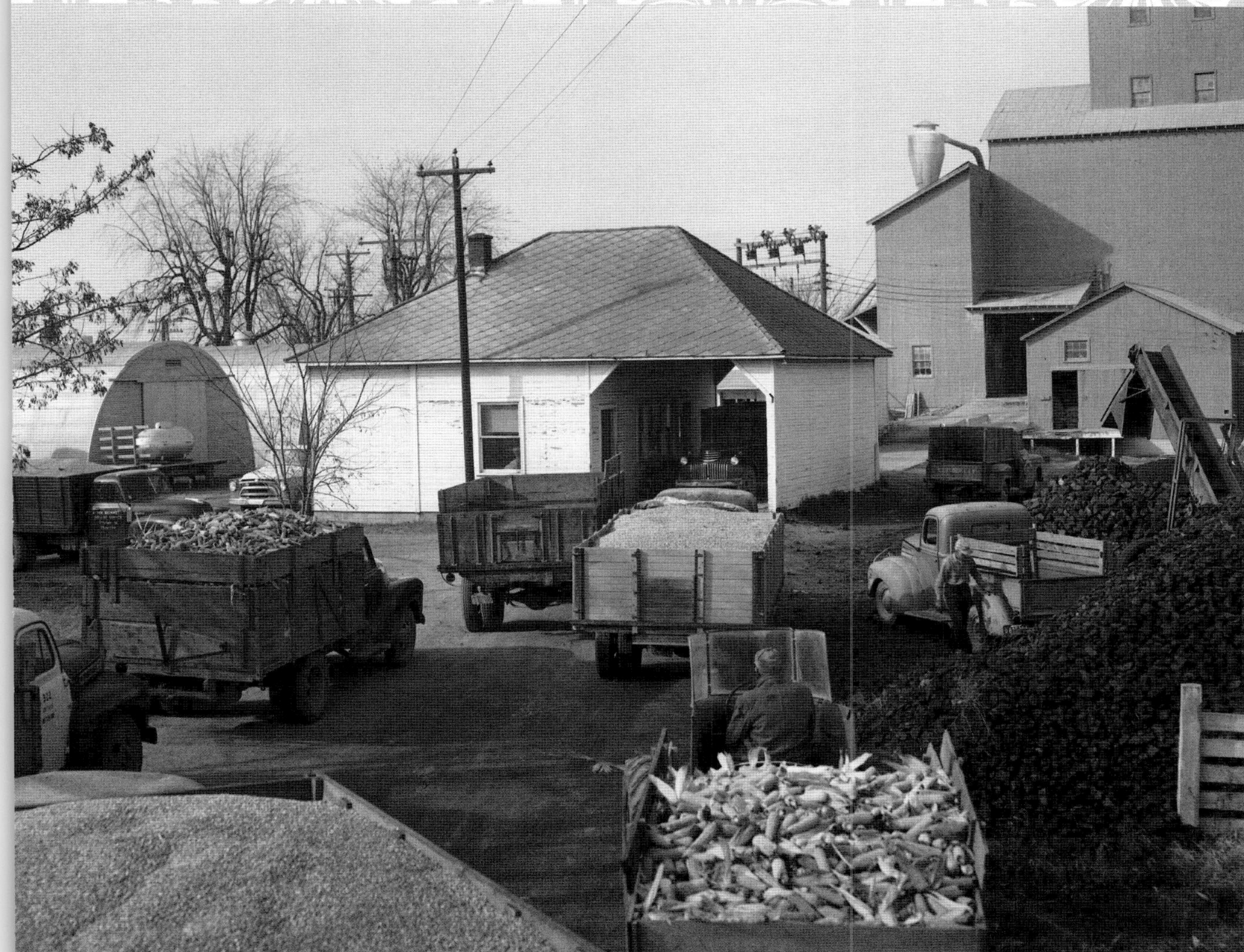

Trucks filled with shelled and unshelled corn wait to be unloaded at the elevator in Montmorenci. (Tippecanoe County, 1960)

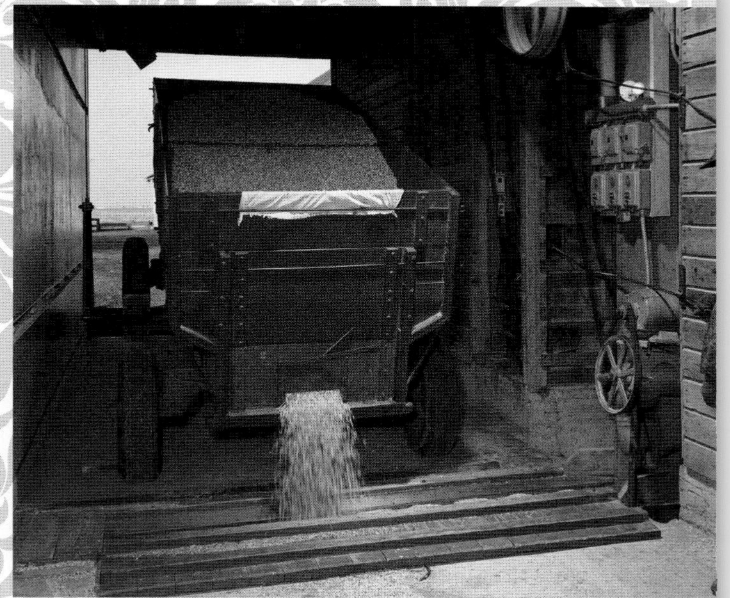

A man dumps shelled corn on a farm owned by Herbert F. Kelley and his son, Walter, in Brookston. (White County, 1963)

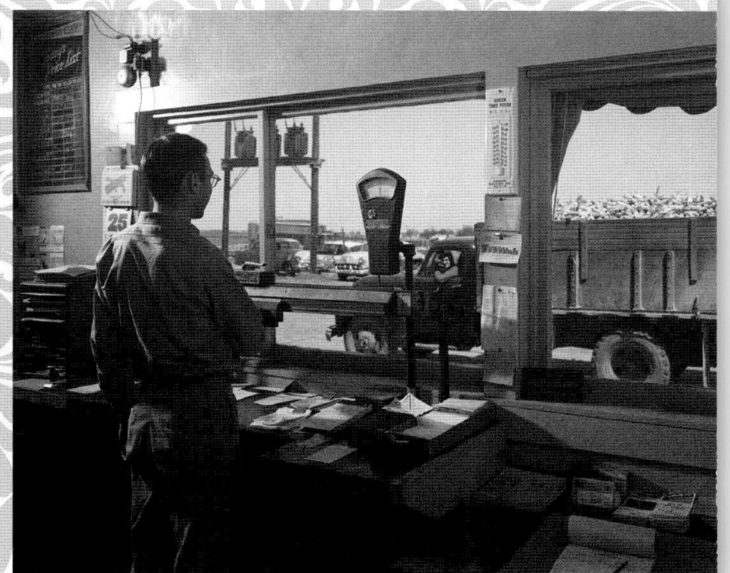

Carl Baker of Delphi has his corn weighed in Camden before it will be ground, mixed, and returned as complete feed by a bulk-feed truck. (Carroll County, 1958)

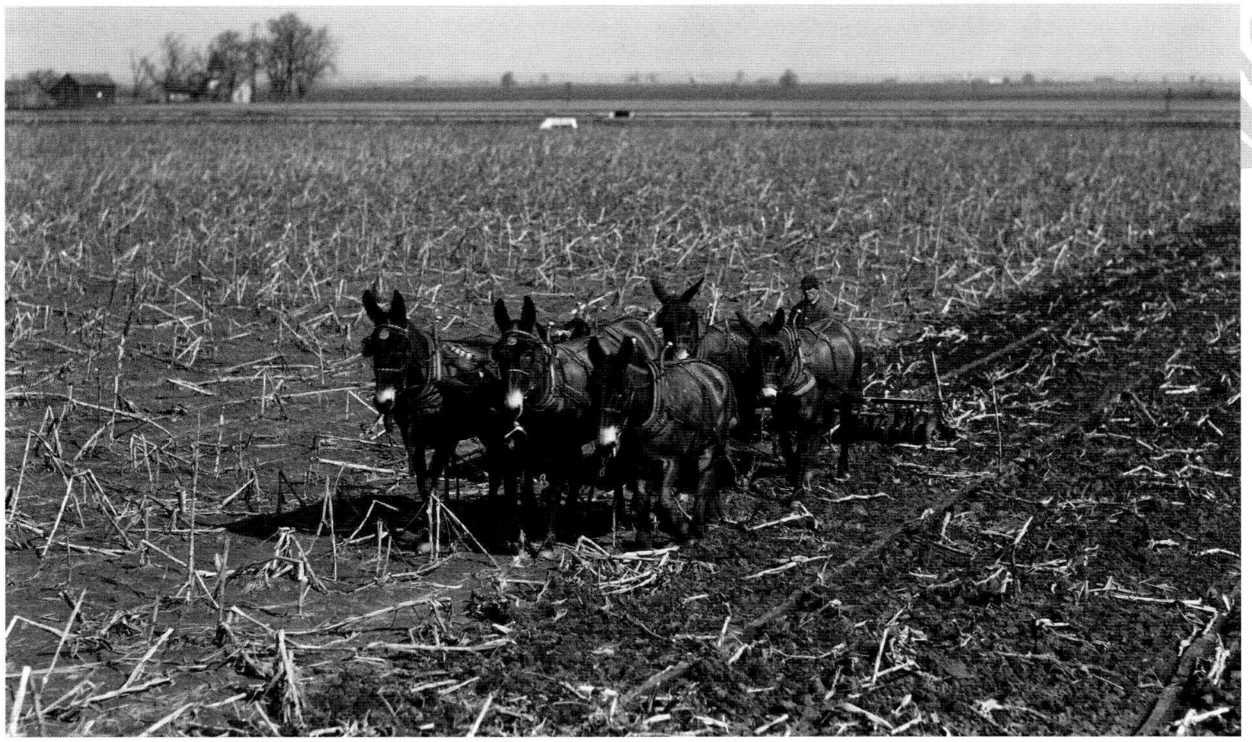

Using a six-mule team to disc under cornstalks on the farm of Jess Andrew from West Point (Tippecanoe County, 1935)

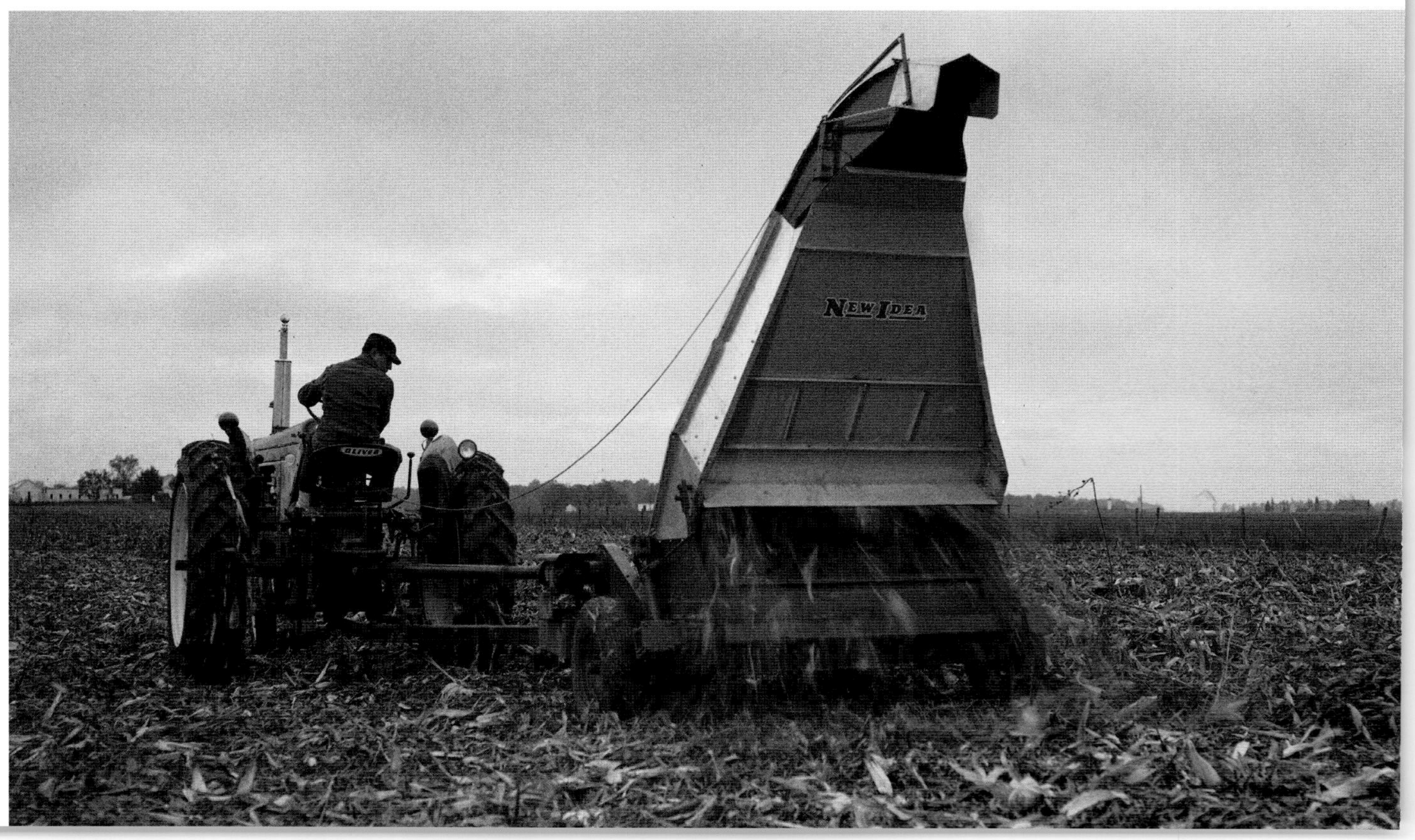

Shredding cornstalks in the field using a New Idea No. 825 field chopper (N.p., 1959)

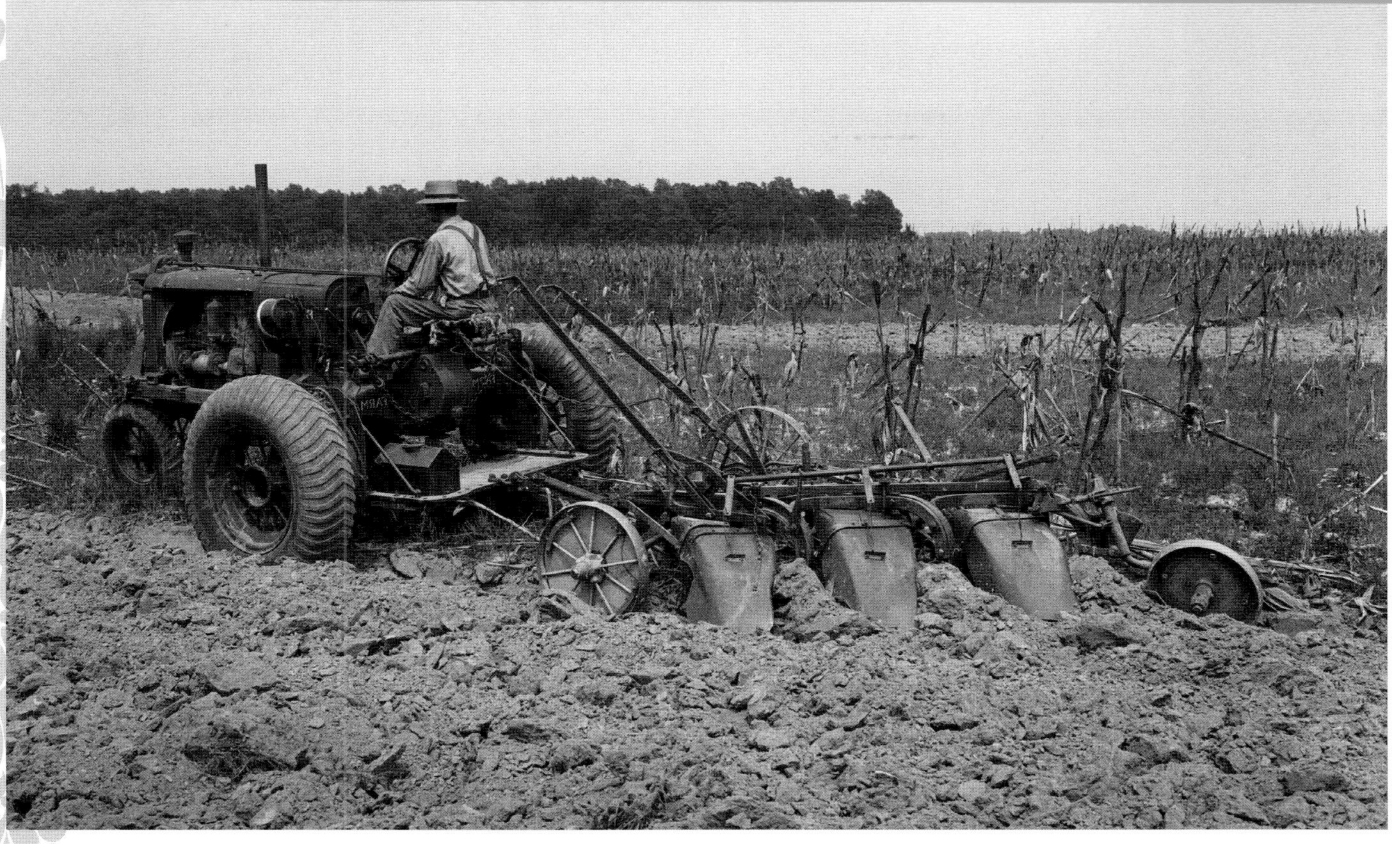

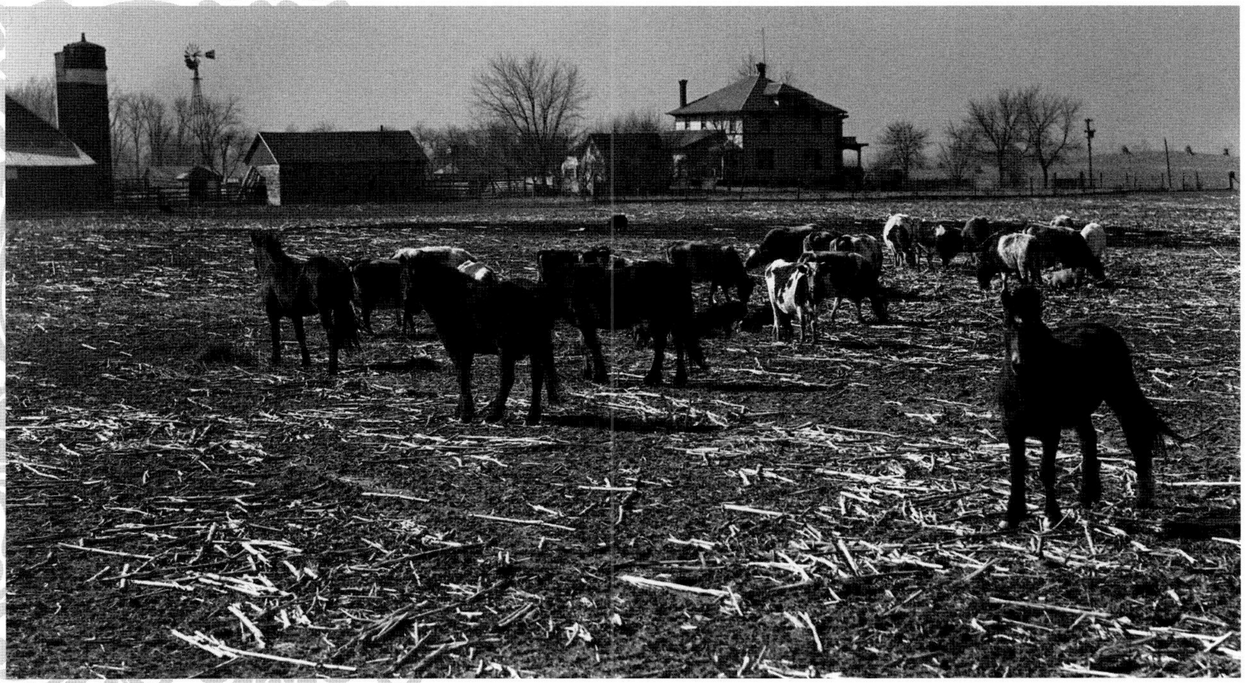

(Top) Plow shields developed by Purdue University are attached to a breaking plow for covering cornstalks. (Tippecanoe County, 1934)

(Bottom) Horses, hogs, and cattle eat cornstalks during the winter on the Owen Skiles farm in Frankfort. (Clinton County, 1939)

Amber Waves of Grain

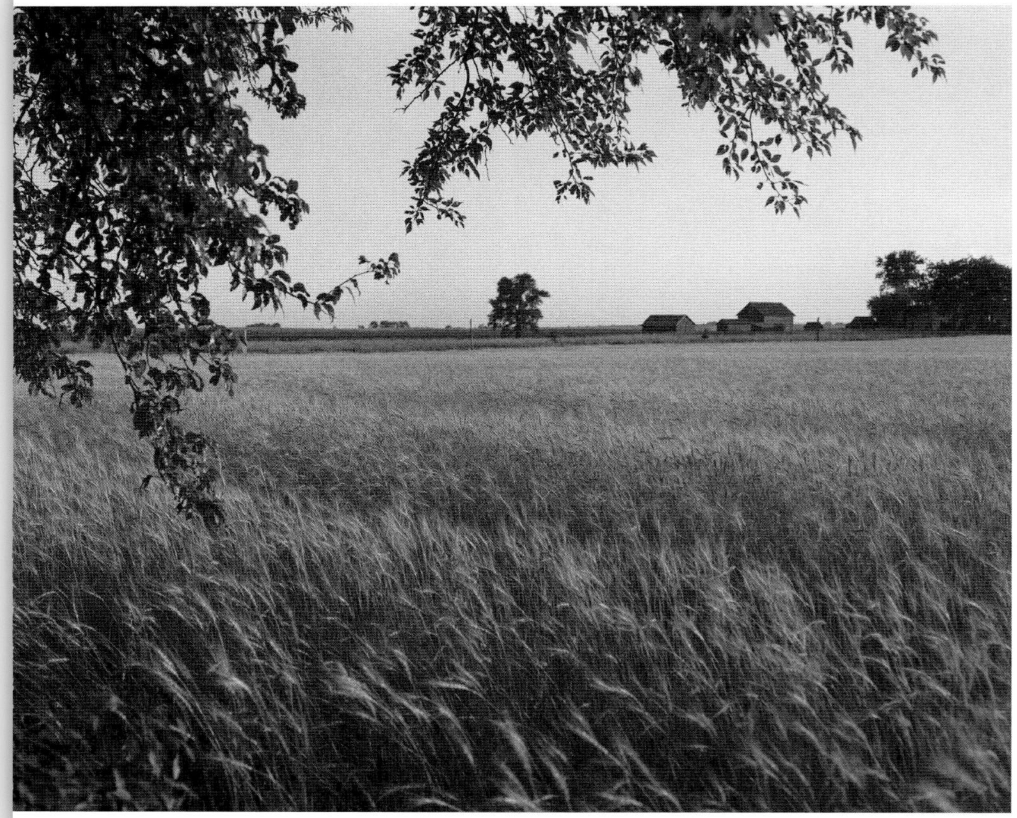

A wheat field in Livingston County, Illinois (1938)

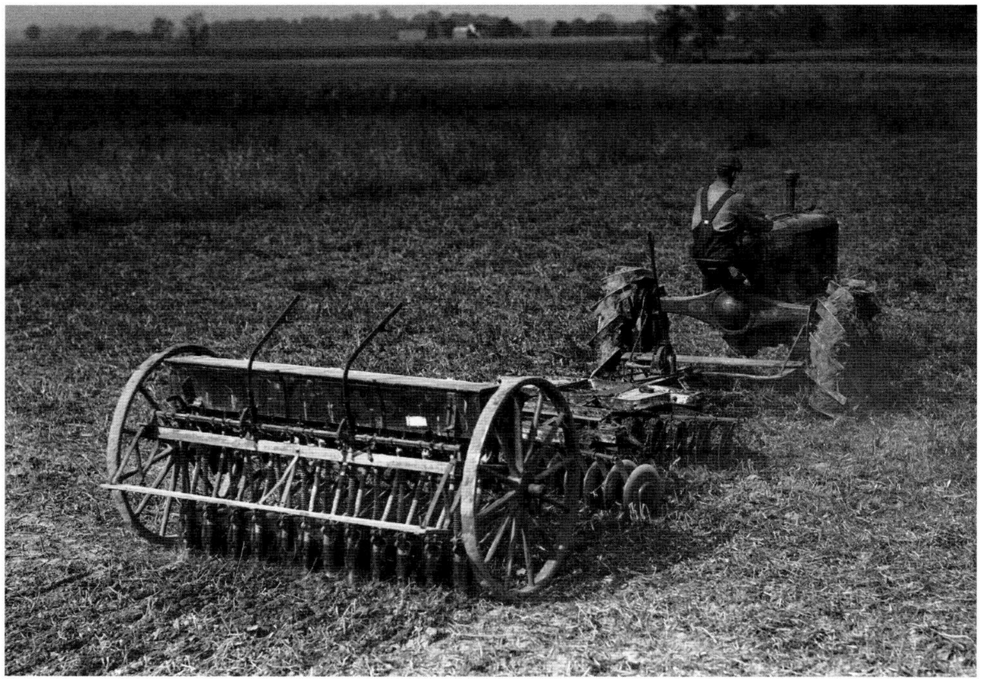

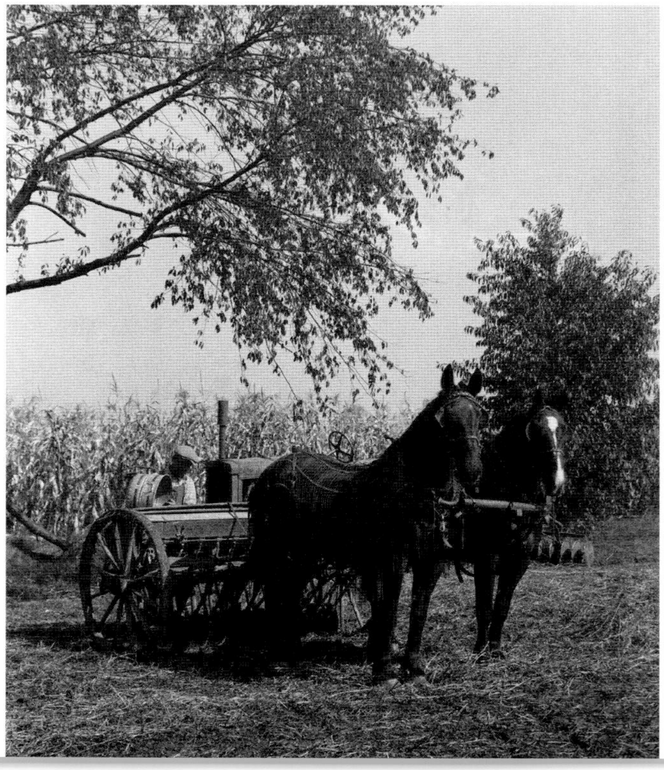

(Top right) This disc and drill are pulled by a steel-wheeled Farmall tractor at the Finis Fouts farm in Deer Creek. The farmer applied 25 pounds of 0-28-12 fertilizer per acre. (Carroll County, 1929)

(Bottom right) A man pours wheat seed into a drill on the Purdue University Soils and Crops Farm near Lafayette. (Tippecanoe County, 1935)

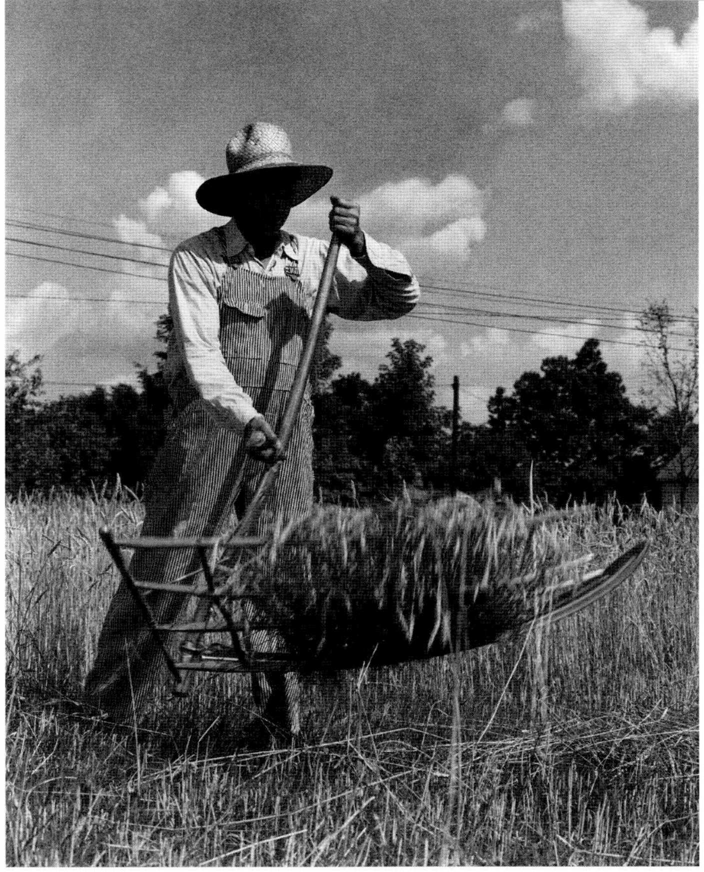

(Top left) Chris J. Yohn from Lafayette cradles wheat. (Tippecanoe County, 1939)

(Bottom left) Alice and Sarah Reinholt from Winamac pitch oats into the threshing machine. (Pulaski County, 1946)

(Right) Edith V. Moore from West Lafayette carries a bundle of wheat under each arm. (Tippecanoe County, 1930)

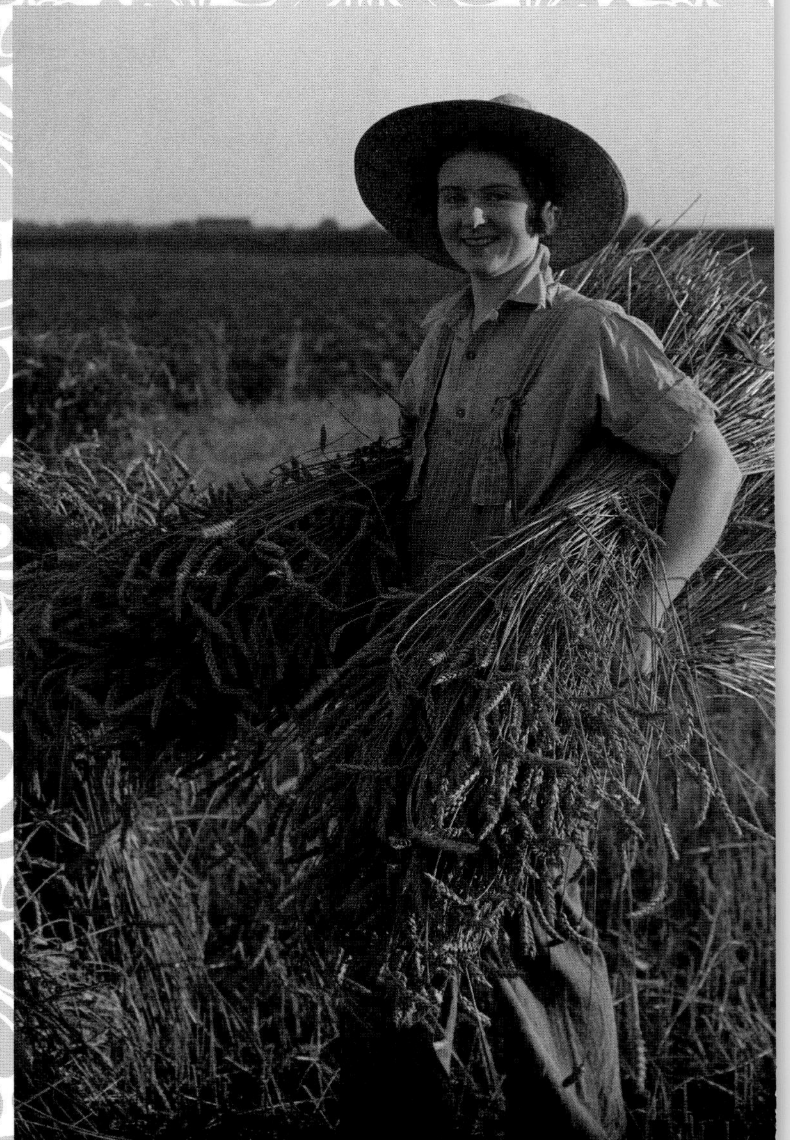

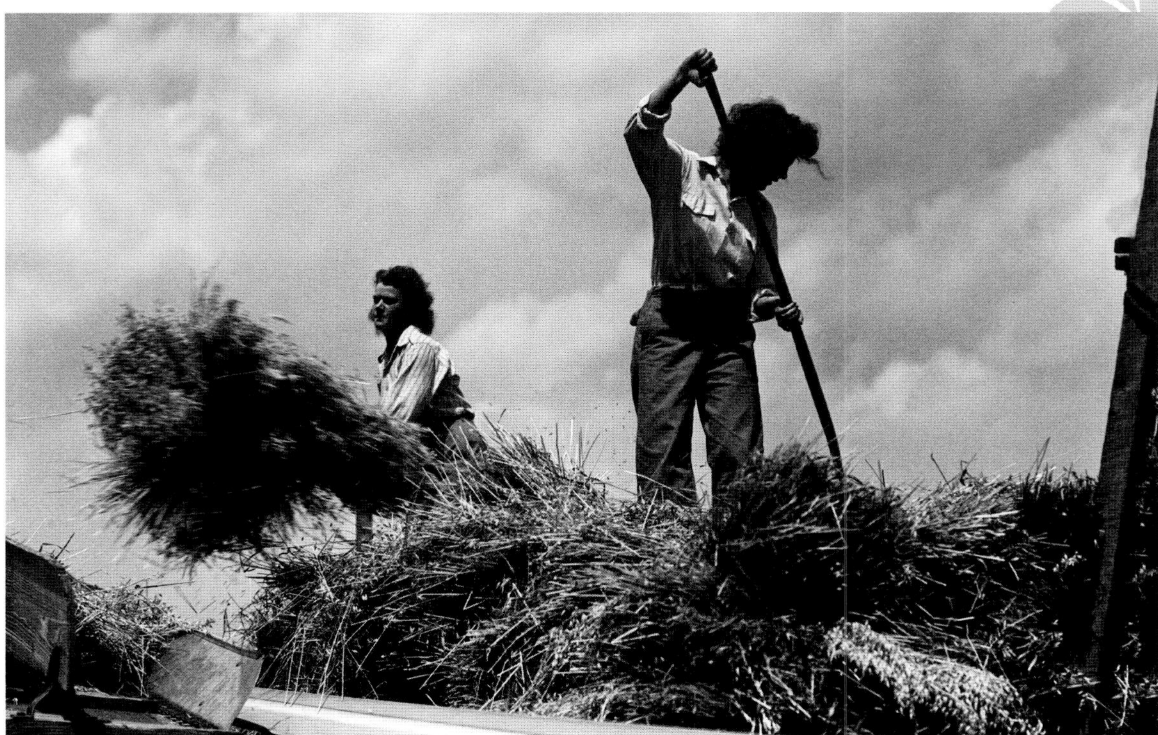

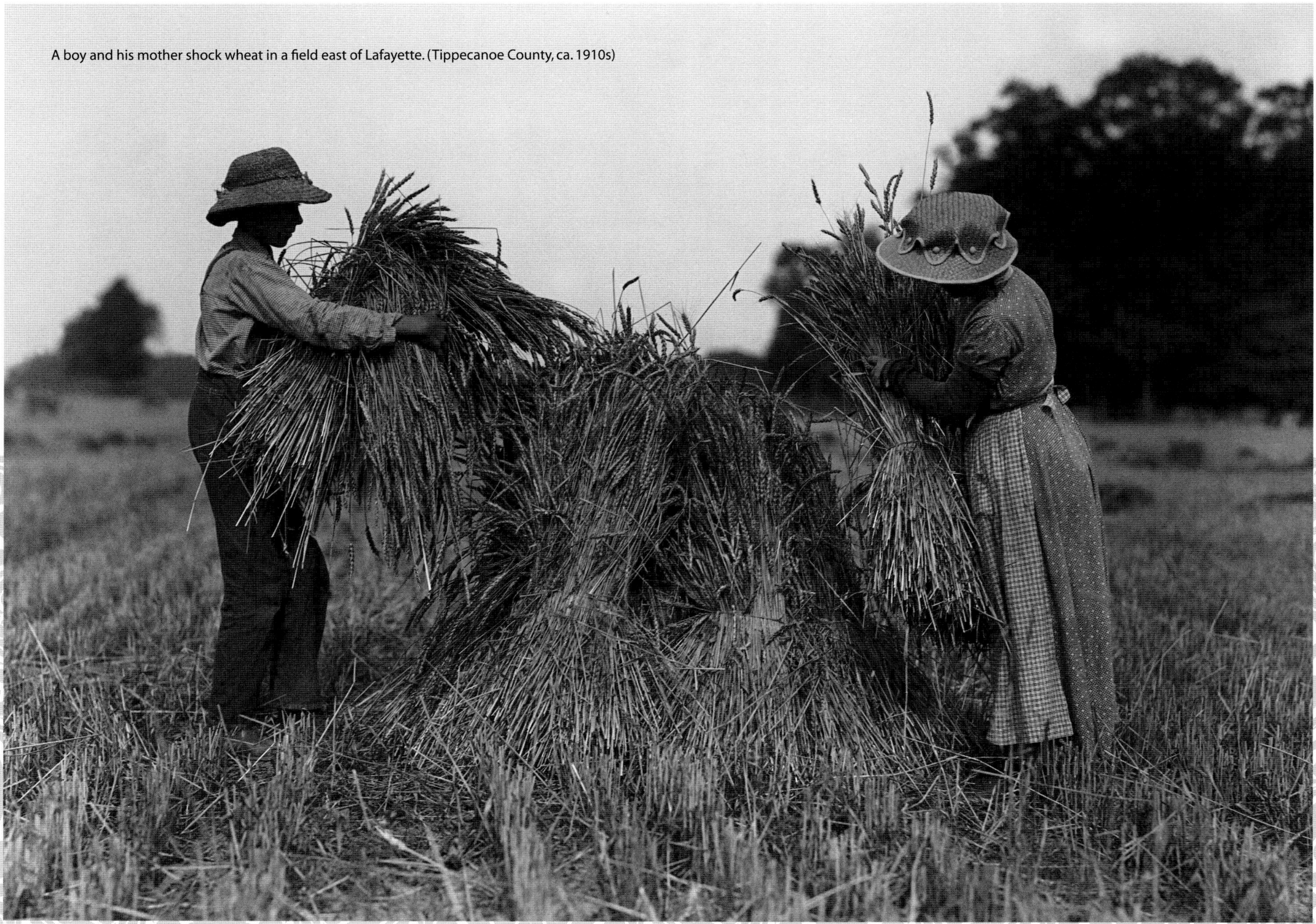

A boy and his mother shock wheat in a field east of Lafayette. (Tippecanoe County, ca. 1910s)

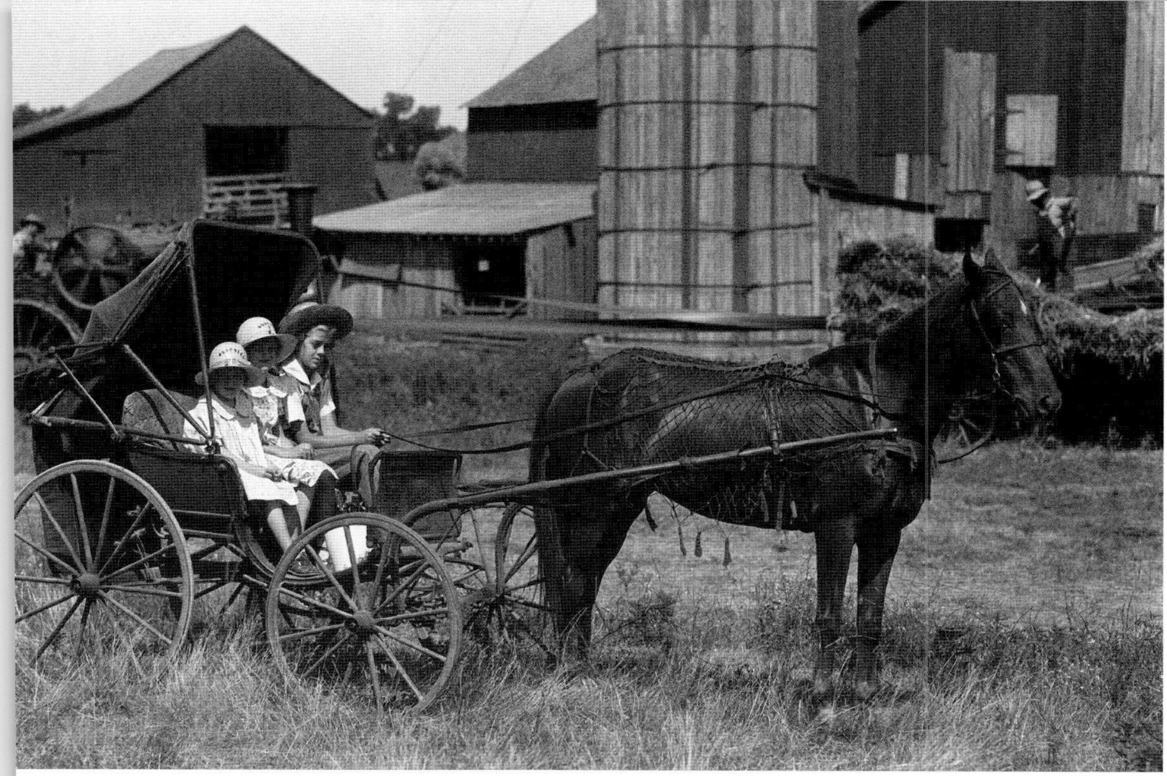

Bringing water to workers at threshing time on the farm of Jess Andrew of West Point (Tippecanoe County, 1932)

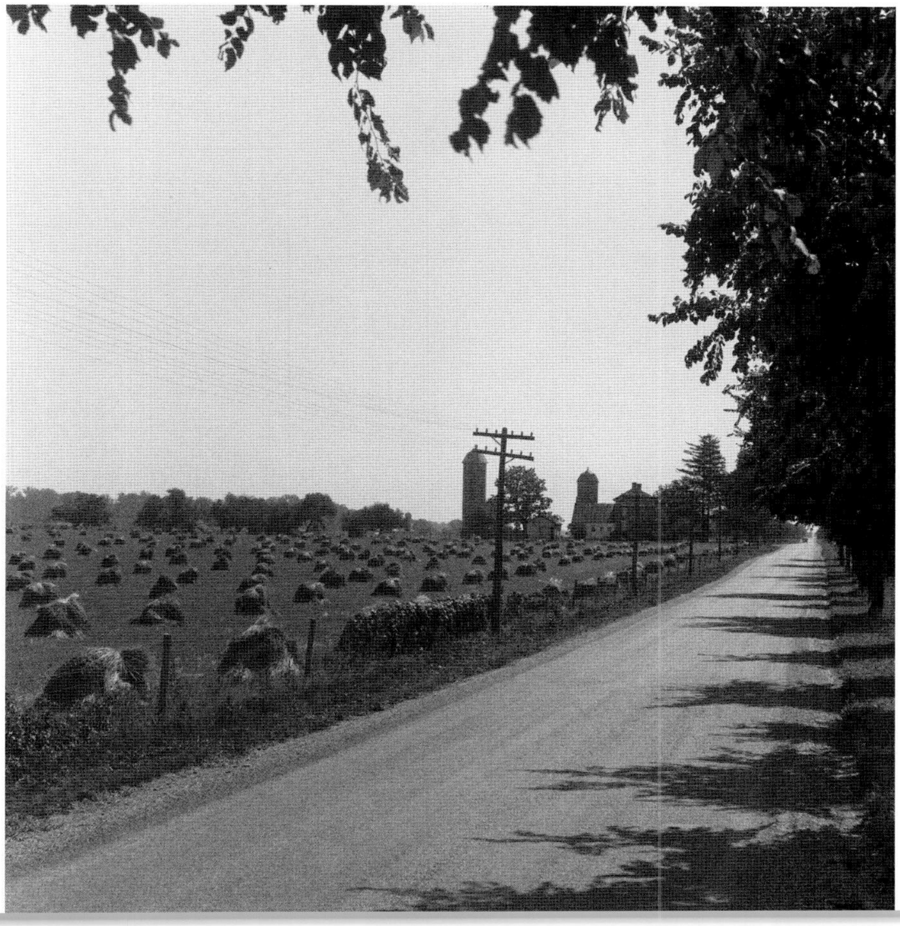

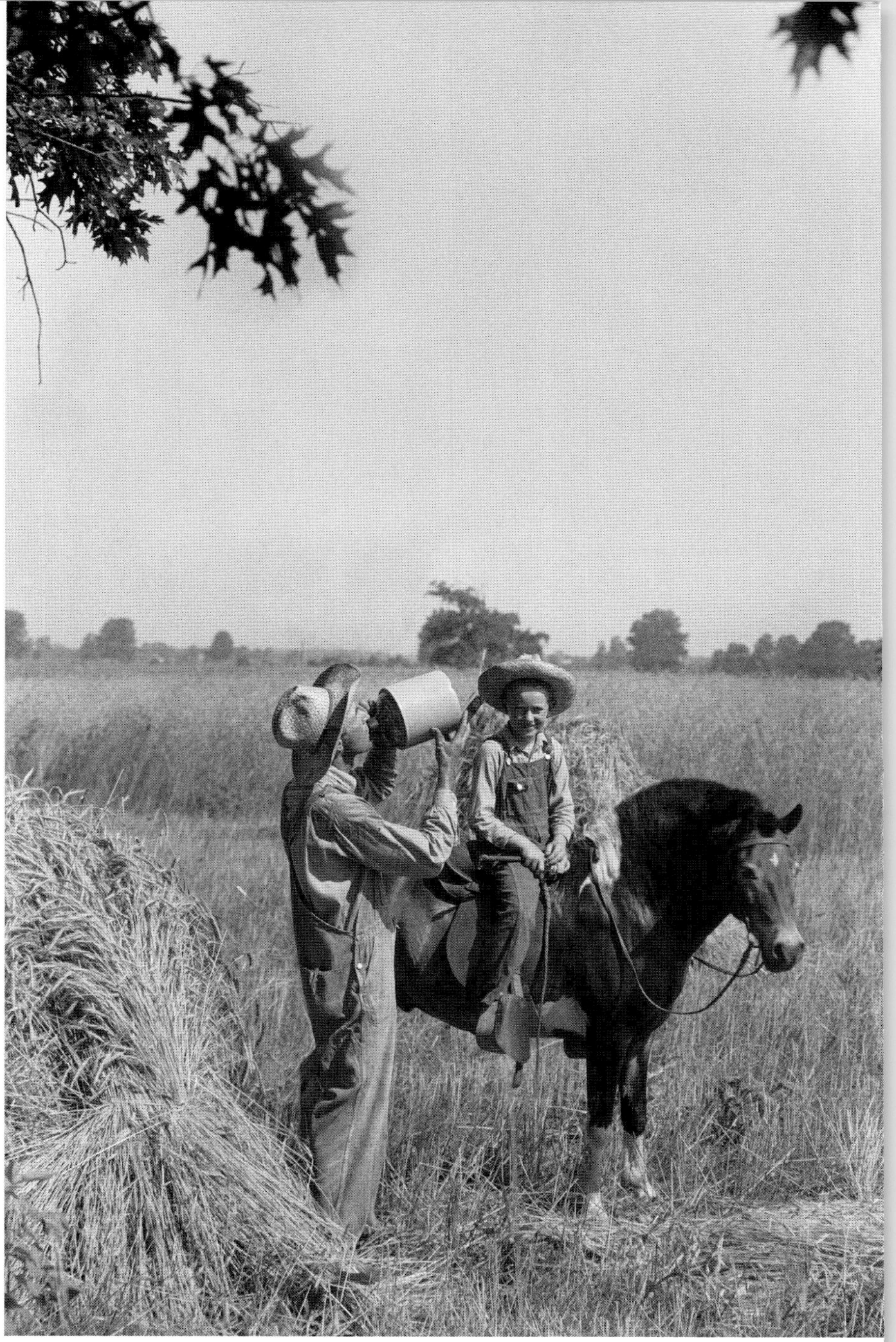

(Above) Benjamin Setzer Jr. (right) from Marshall, Illinois, delivers water to William Wyatt, a hired hand on the farm owned by Setzer's father. (Clark County, IL; 1935)

(Bottom left) A field of shocked wheat on a farm operated by Tansey and McGill (Clinton County, 1934)

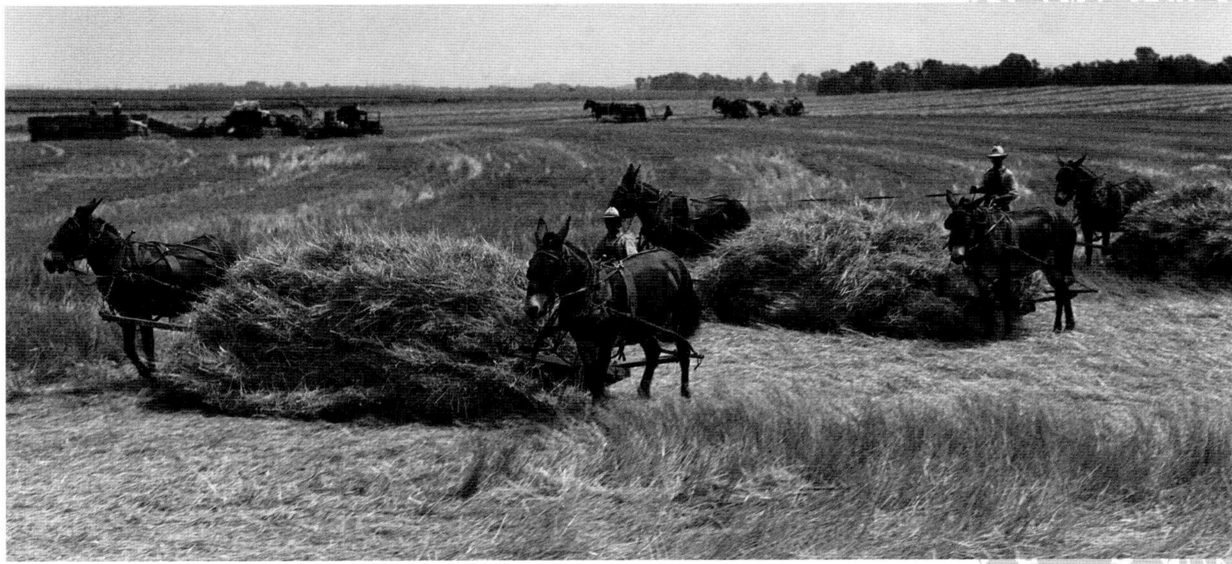

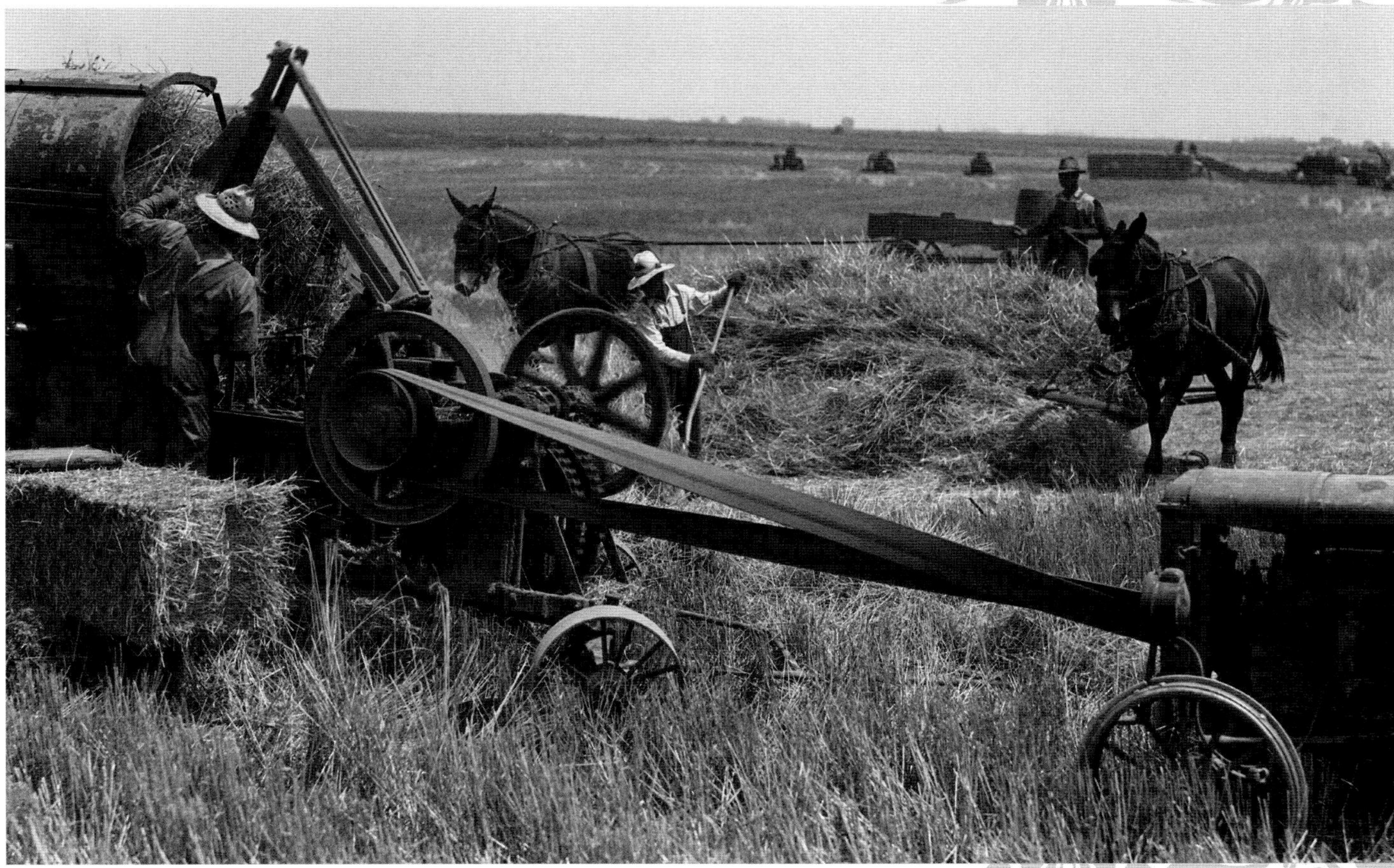

(Top and bottom) Mule teams pulling three bull rakes gather wheat straw on the Hoopeston Canning Company Farm near Hoopeston, Illinois. Then workers bale the straw using equipment powered by a belt that is driven by a tractor engine. (Vermilion County, IL; 1933)

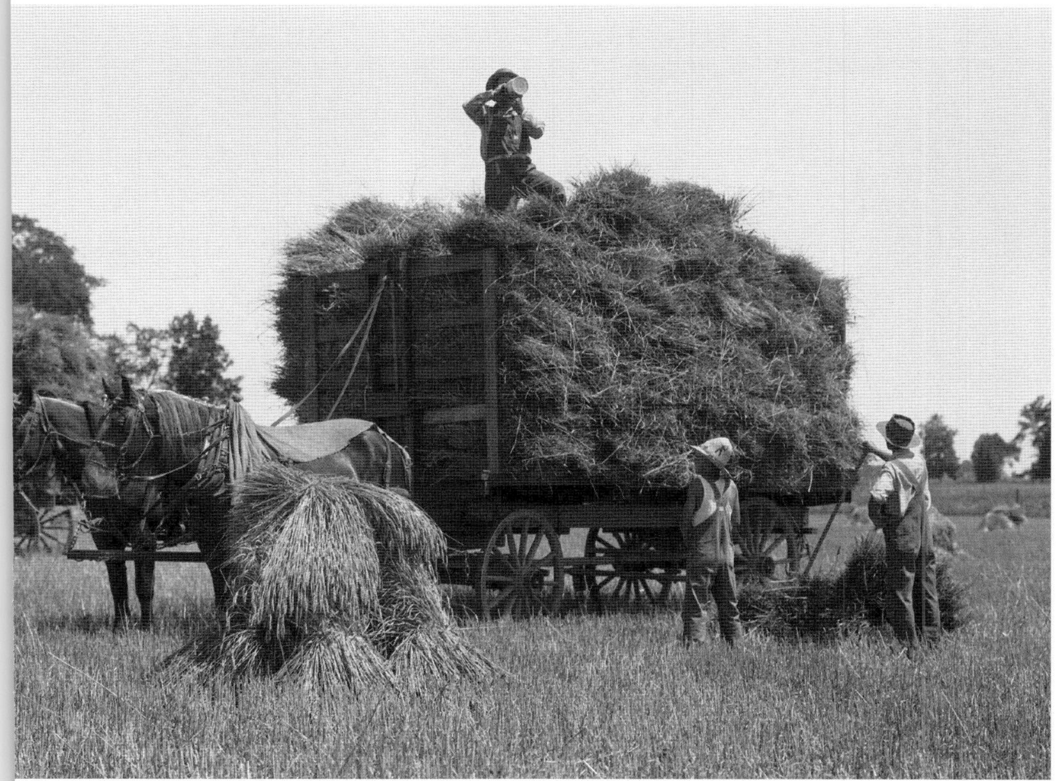

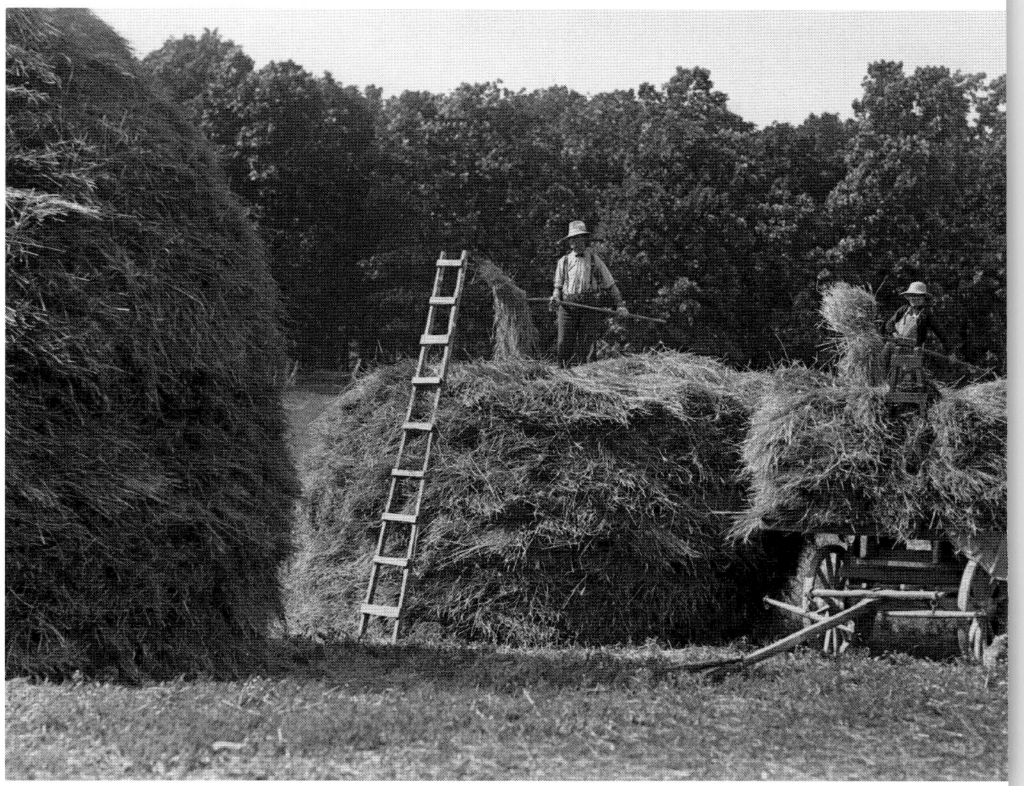

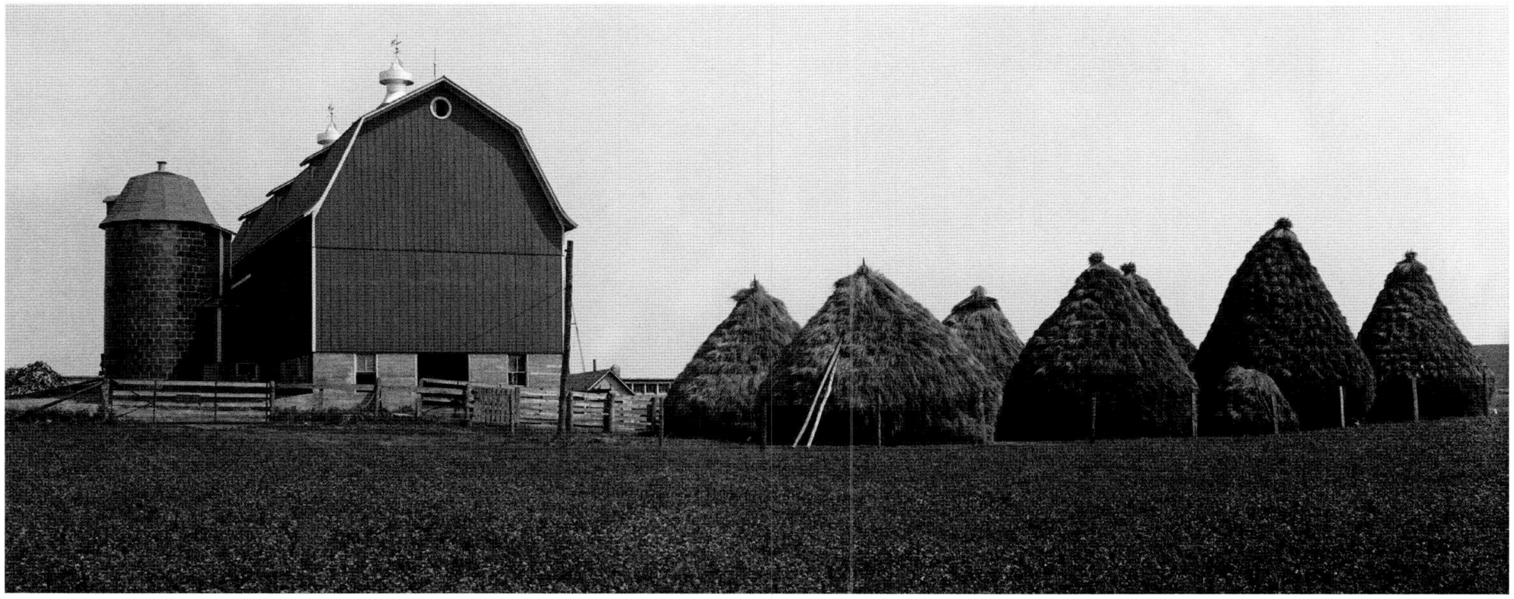

(Top left) Workers take a few minutes to cool off and get a drink in the field of M. G. Wilson of Kokomo. (Howard County, 1932)

(Top right) These men stack wheat bundles by hand in a barn lot. (Tippecanoe County, ca. 1912)

(Bottom) Stacks of small grain in a barn lot await threshing on the farm of Julius Buckholz in Westfield, Wisconsin. (Marquette County, WI; 1922)

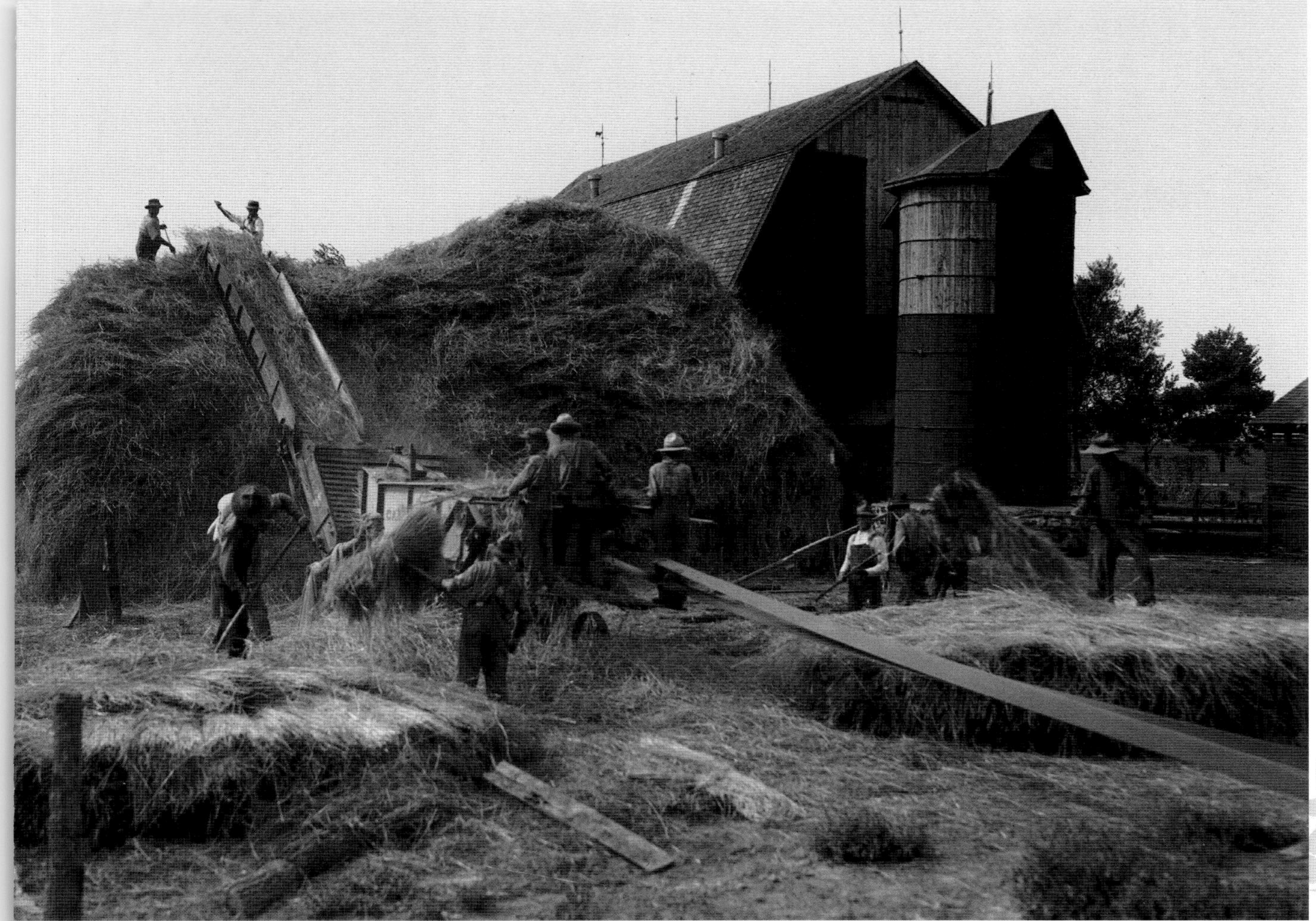

A work crew threshes large stacks of grain. (Wisconsin, 1919)

These men use a steam engine to power a separator to thresh clover seed. (Indiana, ca. 1914)

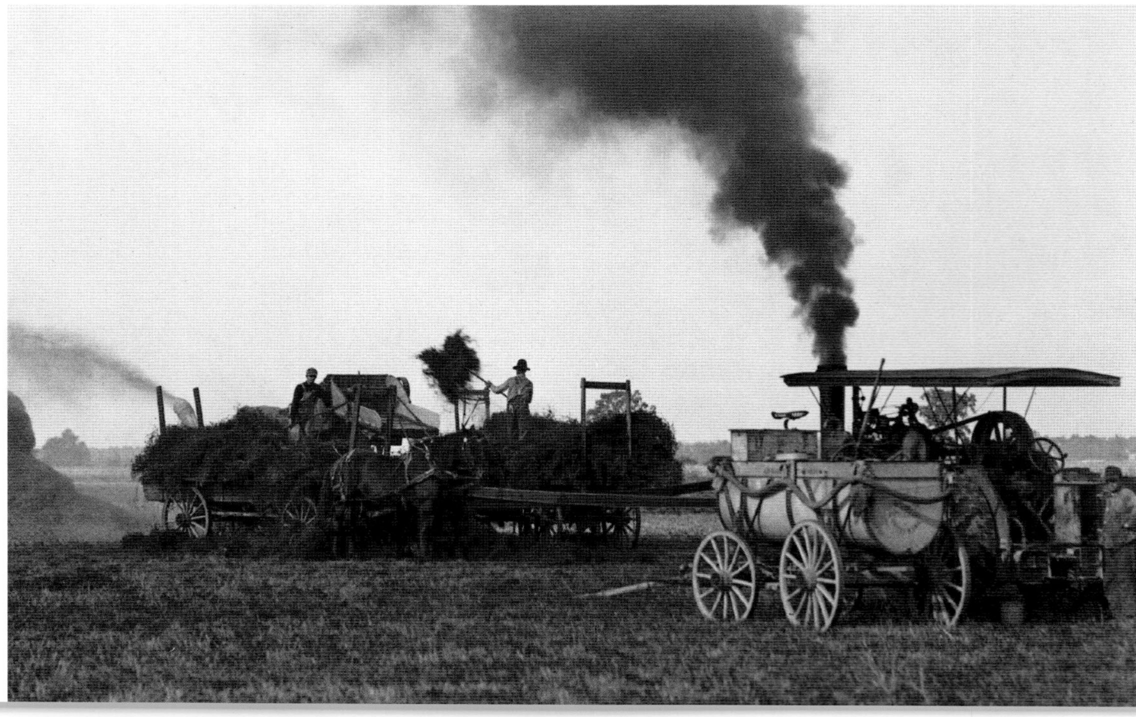

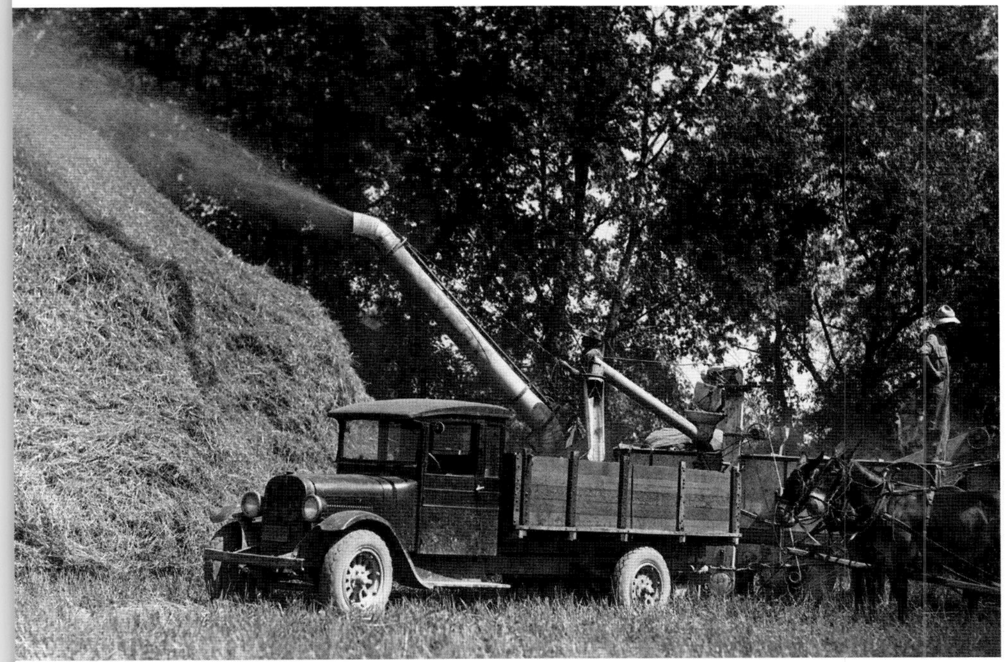

While threshing wheat on the Frank Mahlke farm in Lafayette, the grain is augered directly into a truck. (Tippecanoe County, 1929)

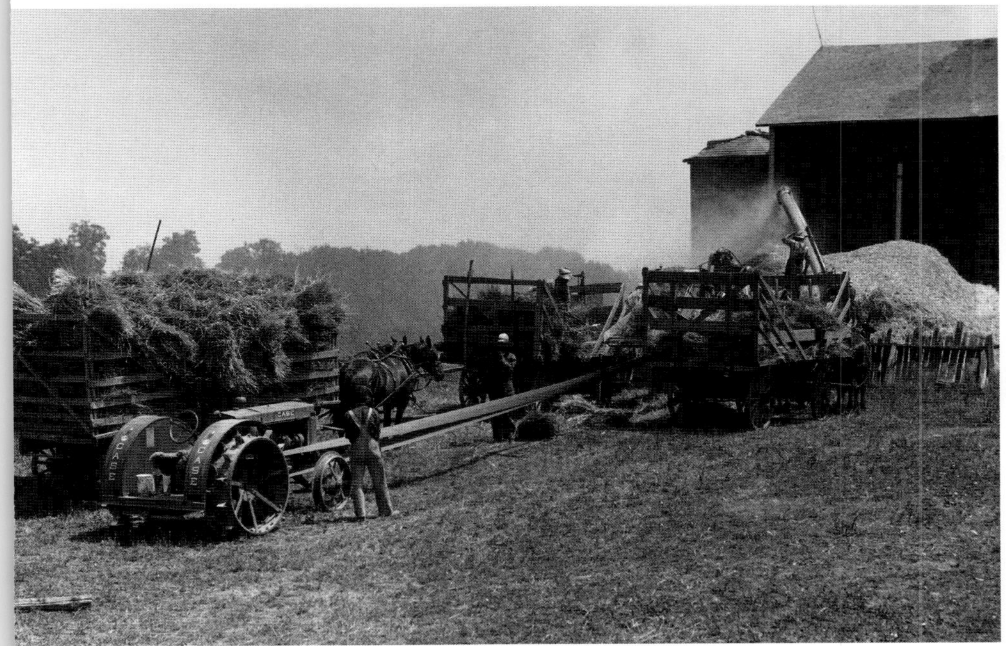

Wheat threshing on the farm of Earl Bayman in South Whitley (Whitley County, 1930)

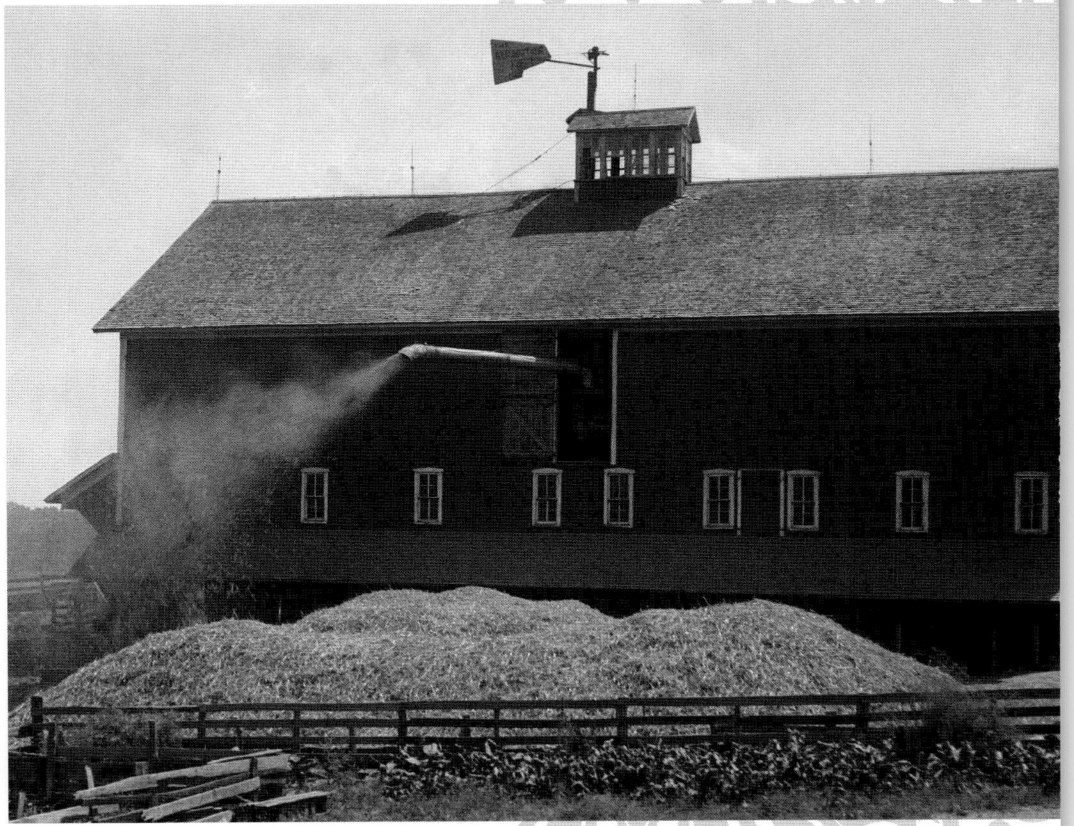

A separator in the barn of P. F. Nye of Goshen blows straw out the door onto a stack. (Elkhart County, 1918)

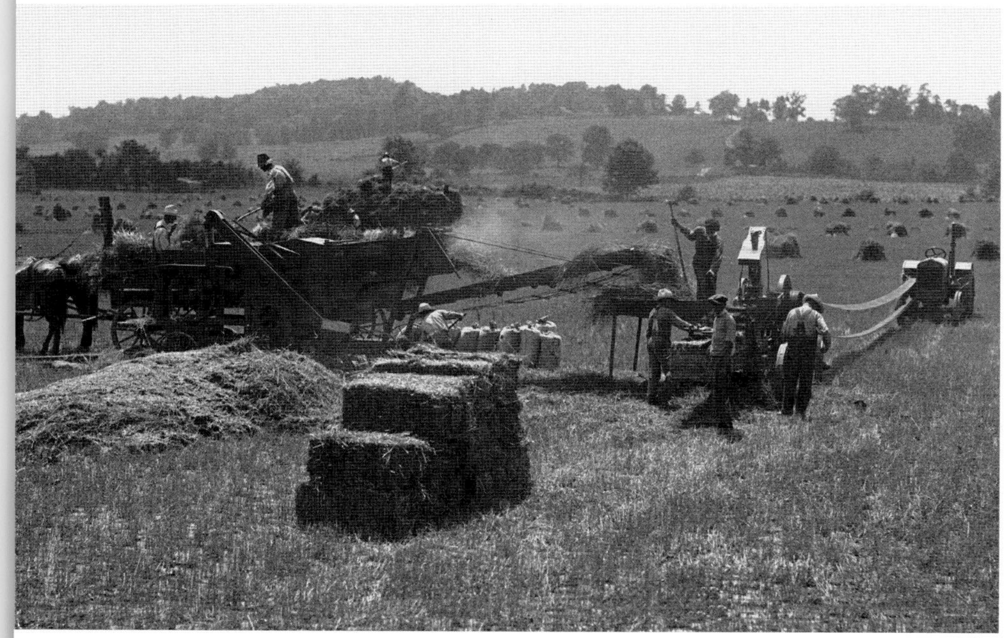

Threshing wheat and baling straw in Bedford at Purdue University's Moses Fell Annex, which is known today as the Feldun Purdue Agricultural Center. (Lawrence County, 1925)

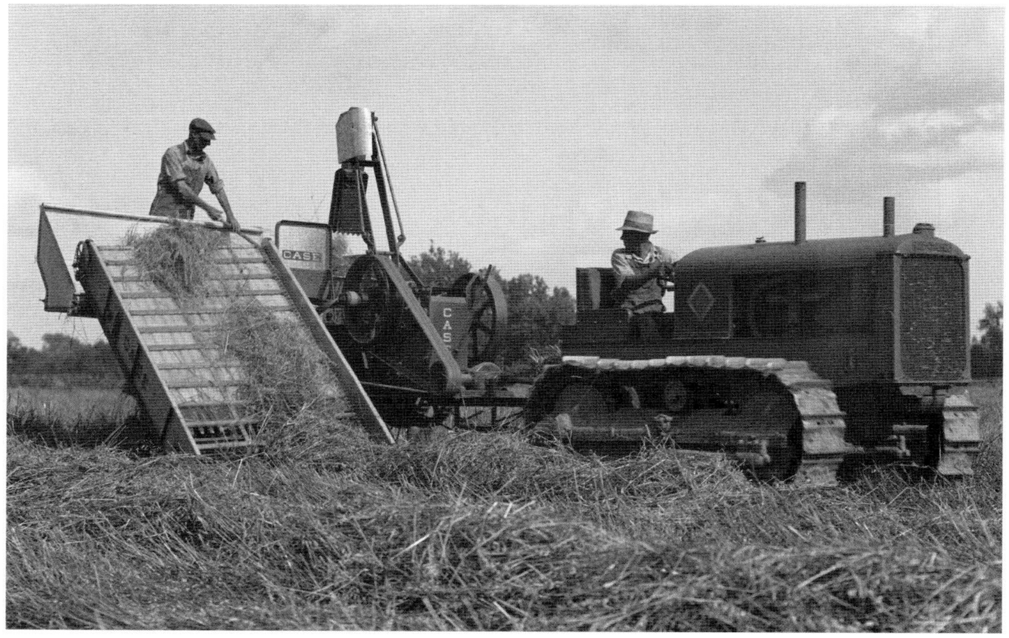

This Allis-Chalmers track-type tractor pulls a Case Pick-up Hay Baler on the farm of Frank B. Shields in Martinsville. (Morgan County, 1932)

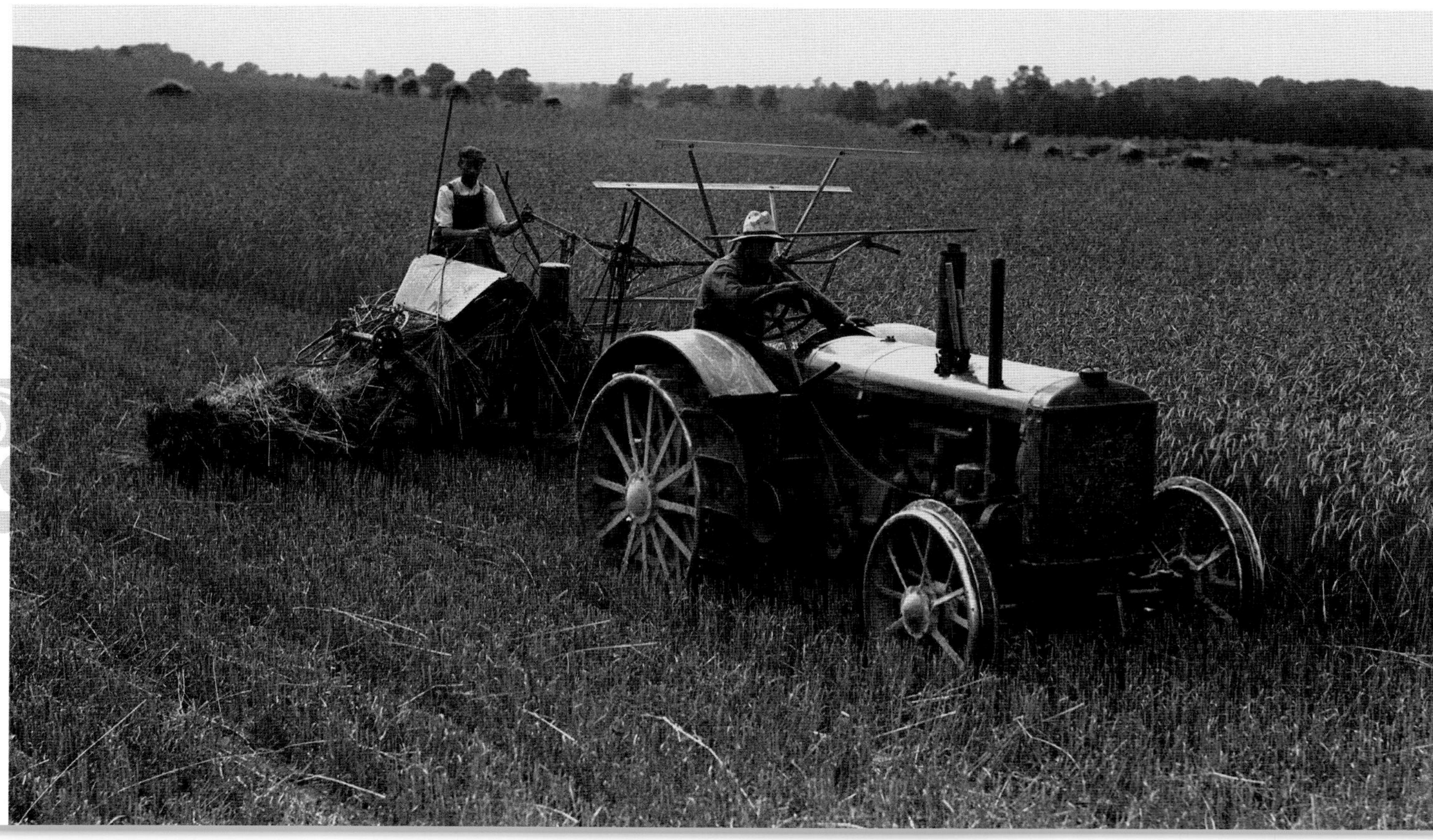

Two men cut wheat on the Lafayette-area farm of Paul Boes. With this Allis-Chalmers tractor and Deering binder, Boes was able to cut 30 acres of wheat per day. (Tippecanoe County, 1929)

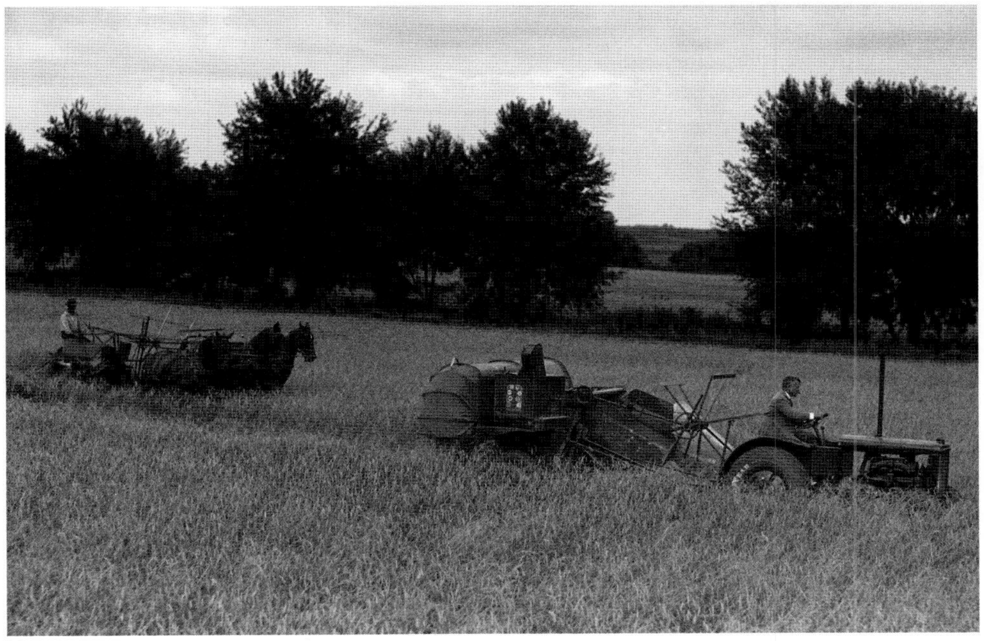

In a competition, a man uses a new Allis-Chalmers combine to harvest wheat while another man follows with a horse-drawn binder. (Indiana, 1933)

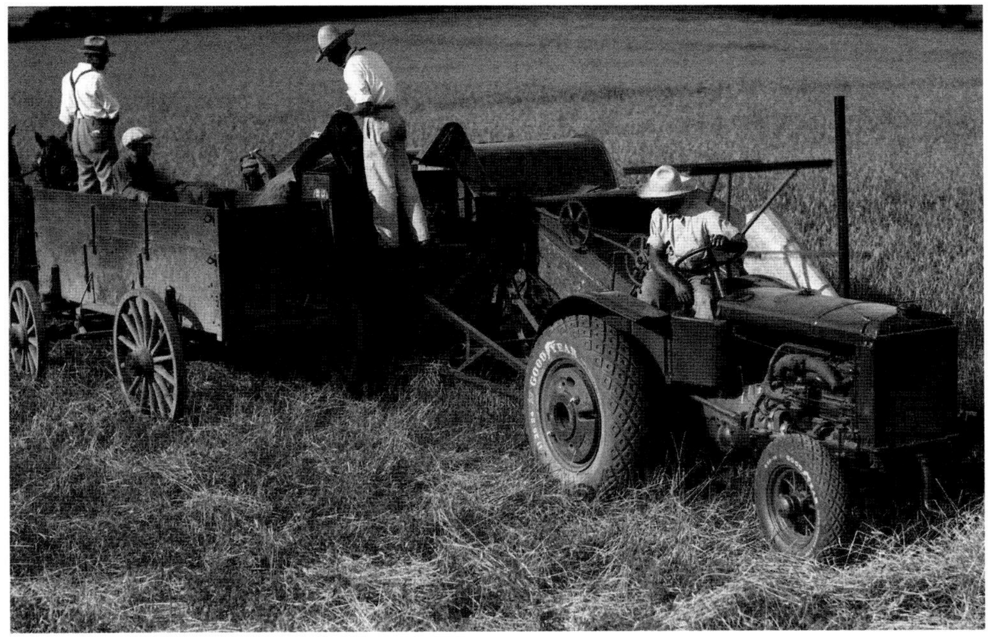

A crew unloads wheat from the new Allis-Chalmers combine into a wagon on the Beaver Dam Farm near LaPorte. (LaPorte County, 1933)

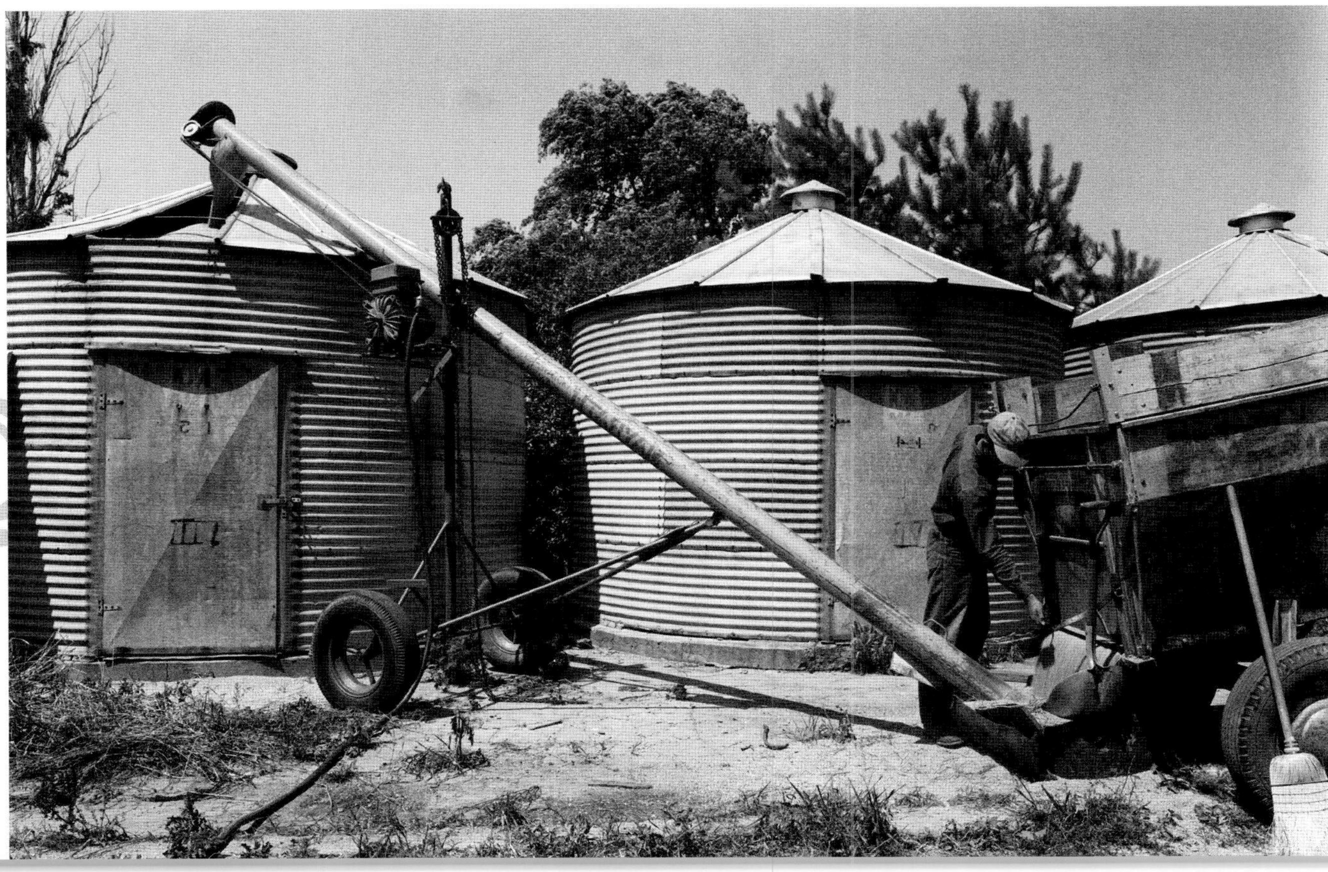

An elevator unloads wheat from a wagon into a bin on the farm of Fred Gutwein and Sons near Francesville. (Pulaski County, 1954)

The Miracle Soybean

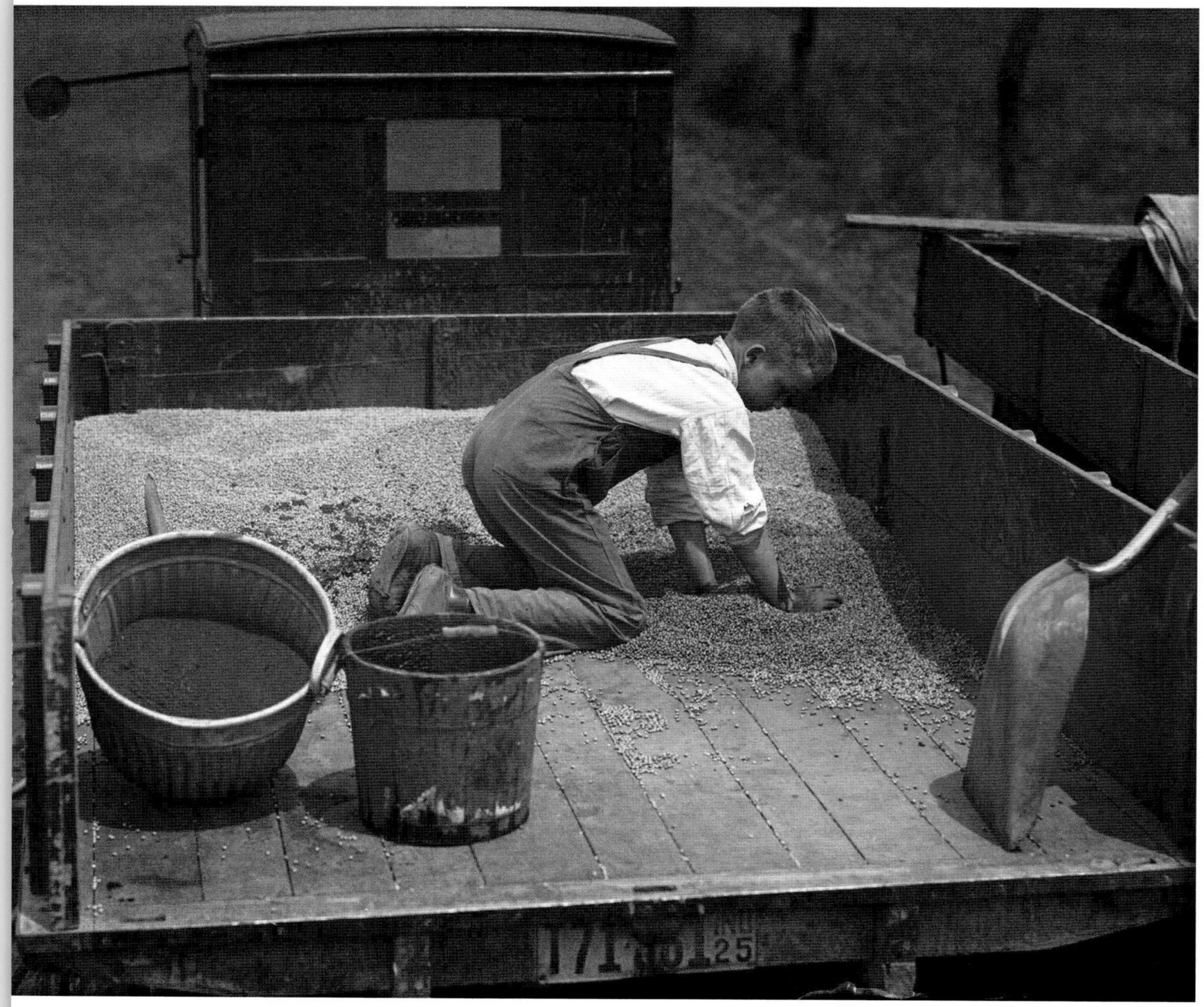

A young man inoculates soybean seed by mixing it with dirt prior to planting on the Charles Meharry farm in Attica. The dirt, which had been taken from a field previously planted to soybeans, contained beneficial bacteria that improved plant health and increased crop yields. (Fountain County, 1925)

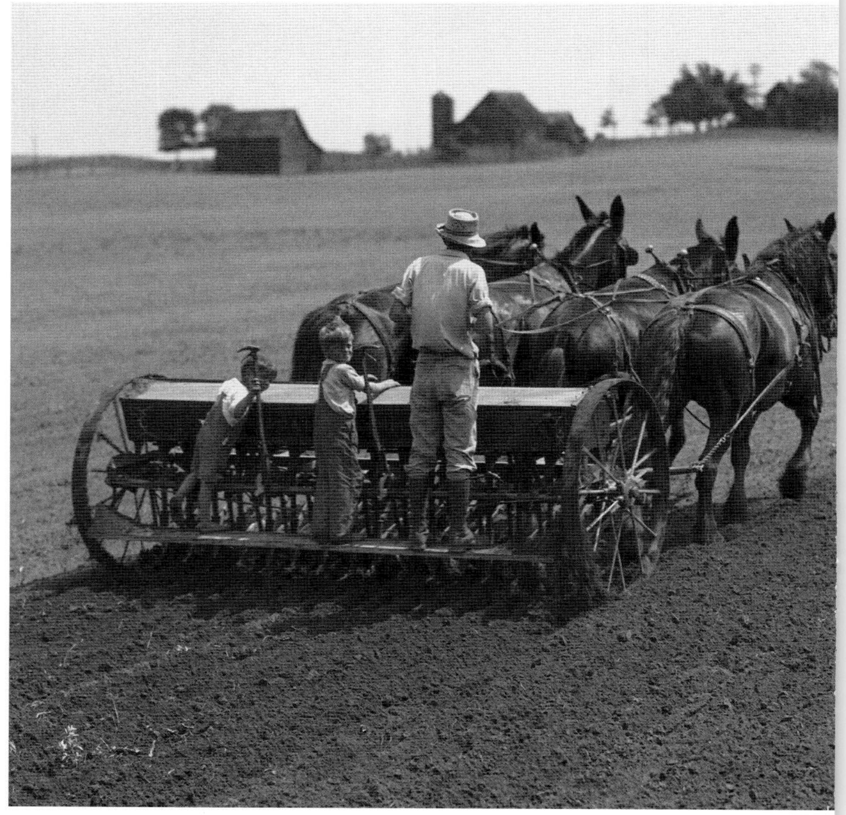

Drilling soybeans with a team of four horses (Indiana, 1932)

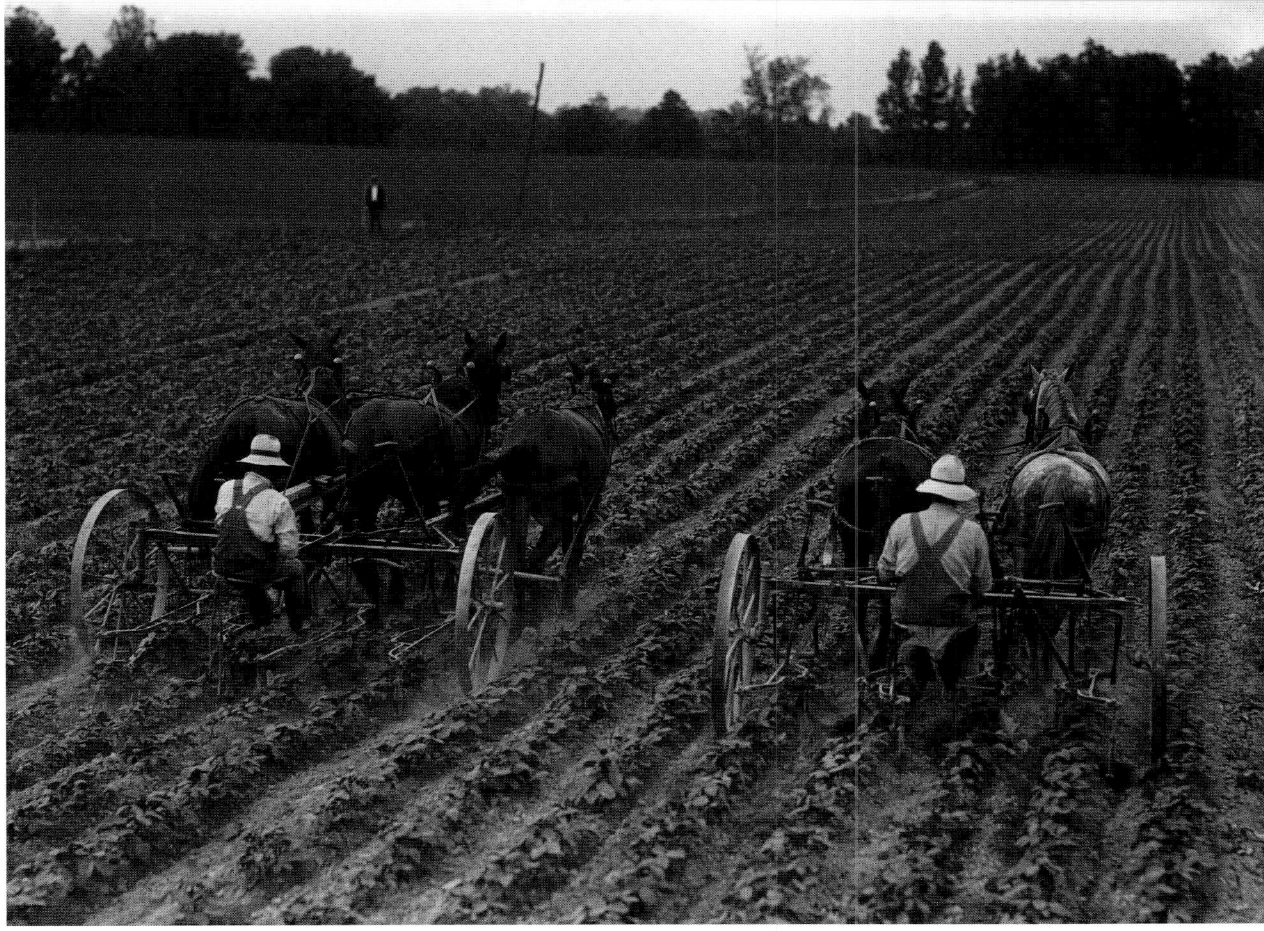

On the Caldwell Farm in Camden, these men cultivate soybeans with a two-row cultivator. At the time, the Caldwell brothers grew 79 acres of soybeans plus 40 acres of soybeans interplanted with corn. (Carroll County, 1923)

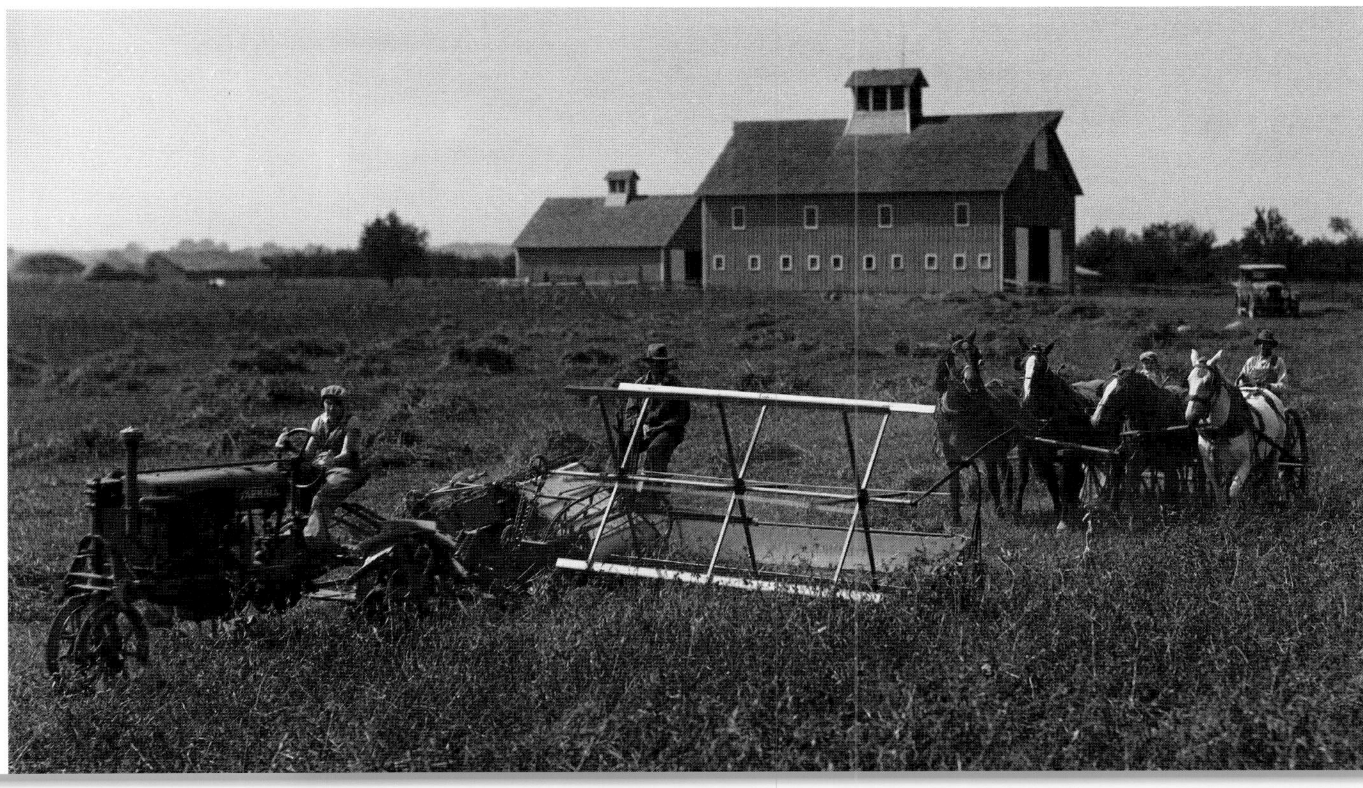

A two-person team cuts soybeans using a Farmall tractor pulling a binder while a man guides a horse-drawn wheat drill behind them on the farm of William Dawson and his son-in-law, Russell Butler, near Lafayette. (Tippecanoe County, 1929)

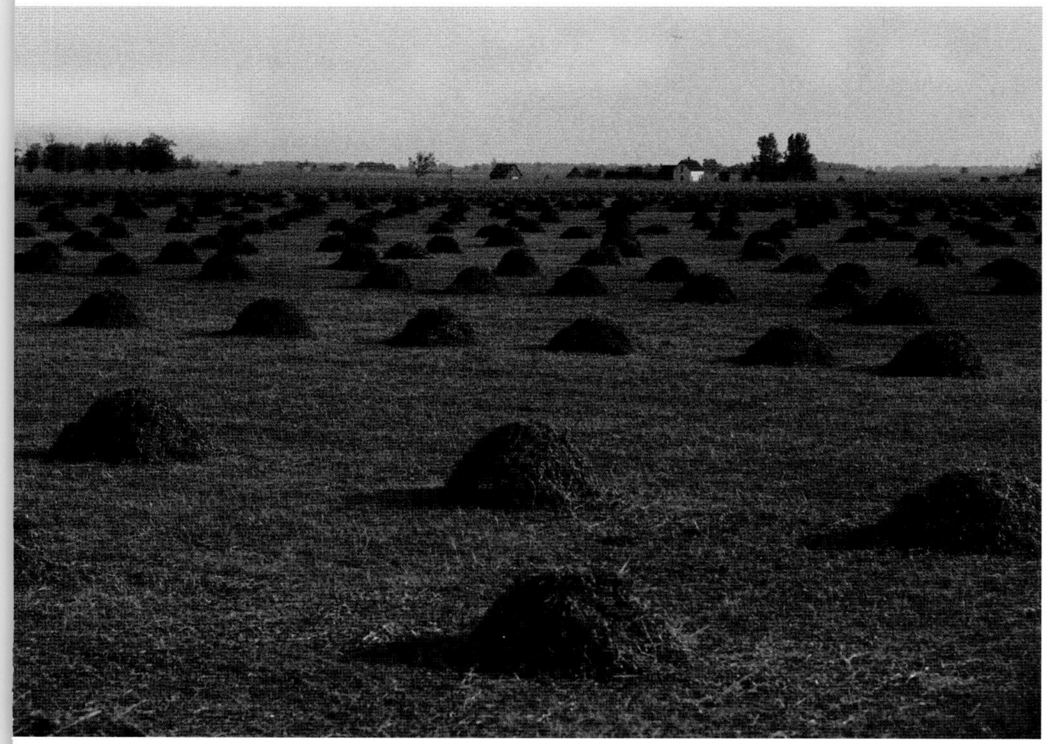

Soybean shocks on the Ora C. Loveless farm in Clarks Hill (Tippecanoe County, 1917)

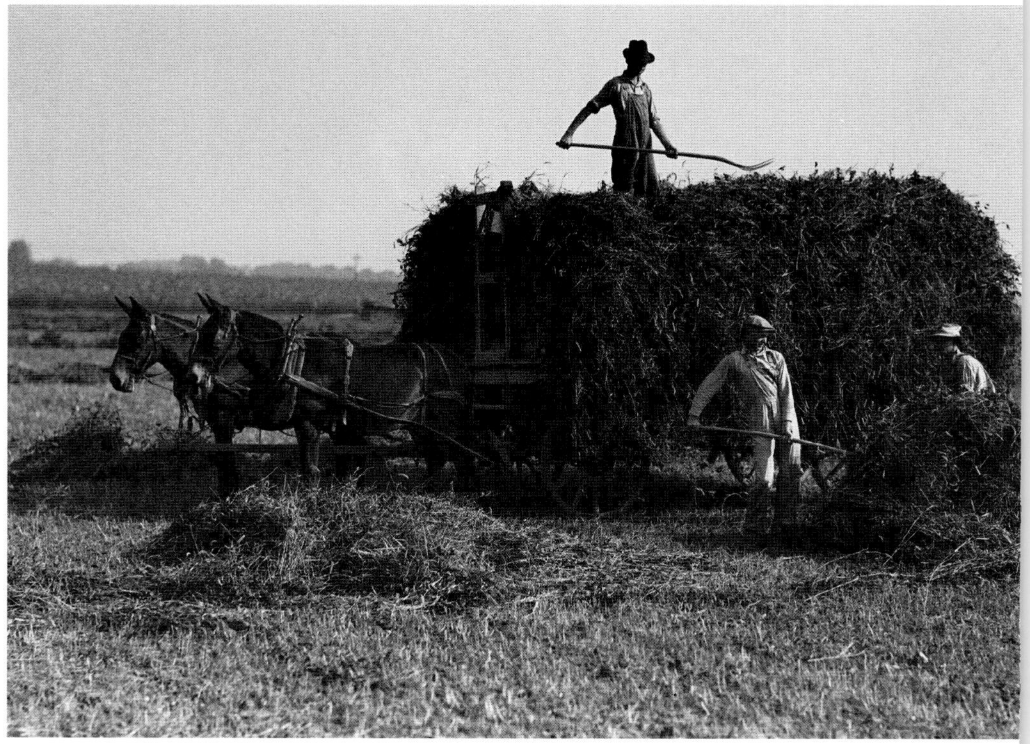

Several workers use pitchforks to load soybean hay onto a wagon on the Lee Marks farm in West Point. (Tippecanoe County, 1929)

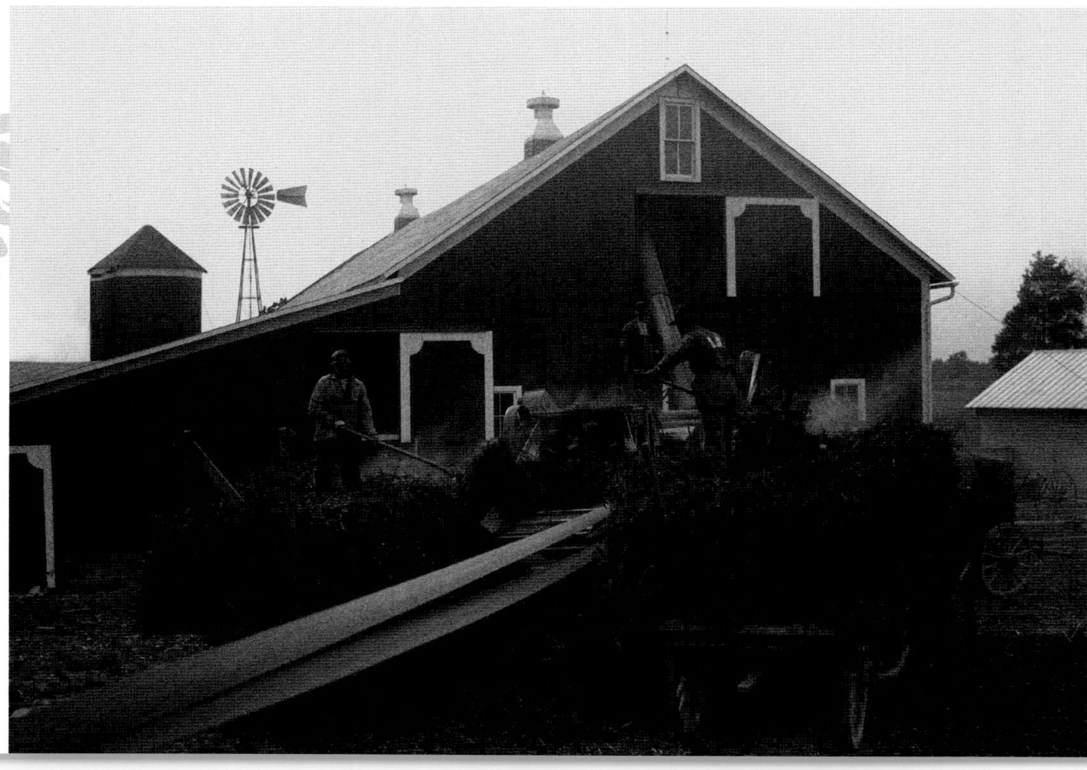

This crew threshes soybeans while blowing stalks into the barn for winter feed and bedding on the Finis Fouts farm near Deer Creek. (Carroll County, 1916)

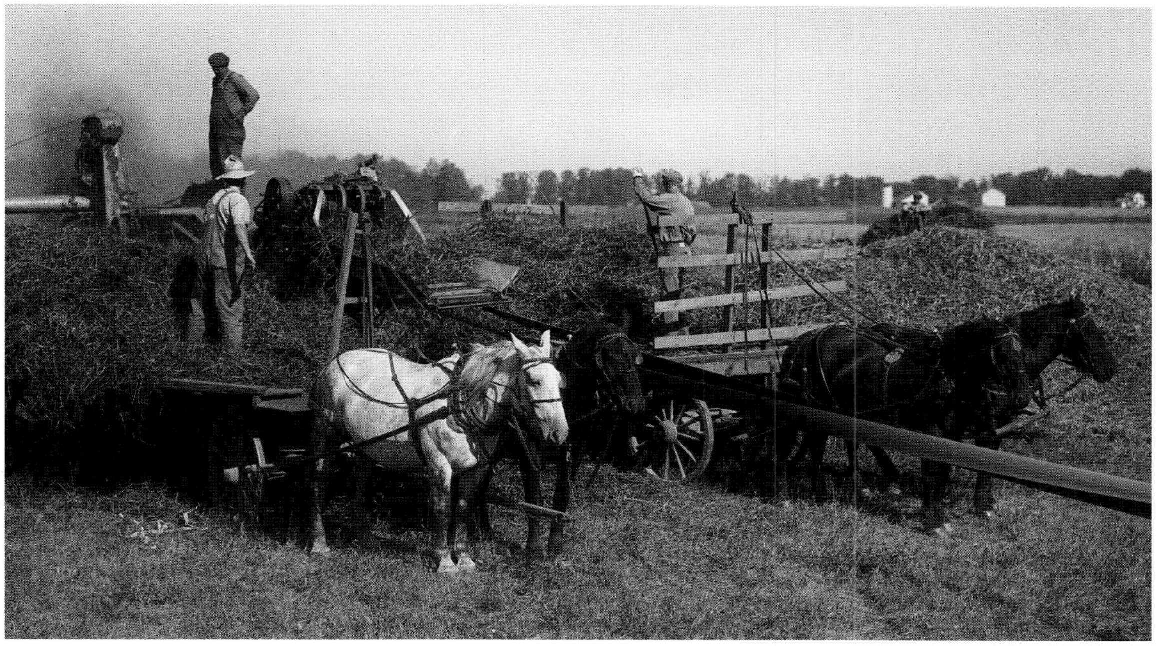

Threshing soybeans on the Lynn Van Natta farm in Battle Ground (Tippecanoe County, 1928)

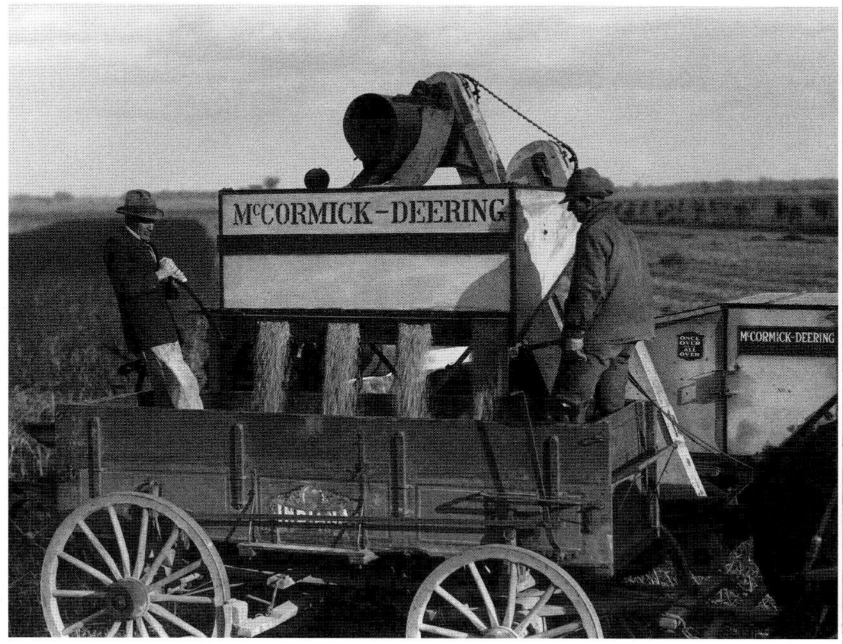

These men dump soybeans from a McCormick-Deering combine into a horse-drawn wagon at the Purdue University Wilson Farm in Lafayette. (Tippecanoe County, 1928)

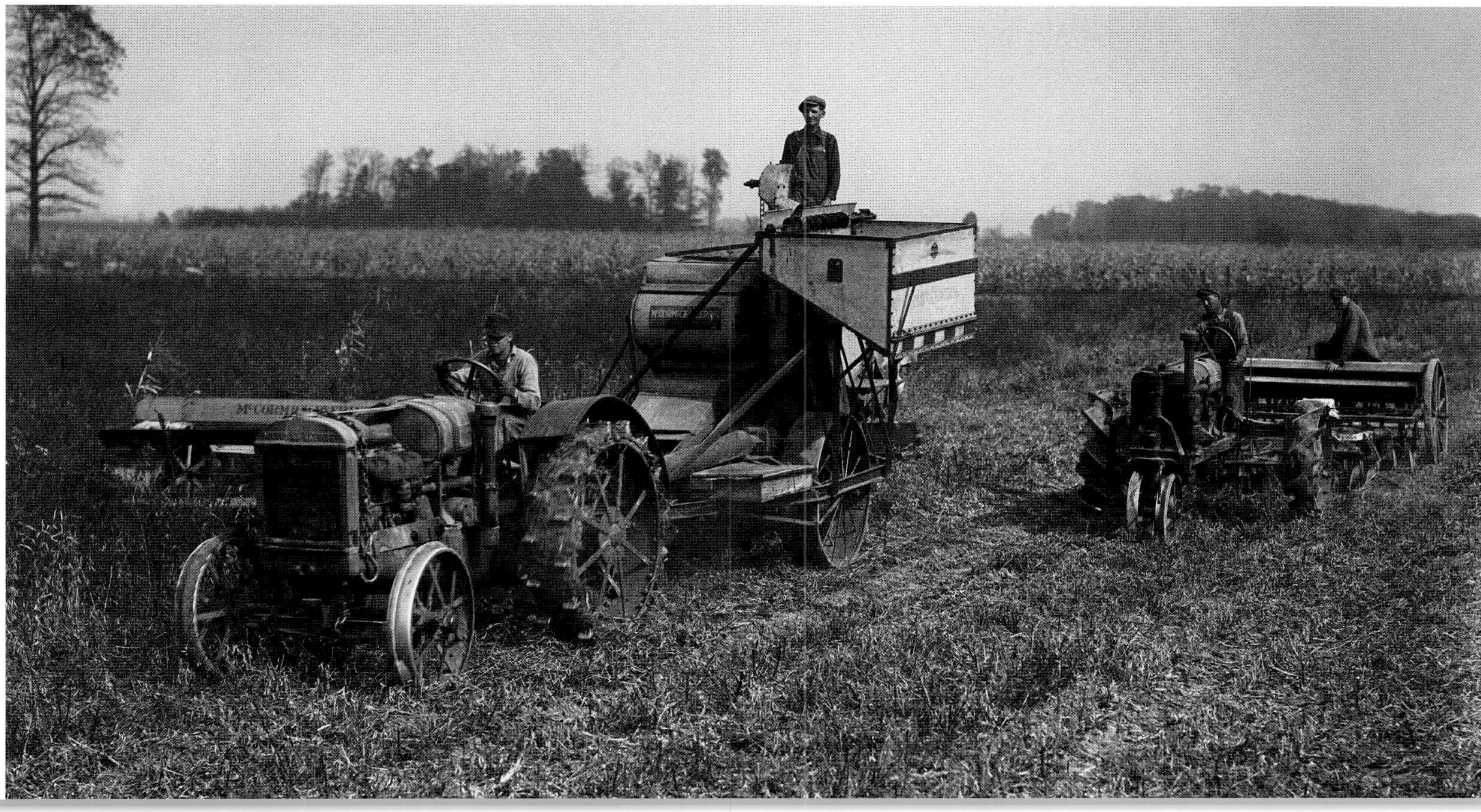

Two men combine soybeans while two others follow with a wheat drill on the farm of Finis Fouts. The introduction of combines allowed farmers to separate soybeans from the plant while still in the field, which made harvesting more efficient. (Carroll County, 1929)

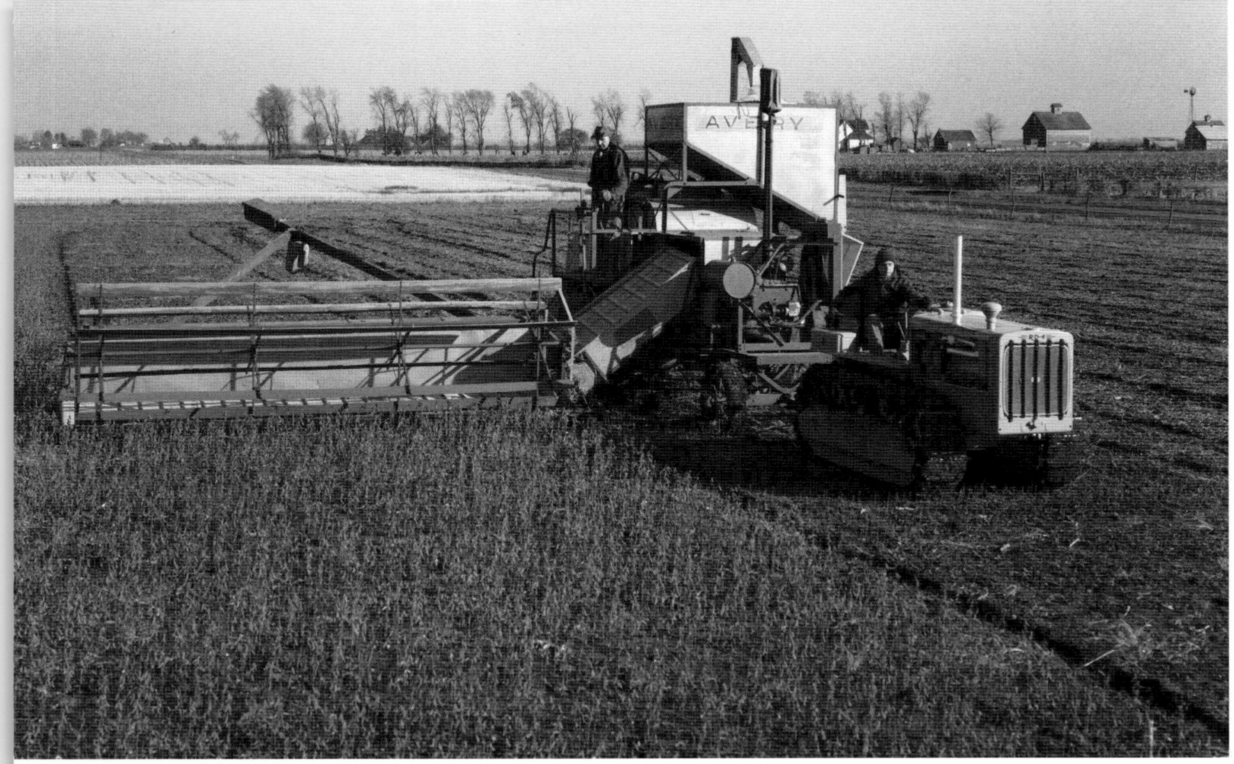

A Caterpillar RD4 at work in a soybean field in Champaign, Illinois (Champaign County, IL; 1936)

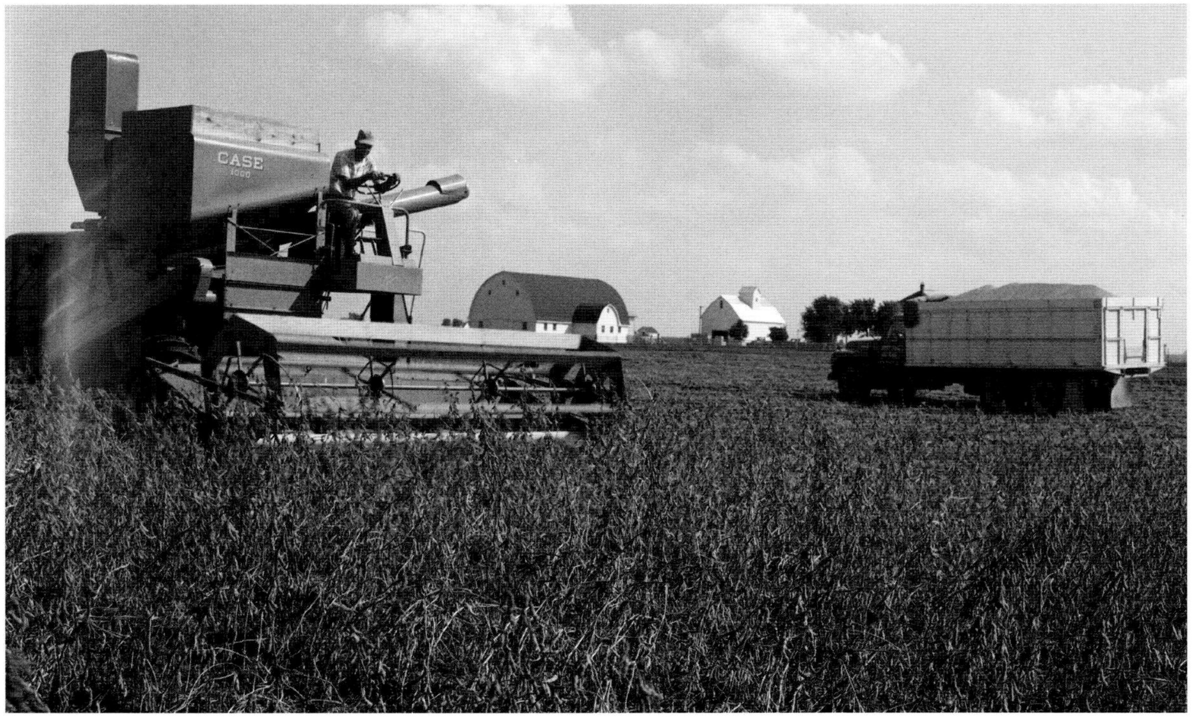

Combining soybeans on the Emmett Lowry farm in Francesville (Pulaski County, 1961)

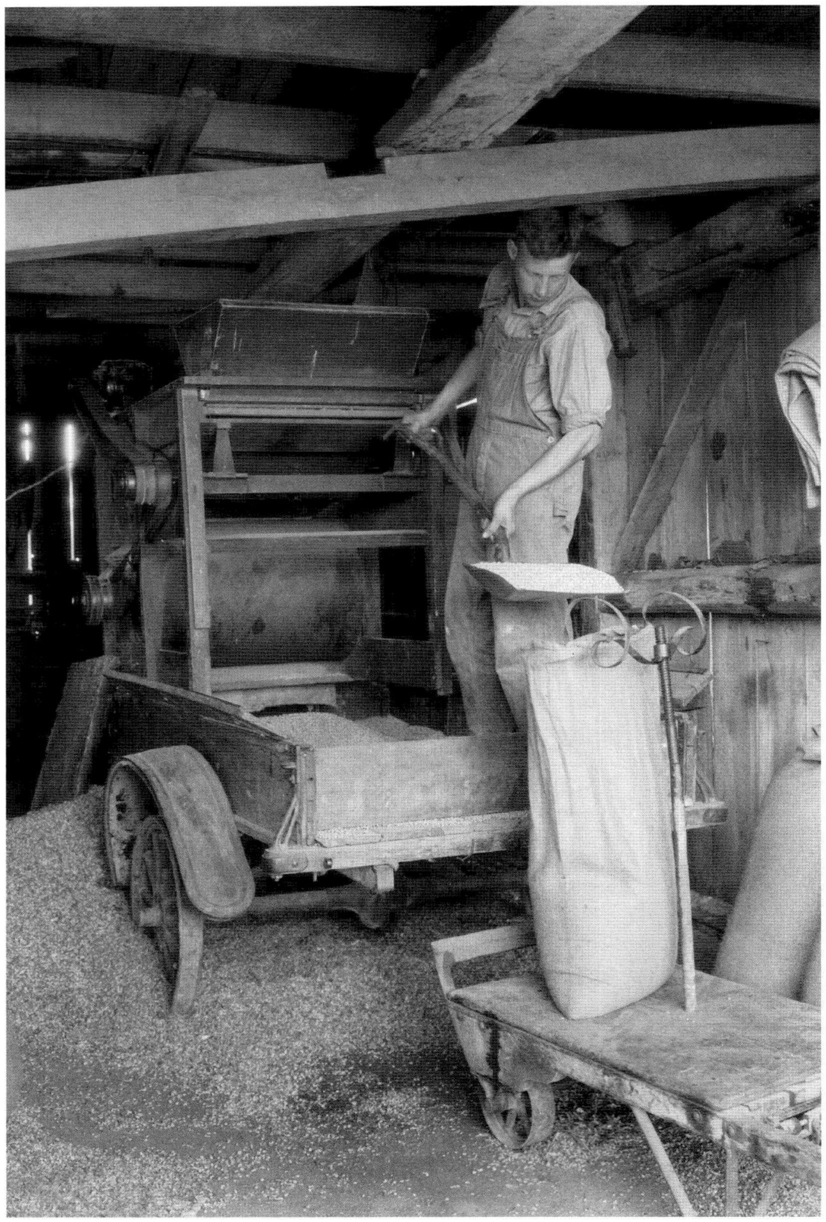

Frederic Fouts uses a homemade stand to hold a sack as he fills it with soybeans grown on the Camden-area farm owned by his father, Taylor Fouts. This simple stand eliminated the need for a second person to hold the sack while filling it. (Carroll County, 1930)

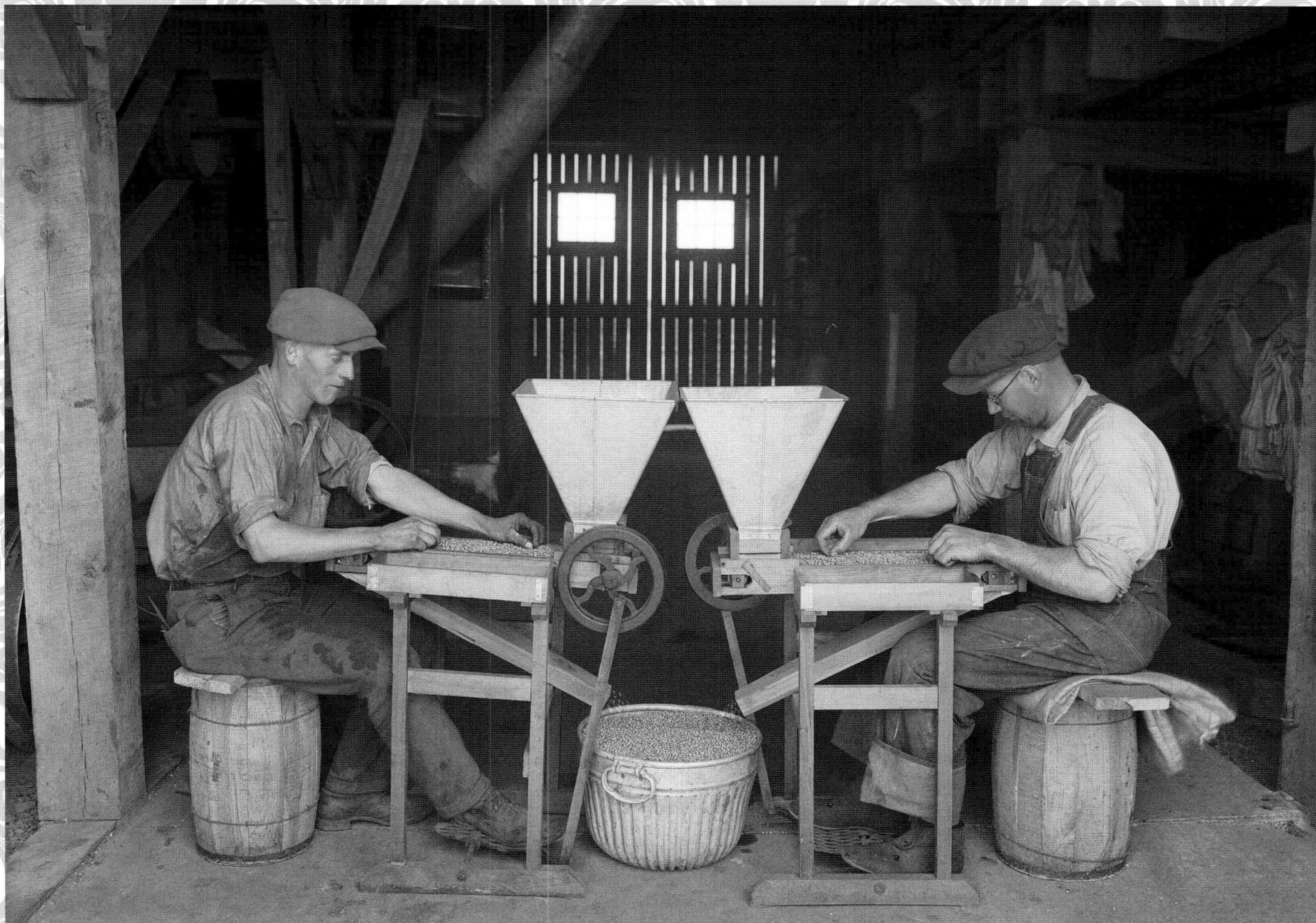

Two men handpick soybeans to be sold for seed on the Charles Meharry farm in Attica. (Fountain County, 1925)

An Apple a Day

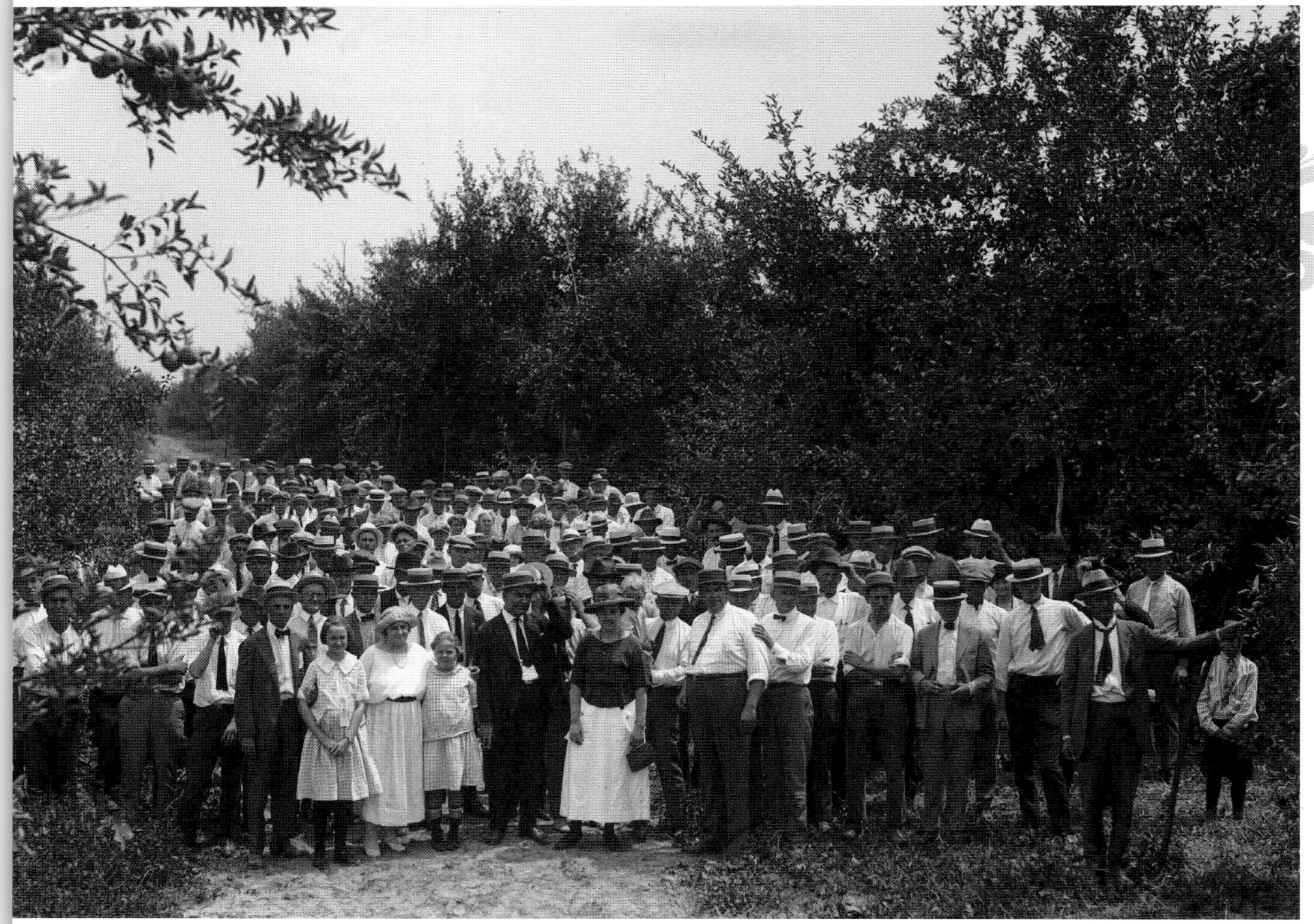

A group assembles for an apple orchard tour at the Simpson Farm in Vincennes. (Knox County, 1922)

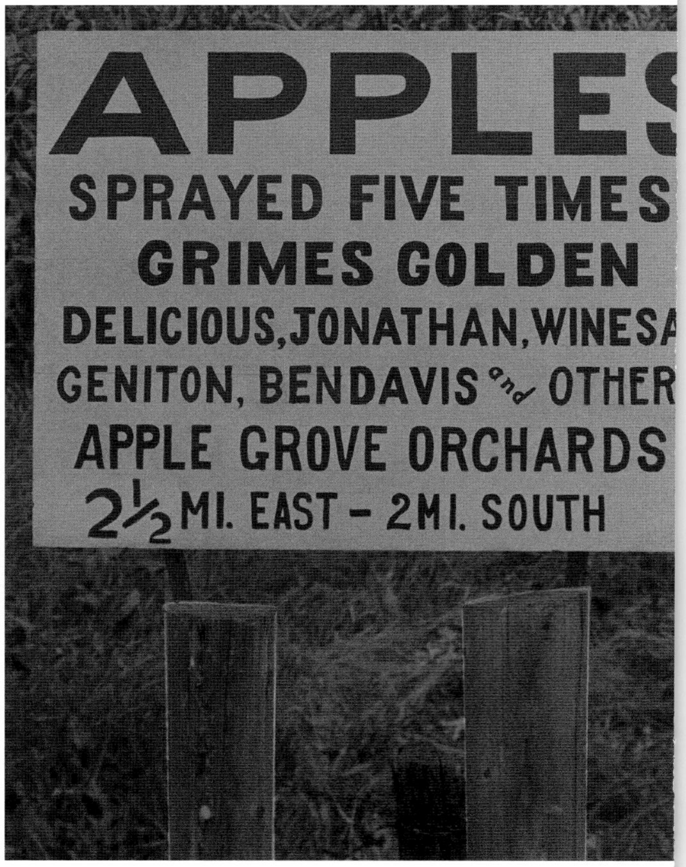

Apple Grove Orchards near Mitchellville, Iowa, advertises apple varieties, noting that they were treated for insects and diseases. (Jasper and Polk Counties, IA; n.d.)

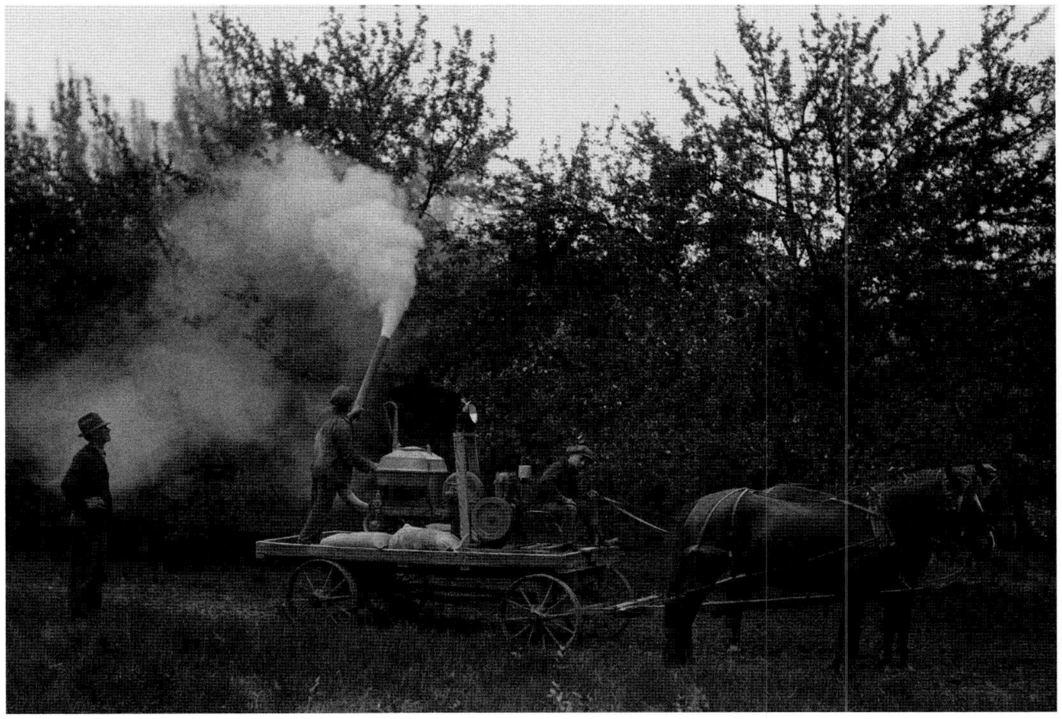

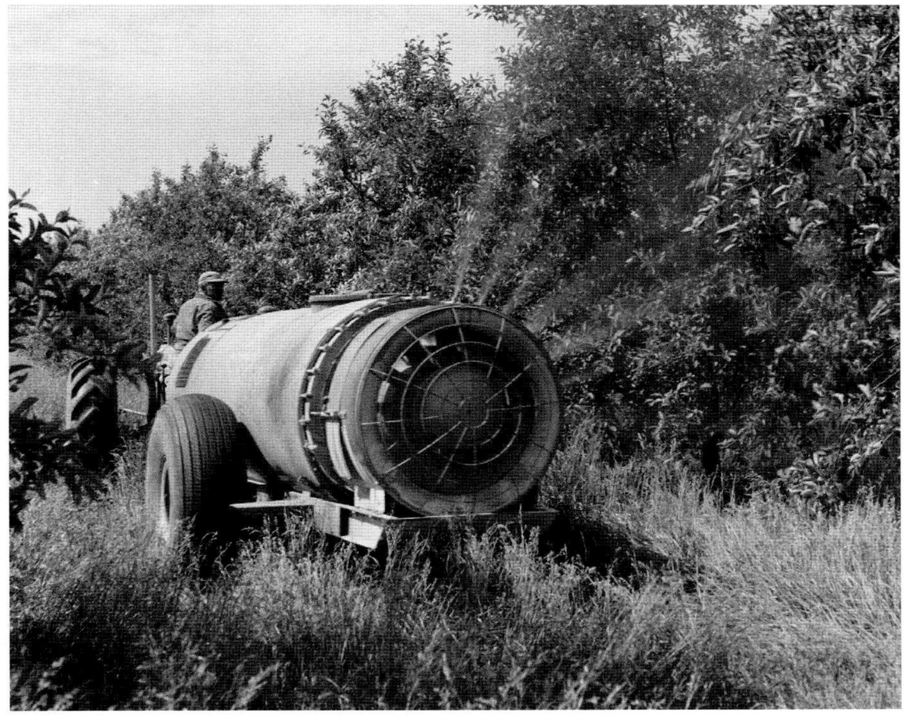

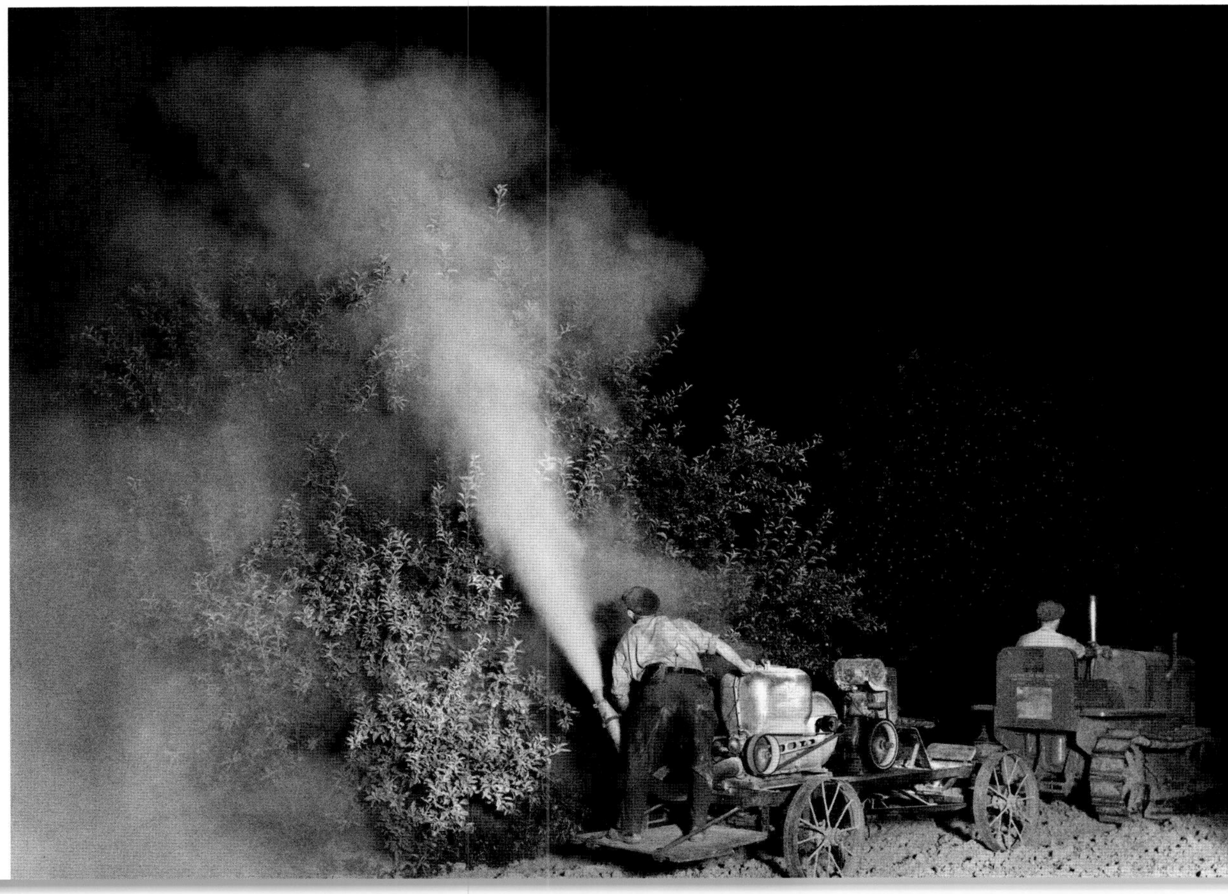

(Top left) Dusting apple trees with sulfur and lead arsenate using a horse-drawn duster at the D. B. Johnson and Son Orchard in Mooresville (Morgan County, 1926)

(Top right) Spraying an apple orchard with advanced equipment (Northern Indiana, 1964)

(Bottom) A dramatic nighttime photograph shows workers dusting the Purdue University Horticulture Farm apple orchard. (Tippecanoe County, 1930)

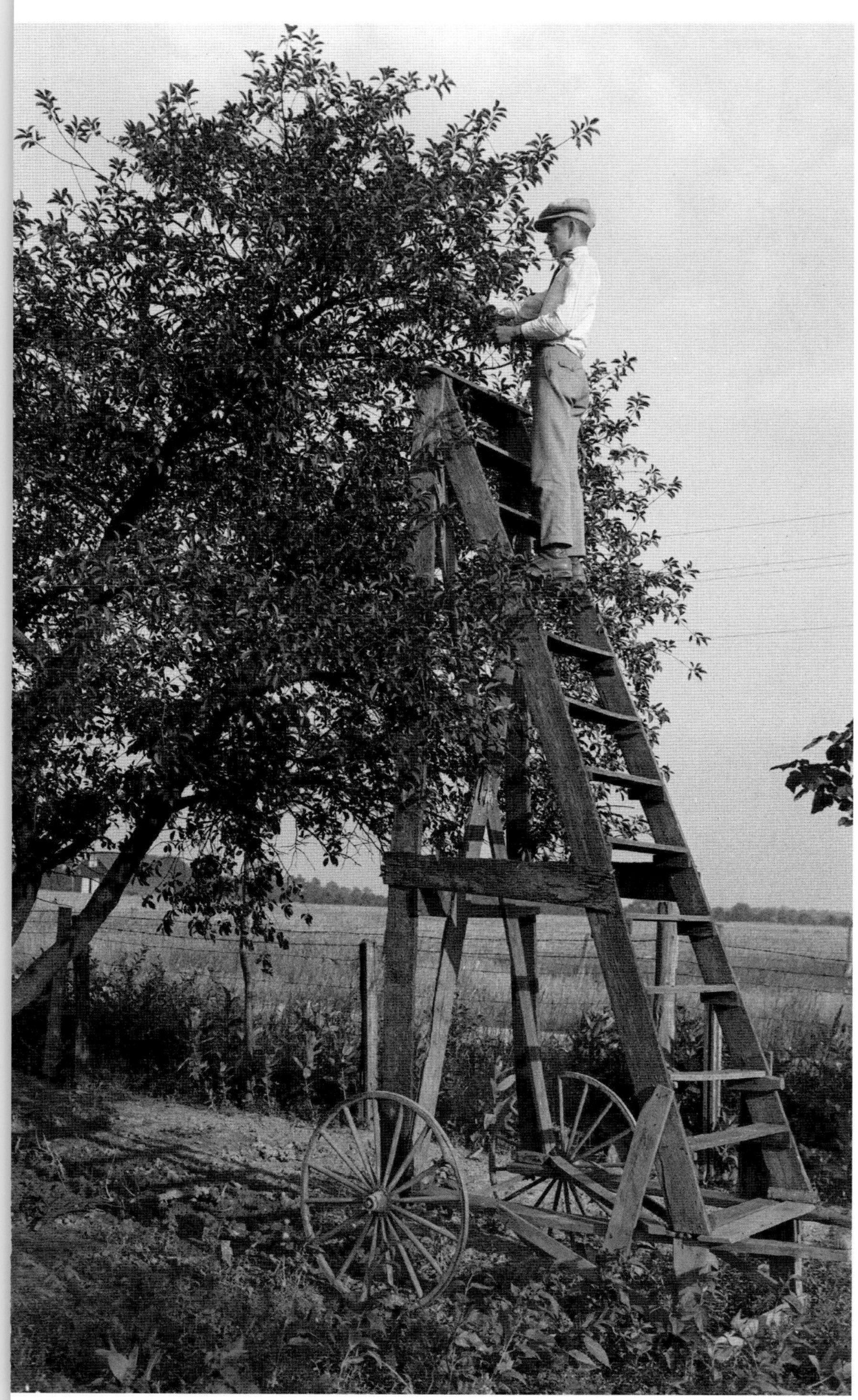

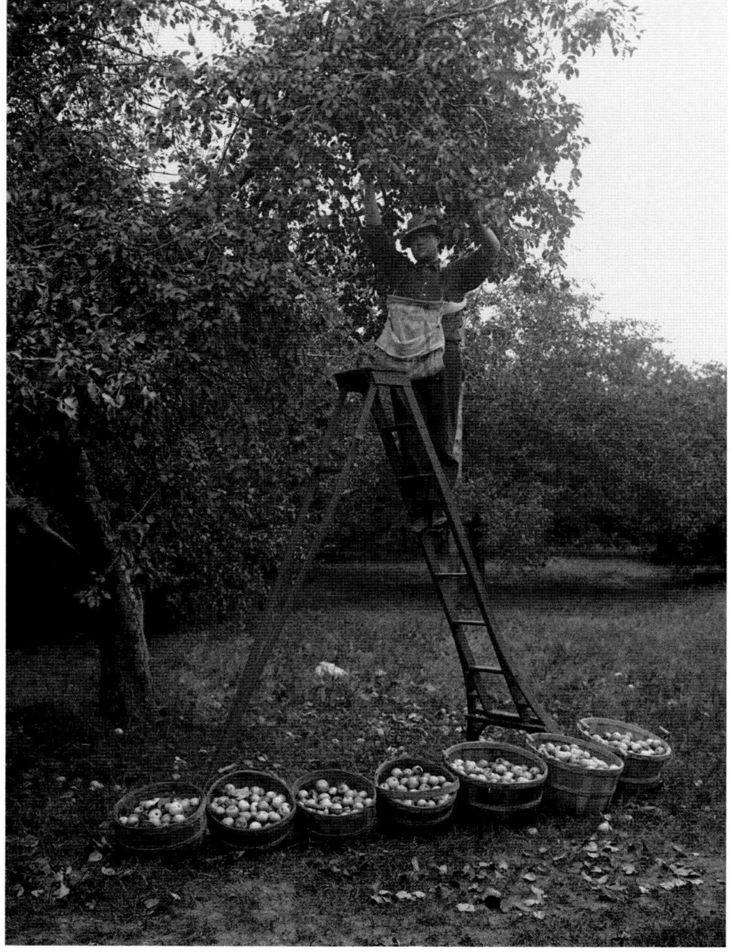

(Far left) This man at the M. Hill farm uses a portable ladder on wheels that can easily be moved from tree to tree. (Marion County, 1924)

(Top right) Handpicking apples at Zion's Orchard in Clarks Hill (Tippecanoe County, ca. 1914)

(Bottom right) Emerald Bundy from Mount Summit places baskets of apples into a Ford sedan with the help of another man. (Henry County, 1924)

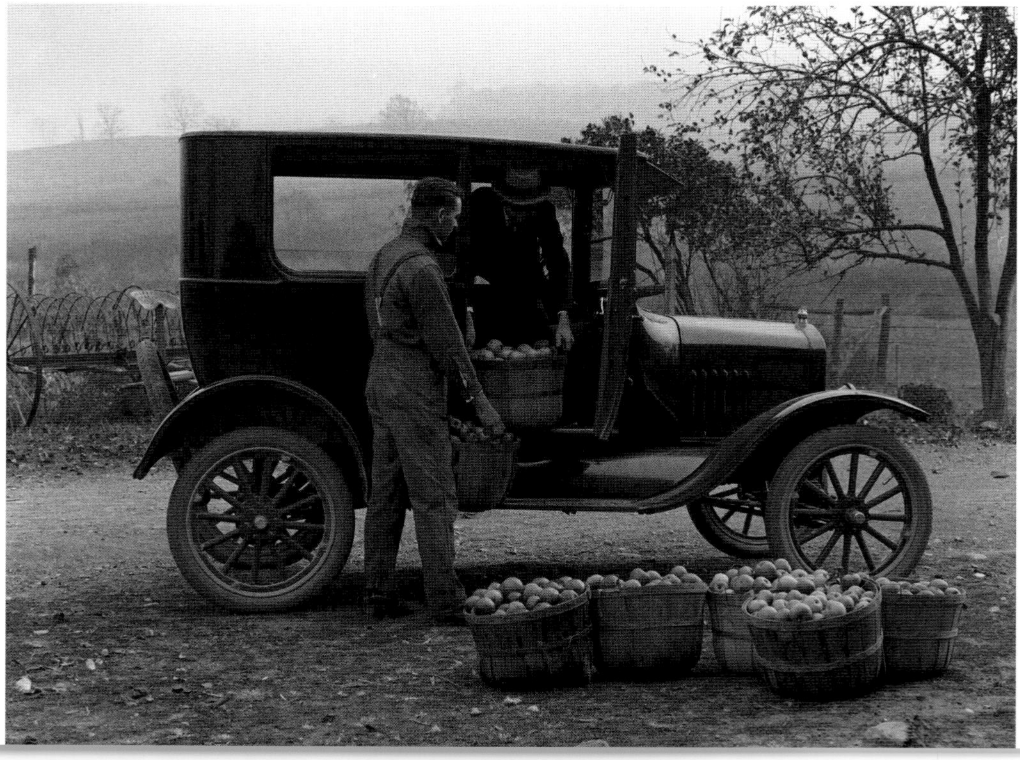

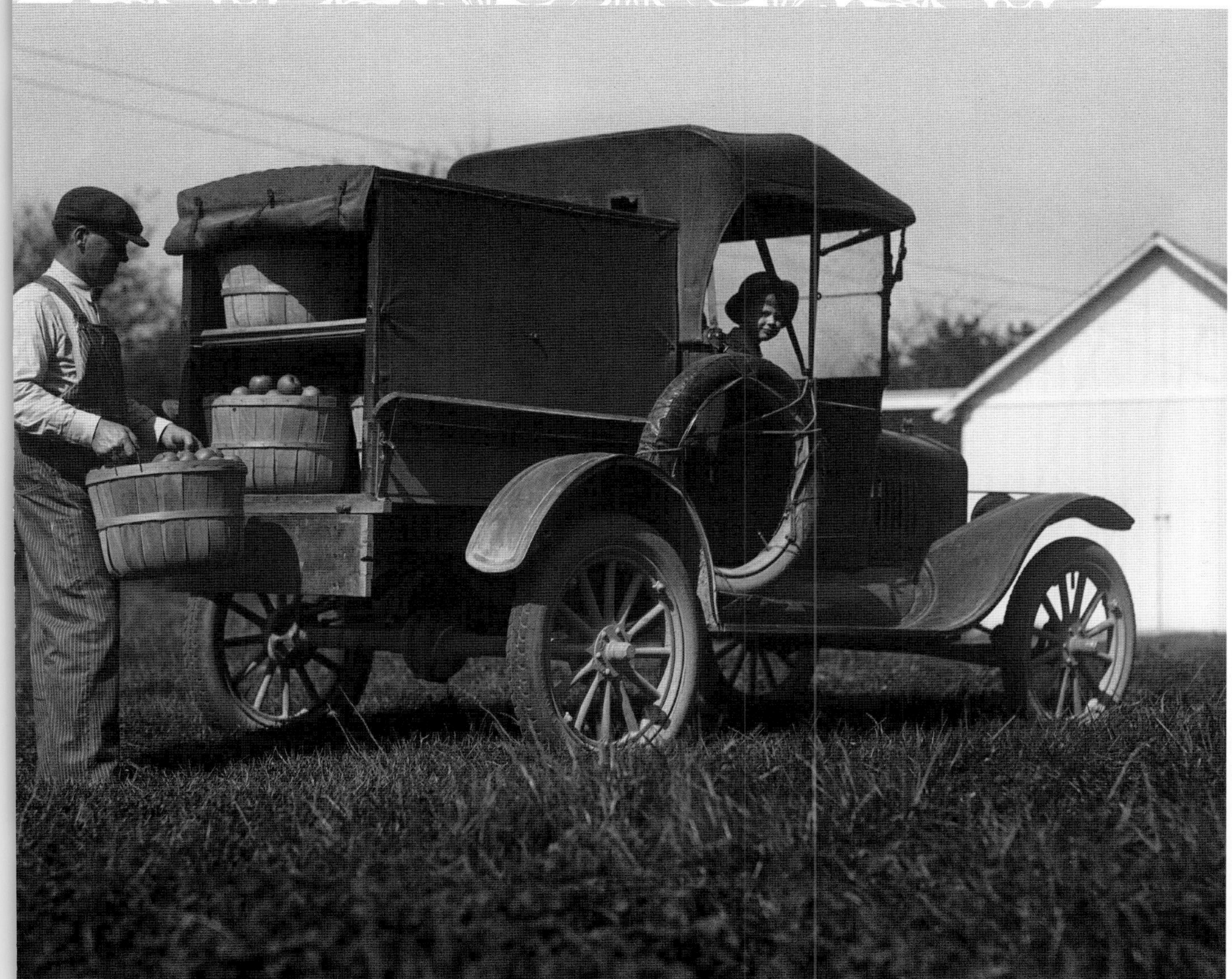

E. E. Brown loads up his Ford truck with apples to take to the local market in Veedersburg. (Fountain County, 1924)

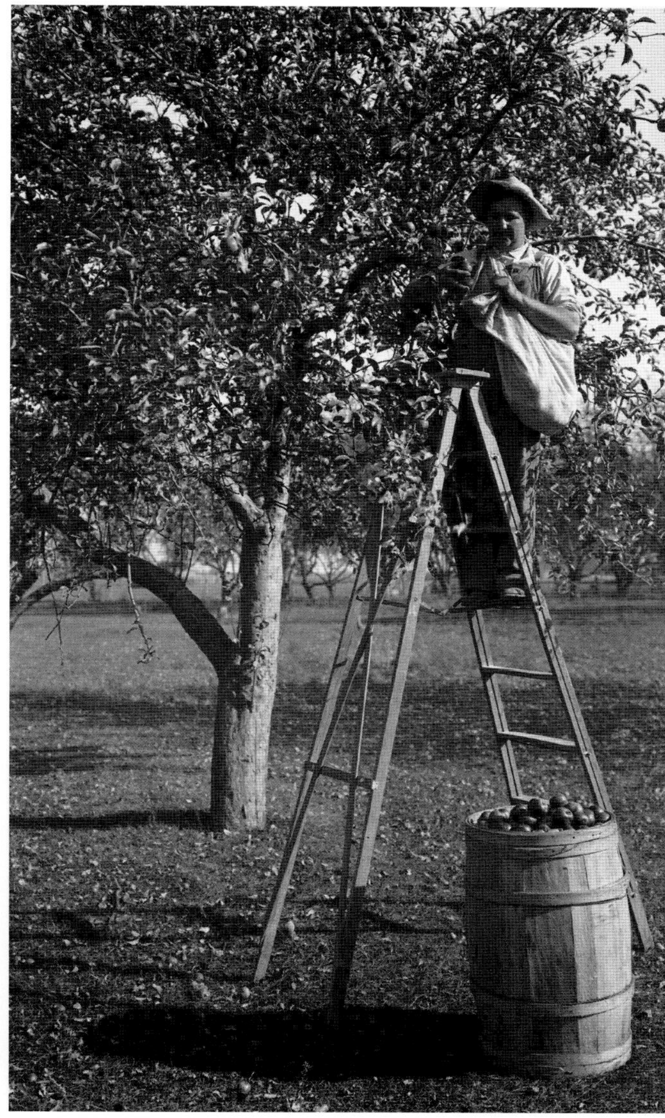

Filling a barrel with apples at an orchard (Indiana, ca. 1932)

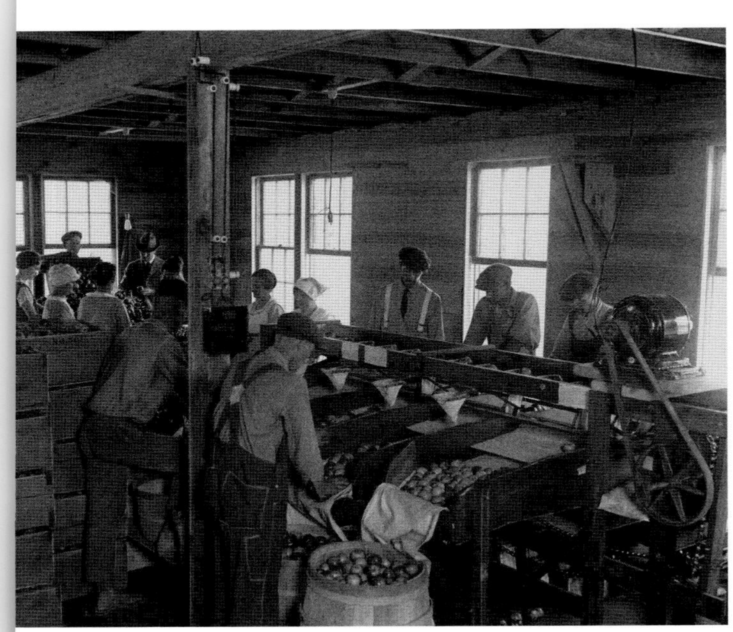

Grading and packing apples at Purdue University's Moses Fell Annex in Bedford (Lawrence County, 1924)

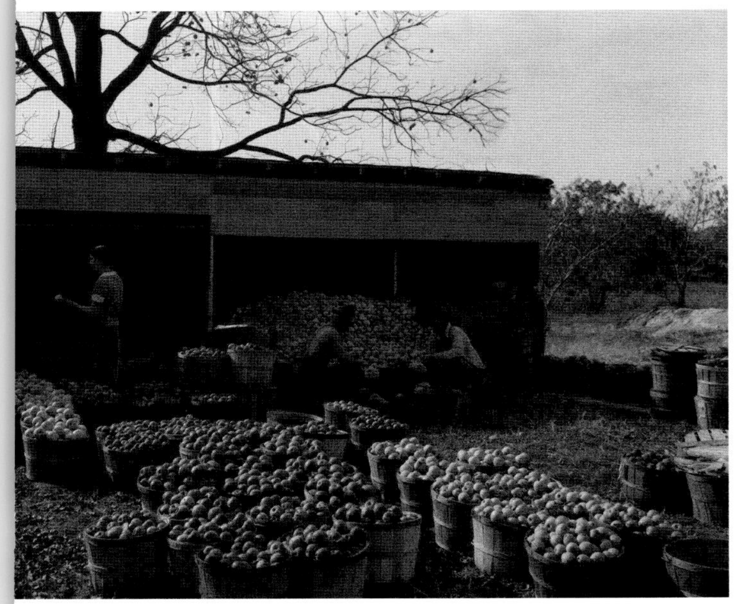

These baskets and bins made up approximately a thousand bushels of apples on the Lafayette farm of Clarence Golden. (Tippecanoe County, 1935)

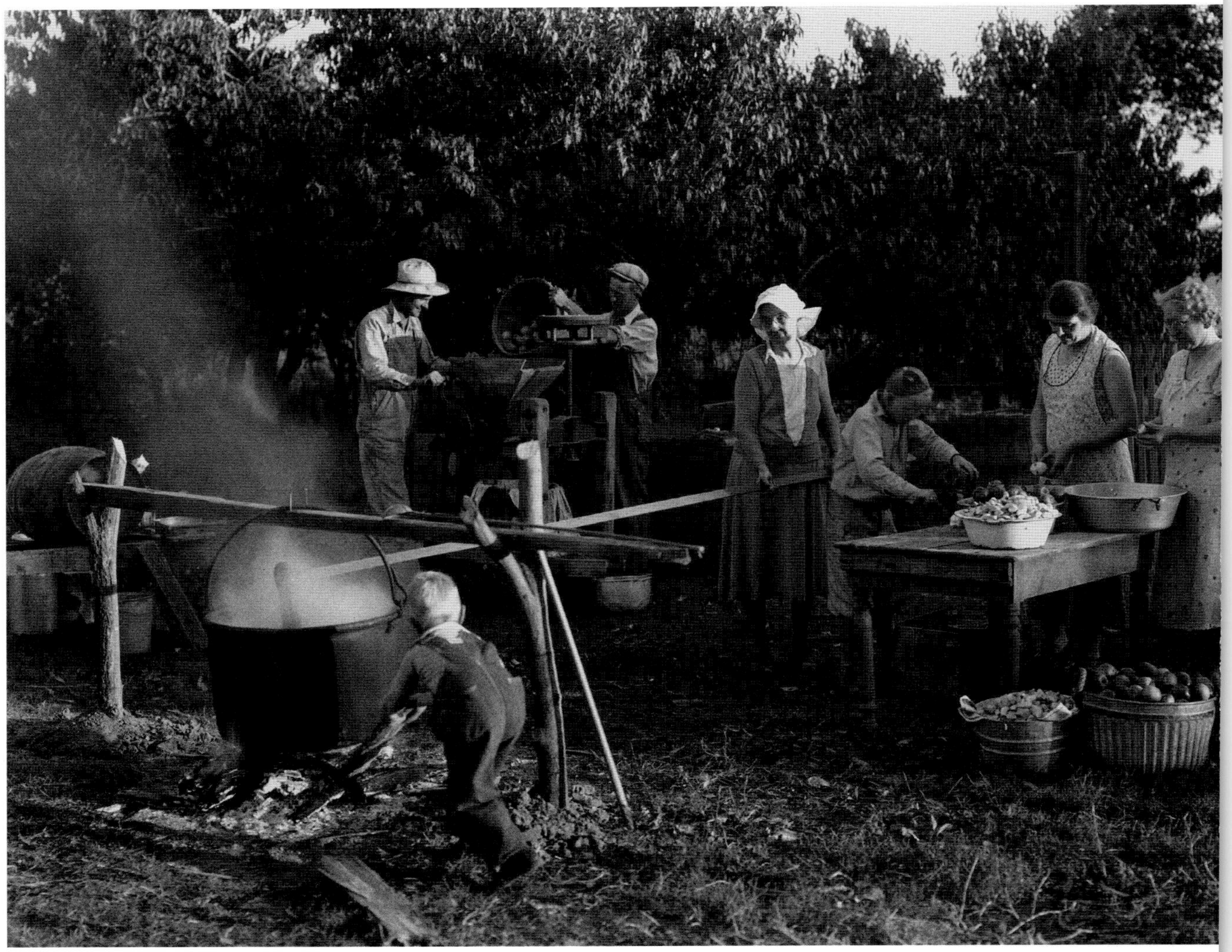

Making apple butter on the farm of Tom Scott in Burrows (Carroll County, 1934)

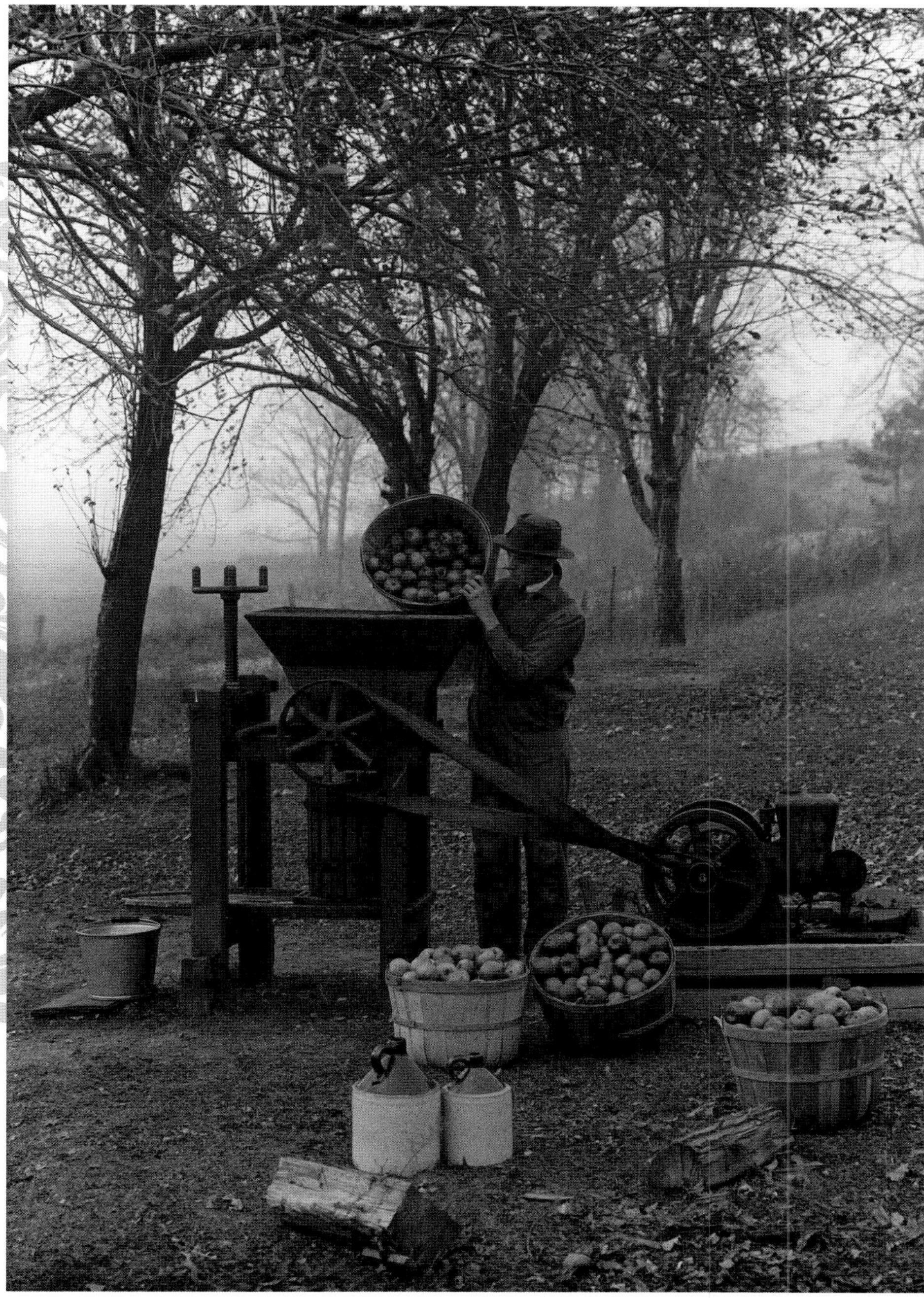

A gasoline motor powers this apple cider press on the farm of Emerald Bundy at Mount Summit. (Henry County, 1924)

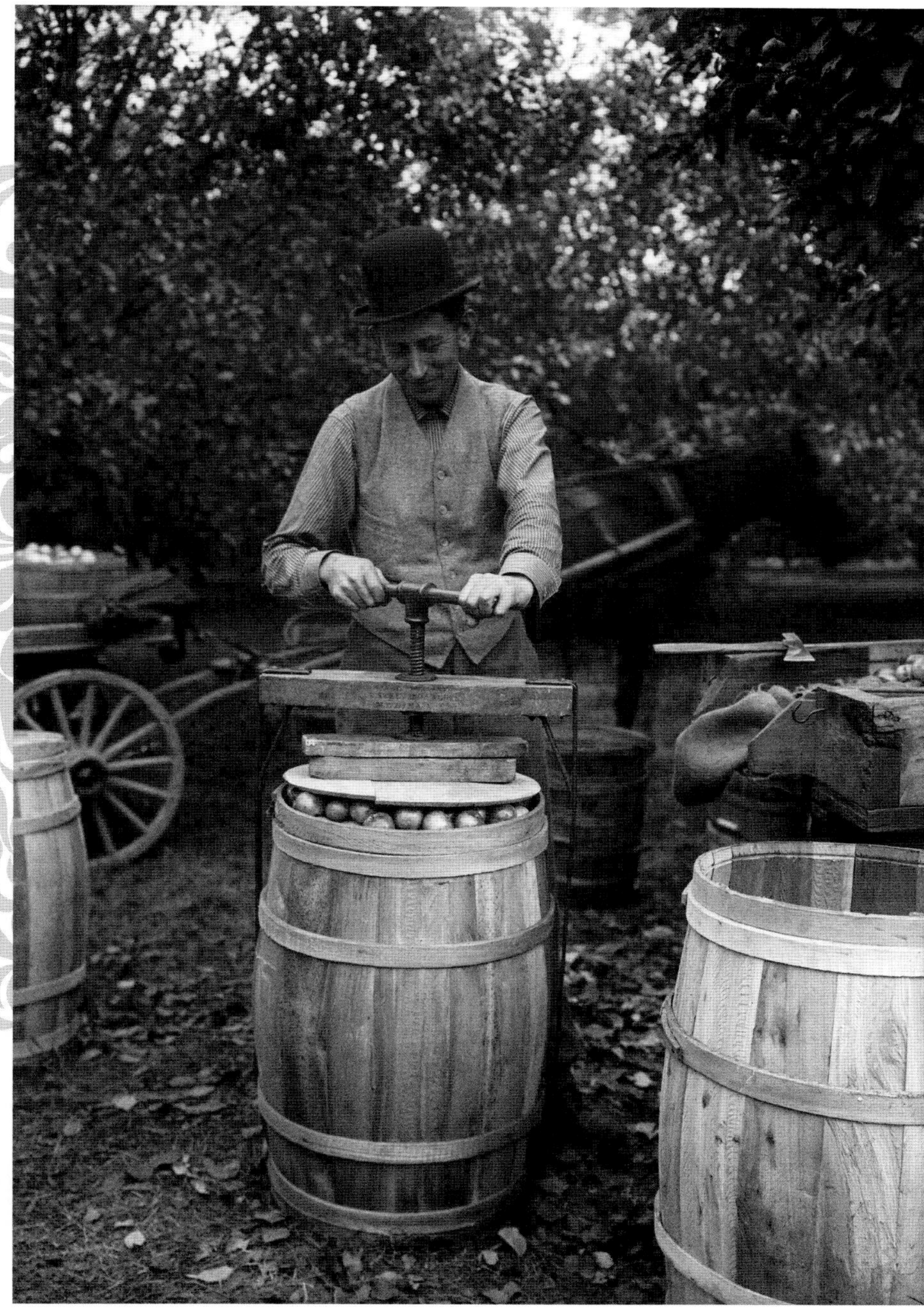

A man packs apples into a barrel at Zion's Orchard in Clarks Hill. (Tippecanoe County, 1915)

A Juicy Peach

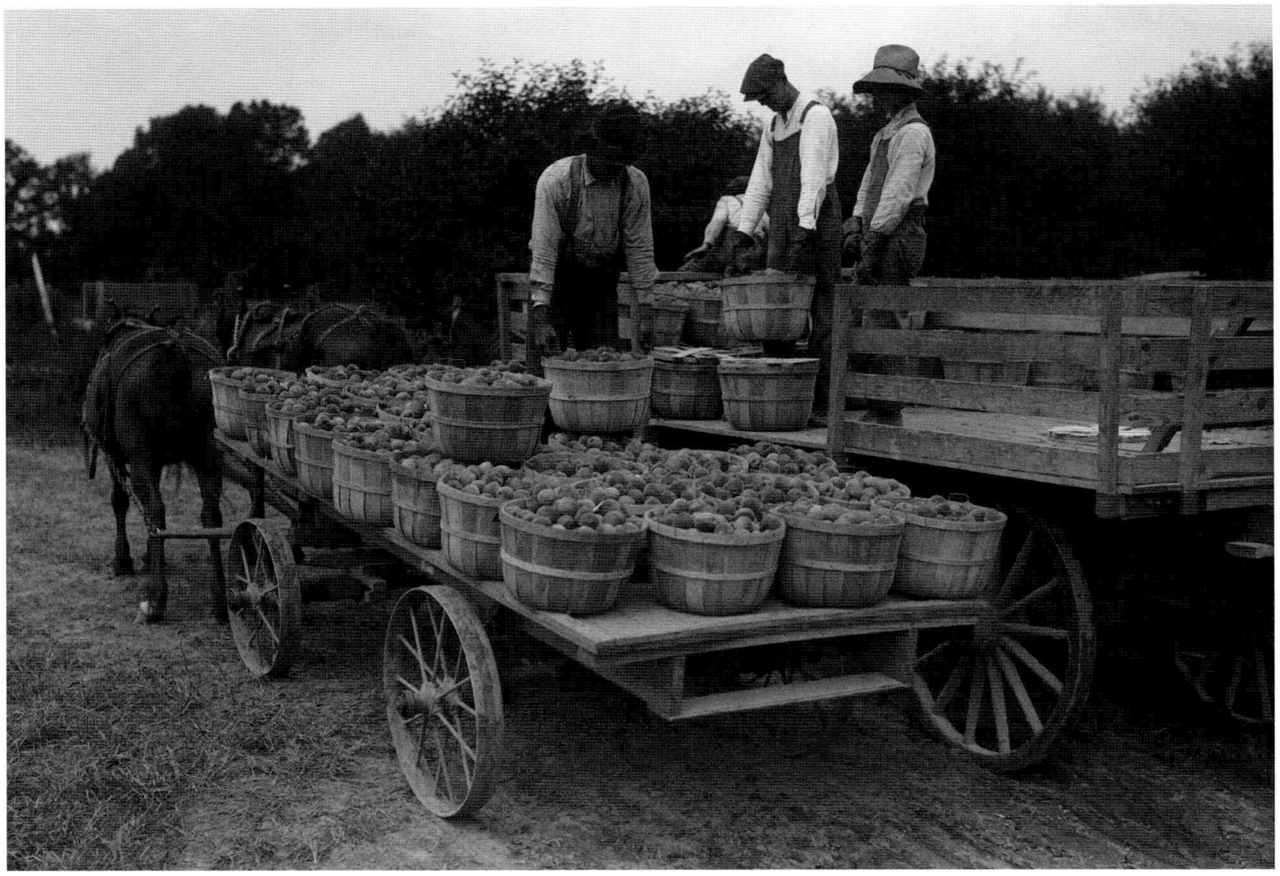

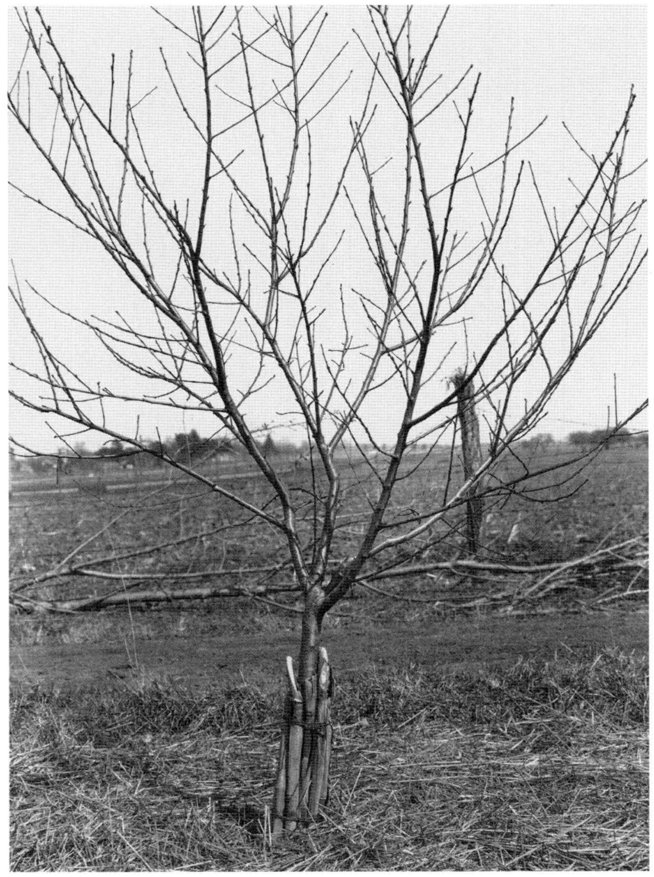

(Top left) Workers load bushel baskets of peaches from an orchard owned by N. Y. Yates of Decker onto a wagon. (Knox County, 1920)

(Top right) T. J. Gwin uses cornstalks tied around the trunk of a young peach tree to protect it from rabbits and mice on his farm in Lafayette. (Tippecanoe County, 1930)

(Bottom) W. S. Miller cultivates a young orchard. Miller was a member of the Geetingsville Spray Ring, a cooperative of orchard owners who paid to purchase and share spraying equipment and materials among their operations. (Clinton County, 1924)

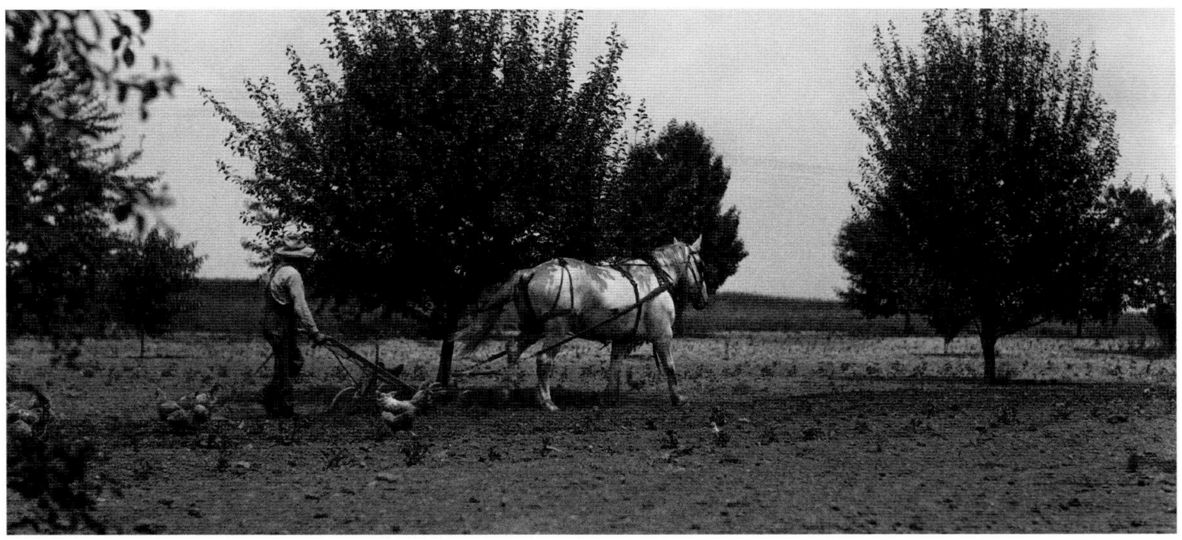

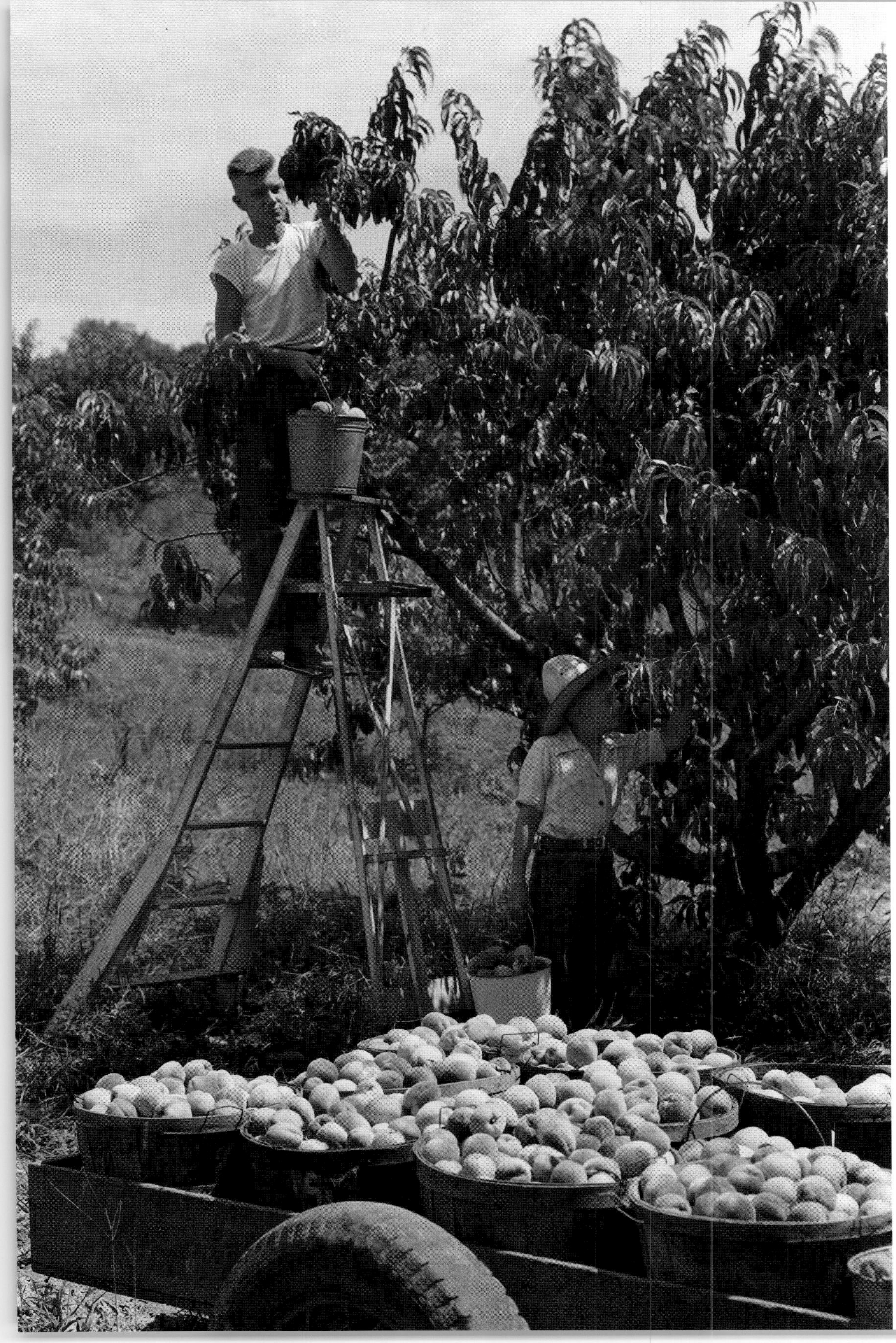

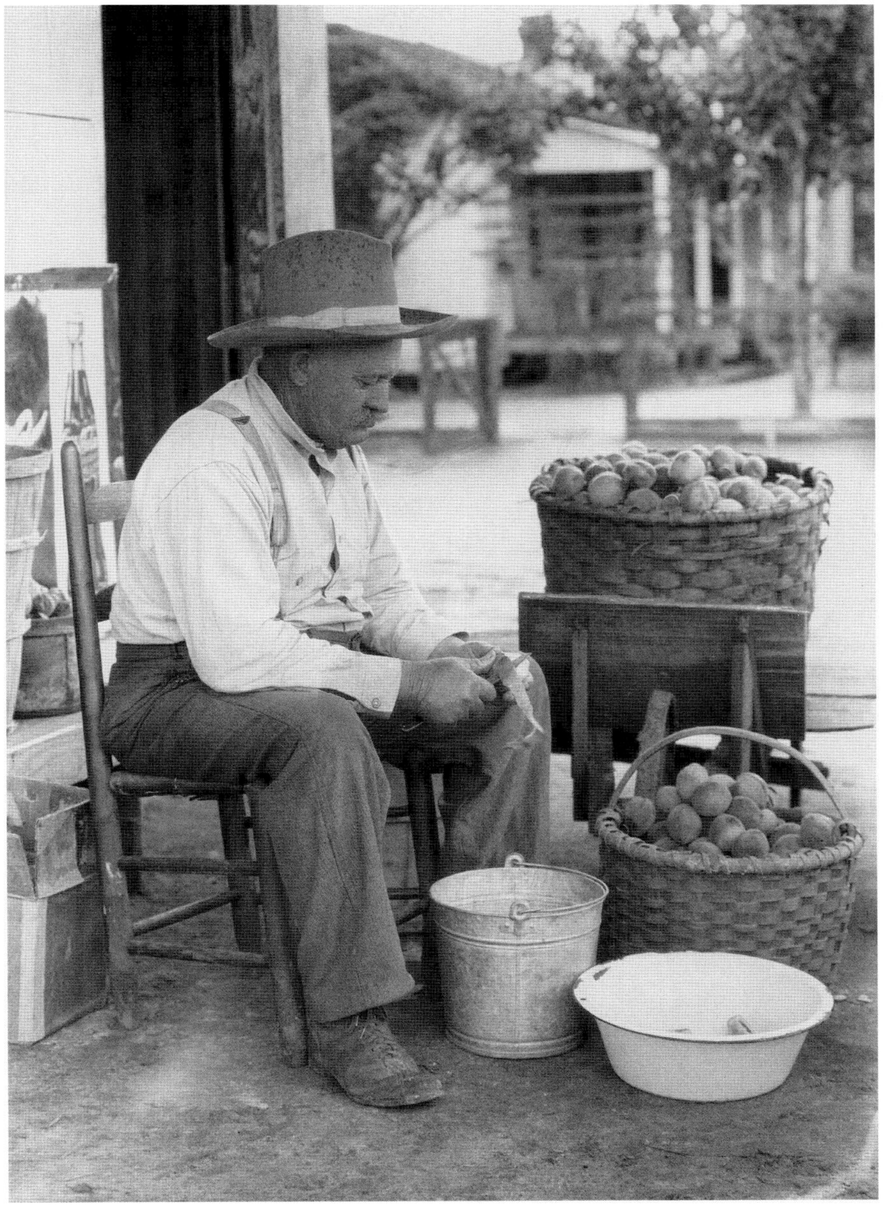

(Left) At the Fry Orchards in Lafayette, Jim (left) and John Allen pick peaches. (Tippecanoe County, 1949)

(Right) Peeling peaches for canning (Georgia, 1928)

Digging Up Spuds for the Winter

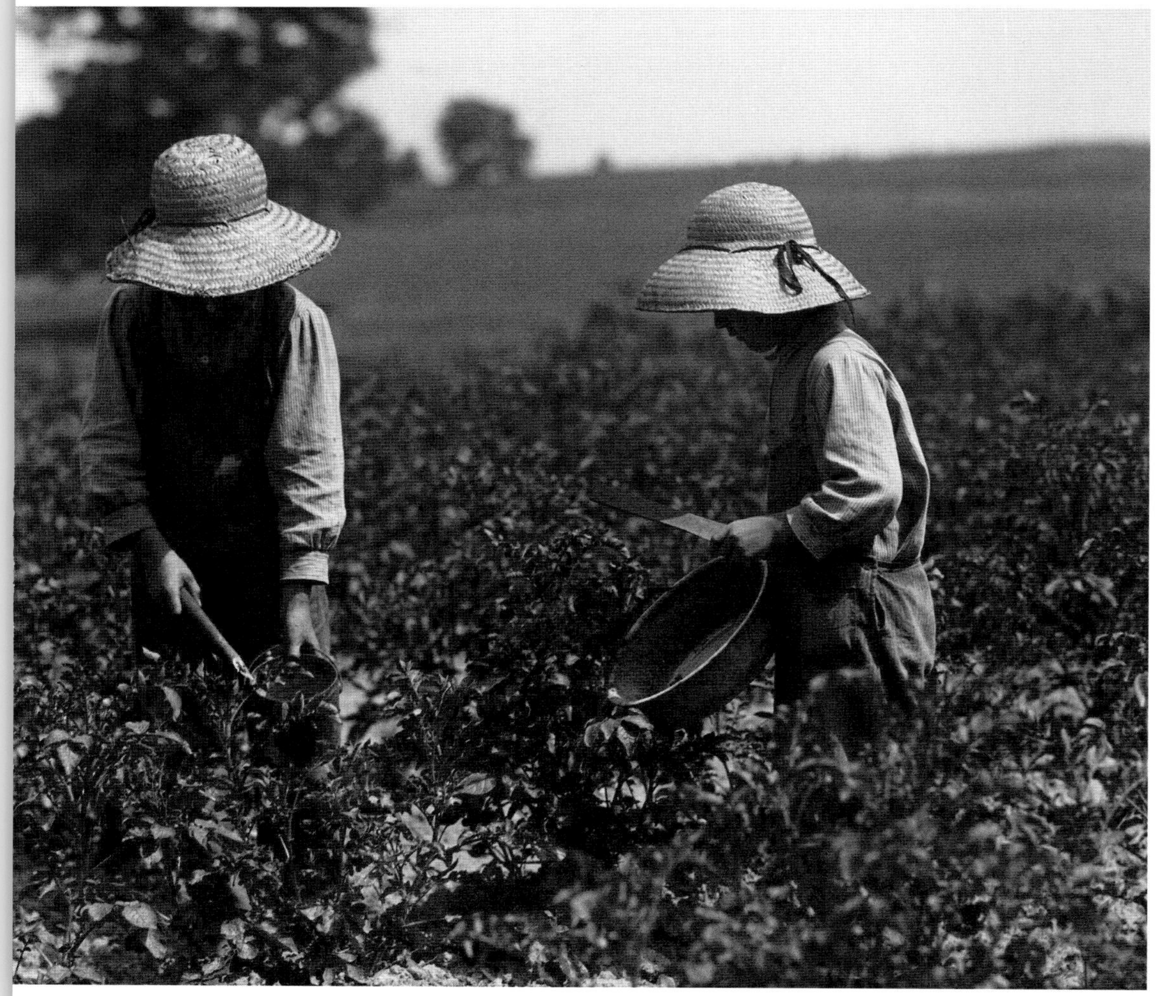

Two boys from the Goshorn family remove insects from a potato crop in Clay City. (Clay County, ca. 1915)

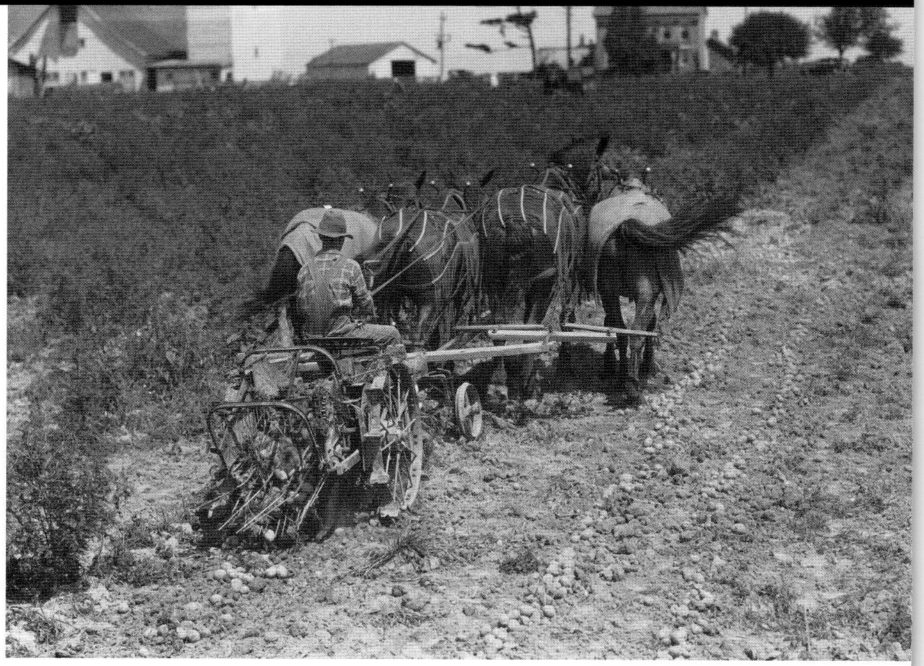

A horse-drawn potato digger on the O. M. Meeker farm in Newton (Newton County, 1929)

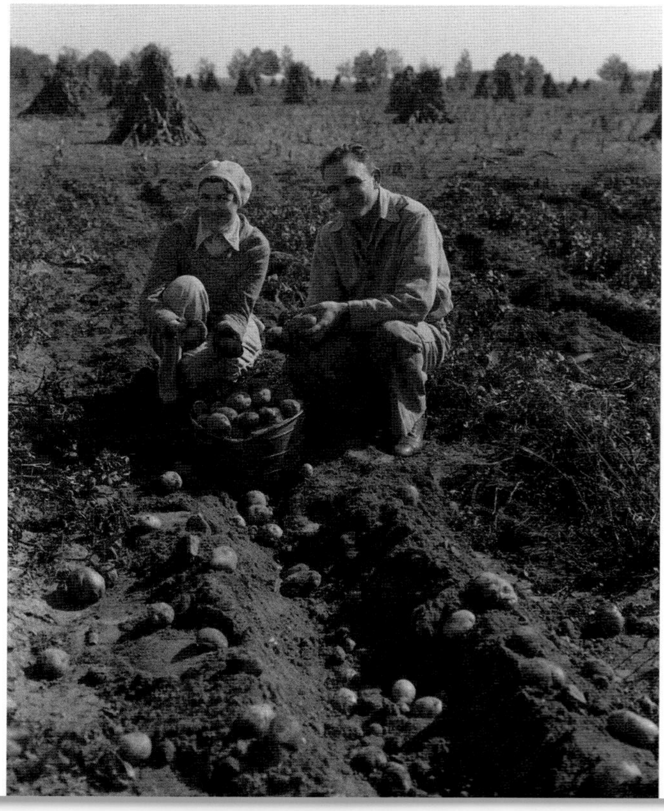

Bernice Beron (left) of North Liberty and Harold Goodling of South Bend pick potatoes. (St. Joseph County, 1934)

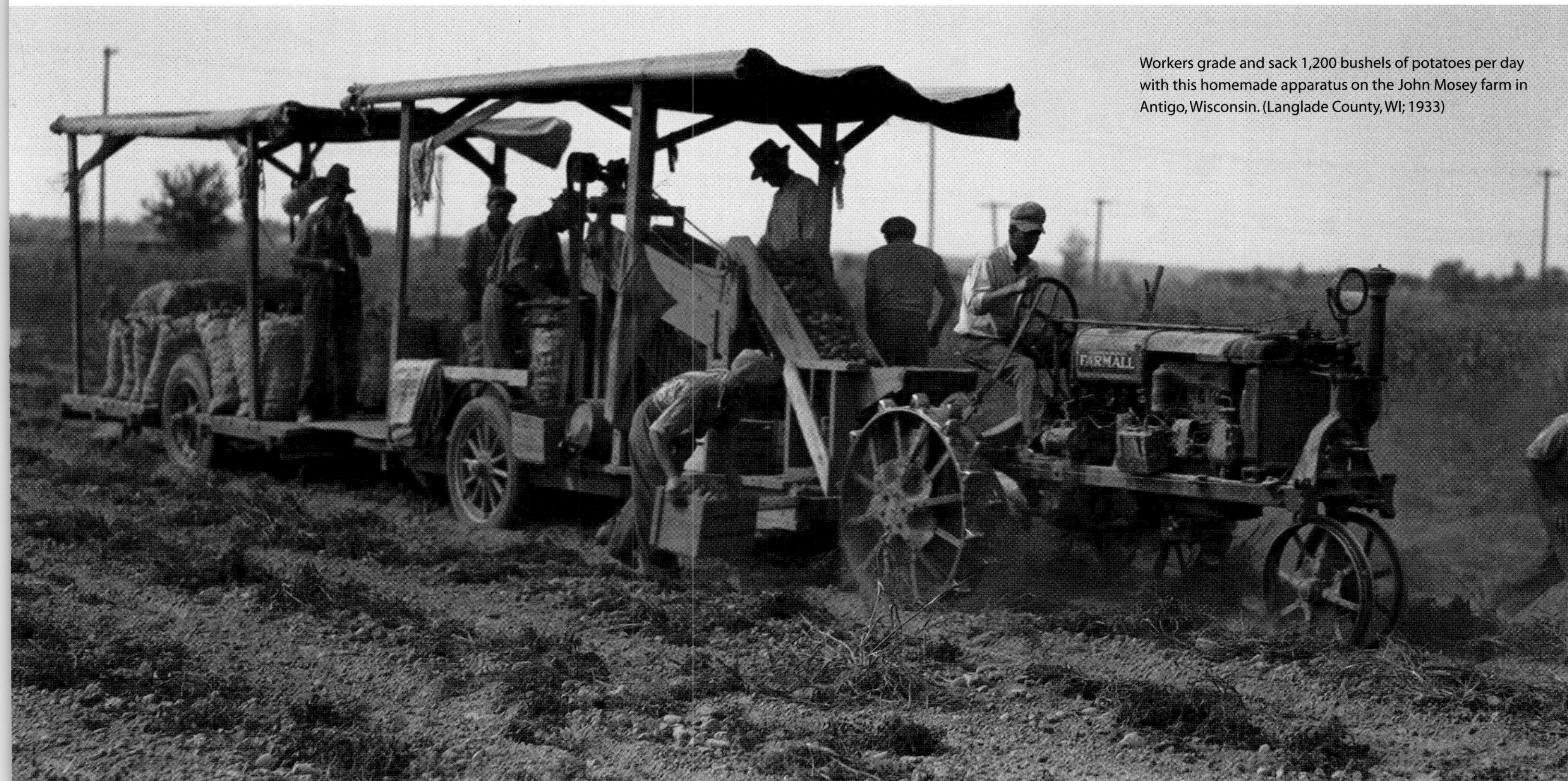

Workers grade and sack 1,200 bushels of potatoes per day with this homemade apparatus on the John Mosey farm in Antigo, Wisconsin. (Langlade County, WI; 1933)

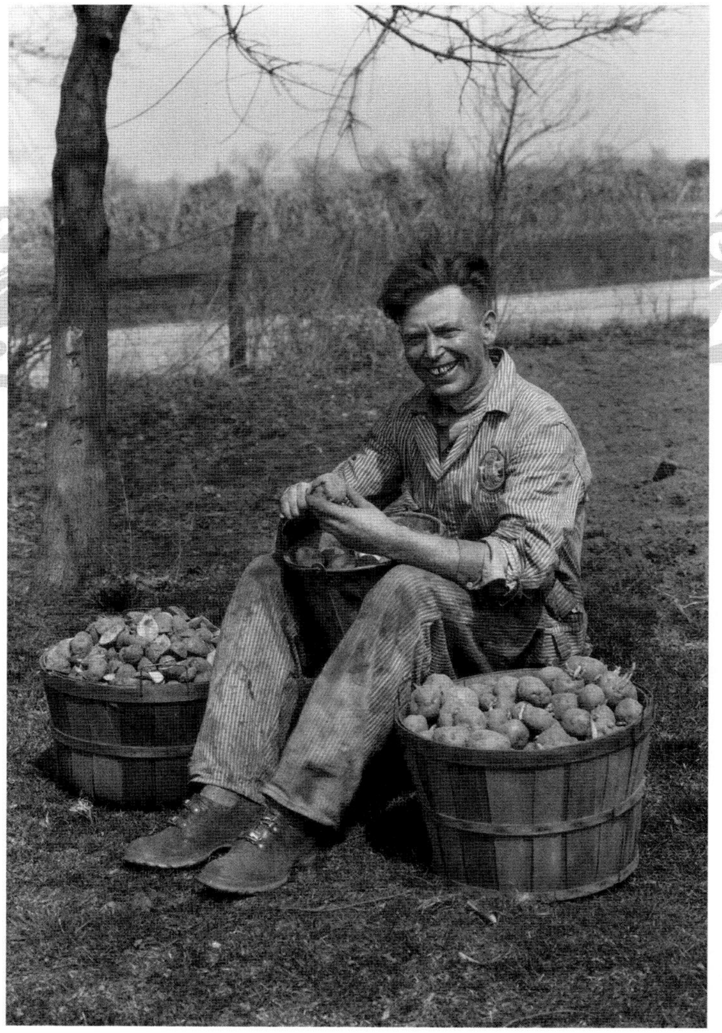

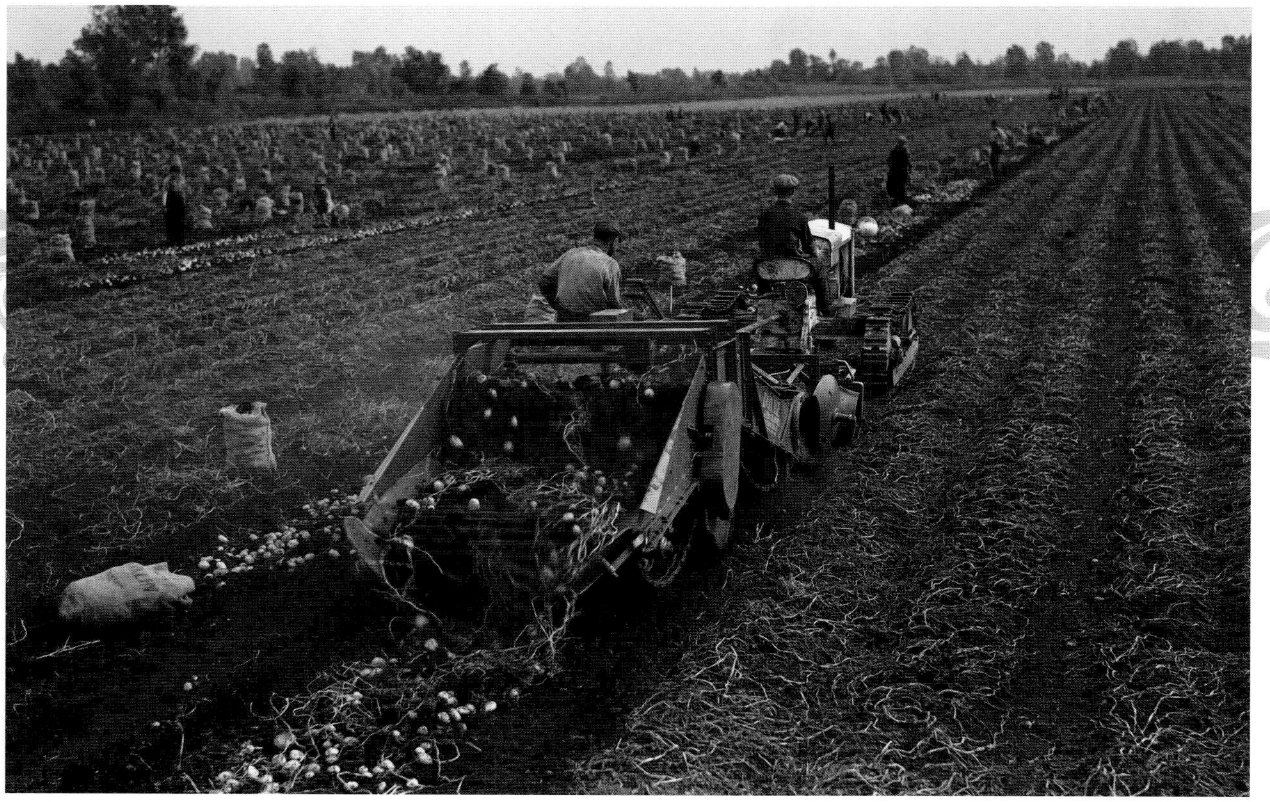

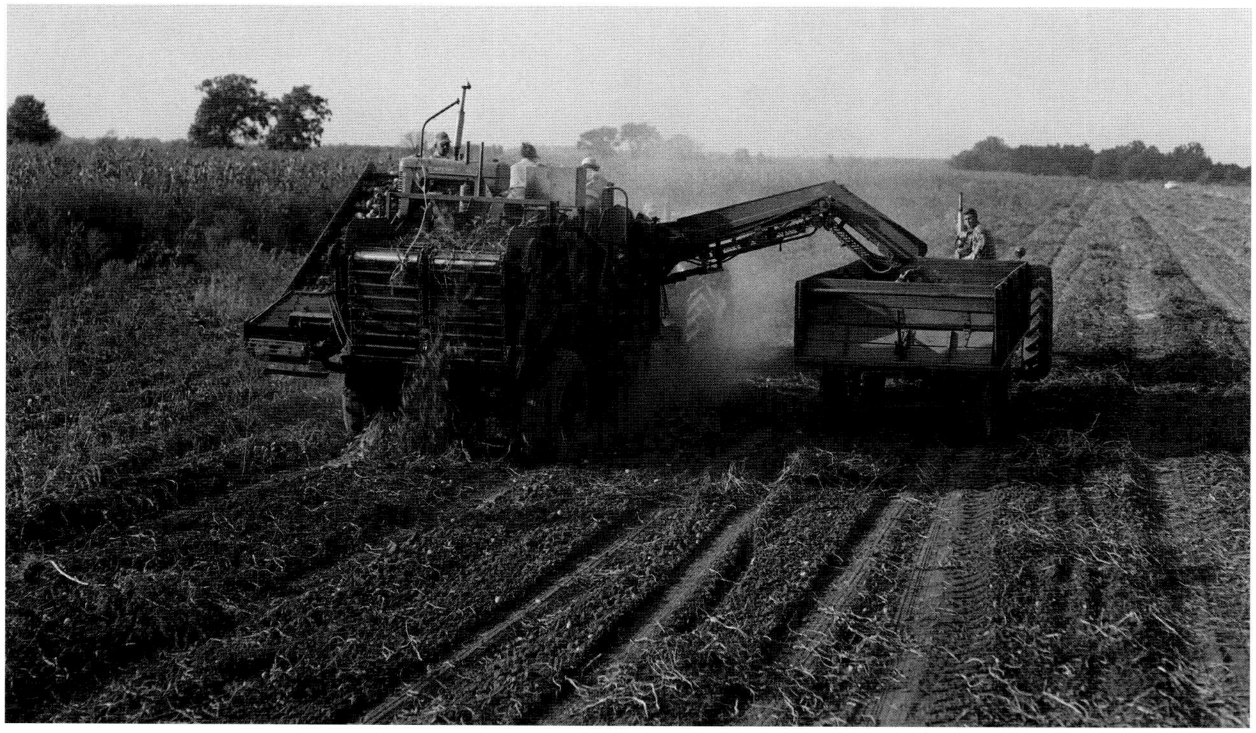

(Above) Ray Willis from Newton cuts Irish Cobbler potatoes for seed. (Newton County, 1931)

(Top right) Digging potatoes on the muck farm of William Gehring (Indiana, 1940)

(Bottom right) A work crew uses this large machine to dig, pick up, and partially clean potatoes. (Sullivan County, 1960)

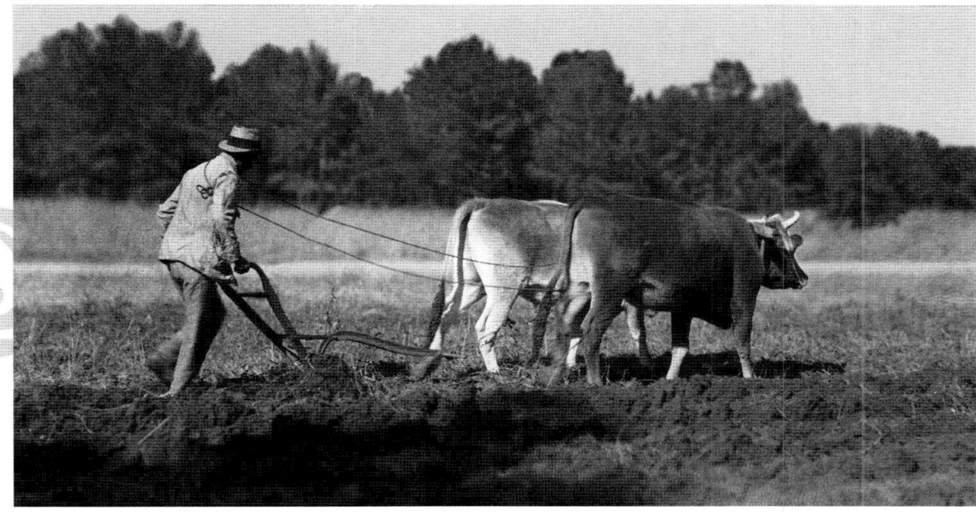

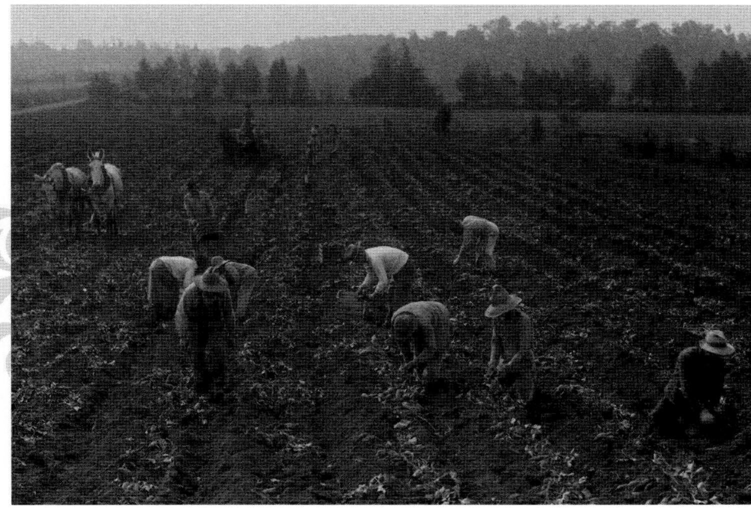

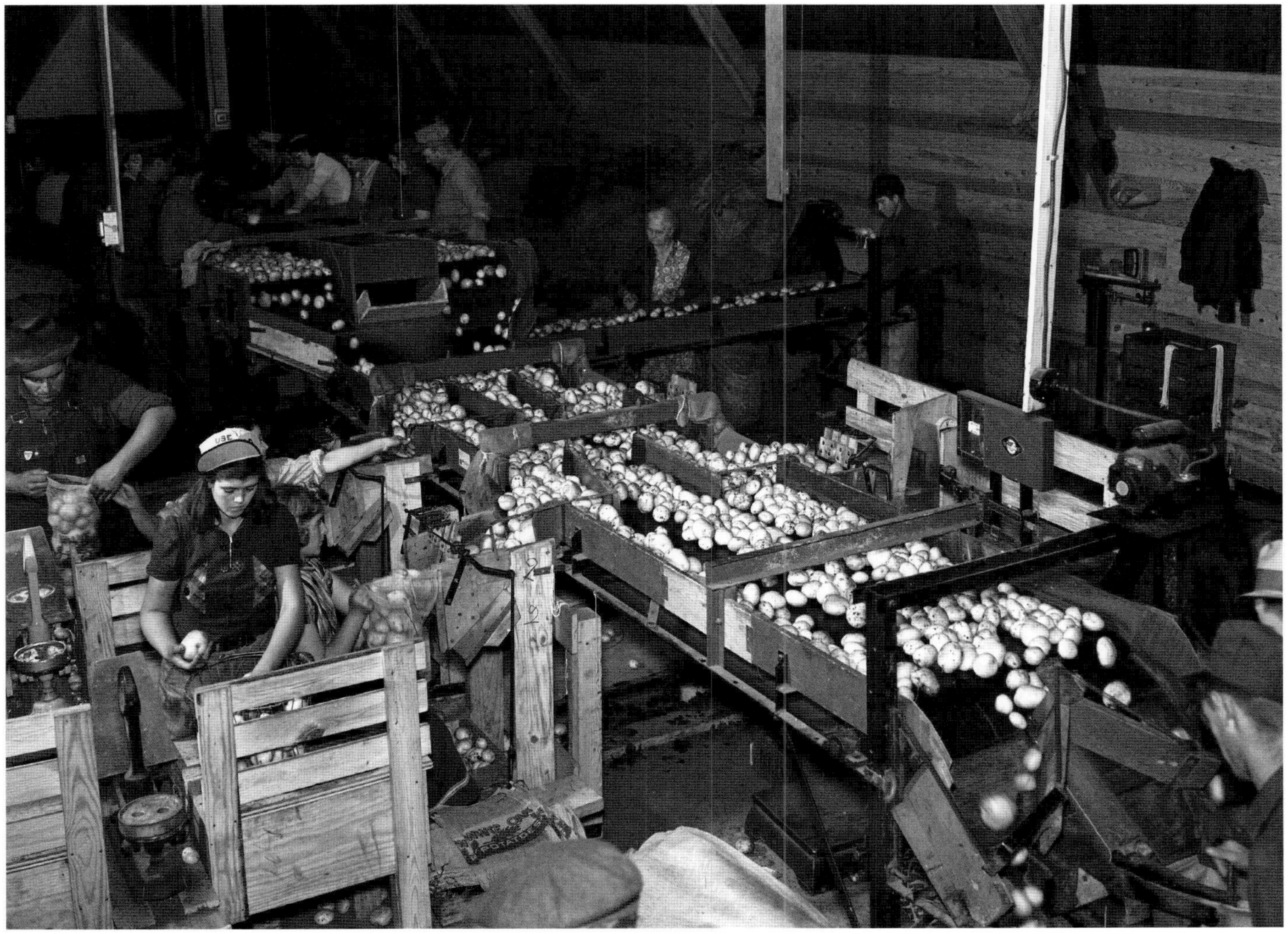

(Top left) A team of oxen pulls a plow that lifts up sweet potatoes in Tallahassee, Florida. (Leon County, FL; 1937)

(Top right) Picking up sweet potatoes in a fertilizer experiment plot on the A. R. Anthony farm in Newton, North Carolina. (Catawba County, NC; 1927)

(Bottom) Sorting, grading, and sacking potatoes in a warehouse on the farm of William Gehring (Indiana, 1940)

Red Ripe Tomatoes

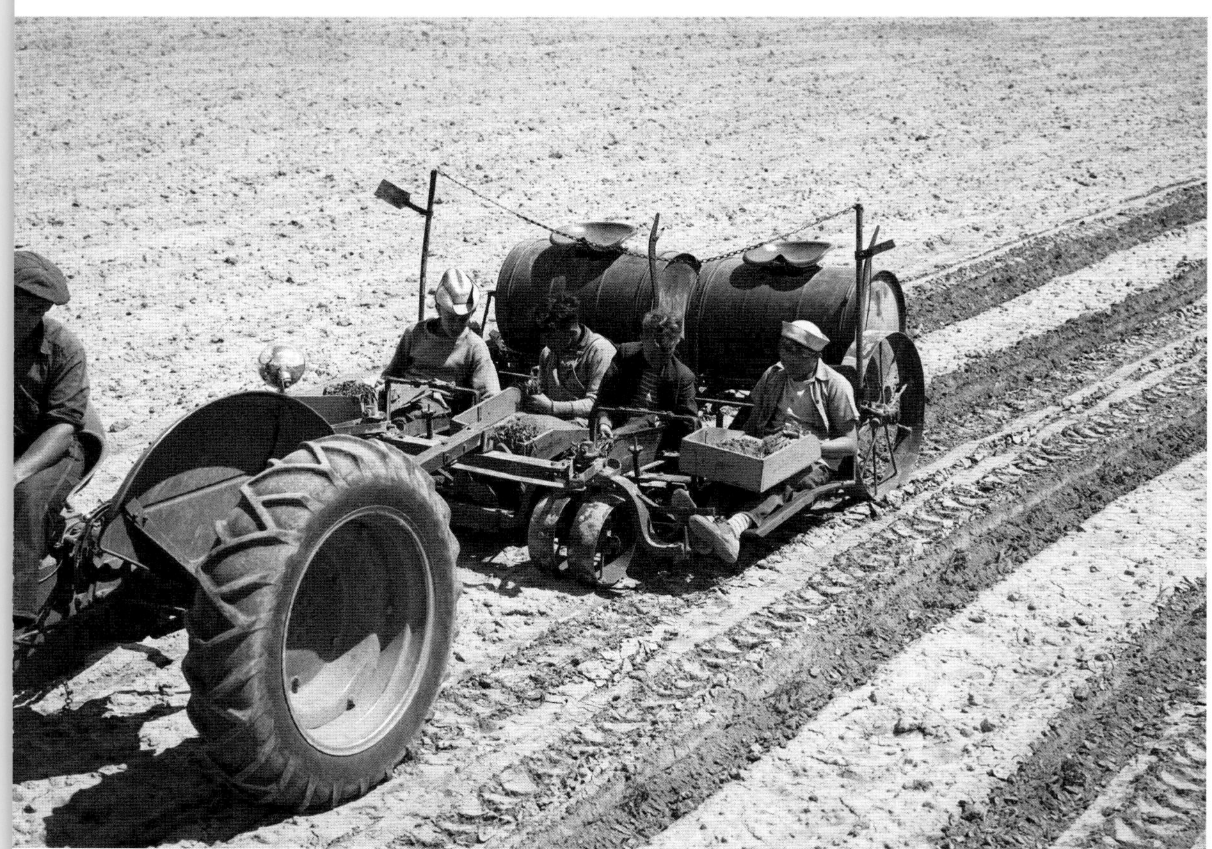

A group of young men sets tomato plants. (Clinton County, 1944)

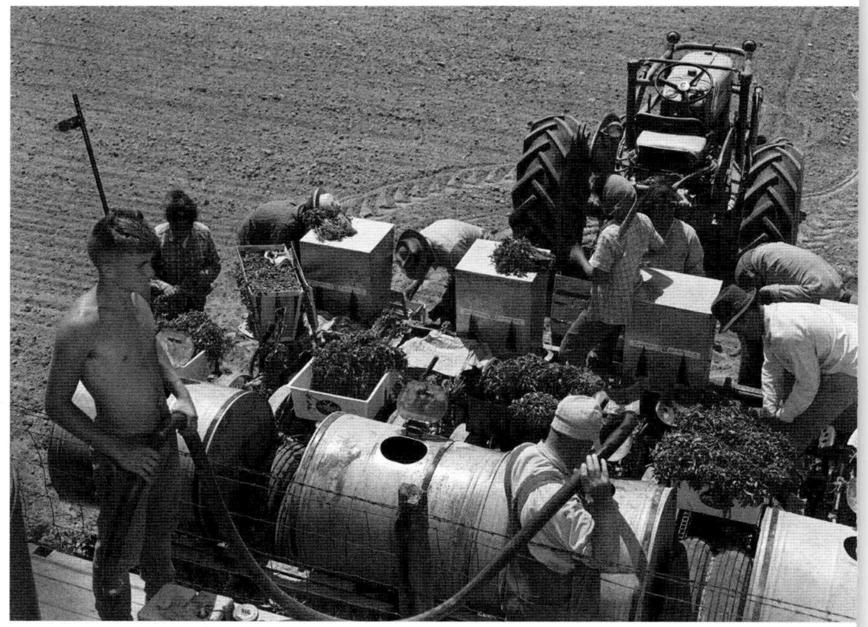

Filling the fertilizer and planter boxes of a four-row tomato planter on the Raymond Jones farm i
Flora (Carroll County, 1960)

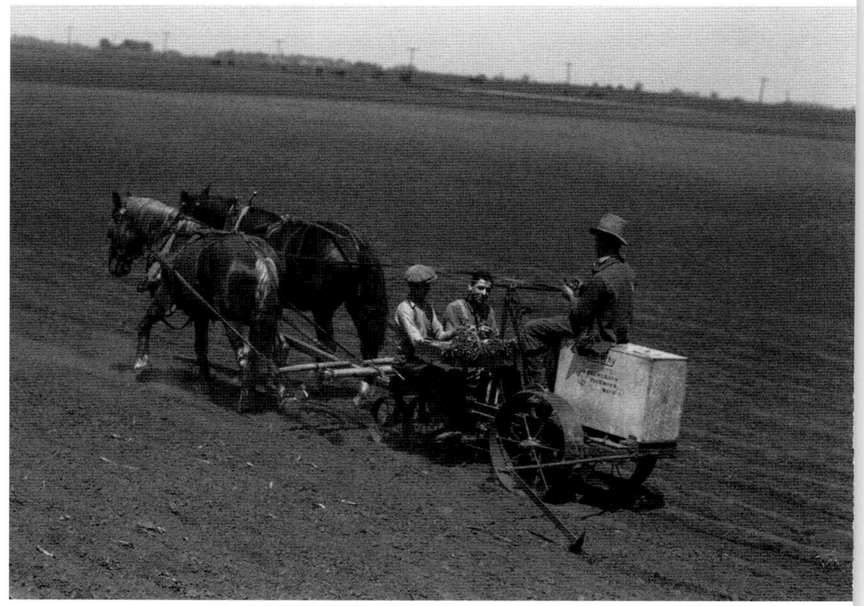

After Joseph Holscher of Fowler contracted with a canning company to grow tomatoes, these workers used a setting machine and plants provided by the company to set six to seven acres of plants per day on his land. Under the agreement, Holscher raised 20 acres of tomatoes, which he fertilized with 2-12-6 fertilizer applied at 300 pounds per acre. (Benton County, 1931)

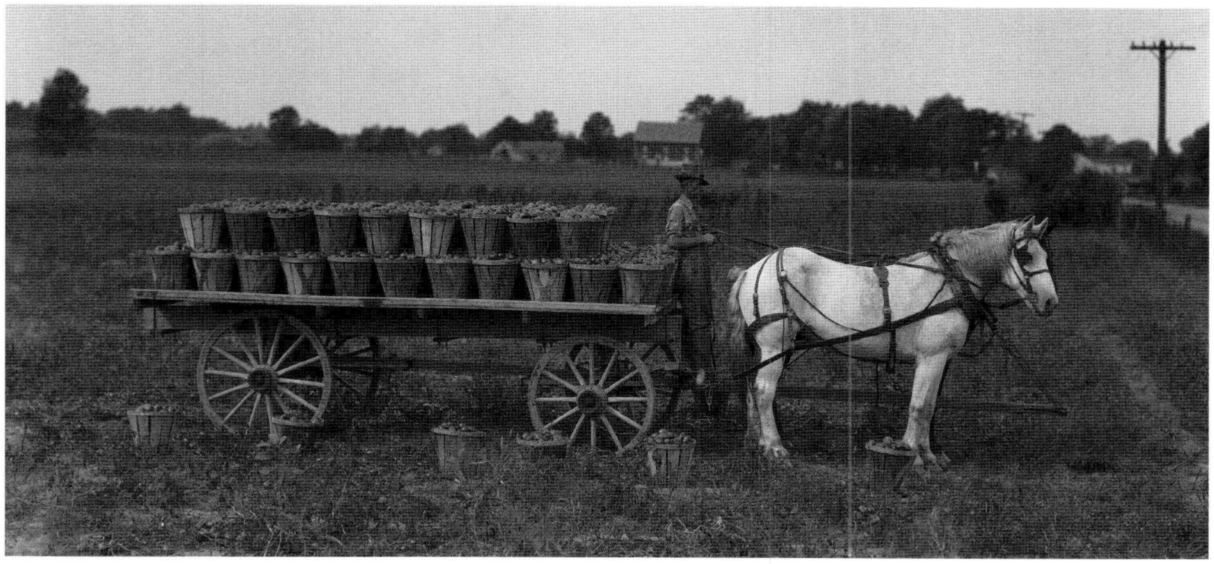

Burn Brewer brings in a load of Greater Baltimore tomatoes from his nine-acre field in Fairmount. (Grant County, 1921)

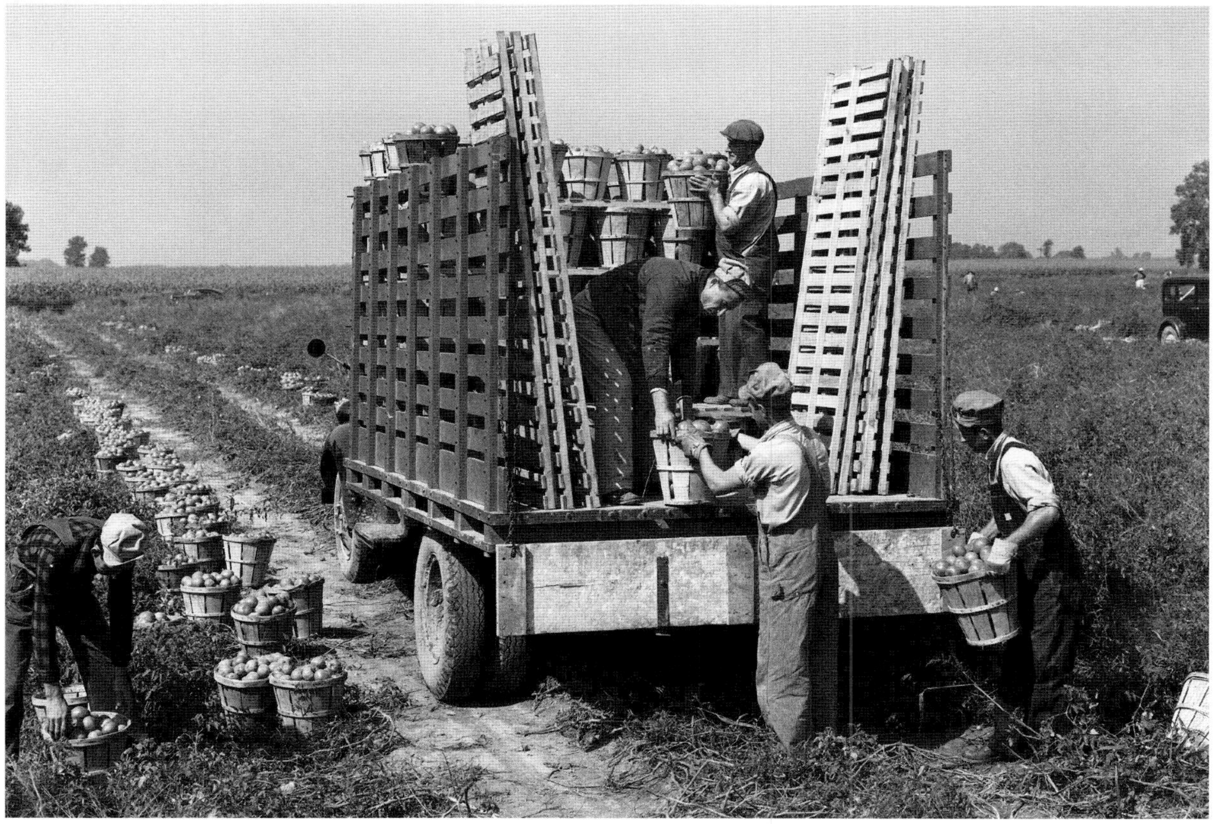

On the farm of Lloyd Shrock in Kokomo, workers load tomatoes into a truck. (Howard County, 1945)

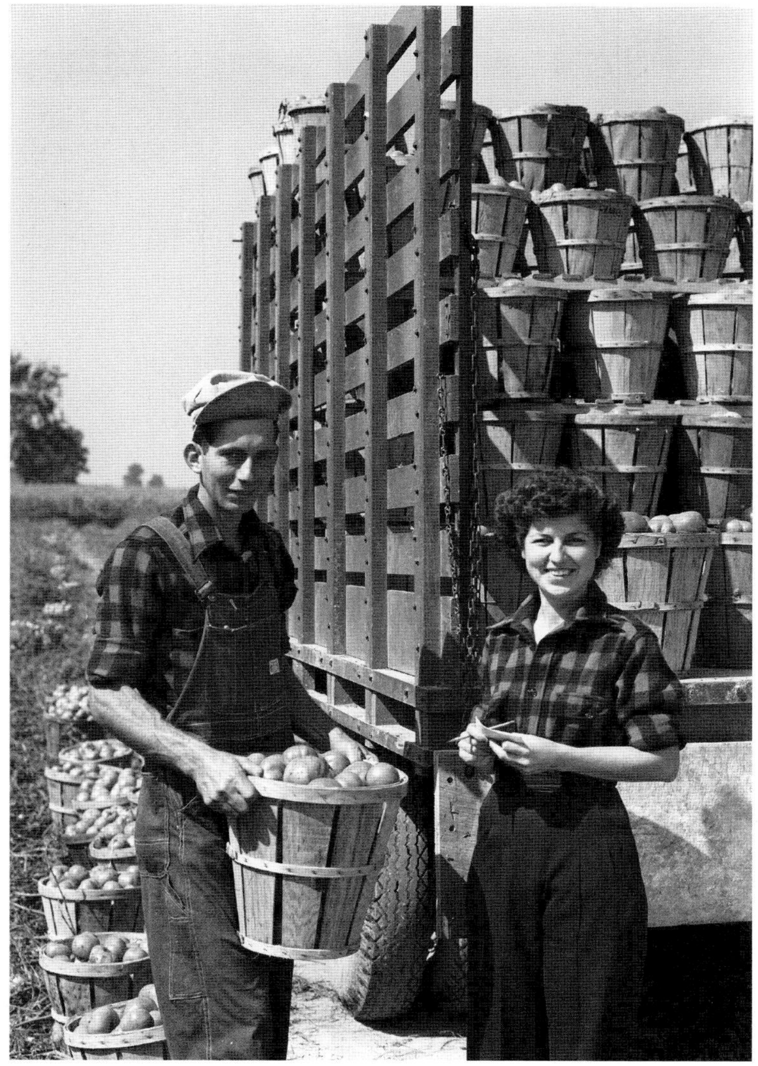

The tomatoes from Lloyd Shrock's Kokomo-area farm are prepared for delivery to a local canning factory. (Howard County, 1945)

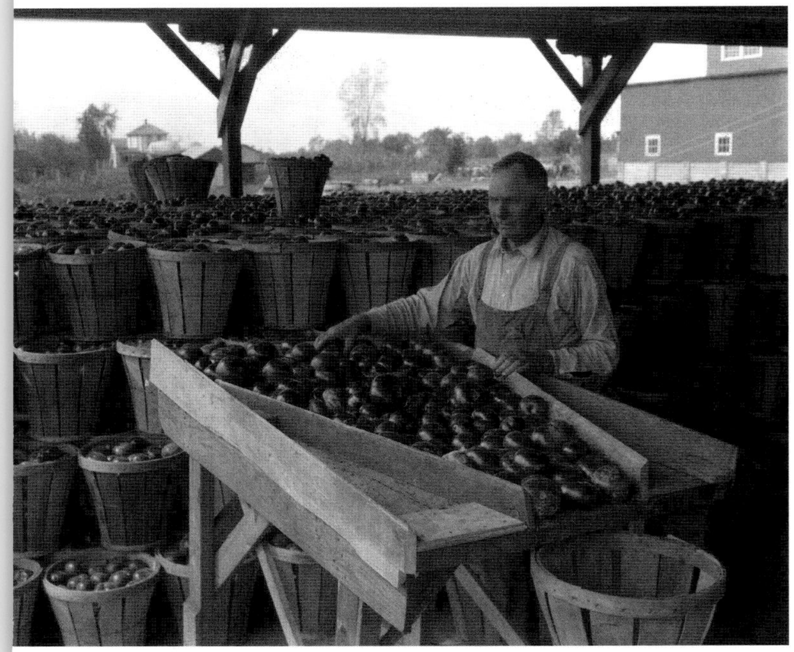

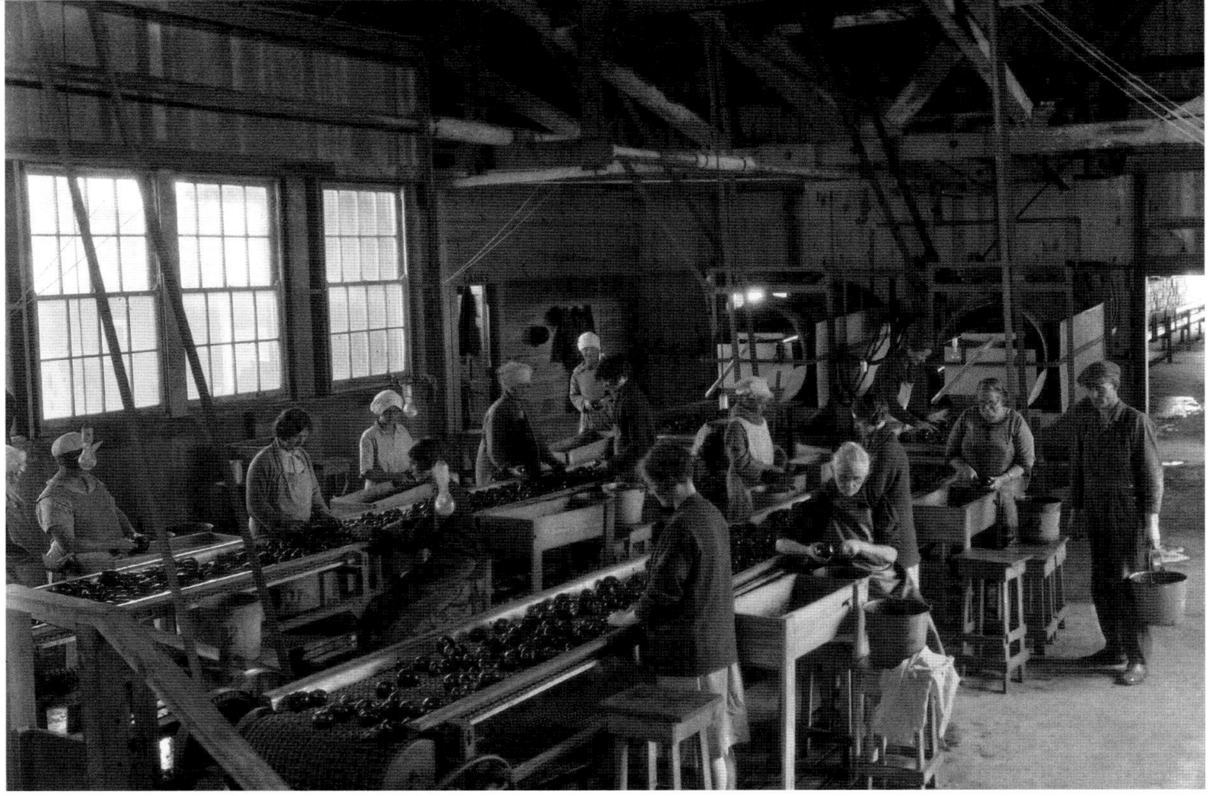

(Top left and right) At the Eaton Canning Company, workers grade and sort tomatoes. (Delaware County, 1927)

(Bottom) Workers at the Klondike Canning Corporation factory place tomatoes into cans. (Tippecanoe County, 1935)

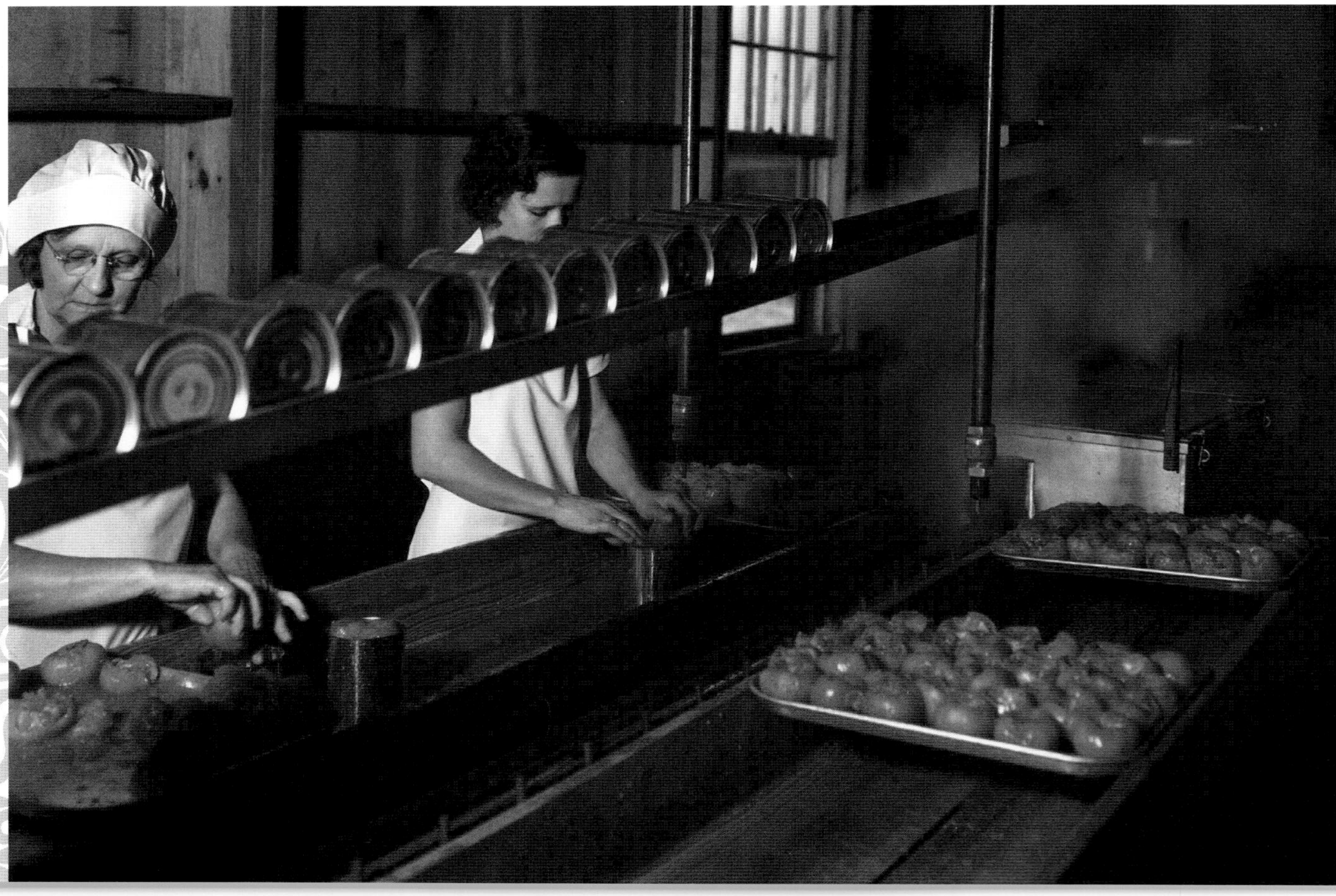

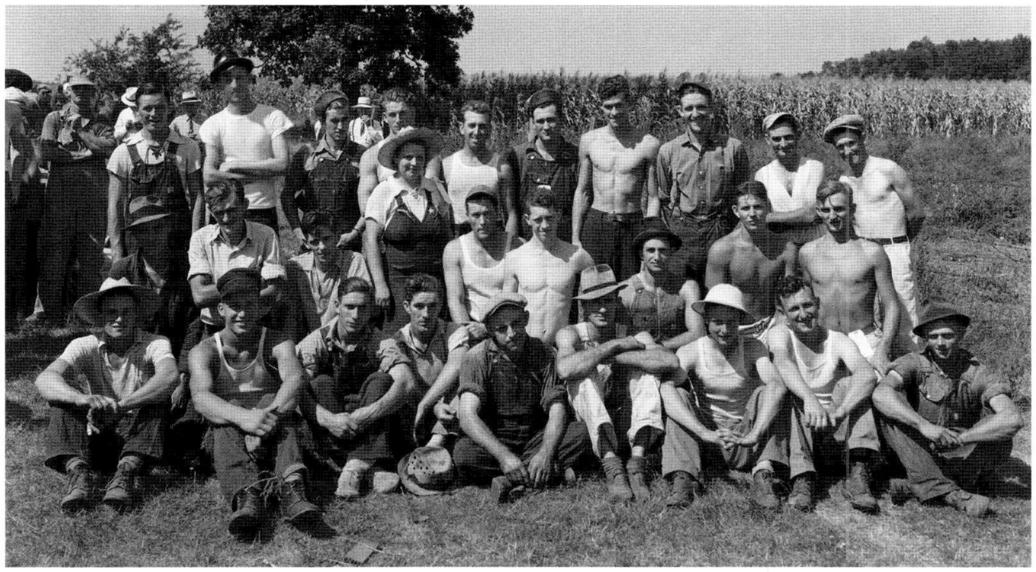

Contestants in the State Tomato Picking Contest and their helpers pose for a photo in Kempton. (Tipton County, 1939)

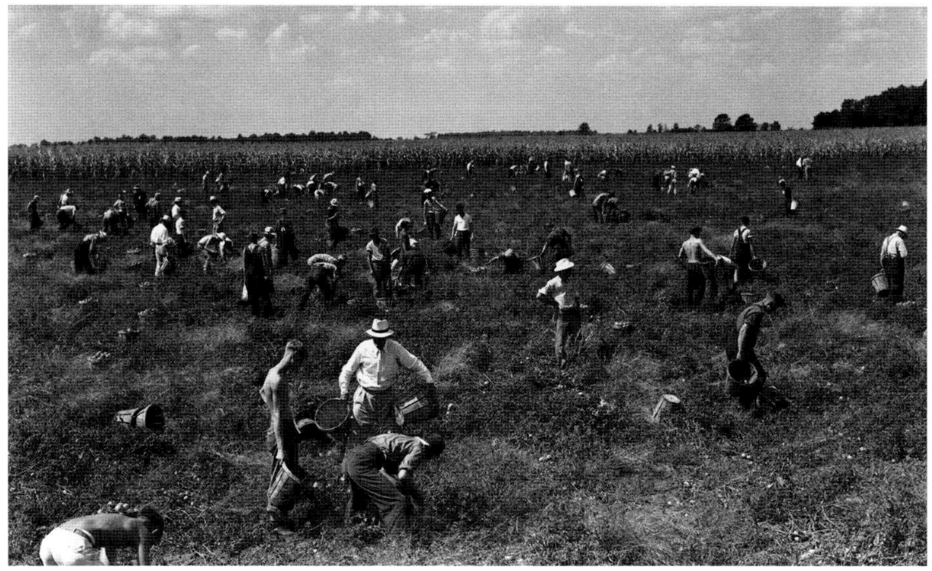

Tomato pickers in action at the Kempton contest (Tipton County, 1939)

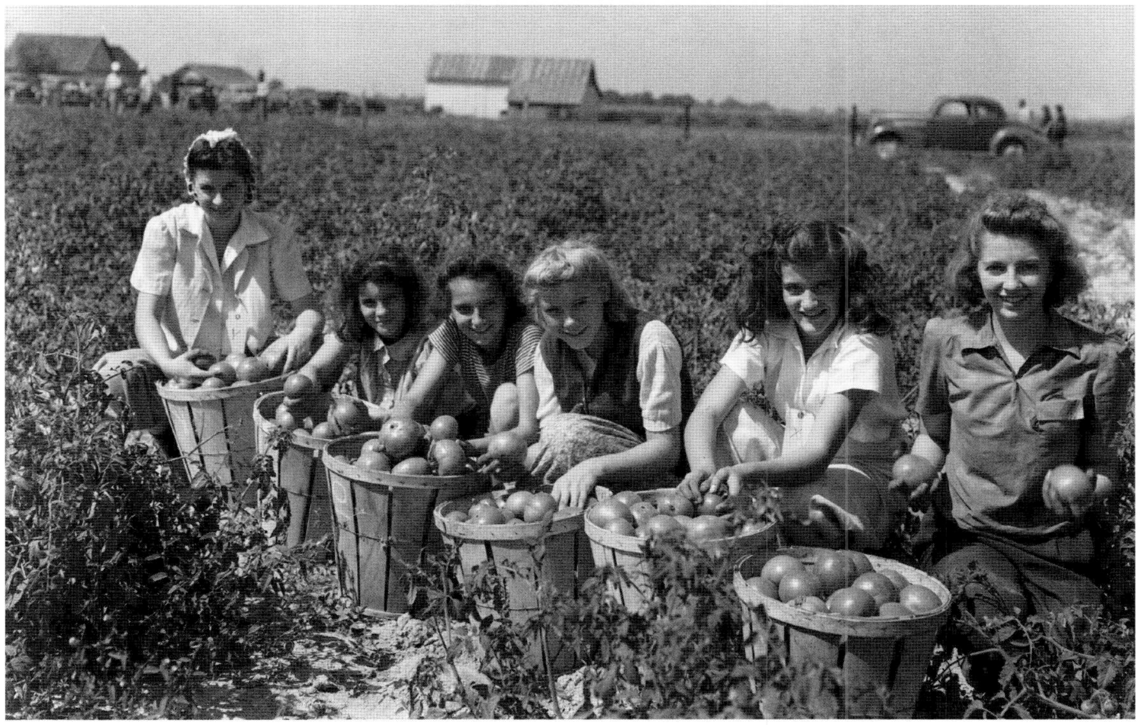

A group of women take part in the Indiana State Amateur Tomato Picking Contest held in Vigo County in connection with the State Tomato Festival. Participants shown here include (from left to right) Gertrude Ellington, Marilyn Bohannon, Barbara Smith, Thelma Ollie, Betty Burge, and Betty Higgins, all of Vigo County. (1943)

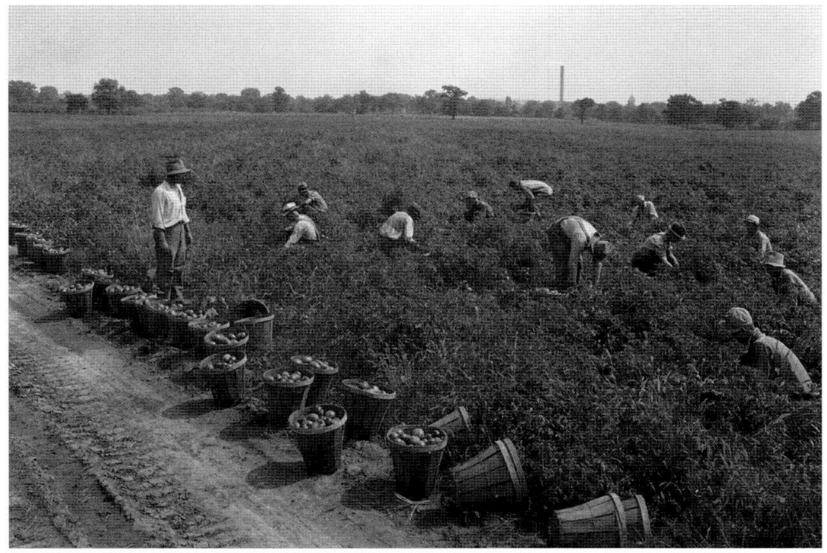

Picking a field of tomatoes on a Purdue University farm (Tippecanoe County, 1935)

Growing Diverse Crops for Market

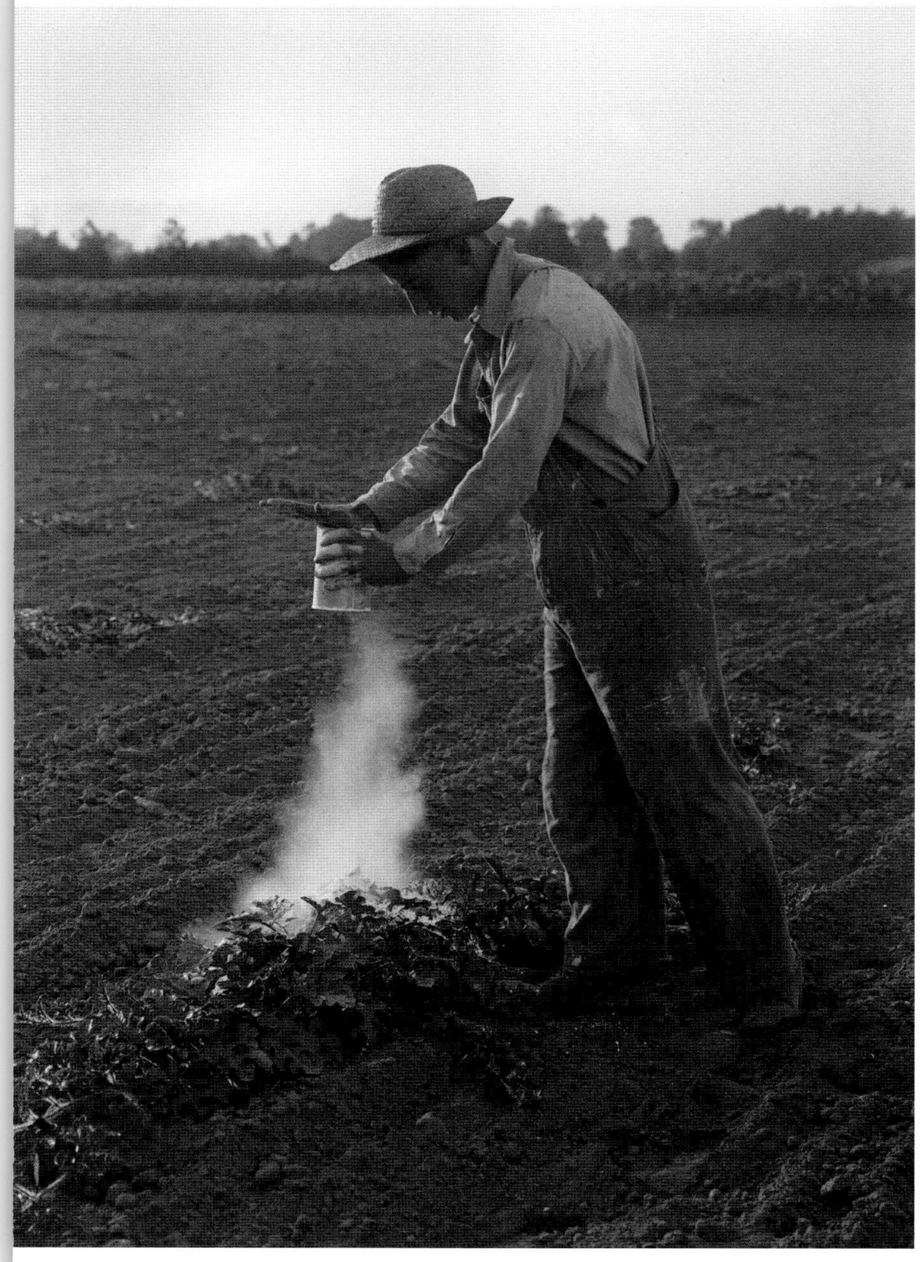

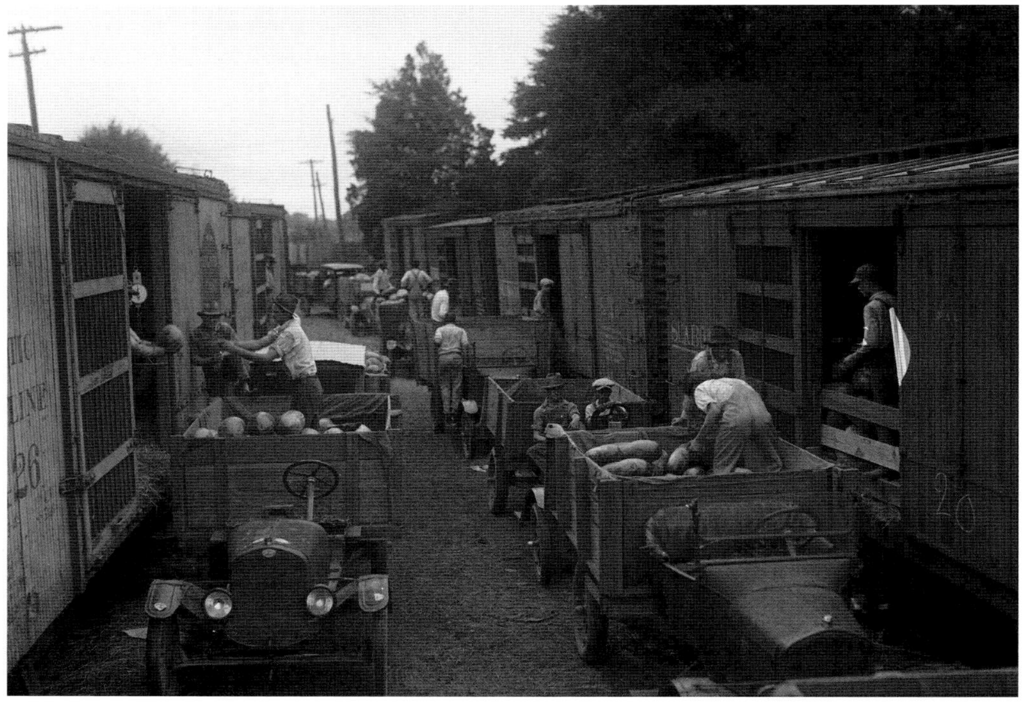

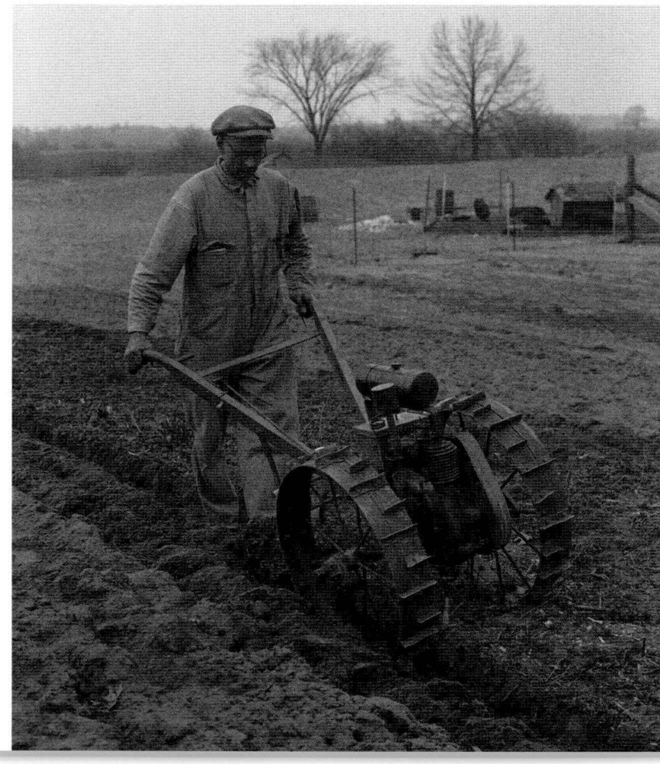

(Far left) Elwood Townsend dusts watermelon plants with insecticide. (Morgan County, 1922)

(Top right) Farmers load watermelons onto railcars for delivery to distant markets. (Georgia, 1928)

(Bottom right) H. W. Fitting of Lafayette plows his truck farm with a garden tractor made by the Walsh Tractor Company of Minneapolis, Minnesota. (Tippecanoe County, 1930)

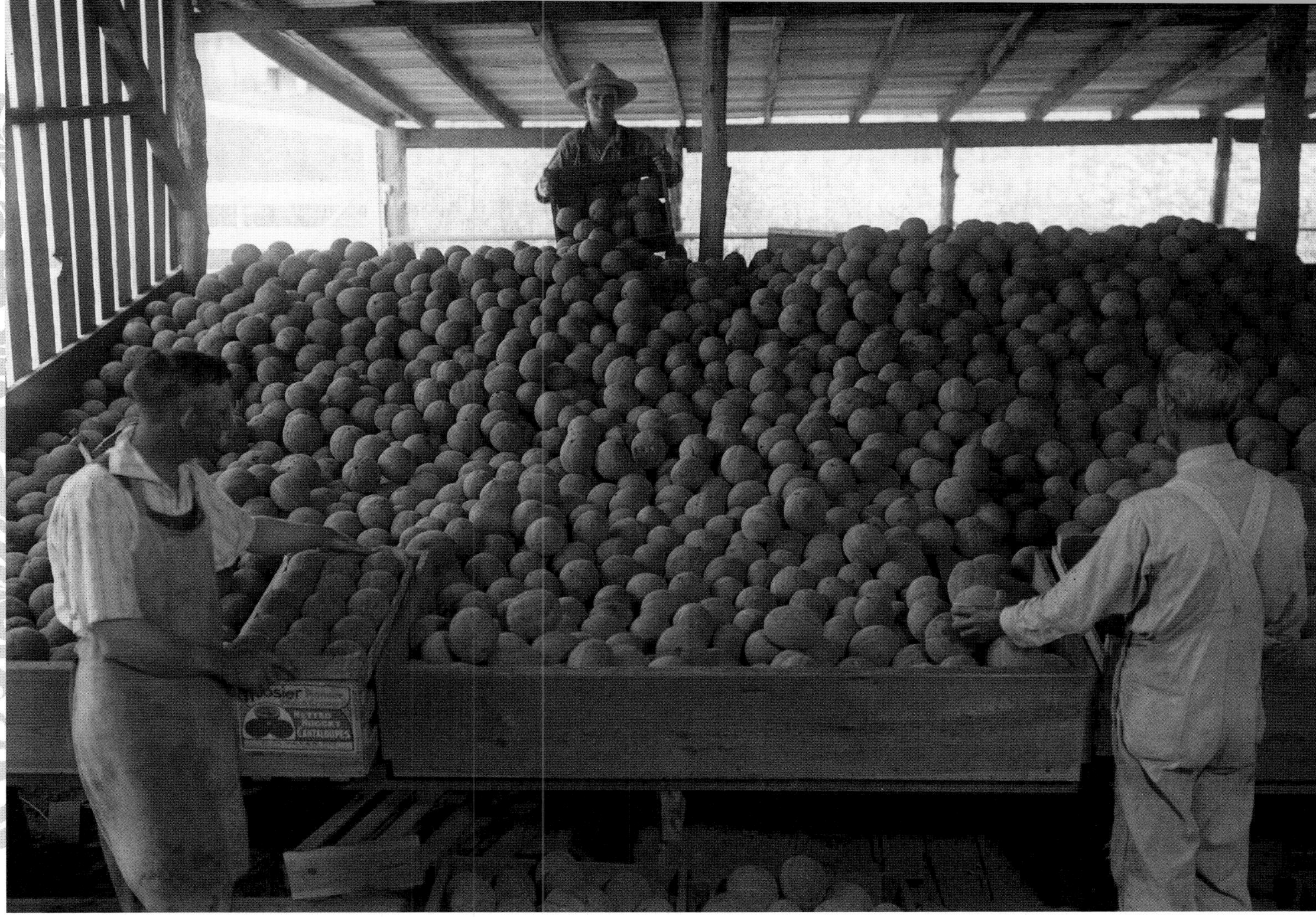

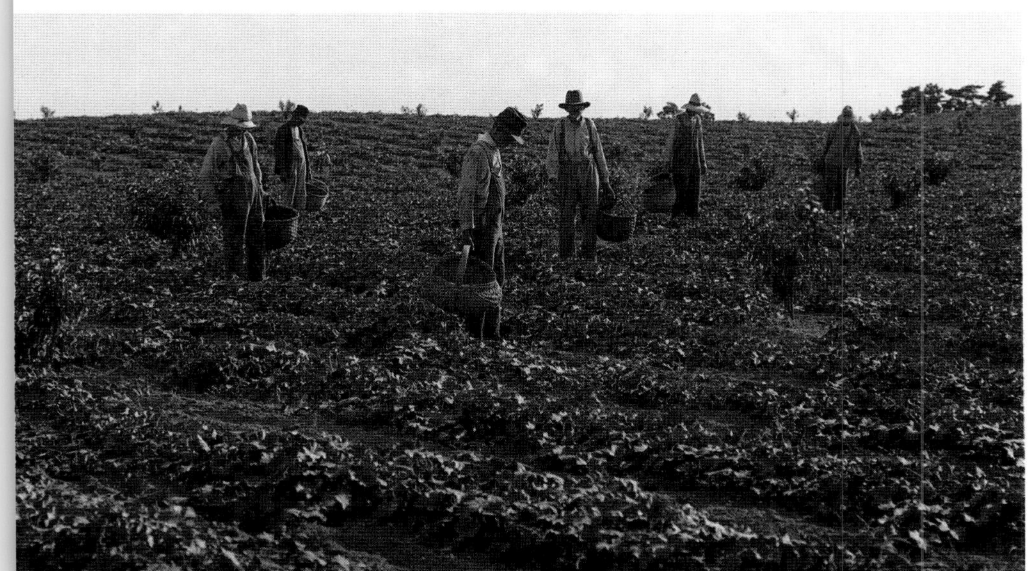

(Top) Workers box cantaloupes in a packing shed at the C. M. Yates farm in Decker. (Knox County, 1920)

(Bottom) Picking cantaloupes on the Lane Farm in Decker (Knox County, 1917)

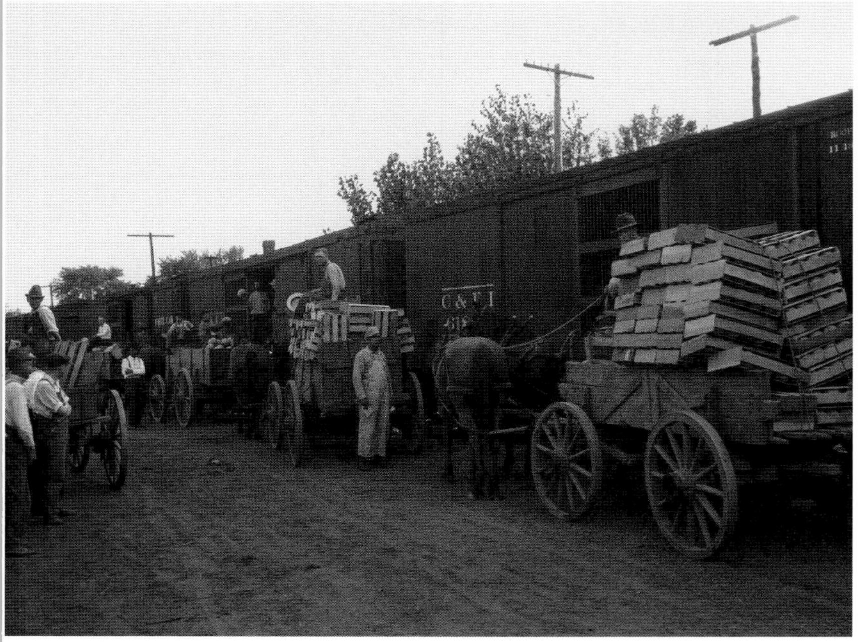

Horse-drawn wagons deliver cases of cantaloupes to the rail yard as part of a cooperative shipping program in Decker. (Knox County, 1922)

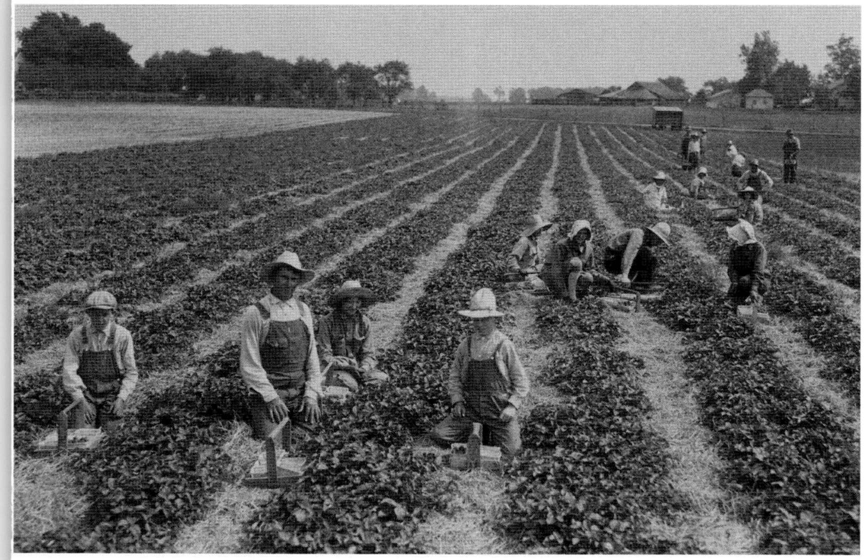

Handpicking strawberries (Indiana, 1931)

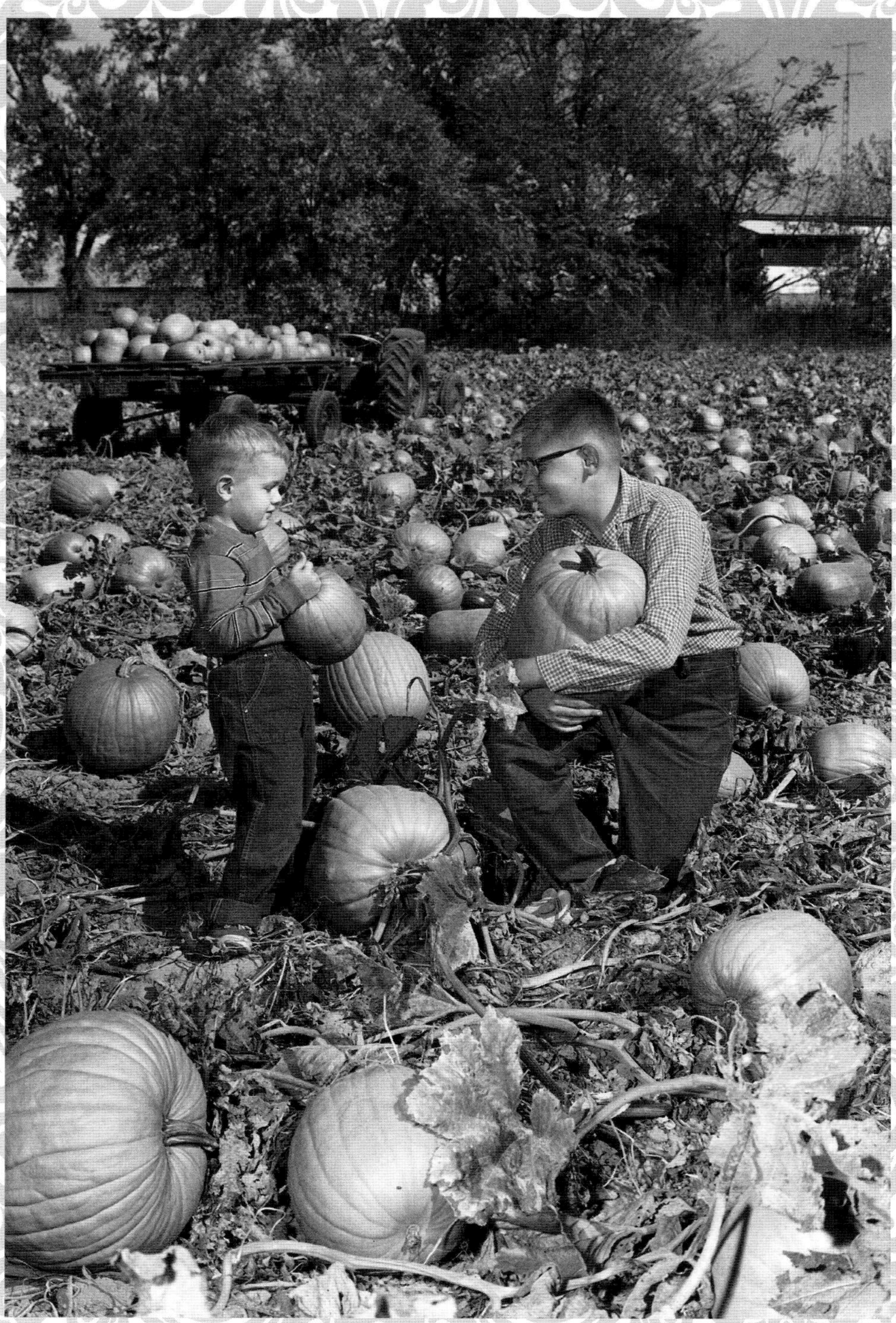

Edward Allen (right) and Joe Vandermark Jr. select the perfect pumpkin on a farm owned by Joe's father. (Tippecanoe County, 1963)

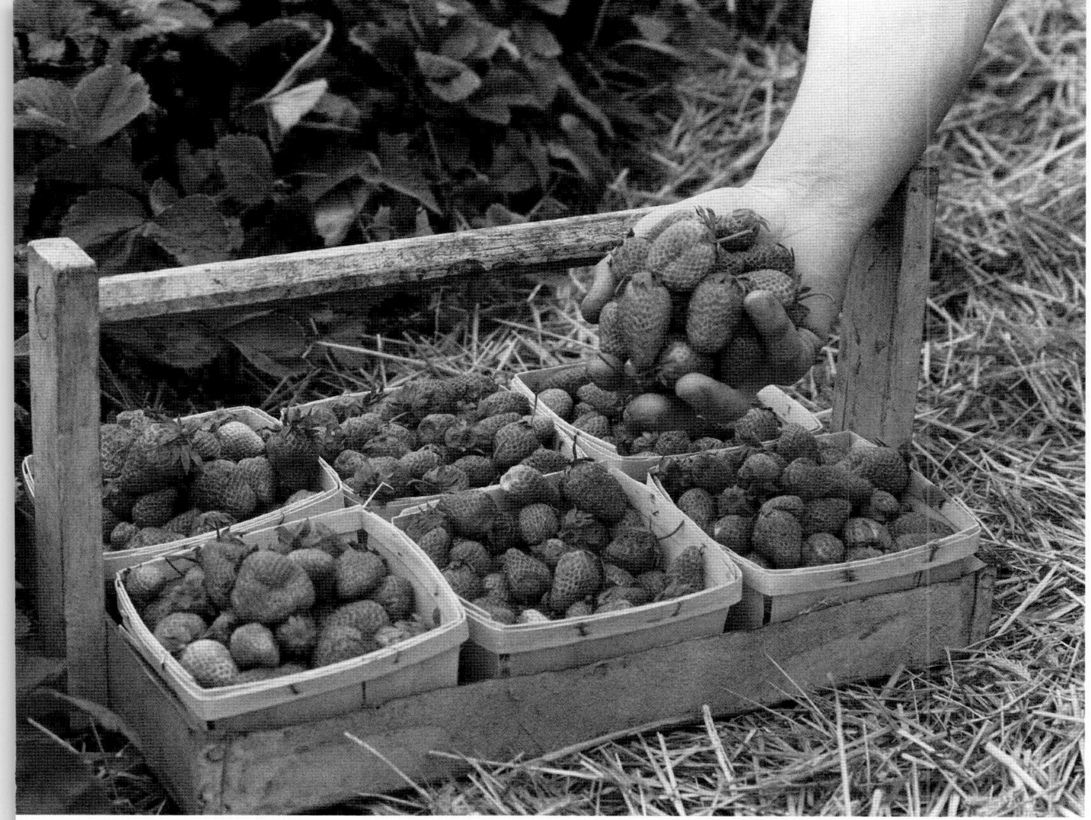

Freshly picked strawberries in a carrier (Indiana, 1931)

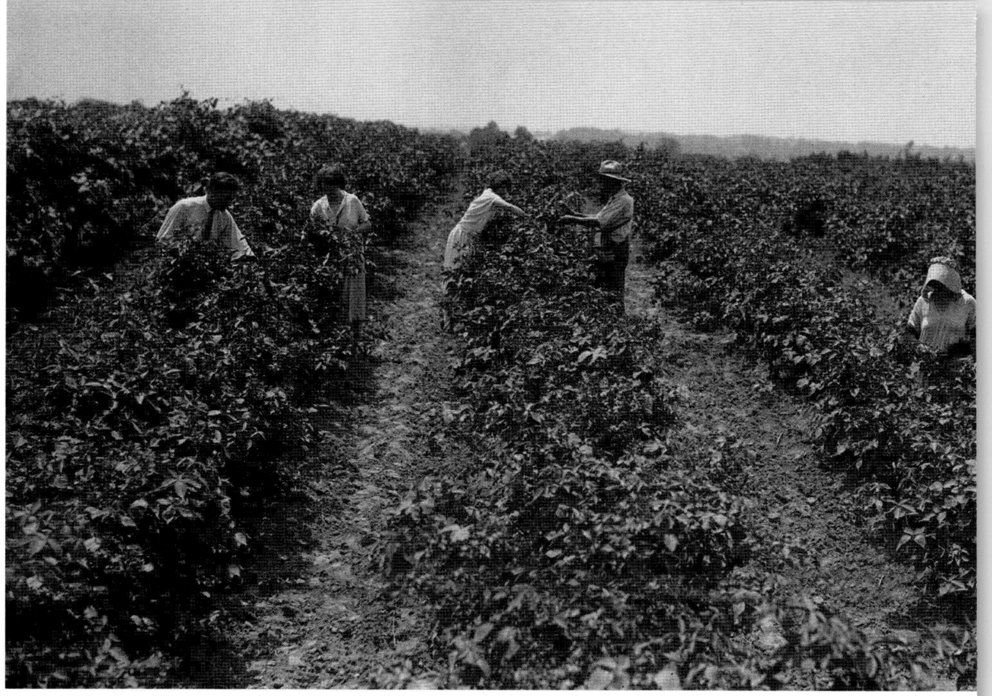

A crew picks early blackberries on the Dan Moore farm in Evansville. Allen wrote, "Dan has already picked and sold 60 gallons from this patch, which is only a fraction of an acre, and they brought an average of 65 cents per gallon. He will have about 60 gallons more before the season closes, but the late ones will not sell quite so high." (Vanderburgh County, 1925)

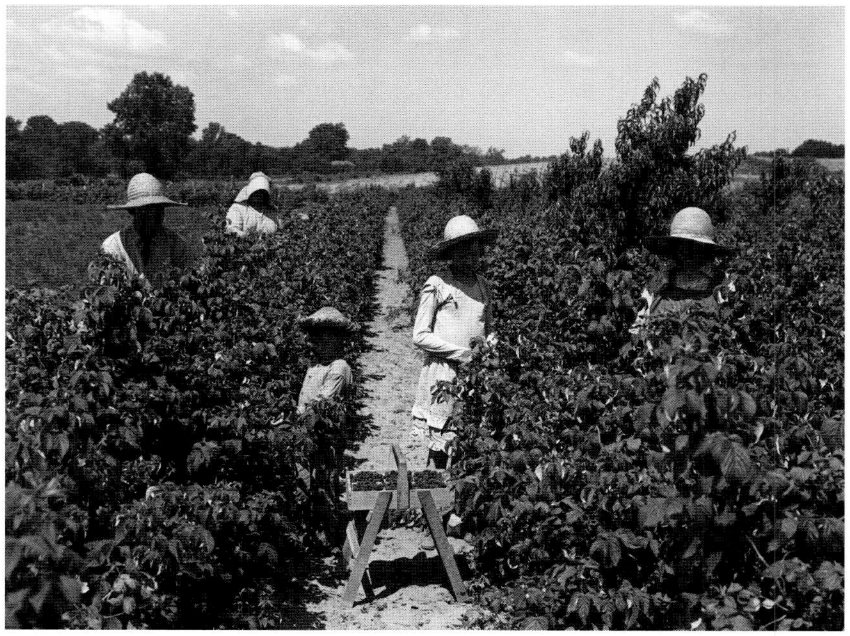

Picking red raspberries during harvest on the H. C. Myers farm (Tippecanoe County, 1927)

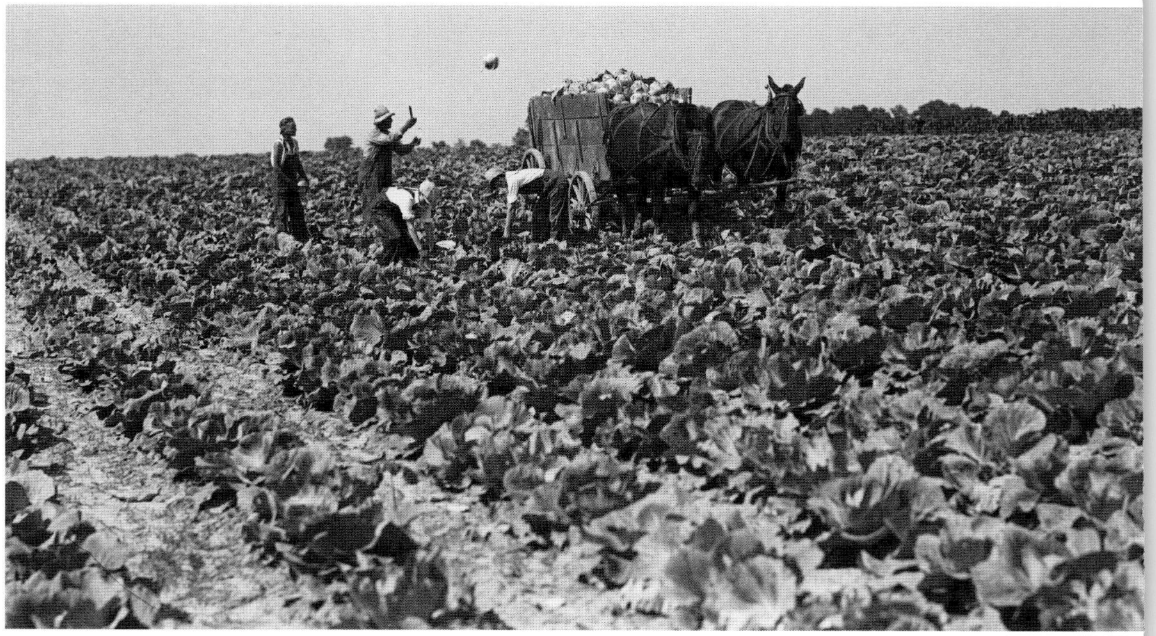

Men cut and pitch cabbages into a wagon on the farm of A. H. Hoeksema in Muscatine, Iowa. (Muscatine County, IA; 1929)

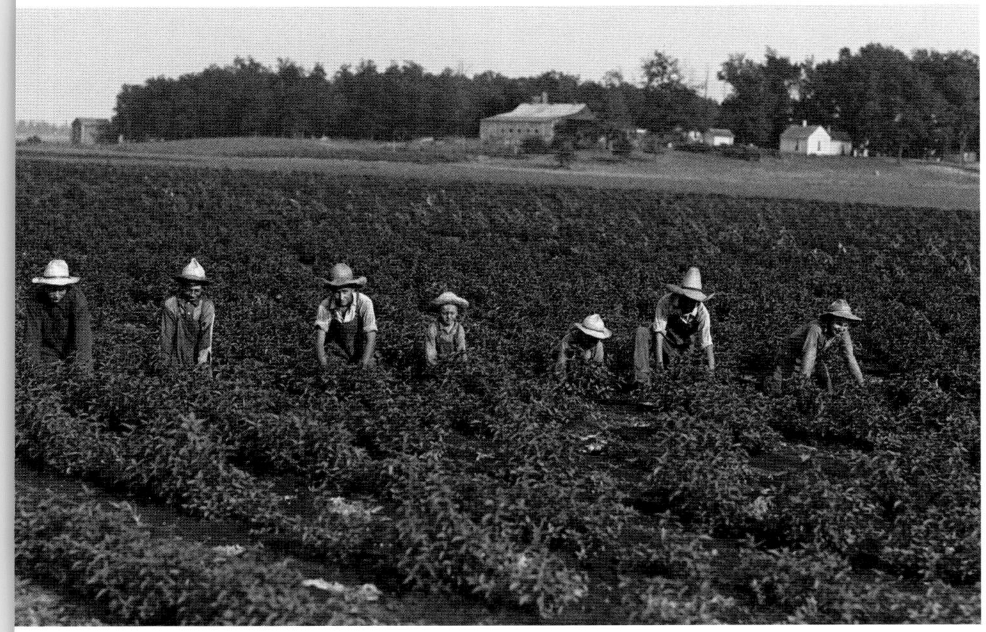

Boys weed a mint field by hand at the H. R. Stauffer farm in Nappanee. (Elkhart County, 1926)

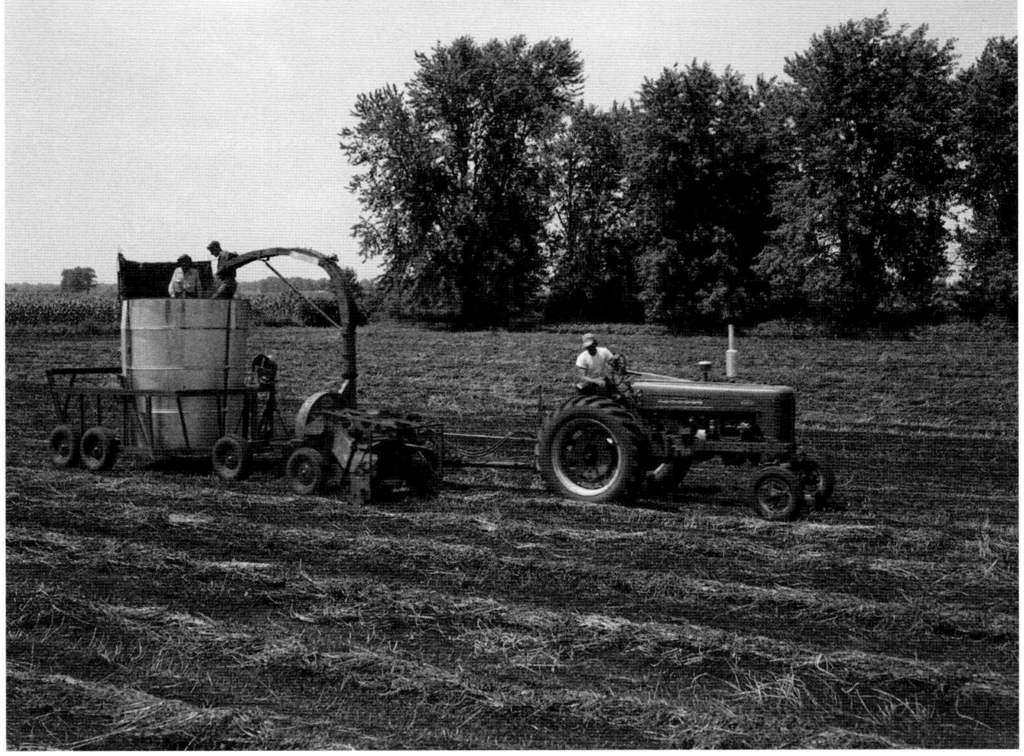

Chopping mint hay into a portable vat before transporting it to a mint still on the Van Johnston farm near North Judson (Starke County, 1957)

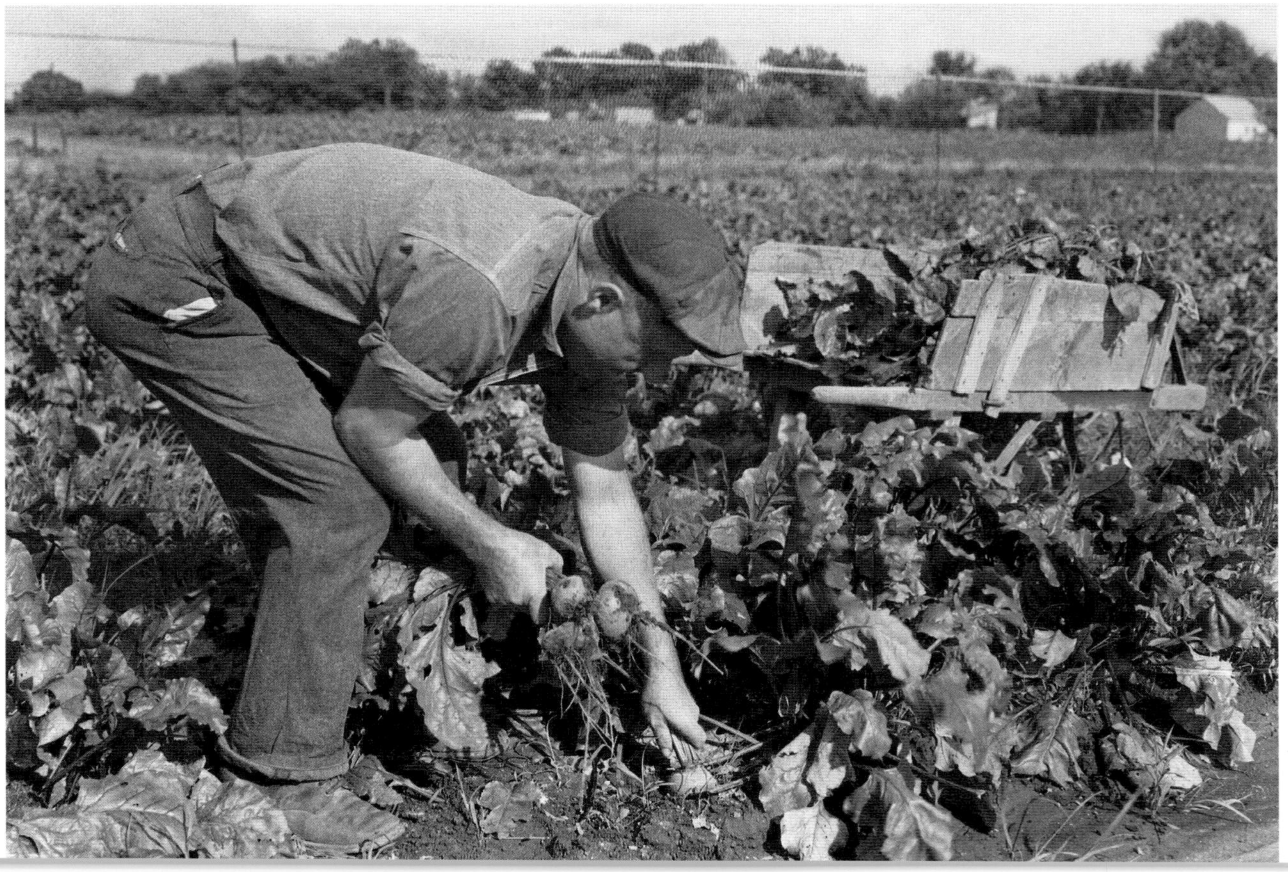

Bill Herdman harvests beets in a field owned by Schilling Greenhouse. (Indiana, 1938)

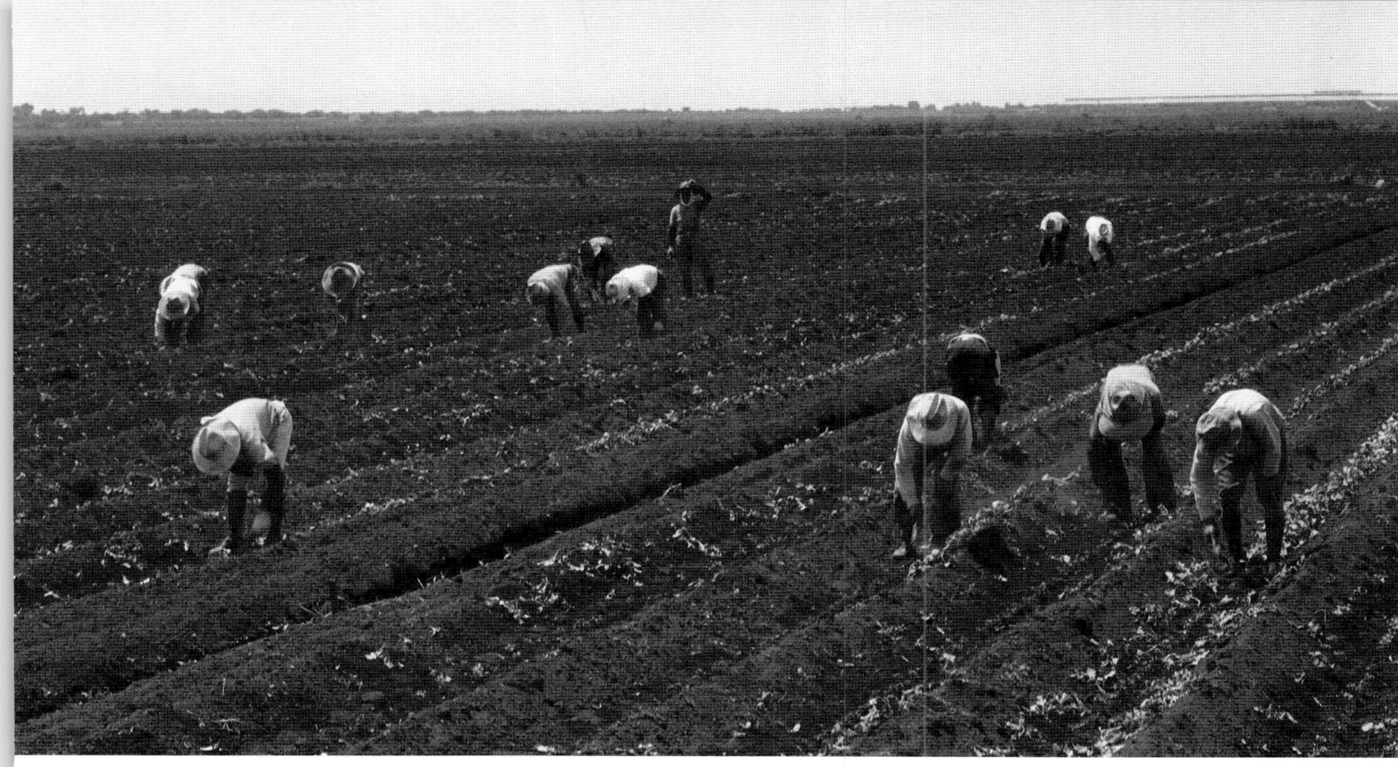

Japanese laborers hand-weed young celery plants on the farm of John C. Maurer and Sons in Stockton, California. (San Joaquin County, CA; 1932)

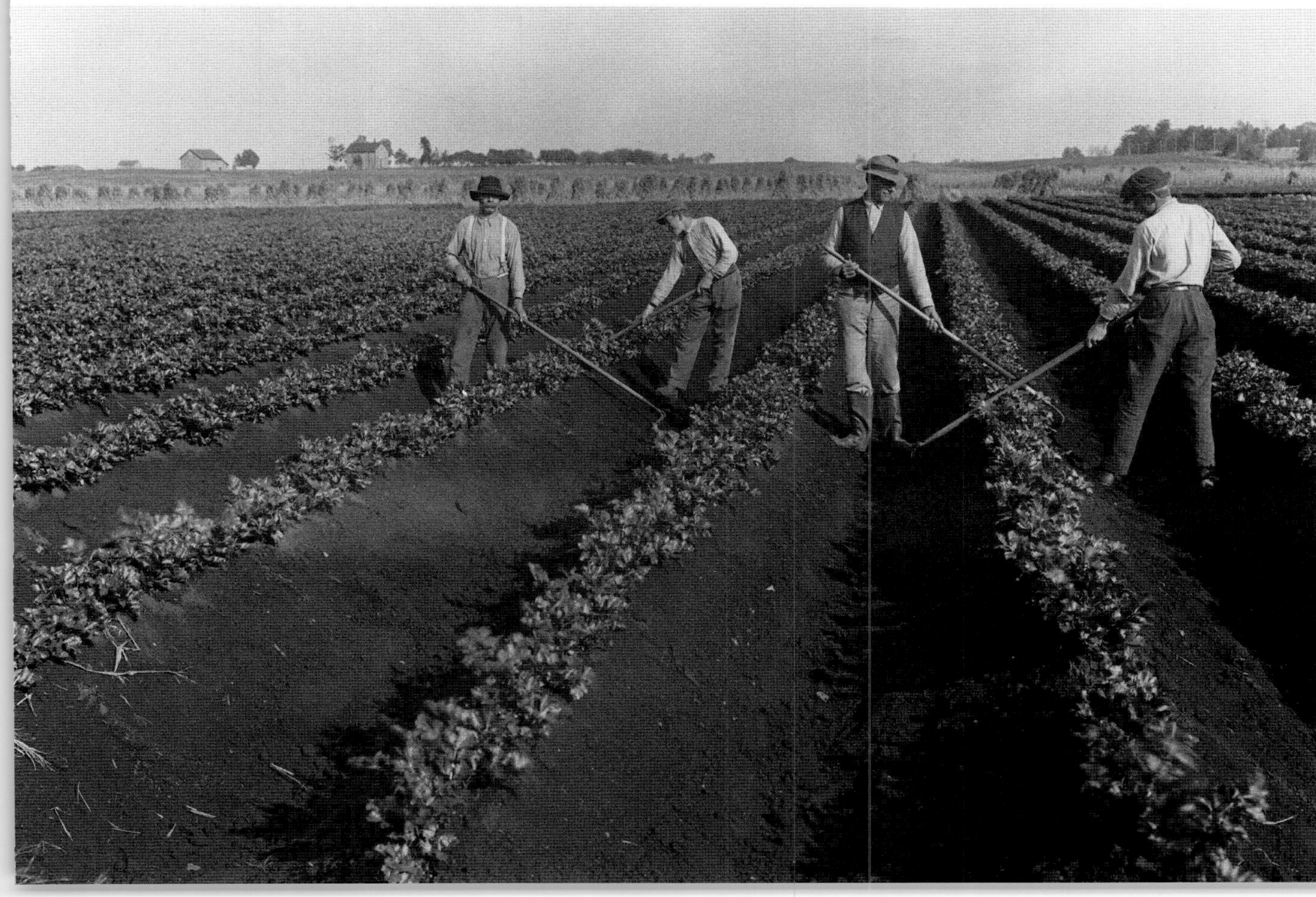

Men pull soil up around plants to blanch celery. (Tippecanoe County, ca. 1915)

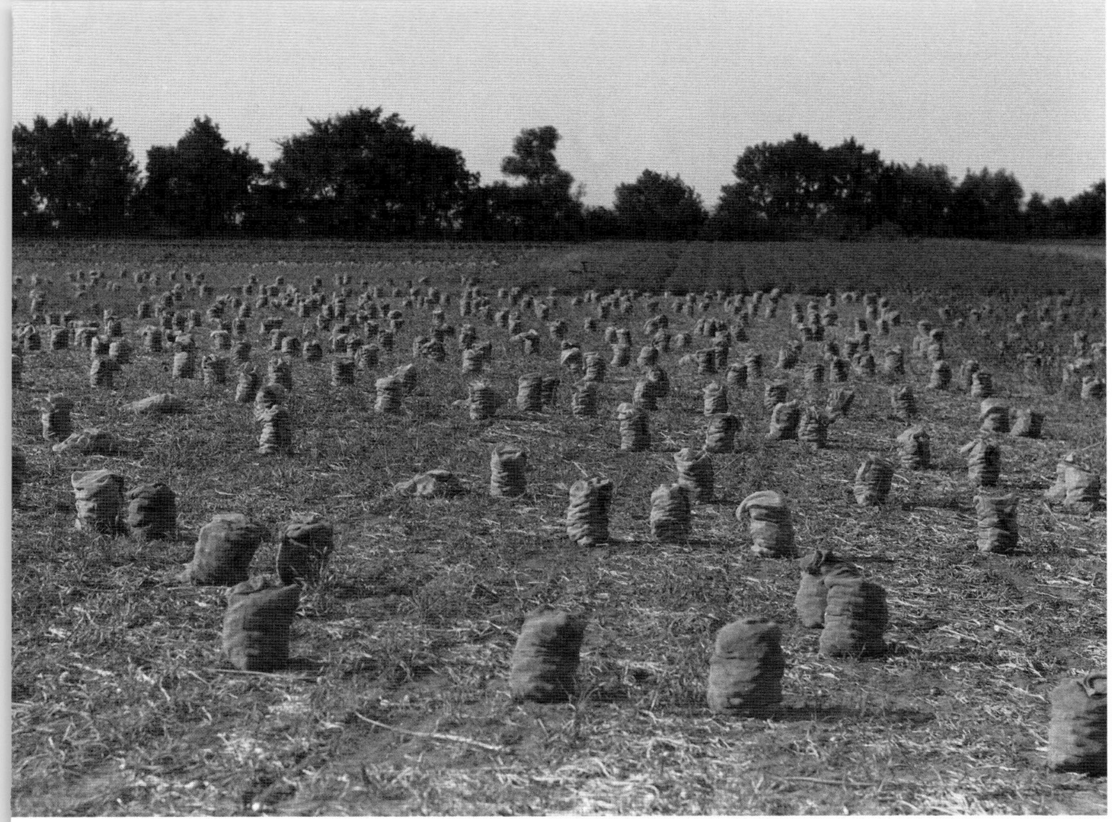

Sacks of onions on the farm of Theodore Kostas from Davenport, Iowa (Scott County, IA; 1929)

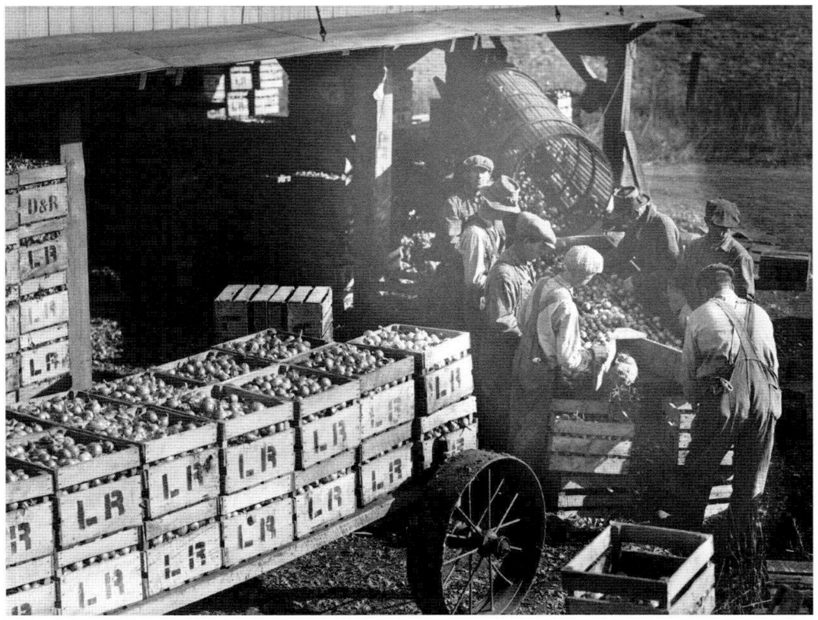

Workers grade and sort onions in the packing shed of Louis Ruderman in LaOtto. (Noble County, 1929)

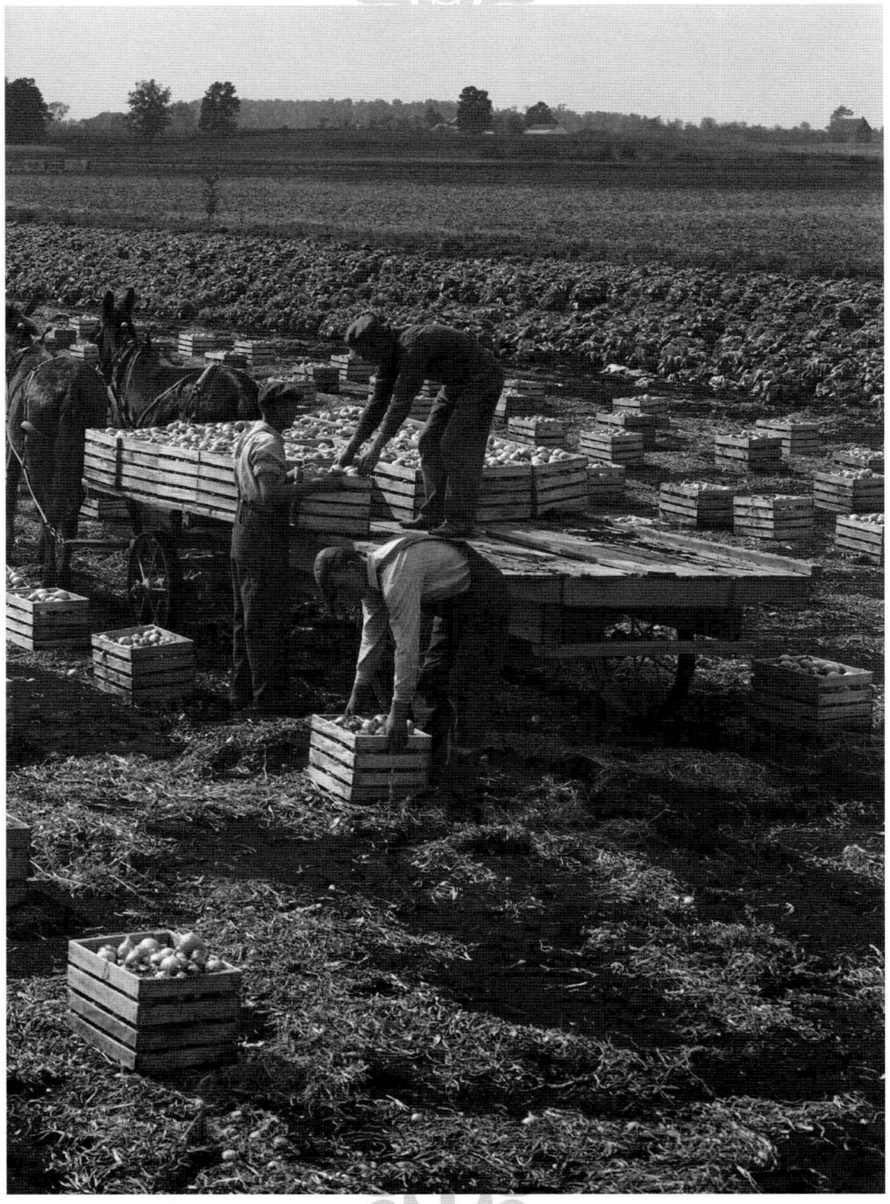

These workers load up crates of onions harvested from the Whit Gast farm in Akron, which produced an average of 1,600 bushels of onions per acre. (Fulton County, 1940)

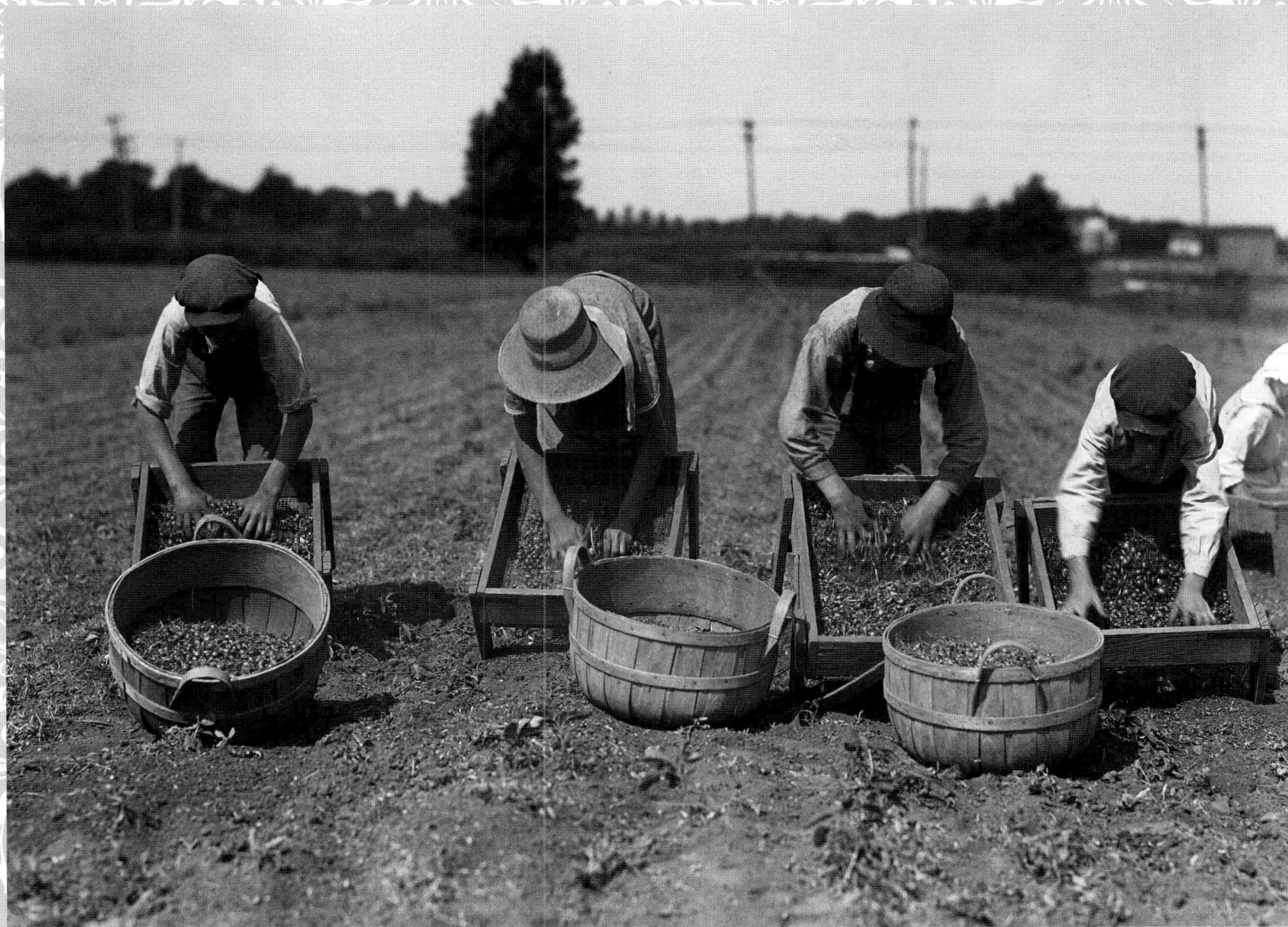

Boys and girls sift dirt out of onion sets at the Wallace Piper farm in Racine, Wisconsin. They could pick and clean 25 bushels per day, earning eight cents per bushel. (Racine County, WI; 1922)

A Fine Tobaccy Crop

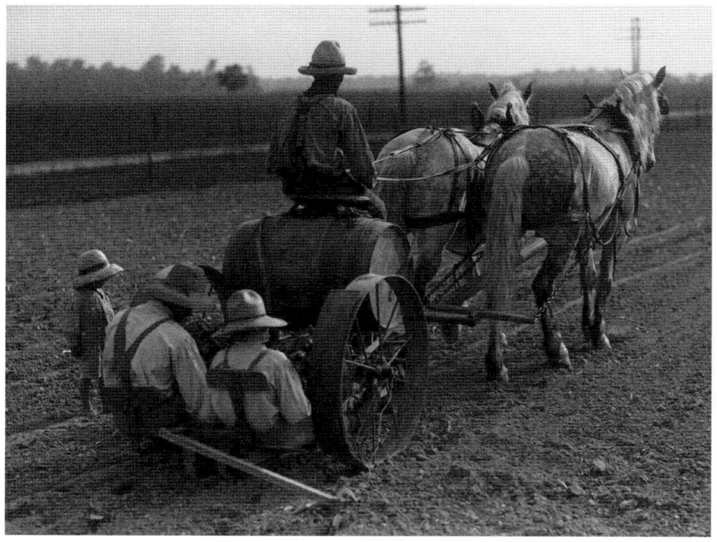

Setting out tobacco plants on the farm of L. Kessler of Lewisburg, Ohio (Preble County, OH; 1916)

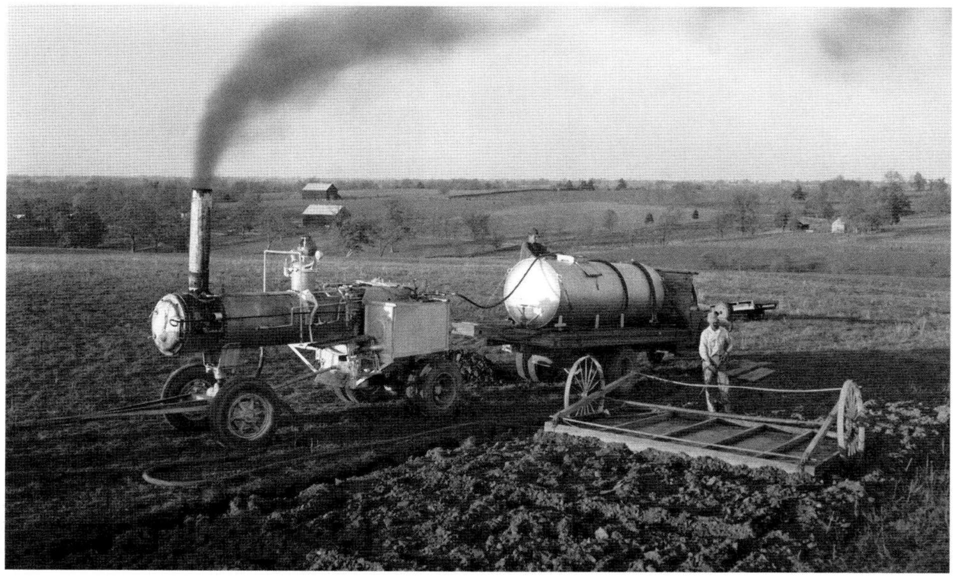

A man uses steam to sterilize a tobacco bed on the farm of W. B. Boston in Versailles, Kentucky. (Woodford County, KY; 1959)

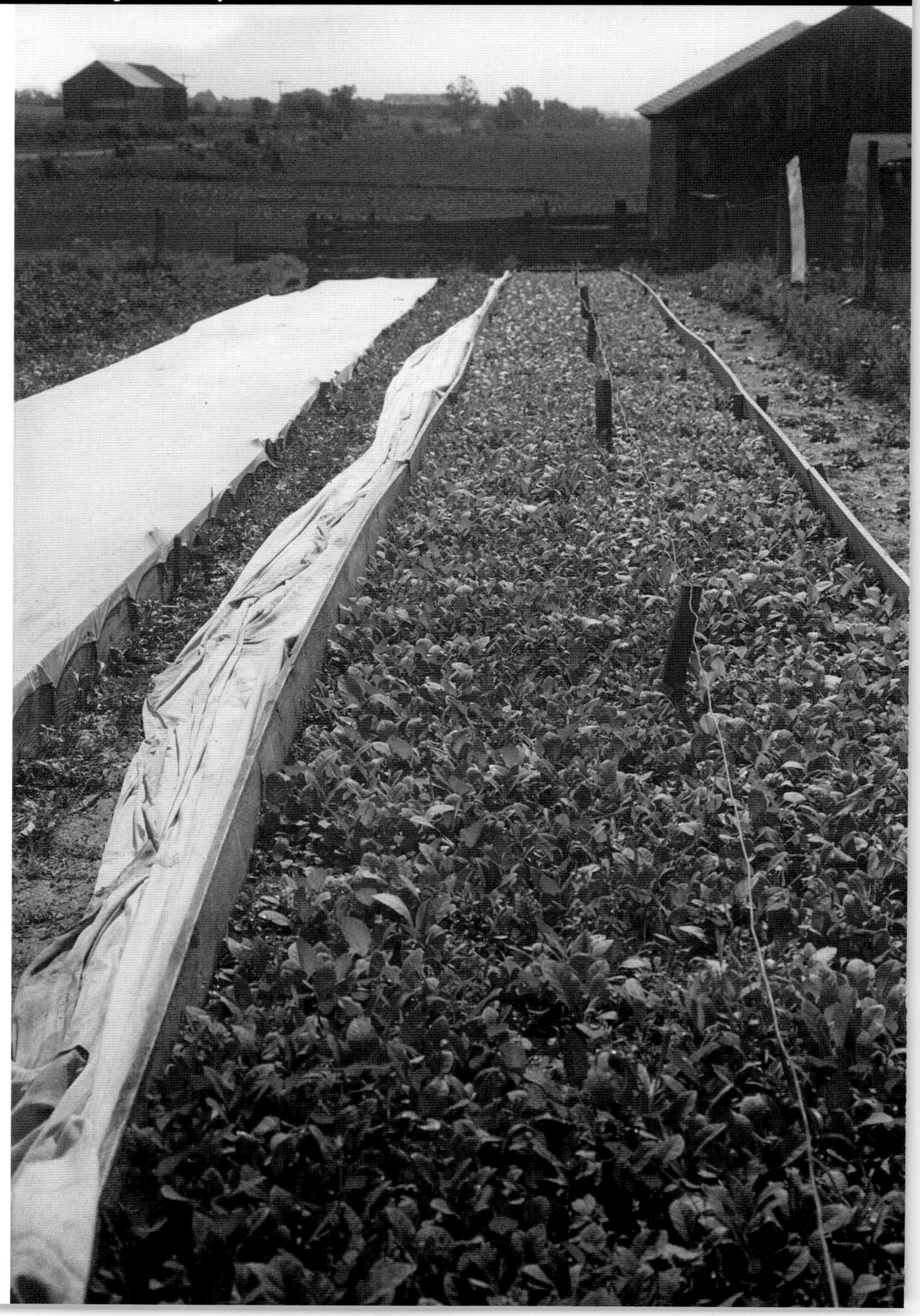

A closeup view of a tobacco bed owned by Peter Yount of Bradford, Ohio (Darke and Miami Counties, OH; 1916)

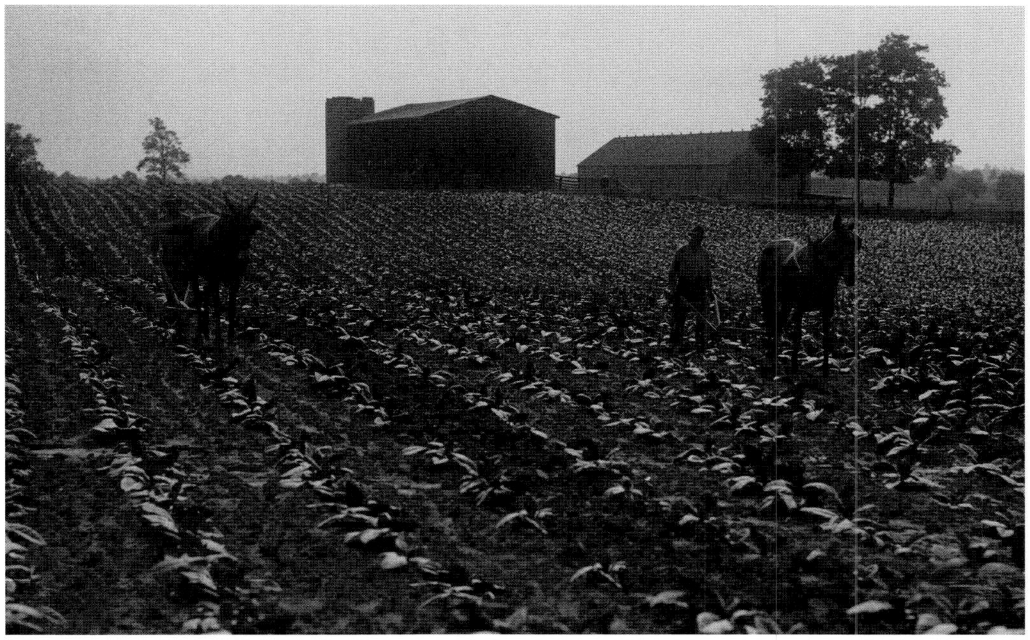

Two men cultivate tobacco on the George Keller farm in Lexington, Kentucky. (Fayette County, KY; 1928)

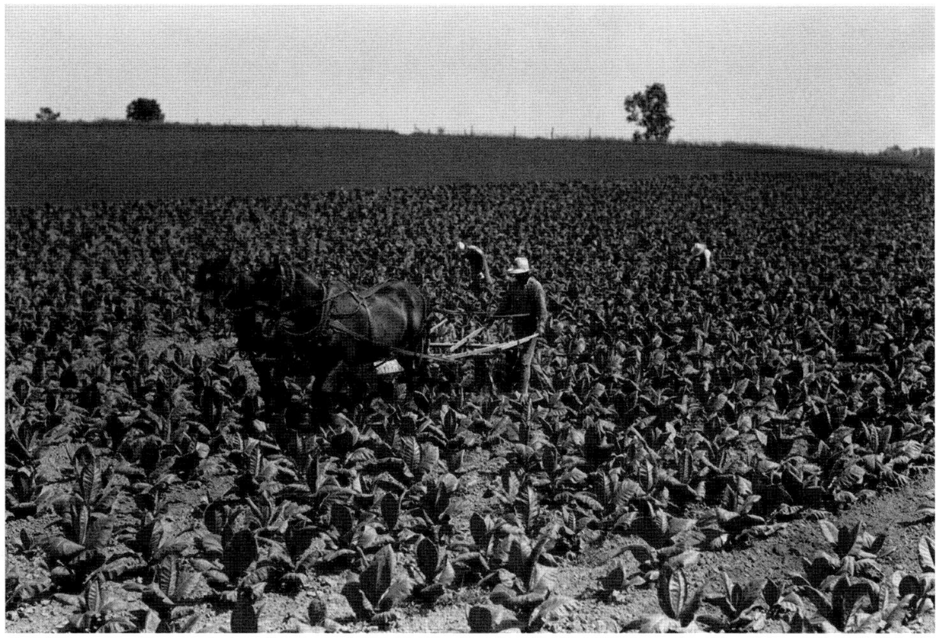

Plowing a field of tobacco on the Ole Harried farm near Albion, Wisconsin (Dane County, WI; 1931)

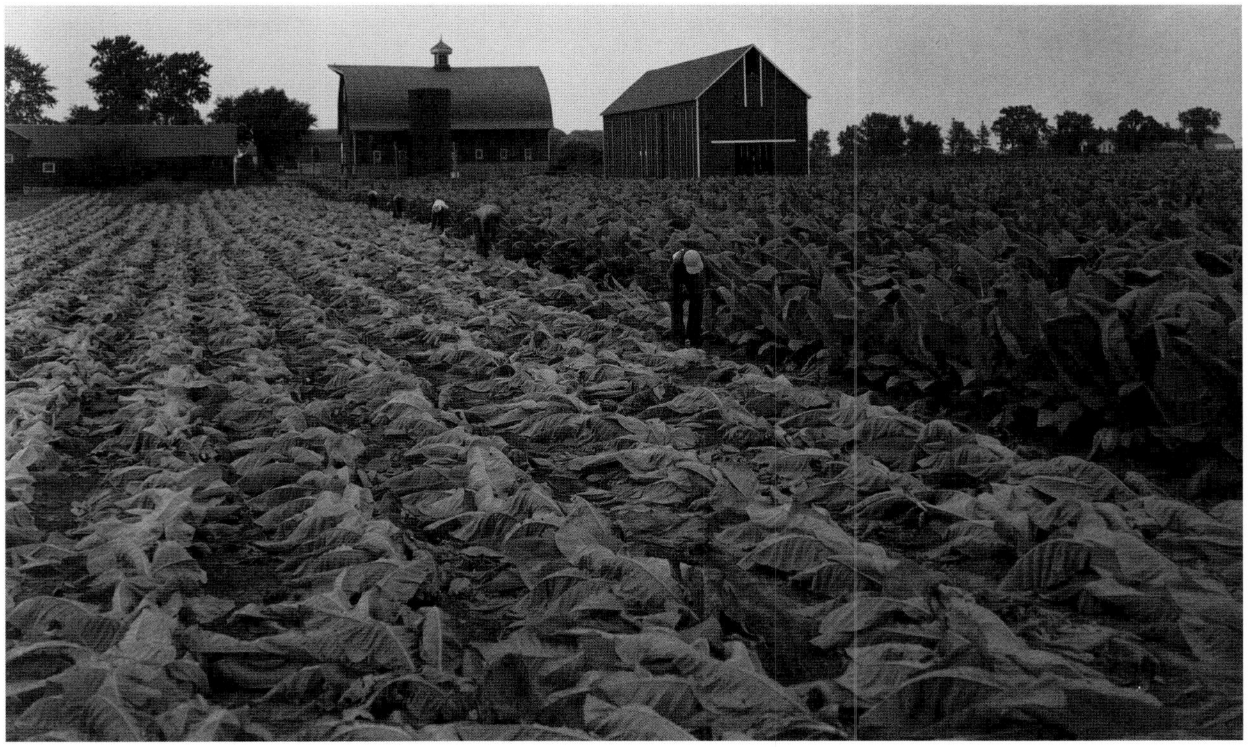

On the Irwin Bliefernicht farm in DeForest, Wisconsin, workers harvest Havana 142 tobacco that was fertilized with manure and supplemented with 2,000 pounds of 3-9-18 fertilizer per acre. (Dane County, WI; 1952)

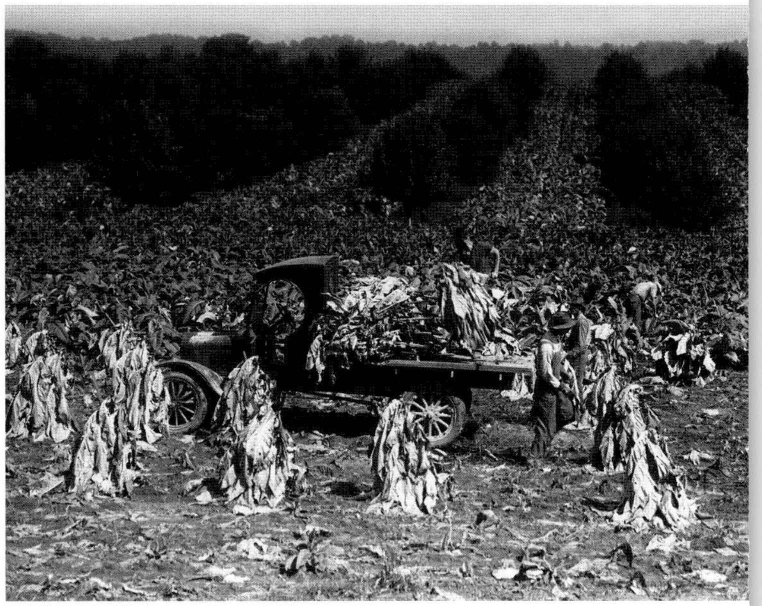

Filling a truck with tobacco harvested from a field owned by J. B. Turner of Mitchell (Lawrence County, 1926)

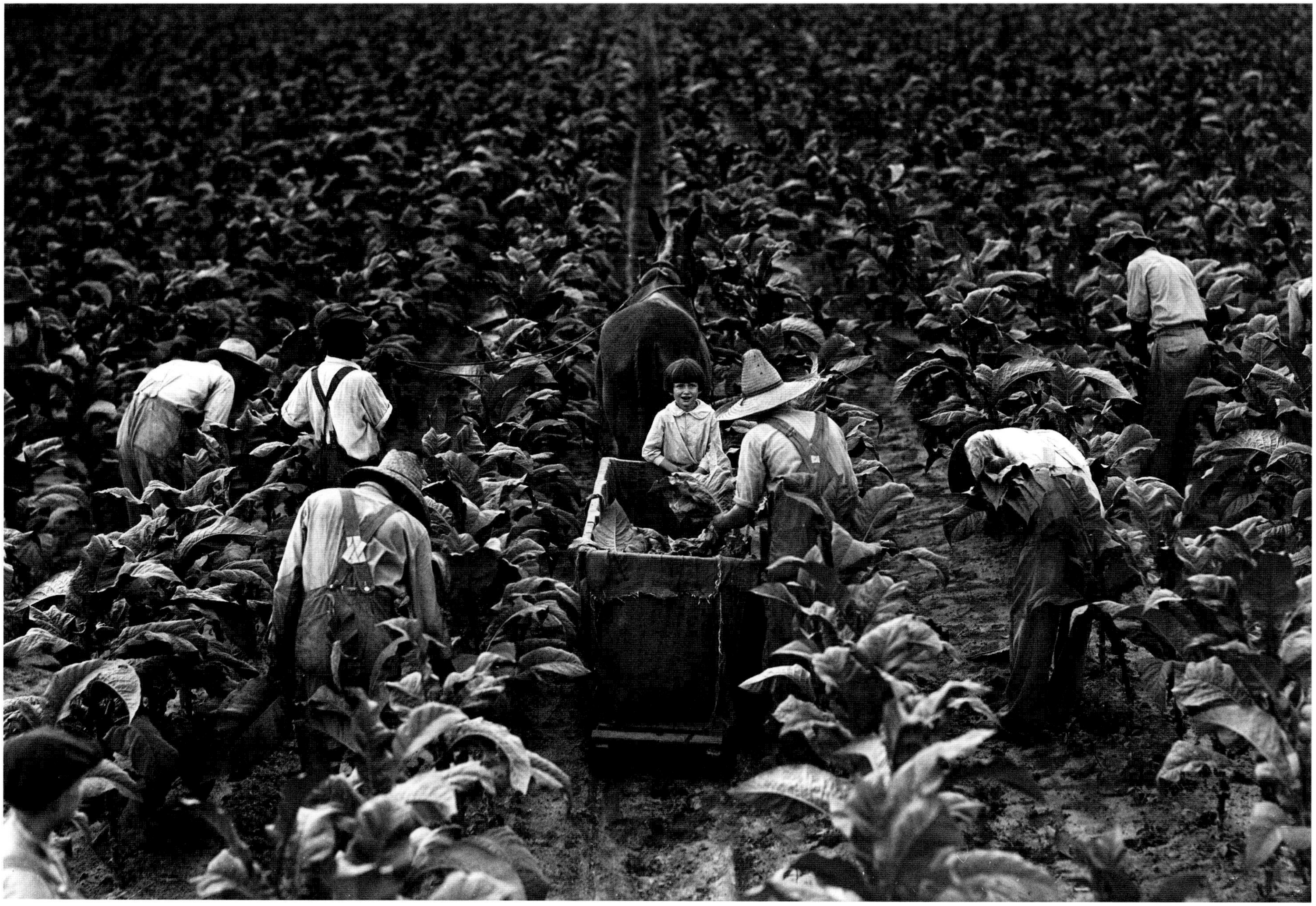

This crew harvests tobacco on the L. C. Weaver farm located in Moultrie, Georgia. (Colquitt County, GA; 1928)

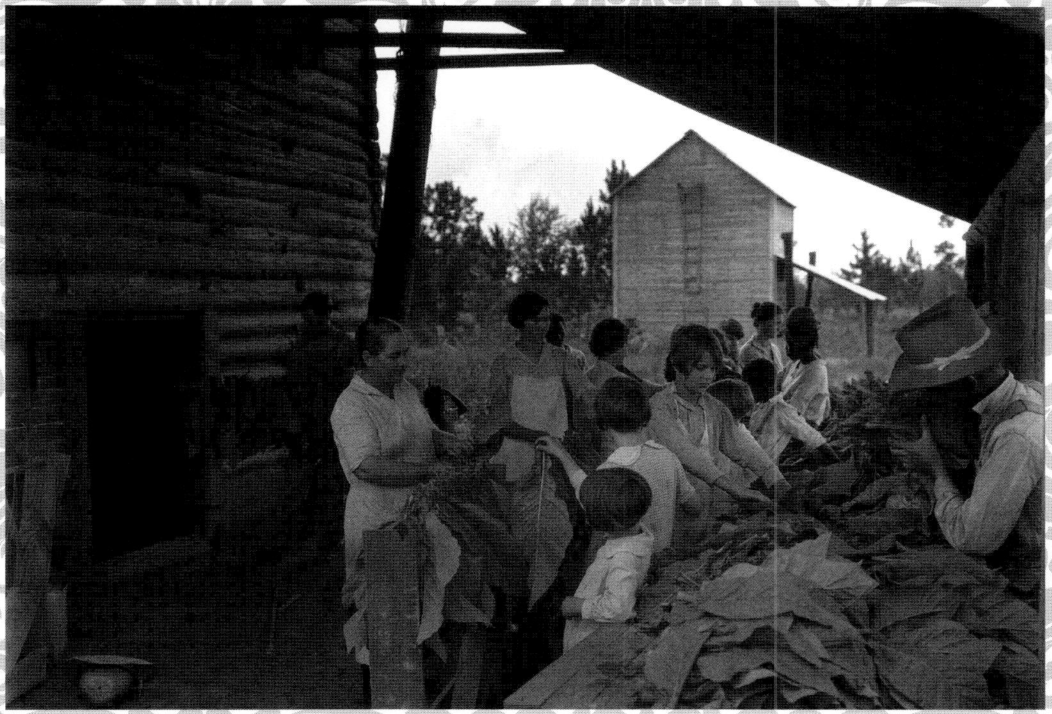

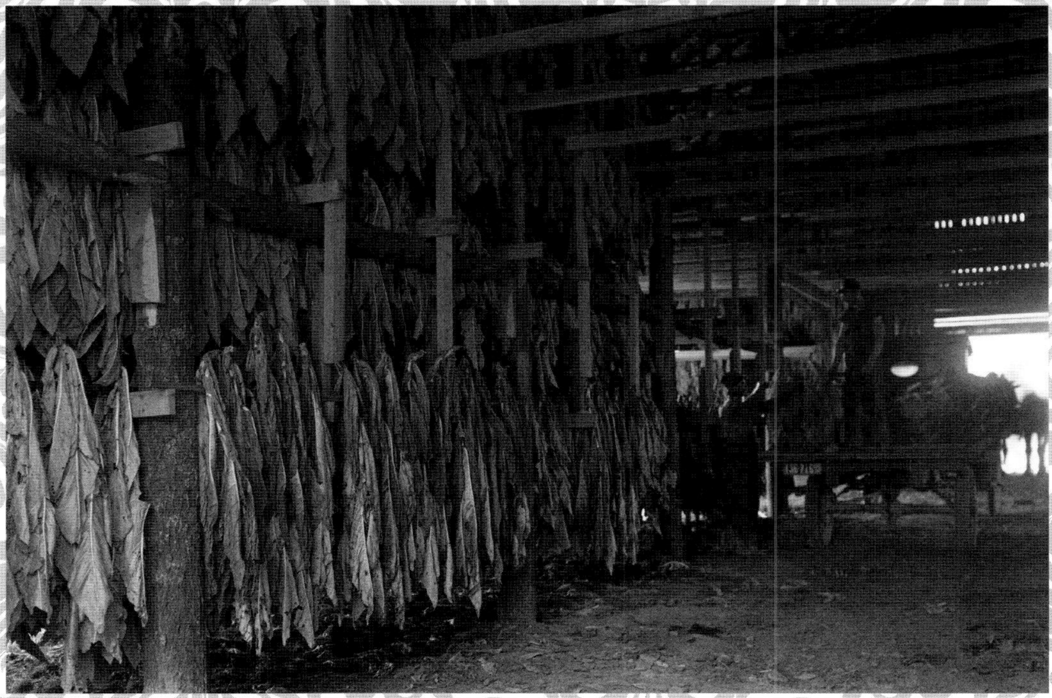

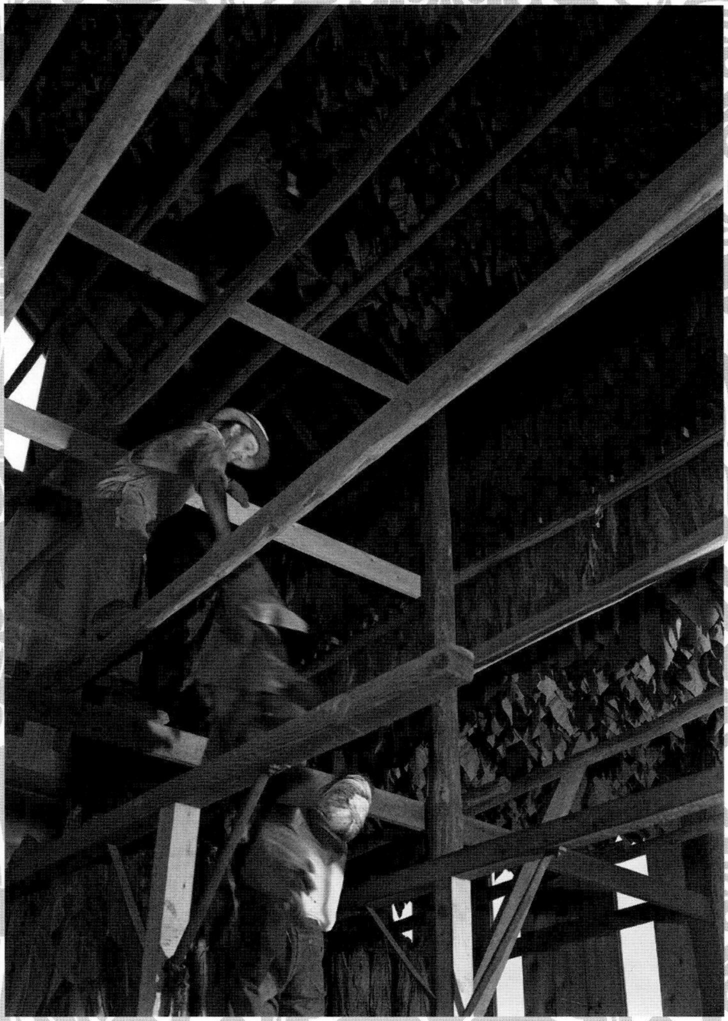

(Top left) Stringing tobacco in preparation for drying on the farm of John G. Norman of Norman Park, Georgia (Colquitt County, GA; n.d.)

(Bottom left) Tobacco hangs in J. B. Turner's drying shed in Mitchell. (Lawrence County, 1926)

(Right) An interior view of a tobacco storage barn in DeForest, Wisconsin (Dane County, WI; 1952)

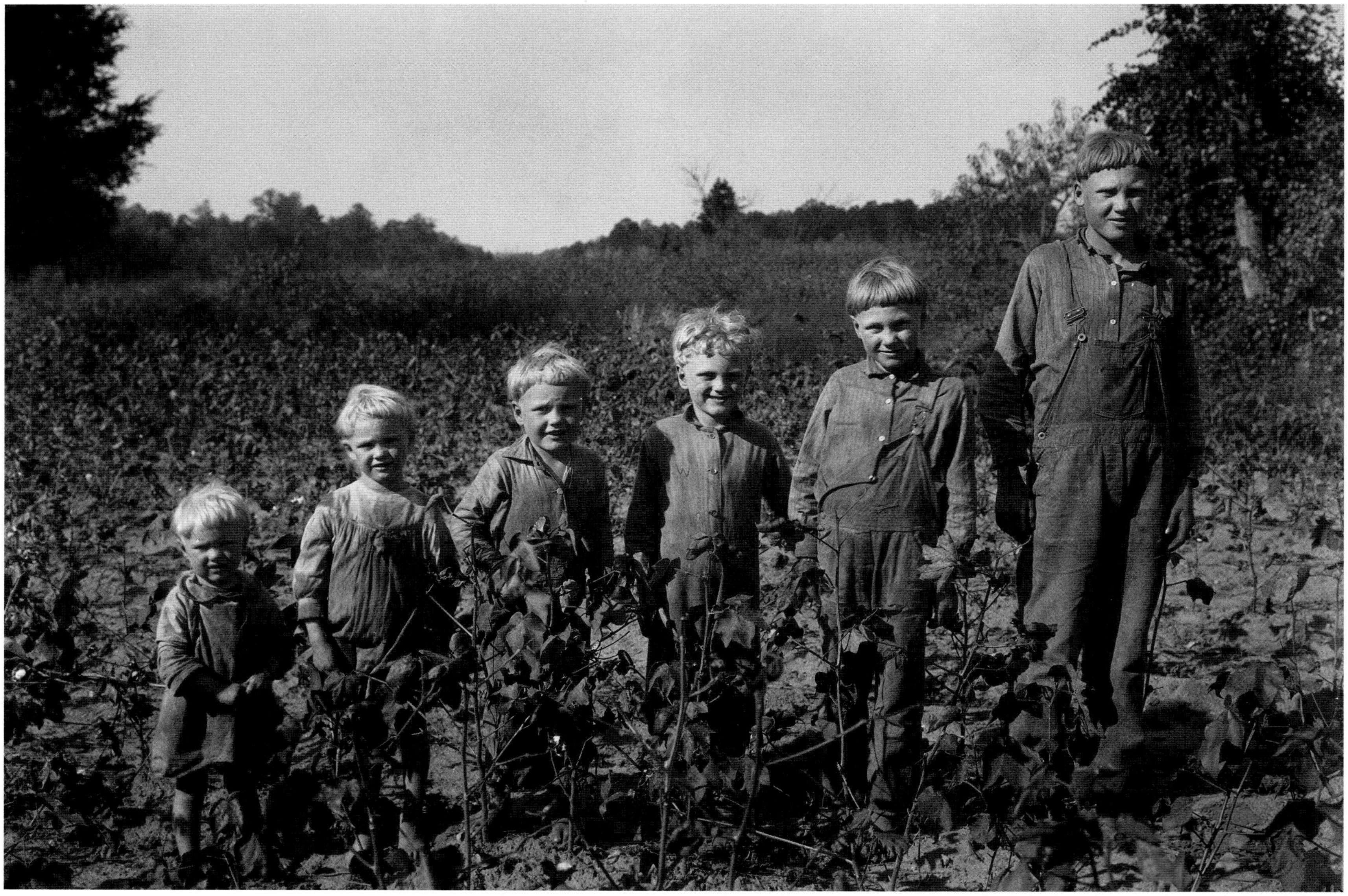

Children pose for a photo in a cotton field. (Georgia or South Carolina, 1927)

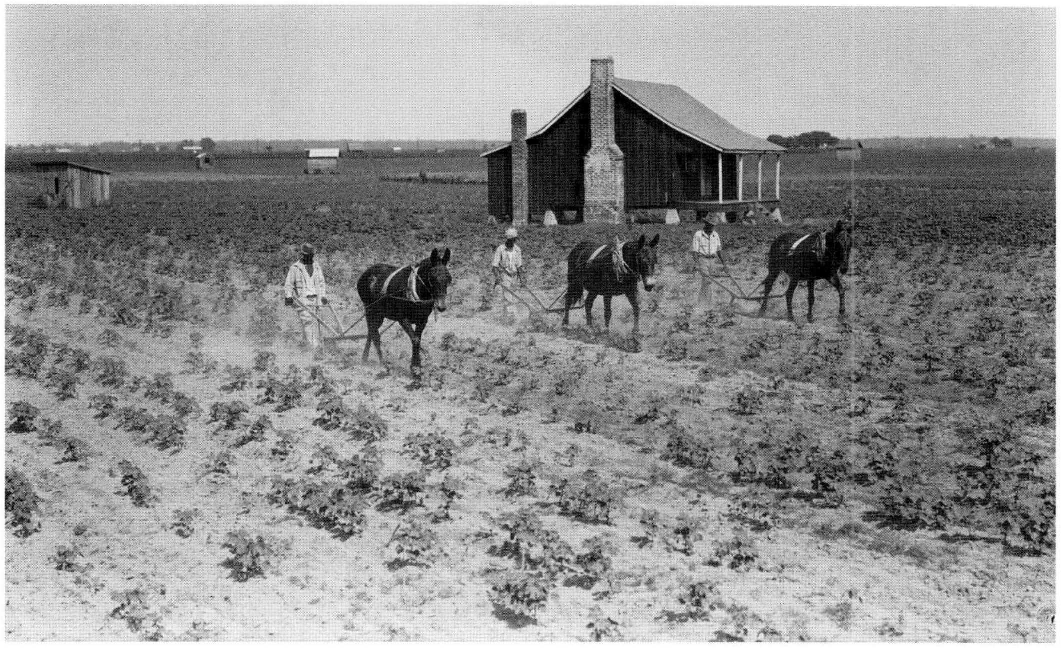

Plowing young cotton with three single-mule teams on the farm of King and Anderson near Clarksdale, Mississippi (Coahoma County, MS; ca. 1932)

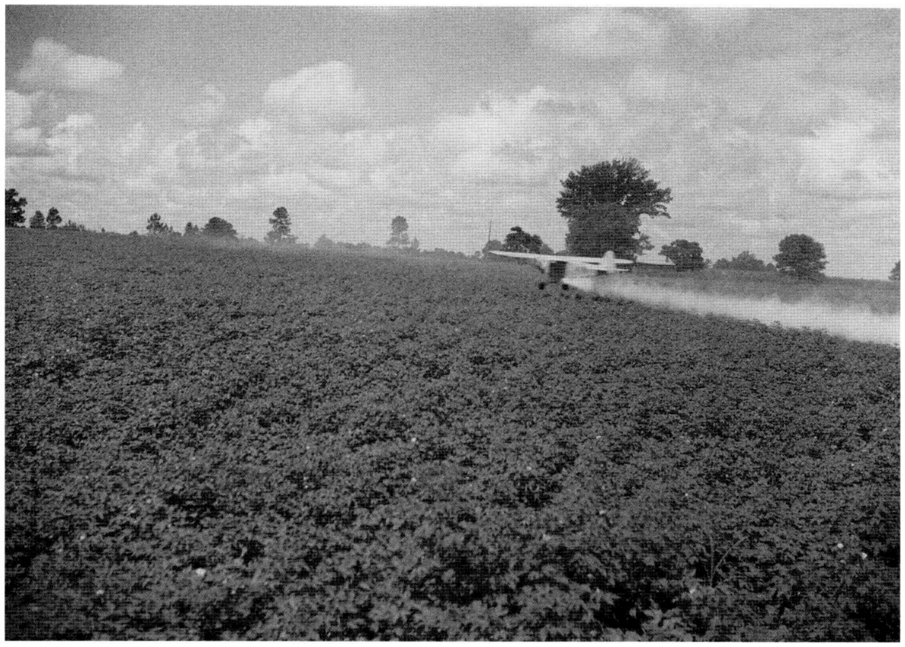

A crop duster treats a cotton field near Hawkinsville, Georgia. (Pulaski County, GA; 1959)

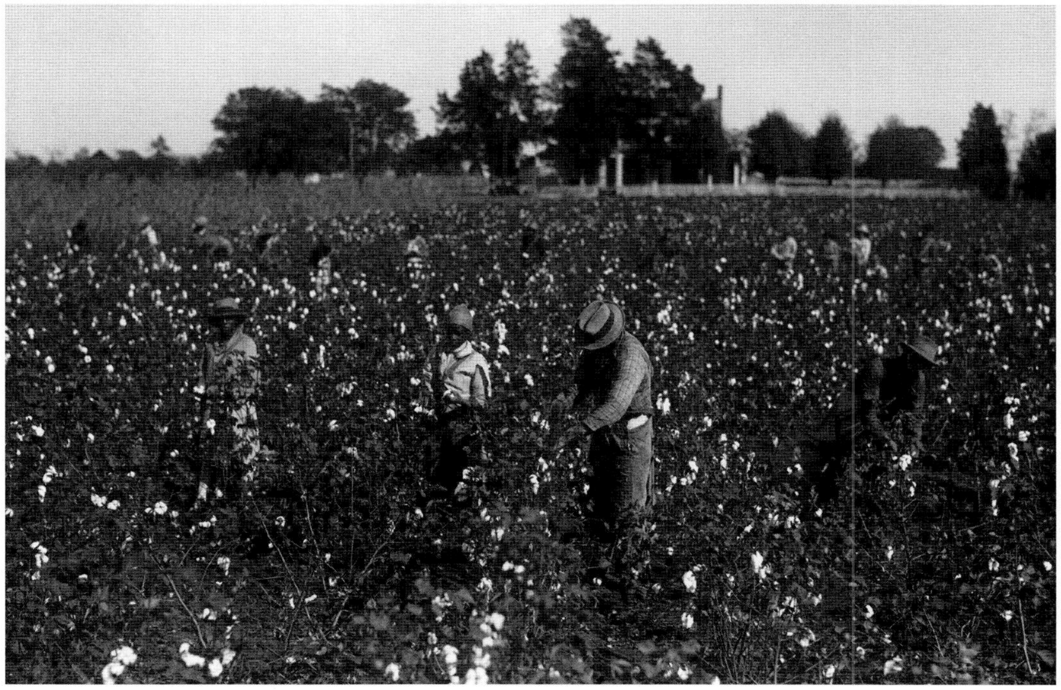

Picking rows of cotton on the farm of John R. Witt in Belle Mina, Alabama (Limestone County, AL; 1927)

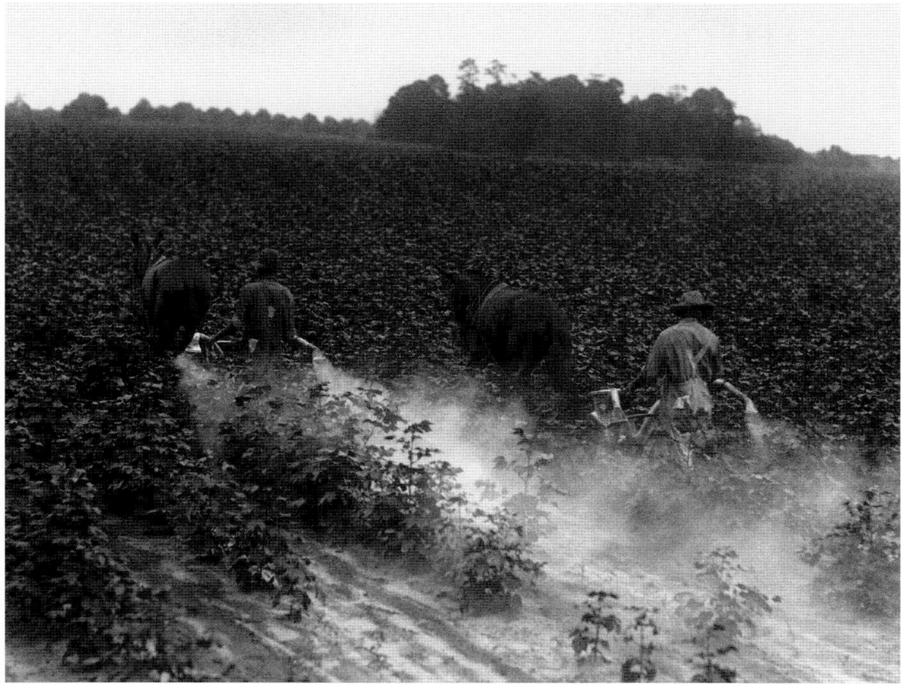

These men use two-row dusters pulled by mules to control insects in cotton on the farm of W. R. Hodges from Dublin, Georgia. (Laurens County, GA; 1928)

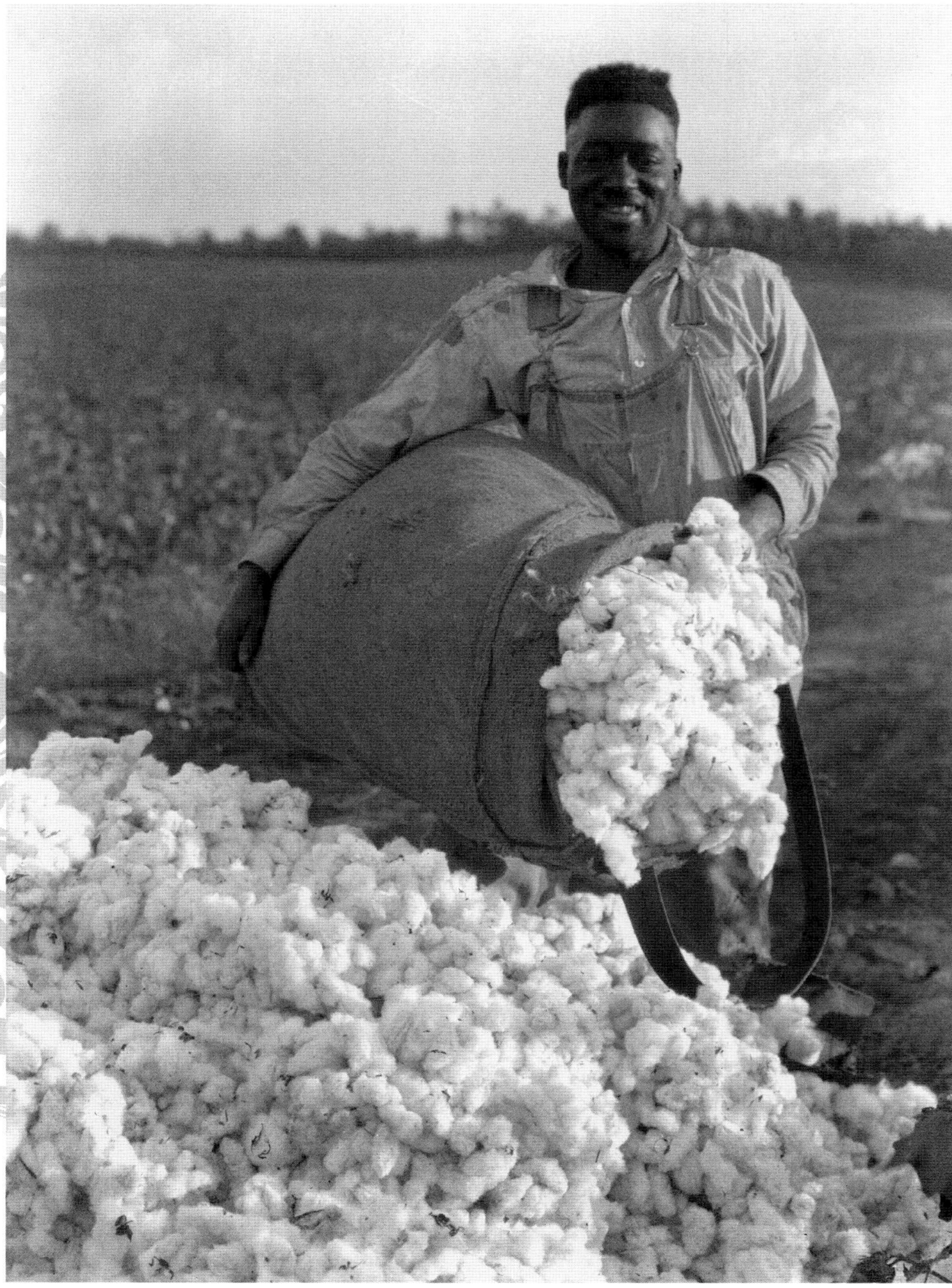

Emptying a bag of cotton on the farm of J. Wade Drake in Anderson, South Carolina (Anderson County, SC; 1927)

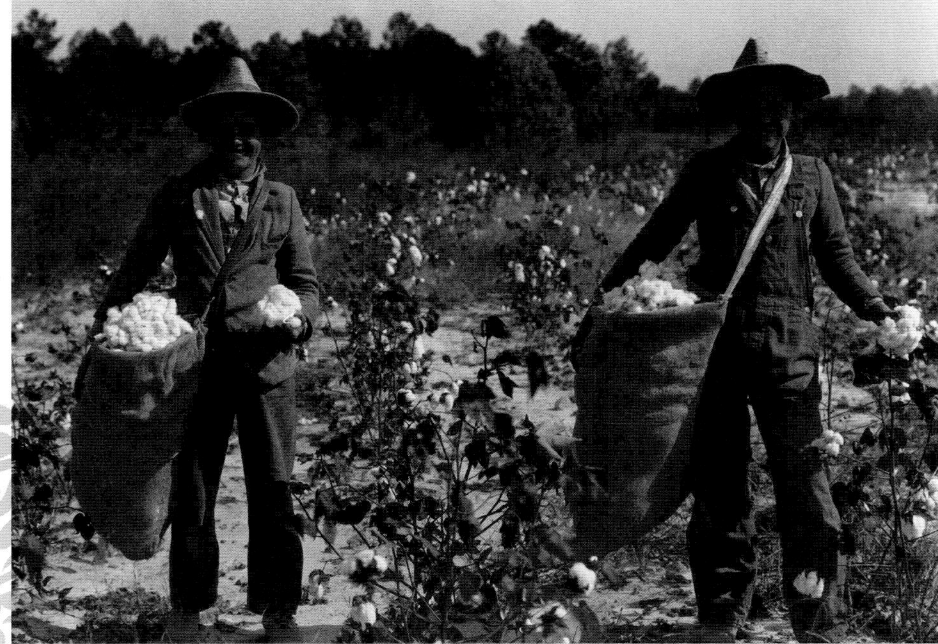

John Robert Knight and George B. Knight Jr. from Grayson, Georgia, pick cotton. Allen wrote, "These boys [along] with one mule do all the work on a 40-acre farm and support their widowed mother. They had out 10 acres of cotton and 15 acres of corn this past season. They aim to feed the surplus corn to hogs and sell the cotton as a cash crop." (Gwinnett County, GA; 1927)

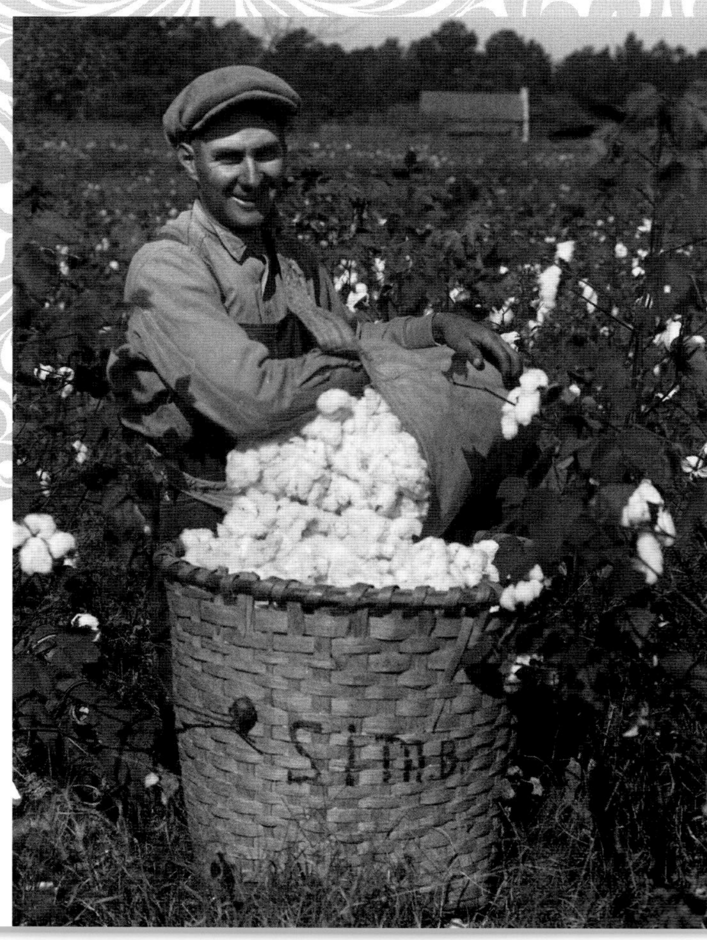

Sim Bennett from Lawrenceville, Georgia, empties his bag of cotton into a basket. (Gwinnett County, GA; 1927)

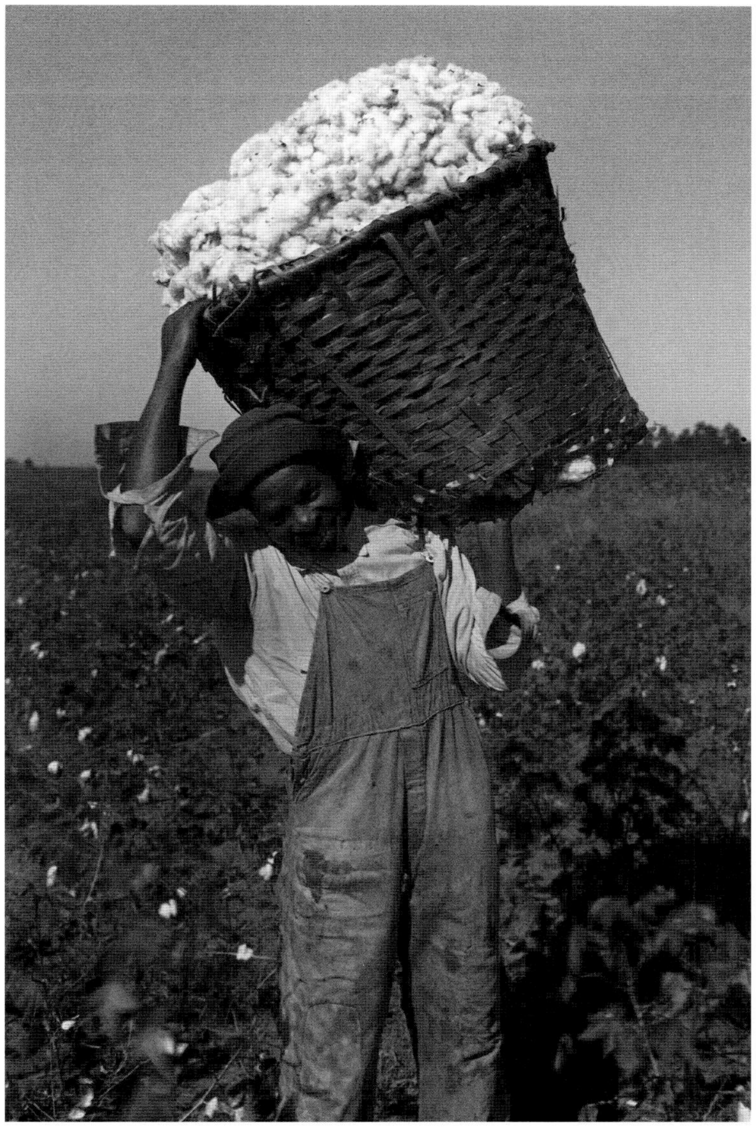

A man carries a basket of picked cotton on his shoulder on the farm of B. L. Redwine in Newnan, Georgia. (Coweta County, GA; 1927)

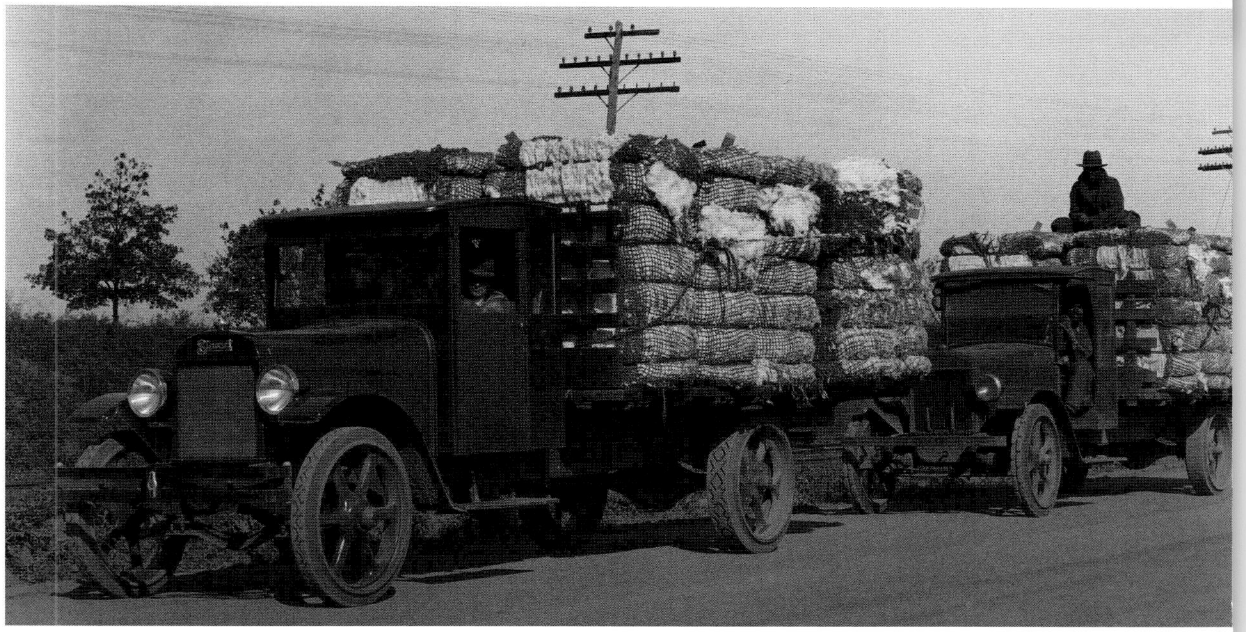

H. E. Childress from Greenville, South Carolina, works with a crew to haul baled cotton to a factory. Each truck is loaded with 5,000 pounds of cotton. (Greenville County, SC; 1927)

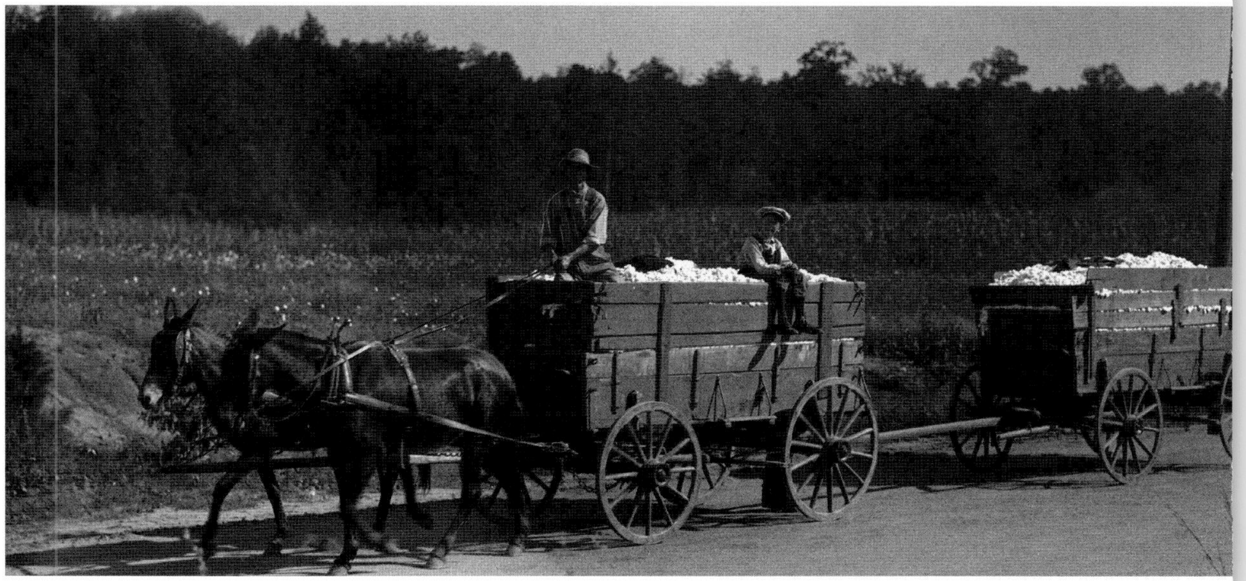

One team of mules pulls two wagonloads of cotton to a cotton gin near Pendleton, South Carolina. (Anderson County, SC; 1927)

Feed for Farm Animals

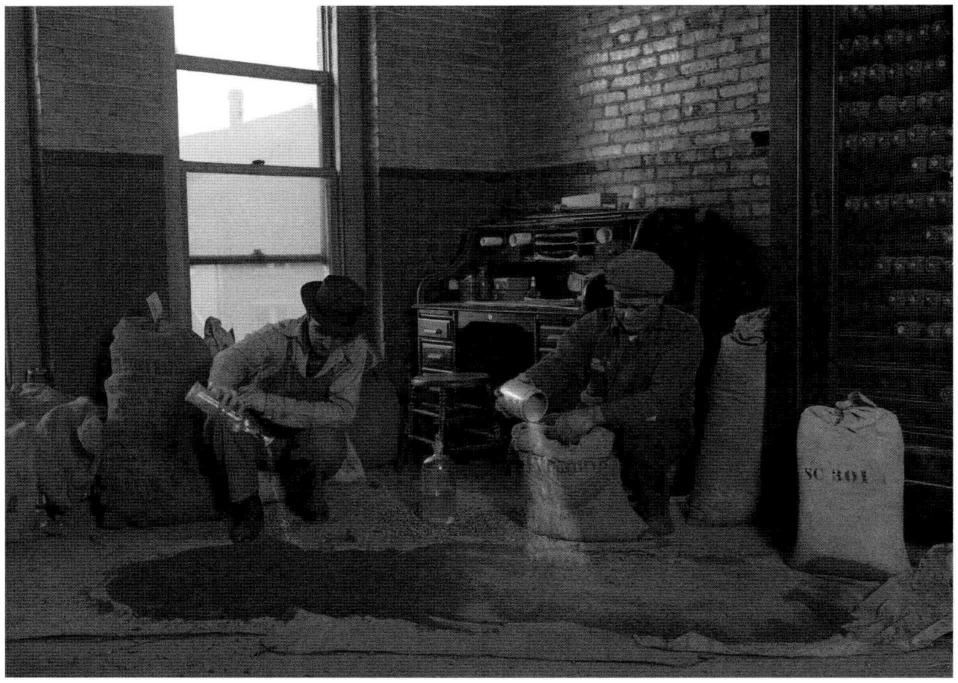

Two men inoculate clover seed with beneficial bacteria by mixing it with dirt at the Purdue University Seed House. (Tippecanoe County, 1926)

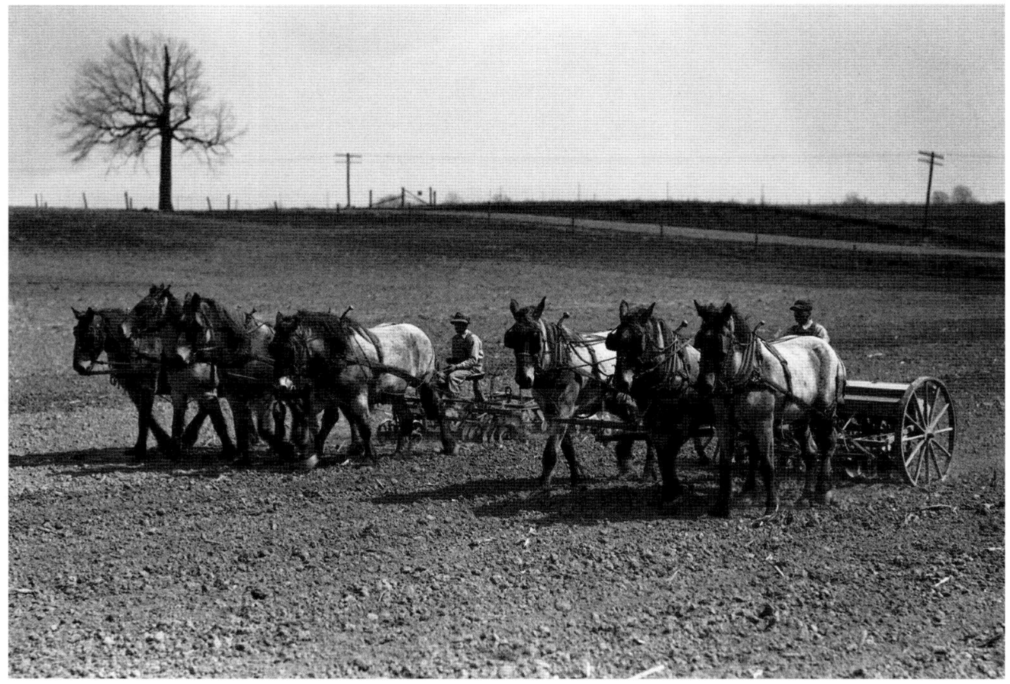

Disking ground and drilling alfalfa seed on the farm of Chauncey Craig near Noblesville (Hamilton County, 1932)

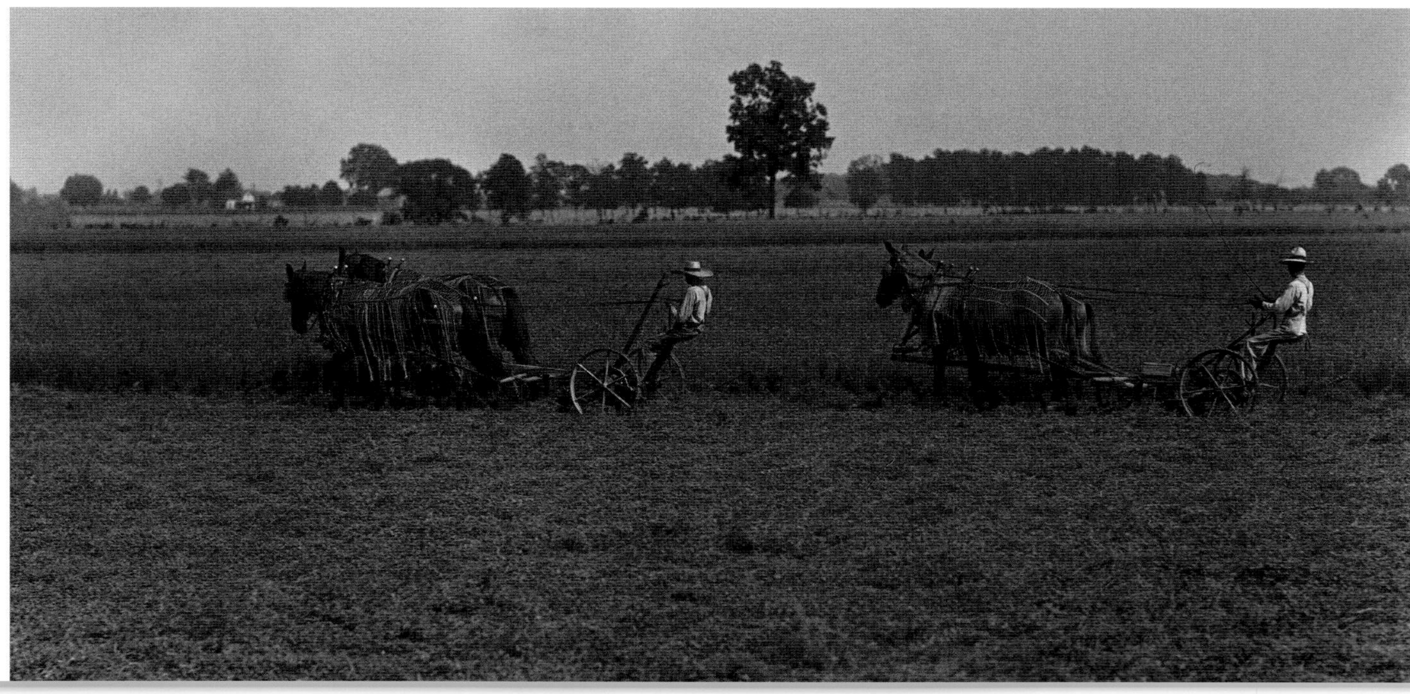

Cutting alfalfa near Elston (Tippecanoe County, early 1910s)

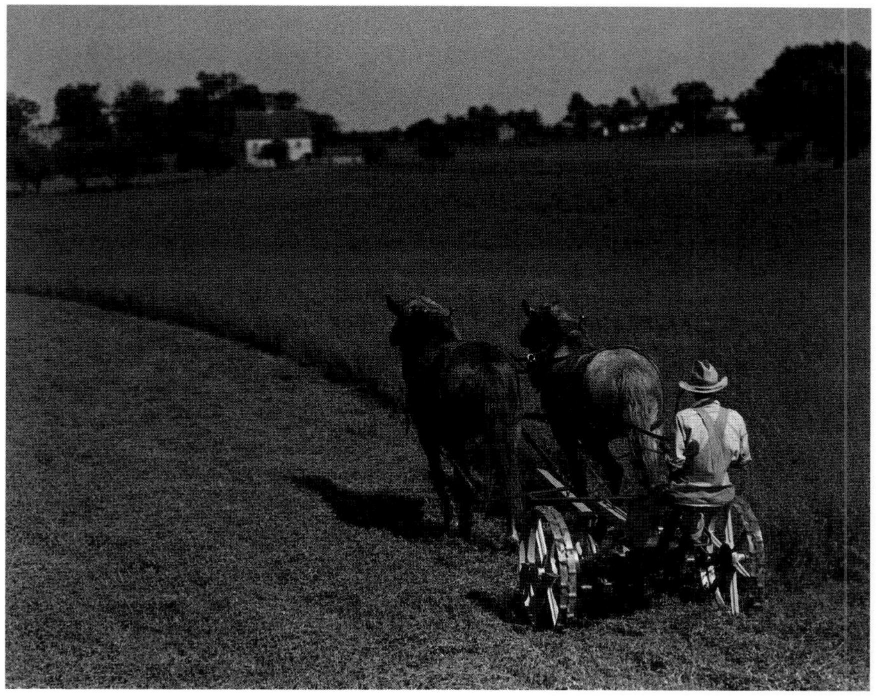

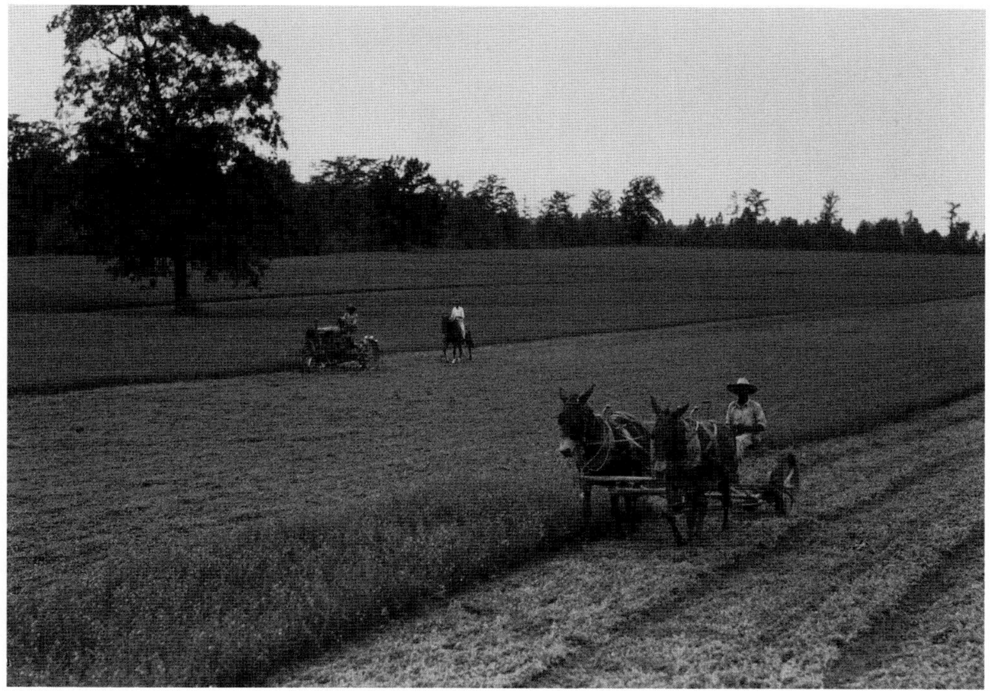

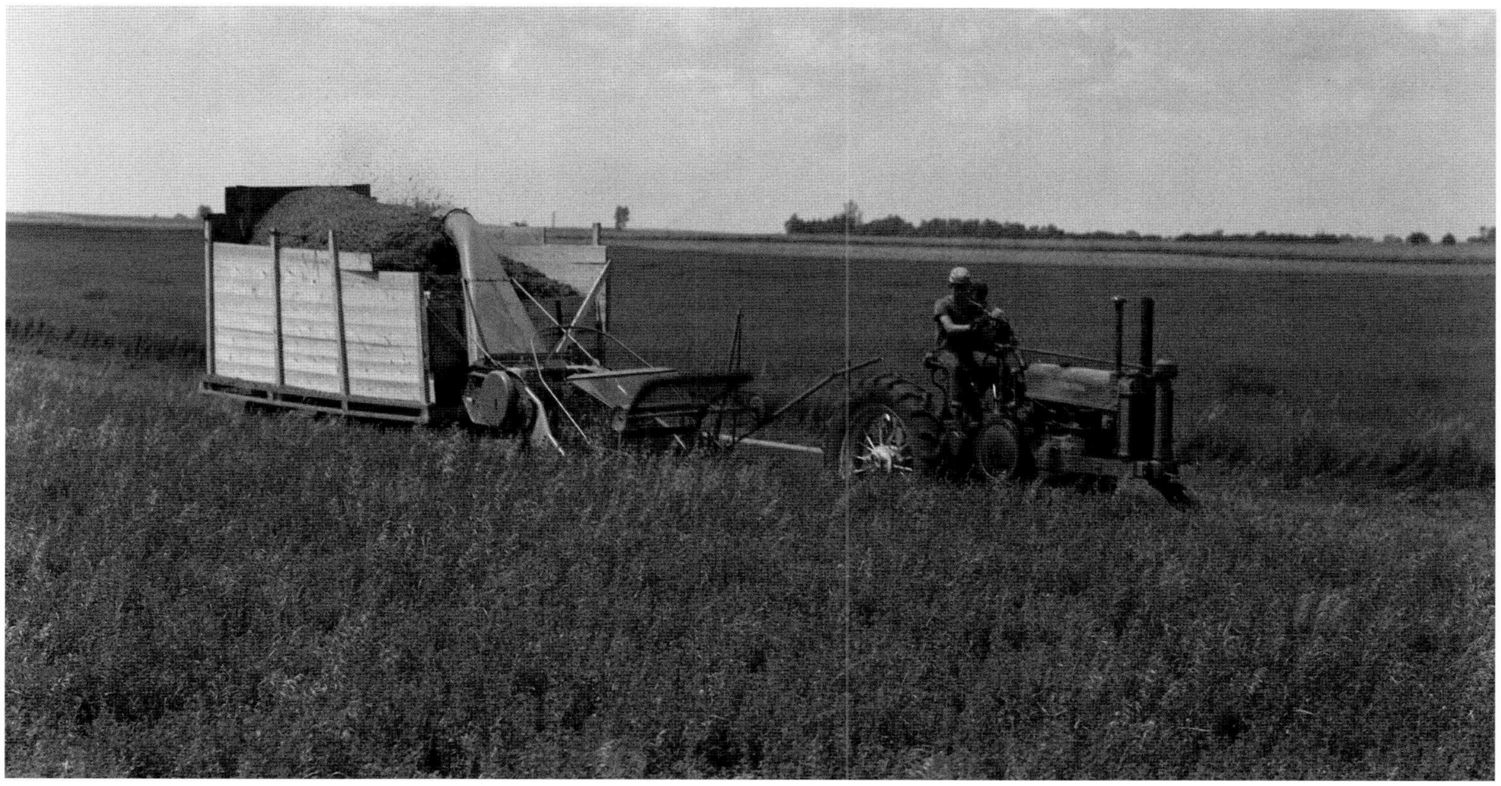

(Top left) A farmer uses a mowing machine to cut alfalfa in the West Point area. (Tippecanoe County, 1939)

(Top right) Cutting alfalfa in a field owned by J. W. Lake and Son in Maud, Mississippi (Tunica County, MS; ca. 1932)

(Bottom) Leo Feller cuts an alfalfa and bromegrass mixture for silage on his farm in Kankakee, Illinois. Feller incorporated molasses pellets with each load to promote fermentation of the silage. (Kankakee County, IL; 1954)

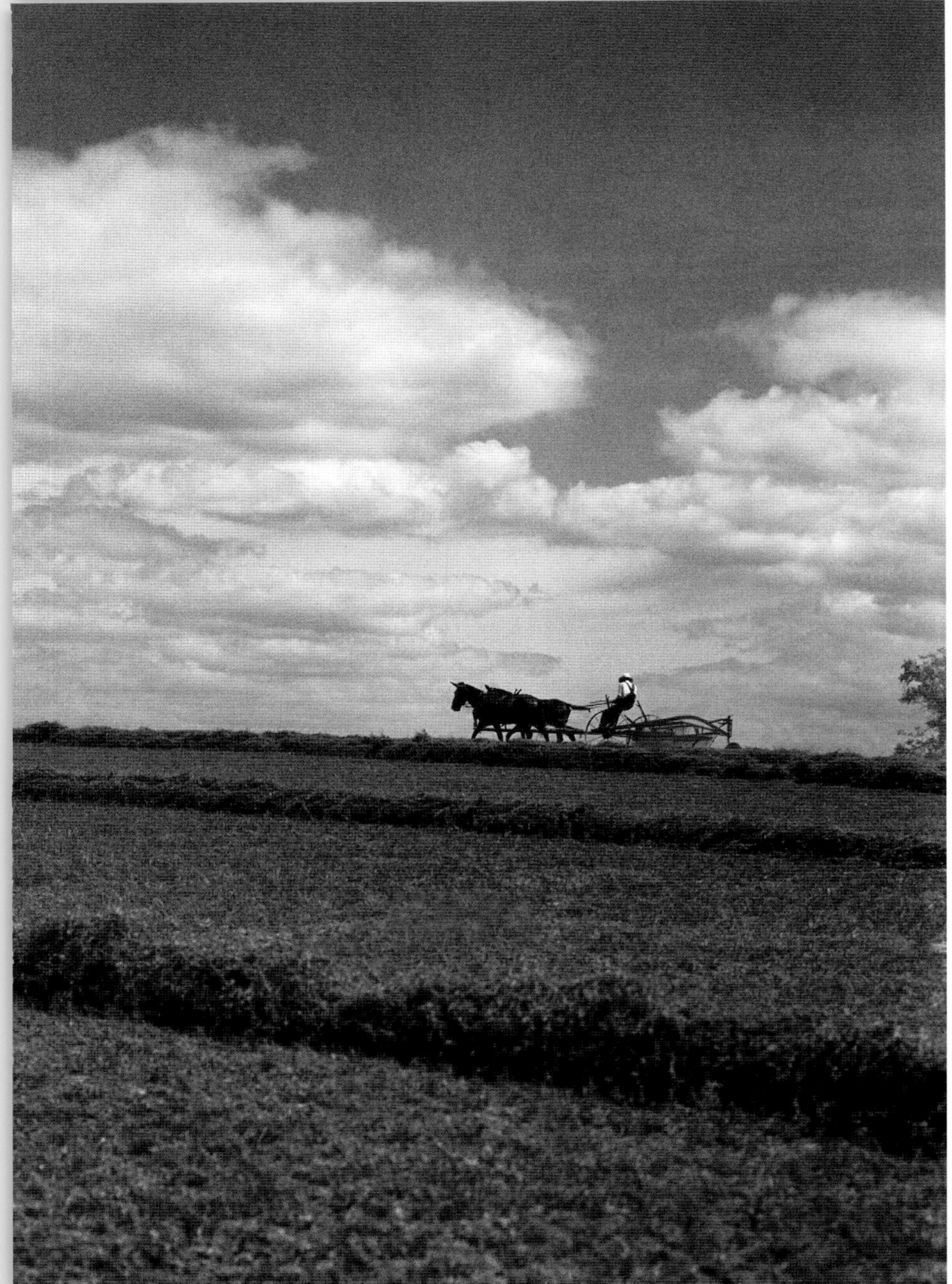

Raking alfalfa with a side-delivery rake on the farm of the Schulze Orchard Company near Oaktown (Knox County, 1935)

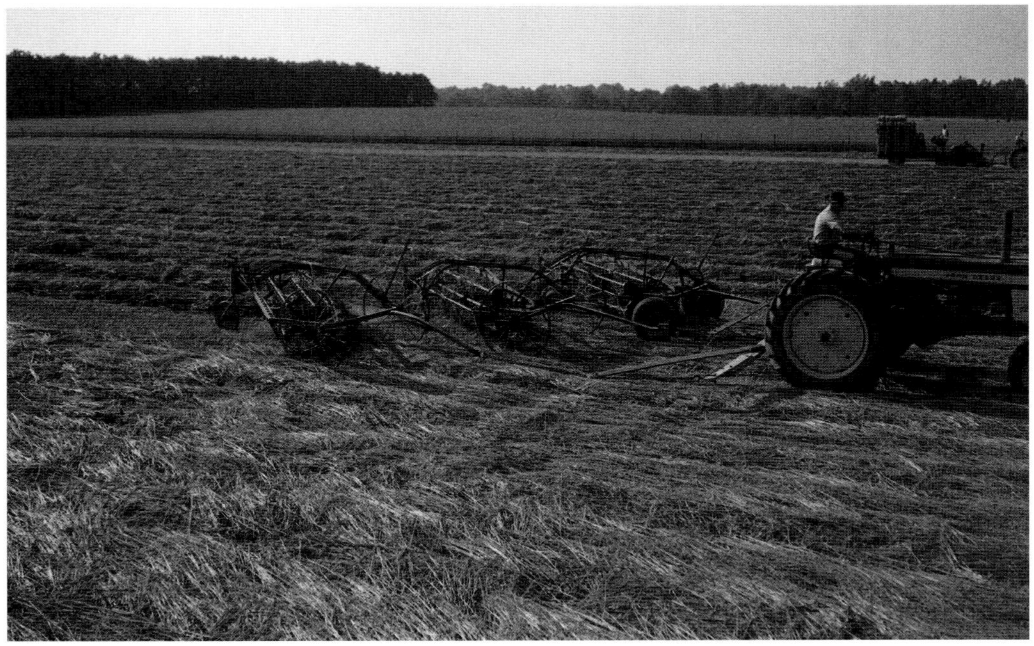

This man uses a tractor to pull three rakes that have been hooked together to put loose hay into rows on the Lawrence Love farm in Eaton. (Delaware County, 1958)

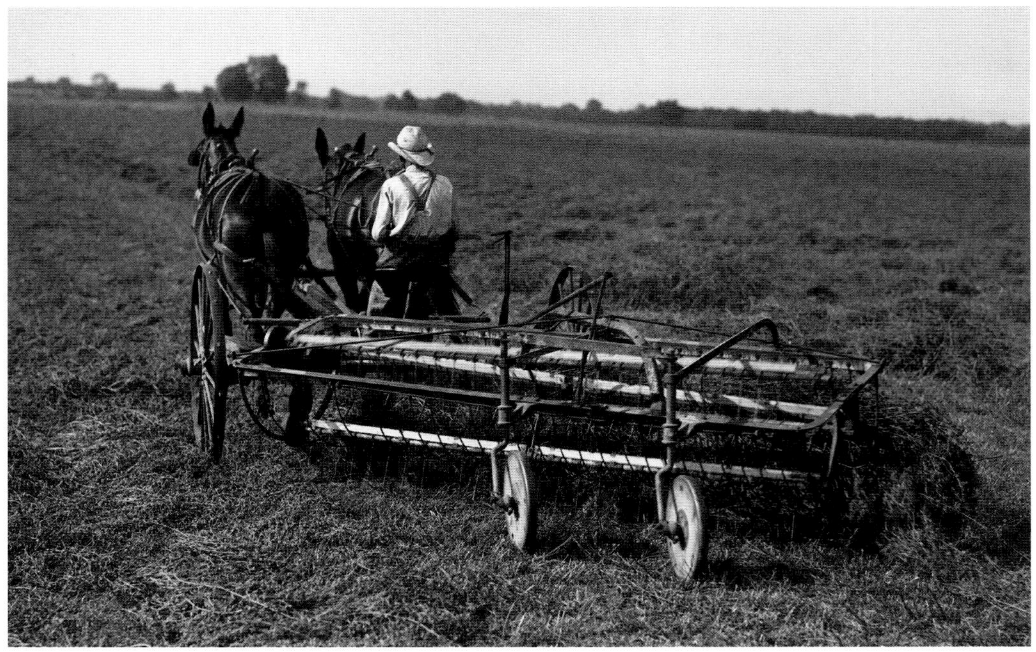

A mule team pulls a side-delivery rake at the Hayland Farms in Taylorville, Illinois. (Christian County, IL; 1920)

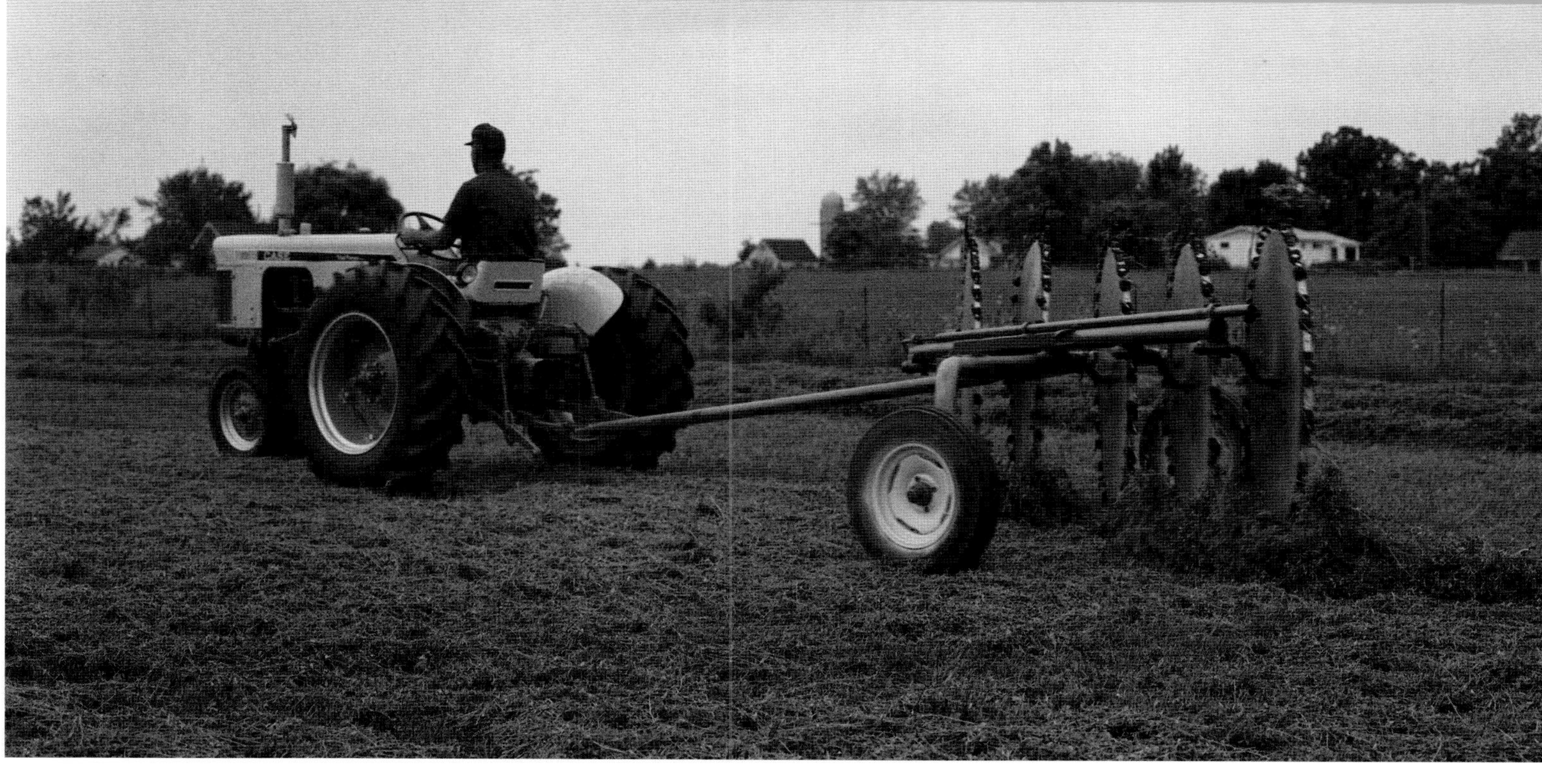

A man rakes hay with a wheel rake near Coldwater, Ohio. (Mercer County, OH; 1965)

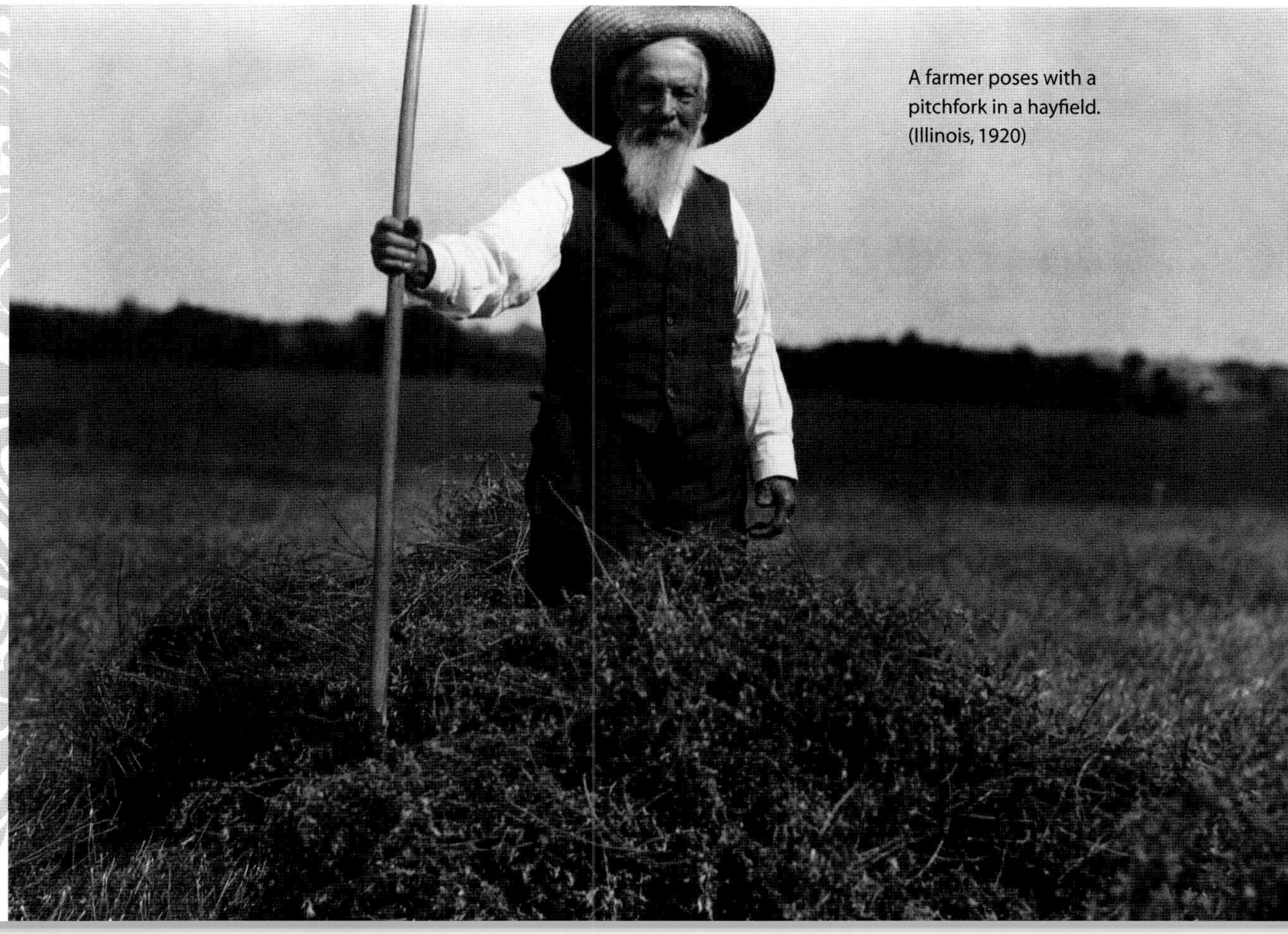

A farmer poses with a pitchfork in a hayfield. (Illinois, 1920)

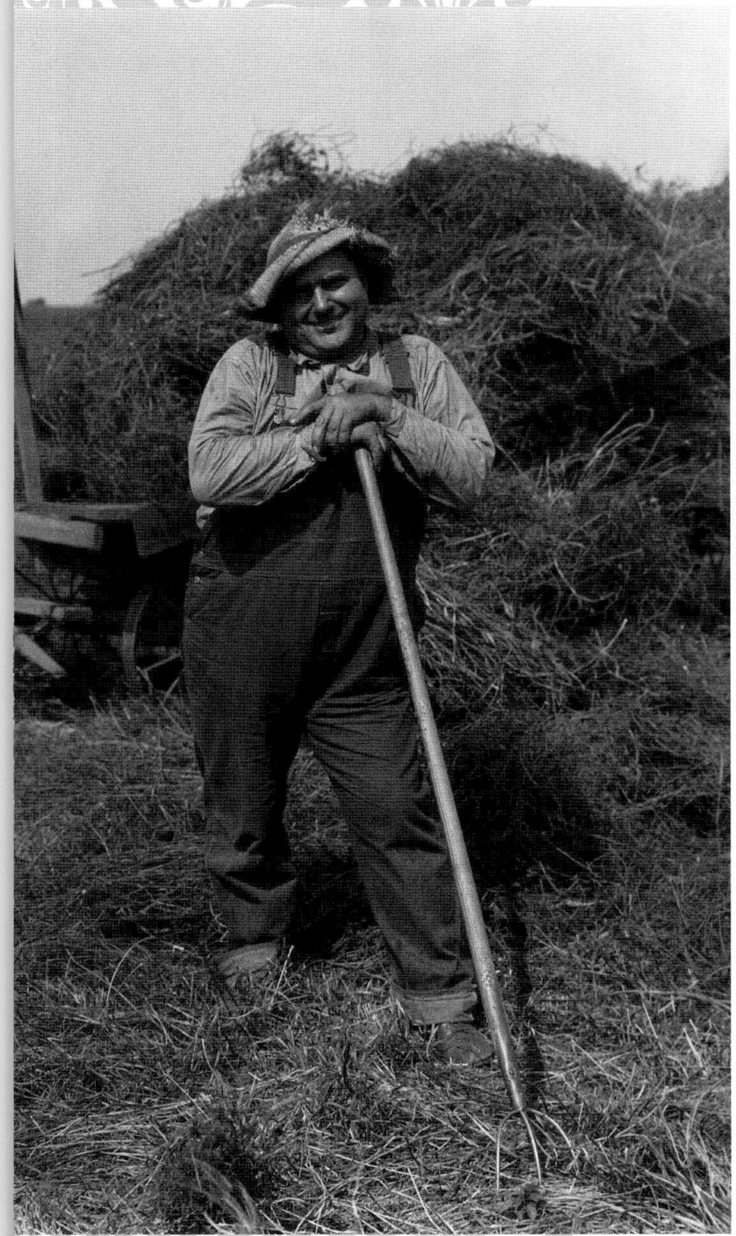

Ray Oswalt, a tenant farmer on the farm of Julius Berlovitz in Romney, pauses for a photograph in a clover field. (Tippecanoe County, 1929)

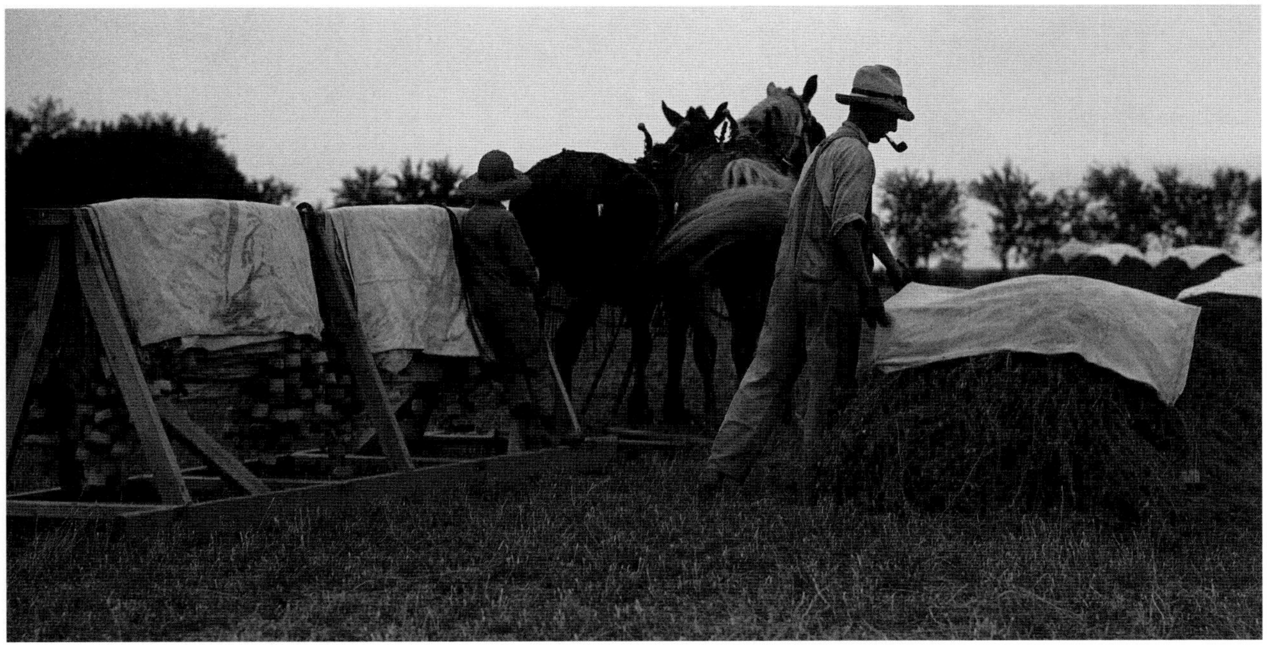

A farmer places canvas covers on hayshocks on the Drew Ten Broeck farm in McLean, Illinois. The horse-drawn rack provided a convenient way to handle the covers. (McLean County, IL; 1917)

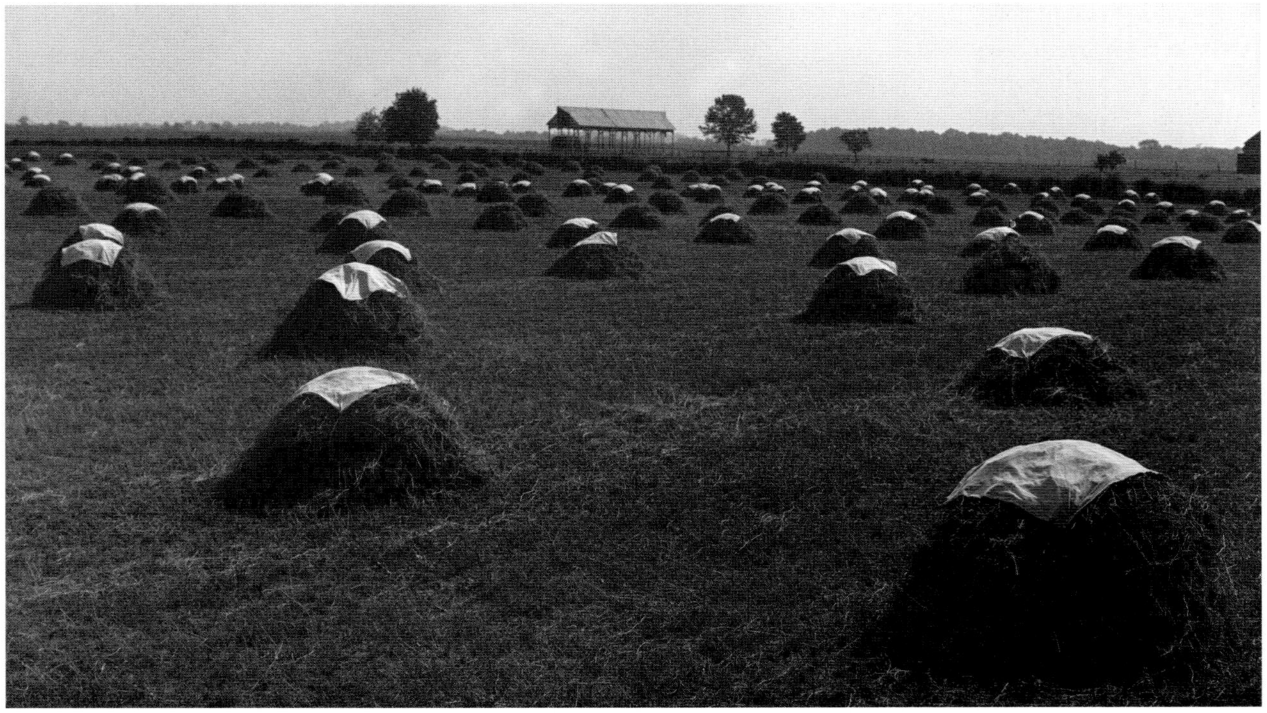

Shocked alfalfa on the farm of J. Crouch and Son (Tippecanoe County, 1914)

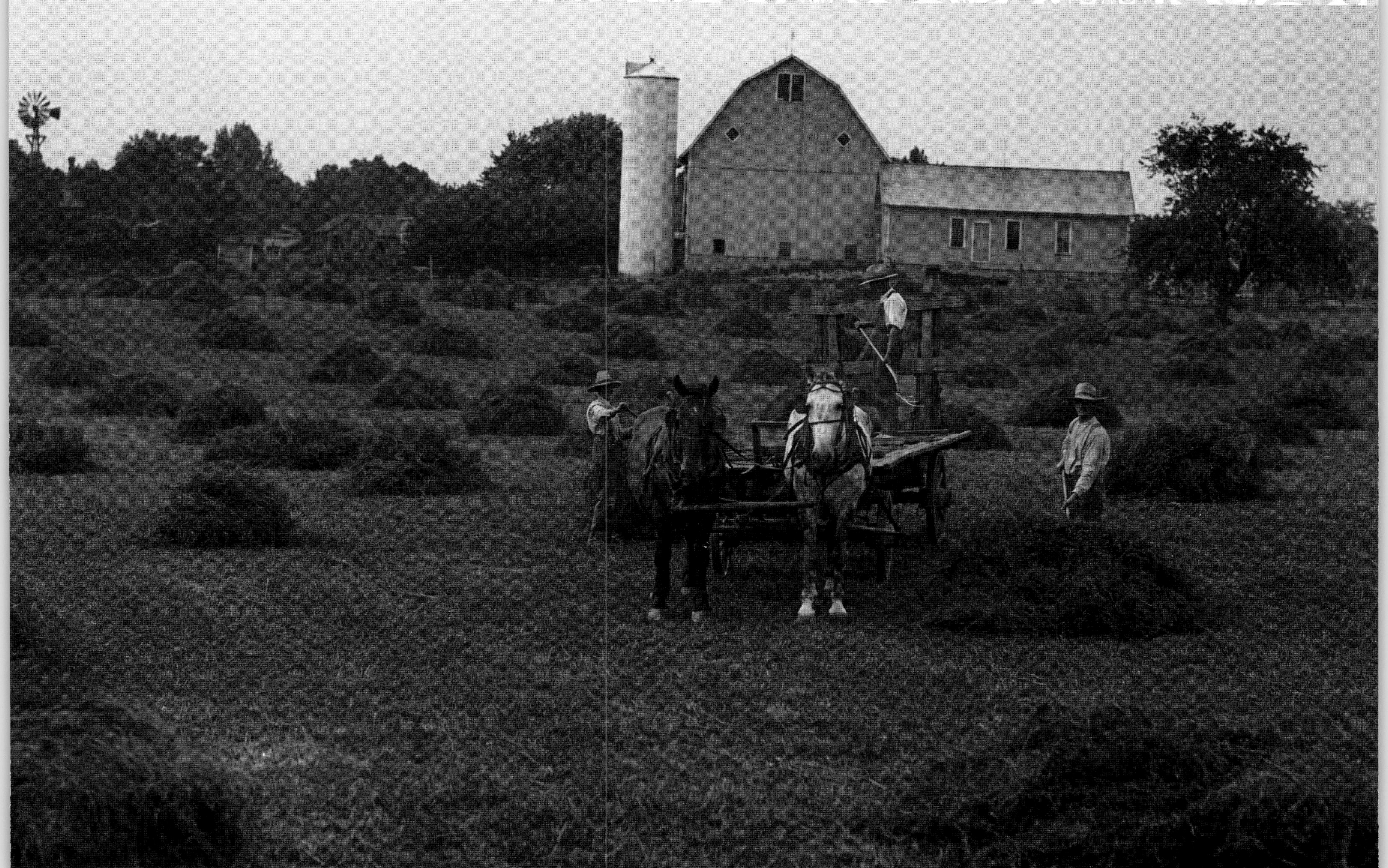

Three men pitch shocks of alfalfa hay onto a wagon on the G. W. Lemler farm in Plymouth. Three cuttings of the crop yielded seven tons per acre. (Marshall County, 1922)

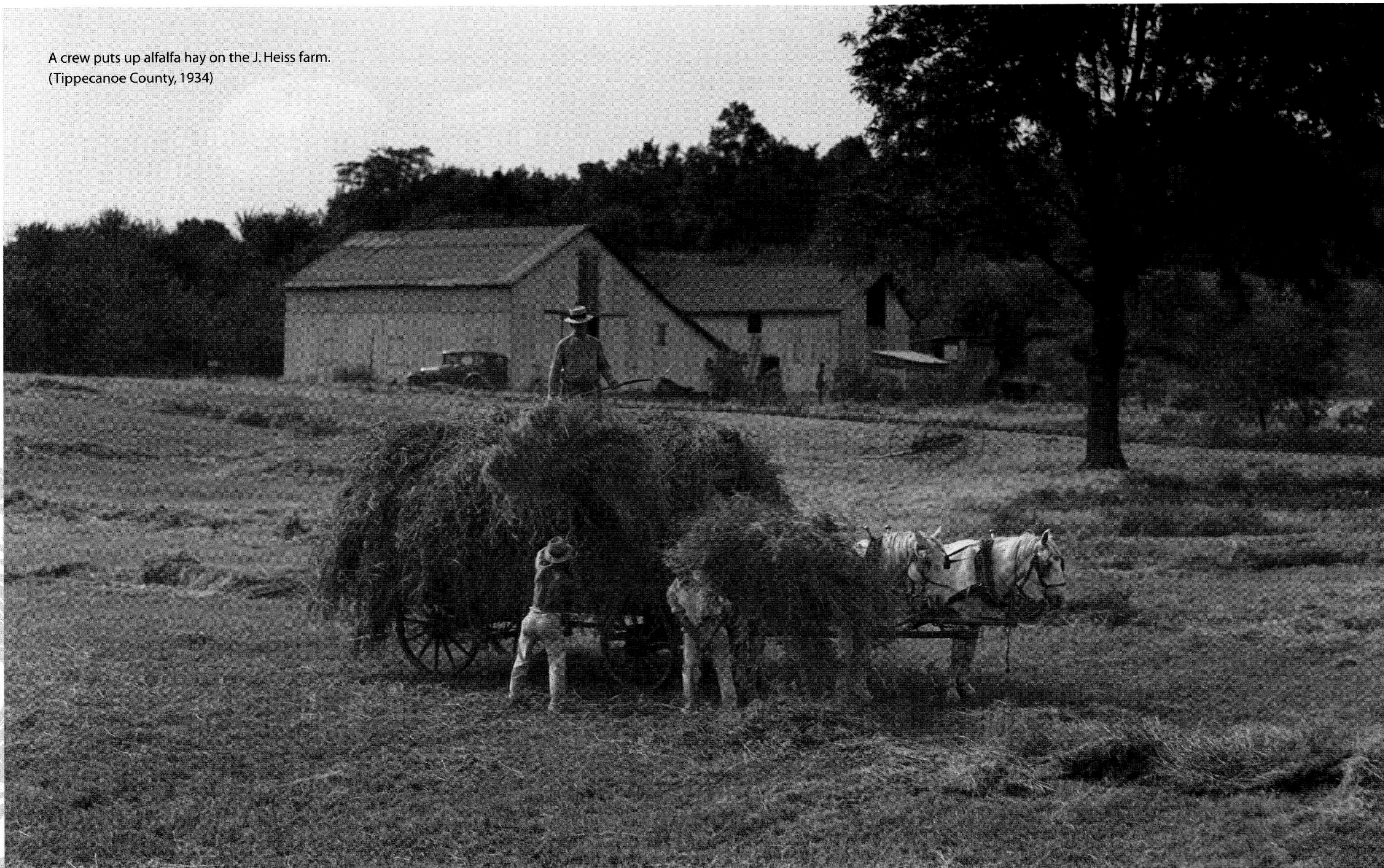

A crew puts up alfalfa hay on the J. Heiss farm.
(Tippecanoe County, 1934)

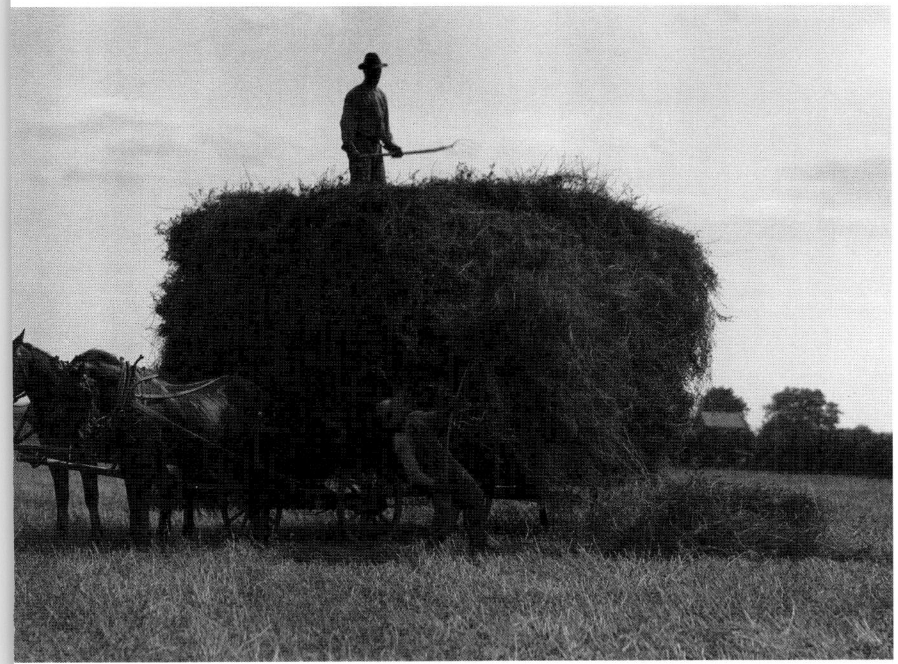

At the Albert Heckle farm in Quincy, Illinois, a man uses a pitchfork and brute strength to move alfalfa hay onto the top of a stack. (Adams County, IL; 1920)

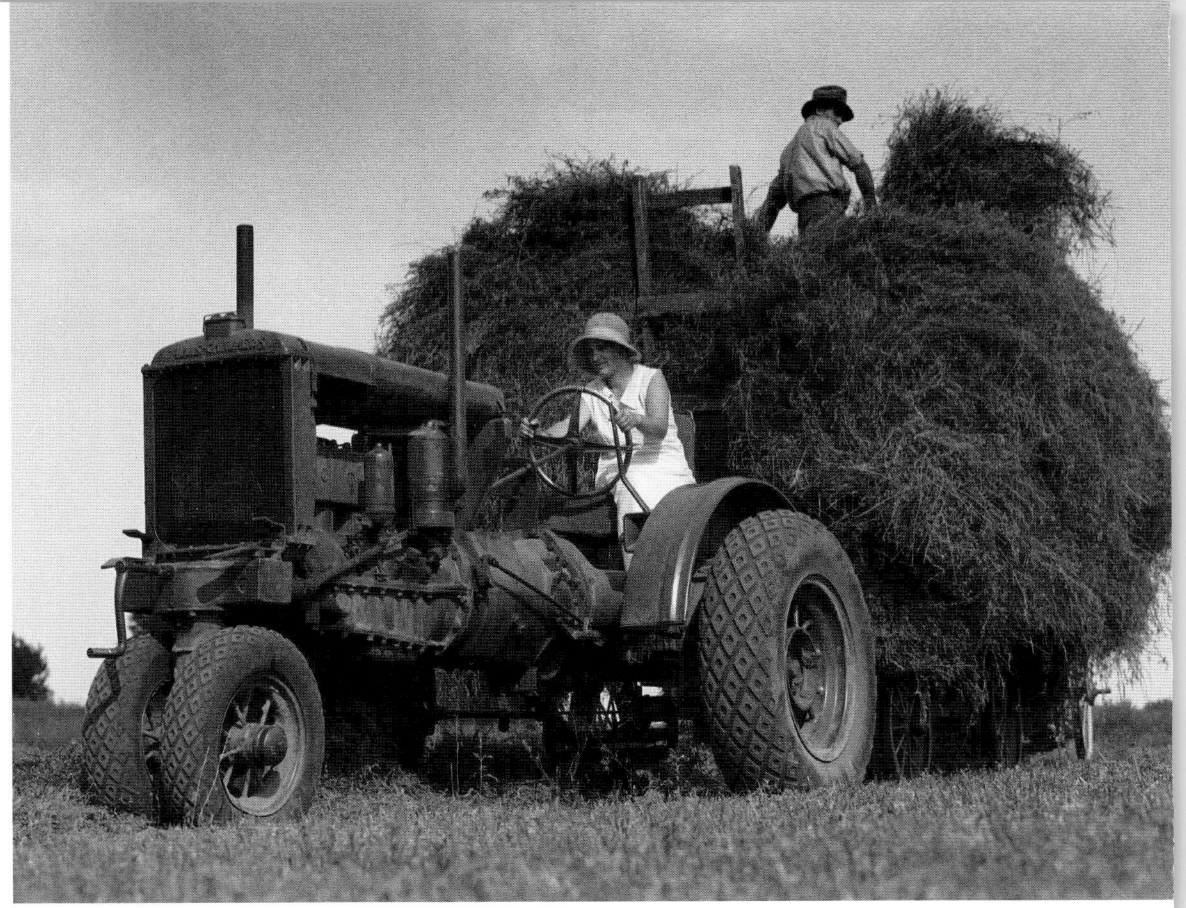

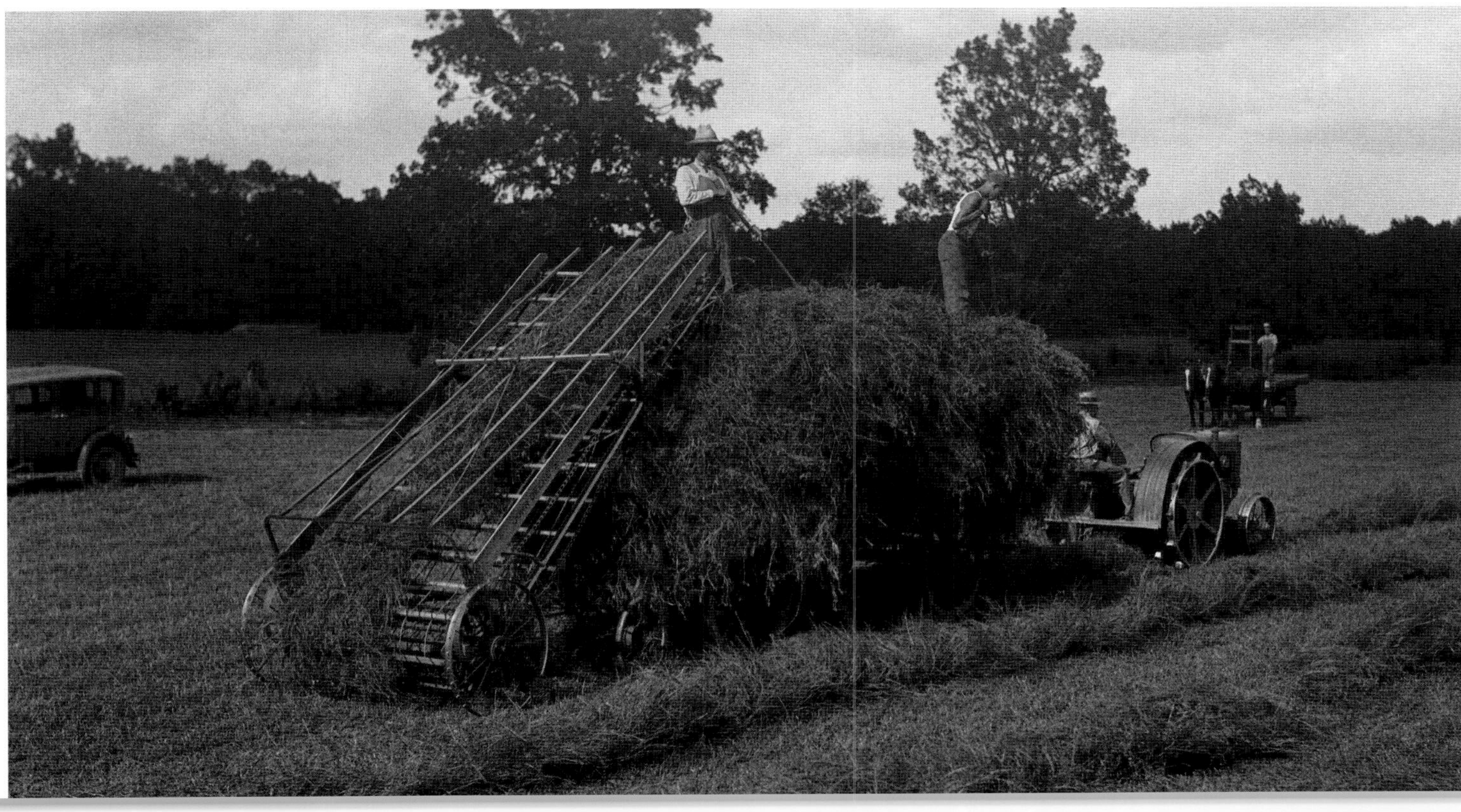

(Top right) Using an Allis-Chalmers tractor with Goodyear tires to work a hayfield on the Albert Grimes farm in Rockford, Illinois (Winnebago and Ogle Counties, IL; 1933)

(Bottom) Loading alfalfa hay high onto a wagon using a hay loader drawn by tractor (Tippecanoe County, 1931)

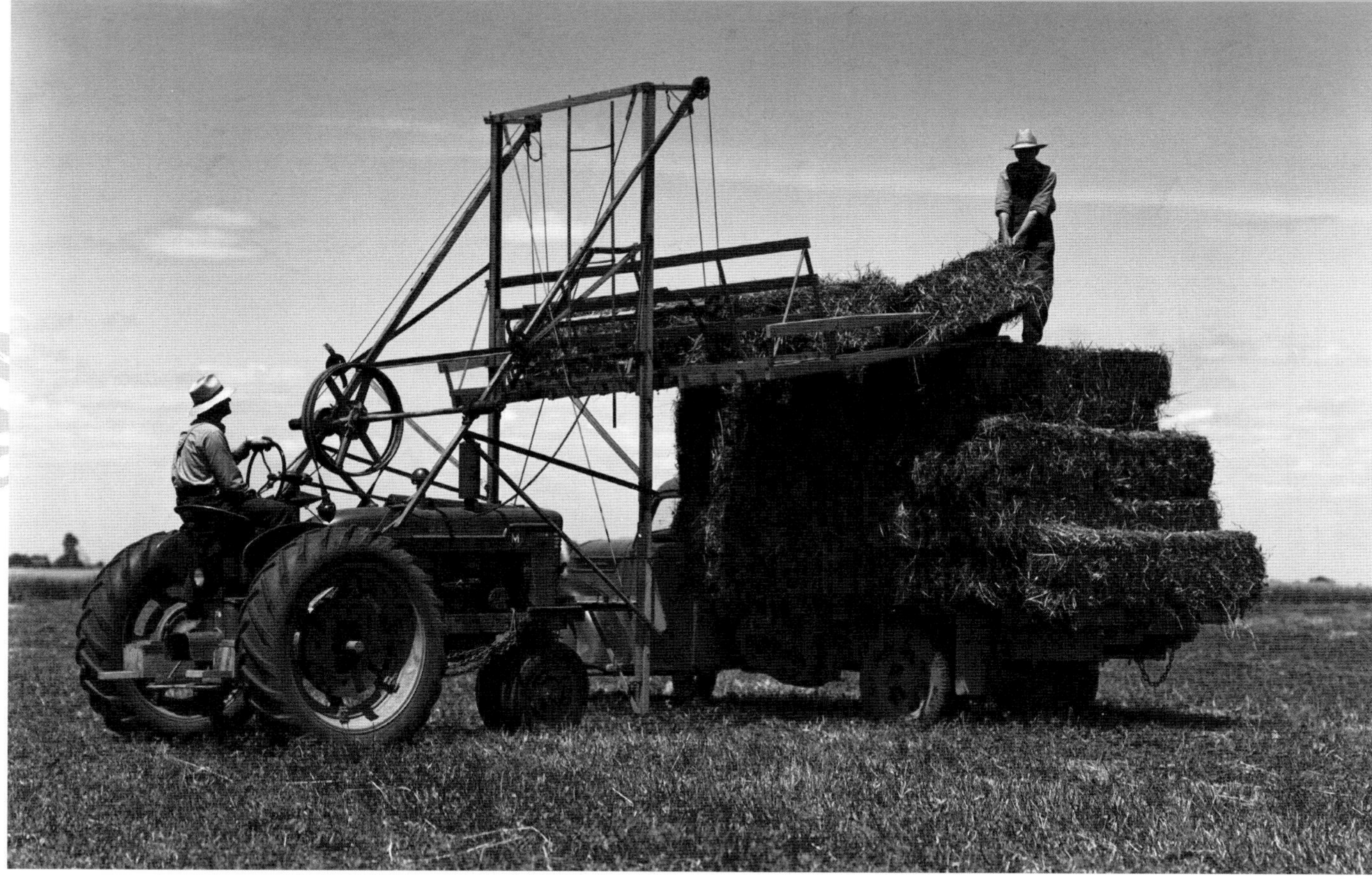

Two men load baled hay onto a wagon with a Gunning Hay Stacker on the farm of Frank Kirkpatrick in Frankfort. (Clinton County, 1945)

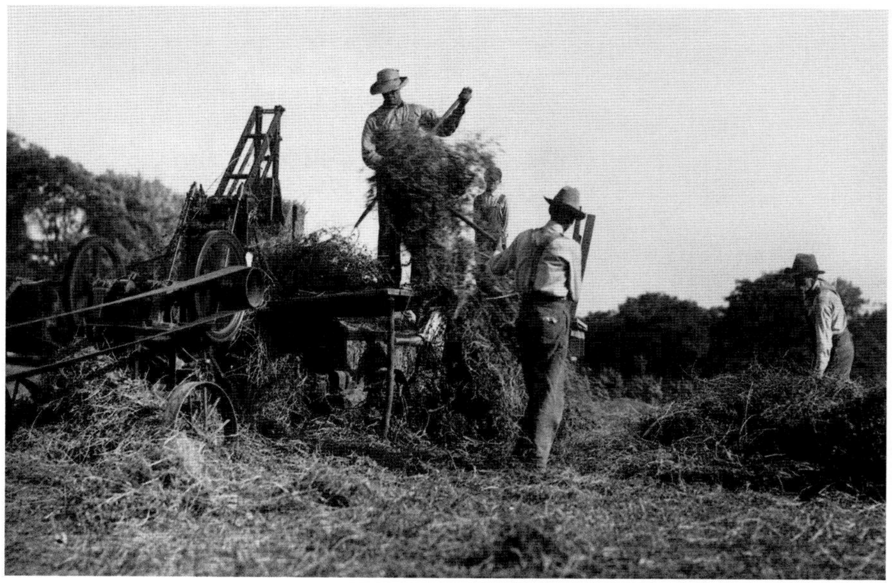

Using a stationary baler in an alfalfa field on the farm of C. D. Greffe near Taylorville, Illinois (Christian County, IL; 1920)

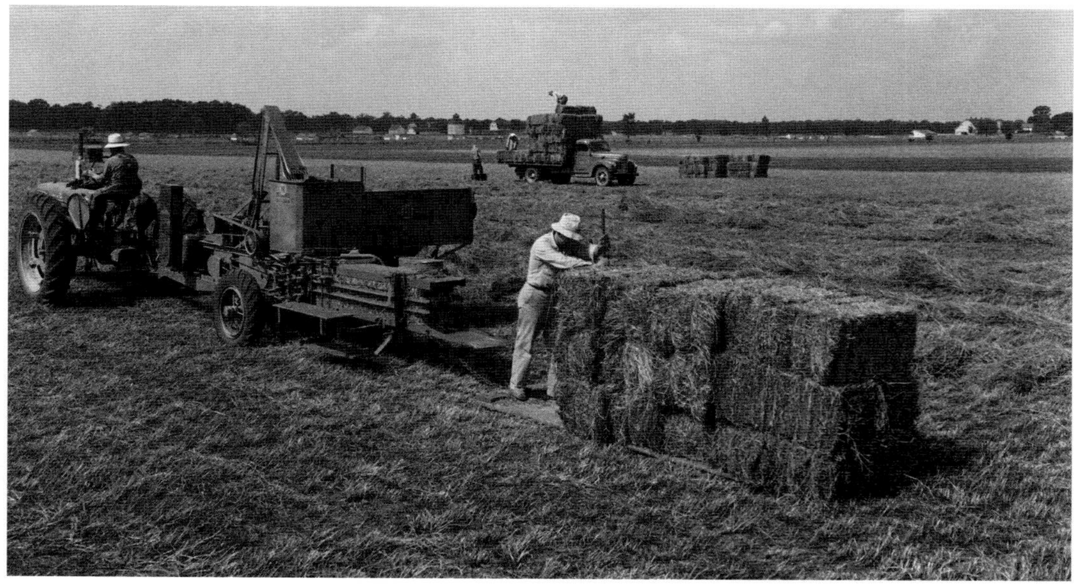

The sled in back of this hay baler made loading and unloading easier for workers on Purdue University's Lynnwood Farm in Carmel. (Hamilton County, 1947)

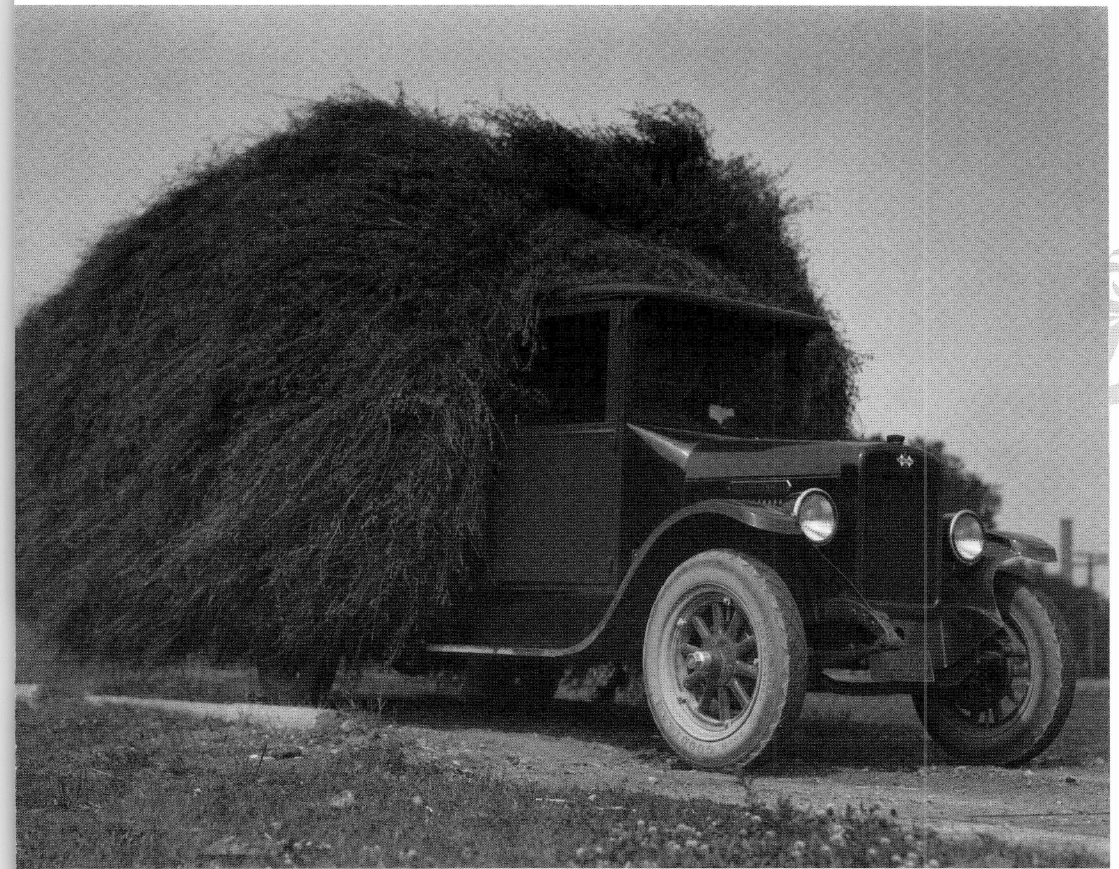

An International truck owned by Roy A. Jones is loaded down with hay. (Tippecanoe County, 1929)

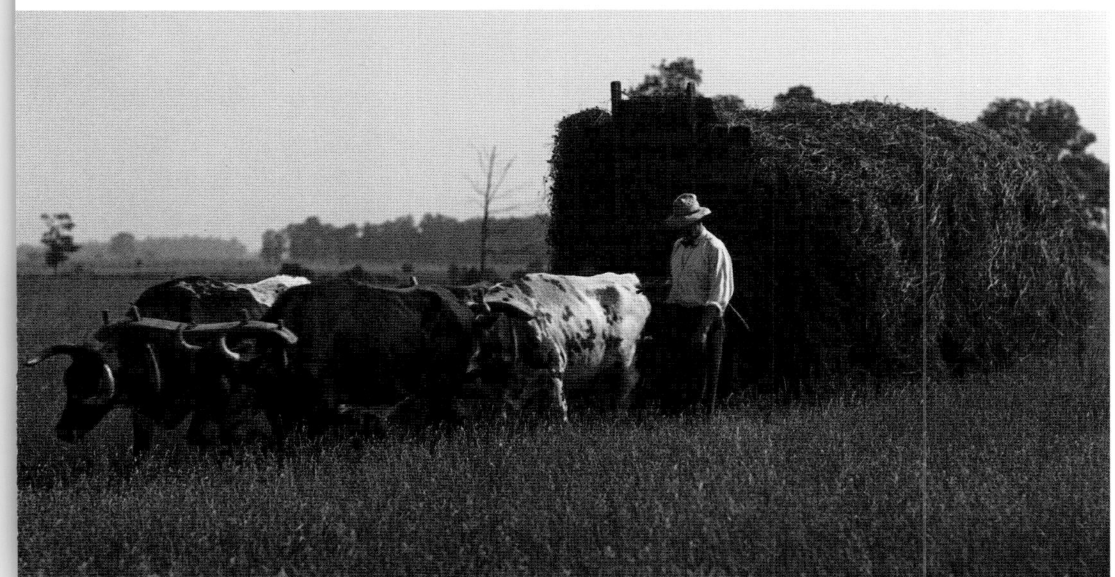

Oxen haul hay from fields belonging to J. E. Green of Muncie. (Delaware County, 1927)

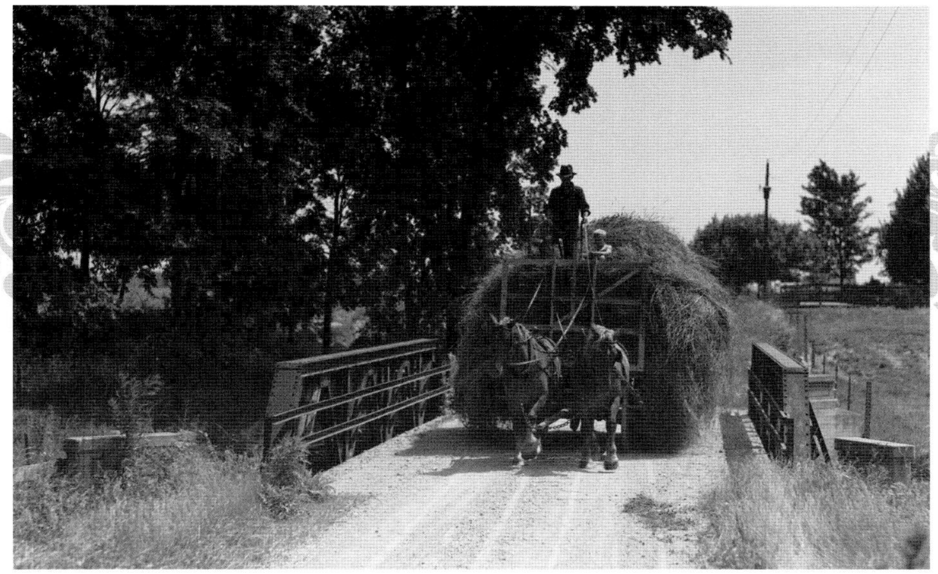

William House drives a team of two Belgian horses to transport hay near his farm in Arcadia. (Hamilton County, 1943)

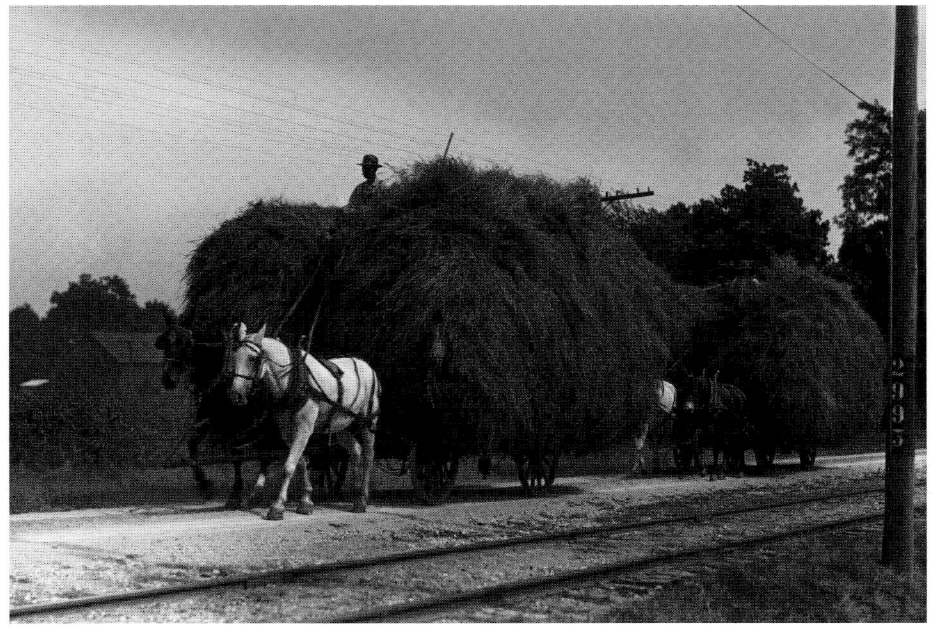

Horses pull two wagonloads of hay near Dayton. (Tippecanoe County, 1913)

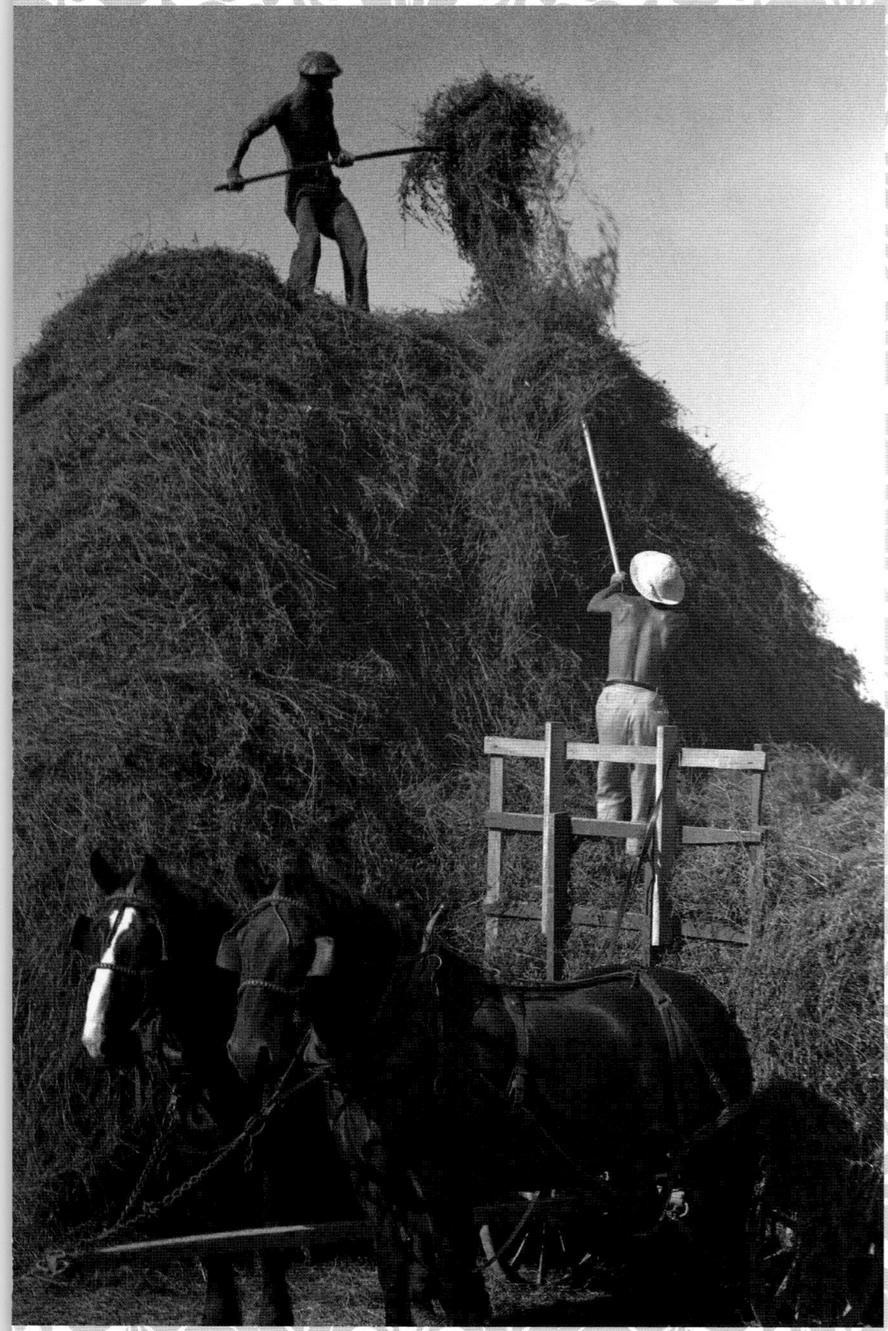

Stacking alfalfa hay on the farm of Charles Wilson of Kennard (Henry County, 1940)

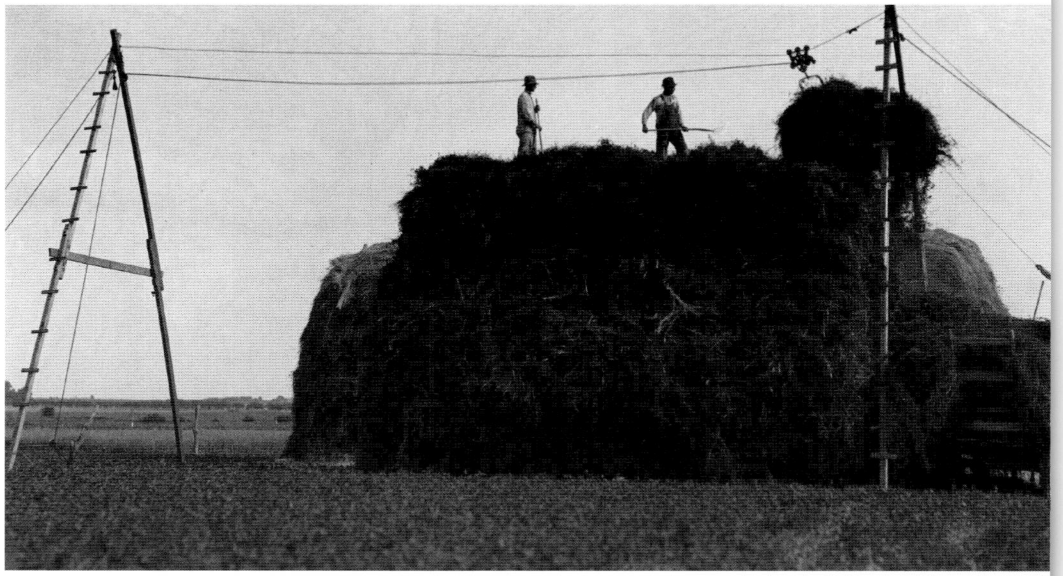

Workers use ingenuity and a special apparatus to build a large haystack at the Deer Park Farm in Utica, Illinois. (LaSalle County, IL; 1918)

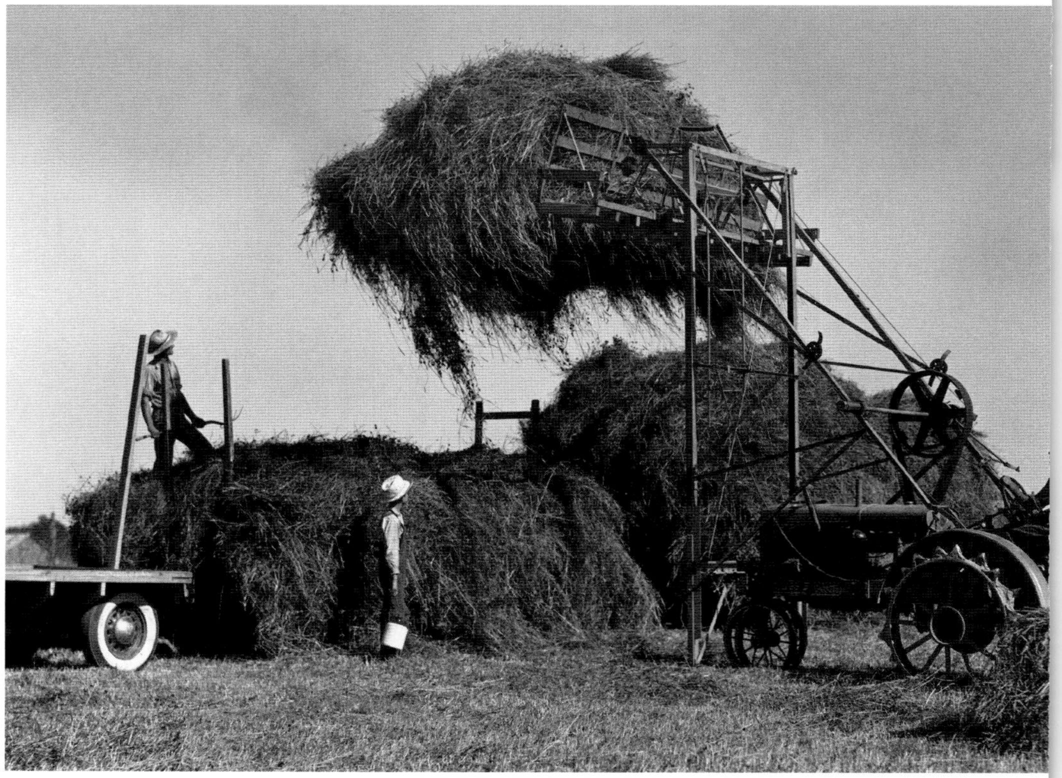

Using a Gunning Hay Stacker to load hay onto a wagon on the farm of Art Kessler (Tippecanoe County, 1944)

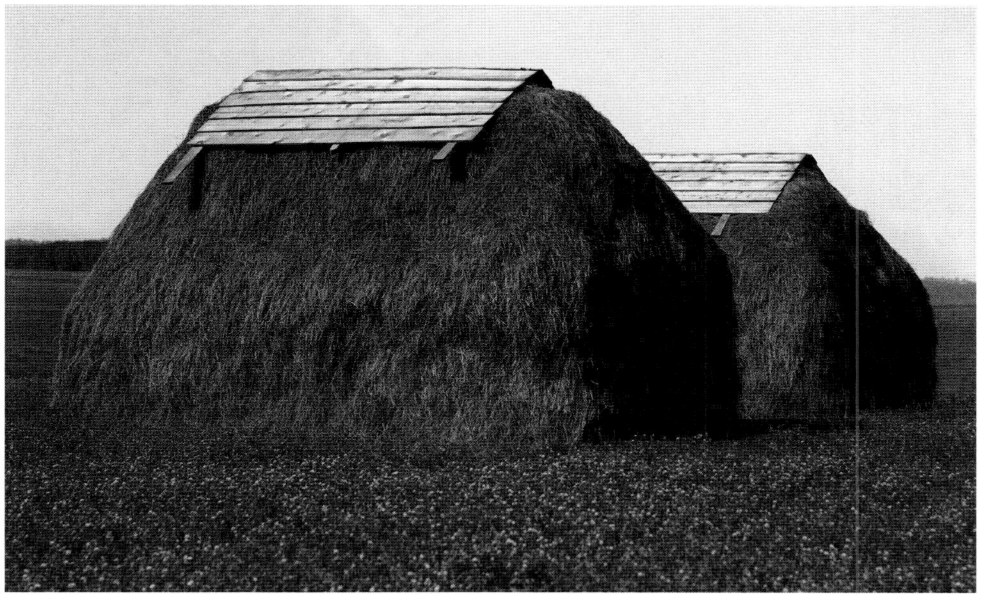

C. R. Fenske in Utica, Minnesota, uses wooden covers on his haystacks. (Winona County, MN; 1919)

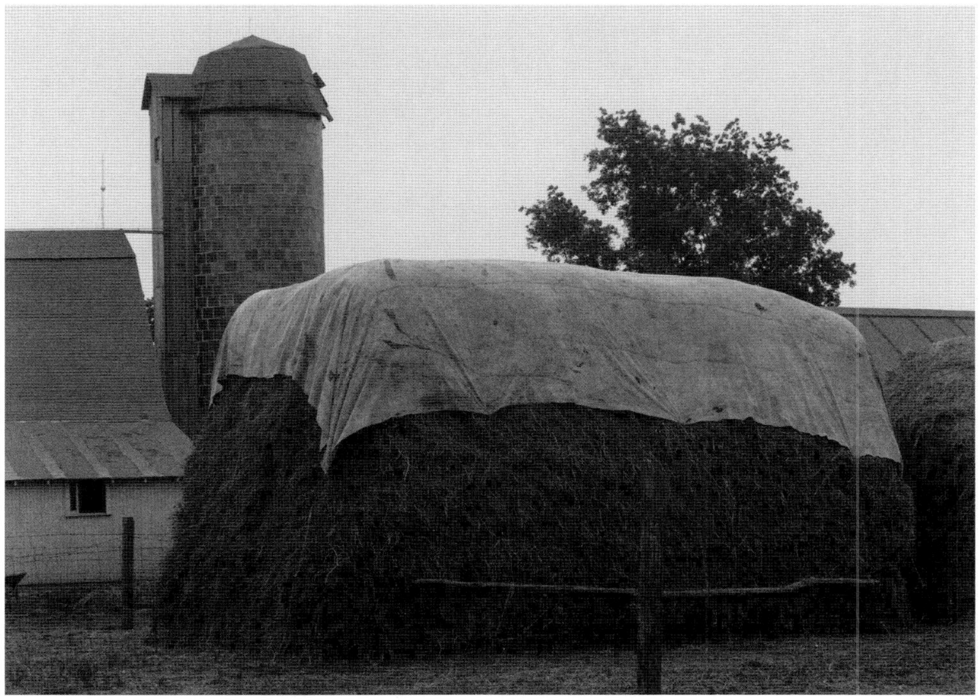

A canvas covers a haystack on the dairy farm of George J. Wiese near Lafayette. (Tippecanoe County, 1931)

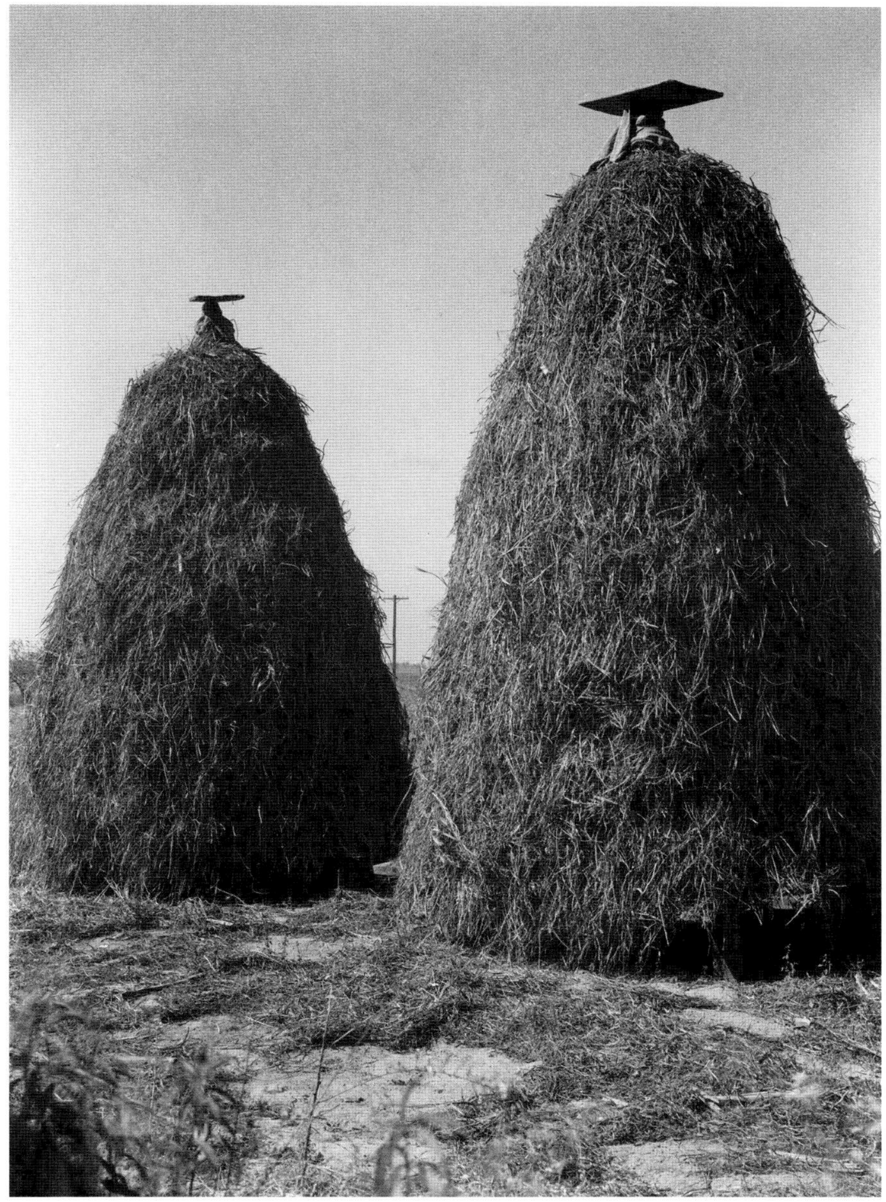

An example of haystacks made in the southern United States (Alabama or South Carolina, 1927)

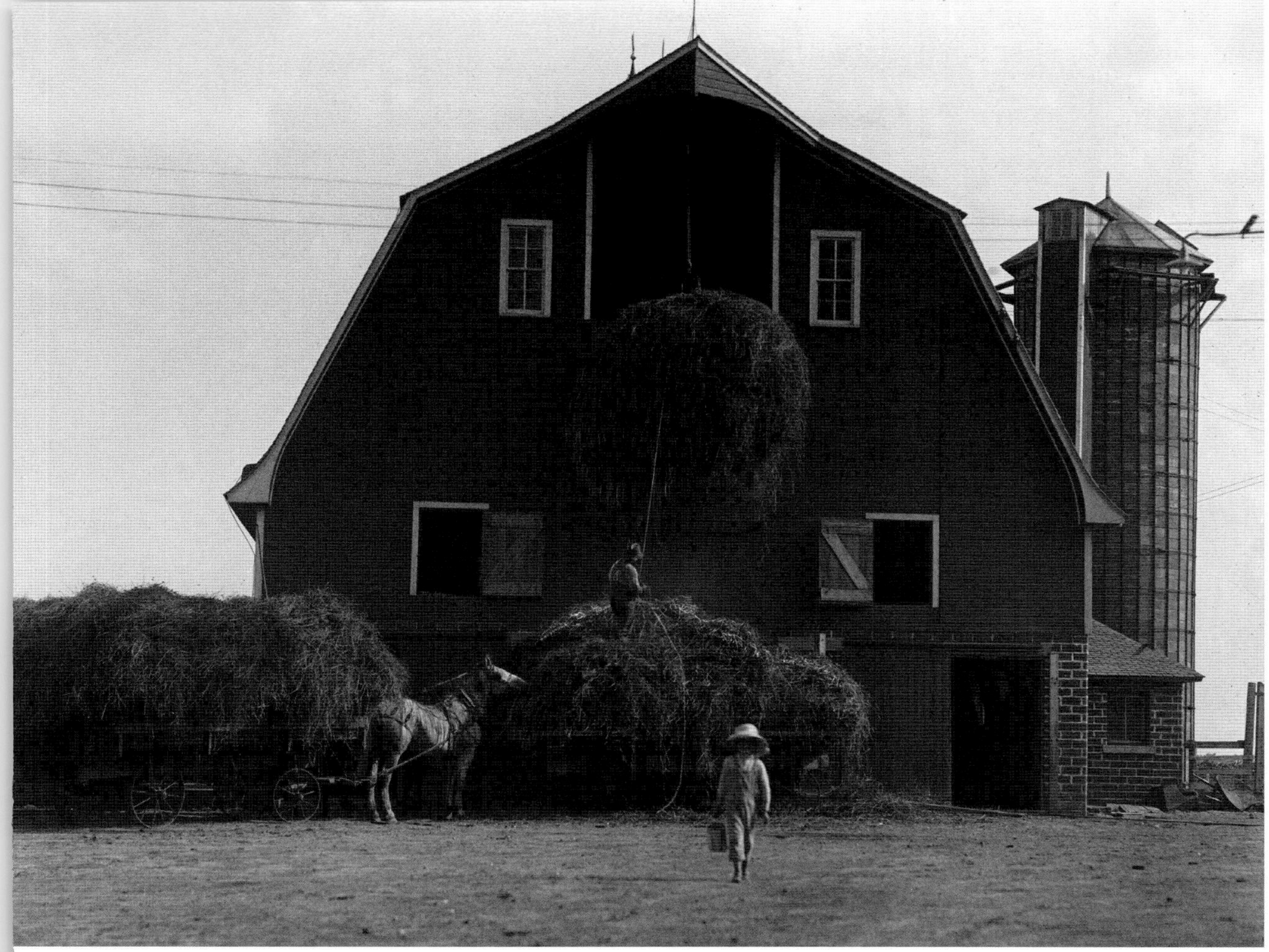

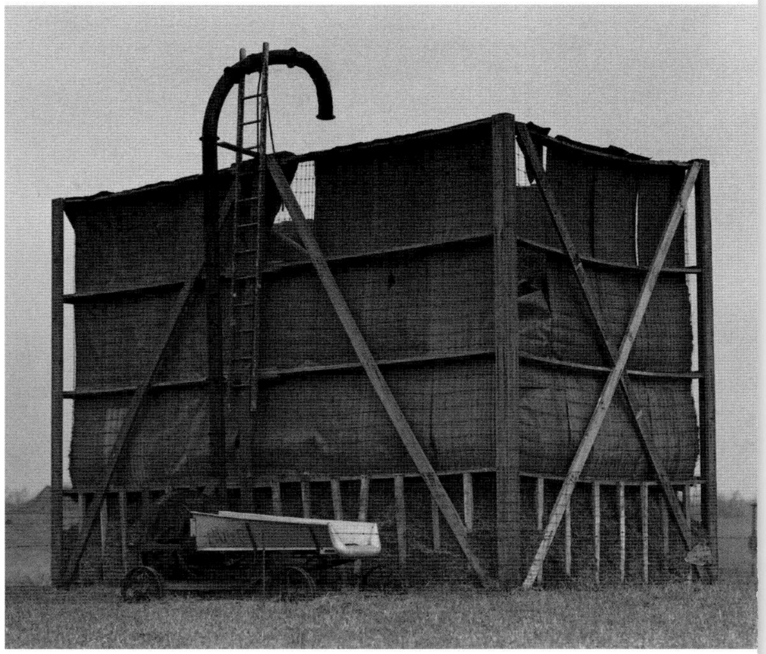

A portable hay shed allowed Bill Jones to feed chopped hay to steers on pasture on his farm near Anderson. (Madison County, 1949)

A farmer hoists hay into the barn on the farm of W. A. Davidson in Brooklyn, Iowa. (Poweshiek County, IA; 1919)

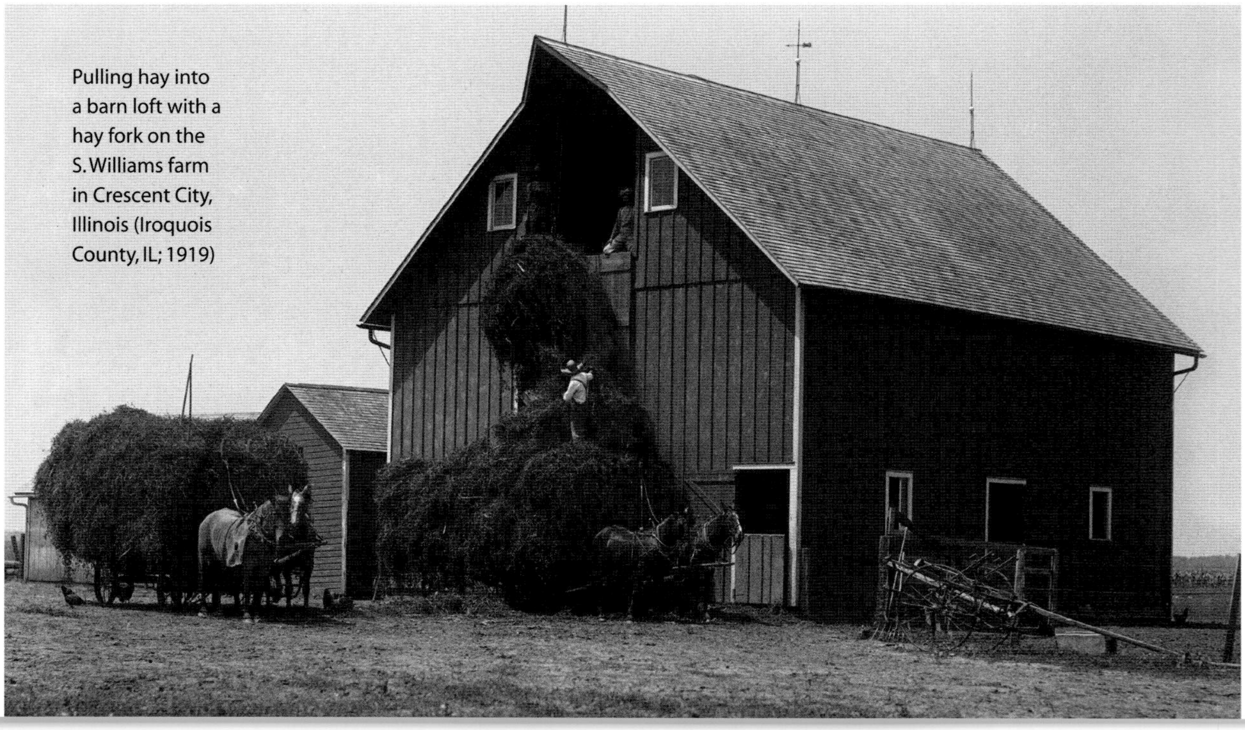

Pulling hay into a barn loft with a hay fork on the S. Williams farm in Crescent City, Illinois (Iroquois County, IL; 1919)

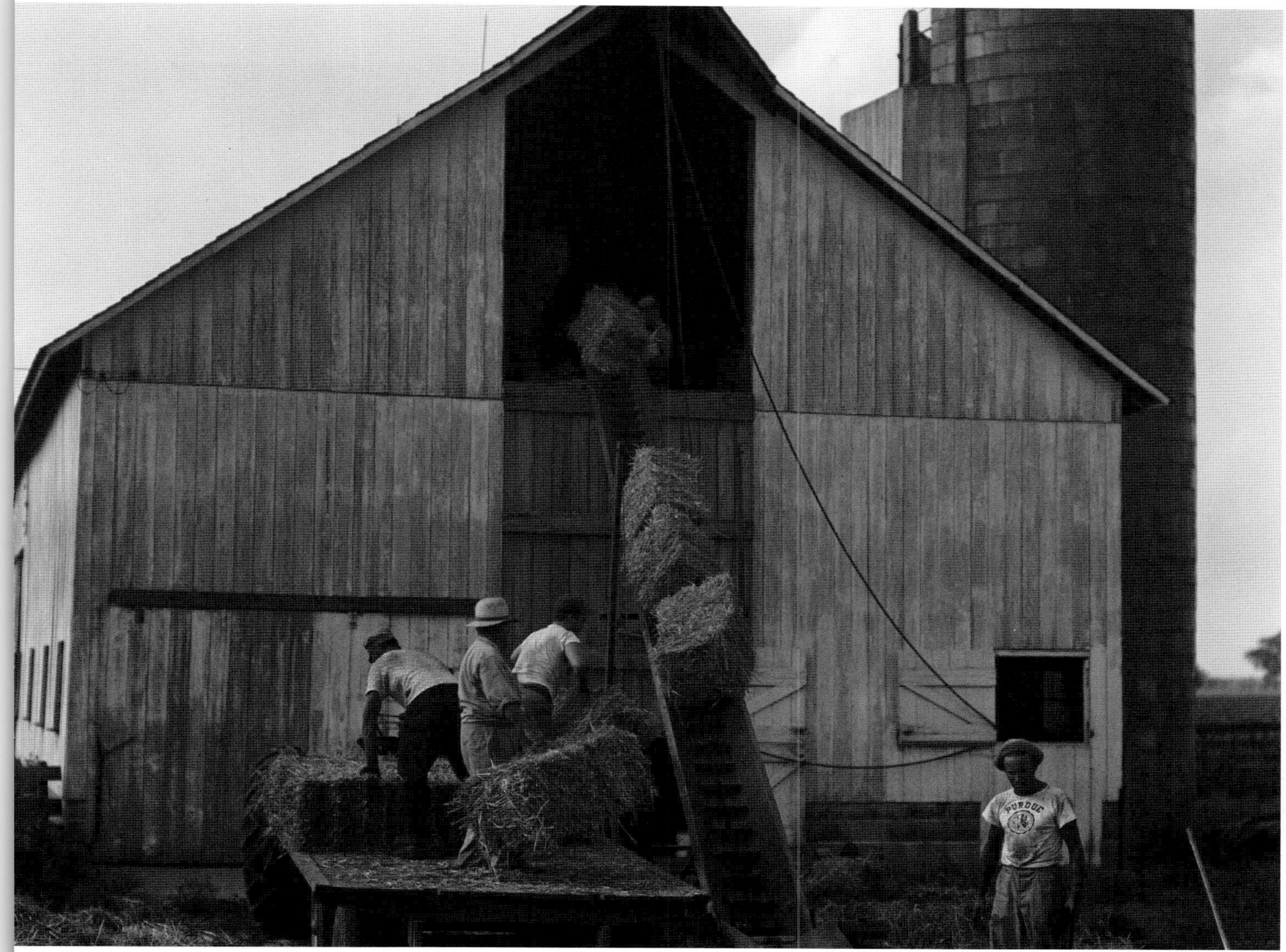

Workers use an elevator to load baled straw into a haymow on a farm near Star City. (Pulaski County, 1946)

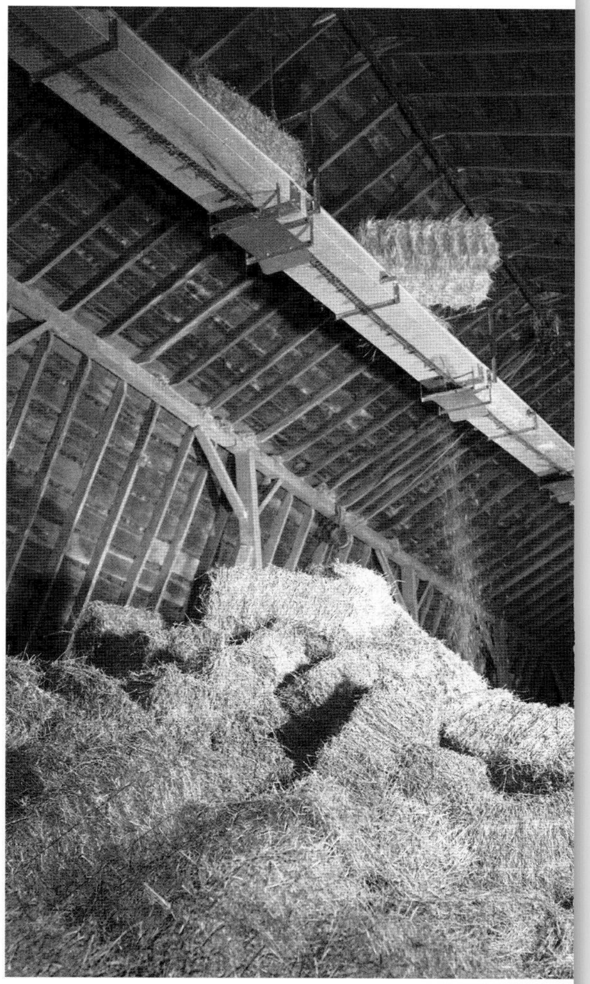

An automated system distributes hay throughout the haymow in this rather large dairy barn on the Alva Braid and Son Farm in Durand, Michigan. (Shiawassee County, MI; 1958)

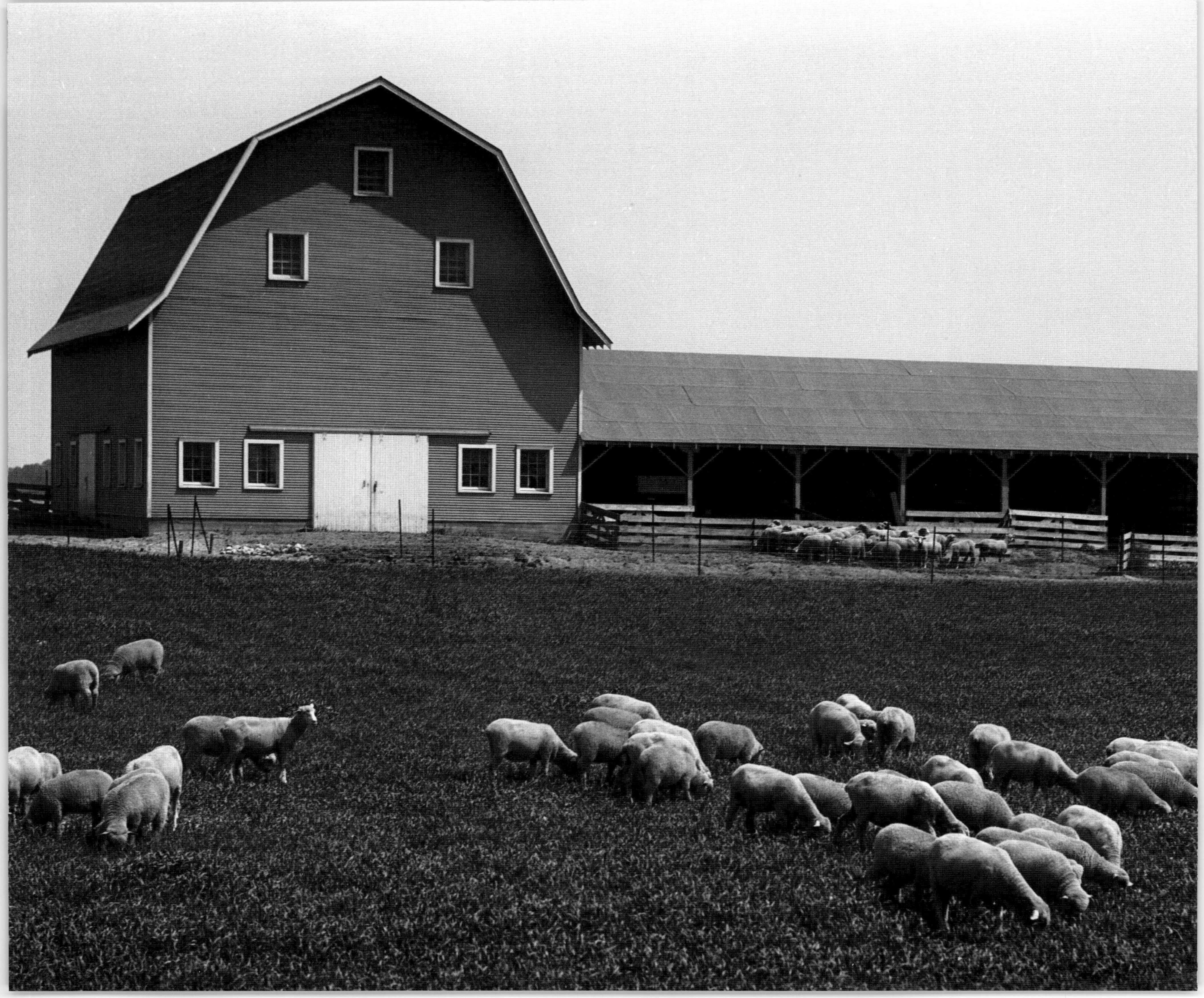

THROUGH THE LENS OF PIONEER PHOTOGRAPHER
J. C. Allen

PART 3 *The Livestock Connection*

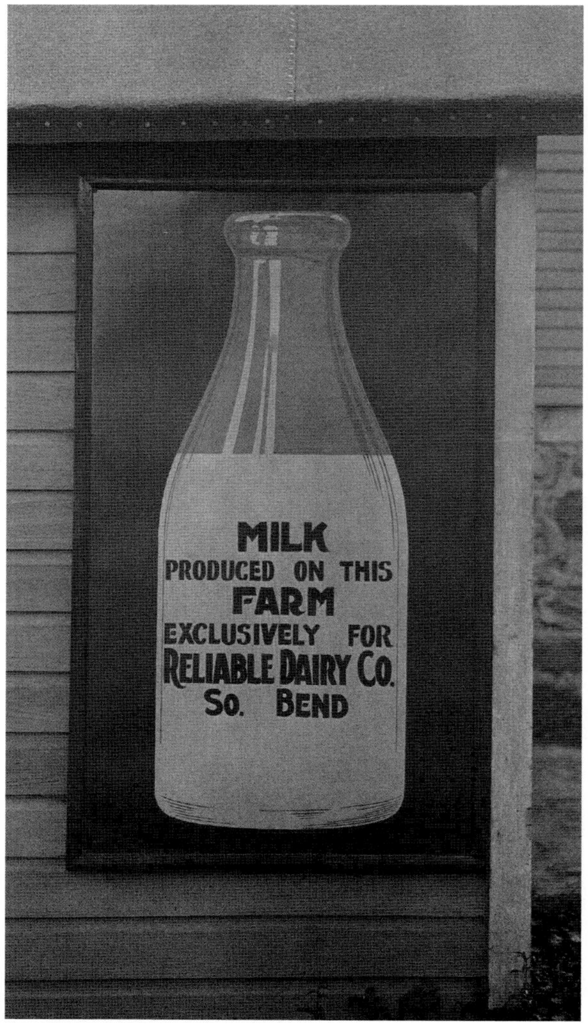

Albert A. Fulk displays a sign at his dairy in South Bend. (St. Joseph County, 1927)

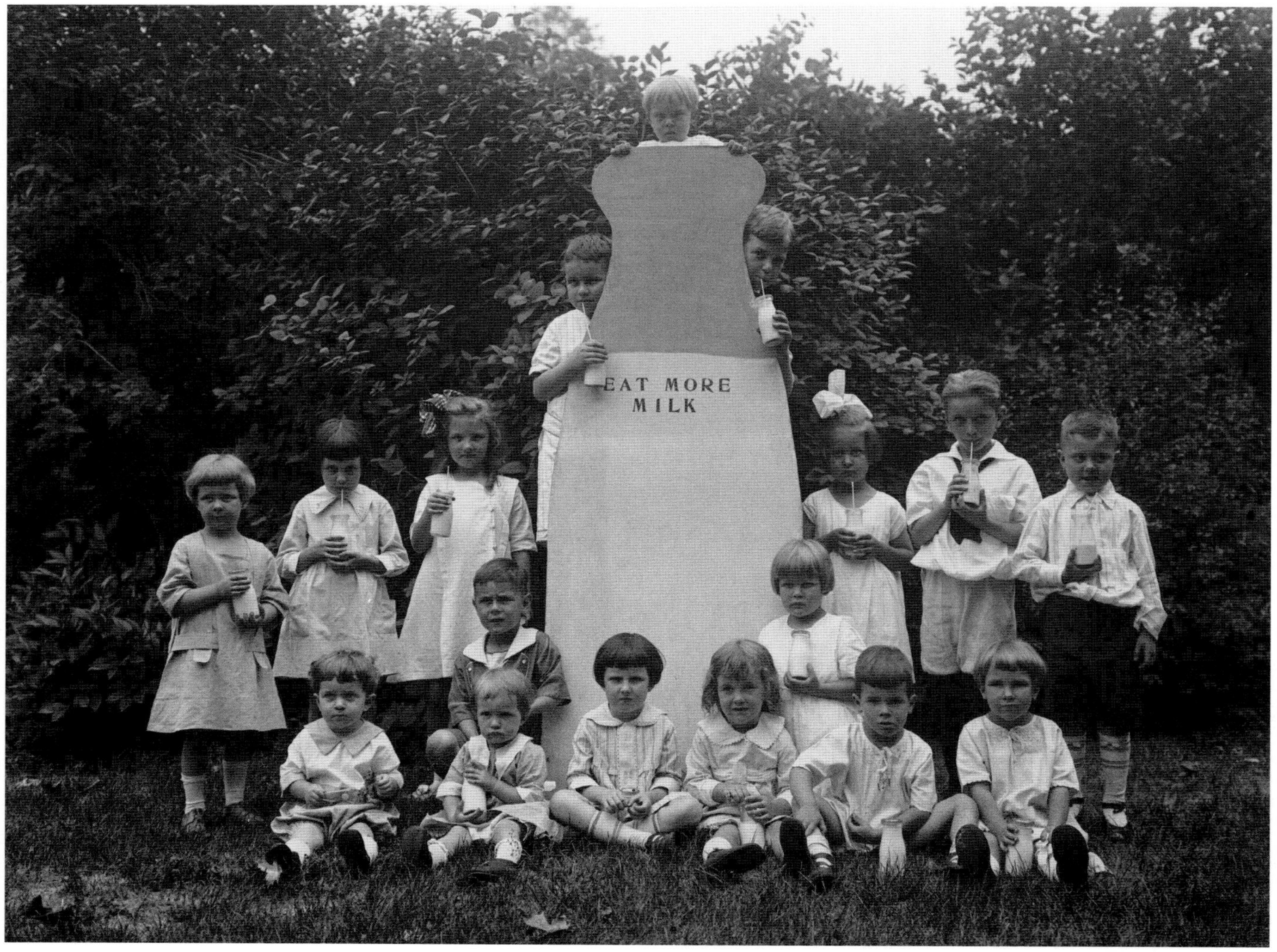

Young children pose around a large plywood milk bottle. (Indiana, 1921)

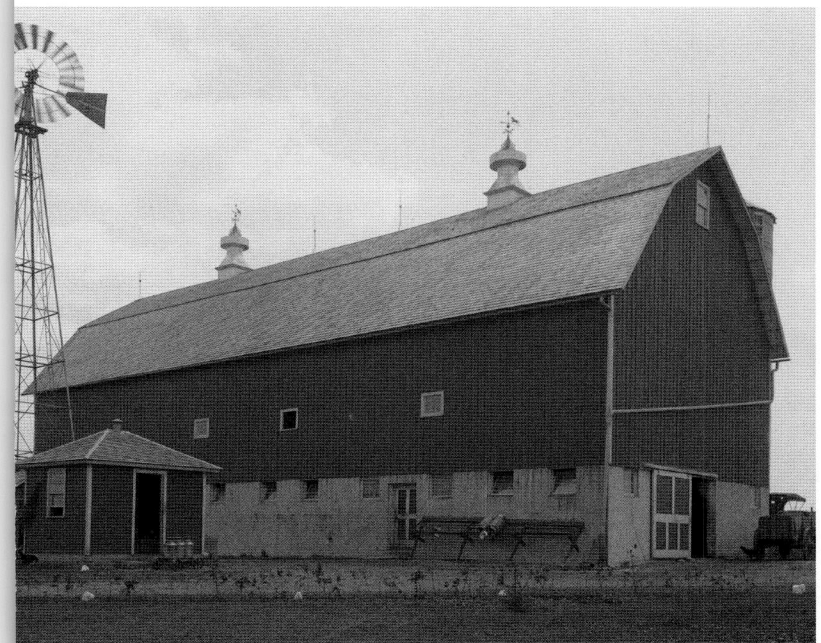

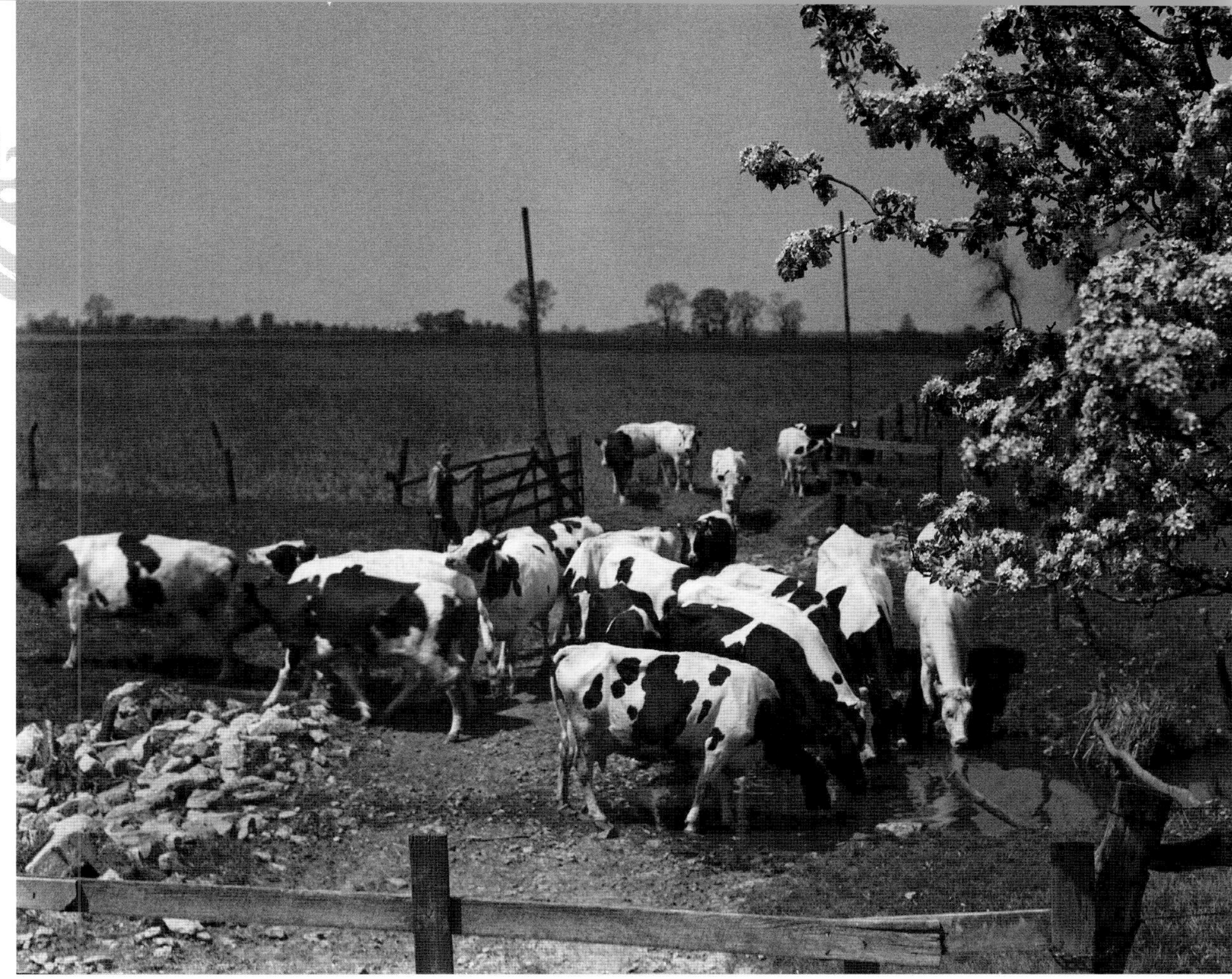

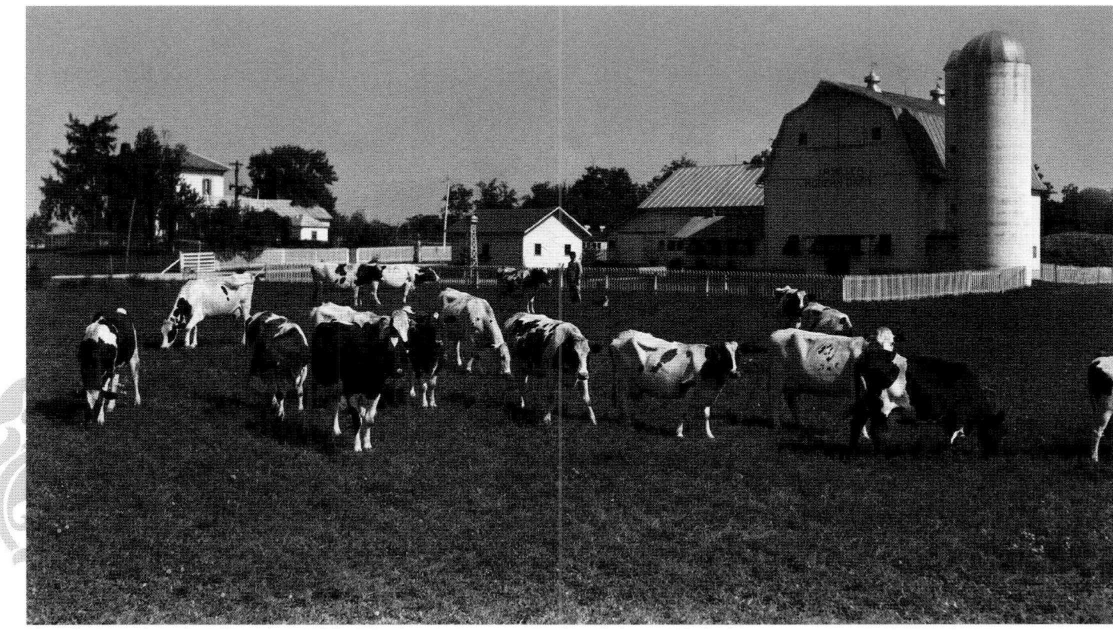

(Top left) A milk house (left foreground), windmill, and dairy barn on the Chris Christianson farm in Geneva, Illinois (Kane County, IL; 1918)

(Top right) A Holstein herd drinks at a small stream on the Oscar Hellmann farm in Worden, Illinois. (Madison County, IL; 1952)

(Bottom) J. R. Miller of Akron won first prize in the Chicagoland's Neatest Dairy Farm Contest with his farmstead and cattle. (Fulton County, 1940)

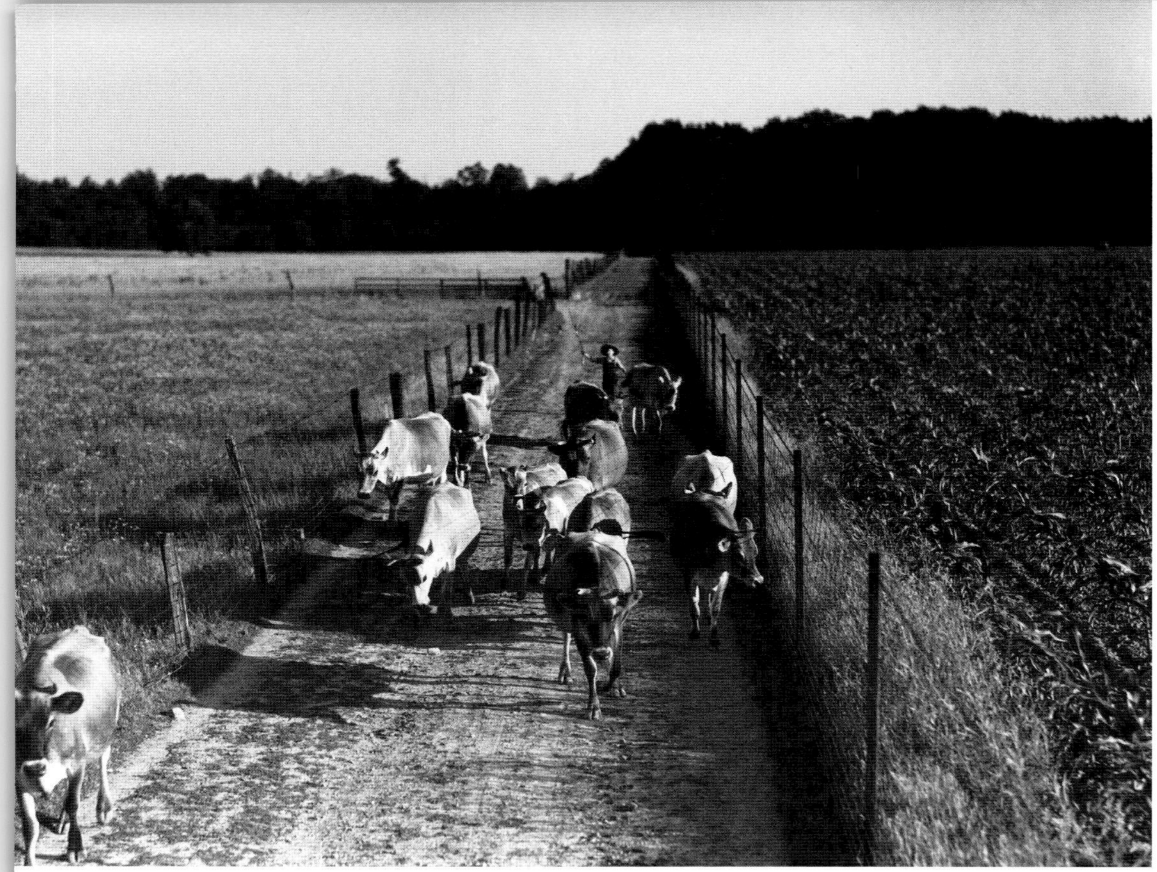

A young boy drives Jersey cattle down a lane on the farm of Howard Richards from New Castle. (Henry County, 1934)

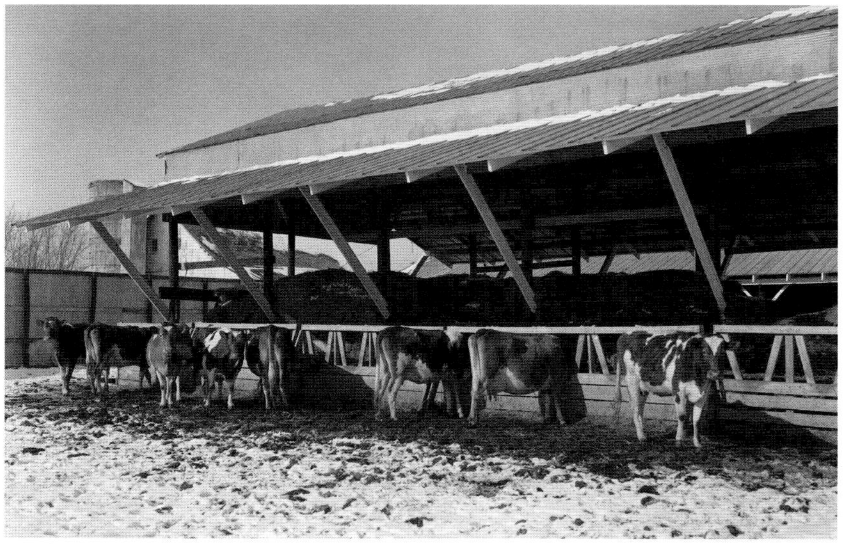

This Guernsey herd uses a self-feeder to eat hay on the Hoagland Farm northeast of Lafayette. (Tippecanoe County, 1955)

Holsteins feed while farmers harvest oats in the field nearby. (Illinois or Wisconsin, 1957)

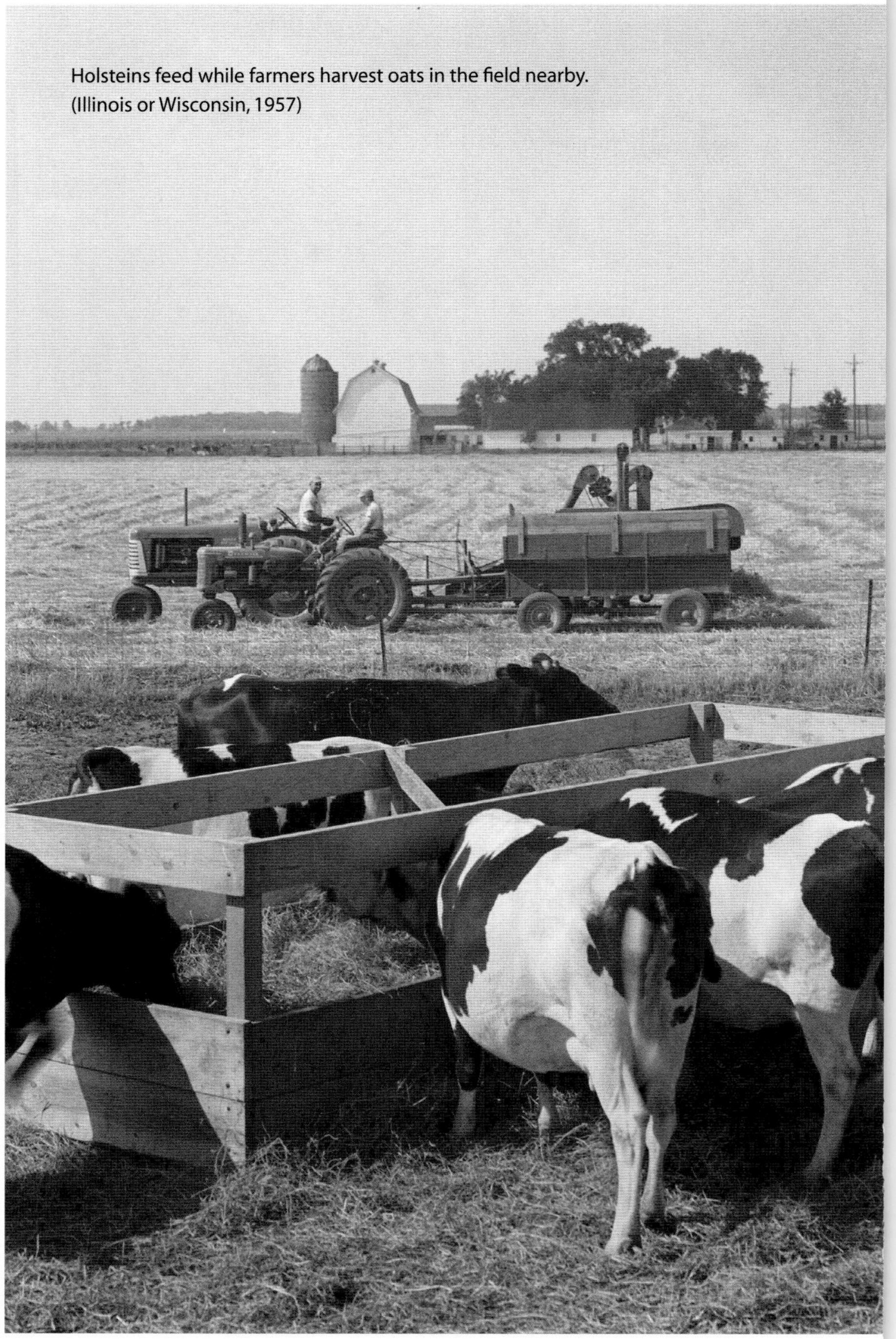

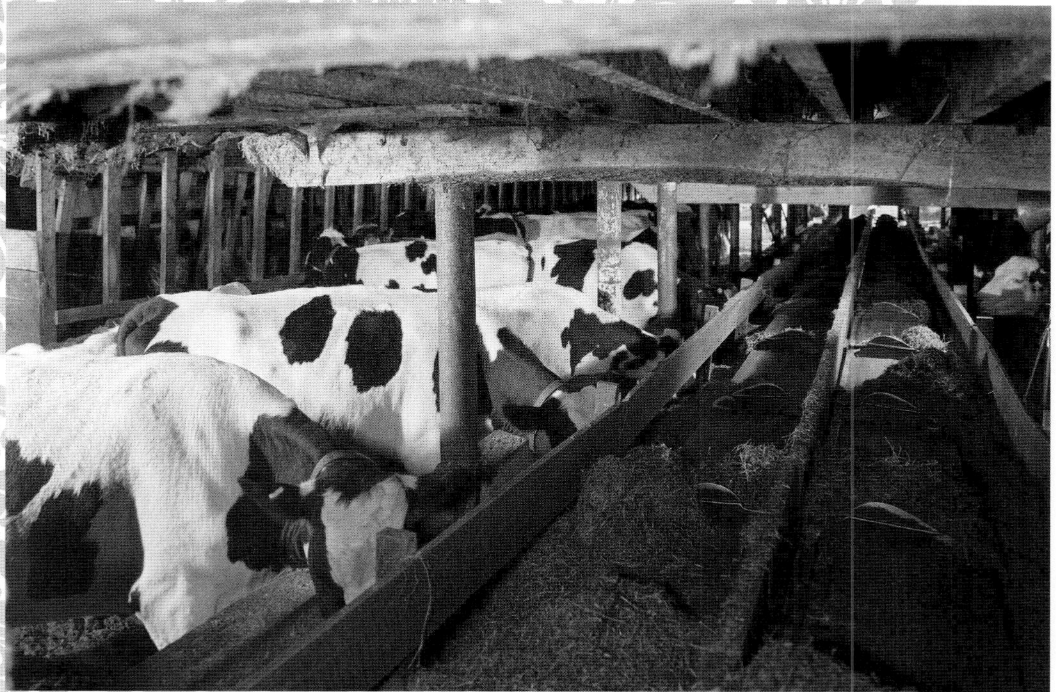

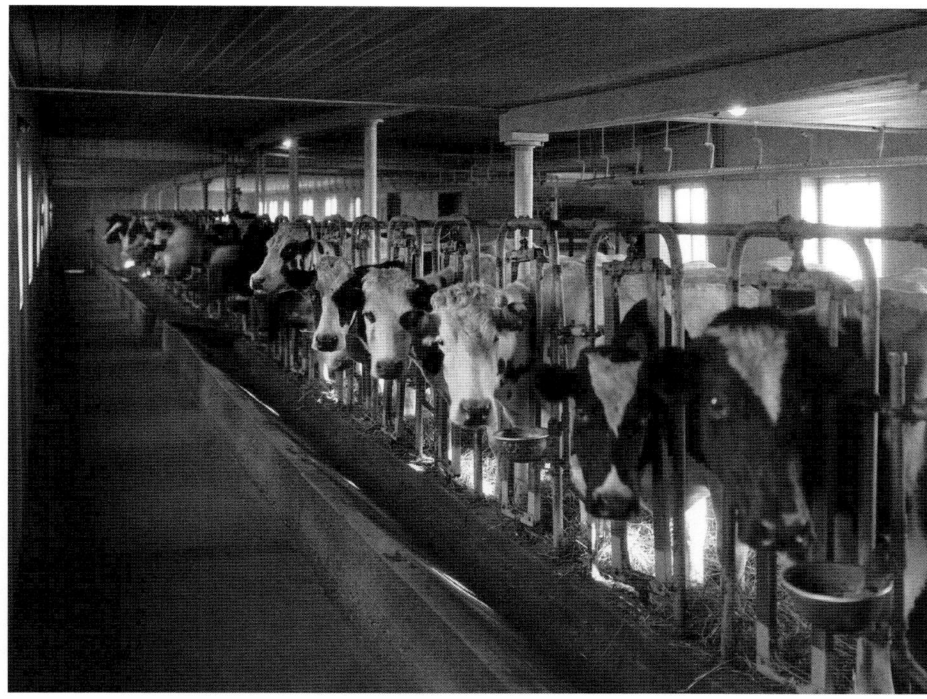

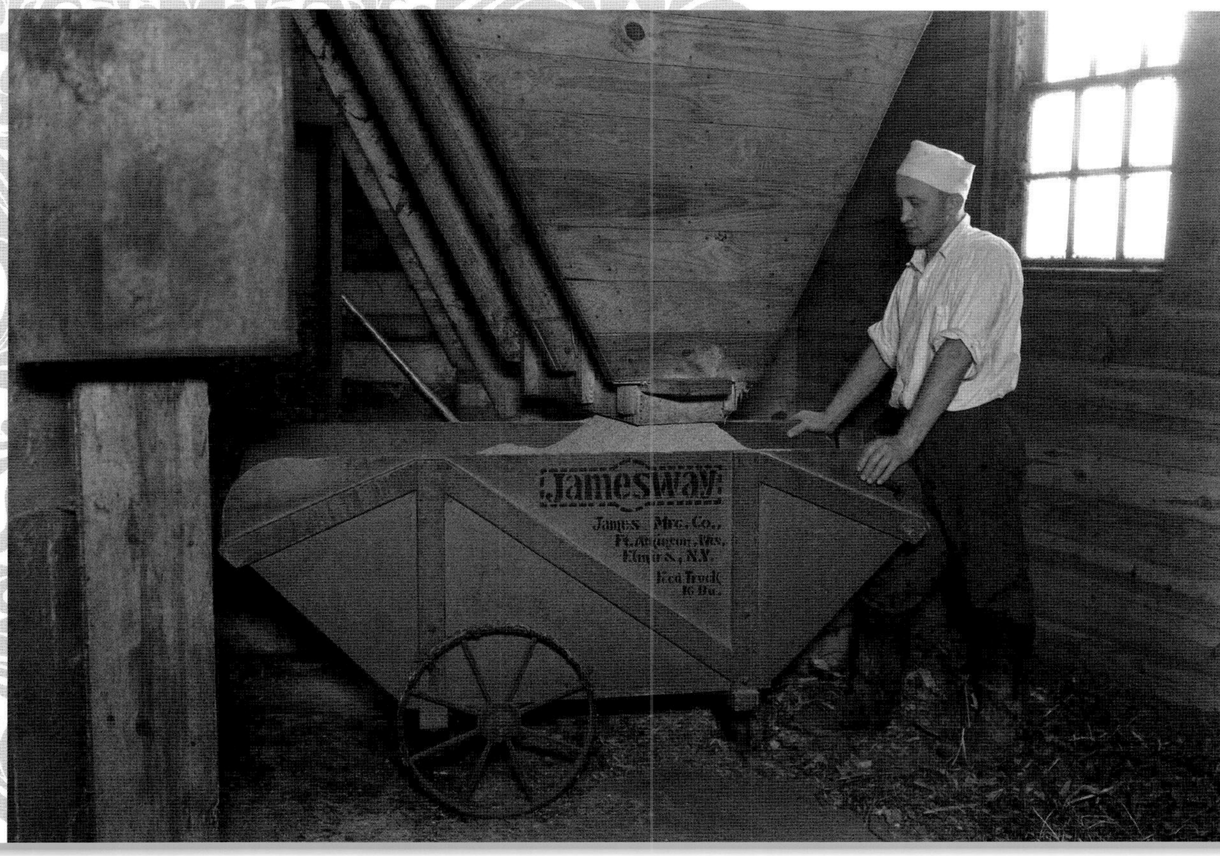

(Top left) An electric belt moves grass silage down a feed trough on the Walter B. Lillie farm near Greenville, Michigan. (Montcalm County, MI; 1956)

(Top right) Holstein cattle inside the barn of Morehouse Dairy in Lafayette (Tippecanoe County, 1932)

(Bottom) Ernest Traumer, manager of the Grand View Dairy Farm in Lafayette, fills his Jamesway Feed Truck with ground feed from the chute of an automatic feed bin. (Tippecanoe County, 1930)

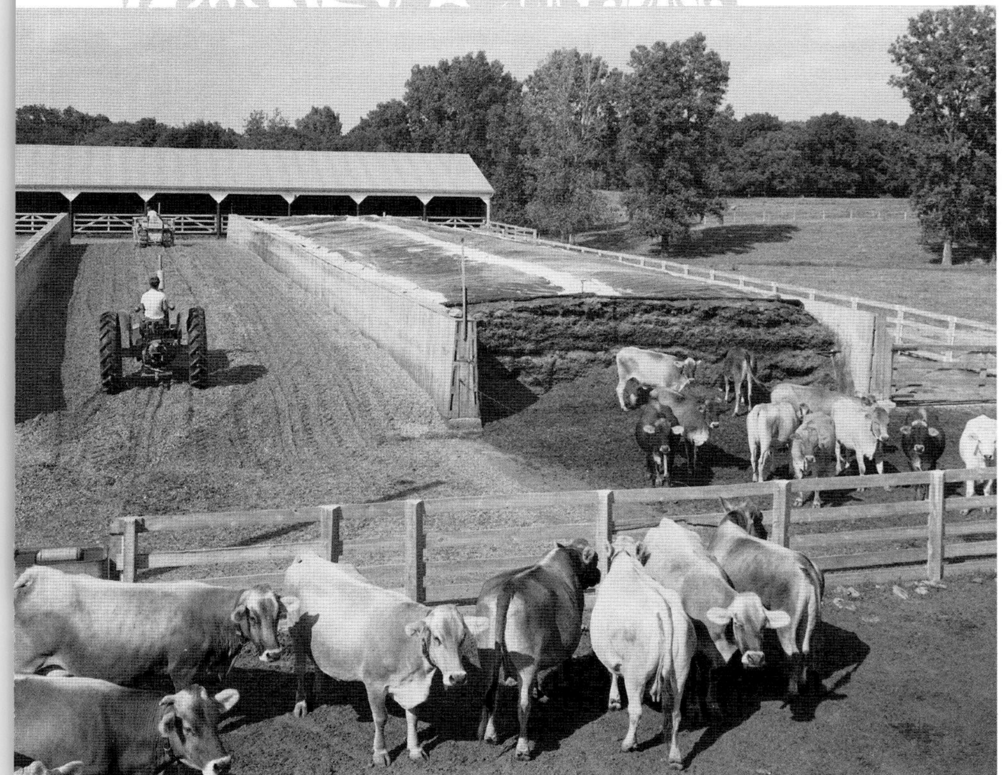

A bunker silo (to right of tractors) holds silage for feeding a Brown Swiss herd at the Conner Prairie Farm in Noblesville. (Hamilton County, 1956)

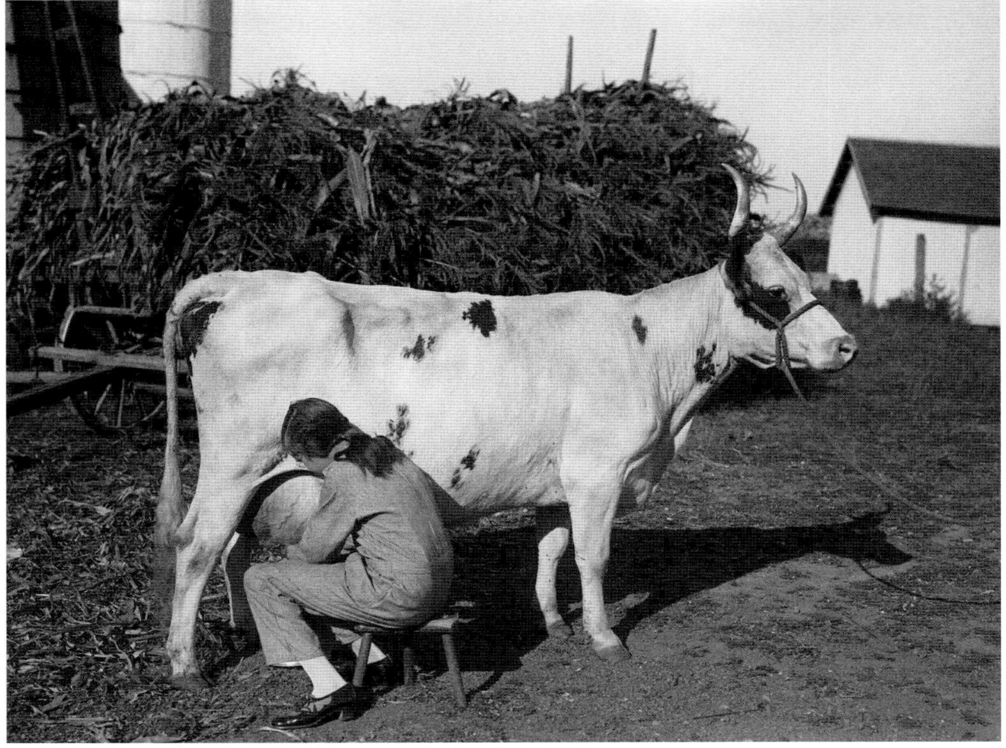

Pauline Eby from Wakarusa hand-milks an Ayrshire cow. (Elkhart County, 1930)

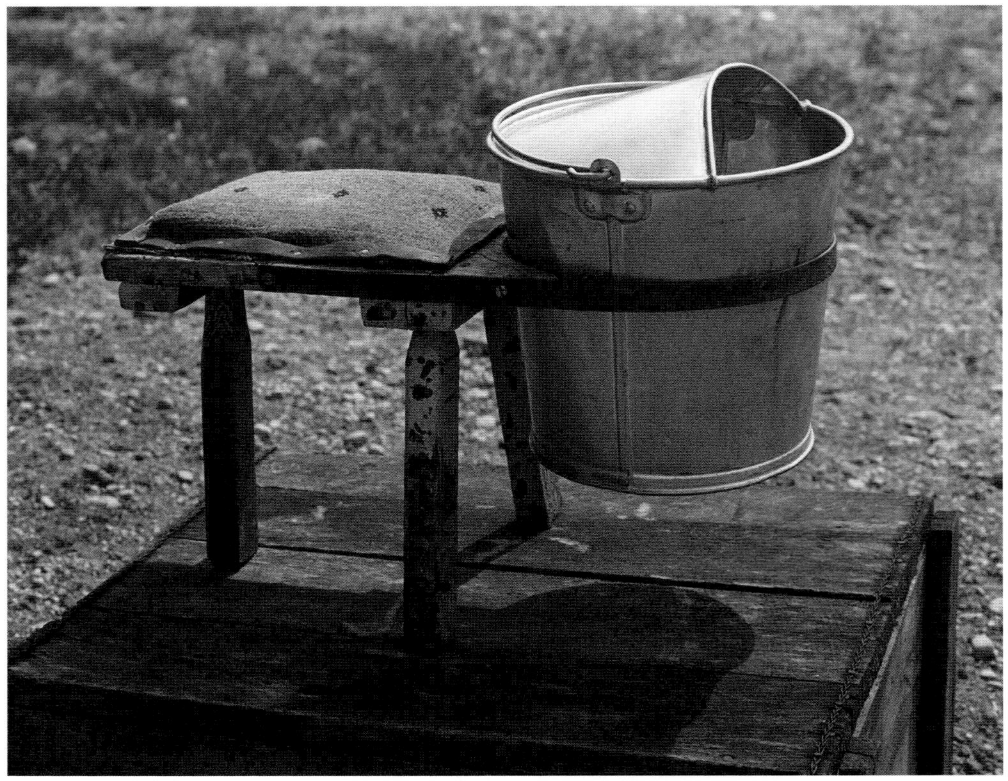

A milking stool from the farm of Dr. F. Dale Barker in Dayton, Ohio (Montgomery County, OH; 1916)

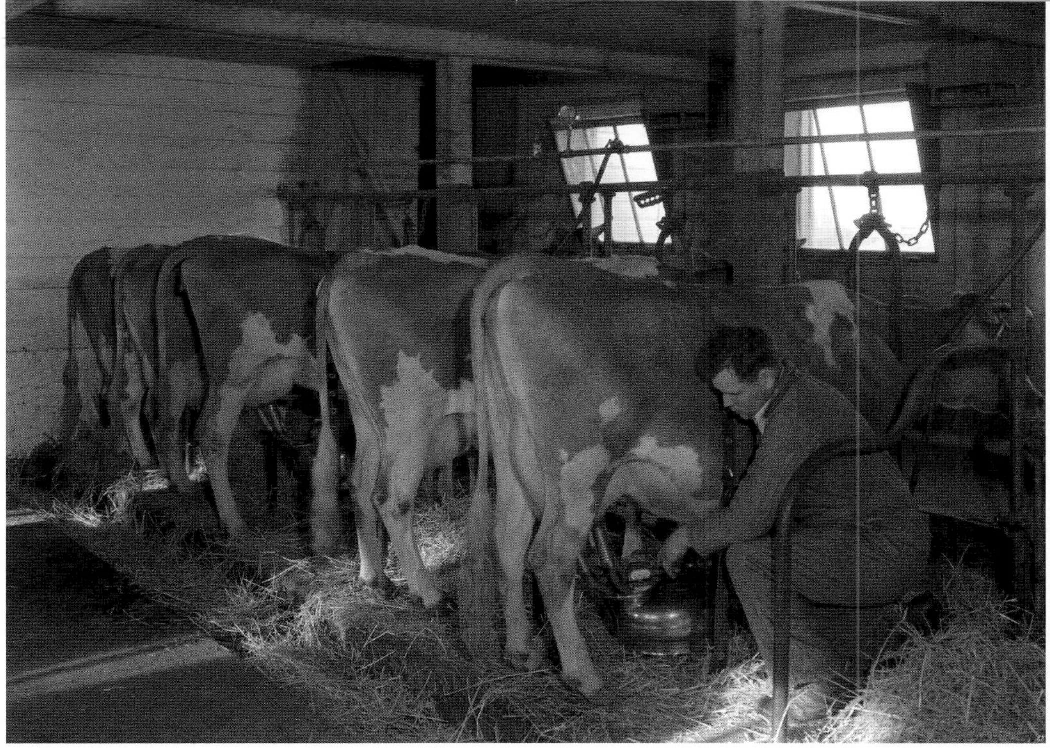

Manager Carl Caldwell of the Bankable Guernsey Farm in Colburn adjusts a Pine Tree milking machine on a cow. (Tippecanoe County, 1931)

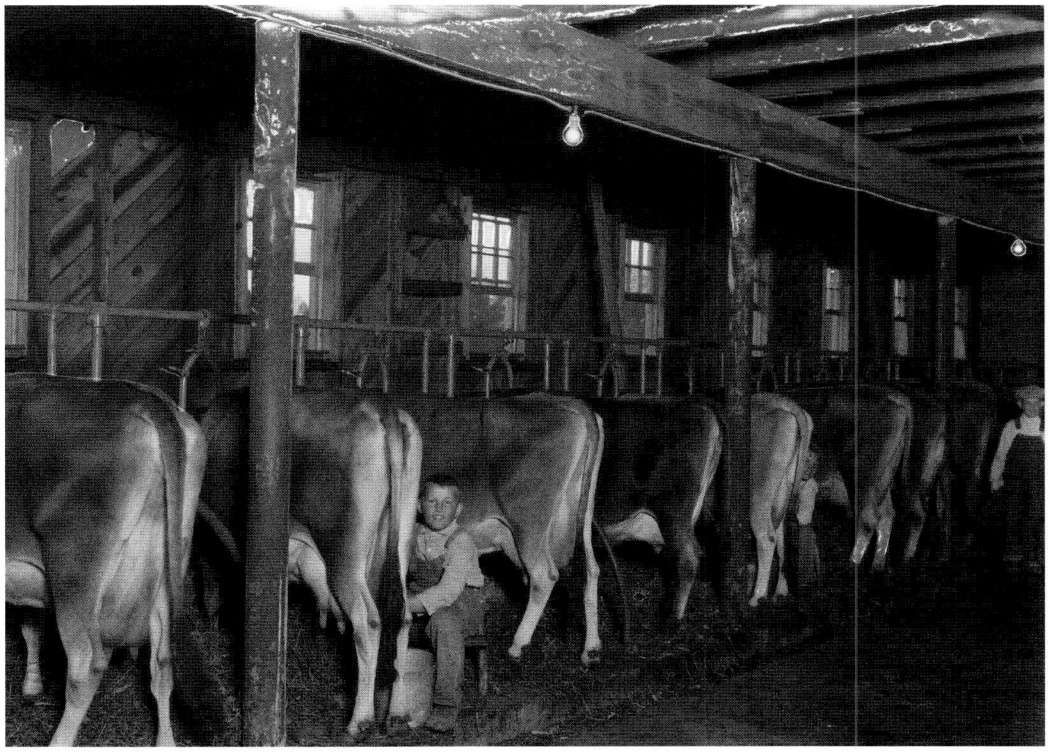

Some boys hand-milk a Jersey herd on the R. S. Edmiston and Sons Farm in Mooresville, North Carolina. (Iredell County, NC; 1927)

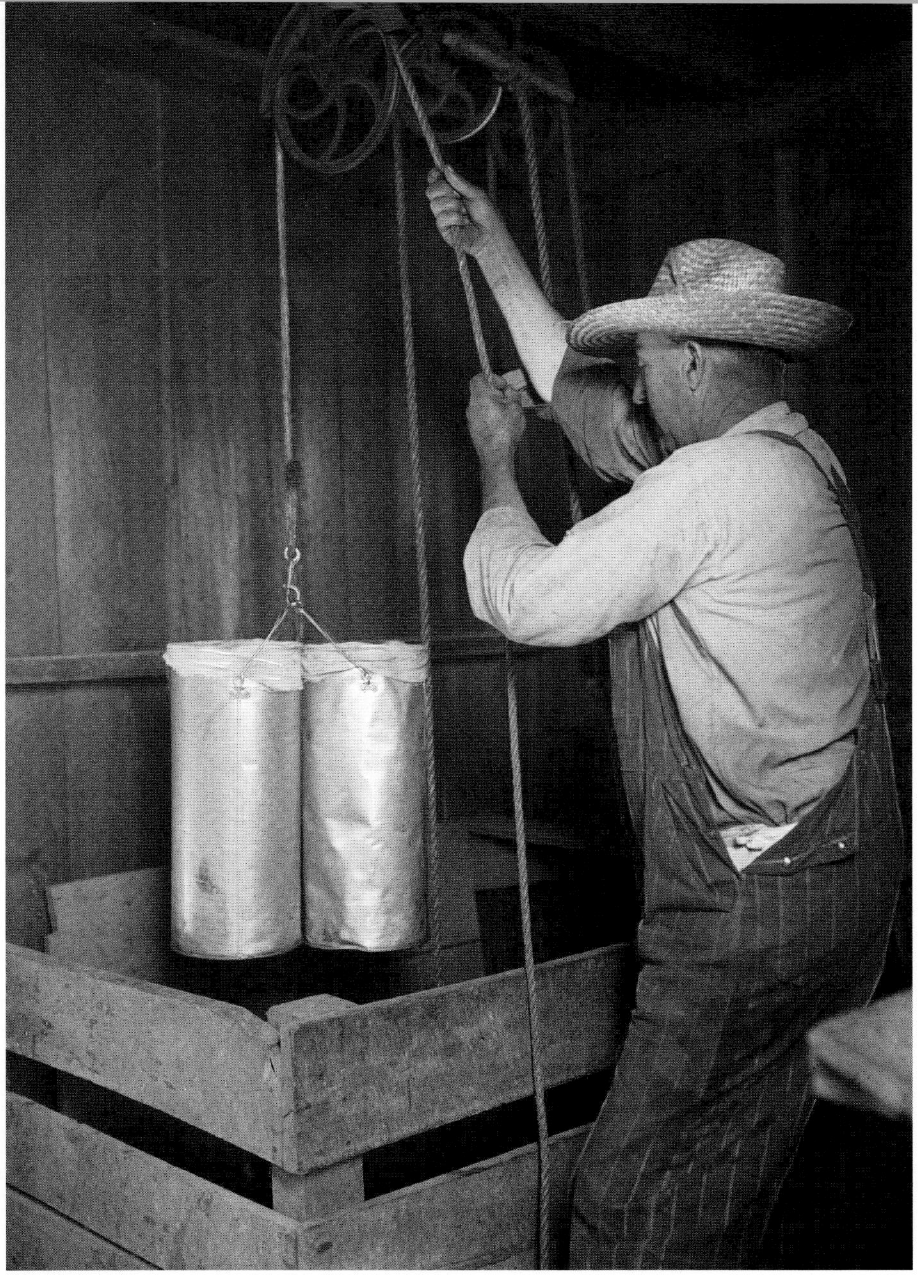

A man lowers milk cans into a well for cooling on the farm of J. T. Carr in Charlestown. (Clark County, 1923)

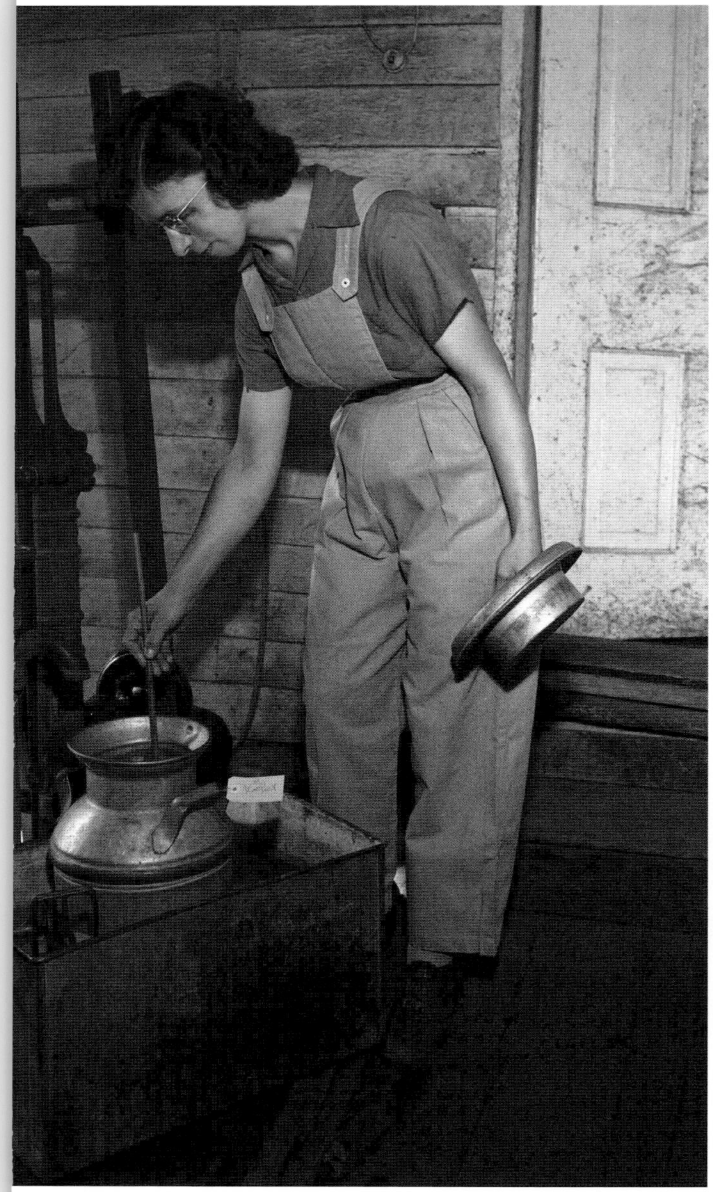

Mary Kochert places cream in a small tank with cold water from the well on the farm of Elizabeth Kochert near Lafayette. (Tippecanoe County, 1944)

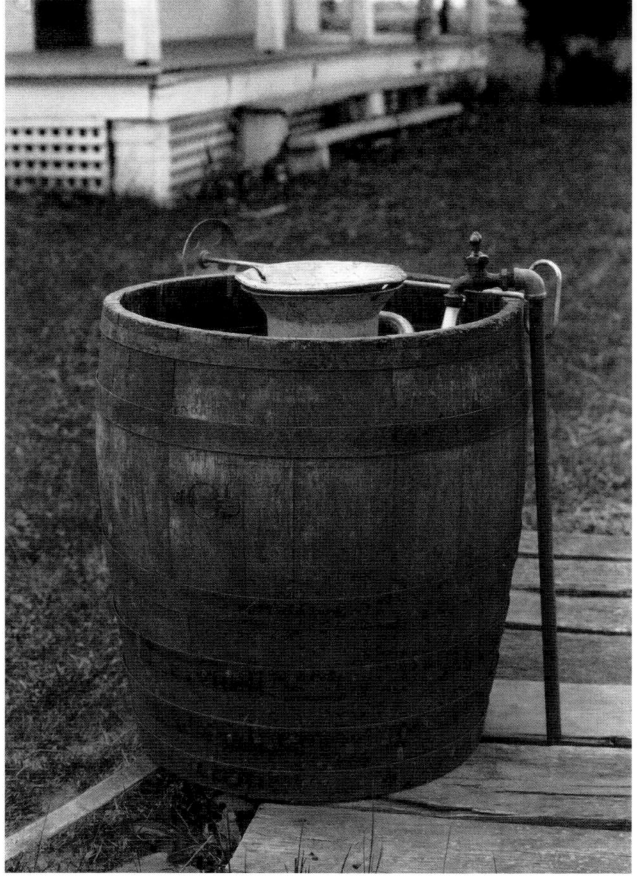

Clark Snider of Muncie used this barrel to cool milk cans on his Muncie-area farm. (Delaware County, 1921)

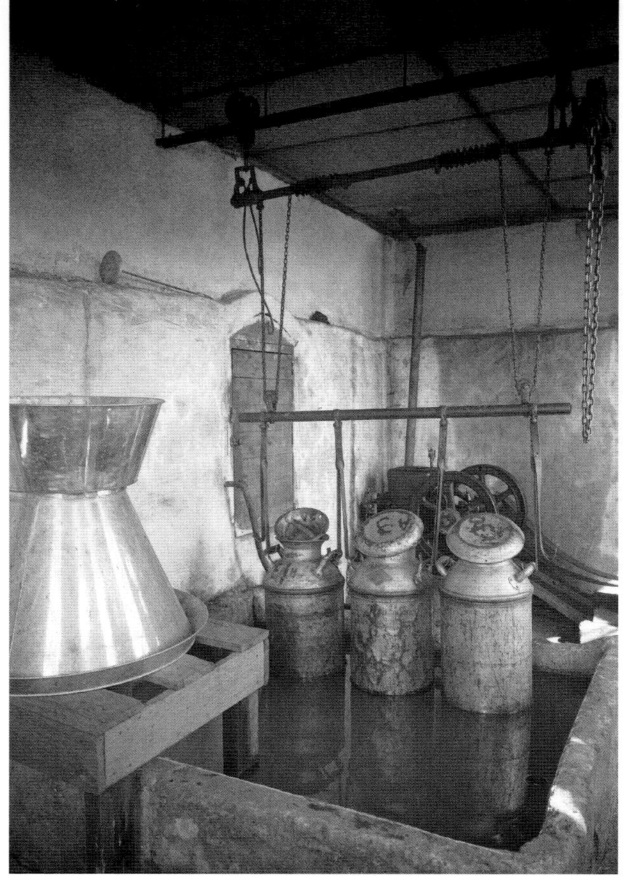

Milk cans are lowered into a cooling tank at the dairy of Dan LaBar in Delavan, Wisconsin. (Walworth County, WI; 1919)

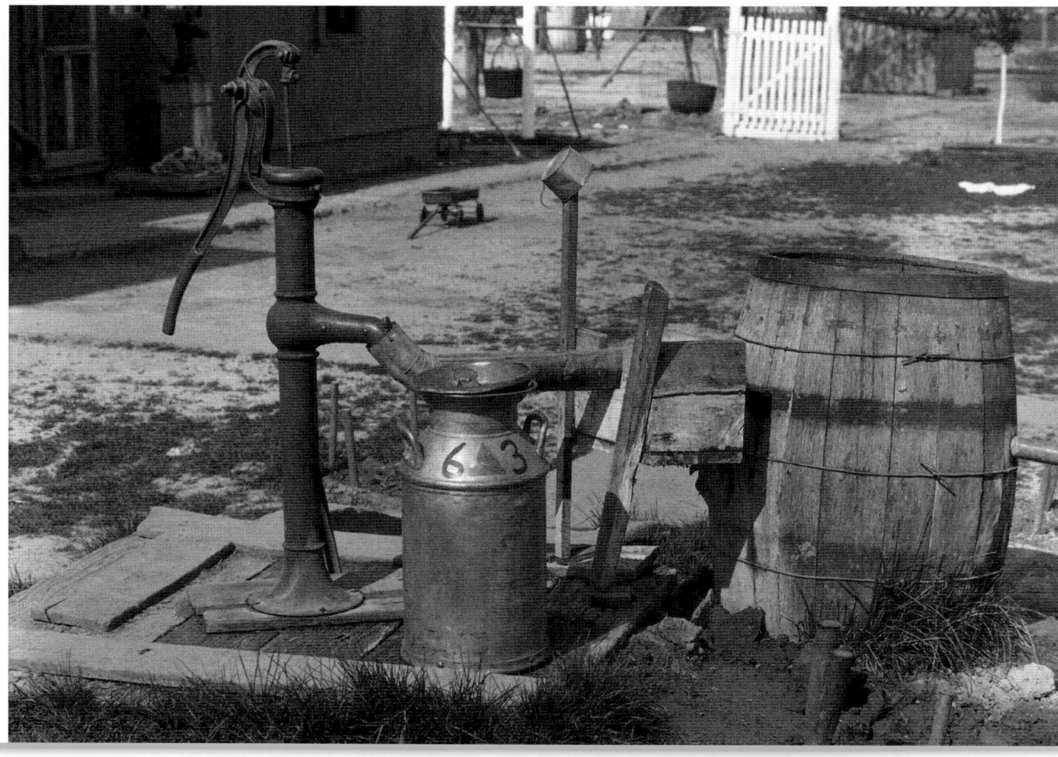

Cream is cooled in a barrel located between the water pump and the stock watering trough on the farm of Frank Stoll. (Daviess County, 1928)

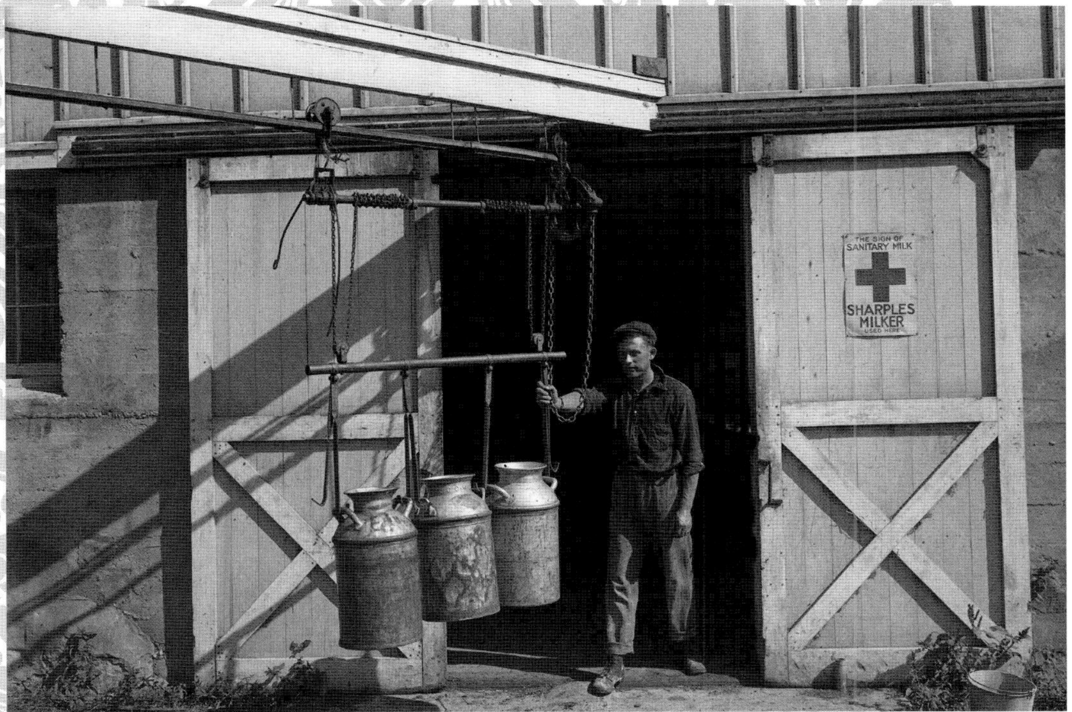

This man uses a rail to move milk cans from the barn to a milk house at the dairy of Dan LaBar in Delavan, Wisconsin. (Walworth County, WI; 1919)

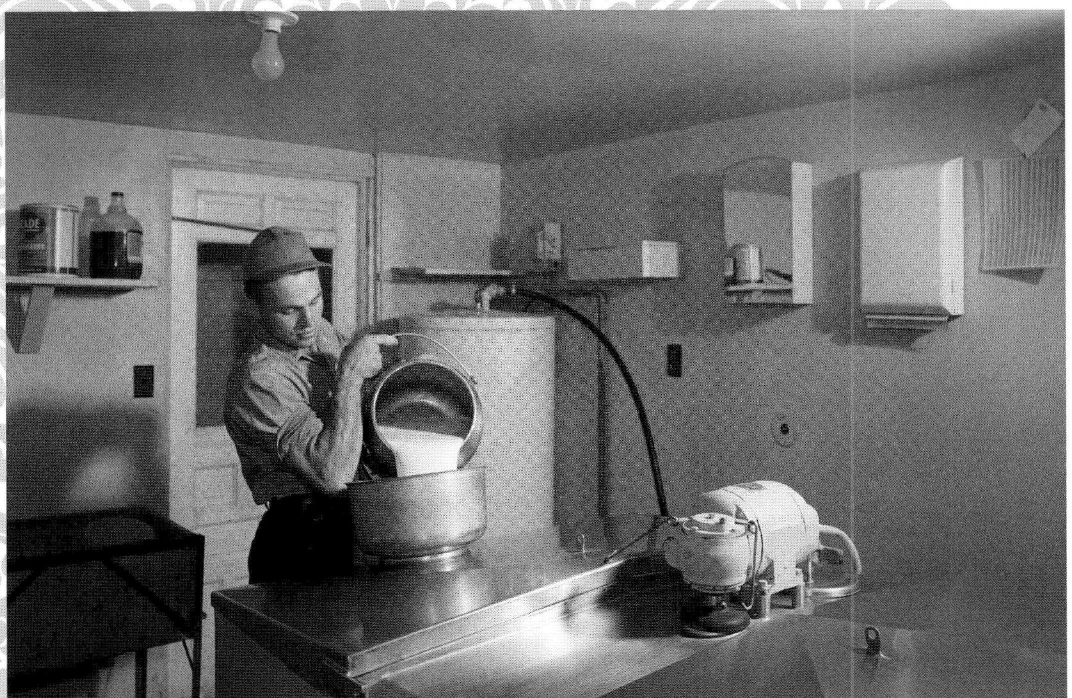

Ralph W. Wirth from Hurley, South Dakota, pours milk into a bulk tank. (Turner County, SD; 1959)

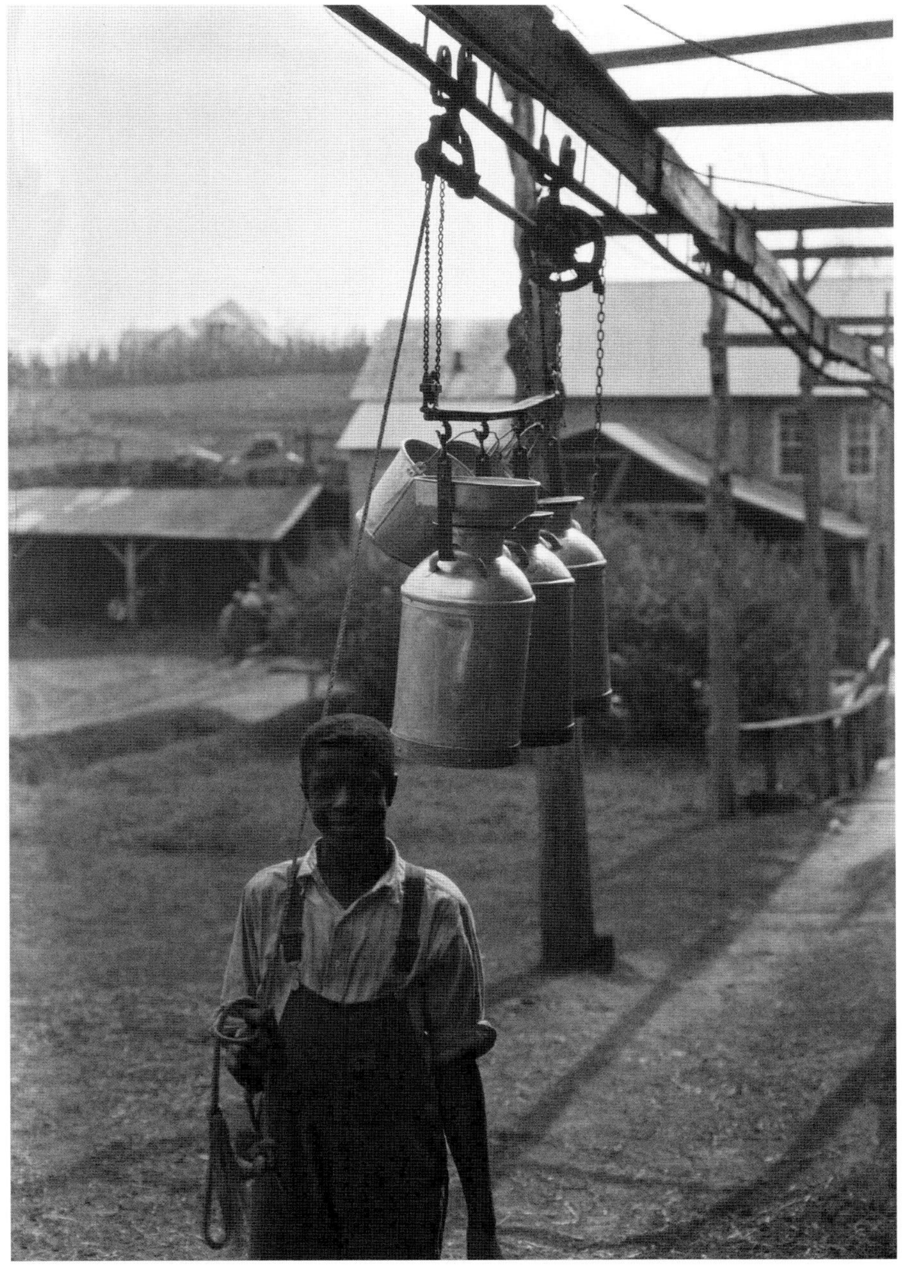

A young man moves milk cans on a track between the barn and the milk house at the dairy of R. L. Shuford in Newton, North Carolina. (Catawba County, NC; 1927)

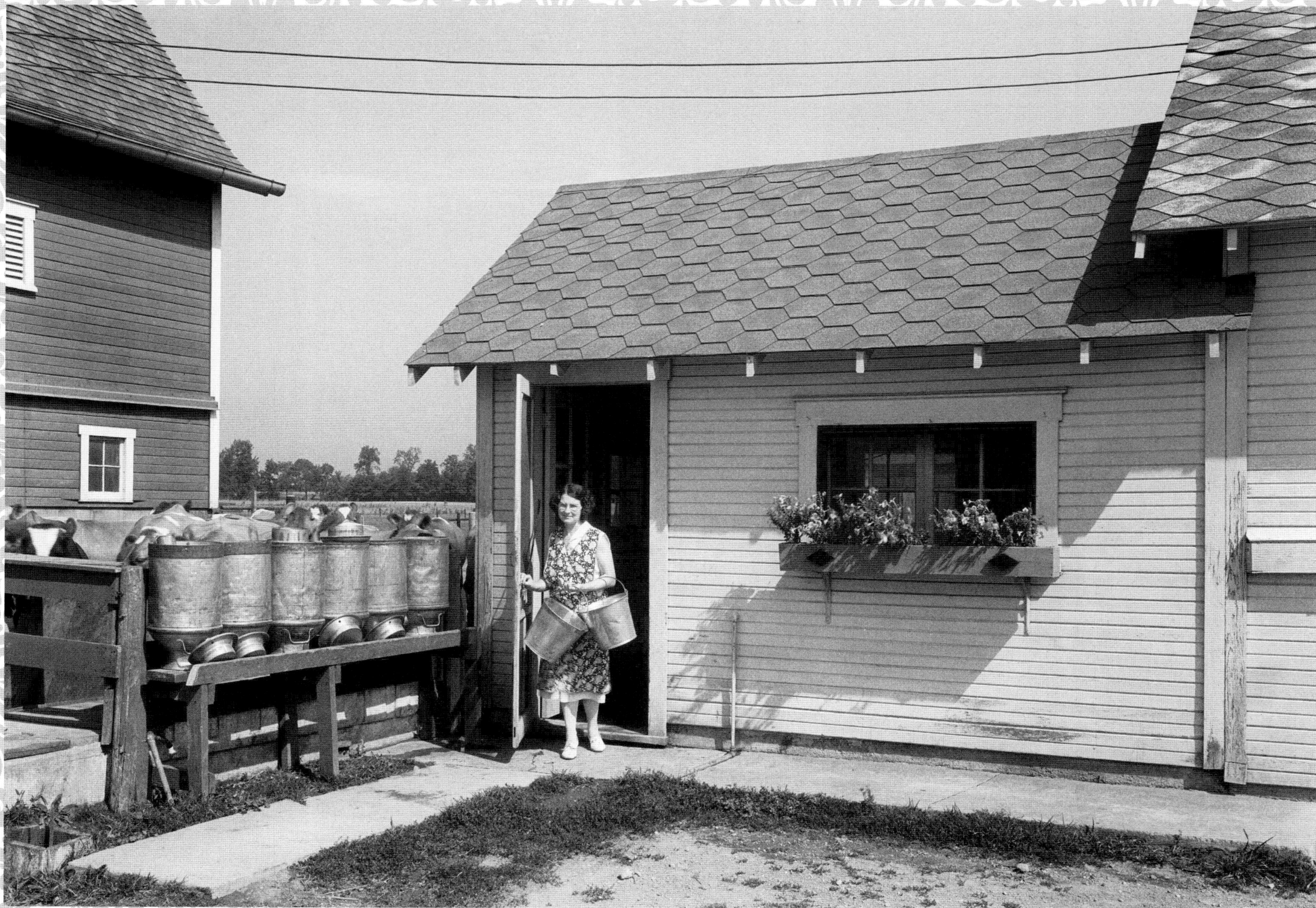

Mrs. Russell Stephenson of Gaston exits a milk house carrying buckets on her arm. (Delaware County, 1934)

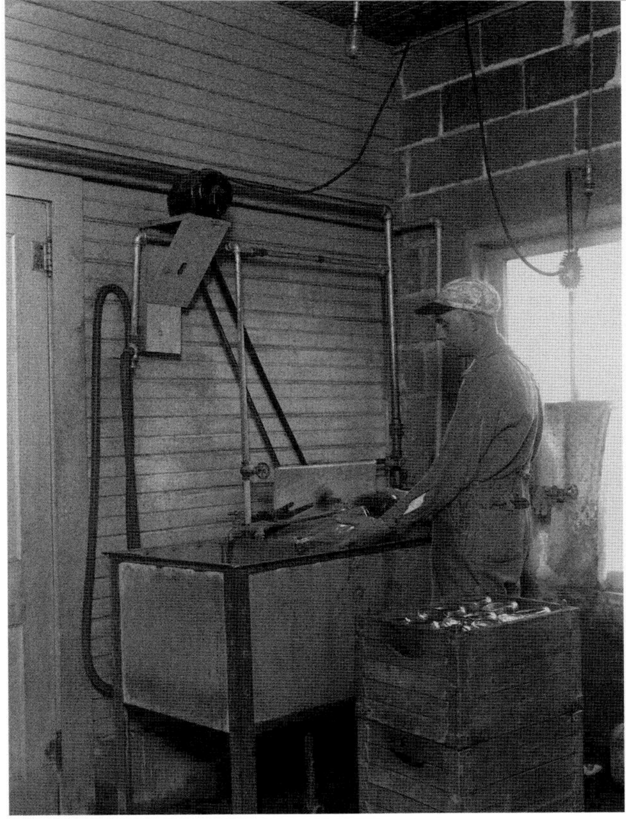

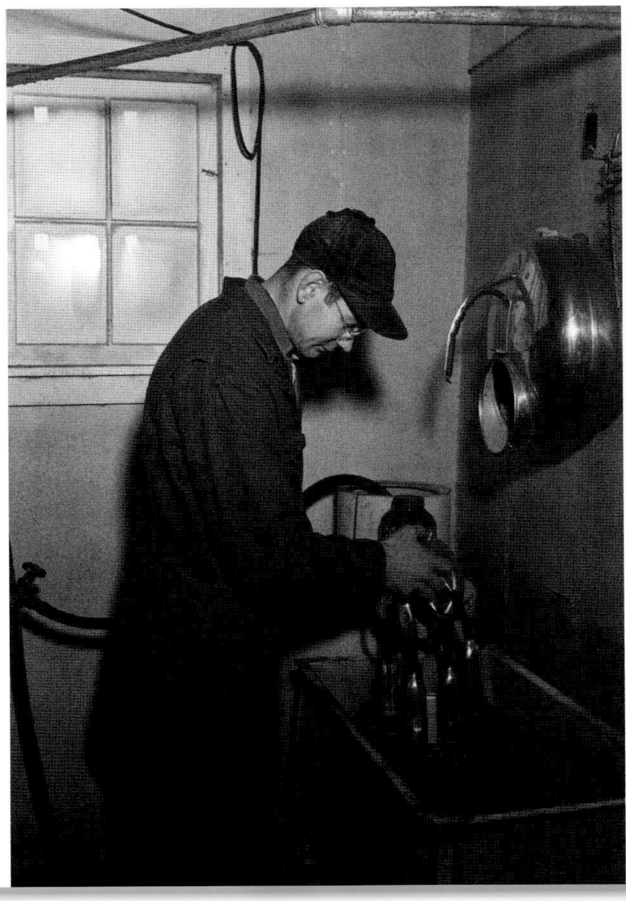

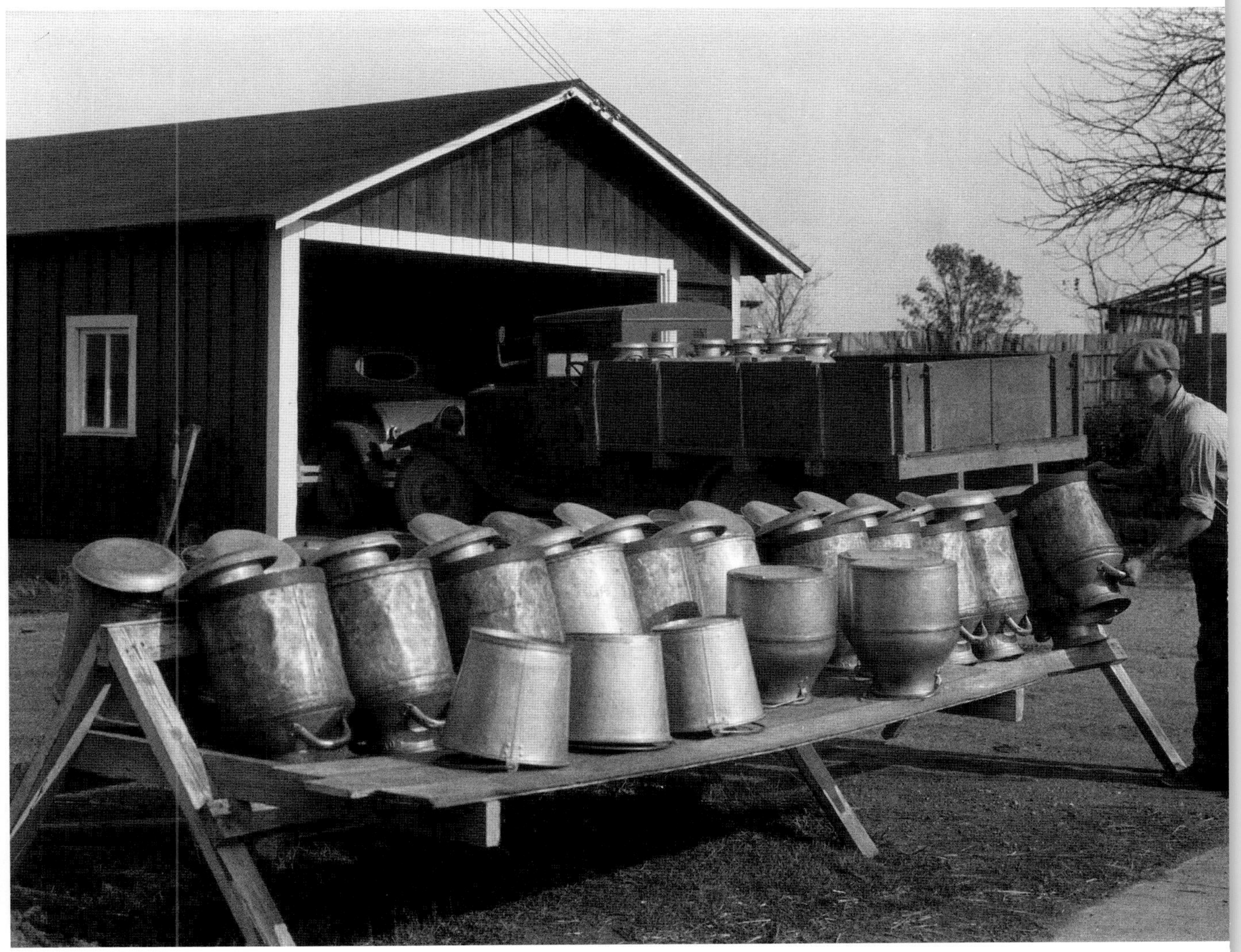

(Top left) A motor-driven milk bottle washer on Mervin Eby's dairy farm in Wakarusa (Elkhart County, 1928)

(Bottom left) Cleaning a milker at the farm of Elmer Longyear from Claypool (Kosciusko County, 1946)

(Right) Dairyman Dick Newman places a milk can on a rack for airing near Culver. (Marshall County, 1930)

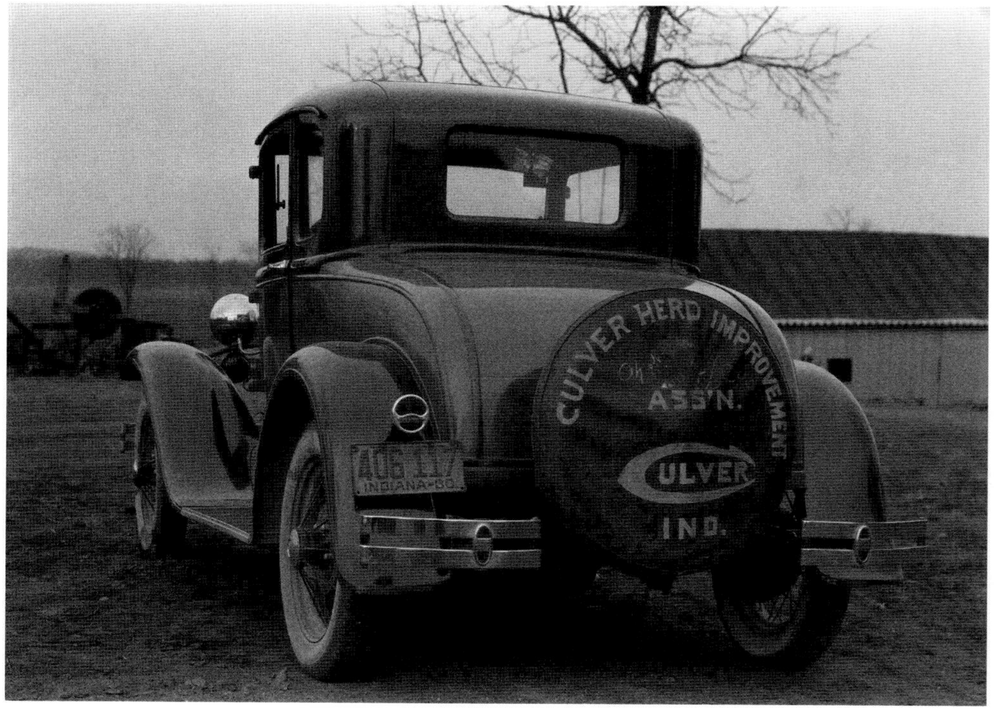

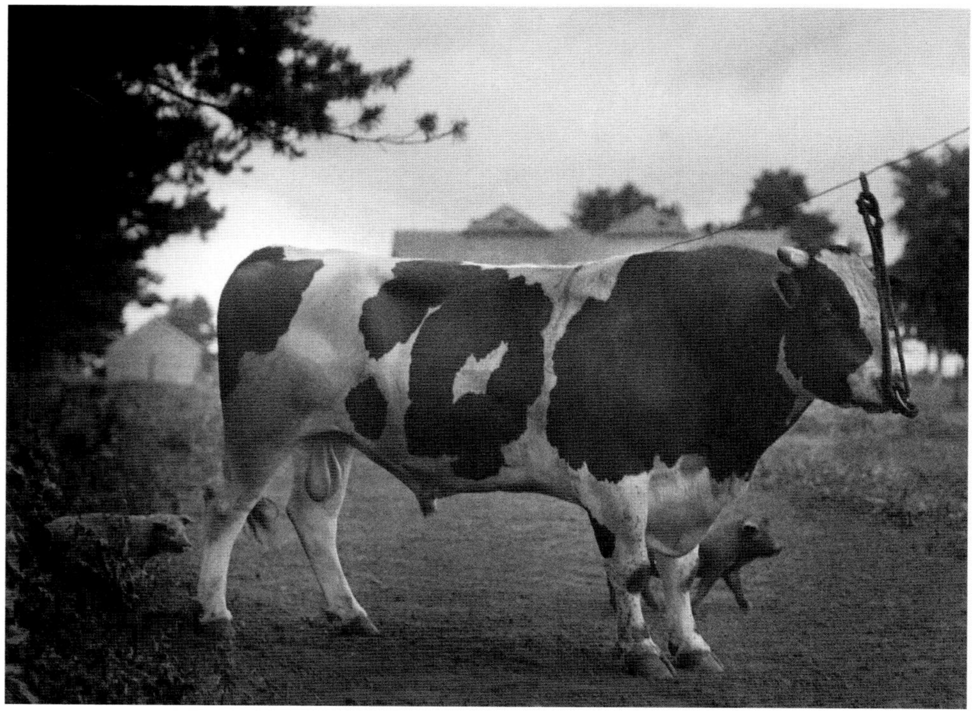

(Top left) A tire cover carries an advertisement for the Culver Herd Improvement Association. (Marshall County, 1930)

(Top right) A Holstein bull is hooked to an exercising wire at the Iowa Agricultural College in Ames, Iowa. (Story County, IA; 1919)

(Bottom) Ira Eby from Wakarusa holds his Ayrshire bull as another man looks on. (Elkhart County, 1930)

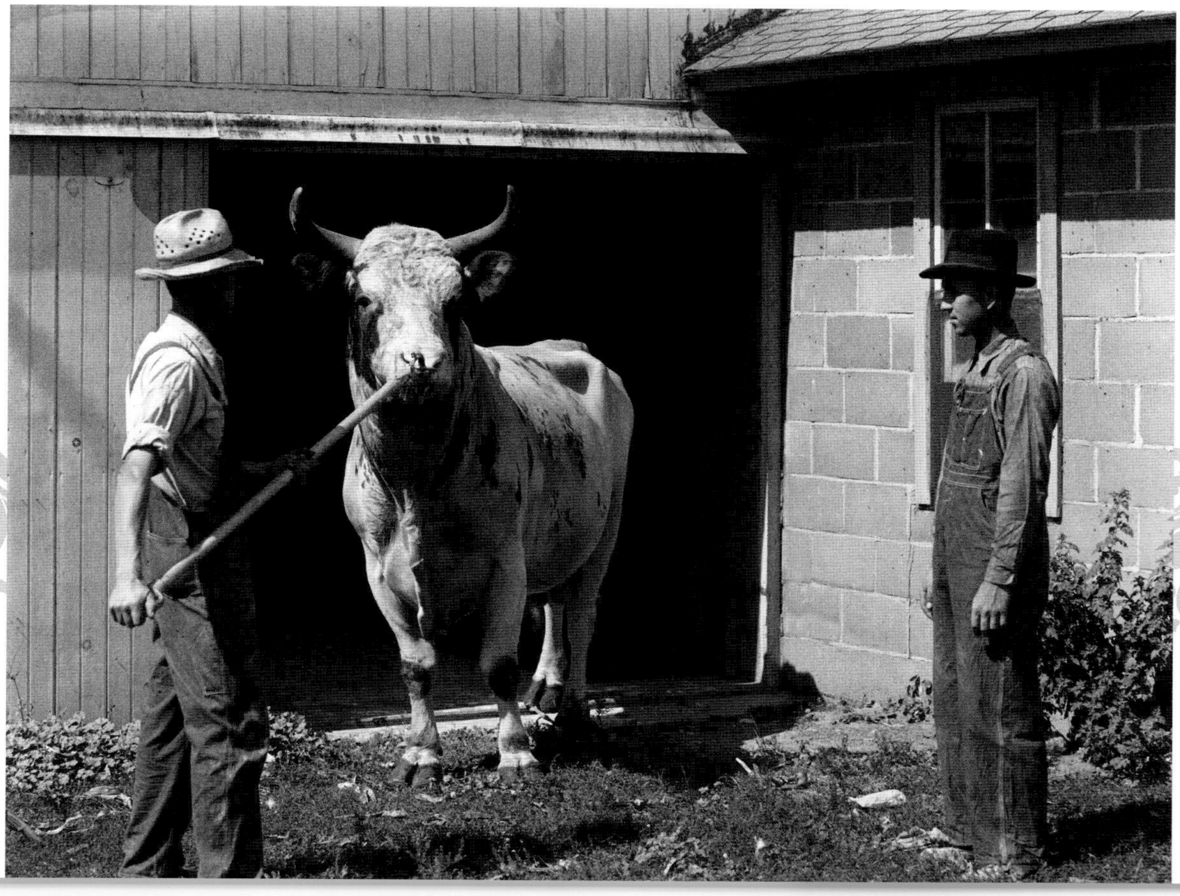

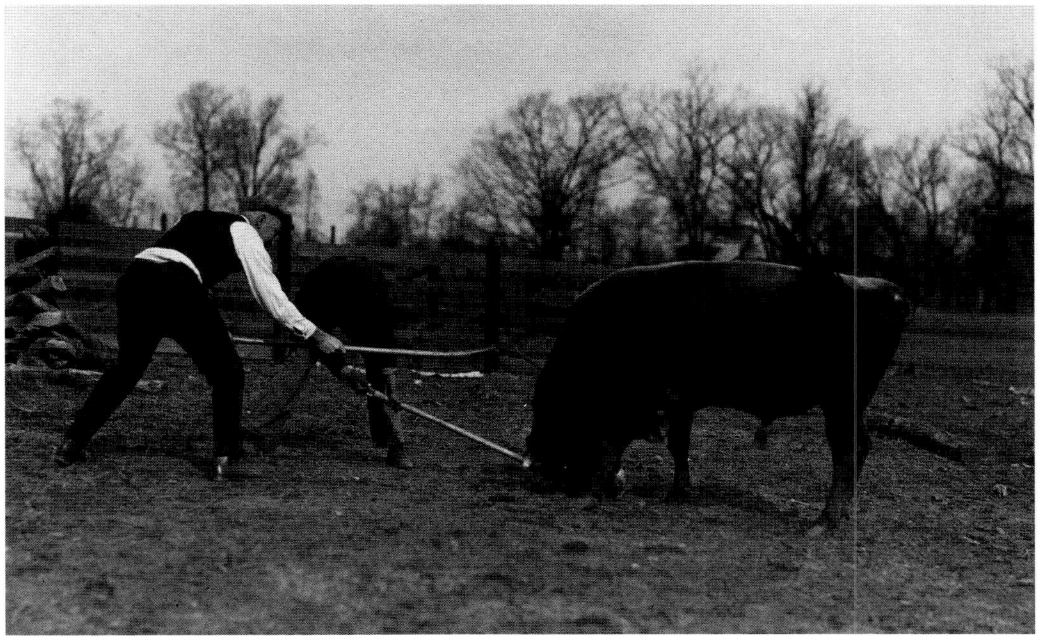

Handling a bull from a Jersey herd on the Spoor Farm near Salem (Washington County, 1932)

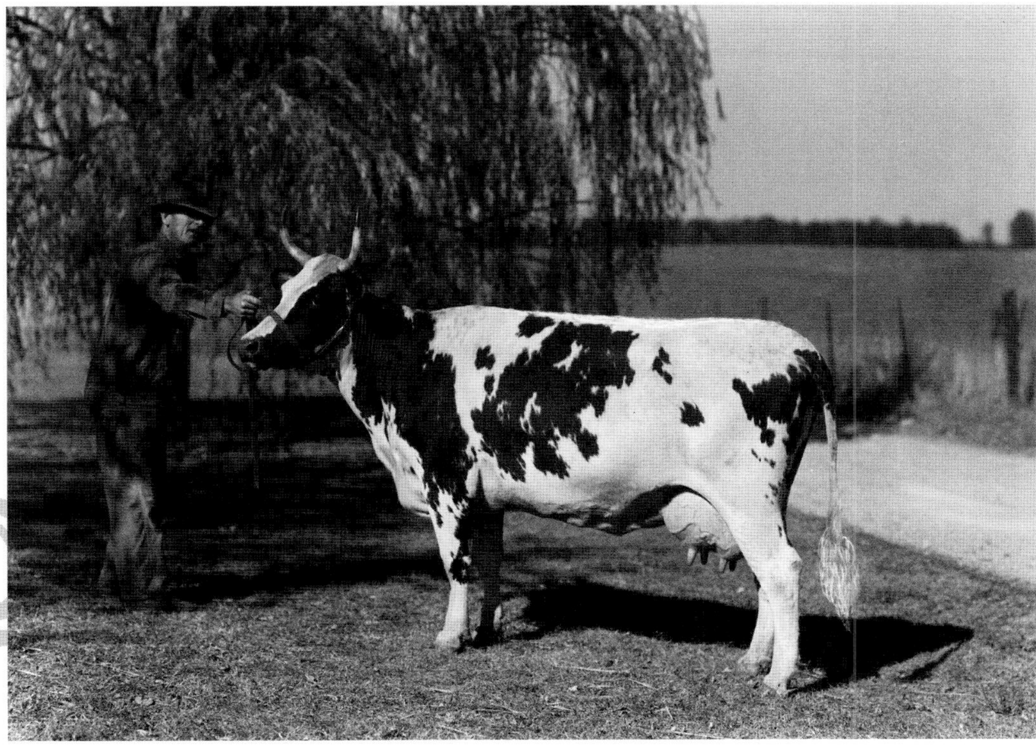

This two-year-old cow owned by Herman Klingel and his mother, Mary Klingel, of Fort Recovery, Ohio, was named a world champion by the Ayrshire Breeders' Association for record milk and butterfat production. (Mercer County, OH; 1946)

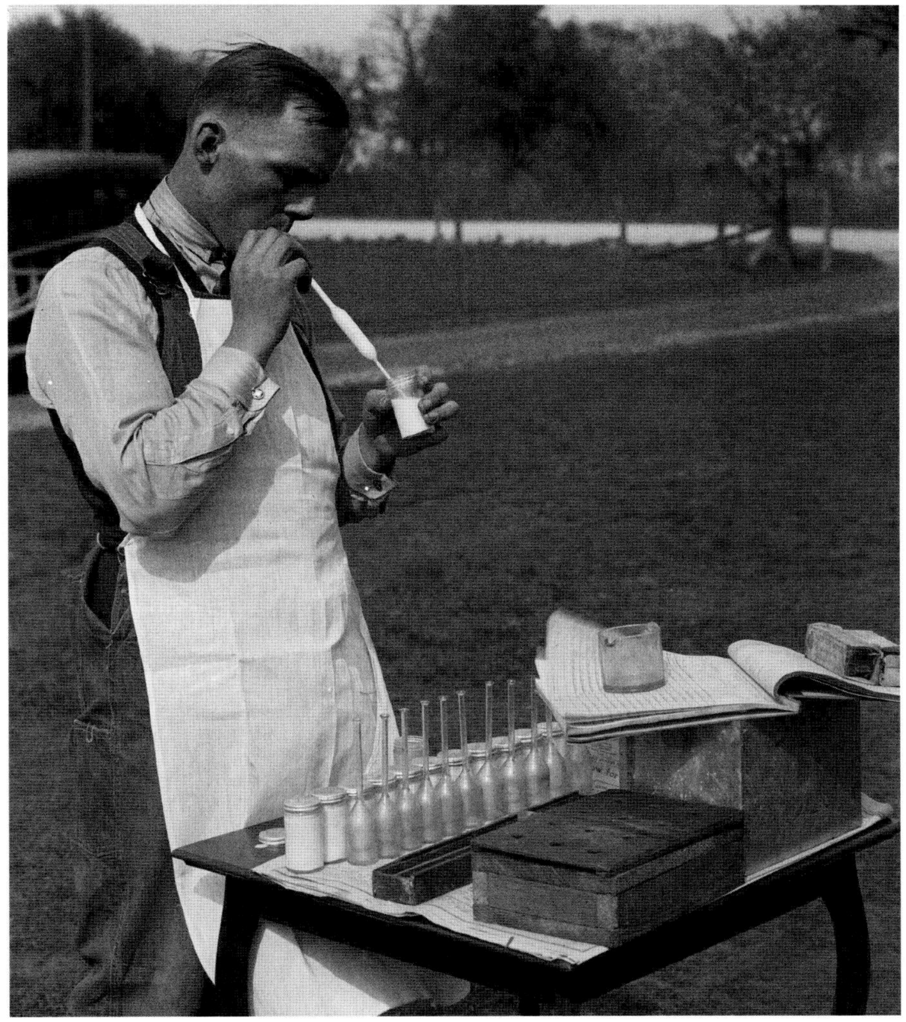

Stewart Rivers, the Lake County cow tester, measures butterfat percentage. Cow testers were hired by Purdue University and paid for by groups of approximately 26 dairy producers to monitor milk and butterfat production on farms and track improvements made through better feeding and genetics. (Lake County, 1923)

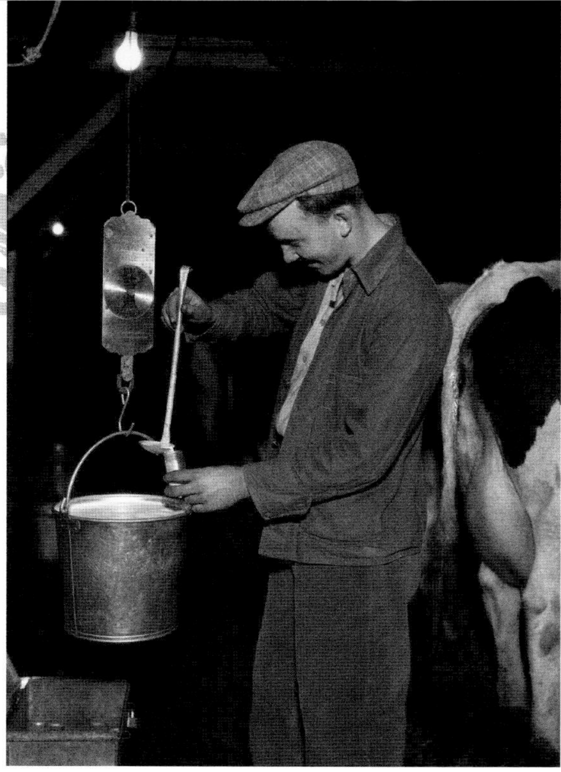

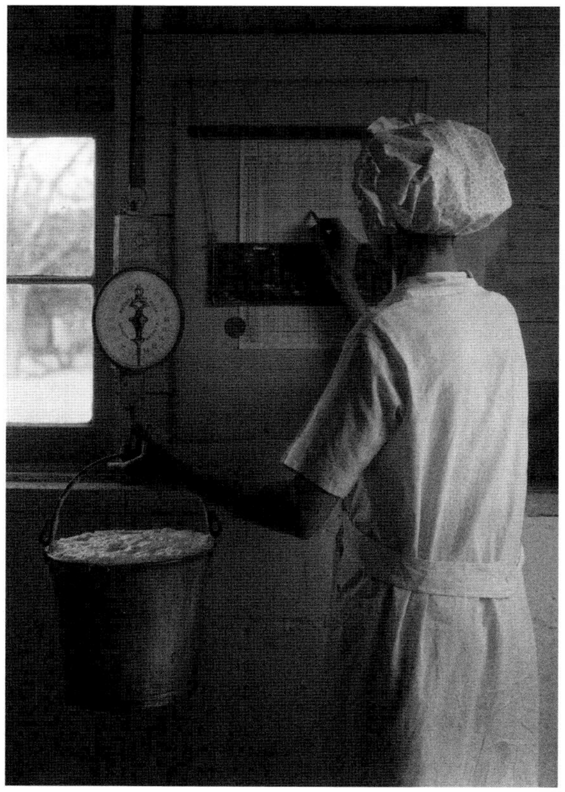

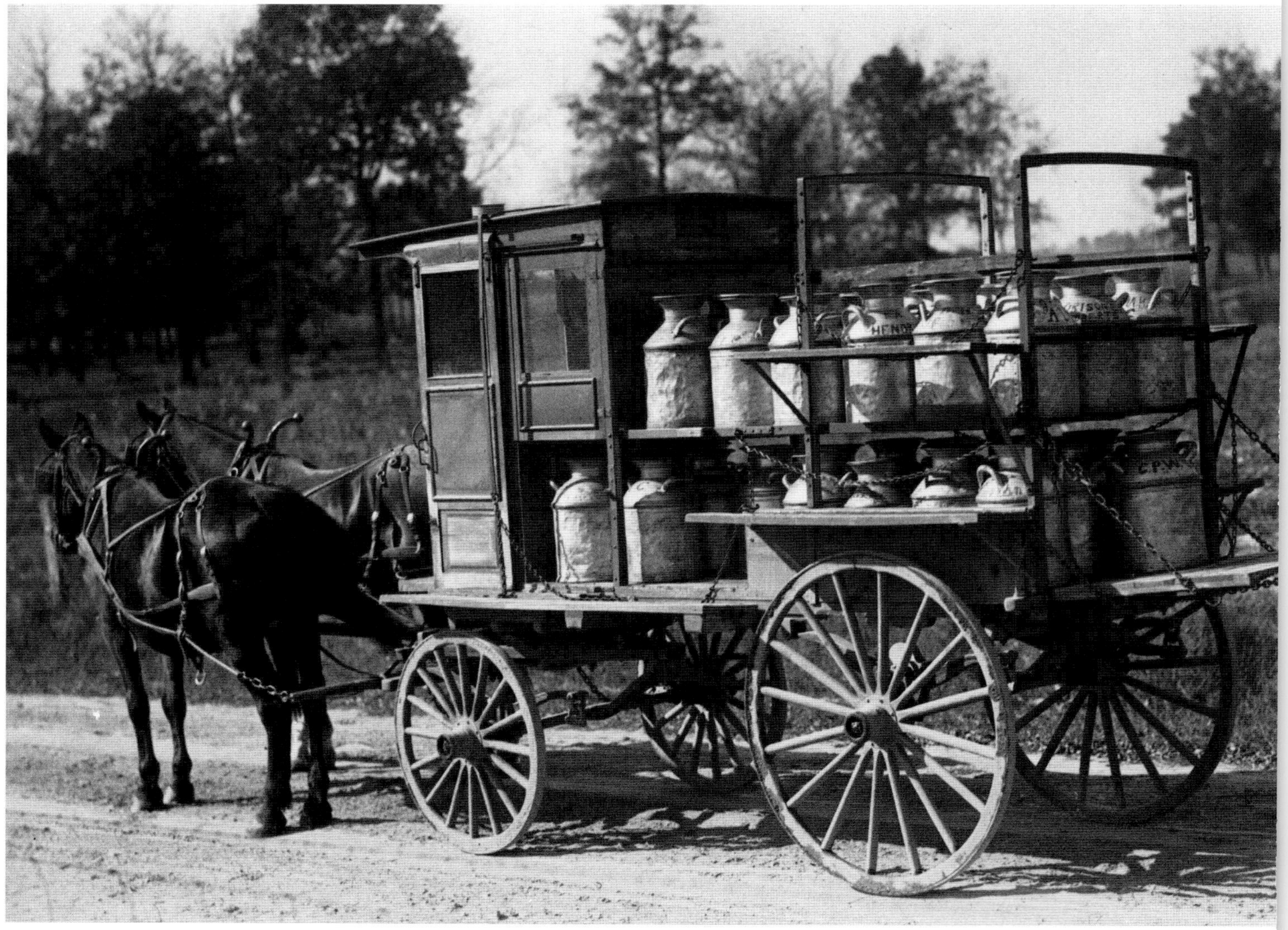

(Top left) Walter E. Amstutz works as the cow tester for the Culver Herd Improvement Association. (Marshall County, 1930)

(Bottom left) Weighing and recording milk production on the dairy farm of C. E. Aikman from Washington (Daviess County, 1928)

(Right) Elwood Baker from Sheridan owned this horse-drawn milk wagon, which could hold up to 90 milk cans. (Hamilton County, 1916)

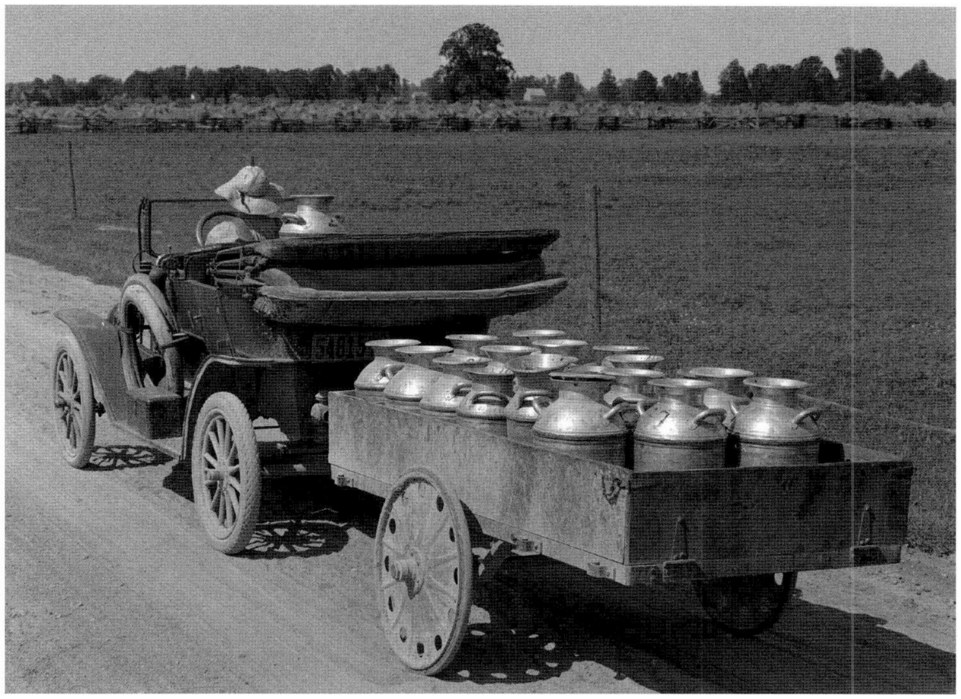

A trailer built for hauling milk cans is attached to a Ford automobile owned by Thomas Rohn from Advance. (Boone County, 1915)

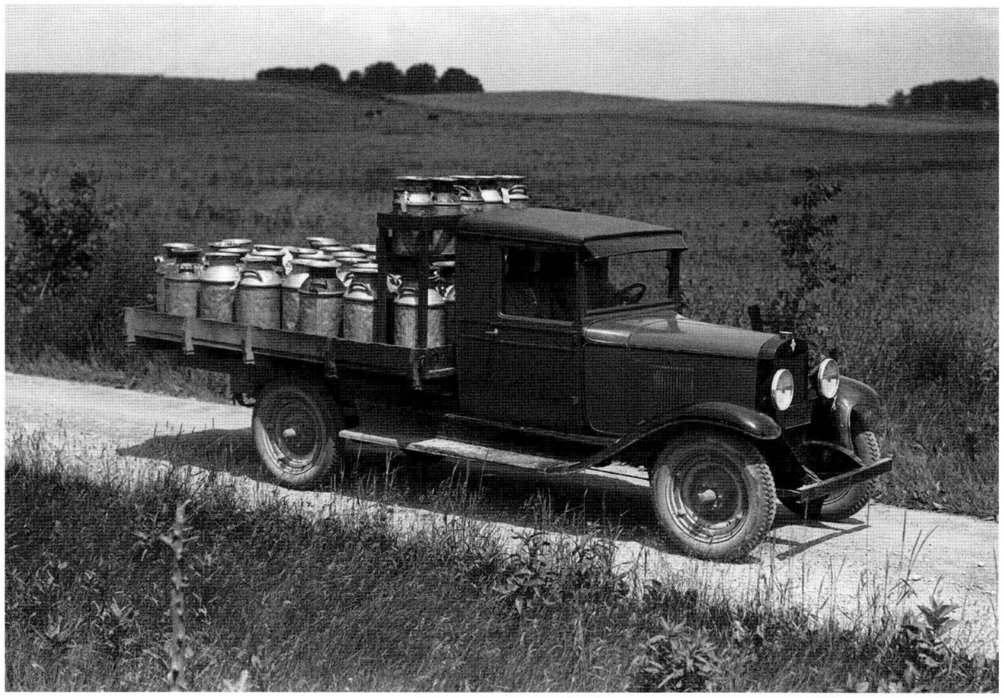

These milk cans are loaded on a truck owned by Simon Van Schepen, who hauled for the Lafayette Ice Cream Company. (Tippecanoe County, 1929)

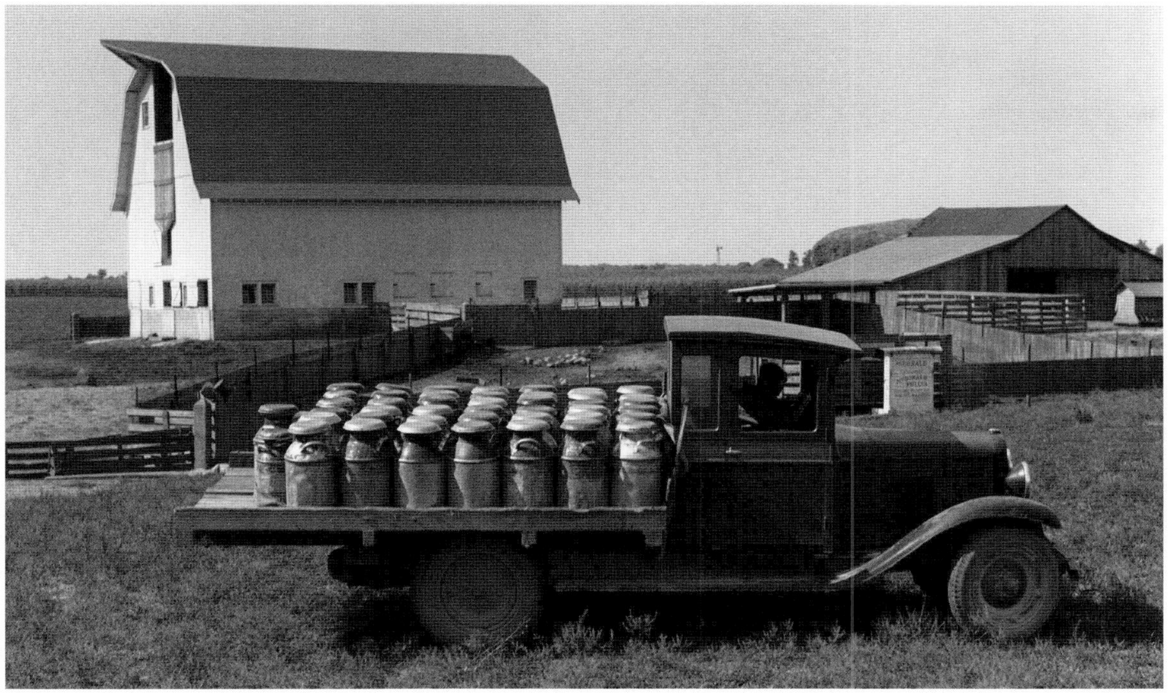

This man collects milk from the farm of Dr. John S. Morrison in Lafayette. (Tippecanoe County, 1930)

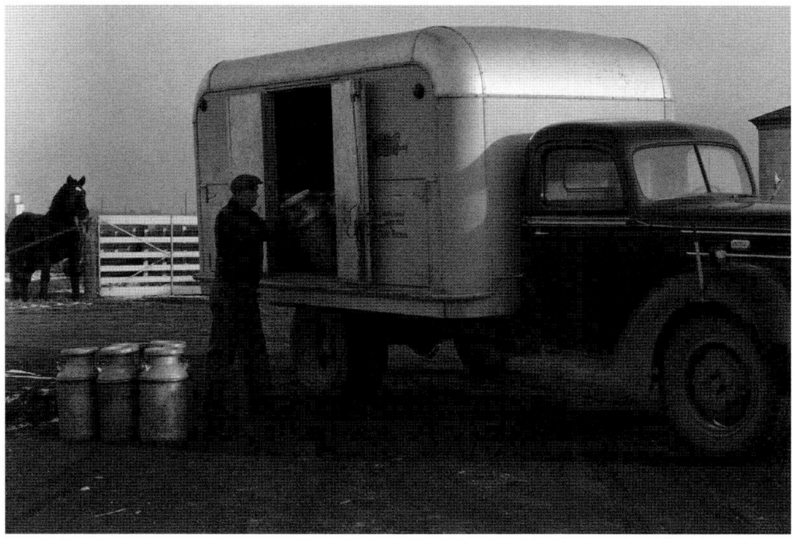

Dale Brown places cans of milk into a truck for delivery to a creamery. (Tippecanoe County, 1941)

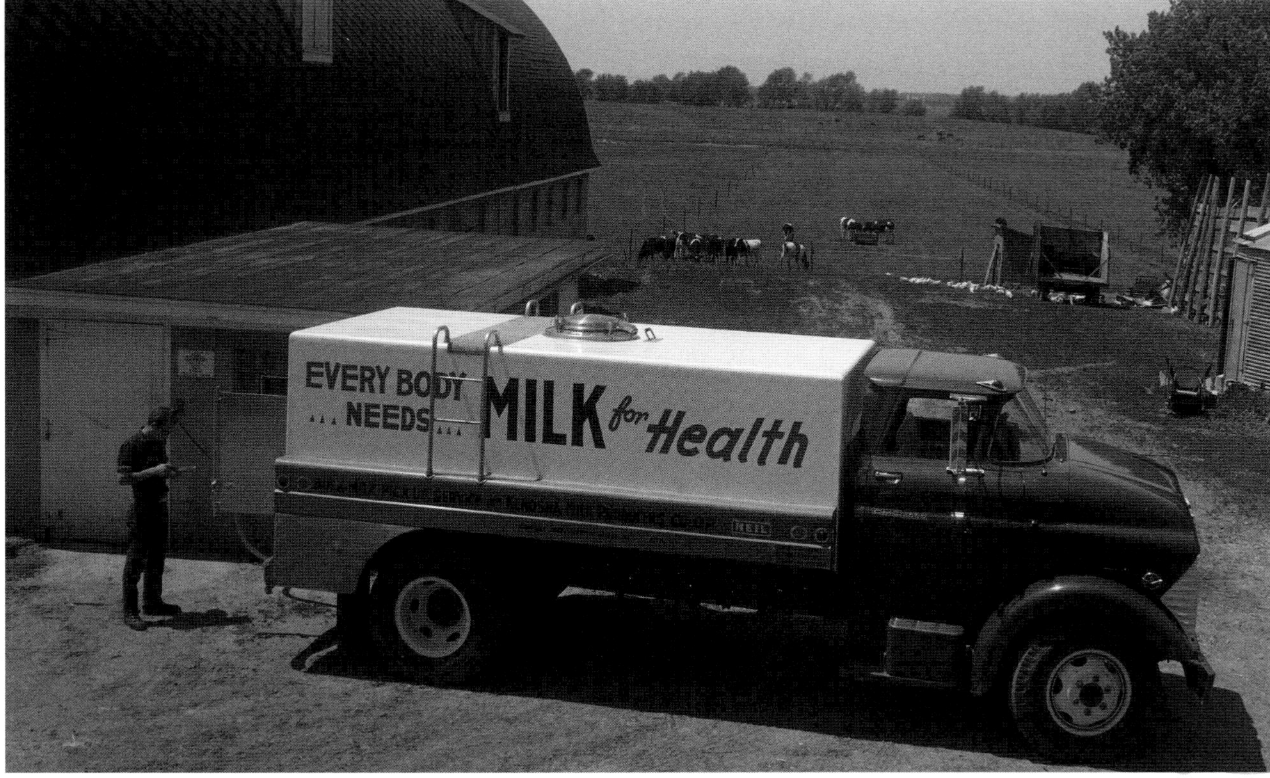

A modern milk truck collects product from a bulk tank on the dairy farm of James M. and Louis Fredrick in Union Grove, Wisconsin. (Racine County, WI; 1963)

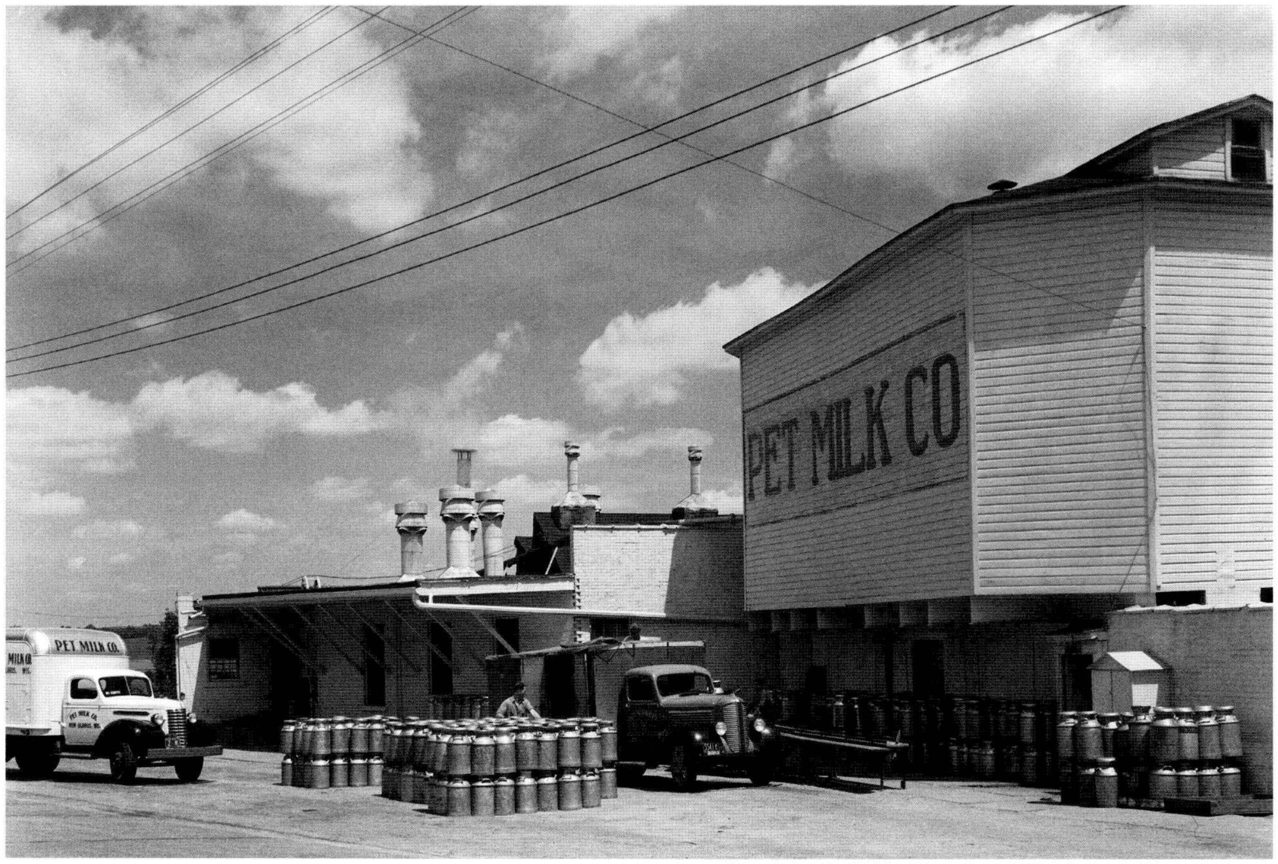

A Pet Milk Company plant at New Glarus, Wisconsin (Green County, WI; 1939)

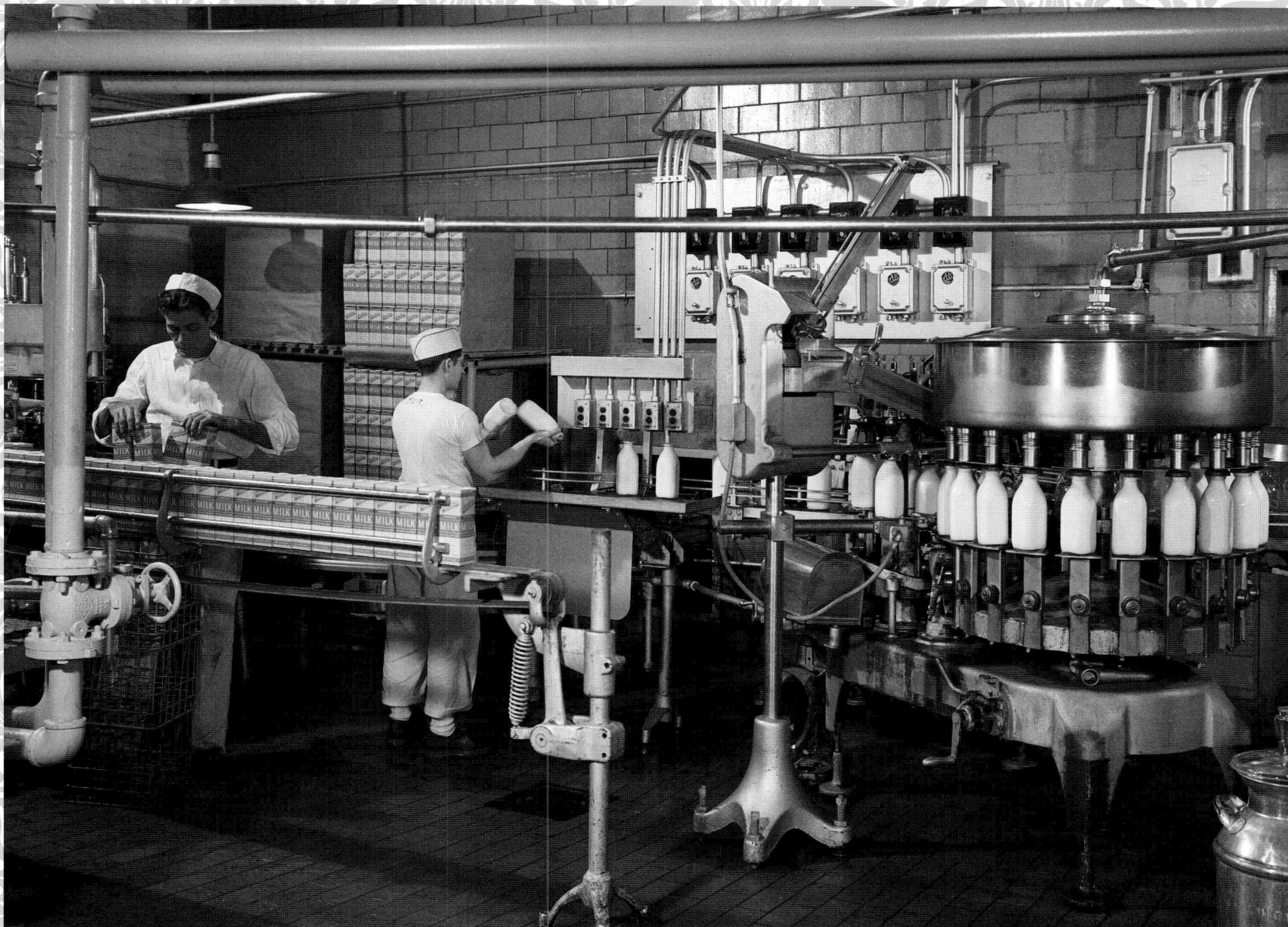

Bottling milk at the Sunshine Dairy in Lafayette (Tippecanoe County, 1949)

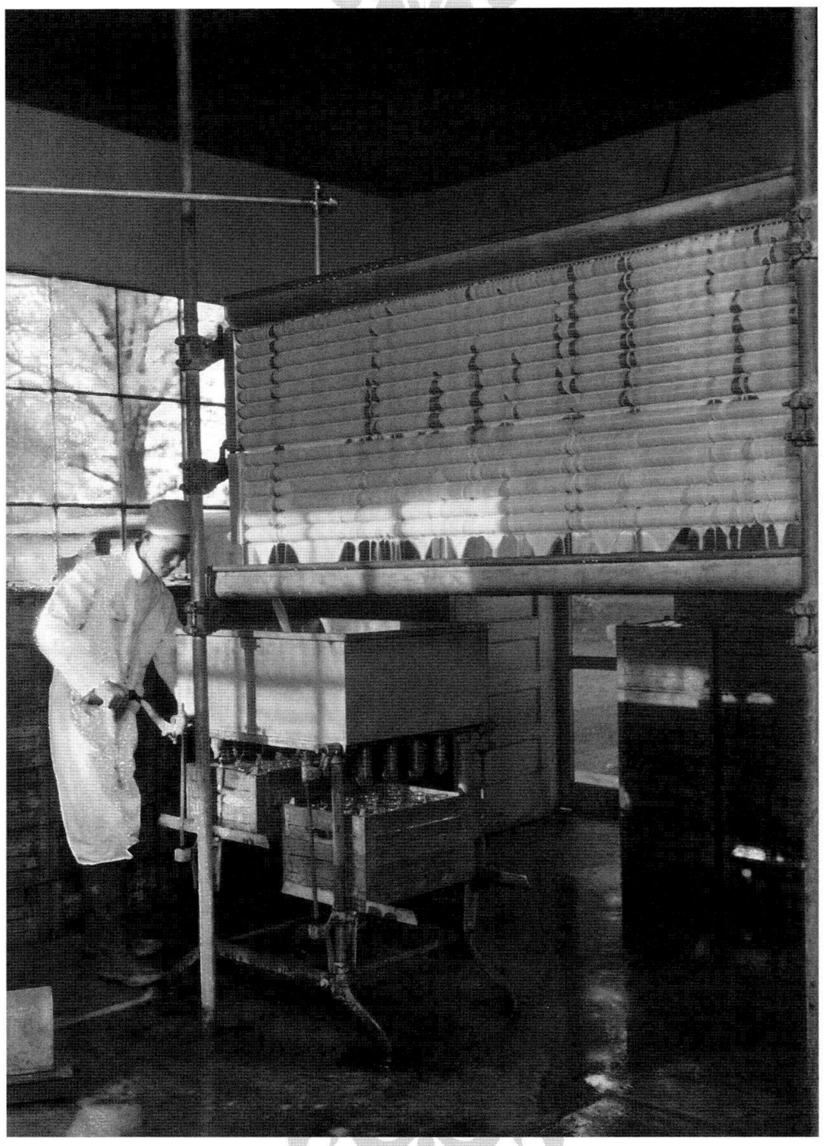

A worker bottles milk at the Morehouse Dairy in Lafayette. A milk cooler is visible in the background. (Tippecanoe County, 1929)

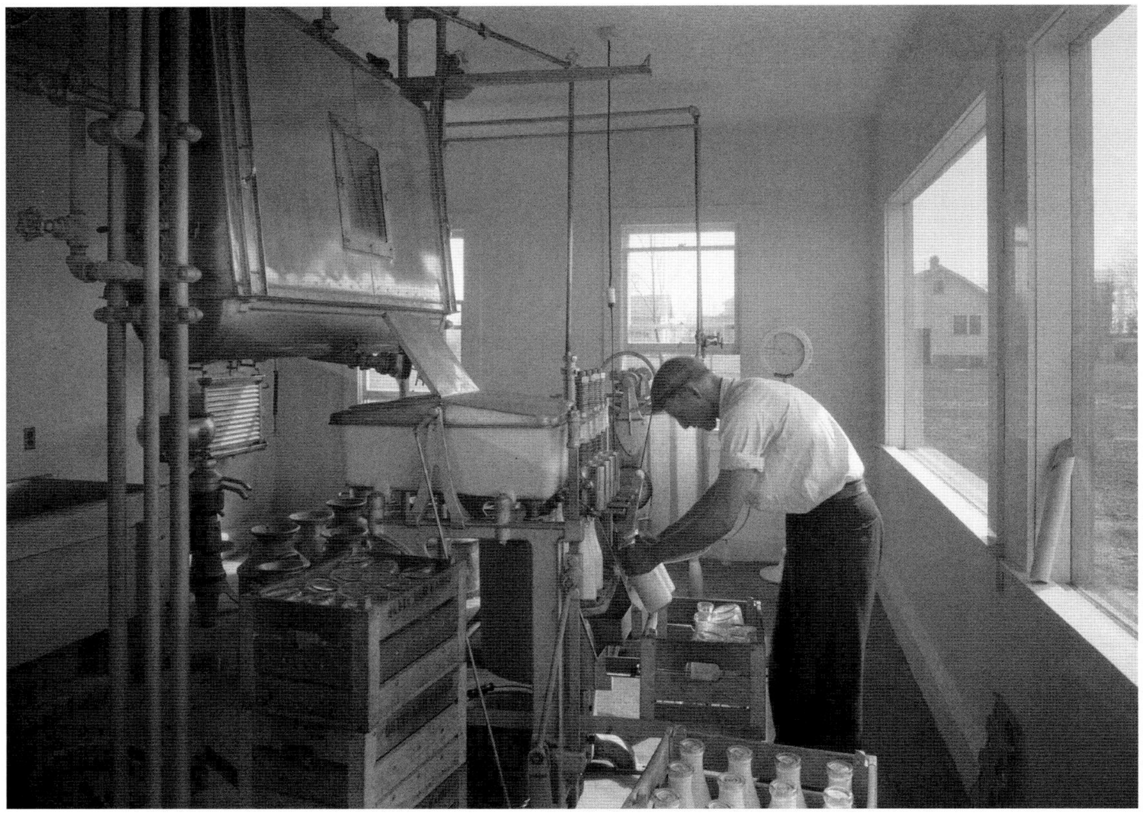

Fred Schornstein, shown here bottling milk, owned the Betty Lou Dairy Farm in Kokomo. (Howard County, 1932)

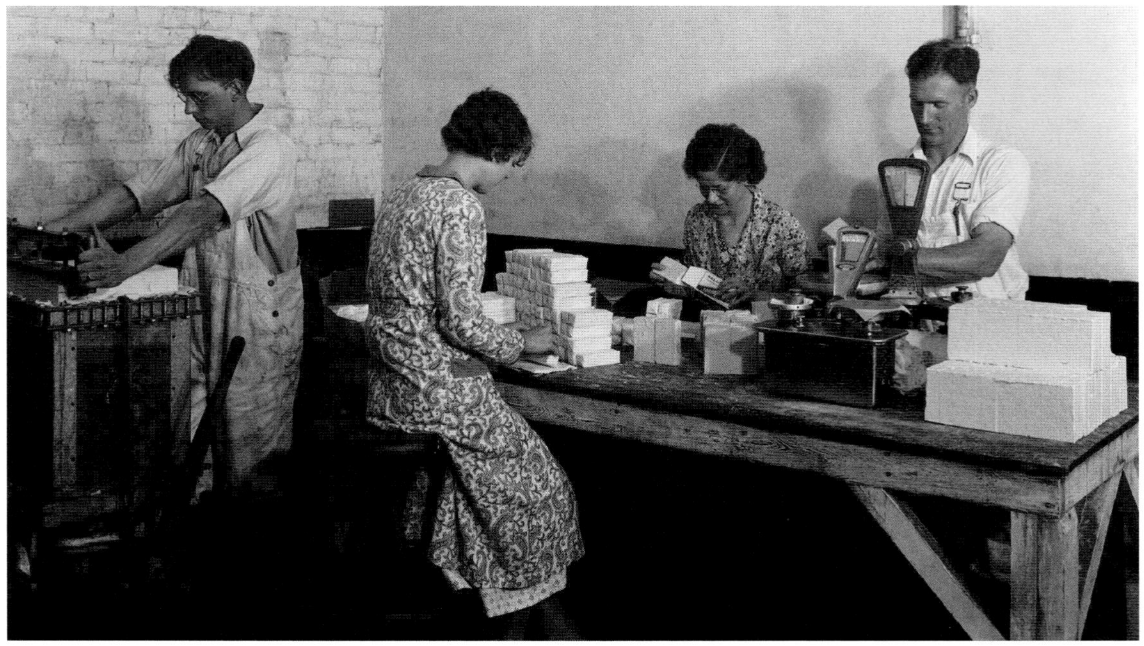

Cutting, weighing, and wrapping butter (Indiana, ca. 1934)

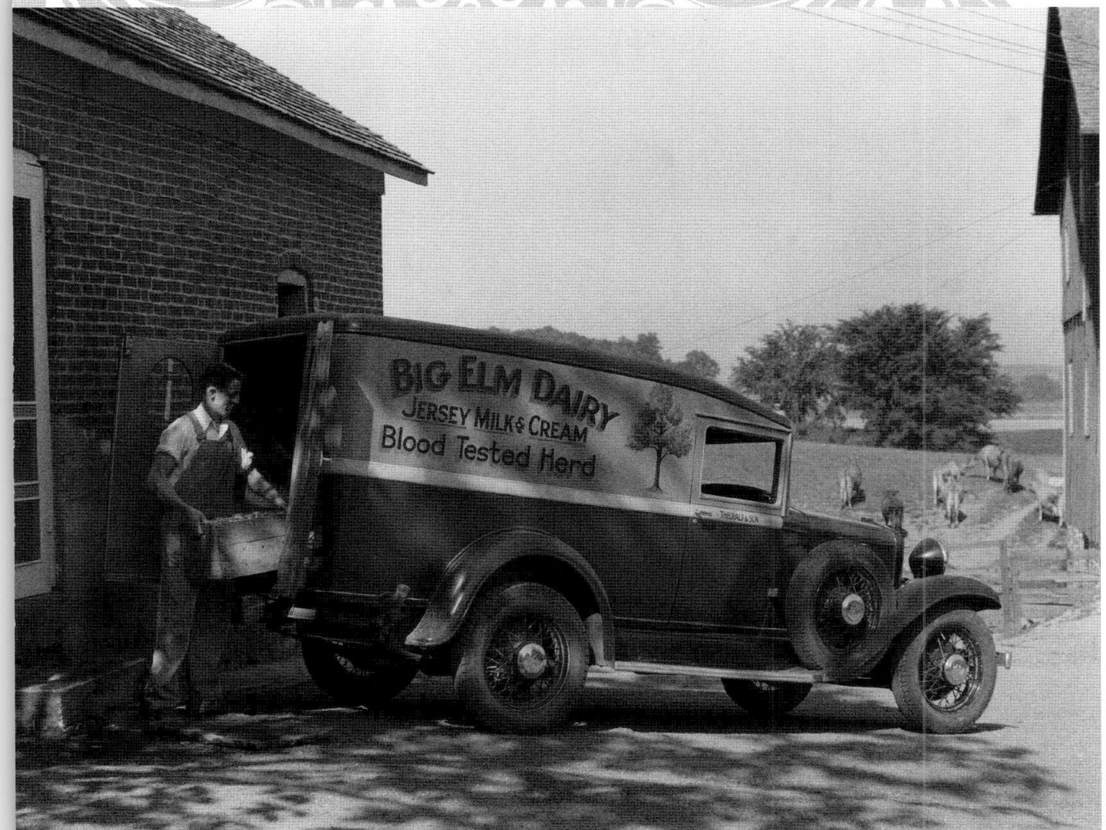

Loading a delivery truck on the Big Elm Dairy of Chancy Theobald in Andrews. (Huntington County, 1934)

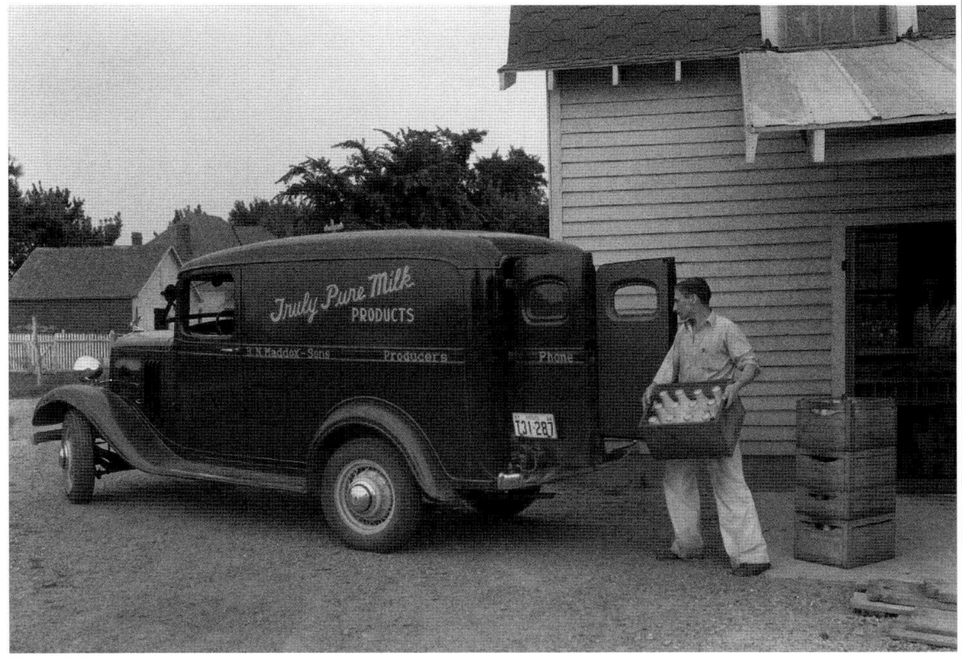

A man places bottles into a milk truck on a dairy owned by R. N. Maddox and Sons in Mayfield, Kentucky. (Graves County, KY; 1934)

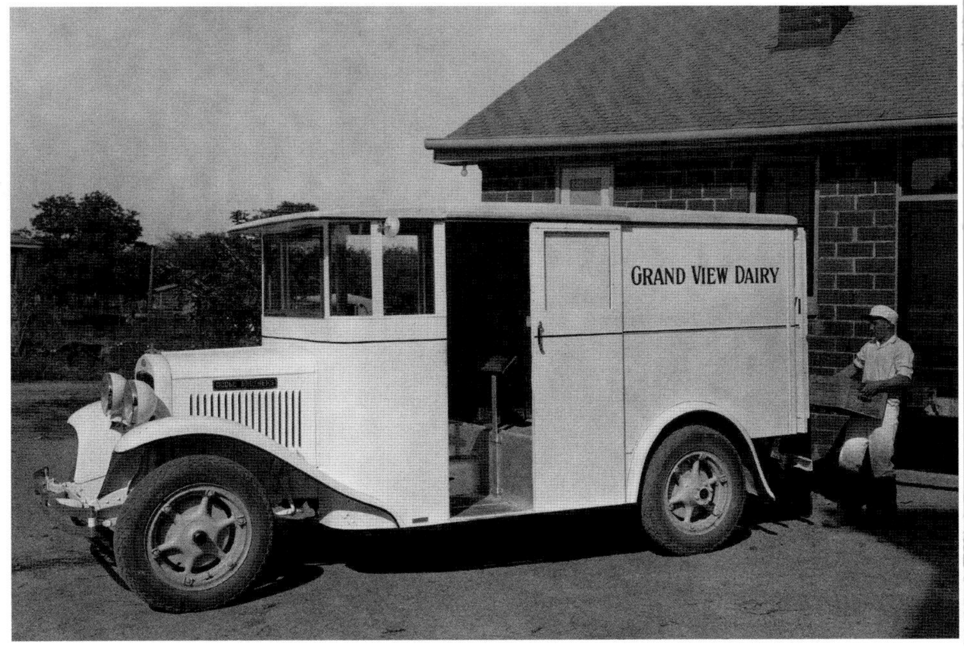

A Grand View Dairy milk truck (Tippecanoe County, ca. 1938)

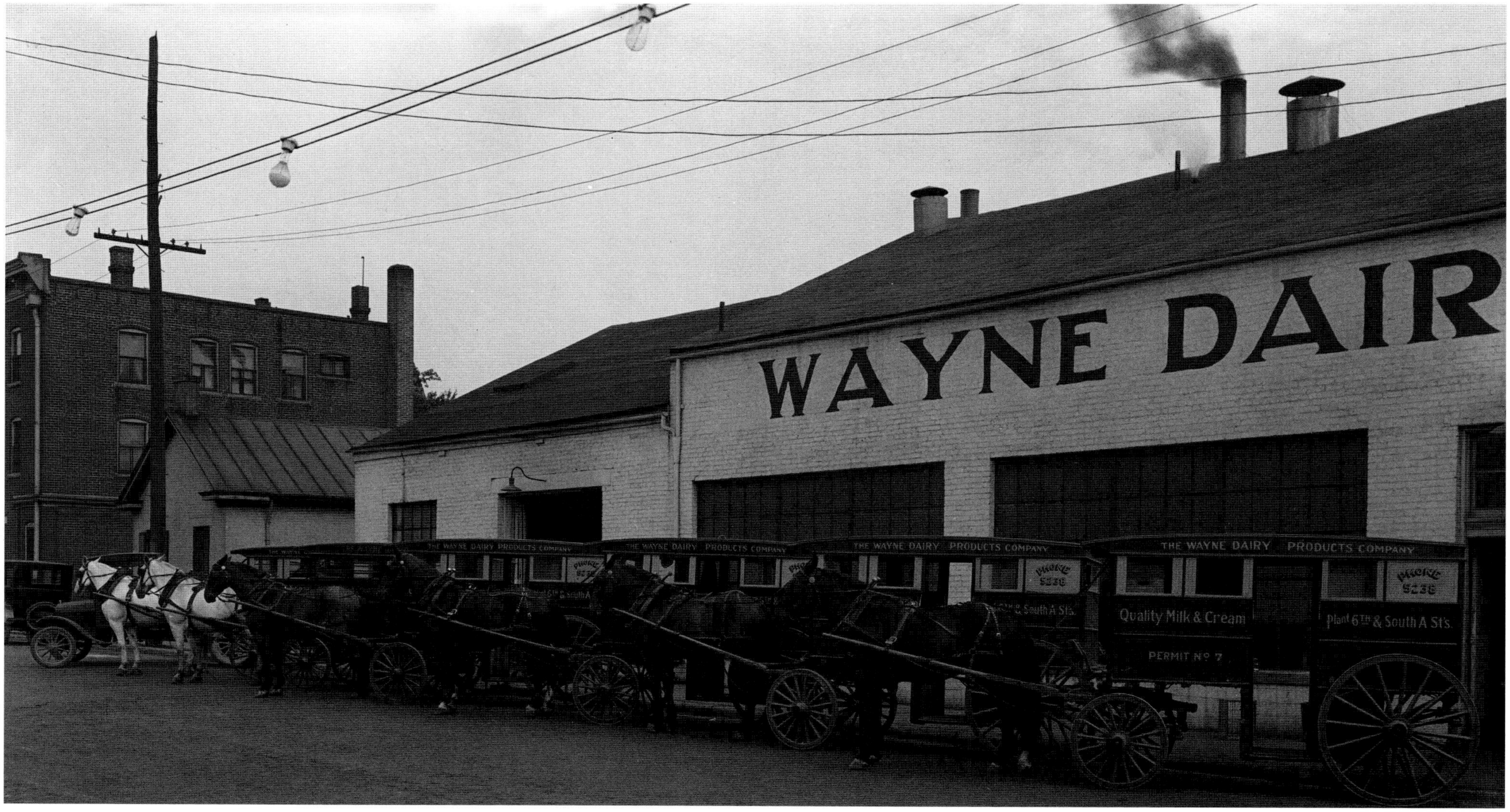

Horse-drawn milk wagons line up in front of the Wayne Dairy Products Company, a farmers' cooperative in Richmond. (Wayne County, 1925)

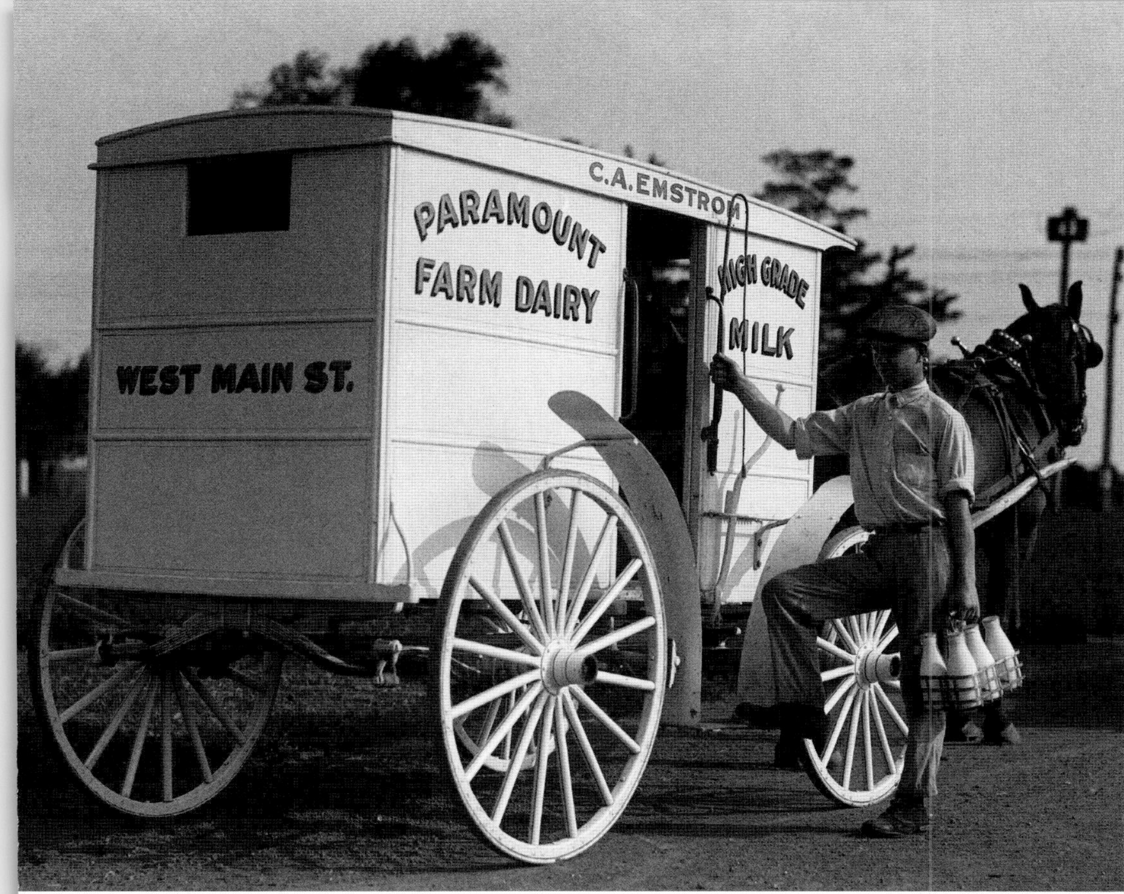

A horse-drawn milk delivery wagon from the farm of C. A. Emstrom in Galesburg, Illinois (Knox County, IL;1919)

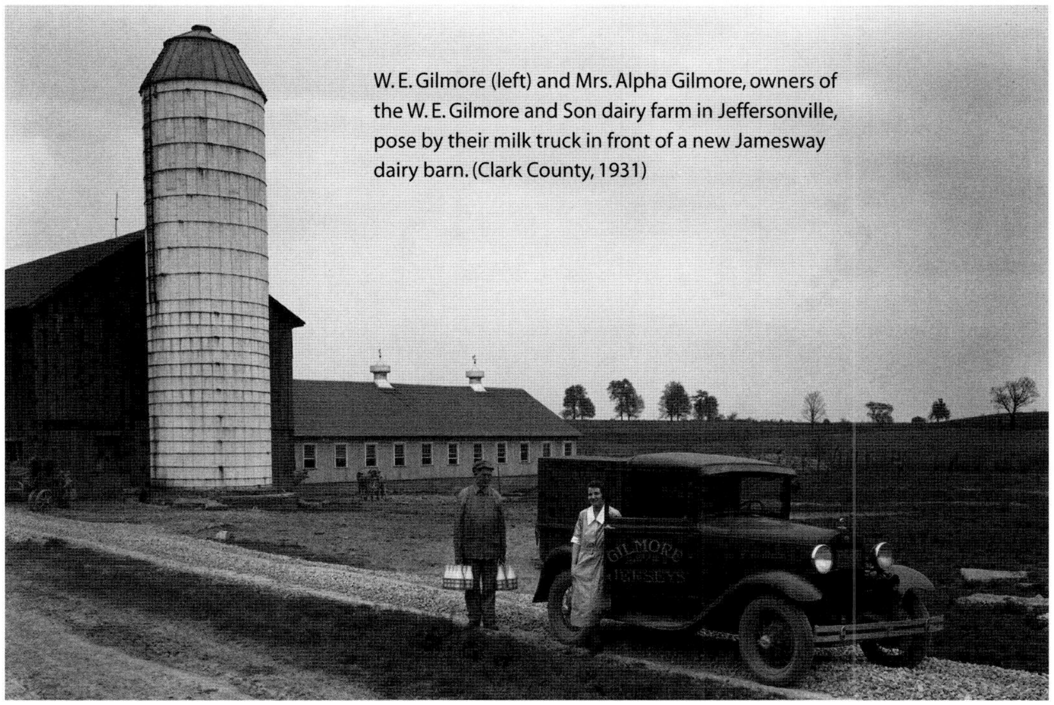

W. E. Gilmore (left) and Mrs. Alpha Gilmore, owners of the W. E. Gilmore and Son dairy farm in Jeffersonville, pose by their milk truck in front of a new Jamesway dairy barn. (Clark County, 1931)

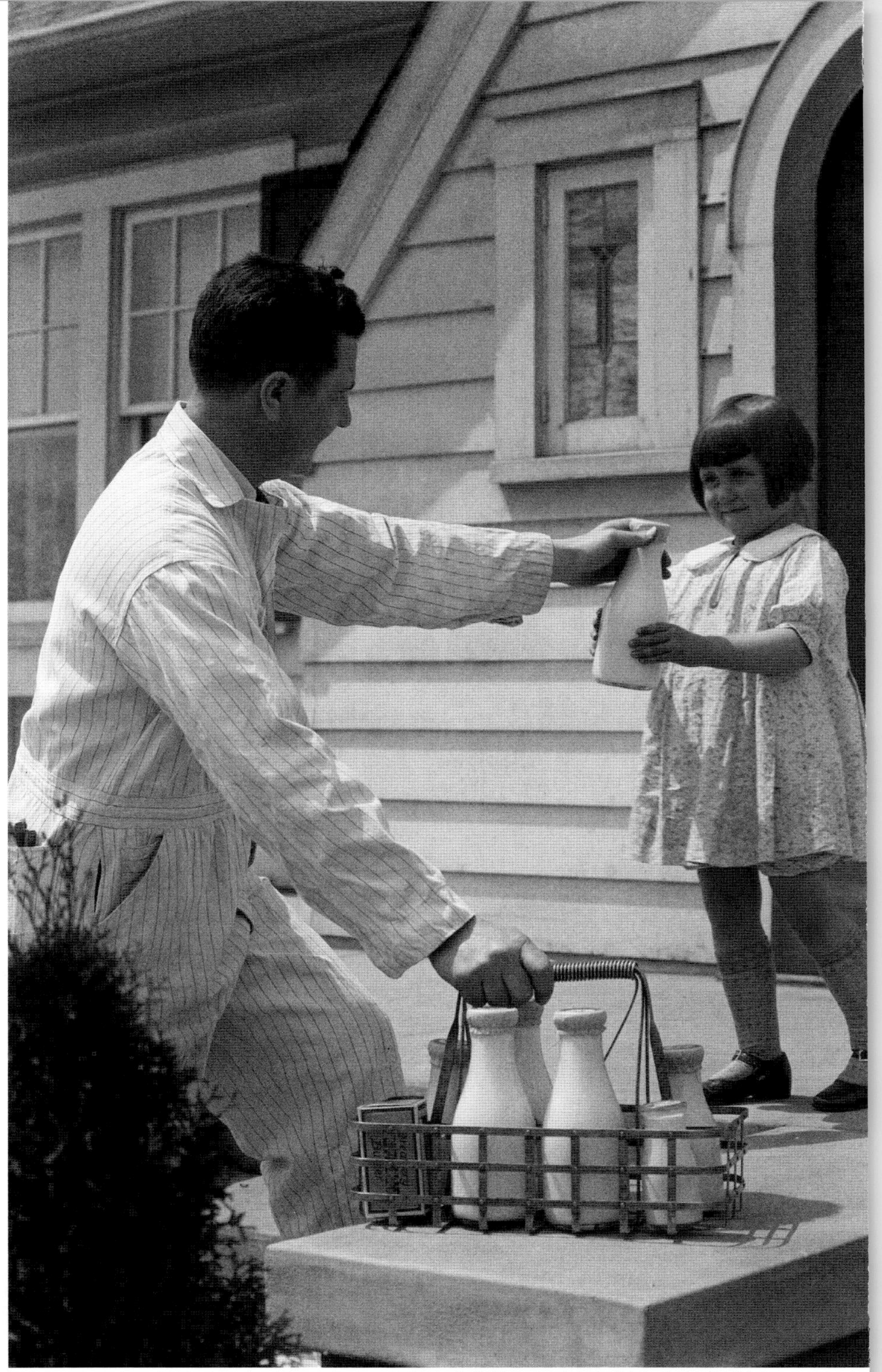

Kenneth Boxell hands a bottle of milk to Doris Kern. (Tippecanoe County, 1933)

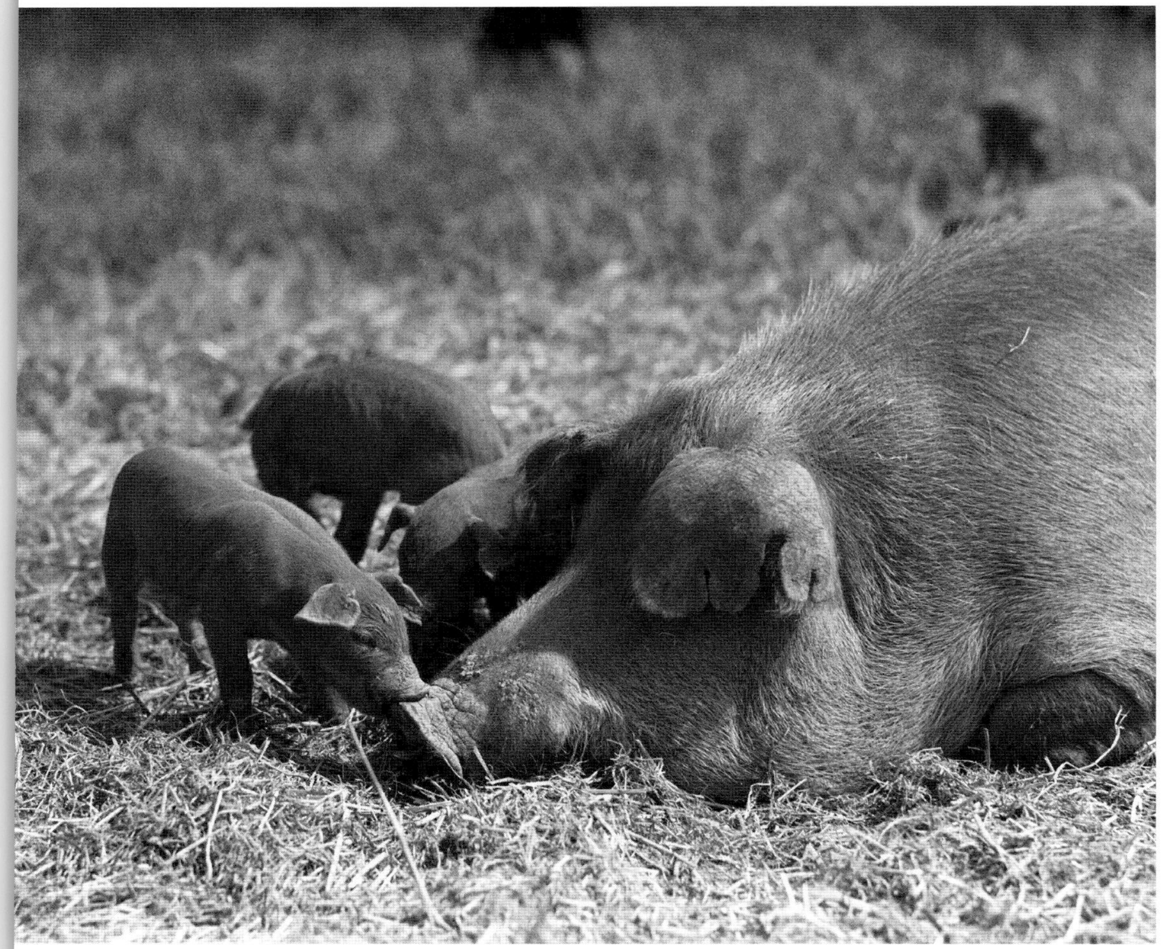

A Duroc piglet chews on a sow's nose. (N.p., 1938)

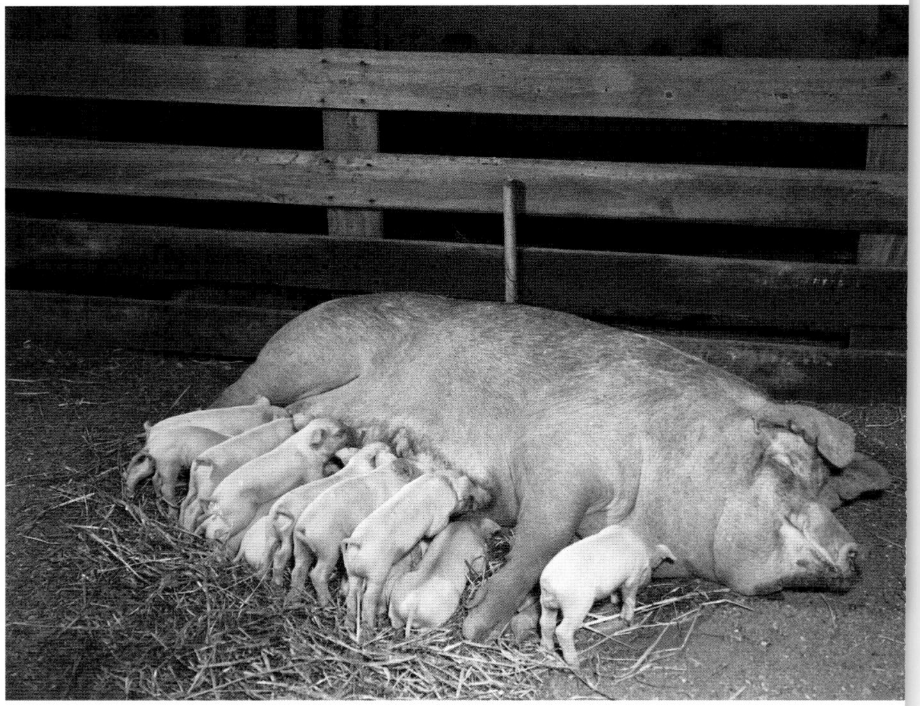

A Chester White sow nurses a large litter of pigs. (Indiana, 1943)

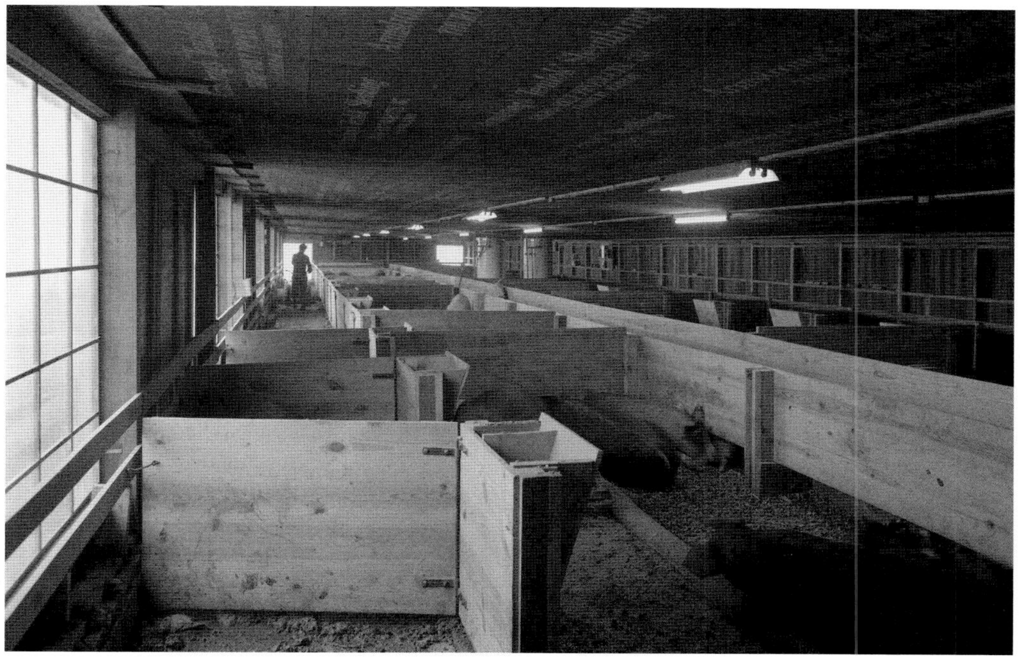

The interior of a modern pig farrowing house on the farm of Don Childress in Linden (Montgomery County, 1954)

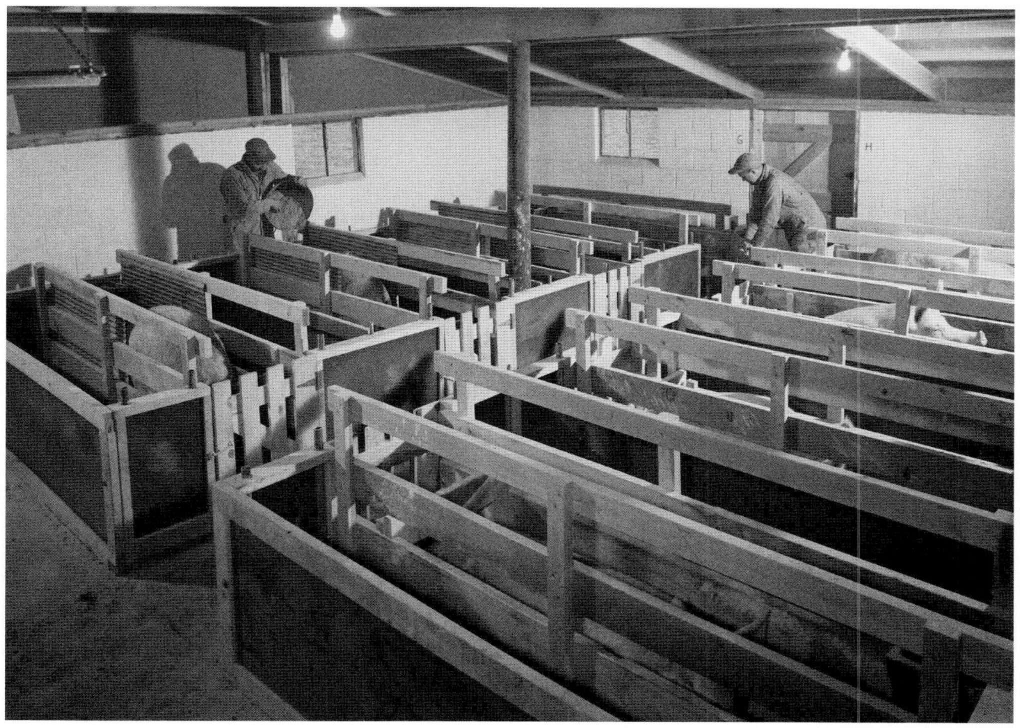

A pig farrowing house on the farm of Bill Congleton in Frankfort (Clinton County, 1956)

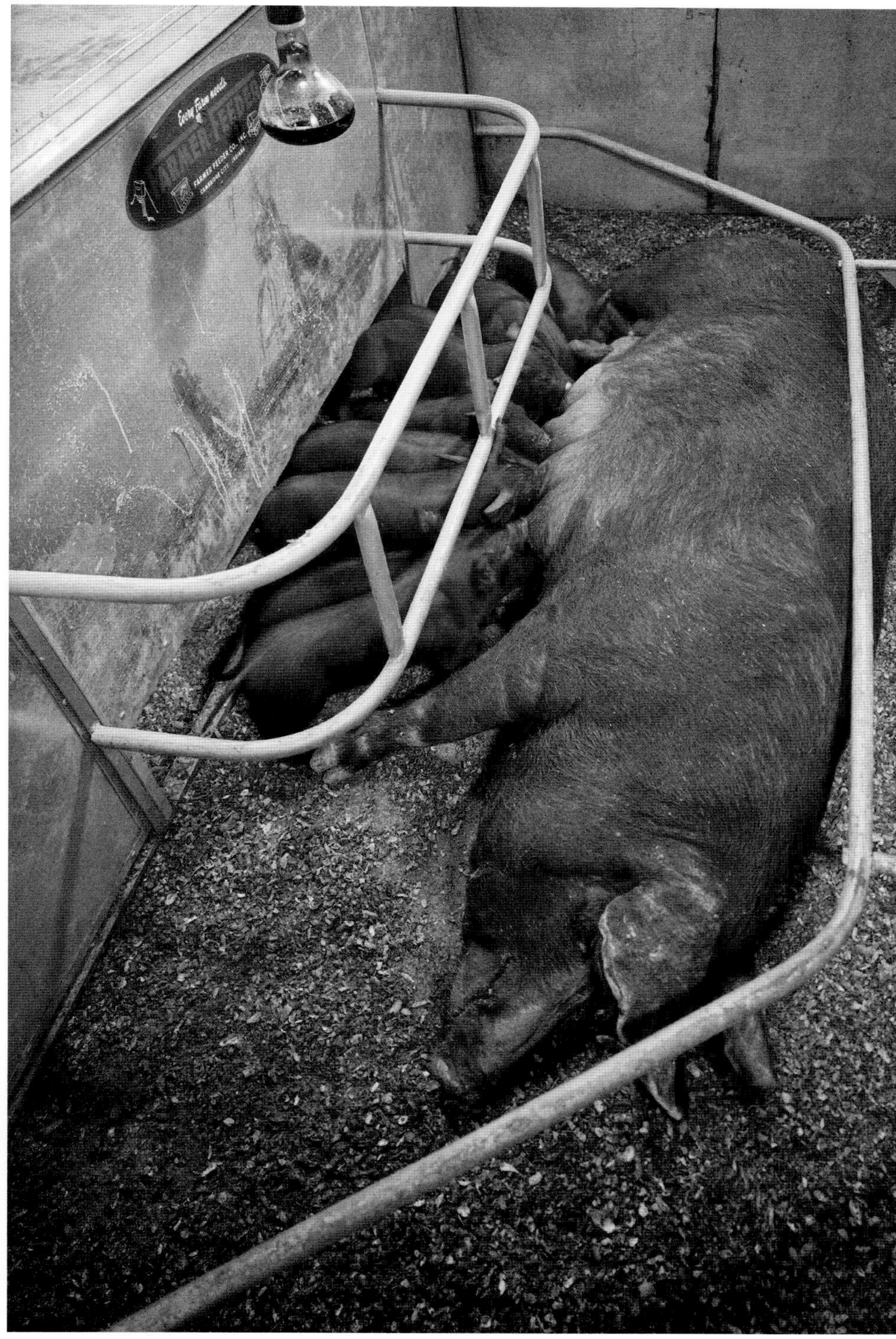

A sow nurses her litter beside a special railing constructed to protect the young pigs at the Conner Prairie Farm at Noblesville. (Hamilton County, 1959)

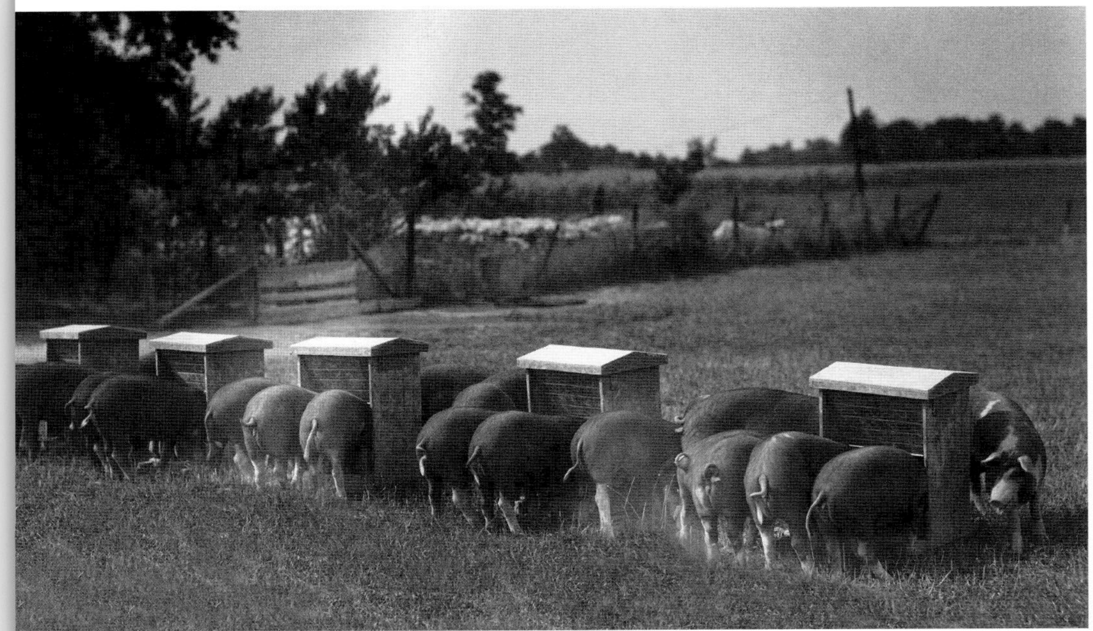

Poland China gilts gather around self-feeders on the Arland and J. Warren Harvey farm. (Hamilton County, 1930)

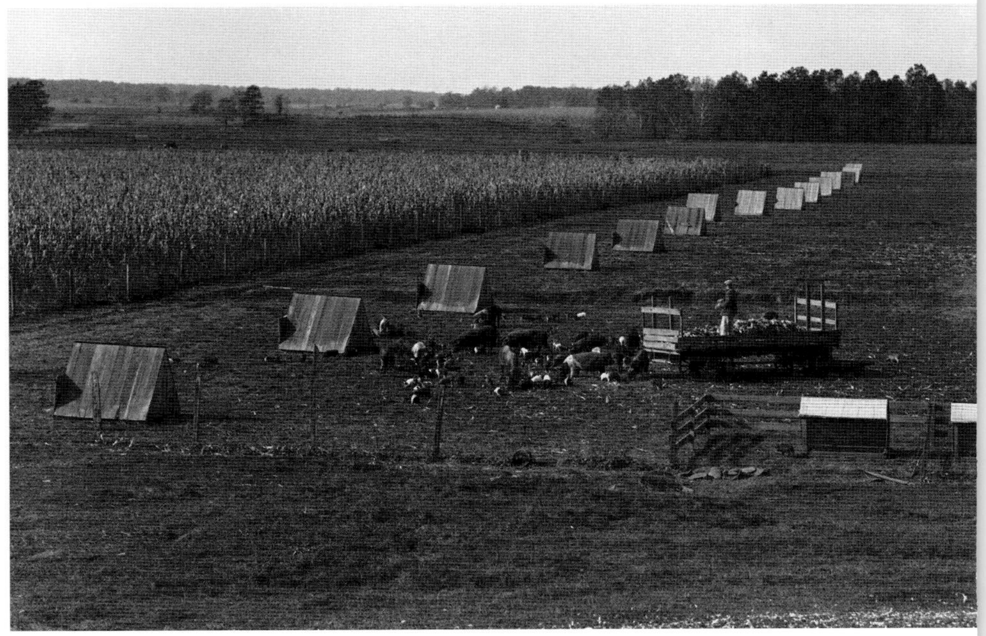

Individual hog houses on grass pasture at the Rainey McCoy farm (Tippecanoe County, 1928)

Cliff Breeder moves a sow and her litter from a central farrowing house to an individual pen and house on pasture. (Tippecanoe or Benton County, 1947)

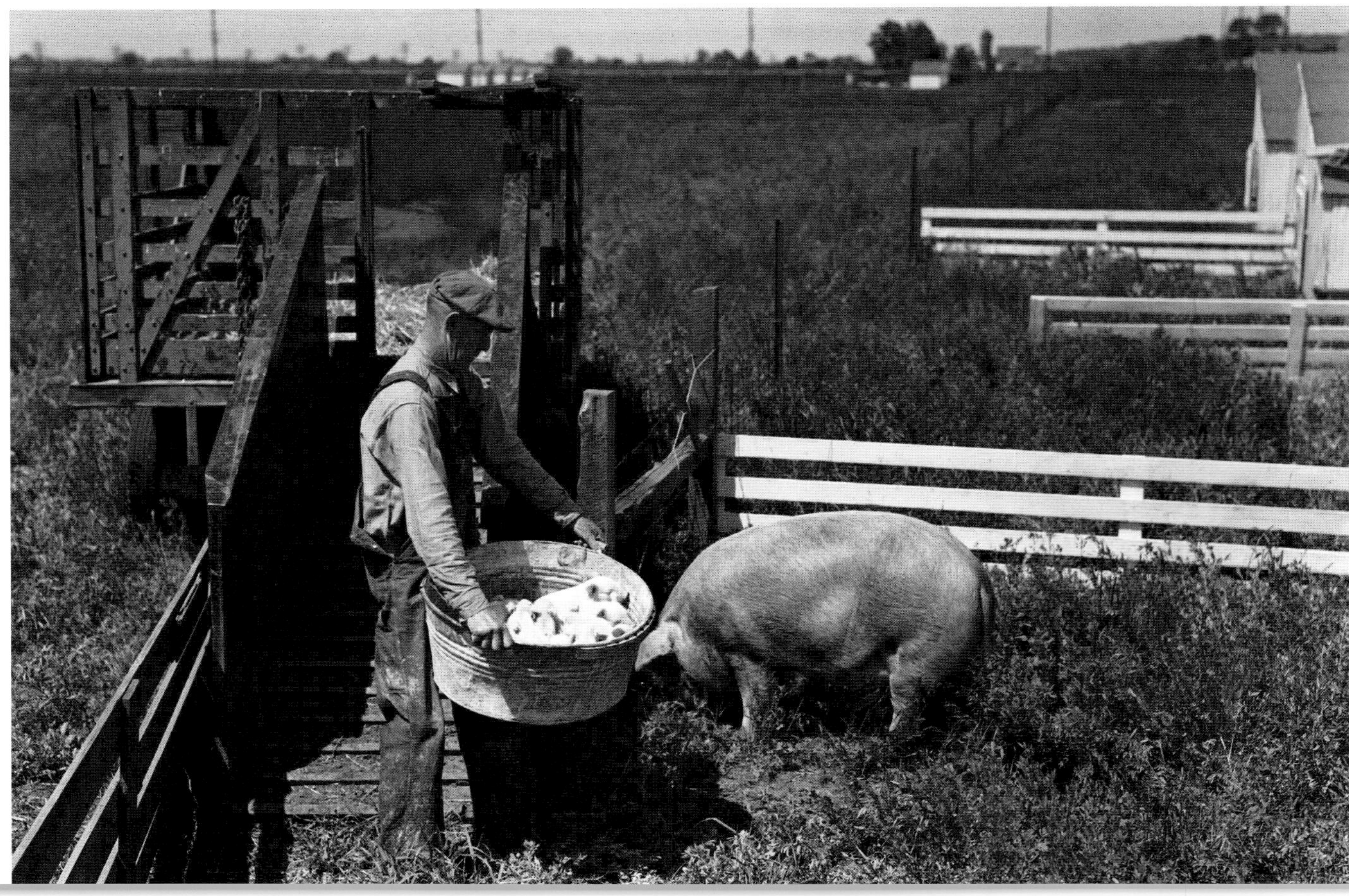

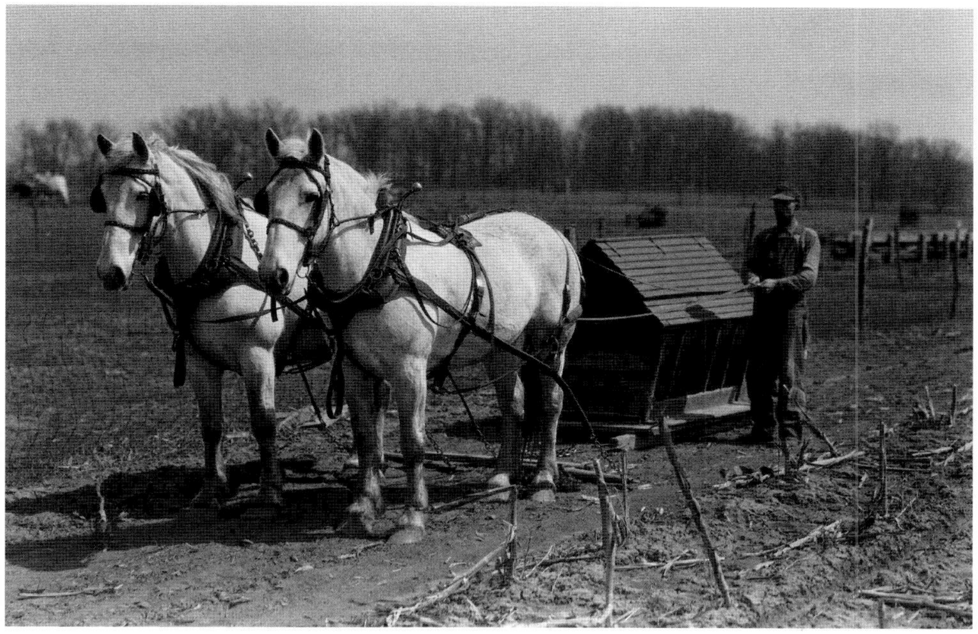

Tenant farmer Bob Schwartz uses a team of white horses to move a hog self-feeder to fresh ground. (Tippecanoe County, 1931)

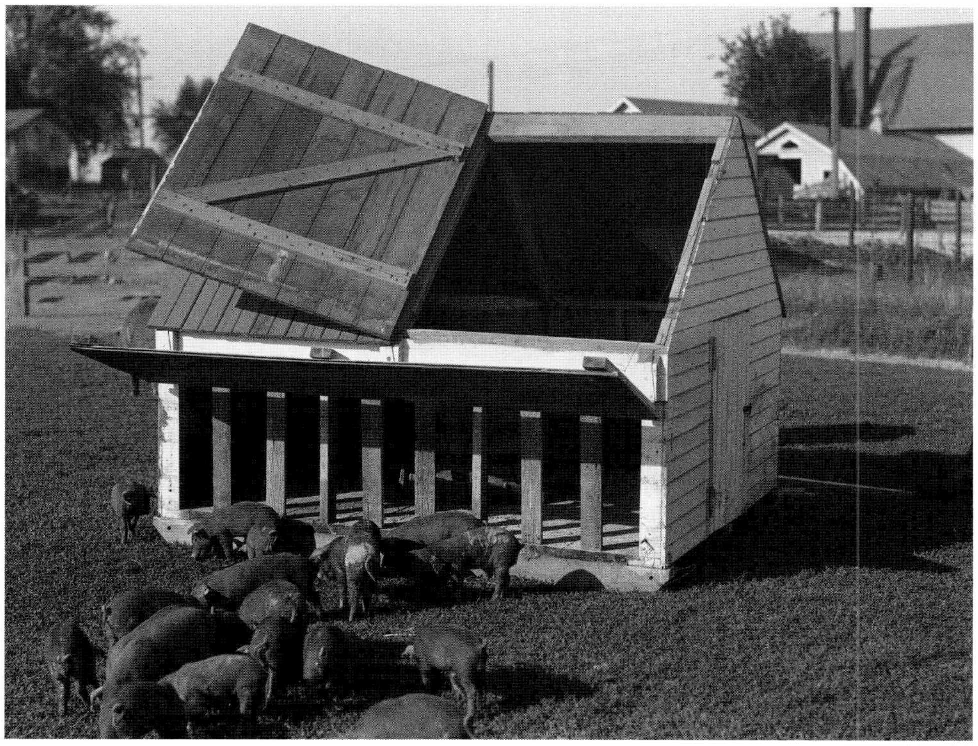

This building, which Allen called a "pig diner," contained a self-feeder. The fencelike structure, known as a creep, limited access to small pigs only. (Indiana, 1933)

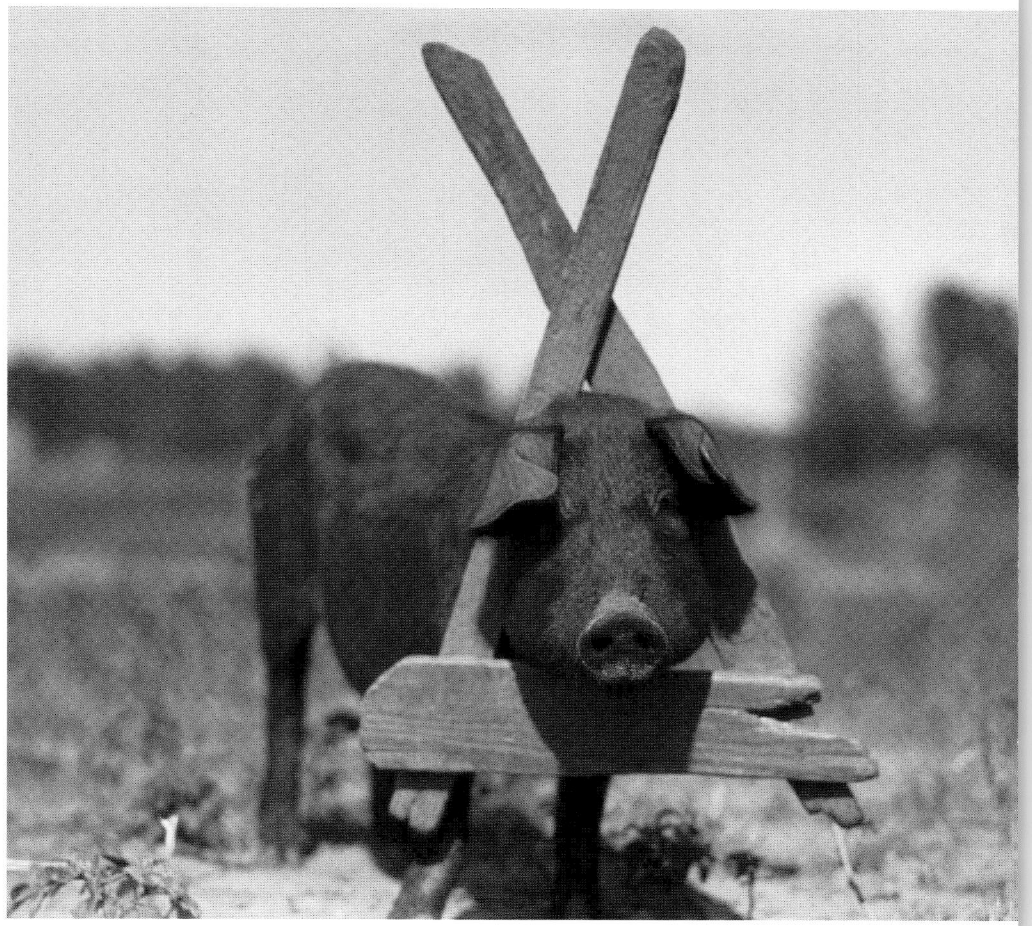

The yoke around the neck of this pig kept it from going through the fence on a Linden-area farm in east Texas. (Cass County, TX; 1933)

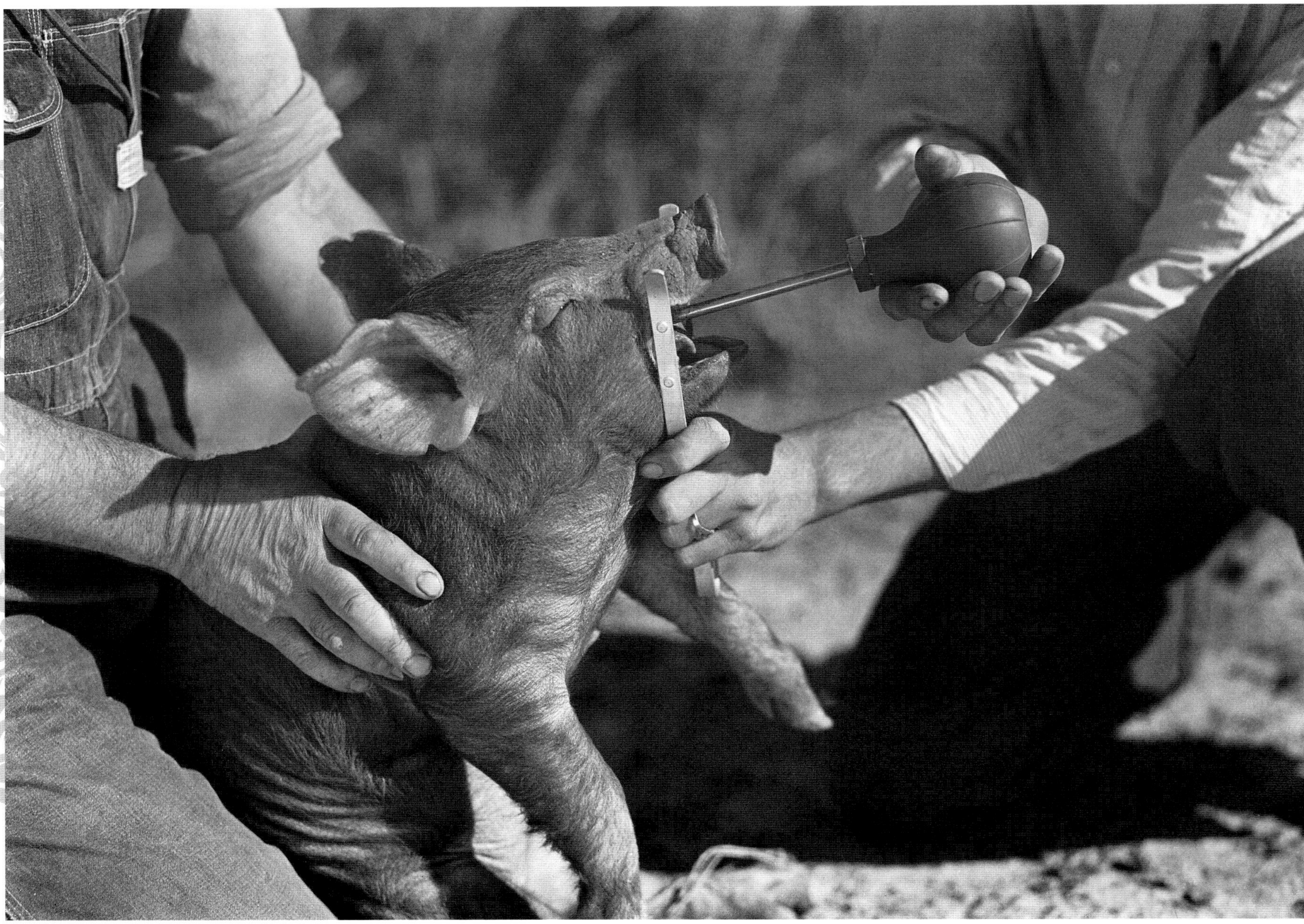

Worming a small pig to kill internal parasites such as roundworms (Indiana, 1938)

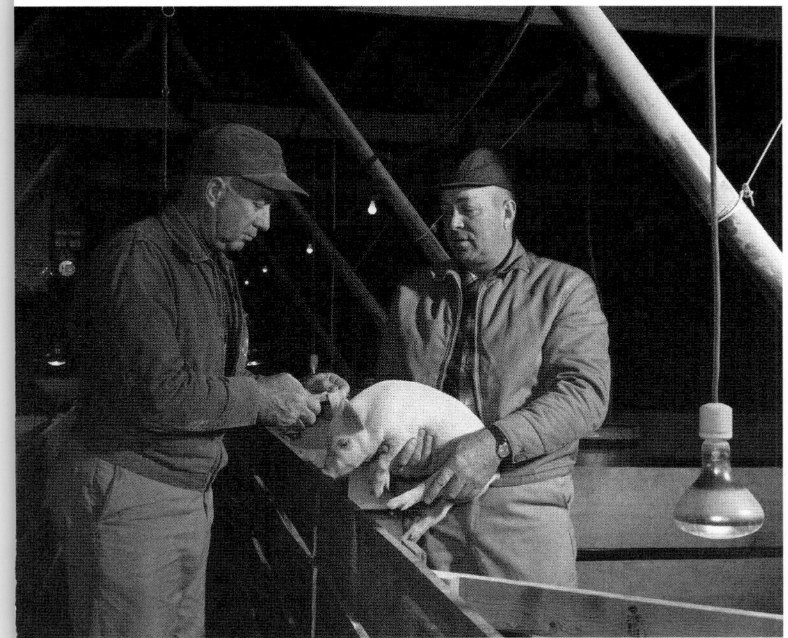

William and Erland Rothenberger from Frankfort notch a pig's ear. (Clinton County, 1966)

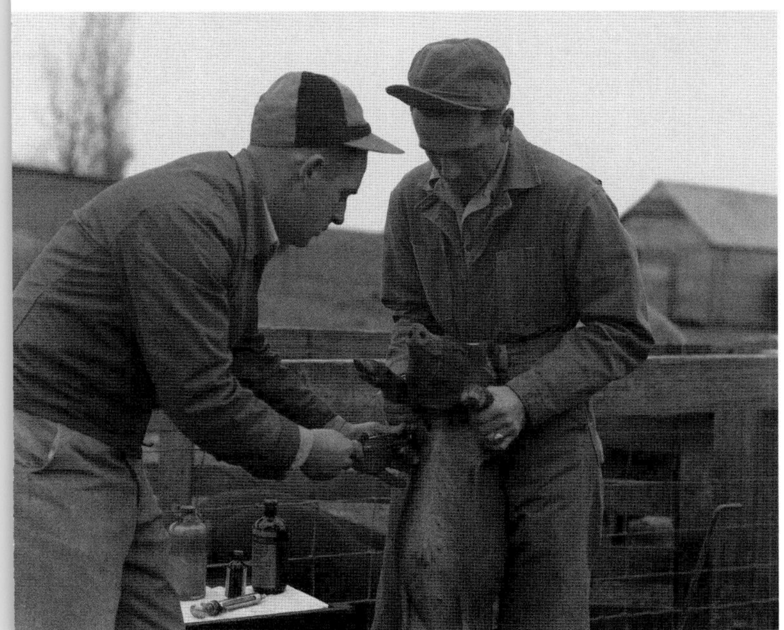

Frank Chilovich (left) vaccinates a pig held by Bob Spangler at the Purina Farm in Gray Summit, Missouri. (Franklin County, MO; 1946)

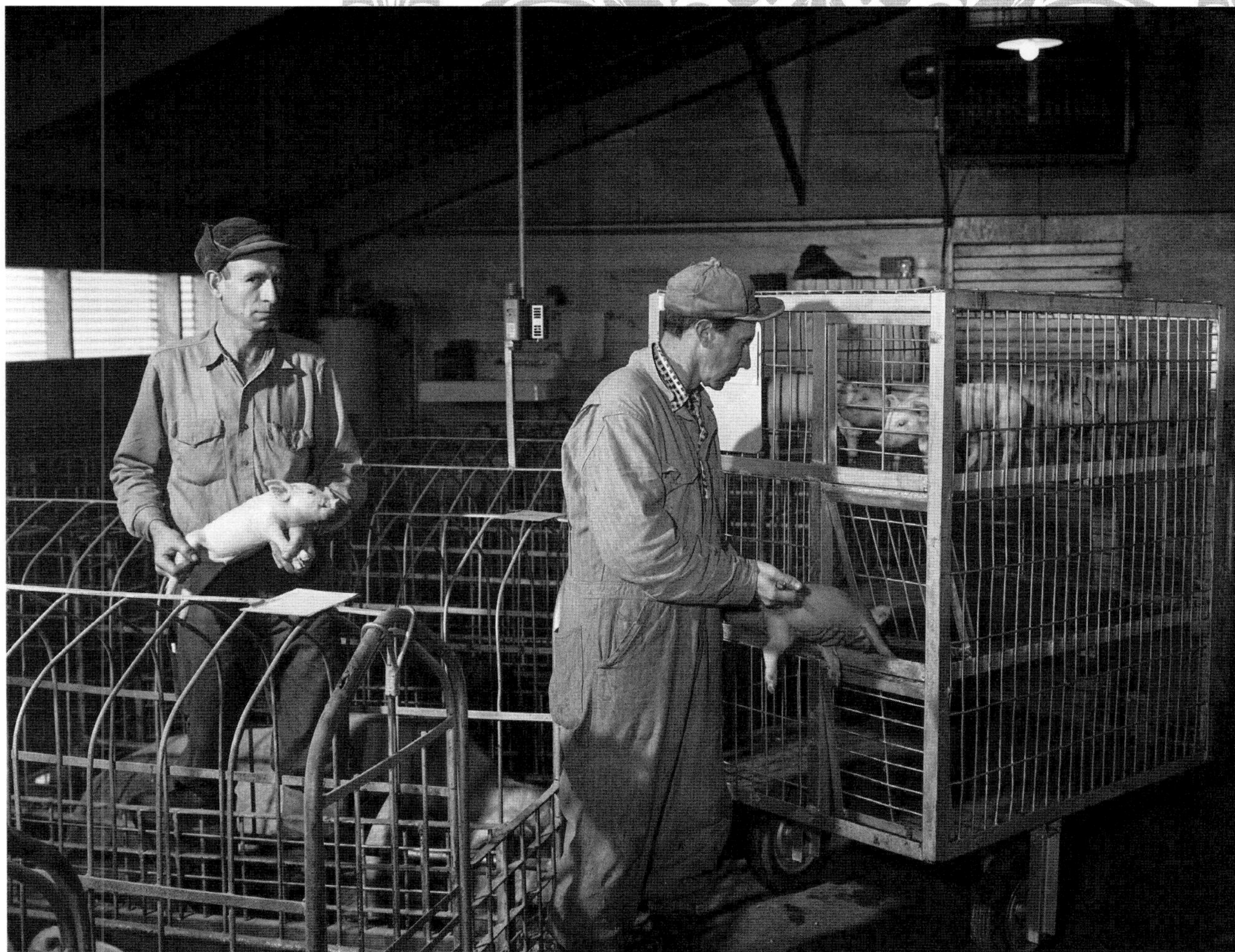

Two men move pigs from a farrowing house to a nursery at Norris Farms in Havana, Illinois. (Mason County, IL; 1958)

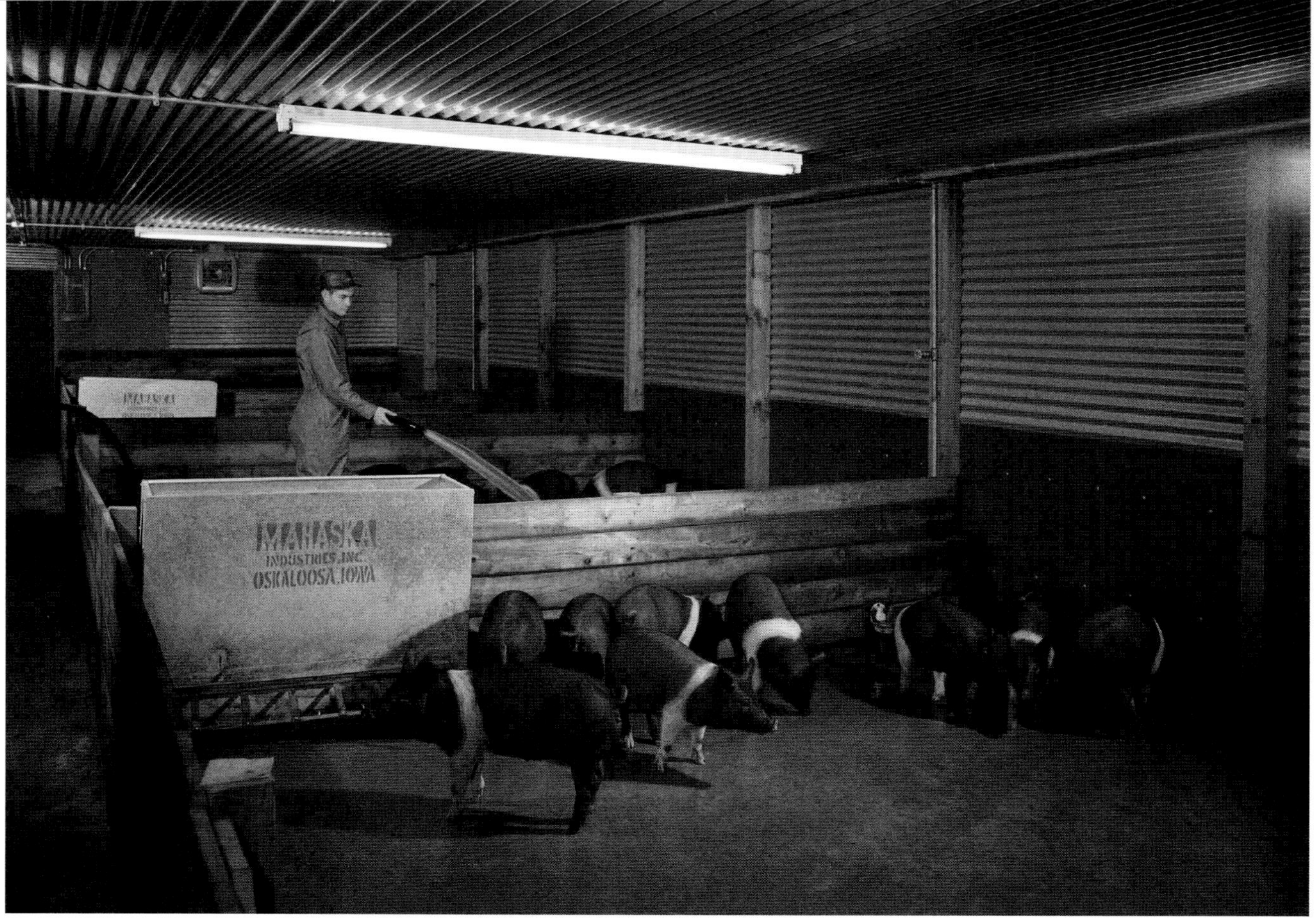

A man cleans manure from concrete inside a pig nursery on the Adams Farm in Tangier. (Parke County, 1961)

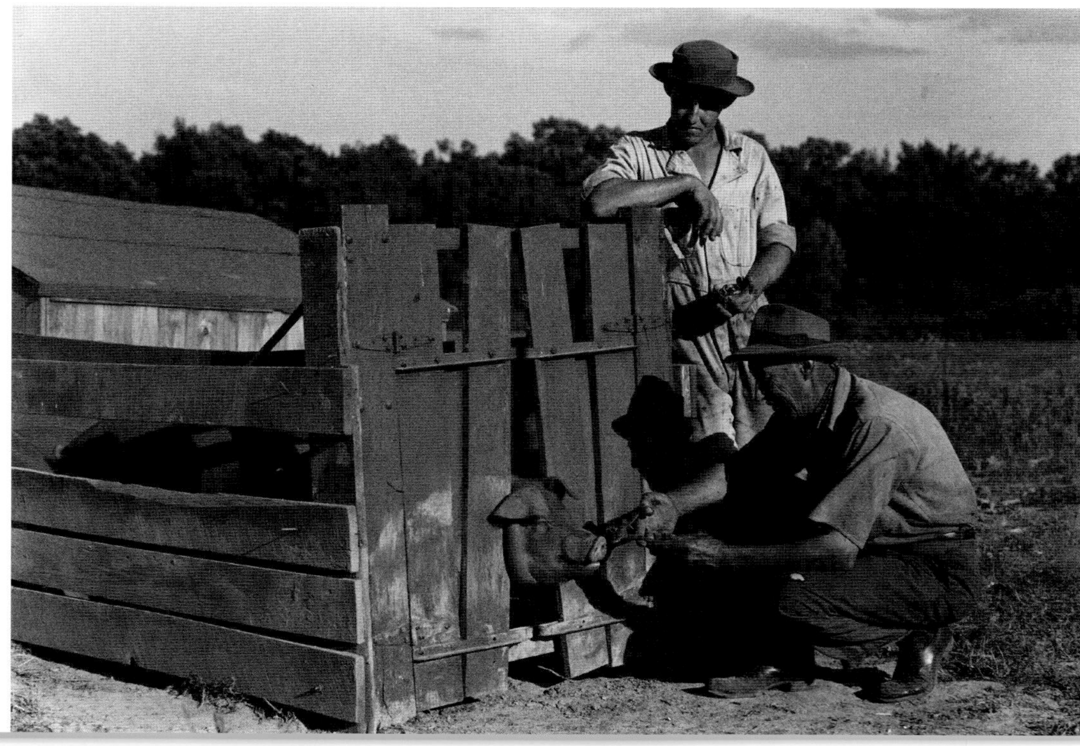

Joe O'Brien places a ring in the nose of a hog while his son looks on in Lafayette. (Tippecanoe County, 1946)

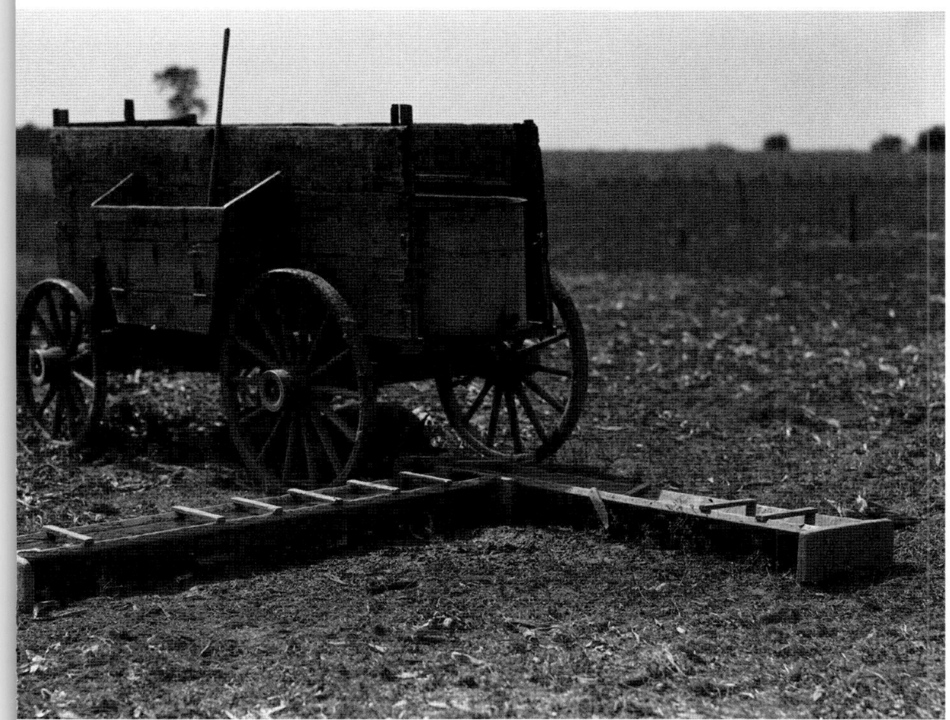

A water storage tank in an old wagon is used for watering hogs on the farm of Edward Funk and Sons in Earl Park. (Benton County, 1932)

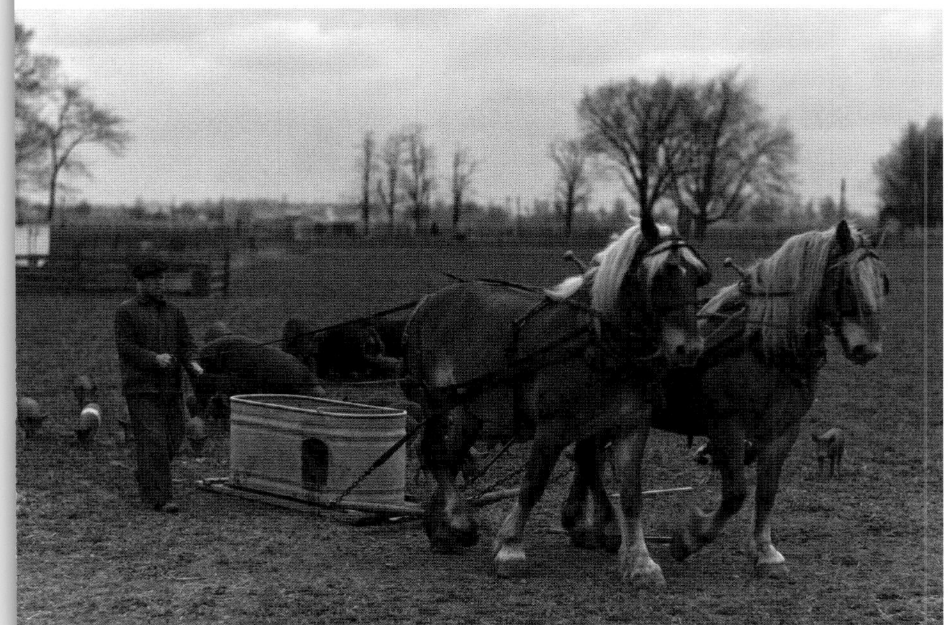

Hauling water for hogs with a team of Belgian mares on the farm of Harvey Hoewischer in Sidney, Ohio (Shelby County, OH; 1943)

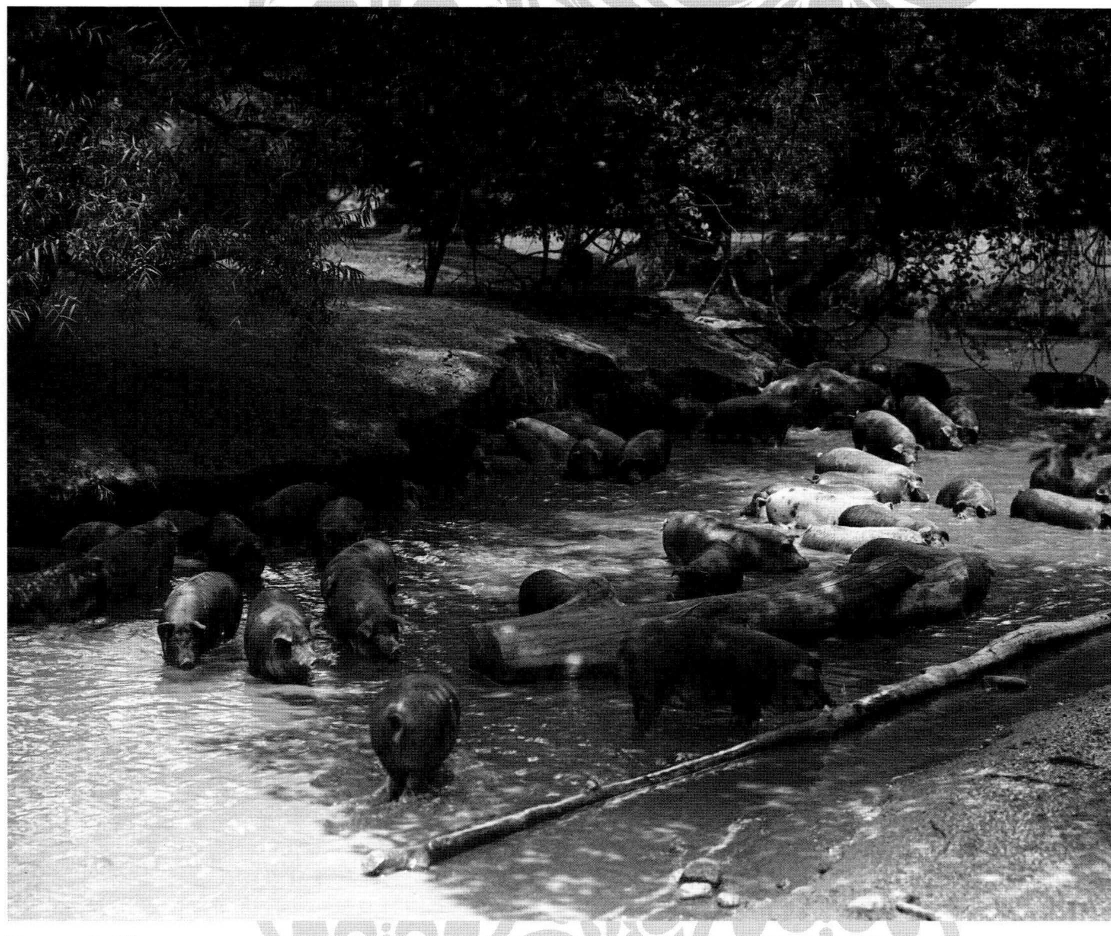

Hogs cool off in a stream on the farm of Lawrence P. Funk in McLean, Illinois. (McLean County, IL; 1917)

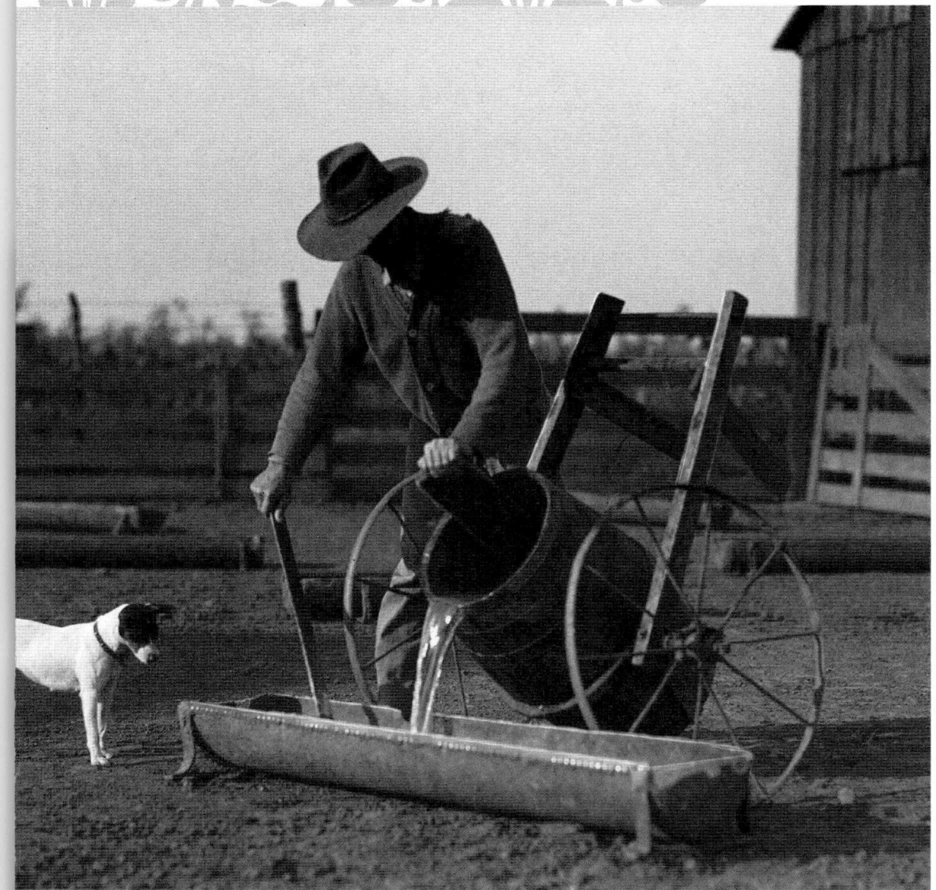

A man uses a barrel retrofitted with wheels to haul water and slop for hogs on the Hollingsworth Farm. (Tippecanoe County, ca. 1913)

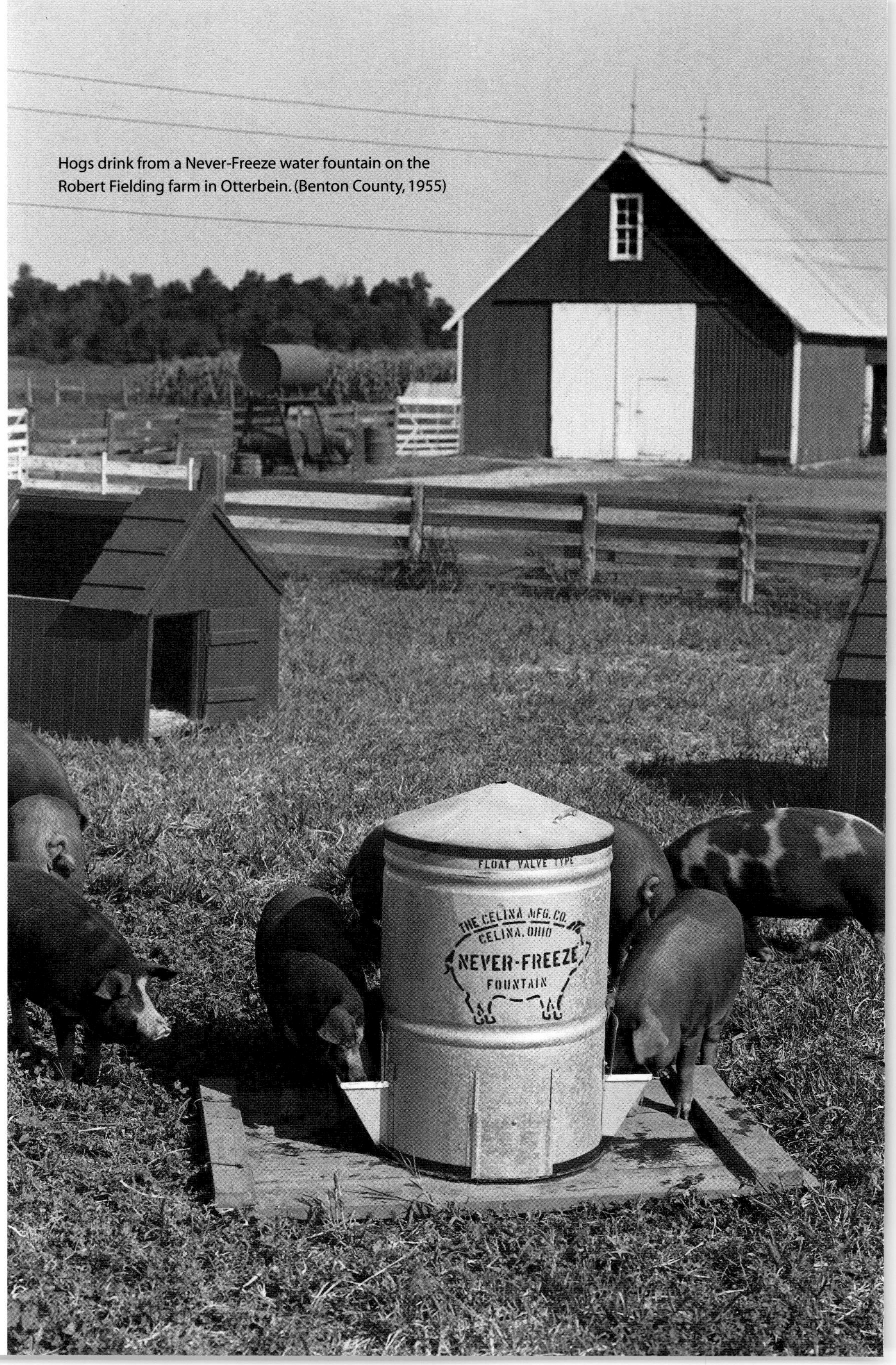

Hogs drink from a Never-Freeze water fountain on the Robert Fielding farm in Otterbein. (Benton County, 1955)

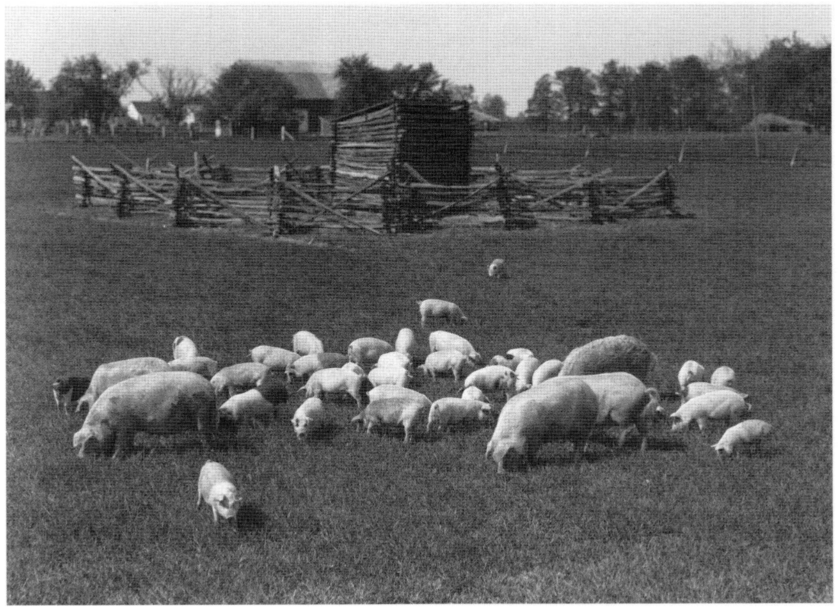

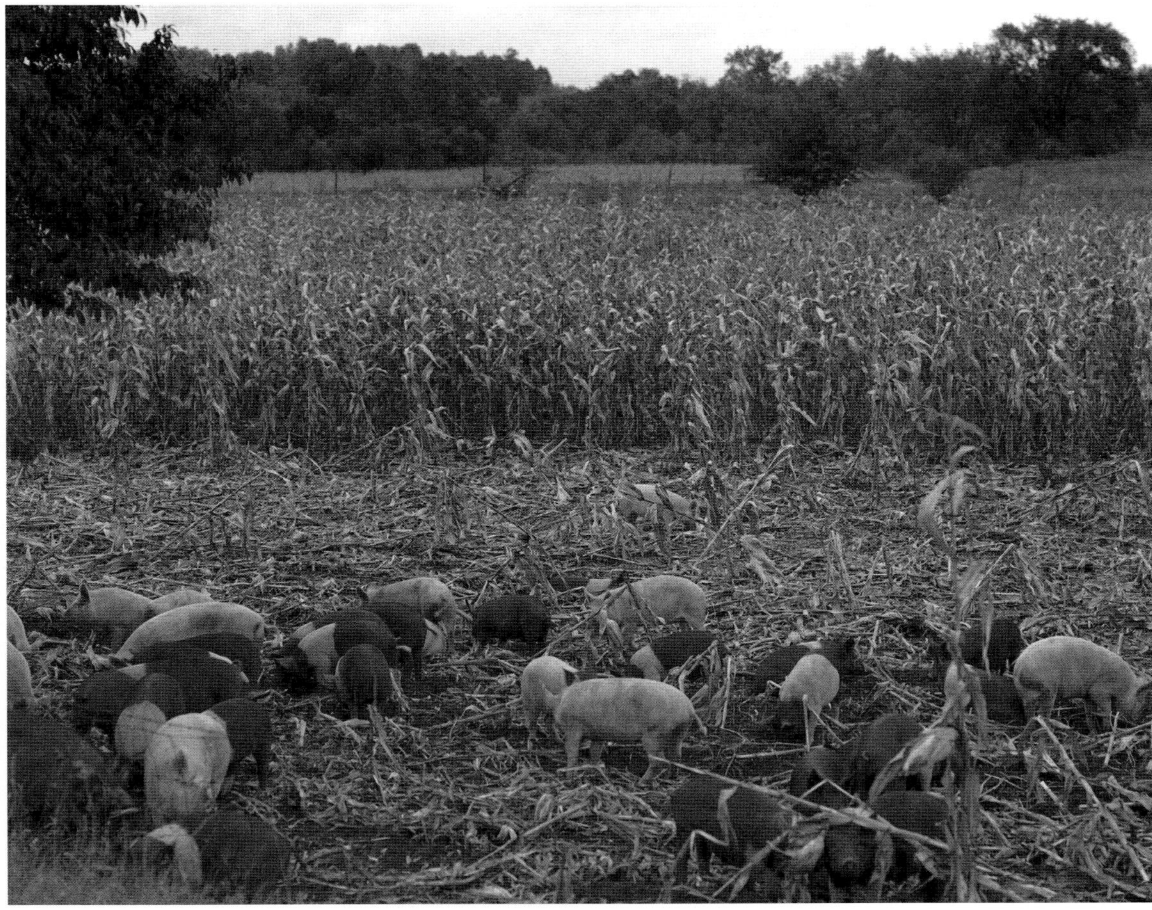

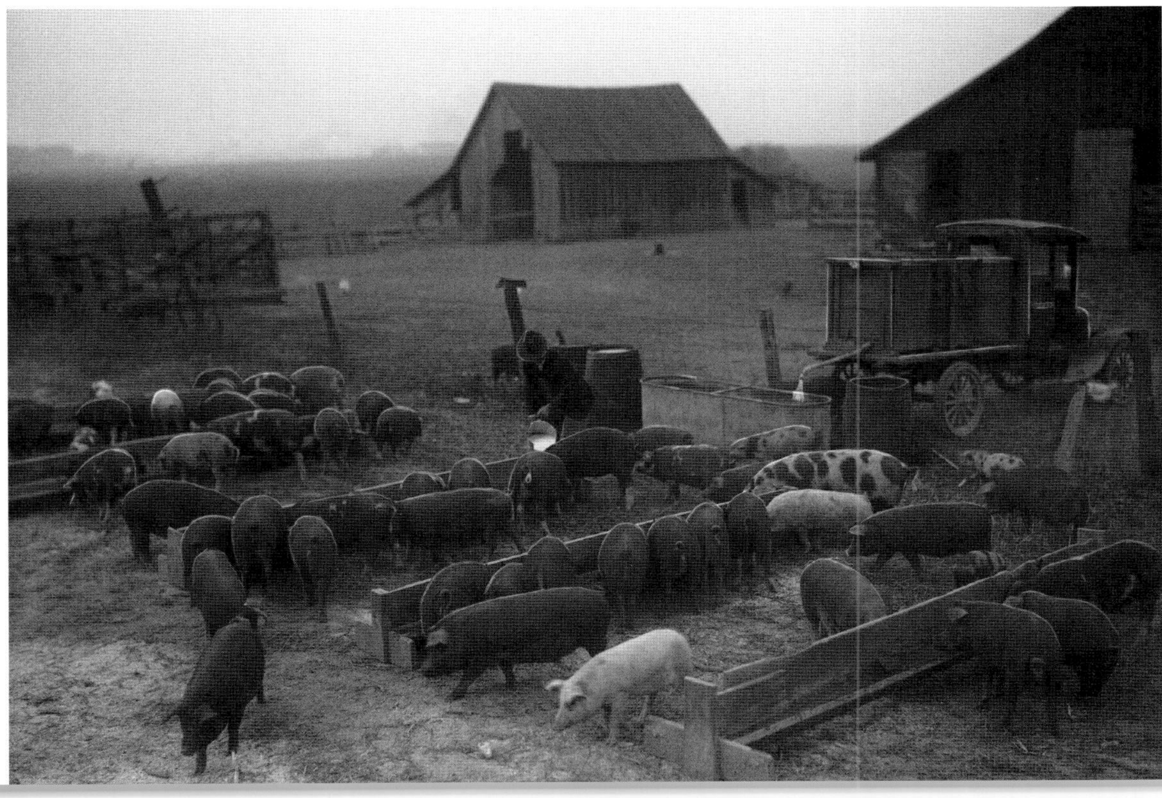

(Top left) Chester White sows graze with young pigs on pasture at the farm of C. Orr in Forest. Allen observed, "[Orr] has five Chester White sows crossed with a Hampshire boar to improve feed and marketing quality of pigs. There are 45 pigs in the bunch. Mr. Orr practices Purdue's recommended Swine Sanitation with excellent results." (Clinton County, 1931)

(Top right) Hogging down corn on the farm of M. Elliott along U.S. Highway 127 near Hudson, Michigan (Lenawee County, MI; 1962)

(Bottom) A man fills troughs with buttermilk from a tank carried on a Ford Model T truck on the Willard Robbins farm in Cayuga. (Vermillion County, 1924)

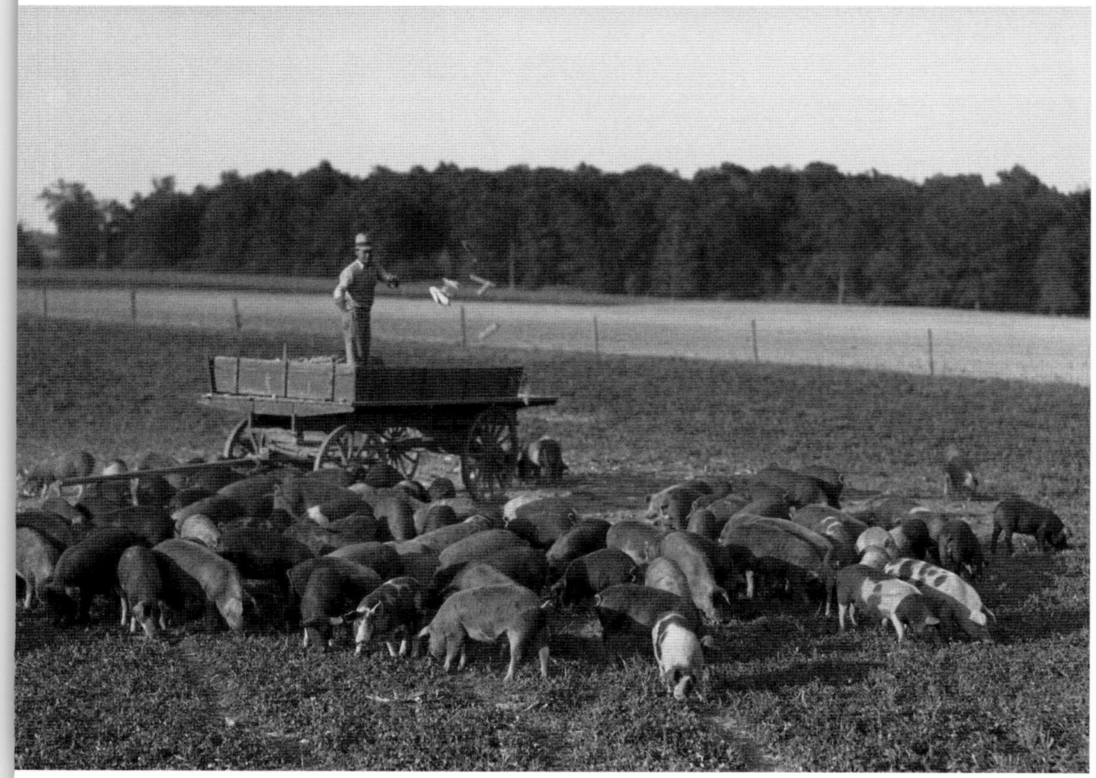

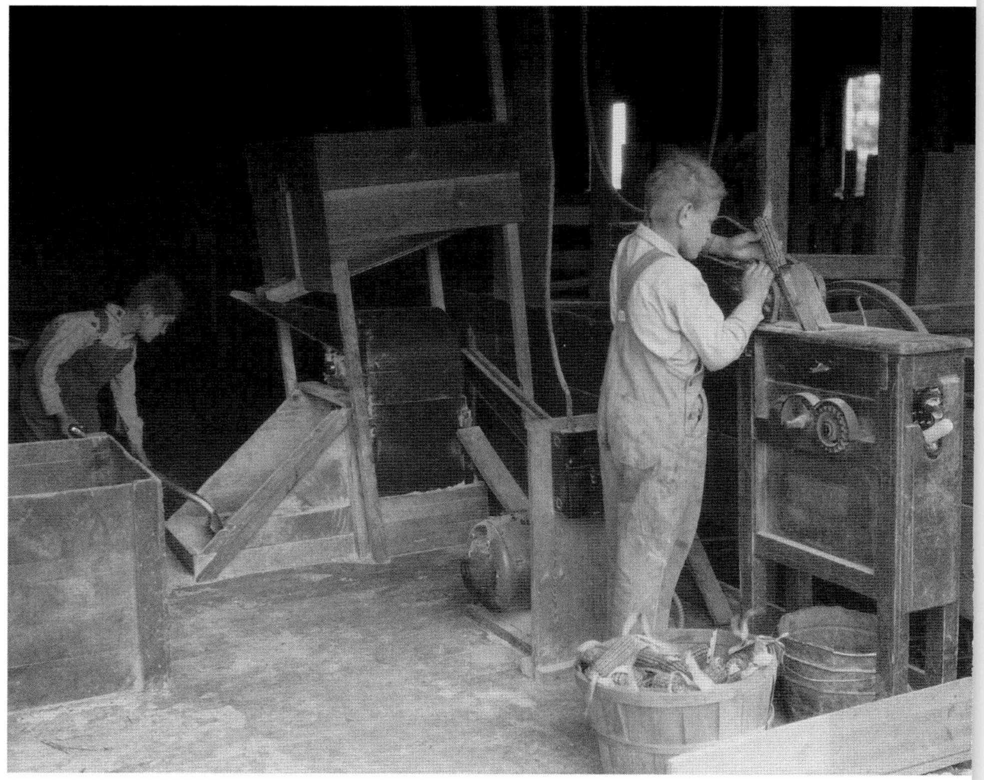

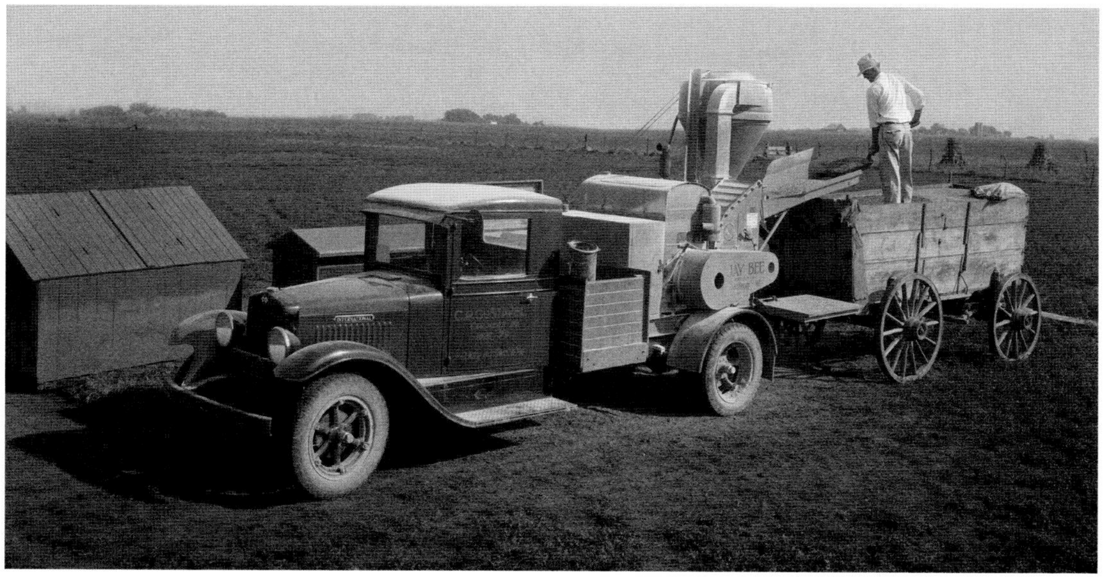

(Top left) A man feeds corn to hogs in a clover field on the Wilson and Wilson Farm near Kokomo. (Howard County, 1934)

(Top right) Grinding feed and shelling corn with an electric motor on the farm of Paul Johnson in Russiaville (Howard County, 1930)

(Bottom) C. R. Gustavel owned this International truck with a portable grinder. (White County, 1931)

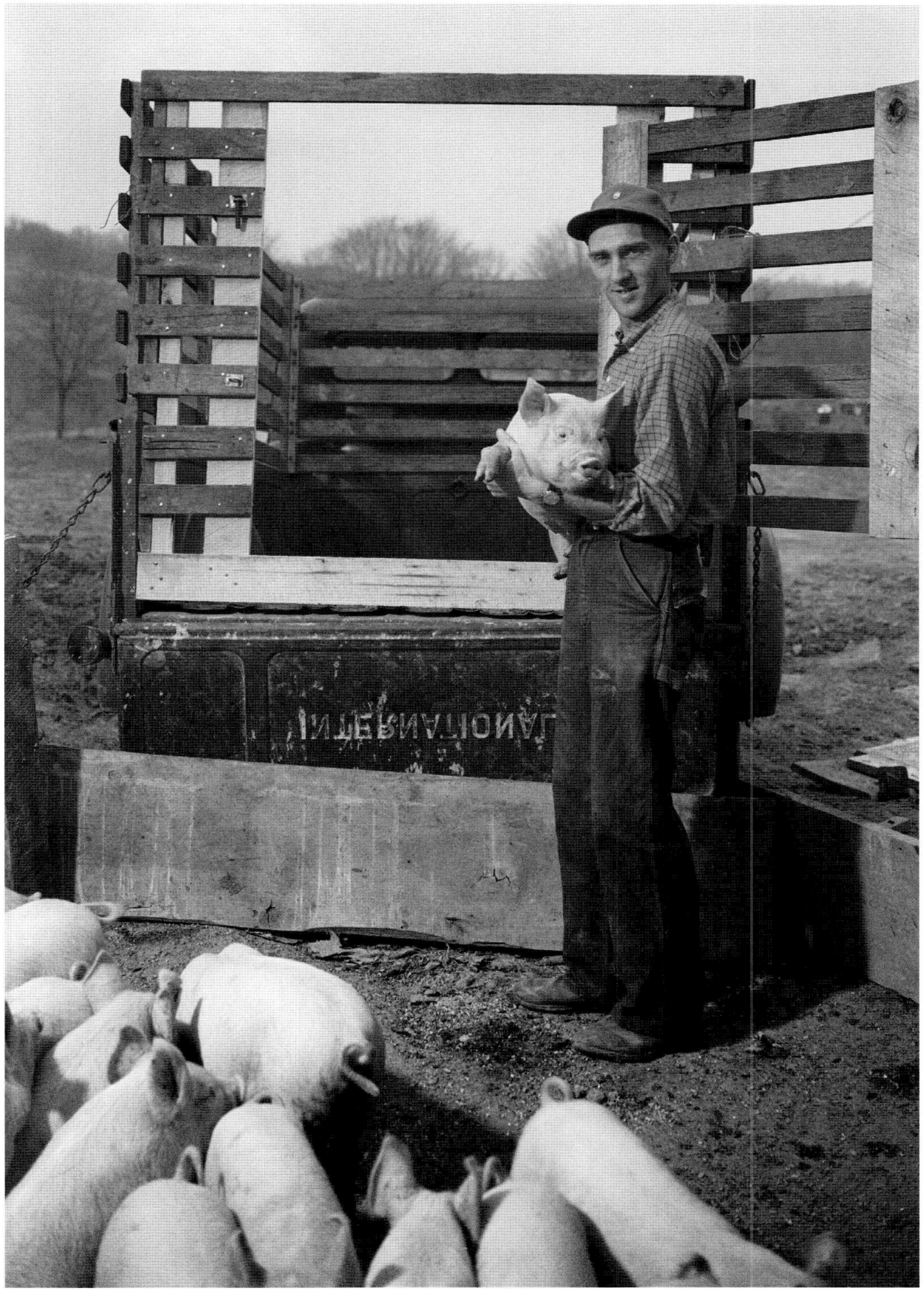

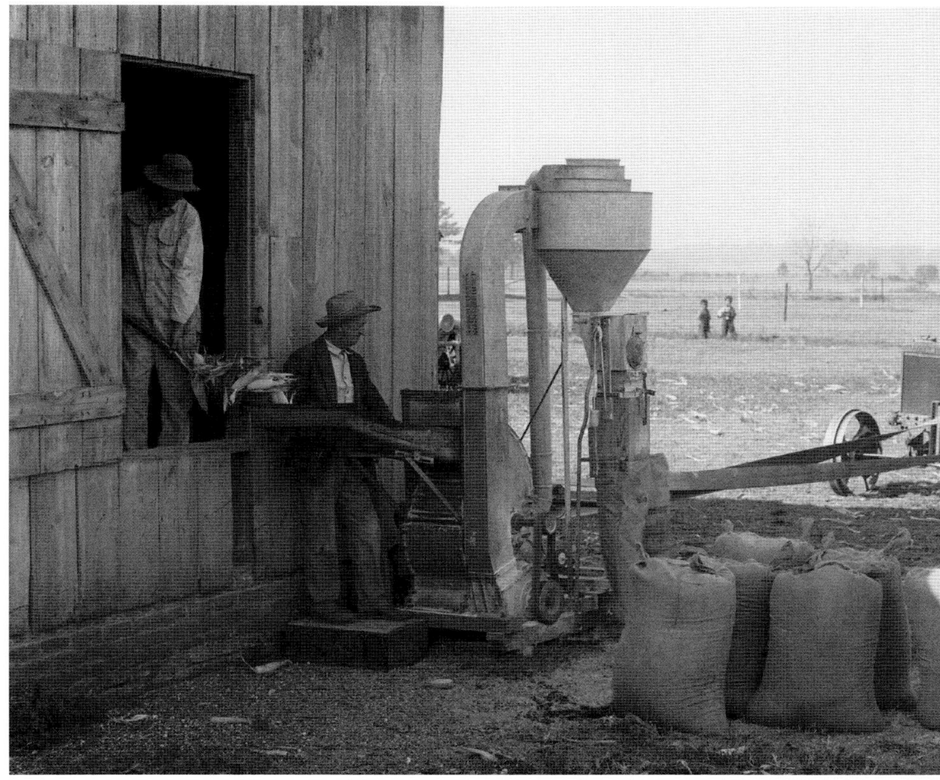

(Left) A man loads feeder pigs at a new hog house on the George L. Cunningham farm in English. (Crawford County, 1959)

(Right) Grinding corn using a tractor and belt on the L. C. Young and Son Farm in Montgomery, Alabama (Montgomery County, AL; 1927)

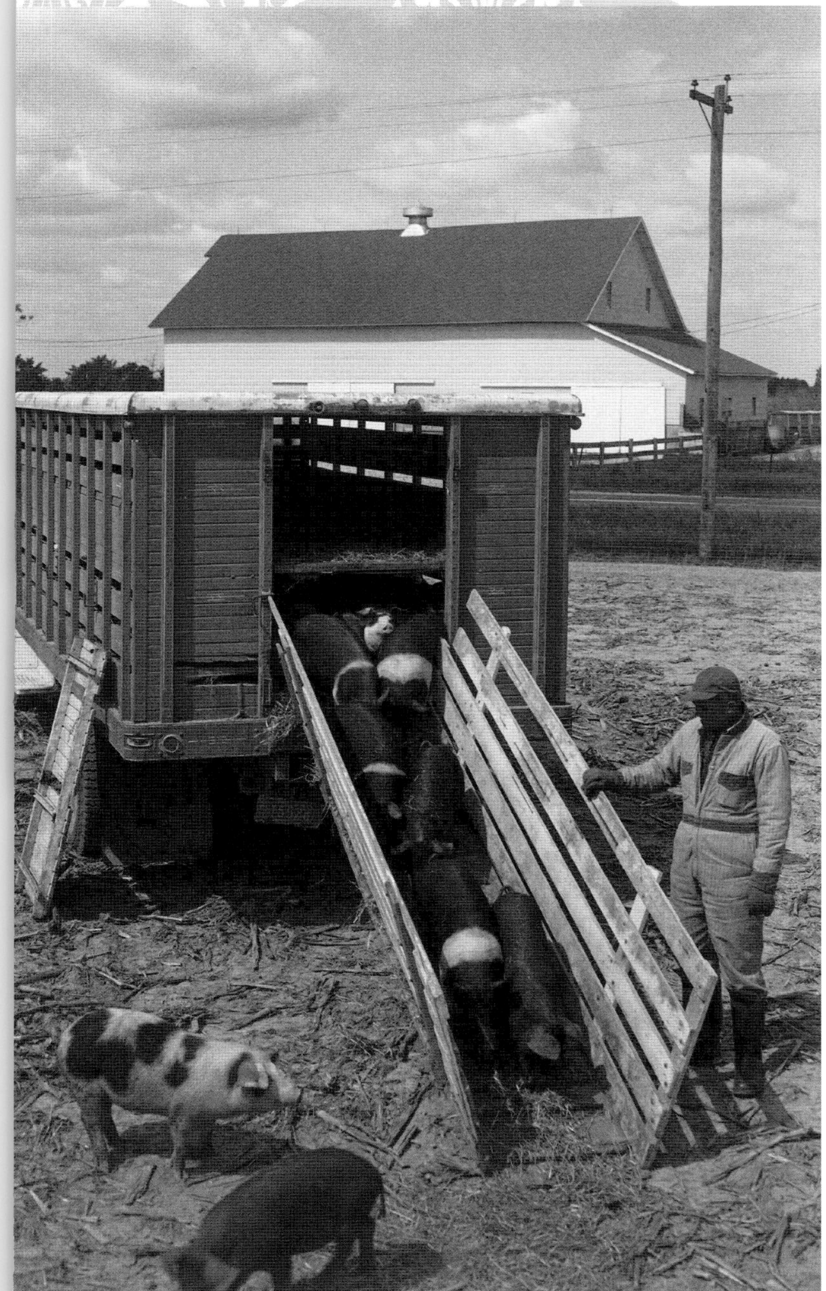

Unloading feeder pigs on the farm of John McCormick in Bringhurst (Carroll County, 1963)

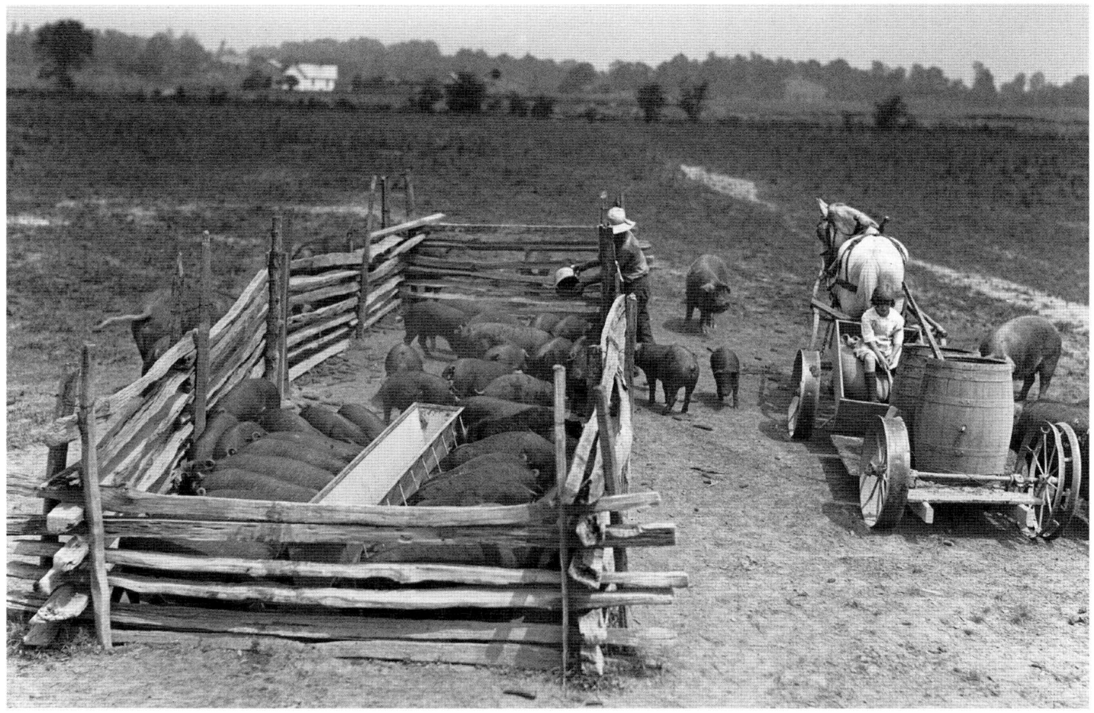

A farmer delivers slop from barrels on a horse-drawn cart to feed pigs inside a split-rail enclosure on the Paul Mitchell farm in Clay City. These animals were fed a semisolid mixture of buttermilk, ground oats, wheat middlings, and corn. (Clay County, 1923)

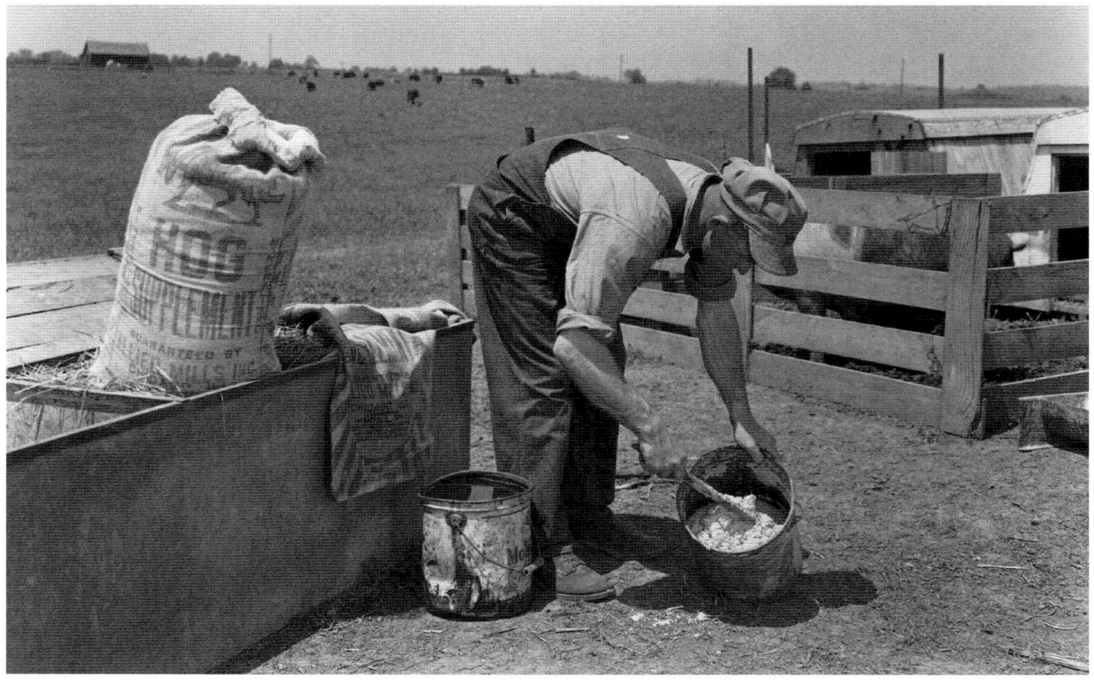

Maurice Lawson mixes bran into slop for brood sows. (Benton County, 1947)

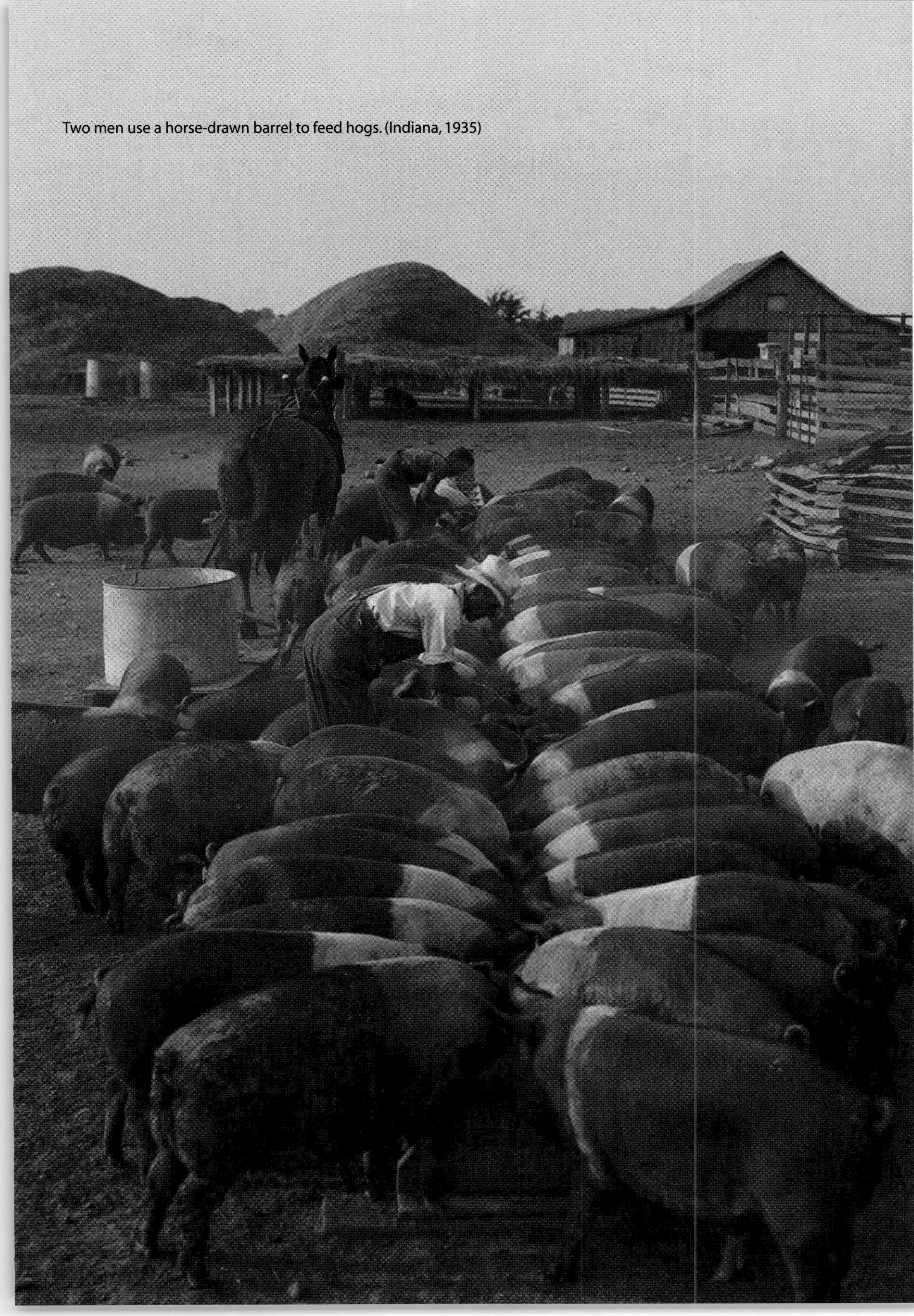

Two men use a horse-drawn barrel to feed hogs. (Indiana, 1935)

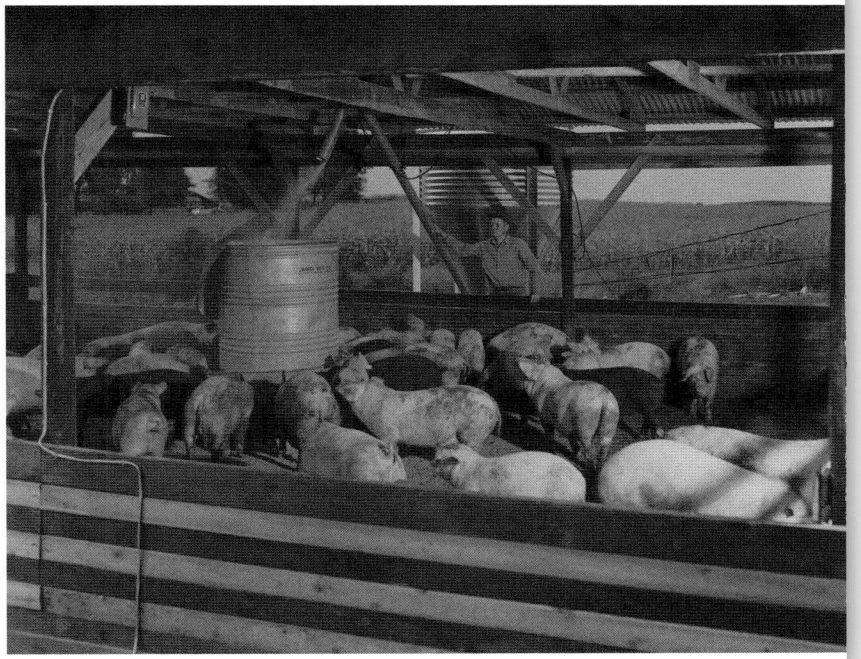

Meat-type hogs raised in confinement on the Noel Shaver farm in Crawfordsville (Montgomery County, 1957)

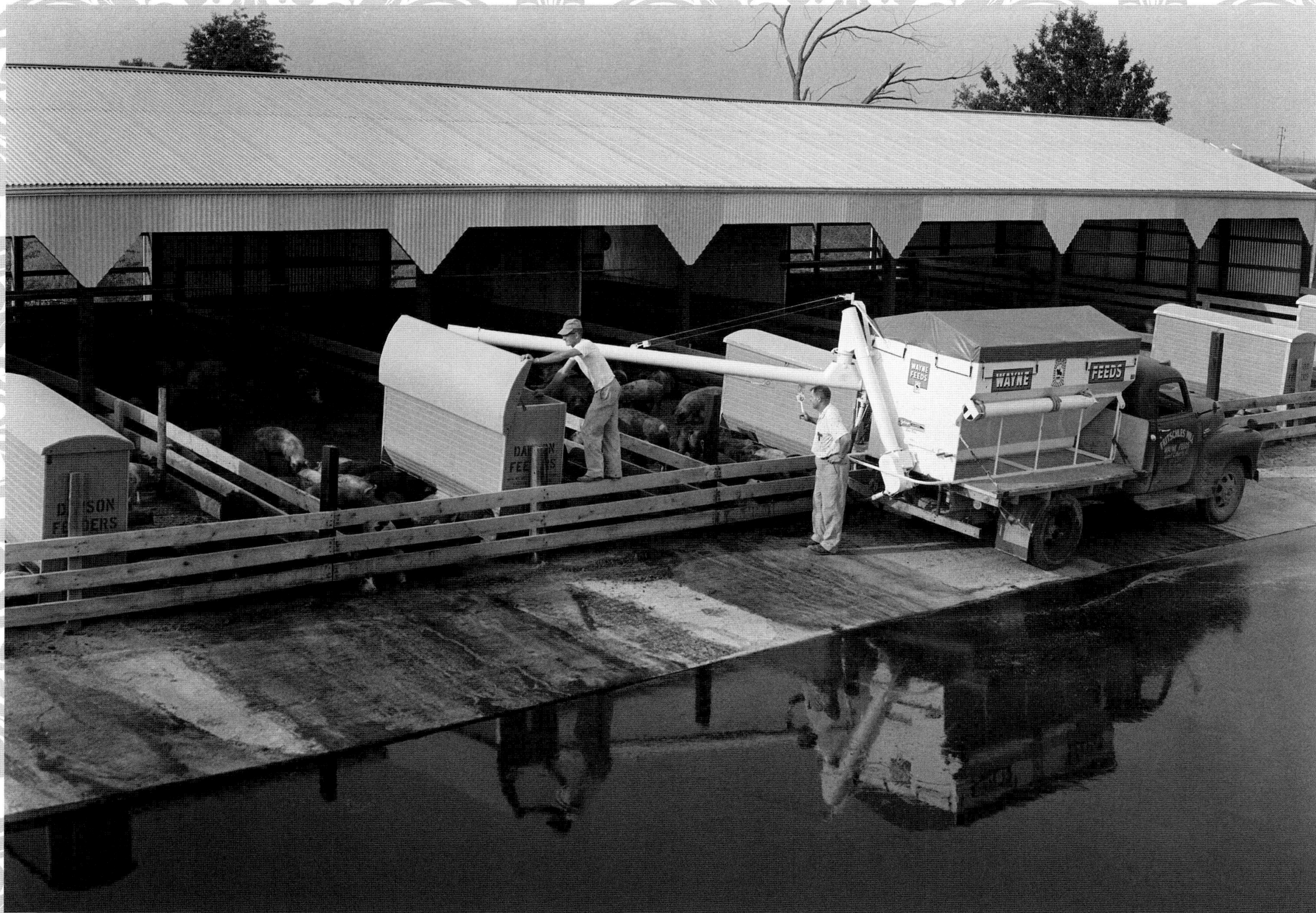

A view of a 400-hog farm owned by Earl Urfer in West Liberty, Illinois (Jasper County, IL; 1961)

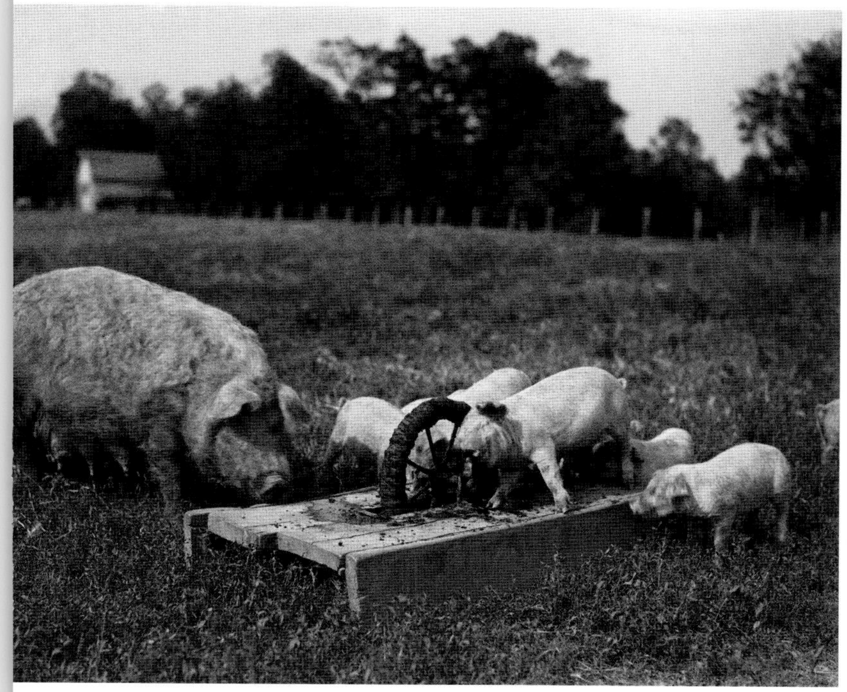

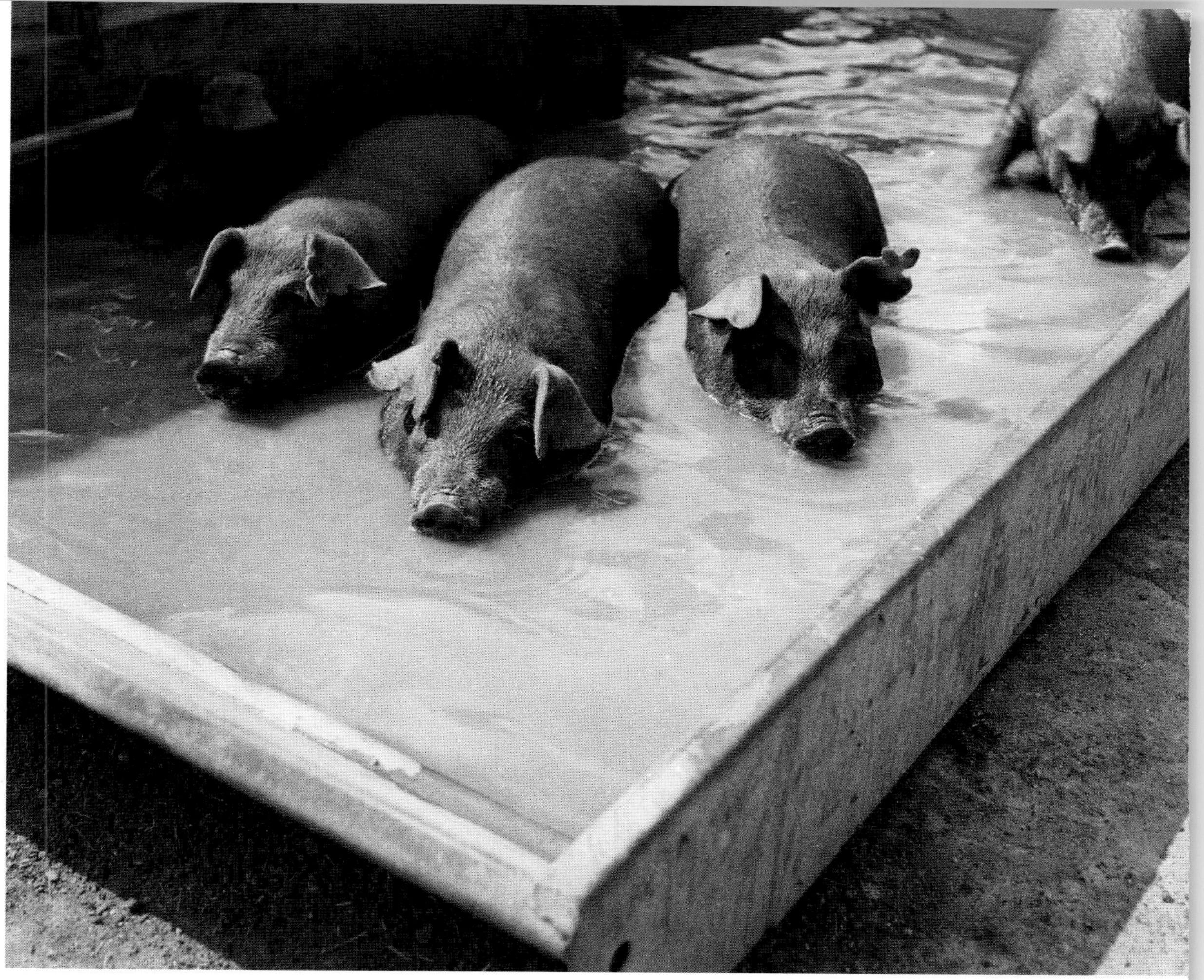

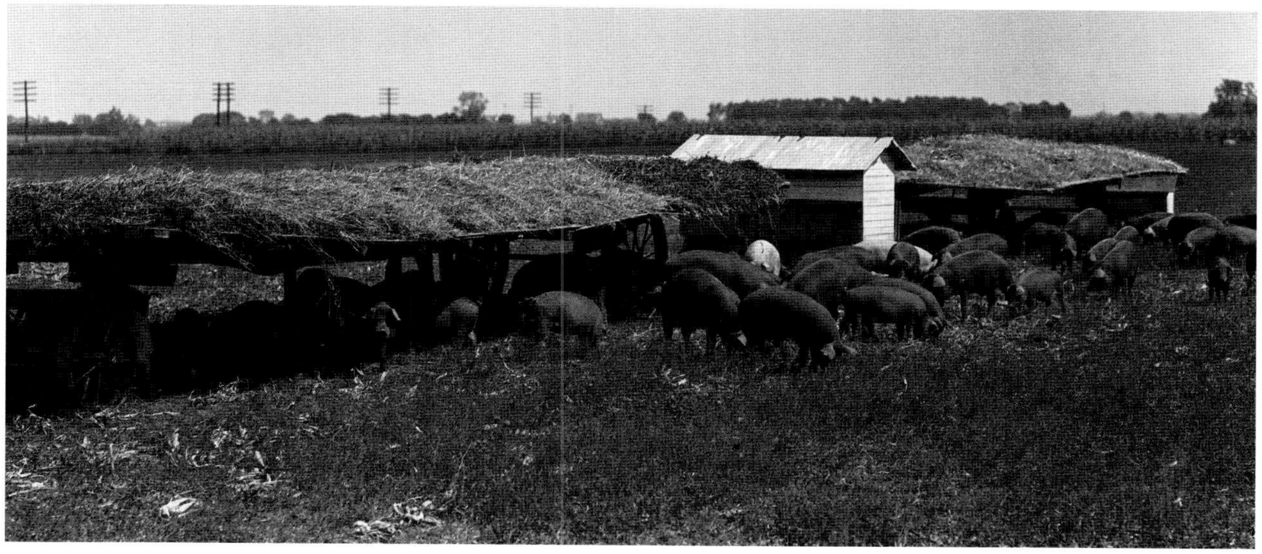

(Top left) Pigs rub against a homemade oiler on the farm of Nelson Stead. (Mercer County, IL; 1920)

(Top right) Hogs cool off in hot weather. (Indiana, 1956)

(Bottom) A portable sunshade mounted on wheels protects hogs on the farm of Edward Funk and Sons in Earl Park. (Benton County, 1932)

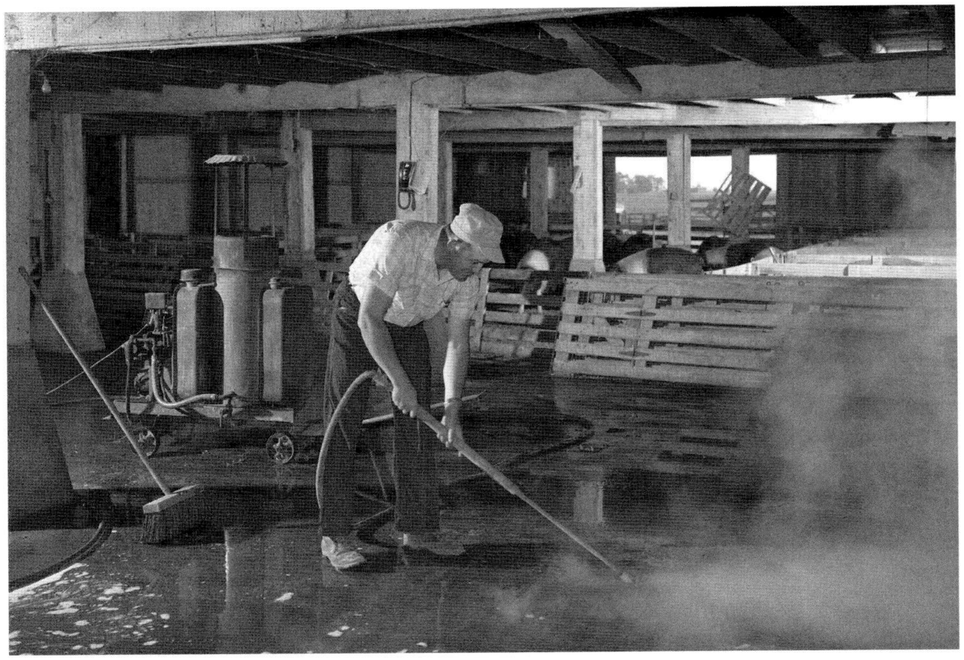

Manford Stewart steam cleans a concrete floor before setting up farrowing pens on the farm of L. L. Stewart and Son of Frankfort. (Clinton County, 1957)

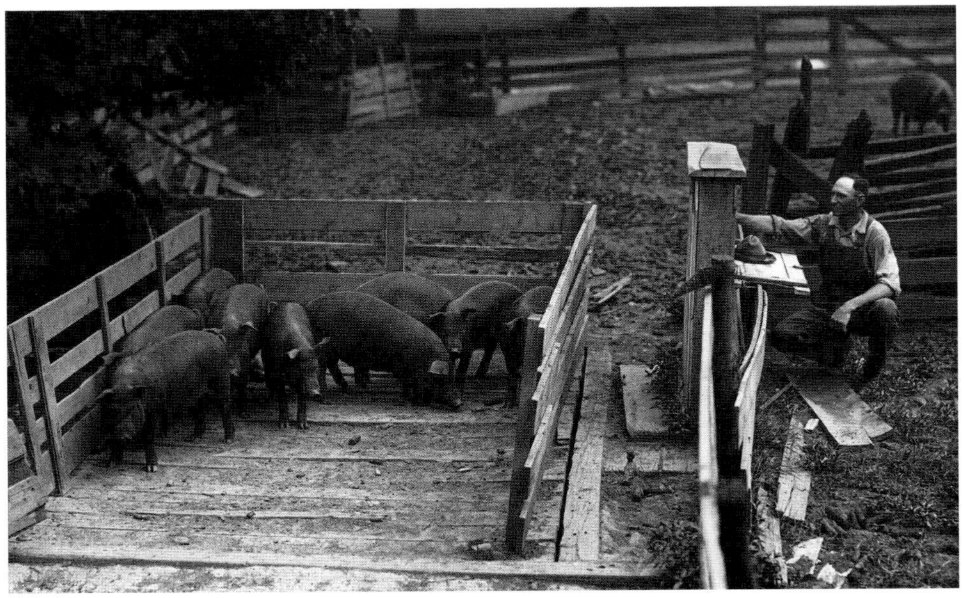

Weighing hogs on farm scales at the E. S. Reutter farm in Fowler (Benton County, 1924)

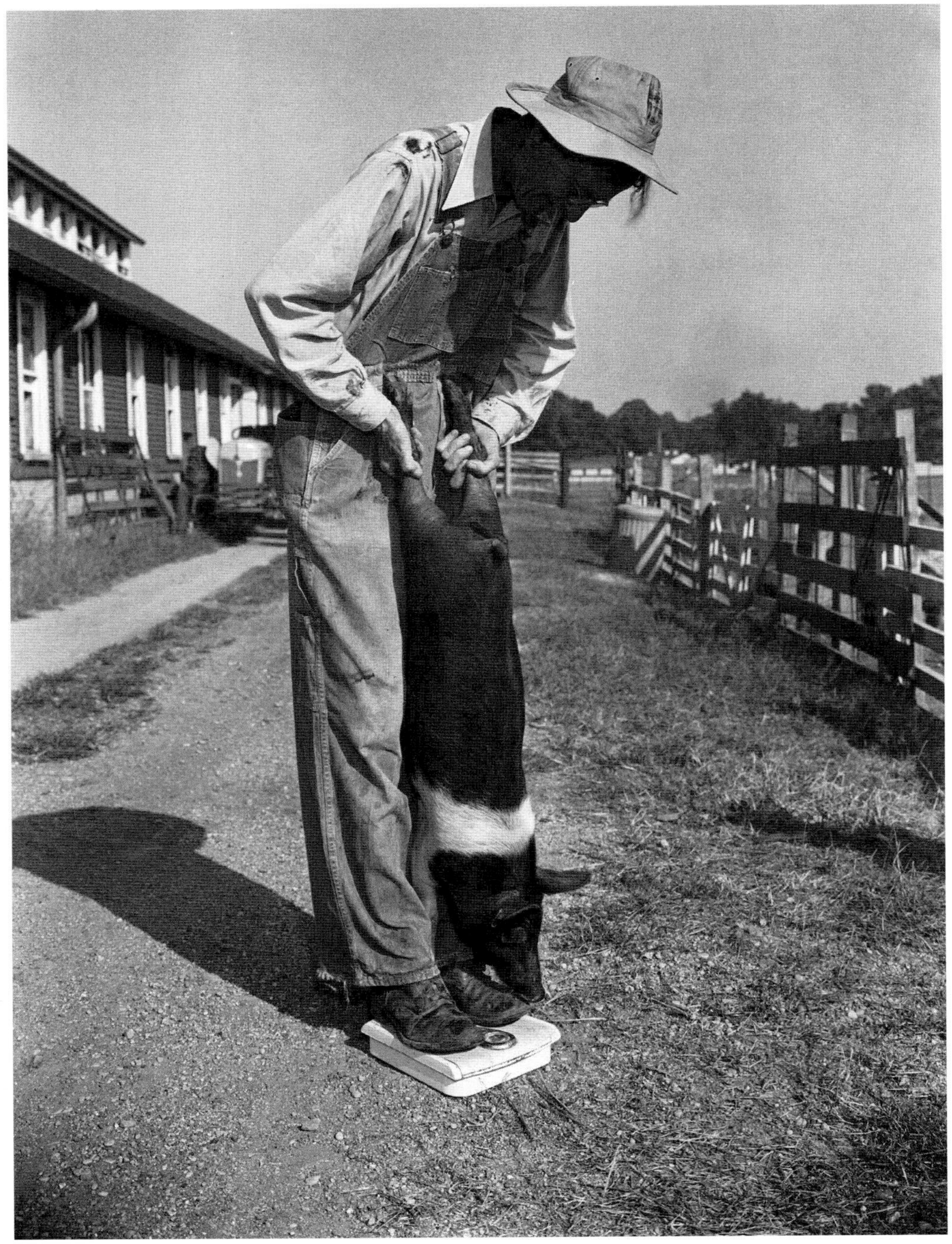

A man uses a bathroom scale to weigh a pig. (Indiana, 1950)

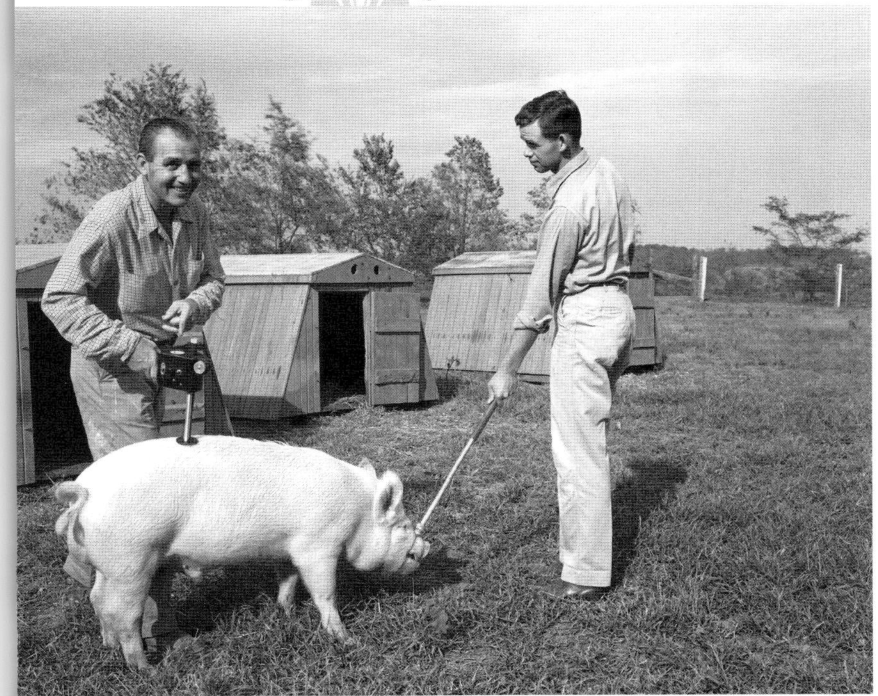

Maurice F. Neville tests the back fat on a Yorkshire boar held by another man. (Tippecanoe County, 1954)

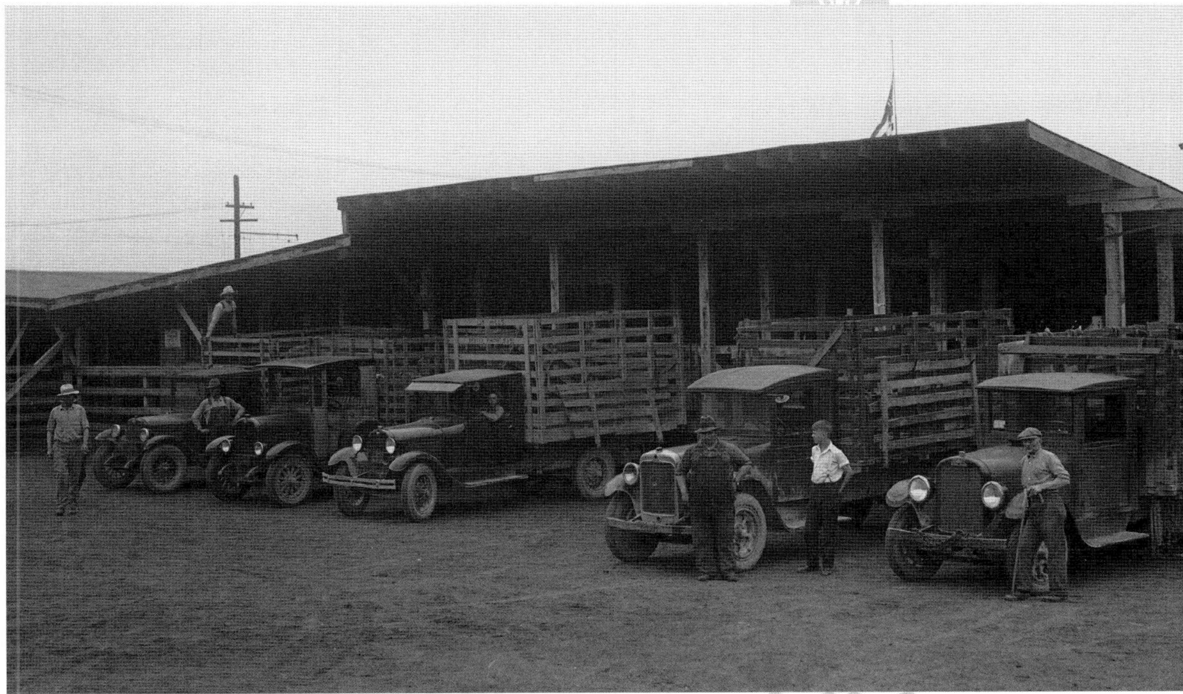

Trucks bring hogs to the Muncie National Stock Yards. (Delaware County, 1929)

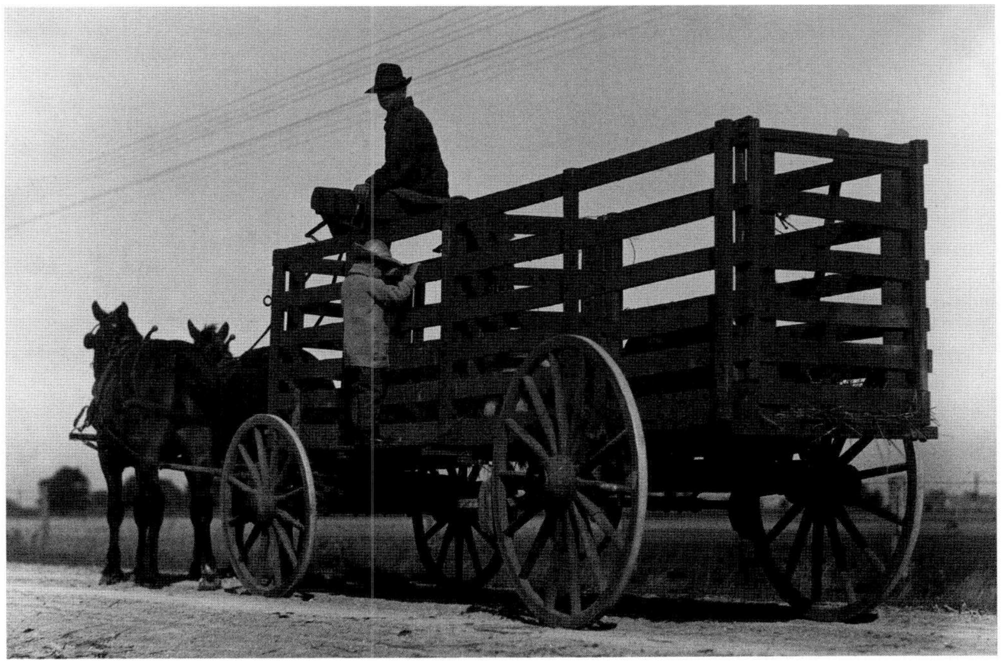

This farmer sold these hogs for nine cents per pound, or nine dollars per hundredweight. (Indiana, 1913)

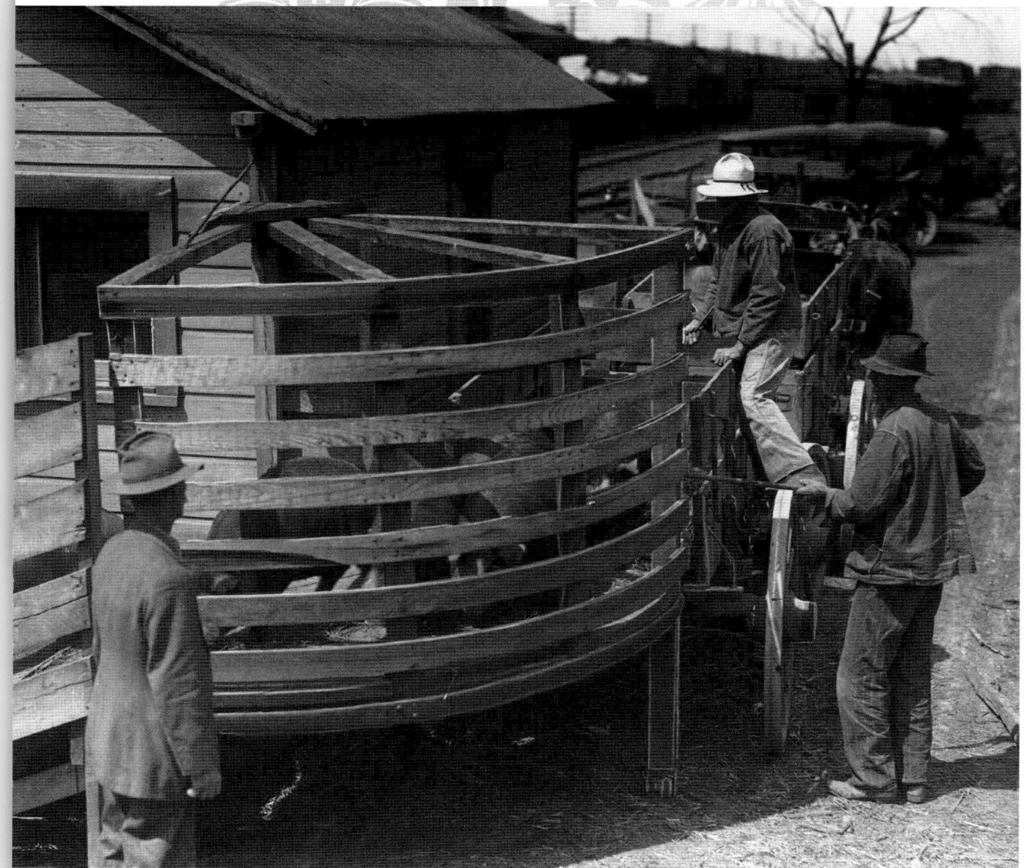

These men prepare to deliver hogs to market using a livestock shipping association in Jennings County. Allen wrote, "Approximately 85 percent of the livestock shipped from this county goes through the cooperative association. The device or chute used for unloading swings on a pivot and turns back against the end of the office while the wagon is being driven in. It is then swung around to the rear of the wagon and the hogs are driven out." (Jennings County, 1923)

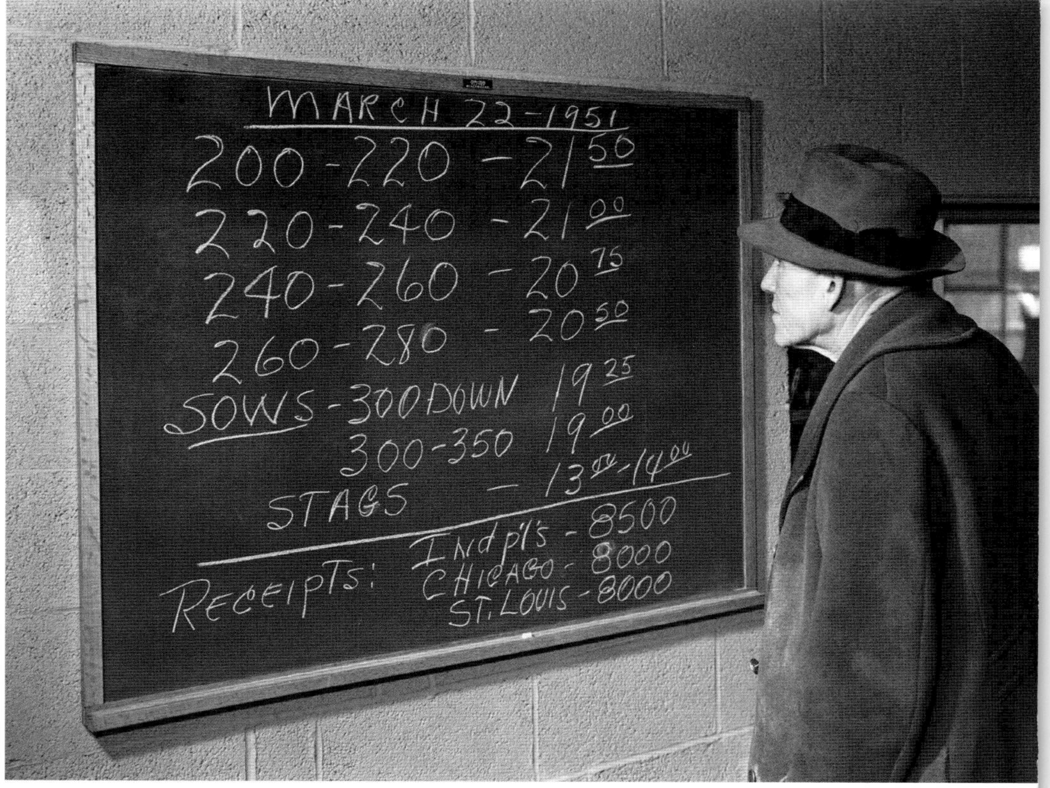

A farmer checks price quotes on a blackboard at the Indiana Farm Bureau Cooperative at the stockyard in Lafayette. (Tippecanoe County, 1951)

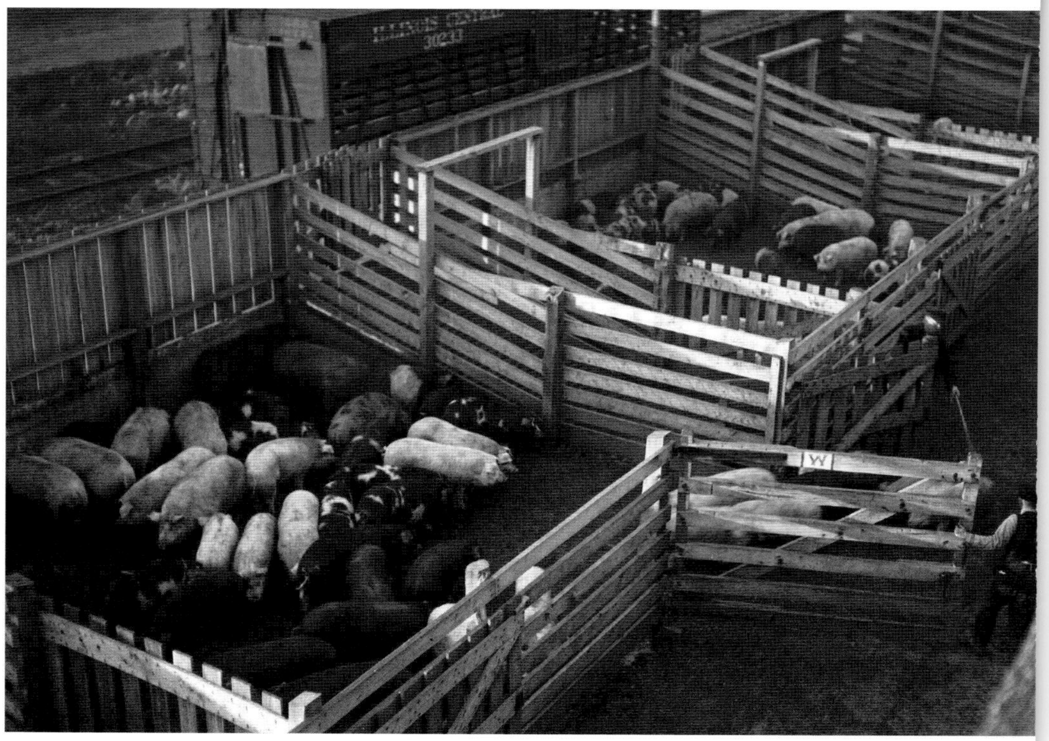

Shipping hogs from the stockyards at Lafayette (Tippecanoe County, 1918)

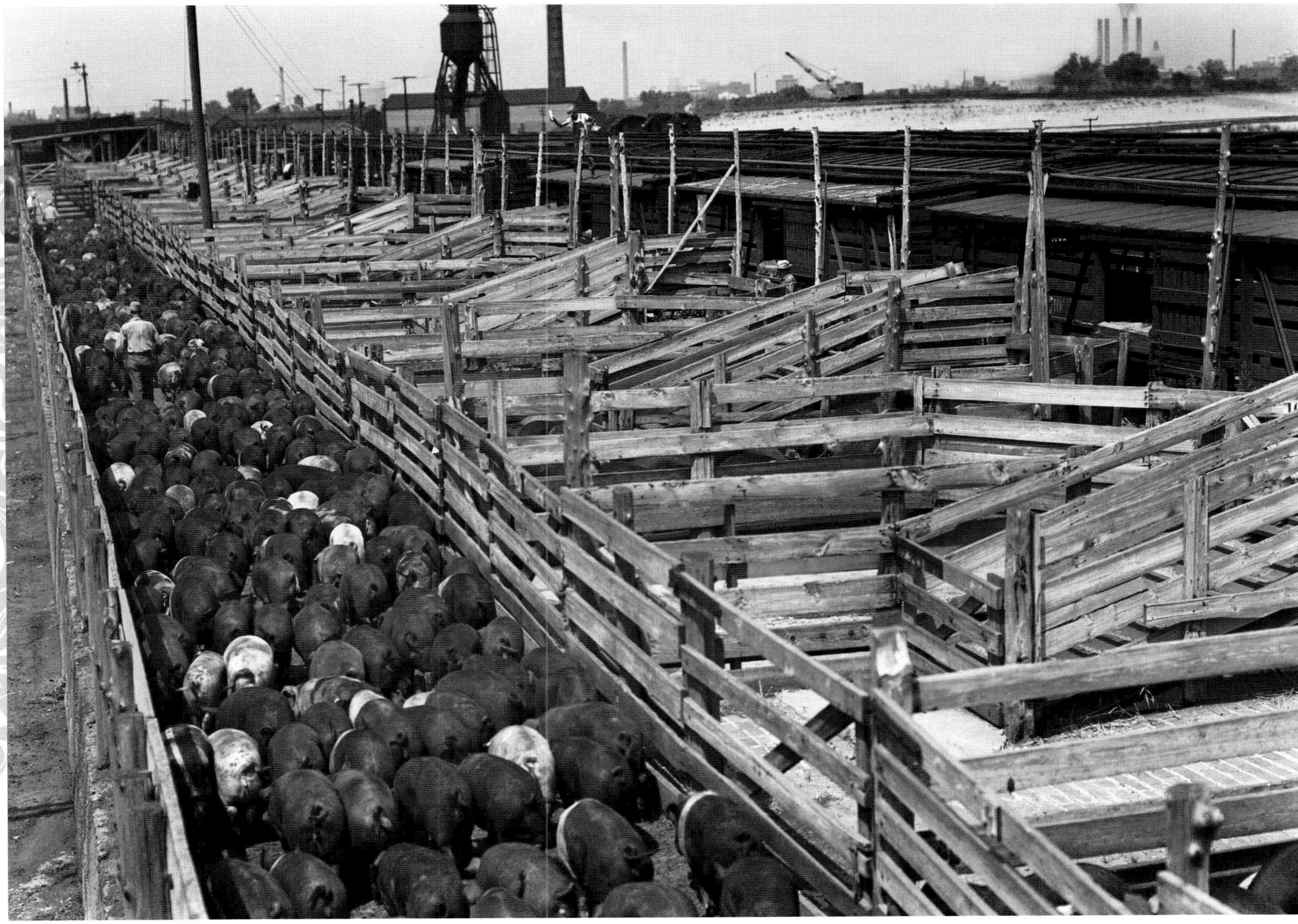

Hogs are loaded into railroad cars at the stockyards in Indianapolis for shipment to buyers in the eastern United States. (Marion County, 1949)

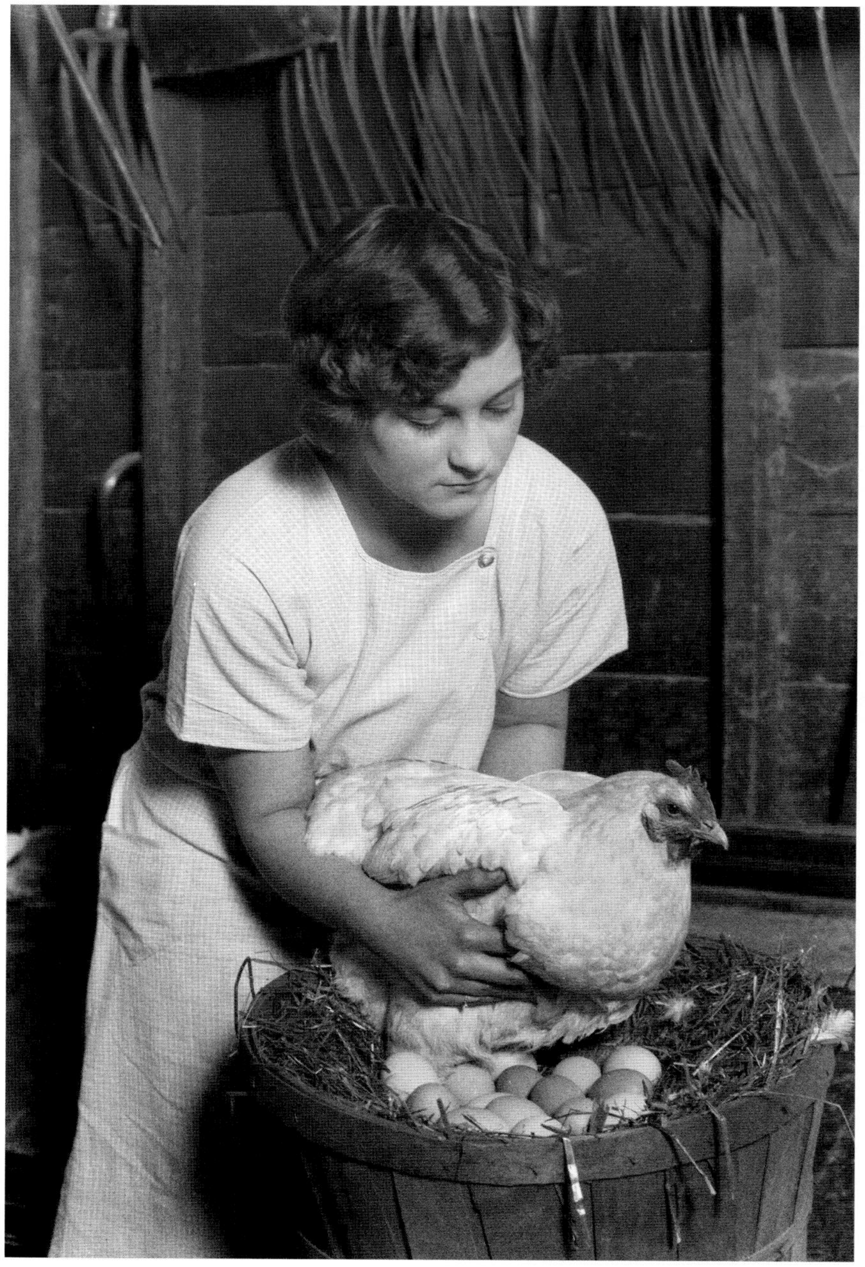

Helen Ryan puts a White Plymouth Rock hen on a nest. (Tippecanoe County, 1928)

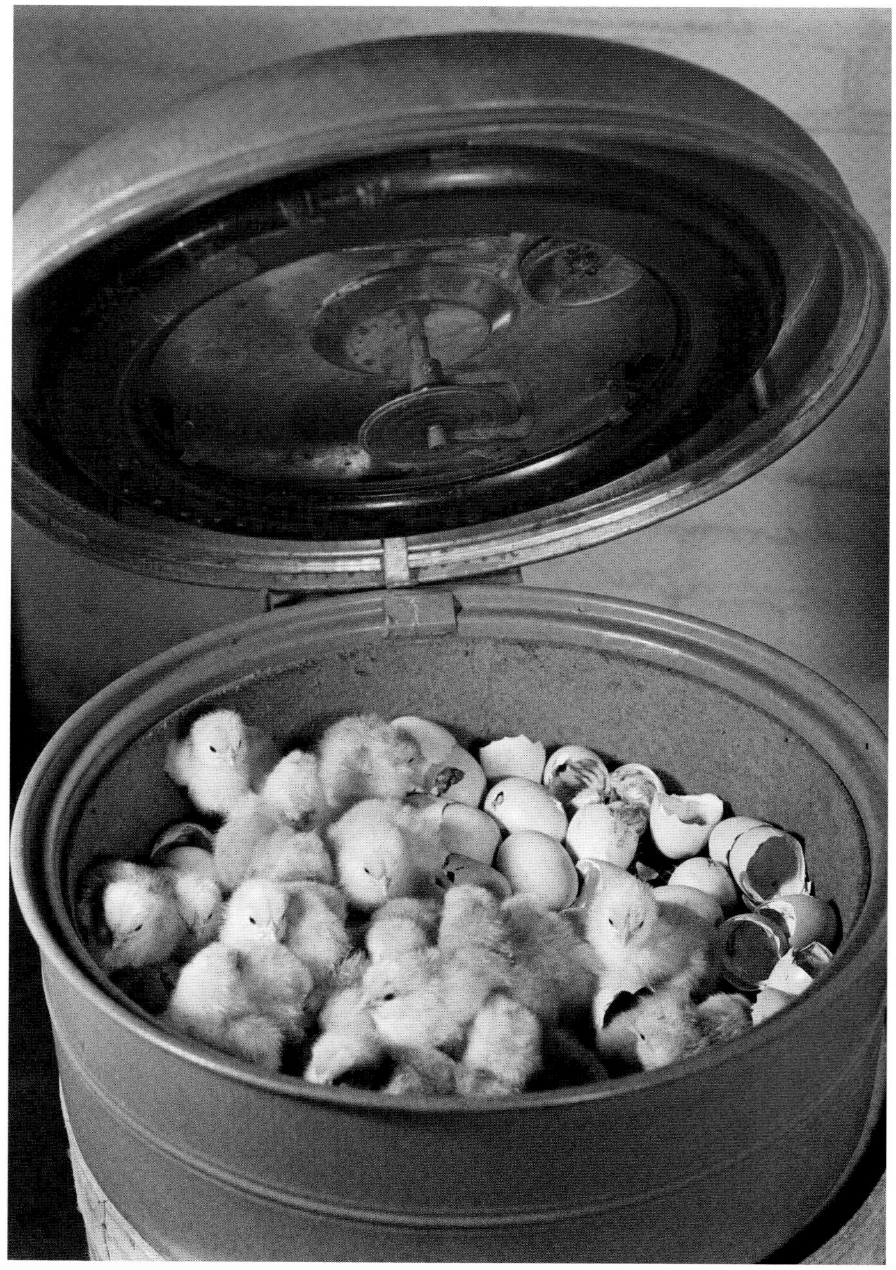

An incubator used in hatching eggs (N.p., 1950)

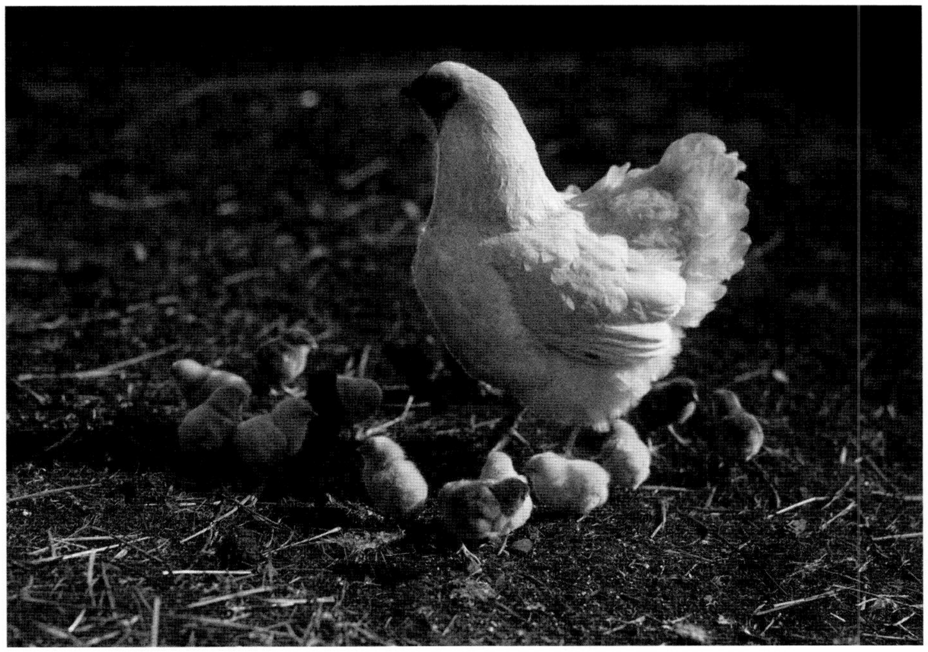

A hen and chicks forage on the farm of Ed Taylor in Montmorenci. (Tippecanoe County, 1918)

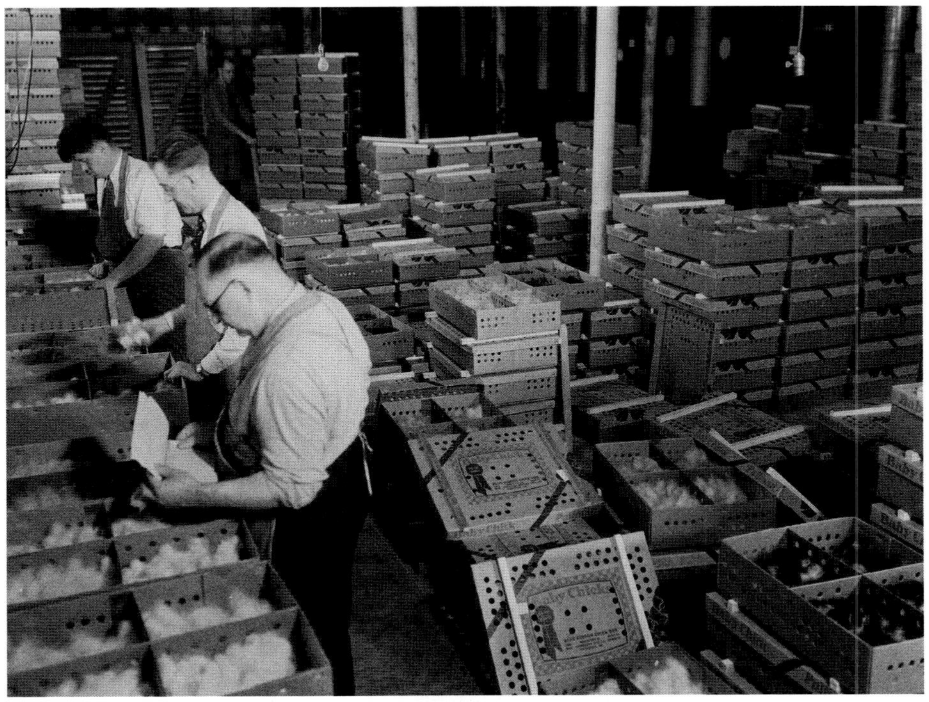

Workers sort and pack baby chicks inside the Wolf Hatchery in Gibsonburg, Ohio. (Sandusky County, OH; 1933)

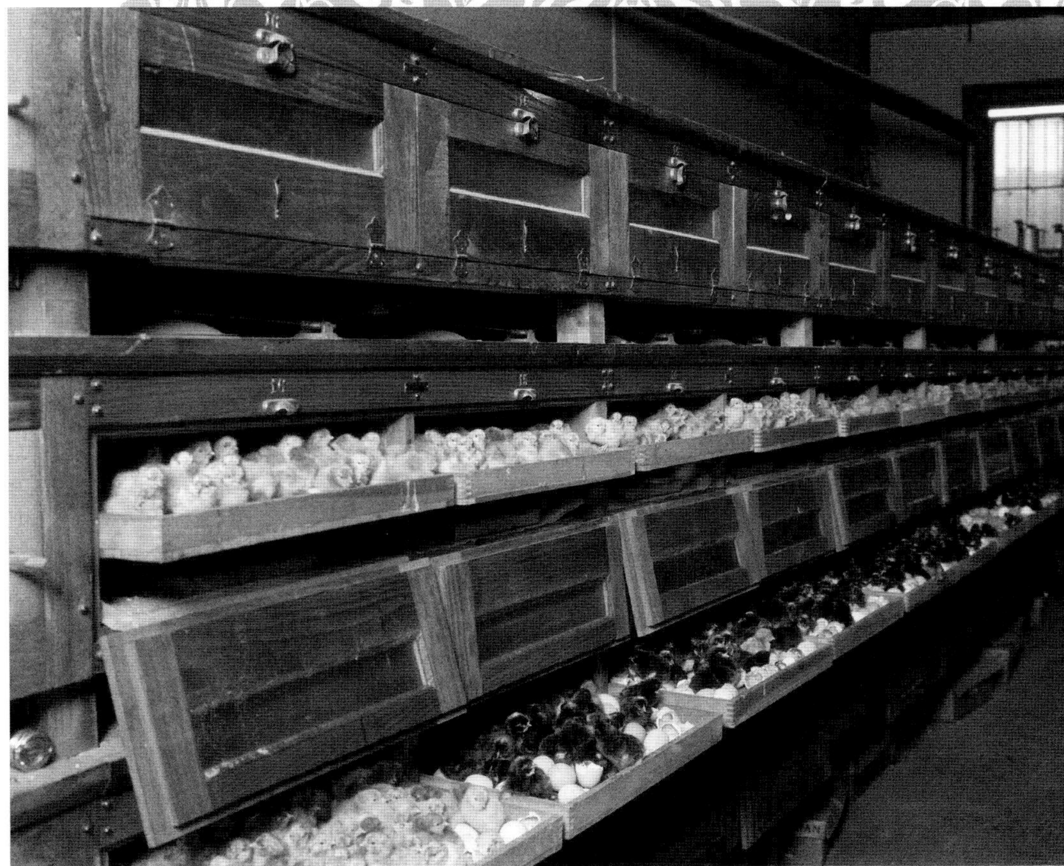

Baby chicks fill incubator trays at the LeRoy Jones Hatchery in West Lafayette. (Tippecanoe County, 1928)

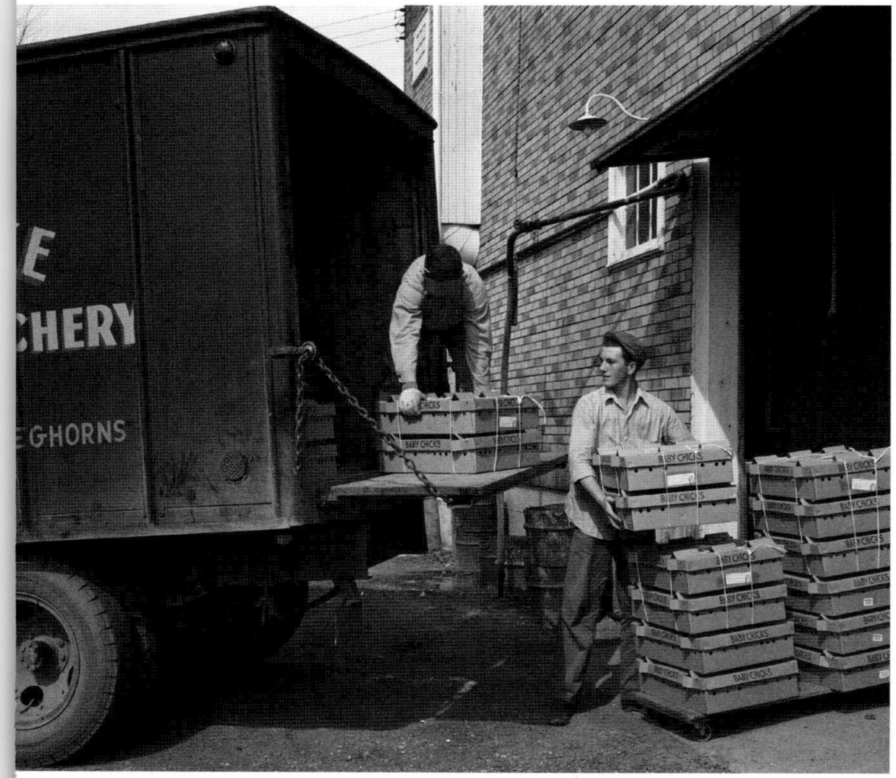

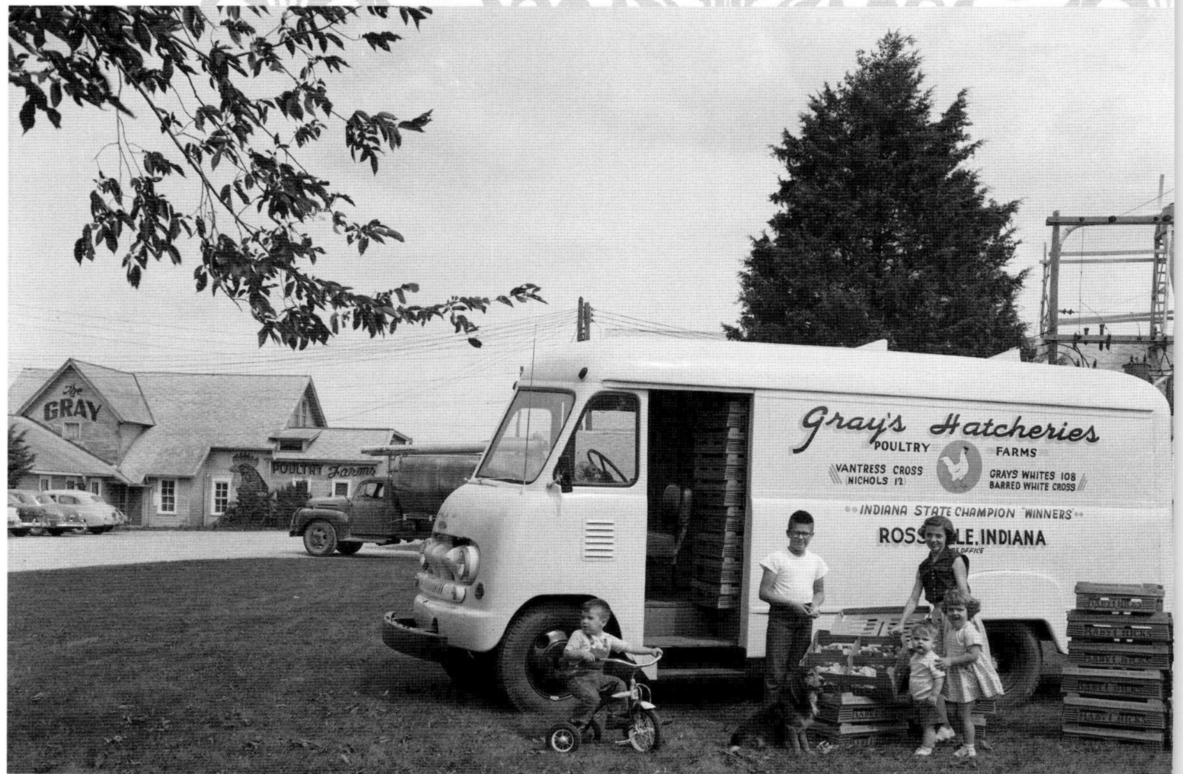

(Top left) Paul Carpenter hands boxes of baby chicks to Robert Carr, who stacks them into a Silver Lake Hatchery truck. (Kosciusko County, 1948)

(Top right) Loading a truck with baby chicks at Gray's Hatcheries in Rossville (Clinton County, 1956)

(Bottom) Gray's Hatcheries in Rossville used this airplane to deliver baby chicks to surrounding areas. (Clinton County, 1952)

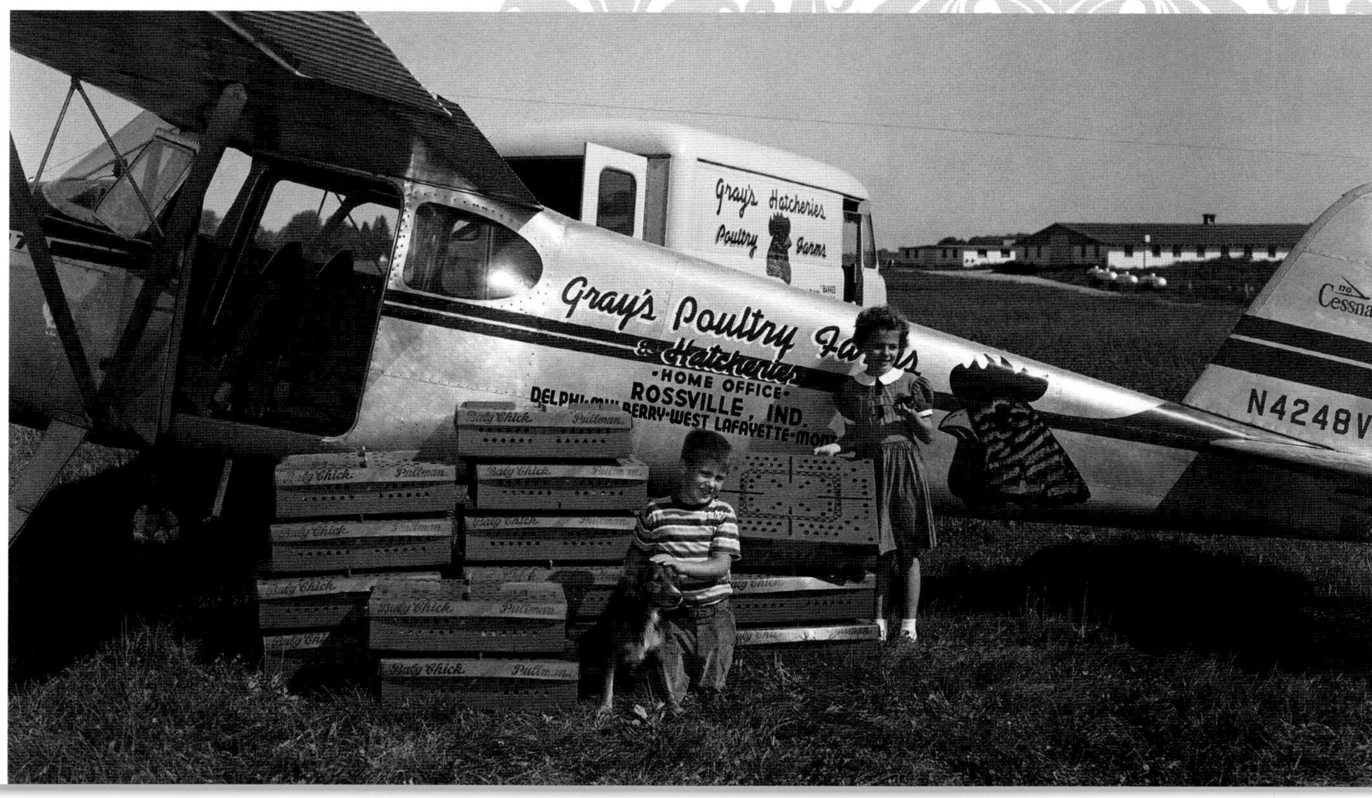

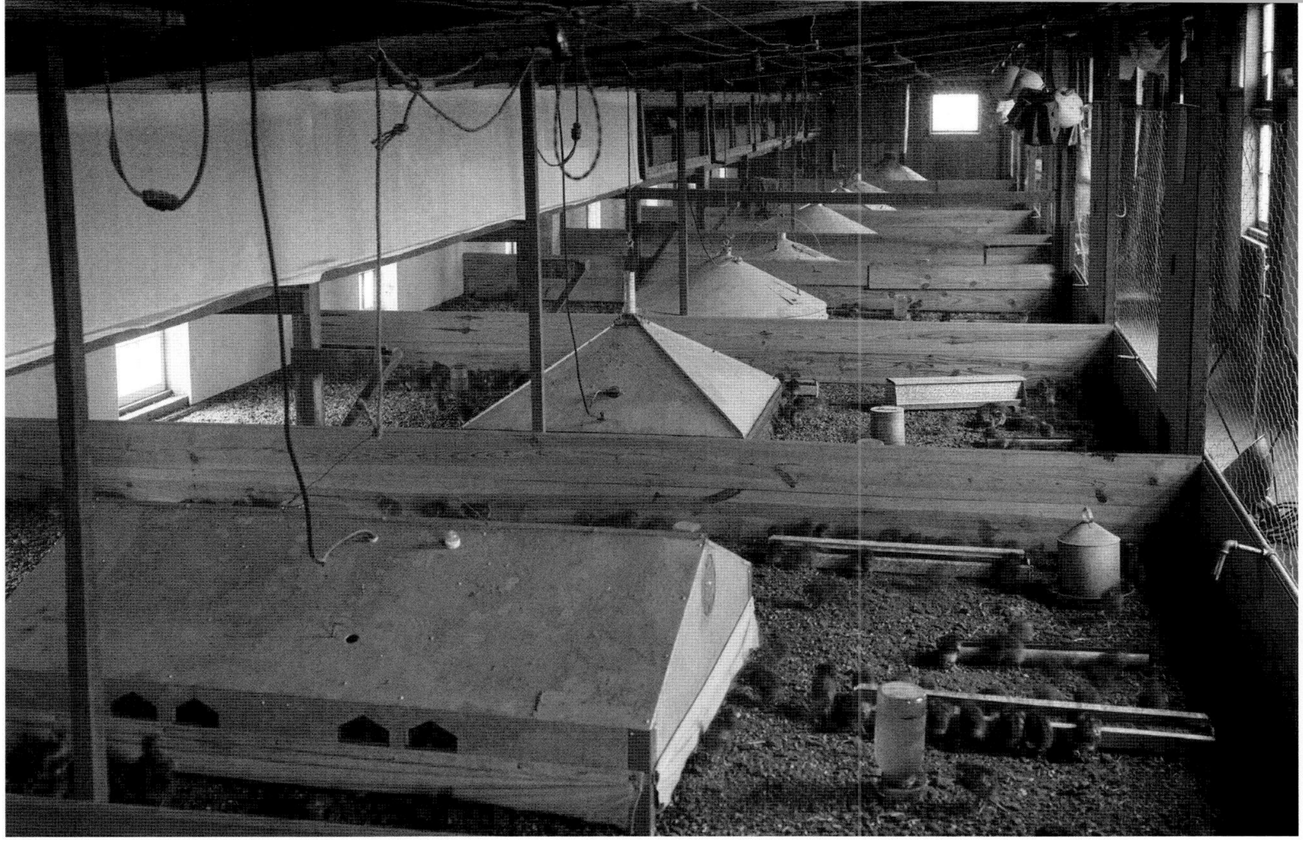

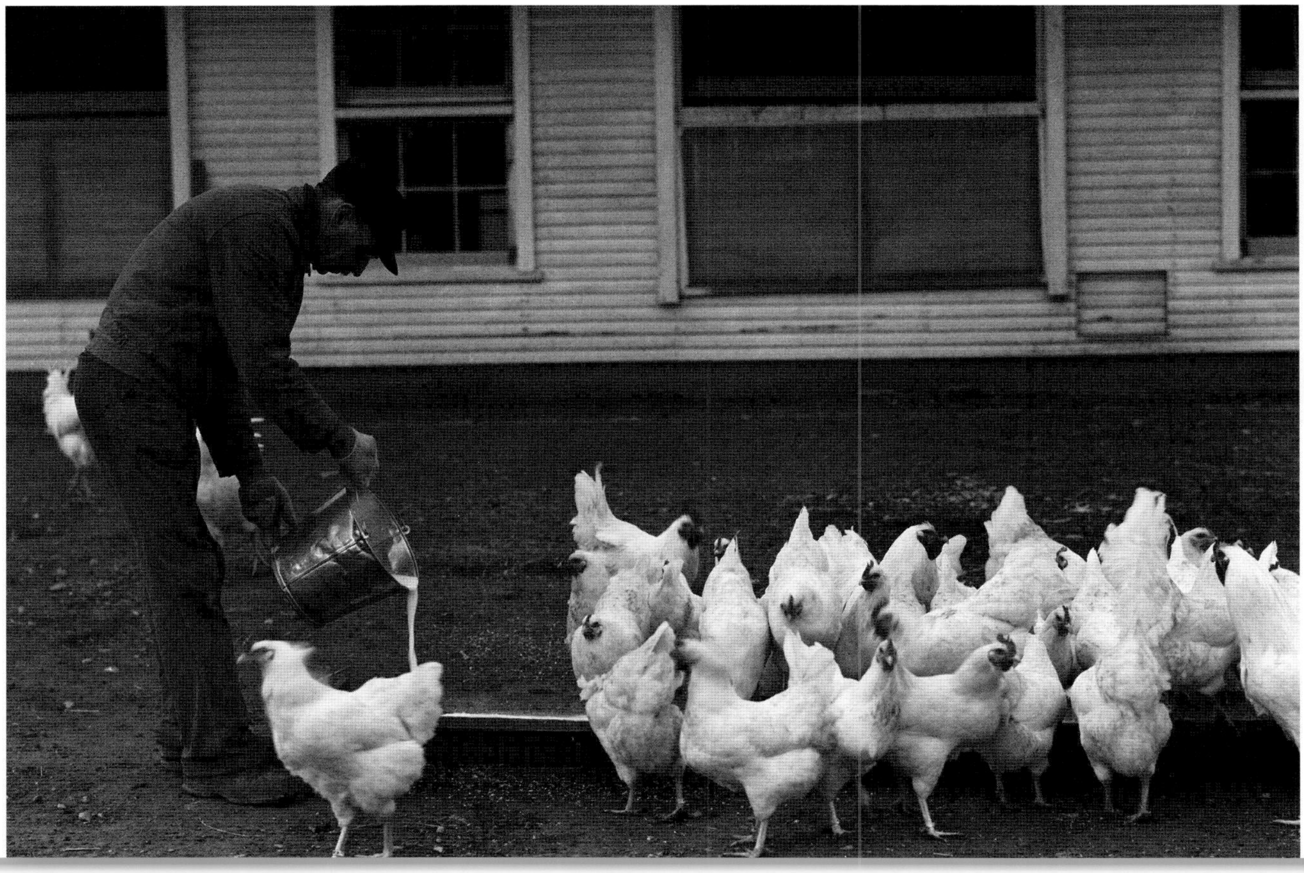

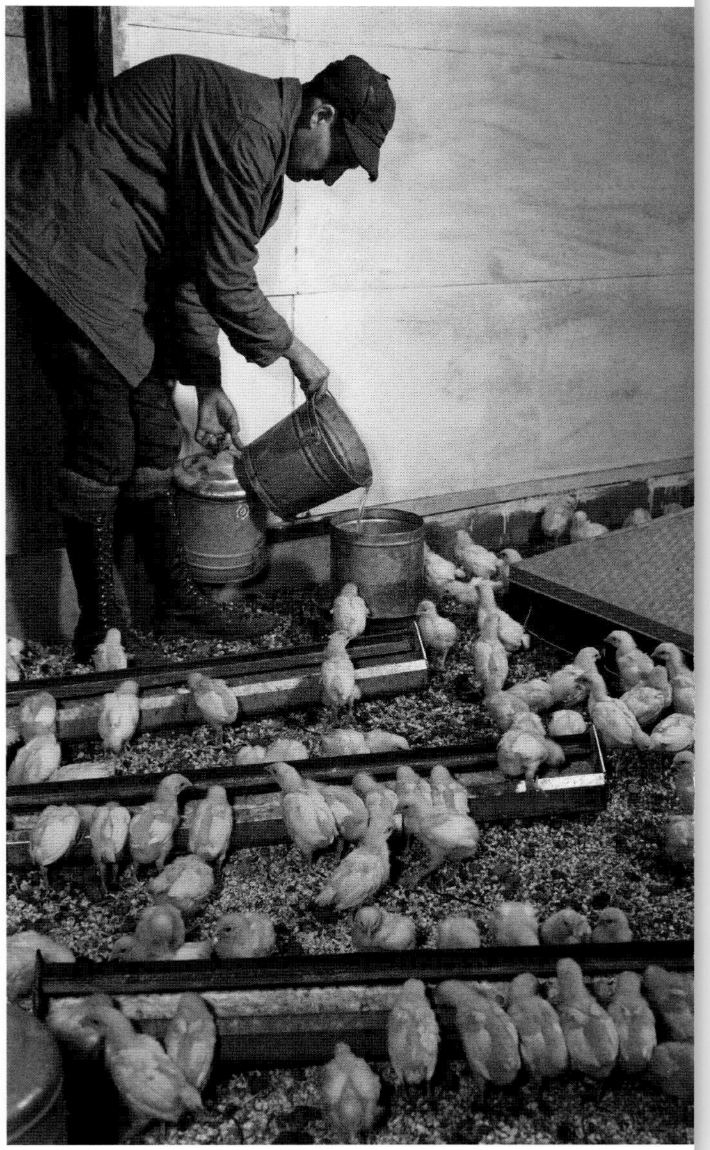

(Top left) Arba Brutus from Pine Village uses electric brooders to produce broilers for the early spring market. (Warren County, 1935)

(Bottom left) Lafayette's Harry Hinckley feeds skim milk to White Leghorn hens. (Tippecanoe County, 1939)

(Right) Frank Blacker of Romney pours fresh water into a fountain for his three-week-old White Plymouth Rock chicks. (Tippecanoe County, 1944)

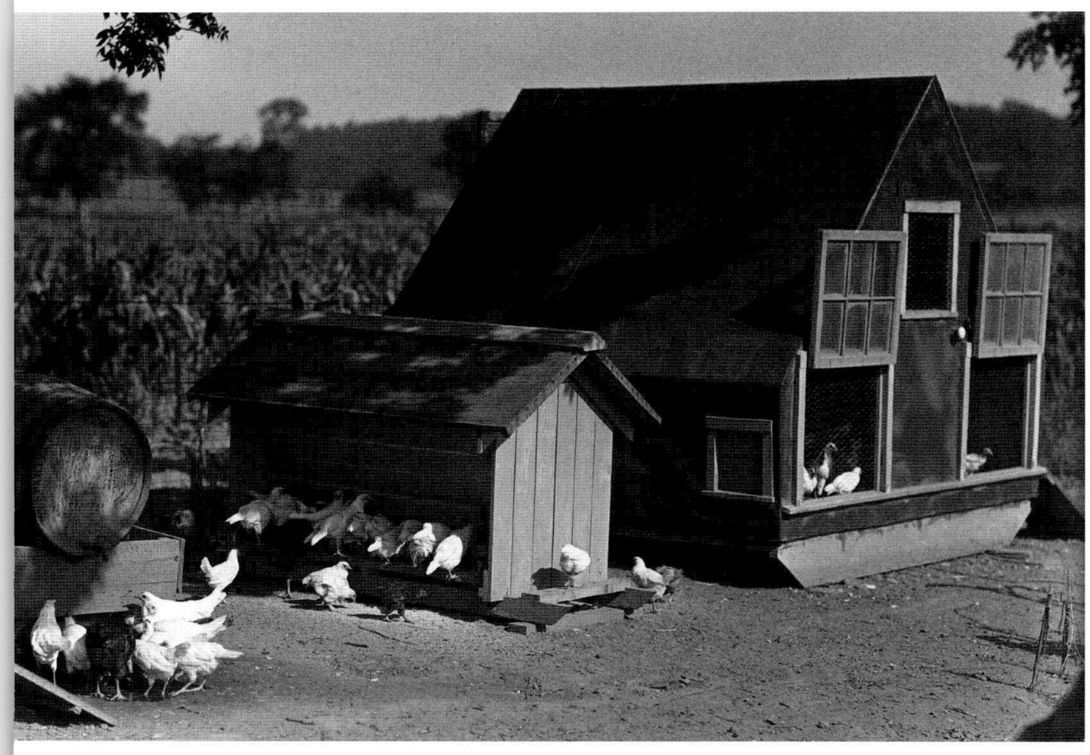

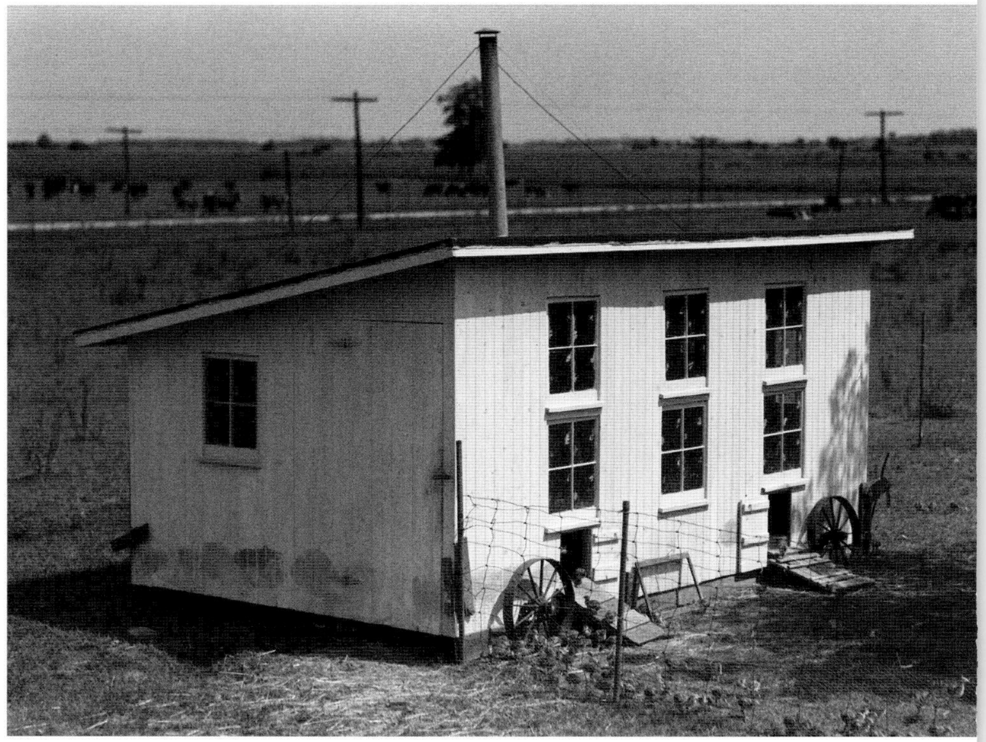

(Top left) A water barrel, feed hopper, and colony house at the Purdue University Poultry Farm (Tippecanoe County, 1913)

(Top right) This turkey brooder house equipped with wheels could be easily moved on the farm of F. L. Spencer in Idaville. (White County, 1934)

(Bottom) White Leghorns on range (Tippecanoe County, 1952)

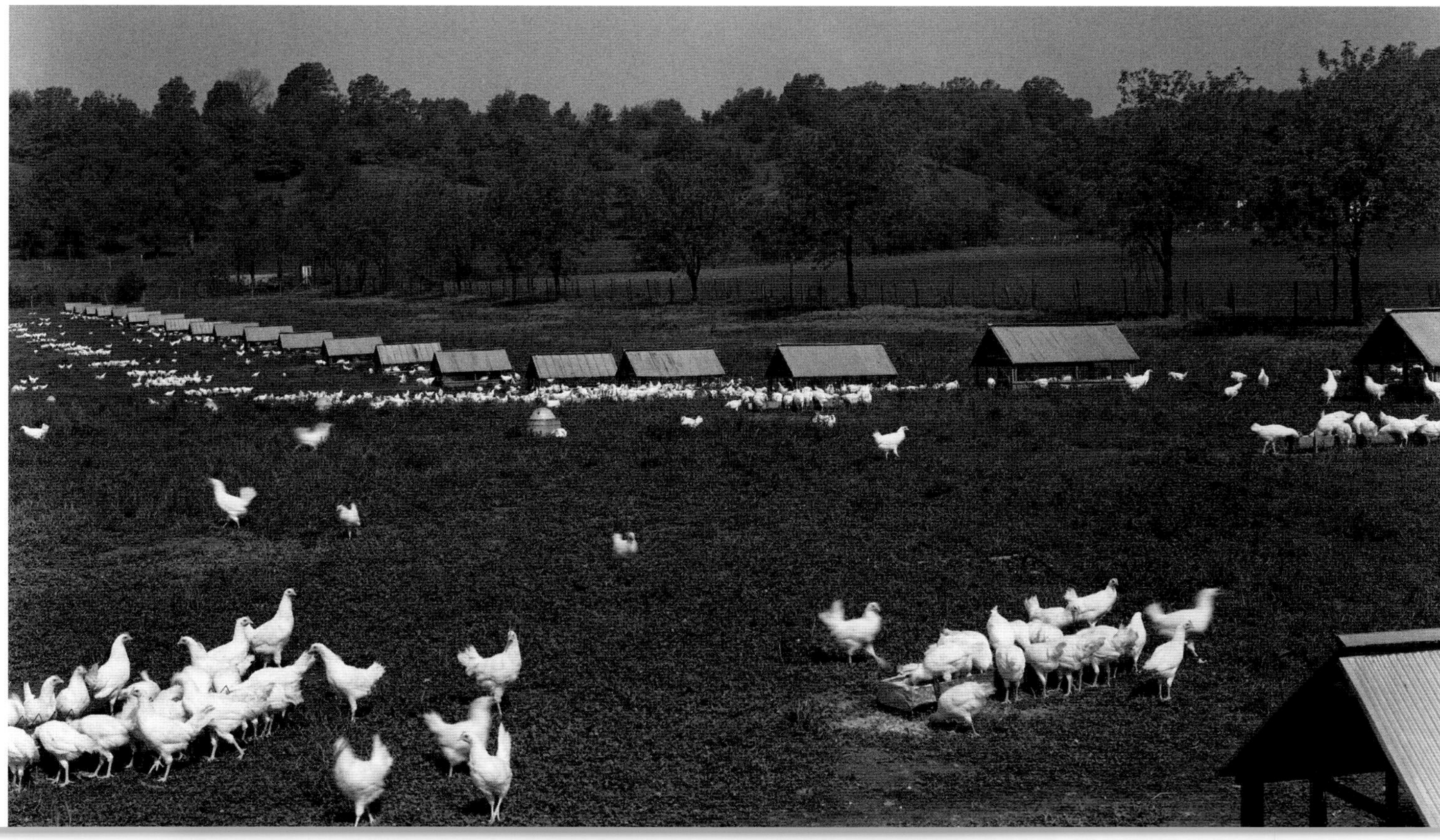

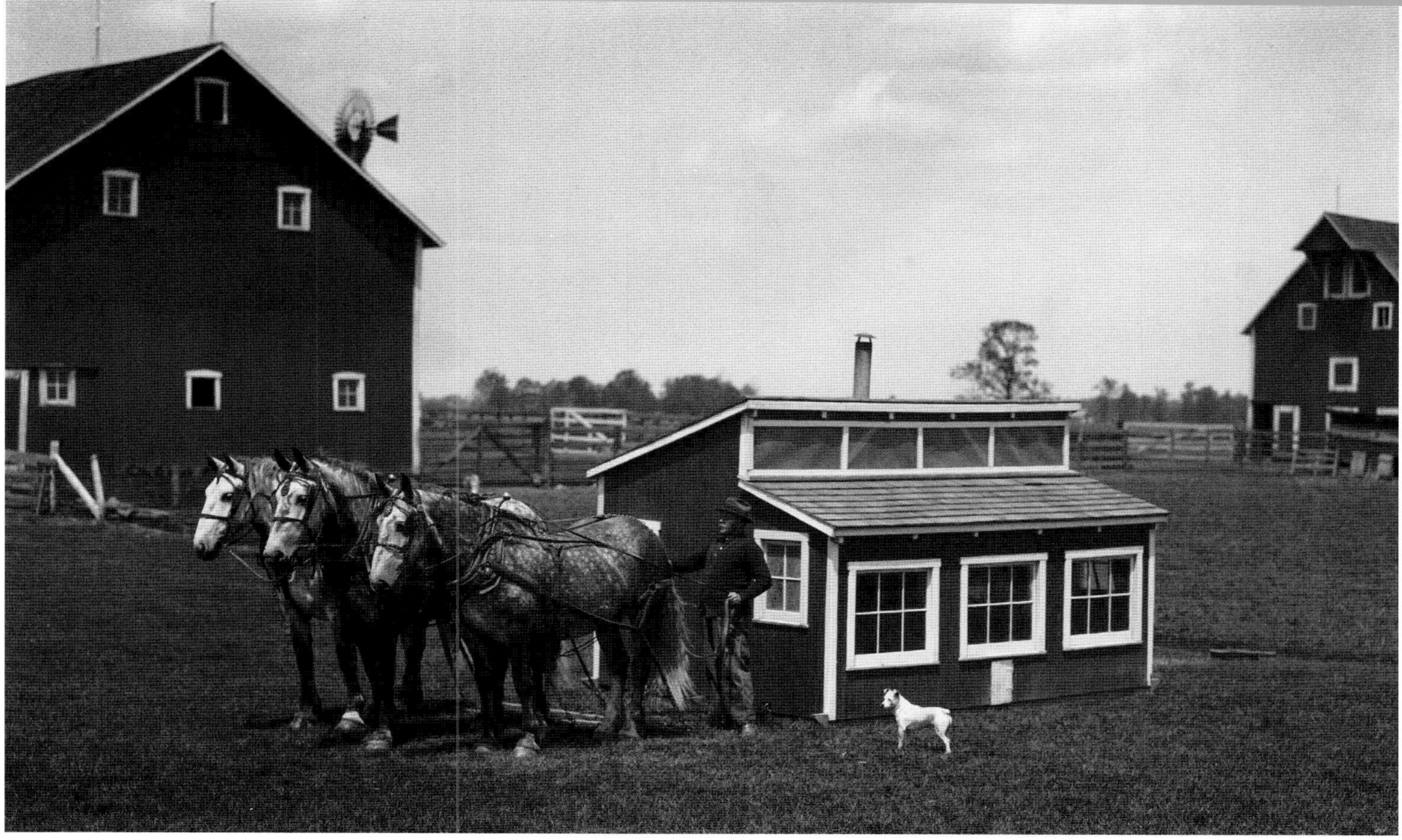

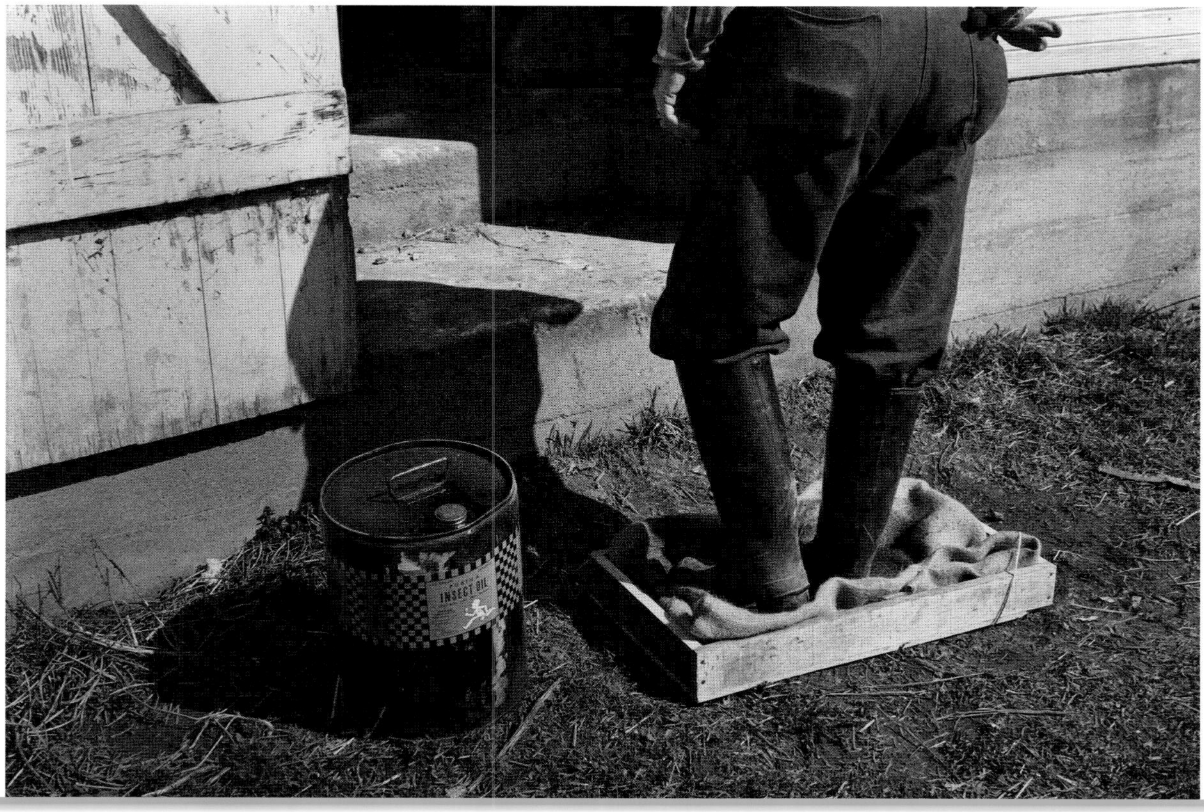

(Top) C. Orr uses his Percheron team to move a brooder house to fresh ground in Forest. (Clinton County, 1931)

(Bottom) Using Purina Insect Oil to clean boots before entering a poultry laying house (Kosciusko County, 1946)

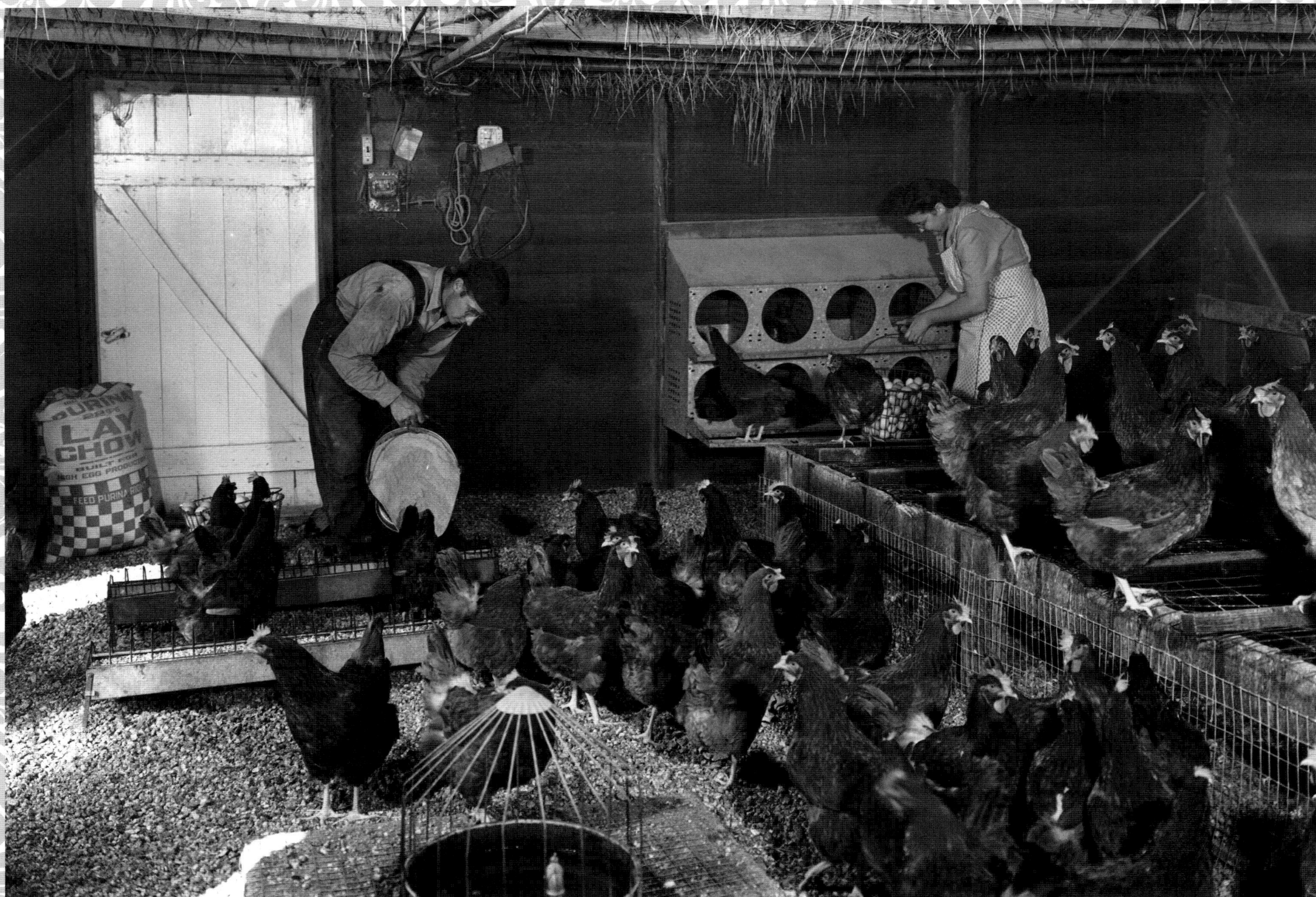

Mr. and Mrs. Kyle DeVault from Romney feed their flock of New Hampshire Red chickens and gather eggs inside their laying house. (Tippecanoe County, 1946)

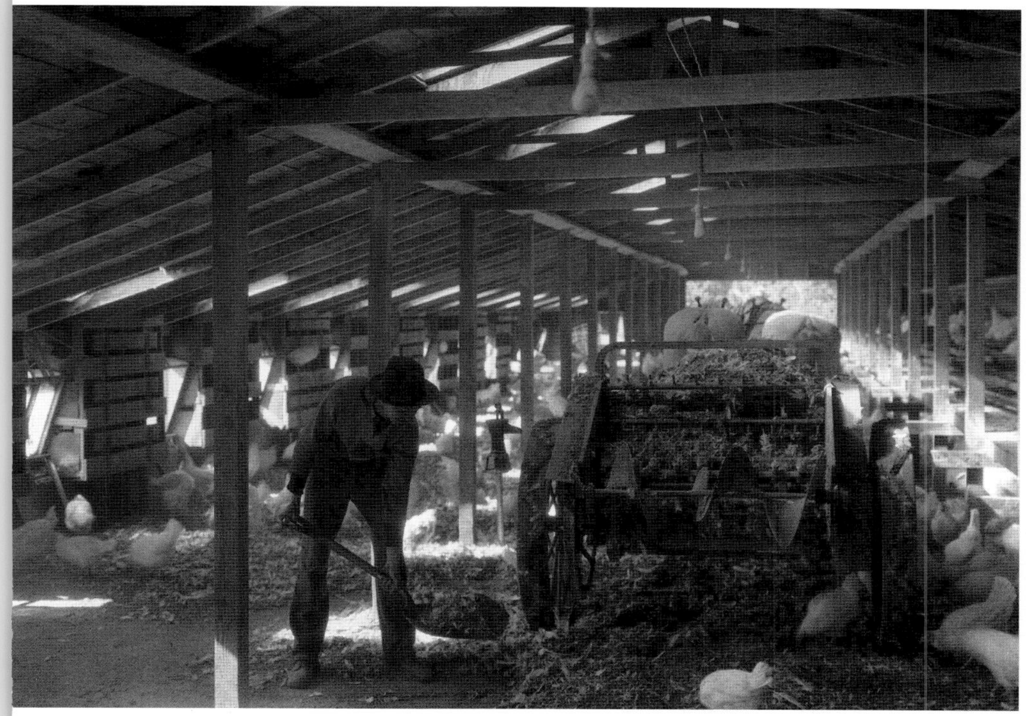

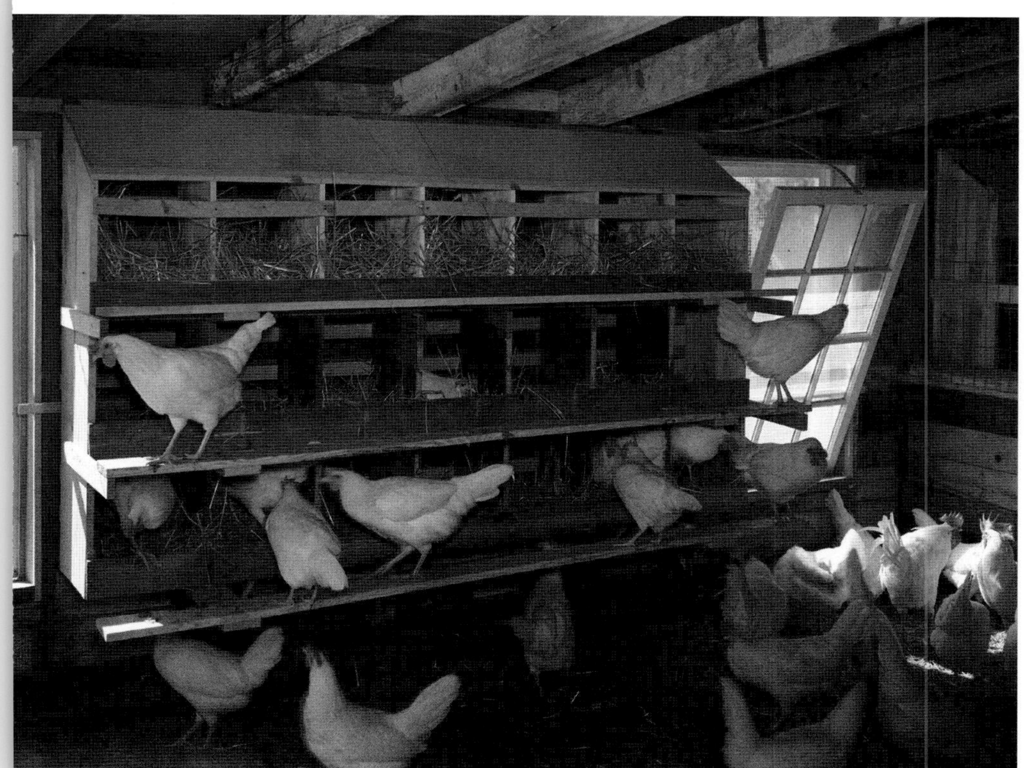

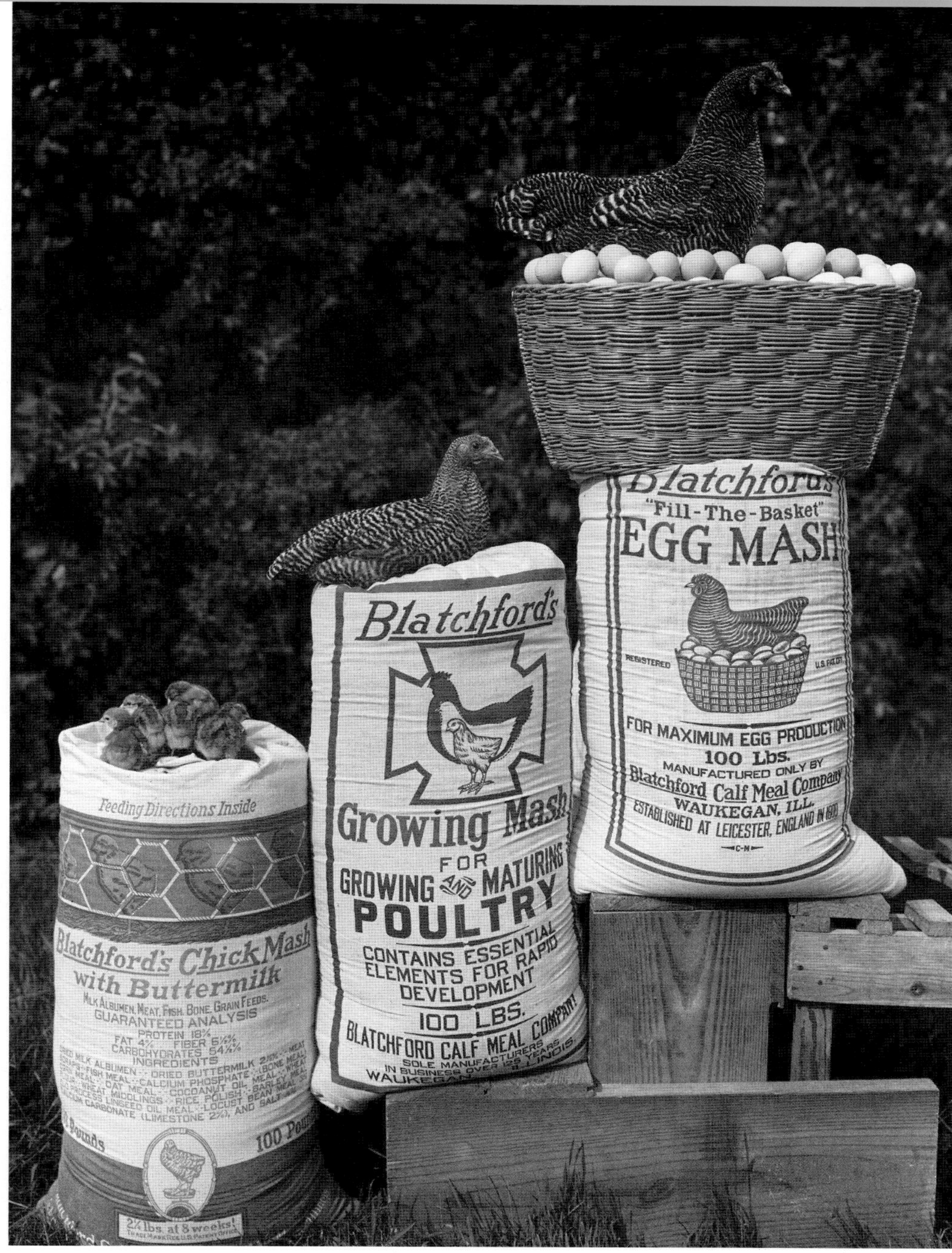

(Top left) A man cleans out one of the poultry houses on the Creighton Brothers Poultry Farm in Warsaw. The large doors at the end of each house allowed a person to drive a manure spreader inside and load it with chicken manure. (Kosciusko County, 1929)

(Bottom left) A laying house with nests on the C. I. Bashore farm in Silver Lake (Kosciusko County, 1939)

(Right) This photo, which Allen took for the Blatchford Calf Meal Company, featured three sacks of chicken feed. (N.p., 1928)

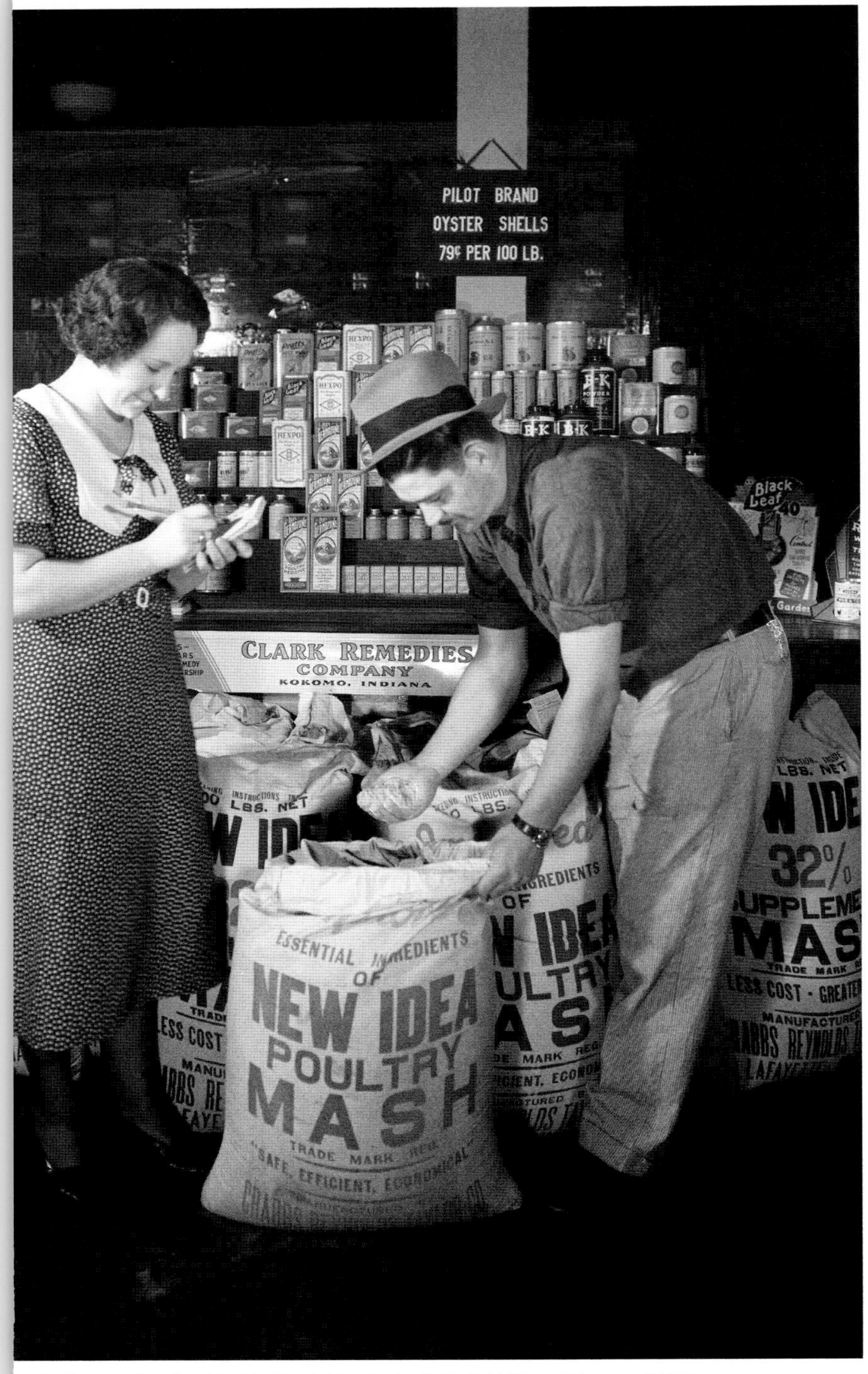

Selling poultry feed inside Gray's Hatcheries in Delphi (Carroll County, 1934)

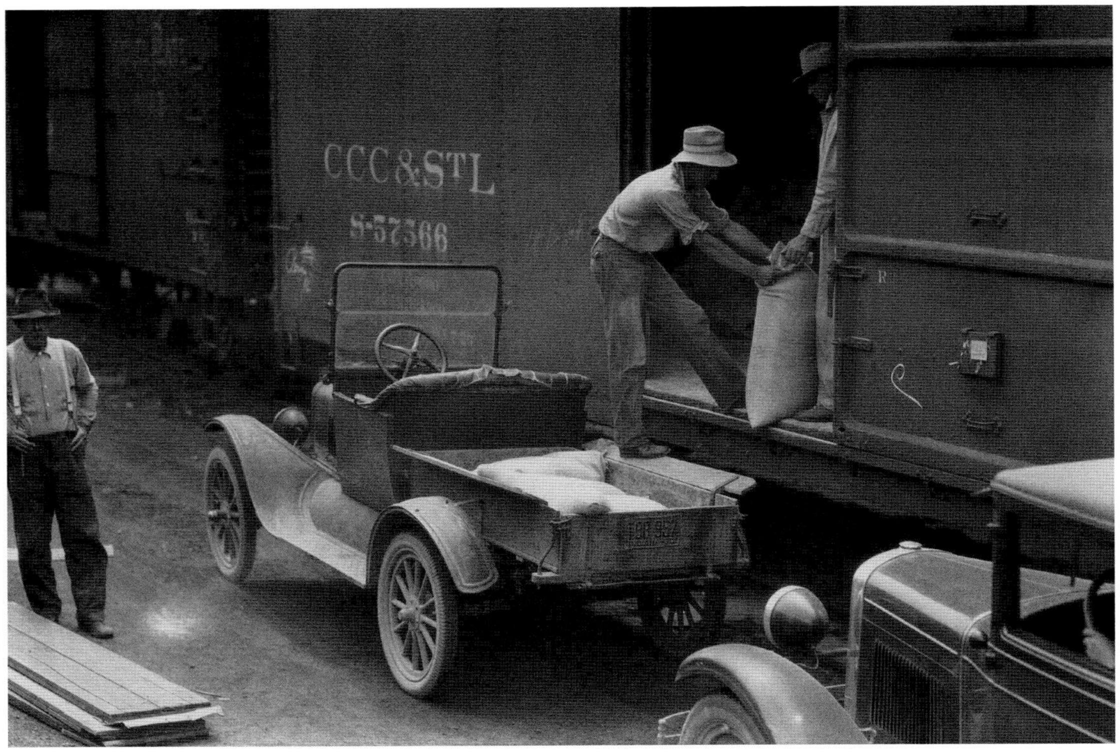

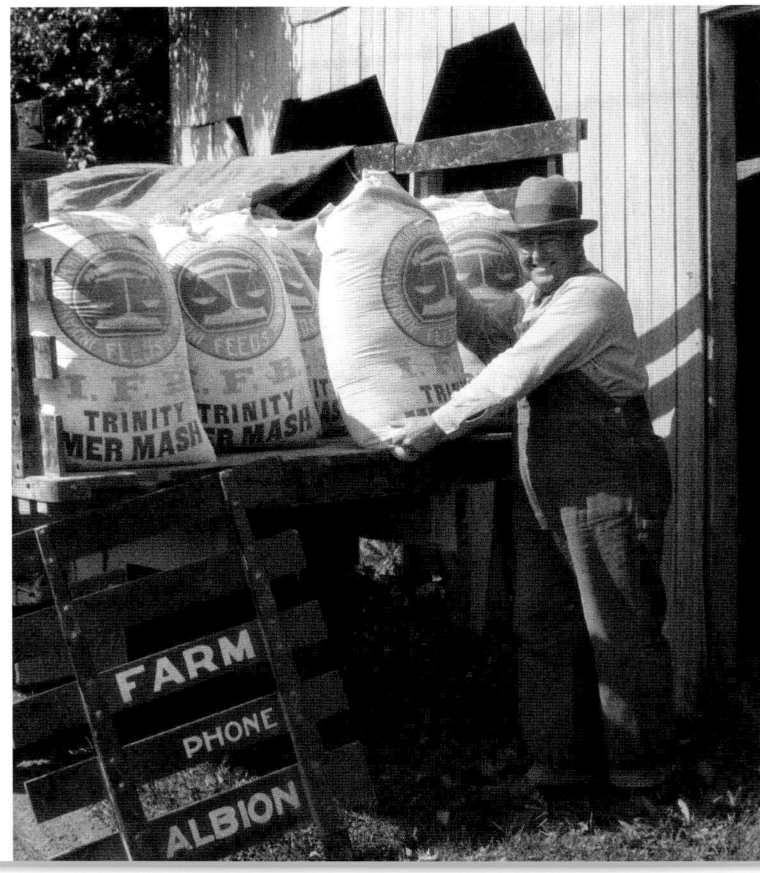

(Top right) Arden Rhodes from Plymouth transfers bags of poultry feed from a railcar at Mentone. (Kosciusko County, 1930)

(Bottom right) James Edwards from Albion unloads feed from a truck. (Noble County, 1930)

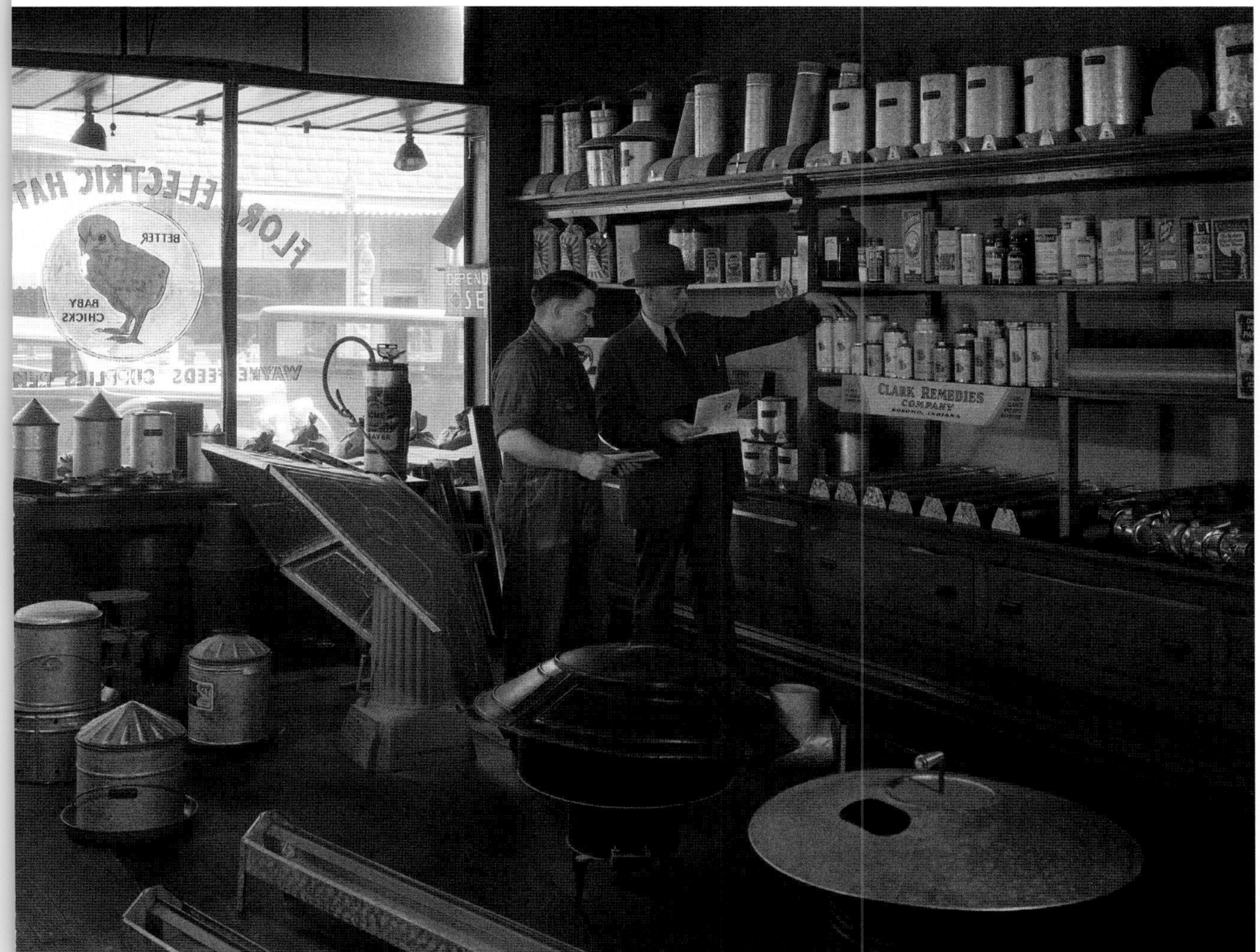

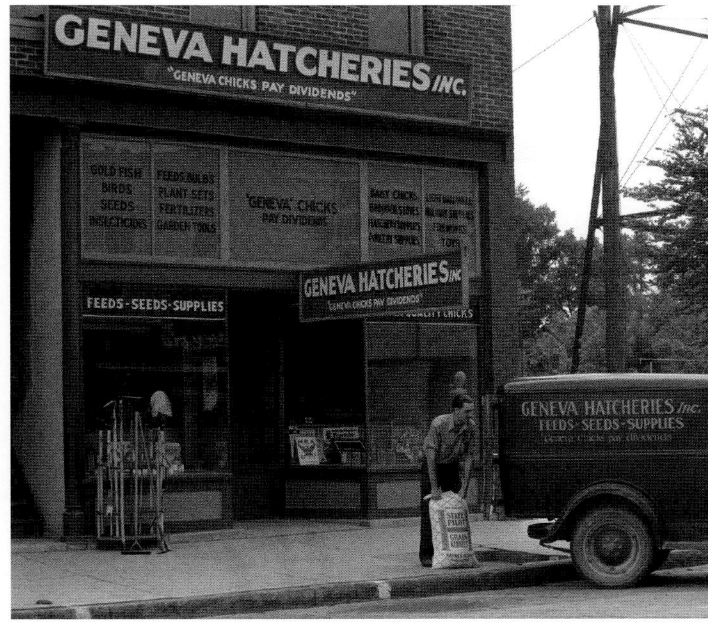

(Left) Poultry medicine and equipment on display at the Flora Electric Hatchery (Carroll County, 1934)

(Right) Loading a delivery truck in front of Geneva Hatcheries (Adams County, 1933)

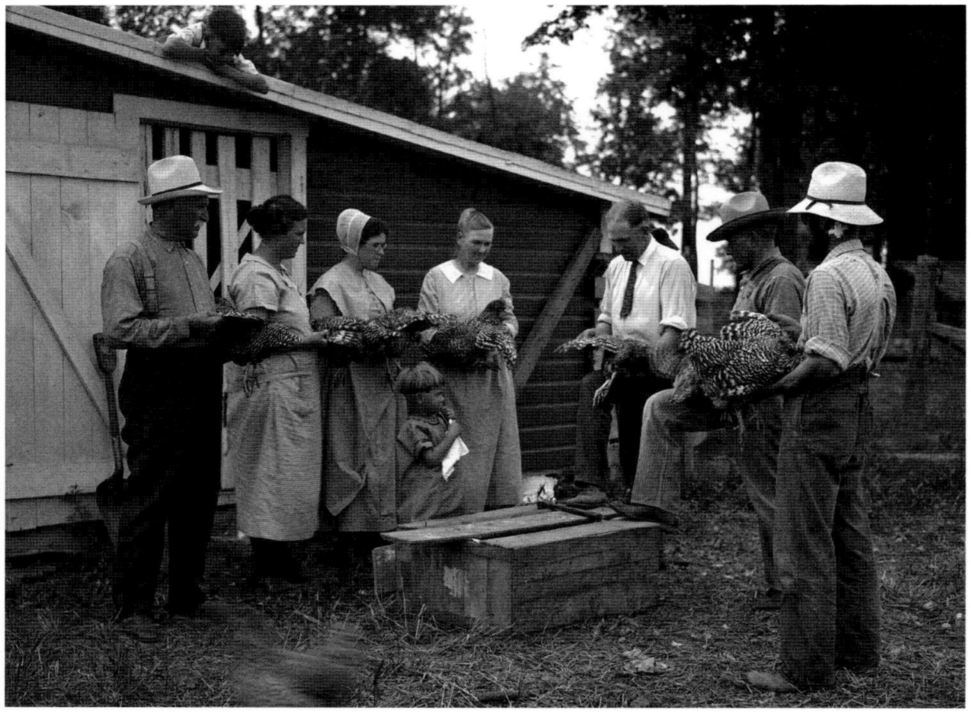

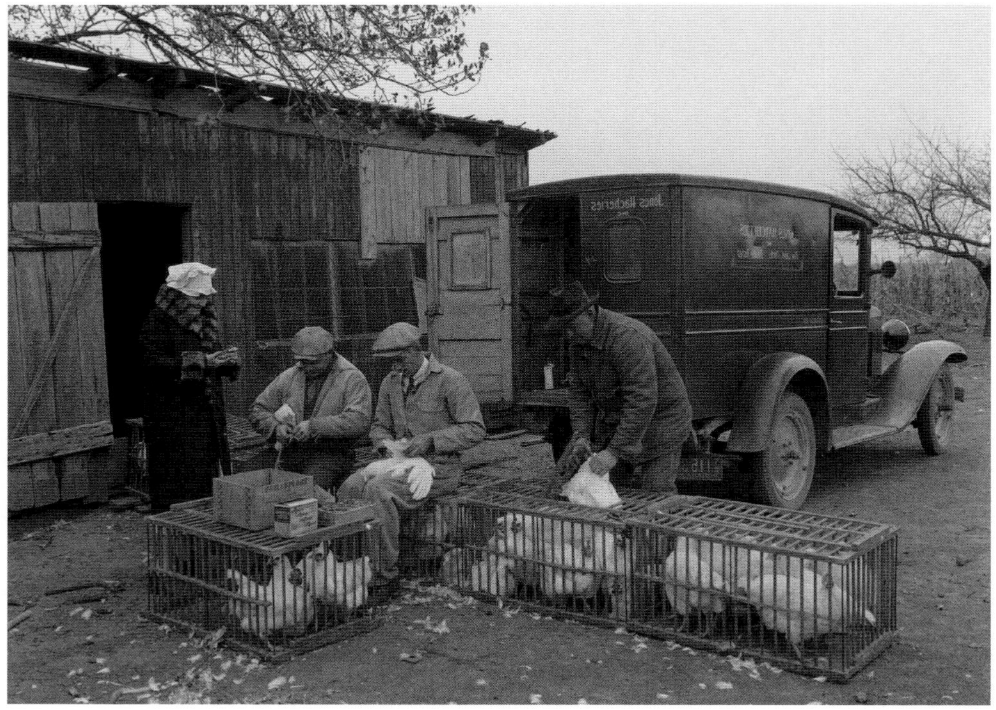

(Top left) Ralph Maggart (third from right), the Carroll County Extension agent, conducts a poultry culling demonstration while a child observes from the roof. (Carroll County, 1934)

(Top right) This group culls underperforming hens and performs blood tests on the flock of Harry Hagerty in Brookston. (White County, 1934)

(Bottom) Students from a vocational agriculture class cull and blood-test a flock owned by the Paul Riley Poultry Farm in West Lafayette. (Tippecanoe County, 1944)

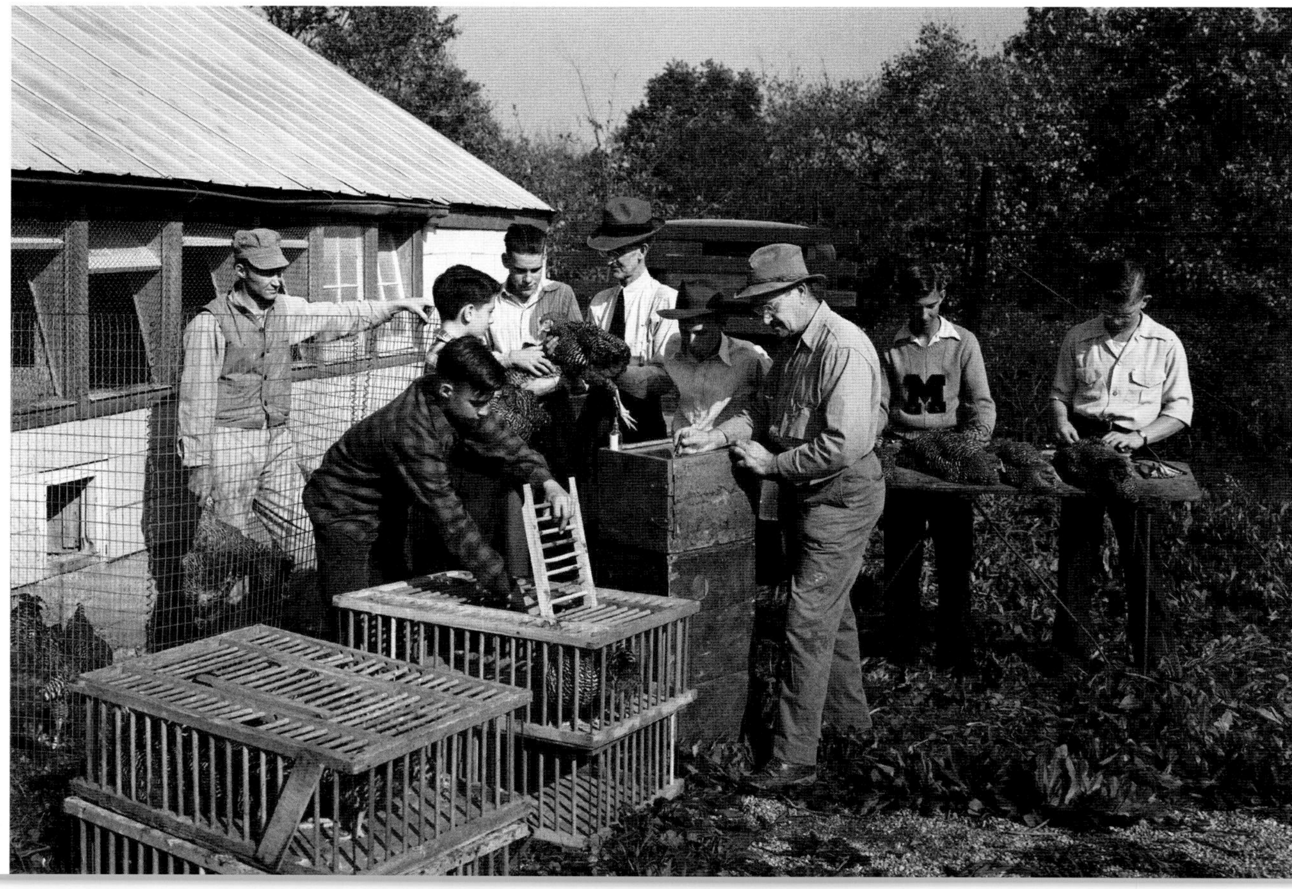

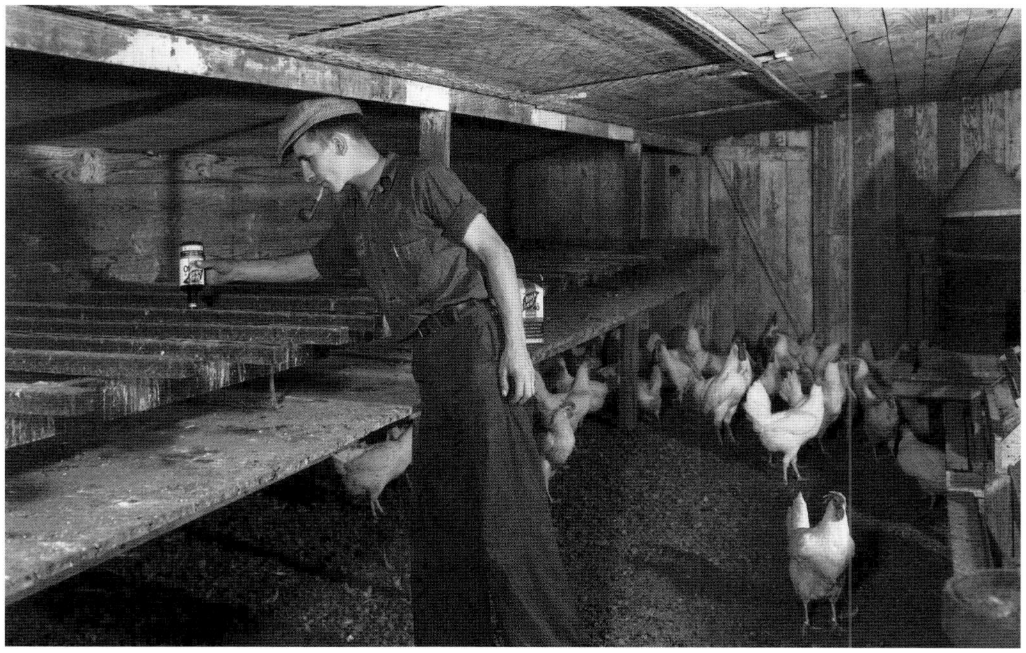

A man puts Black Leaf 40, an insecticide, on chicken roosts on the Paul Riley Poultry Farm. (Tippecanoe County, 1935)

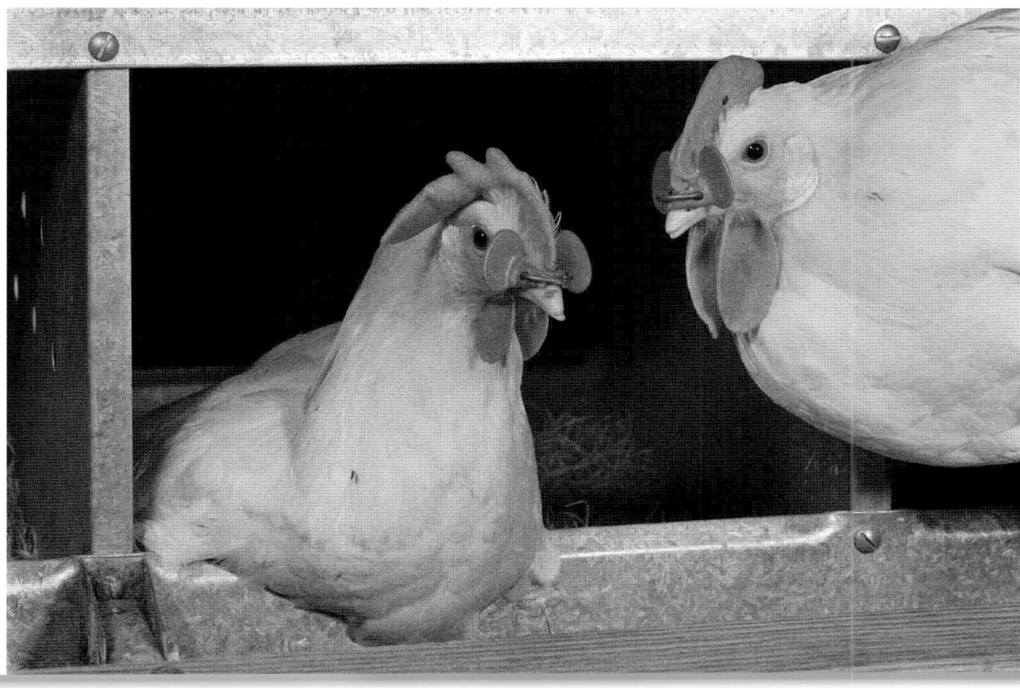

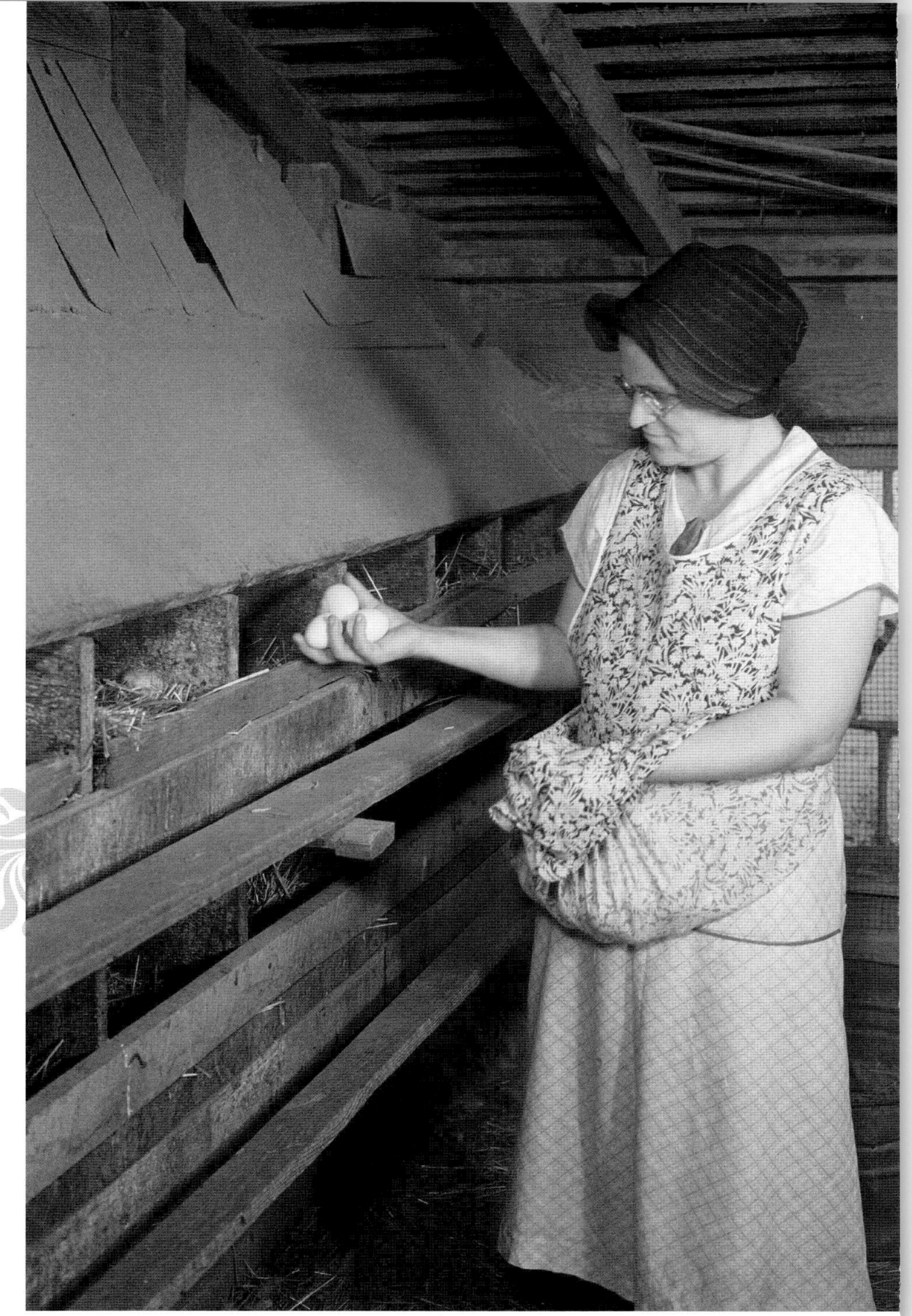

Mrs. William Y. Hartsough of North Manchester gathers eggs in her apron. (Wabash County, 1934)

These eyeglasses for hens helped prevent pecking among White Leghorns on the Arbor Acres Farm in Thorntown. (Boone County, 1963)

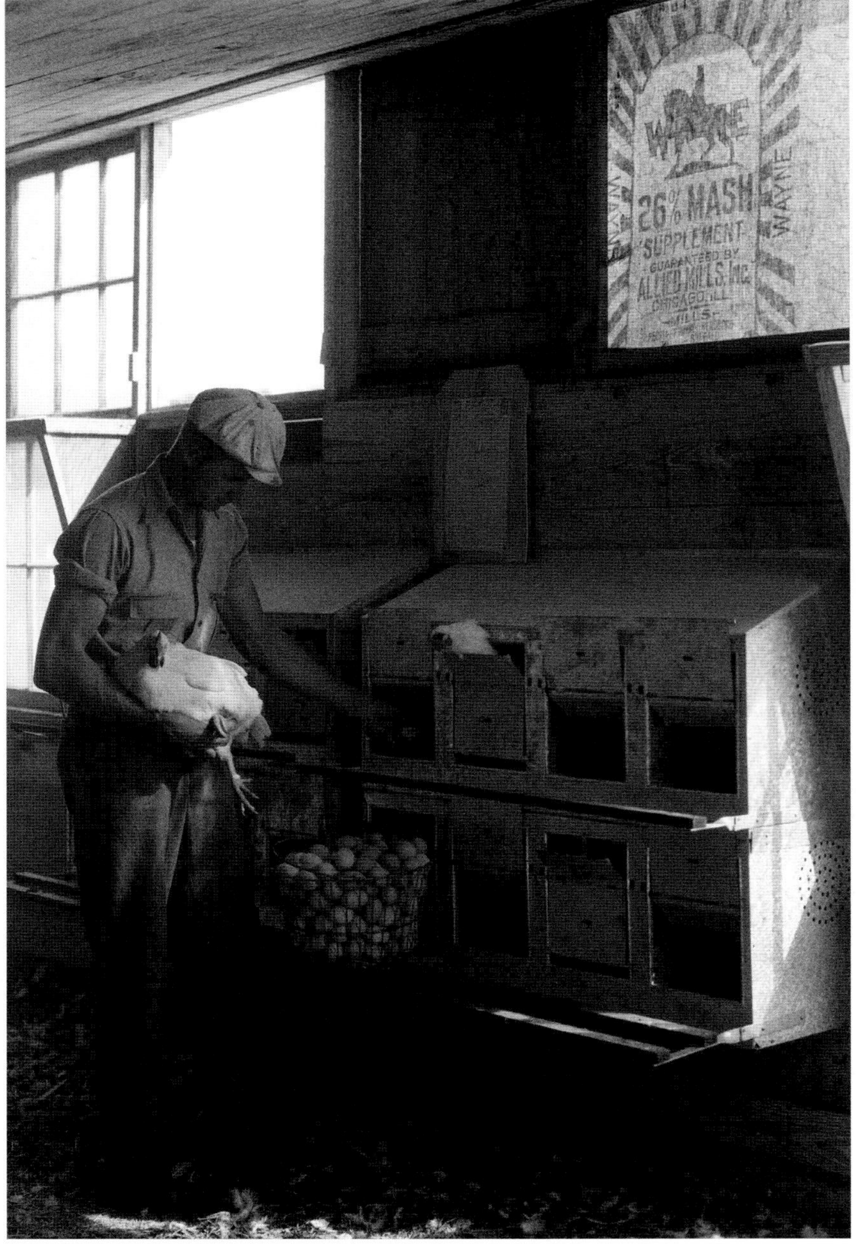

Trap nesting on the farm of Paul Riley in West Lafayette (Tippecanoe County, 1934)

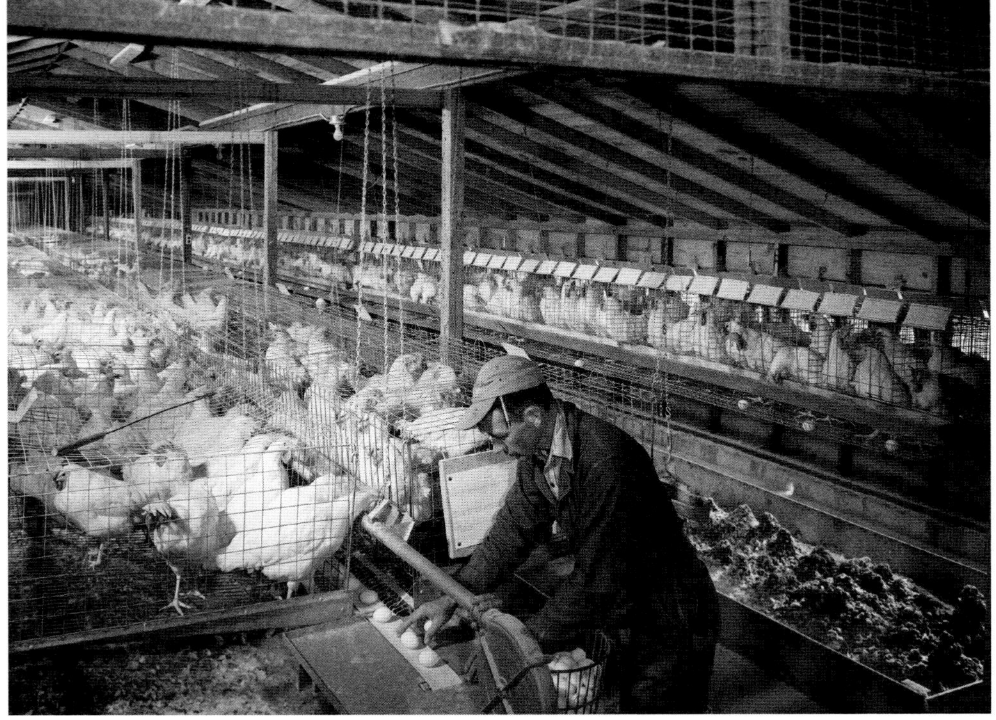

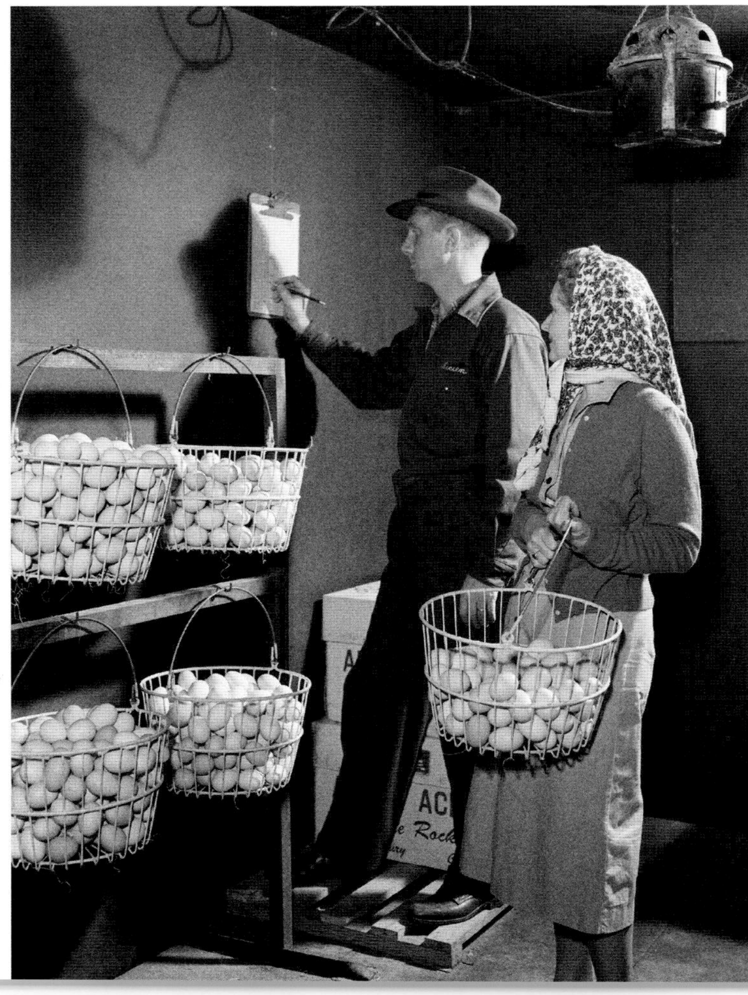

(Top right) Clyde Hicks Jr. from Warsaw picks up eggs as they roll down from special laying cages on the Creighton Brothers Poultry Farm in Warsaw. (Kosciusko County, 1957)

(Bottom right) Jerry Jones keeps detailed records on egg production while Edith Jones looks on. (N.p., 1956)

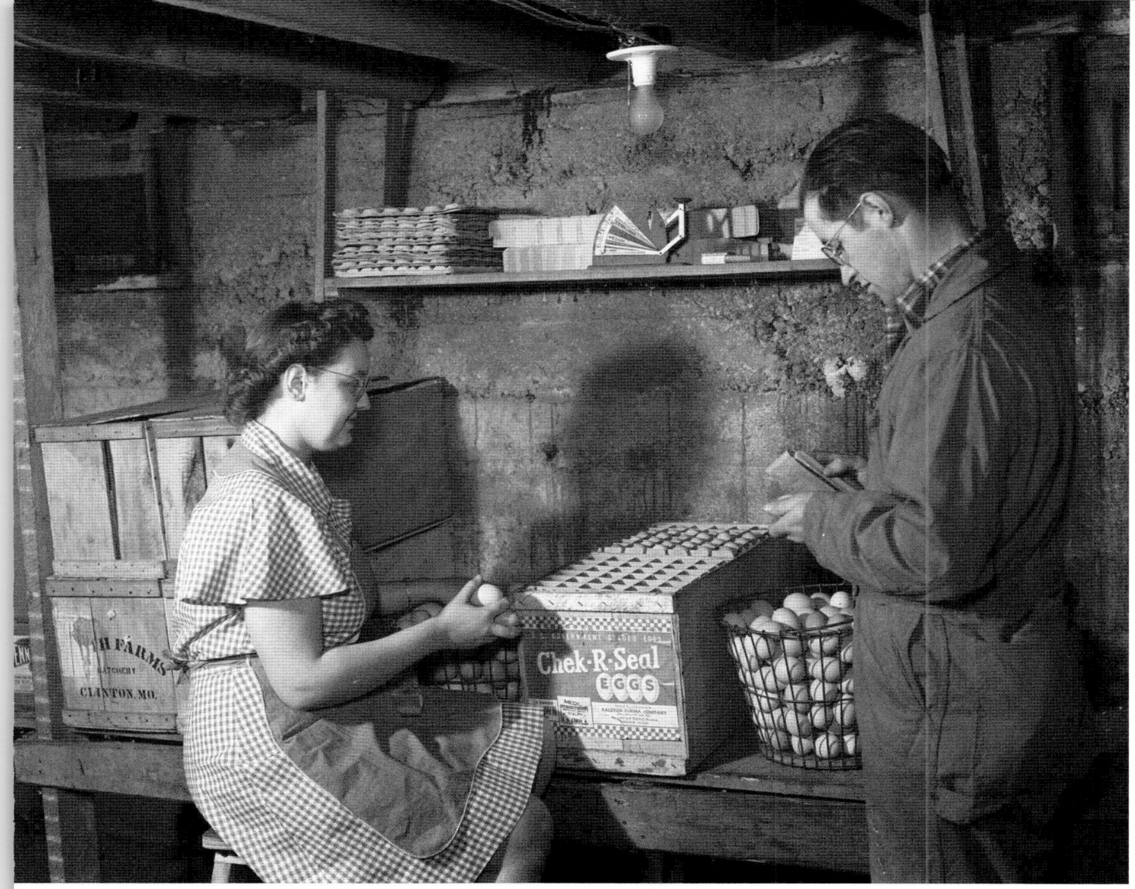

Mr. and Mrs. Kyle DeVault grade and pack eggs in the basement of their Romney home. (Tippecanoe County, 1947)

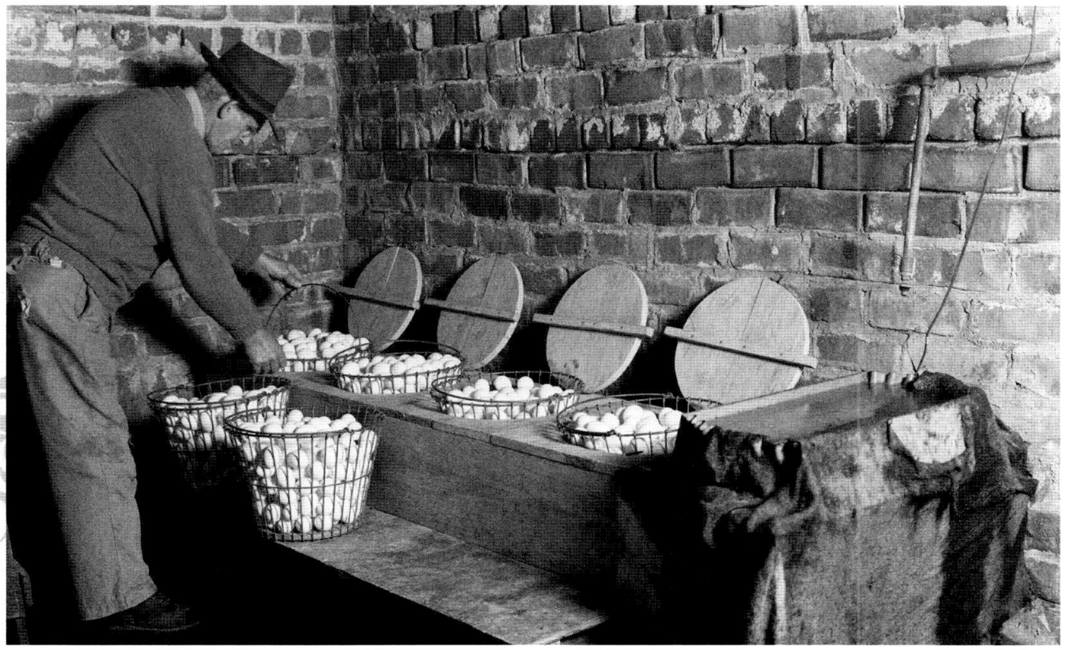

This man cools eggs to keep them fresh on the poultry farm of Lee Foster and Son in Attica. (Fountain County, 1949)

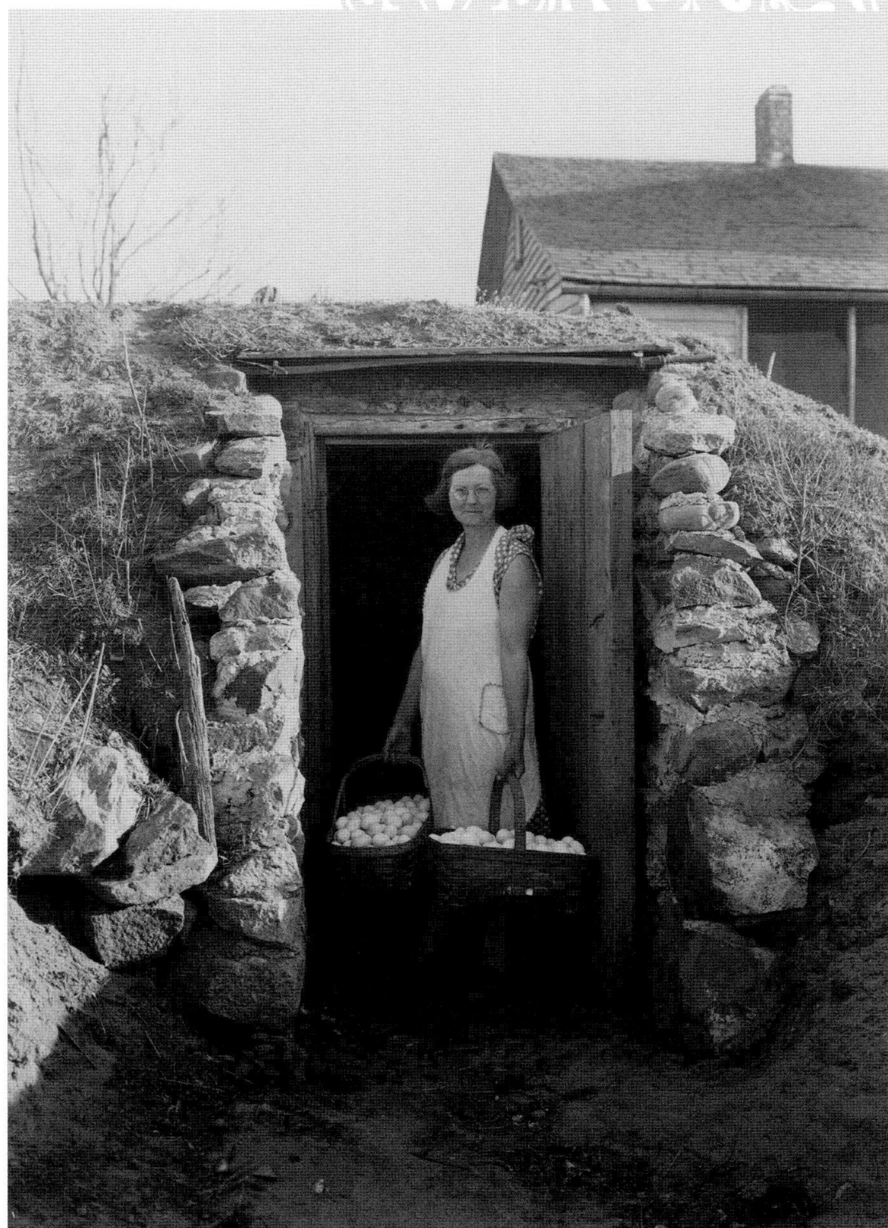

Near Denver, Mrs. Perry Spangler prepares to store eggs in a root cellar until they are shipped. (Miami County, 1934)

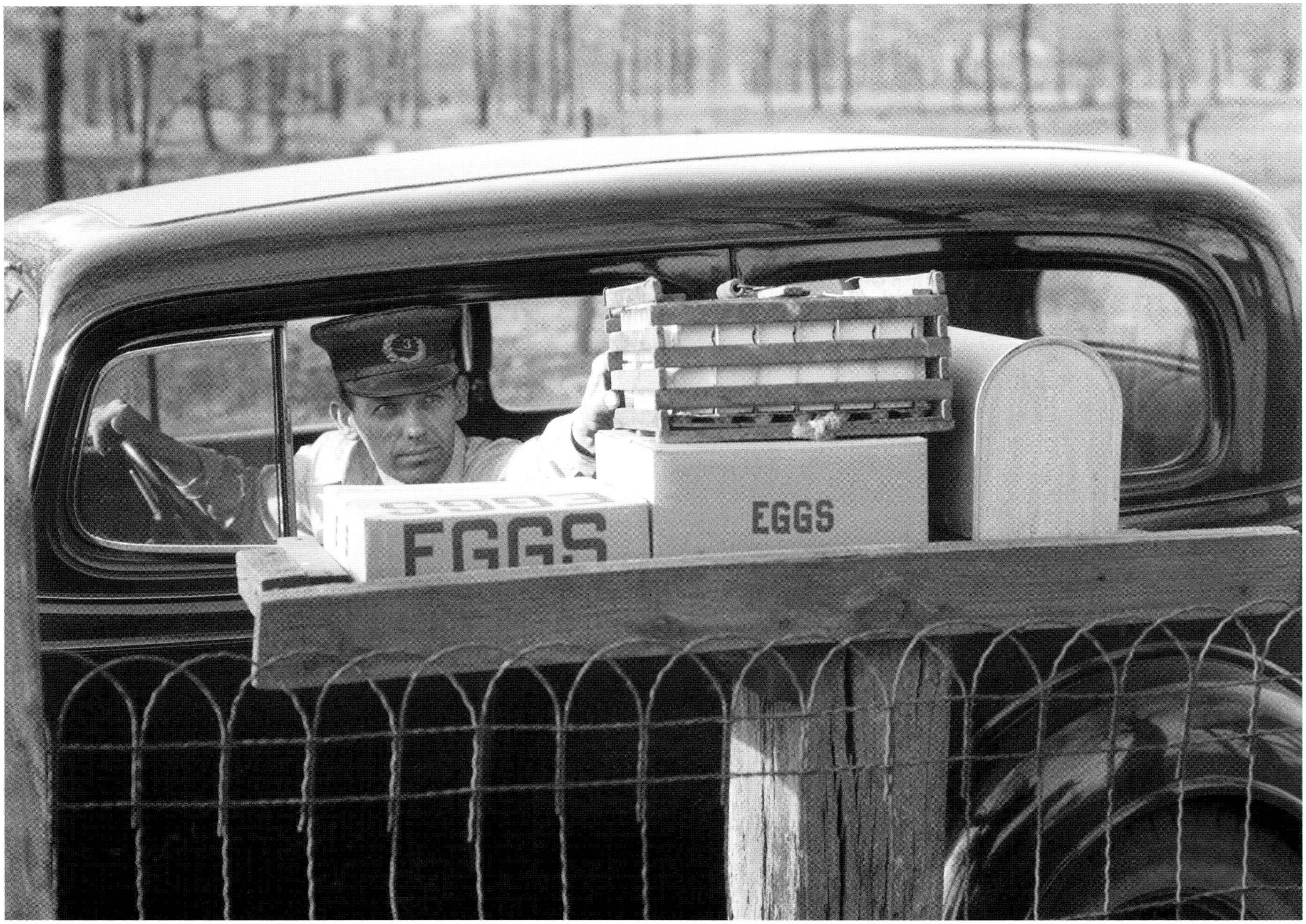

A rural mail carrier gathers eggs for shipment to market at a farm mailbox near Knox. (Starke County, 1934)

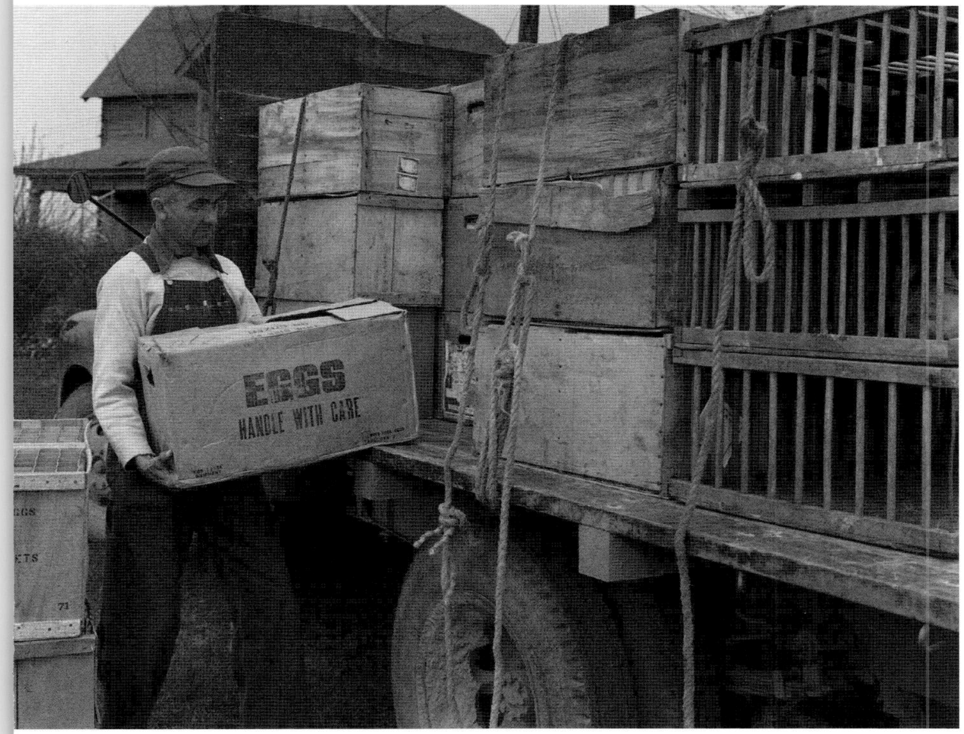

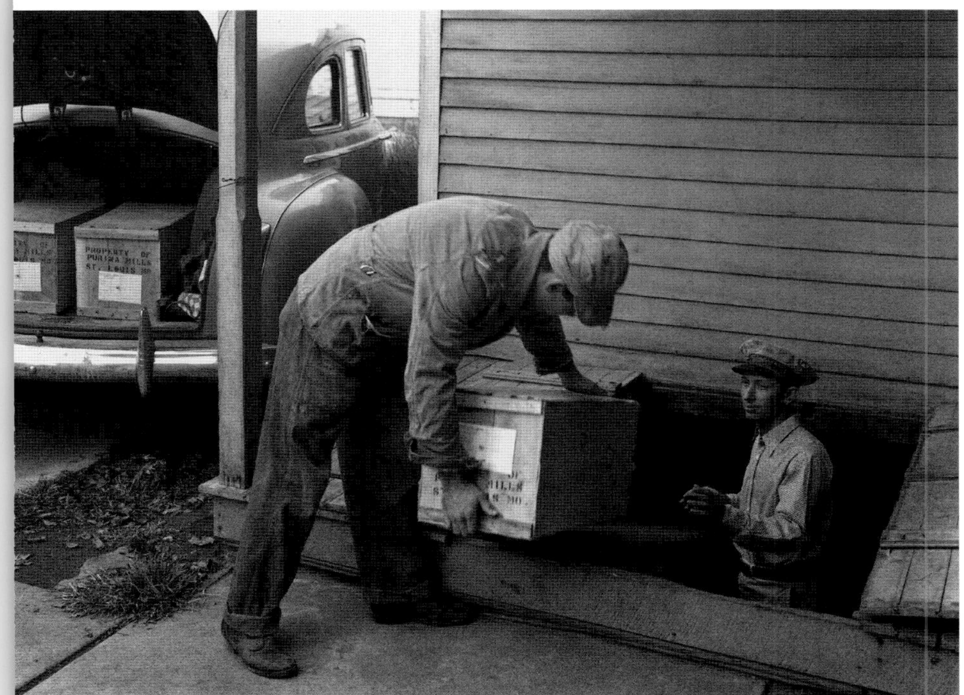

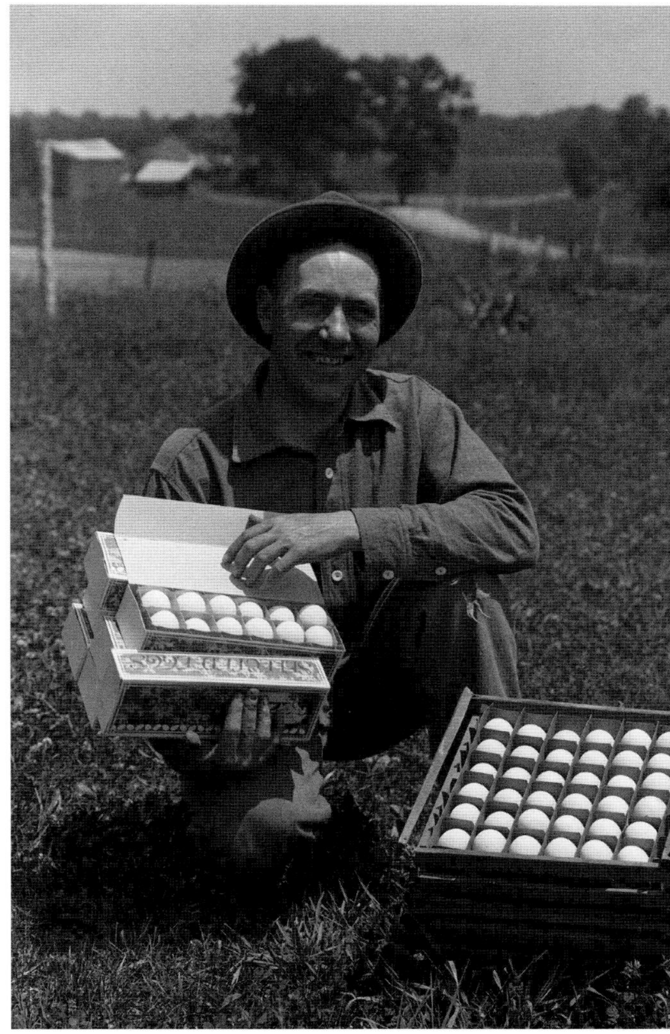

(Top left) A driver for Pickett and Sons, poultry and egg buyers from Sheridan, loads a case of eggs onto his truck. (Hamilton County, 1951)

(Bottom left) John Charles Burdell delivers eggs to a man at a dealer's store in St. Jacob, Illinois. (Madison County, IL; 1946)

(Right) Paul Jones from Columbus displays his special cartons and carriers for marketing fresh eggs. (Bartholomew County, 1925)

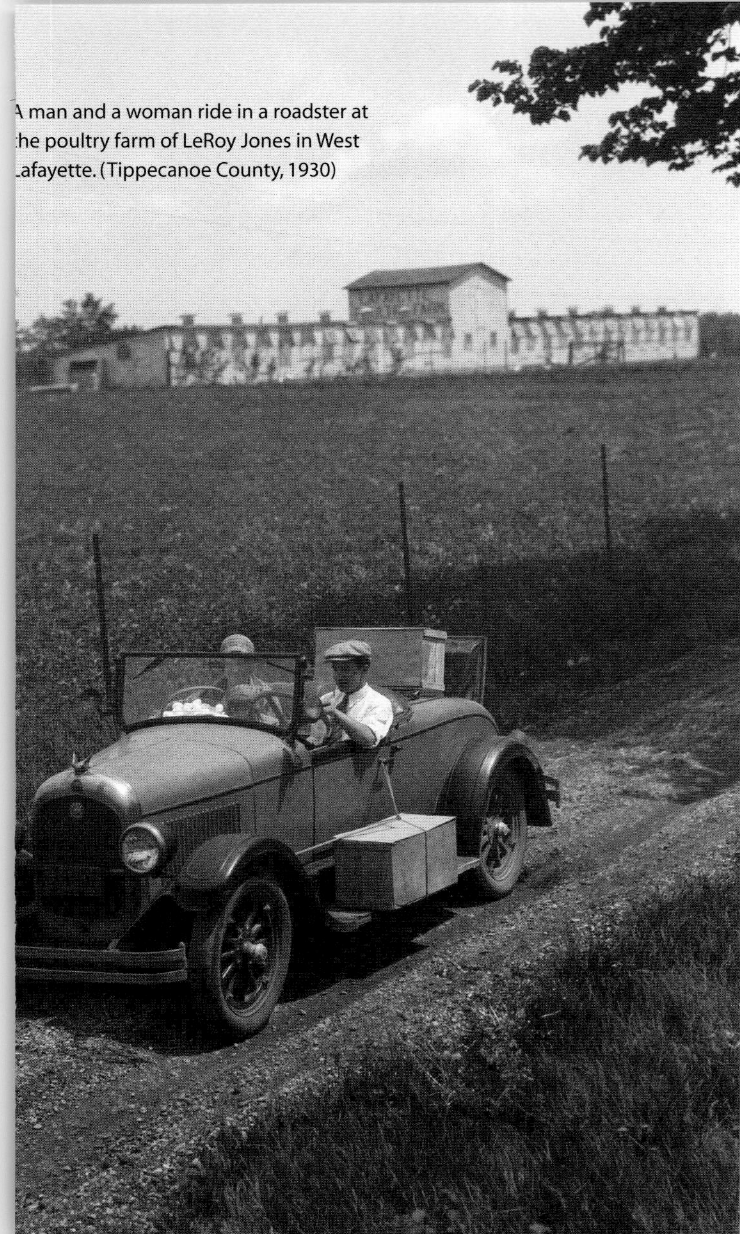

A man and a woman ride in a roadster at the poultry farm of LeRoy Jones in West Lafayette. (Tippecanoe County, 1930)

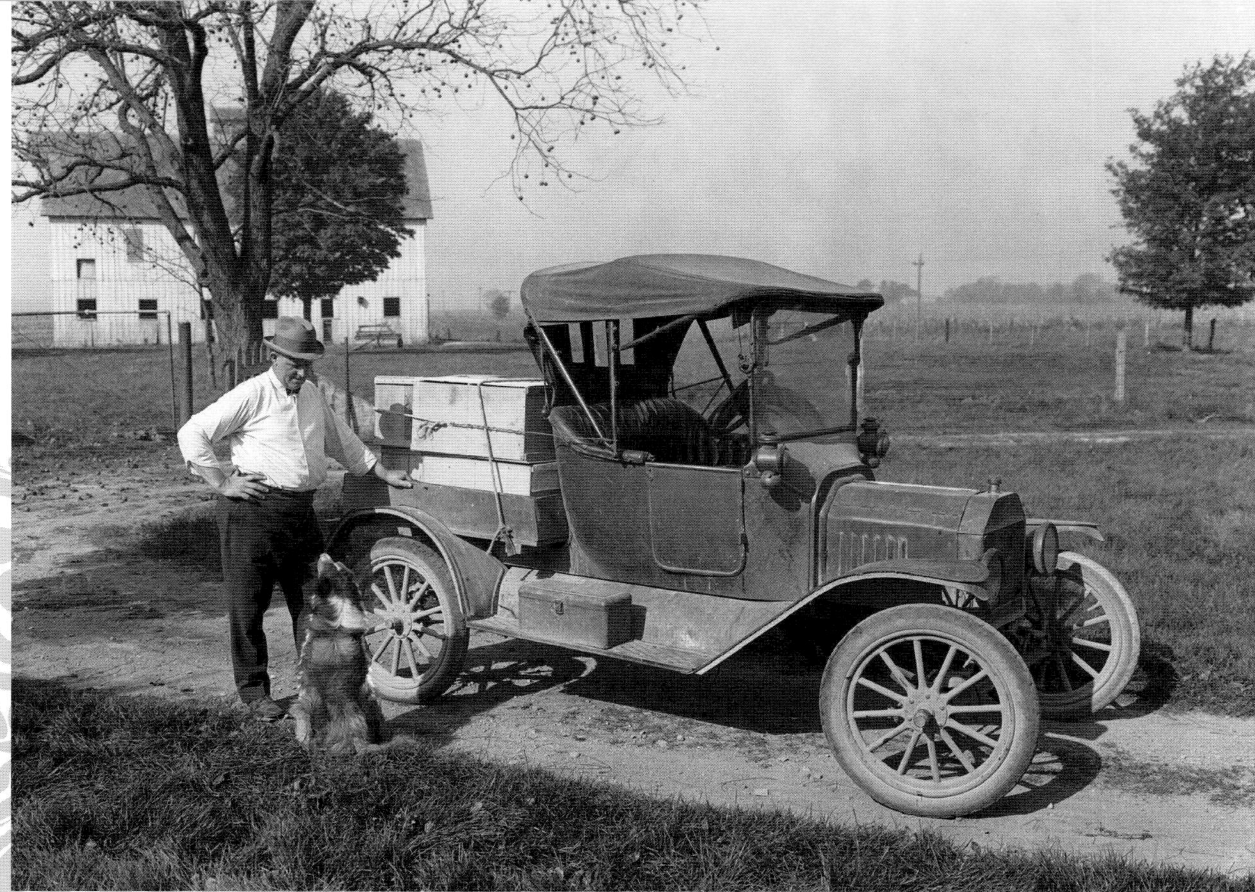

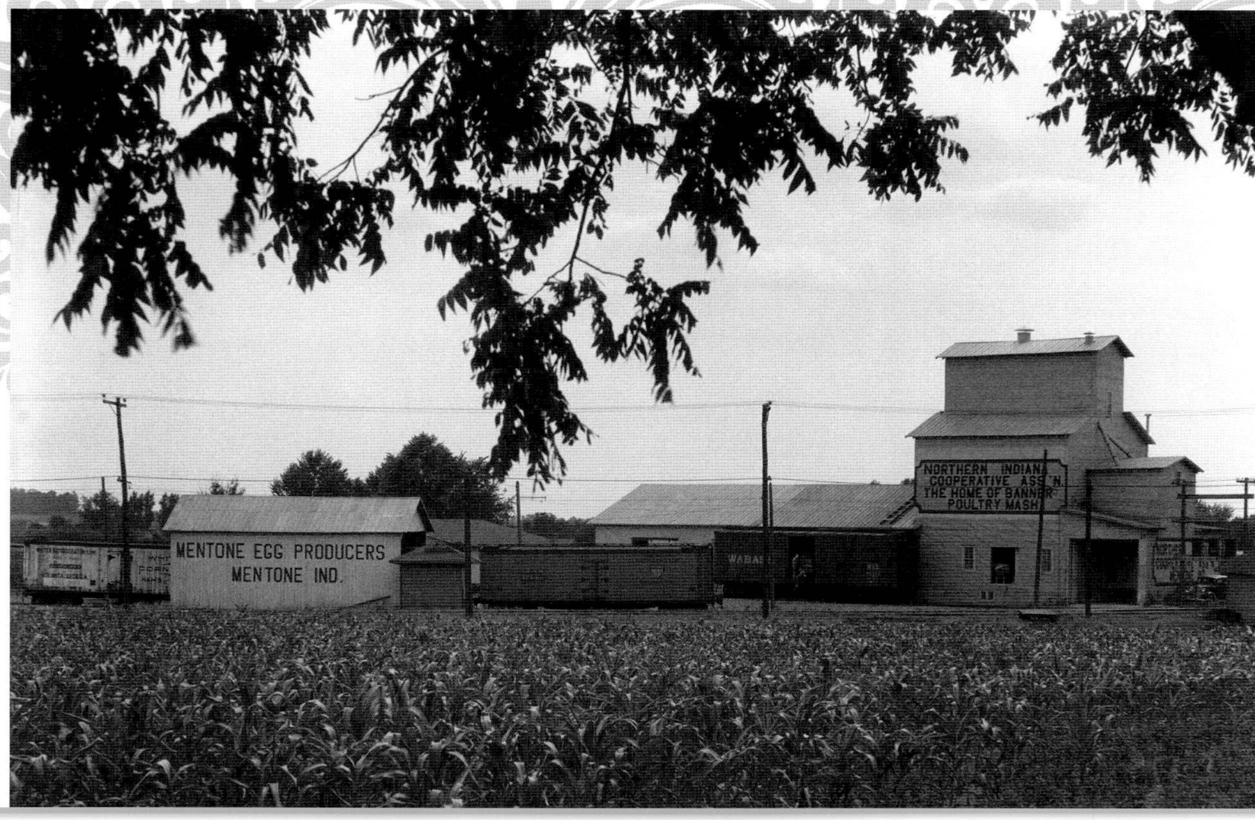

(Top right) Attica-area poultry farmer J. L. Foster prepares to use his Ford pickup truck to deliver eggs for shipment to New York markets. (Fountain County, 1924)

(Bottom right) Members of the Mentone Egg Producers cooperative used these buildings to prepare egg shipments and mix poultry feed. (Kosciusko County, 1934)

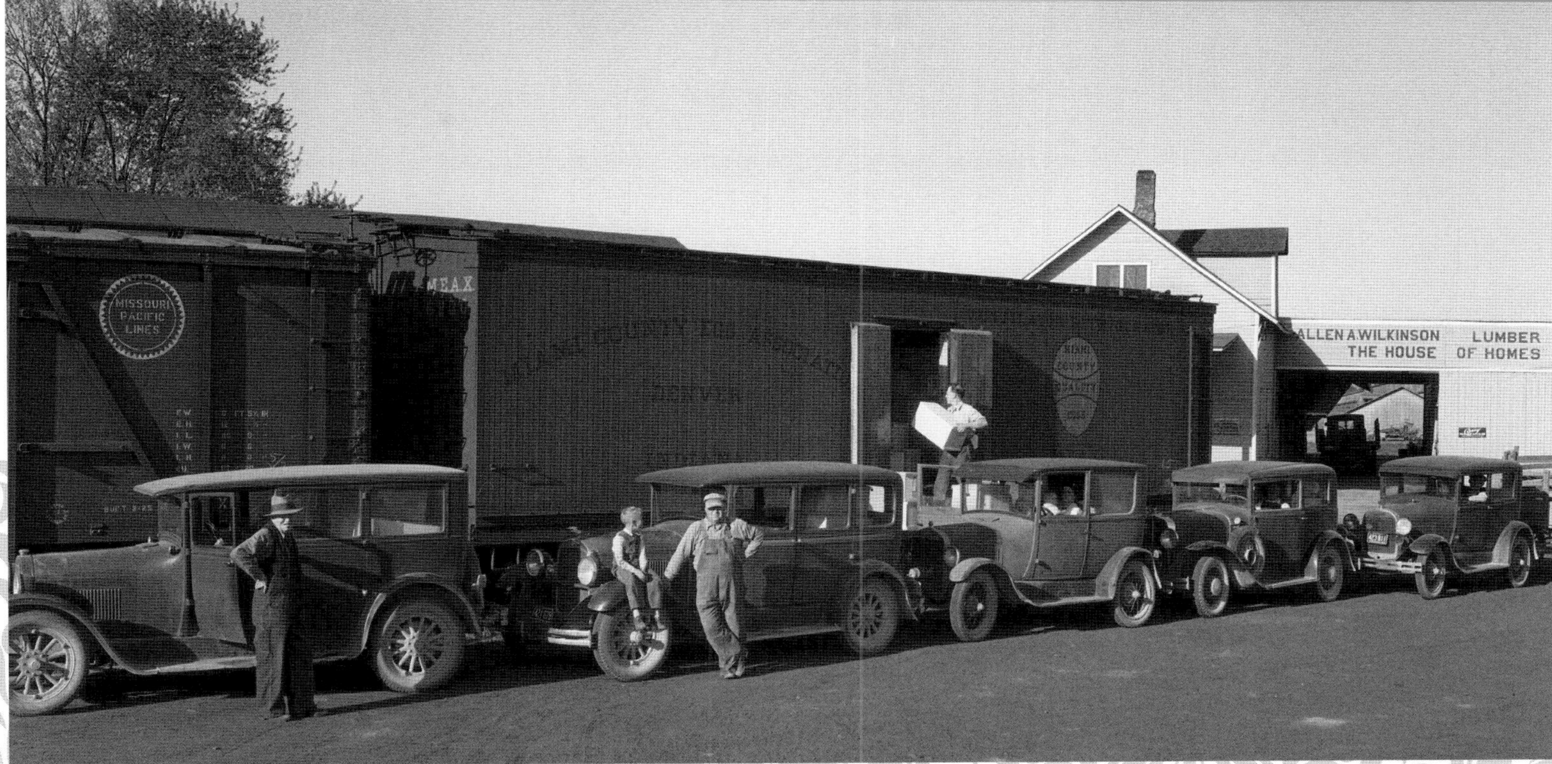

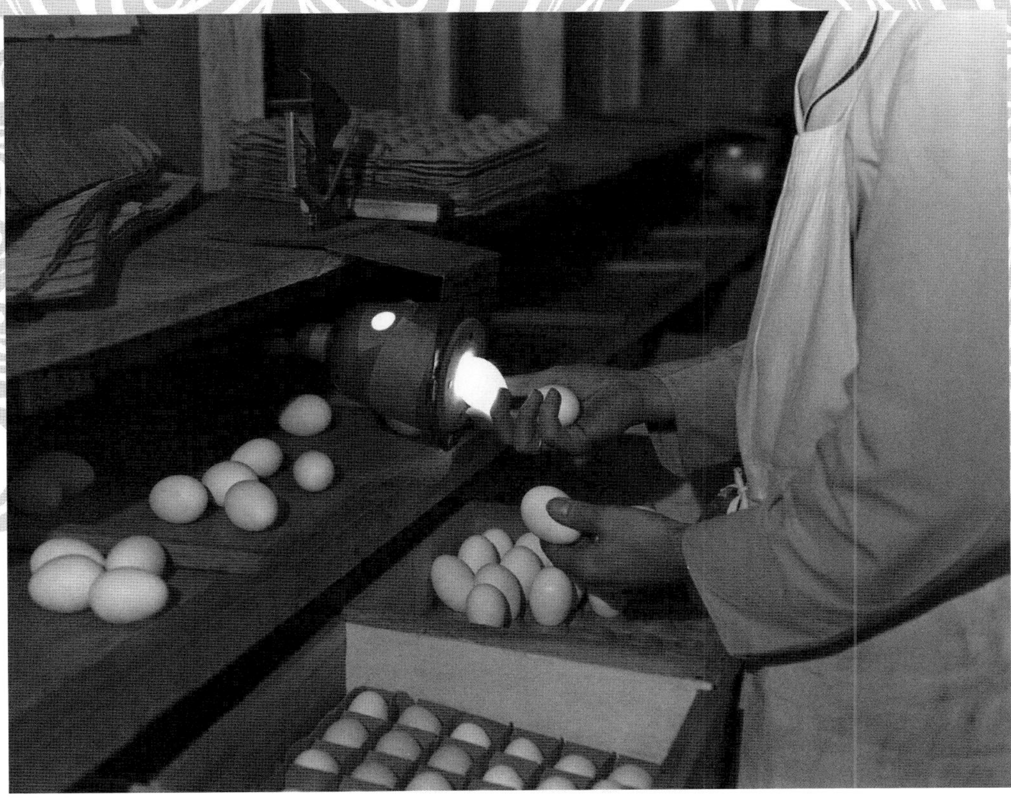

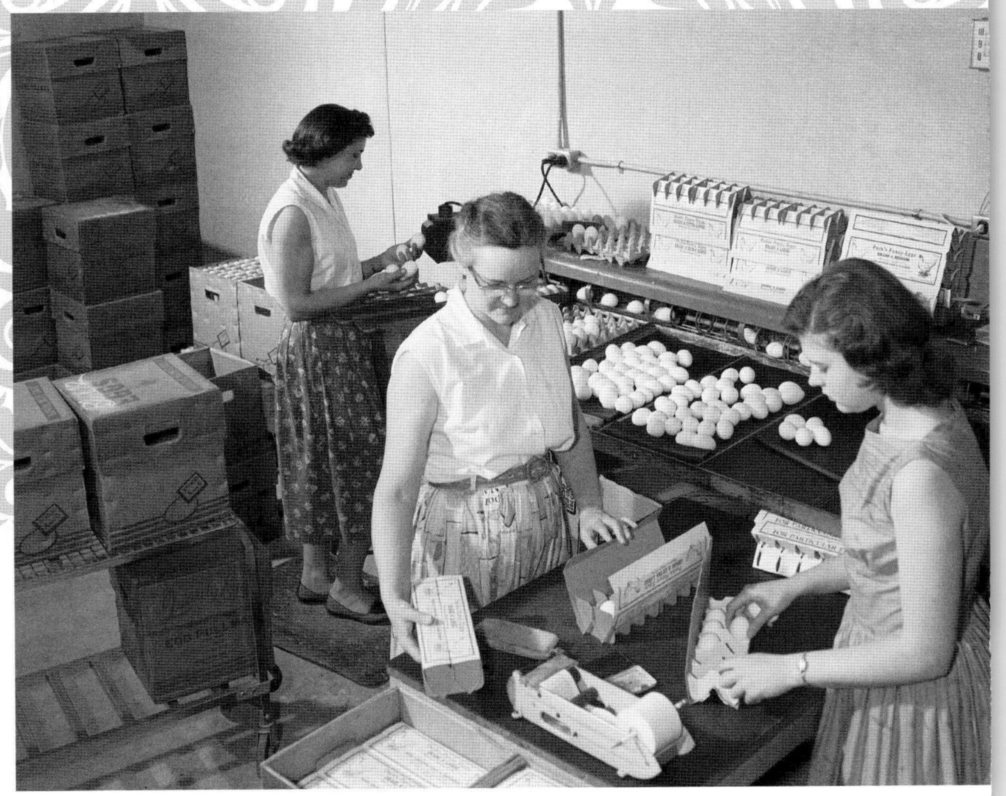

(Top) Farmers deliver eggs to a refrigerated railca
at Denver for cooperative shipping by the Miami
County Egg Association. (Miami County, 1934)

(Bottom left) A worker candles eggs to check for fertilization at the marketing division of St. Louis Egg. (Missouri, 1946)

(Bottom right) Three women candle, cradle, and
pack eggs at the Park Leghorn Farm in Warrensbu
Illinois. (Macon County, IL; 1959)

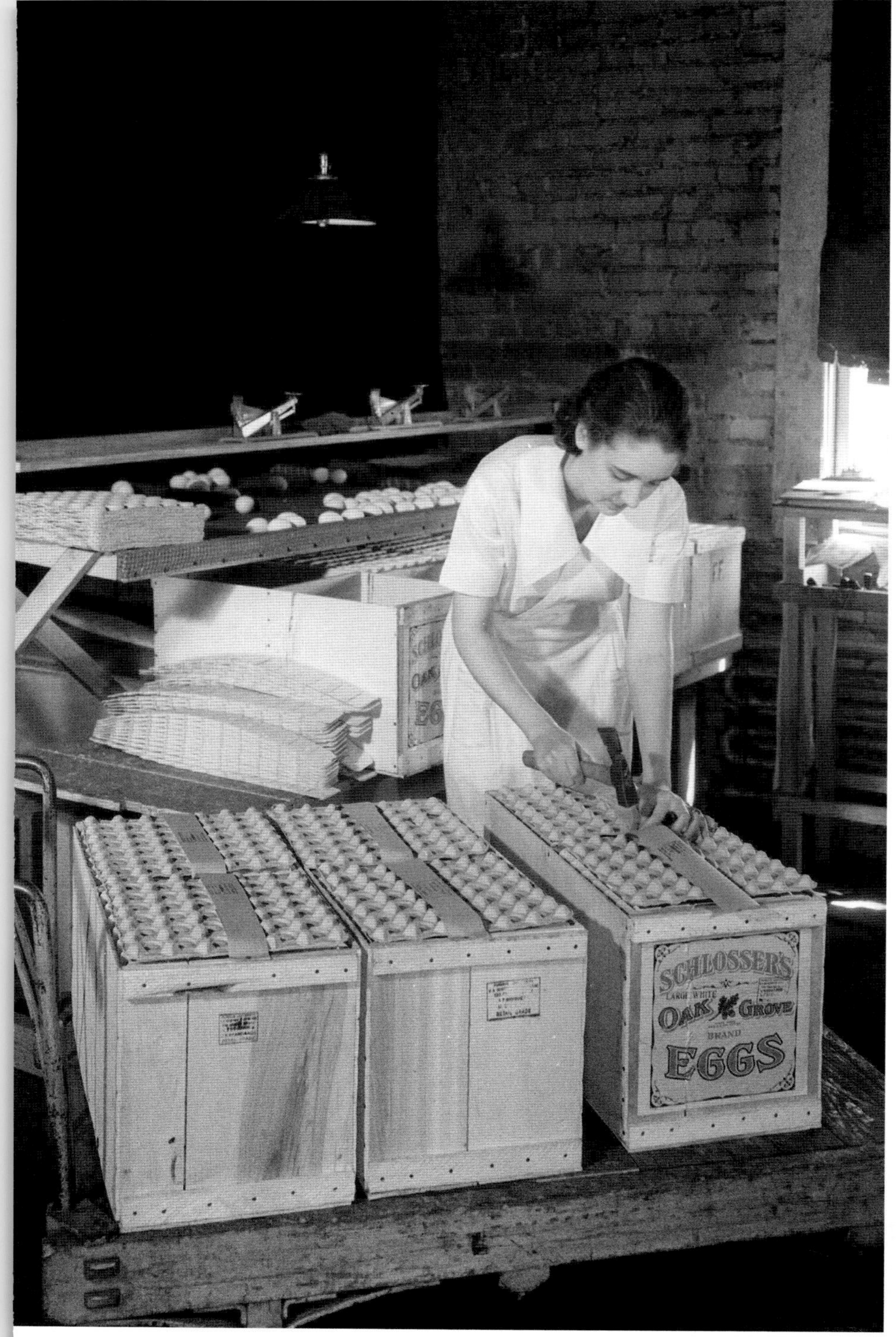

A woman places a government seal on a case of eggs at the Schlosser Brothers packing plant at Frankfort. (Clinton County, 1934)

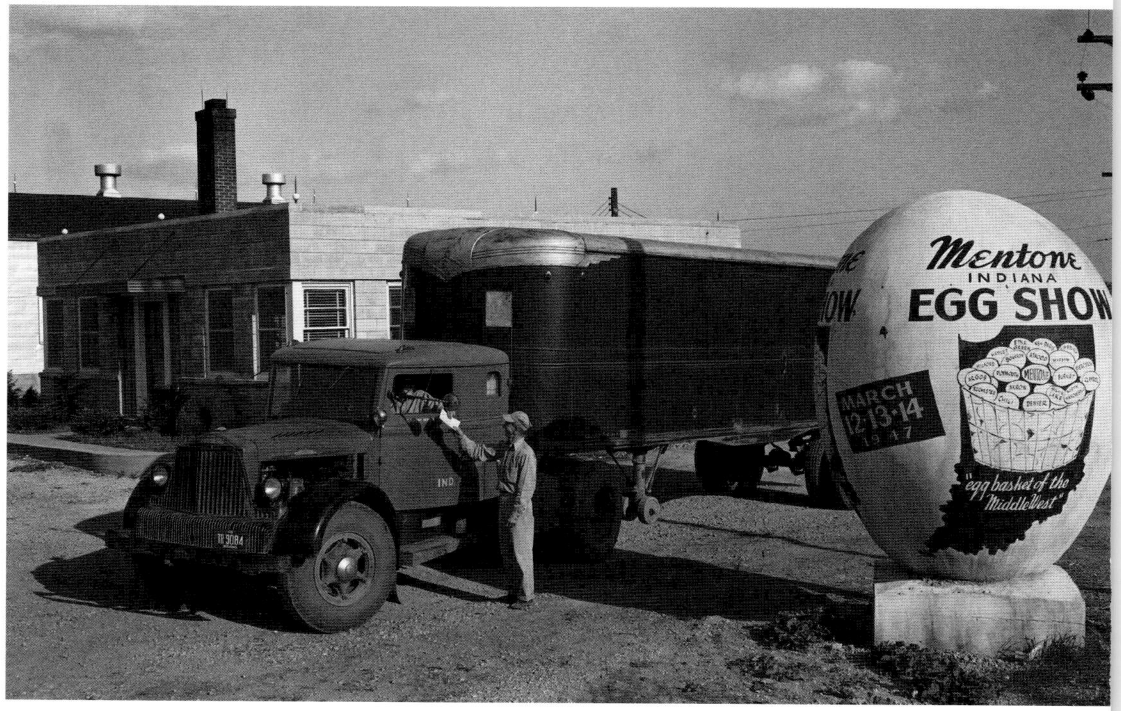

A truck driver prepares to deliver eggs produced in the Mentone area to the New York City market. (Kosciusko County, 1948)

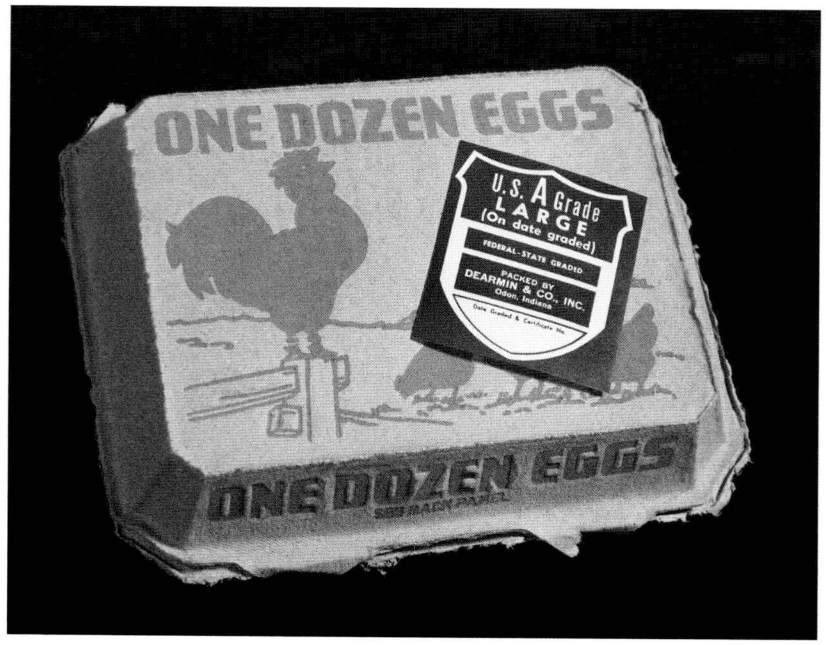

An egg carton with a U.S. Grade A label (Tippecanoe County, 1949)

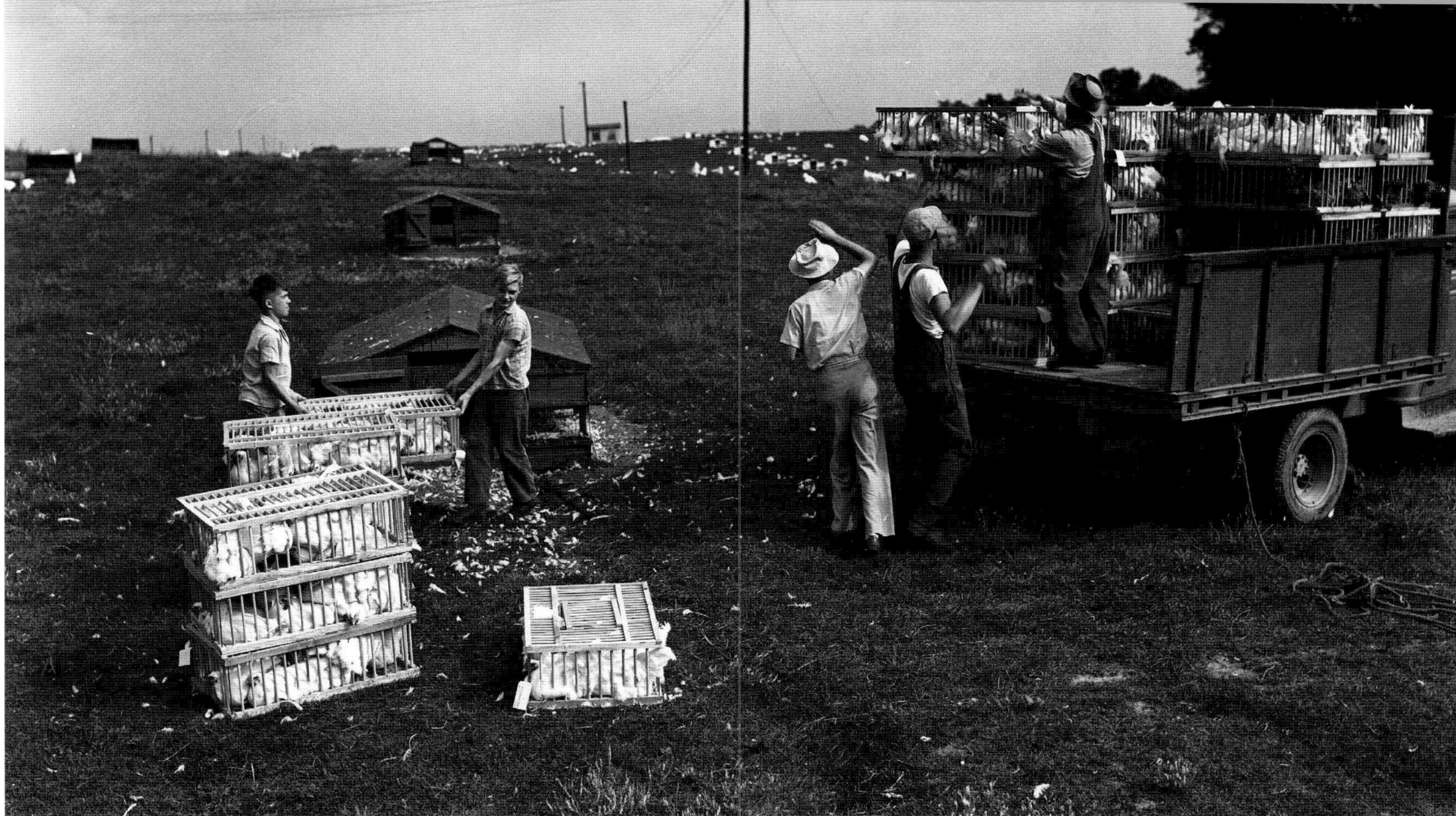

Loading a truck with coops of White Plymouth Rock chickens brought in from the range on the Paul Riley Poultry Farm in West Lafayette (Tippecanoe County, 1945)

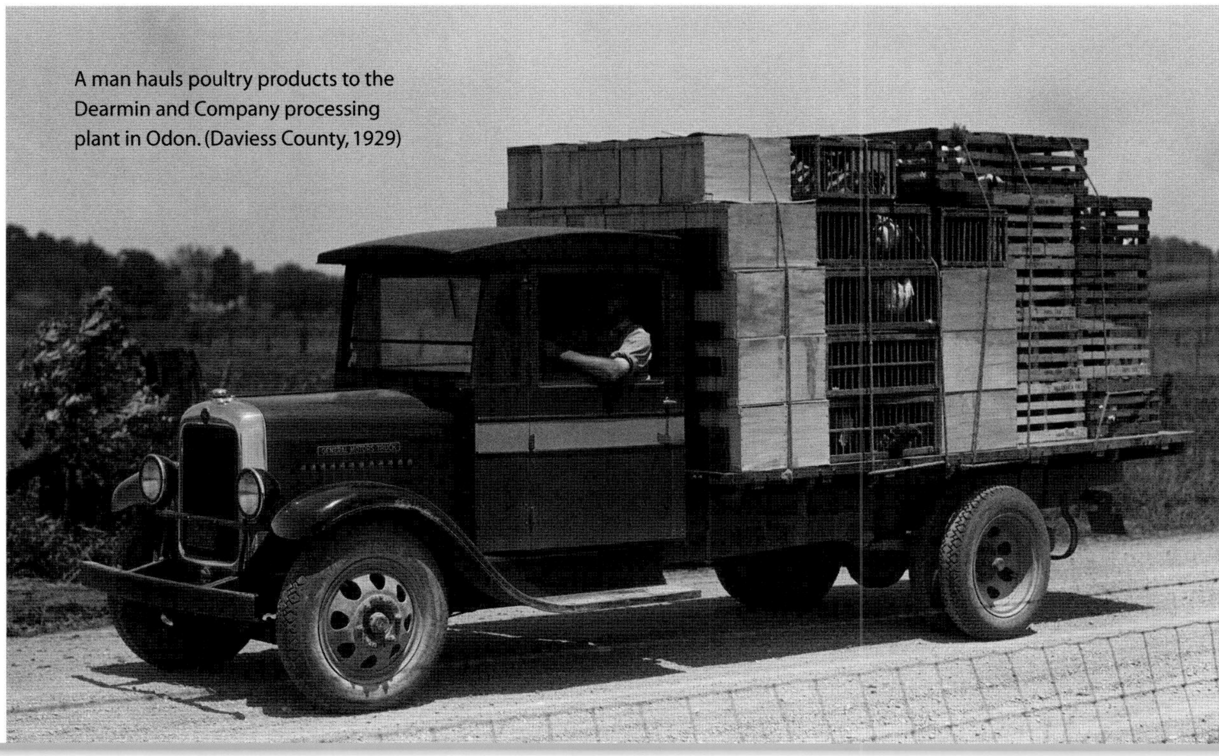

A man hauls poultry products to the Dearmin and Company processing plant in Odon. (Daviess County, 1929)

Wool and Ewe

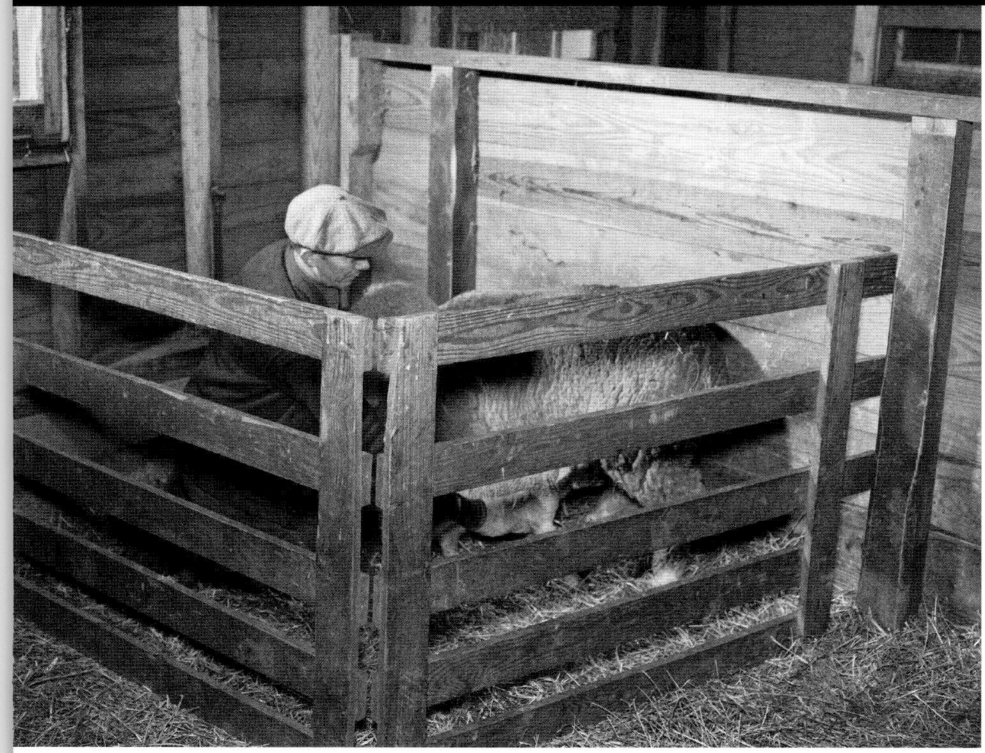

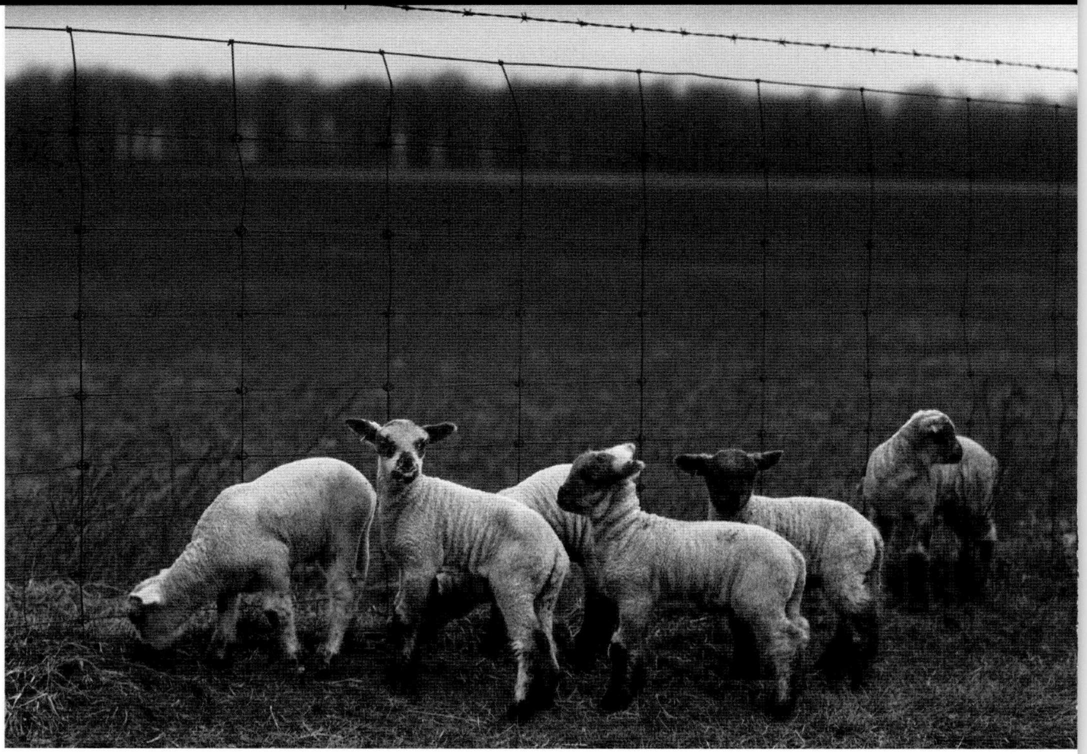

A lambing pen inside a sheep barn on the Purdue University Livestock Experimental Farm (Tippecanoe County, 1940)

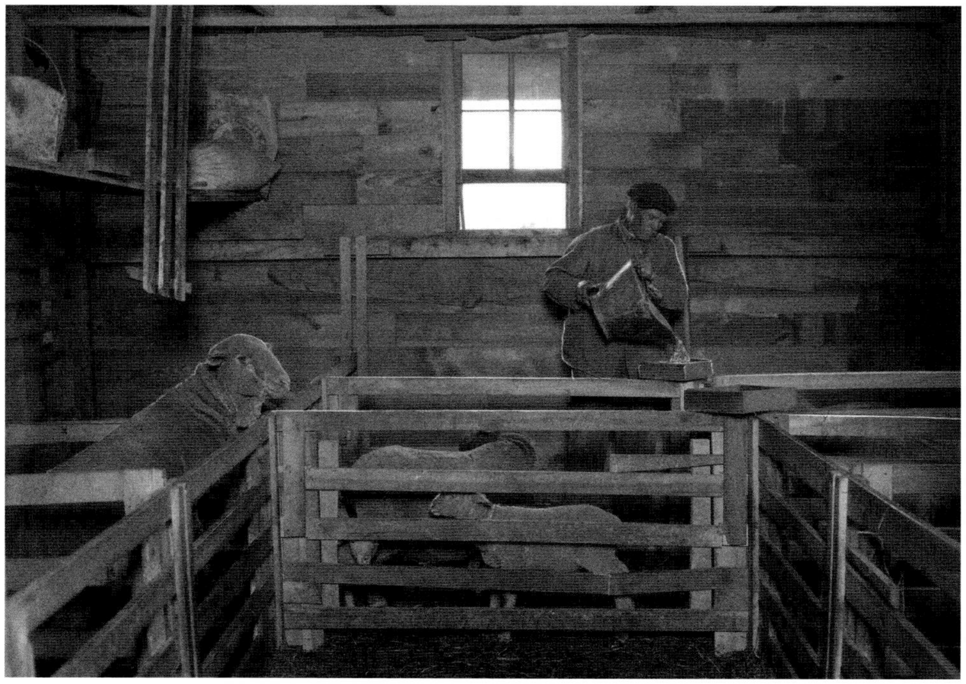

(Above) Lambs at the right age for tail docking (Tippecanoe County, 1934)

(Bottom) A man tends to lambs in pens in the sheep barn of W. W. Miller of Columbia City. Allen noted, "This pen has a double wall with building paper between, and when it was six below outside last winter, the temperature on the inside, with nine ewes in the pens, was 27 above. The size of the pen is 14 by 16 feet and furnishes places for 12 ewes and their lambs." (Whitley County, 1924)

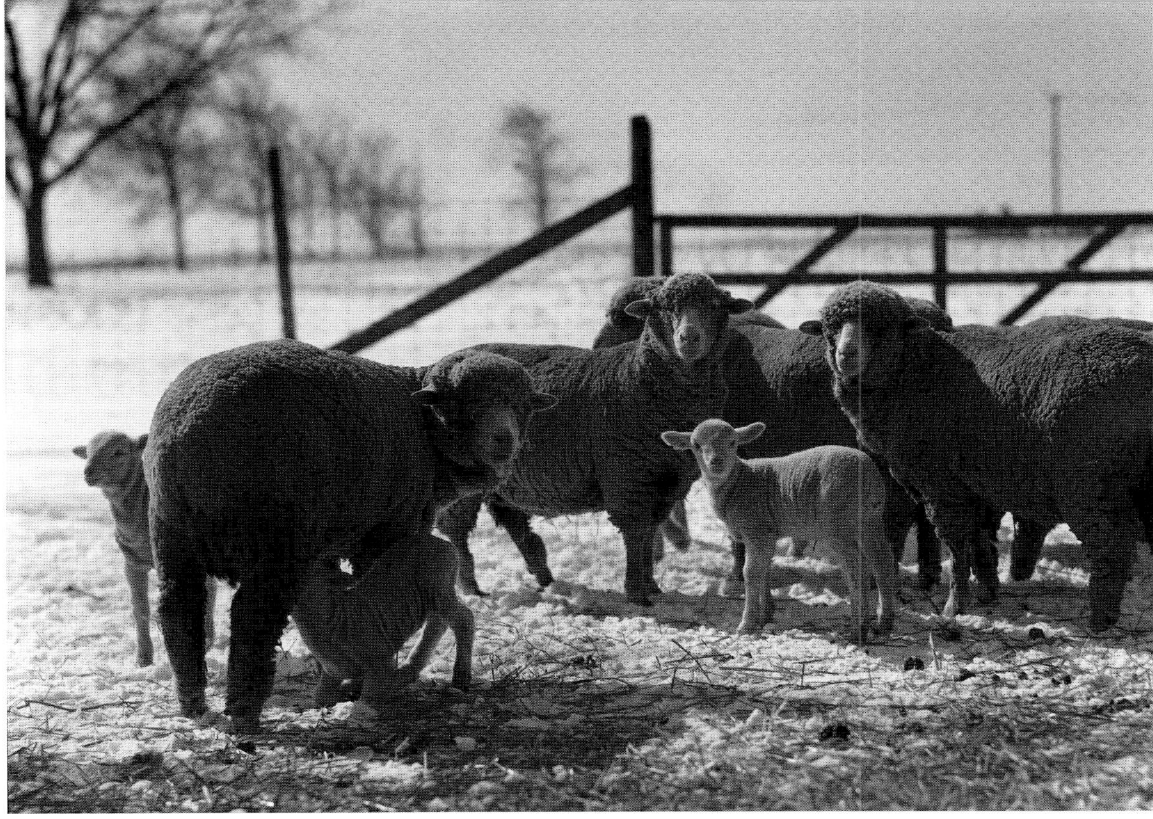

A flock of Rambouillet sheep that was started before 1900 at a Purdue University farm (Tippecanoe County, 1950)

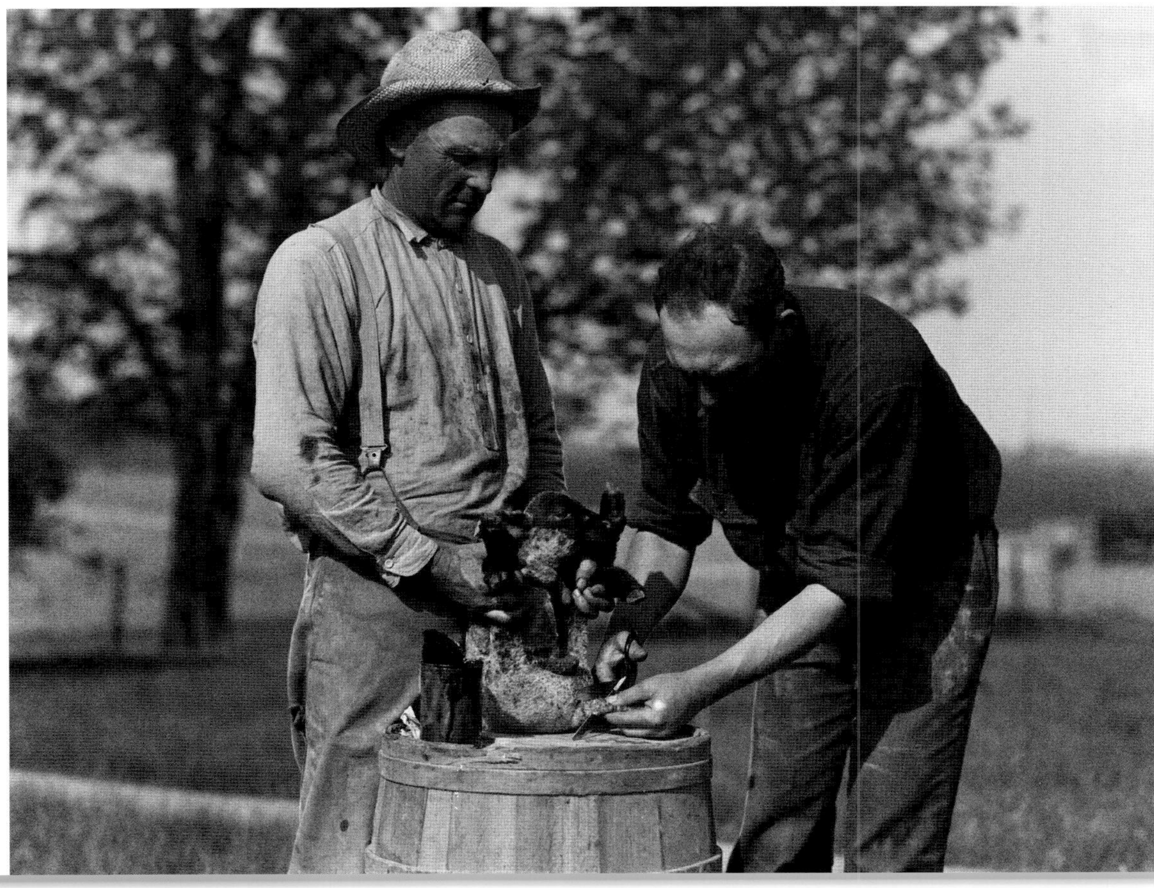

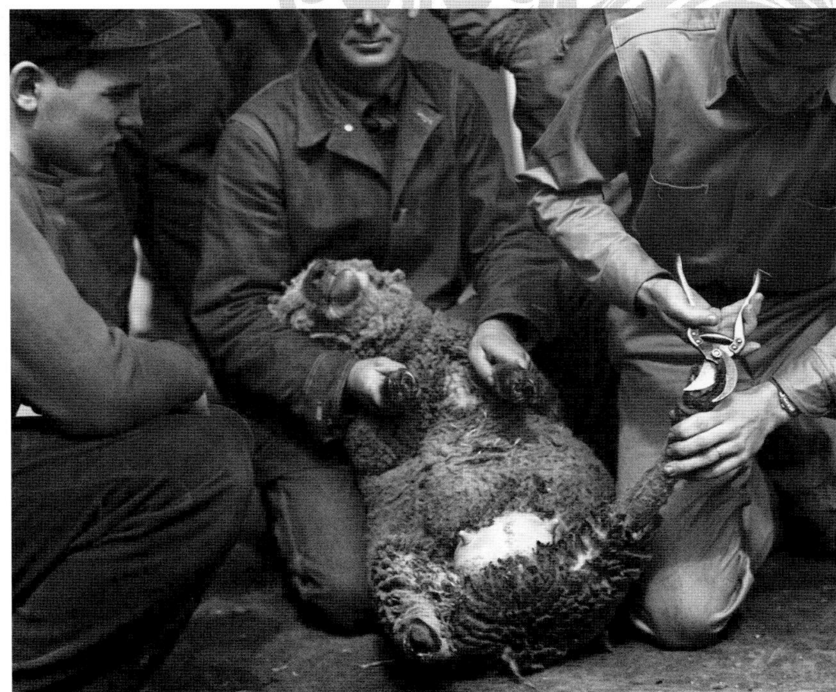

Some young men trim the feet of a Southdown ewe on the farm of Harry McMillen in Romney. (Tippecanoe County, 1944)

Jefferson County Agricultural Extension Agent R. N. Thomas (right) docks the tail of a lamb held by Madison-area farmer Bert Scott. (Jefferson County, 1925)

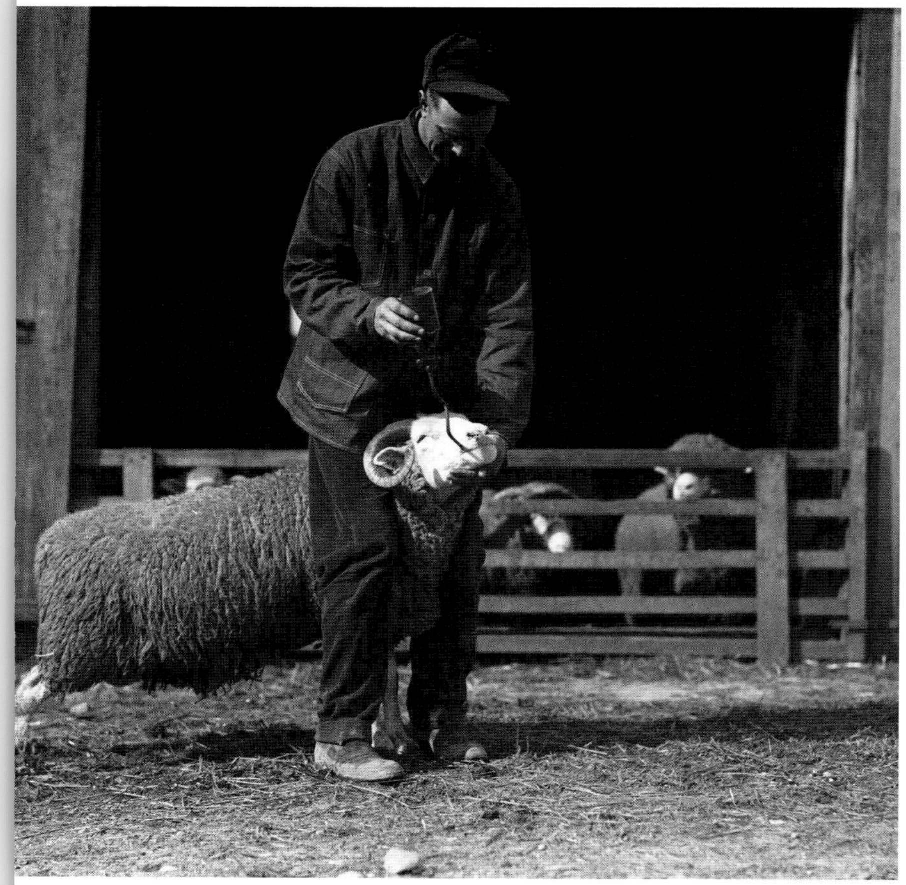

Lindley W. Baker from Cambridge City administers oral medication for stomach worms to a ewe. (Wayne County, 1927)

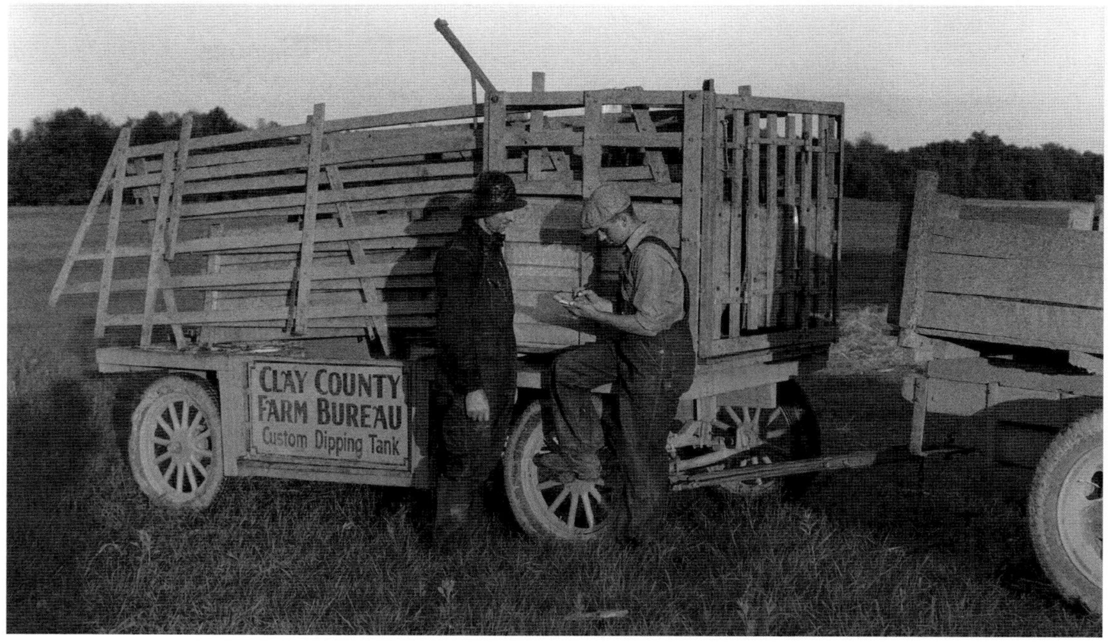

(Above and below) The Clay County Farm Bureau managed this mobile sheep dipping operation, charging Charles Romas from Brazil 10 cents per head to treat his flock for external parasites. (Clay County, 1935)

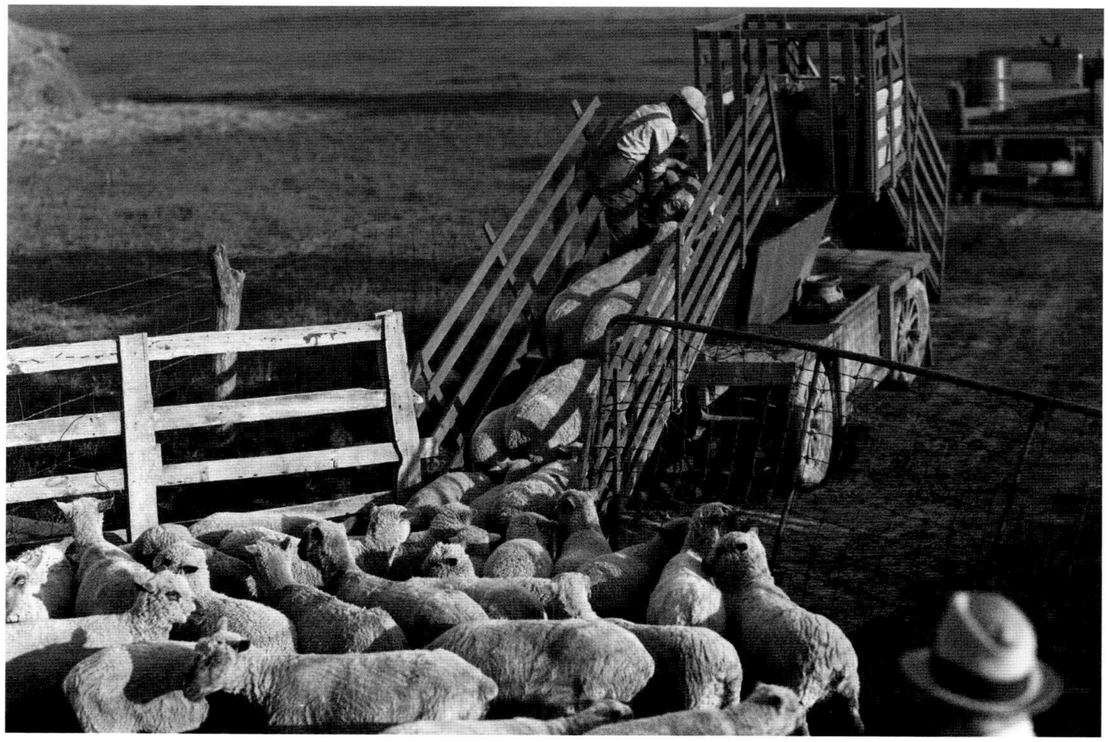

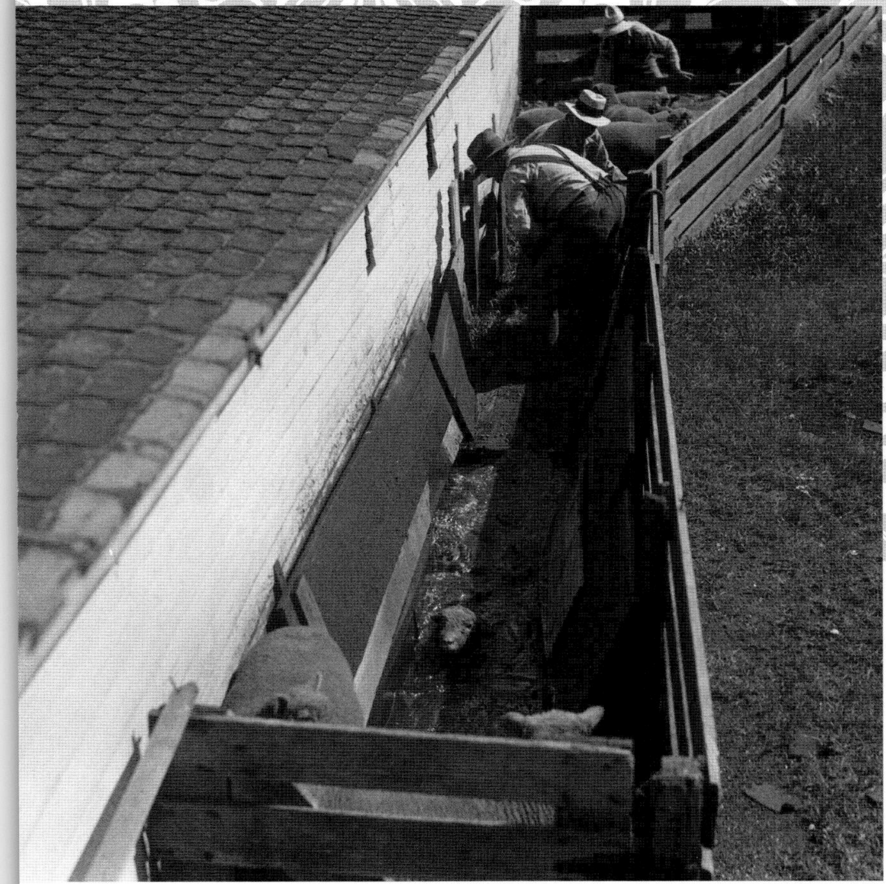

Sheep swim through a dipping tank on the Purdue University campus farm in West Lafayette. (Tippecanoe County, 1931)

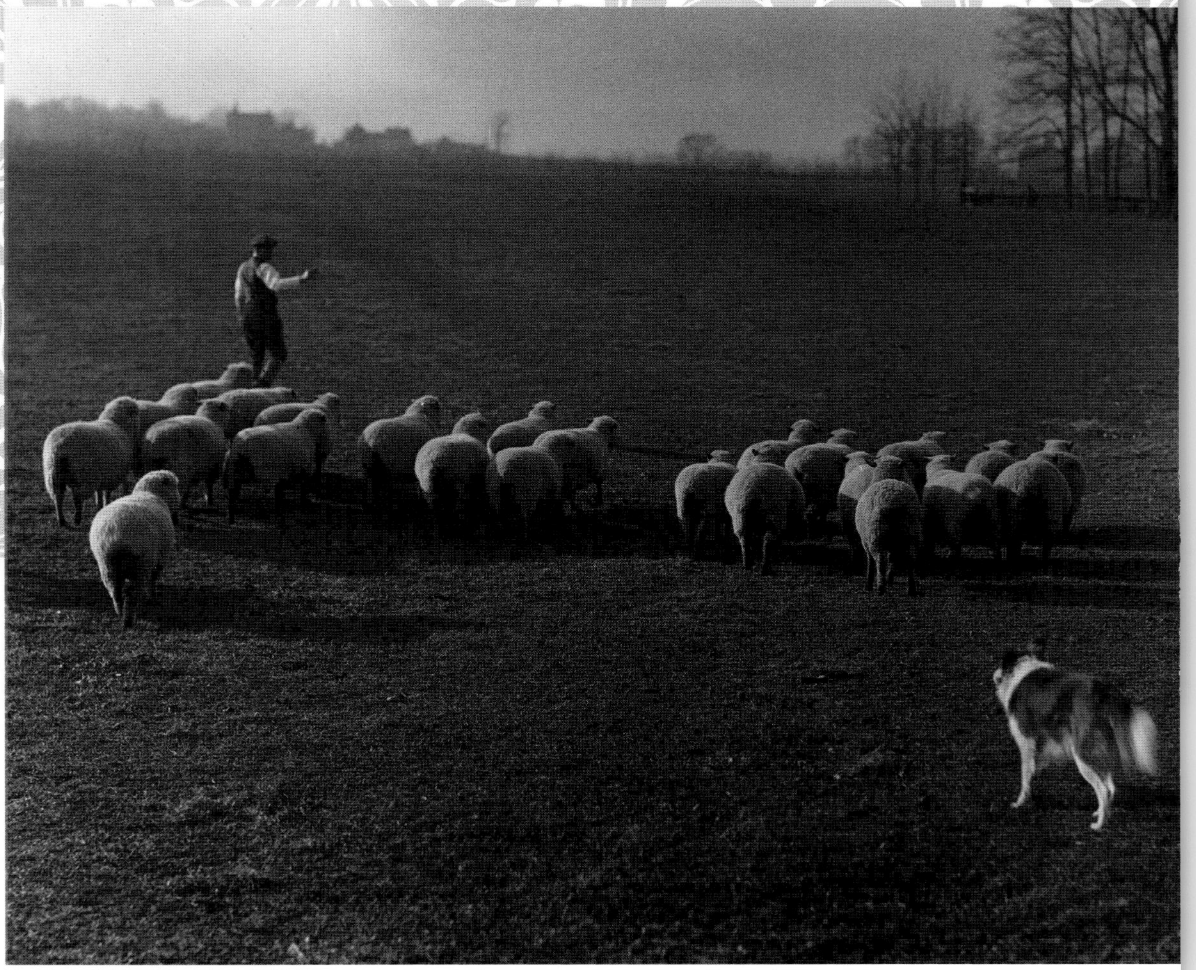

A collie works a Shropshire flock on the farm of Jess Andrew. (Tippecanoe County, ca. 1910)

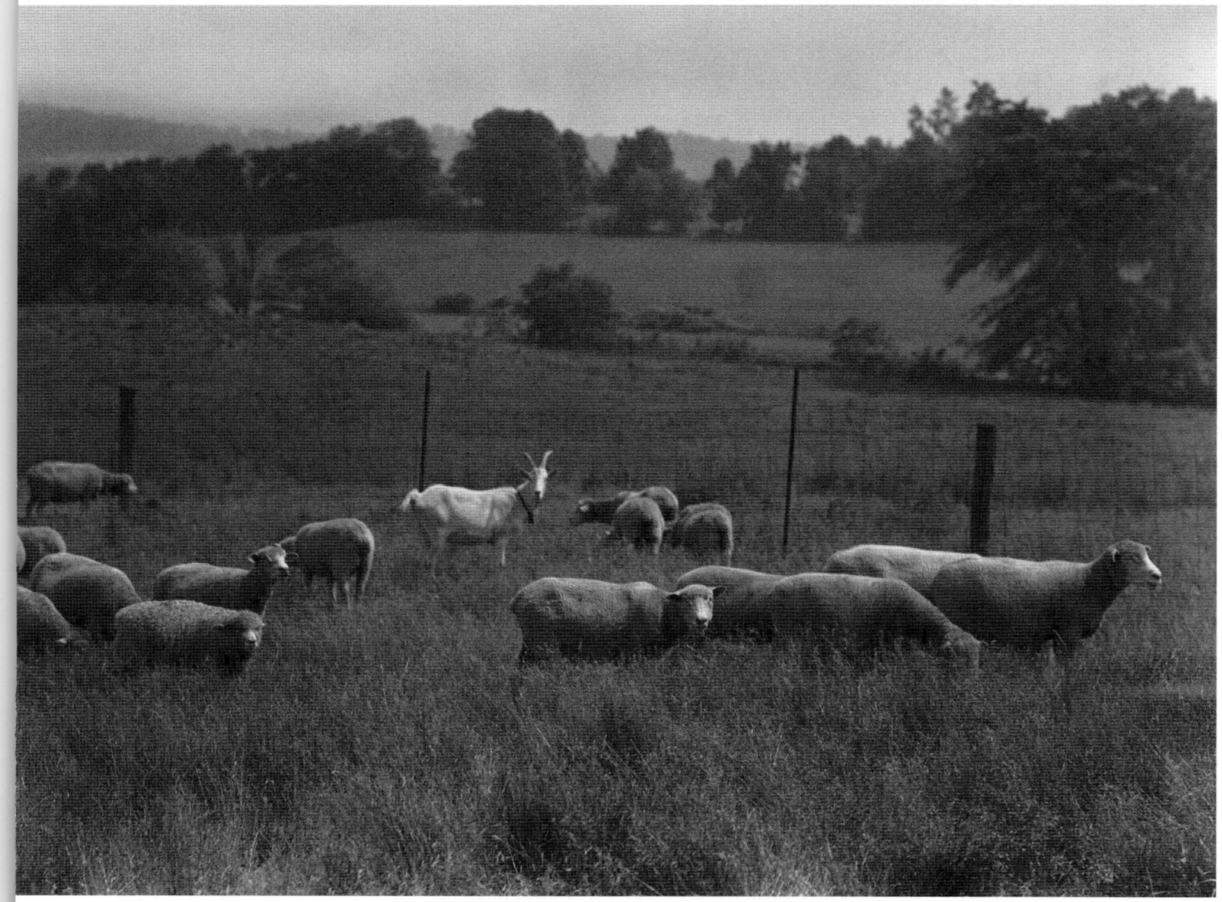

A goat (back center) stands guard in a flock of sheep. (Tippecanoe County, 1953)

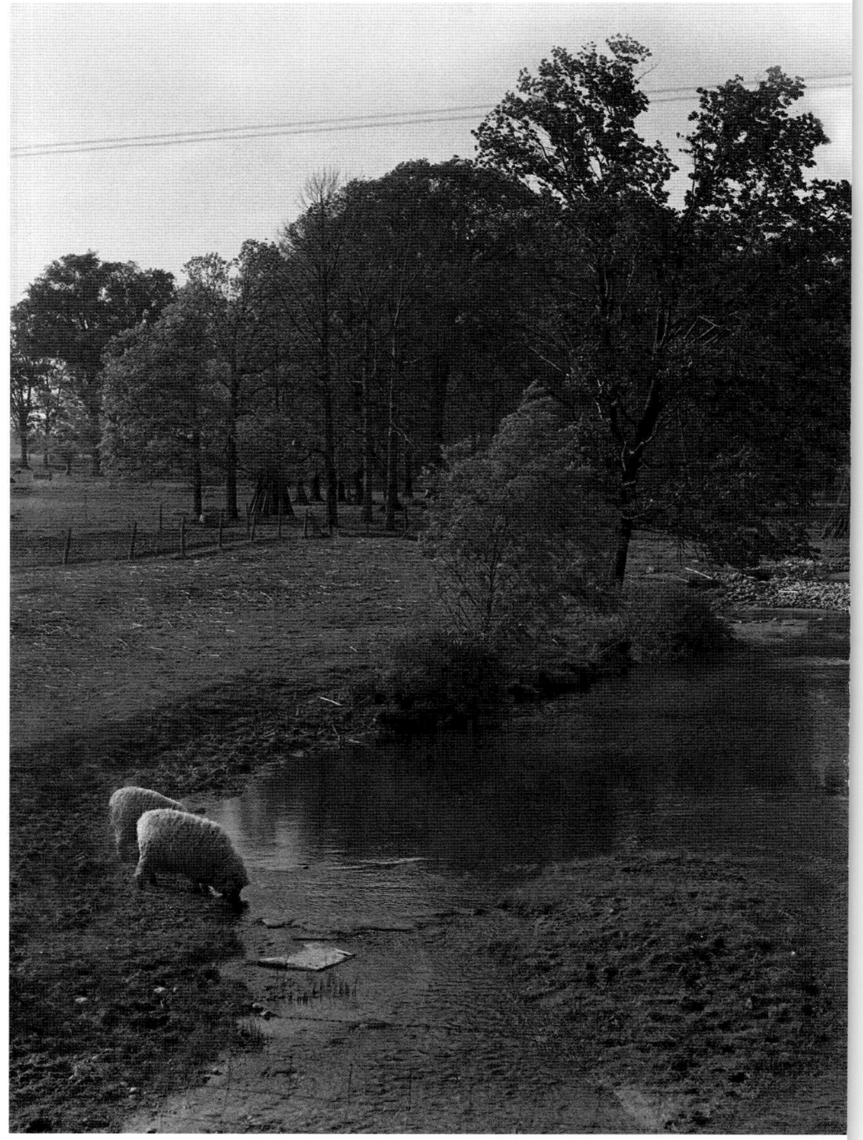

Sheep drink from a stream. (Indiana, 1941)

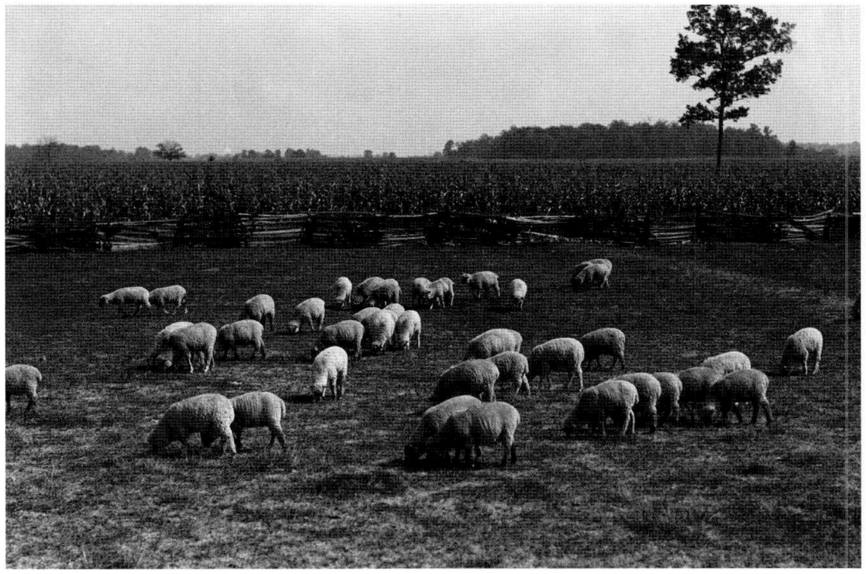

Western lambs graze on bluegrass pasture on the Noah Fouts farm in Deer Creek. (Carroll County, 1914)

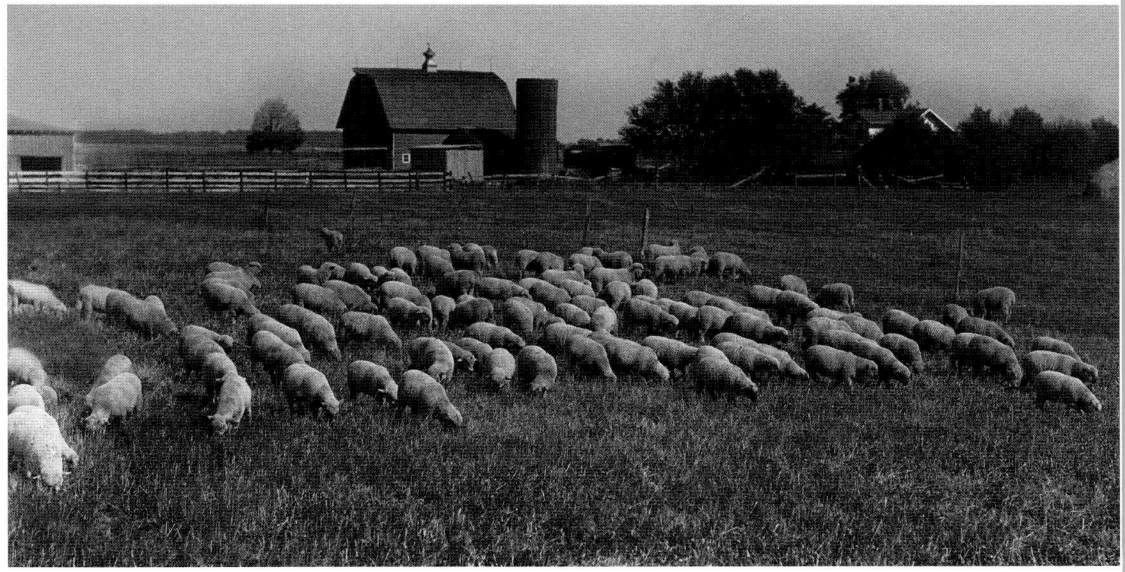

On the Lynn Van Natta farm in Battle Ground, a flock of Western lambs feeds on a clover pasture. (Tippecanoe County, 1928)

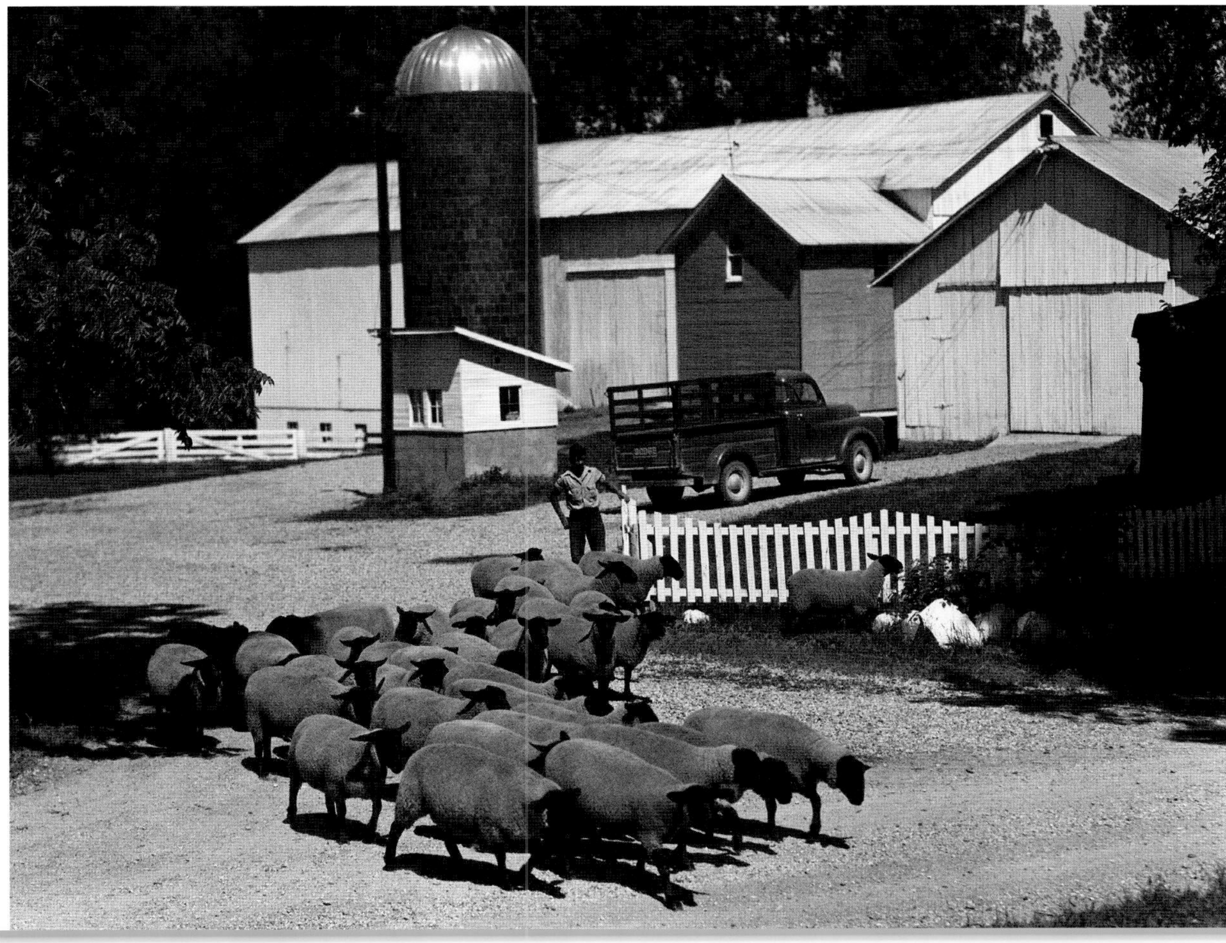

A man moves some Suffolk sheep to pasture on the farm of Fred Dickes in Huntertown. (Allen County, 1953)

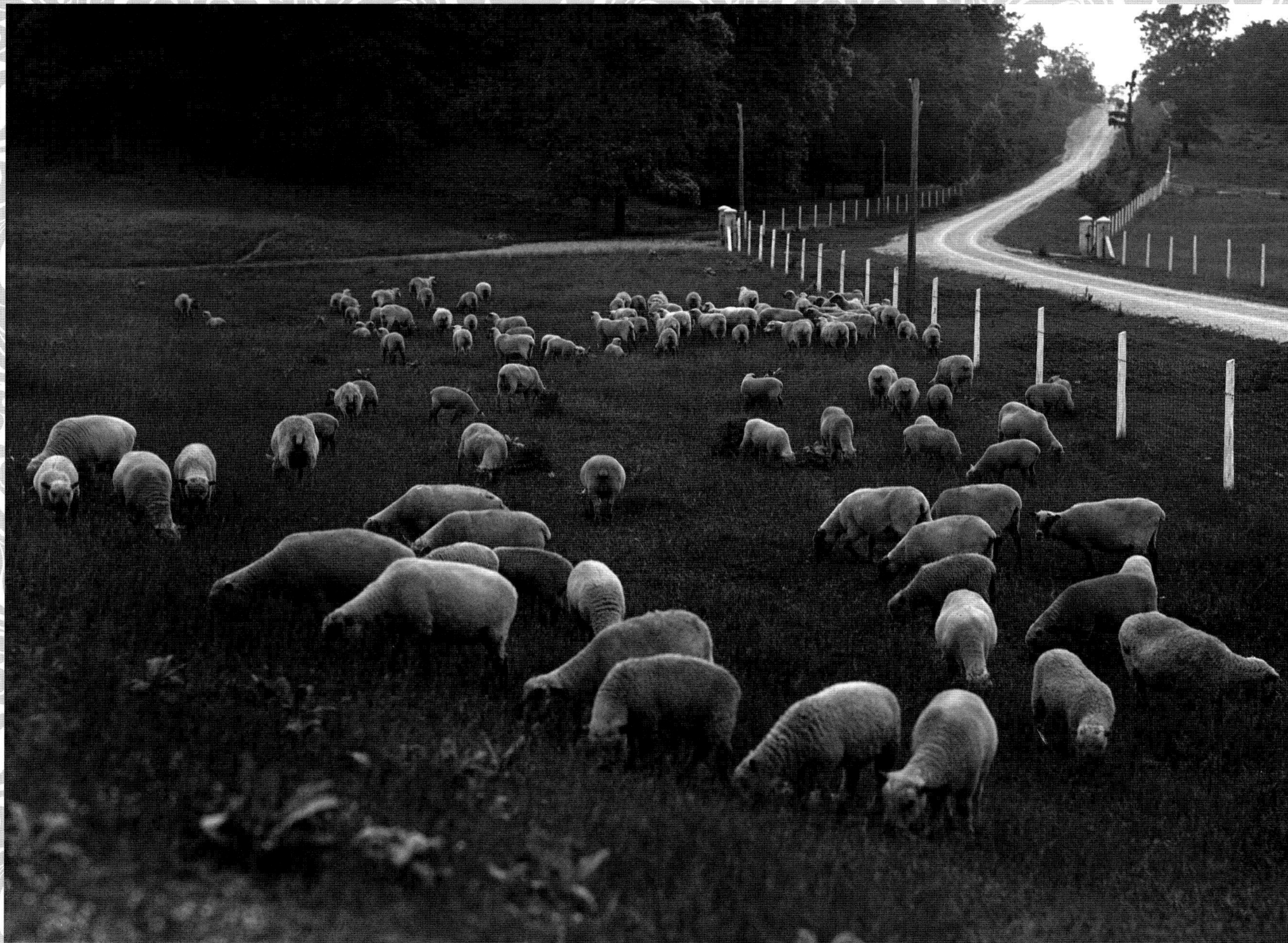

A flock of sheep grazes on the Shadeland Stock Farm with a winding roadway in the background. (Tippecanoe County, 1915)

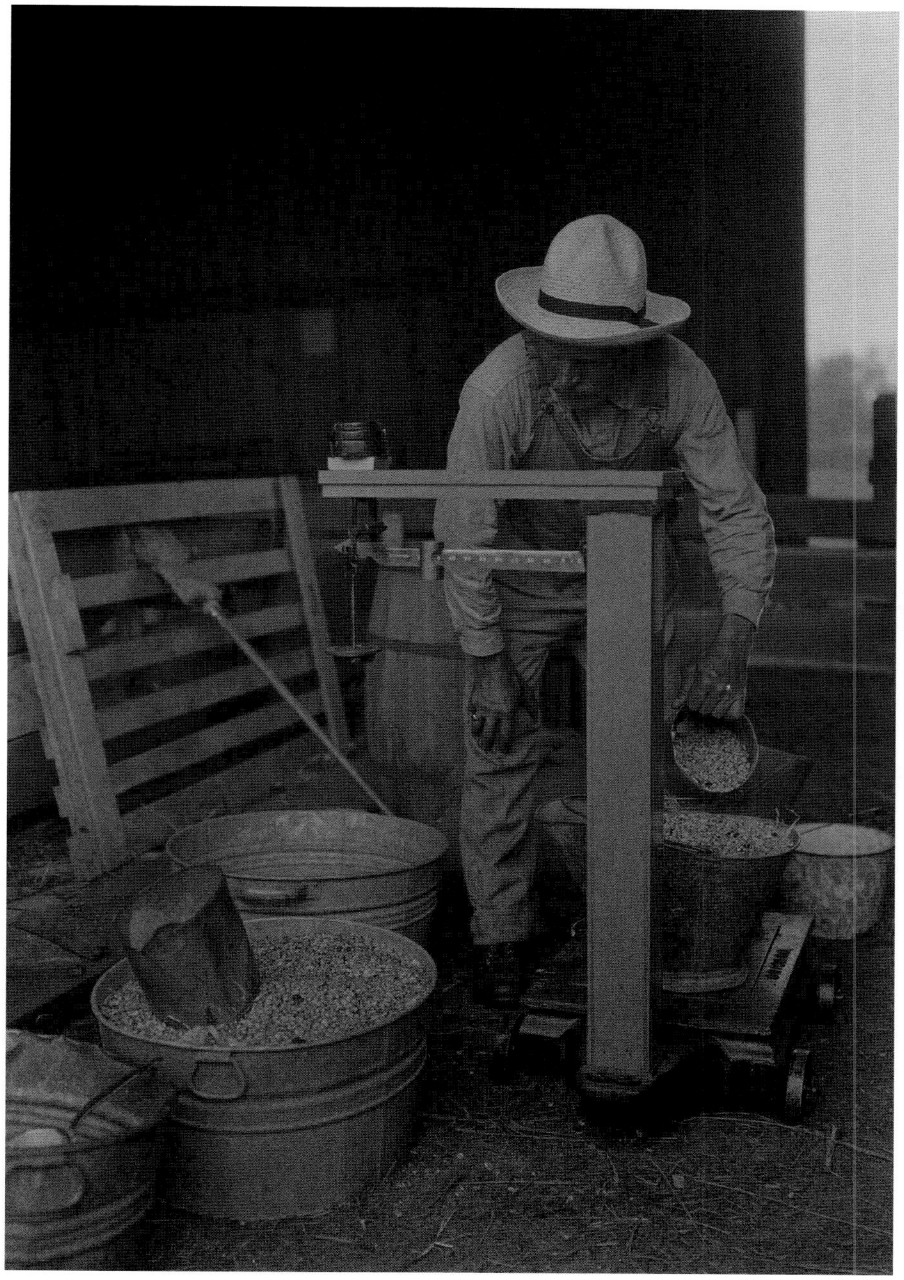

W. W. Miller weighs feed for his sheep. (Whitley County, 1921)

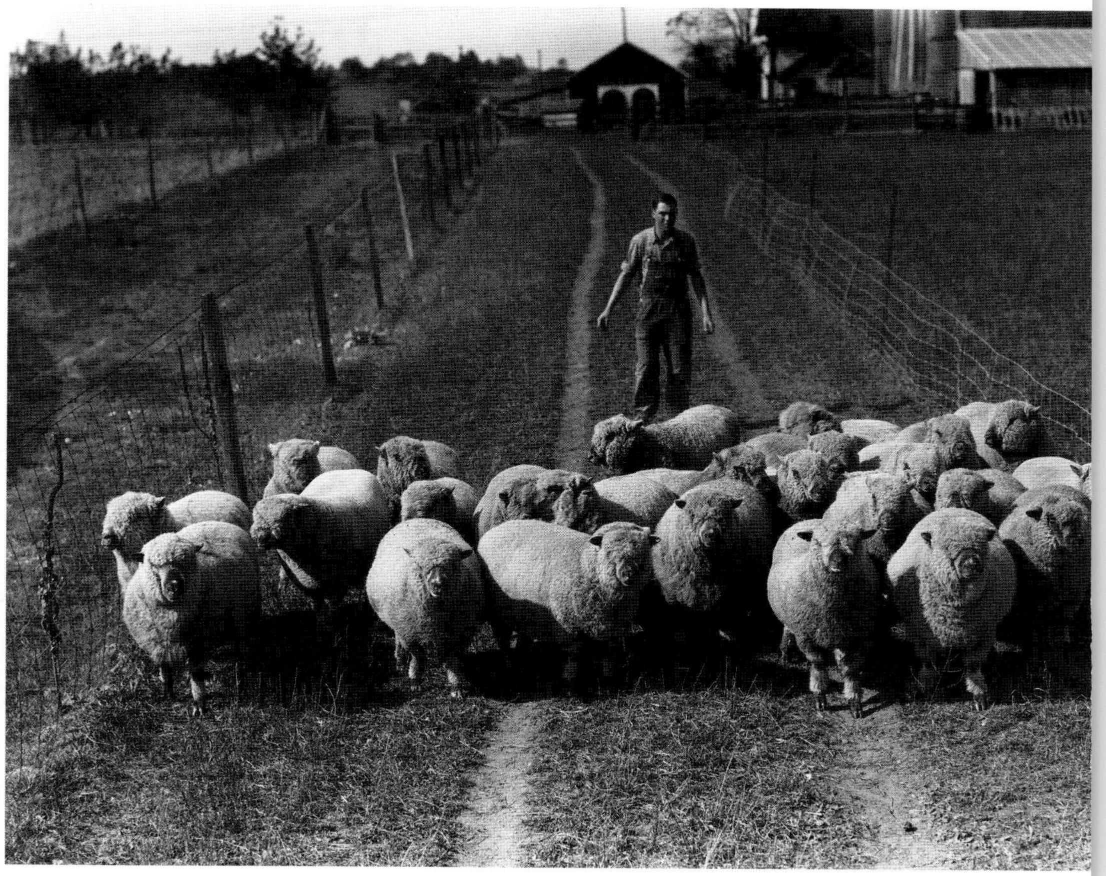

Clair Gilbert herds his flock of Shropshire ewes. (Steuben County, 1929)

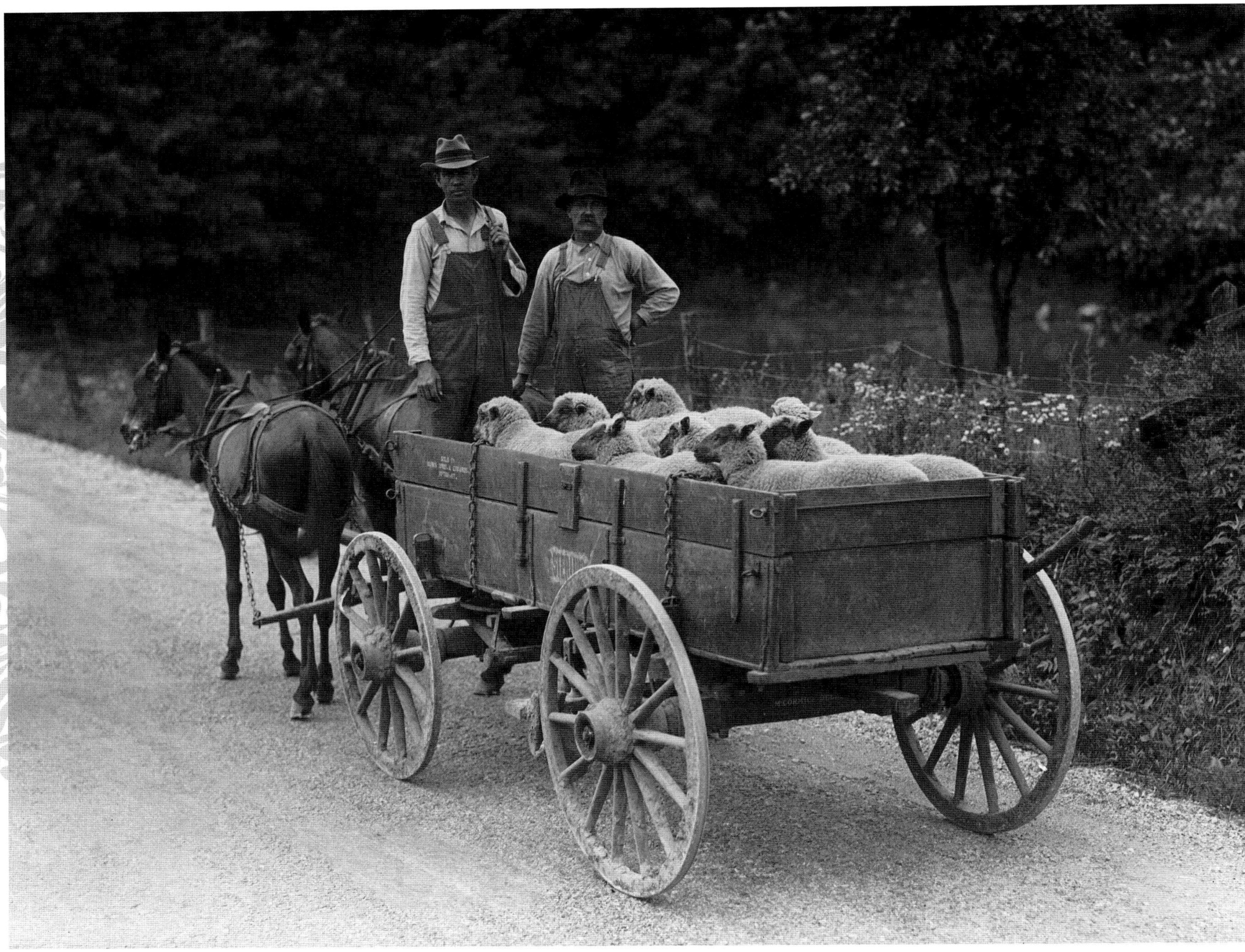

C. M. Hodges and W. G. Sigo use this horse-drawn wagon to transport a load of lambs to market in Bonnieville, Kentucky. (Hart County, KY; 1928)

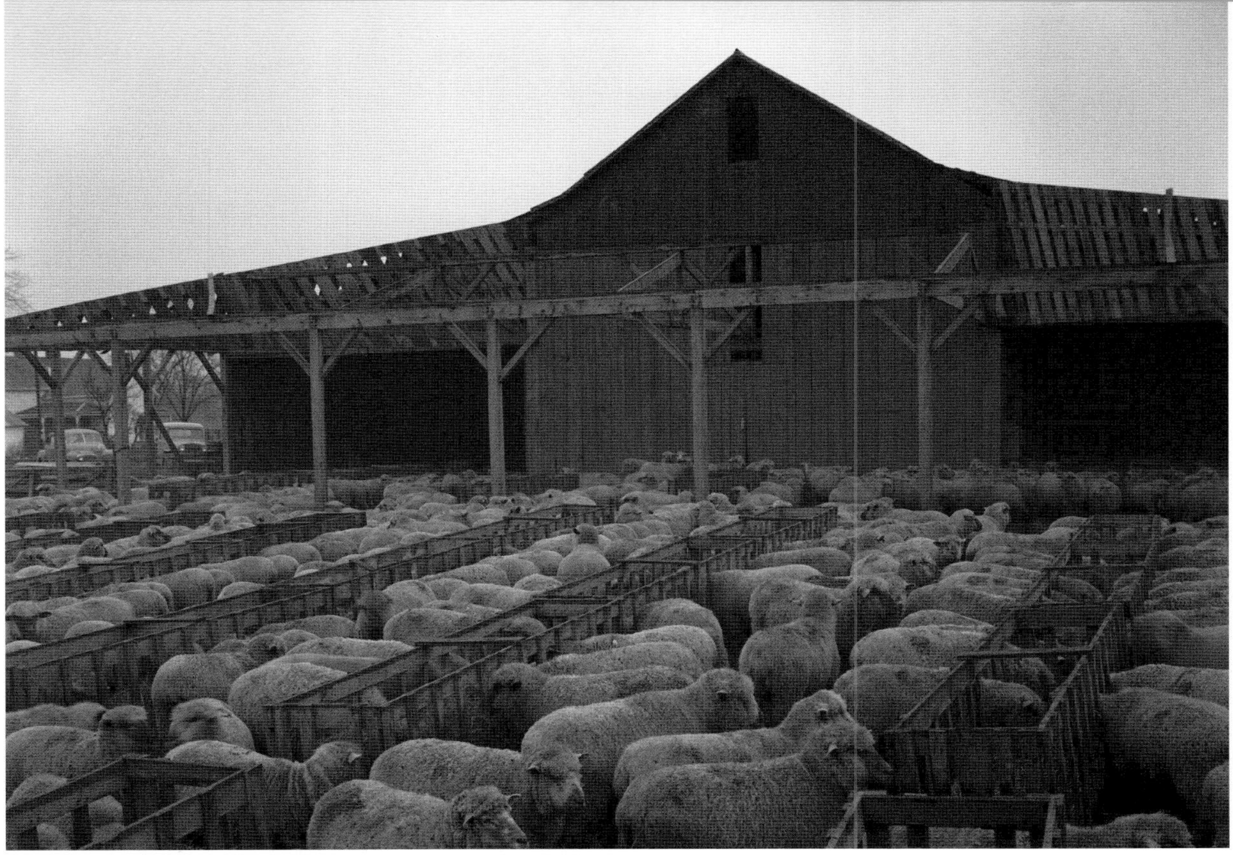

Western ewes feed at troughs at Meadowbrook Farms in Sullivan. (Sullivan County, 1949)

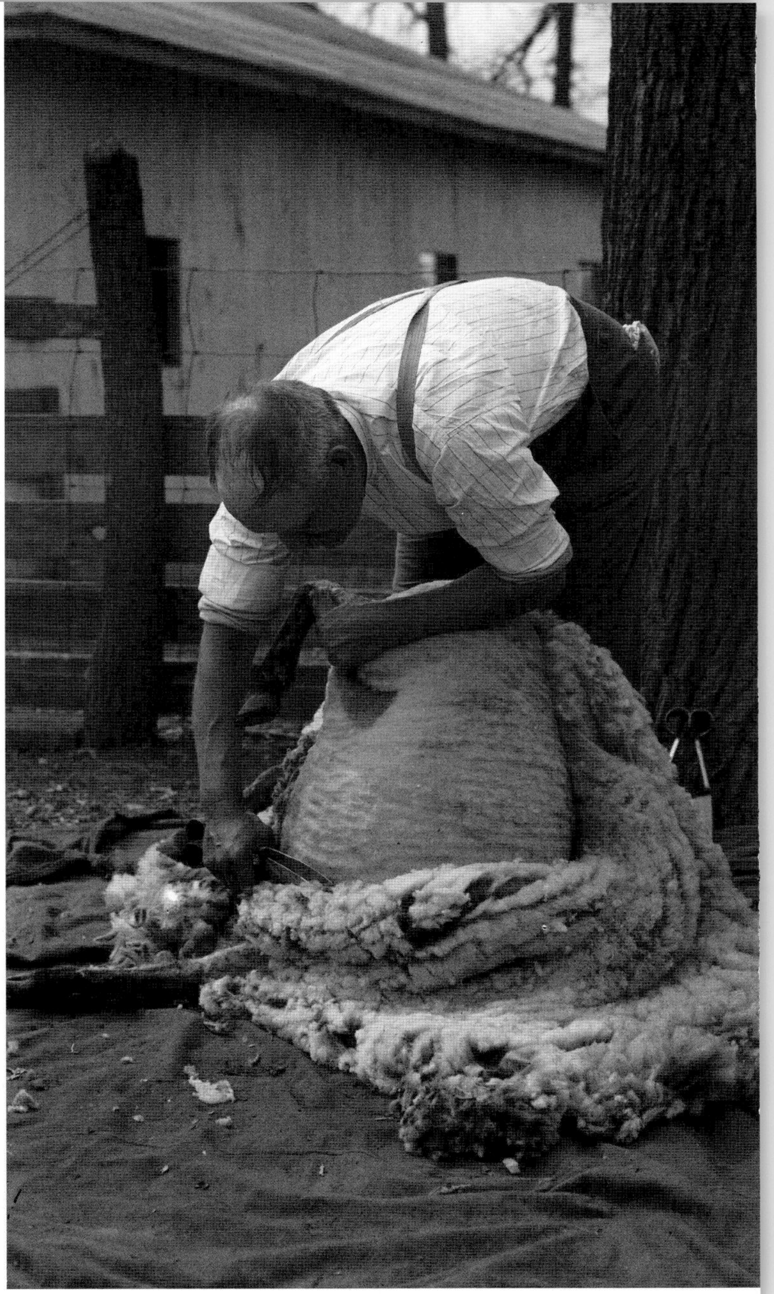

Shearing a sheep with clippers on the Jess Andrew farm in West Point (Tippecanoe County, 1915)

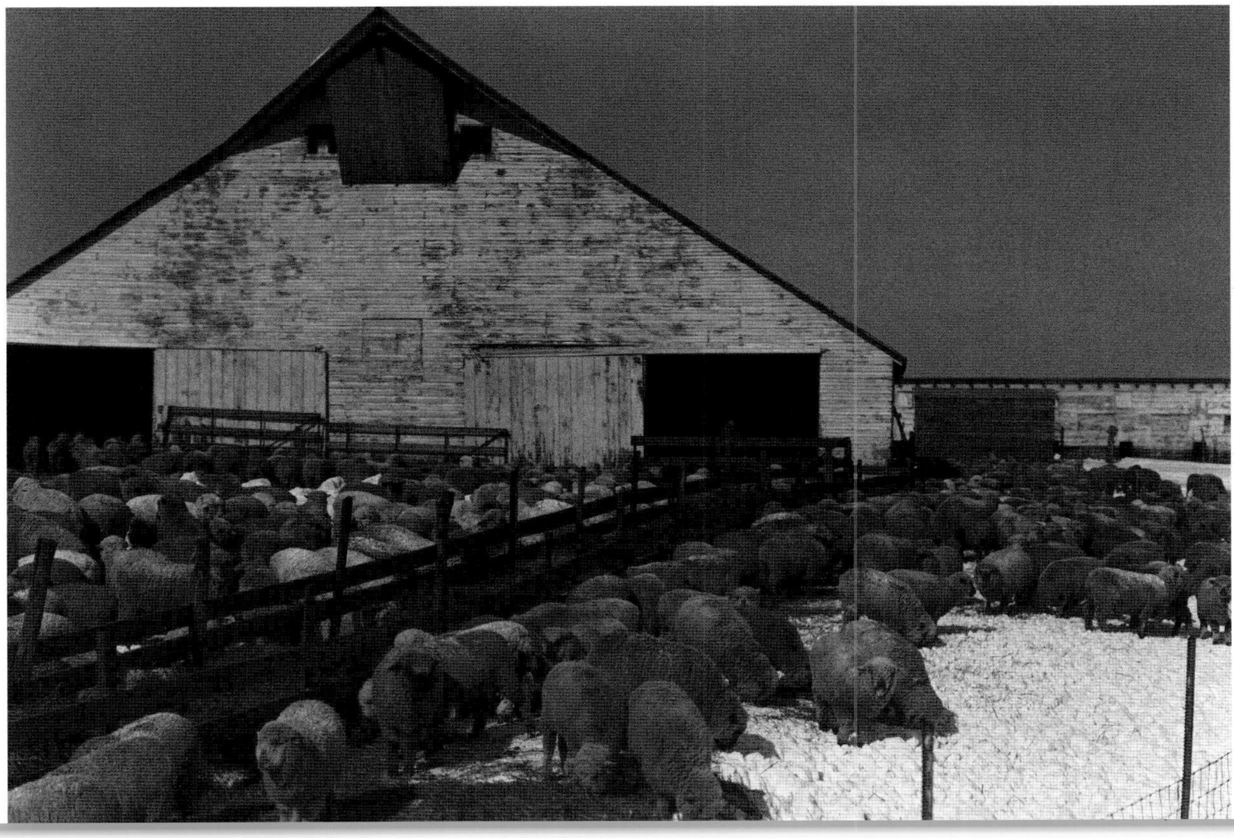

Approximately 625 lambs feed on the farm of Ona Myers in Brookston. (White County, 1944)

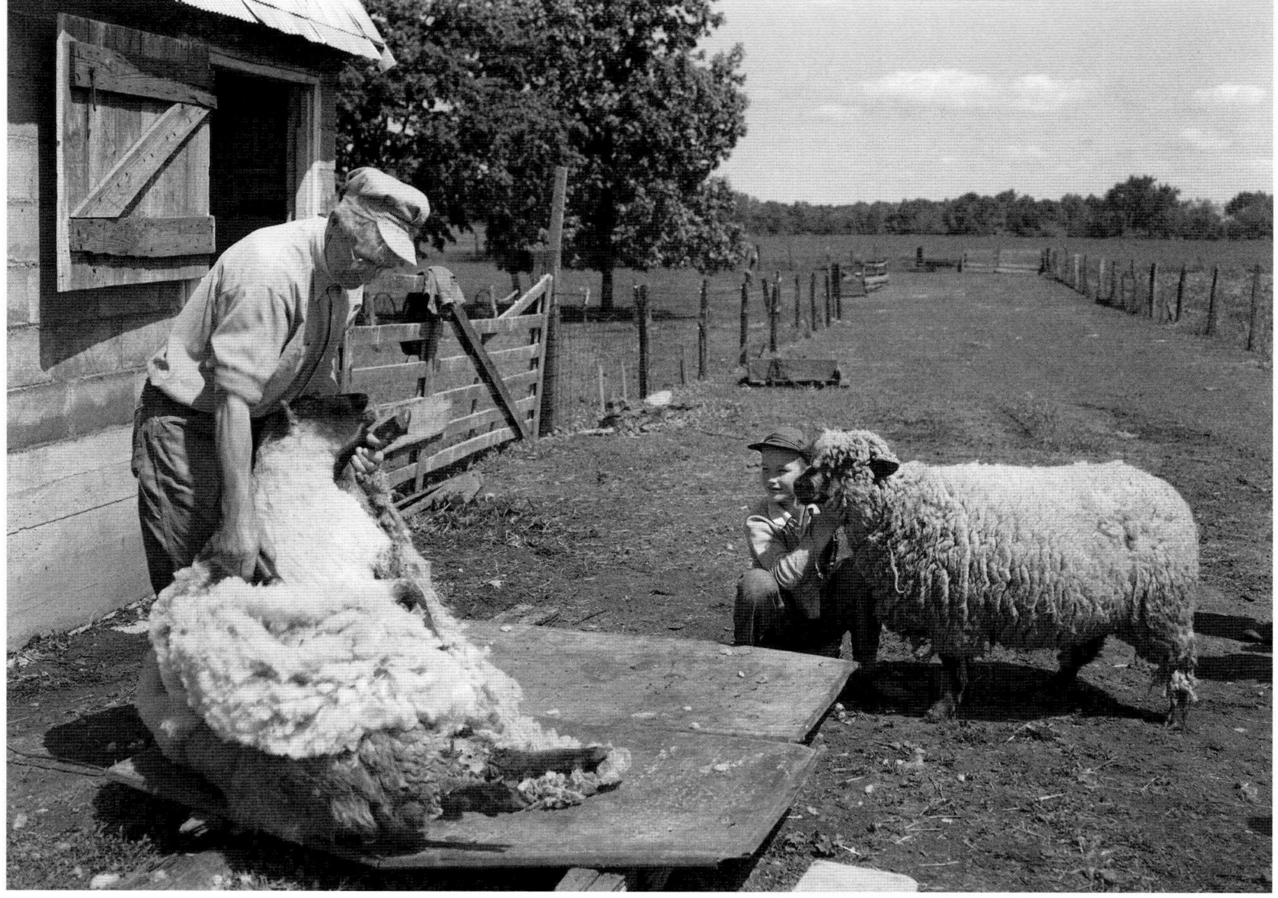

Earl Lumley (left) shears a Shropshire ewe with electric clippers while his son George looks on at the Samuel Homer Honn farm in Lafayette. (Tippecanoe County, 1953)

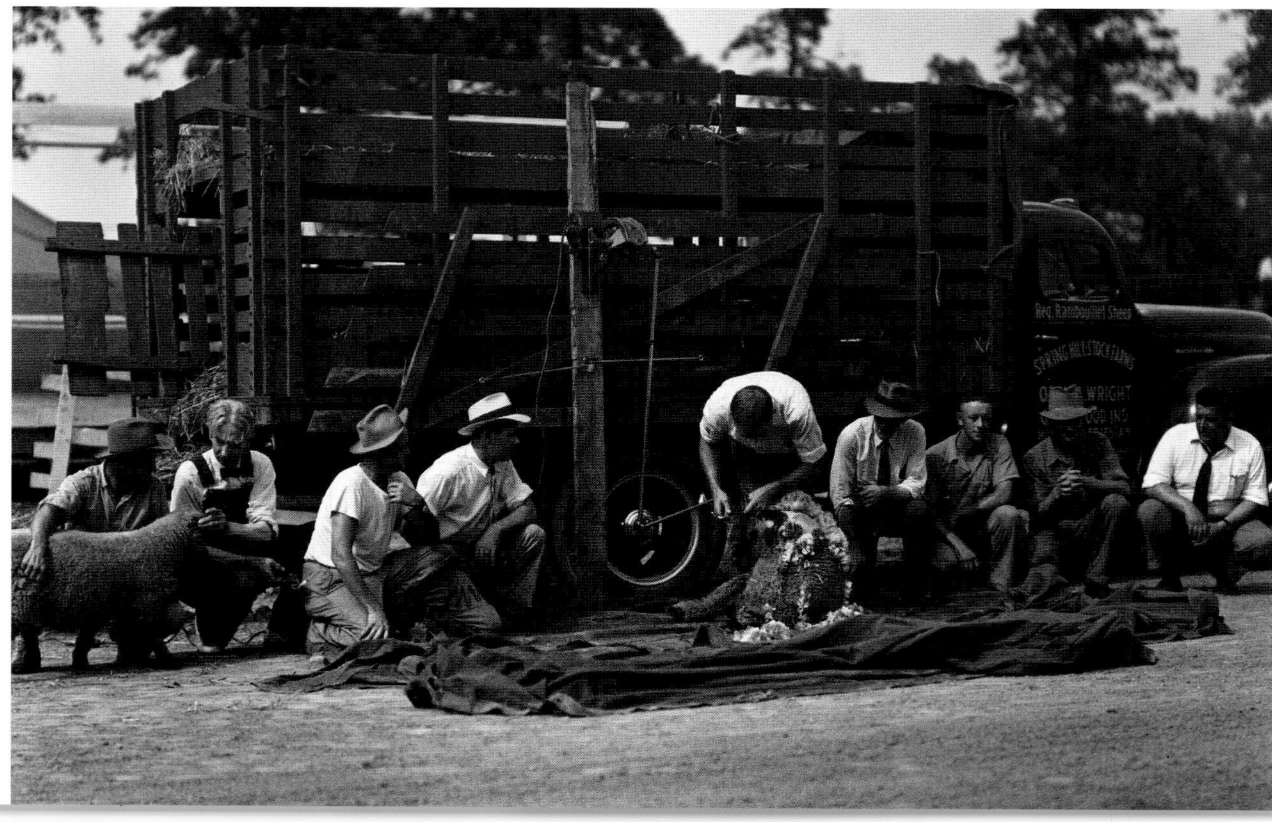

These men watch a sheep shearing demonstration at the Delaware County Fair in Muncie. (Delaware County, 1945)

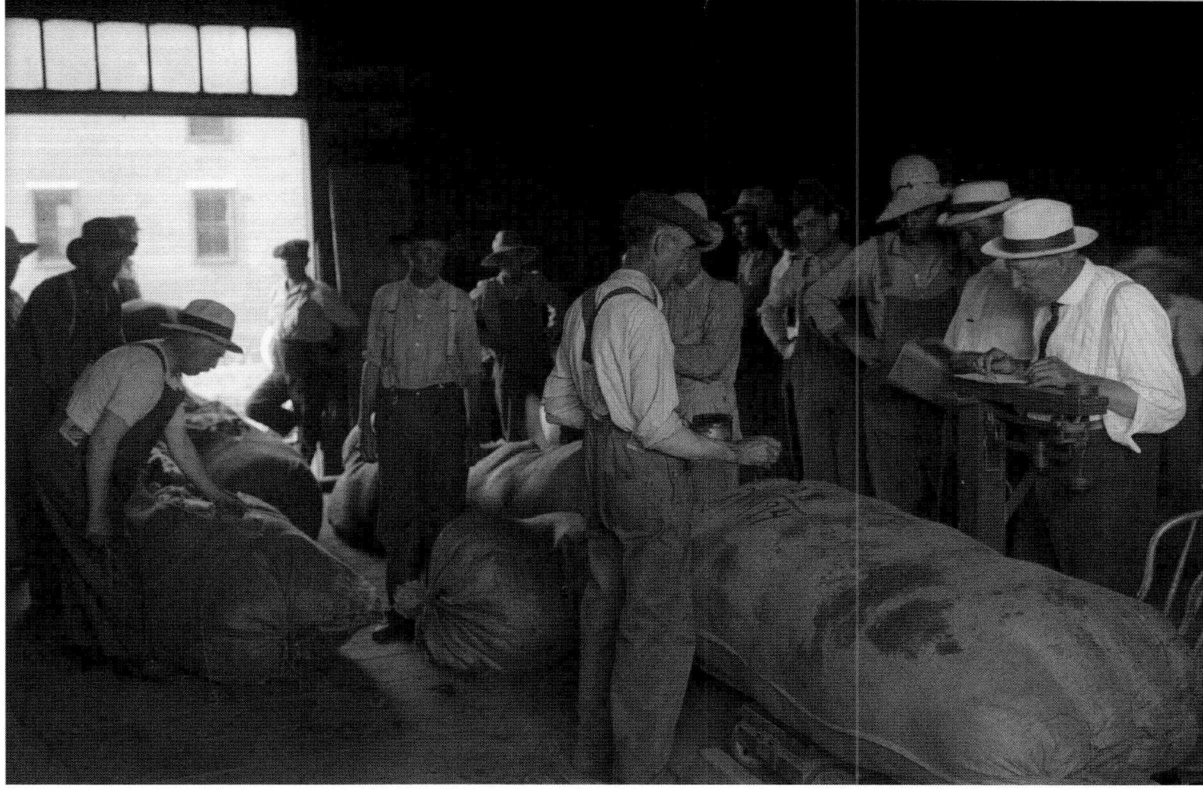

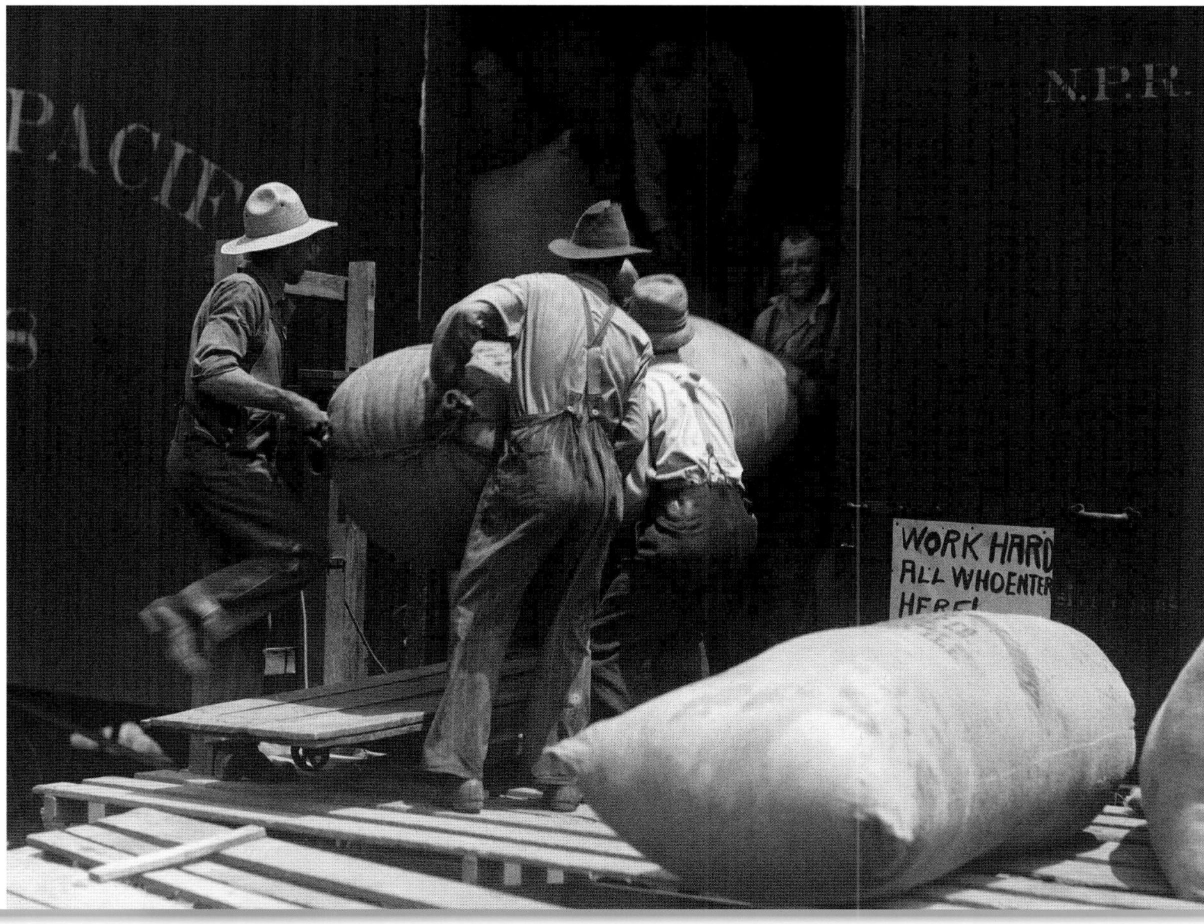

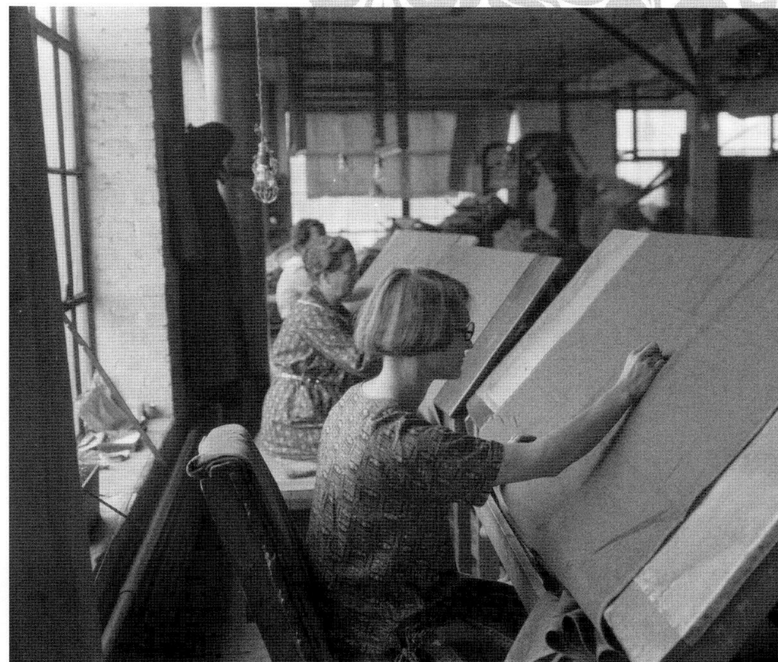

(Top left) Grading and weighing wool in a warehouse operated by the Jefferson County Wool Growers Association in Fairfield, Iowa (Jefferson County, IA; 1919)

(Bottom left) Men at a farmers' cooperative load bags of wool into a railcar. (Shelby County, IL; 1920)

(Right) Women pick foreign matter out of woolen cloth in Columbia City mills. (Whitley County, 1924)

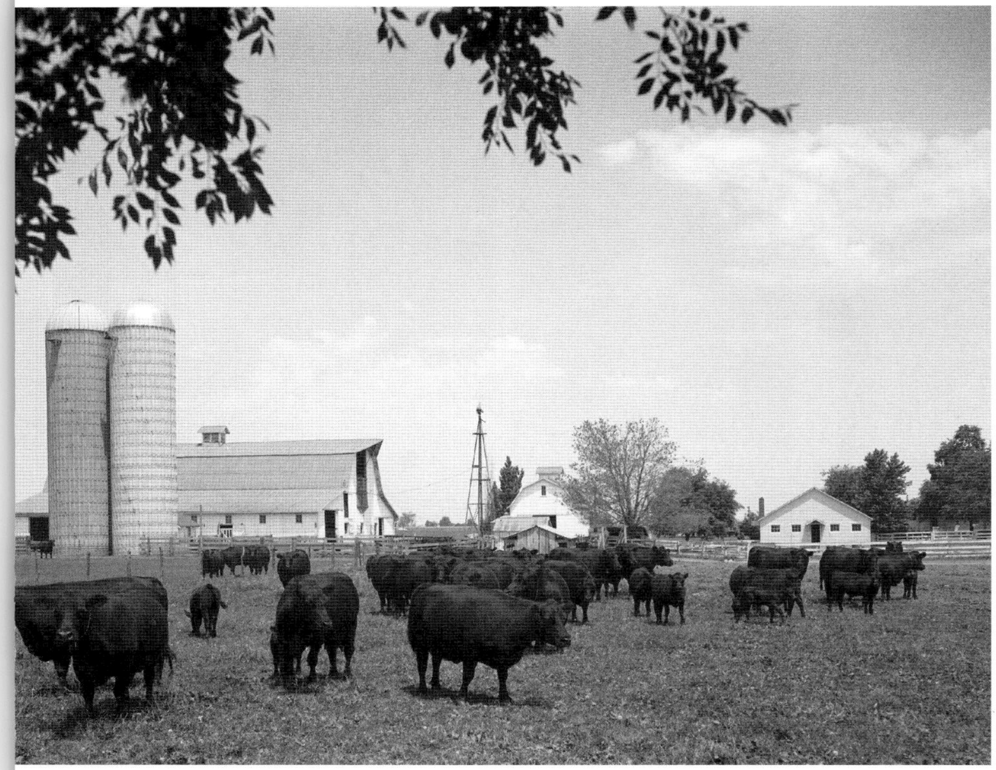

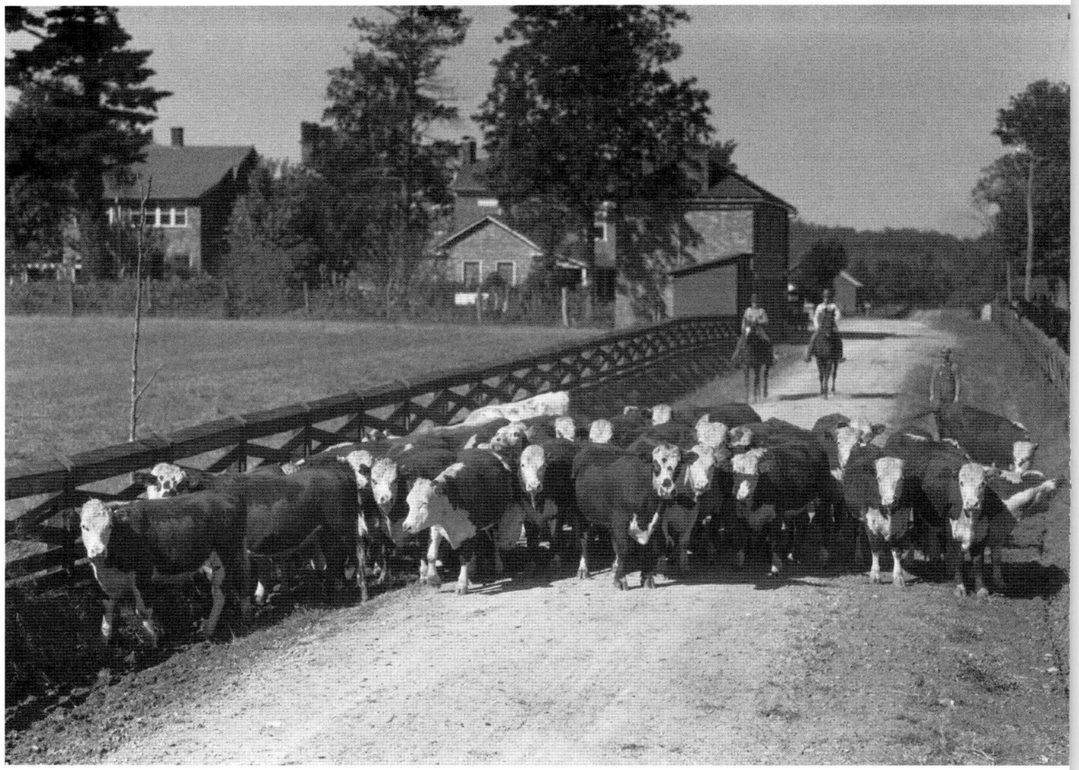

(Top left) An Angus herd, barn, and silos on the farm of Everett Rincker in Strasburg, Illinois (Shelby County, IL; 1959)

(Top right) Driving Hereford steers down a road on the O. S. Bond farm near Auburn, Kentucky (Logan County, KY; 1933)

(Bottom) An Angus cow with a calf on the Aetna Life Insurance Company Farm near Waveland (Montgomery County, 1934)

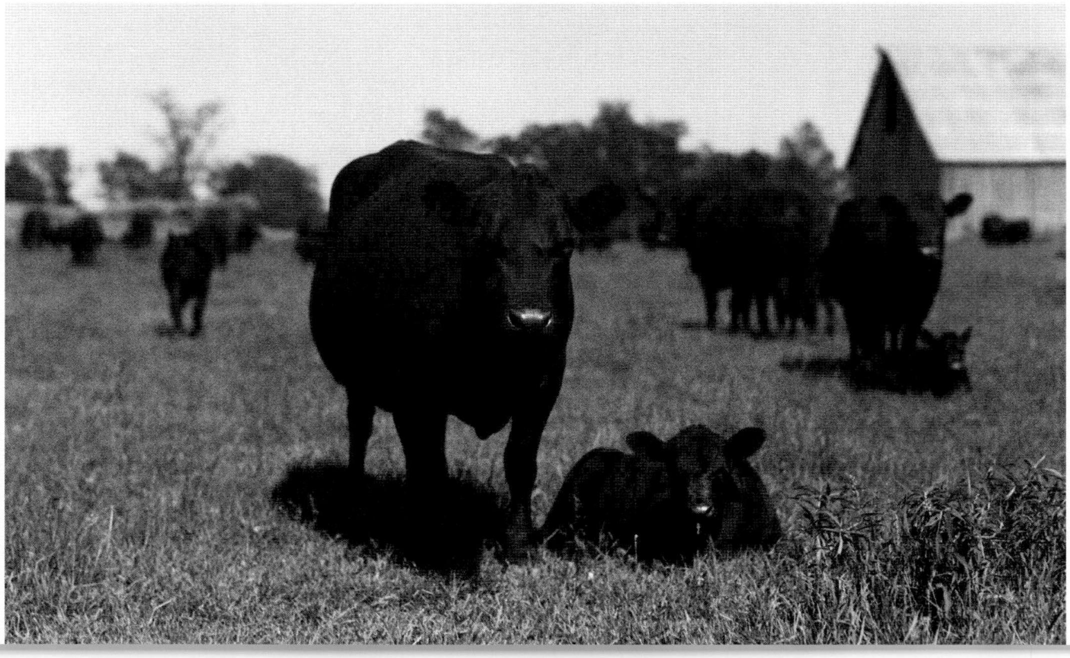

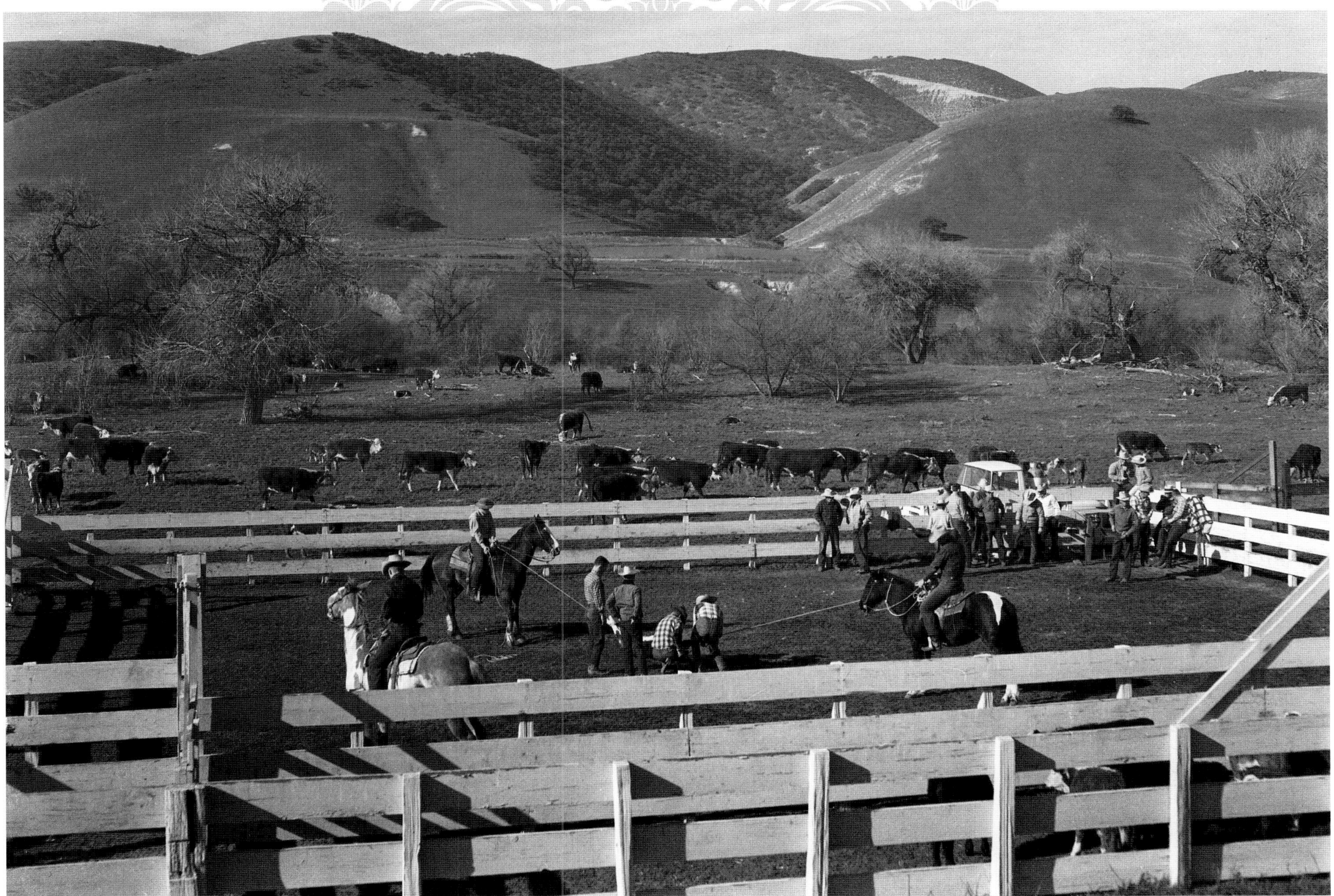

Branding calves on the Michel Orradre Ranch in San Andreas, California (Calaveras County, CA; 1961)

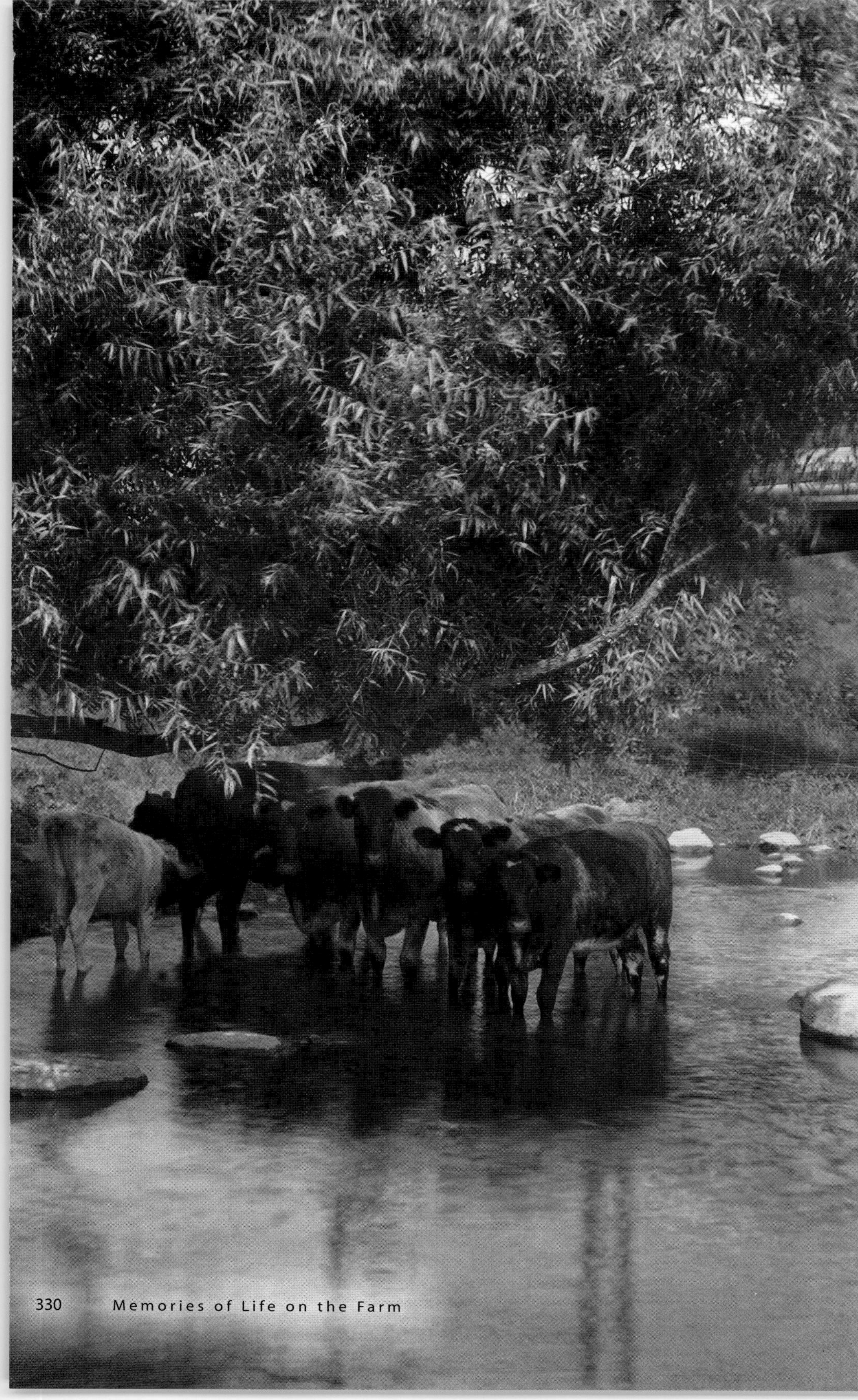

(Left) Shorthorn cattle stand in a stream under a shade tree just west of Brookston. (White County, 1935)

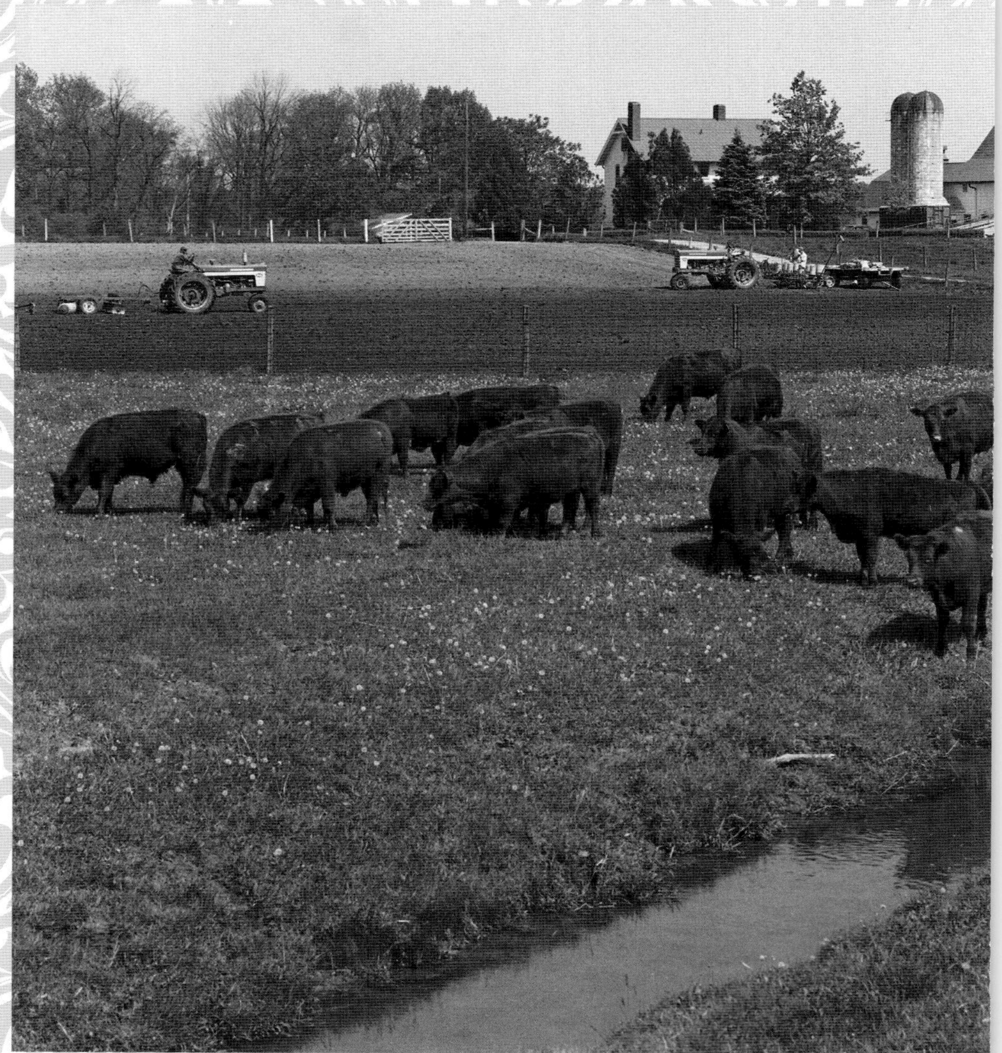

(Above) A view of Purdue University's Lynnwood Farm in Carmel in May (Hamilton County, 1960)

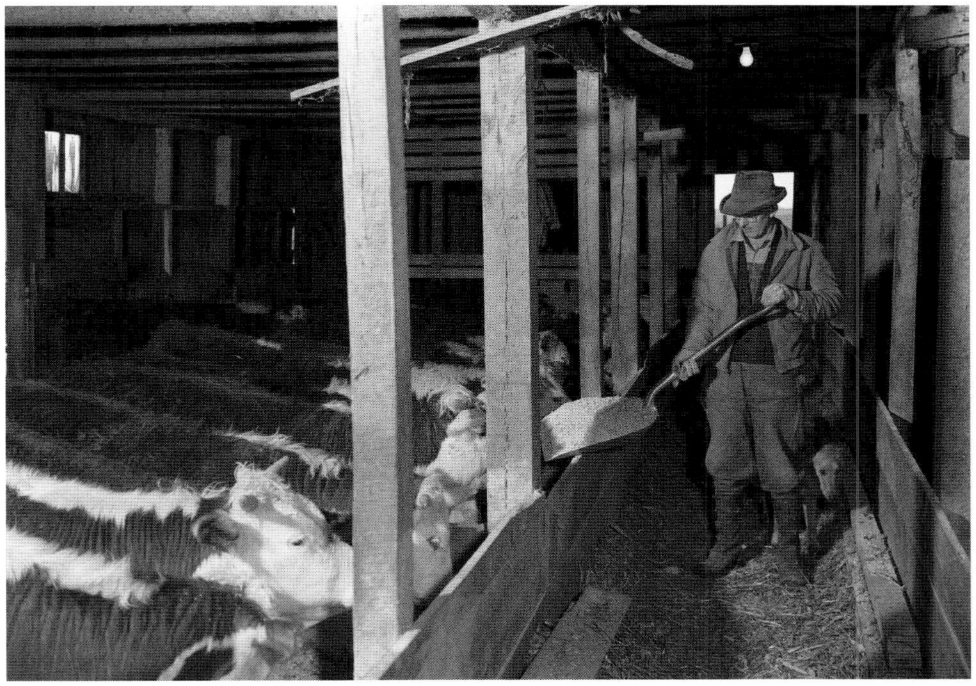

Harry Overpeck feeds cattle on photographer John C. Allen's farm near Linden. (Montgomery County, 1950)

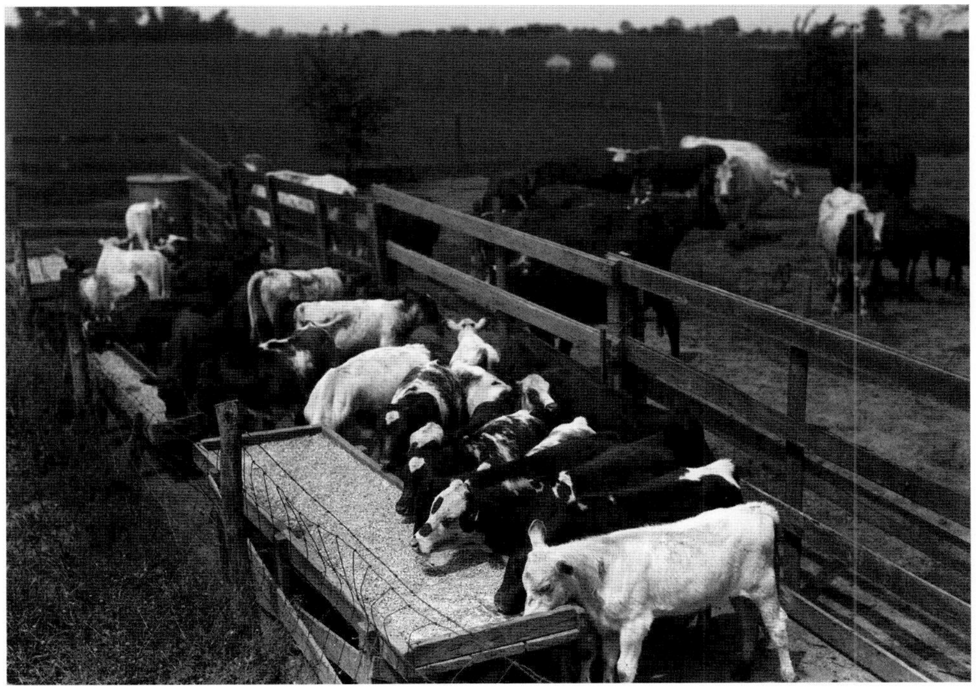

Young calves feed on the farm of Dr. John S. Morrison in Lafayette. (Tippecanoe County, 1930)

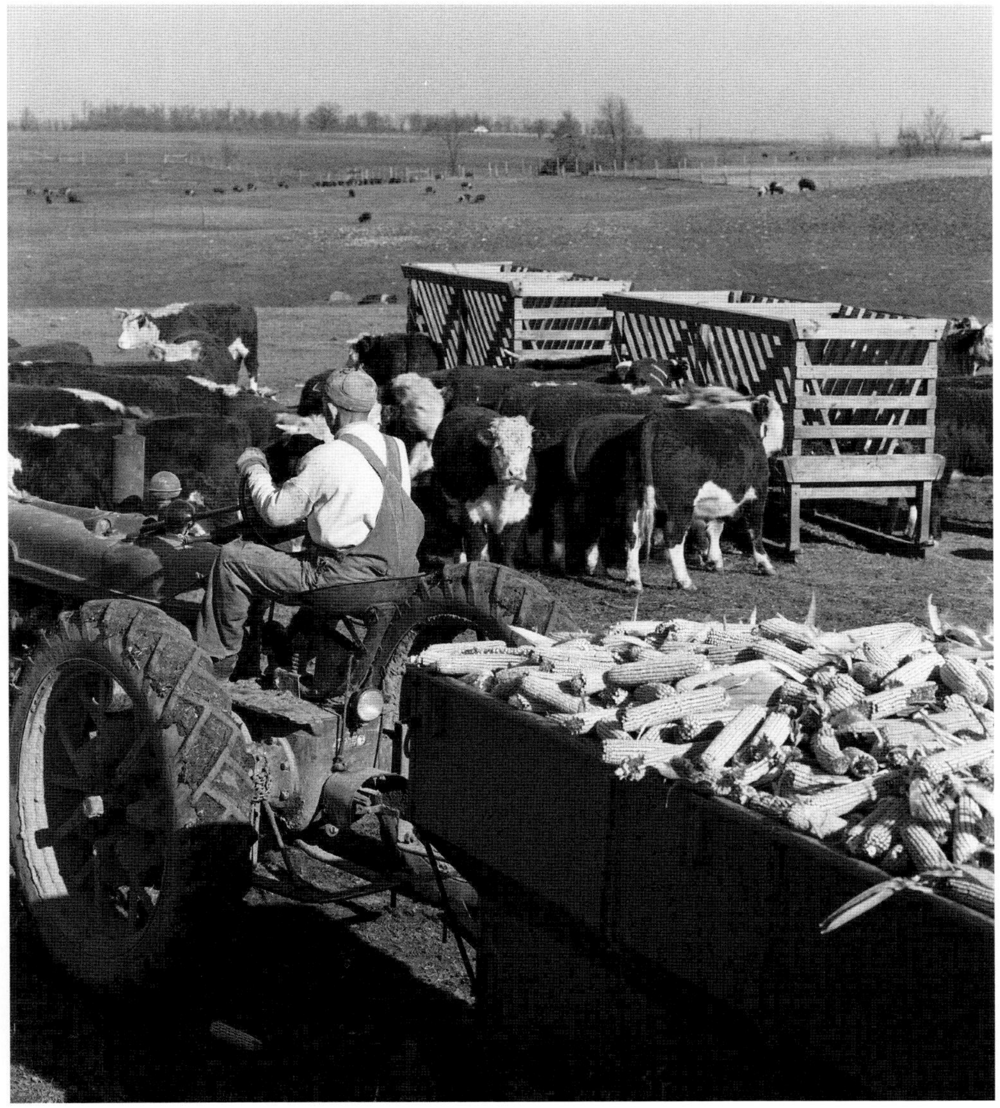

Maurice Lawson hauls a load of corn to his livestock in Otterbein. (Benton County, 1948)

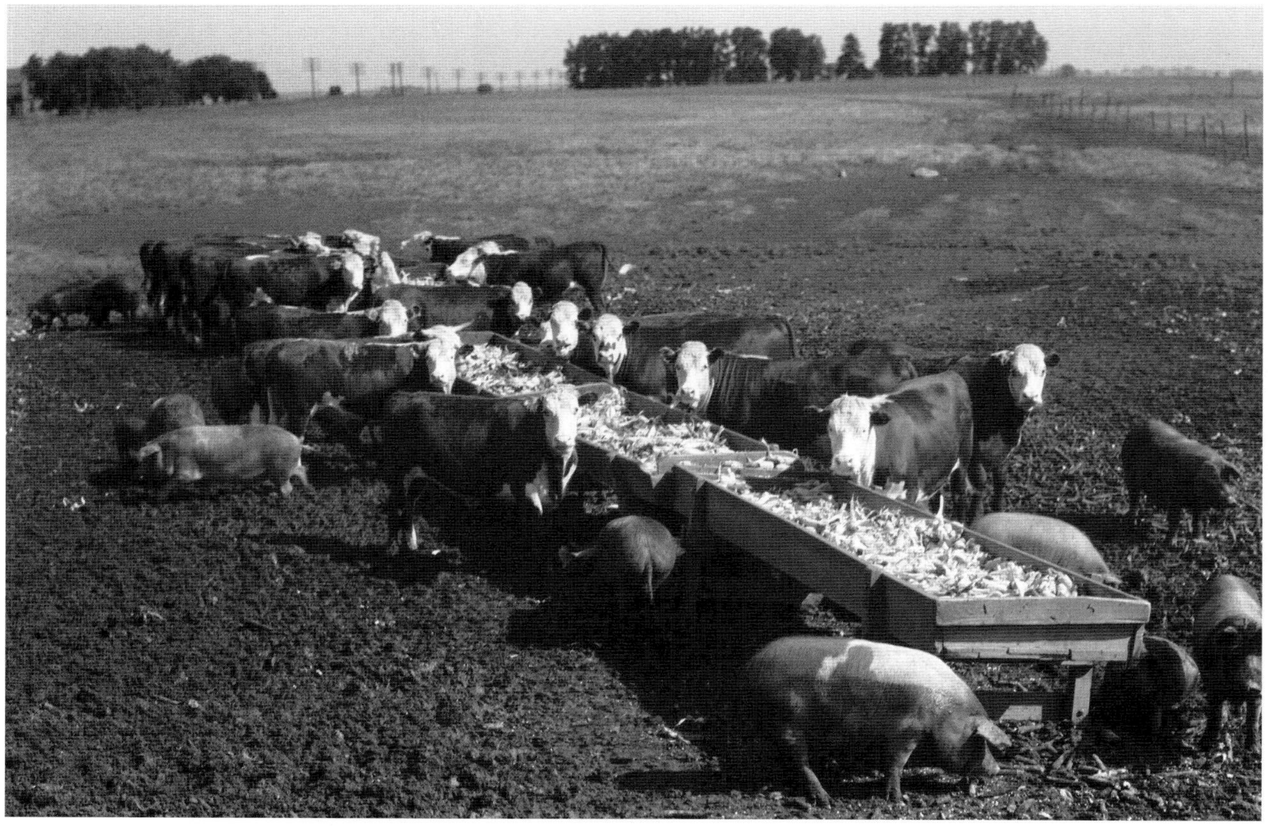

Hereford steers and some hogs gather around a feed trough on the farm of James Kirkpatrick of Fowler, who Allen noted had fed cattle for 43 years when this photo was taken. (Benton County, 1931)

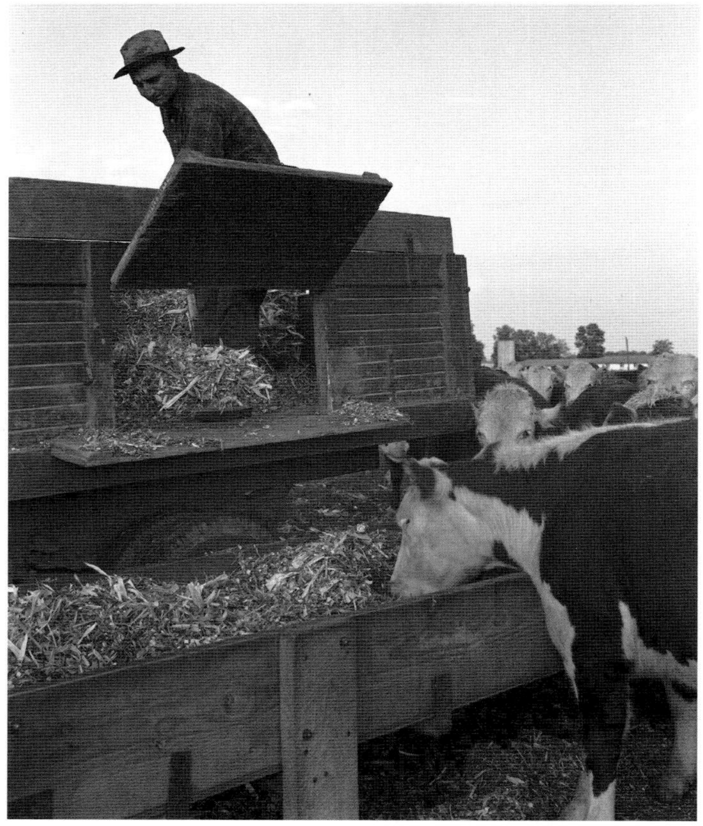

A man feeds silage to Hereford cattle from a modified truck bed on the Morton Brothers Farm in Lebanon. (Boone County, 1945)

A farmer places grass silage into feed bunks for Wayne Darland's cattle herd in Lafayette. (Tippecanoe County, 1955)

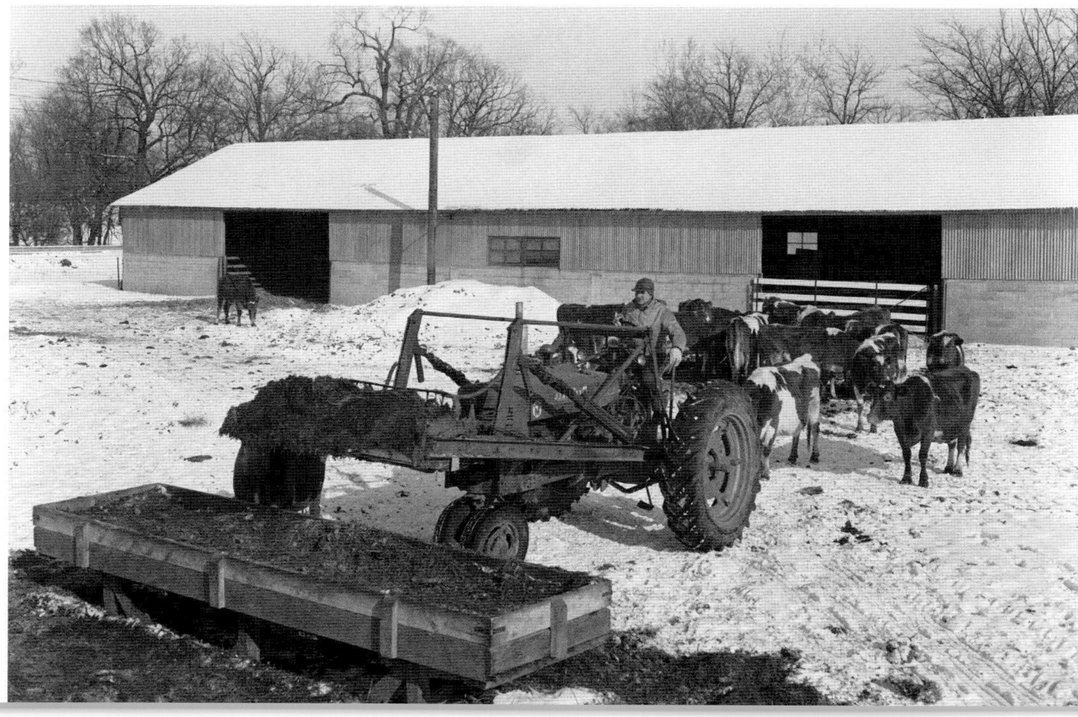

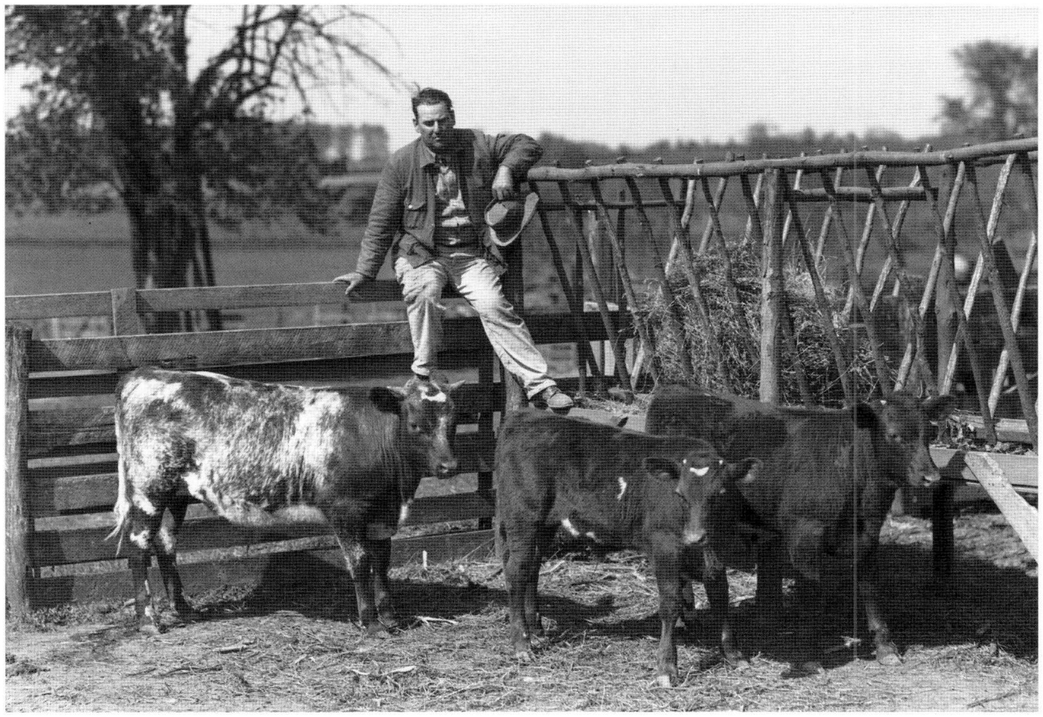

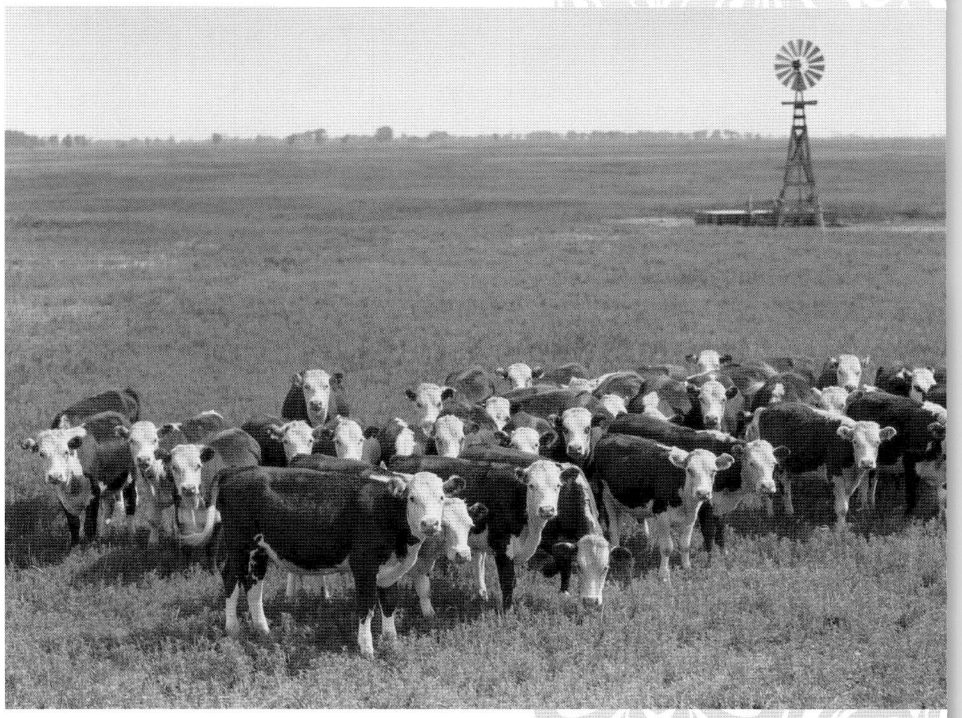

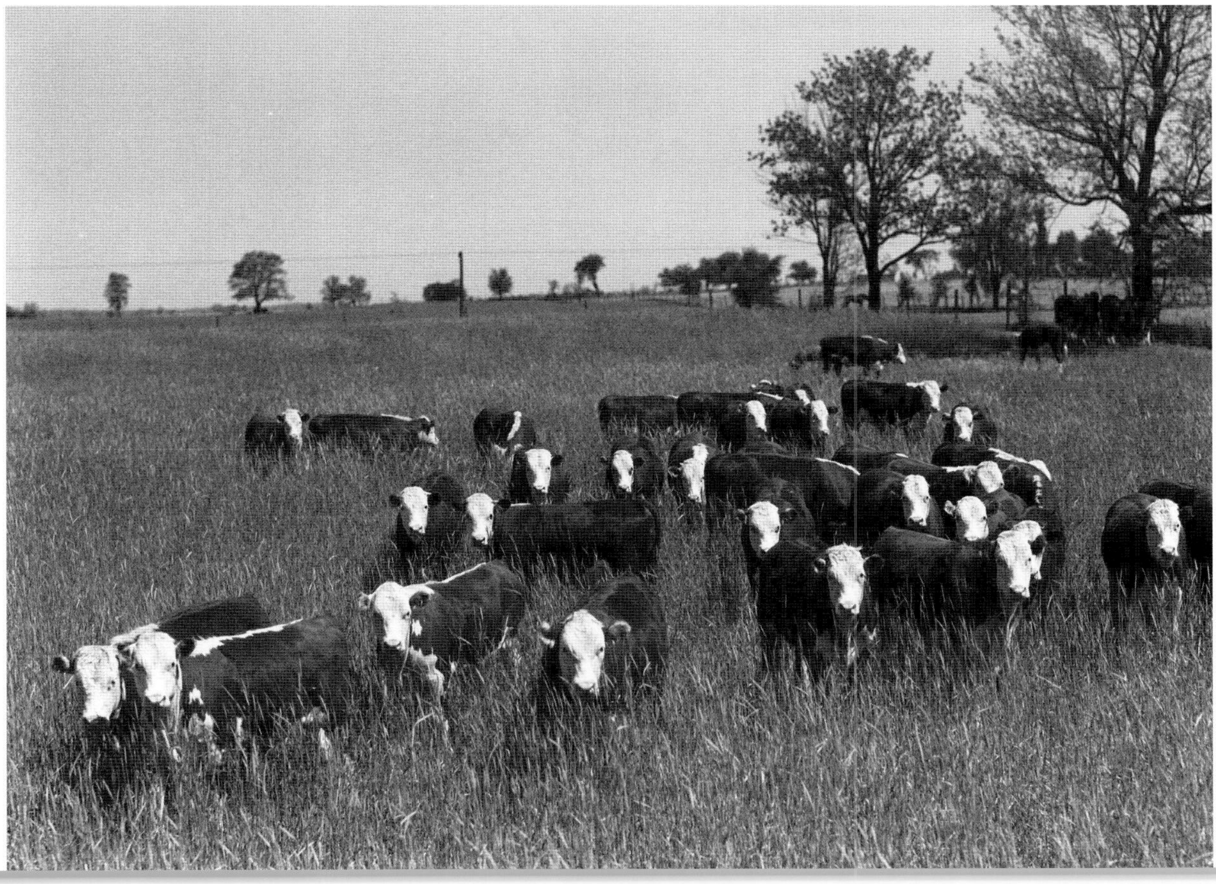

(Top left) Lee Flora of Milton poses with three of his Shorthorn calves. (Wayne County, 1935)

(Top right) Hereford steers gather on a grass pasture with a windmill in the background on the farm of Lewis Mills near Raymond, Kansas. (Rice County, KS; 1932)

(Bottom) A herd of Herefords on a rye pasture on the W. H. Bowman and Son Farm in Rockfield (Carroll County, 1934)

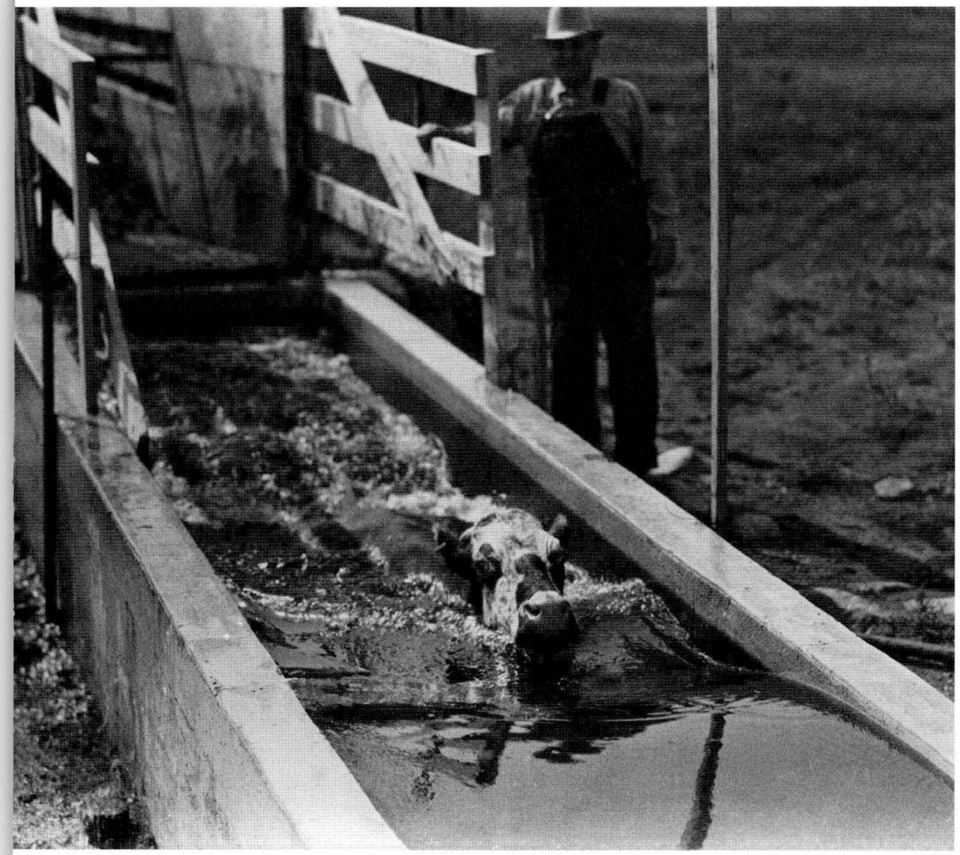

A cattle dipping tank on the Walter Bones Stock Farm in Parker, South Dakota (Turner County, SD; 1939)

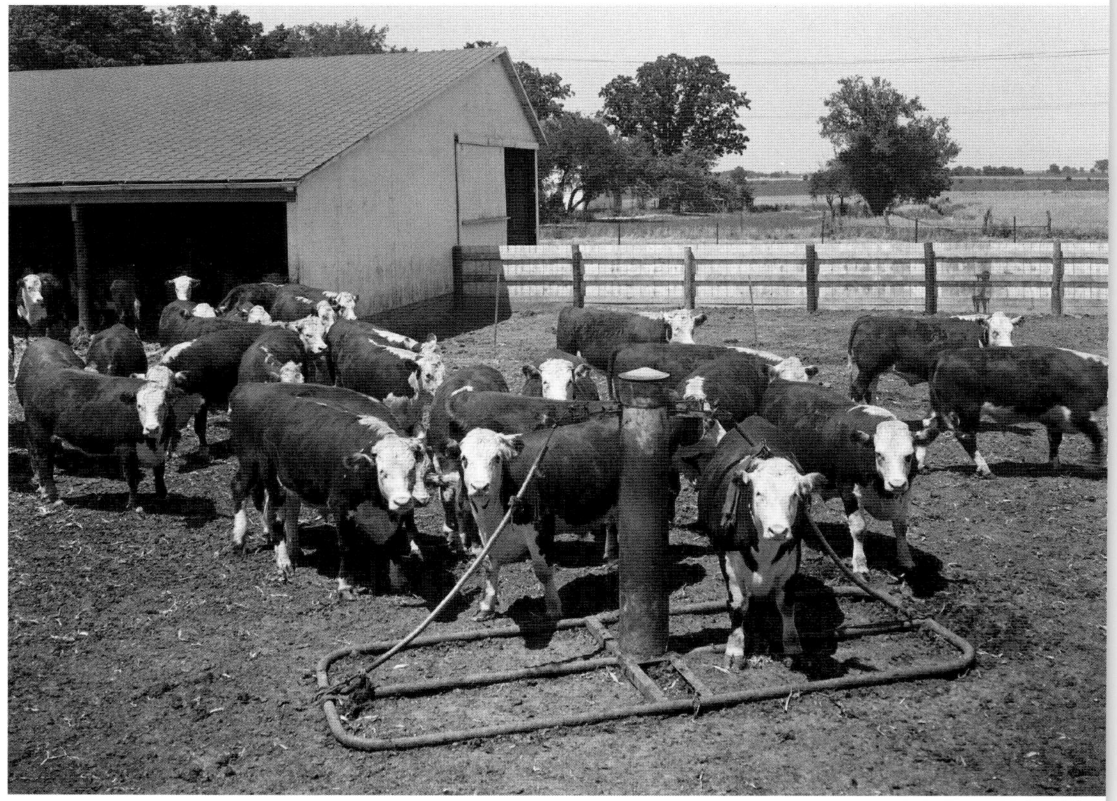

Herefords in a feedlot stand around a livestock oiler on the J. M. Grove farm near Plano, Illinois. (Kendall County, IL; 1955)

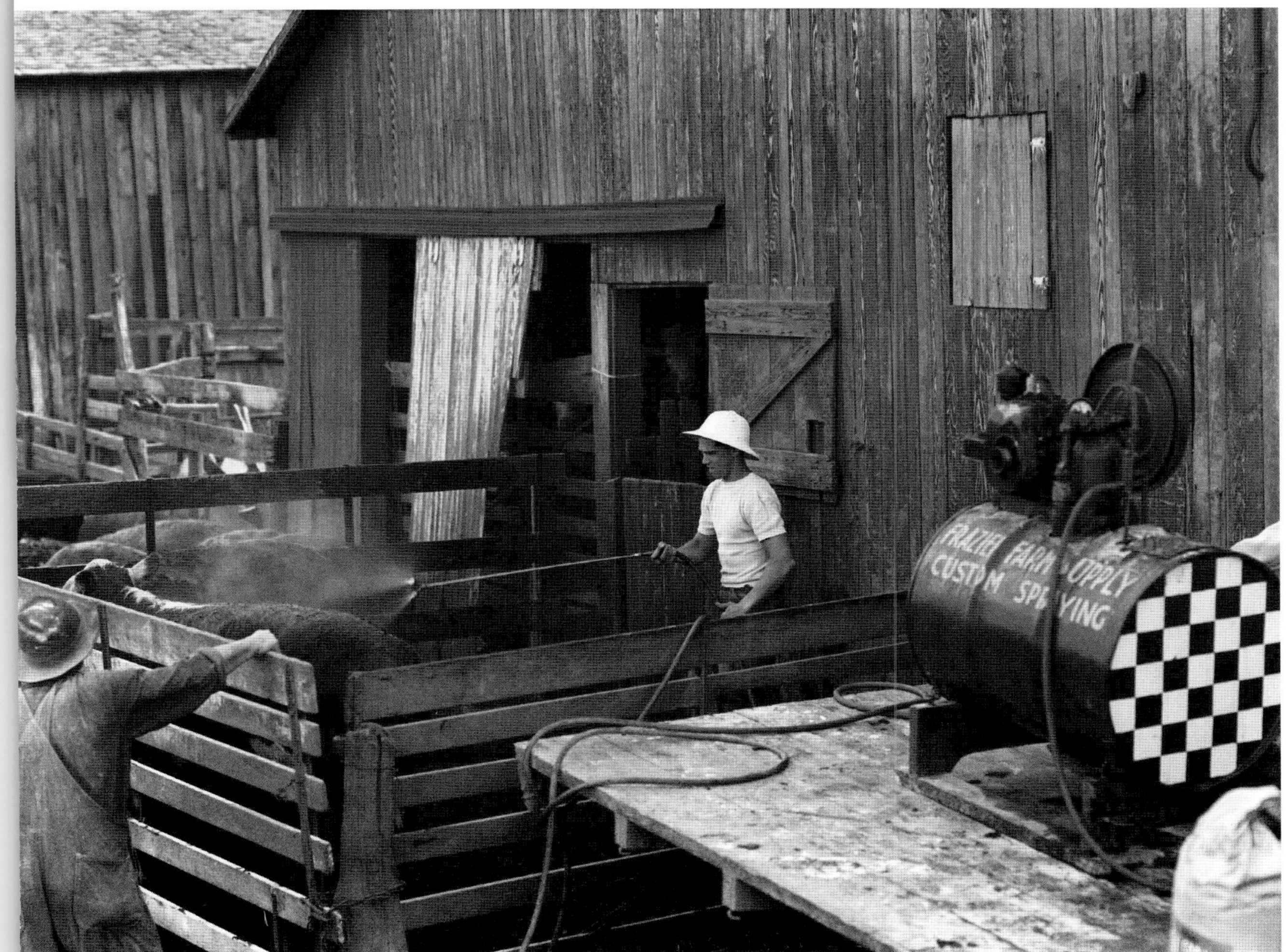

Spraying cattle with DDT for fly control on the Kenneth Royer farm near Linden (Montgomery County, 1947)

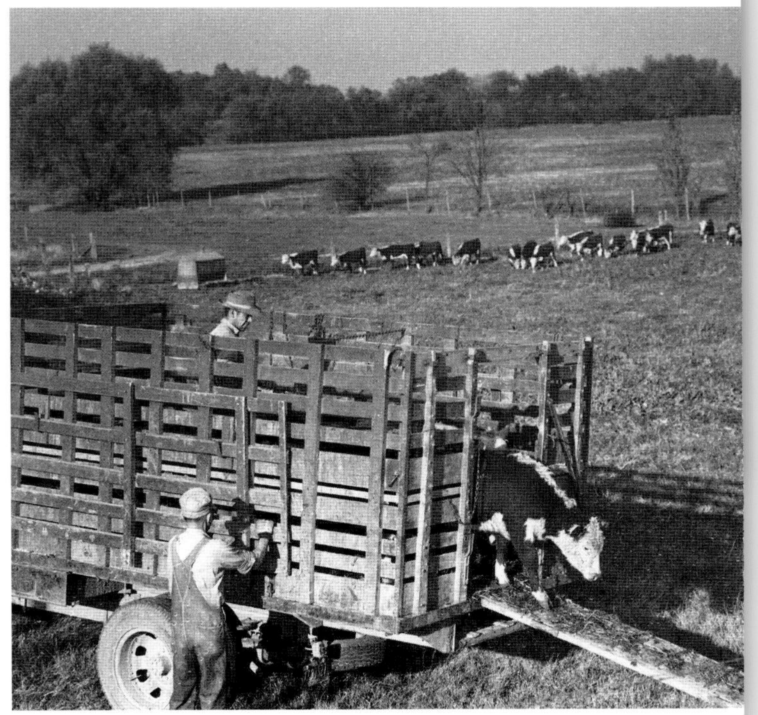

Men unload Herefords on the farm operated by William Miller in Frankfort. (Clinton County, 1948)

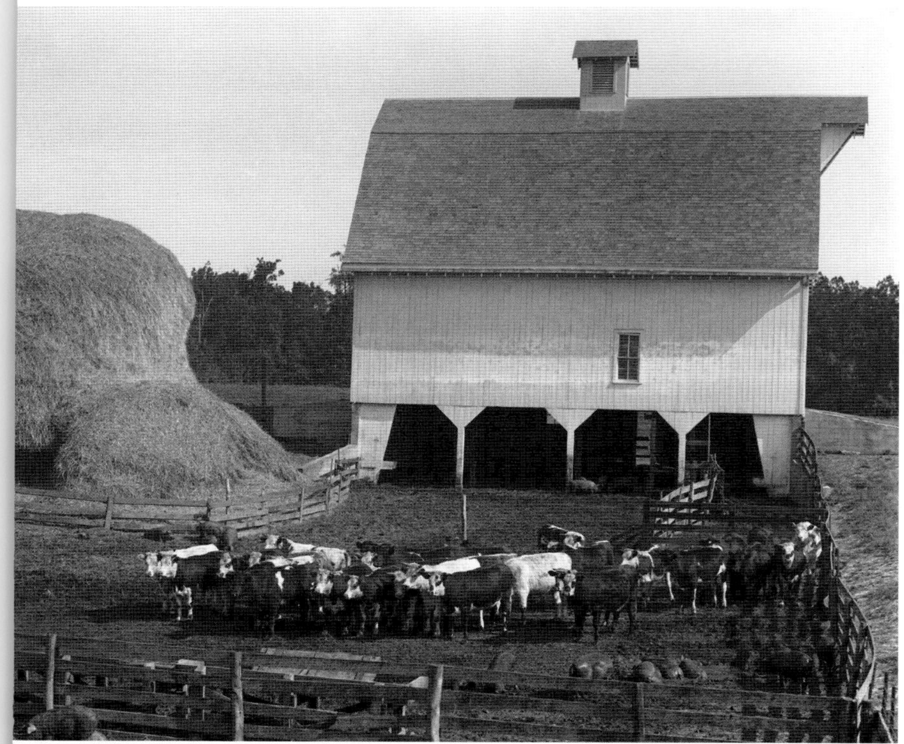

Cattle in front of a feeding shed and large haystack on the Lafayette-area farm of Dr. John S. Morrison (Tippecanoe County, 1931)

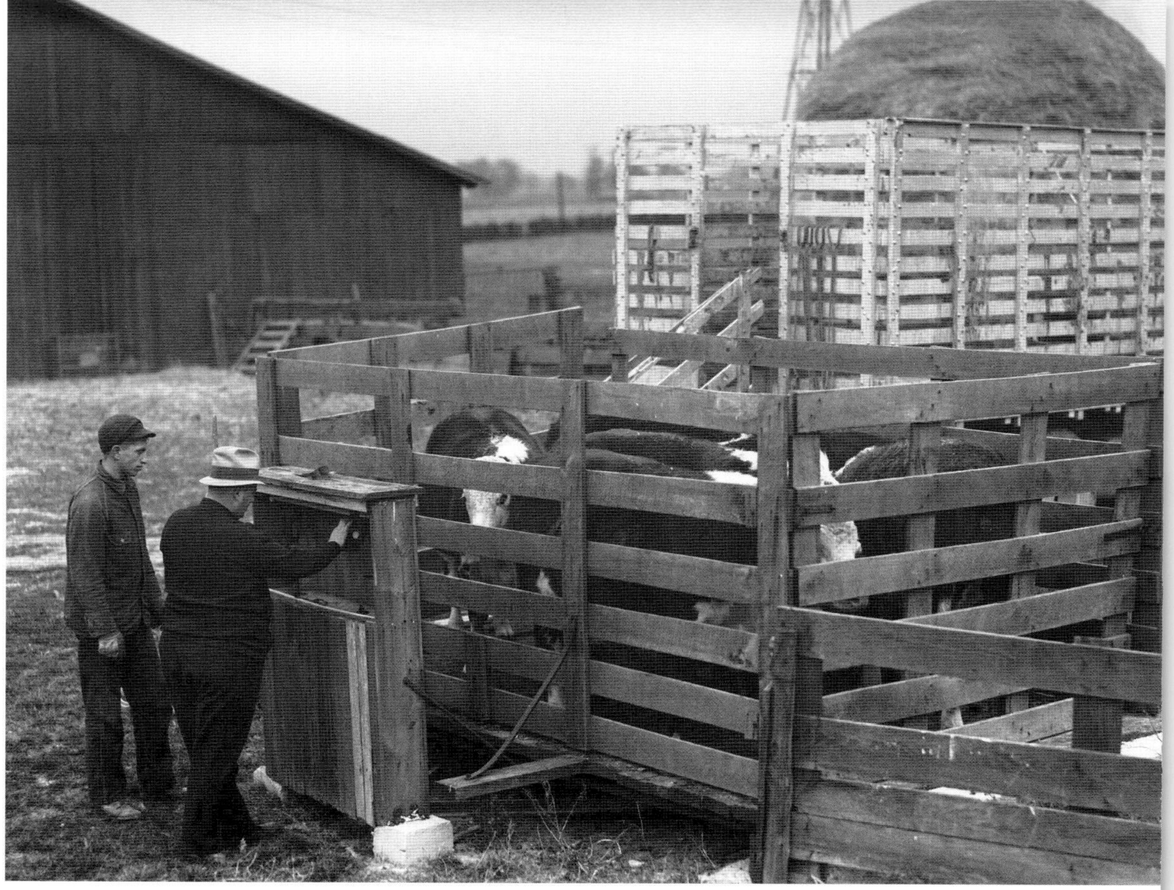

These men weigh Hereford steers on the farm of Dorsey Beal near Bucyrus, Ohio, before sending them to market. (Crawford County, OH; 1934)

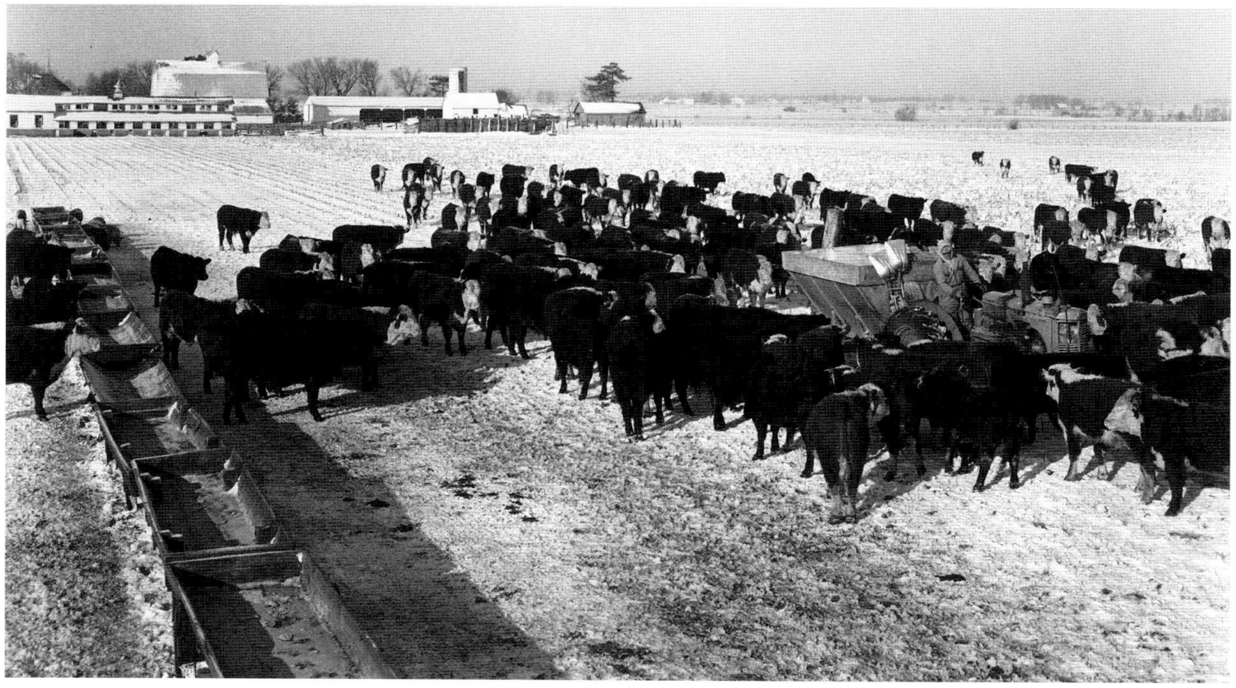

Feeding beef cattle in the winter on the farm of Paul Kennedy and Son in Templeton (Benton County, 1962)

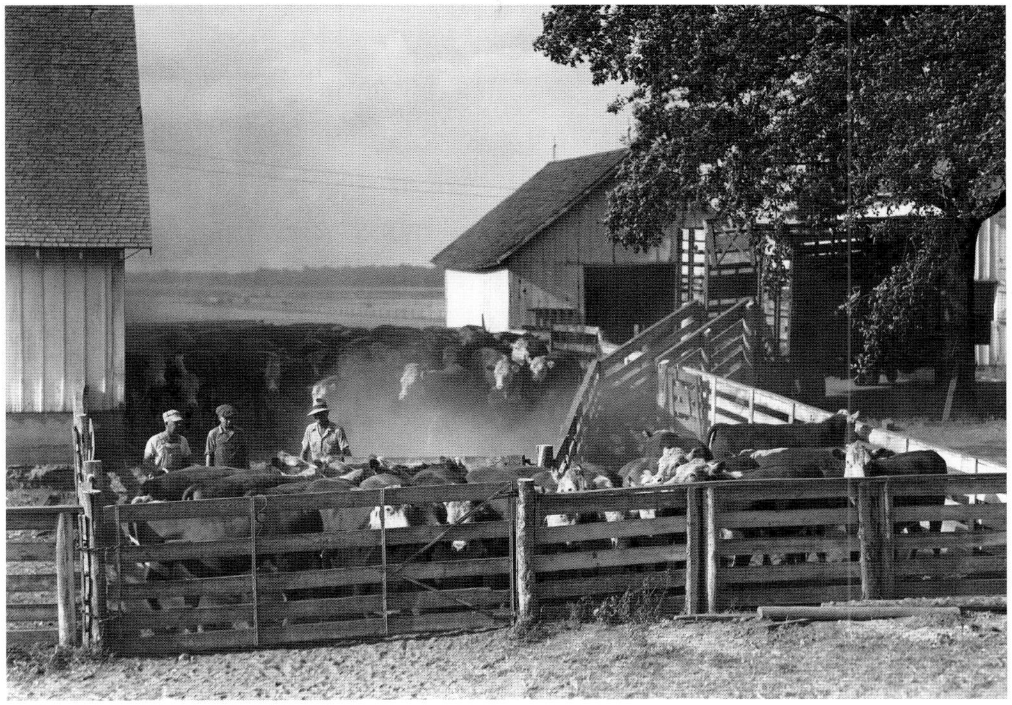

Loading a truck with cattle headed to market on the Warren North farm near Brookston (White County, 1944)

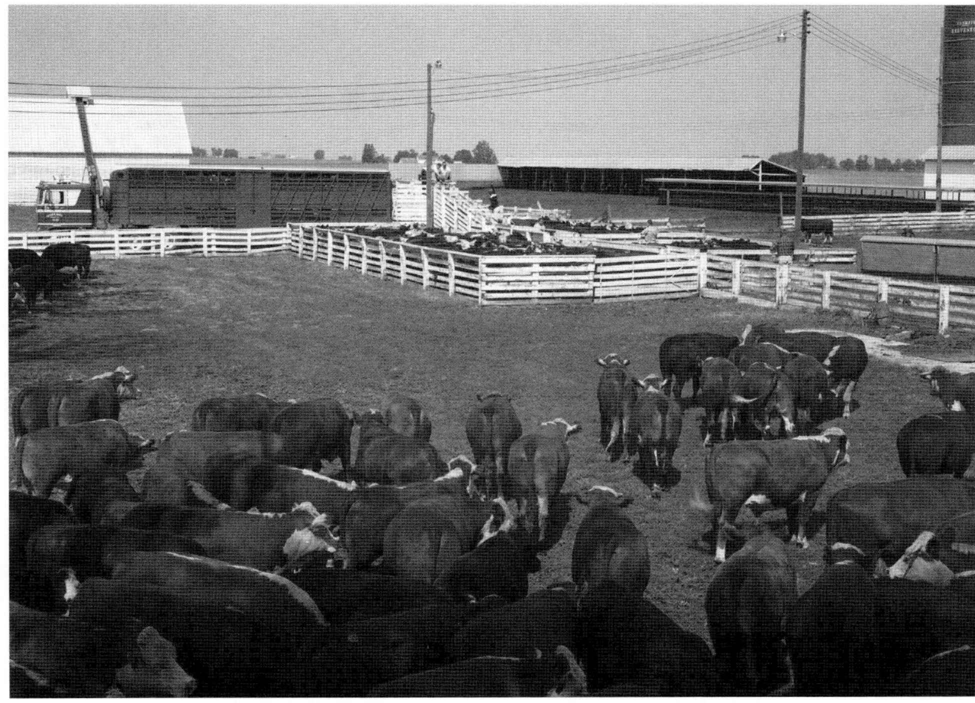

Sorting and shipping cattle on the Warren North farm near Brookston (White County, 1967)

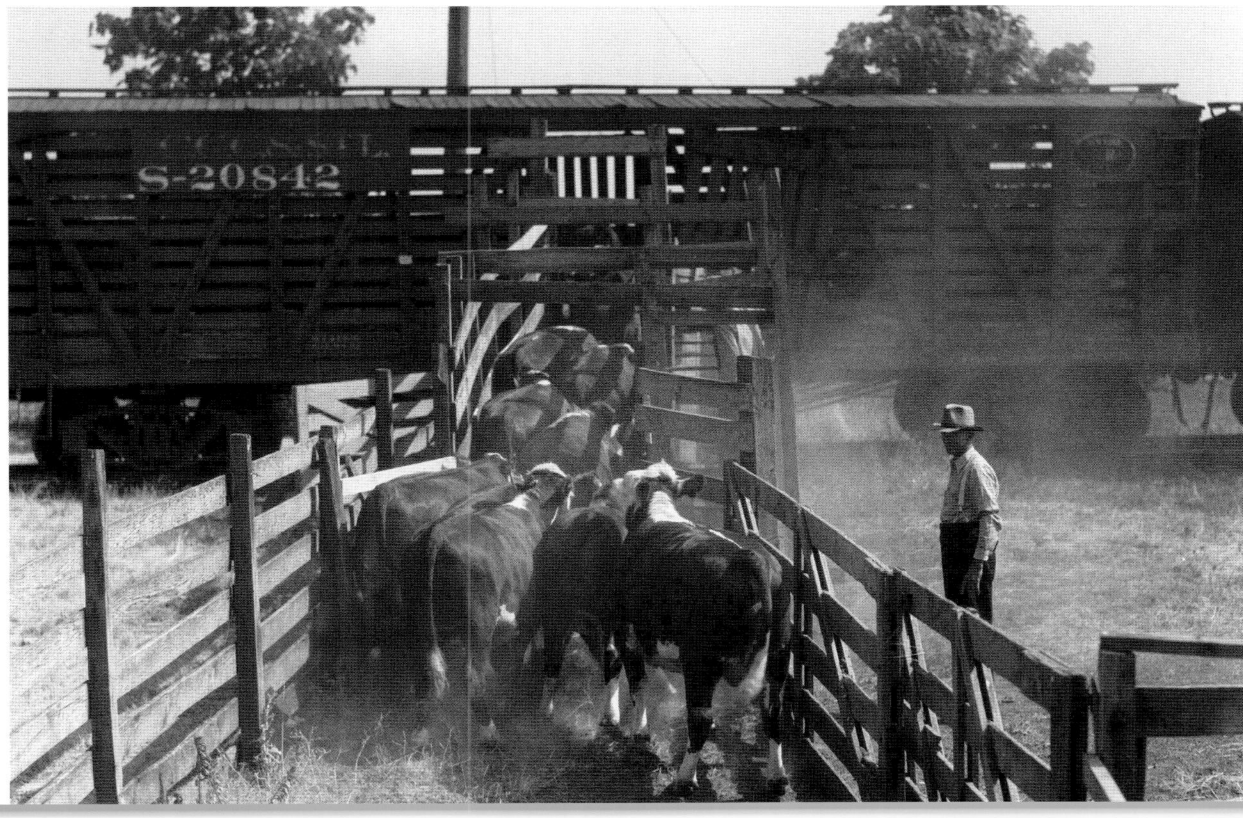

Loading steers for market at the Purdue University campus farm in West Lafayette (Tippecanoe County, 1933)

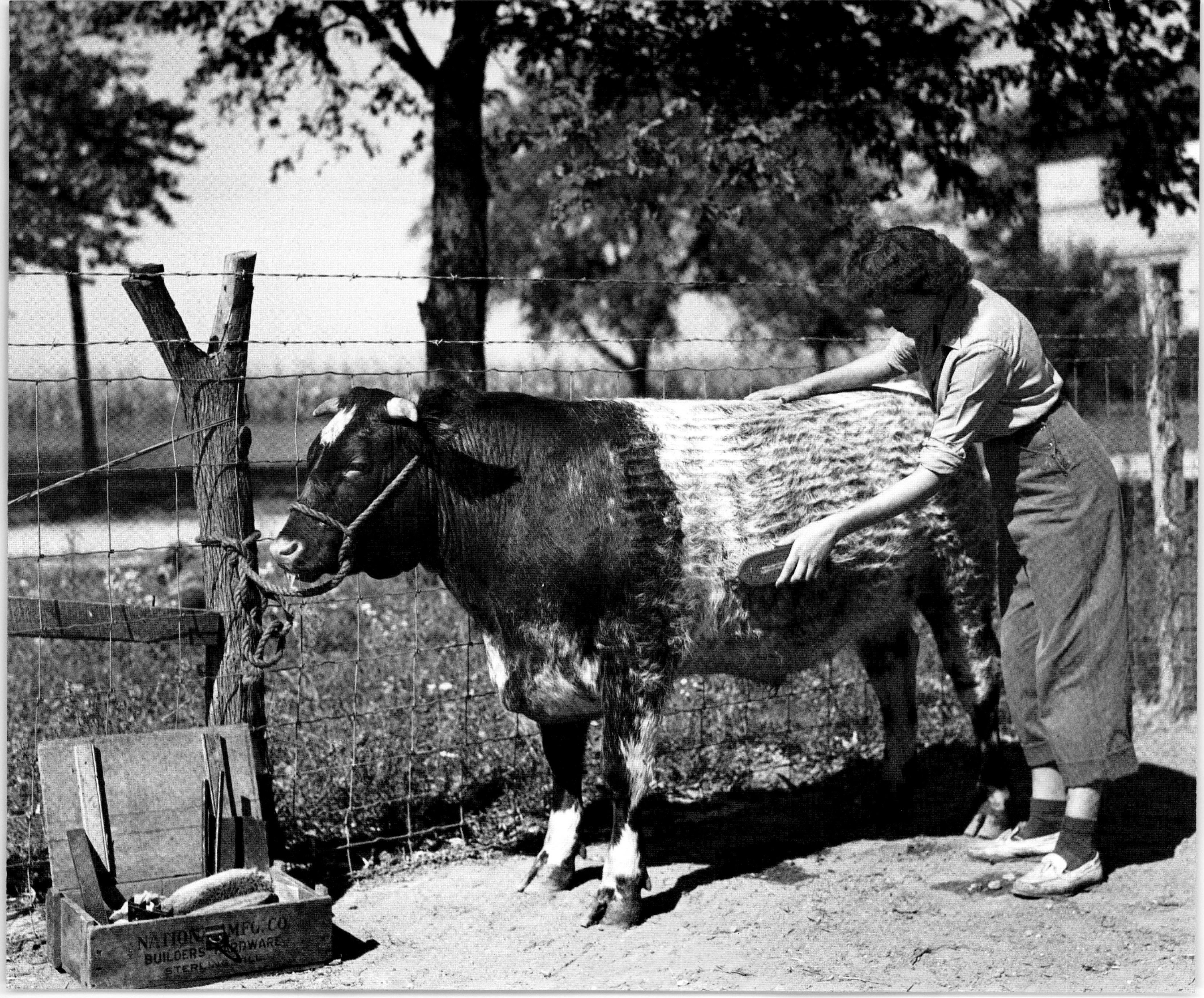
NATION MFG. CO.
BUILDERS RDWARE
ILL.

THROUGH THE LENS OF PIONEER PHOTOGRAPHER
J. C. Allen

PART 4 *The Purdue University Influence*

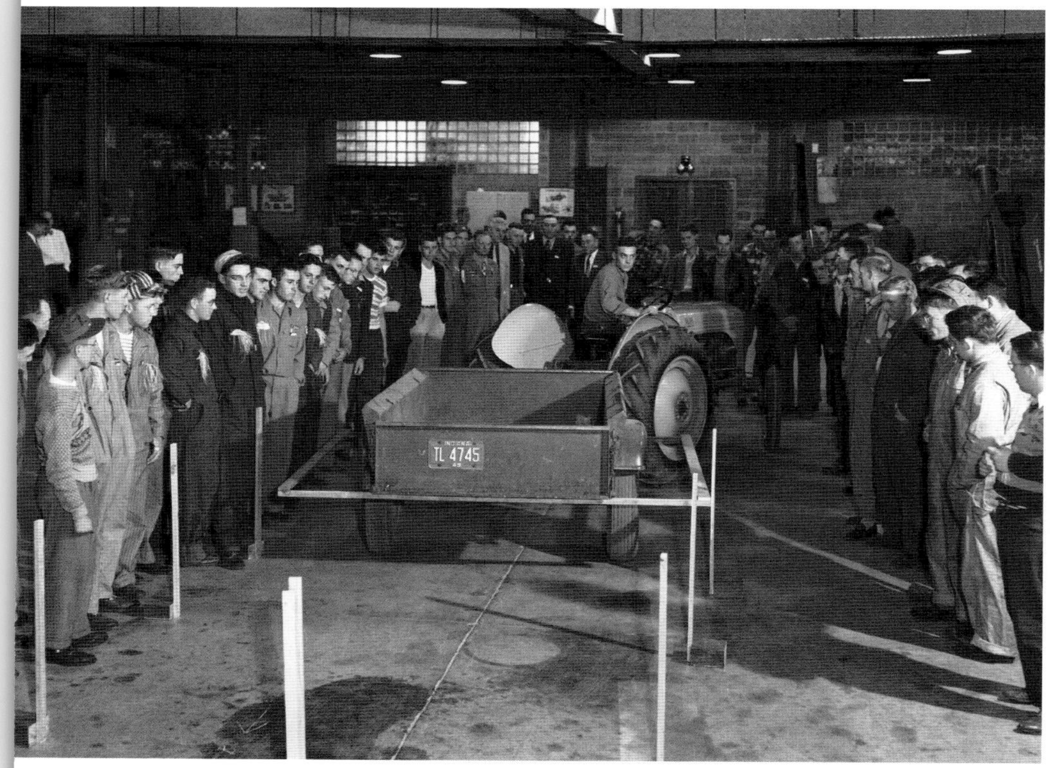

The Department of Agricultural Engineering at Purdue University hosts a tractor school. (Tippecanoe County, 1949)

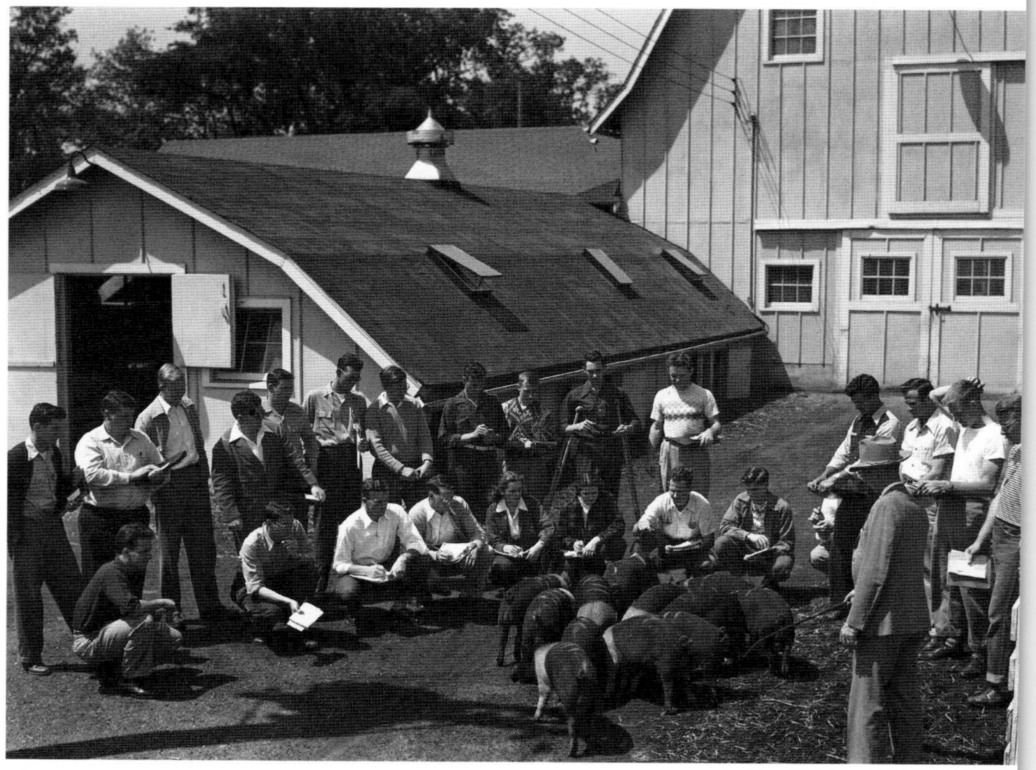

Purdue agriculture students study pigs at the Purdue University campus farm in West Lafayette. (Tippecanoe County, 1948)

Purdue agronomy students work at a corn plot. (Tippecanoe County, 1934)

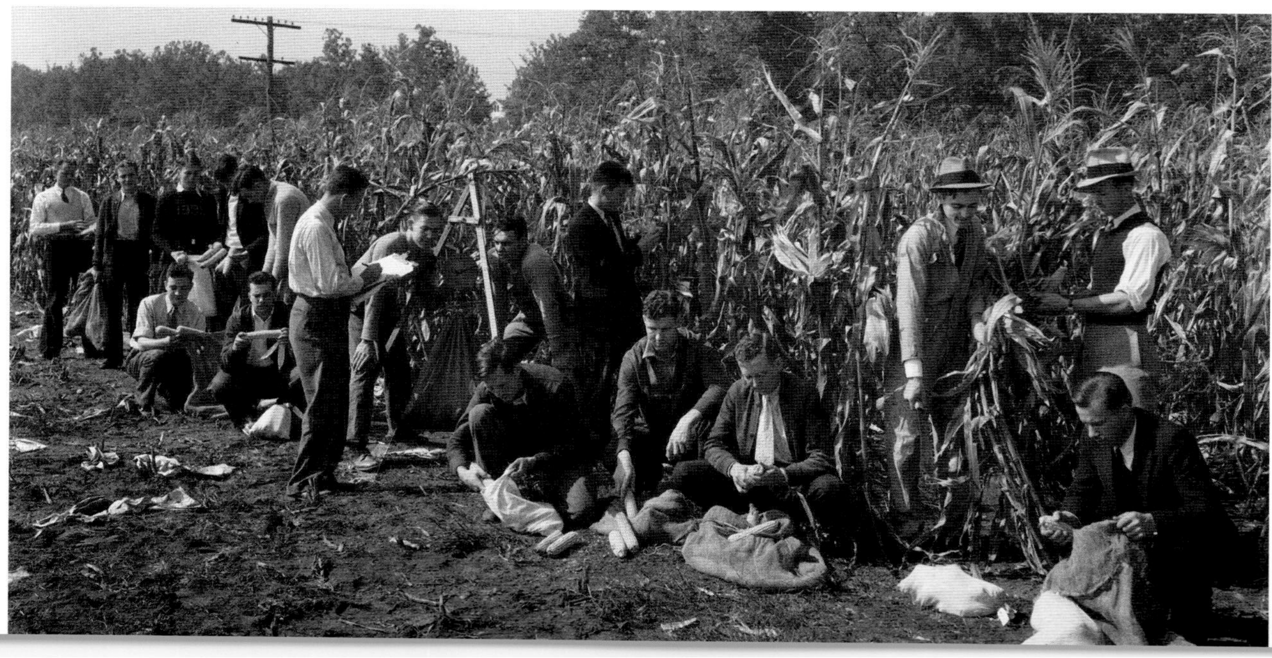

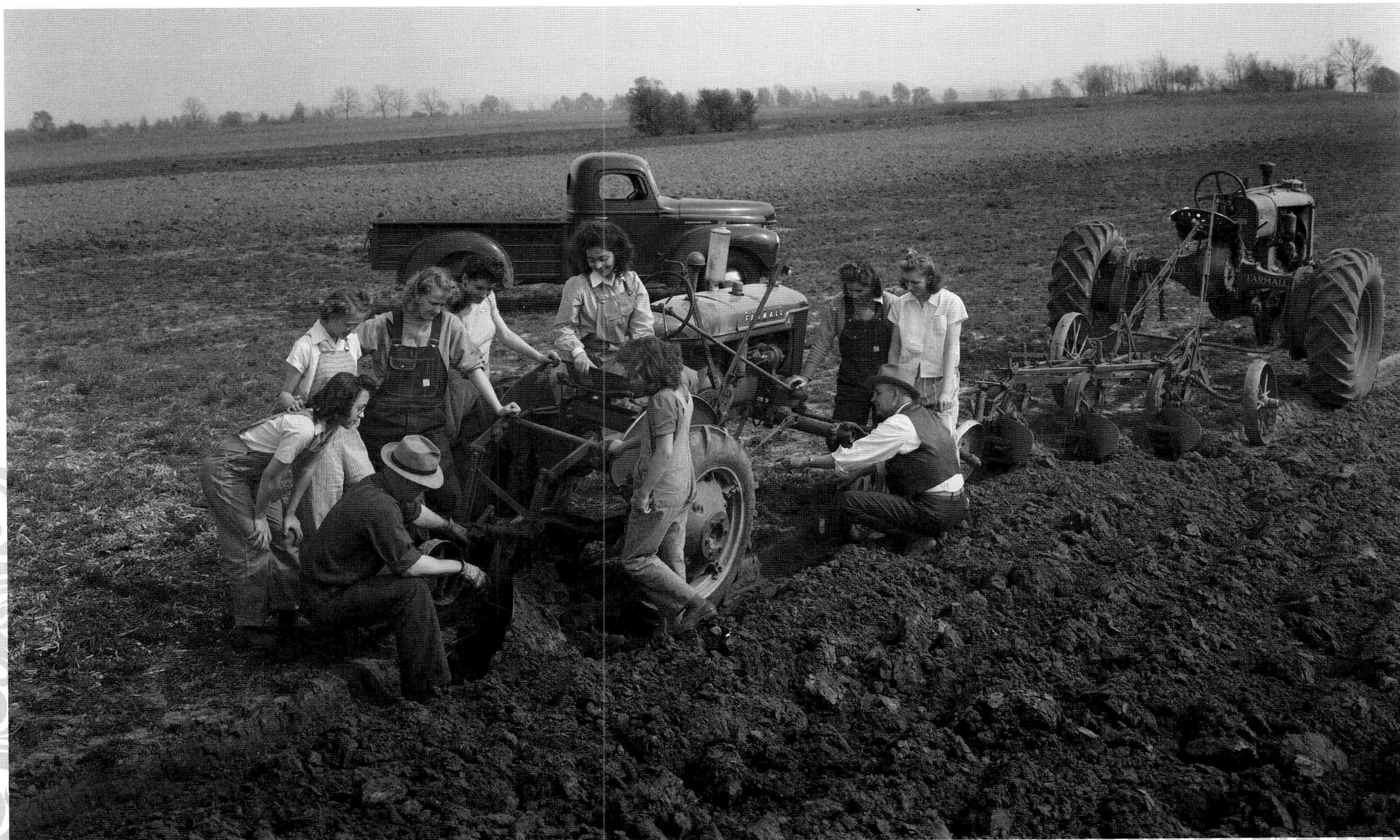

Young women learn about plowing operations at Purdue University. (Tippecanoe County, 1942)

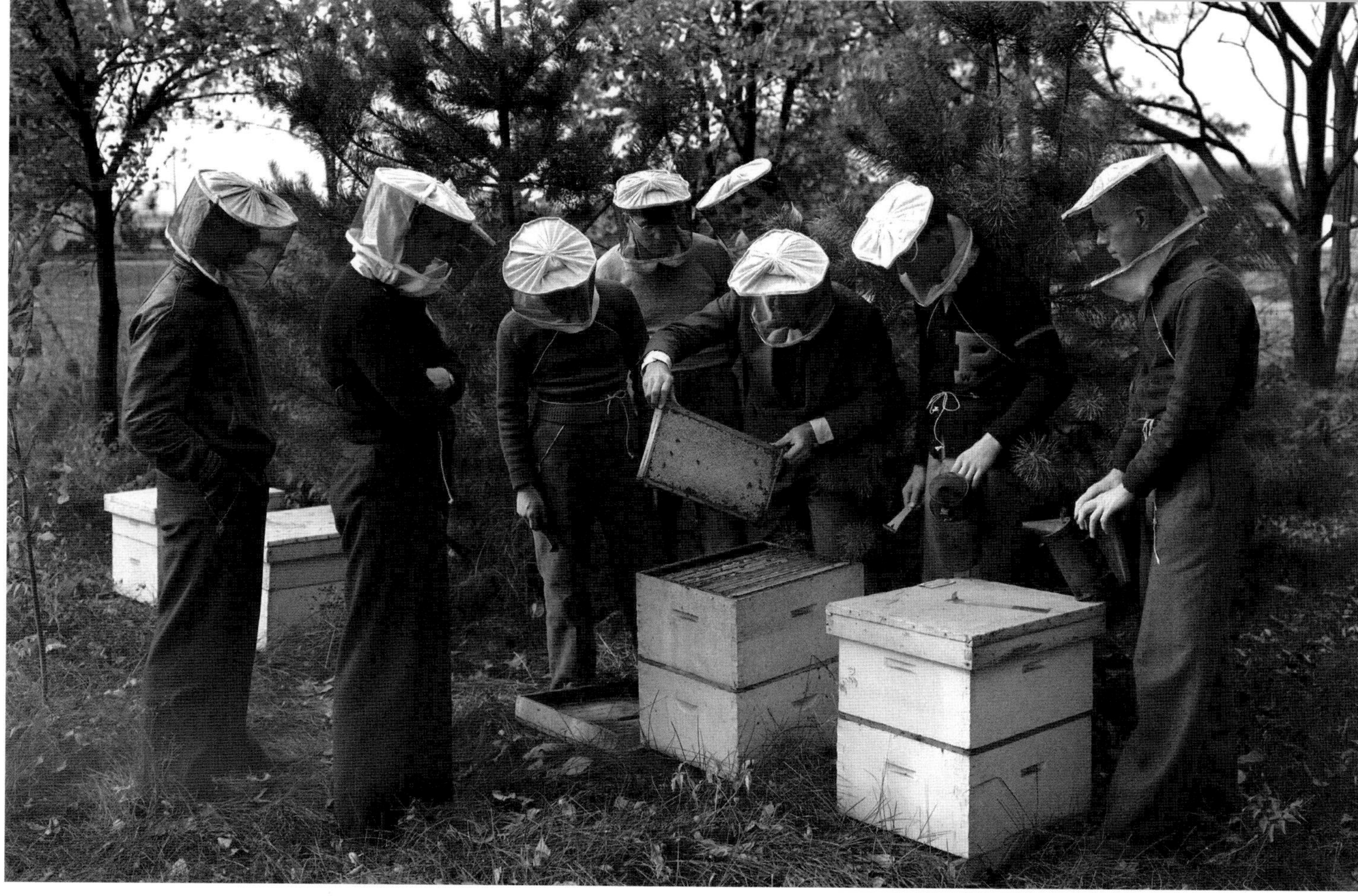

Purdue students work with bees as part of their studies. (Tippecanoe County, 1939)

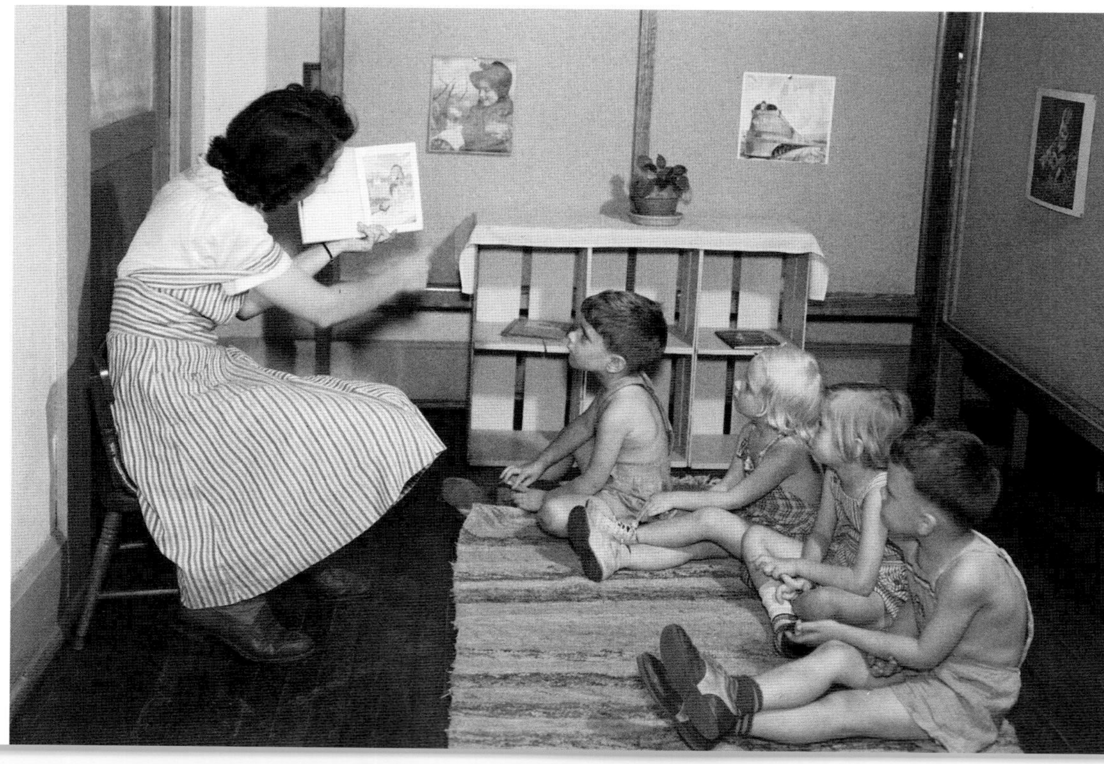

A Purdue student from the School of Home Economics works with a group of young children. (Tippecanoe County, 1942)

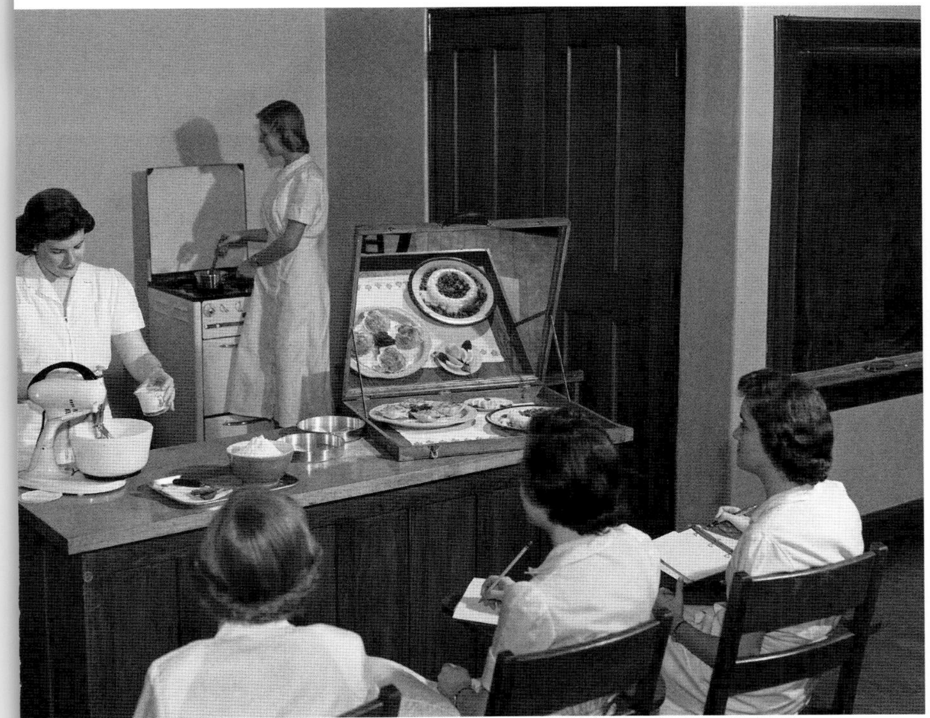

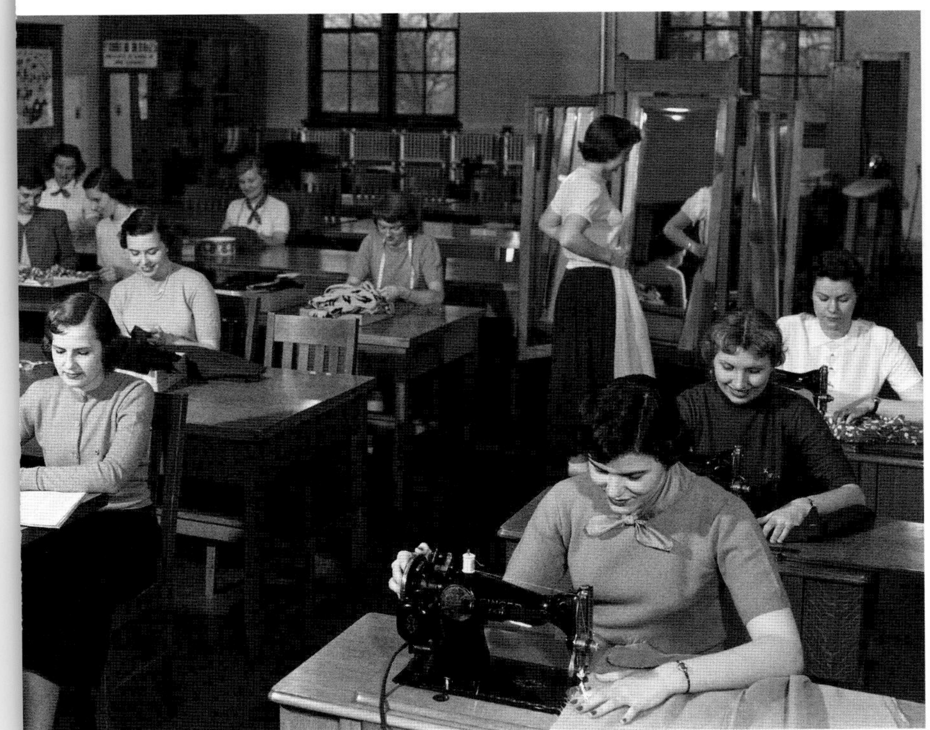

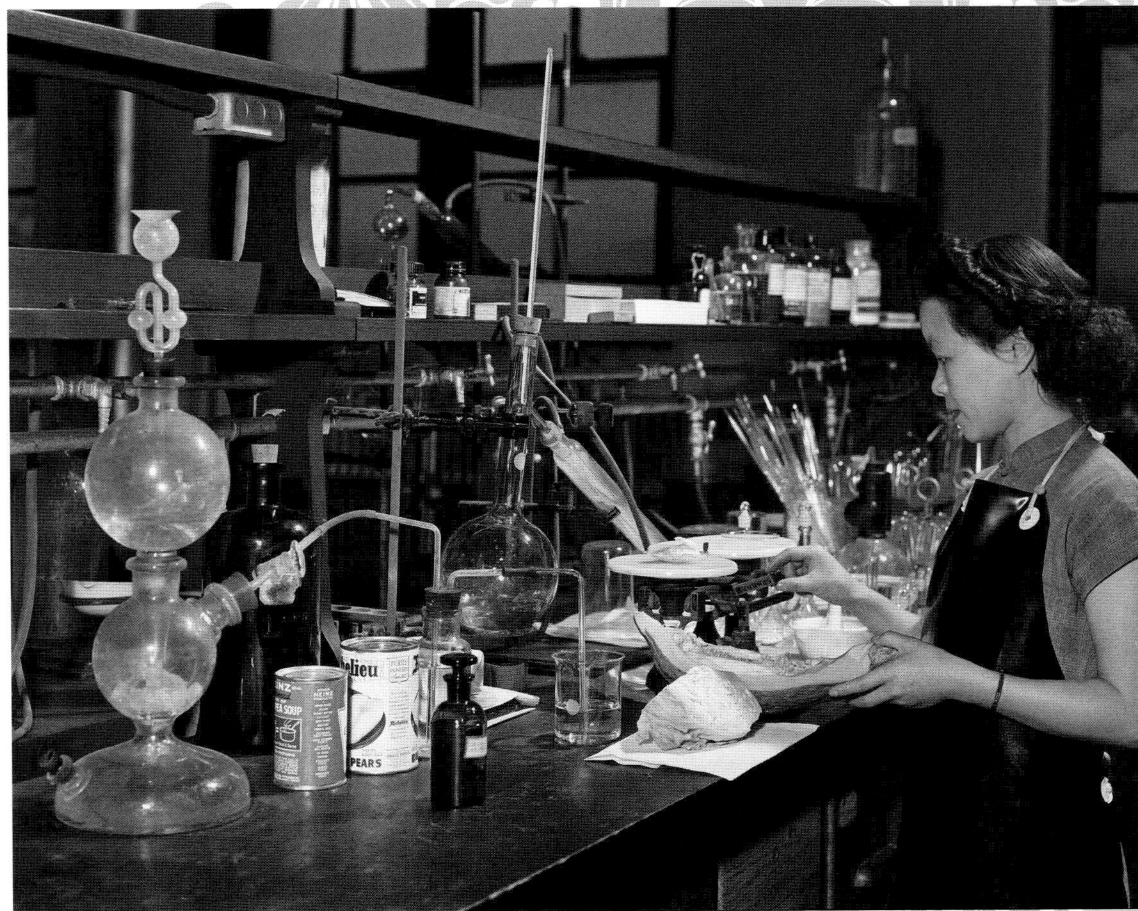

Allen took these photographs of various classrooms at Purdue for a home economics booklet. (Tippecanoe County, 1950)

Research to Improve Farm Productivity

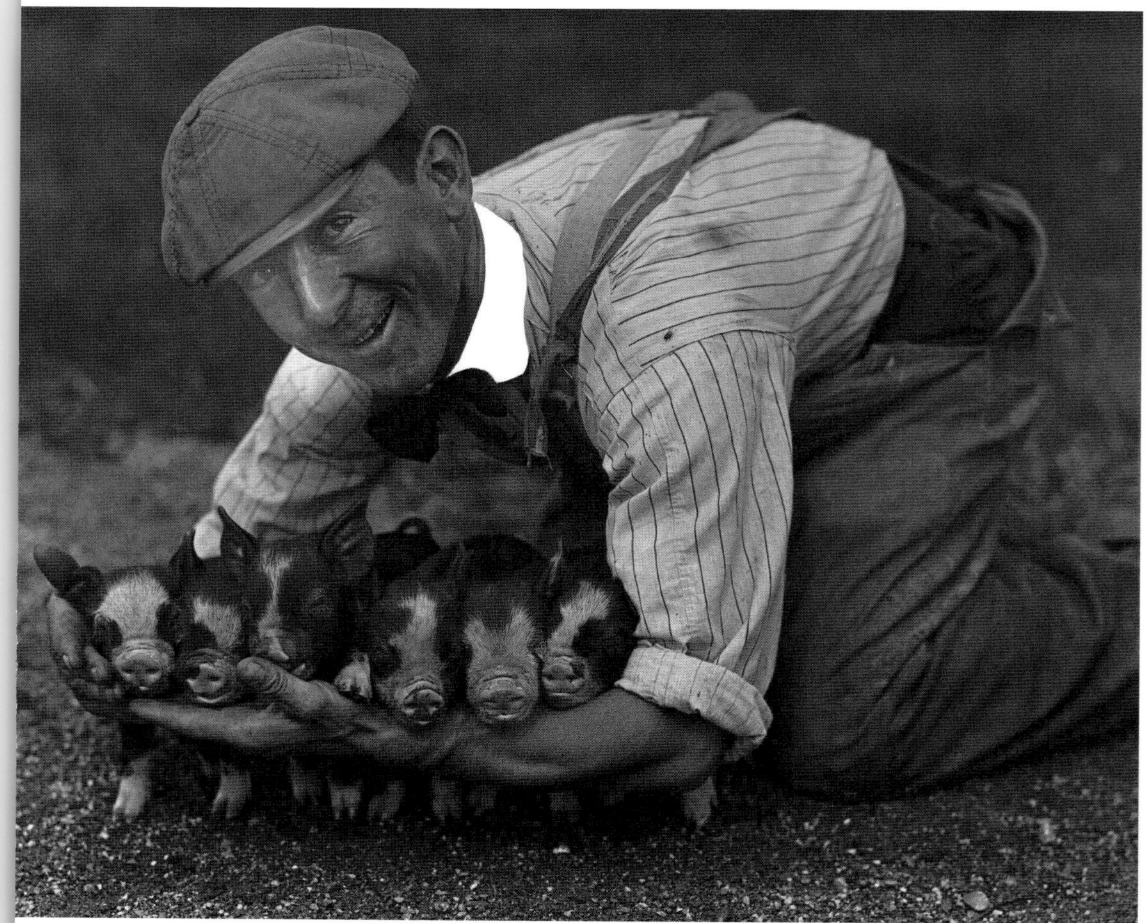

Purdue herdsman "Shorty" Rogers displays an armful of Berkshire pigs. (Tippecanoe County, 1914)

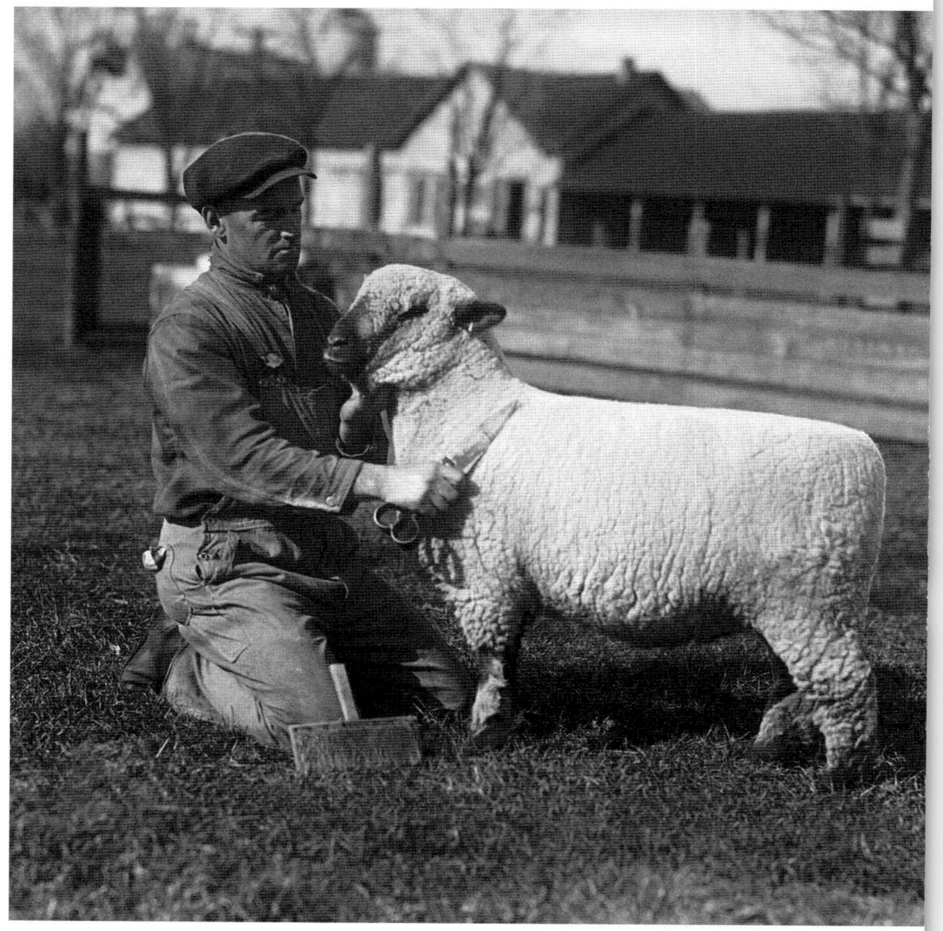

Sam Breese, the Purdue shepherd, uses hand shears and a wool card to prepare the wool of this Shropshire sheep prior to judging. (Tippecanoe County, 1923)

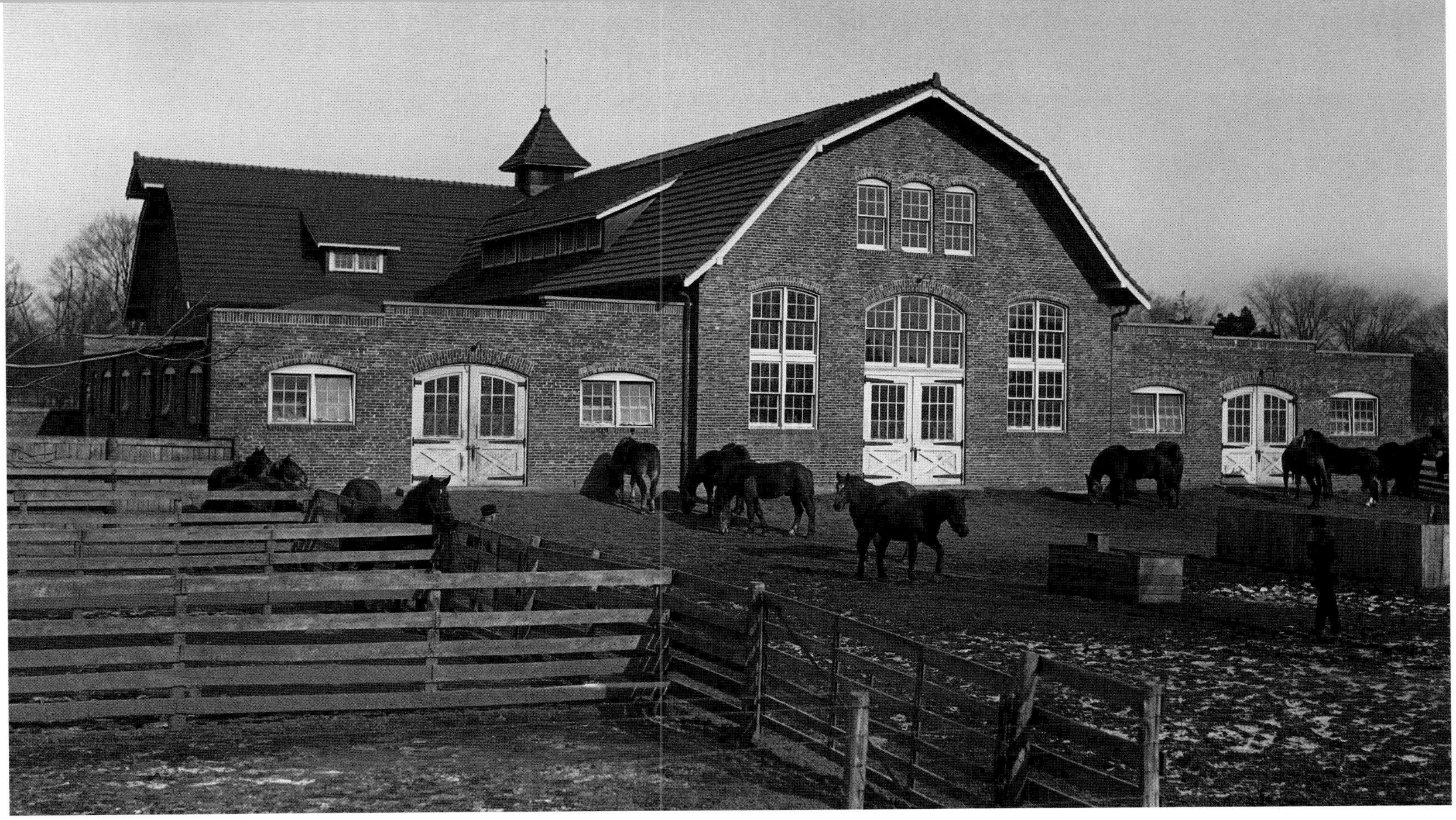

The horse barn on the Purdue University campus farm (Tippecanoe County, 1924)

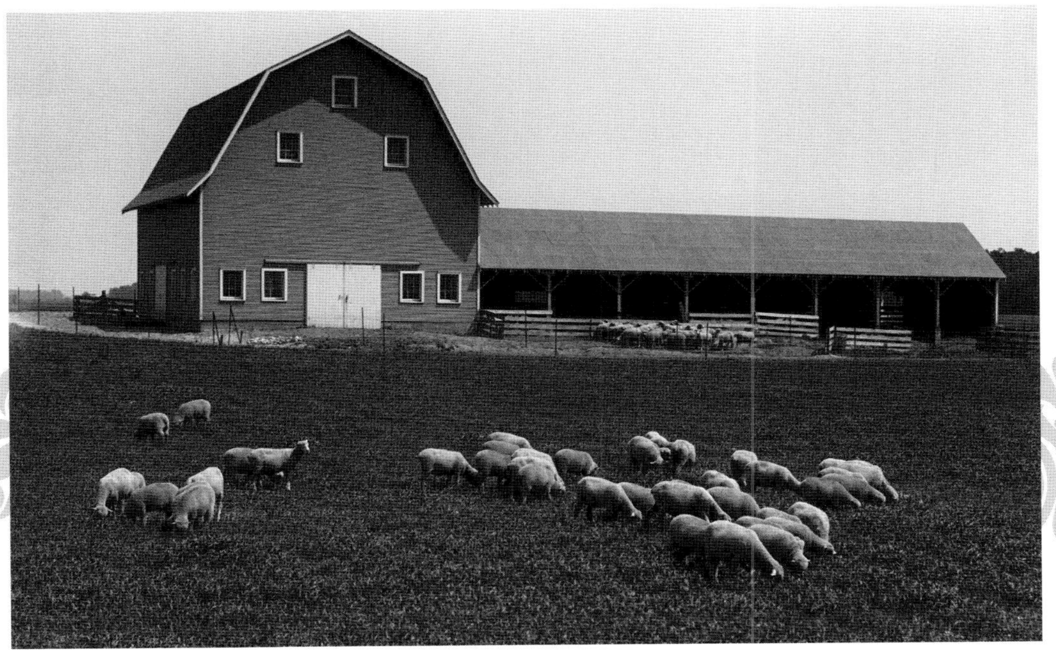

Sheep forage on oats in front of a new sheep barn at the Purdue University Livestock Experimental Farm in West Lafayette. (Tippecanoe County, 1927)

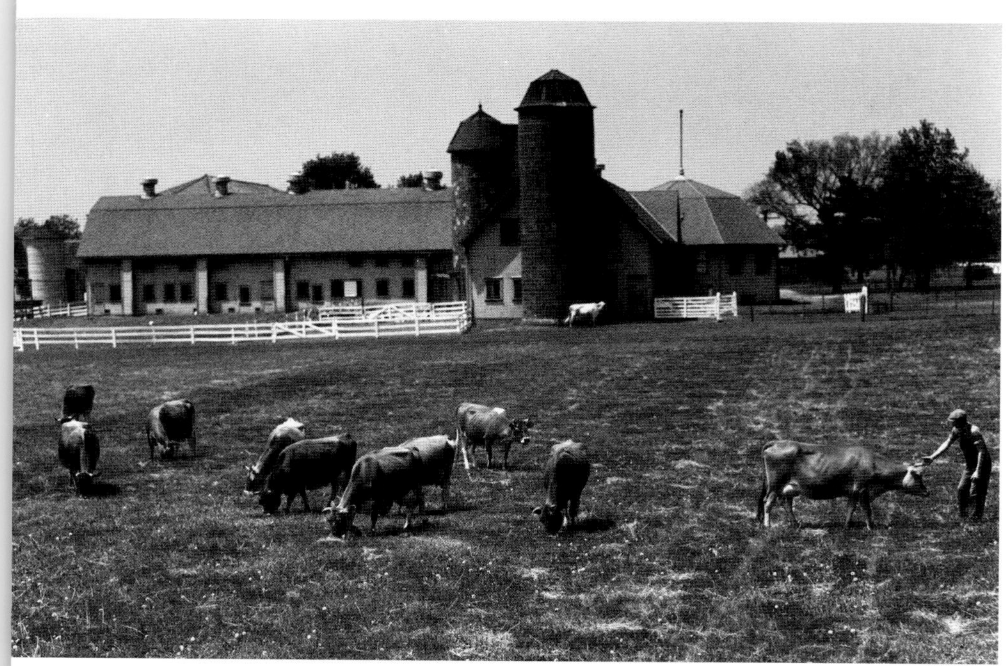

A herd of Jersey cows grazes in front of the dairy barns on the Purdue University campus farm in West Lafayette. (Tippecanoe County, 1935)

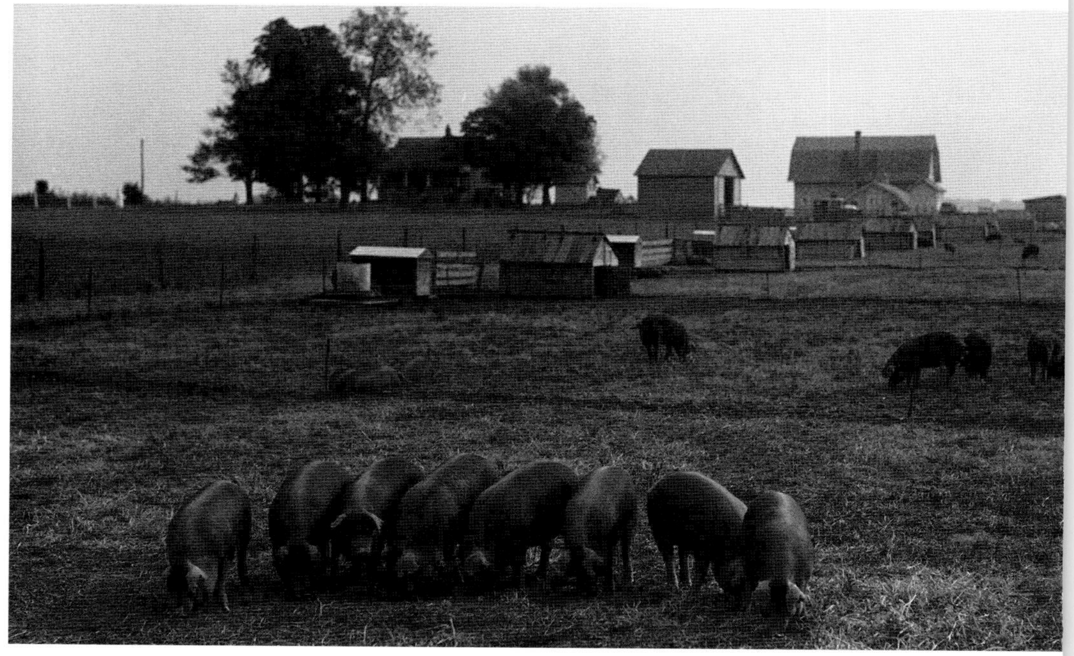

One lot of hogs at the Purdue University Livestock Experimental Farm (Tippecanoe County, 1928)

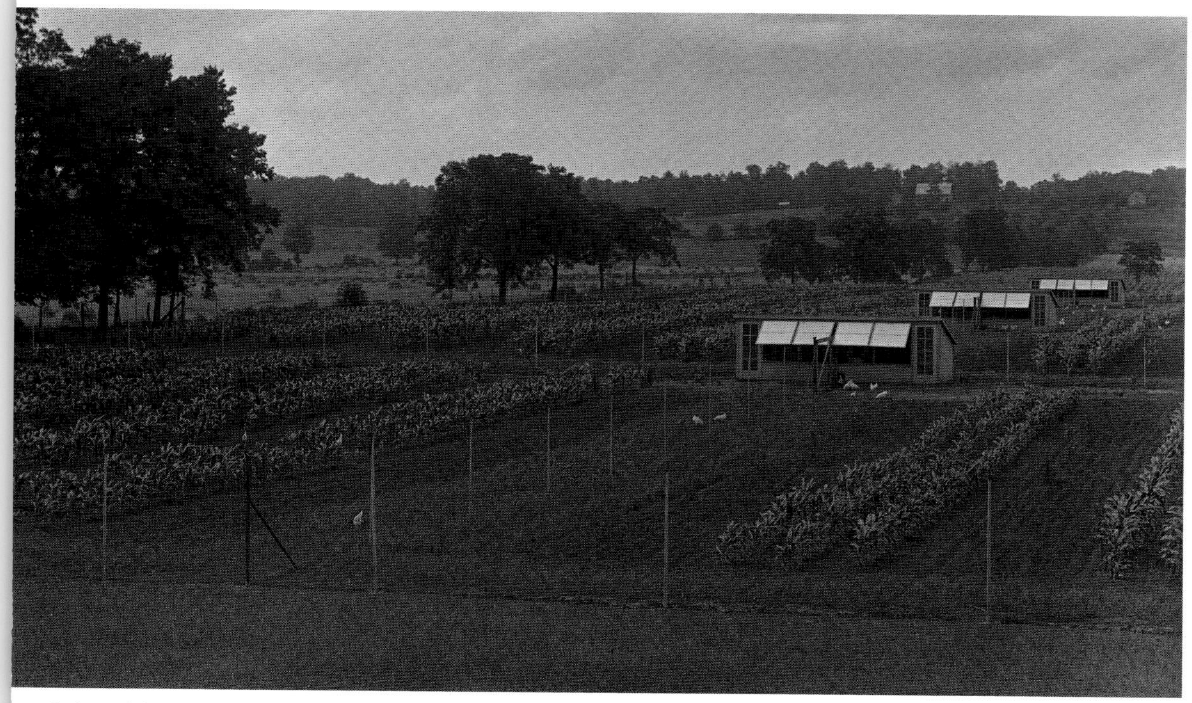

A view of the Purdue University Poultry Farm, which once stood where modern-day dormitories now stand on the west side of the West Lafayette campus (Tippecanoe County, 1912)

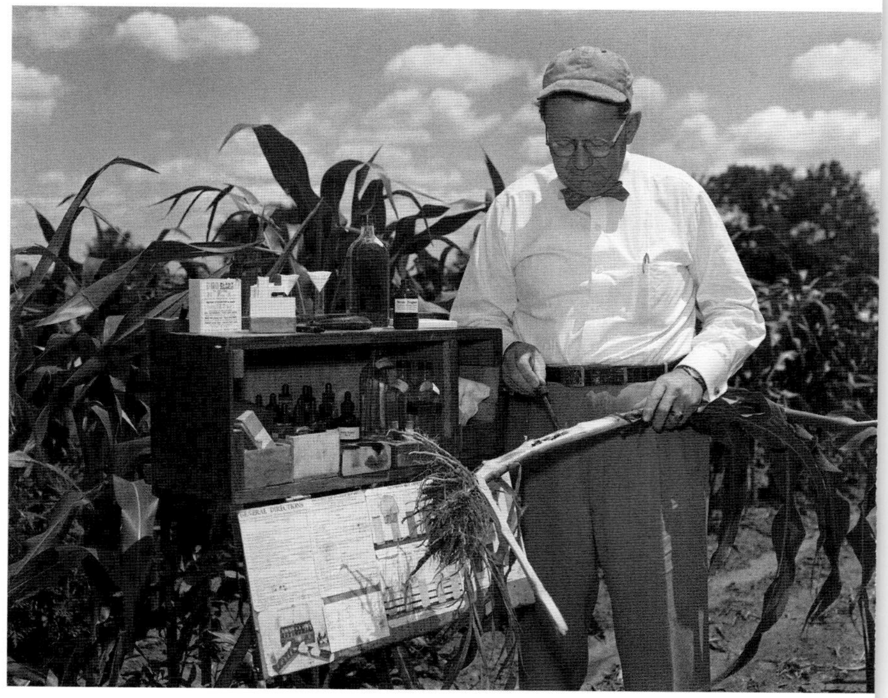

Dr. George Scarseth from the Purdue Department of Agronomy uses a kit to test for nutrients in corn plants and soil. (Tippecanoe County, 1953)

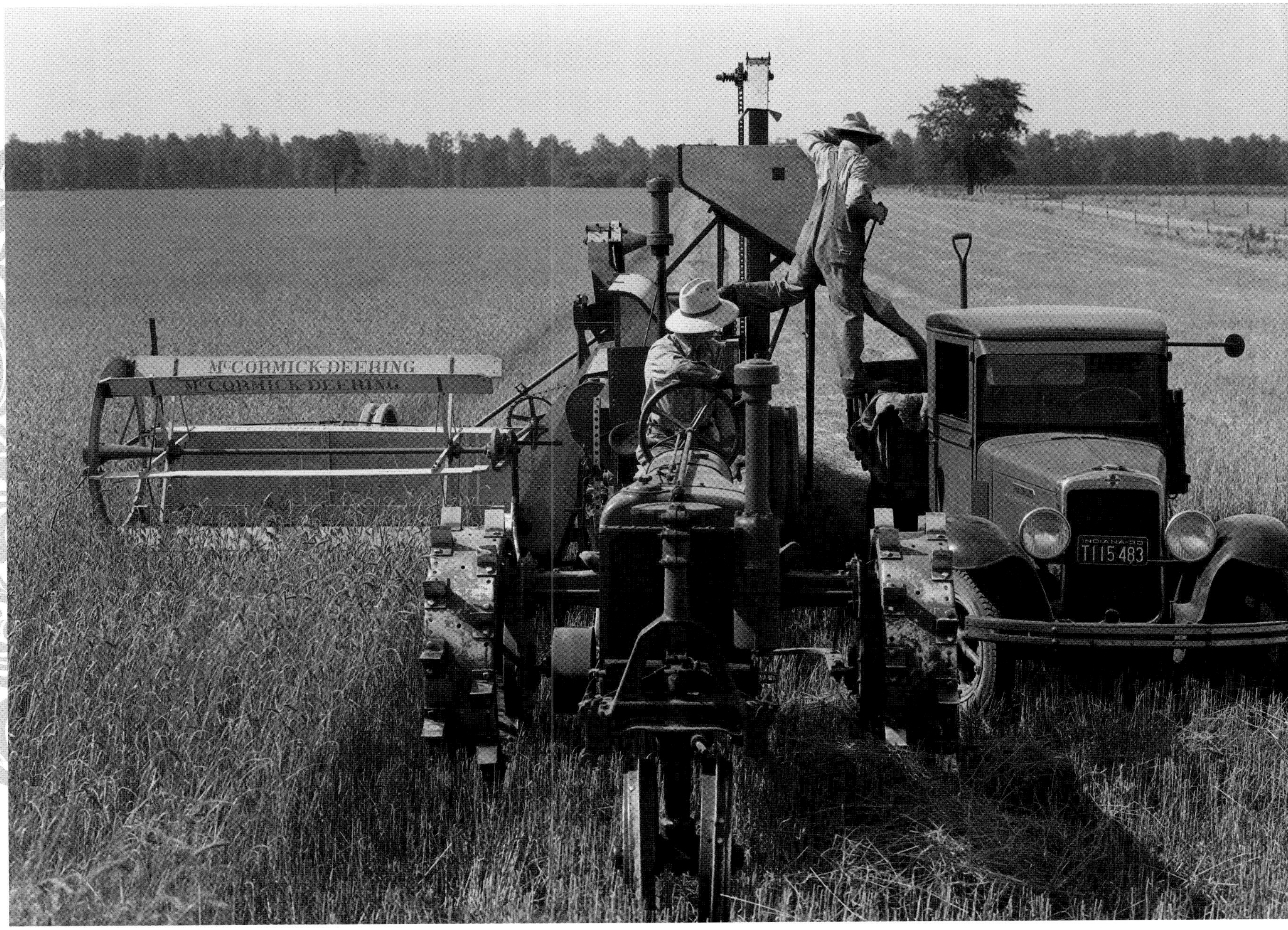

A worker empties grain from an International Harvester combine into a truck on the Purdue University Wilson Farm. (Tippecanoe County, 1933)

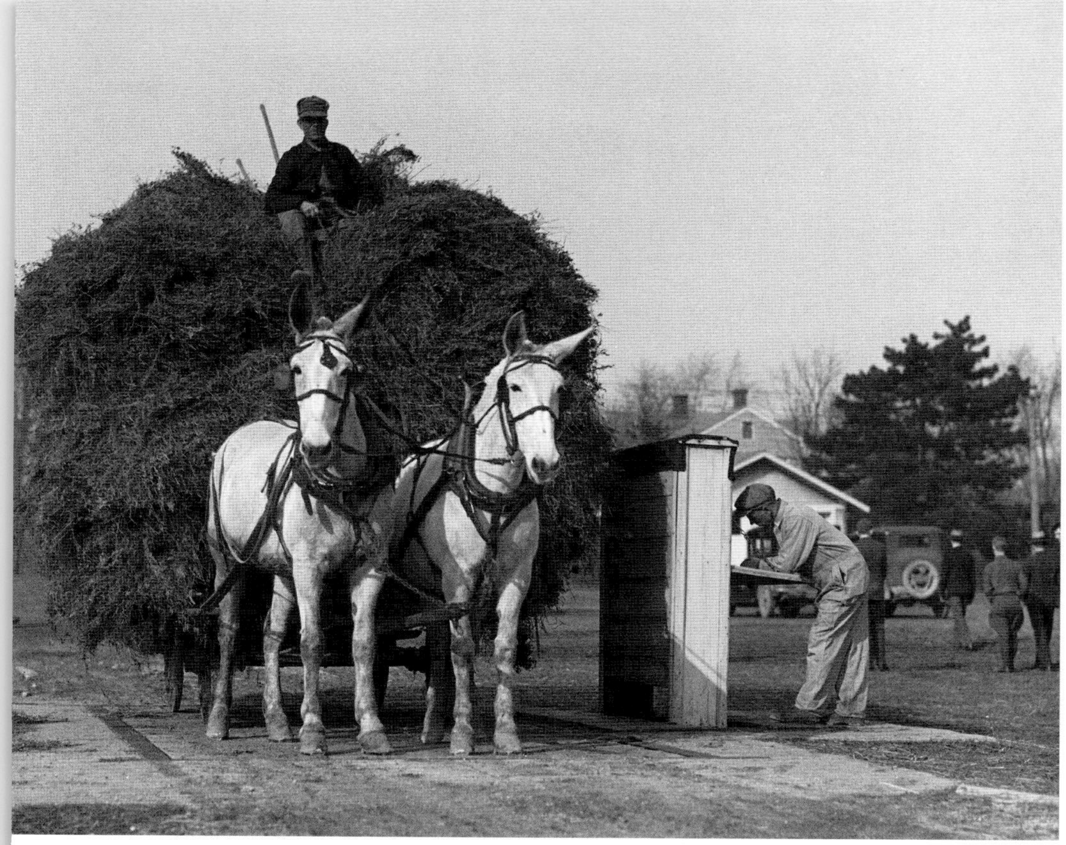

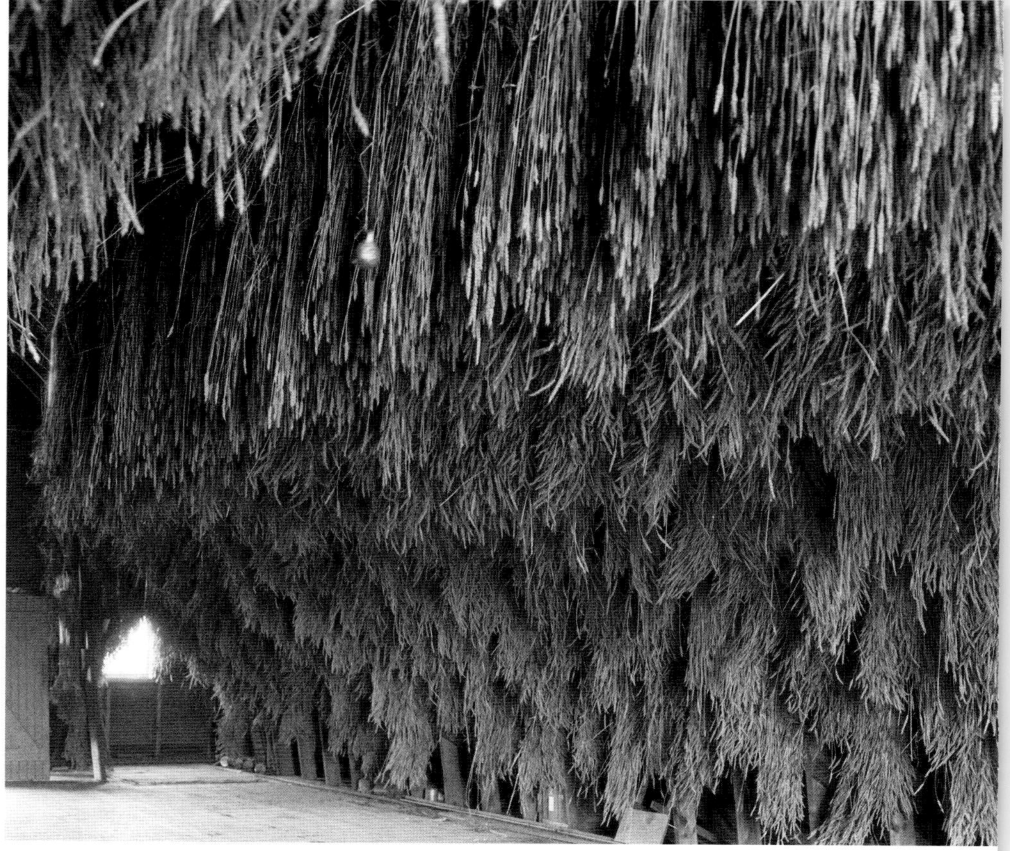

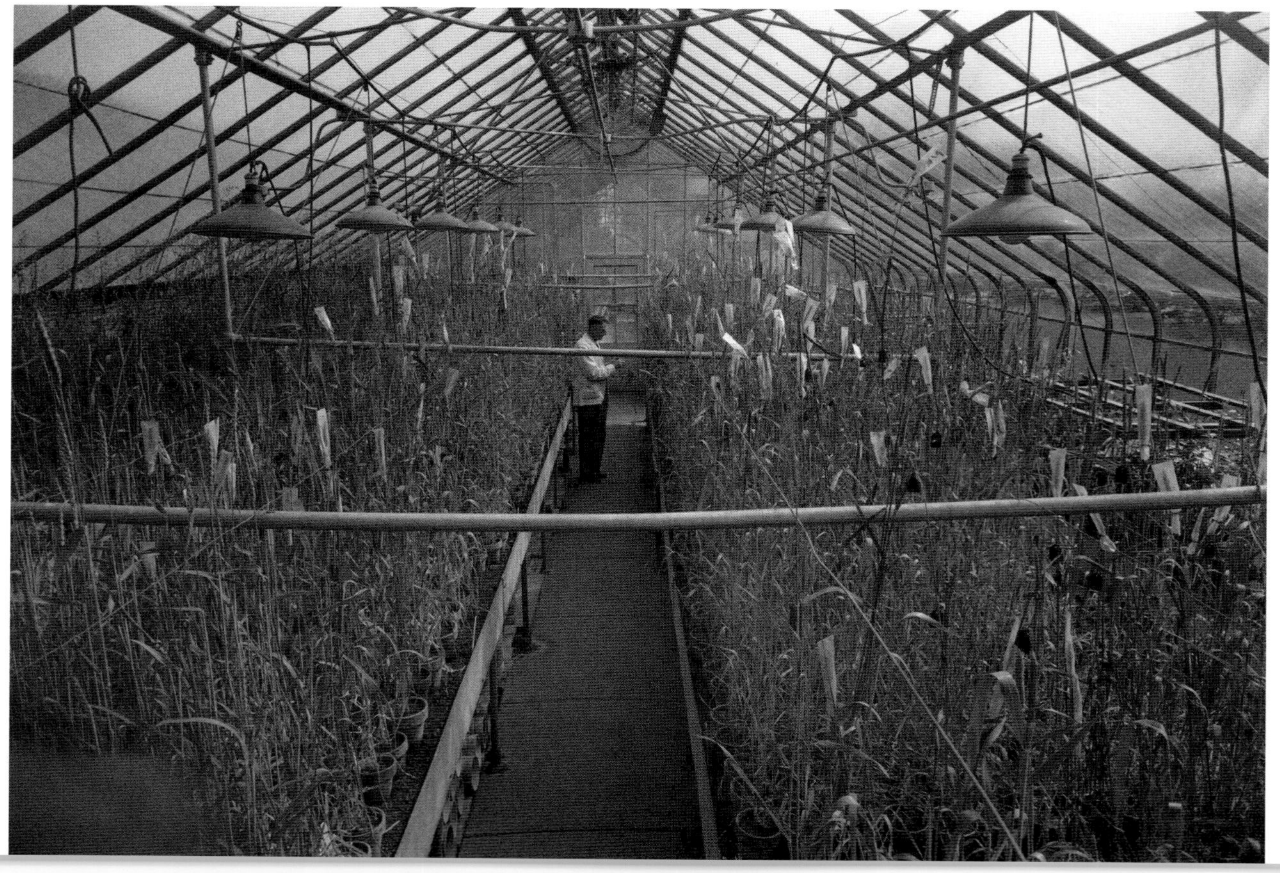

(Top left) Weighing a load of hay pulled by a team of white mules on the Purdue University campus farm. (Tippecanoe County, 1930)

(Top right) Wheat drying at the Purdue Seed House (Tippecanoe County, 1935)

(Bottom) Experimental wheat crosses at a Purdue greenhouse (Tippecanoe County, 1951)

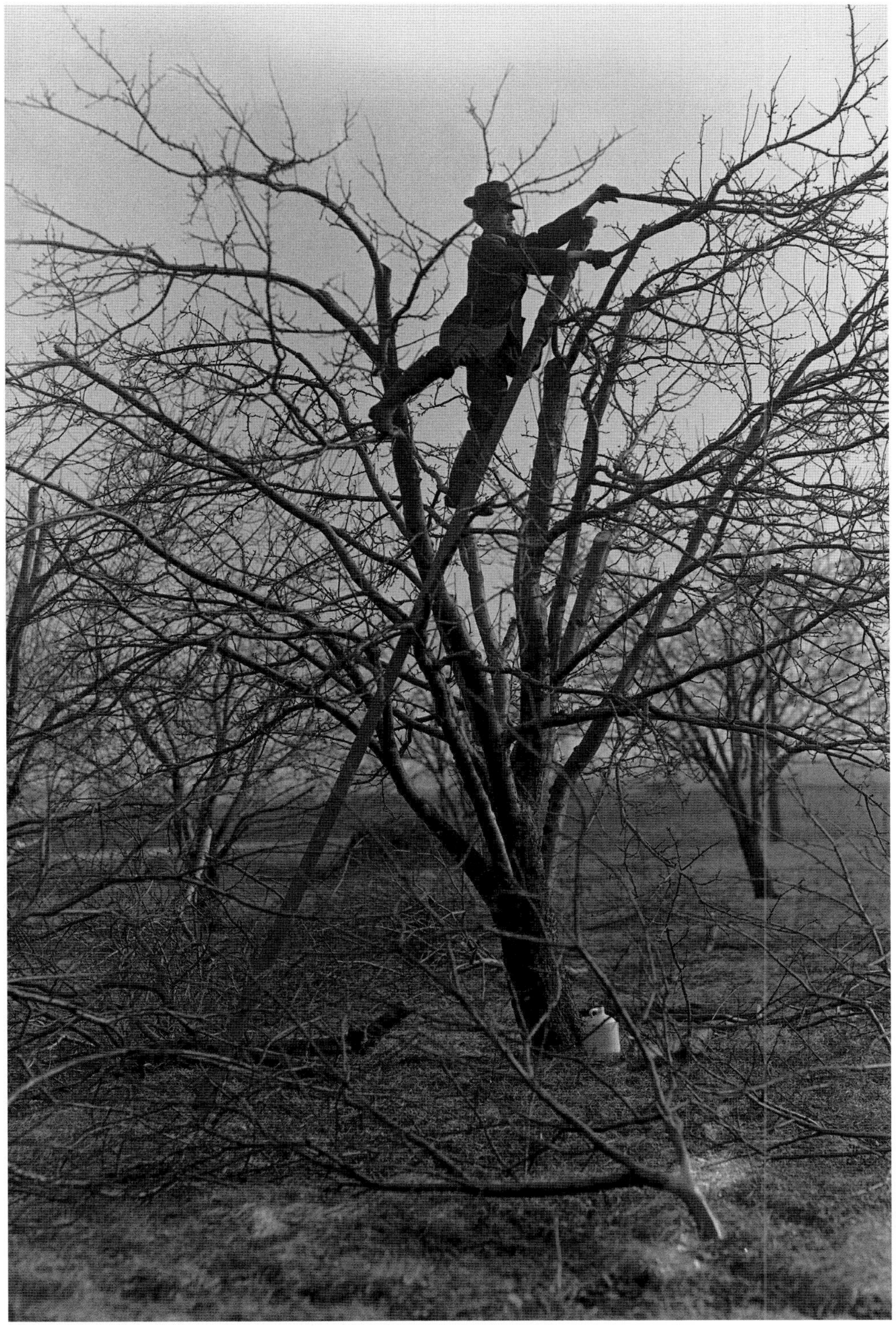

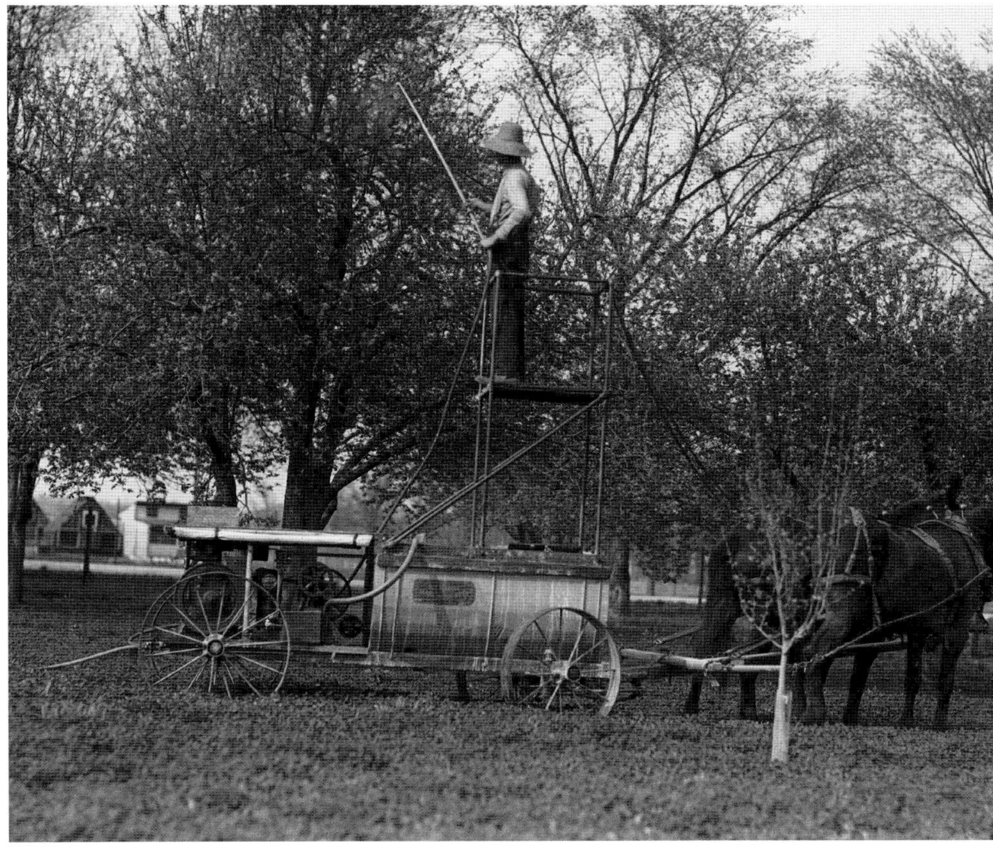

Horses pull spraying equipment used by this worker at the Purdue University orchard. (Tippecanoe County, 1915)

A man prunes an apple tree in the Purdue orchard. (Tippecanoe County, 1912)

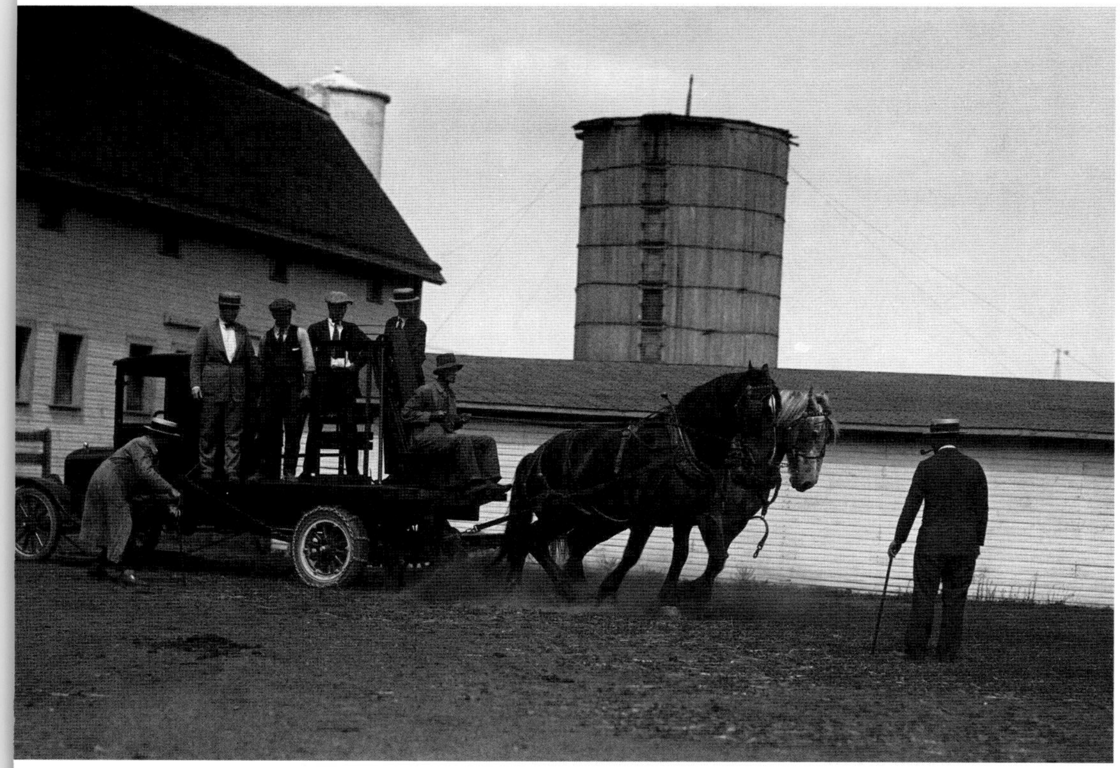

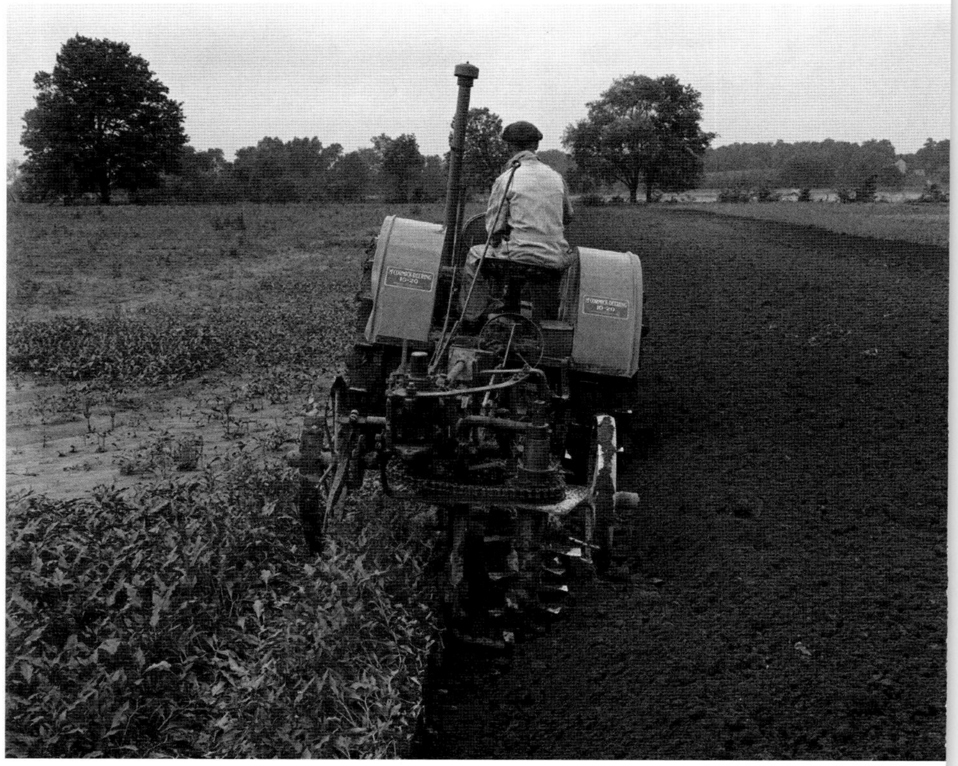

(Top left) Members of the Purdue staff used a dynamometer such as the one on this Ford truck to judge horse pulling contests held at county fairs and the Indiana State Fair. (Tippecanoe County, 1926)

(Top right) Professor William Aitkenhead from the Purdue Department of Agricultural Engineering equipped a plow with a special pulverizing device that he designed and developed to prepare soil for planting. (Tippecanoe County, 1927)

(Bottom) Purdue agricultural engineering scientists developed plow shields that were designed to bury standing cornstalks in order to reduce corn borer populations. (Tippecanoe County, 1934)

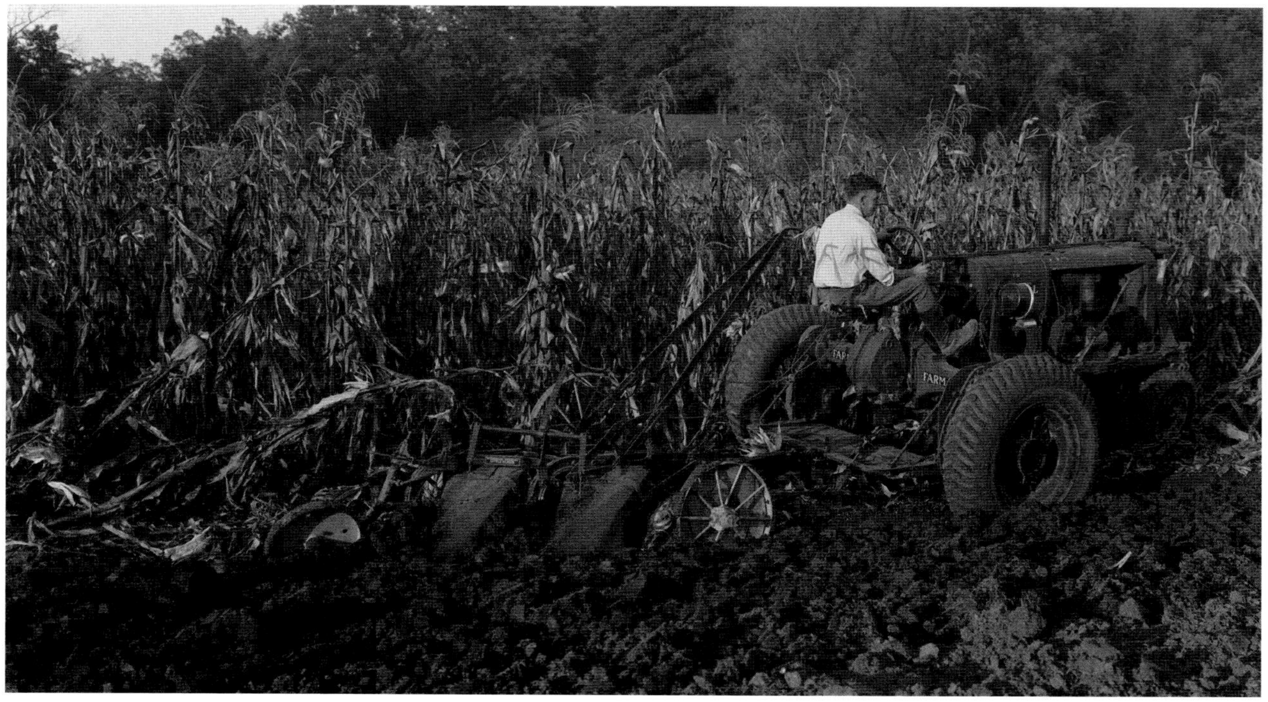

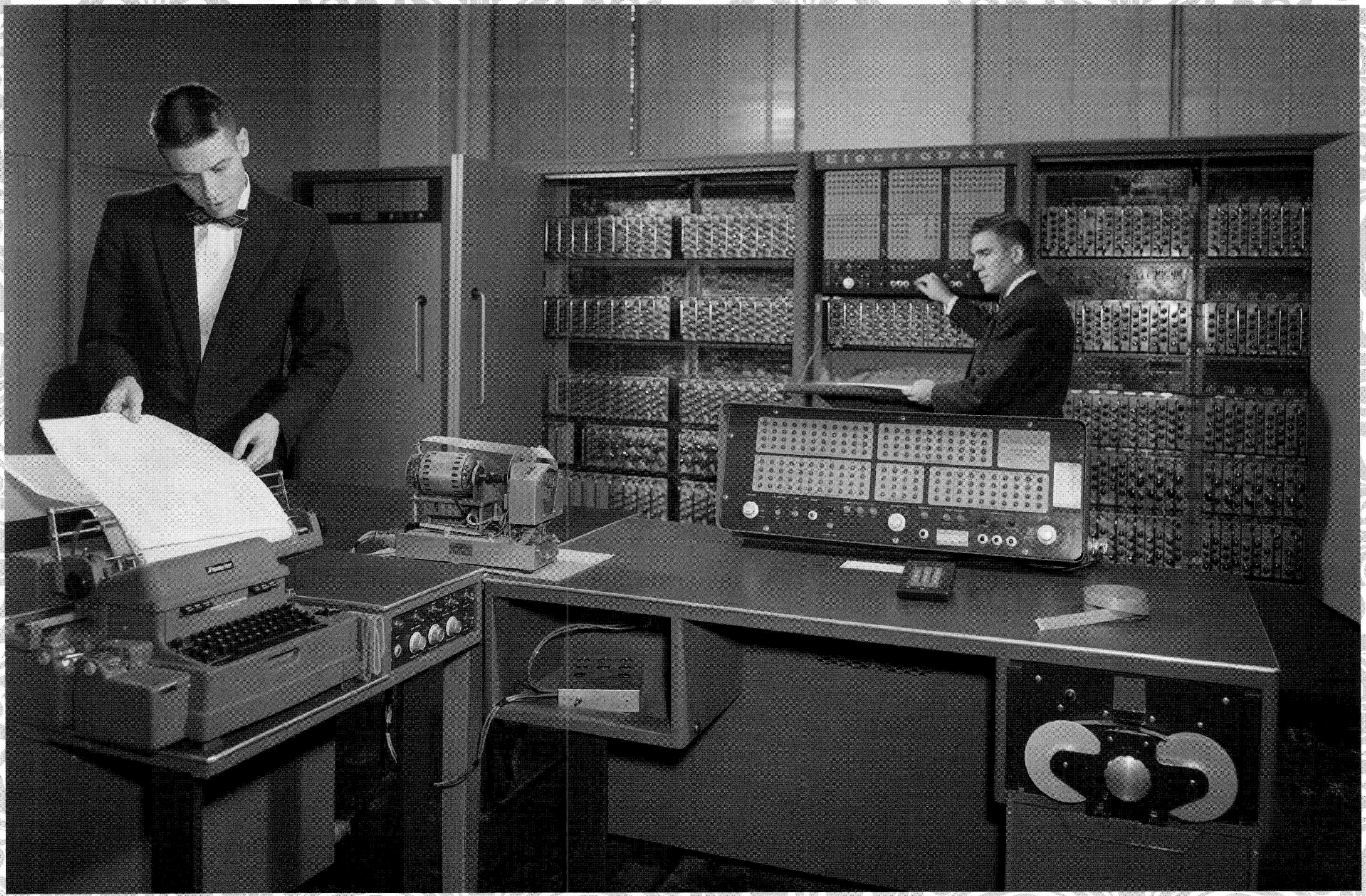

Milton M. Snodgrass (left) and Charles E. French of the Purdue Department of Agricultural Economics work with an "electric brain," a computer used to process large amounts of research data. (Tippecanoe County, 1956)

Education through Outreach

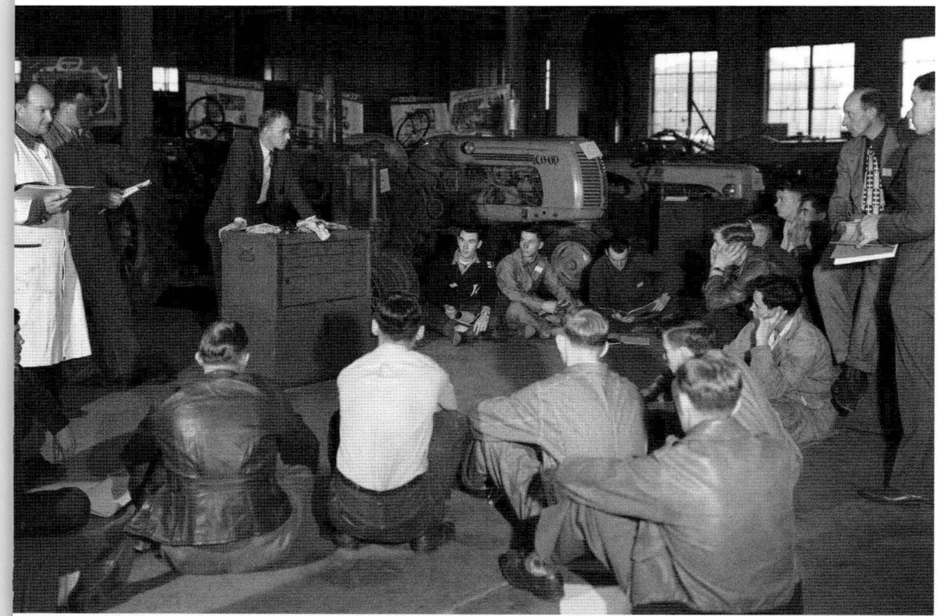

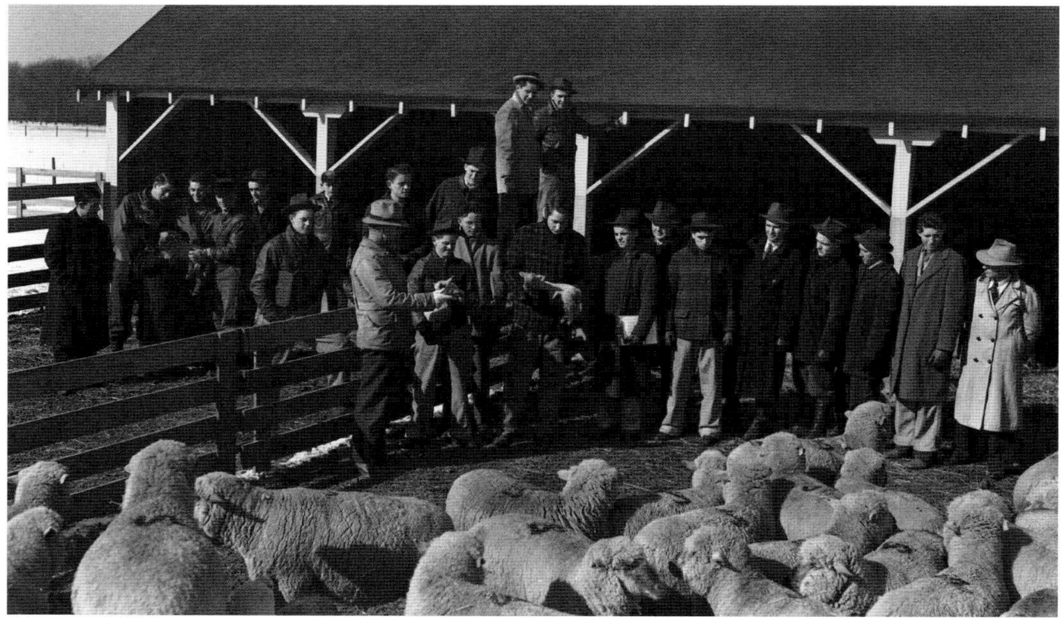

(Top left) Members of a 4-H club attend a workshop on tractor maintenance at Purdue. (Tippecanoe County, 1948)

(Top right) Students attending a Winter Short Course in agriculture inspect sheep at the Purdue University campus farm. (Tippecanoe County, 1942)

(Bottom) Winter Short Course student Robert Risch from Vincennes inspects sealed milk bottles at the creamery in Purdue's Smith Hall. (Tippecanoe County, 1944)

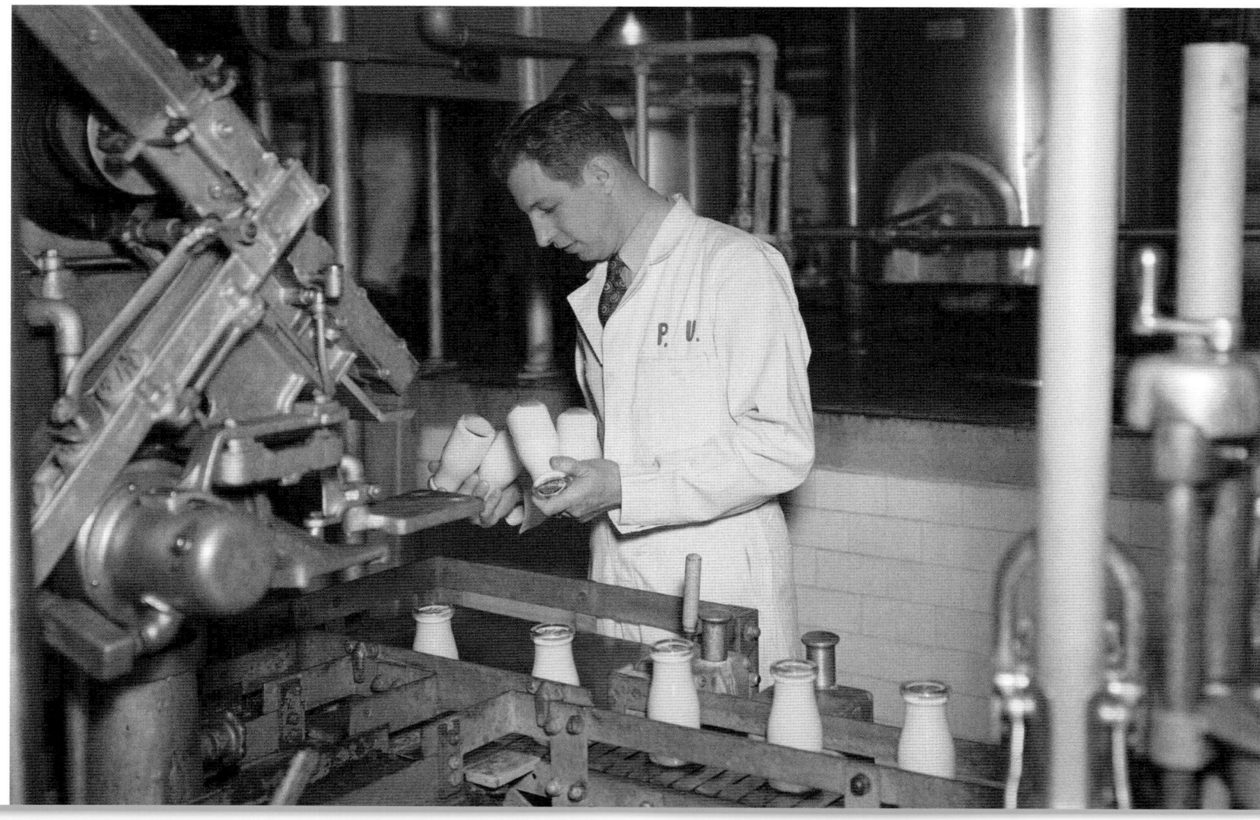

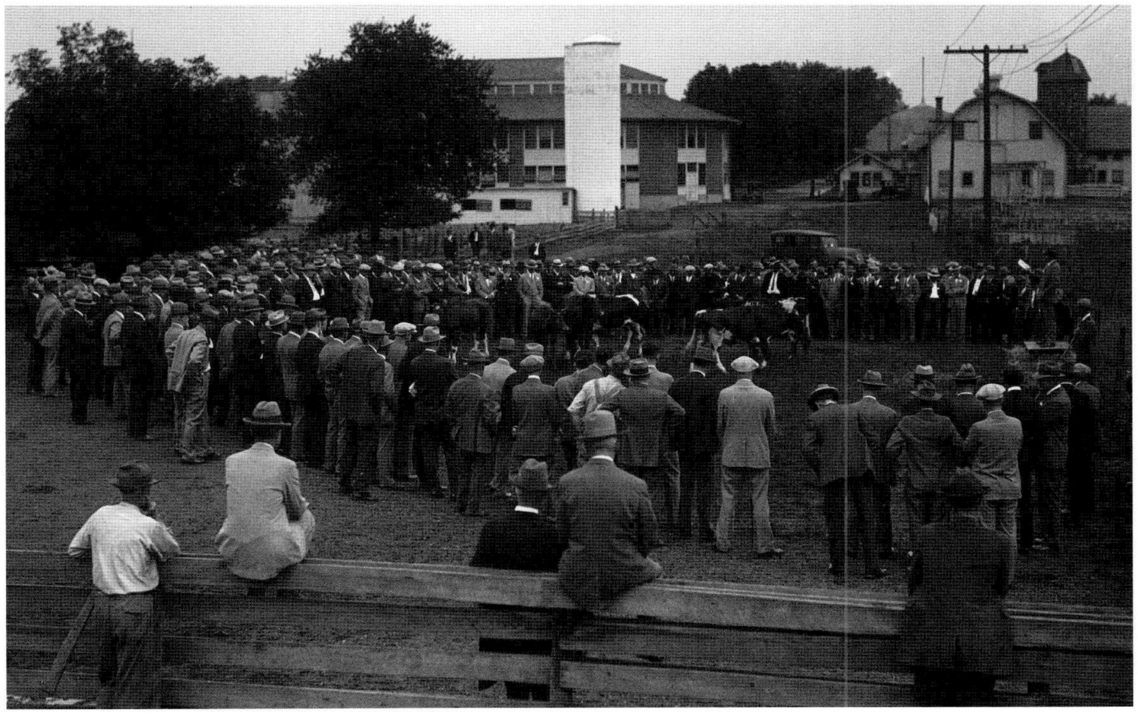

Members of the Indiana Bankers Association make an annual trip to Purdue to learn more about livestock. (Tippecanoe County, 1926)

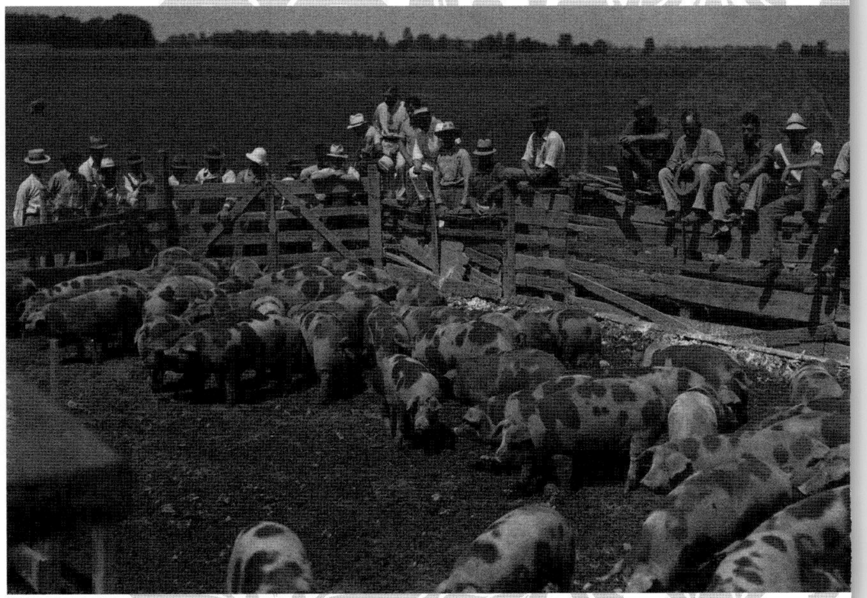

Farmers attend an event led by John Schwab, a Purdue Extension swine expert, to discuss Spotted Poland China hogs on the Cecil Sweigart farm near Cowan. (Delaware County, 1938)

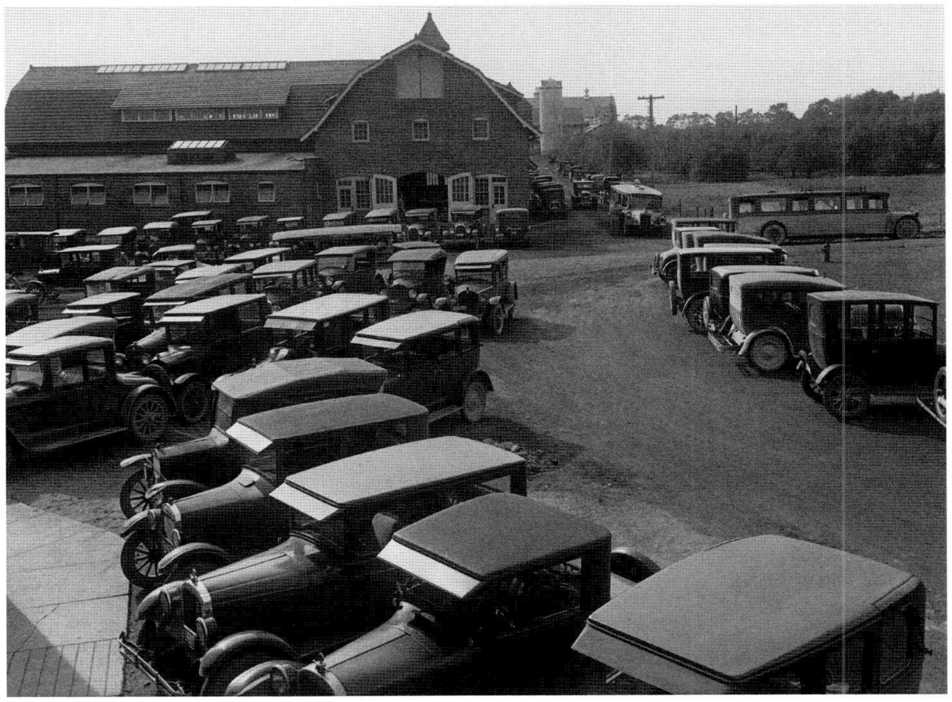

Automobiles parked on the Purdue campus farm during a Swine Day program (Tippecanoe County, 1927)

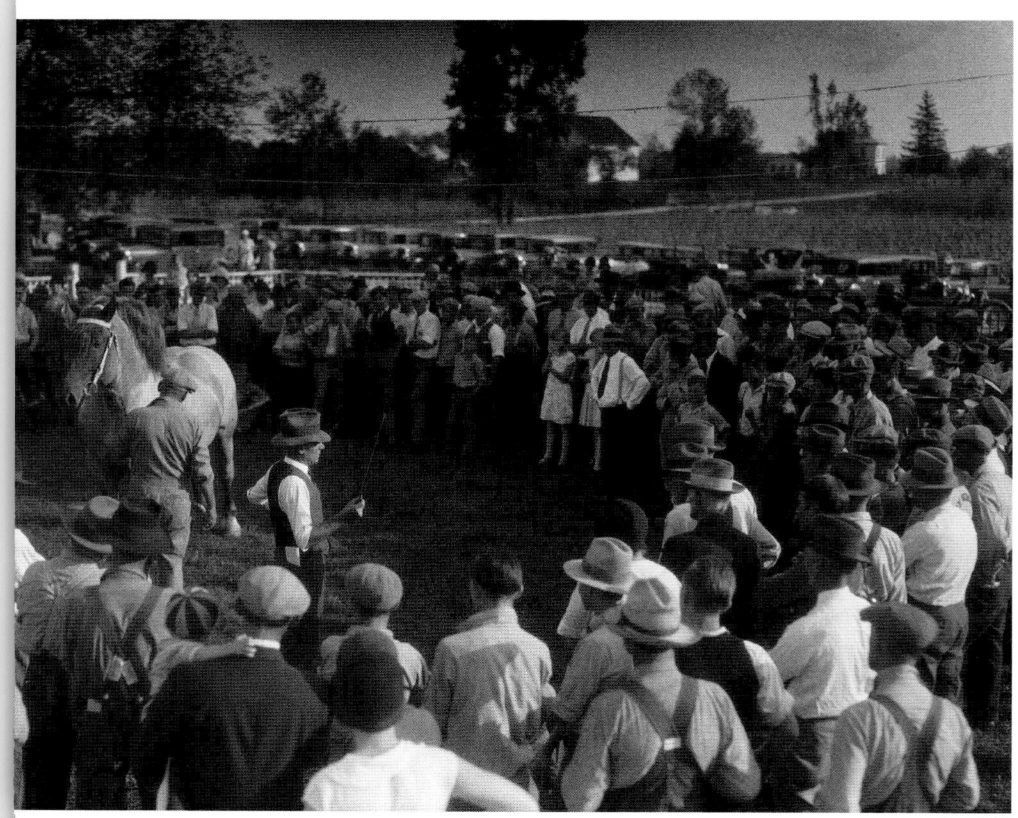

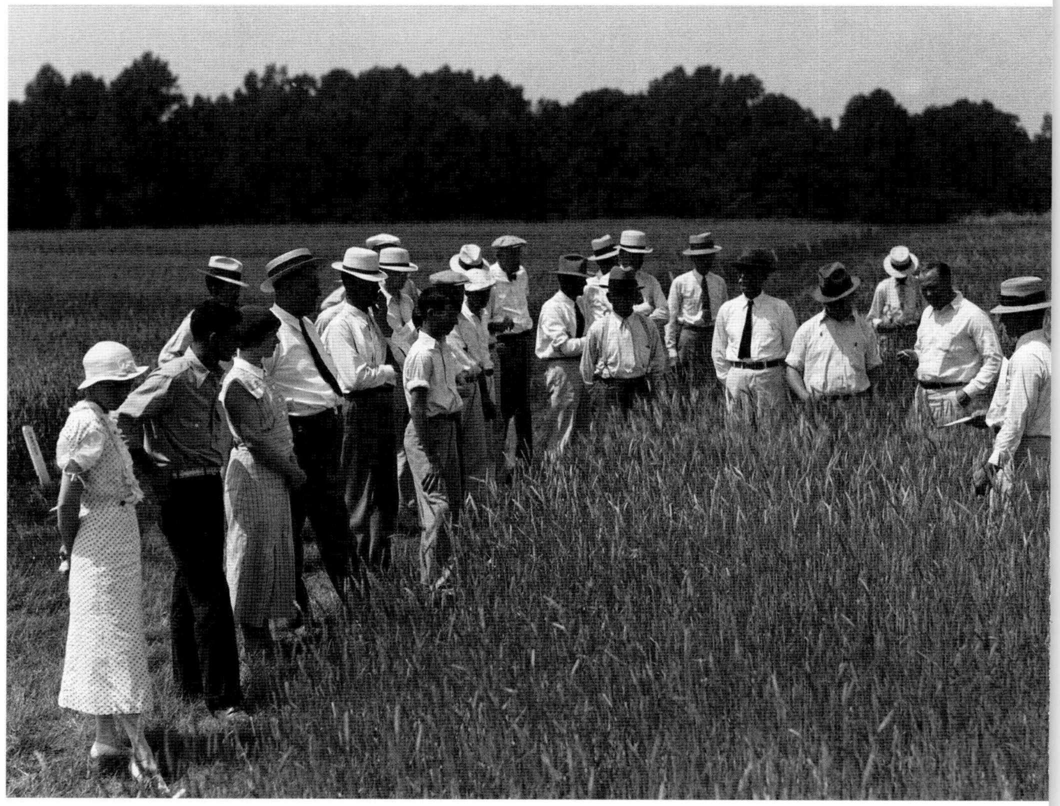

(Top left) Approximately 250 farmers attend the fourth annual show of the Hoosier Gold Medal Colt Club at Purdue University's Herbert Davis Forestry Farm in Farmland. (Randolph County, 1930)

(Top right) A small group examines wheat variety plots at Purdue's Moses-Fell Annex in Bedford during an annual picnic. (Lawrence County, 1934)

(Bottom) Farmers listen to a discussion of pasture crops at the Miller-Purdue Experimental Farm in Upland. (Grant County, 1945)

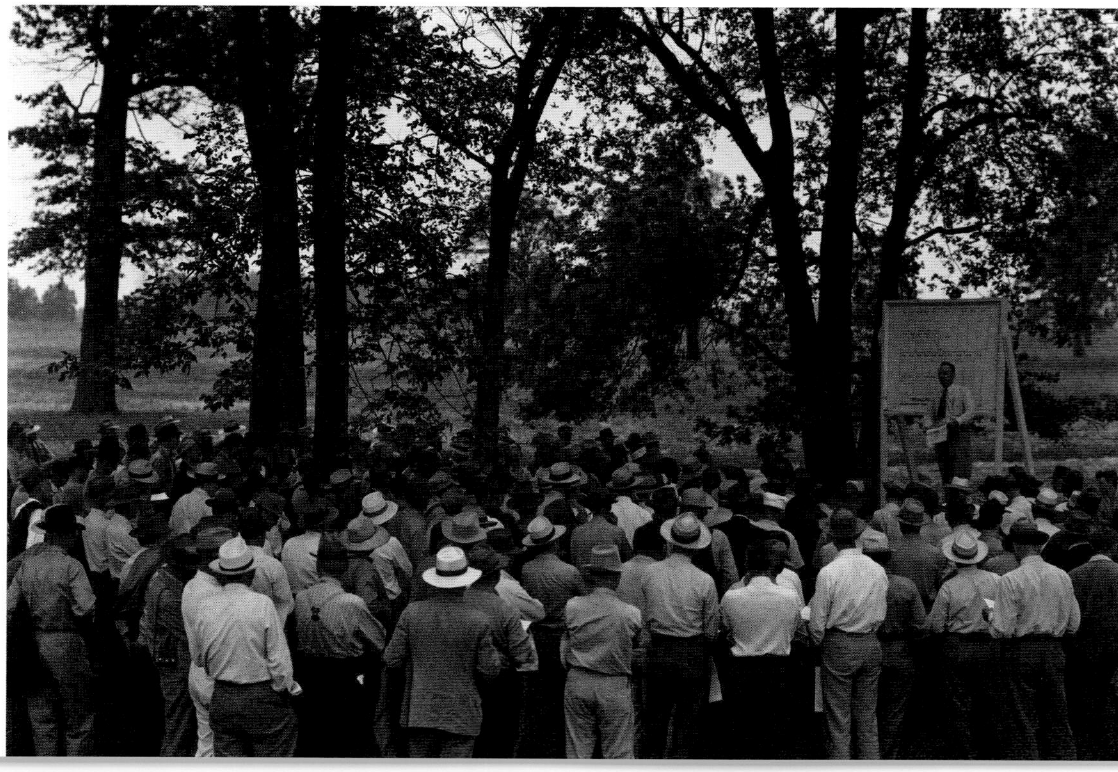

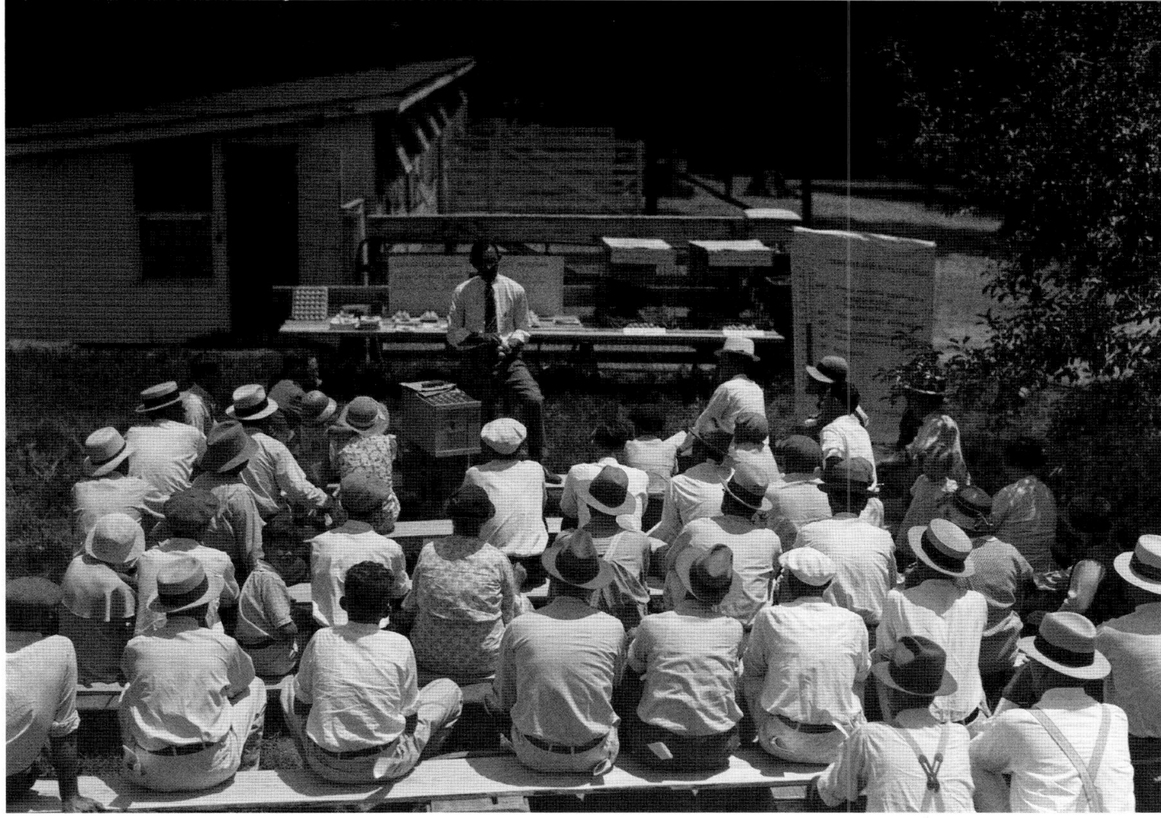

Poultry producers catch up on the latest in poultry science at Purdue's Moses-Fell Annex in Bedford. (Lawrence County, 1934)

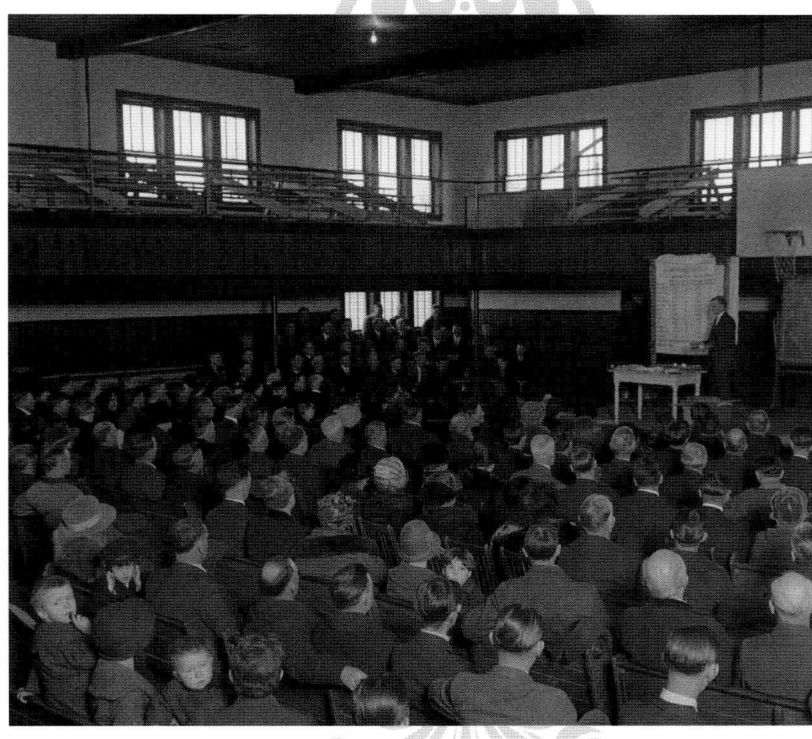

At a Farmers' Short Course in Kentland, LeRoy "Chick" Jones, a poultry Extension specialist at Purdue, discusses how to market eggs for the greatest profit. (Newton County, 1925)

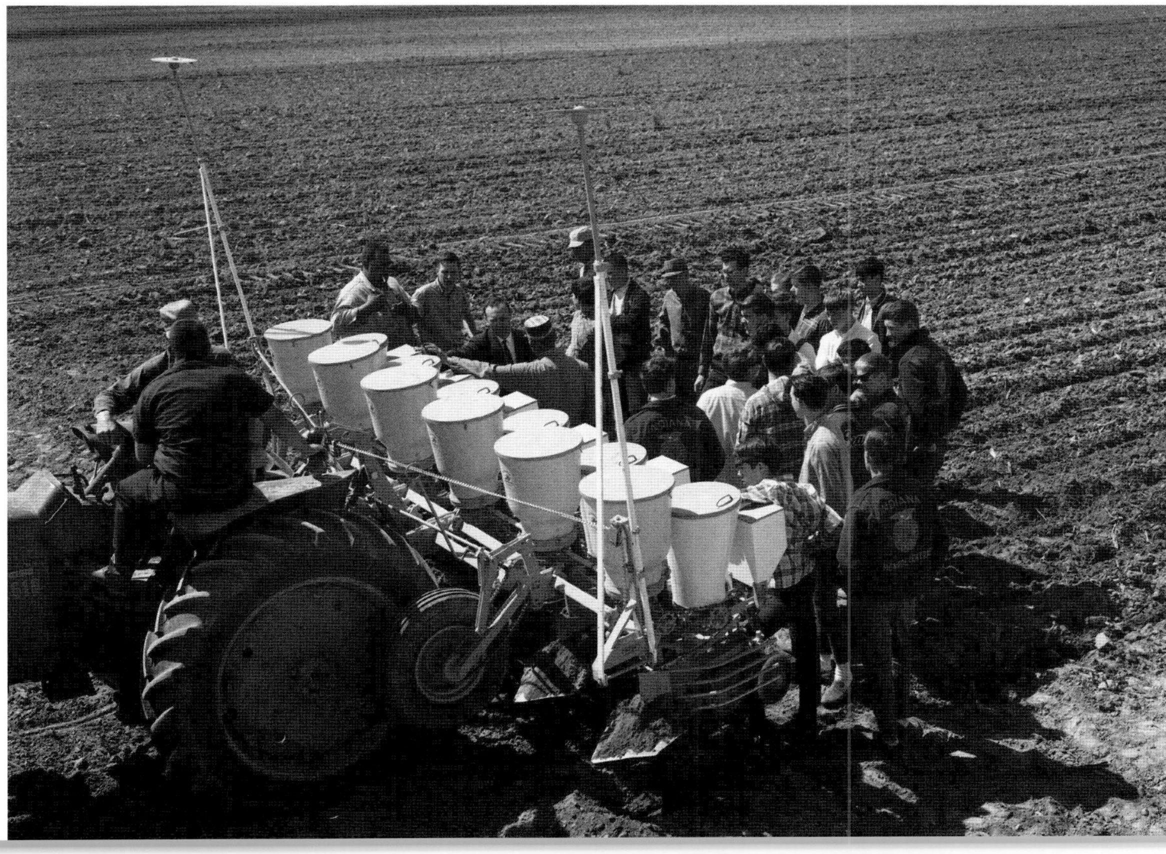

Planting corn with minimum tillage during a field day at the Purdue Herbert Davis Forestry Farm in Farmland (Randolph County, 1966)

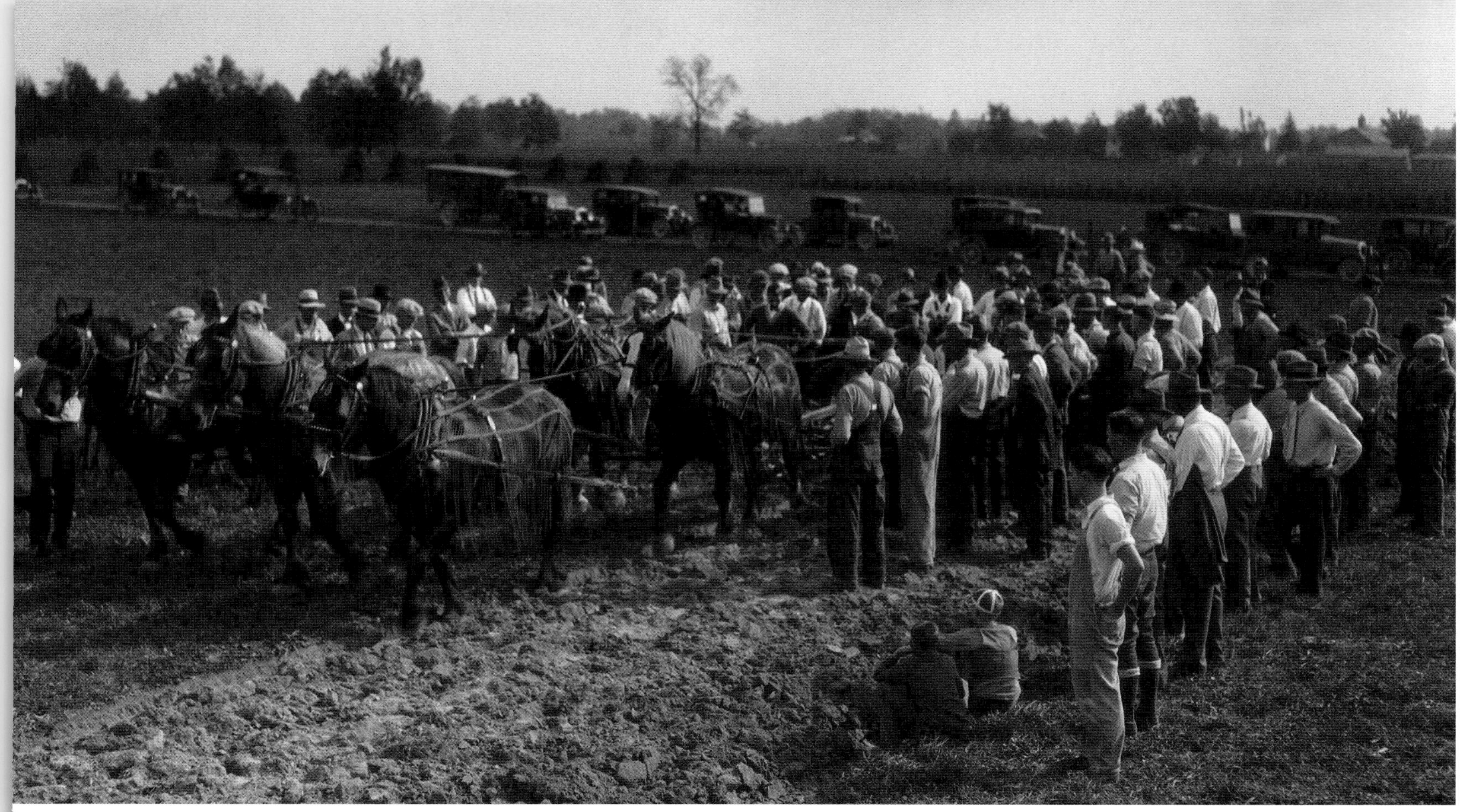

Approximately 200 farmers attend the five-horse-hitch plowing demonstration, a popular event at Purdue's Herbert Davis Forestry Farm in Farmland. (Randolph County, 1930)

A very large crowd listens as Bill Rothenberger tells how he arranged his cattle and hog feeding equipment during the Farm Management Tour on the George Rothenberger and Sons Farm near Frankfort. (Clinton County, 1946)

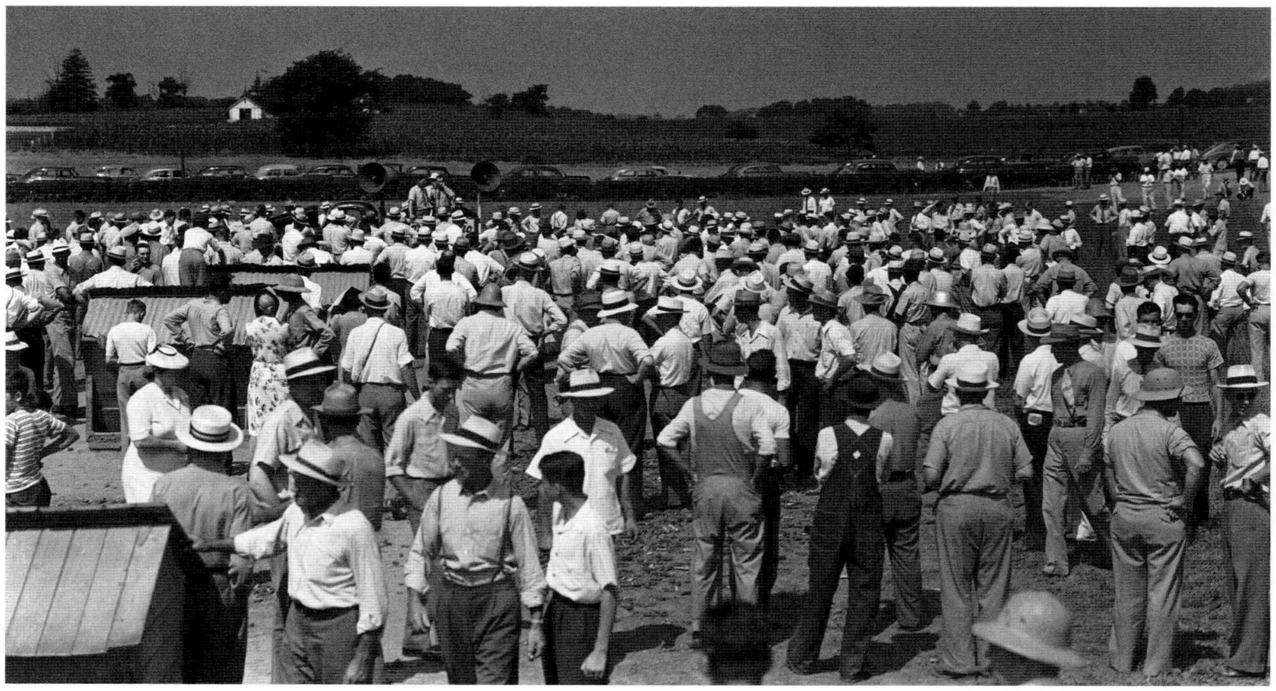

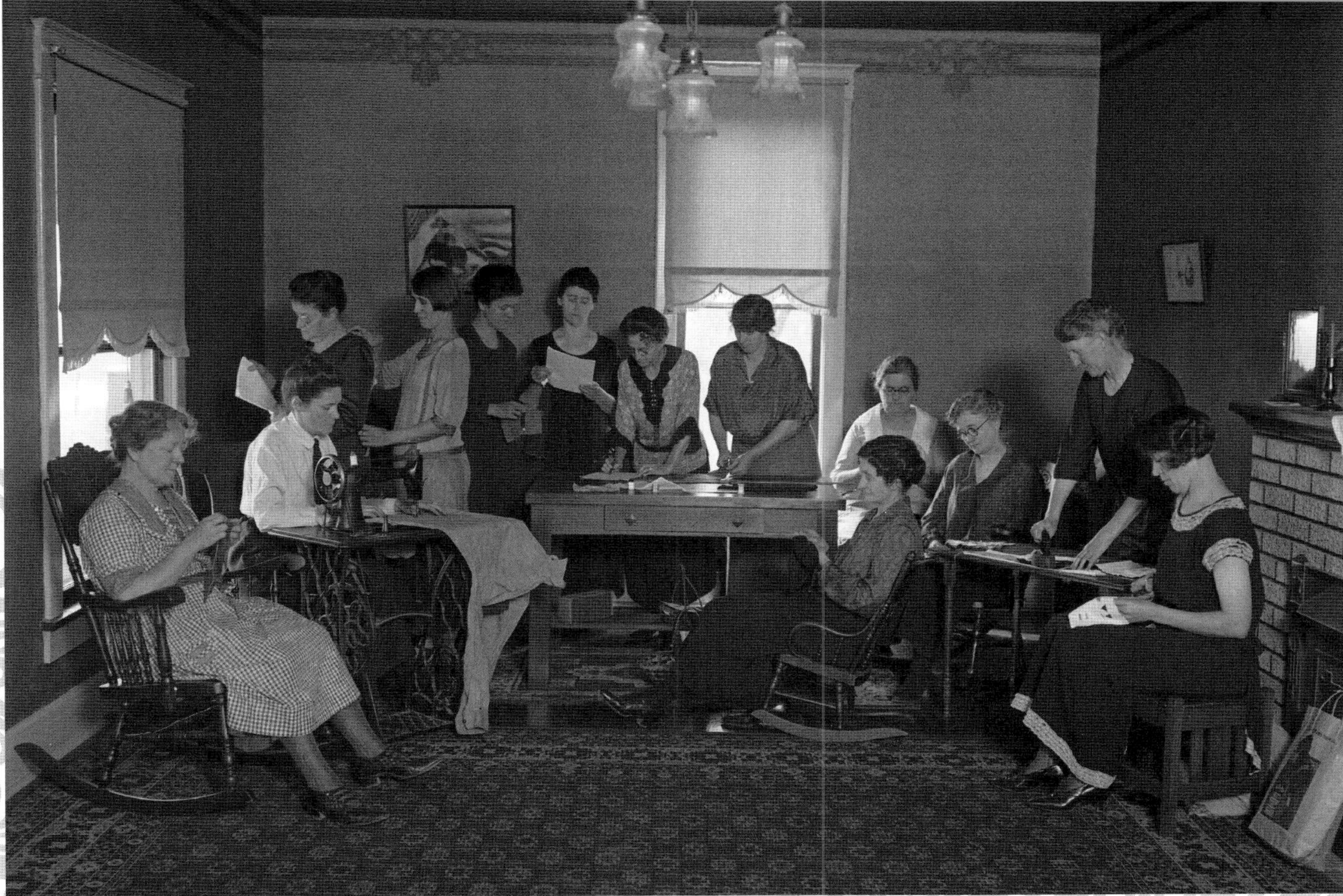

Women attend a regular training meeting of the Helt Township Ladies Sewing Class. (Vermillion County, 1924)

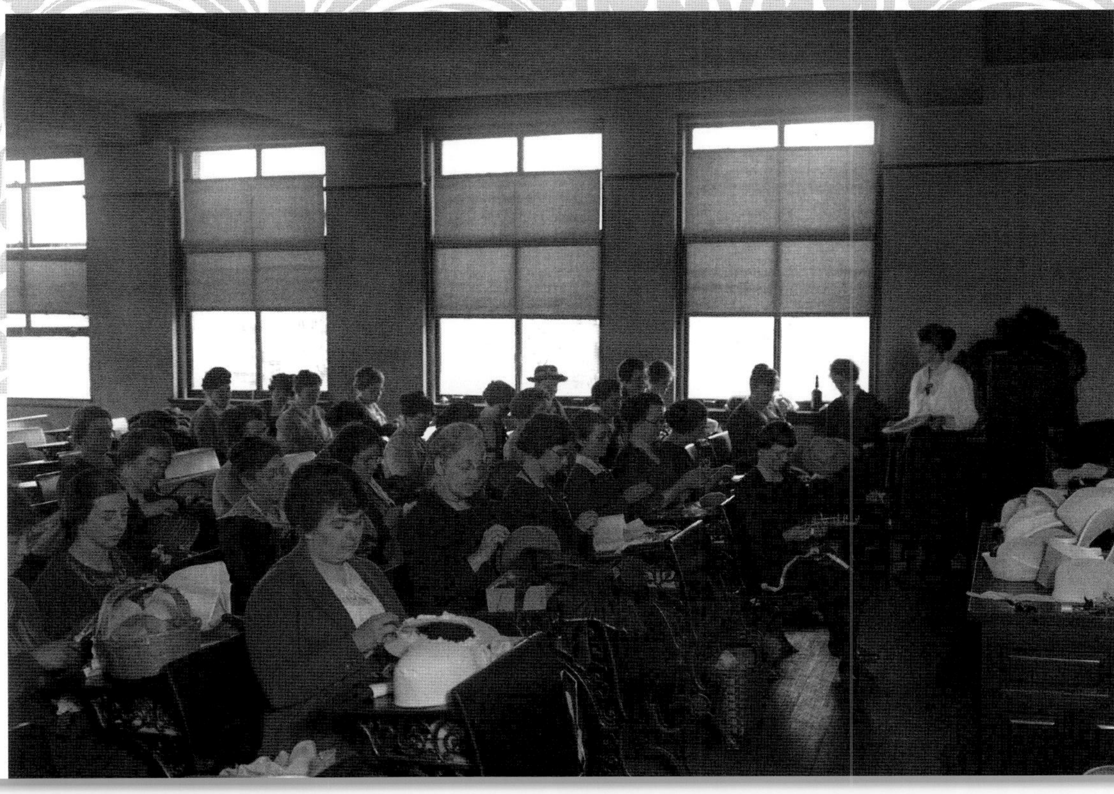

Purdue instructors teach women how to make hats at a millinery school in Crawfordsville. (Montgomery County, 1922)

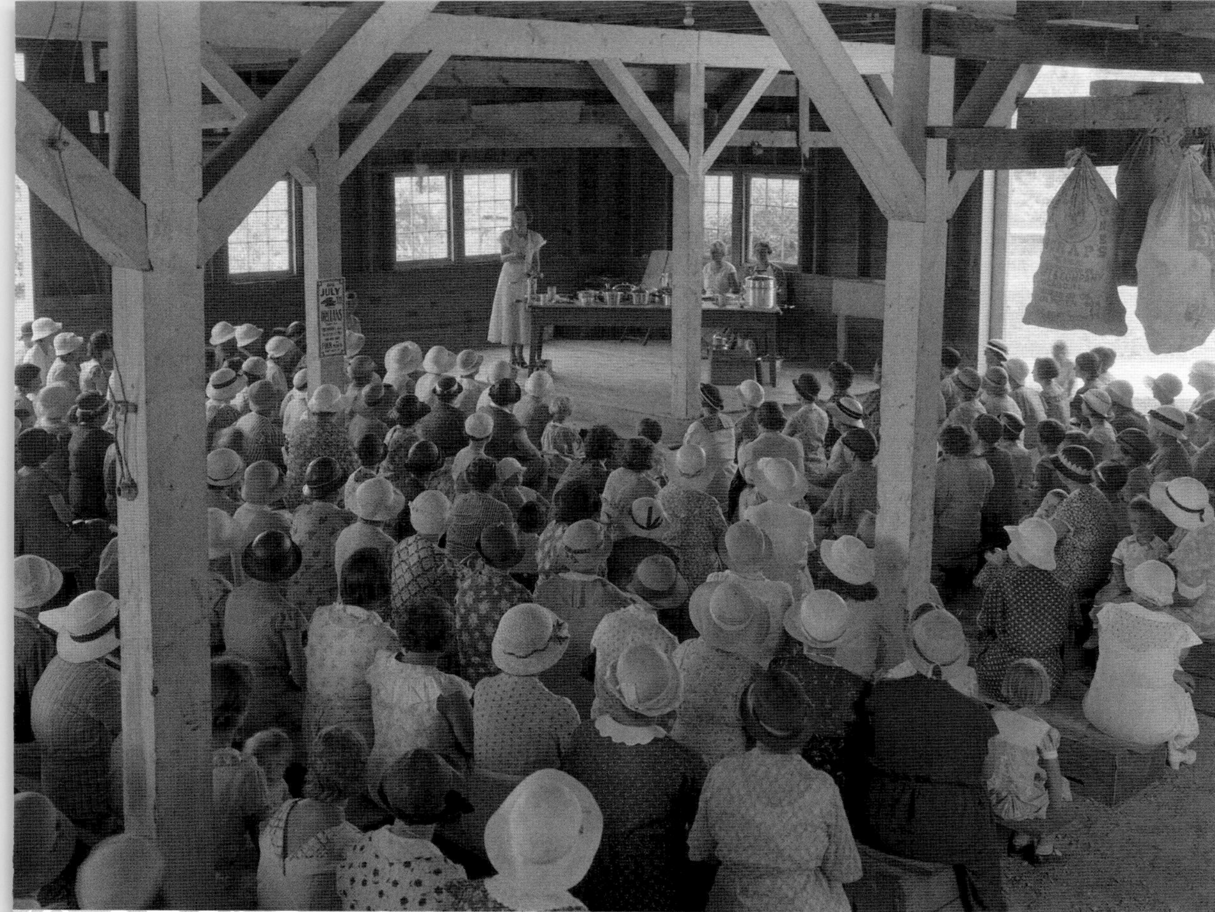

A home economics demonstration group meeting at Purdue's Moses-Fell Annex in Bedford (Lawrence County, 1934)

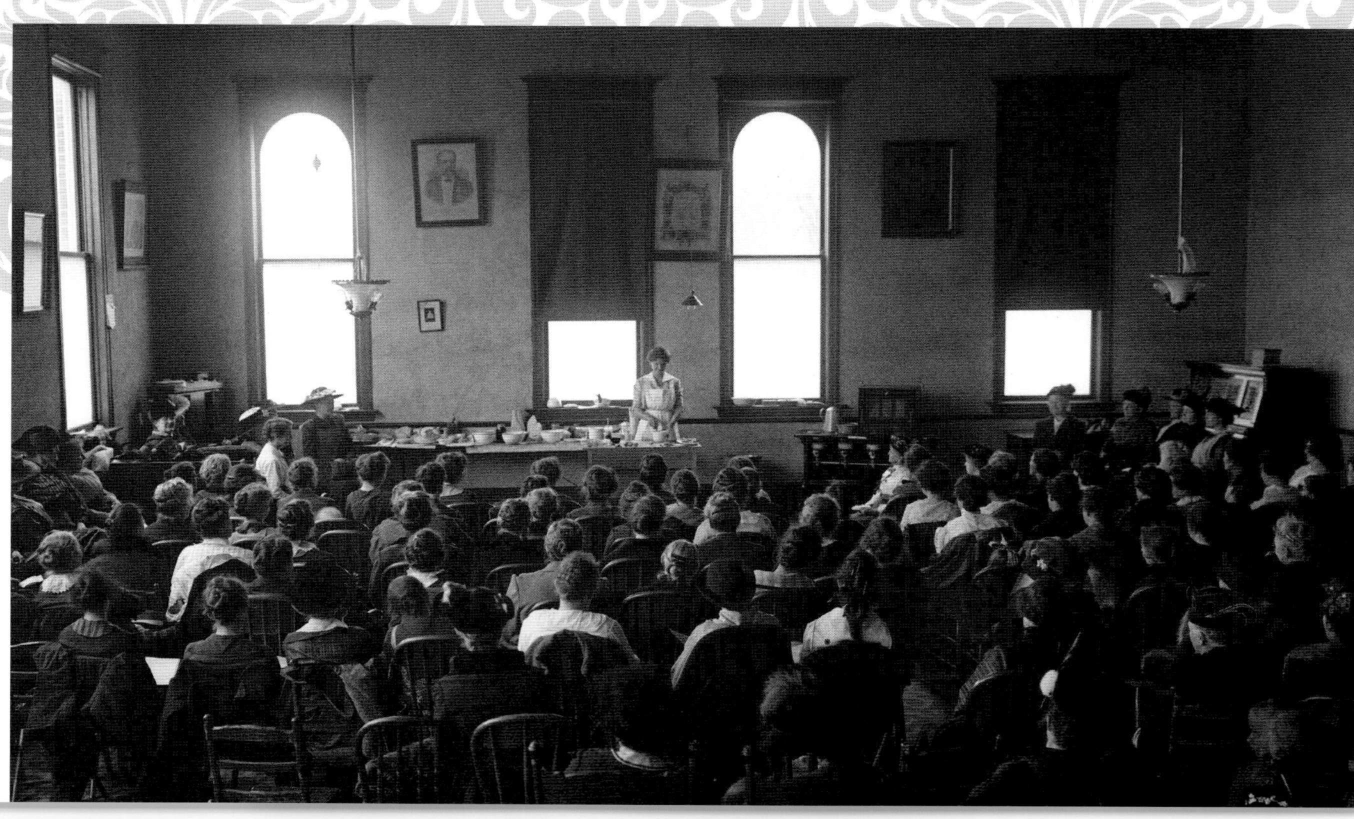

A cooking demonstration at Oxford (Benton County, 1916)

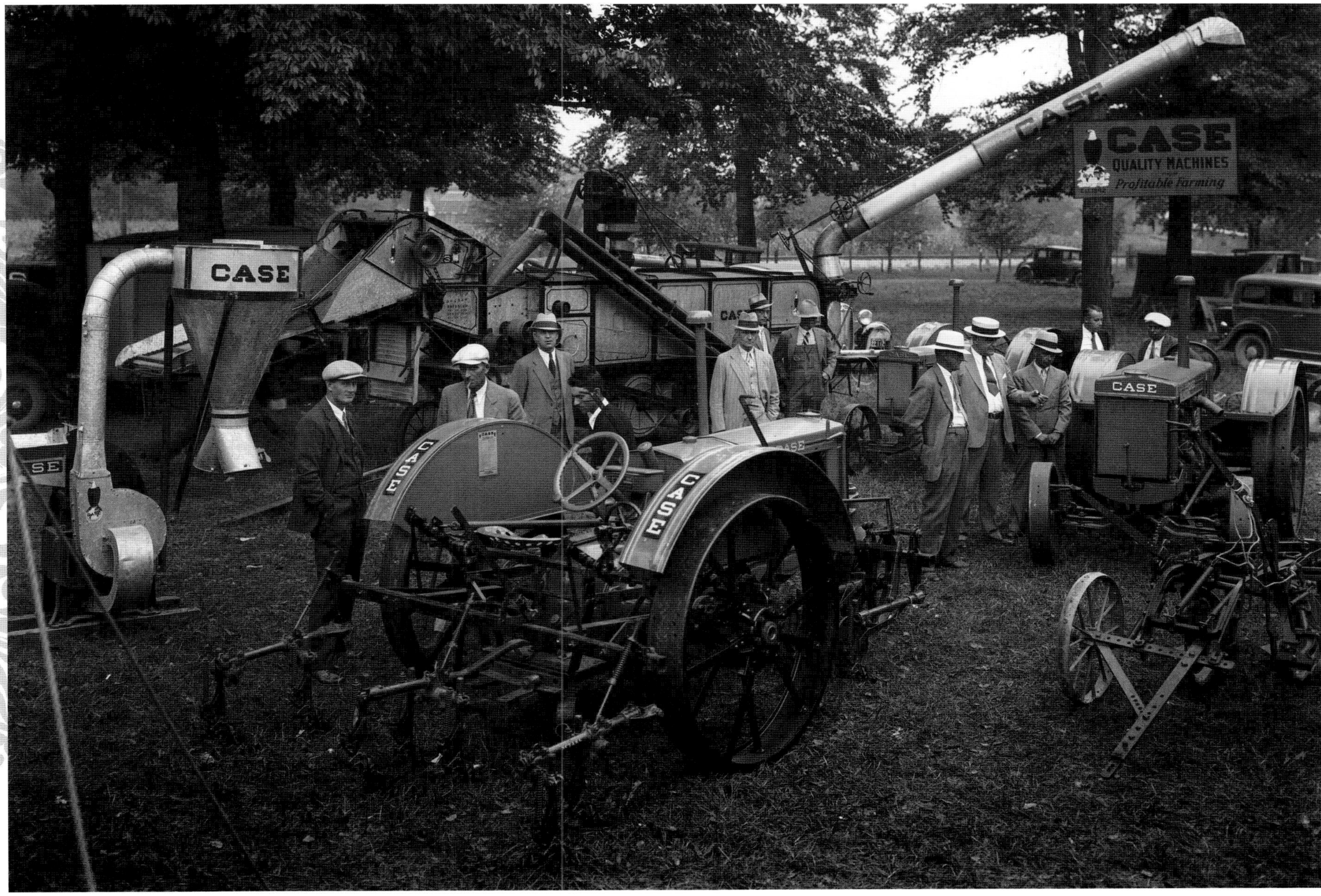

A Case equipment exhibit at the Miami County Fair (Miami County, n.d.)

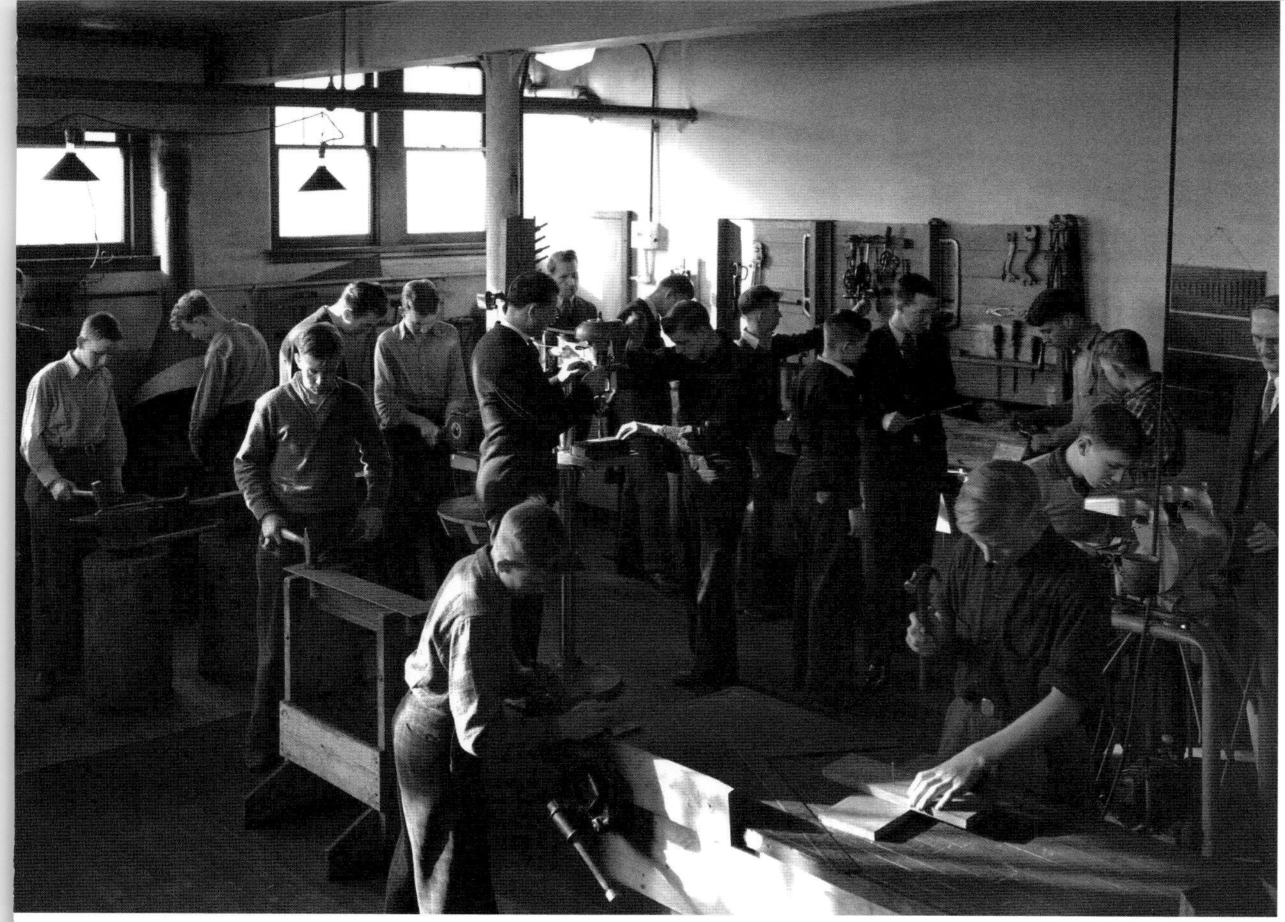

Students learn how to use shop tools at Montmorenci High School. (Tippecanoe County, 1939)

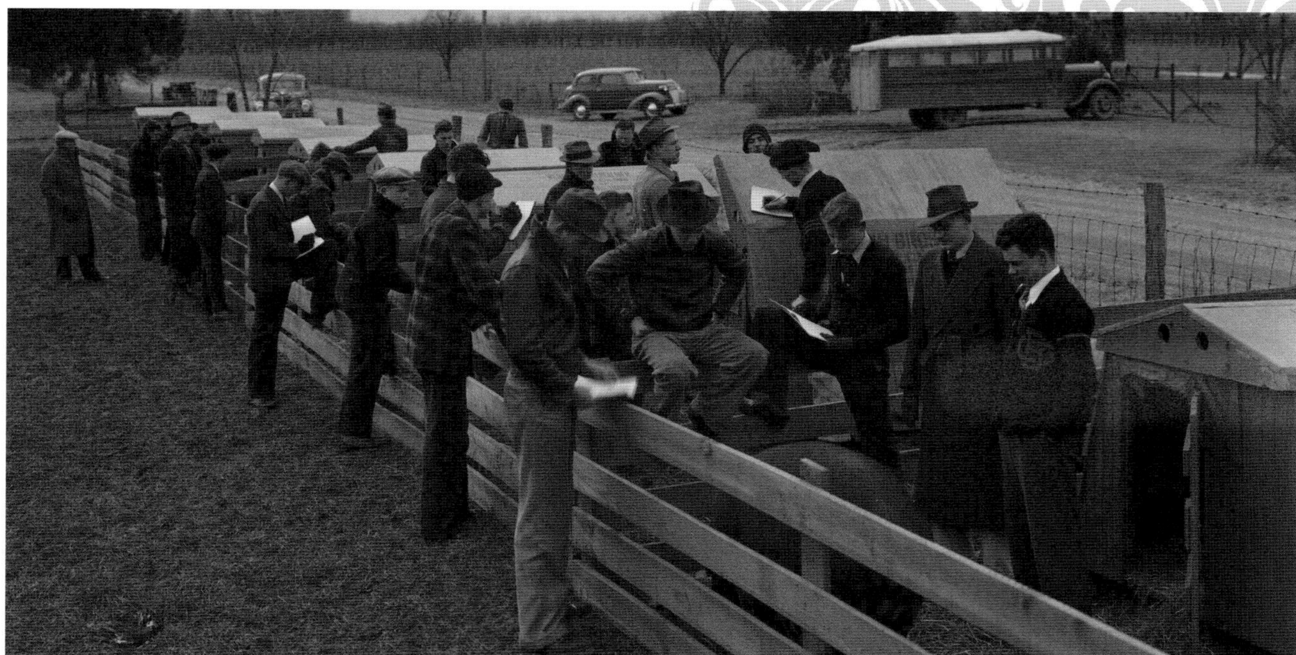

A vocational agriculture class from Battle Ground inspects the layout of a hog operation. (Tippecanoe County, 1939)

(Top left and right) The Purdue Agricultural Extension Service operated this fresh fruit and vegetable marketing trailer, which included a produce display. (Tippecanoe County, 1949)

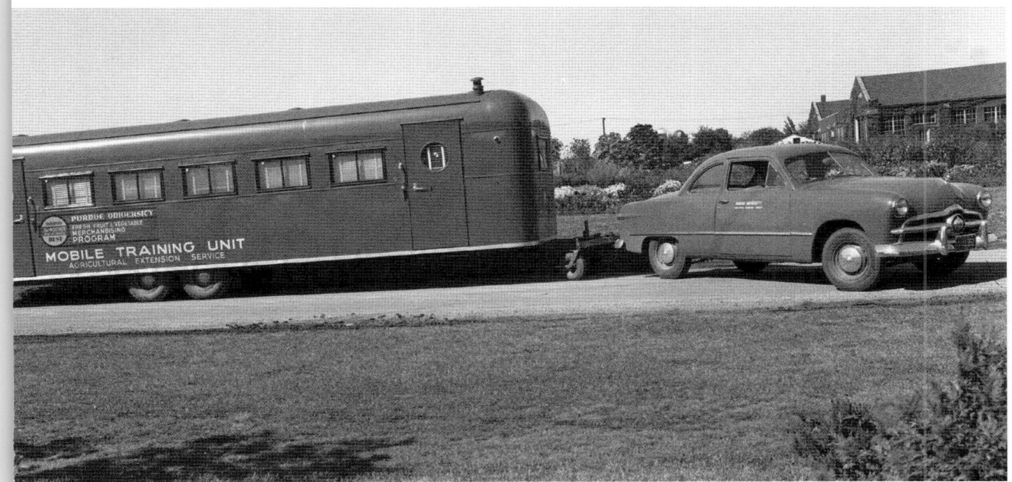

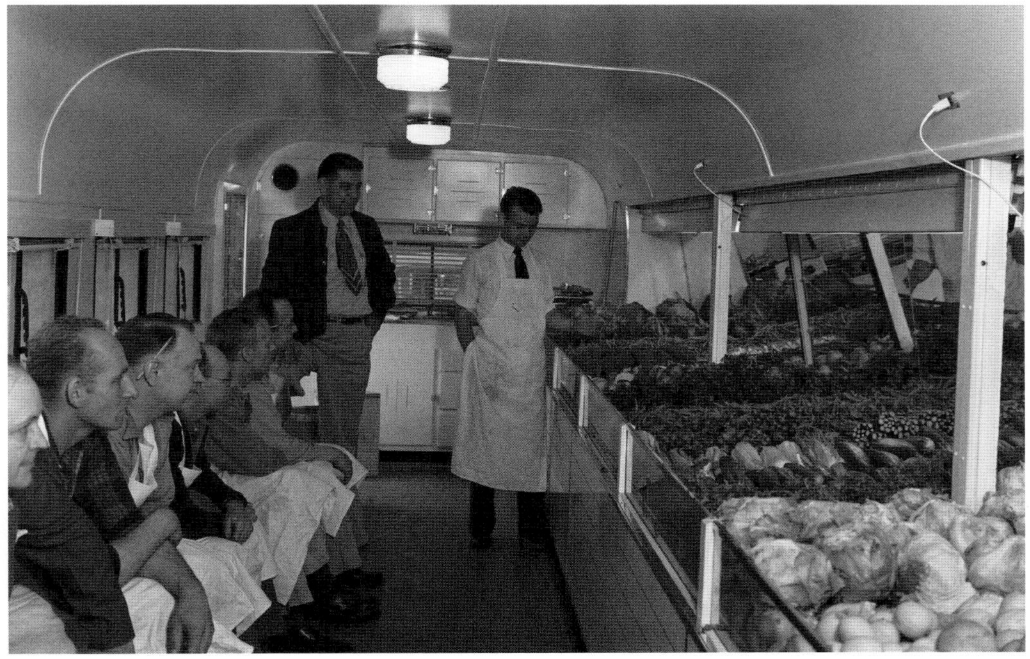

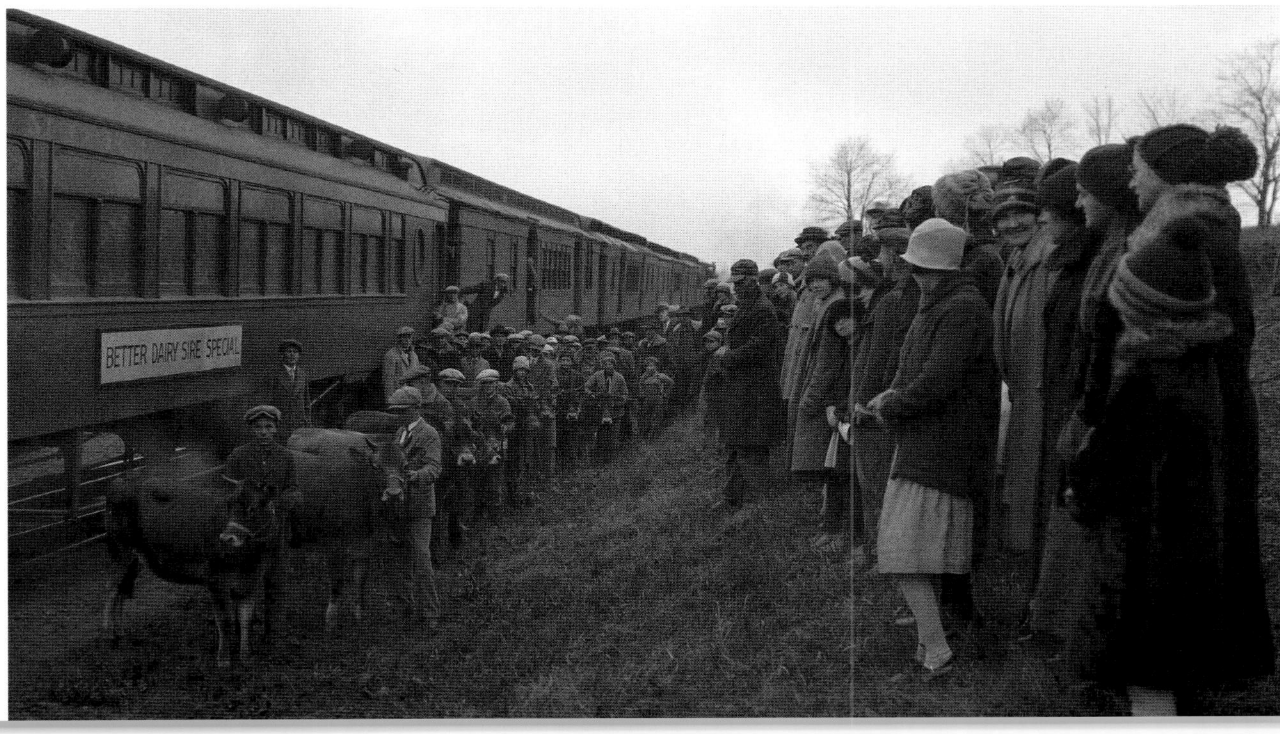

The Better Dairy Sire Special education train from Purdue Extension stops at Lexington, where Jersey calves entered in the Boys' and Girls' Calf Club show draw the interest of the crowd. (Scott County, 1927)

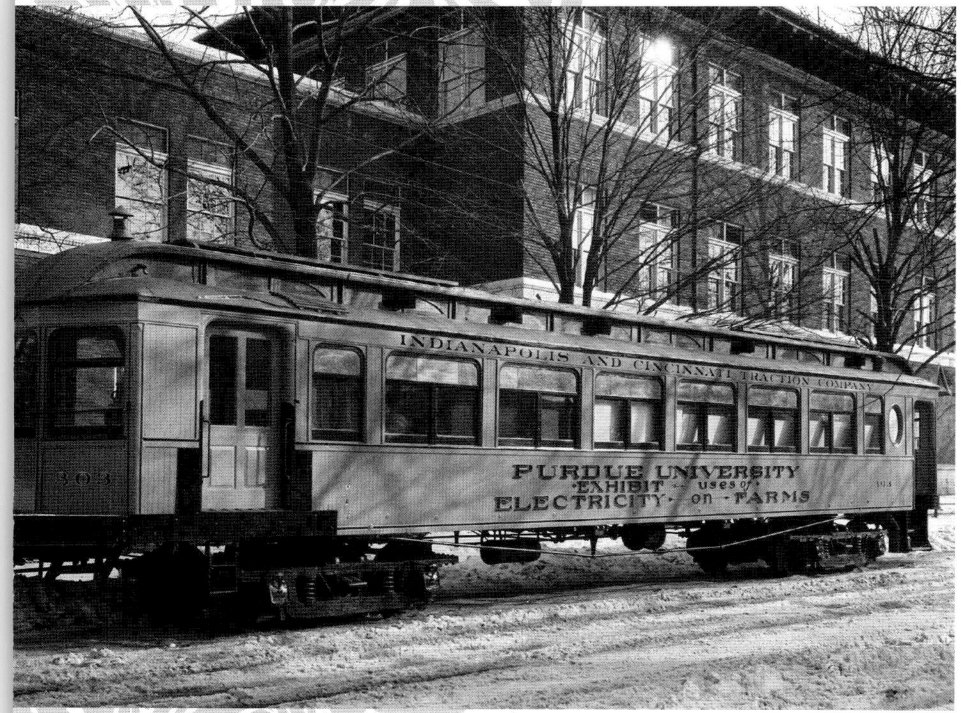

An Indianapolis and Cincinnati Traction Company interurban car becomes an electricity exhibit developed by the Purdue University Agricultural Experiment Station in cooperation with electric light and power companies. (Tippecanoe County, 1927)

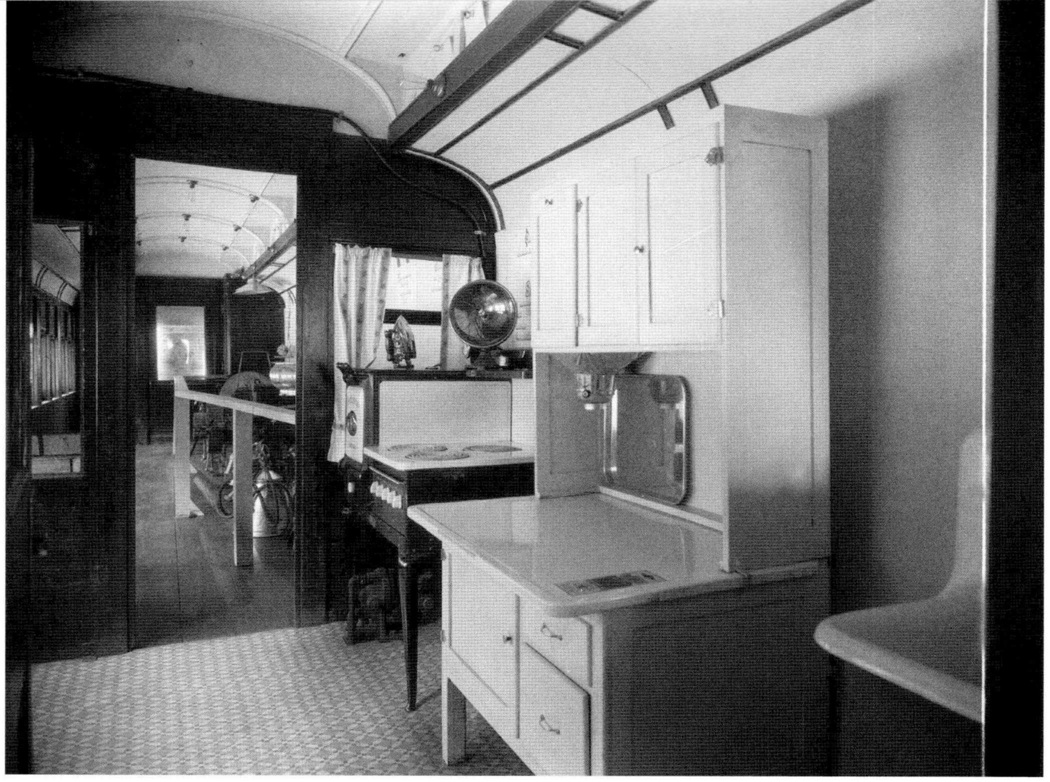

(Above and below) The interior of the electricity exhibit featured all kinds of labor-saving electrical equipment for the farm and home. For eight weeks, it ran on nearly all of Indiana's electric lines, making short stops in almost every part of the state. (Tippecanoe County, 1927)

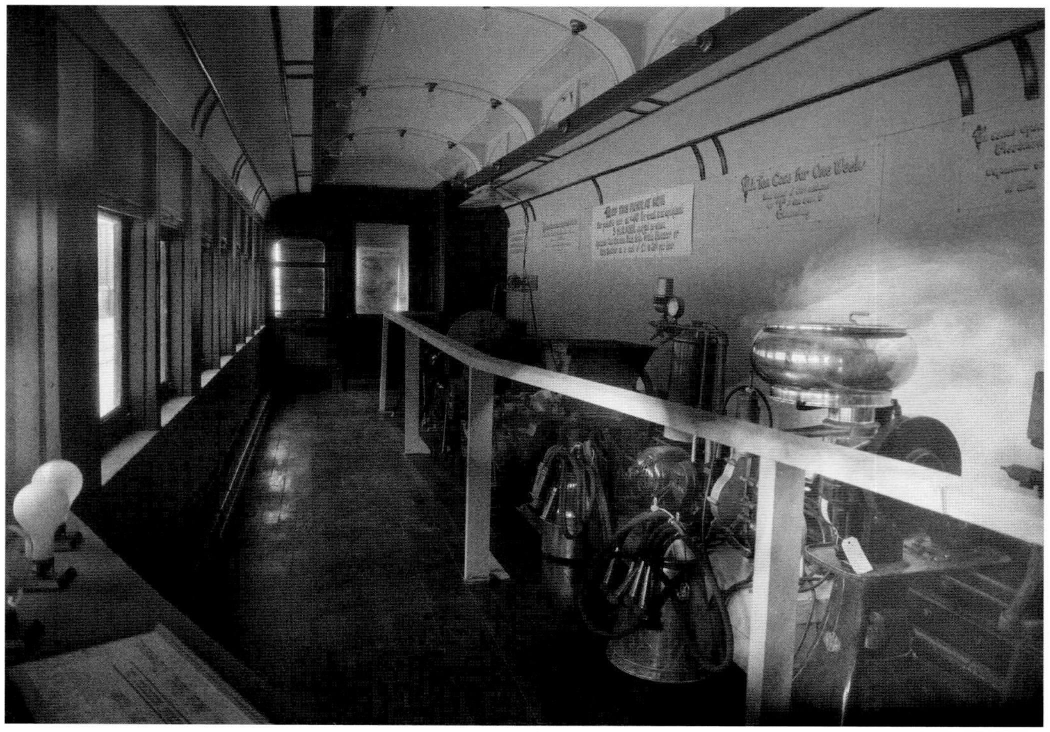

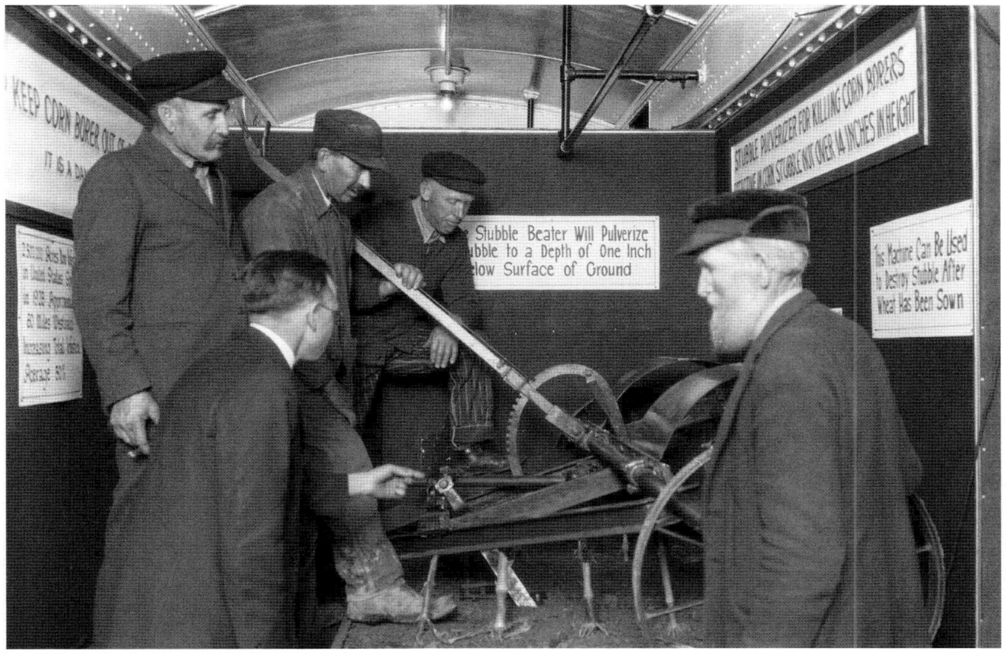

During a stop in Portland, farmers examine a stubble beater on the Purdue Agricultural Extension Service Corn Borer Exhibit Train, which ran on the Pennsylvania Railroad. Purdue staff on board taught farmers just outside the corn borer area how to retard or eliminate this approaching pest. (Jay County, 1927)

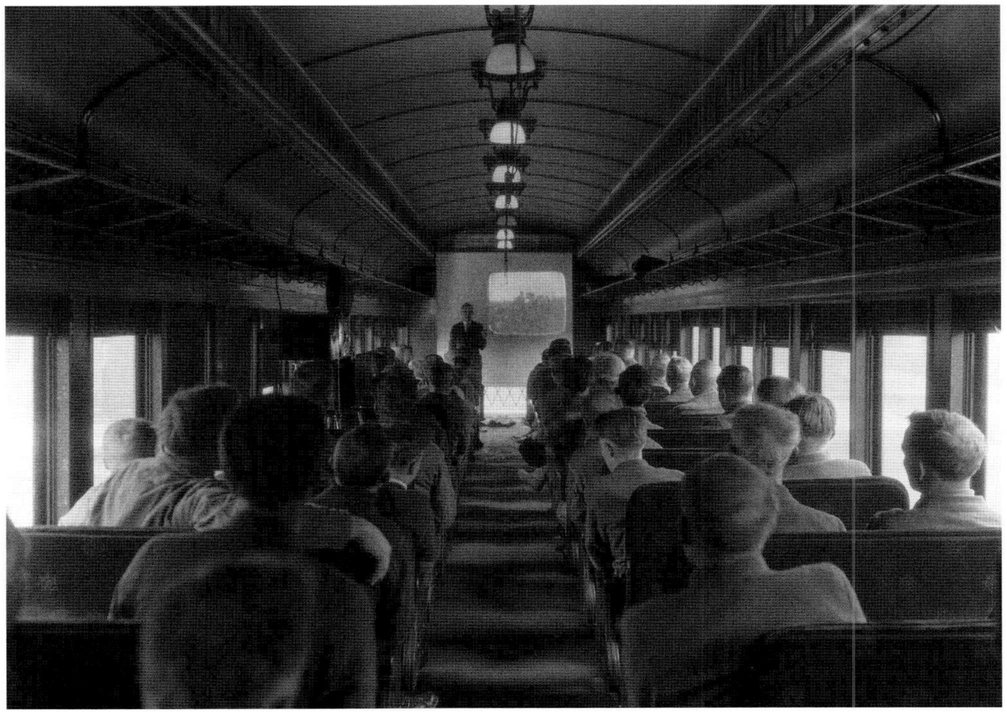

Inside the lecture car on the corn borer train, speakers used lantern slides and moving pictures to show farmers what they needed to do to protect their corn crop. (Jay County, 1927)

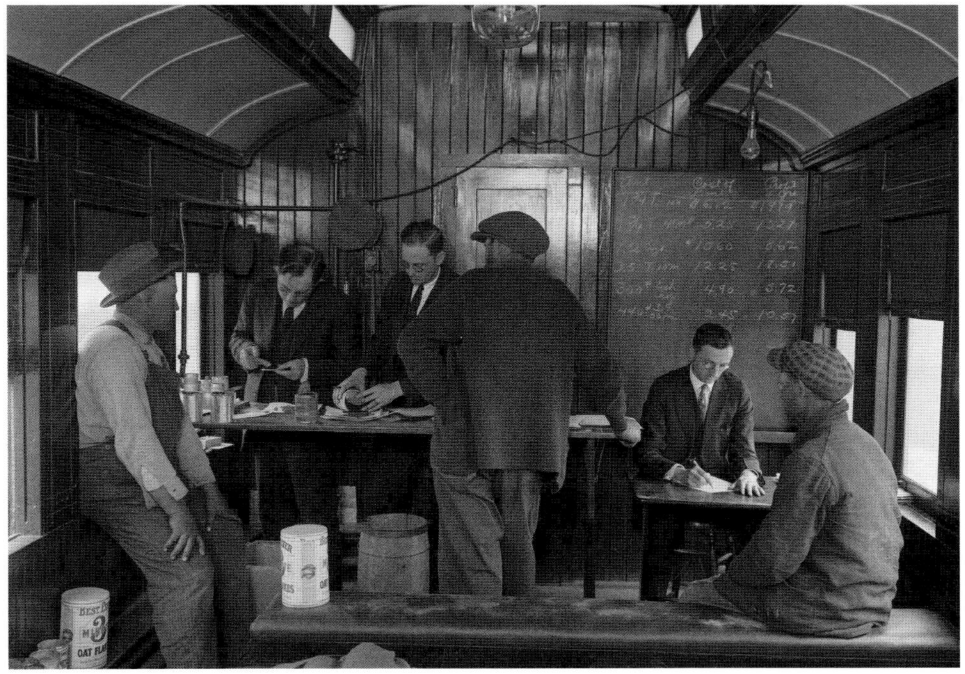

Purdue University specialists test the soil farmers brought to the Limestone Special education train at a stop near Union Center. Each farmer who visited the train received enough free lime to improve the pH of one acre of soil on their farms. (LaPorte County, 1925)

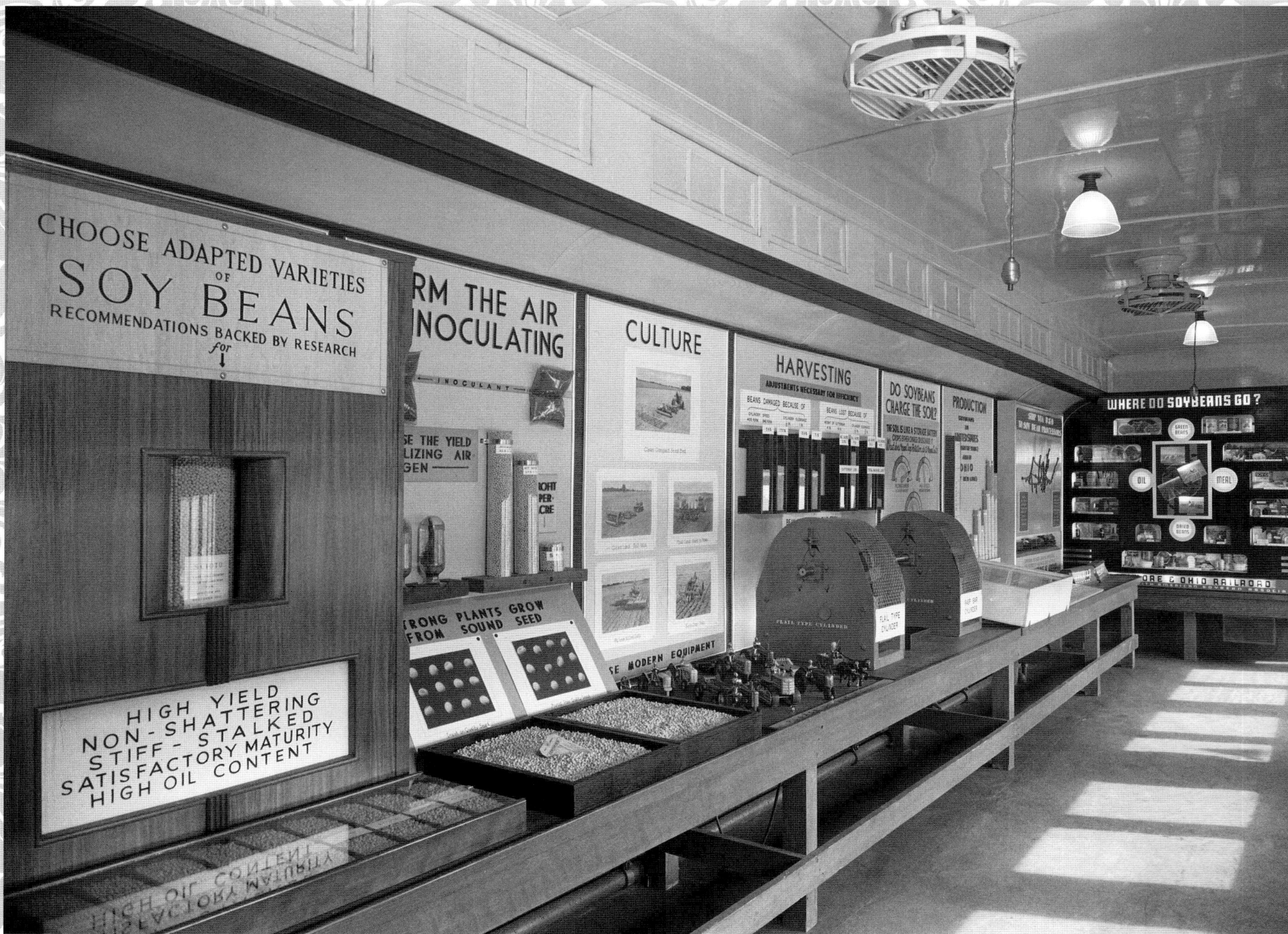

Exhibits on the Soybean Special education train focused on how to grow, care for, and harvest soybeans. The University of Illinois, Ohio State University, and Purdue University helped sponsor the train, which ran on Baltimore and Ohio Railroad lines. (Wood County, OH; 1941)

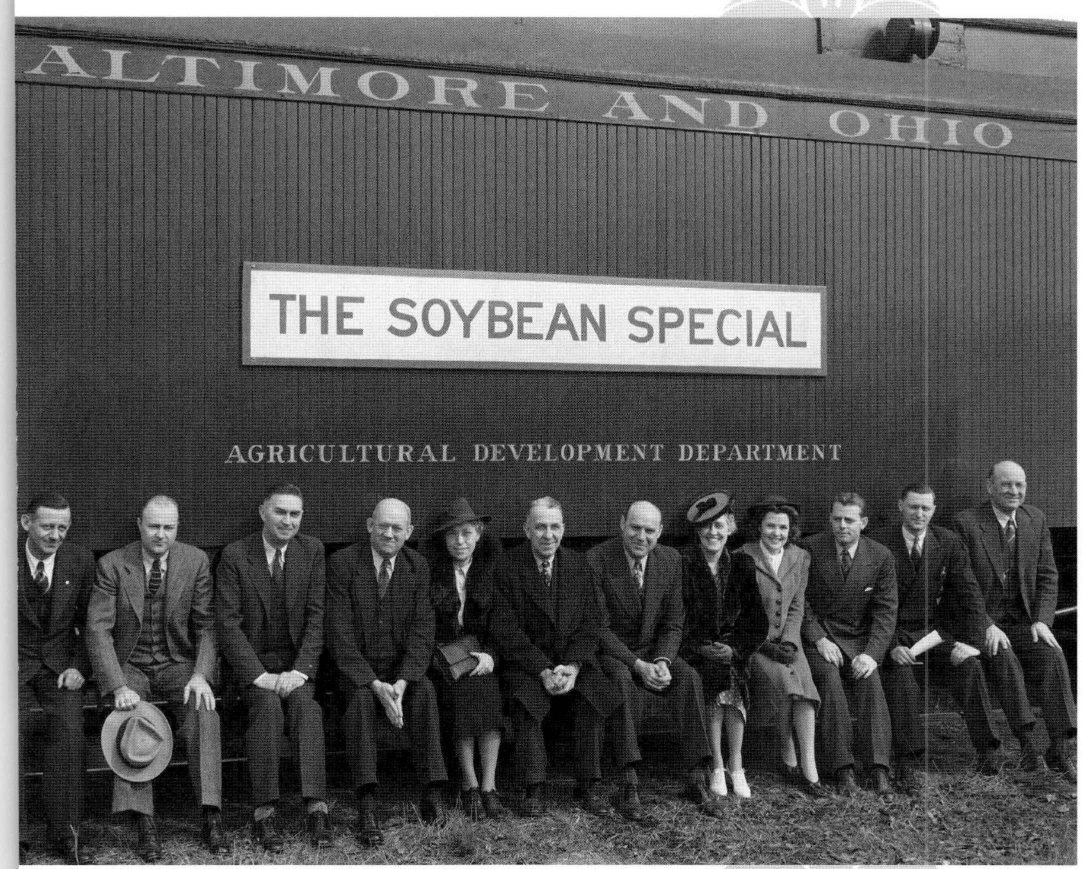

The staff of the Soybean Special poses for a photo in Bowling Green, Ohio. (Wood County, OH; 1941)

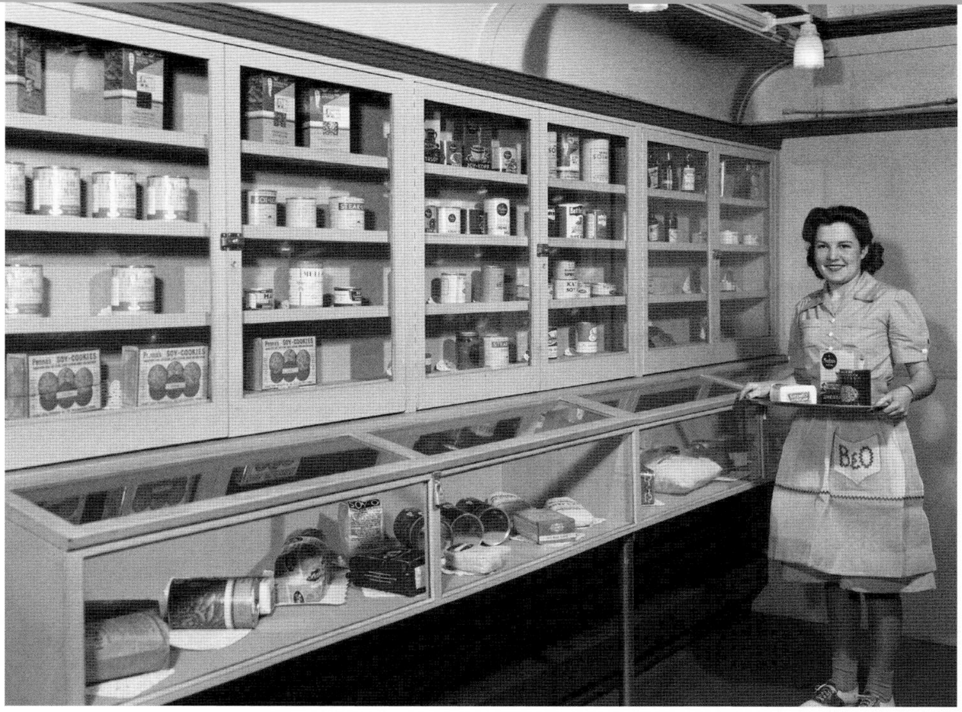

Mary Mowry, a 4-H Club member, stands beside the exhibit of soybean food products on the Soybean Special education train. (Wood County, OH; 1941)

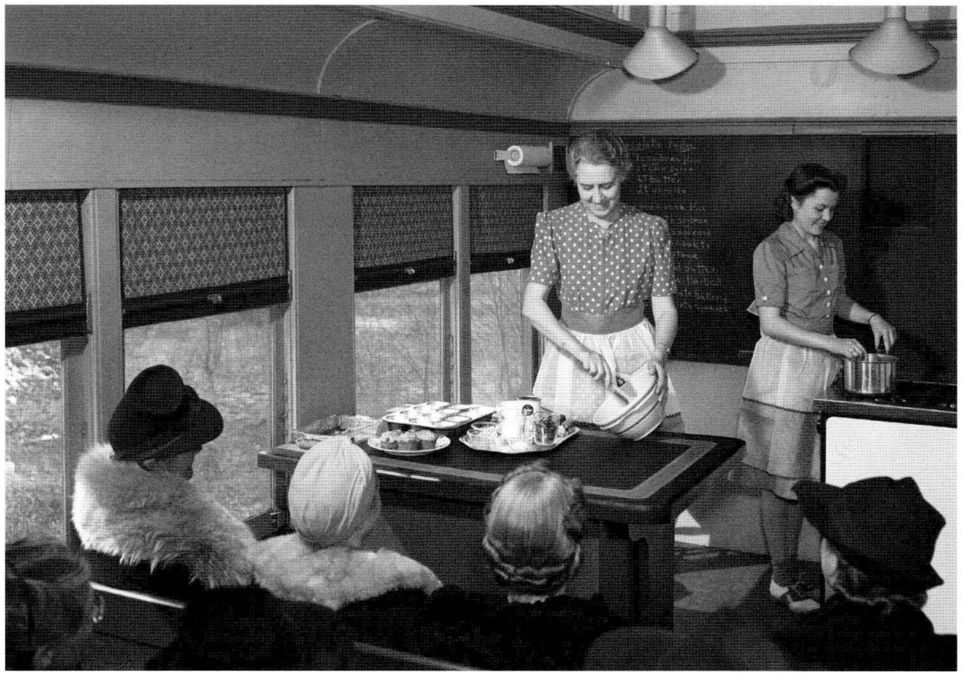

Aboard the Soybean Special, Anita Beadle Vogler (standing at left), former nutrition specialist at Purdue University, presents a cooking demonstration as Mary Mowry (standing at right) assists her. (Wood County, OH; 1941)

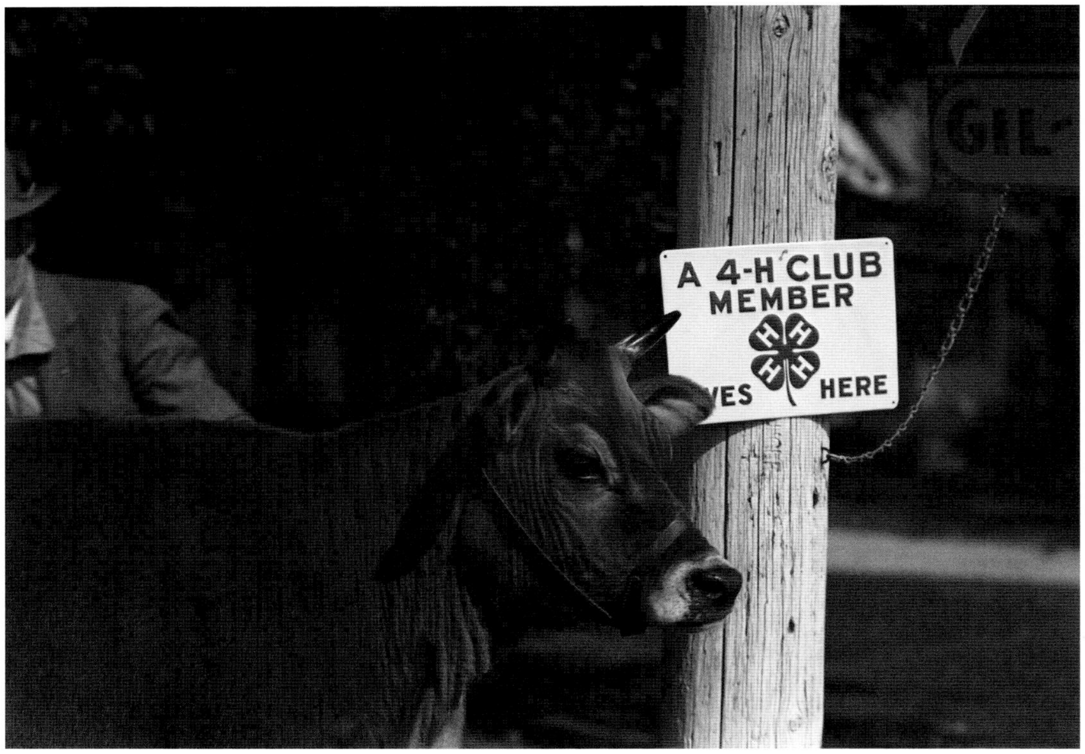

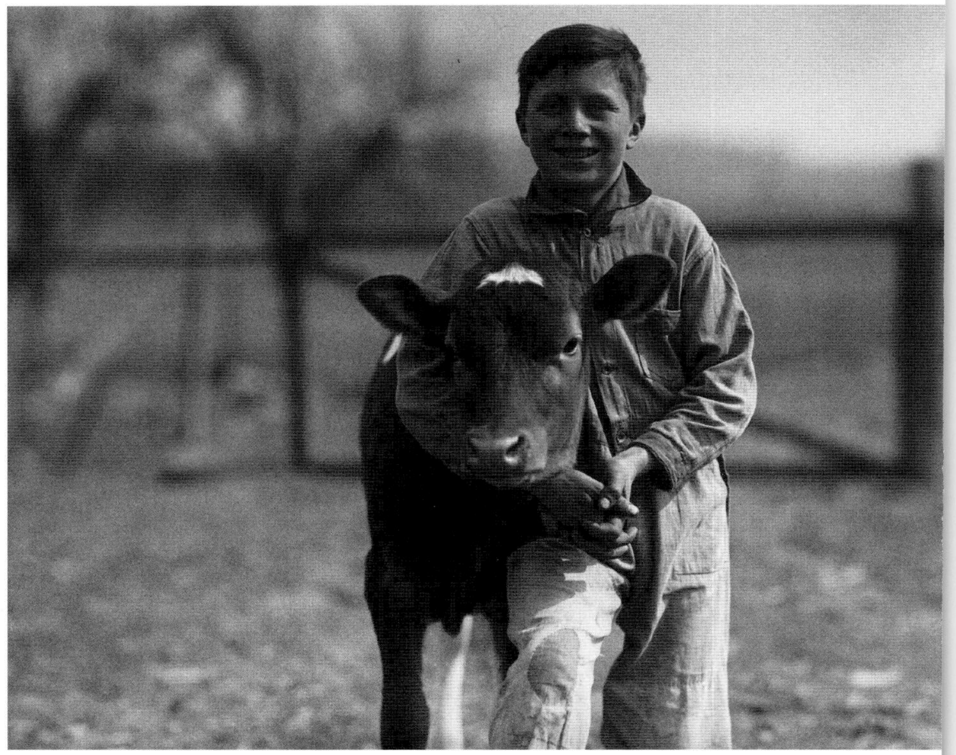

(Top left) A Jersey heifer stands beside a 4-H club sign. (N.p., 1948)

(Top right) Twelve-year-old Everett Eby takes care of Guernsey calves like this one on his father's farm in Elkhart. (Elkhart County, 1926)

(Bottom) Wanda Mansfield from Eaton poses with her Jersey heifer at the Indiana State Fair. (Marion County, 1930)

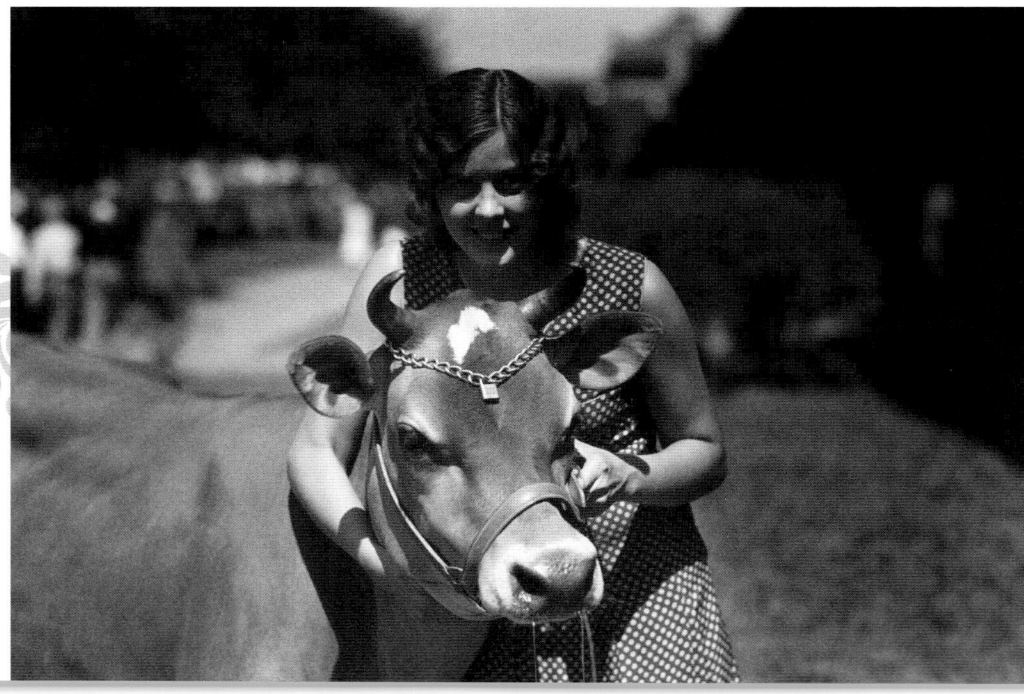

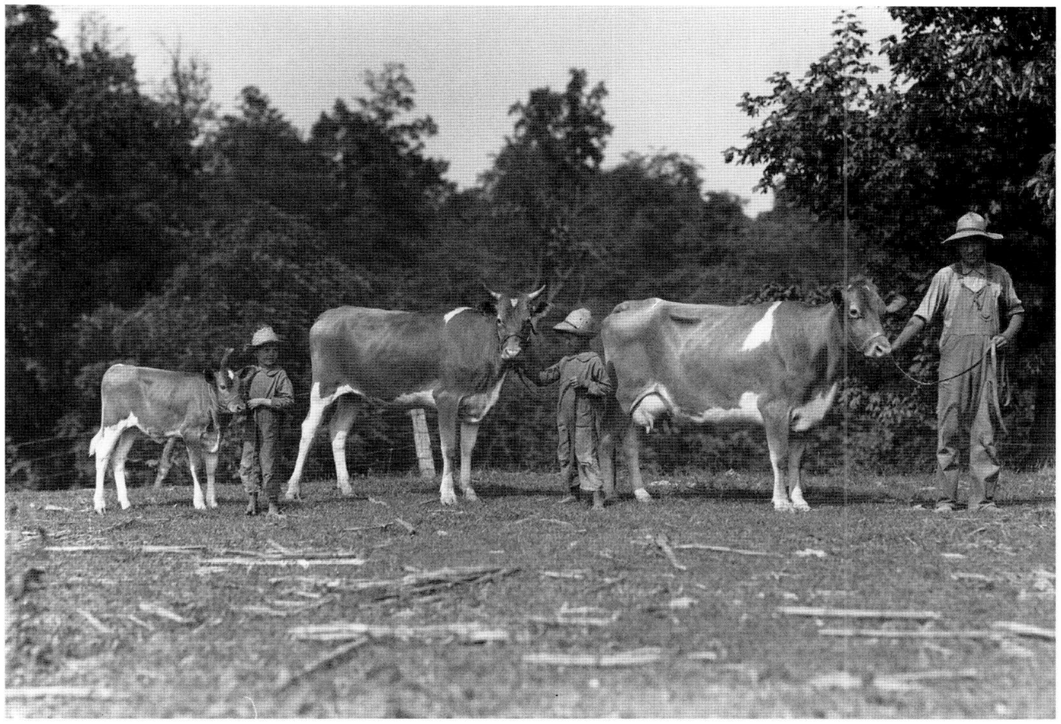

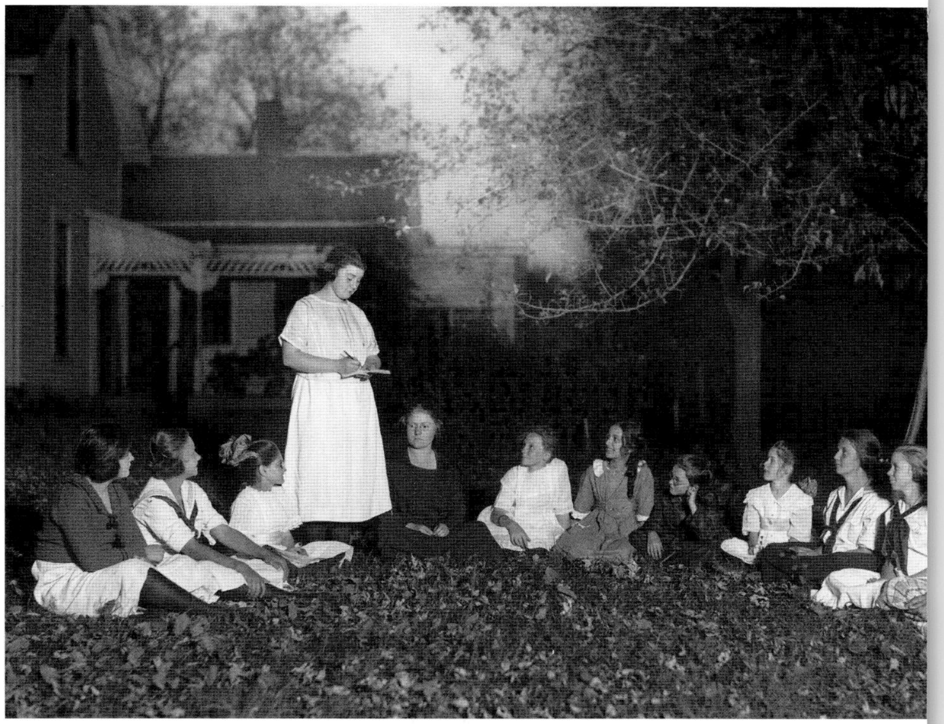

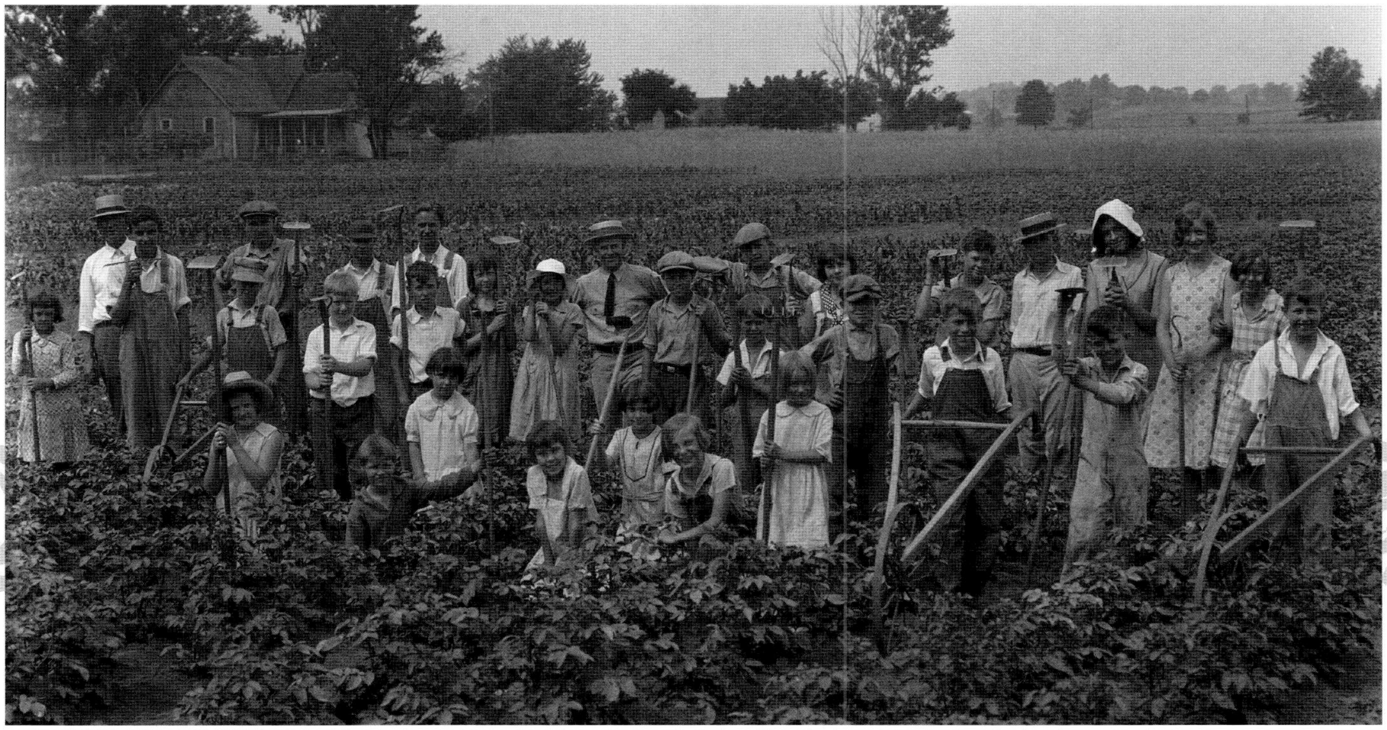

(Top left) Raymond Heald from Plainfield holds his original Hendricks County Calf Club heifer, while his brothers Herman and Morris hold her offspring. (Hendricks County, 1923)

(Top right) Leota Jones (standing), president of the Avon Girls' Club, proudly reports that her club finished "100 percent strong," which meant all of the members completed their projects. (Hendricks County, 1920)

(Bottom) Thirty-four members of the Worthington Garden Club pose in front of their garden. (Greene County, 1931)

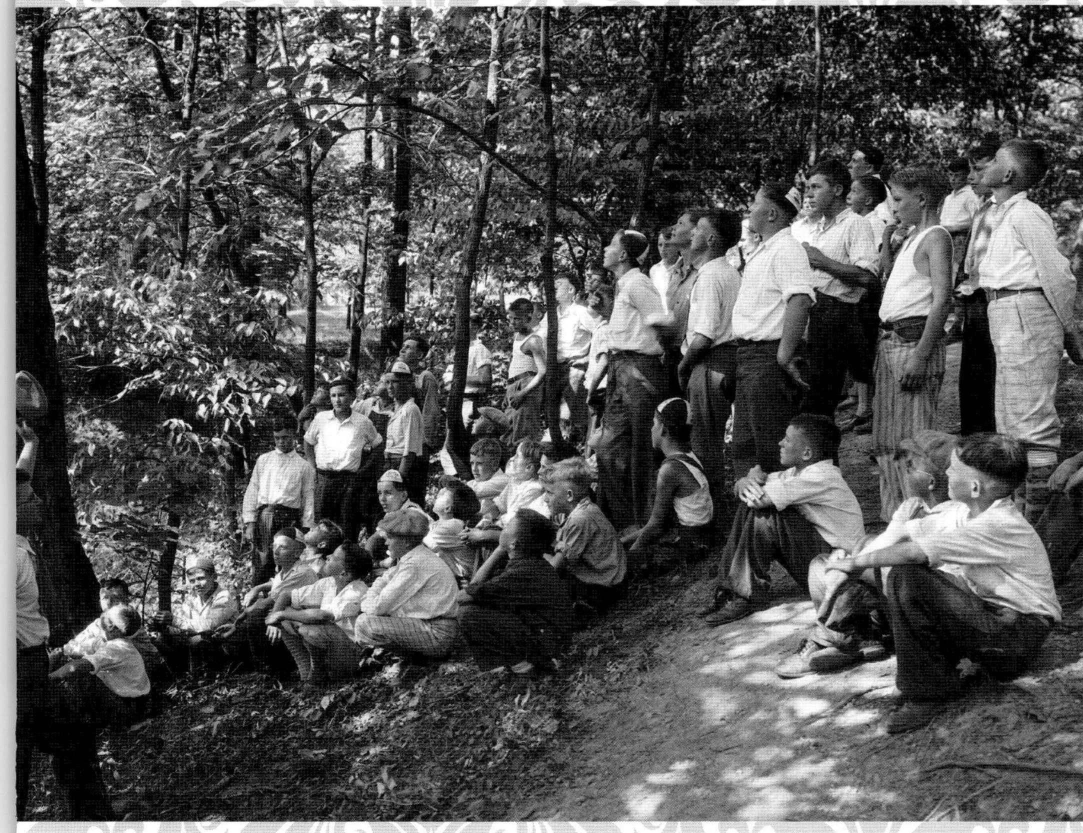

Purdue horticulturist Roy Hull talks to 4-H members about native trees of Indiana at the Richmond Club Conference. (Wayne County, 1931)

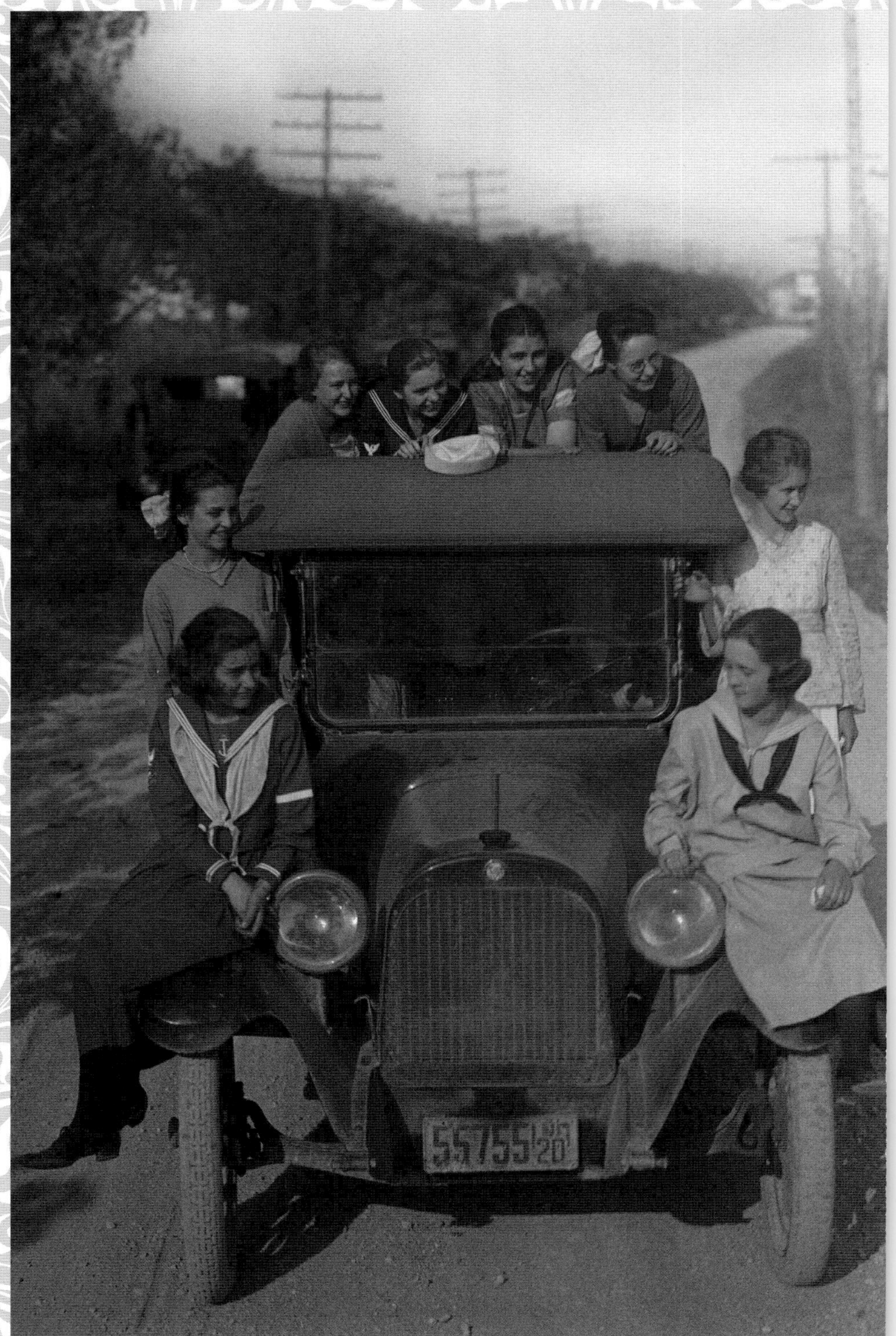

Members of a 4-H club fill a touring car owned by the home demonstration agent at Danville. (Hendricks County, 1920)

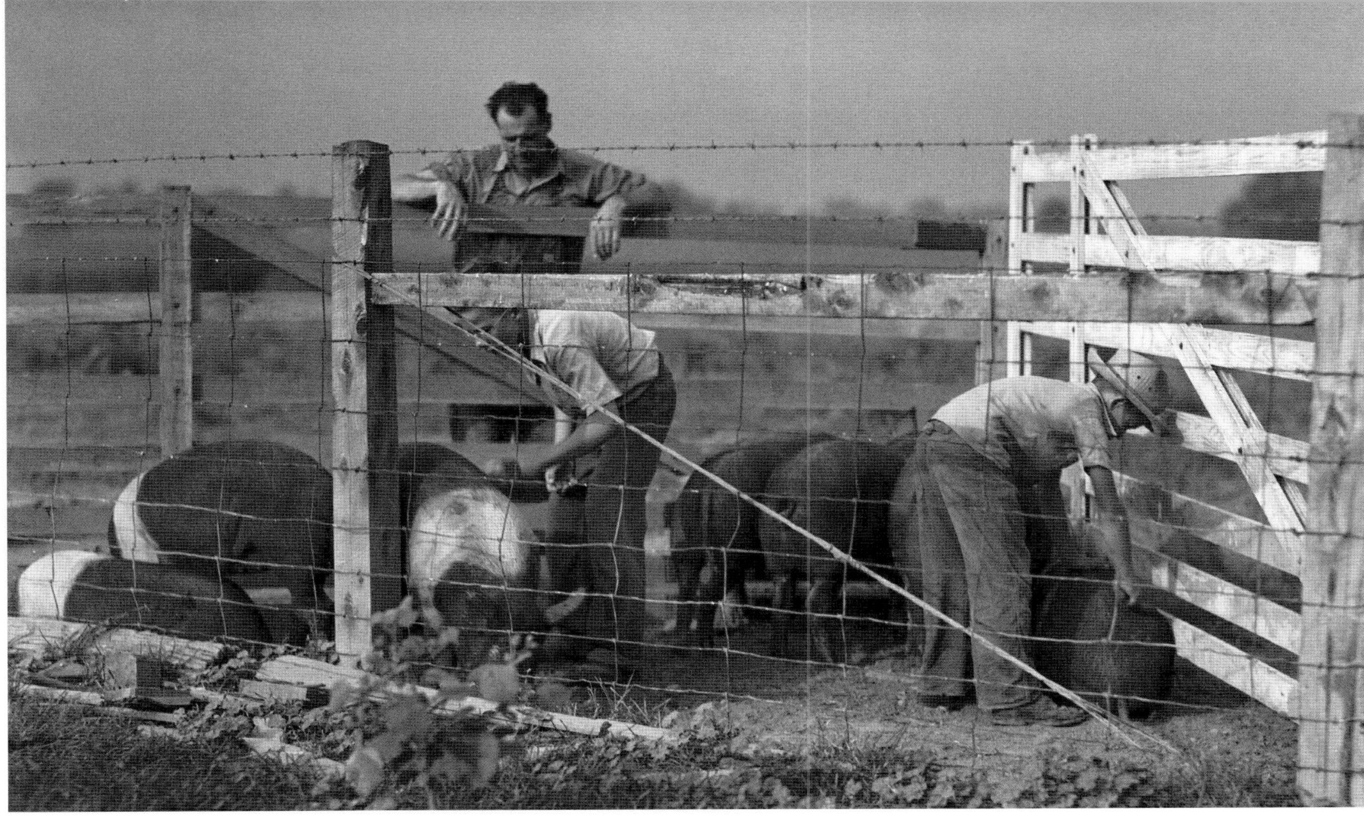

Louis Meyer watches his sons work with Hampshire hogs for their 4-H club project. (Rush County, 1943)

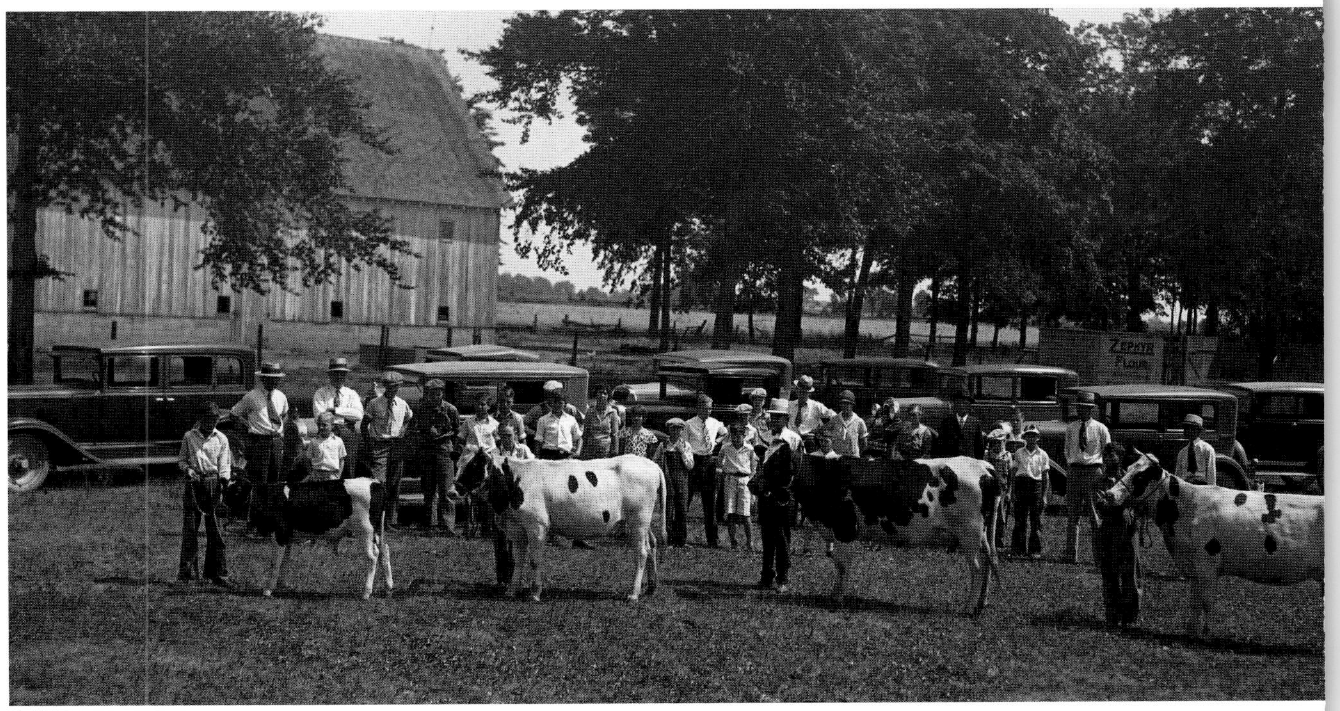

During a tour by the Madison County Dairy Calf Club, members inspect Holsteins that were owned by Max Forrest near Summitville. (Madison County, 1930)

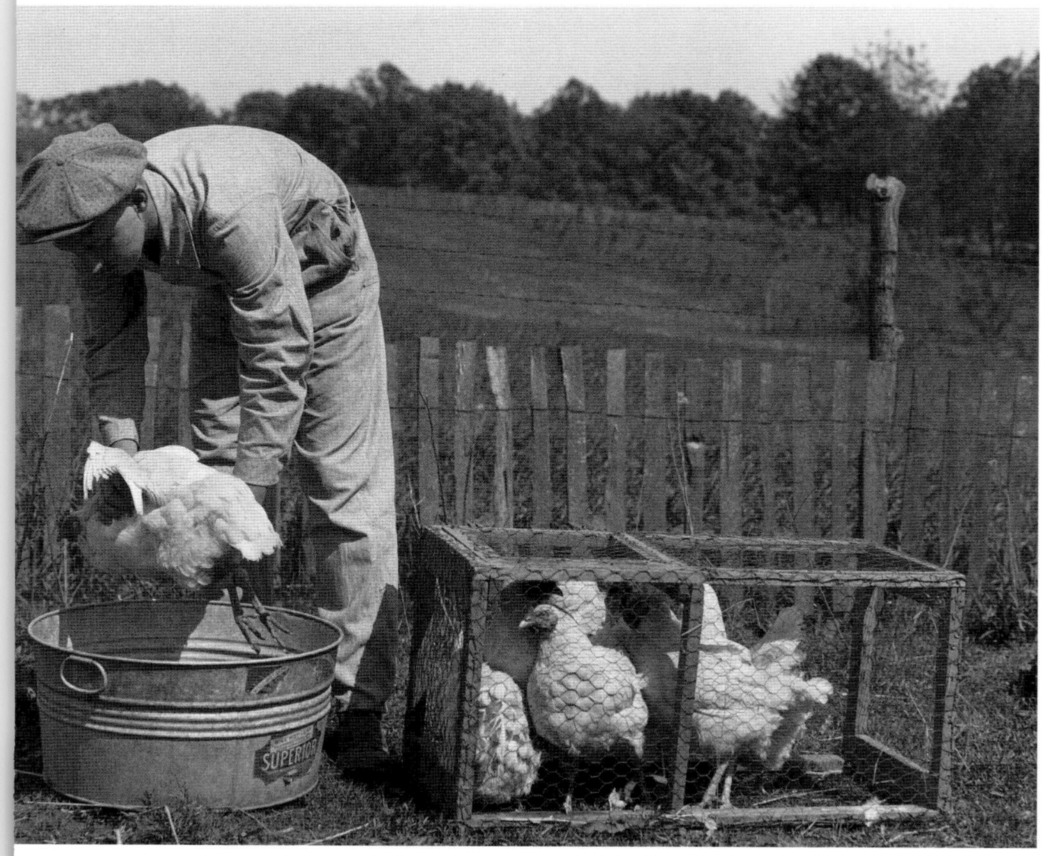

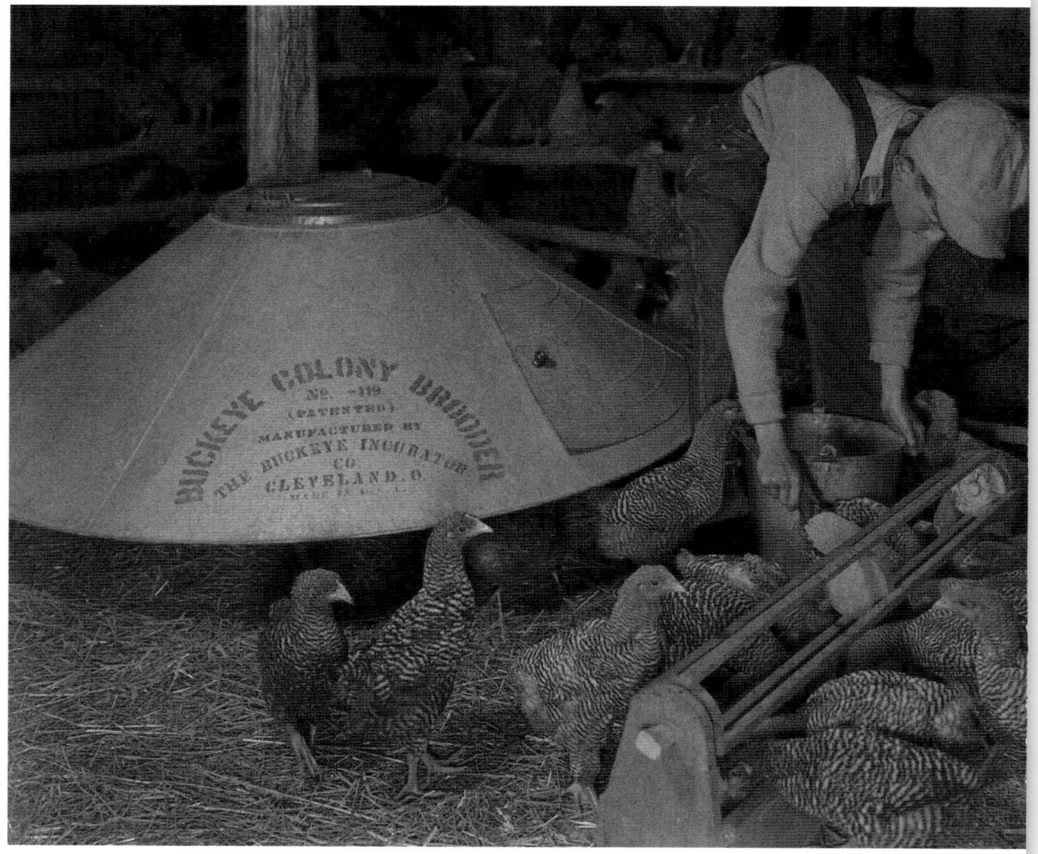

(Top left) Glendon Sendmeyer of Poland dips his chickens to rid them of lice as part of his 4-H club project. (Clay County, 1935)

(Top right) Paul Meyer, a 4-H club member from Bowling Green, feeds his chickens in a brooder house. (Clay County, 1935)

(Bottom) Two 4-H club members from Knightstown—Russell (right) and Lowell Hardin (center)—inspect their corn plot with club leader Bob Amick. (Henry County, 1935)

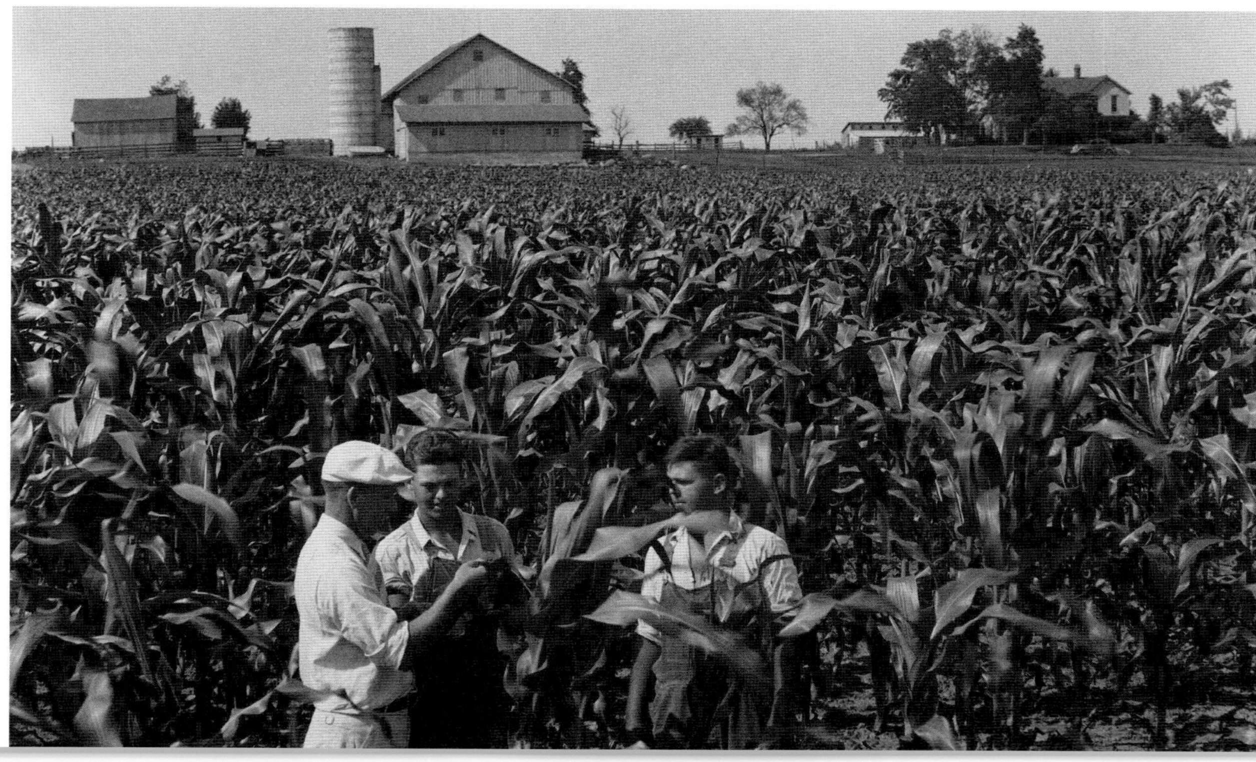

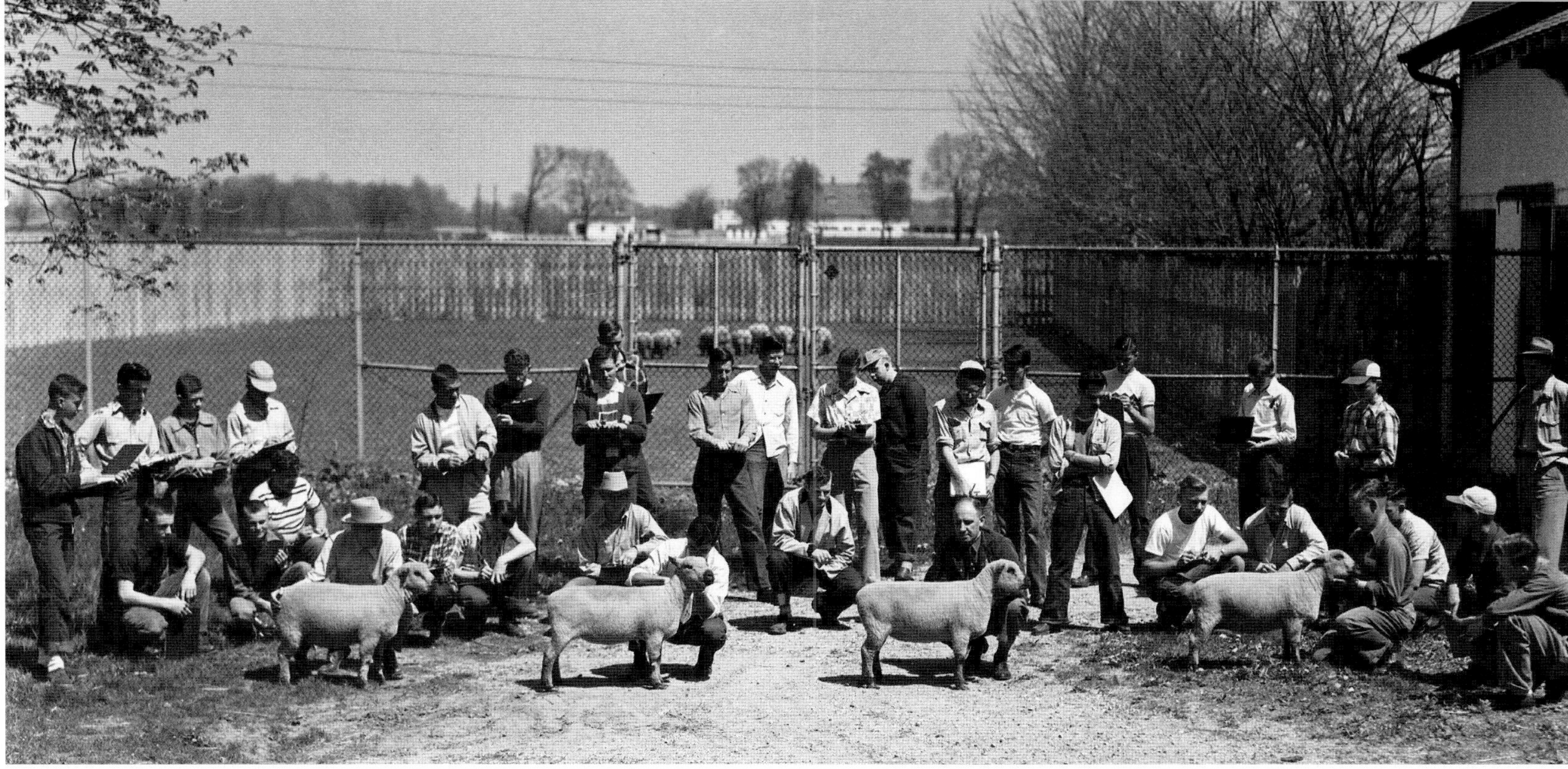

Members of 4-H clubs participate in district judging contests for sheep (top), hogs (bottom left), and cattle (bottom right) at the Purdue University Lynnwood Farm in Carmel. (Hamilton County, 1952)

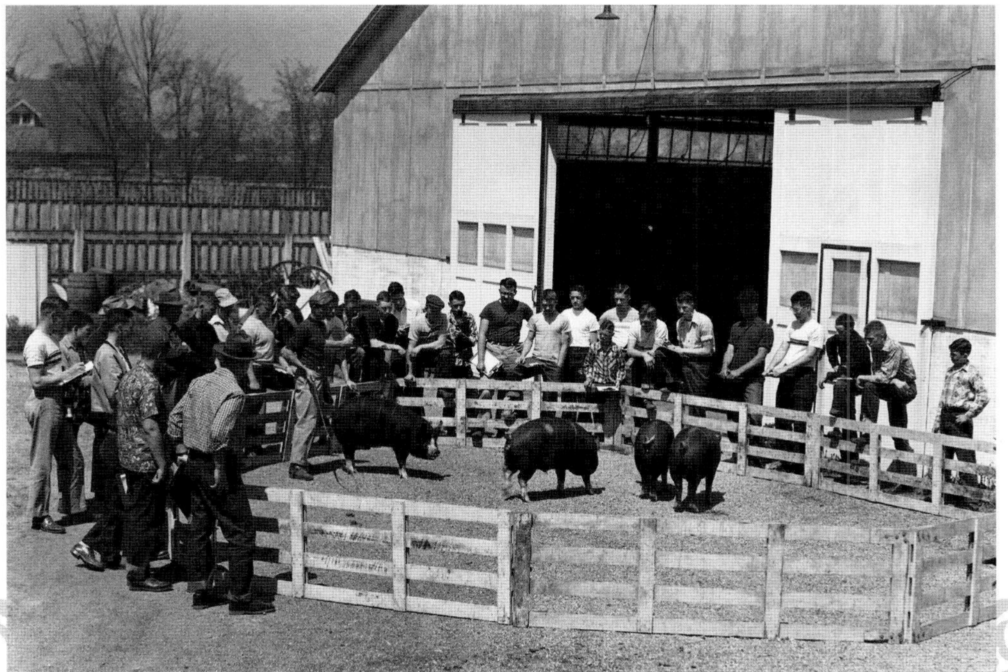

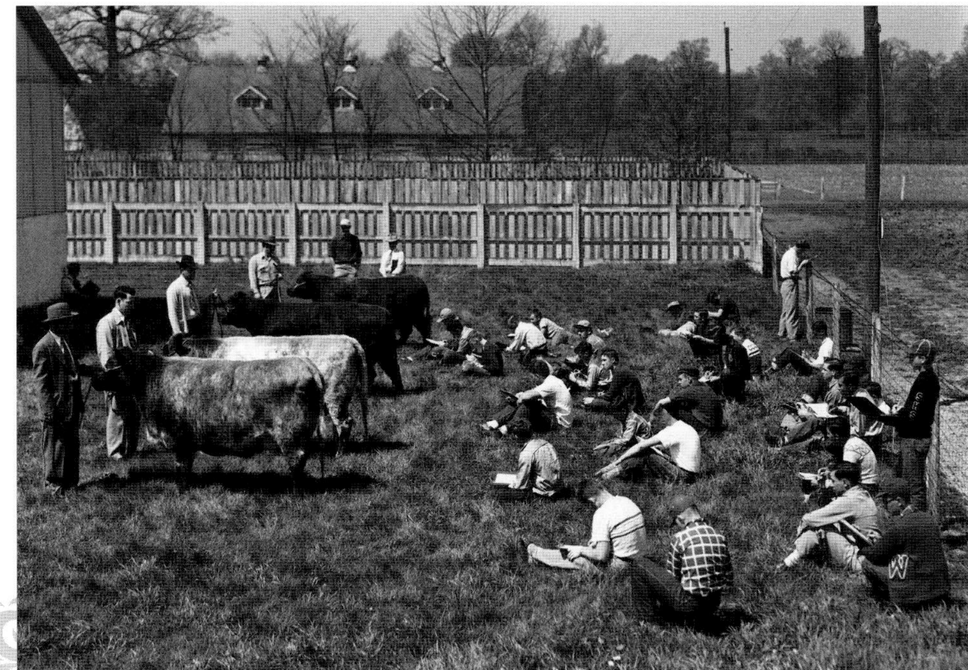

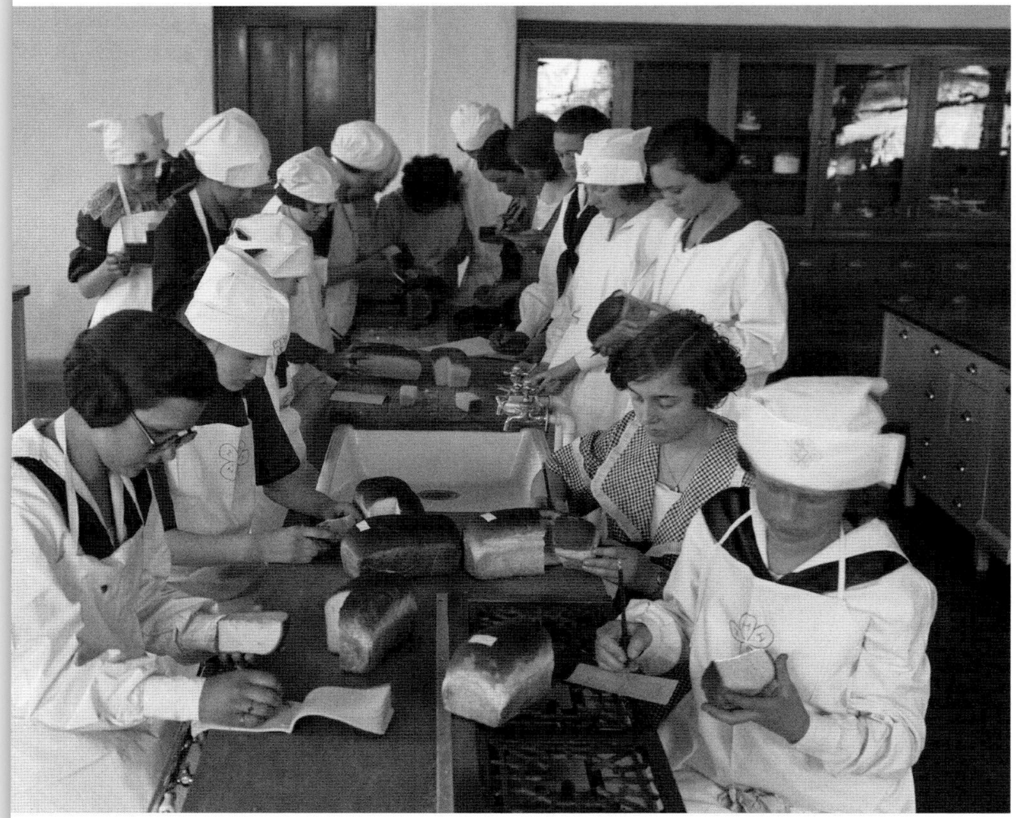

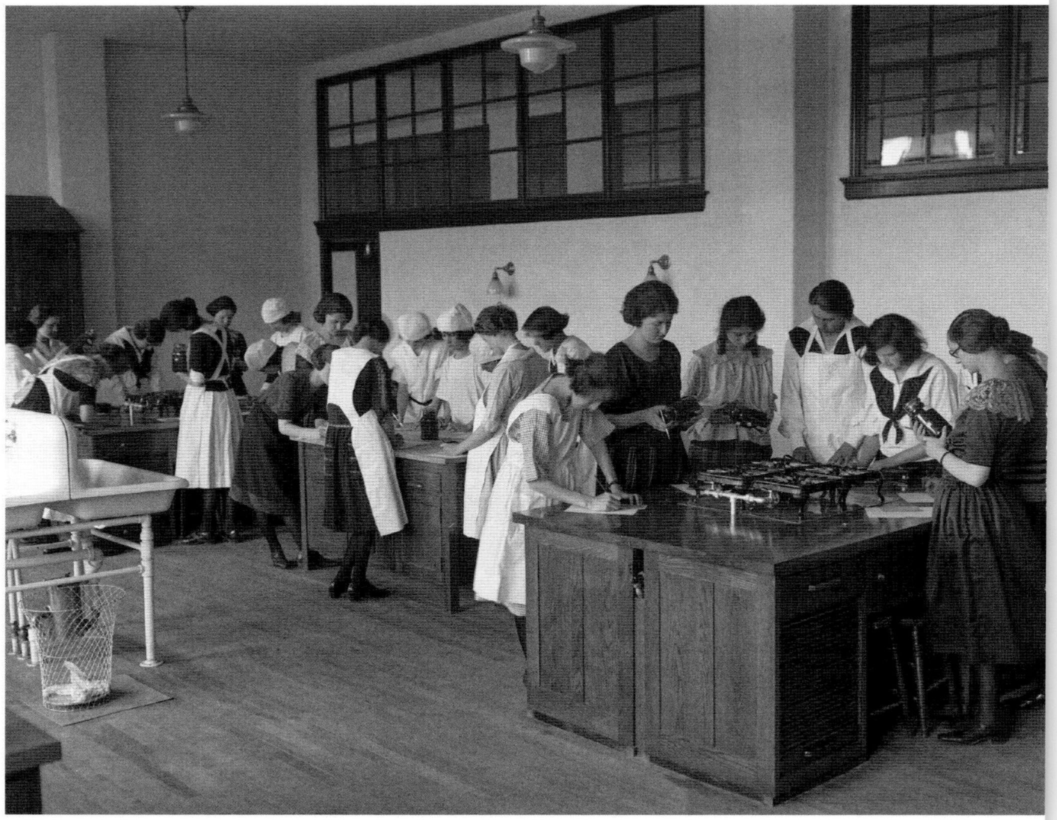

Participants at the 4-H Club Round-Up at Purdue judge bread (left) and canned products (right). (Tippecanoe County, 1923)

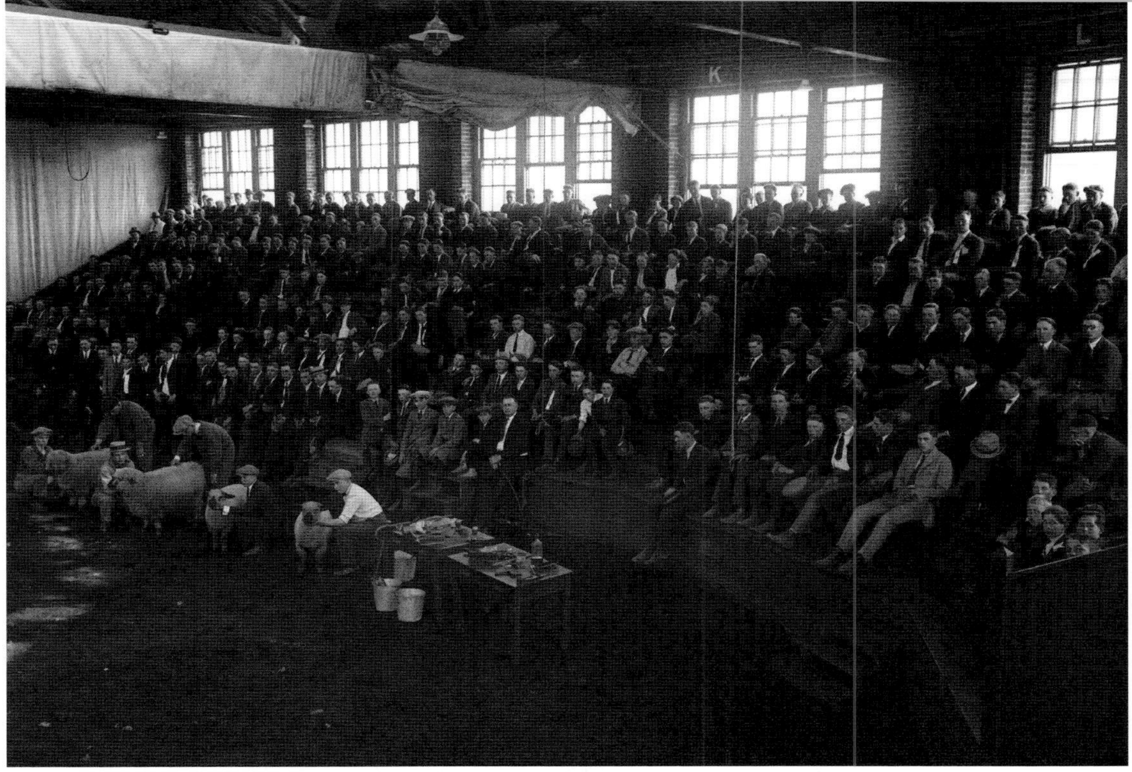

A demonstration on sheep fitting, a process to prepare the animals for competition, takes place inside the Purdue Livestock Judging Pavilion on the campus farm during 4-H Round-Up. (Tippecanoe County, 1923)

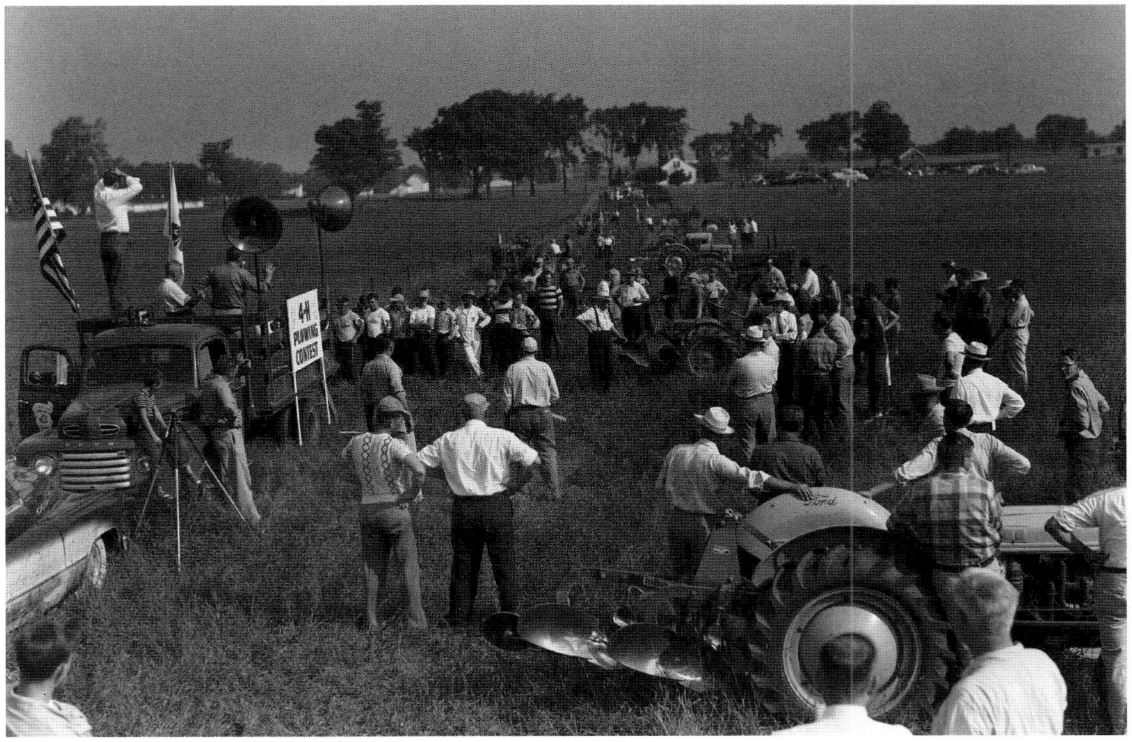

Participants gather for a plowing contest held in connection with a 4-H club show in East Lansing, Michigan. (Ingham County, MI; 1951)

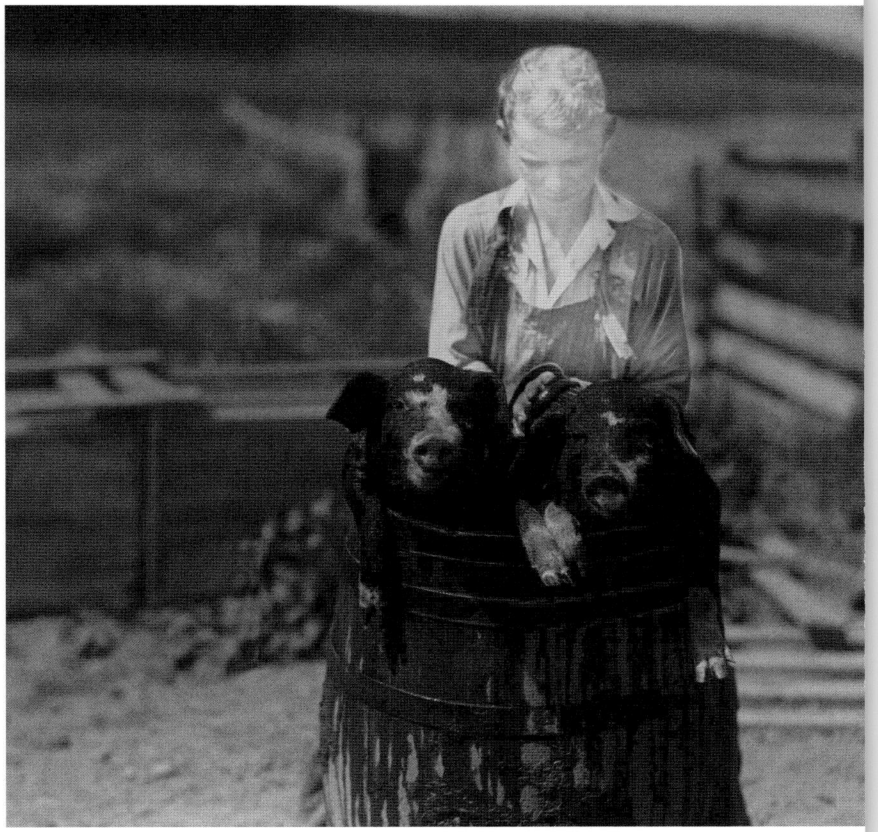

Sixteen-year-old Hollis Williams from Knightstown washes pigs in a barrel for a club project. Allen wrote, "Pigs jump around until most of dirt is washed off, then [he] . . . finishes with a brush." (Henry County, 1924)

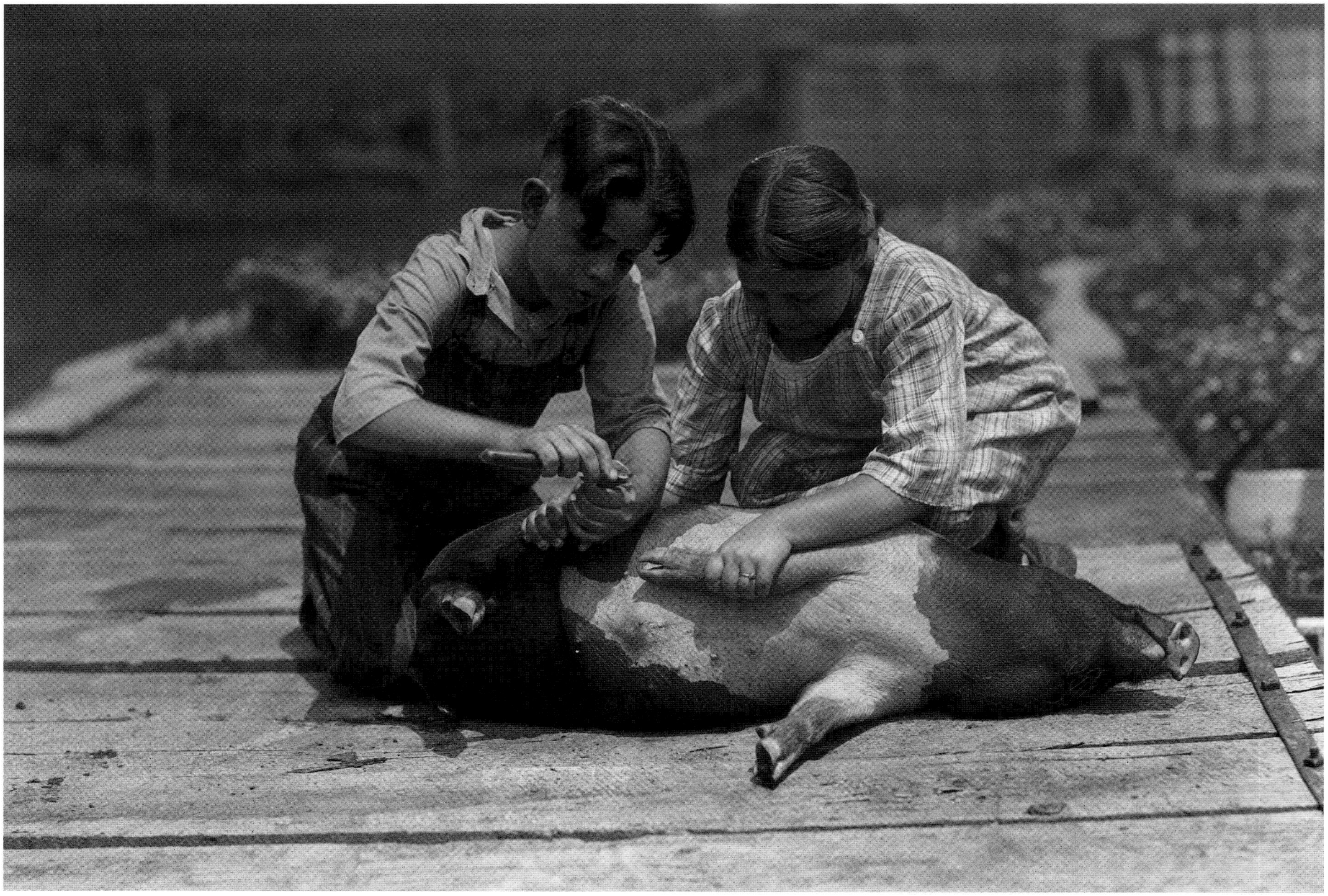

Lloyd Spangler (left) and his sister, Florence, trim the feet of a 4-H club pig in Oxford. (Benton County, 1923)

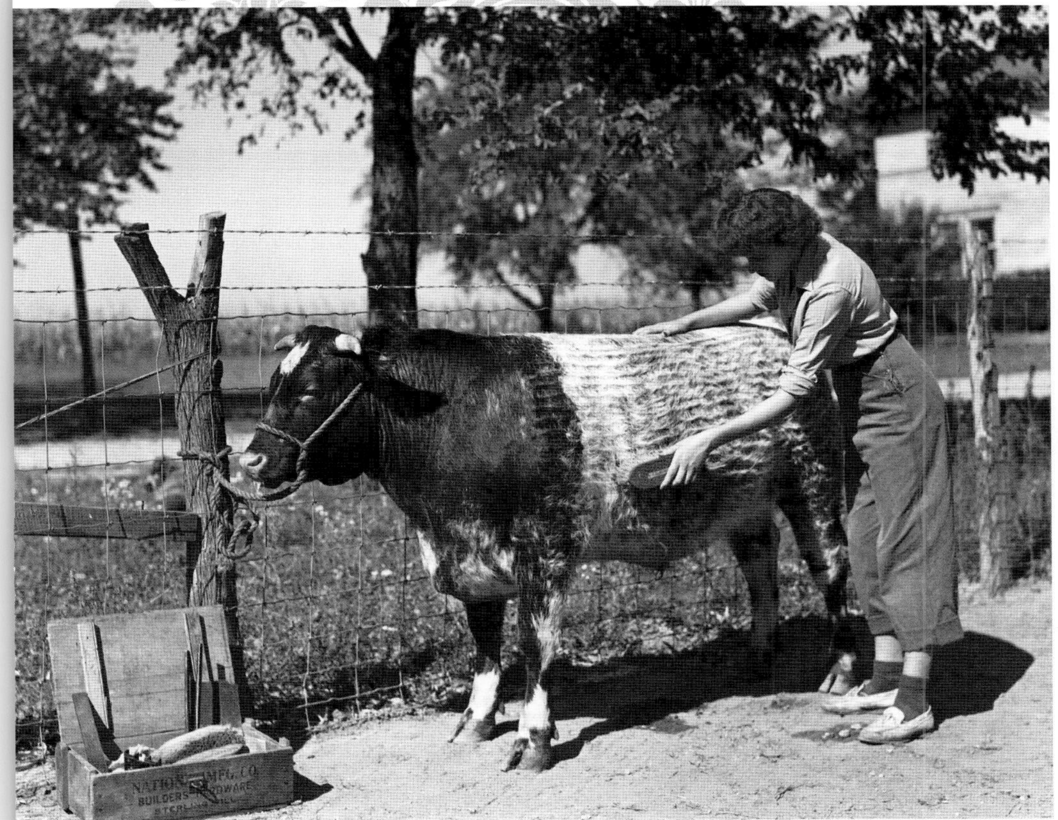

Margaret German from Romney washes her 4-H steer. (Tippecanoe County, 1946)

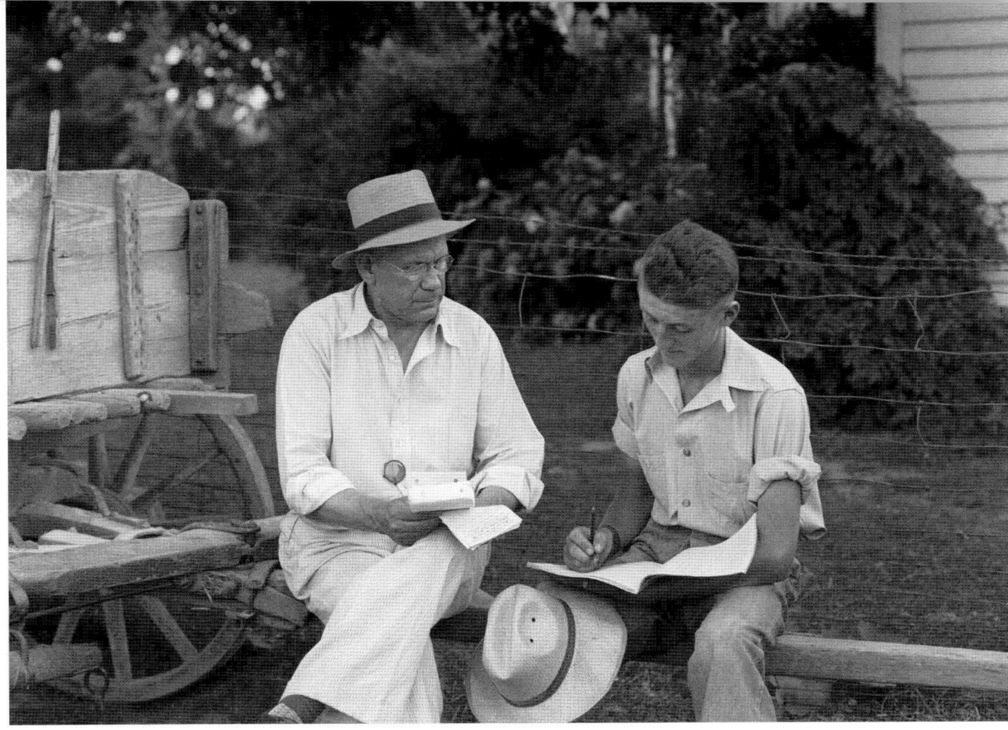

John Stewart (right) from Greensburg works on his 4-H records under the supervision of club leader A. W. McCracken. (Decatur County, 1935)

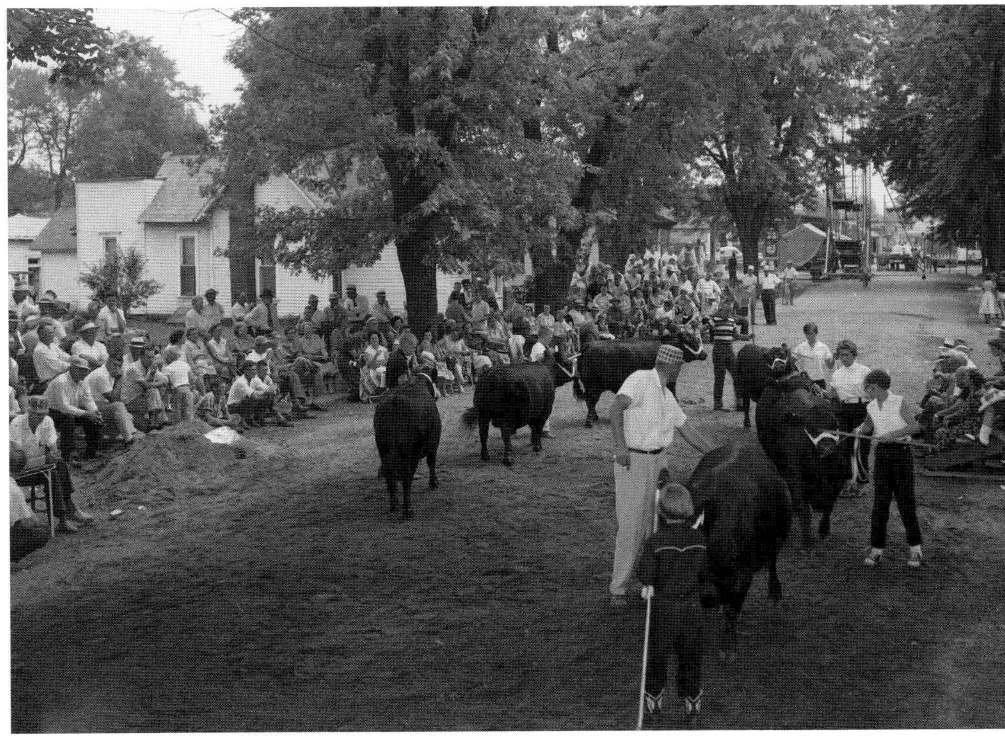

A 4-H cattle show held in connection with a street fair at Russellville (Putnam County, 1956)

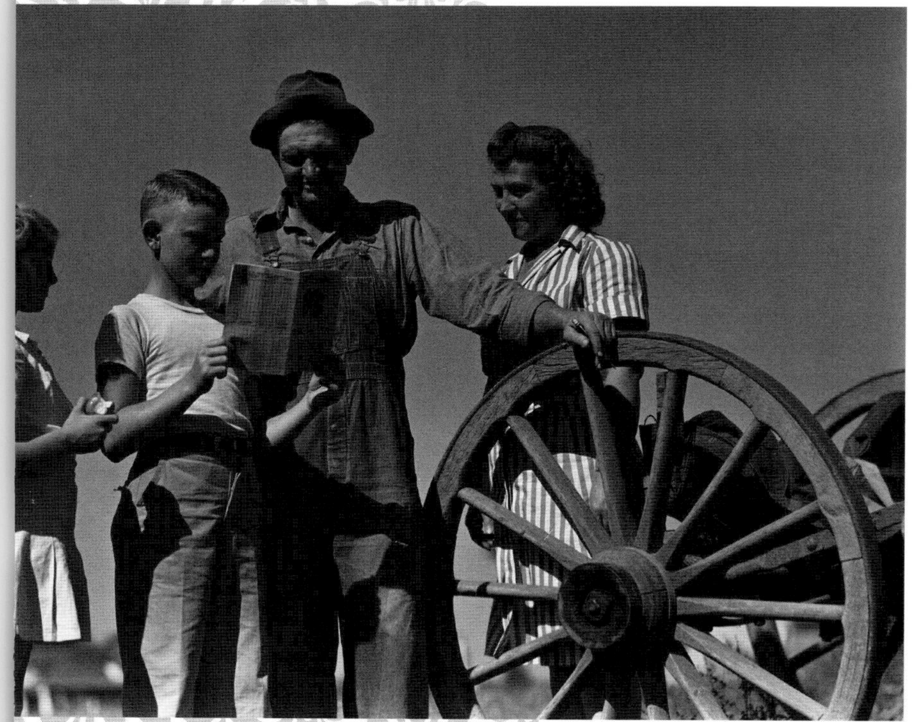

The William Gilliland family from Crawfordsville looks at a 4-H club record book. (Montgomery County, 1946)

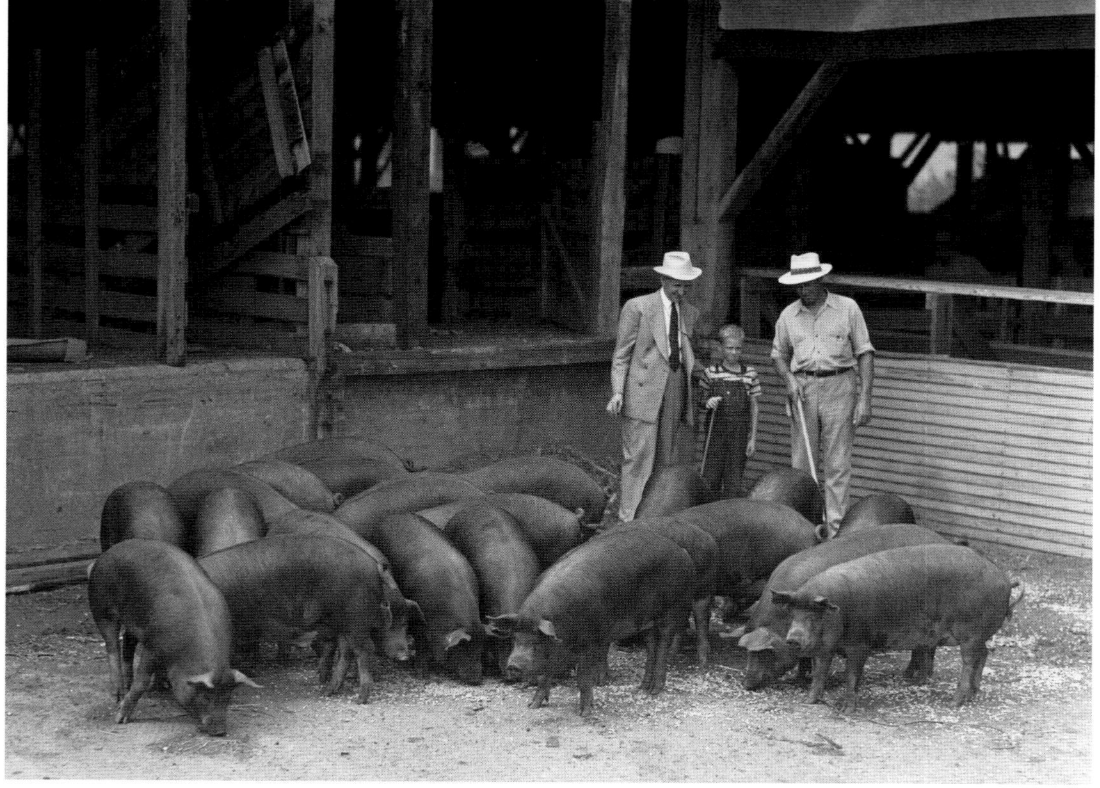

Indiana Governor Henry Schricker (left) congratulates Richmond's LeRoy Turner (right) for winning first place at the eighth annual Hoosier Ton Litter Show. (Wayne County, 1943)

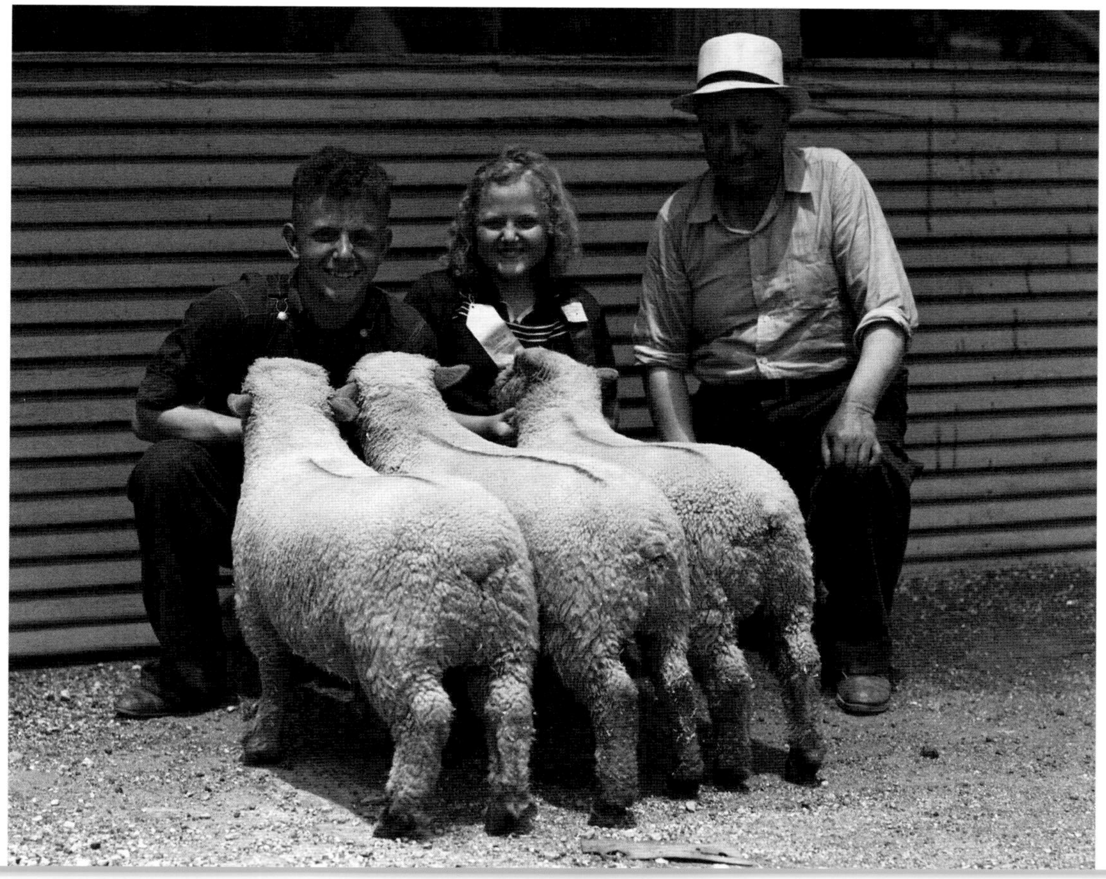

Robert Strange Jr. (left) from Rockville poses with his first-prize pen of three lambs at the Central Indiana 4-H Lamb Club Show alongside his sister and father. (Parke County, 1939)

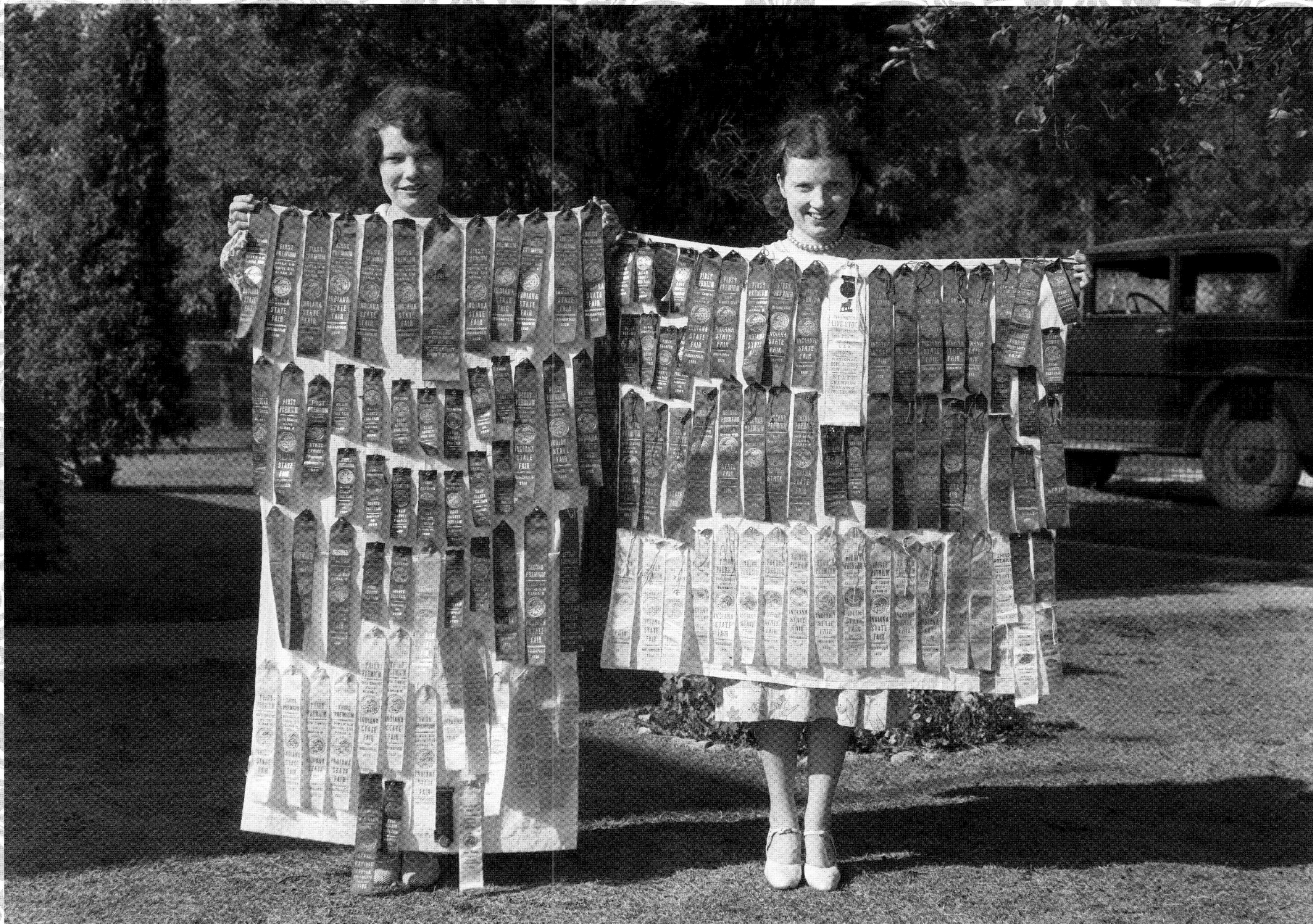

Lorene Krammes and Joenita Burd from Manilla display their collection of 4-H ribbons, which they primarily won from canning competitions at county and state fairs. (Rush County, 1931)

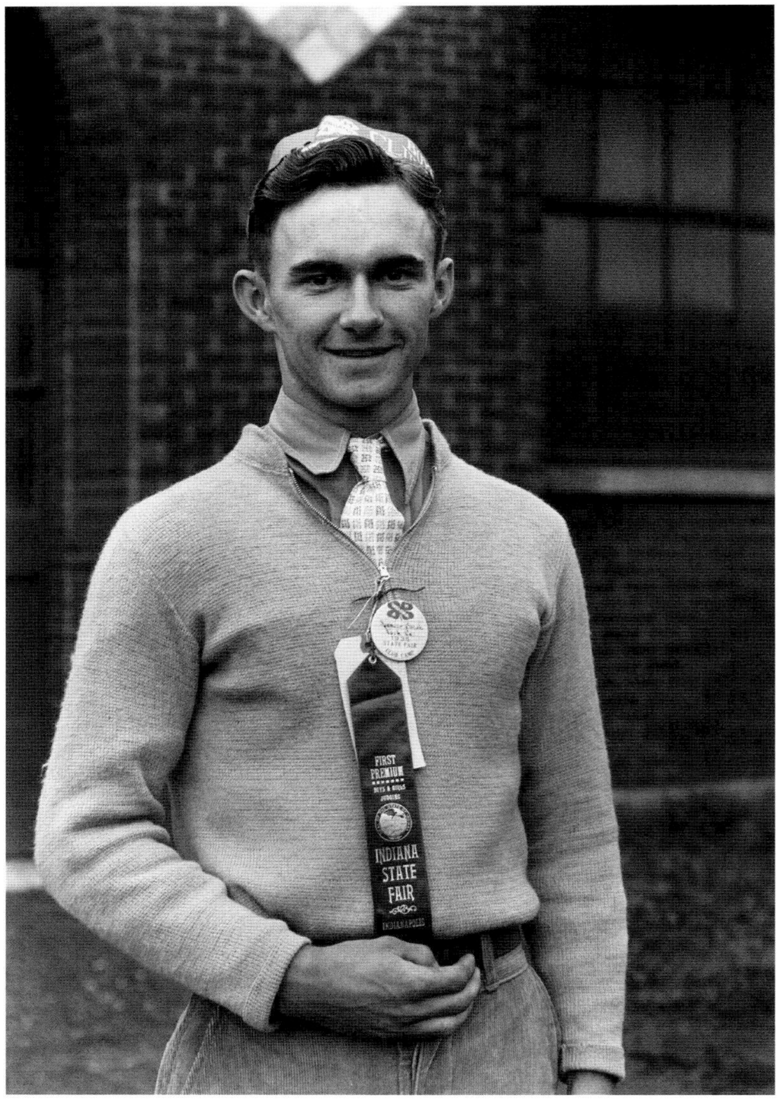

(Above) Spencer Bogue of Rushville scored 660 points out of a maximum 750 to win high individual honors in the Livestock Judging Contest at the Indiana State Fair. He also received a $100 scholarship to Purdue University given by Kingan and Company, a meatpacking plant in Indianapolis. (Marion County, 1935)

(Top right) Frances Flora (left) from Elwood sells her 4-H club poultry to a representative of the Fear Campbell Company of Elwood. (Madison County, 1936)

(Bottom right) Following a 4-H club lamb show at the stockyards in Indianapolis, this first-prize pen of three lambs sold for 20 cents per pound. (Marion County, 1934)

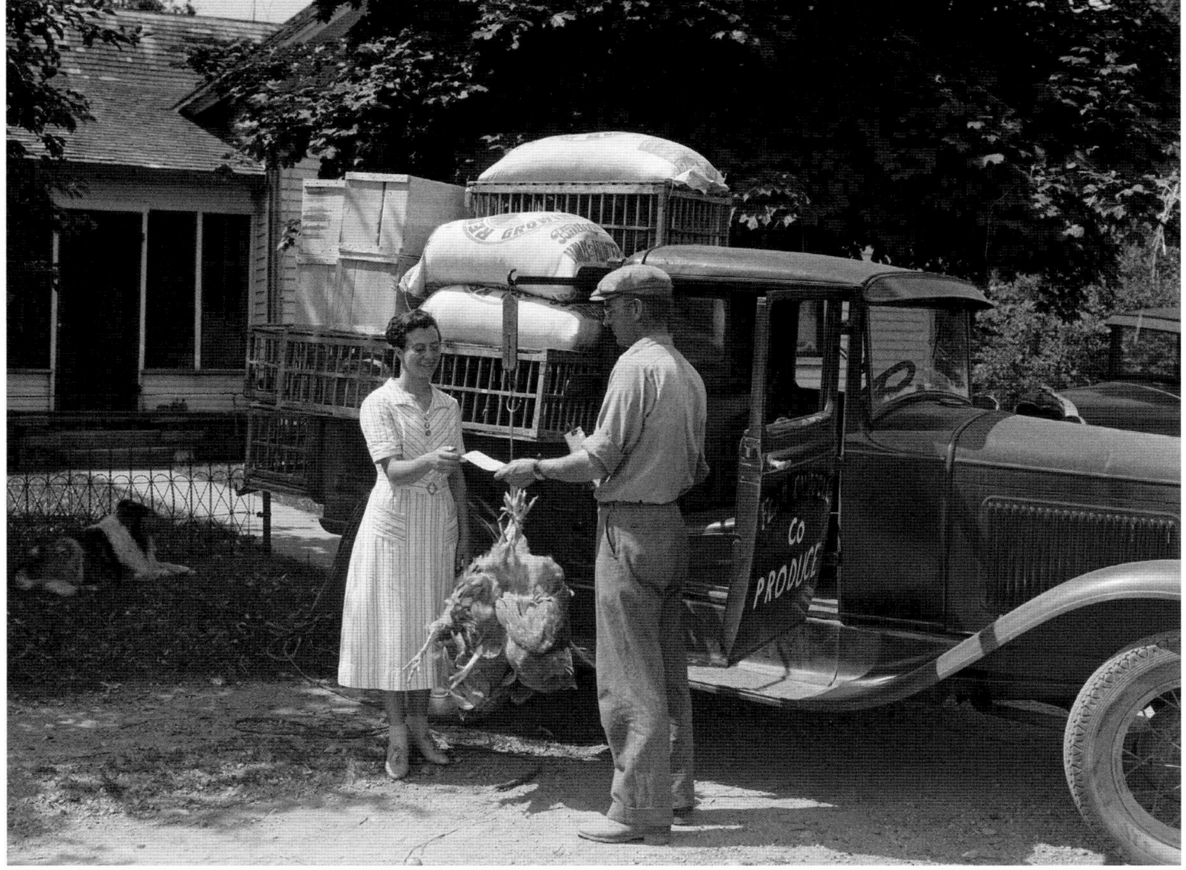

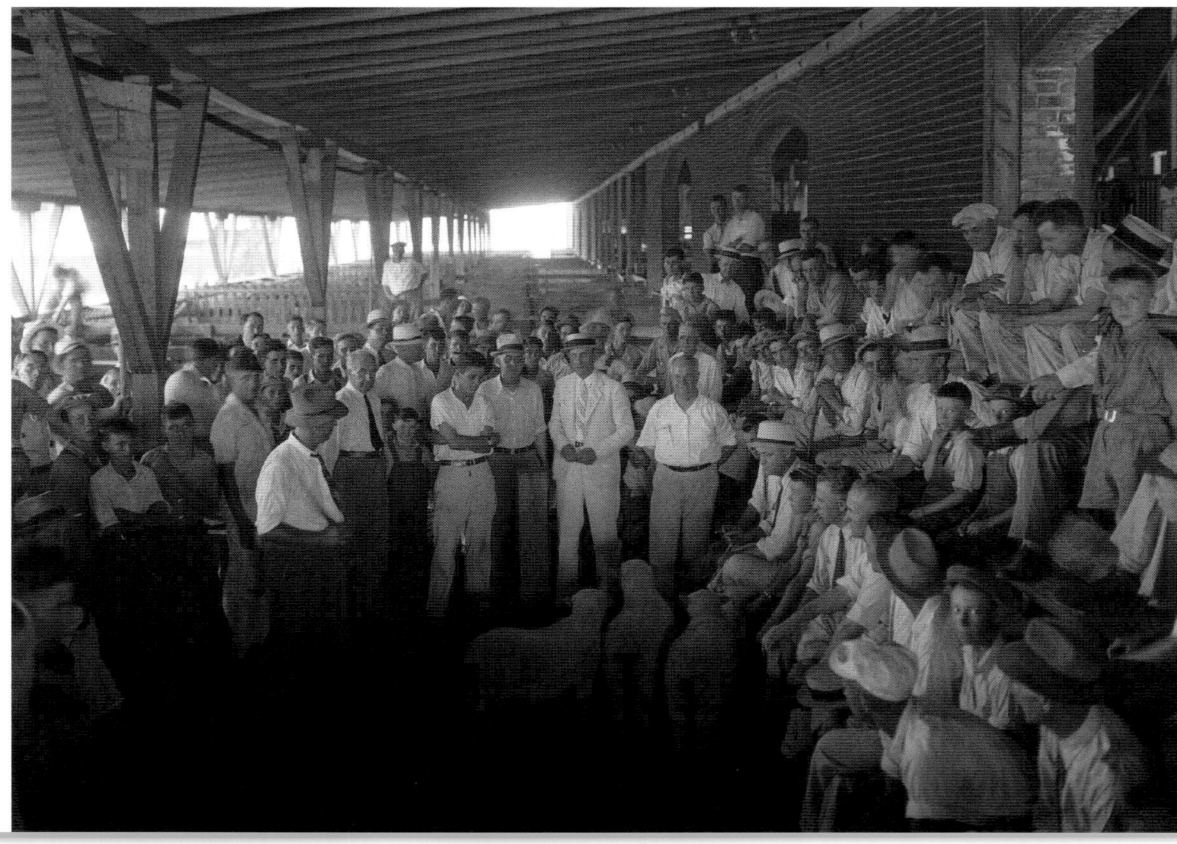

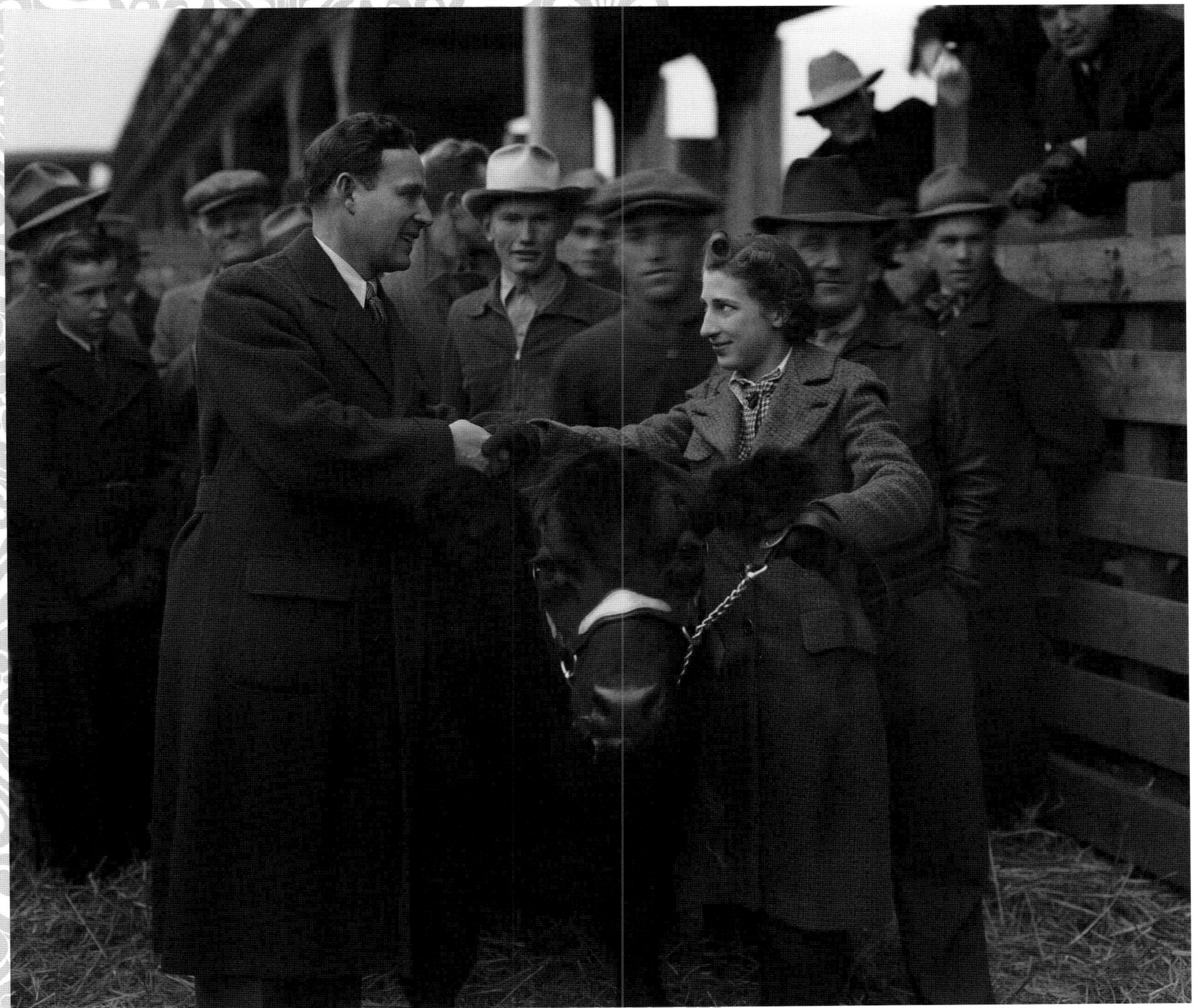

Horace Millhone (left) of Firestone Tire and Rubber Company shakes hands with Irene Brown from Aledo, Illinois, after purchasing her grand champion Angus steer at the International Live Stock Exposition in Chicago. (Cook County, IL; 1938)

NOTES

JCA refers to John Calvin Allen.

The Early Life of John C. Allen

1. James T. Allen, letter of pension claim to John Black, 24 June 1888.
2. James T. Allen, letter of pension claim to John Black, 24 June 1888.
3. Joanne Milkereit, "James Thomas Allen" (unpublished manuscript, Sept. 2007), printout, 1; James T. Allen, letter of pension claim to John Black, 24 June 1888.
4. Milkereit, 1; Mary T. Gibbens, letter to John Allen Benham (grandson of JCA), 1 Sept. 1987.
5. Milkereit, 1.
6. Milkereit, 1; "Died at Clay City," *Terre Haute (IN) Evening Gazette,* 16 Dec. 1897.
7. JCA, unpublished autobiography, ca. 1971, printout, 6.
8. JCA, unpublished autobiography, 6; John Allen Benham (grandson of JCA), email message to Frederick Whitford, 19 June 2017.
9. JCA, unpublished autobiography, 6.
10. JCA, unpublished autobiography, 6.
11. JCA, unpublished autobiography, 6.
12. JCA, Indiana Soldiers' and Sailors' Orphans' Home application for admission, Knightstown, IN, 25 May 1888.
13. JCA, unpublished autobiography, 6.
14. JCA, unpublished autobiography, 6.
15. JCA, unpublished autobiography, 6.
16. JCA, unpublished autobiography, 7.
17. JCA, unpublished autobiography, 7.
18. JCA, unpublished autobiography, 7.
19. JCA, unpublished autobiography, 7.
20. JCA, unpublished autobiography, 6.
21. JCA, unpublished autobiography, 6.
22. JCA, unpublished autobiography, 6.
23. JCA, unpublished autobiography, 7.
24. JCA, unpublished autobiography, 7.
25. JCA, unpublished autobiography, 7.
26. JCA, unpublished autobiography, 7.
27. JCA, unpublished autobiography, 7.
28. JCA, unpublished autobiography, 8.
29. JCA, unpublished autobiography, 8.
30. "Famed Illustrator J. C. Allen Dies," *Lafayette (IN) Journal and Courier,* 21 July 1976, A-4.
31. JCA, unpublished autobiography, 8.
32. [John] Al[len] Benham, "Rural Life Photographer," *Graflex Journal*, no. 1 (2017), 4.
33. R. L. Reeder, "'Have Camera, Will Travel'—and He Did, for Photos by J. C. Allen & Son," *ACE Quarterly* (published by the American Association of Agricultural College Editors), Jan.–Mar. 1976, 10.
34. Reeder, 10; Purdue University Agricultural Alumni Association Certificates of Distinction biography of JCA, 1963.
35. "Purdue's Head Camera Man," *Purdue Alumnus,* May 1933, 19.
36. JCA, unpublished autobiography, 8.
37. JCA, unpublished autobiography, 8.
38. JCA, unpublished autobiography, 8.
39. Chester P. Allen, "Summer Days at Grandma's Farm" (unpublished manuscript on JCA, n.d.), printout, 1.
40. Chester P. Allen, 1.
41. JCA, unpublished autobiography, 8.
42. JCA, unpublished autobiography, 8.
43. JCA, unpublished autobiography, 8.
44. JCA, "I Take Rural Pictures," *Hoard's Dairyman,* 10 Mar. 1959, 275.
45. JCA, unpublished autobiography, 8.

The Photographer Emerges

1. John Allen Benham, "City Directory" (unpublished notes on various city directory entries for JCA, n.d.), handwritten document.
2. Carl Eiche, "J. C. Allen—Chief Photographer," *The Byline* (newsletter of the American Agricultural Editors' Association), ca. Mar. 1977, 4.
3. Purdue University Agricultural Alumni Association Certificates of Distinction biography of JCA, 1963; "Purdue Ag Alumni Honor Four Persons for Outstanding Farming Service," *Lafayette (IN) Journal and Courier,* 19 Jan. 1963, 2.
4. John H. Skinner, letter to Frank G. King, 23 Aug. 1921.
5. JCA, unpublished autobiography, ca. 1971, printout, 8; R. L. Reeder, "'Have Camera, Will Travel'—and He Did, for Photos by J. C. Allen & Son," *ACE Quarterly* (published by the American Association of Agricultural College Editors), Jan.–Mar. 1976, 10.
6. JCA, unpublished autobiography, 8.
7. John H. Skinner, letter to H. J. Reed, 8 July 1910; JCA, Purdue University transcript, 1912.
8. Eiche, 4; Reeder, 10.
9. John O. Allen, unpublished manuscript on JCA, n.d., printout, 1.
10. Reeder, 10.
11. John H. Skinner, letter to Arthur Goss, 27 May 1915.
12. JCA, "I Take Rural Pictures," *Hoard's Dairyman,* 10 Mar. 1959, 275.
13. JCA, unpublished autobiography, 8.
14. John Ashton, letter to John H. Skinner, 27 Feb. 1934.
15. John Ashton, letter to John H. Skinner, 27 Feb. 1934.
16. John H. Skinner, letter to John Ashton, 5 Mar. 1934.
17. JCA, unpublished manuscript on being a photographer, ca. 1940, printout, n.p.
18. JCA, "I Take Rural Pictures," 274.
19. JCA, unpublished manuscript on being a photographer, n.p.
20. Reeder, 10.
21. "J. C. Allen and Son," *Purdue Alumnus,* June/July 1949, 10.
22. Reeder, 10.
23. Reeder, 10.
24. Reeder, 10.

Crafting a Business One Photograph at a Time

1. John H. Skinner, letter to JCA, 26 July 1926.
2. Carl Eiche, "J. C. Allen—Chief Photographer," *The Byline* (newsletter of the American Agricultural Editors' Association), ca. Mar. 1977, 4.
3. JCA, "I Take Rural Pictures," *Hoard's Dairyman,* 10 Mar. 1959, 274.
4. JCA, unpublished manuscript on being a photographer, ca. 1940, printout, n.p.
5. JCA, unpublished manuscript on being a photographer, n.p.
6. John O. Allen (grandson of JCA), in discussion with authors, 21 July 2013.
7. JCA, unpublished manuscript on being a photographer, n.p.

Crafting a Business One Photograph at a Time *(continued)*

8. JCA, unpublished manuscript on being a photographer, n.p.
9. JCA, unpublished manuscript on being a photographer, n.p.
10. JCA, unpublished manuscript on being a photographer, n.p.
11. JCA, unpublished manuscript on being a photographer, n.p.
12. John O. Allen (grandson of JCA), in discussion with authors, 21 July 2013.
13. "J. C. Allen and Son Represent Six Decades of Farm History," *Indiana General News,* 1 May 1970, 2.
14. John O. Allen (grandson of JCA), in discussion with authors, 21 July 2013.
15. JCA, unpublished manuscript on being a photographer, n.p.
16. JCA, "I Take Rural Pictures," 274.
17. John Allen Benham (grandson of JCA), email message to Frederick Whitford, 19 June 2017.
18. Anne Schowe, "The Schowe House at Purdue University: A Brief History" (unpublished manuscript, Mar. 2011), printout, 1.
19. "J. C. Allen and Son," *Purdue Alumnus,* June/July 1949, 11.
20. JCA, letter to Ralph Poulton, 31 Oct. 1950.
21. John H. Skinner, letter to JCA, 31 May 1928.
22. JCA, unpublished autobiography, 9.
23. JCA, unpublished autobiography, 9.
24. "Purdue's Head Camera Man," *Purdue Alumnus,* May 1933, 19.
25. John O. Allen, unpublished manuscript on JCA, n.d., printout, 1.
26. John O. Allen, unpublished manuscript on JCA, 1.
27. JCA, unpublished manuscript on being a photographer, n.p.
28. JCA, "I Take Rural Pictures," 274.
29. "J. C. Allen and Son Represent Six Decades of Farm History," *Indiana General News,* 1 May 1970, 2.
30. John O. Allen (grandson of JCA), in discussion with authors, 20 Sept. 2017.
31. John H. Skinner, letter to Harry J. Reed, 19 Dec. 1941; Harry J. Reed, letter to JCA, 20 Feb. 1942.
32. John O. Allen (grandson of JCA), in discussion with authors, 20 Sept. 2017.
33. JCA, "I Take Rural Pictures," 274.
34. Olivia Maddox, "Moments in Time," *Purdue Agricultures Magazine* 6, no. 1 (Fall 2002): 15.
35. JCA, "I Take Rural Pictures," 274.
36. "Famed Illustrator J. C. Allen Dies," *Lafayette (IN) Journal and Courier,* 21 July 1976, A-4.
37. John O. Allen (grandson of JCA), in discussion with authors, 20 Sept. 2017.
38. JCA, unpublished autobiography, 9; "Purdue's Head Camera Man," *Purdue Alumnus,* May 1933, 19.
39. "Purdue's Head Camera Man," *Purdue Alumnus,* May 1933, 19; John O. Allen, unpublished manuscript on JCA, 2.
40. "John C. Allen and Son," *Purdue Alumnus,* June/July 1949, 10.
41. R. L. Reeder, "'Have Camera, Will Travel'—and He Did, for Photos by J. C. Allen & Son," *ACE Quarterly* (published by the American Association of Agricultural College Editors), Jan.–Mar. 1976, 12.
42. John O. Allen (grandson of JCA), in discussion with authors, 20 Sept. 2017.
43. JCA, unpublished autobiography, 9.
44. John Allen Benham (grandson of JCA), email message to Frederick Whitford, 19 June 2017.
45. JCA, unpublished autobiography, 9.
46. John Allen Benham (grandson of JCA), email message to Frederick Whitford, 19 June 2017.
47. John O. Allen (grandson of JCA), in discussion with authors, 20 Sept. 2017.
48. Purdue University Agricultural Alumni Association Certificates of Distinction biography of JCA, 1963; "Purdue Ag Alumni Honor Four Persons for Outstanding Farming Service," *Lafayette (IN) Journal and Courier,* 19 Jan. 1963, 2.
49. Purdue University, Board of Trustees meeting minutes, 16 Apr. 1952, 37.
50. JCA, unpublished autobiography, 9.
51. Reeder, 10.
52. Eiche, 4.
53. Eiche, 4.
54. Reeder, 10.
55. John Allen Benham (grandson of JCA), email message to Frederick Whitford, 19 June 2017.
56. "Famed Illustrator J. C. Allen Dies," *Lafayette (IN) Journal and Courier,* 21 July 1976, A-4.
57. John O. Allen (grandson of JCA), in discussion with authors, 20 Sept. 2017.

A Lasting Legacy

1. John O. Allen (grandson of JCA), in discussion with Frederick Whitford, 26 June 2017.

SOURCES

JCAS Located in the J. C. Allen and Son Inc. Photographs and Negatives Collection, Purdue University Archives and Special Collections, West Lafayette, Indiana

Allen, Chester P. "Summer Days at Grandma's Farm." Unpublished manuscript on John C. Allen, n.d. Printout. **JCAS**

Allen, James T. Declaration for Invalid Army Pension, 15 June 1885. National Personnel Records Center, Military Personnel Records (NPRC-MPR), National Archives, Washington, D.C. **JCAS**

———. Letter of pension claim to John Black, 24 June 1888. National Personnel Records Center, Military Personnel Records (NPRC-MPR), National Archives, Washington, D.C. **JCAS**

Allen, John C. Correspondence, 1950. **JCAS**

———. Indiana Soldiers' and Sailors' Orphans' Home application for admission, Knightstown, IN, 25 May 1888. **JCAS**

———. "I Take Rural Pictures." *Hoard's Dairyman,* 10 Mar. 1959.

———. Purdue University transcript, 1912. Purdue University Office of the Registrar, West Lafayette, IN.

———. Unpublished autobiography, ca. 1971. Printout. **JCAS**

———. Unpublished manuscript on being a photographer, ca. 1940. Printout (accession ID 20111025). **JCAS**

Allen, John O. Unpublished manuscript on John C. Allen, n.d. Printout. **JCAS**

Ashton, John. Letter to John H. Skinner, 27 Feb. 1934. **JCAS**

Benham, John Allen. "City Directory." Unpublished notes on various city directory entries for John C. Allen, n.d. Handwritten document. **JCAS**

———. Email message to Frederick Whitford, 19 June 2017. **JCAS**

———. "Rural Life Photographer," *Graflex Journal,* no. 1 (2017): 4–7. **JCAS**

Eiche, Carl. "J. C. Allen—Chief Photographer." *The Byline* (newsletter of the American Agricultural Editors' Association), ca. Mar. 1977, 4. **JCAS**

Gibbens, Mary T. Letter to John Allen Benham (grandson of John C. Allen), 1 Sept. 1987. **JCAS**

Indiana General News. "J. C. Allen and Son Represent Six Decades of Farm History," 1 May 1970.

Lafayette (IN) Journal and Courier. "Famed Illustrator J. C. Allen Dies," 21 July 1976.

———. "Purdue Ag Alumni Honor Four Persons for Outstanding Farming Service," 19 Jan. 1963.

Maddox, Olivia. "Moments in Time." *Purdue Agricultures Magazine* 6, no. 1 (Fall 2002), 12–17.

Milkereit, Joanne. "James Thomas Allen." Unpublished manuscript, Sept. 2007. Printout. **JCAS**

Price, Todd. "Telling a Story Through a Lens: The J. C. Allen Photo Collection." *Midwest Open Air Museums Magazine* 38, no. 1–2 (Summer 2017), 20–25. **JCAS**

Purdue Alumnus. "J. C. Allen and Son," June/July 1949, 10–11.

———. "Purdue's Head Camera Man," May 1933, 19.

Purdue University. Agricultural Alumni Association Certificates of Distinction biography of John C. Allen, 1963. **JCAS**

———. "Agriculture in Focus: Through the Lens of J. C. Allen" (announcement for Nov.–Dec. 2009 exhibit at the Purdue University Archives and Special Collections), 13 Nov. 2009. **JCAS**

———. Board of Trustees meeting minutes, 16 Apr. 1952. Purdue University Board of Trustees Meeting Minutes, 1865–Present, Purdue University Archives and Special Collections, West Lafayette, IN.

———. *County Extension Annual Reports, Newton County,* 1918. College of Agriculture Extension Reports Collection, 1913–1972, Purdue University Archives and Special Collections, West Lafayette, IN.

Reed, Harry J. Correspondence, 1942. College of Agriculture Administration records, 1869–2009, Purdue University Archives and Special Collections, West Lafayette, IN.

Reeder, R. L. "'Have Camera, Will Travel'—and He Did, for Photos by J. C. Allen & Son." *ACE Quarterly* (published by the American Association of Agricultural College Editors), Jan.–Mar. 1976, 9–12. **JCAS**

Schowe, Anne. "The Schowe House at Purdue University: A Brief History" (unpublished manuscript, Mar. 2011). Printout. **JCAS**

Shumard-Shelton, Lisa. "Photo Exhibit Showcases History of Farming." *Indiana AgriNews,* 29 Aug. 2003, A1 and A6. **JCAS**

Skinner, John H. Correspondence, 1910–41. College of Agriculture Administration records, 1869–2009, Purdue University Archives and Special Collections, West Lafayette, IN.

Terre Haute (IN) City Directory. Cincinnati, OH[?]: Chas. O. Ebel and Co., 1892.

———. N.p.: W. F. Curtis, 1888.

Terre Haute (IN) Evening Gazette. "Died at Clay City," 16 Dec. 1897. **JCAS**

INDEX

Cities and towns are listed by county.
JCA refers to John Calvin Allen.
Page numbers for main sections are in bold.